GEOTECHNICAL ENGINEERING ANALYSIS AND EVALUATION

McGRAW-HILL SERIES IN GEOTECHNICAL ENGINEERING

Legget and *Karrow* **Handbook of Geology in Civil Engineering** (1983)
Hunt **Geotechnical Engineering Investigation Manual** (1984)
Hunt **Geotechnical Engineering Analysis and Evaluation** (1986)

GEOTECHNICAL ENGINEERING ANALYSIS AND EVALUATION

ROY E. HUNT

Consulting Engineer

McGraw-Hill Book Company

*New York St. Louis San Francisco Auckland
Bogotá Hamburg Johannesburg London Madrid
Mexico Montreal New Delhi Panama Paris
São Paulo Singapore Sydney Tokyo Toronto*

Library of Congress Cataloging in Publication Data

Hunt, Roy E.
Geotechnical engineering analysis and evaluation.

Includes bibliographies and index.
1. Engineering geology. 2. Soil mechanics. I. Title.
TA705.H85 1986 624.15 85-2946
ISBN 0-07-031310-5

1 2 3 4 5 6 7 8 9 0 DOC/DOC 8 9 8 7 6 5

ISBN 0-07-031310-5

The editors for this book were Joan Zseleczky and Geraldine Fahey, the designer was Naomi Auerbach, and the production supervisor was Sally Fliess. It was set in Melior by University Graphics, Inc.

Printed and bound by R. R. Donnelley & Sons, Inc.

CONTENTS

FOREWORD

The scope of this book embraces all of applied soil mechanics, rock mechanics, and engineering geology. Most volumes of such breadth are collections of contributions by many authors hewing to a line laid out by an editor, but seldom achieving unity of presentation, not to mention purpose or approach. Roy Hunt has taken a more difficult path. He has sought a mastery of each of the elements of his book, leaving little of the pertinent literature unexplored, but of far greater importance he has endowed the book with consistency and with a unifying philosophy. That philosophy, grown out of his own experience, is the pervasive influence of geology that cannot be ignored or discounted by the successful geotechnical engineer.

It is a book to be used. The practitioner will soon find that it contains in convenient form all that he needs to start the analysis and evaluation of the problem with which he is concerned, and in many instances all that he needs to bring the problem to a solution. It would be no surprise if this book and its companion volume on investigation would soon show more wear and tear than any other technical books in his library.

It is not a book for the beginner in soil or rock mechanics. Instead, it is for the professional who is already trained in the fundamentals, who knows generally how to approach his problems, but who wants details of procedures or who wants to be sure that he is aware of recent developments.

The author is well aware that the price of certainty—if, indeed, certainty can be attained in geotechnical design and construction—is overconservatism. Yet, failure to anticipate and guard reasonably against the whims of nature is irresponsible and, in the event of a catastrophe that might have been anticipated, indefensible. How to judge where to find the middle ground that represents sound practice is made easier and clearer by the author's consistent attention to what he aptly calls the geologic interface, the one element of every geotechnical design most likely to be misjudged and to lead to grief. The reader who is overly confident in conclusions drawn from his tests and calculations will find in this book the basis for an improved practice.

Ralph B. Peck

PREFACE

The most significant aspects of geotechnical engineering analysis and evaluation pertain to foundations, slopes, embankments, pavements, retaining structures, and underground excavations. Design and construction of such works must deal with an infinite variety of soil formations and rock masses often composed of materials that are far from "ideal" with regard to precise analytical prediction of their response to some construction activity, which is usually desired in terms of potential deformations or rupture. The possibility of a variety of conditions involving less than ideal materials leads to many uncertainties in geotechnical engineering and the practitioner must be prepared to cope with these uncertainties in a confident manner by applying the library of knowledge available, preferably without undue conservatism.

Theoretical concepts and analytical methods are used for guidance but the successful solution to any problem fundamentally requires the application of a large amount of judgment which is based on experience—not only that of the individual but also that of the entire profession. This "global" experience can be said to be incorporated in a catalog of *standards* or *norms* as applied to investigations, analysis, and the formulation of solutions to the engineering problems. Even in locations where procedures are governed by codes, judgment is required, and even then many situations will be encountered which are not covered by code.

There is of course, a large gray area as to what constitutes a standard, because standards vary with the locales, countries, or regions in which they have developed under particular sets of geologic conditions. Caution is required when the practitioner varies from the standards of the project's locale in reaching a solution to a problem, because if difficulties do occur that result in excessive cost overruns, extras, or poor performance related to geotechnology, the practitioner is quite likely to be found at fault during litigation, even if the guilt is not his or hers.

The mature and experienced practitioner would consider varying from local standards if it were in the best interest of the client, but unfortunately the possibility of litigation is a deterrent always casting a forboding shadow. Geotechnical engineering would never progress, however, if new concepts were not applied, and it is the practitioner's responsibility to be aware of new developments, to assess their validity and, if confident about them, to pass this confidence to the client and obtain concurrence for application.

There are generally two ways in which standards new to a particular locale or country may be accepted and adopted. The most common is based on long experience developed from application in some other country or region; the other begins with model or small-scale testing and evolves with full-scale testing under a variety of conditions. A new concept not yet thoroughly proven may be applied to some particular situation if a fair degree of conservatism is maintained and some form of backup provided.

Consulting engineering practice is a very serious affair that frequently involves uncertainties and substantial risk. Malperformance (proven or otherwise) can result in heavy compensation to the damaged or aggrieved parties, or, at times, loss of practice, or, in some countries, even prison. The majority of problems are routine, but in geotechnical engineering difficult and unusual conditions frequently occur as surprises, which means that the practitioner must always be on guard. Standards should be applied but they should be supplemented by new technology when proven and warranted in order to provide better definition of geologic conditions, more reliable data on properties, improved analytical techniques, and more satisfactory construction procedures. The objective is not only to avoid excessive conservatism and project costs, but to be certain that the degree of conservatism selected is not really a myth, for if it is, troubles are likely.

Project types and their attendant geotechnical problems often manifest themselves in an erratic pattern. The consultant may be involved with only routine foundation problems for a long period, for example, then suddenly be faced with an anchored bulkhead or a slope stability or a laterally loaded pile problem. For approaches to solutions the practitioner is faced with the task of searching through various texts, journals, and other technical publications. The primary objective of this book is to provide a ready reference that presents the standard approaches and recent innovations for resolving geotechnical problems in both soil and rock as applied by practitioners generally on an international basis.

Regardless of individual interest, the reader is urged to review Chapter 1—INTRODUCTION—because it provides an overview of several important aspects of geotechnical engineering that deserve special attention, including the geologic interface, the limitations of the technology, and new technology. Article 1.3—Limitations of the Technology—is considered by the author to be particularly important since too many students and practitioners still have the often wrong impression that problems are resolved merely by imputting parameter data into some mathematical equation or other relationship without considering the limitations involved in obtaining truly representative data, the limitations of the theory behind the equation, or the basis for the relationship and its application.

The author is very appreciative of the suggestions of a number of engineers who reviewed various parts of the text. In particular, he is indebted to H.O. Ireland, formerly of the University of Illinois (Chapters 3, 4, 6, and parts of 9), Ernest Jonas of TAMS (Chapter 5 and part of 9), Berry Grubbs of TERRAMAR (Chapter 7), N.R. Morgenstern of the University of Alberta (Chapter 8), and especially to Ralph B. Peck, formerly of the University of Illinois, who reviewed the entire preliminary and final manuscripts in exhausting detail. Appreciation and thanks are warranted also for the individuals, publishers, and manufacturers who were the source of many of the figures and tables included. Photos and other illustrations not credited to some source are those of the author. All of the manuscript was typed by the author whose only regret is not having had a word processor available.

Roy E. Hunt

GEOTECHNICAL ENGINEERING ANALYSIS AND EVALUATION

CHAPTER ONE

INTRODUCTION

1.1 GEOTECHNICAL ENGINEERING

1.1.1 THE SCOPE OF GEOTECHNICAL ENGINEERING

The modern practice of geotechnical engineering, a component of civil engineering, can be considered to include the principles of both soil and rock mechanics and engineering geology. As applied to engineered development and construction, the concepts of geotechnical engineering can be separated into several major problem groups as follows:

Site preparation as required to make a location ready for construction, including methods to improve natural ground conditions, and the effect of development or construction on adjacent, existing structures.

Foundations to support structures, which are divided into two classes, shallow and deep.

Slope stability as related to natural and cut slopes, surface excavations, earth dam embankments, and embankments over soft ground.

Surface retaining structures as required for excavations for buildings and other works, or to retain waterfront earth masses or structures or water bodies.

Closed or underground excavations made for tunnels, caverns, or mines, which may or may not require synthetic support.

1.1.2 THE PRACTICE OF GEOTECHNICAL ENGINEERING

Basics

The solutions to all of these foregoing problems require an evaluation of the response of the geologic environment to the new conditions imposed by the construction activity. The constituent soils and rocks respond to new forces by deforming without rupture, until finally, under a particular set of stress conditions, failure by rupture occurs. Groundwater conditions play a most significant role in all problems, whether they involve stability against failure, compression or consolidation, or earth pressures. Evaluations may be based on theoretical or empirical concepts in concert with observational techniques or, at times, solely on observational techniques which are rooted in experience and judgment.

Engineering Theory and Geologic Reality

Successful solutions, in any case, are dependent upon accurate definition of geologic conditions including stratigraphic relationships, mass defects or discontinuities, material properties, and groundwater conditions. The writings and philosophy of Dr. Karl Terzaghi are legend, but perhaps most important was his constant warning of the need to recognize when to start mistrusting equations and to rely exclusively on observations because of the radical differences between engineering theory and geologic reality. In many of his writings he emphasized the significance of comprehensive geologic study for engineering works. His lectures were presentations in geology and geomorphology, how they relate to a problem, and how the engineering science of soil mechanics has application.

Terzaghi's views of the mutually supporting roles of geological observations, physical testing to measure properties, and theory in the solution of problems of both soil and rock mechanics were summarized in a paper in (1963):[1]

Within the last few decades, in both soil and rock mechanics, a great number of new techniques for testing and field observations have been developed. As a consequence, knowledge of the engineering properties of the materials encountered in subsurface engineering has rapidly increased. In both domains, the successful application of theoretical procedures to the solution of engineering problems is limited to those instances in which the following two conditions are satisfied:

1. Within the range of influence of the proposed engineering operations, the boundaries between materials with significantly different engineering properties should be reasonably well defined. The boundaries divide the subsoil or bedrock into zones. Within each zone, the soil or rock should be sufficiently homogeneous to permit its replacement without serious error by an ideal, perfectly homogeneous substitute, whose engineering properties can be defined by a few numerical values.
2. The number of tests performed should be great enough to permit reliable evaluations of the statistical average of the significant engineering properties of the real material located within each zone. If, in the realm of soil mechanics, the first requirement is satisfied, there is no serious obstacle against compliance with the second requirement, because sampling and soil testing are relatively inexpensive operations. On the other hand, in the field of rock mechanics, the cost of the second requirement may be, and often is, prohibitive, provided the requirement can be satisfied at all. For instance, the cost of securing reliable average values of the deformation characteristics of jointed rock masses is commonly prohibitive, and the determination of the resistance of such masses against failure by shear is still impossible.

Although progress continues in geotechnical engineering, especially in the applications of mathematical relationships, these concepts of Terzaghi are still essentially valid.

Consultant's Obligations

Safety with economy of construction is the obligation of the consulting engineer to his client. Inherent in this obligation is a thorough but not overly extensive subsurface investigation, the extent of which depends upon the type, purpose, and importance of the proposed structure; the geologic conditions; and the failure risk. It is the consultant's responsibility to know when more extensive analysis will lead to construction savings that exceeded the extra study costs required for additional data accumulation. Added construction costs do not necessarily mean safer construction; they can also signify designing for unknown factors because of a lack of thorough study.

Book Scope and Purpose

In scope, this volume and the companion volume [Hunt (1984)[2]] have attempted to follow the pattern of Terzaghi in stressing the importance of geology in geotechnical engineering. The companion volume presents the aspects of the geological environment of significance in engineering works. This volume presents the application of theory and empirical approaches to resolve problems in soil and rock mechanics for various types of construction while relating solutions to geologic conditions. It must be recognized that geotechnical engineering remains very much an art, as well as a science, and there still are many limitations to the applications of the theories and concepts that have been developed. The application of experience and judgment is the most important element of any geotechnical study.

A major purpose of this volume and the companion volume is to serve as reference manuals covering the significant aspects requiring consideration and evaluation for the successful completion of engineered development and construction as related to the geologic environment. Information is presented either to enable solutions to problems to be determined or to locate the solution through source references.

Units

This book is written primarily in the English system of units for the American practitioner, although the metric system popular in South America and Europe and the International system or SI growing in use in Europe and Canada are also included in various figures, tables, and examples. Practitioners now must be familiar with all three systems.

The English system remains important because of the wealth of information on soil properties available from investigations, back analysis of failures, etc., which are essential for preliminary evaluations and for checking the reasonableness of new data obtained for a particular problem. Conversion from English to metric requires knowing only a few simple relationships, and for forces and pressures merely remembering that 1 tsf essentially equals 1kg/cm^2, but units such as psi, ksf, and psf are confusing to those who deal only in the metric system. Most of the world, however, generally is familiar with the English system. The metric and SI systems are essentially the same except for densities, forces, and pressures for which the conversion factors are not simple numbers.

The author prefers the metric system since it is based on a system of 10's and one is less apt to make mistakes in computations. For example, 1 kg/cm^2 equals 10 t/m^2, the density of water is 1 t/m^3 and that of the average soil is 1.8 t/m^3 (112 pcf/62.8), and it is a simple matter to determine overburden pressures and pore-water pressures and relate them to foundation pressures. Conversion tables for the three systems are given in Appendix A.

1.1.3 HISTORICAL NOTE

Early Concepts

For many centuries knowledgeable persons involved with building have been aware that geologic conditions must be considered in order to realize a structure that will not settle, tilt, or collapse. Early construction was based on the builder's experience, and it was not until 1776 when Coulomb presented his earth-pressure theories that analytical methods were introduced. Working with retaining walls, he demonstrated that when a wall is permitted to tilt, an unstable wedge of earth develops behind the wall, bounded by the wall and a failure surface. He related the early pressures imposed on the wall to the weight of the mass in the wedge and the shearing resistance along the failure surface given in terms of internal friction. In 1871, Otto Mohr introduced a general strength theory of rupture for ideal plastic materials, which, combined with Coulomb's work, developed into the well-known expression for geologic materials:

$$s = c + \sigma_n \tan \phi \qquad (1.1)$$

Soil and Rock Mechanics and Geotechnical Engineering

It was not until the work of Terzaghi, however, that *soil mechanics* became recognized as a major discipline of civil engineering. He used the term, along with the term *engineering geology*, in his pioneer work *Endbaumechanik auf bodenphysikalischer Grundlage* published in 1925,[3] and indeed, considered the boundaries between the two disciplines "to be rather artificial." International acknowledgment of the discipline arrived with the *First International Conference on Soil Mechanics and Foundation Engineering* held at Harvard University in June 1936. A major milestone was the publication in 1943 of Terzaghi's *Theoretical Soil Mechanics*,[4] which remains today as an important reference work. This was followed by *Soil Mechanics in Engineering Practice* [Terzaghi and Peck (1948)[5]] and *Fundamentals of Soil Mechanics* [Taylor (1948)[6]] and subsequently many works by other investigators.

Rock mechanics was a discipline developed and practiced for many years by engineers and geologists in the mining and petroleum industries. It did not become recognized as a part of civil engineering until the early 1960s, achieving general recognition at the *First Congress of the International Society for Rock Mechanics* held in Lisbon in 1966. The term *geotechnique* was launched with the publication of that name by the Institution of Civil Engineers of Great Britain in 1948, and although used in Europe and South America for many years, it was not recognized by the American Society of Civil Engineers until

1974, when it adopted *geotechnical engineering* as a division title. The term is now generally accepted to include soil mechanics, foundation engineering, and rock mechanics, and implies a strong dependency on engineering geology (or geological engineering).

The broad scope of the numerous contributions of Terzaghi to geotechnical engineering is truly astounding; many of them remain the basis today for the practice of many elements of this particular art and science. Some of his most significant contributions include the concept of effective stress (1936)[7] and the theory of consolidation (1943),[4] both of which he recognized and published on during the early 1920s, and developments in the concept of the modulus of subgrade reaction (1955).[8] He developed criteria for the support of rock tunnels based on empirical evaluations of rock quality (1946)[9] and was one of the first practitioners to recognize the significance of seafloor instability (1956).[10] These publications, representing a very few of the 278 which he produced, demonstrate the wide range of his interest and contributions.

1.2 THE GEOLOGIC INTERFACE

1.2.1 THE COMPANION VOLUME

Scope

The geologic environment, as it pertains to development and construction, is described in the companion volume [Hunt (1984)[2]] which covers the techniques of exploration, measurement of properties, instrumentation, rock types and rock masses, soil types and soil formations, surface and subsurface water, and geologic hazards (slope failures, ground subsidence, collapse and heave, and earthquakes).

Rock Masses

It is important for the engineer to have a knowledge of the various rock types in terms of petrography, origin, classification, and the characteristic landforms developed from weathering, as well as their characteristic structures as deposited or formed. Processes such as tectonic activity and weathering alter the structure and physical properties resulting in folding, faulting, jointing, and general decomposition and other forms of alteration, providing them with new characteristics completely different from those resulting from original formation.

From the engineering viewpoint, the result is the *in situ rock mass*, consisting generally of intact blocks of various mineral compositions, sizes, and degrees of weathering, divided by planes or zones representing weaknesses identified variously as joints, faults, foliation shears, bedding, etc. The various rock types, when fresh and unweathered, have characteristic properties upon which initial judgments can be based, but these properties are very much affected by in situ conditions, primarily fracturing, decomposition, and groundwater.

Soil Formations

Rock masses disintegrate and decompose under weathering processes resulting in residual soils or loose particles. These are transported by various agents including gravity, water, wind, and glaciers to form new deposits. Soils can be classed by origin as residual, colluvial, alluvial, eolian, and glacial, and subclassed by mode of deposition such as streambed, lacustrine, estuarine, coastal, marine, dune, and loess. Climate plays a major role in the development of the various formations, just as it does in the decomposition of rock masses.

This system of soil classification is important since soils of similar origins and modes of deposition or occurrence have engineering characteristics of compressibility, strength, and permeability that are very similar regardless of geographic location and, in many cases, regardless of climate. The identification of the class of soil formations, therefore, provides information on which to base preliminary judgments regarding the problems to be anticipated for design and construction.

Surface and Subsurface Water

From the engineering viewpoint, the significance of surface water lies in flooding, erosion, and water supply. Subsurface water impacts on excavations and retaining structures, as well as on the physical properties of soils and rocks, and provides a source of water supply. Pollution control and control of groundwater in excavations and slopes and through and around water-retention structures are important elements of engineering studies.

Geologic Hazards

Several events, referred to as the "geologic hazards," occur periodically; some result in loss of life and all cause costly damage to engineering works. Many are natural, but some are the result of human activities. In addition to flooding and erosion, these include slope movements, ground subsidence and collapse, and earthquakes and volcanos. The professional involved in development and construction activity should be aware of these hazards and the natural conditions under which they are likely to occur.

Some, such as earthquakes, ground subsidence, and collapse, are limited to particular geographic locations, whereas others, such as slope failures and flooding, can occur anywhere that physiographic conditions are suitable. The activities of some are preventable, those of many are not, while still others can be provided for in development planning. It is prudent to avoid development in an area with a high risk of occurrences of catastrophic proportions, since predictions of them are often of low reliability.

1.2.2 GEOLOGIC ACCIDENTS: NATURAL OCCURRENCE

In any region where development is proposed, the geologic accidents that may possibly occur as natural events must be identified, the potential risk in terms of the probability of occurrence and the magnitude of the event must be assessed, and the feasibility of providing adequate protection must be evaluated. Where potentially catastrophic consequences may result, the areas of possible effect should be avoided. These may include floodplains, areas on or near a fault zone or on or near a slope, areas adjacent to volcanos, ground over shallow caverns in limestone, or areas exposed to severe storm erosion.

Floods

Floods result from heavy runoff from rainfall and melting snows, the bursting of natural dams formed by landslides, and flood tides and tsunamis along shorelines. The incidence of occurrence is related directly to topographic and climatic conditions. Flood levels and occurrence on land are predicted on the basis of analysis of existing rainfall and runoff data, which in geologic time are often of relatively short duration, and predictions may or may not be reliable, depending upon the adequacy of the data. Predictions based on the determination of the geologic limits of the natural floodplain may be more reliable in many situations.

Erosion

Erosion results from runoff, stream activity, and coastal wave and current forces. It is a major factor in decreasing slope stability and in the siltation of water bodies when it occurs in upland areas. The incidence of occurrence on land is related to vegetation, topography, climate and geology, and the susceptibility to erosion is apparent from an examination of these features, but the actual occurrence of severe erosion depends on storm intensity and the resulting activity of flowing water for a given set of conditions.

Slope Failures

Landslides and other slope failures are the natural result of decomposition of geologic materials, steepening of slopes by erosion, and alteration of inclinations by tectonic movements, earthquakes, rainfall, and snowmelt. The incidence of occurrence is related primarily to topography, geology, and climate. Seafloor mass movements result from earthquakes, hurricane wave forces, and sedimentation loadings.

Certain geologic conditions have a probability of susceptibility to *landsliding*, but natural failure depends for the most part on ground saturation and rainfall intensity. Therefore, actual prediction of failure can be uncertain, even when small movements are already occurring. The probability of occurrence of *other forms* of mass movements (i.e., falls, avalanches, and flows) is often difficult to establish, although the possibility of failure can be identified on the basis of geologic, topographic, and climatic conditions. In many instances, slope failures, as well as erosion and flooding, are neither preventable nor controllable, even with the expenditure of very large sums of money.

Ground Subsidence and Collapse

The occurrence of ground subsidence and collapse as natural events generally is limited to cavernous rocks such as limestone. Climate is the major factor. Cavern growth by solution from groundwater proceeds extremely slowly, and collapse from solution occurs infrequently. If a region with cavernous rock undergoes a long dry period causing a substantial lowering of the groundwater level, however, overburden pressures on the cavern roof are substantially increased and the incidence of collapse becomes high. Collapse of a cavern may result only in a surface depression, but on occasion causes a very large hole many meters across. The potential for collapse is readily predictable, but not the time.

Earthquakes

Earthquakes have a great effect on the land, resulting in warping, faulting, and ground shaking. In addition to the effect on structures, ground shaking may result in liquefaction and subsidence of certain soil types. The effect on bodies of water is the creation of tsumanis and seiches. The incidence of earthquakes and tsunamis is characteristically geographic.

Earthquake occurrence has some degree of predictability based on historic records, and in recent years through the analysis of certain geophysical data, but when quakes will occur and their magnitude are not predictable. They are controllable only in terms of ground and structure response. The liquefaction hazard can be reduced, and structures can be designed to withstand dynamic forces. The magnitude of such forces and their characteristics are predictable only in a general manner, and never with complete certainty. Safety is provided by conservatism in design and the prudent location of important structures.

Volcanos

Volcanic activity can result in disastrous lava flows and enormous dust clouds. The precipitation of the latter can alter or destroy natural vegetation and result in changes in the geologic environment and can result in extensive formations of expansive clays. The incidence of volcanic eruption is essentially unpredictable, but the possibility of the occurrence is usually obvious based on topographic features and historic records.

1.2.3 GEOLOGIC ACCIDENTS: CAUSED BY HUMAN ACTIVITIES

All of the hazards described in the foregoing article that result in natural geologic accidents also can be the result of human activity, many with a higher rate of occurrence than would occur naturally. It is tragic that there is not a greater awareness of the potential impact of human activities on natural conditions, an impact which commonly results in flooding, erosion, landsliding, ground subsidence and collapse, and to a lesser extent even earthquakes.

Floods

Humans expose themselves to flooding by construction in the floodplain. They often attempt to provide protection by placing fill in the floodplain to create higher ground, safe from inundation, but the result of placing the fill and the additional fill that soon follows is to reduce the natural storage capacity of the valley and to raise the flood level.

A very significant cause of flooding in many areas is the depletion of groundwater, oil, or gas which results in *land subsidence* over large areas, often measurable in meters. Coastal areas and interior valleys in arid climates in particular are susceptible. Solutions to the problem are expensive, involving either a prohibition of withdrawal or extensive dike protection at times combined with pumping. Subsidence from oil and gas extraction may be arrested by water injection while the mineral products are being withdrawn.

Erosion and Siltation

The incidence of erosion is increased substantially by stripping natural vegetation from slopes to create farm or grazing lands or by grading and steepening slopes, thereby increasing runoff velocity. Increased erosion means increased siltation in water bodies with deleterious effects on marine life and decreased flood storage capacity.

Groundwater Pollution

Groundwater, a valuable natural resource, can become polluted from disposal of liquid or solid wastes improperly treated either in processing or in ground storage or from saltwater intrusion caused by overpumping of an adjacent freshwater body.

Slope Failures

Slope failures, when they damage developments, are the result primarily of cuts made in the slope, although the removal of vegetation, the placement of hillside fills, and changing the natural drainage to cause water to enter the slope all are contributing factors. The predictability of failure, which depends primarily on precipitation, is as difficult as with natural slopes, but any changes in natural slope conditions requires consideration of measures to increase stability to preclude failure.

Ground Subsidence and Collapse

Ground subsidence and collapse are frequently the results of human activities. Subsidence, the gradual but perceptible lowering of the ground surface, can result in flooding as mentioned, in faulting where large areas are affected, or in structural damage where the subsiding area is relatively small but differential deflections are large. It can result from a number of activities.

In the coal industry, mining large areas underground using the longwall technique can result in a gradual subsidence as the mine opening closes. When the room-and-pillar technique is used, collapse of the mine and the resulting surface subsidence usually occur many years after the mine is closed. Weathering and spalling cause a decrease in pillar strength and eventual rupture of one pillar followed by that of other pillars in sequence under increased roof loads. Local subsidence results from saturation of collapsible soils such as loess or valley alluvium in an arid climate, dewatering for excavations, tunneling in soft ground, and excessive deflection of retained excavation walls. Regional subsidence occurs as a result of fluid extraction as discussed under Floods.

The major cause of ground collapse resulting in holes at the surface is groundwater withdrawal in limestone or other soluble rocks, which not only accelerates solution but by lowering the water table increases overburden pressures on the cavern roofs resulting in raveling of soils into the cavern or in roof collapse. In the subsidence and collapse phenomenon, groundwater withdrawal in soluble rocks is the most dangerous condition.

Earthquakes and Faulting

Earthquakes can be induced by deep-well injection and possibly by fluid extraction. The failure of the Baldwin Hills Reservoir has been postulated to have been caused by displacement along a fault beneath the embankment resulting from oil extraction from the nearby Inglewood oil field [Jansen et al. (1967)[11]]. Earthquakes also are known to be induced by loadings from reservoirs, generally behind dams greater than 100 m in height. The weight of the reservoir and the high pore pressures in the underlying rock masses may cause displacements along buried rock structures resulting in an earthquake. Groundwater extraction can result in surface faults called "growth faults," common to the Houston, Texas, area, for example.

1.2.4 CONSTRUCTION FAILURES: COMMON CAUSES

Once the major geologic hazards in a particular location have been identified and evaluated and the site found to be suitable, there follows the necessity of gathering detailed information for the formulation of design and construction plans. If the information is inadequate or improperly interpreted, construction delays and extra costs may be incurred, and poor performance of structures will result which may lead to damage or even collapse. Various types of construction have their own particular sensitivities to geologic conditions which should be kept in mind during investigation.

Surface Structures

Surface structures on shallow foundations undergo excessive settlements when the compressibility of the soil or rock support is underestimated. Undetected buried strata of weak soils are a common cause of undesirable building settlements. Surface structures and pavements

FIG. 1.1 Wall cracking in warehouse, Queens, New York, 1960. Walls and columns are pile-supported; lightly loaded floors are fill-supported. Soil conditions of miscellaneous fill over thick organic soils are subjecting piles to both downdrag and high lateral forces.

FIG. 1.2 Rupture of corner column of building in Fig. 1.1.

must be designed in consideration of the effects of ground heave from frost action or swelling soils, the latter condition a major cause of distortion in structures. Structures imposing high live loads, such as grain-storage silos or storage tanks, can suffer failure by ground rupture, if not properly assessed. Deep foundations commonly suffer excessive settlements because of either inadequate penetration or the neglect of downdrag forces from negative skin friction during analysis (Figs. 1.1 and 1.2).

Slopes

Failures of cut slopes occur when the slope is cut too steep for the geologic conditions, and retention and internal drainage control are inadequate (Fig. 1.3). Investigation may not disclose a potential weakness plane, such as a clay or mylonite seam or a relict slickensided surface, dipping and daylighting downslope. Many slopes reside naturally at a stability with a safety factor near unity, and any increase in the slope angle by cutting can be expected to decrease stability to an unsafe condition for a given set of climatic and groundwater conditions.

Embankments

Embankment failure can take the form of slope failure or foundation failure. Embankments for roadways or other facilities where it is necessary to raise grade most commonly fail by rupture when placed over soft ground in lifts that are too thick and when placed too rapidly for the shear strength available in the foundation soils (Fig. 1.4). Earth dam embankments fail most commonly from overtopping due to inadequate emergency spillway capacity (determined by hydrological studies), from toe seepage, and from piping through the embankment or the foundation. Embankment design to provide stability and control seepage must be tailored to the borrow materials available for construction. Abutment and foundation seepage control relates to an adequate assessment of the geology.

The foundation stability of concrete dams is related to rock-mass conditions including the orientation of major weakness planes as related to the direction of thrust, the stability against overturning, and the potential for high seepage pressures to develop.

Surface Retaining Structures

Failure of surface retaining structures for excavations or waterfront development usually occurs because of improper assessment of earth pressures or because of improper construction techniques. These techniques are important in all works, but in retaining structures a small

FIG. 1.3 Landslide and structural failure of 6-m-high cantilever wall, BR 101, Mangaratiba, R.J., Brazil. Single row of anchors penetrating into rock stabilized the base. Residual soil slope rises over 100 m. Failure occurred in 1974, and as of 1981 stabilization was still under construction with a series of walls installed at various levels along the slope.

FIG. 1.4 Mud waves and lateral displacement of large area at the toe of a failed 20-m-high fill placed over soft soils. Failure occurred after a period of heavy rains which increased fill weight. Mud waves extend for over 100 m and occurred in a relatively shallow deposit of soft soils overlying strong residual soils.

change in construction procedure can have a much greater detrimental effect on stability than it will in the case of an earth or concrete dam, for example. Retaining structures may also fail due to inadequate penetration below the excavation bottom, which can permit vertical deflection of the wall, or bottom heave or uplift, depending upon geologic conditions. Inadequate penetration of waterfront structures can permit removal of support by scour.

Tunnels

Tunnel failures normally occur during construction when unforeseen geologic conditions are encountered such as running, squeezing, or swelling ground. Designs normally provide for adequate final support and relatively few tunnels perform poorly after construction. During construction, however, the practices employed for the geologic conditions encountered are the basis for good tunneling performance as measured by costs, schedules, and the protection of workers' lives.

1.3 LIMITATIONS OF THE TECHNOLOGY

1.3.1 BASIC FACTORS

Problem Categories

Geotechnical problems may be divided into three general categories related to physical properties: deformation without rupture, failure by rupture, and groundwater movement and seepage pressure. The latter condition is also significant in deformation and rupture.

Deformation

Deformations in geologic materials can be elastic (all strain is recoverable upon removal of applied stress), plastic (strain is nonrecoverable), a combination of both, or occasionally viscous. They can take the form of compression, expansion, or shear in soil or rock or bending in rock. Deformations are evaluated on the basis of elastic theory or elastic strain concepts, the theory of consolidation, or empirical relationships. The magnitude and distribution of stresses from applied loads which result in deformations are usually determined on the basis of elastic theory.

Strength

Under certain conditions of applied and confining stresses, failure by rupture occurs when the shear strength of a material is exceeded. Shear strength is not a definitive value for a given material, but rather is influenced by a number of factors. Relationships among these factors have been established by a number of theories and tests. Shear-strength relationships also form the basis for earth-pressure theories. In all cases, pore pressures resulting from groundwater conditions play a significant role.

Permeability

The movement of water through the subsurface is described by Darcy's law which is based on specific and limiting conditions.

Project Execution

The objectives of a geotechnical investigation are to define the response of the geologic materials in terms of deformation, strength, and groundwater conditions to the new conditions of stresses and other factors to be imposed by the construction. The successful execution of a project requires an understanding not only of the limitations of the theories and concepts available for analysis but also of the limitations to defining the elements of the geologic environment and the engineering properties of the materials.

The necessity for experience and the application of mature judgment cannot be overemphasized. When the uncertainties are recognized and properly evaluated they are provided for by adequate margins of safety.

1.3.2 GEOLOGIC MATERIALS AND THEORY

Ideal Materials

Geotechnics is often referred to as both art and science. *Art* refers to the necessity of applying substantial judgment based on practical experience to most problems in geotechnical engineering because the theoretical concepts upon which the *science* is based are by and large developed for ideal materials.

In a geomechanical sense an ideal material has the characteristics of isotropy, homogeneity, and continuity *(see Art. 4.1.2 for definitions)*. It may also be elastic or plastic. Hard, crystalline rock masses, free of discontinuities, have properties that are elastic or nearly so. Other geologic materials may have elastic properties under low stress levels, but in general combinations of elastic and plastic properties are common. Besides certain rock masses, ideal geologic materials can be considered to include uniform sand deposits and some alluvial clay deposits. These materials, although not elastic, can be essentially isotropic, homogeneous, and continuous.

Nonideal Materials

Deposits and formations of geologic materials are variable and irregular, and the presence of an ideal material to any significant extent is the exception rather than the rule. As mentioned, crystalline rocks free of defects may approach an elastic material, even though many types are not truly homogeneous. Fracturing causes all rock masses to be discontinuous. The detrital sedimentary rocks are commonly anisotropic, as are foliated metamorphic rocks. Decomposition from weathering produces anisotropy and nonhomogeneity in all rock types.

Residual soils are generally anisotropic and nonhomogeneous. Alluvial deposits are characteristically stratified and, therefore, strongly anisotropic, although particular beds may be quite homogeneous and even isotropic. Eolian deposits are characteristically anisotropic but homogeneous.

Complicating the development of geologic profiles from those originally formed or deposited are the continuous changes brought about by erosion, deposition, solution, cementation, tectonic activity, and the natural processes of prestressing. The result in a given location can be a profile highly variable in depth and lateral extent.

Groundwater Flow

The movement of water through geologic materials is defined by Darcy's law in terms of the coefficient of permeability, valid only for the case of laminar flow of an incompressible fluid through a saturated medium. Air or gas trapped in the material, or conditions of turbulent flow, invalidate the conditions for validity. Laminar flow occurs in soils ranging from uniform coarse sands to clays, and in some fractured rock masses where openings are not overly large. Turbulent flow occurs in very coarse sands and gravels and in fractured rock with large openings. The coefficient of permeability also is very much affected by anisotropy and layering, usually resulting in values in the horizontal direction higher than those in the vertical direction.

1.3.3 FIELD AND LABORATORY INVESTIGATION

Geologic Sections

The initial objective for a given project is to determine the representative stratigraphy and groundwater conditions beneath the site, presented in the form of geologic sections supported by other data such as test boring logs or geophysical sections. Most explorations, such as test borings (Fig. 1.5), use of the cone penetrometer (Fig. 1.14), or test pits, provide data only at a point, whereas geophysical methods, test trenches, and adits provide data along a line. Geophysical data are remotely sensed and subject to many limitations, but are useful for interpretations between borings or other data bases. Test trenches and adits are relatively costly but are necessary in certain situations such as the use of trenches to explore fault zones or the use of adits to define rock conditions for a concrete dam abutment.

In many cases, several exploration methods are employed to obtain a data base, and then experience and judgment are applied to extrapolate and interpret the data to obtain geologic sections truly representative of subsurface conditions. Three-dimensional sections, as illustrated by the example given as Fig. 5.2, are extremely useful in complex conditions. Adequate interpretation requires an understanding of geologic processes, and even then all pertinent features may not be disclosed.

FIG. 1.5 The rotary drill rig used for the SPT, undisturbed sampling, and rock coring hasn't changed much in 30 years.

Property Measurements

Engineering properties are determined indirectly from correlations with empirical tests, directly by laboratory tests on specimens retrieved undisturbed from test borings or pits, or directly from tests made in situ.

Indirect Tests

The standard penetration test (SPT) and the cone penetrometer test (CPT) provide data from which correlations are made to obtain estimates of strength and compressibility. Although theoretically more versatile than the SPT, because both point resistance and shaft friction are measured during testing, the CPT is limited to relatively weak materials and does not provide a sample of the materials for examination. "Representative" samples are recovered from the SPT (Fig. 1.6), but the test is purely empirical and re-

FIG. 1.6 Split barrel or spoon sample with thinly layered silts and silty sands after an SPT. The basis for most foundations designed in the United States and South America is the SPT.

sults are very much influenced by operational procedures and equipment conditions. Wide variations in values in a given soil formation are common, and the data must be applied with sound judgment and an understanding of the test limitations.

Laboratory Testing

In the laboratory, engineering properties are measured by tests performed on samples retrieved from the field. Materials from which undisturbed samples can be obtained in test borings generally are limited to soft to stiff intact clays and sound to intact rock. Mixtures with gravel particles (residual soils, glacial till), strong soils such as fissured clays, and decomposed and fractured rock are usually very difficult to sample undisturbed in test borings. In strong cohesive soils, undisturbed block samples (Fig. 1.7) are retrieved from test pits. Truly undisturbed samples of cohesionless sands and silts can hardly be obtained under any conditions. Laboratory tests on these materials are performed on specimens fabricated at the in situ density.

In Situ Testing

Testing in situ also involves many limitations affecting the adequacy of the data obtained. In rock masses, excavation and test preparation can change natural conditions, and the test may not stress a representative mass, as is often the case with plate-jacking tests, for example. Pressuremeter testing in soil and rock normally stresses the tested mass in a direction perpendicular to the direction of applied foundation stresses, as does vane shear testing in soils. Results from these tests, therefore, are affected strongly by material anisotropy.

Because of scale effects, the only tests yielding truly representative data are full-scale load tests applied to in situ conditions, and even these generally are not representative of changes that occur as a long-term function of time, such as changing groundwater conditions or time-related deformations. Determination in situ of the permeability coefficient is very much affected by layering and other aspects of ground stratigraphy, as well as the degree of saturation and the grain size distribution of the materials.

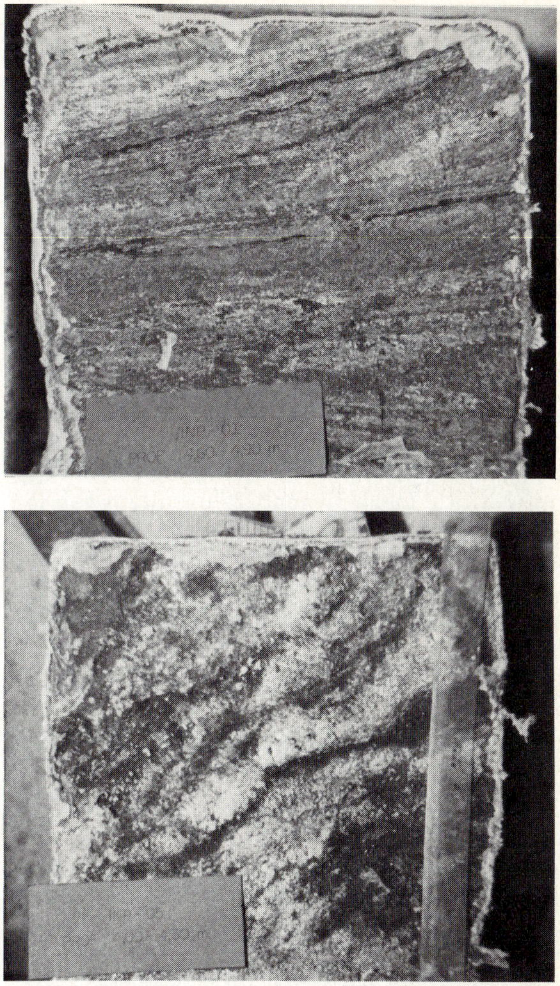

FIG. 1.7 Block samples, 30 × 30 × 30 cm, of saprolite derived from gneiss, São Paulo state, Brazil. The test pit was stopped at a depth of 5 m because facilities were not available for shoring the walls or for penetrating a soft rock layer, but the formation extends to depths of over 20 m. The stability of a 40-m-high cut to be made at this location will be very much affected by the weakness planes.

1.3.4 ANALYTICAL LIMITATIONS

Strength and Rupture

Limiting Equilibrium

Problems involving the evaluation of failure by rupture, such as for foundations, slopes and retaining structures, are commonly analyzed by limit equilibrium methods. At limiting equilibrium the forces acting to cause failure are considered to be in balance with the forces acting to resist failure. The basic considerations common to all limit equilibrium analyses are the location and shape of the potential critical-failure surface; the magnitude of the forces acting upon it to cause failure, which can include the weight of the overlying earth mass, seepage forces, and imposed loads; and the magnitude of the shearing resistance available along the failure surface.

This approach does not permit an assessment of the stress distribution in the ground nor of the deformations that occur. It also assumes that shear strength is mobilized simultaneously at every point along the failure surface and does not consider that failure is often progressive. The most critical factor in analysis usually is the selection of representative values for strength, a procedure which has its difficulties as indicated in the preceding article. Limiting equilibrium analysis is valid for many problems, however, because it usually defines the condition at which collapse will occur, yielding a conservative evaluation to which safety factors are applied for design.

Critical Factors

In all analysis of failure by rupture the most critical factor is the selection of representative values for material strength, which is affected by a large number of variables. For conditions of static loadings, these include magnitude, direction and rate of load application, seepage forces, presence of existing failure surfaces or surfaces along which some past displacement has occurred, distribution of the normal stress along the failure surface, and confining stresses. An additional factor can be the effect of repetitive or cyclic loads from earthquakes.

Deformation

Imposed Stresses

Most practical problems involving foundations for structures are concerned with deformation that occurs under stresses less than those producing rupture, in the form of either compression or expansion of geologic materials. The evaluation of the compression that will occur under an applied load requires first the determination of the magnitude of stresses that will be imposed. Stress distributions are described on the basis of elastic theory considering geologic materials as ideal, elastic media, although this is seldom true *(see Art. 1.3.2)*.

General theories have been developed for homogeneous, elastic, isotropic bodies; for thinly layered masses; for rigid strata over weaker ones; for rigid strata underlying weaker ones; or for single-pile foundations. The distribution of stresses associated with a rigid foundation as compared to a flexible one, as well as those associated with a pile group, are not well defined. Stress distribution in discontinuous rock masses has been approached only in a very general empirical manner, and on important projects strains are measured by instrumentation from which stress distributions are estimated.

Settlement Analysis

The rate of compression of saturated, intact (nonfissured clays) is analyzed on the basis of the *theory of consolidation*, which assumes that Darcy's law is valid as pore pressures dissipate under an applied load and compression occurs. Analysis provides a reasonably accurate measurement of the time rate of consolidation for certain materials. Some clays, and especially clays with organic matter, however, undergo a large amount of secondary compression during which settlement occurs but pore pressures have not entirely dissipated. Their behavior, therefore, is not within the concepts of consolidation theory. Analyses of time rates for these materials are considered to provide only gross estimates.

The total compression, which does not depend on the validity of Darcy's law, can be predicted reliably for many materials. These include cohesionless granular soils, strong granular cohesive materials with large particles, rock masses, and materials not fully saturated. In these materials the assessment of compressibility can be based, in a general manner, on deformation moduli assigned to the material. Moduli provide information only on the total amount of "short-term" deformation that will occur, and not on any time-related aspects, including creep.

Moduli often are selected to define the strain that will occur under a given stress condition in the nonelastic range. Elastic moduli have validity only when applied to the elastic deformation range, which for most geologic materials exists only at very low stress levels. As mentioned in Art. 1.3.3, obtaining moduli values from laboratory or in situ tests that are truly representative is often difficult because of problems inherent in procuring undisturbed samples, in disturbing the ground in preparation for testing, in the size of the tested area as compared to the actual area stressed by the structure, or because of anisotropic effects.

Soil-Structure Interaction

Another group of deformation problems involves soil-structure interaction. Mat and continuous foundations, flexible retaining structures, and laterally loaded piles may be designed on the basis of the moments and shears that develop in the structure from deflections resulting from soil deformation. The deformations of soil and structure are interrelated.

Some solutions are based on the beam- or plate-on-an-elastic subgrade theory, in which soil deformations are defined either in terms of a modulus of subgrade reaction or an elastic modulus. The validity of solutions depends on how representative are the values of the load-deformation characteristics of the soils. Solutions for flexible retaining structures generally are based on either theoretical or empirical distributions of earth pressures which vary with soil type and strength properties.

Foundations supported on soils and subjected to dynamic forces from earthquakes are also soil-structure interaction problems *(see Art. 1.3.5)*.

Finite Element Method

The finite element method *(see Art. 1.4.3)* is finding wide application to problems dealing with the evaluations of deformations resulting from changing stress conditions as in the case of soil-structure interaction for static and dynamic forces, in rock masses supporting concrete dam foundations, or in open or closed excavations in rock masses. Time-deformation relationships can be analyzed but in all cases the critical aspect of validity is representative definition of material properties and the geometry of the geologic regime, usually a difficult and expensive task, especially in rock masses.

Groundwater Movement and Seepage Forces

The flow of groundwater through geologic materials can be represented by flow nets from which pore-water pressures, seepage forces, and flow quantities can be determined, but the validity of the solution is based on the applicability of Darcy's law, and on employing representative values for the permeability coefficient for all significant strata. The application of flow-net analysis in stratified soils can be somewhat complex, and can be done in rock masses only in a general manner. Flow-net analysis, when properly performed, yields reliable results and is particularly important for problems involving earth dams, concrete dam foundations, slopes, rigid retaining walls, and open excavations.

1.3.5 PRACTICAL SITUATIONS

Foundations

Design Criteria

Foundations for structures are placed in two general categories: shallow and deep. In all cases the objective is to provide support to a structure which will restrict settlement to within tolerable limits and will avoid failure by ground rupture. For shallow foundations, as shown in Fig. 1.8, the approach is to evaluate soil compressibility or expandability and to select an allowable bearing value based on foundation depth and shape. In cases where high live loads will be applied, the stability against shear failure is important.

In deep pile or pier foundations, the approach is to determine the ultimate capacity for each unit on the basis of soil strength by considering both end-bearing and shaft friction, and then to apply a safety factor to obtain a supporting value that will limit deflections. In some cases settlement analysis is performed for pile groups, particularly where they will gain support from or will impose loads on compressible clays.

Shallow Foundations

The compressibility of many geologic materials, including sands, residual soils, decomposed rock, and fractured rock masses, can be evalu-

FIG. 1.8 Shallow square footings bearing in a very still clayey residual soil.

ated only in a general manner since undisturbed sampling is either difficult or impossible. On routine projects in all soils except clays, the usual procedure is to select an allowable bearing value based on either the SPT or the CPT. Penetration tests in clays are not considered to provide reliable data upon which to base estimates of settlement.

Where refined estimates of settlement are required for materials from which samples cannot be obtained in the undisturbed state suitable for laboratory testing, analysis of compression is based on soil moduli as discussed in Art. 1.3.4. The variations from reality of estimates of settlement will depend on the variations from elastic theory, both in terms of physical properties of the tested materials, and the distribution of the stresses imposed by the structure. The most reliable basis for estimating deformations from compression of these materials is from expensive full-scale load tests that bracket all significant conditions. Testing must stress the same zone, to the same magnitude or higher, and with the same load orientation as the proposed foundation, to be meaningful. Moduli only provide estimates of essentially instantaneous settlement under a given stress condition and do not provide information on time factors.

The SPT and other in situ tests to measure moduli are not suitable for soft to stiff intact saturated clays since compression in these materials is very much a time-related phenomenon. Analysis is performed on the basis of the theory of consolidation. In organic materials the magnitude and time rate can be assessed with some degree of accuracy only by instrumented large-scale load tests.

Foundation design in expansive materials is based first on identifying their presence and potential swell pressures and activity. Then, if the potential for expansion of detrimental magnitude is judged to exist, designs are prepared either to isolate the foundations and structures from the expansive or active zone or to permit deflections without excessive structural distortion.

Deep Foundations

Deep foundations consisting of either driven or bored piles acquire their supporting capacity from end bearing, shaft friction, or both; caissons obtain their capacity primarily or totally in end bearing. Depending upon their spacing, deep foundations may act either as single units or as a group.

Theory provides the basis for estimating the capacity in end bearing or in shaft friction for single units from shear strength, but empirical relationships are employed since several factors are not well understood. It has been found, for example, that values for end bearing and shaft friction in sands increase to a particular depth with respect to overburden pressure, termed the "critical depth" (generally considered to be 10 to 20 pile diameters), beyond which the effective stress adjacent to the pile is not necessarily equal to the effective overburden pressures. In addition, piles deriving support from both end bearing and shaft friction are difficult to analyze with precision because pile settlement is required to mobilize shaft friction and the percentage of available friction is related directly to the amount of settlement occurring. The problems of downdrag on single piles from negative shaft friction and deflection under laterally applied loads are relatively well understood, although long-term creep deformations are not well defined.

Another area where substantial data are lacking is that of values for shaft friction for bored piles in strong soils and decomposed or fractured rock. Such values become critical where deep penetration is required for high-capacity piles supported by poor-quality rock where end-bearing values are relatively low. In any event, it is possible to perform load tests on single units to confirm an allowable capacity that will restrict deflections to desirable limits. In very long piles elastic compression of the unit itself must be considered.

Rigorous methods are not available for analysis of pile action in a group, either under vertical or lateral loadings. The efficiency of a pile in a group depends on several factors including the type of material providing support to the pile, the installation method (bored, driven, jacked), whether or not the pile cap is bearing on the surface, and pile spacing. Since driving closely spaced piles in sand densifies the materials, group capacities greater than the sum of the capacities of single piles are possible, whereas in clays affected by remolding, group capacity may be only two-thirds of the sum of the capacities of the single piles. Evaluation of group capacities usually is based on empirical relationships since load tests on groups, and particularly on large groups, is a relatively costly procedure.

Settlement analyses and general stability analyses are performed, for pile groups supported in or stressing a clay stratum, by the equivalent pier method which is essentially an empirical concept. Stresses and deformations resulting from foundations placed at some significant depth below the surface are not well understood because of the effects of confinement on soil properties and stress distribution.

Slope Stability

Analytical Factors

The analysis of the stability of a natural or cut slope, as approached by the limit equilibrium method, requires the determination of the shape and location of the potential failure surface and the shearing resistance that will be mobilized along it. The potential failure surface is readily determined in uniform sand or normally consolidated clays, and in stratified soil or rock formations where a stratum or discontinuity represents an obvious weakness plane.

It is assumed that the available shear strength will be mobilized simultaneously along the entire failure surface. The critical factors are the selection of representative strength parameters and the assumptions for the maximum seepage pressures which may prevail at some future time. Seepage pressures are difficult to assess with certainty as they depend on ground saturation and rainfall conditions. They may serve to reduce the normal stress directly and, therefore, the frictional component of strength resisting rupture, and they are the primary cause of slope failures. Other factors which change slope conditions with time and are difficult to assess are erosion or tectonic activity causing a slope to steepen, decomposition weakening slope materials, and dynamic loadings from earthquakes.

If a cut is to be made, evaluations are performed to determine the stable slope angle, which in high cuts is obtained by benching (as shown in Fig. 1.9), and drainage. If stability is in question, then some form of retention is required.

Nonanalyzable Conditions

Many forms of slope movement cannot be analyzed in the present state of the art, including most types of falls, avalanches, flows, and progressive slides, although the potential for their occurrence can be recognized. Avalanches and flows occur suddenly and can begin at any point on a slope. Progressive slides, common to residual soils and stiff, fissured clays, and failure by lateral spreading such as may occur in glacial marine and lacustrine soils, also cannot be analyzed since the location of the critical failure surface where rupture would begin cannot be determined.

FIG. 1.9 Slope stabilization by benching in residual soils and decomposed rock at the beginning of the excavation for a 300-m-deep open-pit mine.

Earth Dams and Embankments

Because earth dams are fabricated structures, the conditions of geometry and strength are relatively well defined compared with natural slopes and stability analysis is performed with greater confidence. Instrumentation provides information on pore pressures during and after construction. Embankments over soft ground usually are analyzed as stability problems but can be evaluated as bearing-capacity problems. The critical factors are the selection of representative strength parameters and construction procedures related to magnitudes and time rates of loading. Measurements of undrained strength are affected by a number of factors which present difficulties in selecting a truly representative value.

Retaining Structures

Rigid and Semirigid Walls

Retaining structures are required to support cut slopes and open excavations if space is not available for a stable cut slope or for reasons of economy. Designs usually are based on earth pressures. For problems involving rigid *gravity walls*, earth pressures can be determined with reasonable accuracy based on theory as long as soil properties are accurately assessed and seepage forces accounted for. An important element in the analysis is the assumption as to whether or not the wall will yield, since wall movement directly affects the magnitude of pressures mobilized.

Anchored curtain walls, such as those illustrated in Fig. 1.10, are usually considered as semirigid walls with movements very much restricted by the tiebacks. Designs are often based on stability analysis rather than earth pressures, and their performance is very much reliant on good construction practices, particularly with regard to anchor installation. The control of subsidence of adjacent ground and the settlement of nearby structures is a necessary consideration.

Flexible Walls

Flexible structures often are used to retain open excavations (Fig. 1.11) or waterfronts. In flexible-wall design the objectives are to find the maximum moment in the wall resulting from the earth-pressure forces, permitting the selection of an adequately stiff wall member, and to determine the forces on the wall support system (struts, anchors, or braces). An additional objective often is to restrict wall movements to avoid back-slope subsidence, as mentioned above.

FIG. 1.10 Slope stabilization with anchored panels (left) and retention with an anchored curtain wall of a 20-m-deep excavation for a 50-story office building, Rio de Janeiro.

FIG. 1.11 Anchored sheet-pile wall for 13-m-deep excavation for the cooling-water intake structure of a nuclear power plant. Irregular heights of the sheets reflect the variation in the rock surface.

The basis for the computations of earth pressure is the classical theories, but the total pressures and their distributions are modified by semiempirical or empirical relationships. The classical or Rankine distributions have been found to be reasonably valid for cantilever walls but not totally applicable in other cases. Anchored bulkheads are evaluated on the basis of the Rankine distributions with empirical modifications, and strut-braced or anchored flexible walls are evaluated based entirely on empirical relationships. In general, studies have shown that wall flexibility causes earth pressures to redistribute and vary significantly from the classical theories.

In designing temporary retaining structures, the unknowns are provided for with adequate safety factors and by adding contingency supports during construction if wall deflections become excessive. The latter provision requires accurate monitoring of wall movements during construction. The assessment of the unknown factors is more critical for permanent structures, such as sheet-pile bulkheads. In this case it has been found that earth pressures in soils with grain sizes smaller than coarse sand can increase with time. Arching may temporarily or permanently reduce the maximum bending moment and result in a significant increase in anchor stresses. These effects are avoided by using backfill as coarse as or coarser than coarse sand. Regardless of the design assumptions, the performance of flexible retaining structures is very much dependent on construction procedures which can change the earth-pressure distribution from that assumed for design.

Backslope Subsidence

Backslope subsidence cannot be assessed by classical methods and traditionally has been estimated on the basis of observational data. Diaphragm walls and continuous sheet-pile walls in soft clays, where backslope subsidence can be a serious problem, have been analyzed by the finite element method to estimate wall deflections and associated subsidence. The analysis incorporates values of soil moduli as well as models of the excavation sequence. The selection of soil moduli to represent various levels of shear stress for various forms of behavior is critical in obtaining realistic results.

Underground Openings

General Considerations

Most underground openings (tunnels, caverns, mines, etc.) require some form of support during and after construction to retain the opening and, at times, to restrict subsidence of the overlying ground surface. Tunnel entrances (Fig. 1.12) often present slope stability problems.

The nature of underground construction is such that it is virtually impossible to determine all geologic conditions in advance of construction with present exploration techniques. Pilot excavations are costly but important, but even with them all significant conditions may not be disclosed. Experienced geologic interpretation is the usual basis for prediction of conditions hazardous to underground construction, particu-

FIG. 1.12 Entrance to cooling-water discharge tunnel for nuclear power plant. Slope is stabilized with a series of low, anchored curtain walls.

larly in rock masses. Construction techniques are the critical elements in the satisfactory performance of underground works.

Important factors that require assessment are material hardness as it pertains to removal and advance, the "stand-up" time of the materials which relates to the amount of tunnel advance possible before placing supports, and the development of earth pressures that require temporary and permanent support. Stand-up time can be assessed by empirical relationships only in a very general manner.

Soil Formations

Earth pressures in tunnels are a problem in soil-structure interaction and cannot be determined precisely by analytical methods. The performance depends on the rigidity of the tunnel lining and soil types, but the pressure distribution is markedly affected during tunneling and installation of the lining. Earth pressures on tunnels during construction may be only 20% of overburden in clays but may increase with time to full overburden pressure and can exceed overburden pressures if swelling occurs. In sands, where arching is important, earth pressures can be much less than overburden and may not increase significantly with time.

Deformation around a tunnel and surface subsidence, as well as loads on linings, can be evaluated by theoretical analysis of the states of stress and strain near a circular cavity employing the finite element method. The validity of such an analysis, however, is limited by the assumptions made concerning soil behavior and tunnel deformation, and can only to a limited extent account for construction methods and procedures and the rate of advance of the heading. Subsidence potential often is estimated on the basis of empirical data, but does not account for accidents caused by blowouts, collapse, or running ground.

Rock Masses

Refined analytical techniques are difficult to apply with certainty to rock masses because of the problems associated with determining accurately all of the geologic parameters including the geometry, the strength, and other properties of the blocks and discontinuities. In addition, the problem is usually three-dimensional and time-dependent. Support determinations for rock masses normally are based on relationships developed from experience.

Underground excavations in rock frequently employ instrumentation to monitor stress-strain relationships and time-dependent deformations. The major factors to be determined and evaluated are rock-mass quality and groundwater conditions. Particularly troublesome conditions are caused by running, squeezing, or swelling ground.

Earthquake Forces

Significance

The significance of earthquake forces with respect to development and construction can be placed into two general categories, the effect on the land and water bodies and the effect on structures [see Chap. 11 of Hunt (1984)[2]]. The dynamic forces impose cyclic loadings which can cause the development of high pore pressures and liquefaction of certain soil types resulting in failures of slopes or level ground, or general ground subsidence with or without liquefaction. Ground warping and fault displacements can occur prior to, during, or following an earthquake. Ground shaking invokes a dynamic response in structures as the seismic waves cause vertical and horizontal displacements in periodic motion. Ground-motion elements of most significance are the horizontal and vertical components of gravity, the frequency content or period, and duration. These factors vary with earthquake magnitude, site distance from the source, local topography, and geologic conditions.

Potential Force Assessment

There are many limitations on an accurate assessment of earthquake forces. Relatively well investigated and established, or definable, are the responses of various soil types and structures to dynamic loadings with various characteristics. Unpredictable, with a high degree of uncertainty, in the present state of the art, are the occurrence and magnitude of a possible earthquake and the characteristics of the ground motion which it may induce at a given location.

The current approach is to perform a statistical evaluation of existing records for a given locale (which in geologic time usually represents a very small sample), select a design earthquake with a characteristic acceleration and frequency content, and then attenuate it through some distance to the site, with the source usually identified as an active or potentially active fault (a "capable" fault). Earthquake magnitudes have been related to length of fault break, and this relationship may be used to increase a potential magnitude over that in the historical records. The limitation in this approach is the determination that a fault is or has been recently active (movement within the past 35,000 years is one criterion generally used to define a "capable" fault), or is considered as a dead or noncapable fault. Determination of fault activity is a difficult and costly procedure, and faults considered "dead" have been found with alarming frequency in recent years to actually be active or the cause of significant earthquakes.

Soil-Structure Interaction

The "design earthquake," determined as described above, is applicable to motions generated in the basement rock. If the structure is supported on soils, the effect of the soils between the bedrock and the structure requires evaluation. Since ground shaking causes structure motion which can feed back into the soil, the problem is one of soil-structure interaction and is relatively complex. Most soils tend to amplify the rock motion, at times by a factor of 2 or 3, and the amplification must be considered in analysis. The excitation from the underlying rock basically applies shearing stresses to the soils. The soil-structure interaction problem has been evaluated by halfspace analysis, but in recent years the approach is to use finite element analyses that account for soil properties including the variation of the shear modulus and material damping with strain. The application of either technique requires substantial judgment since there are limitations inherent in both, as well as uncertainties in determining soil properties and ground-motion characteristics.

Slopes and Embankments

For many years slopes and embankments have been evaluated with pseudostatic analysis in which a horizontal earthquake force is included in the standard limit equilibrium stability analysis. The procedure has been found unreliable in many cases and is now used more often as a first approximation. Dynamic analysis, similar to soil-structure interaction analysis, is considered to yield more reliable results but still contains uncertainties.

1.4 NEW TECHNOLOGY

1.4.1 PROGRESS

During the early development stages of a science, major breakthroughs occur periodically as new theories and concepts arrive to fill major gaps in knowledge, as happened, for example, with the theory of consolidation in soil mechanics. As the science matures with the slowly acquired experience gained from application, however, the accumulation of knowledge is seldom augmented with some astounding new development, but instead grows gradually step by step. Such has been the case with geotechniques over the past three or four decades. The concepts developed in the early years still remain by and large the basis for analysis and design used today, although in some cases their rate and scope of application have broadened. For example, the application of deformation moduli for evaluating compression is finding wide use in recent years, yet Terzaghi dealt with its application in detail in his *Theoretical Soil Mechanics*, published in 1943.[4]

The most significant development in analytical methods over the past decade or so has been associated with computer technology and the finite element method. Laboratory testing techniques have undergone relatively little change except for the measurements of dynamic properties and some modification in strength testing techniques in the triaxial apparatus. Most of the new developments in the field and laboratory are associated with exploration, in situ testing, and instrumentation. The last development not only provides basic data required for analytical evaluations, but also provides the means for monitoring and confirming the conclusions drawn concerning ground and structural response to applied loads.

1.4.2 EXPLORATION, IN SITU TESTING AND INSTRUMENTATION

In the broad sense exploration is performed to define the geologic environment and to obtain samples for laboratory testing. It is also closely allied with in situ testing to measure material properties as they exist in the ground. Instrumentation can be installed to measure in situ properties to obtain data for analysis or to monitor changes induced by construction. Exploration and instrumentation are described in detail in Hunt (1984)[2] as are property measurements, which are also discussed in subsequent chapters in this volume. The significant new developments are summarized here. Many are not truly new to the past decade or so, but their application is being recognized as a standard procedure. This phase of a geotechnical program can be considered as the most important, since the results of rigorous analysis are only as good as the input.

Geotechnical investigation can be divided into four general areas: determination of the surficial geology and interpretation of subsurface conditions; exploration to define the stratigraphy and to recover samples of the materials for identification and laboratory property testing; in situ measurements of properties; and instrumentation to obtain property data and to monitor changing conditions of deformation and stress.

Engineering Geology Mapping

In a preliminary study phase, the interpretation of remote-sensing imagery and the preparation of engineering geology maps (Fig. 1.13) illustrating surficial and shallow surficial conditions is

FIG. 1.13 Engineering geology map, prepared from interpretation of remote-sensing imagery and shown here plotted on an aerial photo, depicts shallow subsurface geology, in this case for a community about to receive a sanitary sewer system. Scale 1:8000.

becoming a standard practice. The availability of satellite and side-looking airborne radar (SLAR) imagery of high quality has permitted the accurate interpretation of geologic conditions over very large areas, leading, for example, to the identification of many new lineaments, often indicative of faulting. The sequential procurement of satellite imagery over a period of many years may yield data on subtle changes in the earth's surface on a large scale.

FIG. 1.14 Truck-mounted CPT rig jacked up to provide full reaction of 15 tons. (Courtesy of Geo-Insitu, São Paulo, Brazil.)

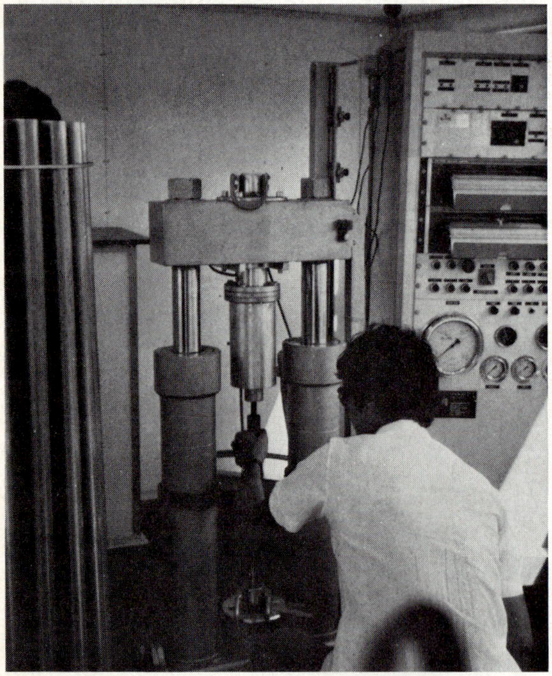

FIG. 1.15 Interior of cab of Fig. 1.14 showing CPT jacking device and readout unit for point resistance and shaft friction versus depth. (Courtesy of Geo-Insitu, São Paulo, Brazil.)

Stratigraphic Sectioning and Sampling

Subsurface stratigraphic sectioning and sampling are still carried out primarily with test borings, but the CPT (Figs. 1.14 and 1.15) and geophysical methods are finding wider application, generally as methods to augment test boring information. The most significant new test boring development is the application of wireline drilling techniques, particularly for offshore exploration (Figs. 1.16 and 1.17). Using this technique, it is possible to recover rock cores as well as drive and push soil samples and to perform CPTs in hundreds of meters of water at penetrations below the seafloor of hundreds of meters. The recovery of core specimens in fractured rock during land operations has been improved by integral coring techniques, and the development of borehole cameras permits the examination of the walls of boreholes and other underground openings.

Geophysical surveys are commonly used offshore, usually including side-scan sonar and high-resolution seismic reflection surveys. The latter detect faults, slide scars, and other features, as well as pictorializing stratigraphy. Strata depths and physical properties, however, can be determined accurately only from test borings, the CPT, or seismic refraction surveys. Seismic refraction surveys commonly are employed on land to define stratigraphy and to aid in material identification through propagation velocity measurements. A very new development showing some promise, especially for the detection of underground voids and cavities, is profiling with radar techniques.

In Situ Property Measurements

The measurement of engineering properties in situ is a rapidly growing trend, particularly in materials difficult to sample undisturbed. Most are used in boreholes. The gamma-gamma probe is used to measure natural density, and the neu-

FIG. 1.16 Wireline drill rig mounted on ship for offshore foundation investigations.

FIG. 1.17 Undisturbed sample being retrieved with a ship-mounted wireline drill rig operating in 100 m of water and drilling to penetrate to 130 m below the seafloor.

tron probe is used to measure natural water content. Crosshole seismic methods are used to measure direct wave velocities from which dynamic deformation moduli are obtained for use in seismic response studies and for the evaluation of rock quality. The pressuremeter is used for the measurement of deformation moduli and limiting strength values, and the Cankometer, a self-boring pressuremeter, has been designed to measure shear modulus, shear strength, pore pressure, and lateral stress. Strain meters and stress cells are installed in boreholes in rock masses to provide measurements of properties such as residual stress.

To measure earthquake ground and structural response an increasingly large network of accelerographs is being installed in the United States and other countries. These strong-motion networks will provide data for more definitive evaluations of the effects of elastic waves on structures and the ground.

Instrumentation

Perhaps the major change in exploration technology has been the recognition of the importance of instrumentation to monitor changes in the ground occurring naturally or as a result of construction activity. In light of the limitations of the theories and concepts of geotechnical engineering, designs can be prepared to the limit of safety (often a fictitious number applied to compensate for the limitations of knowledge), and then the adequacy of the design checked by instrumentation in terms of deformations, stresses, and loads.

Precise surveying, in a relatively small time interval, of slope movements and other deformations, is possible with the laser geodimeter. Mechanical and electrical extensometers are installed to monitor deflections and strains under a wide range of conditions including settlement of structures, ground compression, movements downslope and across faults, movements in rock slopes and around tunnel openings, and longitudinal strains in earth dams and between the dam and the abutments. Inclinometers measure lateral movements in soils, and deflectometers measure lateral movements in rock masses. The acoustical emission device shows promise for monitoring changing stress conditions in any deforming body and forewarning of possible catastrophic events such as collapse of a slope or a mine, movement along faults, and leakage in dams. In situ pressures and stresses are measured with piezometers, stress and load cells, strain gauges and meters, and stress meters. With improvements in electronics and the miniaturization of equipment, the development of new instrumentation has almost unlimited horizons.

1.4.3 ANALYTICAL TECHNIQUES

A Review

The more significant changes in analytical techniques have resulted from the development of computer technology, most importantly in soil-structure interaction problems and the application of the finite element method. Static soil-structure interaction problems solved on the basis of beam- or plate-on-an-elastic-subgrade theory include continuous and mat foundations, flexible sheet-pile and diaphragm-type retaining walls, and laterally loaded piles. In the solution, soil properties are defined in terms of moduli established for various amounts of deflection, and the corresponding moments, shears, and deflections that develop in the structure under applied load are determined either by difference equations or the finite element method. Interaction between soil and structure also occurs when a heavy structure is subjected to dynamic forces, and is evaluated by the FEM as described previously.

Other significant developments include soil-suction theory as applied to partially saturated soils such as expansive clays and soils in slopes, load-transfer mechanisms for piles in soils and socketed in rock, and piles for offshore structures under cyclic, lateral loads from sea waves.

The Finite Element Method (FEM)

Applications

The most important advance in the analytical aspects of geotechnical engineering is the application of the finite element method (FEM). It is used for the evaluation of displacement, force, strain, or stress fields starting from any initial boundary force or displacement field. In the analysis, anisotropic conditions, as well as stress and time-dependent constitutive relationships and construction sequence, can be considered.

In soil mechanics the technique is used primarily in soil-structure interaction problems as mentioned in the preceding paragraph. In rock mechanics problems it has been used to evaluate stresses and deformations beneath concrete dam foundations, in slopes, and in rock with residual stresses and to evaluate the deformation and potential for collapse of underground openings including tunnels, caverns, and mines. The FEM is a very powerful analytical tool, but the results are only as accurate as the assumptions made regarding the geometry of the soil or rock formations and the physical properties assigned to the materials.

Description of Method

The objective of the finite element method is the numerical solution of problems in the mechanics of "continua," or uninterrupted structures. In civil engineering, complex structures are analyzed by analogues. A truss, for example, is idealized as a number of pin-connected, axially loaded straight members composing an assembly of elements interconnected at nodal points. The behavior of the assembly is assumed to be a function of the force-displacement characteristics of the individual elements.

In the FEM, analogous to the truss problem, a continuum is subdivided into a discrete number of elements, interconnected at a finite number of nodal points. At each point imaginary forces, equal in magnitude to some integral function of the stresses along element boundaries, are assumed to act. These forces are related to nodal displacement by means of specific stiffness matrices. This coupling allows the evaluation of displacement, force, strain, or stress fields resulting from any initial boundary force or displacement field. Mathematically, the method is somewhat similar to matrix analysis of structures. Since it is assumed that the material behaves as a continuum, however, the FEM is *not suitable* for modeling progressive failure or the generation of failure surfaces.

Material Properties Input

The stiffness, or stress-strain matrix, is accounted for by material properties relating either normal stresses and normal strains or shear stresses and shear strains. In a *linear elastic* model, the terms of the stiffness matrix are defined by any two elastic constants, normally Young's modulus and Poisson's ratio. It is usually more precise to consider the materials to be *nonlinear* and to use elastic constants which vary as functions of stress or strain level, such as the tangent or secant modulus. Drnevich (1975)[12] describes the use of the constrained modulus and the shear modulus and cites the relatively simple definitions obtained for all elements of the matrix, as well as the major advantage that the compression and shear moduli are relatively independent of the relationships of elasticity. When nonelastic moduli are used, however, a knowledge of in situ stress conditions is required. Initial stresses and likely relationships between the coefficient of lateral earth pressure (K_o) and the other soil parameters are important.

1.4.4 CONSTRUCTION TECHNIQUES

New construction techniques have been developed mainly in the areas of site preparation and ground improvement; deep foundations for offshore structures; the use of soil slurry in trenches, walls, and tunnels; and the use of tunnel boring machines (TBMs). As was the case with a number of important exploration methods, many of the new developments in construction have been used for many years in Europe and Japan before finding acceptance in the United States.

Ground Improvement

New techniques in ground improvement for the support of structures are numerous including surcharging in connection with wick drains, dynamic compaction, ground reinforcing with columns formed in situ, and geotextiles. *Wick drains* are rapidly replacing sand drains as a means of decreasing surcharge time. Composed of filter materials contained on rolls, the wicks are threaded rapidly into the ground, usually without jetting. Economies are realized over the use of sand drains due to the rapid installation without the use of water or the necessity for supplying sand of a narrow gradation.

Dynamic compaction is a relatively simple operation performed by a standard crane dropping a heavy weight from a substantial height to impact on the ground surface. It has been used to compact in situ various poor soils including fills of silty clay, dredged sands, and random materials, in many cases underlain by weak organic soils, to achieve a degree of support suitable for lightly loaded structures on shallow foundations.

Sites with poor soils are also being prepared for support of relatively light structures by reinforcing the ground with *columns formed in situ* of stone, sand, or stabilized soil. The methods are particularly useful for underwater sites or locations with very shallow groundwater. *Geotextiles* are finding a wide range of uses in ground reinforcing, as beneath roadway embankments over soft ground, in embankments and other fabricated slopes, in foundation fills and reinforced earth walls, and in culvert bedding and cover, in addition to being used as filter materials in drains or as impermeable liners.

Rollcrete is a low-cement-content aggregate mixture prepared in a batch plant, hauled to the site by earthmoving equipment, and compacted in layers by vibratory rollers in an operation similar to fabricating a compacted earth fill. It is being used to construct dams as an alternate to rock fill and mass concrete and as backfill in large excavations for important structures. Strengths have been found to be higher than those of mass concrete with equivalent cement content and placed conventionally.

Deep Foundations

Significant innovations have been incorporated in the installation of piles by driving and piles for offshore structures. *Dynamic testing* during pile driving provides a means for evaluating pile capacity, hammer performance, driving stresses and pile damage during installation and furnishes data for estimating soil strength parameters and distribution of side friction and end bearing. It is rapidly becoming standard procedure for offshore structures where static load tests are extremely costly and difficult to perform.

Piles for offshore platforms commonly are of large diameter and thick walled, often driven to penetrations of several hundred feet to achieve axial capacities from 1000 to 4000 tons. *Hammers* with extremely high-rated energies and hammers suitable for driving underwater have been developed to achieve required penetrations. An *insert* or *grouted pile*, a combination of a driven and a bored pile and grouting, has been developed for conditions where the required penetration cannot be achieved by driving, as in tropical waters where cemented strata are interbedded with weak soils. Such piles also find application on land in similar subsurface conditions where high capacities are required and rock is at prohibitive depths.

Open and Closed Excavations

Anchored concrete curtain walls are increasing in application for the retention of hillside cut slopes and fills and open excavations. High capacities and good control of backslope subsidence are possible, and open excavations remain unencumbered by bracing systems. Excavation walls are also being constructed with the *slurry trench technique* which provides excellent groundwater control during construction. Displacing the slurry with tremie concrete after a reinforcing cage has been installed and anchoring it with tiebacks during excavation provide for a permanent wall of high quality. Additional strength is provided for deep walls by the installation of soldier piles during the tremie operation [the soldier-pile-tremie-concrete (SPTC) wall].

Tunneling operations in saturated granular soils and in soft clays have very much improved with the use of the *slurry shield* or *slurry-face tunnel boring machine (TBM)*, and in soils above the water table or with relatively low permeabilities below the water table with the *earth pressure balance (EPB) shield*. In addition to relatively high advance rates, surface subsidence from tunneling is effectively minimized. In rock masses, tunneling is also being achieved with increasing frequency with TBMs or with the "new Austrian tunneling method" which employs shotcrete and rock bolts as a support system.

1.5 SYNOPSIS

The preceding discussion provides an overview of the significance of geology in geotechnical engineering and the present state of the art. The companion volume [Hunt (1984)[2]] covers the elements of geology of importance to the engineer including historical and structural geology, geomorphological relationships for various rock and soil formations, descriptions and classifications of rock masses and soil formations, the occurrence of surface and subsurface water, and the geologic hazards—slope failures, ground subsidence and collapse, and earthquakes.

Throughout the companion volume geologic conditions are related to their engineering significance, including descriptions of engineering classification systems for rock masses and soil formations, the characteristic engineering properties of the various classes and types of geologic materials, and the engineering aspects of surface and subsurface water. To make discussions on the geologic environment more meaningful as it is related to geotechnical engineering, it was necessary to introduce early in the volume the engineering properties of strength, deformation, and permeability. A knowledge of these properties is necessary for an understanding of the engineering implications of slope stability, ground subsidence and collapse, and earthquakes. Because the major tie between engineering geology and geotechnical engineering is through exploration, this subject was covered in detail.

Following this chapter, this volume covers permeability, strength, and deformation of geologic materials and then presents the application of these properties to site preparation and ground improvement; shallow and deep foundations; slope and embankment stability analysis; retaining structures for slopes, open excavations, and waterfront structures; and underground openings such as tunnels and mines.

The approach throughout these two volumes is to treat soil and rock mechanics together since, although the material properties may very much differ in general, there is a good degree of overlap, and the analytical concepts are very similar. Earth dams and pavements are not treated as separate chapter headings, but the important elements considered in their analysis and design are covered in the chapters dealing with permeability, site preparation, ground improvement, and slopes and embankments. Soil and rock dynamics and earthquake loadings also are covered in various chapters.

REFERENCES

1. Terzaghi, K. (1963) Discussion of "An Approach to Rock Mechanics," by K. John, 1962, *Proc. ASCE, J. Soil Mech. and Found. Engrg. Div.*, Vol. 89, No. SM1, pp. 295–300.
2. Hunt, R. E. (1984) *Geotechnical Engineering Investigation Manual*, McGraw-Hill Book Co., New York.
3. Terzaghi, K. (1925) *Erdbaumechanik auf bodenphysikalischer Grundlage*, Deuticke, Vienna.
4. Terzaghi, K. (1943) *Theoretical Soil Mechanics*, John Wiley & Sons, New York.
5. Terzaghi, K. and Peck, R. B. (1948) *Soil Mechanics in Engineering Practice*, John Wiley & Sons, New York.
6. Taylor, D. W. (1948) *Fundamentals of Soil Mechanics*, John Wiley & Sons, New York.
7. Terzaghi, K. (1936) "The Shearing Resistance of Saturated Soils and the Angle Between Planes of Shear," *Proc. 1st Intl. Conf. Soil Mechs. and Found. Engrg.*, Cambridge, Mass., Vol. 1, pp. 54–56.
8. Terzaghi, K. (1955) "Evaluations of Coefficients of Subgrade Reaction," *Geotechnique*, London, Vol. 5, No. 4, pp. 297–326.
9. Terzaghi, K. (1946) "Rock Defects and Loads on Tunnel Supports," Sec. 1 in *Rock Tunneling With Steel Supports*, Proctor and White, eds., Commercial Shearing and Stamping Co., Youngstown, Ohio, pp. 17–99.
10. Terzaghi, K. (1956) "Varieties of Submarine Slope Failures," *Proc. Texas Conf. on Soil Mechs. and Found. Engrg.*, No. 8, Austin, Tex.
11. Jansen, R. B., Dukleth, G. W., Gordon, B. B., James, L. B., and Shields, C. E. (1967) "Earth Movement at Baldwin Hills Reservoir," *Proc. ASCE, J. Soil Mechs. and Found. Engrg. Div.*, Vol. 93, No. SM4, July.
12. Drnevich, V. P. (1975) "Constrained and Shear Moduli for Finite Elements," *Proc. ASCE, J. Geotech. Engrg. Div.*, Vol. 101, No. GT5, May, pp. 459–473.

BIBLIOGRAPHY

Casagrande, A. (1964) "Role of the Calculated Risk in Earthwork and Foundation Engineering," *Terzaghi Lectures, 1963–1972*, ASCE, New York, 1974, pp. 72–138.

Clough, G. W. (1974) "Finite Element Analysis of Ground Support Systems," Lecture presented at New York Met. Section, ASCE, Found. and Soil Mech. Sem.

Deere, D. U. (1975) "General Review," Session 6—Summary and Discussion, *Proc. 16th Symp. on Rock Mechs., Design Methods in Rock Mechanics*, ASCE, New York, 1977.

Legget, R. F. (1979) "Geology and Geotechnical Engineering," *Proc. ASCE, J. Geotech. Engrg. Div.*, Vol. 105, No. GT3, March, pp. 342–391.

Peck, R. B. (1974) "Conference Overview," *Proc. ASCE, Conf. on Analysis and Design in Geotech. Engrg.*, Austin, Tex., Vol. II, pp. 127–134.

Peck, R. B. (1977) "Case Histories in Soil Mechanics," *Proc. 9th Intl. Conf. Soil Mech. and Found. Engrg.*, Tokyo, Case History Vol., published by The Japanese Soc. of Soil Mechs. and Found. Engrg., 1981, p. 1–14.

Seed, H. B. (1974) "The Role of Computers in Analysis and Design," *Proc. ASCE, Conf. on Analysis and Design in Geotech. Engrg.*, Austin, Tex., Vol. II, p. 135.

Terzaghi, R. D. and Voight, B. (1979) "Karl Terzaghi on Rock Slides: The Perspective of Half a Century," *Rock Slides and Avalanches*, Vol. 2, *Engineering Sites*, B. Voight, ed., Elsevier Scientific Pub. Co., Amsterdam, Chap. 2.

Wroth, C. P. (1976) "Review of Earth Pressure Theory and Earth Pressure Behavior," *Geotechnical Lecture Series on Lateral Earth Pressure*, Boston Soc. Civ. Engrs.

PART I

ENGINEERING PROPERTIES OF GEOLOGIC MATERIALS

CHAPTERS

PURPOSE AND SCOPE

Part I covers the engineering properties of soil and rock, including both static and dynamic response to applied forces. The engineering properties are covered in detail in the companion volume [Hunt (1984)[1]] from the aspects of laboratory and field measurements of quantities. Sections on hydraulic properties are broader in scope in the companion volume, covering flow systems and analysis and groundwater and seepage control, in addition to testing.

The chapter on Basic, Index, and Hydraulic Properties is given in Part I as a general summary of the companion volume to provide a ready reference to terminology and relationships as they pertain to the practical problems presented in later parts of this volume. The chapters on Strength and Deformation are approached more from the aspects of theory and application of properties to problem solving than from the aspects of testing.

CHAPTER TWO
BASIC, INDEX, AND HYDRAULIC PROPERTIES

2.1 INTRODUCTION

The engineering properties of geologic materials are considered to include the hydraulic and mechanical properties (strength and deformation), as well as the basic and index properties. Laboratory and field methods of measurement of the engineering properties are covered in detail in the companion volume [Hunt (1984)[1]] and are presented in this and the subsequent two chapters more from the perspective of application than from testing.

Basic properties include the fundamental characteristics of geologic materials and are used primarily for identification and correlations between similar materials regarding response to some construction activity. Some are used in engineering calculations.

Index properties define certain physical characteristics used basically for classifications, but also for correlations with engineering properties.

Hydraulic properties, expressed in terms of permeability, pore pressures, or seepage forces, are engineering properties. They involve the flow of fluids through geologic media.

The classification and description of rocks and soils are presented in detail in the companion volume.

2.2 BASIC AND INDEX PROPERTIES

2.2.1 GENERAL

A summary of the laboratory tests performed to obtain measurements of the basic and index properties, applications of the property measurements, and references covering the performance of the tests are given in Table 2.1. In situ testing methods are summarized in Table 2.2.

Basic and index test data are required for material identification and classification and as input for engineering analysis. In addition, correlations based on basic and index properties with data obtained from other investigations in which extensive testing was performed or engineering properties were determined by back analysis of failures provide data for preliminary studies and analysis as well as a check on the reasonableness of data obtained during investigation.

2.2.2 INTACT ROCK SPECIMENS

Intact specimens of fresh to slightly weathered rock, free of major weakness planes representing discontinuities, are normally tested in the laboratory to obtain measurements of volume-weight relationships, hardness, durability, reactivity, strength, and sonic velocities.

TABLE 2.1
BASIC AND INDEX PROPERTIES: LABORATORY TESTS

Material	Property/test	Applications	Reference
Intact rock specimens	Basic properties	Correlations, analysis	
	Specific gravity	Mineral identification	
	Porosity	Property correlations	ISRM (1981)[2]
	Density	Material and property correlations	
		Engineering analysis	ISRM (1981)[2]
	Hardness	Material correlations	ISRM (1981)[2]
		Tunneling machine excavation evaluation	
	Durability		ASTM C131
	LA abrasion	Evaluation of construction aggregate quality	
	British crushing		
	Slaking	Weathering proneness	ISRM (1981)[2]
	Reactivity	Reaction between cement and aggregate	ASTM C227, C289, C295
	Sonic velocities	Computations of dynamic properties	ASTM D2845
	Index properties	Classification/correlations	
	Uniaxial compression	See rupture strength	ASTM D2938
	Point-load test	See rupture strength	Broch and Franklin (1972)[3]
Soils	Basic properties	Correlations, classification	
	Specific gravity	Material identification	ASTM D854
		Void ratio computation	
	Moisture or water content	Material correlations in the natural state	ASTM D2216
		Computations of dry density	
		Computations of Atterberg limits	
	Density: natural (unit weight)	Material correlations	
		Engineering analysis	
	Density: maximum	Relative density computations	ASTM D2049
		Moisture-density relationships	
	Density: minimum	Relative density computations	ASTM D2049
	Optimum-moisture density	Moisture-density relationships for field compaction control	ASTM D698, D1557
	Sonic velocities	Computations of dynamic properties	*See Art. 3.3.5*
	Index properties	Correlations, classification	
	Gradation	Material classification	
		Property correlations	ASTM D422
	Liquid limit	Computation of plasticity index	
		Material classification	ASTM D423
		Property correlations	
	Plastic limit	Computation of plasticity index	
		Field identifications	ASTM D424
	Shrinkage limit	Material correlations	
	Organic content	Material classification	ASTM D427 Arman (1970)[4]

Volume-Weight Relationships

Included in volume-weight relationships are specific gravity (SG or G_s), density (ρ or γ), and porosity (n) as defined in Table 2.3. Bulk density normally refers to the field, or in situ, density.

The specific gravities of most rock-forming materials range from 2.65 to 2.8, although heavier minerals such as hornblende, augite, or hematite vary from 3 to 5 and higher [see Table 5.5 of Hunt (1984)[1]]. Porosity depends largely on rock

TABLE 2.2
BASIC AND INDEX PROPERTIES: IN SITU MEASUREMENTS

Material	Tool or method	Applications	Reference
Rock masses	Basic properties		
	Gamma-gamma borehole probe	Continuous measure of density	ISRM (1981)[2]
	Neutron borehole probe	Continuous measure of porosity	ISRM (1981)[2]
	Index properties		
	Rock coring	Measure the RQD (rock quality determination) used for various empirical correlations	Deere (1963)[5]
	Seismic refraction	Estimated rippability on the basis of P-wave velocities	Hunt (1981)[1]
Soil formations	Basic properties		
	Gamma-gamma borehole probe	Continuous measure of density	Crosby et al. (1981)[6]
	Neutron borehole probe	Continuous measure of moisture content	Crosby et al. (1981)[6]
	Sand-cone density apparatus	Measure surface density	ASTM D1556
	Nuclear density moisture meter	Surface measurements of density and moisture	ASTM D2922
	Balloon apparatus	Measure density at surface	ASTM D2167

TABLE 2.3
VOLUME-WEIGHT RELATIONSHIPS FOR INTACT ROCK SPECIMENS*

Property	Symbol	Definition	Expression	Units
Specific gravity (absolute)	G_s	The ratio of the unit weight of a pure mineral substance to the unit weight of water at 4°C. $\gamma_w = 1$ g/cm³ or 62.4 pcf.	$G_s = \gamma_m/\gamma_w$	
Specific gravity (apparent)	G_s	The specific gravity obtained from a mixture of minerals composing a rock specimen.	$G_s = \gamma_m/\gamma_w$	
Density	ρ or γ	Weight W per unit volume V of material.	$\rho = W/V$	t/m³
Bulk density	ρ	Density of rock specimen from field. (Also g/cm³, pcf.)	$\rho = W/V$	t/m³
Porosity	n	Ratio of pore or void volume V_v to total volume V_t.	$n = V_v/V_t$	%
		In terms of density and the apparent specific gravity.	$n = 1 - \rho/G_s$	%

*From Hunt (1984).[1] Reprinted with permission of McGraw-Hill Book Co.

origin. Slowly cooling igneous magma results in relatively nonporous rock, whereas rapid cooling associated with escaping gases yields a porous mass. In sedimentary rocks porosity depends on the amount of cementing materials present and on the size, grading, and packing of particles. Densities of fresh intact rock, for the most part, do not vary greatly unless they contain significant amounts of the heavier minerals. Typical values for density are given in Table 3.12. Permeability of intact rock is related to porosity, although normally the characteristics of the in situ rock govern rock-mass permeability. There are strong correlations among density, porosity, and strength.

Hardness

Hardness is the ability of a material to resist scratching or abrasion. Correlations can be made with density, uniaxial compressive strength, and sonic velocities, with the rate of advance for tunneling machines and other excavation methods, and with rock quality for foundation support. The predominant mineral in the rock specimen and the degree of weathering decomposition are the controlling factors. The Schmidt hammer rebound hardness test [ISRM (1981)[2]] provides very useful data.

Durability

Durability is the ability of a material to resist degradation by mechanical or chemical agents. It is the factor controlling the suitability of rock material used as aggregate for roadway base course or in asphalt or concrete. Slake durability provides an index to the potential for swelling, weakening, or disintegration when a rock is exposed to weathering processes in excavations. The predominant mineral in the specimen, the microfabric, and the degree of decomposition are controlling factors.

Reactivity: Cement-Aggregate

A reaction between soluble silica in the aggregate from crushed rock and the alkali hydroxides derived from portland cement can produce abnormal expansion and cracking of mortar and concrete, often with severely detrimental effects to pavements, foundations, and concrete dams. There is often a delay of about 2 to 3 years after construction, depending upon the aggregate type used, for deterioration to become apparent.

Sonic Velocities

Compression or P-wave velocities determined by ultrasonic testing in the laboratory are compared with field P-wave velocities obtained by seismic direct methods (such as crosshole tests) to obtain measures of rock quality in terms of the *velocity ratio* (V_F/V_L) or the *velocity index* $(V_F/V_L)^2$. The field velocities reflect the influence of rock-mass defects on the sonic waves; the closer the velocity ratio or velocity index approaches to unity, the higher is the rock quality as it approaches sound intact rock *(see Arts. 3.3.5 and 4.1.4)*. Shear-wave velocities are used to calculate the dynamic moduli of intact rock (Table 4.2).

Uniaxial Compression and Point-Load Strength

These strength values are used mainly for correlations and in rock classification systems [see Art. 5.2.7 in Hunt (1984)[1]].

2.2.3 ROCK MASSES

Properties of rock masses divided by systems of discontinuities into blocks ranging from fresh to decomposed rock are best measured in situ. Mass density is the basic property. Sonic-wave velocities and the rock-quality designation (RQD) are used as index properties to rock quality and for various correlations [Hunt (1984),[1] p. 217].

Rippability

Rippability refers to the ease of excavation by construction equipment. Since it is related to rock quality in terms of hardness and fracture density, which may be measured by seismic refraction surveys, correlations have been made between rippability and seismic P-wave velocities as given in Table 5.1. If the material is not rippable by a particular piece of equipment, then jackhammering or drilling and blasting are required.

2.2.4 SOILS

General

The basic and index properties of soils generally are considered to include volume-weight and

TABLE 2.4
VOLUME-WEIGHT RELATIONSHIPS FOR SOILS*

Property		Saturated sample (W_s, W_w, G_s are known)	Unsaturated sample (W_s, W_w, G_s, V are known)	Illustration of sample
Volume components	Volume of solids V_s	$\frac{W_s}{G_s\gamma_w}$†		
	Volume of water V_w	$\frac{W_w}{\gamma_w}$‡		
	Volume of air or gas V_a	zero	$V - (V_s + V_w)$	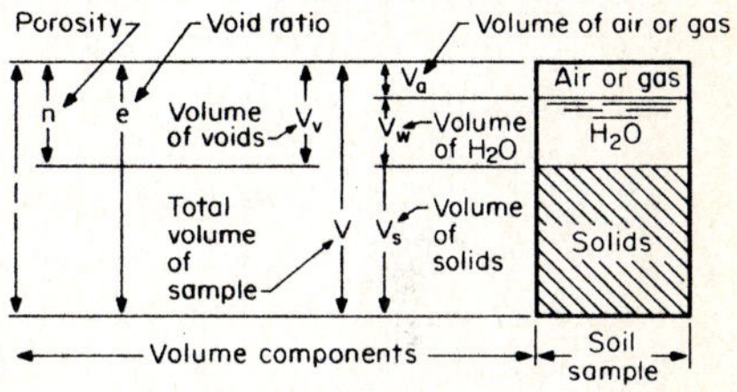
	Volume of voids V_v	$\frac{W_w}{\gamma_w}$‡	$V - \frac{W_s}{G_s\gamma_w}$	
	Total volume of sample V	$V_s + V_w$	Measured	
	Porosity n	$\frac{V_v}{V}$ or $\frac{e}{1+e}$		
	Void ratio e	$\frac{V_v}{V_s} = \frac{G_S\gamma_w}{\gamma_d} - 1$		$e = wG_s/100$ (saturated soil)
Weights for specific sample	Weight of solids W_s	Measured		
	Weight of water W_w	Measured		
	Total weight of sample W_t	$W_s + W_w$		
Weights for sample of unit volume	Dry-unit weight γ_d	$\frac{W_s}{V_s + V_w}$	$\frac{W_s}{V}$	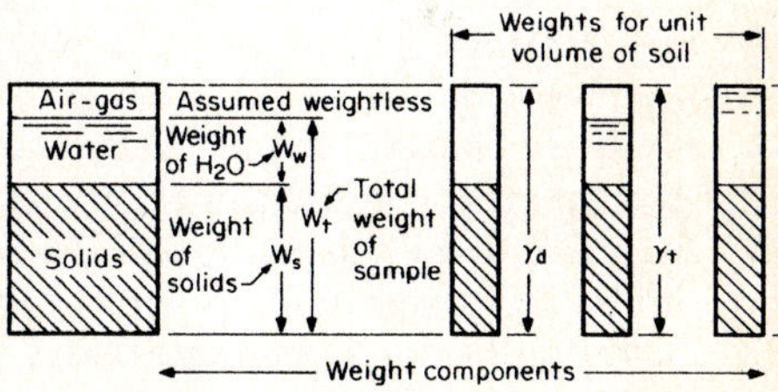
	Wet-unit weight γ_t	$\frac{W_s + W_w}{V_s + V_w}$	$\frac{W_s + W_w}{V}$	
	Saturated-unit weight γ_s	$\frac{W_s + W_w}{V_s + V_w}$	$\frac{W_s + V_v\gamma_w}{V}$	
	Submerged (buoyant) unit weight γ_b	$\gamma_s - \gamma_w$‡		
Combined relations	Moisture content w	$\frac{W_w}{W_s}$		
	Degree of saturation S	1.00	$\frac{V_w}{V_v}$	$\gamma_d = \frac{\gamma_t}{1 + w}$ $\gamma_s = \gamma_d + \gamma_w\left(\frac{e}{1 + e}\right)$
	Specific gravity G_s	$\frac{W_s}{V_s\gamma_w}$		

*After NAVFAC (1971).[7]

†γ_w is unit weight of water, which equals 62.4 pcf for fresh water and 64 pcf for sea water (1.00 and 1.025 g/cm^3). Where noted with ‡ the actual unit weight of water surrounding the soil is used. In other cases use 62.4 pcf. Values of w and s are used as decimal numbers.

moisture-density relationships, relative density, gradation, plasticity, and organic content.

Volume-Weight Relationships

Definitions of the various volume-weight relationships for soils are given in Table 2.4. Commonly used relationships are void ratio e, soil unit weight (also termed density or mass density and reported as total or wet density) γ_t, dry density γ_d and buoyant density γ_b, moisture (or water) content w, saturation degree S, and specific gravity of solids G_s. The natural density is an important element in most geotechnical analyses and, except for cohesive materials from which undisturbed samples may be procured, its accurate measurement often is difficult to obtain, especially in clean granular soils below the water table.

Moisture-Density Relationships (Soil Compaction)

Optimum moisture content and *maximum dry density* relationships are commonly used to specify a standard degree of compaction to be achieved during the construction of a load-bearing fill, embankment, earth dam, or pavement. Specification is in terms of a percent of maximum dry density, and a range in permissible moisture content often is specified as well (*see Art. 5.2.2*). The density of a soil can be increased by mechanical compaction. If the moisture content is increased in increments, the density will also increase in increments under a given compactive effort, until eventually a peak or maximum density is achieved for some particular moisture content. The density thereafter will decrease as the moisture is increased.

Relative Density D_R

Relative density refers to an in situ degree of compaction, relating the natural density of a cohesionless granular soil to its maximum density (the densest state to which a soil can be compacted, D_R = 100%) and the minimum density (the loosest state that dry soil grains can attain, D_R = 0%). The relationship is illustrated in Fig. 2.1, which can be used to find D_R when γ_N (natural density), γ_D (maximum density), and γ_L (loose density) are known. D_R may be expressed as

$$D_R = \frac{1/\gamma_L - 1/\gamma_N}{1/\gamma_L - 1/\gamma_D} \tag{2.1}$$

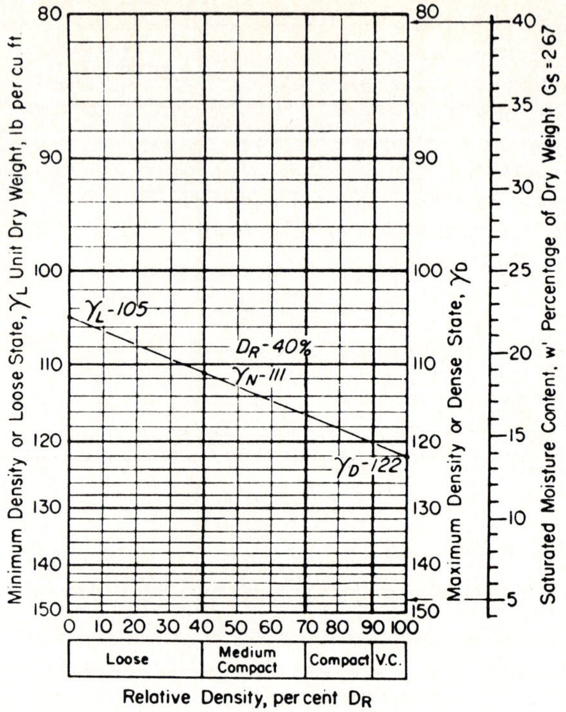

FIG. 2.1 Relative density diagram. [From Burmister (1948).[8] Copyright ASTM, 1916 Race Street, Philadelphia, PA 19103. Reprinted with permission.]

D_R is used for classification of the degree of in situ compactness as given in Fig. 2.1 or, more commonly, to classify in situ density as follows: very loose (0 to 15%), loose (15 to 35%), medium dense (35 to 65%), dense (65 to 85%), and very dense (85% to 100%). Correlations with N values of the SPT are given in Table 3.5. Measurements of D_R are limited to material with less than about 35% nonplastic soil passing the no. 200 sieve because fine-grained soils falsely affect the loose density.

Void ratio and unit weight are directly related to D_R and gradation characteristics. Permeability, strength, and compressibility are also related directly to D_R and gradation characteristics.

Gradation (Grain Size Distribution)

Gradation refers to the distribution of the various grain sizes in a soil specimen plotted as a function of the percent by weight passing a given sieve size. A parameter often used for correlations is the *coefficient of uniformity* C_u defined as the ratio between the grain diameter of 60% finer to the grain diameter corresponding to the 10% finer line, or:

$$C_u = D_{60}/D_{10} \tag{2.2}$$

Gradation curves from cohesionless granular soils may be used to estimate γ_D and γ_L, and if γ_N or D_R is known, estimates can be made of the void ratio e, porosity n, friction angle ϕ, and the coefficient of permeability k. Burmister (1948)[8] developed a system for classifying gradation curve characteristics based on curve shape and the range of sieve sizes encompassed [see Hunt (1984),[1] p. 146], S shapes are the most common, characteristic of well-sorted (poorly graded) sands deposited by flowing water, wind, and wave action. Gradation characteristics are useful as a basis for estimating a number of parameters. General relationships for estimating maximum compacted densities and minimum densities and the influence of grain angularity are given in Fig. 2.2. Estimations of k values may be made from Figs. 2.5 and 2.6.

Gradation relationships are used as the basis for soil classification systems, along with plasticity relationships, such as the *Unified System* given as Table 2.5.

Plasticity

Atterberg limits, which include the liquid limit, the plastic limit, and the shrinkage limit, are used to define the plasticity characteristics of clays and other cohesive materials.

Liquid limit (LL) is the moisture content at which a soil passes from the liquid to the plastic state as moisture is removed. At the LL, the undrained strength $s_u \approx 0.03$ tsf. The LL correlates approximately with the compression index C_c for normally consolidated clays of low to moderate sensitivity as follows:

$$C_c \approx 0.009(\text{LL} - 10\%) \tag{2.3}$$

Plastic limit (PL) is the moisture content at which a soil passes from the plastic to the semisolid state as moisture is removed.

Plasticity index (PI) is defined as PI = LL − PL.

Shrinkage limit (SL) is the moisture content at which no more volume change occurs upon drying.

Liquidity index (LI) is used for correlations and is defined as:

$$\text{LI} = \frac{w - \text{PL}}{\text{LL} - \text{PL}} = \frac{w - \text{PL}}{\text{PI}} \tag{2.4}$$

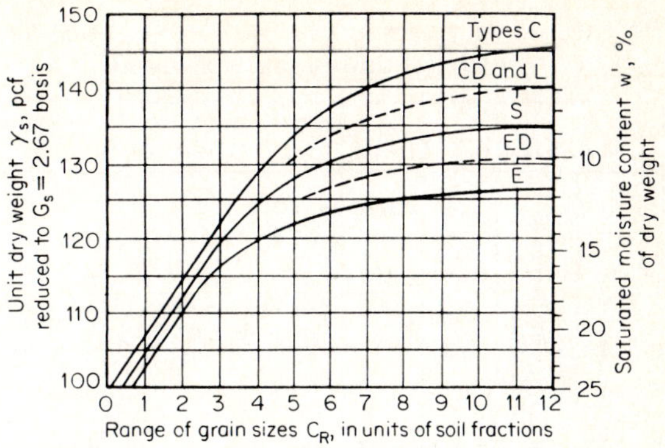

(a) Maximum compacted densities: D_R obtained by vibrations = 100%.

(b) APPROXIMATE MINIMUM DENSITIES, 0% D_R			
	Decrease in density, pcf		
Range in grain sizes, C_R	**Coarser soils**		**Finer soils**
1 to 3	10	to	20
3 to 5	20	to	25
5 and greater	25	to	30+

(c) APPROXIMATE INFLUENCE OF GRAIN SHAPE ON DENSITY	
Grain shape	**Change in density, pcf**
	−
Very angular	10 to − 15
Subangular	0 to normal
Rounded or waterworn	+2 to +5
0.5% mica	−2 to −5

FIG. 2.2 Maximum compacted densities, approximate minimum densities, and influence of grain shape on density for various gradations. [From Burmister (1948).[8] Copyright ASTM, 1916 Race Street, Philadelphia, PA 19103. Reprinted with permission.]

TABLE 2.5
UNIFIED SOIL CLASSIFICATION SYSTEM (ASTM D2487) [After USAWES (1967)[17]]

Major Divisions			*Group Symbols*	*Typical Names*	*Laboratory Classification Criteria*		
Coarse-grained soils (More than half of material is larger than No. 200 sieve size)	Gravels (More than half of coarse fraction is larger than No. 4 sieve size)	Clean gravels (Little or no fines)	GW	Well-graded gravels, gravel-sand mixtures, little or no fines	Determine percentages of sand and gravel from grain-size curve. Depending on percentage of fines (fraction smaller than No. 200 sieve size), coarse-grained soils are classified as follows: Less than 5 per cent: GW, GP, SW, SP; More than 12 per cent: GM, GC, SM, SC; 5 to 12 per cent: *Borderline* cases requiring dual symbols[b]	$C_u = \dfrac{D_{60}}{D_{10}}$ greater than 4; $C_c = \dfrac{(D_{30})^2}{D_{10} \times D_{60}}$ between 1 and 3	
			GP	Poorly graded gravels, gravel-sand mixtures, little or no fines		Not meeting all gradation requirements for GW	
		Gravels with fines (Appreciable amount of fines)	GM[a] d / u	Silty gravels, gravel-sand-silt mixtures		Atterberg limits below "A" line or P.I. less than 4	Above "A" line with P.I. between 4 and 7 are *borderline* cases requiring use of dual symbols
			GC	Clayey gravels, gravel-sand-clay mixtures		Atterberg limits below "A" line with P.I. greater than 7	
	Sands (More than half of coarse fraction is smaller than No. 4 sieve size)	Clean sands (Little or no fines)	SW	Well-graded sands, gravelly sands, little or no fines		$C_u = \dfrac{D_{60}}{D_{10}}$ greater than 6; $C_c = \dfrac{(D_{30})^2}{D_{10} \times D_{60}}$ between 1 and 3	
			SP	Poorly graded sands, gravelly sands, little or no fines		Not meeting all gradation requirements for SW	
		Sands with fines (Appreciable amount of fines)	SM[a] d / u	Silty sands, sand-silt mixtures		Atterberg limits above "A" line or P.I. less than 4	Limits plotting in hatched zone with P.I. between 4 and 7 are *borderline* cases requiring use of dual symbols
			SC	Clayey sands, sand-clay mixtures		Atterberg limits above "A" line with P.I. greater than 7	
Fine-grained soils (More than half material is smaller than No. 200 sieve)	Silts and clays (Liquid limit less than 50)		ML	Inorganic silts and very fine sands, rock flour, silty or clayey fine sands, or clayey silts with slight plasticity	Plasticity Chart (see below)		
			CL	Inorganic clays of low to medium plasticity, gravelly clays, sandy clays, silty clays, lean clays			
			OL	Organic silts and organic silty clays of low plasticity			
	Silts and clays (Liquid limit greater than 50)		MH	Inorganic silts, micaceous or diatomaceous fine sandy or silty soils, elastic silts			
			CH	Inorganic clays of high plasticity, fat clays			
			OH	Organic clays of medium to high plasticity, organic silts			
Highly organic soils			Pt	Peat and other highly organic soils			

Plasticity Chart

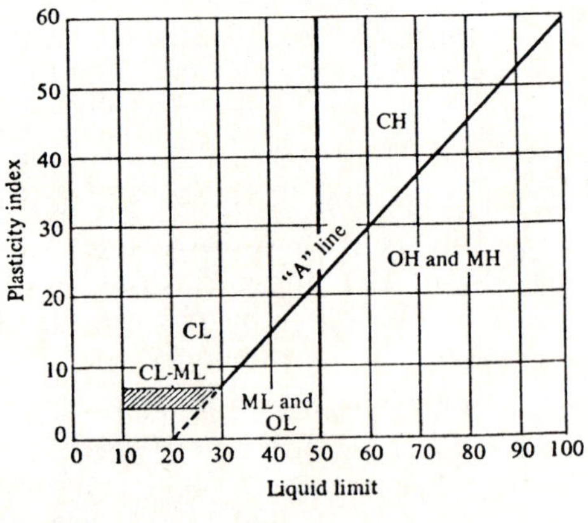

[a]Division of GM and SM groups into subdivisions of d and u are for roads and airfields only. Subdivision is based on Atterberg limits; suffix d used when L.L. is 28 or less and the P.I. is 6 or less; the suffix u used when L.L. is greater than 28.

[b]Borderline classifications, used for soils possessing characteristics of two groups, are designated by combinations of group symbols. For example: GW-GC, well-graded gravel-sand mixture with clay binder.

A plot of PI vs. LL provides the basis for cohesive soil classification as shown on the plasticity chart (Fig. 2.3) and for various correlations. In general, the higher the value of PI and LL, the greater is the tendency of the specimen to shrink upon drying and to swell upon wetting (except for organic soils). The relationship between the natural moisture content and the PL and LL is an indication of the soils' *consistency*, which is related to strength and compressibility (see Table 3.7 for strength relationships). The *liquidity index* expresses this relationship quantitatively. For example, where the natural water content approximates LL, LI $\approx$ 1 and $U_c \approx 0.3$ to 1.0 tsf and where the natural water content approximates PL, LI $\approx$ 0 and $U_c \approx 1$ to 5 tsf.

Organic Content

Organic materials are found as pure organic matter or as mixtures with sand, silt, or clay. Their engineering characteristics relate to the percentage of organic matter in a specimen.

2.3 HYDRAULIC PROPERTIES

2.3.1 GROUNDWATER

Engineering Significance

Groundwater can be encountered in various modes of occurrence as classified by geologists. The static water table or the phreatic surface has been defined as the level within a body of subsurface water at which water pressures are equal to atmospheric pressure, or at which the neutral stress u_w in the soil equals zero. Above the "water table" in the *upper zone* water may be found as gravity or vadose water, hygroscopic moisture, pellicular water, or perched water or in the capillary fringe. The latter two conditions may represent saturation. Below the water table in the *saturated zone* may be found formations termed as aquitards, aquicludes, aquifuges, aquifers, or connate water. [For descriptions see Art. 8.3.1. of Hunt (1984).[1]]

Engineers are primarily concerned with aquifers, which can be artesian, aquitards, and the capillary zone. Water flowing into excavations from aquifers must be controlled to maintain dry working conditions and to reduce pressures on retaining structures. Water flowing through, around, or beneath dams or other retaining structures requires control to prevent excessive losses, seepage pressures, and piping. Aquitards, represented by saturated silts, can allow seepage into excavations and may constitute weakness zones in excavation walls and zones susceptible to "quick" conditions in the bottom of excavations or in slopes.

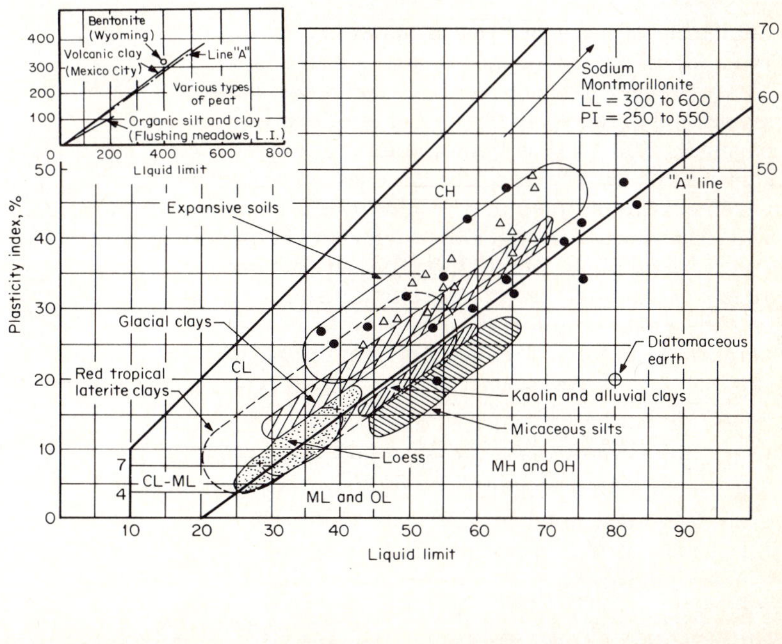

FIG. 2.3 Plasticity chart for Unified Classification System (see Table 2.5). [From Hunt (1984).[1] Used with permission of McGraw-Hill Book Company.]

Artesian conditions can cause boiling or piping and uplift in excavation bottoms which may result in sudden failure. Underconsolidation of a soil stratum results if artesian pressures prevent drainage and consolidation causing the soil to remain in a very loose or very soft state. The condition is found in river valleys and other lowlands where buried aquifers extend continuously into nearby hills. Most importantly, the engineer must be aware that groundwater conditions are transient and that those encountered during investigation will not necessarily be those existing during construction or at some later date.

Associated Phenomena

Permeability is the capacity of a material to transmit water. Flow through a geologic medium is quantified by a material characteristic termed the *coefficient of permeability* k, expressed in terms of Darcy's law, valid for laminar flow in a saturated, homogeneous material, as

$$k = \frac{q}{iA} \quad \text{cm/s} \qquad (2.5)$$

where q = quantity of flow per unit of time, cm^3/s
i = hydraulic gradient, i.e., the head loss per length of flow h/L (a dimensionless number)
A = area, cm^2

Capillarity is the tendency of water to rise in "soil tubes" or connected voids to elevations above the groundwater table. It provides the moisture that results in heaving of buildings and pavements from the volume increase of expansive soils or from freezing, or the destruction of pavements from "pumping" under wheel loads where pavement support is provided by fine-grained soils. The potential for the detrimental effects of capillary rise is a function of soil type and the depth to the static water table or to a perched water table. Rating criteria for drainage, capillarity, and frost heave in terms of soil type are given in Table 2.6.

Piping refers to two phenomena: (1) water seeping through fine-grained soil, eroding the soil grain by grain and forming pipes or tunnels and (2) water under pressure flowing upward through a granular soil with a head of sufficient magnitude to cause soil grains to lose contact and capability for support. Also termed *boiling* or *liquefaction*, piping is the cause of a "quick" condition (as in quicksand) in which the sand essentially liquifies.

"Cyclic" liquefaction refers to the complete loss of supporting capacity when dynamic earthquake forces cause a sufficiently large temporary increase in pore pressure in the mass [see *Arts. 3.3.5, and 11.3.3 of Hunt (1984).*[1]]

2.3.2 SUBSURFACE FLOW

The Hydraulic Gradient

Hydrostatic conditions refer to pressures in fluids when there is no flow. The pressure P at depth h in water equals the unit weight of water γ_w times the depth plus atmospheric pressure P_a, and is equal in all directions, i.e., $P_v = P_h$.

Groundwater flow occurs when there is an imbalance of pressure from gravitational forces acting on the water, and it seeks to balance the pressure.

Hydraulic gradient and *permeability* are the two factors upon which groundwater movement is dependent. The *hydraulic gradient* between two points on the water table is the ratio between the difference in elevation of the two points and the distance between them. It reflects the friction loss as the water flows between the two points.

Flow-condition nomenclature is illustrated in Fig. 2.4:

- *Static condition* refers to no flow, and in Fig. 2.4*a*, water will rise to the same piezometric level in any tube extending from the inclined sand-filled glass tube.
- *Pressure surface* is created when flow is allowed to occur. The levels in the tubes drop as shown on Fig. 2.4*b*.
- *Hydraulic head* h is the difference in water-level elevation between the two tubes, or the head loss during flow.
- *Hydraulic gradient* i is the ratio of the hydraulic head h to the length of flow path L expressed as

$$i = h/L \qquad (2.6)$$

TABLE 2.6
TENTATIVE CRITERIA FOR RATING SOILS WITH REGARD TO DRAINAGE, CAPILLARITY, AND FROST HEAVING CHARACTERISTICS*

Fineness identification†	"Trace fine sand"	"Trace silt"	"Little silt" (coarse and fine)	"Some fine silt" "Little clayey silt" (fissured clay soils)	"Some clayey silt" (clay soils dominating)
Approx. effective size, D_{10} mm‡	0.4 0.2	0.2 0.074	0.074 0.02	0.02 0.01	0.01
Drainage	Free drainage under gravity excellent	Drainage by gravity good	Drainage good to fair	Drains slowly, fair to poor	Poor to impervious
Approx. range of k, cm/s	0.5 0.10 0.2	0.020 0.04	0.0010 0.006	0.0002 0.0004	0.0001
	← Deep wells	← Well points successful →			
Capillarity Approx. rise in feet, H_c	Negligible 0.5	Slight 1.5 1.0	Moderate 7.0 3.0	Moderate to high 15.0 10.0	High 25.0
Frost heaving susceptibility Groundwater within 6 ft or $H_c/2$	Non-frost-heaving	Slight	Moderate to objectionable	Objectionable	Objectionable to moderate

*Criteria for soils in a loose to medium-compact state. From Burmister (1951).[9]

†Fineness classification is in accordance with the ASEE Classification System.

‡Hazen's D_{10}. The grain size for which 10% of the material is finer.

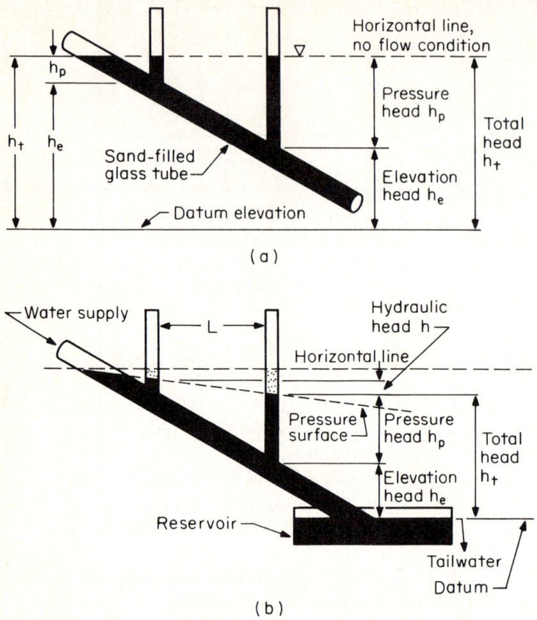

FIG. 2.4 Diagrams illustrating the hydraulic gradient of groundwater: (*a*) no-flow condition and (*b*) flow creates an hydraulic gradient. [From Hunt (1984).[1] Used with permission of McGraw-Hill Book Company.]

- *Pressure head* h_p is the height to which water will rise in the vertical tube from the point of interest or reference (also referred to as piezometric head).
- *Elevation head* h_e is the height of the point of interest or reference with respect to some arbitrary datum.
- *Tailwater elevation*, as in the case of a lake or pool where the elevation is constant, is the reference datum for most seepage problems.
- *Total head* h_t equals the pressure head plus the elevation head.
- *Steady-state condition* usually means a state of constant flow, no acceleration or deceleration or change in piezometric levels.

Permeability

Permeability has been defined in the previous article as the capacity of a material to transmit water, expressed in terms of its coefficient k, which gives the superficial or discharge velocity per unit of gradient, as if the flow occurred through the total volume of the medium, not only the void area.

Values for k are determined from laboratory or in situ tests as summarized in Table 2.7. For many practical problems, sufficiently accurate values for k may be obtained from charts and tables which may provide correlations with gradation characteristics such as given in Table 2.8 and Fig. 2.5, or with gradation characteristics and relative density as given in Fig. 2.6.

Stratification affects permeability significantly and granular formations are commonly layered. Water will flow much more freely in clean sand or gravel beds than through silty or clayey beds in stratified soils. This often causes horizontal permeabilities to be greater than vertical permeabilities by factors of 10 or 100 or more. An evaluation of the effect of stratification is given in Fig. 2.7.

"Secondary permeability" is a term given by Terzaghi to *rock-mass permeability*. In most practical situations it is primarily a function of joint conditions including spacing, aperture width, and the nature of the filling, or of cavity size and distribution. Because of the great irregularity of these features, including their normally discontinuous nature in most rock masses, an estimation of rock-mass permeability can be subject to substantial error. Where rock is essentially intact, values are extremely low for most rock types, except for porous rocks such as some sandstones (argillaceous or highly cemented sandstones may have low permeabilities). Some typical values are given in Table 2.9.

Pore-Water Pressures

Pore-water pressure (u or u_w) is the pressure existing in the water in the pores, or void spaces, of a saturated soil element.

Cleft-water pressure refers to the pressures existing on the water in saturated joints or other fractures in rock masses.

Excess hydrostatic pressure is that pressure capable of causing flow of water ($h\gamma_w$ in Fig. 2.8*b*).

No-Flow and No-Applied-Stress Condition

For the condition of no flow and no applied stress, pore-water pressures are equal to the unit

TABLE 2.7
TESTS TO MEASURE PERMEABILITY COEFFICIENT

Material	Test		Application
		IN SITU TESTS	
Rock masses	Constant-head test	In boreholes to measure k in heavily jointed rock masses.	Free-draining materials. Requires ground saturation.
	Falling-head test	In boreholes to measure k in jointed rock masses. Can be performed to measure k_{mean}, k_v or k_h.	Slower draining materials or below water table.
	Rising-head test	Same as for falling-head test.	Same as for falling-head test.
	Pumping tests	In wells to determine k_{mean} in saturated uniform formations.	Not representative for stratified formations. Measures average k for entire mass.
	Pressure testing	Measures k_h in vertical borehole.	Requires clean borehole walls. Pressures can cause joints to open or to clog from migration of fines.
Soil formations	Constant-head test	In boreholes or pits to measure k in free-draining soils.	Free-draining soils. Requires ground saturation.
	Falling-head test	In boreholes in slow-draining materials, or materials below GWL. Can be performed to measure k_{mean}, k_n, or k_h.	Slow-draining materials or below water table.
	Rising-head test	Similar to falling-head test.	Similar to falling-head test.
	Pumping tests	In wells to measure k_{mean} in saturated uniform soils.	Results not representative in stratified formations.
		LABORATORY TESTS	
Soils	Constant-head test		k in free-draining soil.
	Falling-head test		k in slow-draining soil.
	Consolidometer test		k in very slow-draining soil (clays).

weight of water times the depth below the free-water surface as shown in Fig. 2.8*a*, expressed as:

$$u = \gamma_w z_w \tag{2.7}$$

Buoyancy pressures refer to the vertical pressures acting on each end of the soil column; on the specimen bottom in Fig. 2.8*a* the buoyancy force equals 30 g/cm^2.

Upward Flow Condition

In Fig. 2.8*b*, a head of 20 cm (seepage force) causes an increase in pore pressure at the base of the soil column supported on a screen to u = 50 g/cm^2. The tailwater is barely overflowing and the 20-cm head has been dissipated in viscous friction in the soil specimen.

Boundary water pressures act on the specimen; on the bottom they are equal to the *buoyancy force* (30 g/cm^2) plus the *seepage force* (20 g/cm^2), and on the top they are equal to the water pressure (10 g/cm^2).

Effective Stresses

The effective stresses ($\bar{\sigma}_v = \sigma_v - u$; see *Art. 3.2.2*) may be found either from boundary forces, considering total soil weight, or from seepage forces,

TABLE 2.8
TYPICAL PERMEABILITY COEFFICIENTS FOR VARIOUS MATERIALS*

	Particle-size range				"Effective" size		Permeability coefficient k		
	Inches		Millimeters						
	D_{max}	D_{min}	D_{max}	D_{min}	D_{20}, in	D_{10}, mm	ft/year	ft/month	cm/s
				TURBULENT FLOW					
Derrick stone	120	36			48		100×10^6	100×10^5	100
One-man stone	12	4			6		30×10^6	30×15^5	30
Clean, fine to coarse gravel	3	¼	80	10	½		10×10^6	10×10^5	10
Fine, uniform gravel	⅜	1/16	8	1.5	⅛		5×10^6	5×10^5	5
Very coarse, clean, uniform sand	⅛	1/32	3	0.8	1/16		3×10^6	3×10^5	3
				LAMINAR FLOW					
Uniform, coarse sand	⅛	1/64	2	0.5		0.6	0.4×10^6	0.4×10^5	0.4
Uniform, medium sand			0.5	0.25		0.3	0.1×10^6	0.1×10^5	0.1
Clean, well-graded sand and gravel			10	0.05		0.1	0.01×10^6	0.01×10^5	0.01
Uniform, fine sand			0.25	0.05		0.06	4000	400	40×10^{-4}
Well-graded, silty sand and gravel			5	0.01		0.02	400	40	4×10^{-4}
Silty sand			2	0.005		0.01	100	10	10^{-4}
Uniform silt			0.05	0.005		0.006	50	5	0.5×10^{-4}
Sandy clay			1.0	0.001		0.002	5	0.5	0.05×10^{-4}
Silty clay			0.05	0.001		0.0015	1	0.1	0.01×10^{-4}
Clay (30 to 50% clay sizes)			0.05	0.0005		0.0008	0.1	0.01	0.001×10^{-4}
Colloidal clay ($-2\mu \leq$ 50%)			0.01	10 Å		40 Å	0.001	10^{-4}	10^{-9}

*From Hough (1957).[10] Reprinted with permission of John Wiley & Sons, Inc.

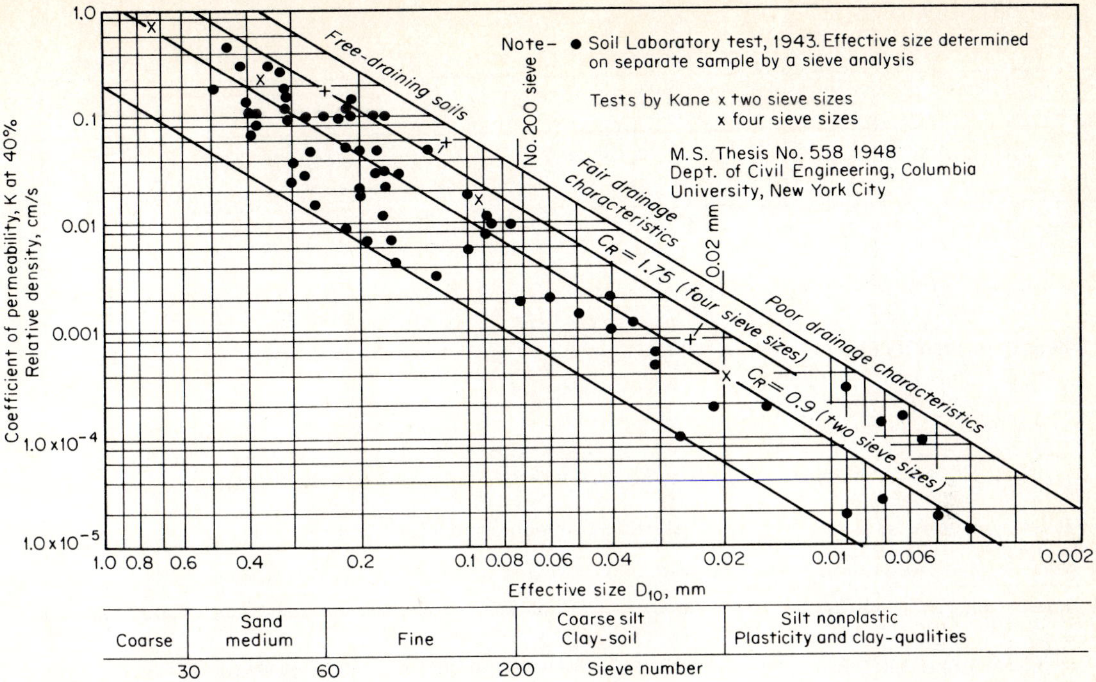

FIG. 2.5 Relationship between permeability and Hazen's effective size D_{10}. Coefficient of permeability reduced to basis of 40% D_R by Fig. 2.6. [From Burmister (1948).[8] Copyright ASTM, 1916 Race Street, Philadelphia, PA 19103. Reprinted with permission.]

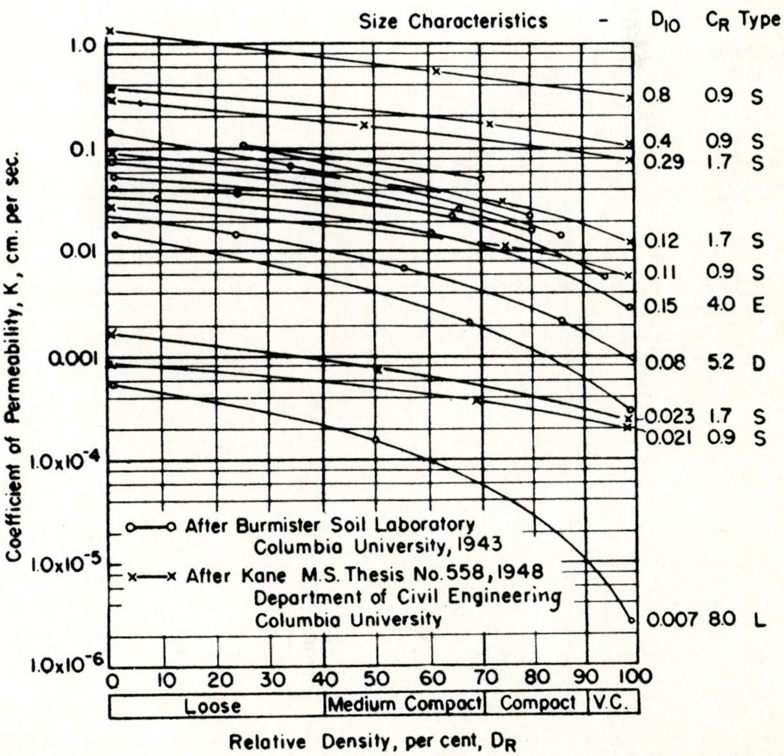

FIG. 2.6 Permeability–relative density relationships. [From Burmister (1948).[8] Copyright ASTM, 1916 Race Street, Philadelphia, PA 19103. Reprinted with permission.]

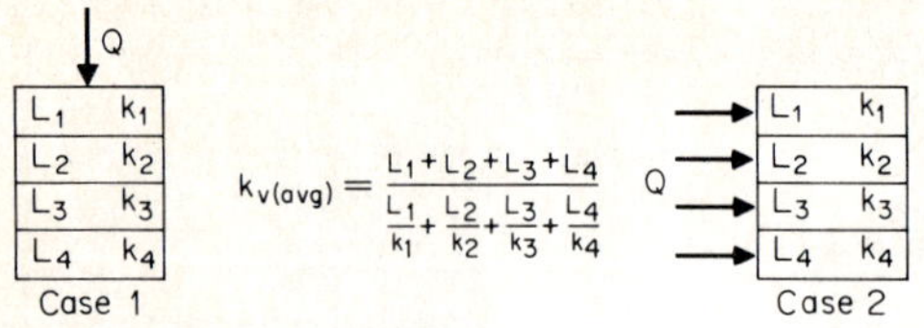

$$k_{v(avg)} = \frac{L_1+L_2+L_3+L_4}{\frac{L_1}{k_1}+\frac{L_2}{k_2}+\frac{L_3}{k_3}+\frac{L_4}{k_4}}$$

$$k_{h(avg)} = \frac{k_1 L_1 + k_2 L_2 + k_3 L_3 + k_4 L_4}{L_1+L_2+L_3+L_4}$$

where Q = quantity of flow
L = flow path length
k = coefficient of permeability

NOTE: The electrical analogy:
If $L_1 = L_2 = L_3 = L_4 = 1$, and $k_1 = 1$, $k_2 = 2$, $k_3 = 3$, $k_4 = 4$, then in case 1, $k_{v(avg)} = 1.9$ and in case 2, $k_{h(avg)} = 2.5$

FIG. 2.7 An evaluation of the effect of stratification on permeability. [From Salzman (1974).[11]]

TABLE 2.9
TYPICAL PERMEABILITY COEFFICIENTS FOR ROCK AND SOIL FORMATIONS*

	k, cm/s	Intact rock	Porosity n, %	Fractured rock	Soil
Practically impermeable	10^{-10} 10^{-9} 10^{-8} 10^{-7}	Massive low-porosity rocks	0.1–0.5 0.5–5.0		Homogeneous clay below zone of weathering
Low discharge, poor drainage	10^{-6} 10^{-5} 10^{-4} 10^{-3}	Sandstone Weathered granite Schist	5.0–30.0	Clay-filled joints	Very fine sands, organic and inorganic silts, mixtures of sand and clay, glacial till, stratified clay deposits
High discharge, free draining	10^{-2} 10^{-1} 1.0 10^{1} 10^{2}			Jointed rock Open-jointed rock Heavily fractured rock	Clean sand, clean sand and gravel mixtures Clean gravel

*After Hoek and Bray (1977).[20]

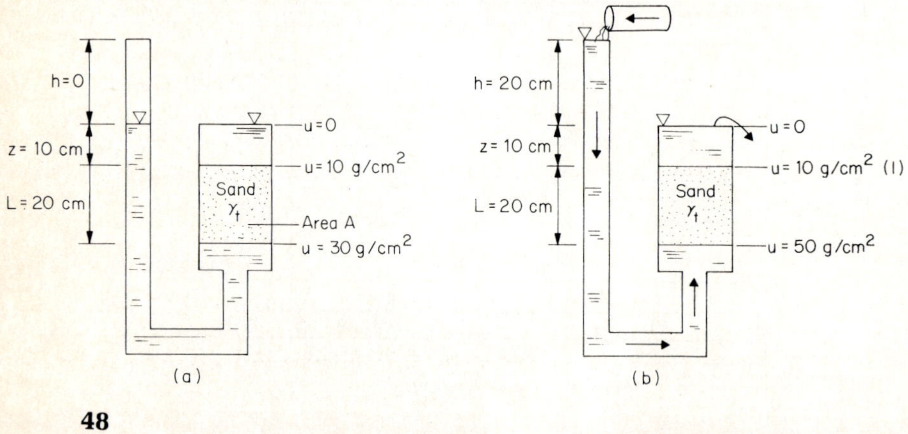

FIG. 2.8 Pore-water pressures. (a) No-flow condition. Bouyancy pressures act on each end of the soil specimen. (b) Upward flow condition. Boundary pressures act on each end of the specimen. At (1) there is 20 g/cm² lost in seepage.

considering submerged weight. At the bottom of the specimen in Fig. 2.8*b*, $\bar{\sigma}_v$ equals either:

1. Total specimen weight ($LA\gamma_t$) plus the overlying water weight ($zA\gamma_w$) minus the pore-water pressure $(h - L - z)A\gamma_w$, or
2. Submerged specimen weight ($LA\gamma_b$) minus the seepage force ($hA\gamma_w$). For $\bar{\sigma}_v = 0$, $LA\gamma_b - LA\gamma_w = 0$ and a "quick" condition exists.

Applied Stresses

Applied stresses cause an increase in pore pressures. Loading a clayey soil causes the process of consolidation *(see Art. 4.3.4)*. The pore water first carries the load; then, as the water drains from the soil, pore pressures dissipate, the voids become smaller, and the load is transferred to the soil skeleton. If the load is applied rapidly, however, the soil has no time to drain, friction is not mobilized, and the substantially lower undrained shear strength prevails. Even if drainage occurs, the frictional component of strength will be reduced by the amount of pore pressure.

Neutral stress refers to the pressure in the pore water because it does not contribute either to compression or to an increase in shearing resistance.

Seepage

Velocity

The average seepage velocity v_s of water flowing through the pores of a saturated soil mass is equal to the discharge velocity ($v_d = ki$) times the ratio $(1 + e)/e$, or the discharge velocity divided by the effective porosity n_e expressed as:

$$v_s = \frac{ki}{n_e} \quad \text{ft/day or cm/s} \tag{2.8}$$

The practical significance of seepage velocity lies in the field evaluation of *k*, using dye tracers and measuring the time for the dye to travel the distance between two observation holes, and in estimating the rate of movement in pollution-control studies.

Pressures and Liquefaction

Seepage pressures *j* are stresses in the soil caused by the flow of water. They are equal to the hydraulic gradient *i* times the unit weight of water expressed as:

$$j = (h/L)\gamma_w = i\gamma_w \tag{2.9}$$

and act in the direction of flow.

Where the flow tends to be upward, as at the toe of a dam, along slopes, or in the bottoms of excavations, the pressure is resisted by the weight of the overlying soil column.

"Quick conditions" (*boiling* or *liquefaction*) occur when the upward seepage gradient increases until the seepage pressure exceeds the submerged weight of the soil, and the soil column is uplifted. The result is complete loss of intergranular friction and the supporting capacity of the soil. (*"Cyclic" liquefaction* occurs under dynamic loadings) *(see Art. 3.3.5)*.

Critical gradient i_{cr} is the hydraulic gradient required to produce liquefaction and equals the ratio of the submerged unit weight of the soil to the unit weight of water.

$$i_{cr} = \frac{\gamma_t - \gamma_w}{\gamma_w} = \frac{\gamma_b}{\gamma_w} \tag{2.10}$$

In coarse to fine sand, i_{cr} is about 0.9 to 1.0, and in lay person's terms the result is "quicksand." The example given in Fig. 2.9 is stable since $i_{cr} = 62/20 = 3.1$.

The *factor of safety (FS)* against liquefaction, usually taken as 3 or more because of the disastrous nature of such a failure, is defined as

$$FS = i_{cr}/i \tag{2.11}$$

Because seepage pressure is directly proportional to the hydraulic gradient, potentially the most dangerous areas for liquefaction are those where the upward gradient is large and the counterbalancing weight is small such as along the toe of slopes or in excavation bottoms below the water table. The analysis of flow through soils and the determination of seepage pressures may be performed with flow nets as described in the following section.

Rock Masses

Cleft-water pressures is the term used to designate the water pressures that develop in rock-mass fractures. When too high, they can result in

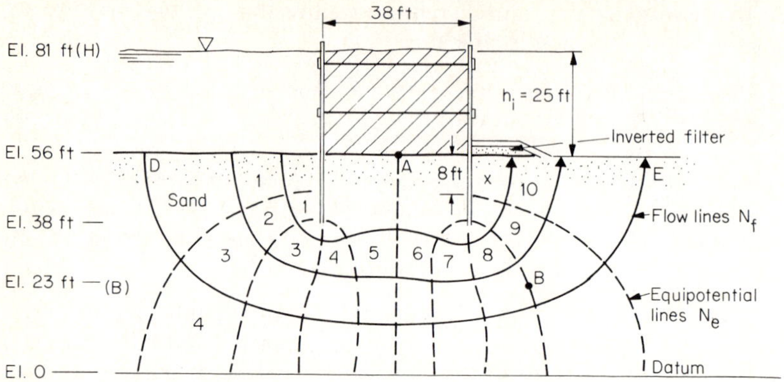

FIG. 2.9 Example of a confined flow net beneath a cofferdam to illustrate analysis to find pressure heads, seepage quantity, and seepage pressure in sands with k = 0.1 ft/min.

1. From the figure $N_f = 4$, $N_e = 10$, $\mathcal{S} = N_f/N_e = 0.4$

2.

Point	Elevation head h_e, ft	Total head h_t, $h_t = H - N_f(h_1/N_e)$	Pressure head h_p, $h_p = h_t - h_e$, ft	Water pressure p, $h_p \times 62.4$, psf
A	56	81 − 5(25/10) = 68.5	12.5	780
B	23	81 − 8(25/10) = 61.0	38.0	2371

3. Piezometer at point *B* will show water level 38.0 ft above point *B*

4. Seepage quantity $q = \mathcal{S}_k h_1 = 0.4(0.1)25 = 1.0\ ft^3/min$ per foot of cofferdam

5. Gradient in region $x = i_x = \Delta h_1/l = 25/10/8 = 0.313$

6. Seepage pressure in zone $x = 0.313 \times 62.4 = 19.5$ pcf

instability of rock foundations for concrete dams, and they are the common cause of slope failures. Seepage can result in the softening of joint fillings, and the development of high pore pressures in the filling material can reduce strength.

Stress changes can significantly affect seepage and permeability in rock masses. Compressive stresses cause closure of joints even under relatively low stress levels, reducing seepage flow, although sufficient closure of other voids to reduce permeability occurs in most rocks only under relatively high stress levels. Tensile stresses can increase permeability and flow, with the increase commonly occurring as a rock slope begins to deflect. The failure of the Malpasset Dam is considered to be the result of tensile stresses increasing under the toe of the concrete arch dam in the foundation gneiss. It has been estimated that the gneiss had a permeability 1000 times greater under the tensile stresses than when in compression.

Flow-Net Analysis

General

Flow through a soil medium may be represented by a flow net: a two-dimensional graphical presentation of flow consisting of a net of flow lines and equipotential lines, the latter connecting all points of equal piezometric level along the flow lines.

Flow nets are used to evaluate:

1. Seepage quantities exiting through or beneath a dam or other retaining structure.
2. Seepage pressures that may result in uplift below a dewatered excavation or at the toe of a dam.
3. Exit gradients and the potential for liquefaction in a dam, slope, or excavation.
4. Pore pressures along potential failure surfaces in a slope.
5. Inflow quantities into a well or other opening in the ground.

There are two general flow conditions encountered in analysis:

Confined flow refers to the case where the phreatic surface is known, as beneath cutoff walls (Figs. 2.9 and 9.48).

Unconfined flow refers to the case where the location of the phreatic surface is not known, as in earth dams (Fig. 8.13) and slopes (Fig. 8.14).

Flow-Net Construction

Flow-net construction is a graphical procedure accomplished by trial and error. The flow zone, bounded by the phreatic surface and an impervious stratum, is subdivided on a scaled drawing of the problem area as nearly as possible into equidimensional quadrilaterals formed by *flow lines* and *equipotential lines* crossing at right angles, as shown in Fig. 2.9.

In the simple case given as Fig. 2.9, a *flow line* is represented as the path along which a particle of water flows on its course from D to E through the saturated sand mass. Each flow line starts at some point along the surface of the sand mass where it has a pressure head h_p; thereafter the flowing particle gradually dissipates this head in viscous friction until it reaches the exit surface. Along each flow line there is a point where the water has dissipated any specific portion of its potential. *Equipotential lines* connect all such points of equal *total head* on the flow lines. Piezometers measure the pressure head, or the total head minus the elevation head.

The basic assumptions in flow-net construction are that Darcy's law is valid and that the soil formation is homogeneous and isotropic. *Anisotropic conditions* resulting from stratification normally cause horizontal permeability to be greater than vertical in natural soil formations, whereas in a zoned earth dam the effect can be complex. In general, at boundaries between zones of materials with different permeability, both the flow lines and the equipotential lines are deflected. Where seepage moves from a zone of higher to one of lower permeability, more area is required for the same quantity of flow per unit of time and the flow lines deflect to make the flow channels wider. Flow channels are narrower in the opposite situation where water seeps to a more pervious zone.

Anisotropic conditions are accounted for in flownet constuction by shrinking the dimensions of the cross section in the direction of the greater permeability. For example, if $k_h > k_v$, the horizontal scale is reduced by multiplying the true distance by $\sqrt{k_v/k_h}$ producing a transformed section, and the flow net is constructed in the ordinary manner. The scale is again expanded to the true section, in which the flow and equipotential lines are no longer perpendicular, and analysis is performed using a weighted coefficient of permeability $k = \sqrt{k_v k_h}$.

Details of flow-net construction can be found in Casagrande (1940),[14] Sherard et al. (1963),[15] and Cedergren (1977).[16]

Analysis

Seepage quantity can be calculated, using the flow net, from the expression:

$$q = \frac{N_f}{N_e} kh \tag{2.12}$$

where N_f = number of flow channels (space between any adjacent pair of flow lines)
N_e = number of equipotential drops along each flow channel
k = coefficient of permeability
h = total head loss (h_1 in Fig. 2.9)
N_f/N_e = $\mathcal{S}$, the shape factor
q = discharge or quantity of flow per unit width (commonly given in ft^3/s per running foot, or m^3/s per running meter)

Examples of computations of seepage quantities for confined flow conditions are given in Figs. 2.9 and 9.48, and for unconfined flow conditions in Fig. 8.13*a*.

Seepage pressure is equal to the hydraulic gradient times the unit weight of water ($p_s = i\gamma_w$) and acts in a direction parallel to the flow lines. In the example of Fig. 2.9, p_s in zone $x = 0.313 \times 62.4 = 19.5$ pcf, which is resisted by the submerged weight of the overlying sand column, for the average sand about 60 pcf; therefore, liquefaction or boiling is not imminent. The inverted filter provides added protection. Other examples are given in Figs. 8.13*a* and 9.48.

Pore-water pressures may be determined from flow nets as illustrated in Fig. 8.13*b*. The application to stability analysis of slopes is illustrated in Fig. 8.38.

Conclusions

Flow nets are useful tools since even a crude flow net will permit fairly accurate determination of seepage quantities and pressures in soils. They can be time-consuming to construct, especially when anisotropic conditions are analyzed, and each time the dimensions are changed, as by increasing the depth of a cutoff wall, new flow nets are constructed.

2.3.3 GROUNDWATER AND SEEPAGE CONTROL

Importance

Groundwater and seepage control is a most important consideration in the stability of natural slopes, dams, and levees; excavations for structures, cut slopes, open-pit mines, tunnels, and shafts; buried structures; pavements; and side-hill fills.

Investigators must consider control both under conditions during construction and under those after it, inasmuch as conditions are likely to be different from those encountered during construction as a result either of natural events or of causes brought about by construction itself.

Necessity

During construction control is required to

- Provide a dry excavation and permit construction to proceed efficiently
- Reduce lateral loads on sheeting and bracing in excavations
- Stabilize "quick" bottom conditions and prevent heaving and piping
- Improve supporting characteristics of foundation materials
- Increase stability of excavation slopes and side-hill fills
- Cut off capillary rise and prevent pumping and frost heaving in pavements
- Reduce air pressure in tunneling operations

After construction control is required to:

- Reduce or eliminate uplift pressures on bottom slabs and permit economies from the reduction of slab thicknesses for basements, buried structures, canal linings, spillways, drydocks, etc.
- Provide for dry basements
- Reduce lateral pressures on retaining structures
- Control embankment seepage in all dams
- Control seepage and pore pressures beneath pavements, side-hill fills, and cut slopes
- Prevent surface and groundwater contamination from pollutants.

Control Methods

Control methods may be placed in three main categories:

1. *Cutoffs and barriers,* which when constructed or installed properly have the potential to seal off flow. Included in this category are liners, blankets and membranes of clay, asphalt, plastic, concrete, etc.; walls of compacted earth, concrete, bentonite slurry, or ice; and grouting, which often is an imperfect cutoff.
2. *Dewatering systems,* which serve to lower the water table and reduce pore-water pressures, or in some cases only to reduce pore pressures. Included are simple sump pumping, wellpoints (single and multistage systems, vacuum, or jet-inductor systems), deep wells, and electroosmosis.
3. *Drains,* which serve to control flow, in some cases lowering the water table, in others reducing pore pressures and seepage forces. They have many forms such as blankets, triangles, trapezoids, and cylinders, or are constructed in trenches.

Other factors:

Filters are provided between zones with significant differences in permeability to control flow velocities and prevent migration of fines and piping. They are most important adjacent to drains where it is desirable to prevent clogging or piping.

Surface treatments are provided to deter or prevent infiltration of water on slopes.

TABLE 2.10§
GROUNDWATER CONTROL METHODS AND COMMON APPLICATIONS

Method	Engineering problem									
	Earth and rock-fill embankments	Dam foundations	Slopes	Retaining structures (slopes)	Open excavations	Closed excavations (tunnels, etc.)	Building basements	Site surcharging	Pavements	Pollution-control systems
Cutoff:										
Liners*	X									X
Earth walls*	X									X
Slurry walls*†‡		X			X	X	X			X
Ice walls†					X	X				
Concrete walls*		X			X	X	X			X
Sheet pile walls†					X					
Grout walls*‡		X			X	X				
Dewater:										
Sump pumps†					X					
Wellpoints†		X			X					X
Deep wells†‡		X	X		X	X				X
Electroosmosis†			X		X	X				
Drain:										
Blanket drains*	X			X			X		X	
Trench drains*		X					X		X	
Triangular drains*	X			X						
Circular vertical drains*			X							
Circular subhorizontal drains*‡			X	X		X				
Drainage galleries†			X			X				
Relief or bleeder wells†‡		X	X					X		

*Permanent installation.

†Construction control.

‡Corrective measure after construction.

§From Hunt (1984).[1]

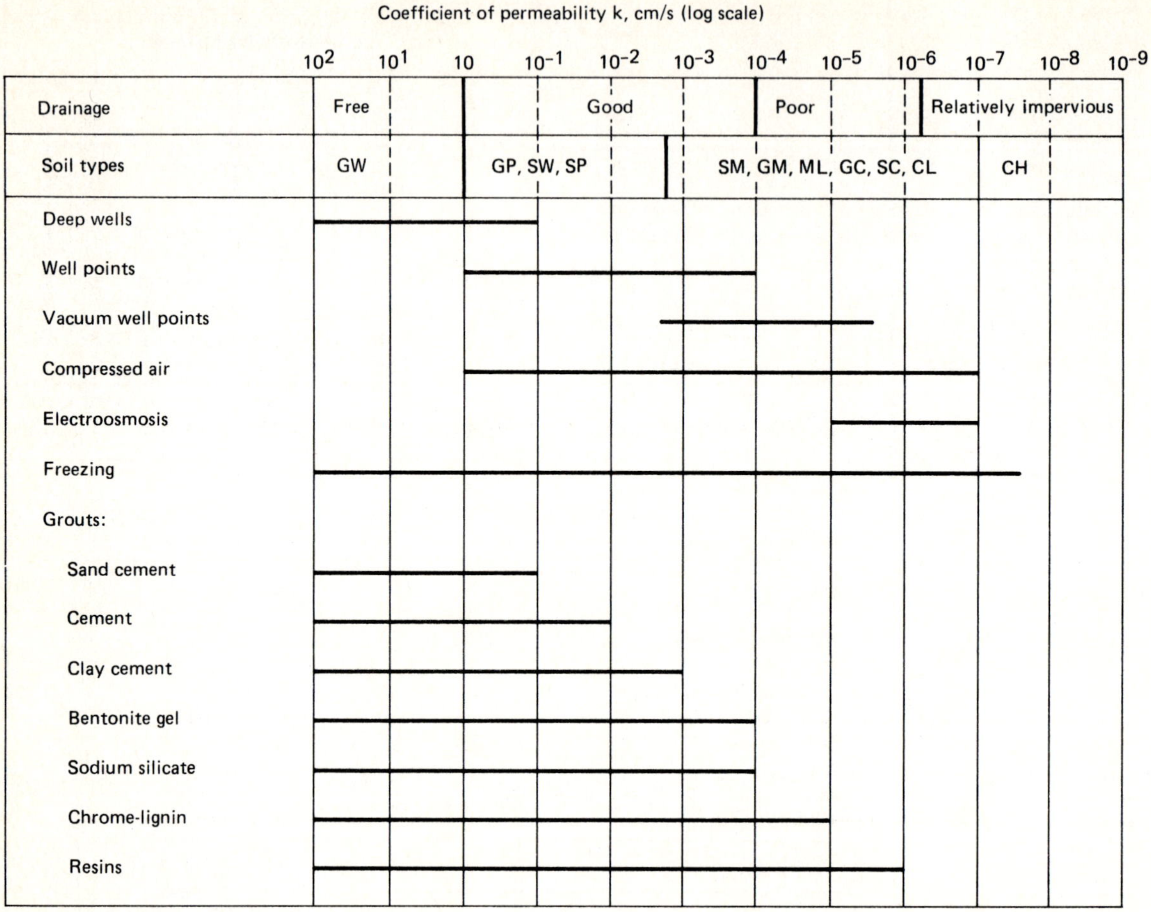

FIG. 2.10 General applicability of some methods for controlling groundwater and seepage in soils as a function of grain-size characteristics. (Other methods not shown include liners, walls, and drains.) [From Hunt (1984).[1] Used with permission of McGraw-Hill Book Company.]

Groundwater and seepage control methods and their common applications are summarized in Table 2.10. A general comparison of many of the various methods in terms of soil-gradation characteristics is given in Fig. 2.10. Not included are liners, walls, and drains which are normally integral parts of a structure. The methods are described in detail in subsequent chapters where they are applicable to a particular construction problem.

REFERENCES

1. Hunt, R.E. (1984) *Geotechnical Engineering Investigation Manual*, McGraw-Hill Book Co., New York.
2. ISRM (1981) *Rock Characterization Testing and Monitoring*, E. T. Brown, ed., publishing for the Comm. on Testing Methods, Intl. Soc. for Rock Mech., by Pergamon Press, Oxford, England.
3. Broch, E. and Franklin, J. A. (1972) "The Point Load Strength Test," *Intl. J. Rock Mech.*, Min. Soc., Vol. 8.
4. Arman, A. (1970) "Engineering Classification of Organic Soils," *HRB No. 30*, Highway Research Board, Washington, D.C., pp. 75–89.
5. Deere, D. U. (1963) "Technical Description of Rock Cores for Engineering Purposes," *Rock Mech. Engrg Geol.*, Vol. 1, pp. 18–22.
6. Crosby, J. W. III, Konstantinidis, B., and Davis, P. (1981) "Geotechnical Applications of Borehole Geophysics," *ASCE J. Geotech. Engrg. Div.*, Vol. 107, No. GT10, Oct., pp. 1255–1267.

7. NAVFAC (1971) *Design Manual DM-7, Soil Mechanics, Foundations and Earth Structures*, Naval Facilities Engineering Command, Alexandria, Va.

8. Burmister, D. M. (1948) *The Importance and Practical Use of Relative Density in Soil Mechanics*, Vol. 48, Amer. Soc. for Test. and Mat., Philadelphia, Pa.

9. Burmister, D. M. (1951) *Identification and Classification of Soils—An Appraisal and Statement of Principles*, ASTM Spec. Pub. 113, Amer. Soc. for Test. and Mat., Philadelphia, Pa., pp. 3–24, 85–91.

10. Hough, K. B. (1957) *Basic Soils Engineering*, The Ronald Press, New York.

11. Salzman, G. S. (1974) "Seepage and Groundwater," Lecture notes prepared for Gilbert Assocs., Joseph S. Ward and Assocs., Caldwell, N.J.

12. Hoek, E. and Bray, J. W. (1977) *Rock Slope Engineering*, Inst. of Min. and Met., London.

13. Lambe, T. W. and Whitman, R. V. (1969) *Soil Mechanics*, John Wiley & Sons, New York.

14. Casagrande, A. (1940) "Seepage Through Dams," *Contributions to Soil Mechanics 1925–1940*, Boston Soc. Civ. Engrs.

15. Sherard, J. L., Woodward, R. J., Gizienski, S. F., and Clevenger, U. A. (1963) *Earth and Earth-Rock Dams*, John Wiley & Sons, New York.

16. Cedergren, H. R. (1977) *Seepage, Drainage and Flow Nets*, John Wiley & Sons, New York.

17. USAWES (1967) *The Unified Soil Classification System*, Tech. Memo No. 3-357, U.S. Army Eng. Waterways Exp. Sta., Corps. of Eng., Vicksburg, Miss.

CHAPTER THREE

STRENGTH OF GEOLOGIC MATERIALS

3.1 SOME BASIC CONCEPTS

3.1.1 FAILURE BY RUPTURE

Failure Modes

Strength is considered to be the ability of a material to resist failure by rupture. When a stress is applied to a geologic material it begins to deform as evidenced by strains. The deformation can take the form of compression, tension, or shear acting in directions where the deformation constraint is minimal. *Rupture* occurs, usually along well-defined failure surfaces, as the magnitude of the applied force is increased at some significant rate, when some characteristic stress level is reached.

There are several modes of failure by rupture as shown on Fig. 3.1. *Soils* fail either in tension or in shear. In most practical problems only the resistance to failure by compressive shear requires consideration, although in some cases tensile stresses causing cracks are significant. Rock failure modes have a greater variety. Shear failure occurs in weak rocks or along joints or in the crushed rock of fault zones. Tensile failure occurs during toppling in slopes or during the exfoliation or sheeting of granite masses. Bending or flexural failure can occur in a slab forming the roof over a mine. Crushing or compressive failure occurs under the drill bit of boring machines or in mine pillars.

Analytical Approaches: Rupture Versus Deformation

Both deformation without rupture and failure by rupture occur under applied stresses; the analytical approach to each phenomenon, however, is substantially different. *Deformations*, such as may occur under a foundation loading, are computed by applying elastic theory to determine the magnitude of stresses imposed on some element in the mass (*see Chap. 4*). *Rupture failure* analysis, in most cases, requires the determination of a failure surface and the forces acting to cause failure as well as the resisting forces, which consist basically of shear strength, acting along the failure surface. The concept of a limiting value that can be reached when the forces acting to cause failure are in balance with the forces acting to resist failure is termed *limiting equilibrium*. Practical examples are given in Fig. 3.55.

Stability against rupture of foundations, slopes, rigid retaining structures, and anchors is analyzed by most practitioners by applying limiting equilibrium concepts. Analysis of flexible retaining structures also requires determining the mo-

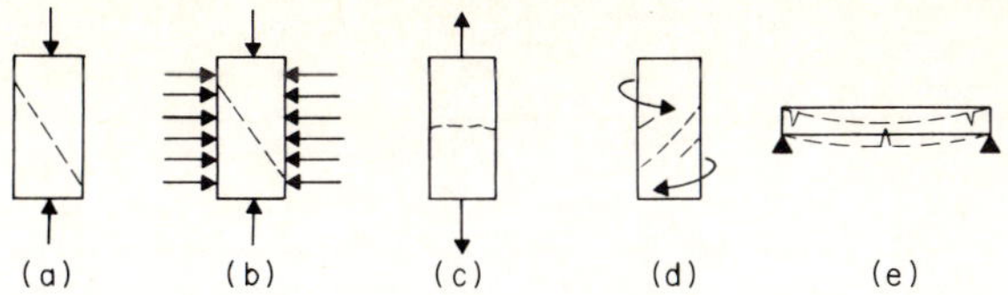

FIG. 3.1 Forms of rupture strength: (*a*) unconfined compression; (*b*) confined compression; (*c*) tension; (*d*) torsion; and (*e*) bending. Parts (*a*) and (*b*) are applied commonly to soil and rock; parts (*c*), (*d*), and (*e*) are applied normally to rock.

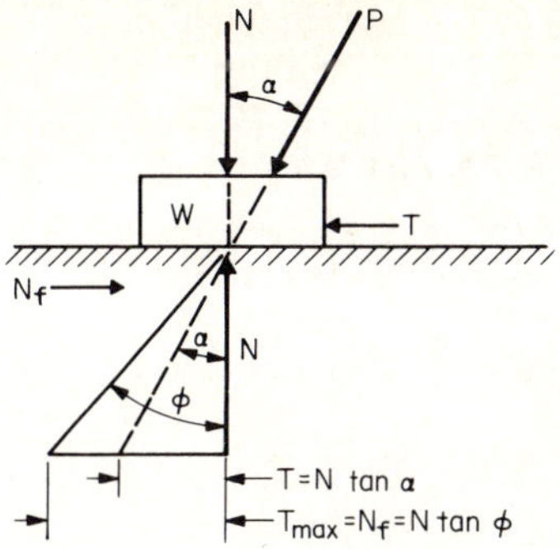

FIG. 3.2 Frictional force *f* resisting shearing force *T*.

ment capacity of the structural member and the support system required to resist deformation caused by lateral earth pressures which are a function of material strength.

Basic Shear Strength Parameters

Shear strength is the measure of rupture strength ordinarily applied in geotechnical problems and is represented by the strength mobilized during the displacement of adjacent elements along a planar or curved surface. *Shear strain* is the angle of displacement between elements during displacement. *Components* of shear strength can be the angle of internal friction ϕ or cohesion c or both. Which of these components act to resist rupture depends on a number of factors including the type of material, the rate of applied loading, the presence of preexisting failure surfaces, and the stress history of the formation. The actual values obtained during laboratory or field testing depend on the test method as well.

Frictional resistance is illustrated in Fig. 3.2. The block with weight *W* resting on a level surface is subjected to driving force *P* which has a normal stress component *N* (which includes the block weight) and a shearing stress component *T*. *P* is applied in increments until slip occurs. The resistance to the shearing stress component is the frictional force N_f, in which *f* is the coefficient of friction. Alpha (α) is the angle of obliquity, or resultant of *N* and *T*, and the angle ϕ represents the angle at which slip occurs, or the maximum angle of obliquity (*see Fig. 3.7e*). The maximum shearing resistance S_{max} is the product of the normal force and the tangent of ϕ or:

$$T_{max} = S_{max} = N \tan \phi \quad \text{kg/cm}^2, \text{kPa, tsf} \tag{3.1}$$

This concept is fundamental to geotechnical engineering; soil or rock strength represented by a frictional force between particles resists a shearing force with components of normal and shearing stresses.

Cohesion c, the other element of strength in geologic materials, is generally considered to be a bonding between surfaces of colloidal particles resulting from electrochemical or other forces. It is *independent* of normal forces, and the strength of a material exhibiting both cohesion and friction may be expressed in the Mohr-Coulomb equation:

$$s = c + \sigma_n \tan \phi \tag{3.2}$$

Cohesion is often given as s_u, the undrained shear strength of soils, for the special case where $\phi = 0$. This case occurs commonly when normally consolidated clays are loaded rapidly to failure and there is no time for drainage of the material to occur.

3.1.2 EARTH PRESSURES

The determination of earth pressures is fundamental to the solution of all stability problems involving soils, as they either cause failure directly or resist the forces causing failure.

Earth Pressure at Rest

The "at-rest" state of stress refers to earth pressures existing in the soil under natural conditions with no influence of construction or changes in loading. *Hydrostatic pressures* act equally in all directions and exist in fluids. The pressure at a given depth is equal to the product of the unit weight of the fluid and the depth, plus atmospheric pressure. *Earth stresses* below the ground surface, however, can be complex as they are influenced by topographic expression. The simplest stress pattern exists beneath a horizontal ground surface and is referred to as *geostatic stress*, as shown in Fig. 3.3.

Vertical stress σ_v on an element of soil or rock at some depth Z is equal to the summation of products of the unit weight and thickness for each distinct stratum thickness z in which the unit weight is expressed in terms of effective stresses as follows:

γ_t = total unit weight above the water table, or
γ_b = submerged (bouyant) unit weight below the water table ($\gamma_b = \gamma_t - \gamma_u$)

Below the groundwater level (GWL) the effective stress, $\bar{\sigma}_v = \gamma_t z - \gamma_w z$ where $\gamma_w z = u$, the total water pressure at a point or the neutral stress (*see Art. 2.3.2*); therefore, $\bar{\sigma}_v = \gamma_t z - u$. Some relationships for determining overburden pressures are given on Fig. 3.4.

Lateral stress is influenced by the material strength and at depth Z is given in terms of the *coefficient of lateral earth pressure at rest* K_o, also known as the *coefficient of lateral stress ratio* expressed as

$$K_o = \bar{\sigma}_v/\bar{\sigma}_h \qquad (3.3)$$

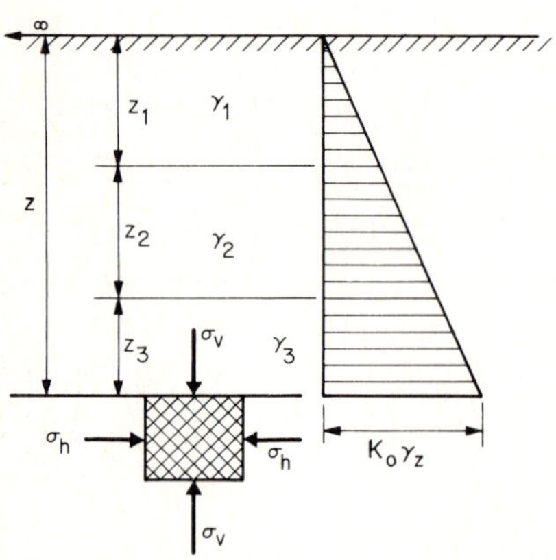

$$\sigma_v = \sum_0^z \gamma_i z_i \qquad \sigma_h = K_o\sigma_v$$

FIG. 3.3 The geostatic stress condition and earth pressure "at rest."

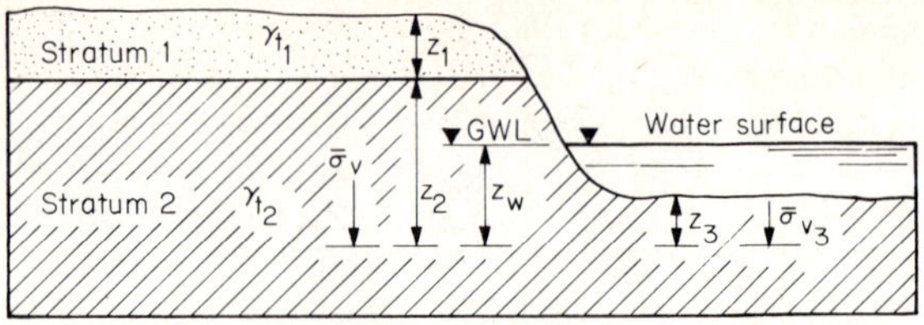

$$\sigma_v = \gamma_{t1}z_1 + \gamma_{t2}z_2 - \gamma_w z_w \qquad \sigma_{v3} = (\gamma_t - \gamma_w)z_3$$

or

$$\sigma_v = \gamma_{t1}z_1 + \gamma_{t2}(z_2 - z_w) + \gamma_b z_w \qquad \sigma_{v3} = \gamma_b z_3$$

FIG. 3.4 Overburden stresses below the water surface.

For a discussion of values, applications, and methods of measurements of K_o see Art. 3.2.3.

Passive and Active Earth Pressures

The "at-rest" state is the natural or in situ condition in which the mass is stable; i.e., no movement is occurring. The passive and active states of stress refer to conditions where movement of the mass occurs and states of limiting equilibrium are reached. If the ground surface is horizontal, a *passive state of stress* is a condition in which a soil mass under horizontal compression exerts its maximum resistance, and the *active state of stress* is a condition in which a soil mass reaches a state of failure when it is permitted to stretch in a horizontal direction (*see Art. 3.2.3*).

Induced and Residual Stresses

Natural events or construction activity can induce stresses that can increase or decrease earth pressures. Stresses are *increased* by deposition, excavation, embankment or foundation loadings, a lowering of the water table, or tectonic activity and metamorphism. Stresses are *reduced* by erosion of material from the surface, by excavation which removes load, or by rising water tables. Tunnels and other underground openings can result in complicated changes about the opening, represented generally by stress decrease in both the vertical and lateral directions (*see Art. 10.1.2*).

Residual stresses remaining in rock masses from tectonic activity, metamorphism, etc. can be many times overburden stresses [see Art. 6.6 of Hunt (1984)[1]]. They are evidenced by slabbing of rock from the walls of steep-sided valleys, the deflection and slabbing of walls and floors in excavations, and rock bursts in deep mine chambers and tunnels. Because rock is more nearly elastic than soil, the strains that occur upon stress release are much greater and occur much more rapidly in rock masses, often explosively. Heavily overconsolidated clays and clay shales usually contain residual stresses which cause heave during excavation.

3.1.3 PRINCIPAL STRESS RELATIONSHIPS

Significance

A shearing force, with components of normal and shearing stresses, is resisted by a frictional force in Fig. 3.2. The maximum shear strength may be expressed in terms of the frictional angle ϕ and the normal stress N, or σ_N. In the ground the magnitudes of these normal and shearing stresses are related to the magnitudes of the confining pressures caused by overburden, applied stresses from construction activity, or some other source such as residual stresses. In stability analysis it is necessary to describe the pertinent stress system; this is achieved by employing methods from engineering mechanics for analyzing stress at a point on a plane passing through an element.

Principal Stresses

A fundamental principle of mechanics holds that three *principal planes* exist through any point in a mass, always intersecting each other at right angles (acting orthogonally). In order of decreasing magnitude of normal stress these planes are known as the *major, intermediate,* and *minor* principal planes, and on each there are no shearing stresses. In Fig. 3.5 O_p represents any point within a stressed mass, or the origin of the planes, and OA and OB are the major and minor principal planes. The intermediate principal plane is that of the sketch.

The *principal stresses* are those acting perpendicular to the principal planes, including the major (maximum) principal stresses σ_1, the minor (minimum) principal stress σ_3, and the intermediate principal stress σ_2. The relationship between the principal stresses and the normal stress and the shear stress acting on any other plane is shown in Fig. 3.5. In geomechanics problems the intermediate principal stress may often be considered to be equal to the minor principal stress.

The Mohr Diagram

Equilibrium requires that the sum of the forces given in Fig. 3.5 equal zero. Therefore, σ_n and τ can be expressed in terms of the principal stresses and the angle θ as:

$$\sigma_n = \frac{\sigma_1 + \sigma_3}{2} + \frac{\sigma_1 - \sigma_3}{2} \cos 2\theta \tag{3.4}$$

$$\tau = \frac{\sigma_1 - \sigma_3}{2} \sin 2\theta \tag{3.5}$$

If points are plotted to represent coordinates of normal and shearing stresses acting on a particular plane for all values of θ given in Eqs. 3.4 and 3.5, their loci form a circle which intersects the horizontal axis at coordinates equal to the major

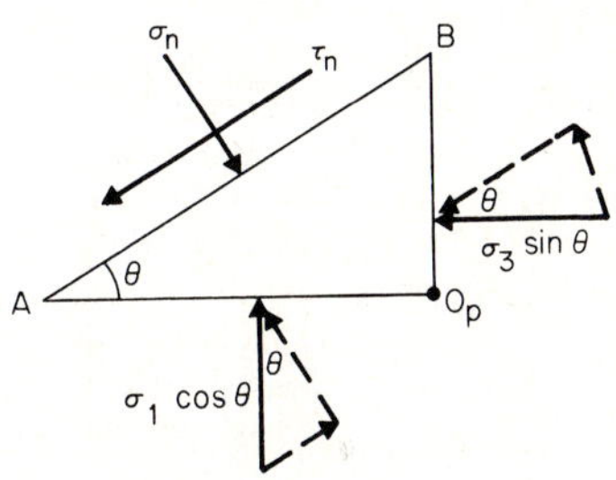

FIG. 3.5 Stresses on a random plane through a point (σ_2 is the plane of the paper).

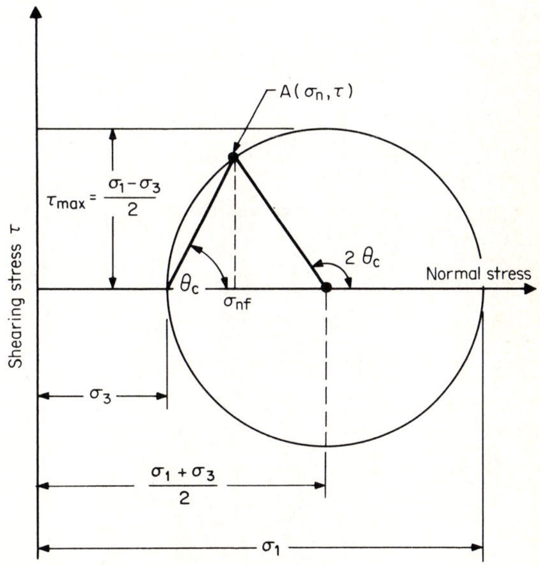

FIG. 3.6 The Mohr diagram relating τ, σ_{nf}, σ_3, and σ_1 ($\theta_c = \theta$ at failure; σ_{nf} = normal stress at failure).

(σ_1) and minor (σ_3) principal stresses. The circle is referred to as the *Mohr diagram*, or *Mohr's circle*, given in Fig. 3.6.

Many fundamental stress relationships may be illustrated by Mohr's circle, as shown in Fig. 3.7. Figure 3.7*a* shows the major and minor principal planes intersecting at the point of origin O_p. The maximum shearing stress τ_{max} equals the circle's radius and occurs on planes with $\theta = 45°$ as shown in Fig. 3.7*b*, and

$$\tau_{max} = \tfrac{1}{2}(\sigma_1 - \sigma_3) \qquad (3.6)$$

Conjugate shearing stresses occur on planes at right angles to each other and are numerically equal but opposite in sign, as in Fig. 3.7*c*. The resultant stress on any plane has a magnitude $\sqrt{\sigma^2 + \tau^2}$ and an obliquity $\alpha = \tan^{-1}(\tau/\sigma)$, as in Fig. 3.7*d*. The maximum angle of obliquity α_{max} (Fig. 3.7*e*) is a line drawn tangent to the circle through the origin. The limiting obliquity is the criterion of slip for frictional forces (Fig. 3.2); therefore, it is the plane of maximum obliquity, and not the plane of principal shear [defined by $\tau = \tfrac{1}{2}(\sigma_1 - \sigma_3)$], that is likely to fail. It may be seen from Fig. 3.7*e* that the plane of maximum shear has a slightly smaller obliquity than that designated as α_{max}. In geomechanics problems, $\alpha_{max} = \phi$.

Field and Laboratory Relationships

Some field relationships between principal stresses and the normal and shearing stresses acting on failure surfaces are shown in Fig. 3.8.

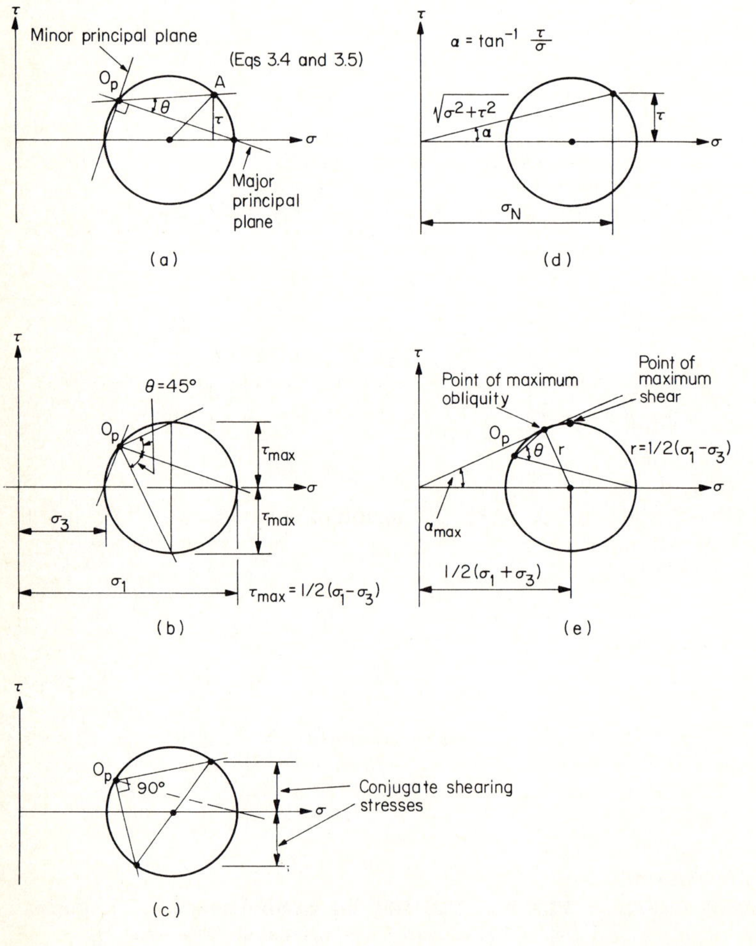

FIG. 3.7 Fundamental stress relationships demonstrated by the Mohr diagram: (*a*) stress condition for Fig. 3.5; (*b*) principal shearing stresses; (*c*) conjugate shearing stresses; (*d*) obliquity and resultant stress; and (*e*) maximum obliquity.

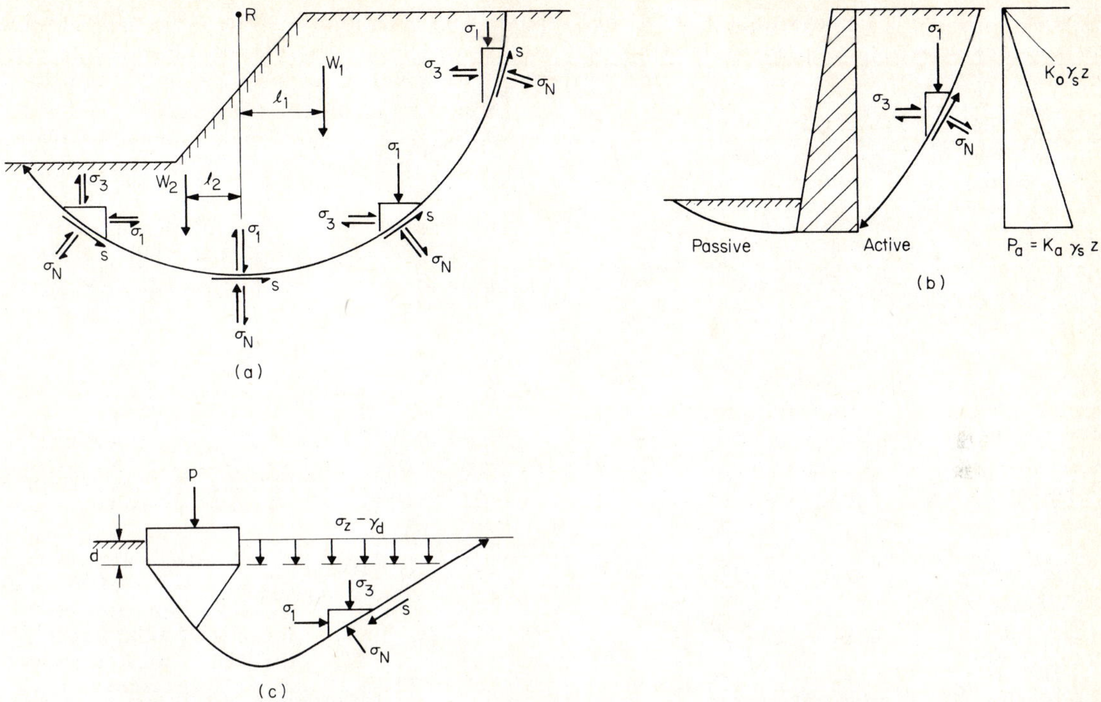

FIG. 3.8 Field relationships between principal stresses and the normal and shearing stresses, and the shearing resistance S: (*a*) rotational slope failure; (*b*) retaining wall; and (*c*) shallow foundation. [Parts (*a*) and (*b*) after Burmister (1953).[2] Copyright ASTM, 1916 Race Street, Philadelphia. PA 19103. Reprinted with permission.]

The normal stress is a function of the overburden stress γ_z which varies with the orientation of the failure surface to the ground surface. It can be vertical ($\sigma_v = \gamma_z$) or horizonal ($\sigma_h = K_o \gamma_z$), but usually is neither as shown in Fig. 3.8. The relationship between the principal, normal, and shearing stresses in a laboratory triaxial test is given in Fig. 3.9*a*. In this case $\sigma_1 = \sigma_3 + \sigma_d$, where σ_3 is a confining pressure and σ_d is a deviator stress. The distribution of compressive and shearing stresses beneath a circularly loaded area during general shear (Fig. 3.8*c*) is shown in detail in Fig. 6.29.

3.2 BASIC STRENGTH RELATIONSHIPS

3.2.1 FAILURE CRITERIA

General

Failure by rupture occurs in a material under an applied stress when the available shearing strength is equal to the shearing stresses. Several commonly accepted basic criteria have been developed to predict the principal stresses at failure and the material strength as follows:

Soil Mechanics (This Article)

- The Mohr criterion of shear failure:

$$\tau = \sigma_n \tan \alpha \tag{3.7}$$

- The Coulomb criterion of failure, also known as the Mohr-Coulomb equation:

$$s = c + \sigma_n \tan \phi \tag{3.8}$$

- The Terzaghi-Coulomb equation giving strength in terms of effective stresses:

$$s = \bar{c} + (p-u) \tan \bar{\phi} = \bar{c} + \bar{\sigma}_n \tan \bar{\phi} \tag{3.9}$$

Rock Mechanics (See Art. 3.4)

- Griffith criterion for brittle failure

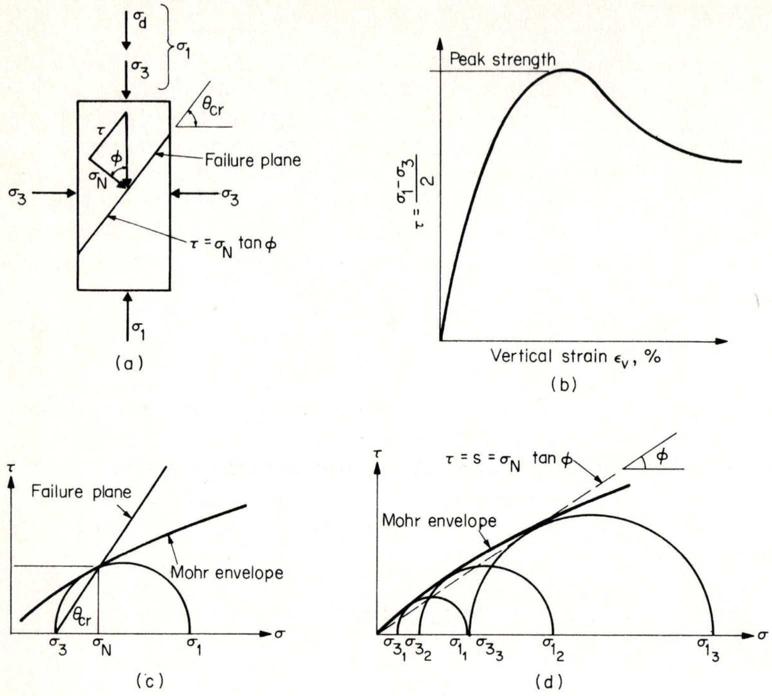

FIG. 3.9 The triaxial compression test and the Mohr envelope: (*a*) stresses on triaxial test specimen; (*b*) stress-strain curve from triaxial test; (*c*) Mohr circle for triaxial compression test at failure; and (*d*) Mohr's envelope and the friction angle for a range of confining pressures.

- Coulomb-Navier criterion:

$$S_s = \tau_{max} = \tau_o + f\sigma_n \tag{3.10}$$

- Also the Terzaghi-Coulomb equation, Eq. 3.9.

Mohr Criterion of Rupture

Mohr's strength theory holds that shearing stress on any plane in Fig. 3.7*d* may be expressed as:

$$\tau = \sigma_n \tan\alpha \tag{3.7}$$

and if α is at its limiting obliquity, the shearing stress is also at its limiting or maximum value, which is the shearing strength. Therefore, the shearing strength (for a cohesionless material) may be written as

$$s = \sigma_n \tan\phi \tag{3.11}$$

where ϕ is the angle of maximum obliquity, resulting in an expression similiar to Eq. 3.1 for a frictional force acting on a free body.

The relationship between the principal stresses (σ_1, σ_3), τ, σ_n, and ϕ may be illustrated in practice by the triaxial compression test. When a soil specimen is confined triaxially and an axial load is applied, rupture will occur when the shear stress on the plane at the angle θ_{cr} to the major principal stress exceeds the shear strength of the material. This is the basis for the Mohr criterion of rupture. Stresses on a laboratory specimen are shown in Fig. 3.9*a*. The confining stress σ_3 is the minor principal stress. (The intermediate principal stress σ_2 can be disregarded in many practical problems and in the triaxial test is the same as σ_3.) The major principal stress, $\sigma_1 = \sigma_3 + \sigma_d$, where σ_d is the applied or *deviator stress*, which is increased until failure. Figure 3.9*b* illustrates the stress-strain curve for a given confining pressure.

The relationship between σ, τ, and the failure plane are given on the Mohr circle, Fig. 3.9*c*. The failure plane is inclined at an angle θ_{cr} to which the major principal stress is acting at the limiting stress condition. As shown in Fig. 3.10, the *theoretical failure plane* lies at an angle $\theta = 45° + \phi/2$; however, the actual or *observed failure plane* may be found to vary from this value. The normal stress σ_n is perpendicular to the failure plane (Fig. 3.9*a*) and is the stress corresponding to the point where Mohr's circle is tangent to Mohr's envelope.

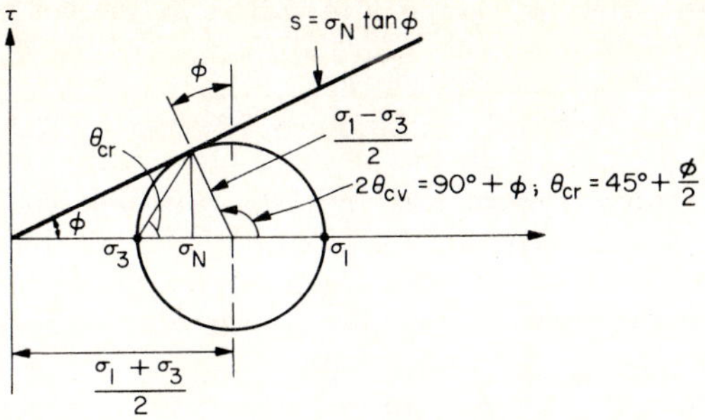

FIG. 3.10 Relationships between ϕ and the principal stresses at failure:

$$\sin \phi = \frac{\sigma_1 - \sigma_3}{\sigma_1 + \sigma_3} \qquad \textbf{(3.12)}$$

$$\frac{\sigma_1}{\sigma_3} = \frac{1 + \sin \phi}{1 - \sin \phi} = \tan^2(45 + \phi/2) = N_\phi$$

NOTE: **For significance of Eq. 3.12 see Art. 3.2.3, Passive and Active Earth Pressures.**
For *p-q* diagrams:

$$\sin \phi = \frac{\sigma_1 - \sigma_3}{\sigma_1 + \sigma_3} = \frac{p}{q} \qquad \textbf{(3.13)}$$

Mohr's envelope is obtained from several Mohr's circles which are plotted for values of σ_1 and σ_3 resulting from tests performed at various confining pressures. In cohesionless soils, a line drawn tangent to the circles goes through the origin at an angle equal to ϕ (Fig. 3.9*d*) which forms the Mohr envelope and represents the limit between a stable condition and failure (Mohr's theory of rupture). The shear strength along this line is defined by $\tau = s = \sigma_n \tan \phi$ (Eq. 3.11). Mohr's envelope drawn for a series of tests performed at a wide range of confining pressures will be slightly curved because of the variation of frictional resistance due to grain interlocking. The proportional increase is higher at lower confining pressures. In practice, the envelope line to define ϕ is selected as the best fit through the origin and tangent to the circles in the region of confining pressures of interest. The curvature of the envelope is greatest for dense granular soils and decreases as a soil becomes looser.

There are several useful relationships between ϕ and the principal stresses at failure. These are given on Fig. 3.10.

Mohr-Coulomb Failure Law

The foregoing discussion applies to stress conditions developing during failure in a dry, cohesionless soil, such as quartz sand. Many materials encountered in practice yield a strength envelope intercepting the vertical axis at some value which is termed *cohesion c* as shown in Fig. 3.11. Cohesion is independent of the normal stress, and the shear strength is expressed by the Mohr-Coulomb failure law

$$s = c + \sigma_n \tan \phi \qquad (3.8)$$

The rational use of the Mohr-Coulomb relationship requires an understanding of the various factors that effect shear strength, for c and ϕ are not constant soil properties but rather empirical coefficients that vary with soil drainage and stress history, test conditions, and other factors (*see Art. 3.3*).

3.2.2 TOTAL AND EFFECTIVE STRESS CONCEPTS

Total Stresses

If no drainage is permitted from a fully saturated soil as load is applied, the stress at failure is carried partially by the pore water and partially by the soil particles, between which there is thus some intergranular stress. The peak "undrained" strength parameters mobilized depend on the soil type:

- Cohesionless soils—The envelope passes through the origin (Figs. 3.12 and 3.13) and strength is expressed in terms of ϕ as in Eq. 3.11.
- Soils with cohesion—The envelope intercepts the shear stress ordinate at a value taken as

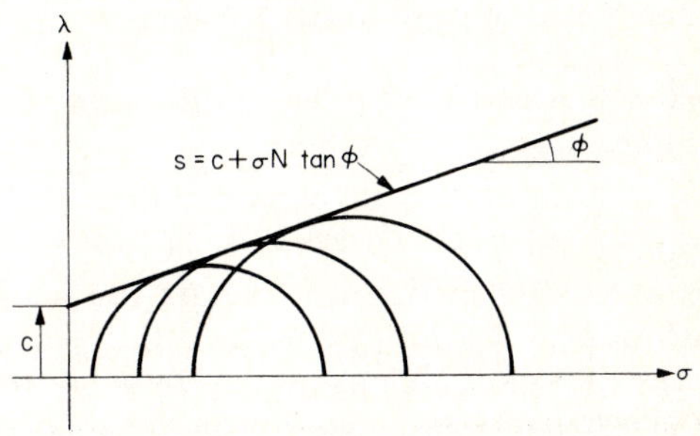

FIG. 3.11 Mohr-Coulomb failure envelope for laboratory test on material exhibiting cohesion.

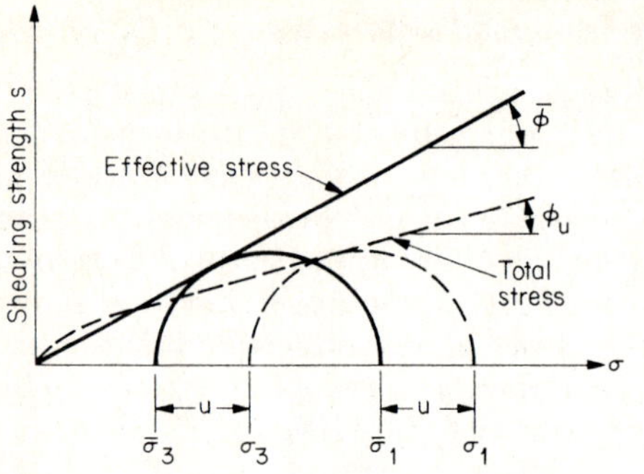

FIG. 3.12 Mohr's envelopes for total and effective stresses from CU triaxial tests on *loose* saturated sands. ϕ_u is termed the consolidated-undrained angle of shearing resistance and $\bar{\phi}$ is the effective angle.

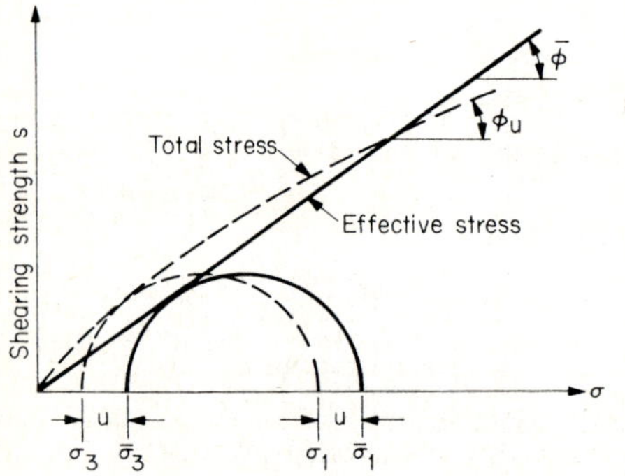

FIG. 3.13 Mohr's envelopes for total and effective stresses from CU triaxial tests on *dense* saturated sands.

cohesion c, as in the case shown in Fig. 3.11, and the strength is expressed by Eq. 3.8.

- Normally to slightly overconsolidated clays—There is no frictional component ($\phi = 0$) and the strength is expressed as s_u, the undrained strength as discussed in Art. 3.3.3.

Effective Stresses and the Terzaghi-Coulomb Equation

If the saturated soil is permitted to drain as load is applied, the stress is initially carried by the pore water, but as drainage occurs the load is transferred to the soil grains, and the *effective stresses* (intergranular stresses) are mobilized. The *effective stress is equal to the total normal stress minus the pore-water pressure,* and the peak "drained" shear strength is expressed by the equation

$$s = (\sigma_n - u) \tan \bar{\phi} = \bar{\sigma}_n \tan \bar{\phi} \qquad (3.14)$$

As shown in Figs. 3.12 and 3.13, ϕ values based on effective stresses are often higher than those based on total stresses as long as the pore pressure u is positive. In dense sands u is negative and the undrained strength is higher; i.e., ϕ is higher. If pore pressures are measured during *undrained* loadings, effective stresses can be computed. For soils with a cohesion intercept (Fig. 3.11), the drained strength is expressed by the *Terzaghi-Coulomb equation*

$$s = \bar{c} + (p - u) \tan \bar{\phi} = \bar{c} + \bar{\sigma}_n \tan \bar{\phi} \qquad (3.9)$$

Effective stresses are also discussed under Arts. 2.3.2 and 3.1.2.

In *partially saturated soils,* strength is controlled by effective stresses, but the effective stress concept cannot be applied directly because of pressures in the air or other gas in the partially saturated voids. For dam embankments, strength should be estimated during the construction phase from tests performed to duplicate in situ conditions as closely as possible in terms of percent saturation, total stress, and the pressure on the liquid phase [Lambe and Whitman (1969)[3]]. Since slope failure in the field often occurs under saturated conditions, however, samples are normally saturated during testing to yield more conservative strength data (see discussion on *soil suction, Art. 3.3.2*). *Apparent cohesion* results from capillary forces in partially saturated, fine-grained granular soils such as fine sands and silts and provides a temporary strength which is lost upon either saturation or drying. An important factor affecting strength in partially saturated soils is negative pore pressures, termed *soil suction* [Brand (1982)[4]], as discussed in Art. 3.3.2.

Pore-Pressure Parameters

Pore-water pressures have been discussed in Arts. 2.3.2 and 3.1.2. They are measured in the laboratory either by the triaxial shear apparatus or by the consolidometer and in the field with piezometers. In the field they may be increased by stresses induced from construction loadings, by the height of groundwater acting on an element at depth in the ground, or by dynamic forces during earthquakes. Their major significance is that they act to reduce the intergranular stress that develops between soil particles and may therefore reduce soil strength.

Pore-pressure parameters are used to express the portion of stress increment carried by the pore fluid in terms of the ratio of the pore pressure increment (Δu) to the total stress increment ($\Delta\sigma$).

Parameter A

Pore-pressure parameter A is expressed for triaxial loading as

$$A = \frac{\Delta u - \Delta\sigma_3}{\Delta\sigma_1 - \Delta\sigma_3} \tag{3.15}$$

or for the normal *undrained* test (CU) where $\Delta\sigma_3 = 0$, as

$$A = \frac{\Delta u}{\Delta\sigma_1} \tag{3.16}$$

Parameter A is the most significant in practice. Values depend on soil type, state of stress, strain magnitude, and time. Typical values are given in Table 3.1 for conditions at failure, but important projects always require measurement by test. High values occur in soft or loose soils. Minus values indicate negative pore pressures, which occur in dense sands and heavily preconsolidated clays as the result of the tendency to increase in volume during shear *(dilatancy)*. Pore pressures are most responsive to applied stress in sensitive clays and loose fine sands which are subject to liquefaction. In soft or loose soils higher shear strain magnitudes result in higher values for parameter A.

TABLE 3.1
TYPICAL VALUES OF PORE-PRESSURE PARAMETER *A* AT FAILURE*

Soil type	Parameter A
Sensitive clay	1.5 to 3.0
Normally consolidated clay	0.7 to 1.3
Overconsolidated clay	0.3 to 0.7
Heavily overconsolidated clay	−0.5 to 0
Loess	−0.2
Very loose fine sand	2.0 to 3.0
Medium compact fine sand	0.0
Dense fine sand to silt	0.3 to 0.5

*After Leonards (1962).[5]

Parameter A is measured in the CU triaxial test (*see Art. 3.3.3*) and is used to estimate the magnitude of initial excess pore pressure produced at a given point in the subsoil by a change in the total stress system. With piezometers used to measure in situ pore pressures, the validity of the estimates can be verified and the loading rate during surcharging or embankment construction can be controlled to avoid failure caused by exceeding the undrained shear strength (*see Art. 5.3.2*).

Parameters B and C

Parameter B is the pore-pressure change corresponding to a total stress change for isotropic loading (three-dimensional), expressed as

$$B = \frac{\Delta u}{\Delta\sigma_3} \tag{3.17}$$

Parameter C is the change in pore pressure corresponding to the change in direct stress occurring when soil is loaded into the odometer (one-dimensional compression), expressed as

$$C = \frac{\Delta u}{\Delta\sigma_1} \tag{3.18}$$

In a *saturated soil* the compressibility of the soil skeleton is almost infinitely greater than the compressibility of the pore water, and thus essentially all of the stress increment applied is carried by the pore fluid in confined or isotropic loading; i.e., $C \approx B \approx 1$. In a *dry soil* the compressibility of the pore air is infinitely greater than the compressibility of the soil skeleton and $C \approx B \approx 0$. In *partially saturated soil* the very high compressibility of air relative to water and the soil skeleton results in low values of B and C until the saturation approaches 100%. Parameter C is little used in practice; parameter B is used as a control factor in specimen saturation under back pressure during triaxial testing (*see Art. 3.3.2*). The pore pressure parameters are discussed in detail in Bishop and Henkel (1962)[6] and Lambe and Whitman (1969).[3]

Stress Path Method (*p-q* Diagrams)

If Mohr's circles are plotted on a single diagram for many states of stress for a given specimen, or for many states of stress for a number of different specimens, the diagram becomes confusing and difficult to interpret. The *stress path method* is used to depict successive states of stress as a soil element is loaded or unloaded through a series of stages. The states of stress are plotted as *stress points*, rather than Mohr's circles, and the points are connected to form a *stress path*.

Stress Points

Stress points are determined from triaxial test and plotted as coordinates given by p and q where p locates the center of a Mohr's circle and q locates the point on Mohr's circle equal to τ_{max} as defined by

$$p = \frac{\sigma_1 + \sigma_3}{2} \quad \text{and} \quad q = \pm \frac{\sigma_1 - \sigma_3}{2} \tag{3.19}$$

q is plus if σ_1 is inclined equal to or less than $\pm 45°$ to the vertical and minus if σ_1 is inclined less than $\pm 45°$ to the horizontal. In most cases for which the stress point representation is used, the principal stress acts on vertical and horizontal planes and p and q may be expressed as

$$p = \frac{\sigma_v + \sigma_h}{2} \quad \text{and} \quad q = \frac{\sigma_v - \sigma_h}{2} \tag{3.20}$$

Plotting a stress point is equivalent to plotting one single point of Mohr's circle; numerically, $q = \frac{1}{2}\sigma_d$, as shown in Fig. 3.14.

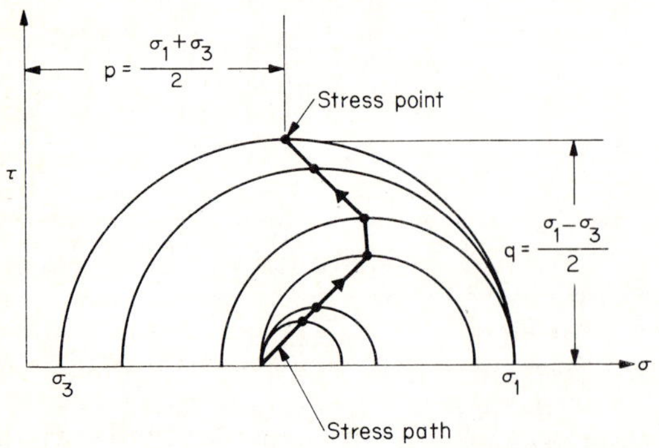

FIG. 3.14 Stress points and stress path plotted from a series of Mohr's circles. [From Lambe (1967).[7]]

Stress Paths

A stress path is obtained by connecting a series of stress points with lines or curves as shown in Fig. 3.14; the series of points can be from a single test, with values of p and q plotted as the test proceeds, or from several tests. The K_f line drawn through several stress points provides a line similar to Mohr's envelope as shown in Fig. 3.15. It makes an angle $\bar{\alpha}$ with the horizontal axis, and since it is based on plots of τ_{max}, the maximum obliquity $\bar{\phi}$ is found from

$$\bar{\phi} = \sin^{-1}(\tan \bar{\alpha}) \tag{3.21}$$

For soils exhibiting cohesion, the K_f line intercepts the vertical axis at $\bar{a}$ and $\bar{c}$ is found from

$$\bar{c} = \bar{a} / \cos \bar{\phi} \tag{3.22}$$

Three types of stress path are shown in Fig. 3.16*a*: σ, the path for total stress; $\sigma - u_s$, the path for total stress minus static pore pressure or minus steady-state pore pressure; and $\bar{\sigma}$, the path

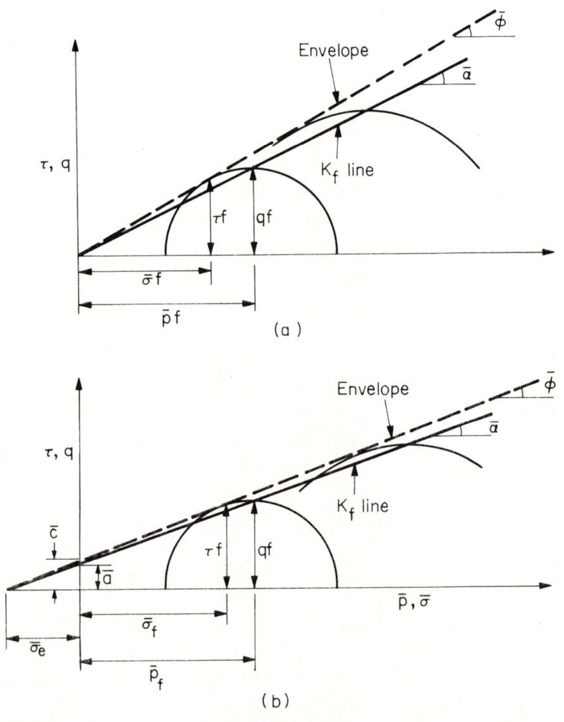

FIG. 3.15 Mohr's envelope and the K_f line for failure conditions: (*a*) cohesionless soil and (*b*) cohesive soil. [From Lambe and Marr (1979).[8]]

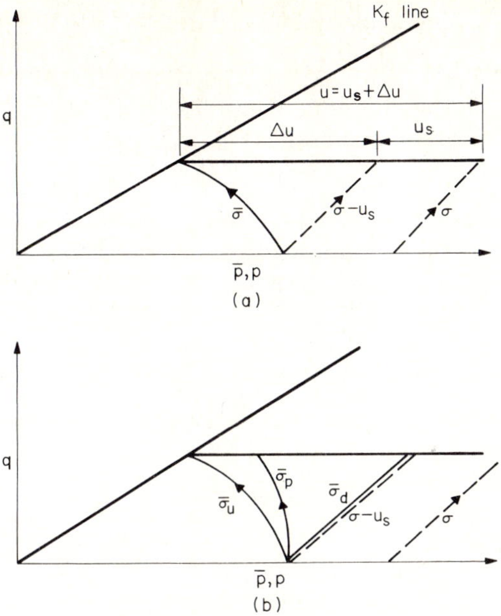

where σ = path for total stress
$\sigma - u_s$ = path for total stress minus static pore pressure or steady-state pore pressure
$\bar{\sigma}$ = path for effective stress
$\bar{\sigma}_u$ = path for effective stress, undrained
$\bar{\sigma}_p$ = path for effective stress, partly drained
$\bar{\sigma}_d$ = path for effective stress, drained
$\bar{\sigma}_d = \sigma - u_s$

FIG. 3.16 Stress paths for total and effective stresses, drained and undrained tests for NC clay. [From Lambe and Marr (1979).[8]]

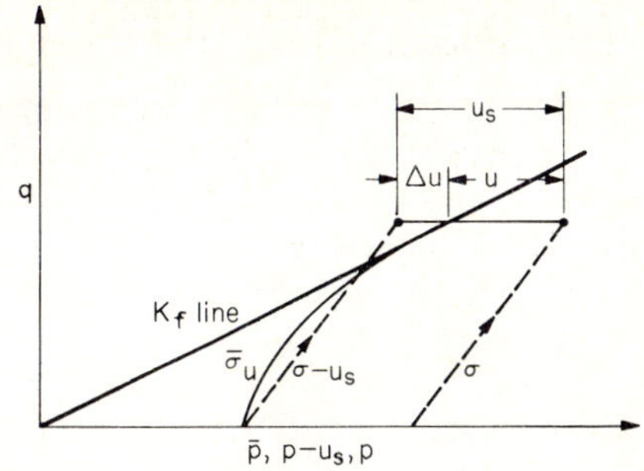

FIG. 3.17 Stress paths for undrained test, OC clay. [From Lambe (1967).[7]]

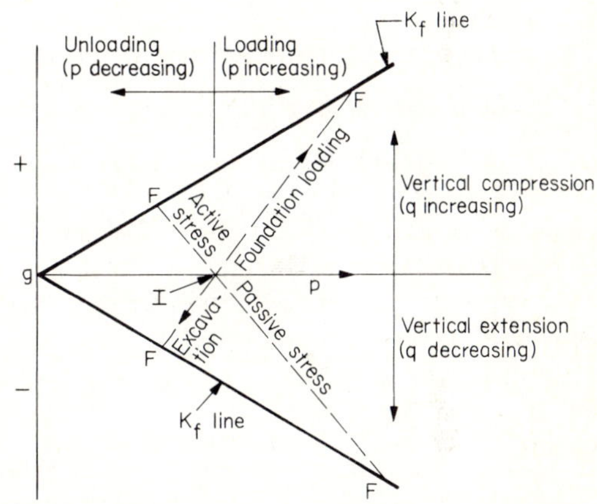

FIG. 3.18 Stress paths for various drained conditions. [From Lambe and Marr (1979).[8]]

for effective stress. Possible effective stress paths for a given total stress are shown in Fig. 3.16*b*. There are two possible extremes: $\bar{\sigma}_u$ for completely undrained conditions and $\bar{\sigma}_d$ for fully drained conditions. Between $\bar{\sigma}_u$ and $\bar{\sigma}_d$ are an infinite number of effective stress paths for $\bar{\sigma}_p$, partial drainage. Figure 3.16*a* illustrates the development of positive pore pressures for an NC clay, whereas Fig. 3.17 illustrates negative pore pressures developing in an OC clay. Pore pressures are related to volume changes during shear; the volume changes are in turn related to the stress history of the material.

Stress paths for four conditions leading to failure are shown in Fig. 3.18. In each case the sample starts from an initial isotropic stress state point *I*. Reducing *p* constitutes "unloading"; increasing *p* constitutes "loading." Increasing *q* results in "vertical compression," and decreasing *q* results in "vertical extension" of the soil element. These four changes are analogous, respectively, to the wall-active case, foundation loading, the wall-passive case, and excavation. The stress path method and some applications can be found in Lambe (1967),[7] Lambe and Whitman (1969),[3] and Lambe and Marr (1979).[8] One simple application is to estimate total or effective stress parameters from a single direct shear or triaxial shear test. Stress points (p, q) are plotted for the latter part of the test and the points connected by the K_f line as shown in Fig. 3.19.

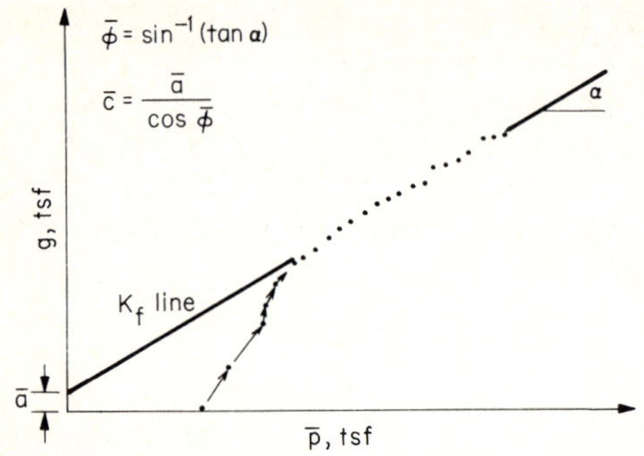

FIG. 3.19 Approximating strength parameters from a single direct shear or triaxial shear test using the stress path method.

3.2.3 AT-REST, PASSIVE AND ACTIVE STRESSES

Coefficient of Lateral Earth Pressure at Rest (K_o)

The "at-rest" state of stress was described in Art. 3.1.2, and the coefficient of lateral earth pressure at rest, or the coefficient of lateral stress ratio, was defined as $K_o = \bar{\sigma}_v/\bar{\sigma}_h$ (Eq. 3.3).

Applications of K_o

In practice K_o has a number of useful applications, including:

- Computations of lateral thrusts against retaining structures where the active state of stress is not mobilized (*see Art. 9.2.3*)
- Reconsolidation of triaxial test specimens according to an in situ anisotropic stress path [Lambe and Whitman (1969)[3]] (*see Art. 3.3.2*)
- Analysis of shaft resistance of pile foundations (*see Art. 7.3.4*)
- Computation of lateral swelling pressures against friction piles in expansive soils [Kassif and Baker (1969)[9]]
- Settlement computations in various situations [Lambe (1964)[10]]
- Analysis of progressive failure in clay slopes [Lo and Lee (1973)[11]]
- Prediction of pore pressures in earth dams [Pells (1973)[12]]

Measurements of K_o

Laboratory measurements are made directly with a semirigid confining ring similar to the consolidometer but with built-in cells to measure lateral and pore-water pressures [Abdelhamid and Krizek (1976)[13]] and indirectly using the consolidometer [Zeevaert (1953)[14]] or the triaxial compression test [Poulos and Davis (1972)[15]]. In situ measurements are made directly with the self-boring pressuremeter or Cankometer [Massarsch and Broms (1976)[16]] or a thin-wall earth-pressure cell of the Glötzl type [Massarsch and Broms (1976)[16]] or indirectly with the CPT [Schmertmann (1977)[17]].

Estimating K_o

Elastic solids—K_o is given in terms of Poisson's ratio ν

$$K_o = \frac{\nu}{1-\nu} \qquad (3.23)$$

Sands and *normally consolidated clays*—K_o is given in terms of ϕ

$$K_o = 1.0 - \sin \bar{\phi} \qquad (3.24)$$

It can be seen that the lateral earth pressure decreases as ϕ increases. Usual value ranges are 0.4 to 0.5, but mechanical compaction adjacent to walls can result in values as high as 1.0 to 1.5 (*see also Table. 9.1*).

Overconsolidated clays—K_o can range from 0.5 to almost 3.0 in the natural state and higher in clays subjected to mechanical compaction (Table 9.1). There appears to be no "standard" procedure for estimating K_o for OC clays although one approach that has received fairly wide acceptance is that presented by Brooker and Ireland (1965)[18] based on the relationship between K_o and OCR (overconsolidation ratio) and PI, as given in Fig. 3.20. This family of curves is based on data from five clays: London clay, Weald clay, Chicago clay, Goose Lake flour, and Bearpaw shale [Mayne and Kulhawy (1982)[19]].

Overconsolidation Ratio (OCR)

The overconsolidation ratio is defined as the ratio of the maximum past pressure p_c to the existing effective overburden pressure $\bar{p}_o$ (or $\bar{\sigma}_v$). In general, for a NC clay, $p_c/\bar{p}_o$ = 0.8 to 1.5 and for an OC clay, $p_c/\bar{p}_o > 1.5$ [Clemence and Finbarr (1981)[20]]. The OCR is used to estimate consolidation in clays [Schmertmann (1977)[17]], for cor-

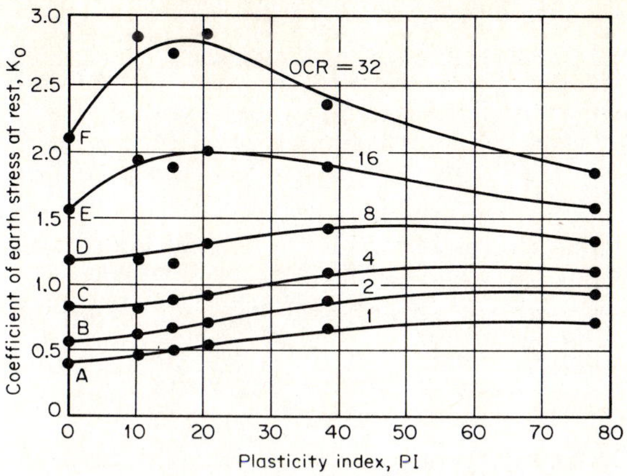

FIG. 3.20 Relationship between K_o, OCR, and PI. [From Brooker and Ireland (1965).[18] Reprinted with permission of National Research Council of Canada.]

relation of strength properties [Terzaghi and Peck (1967)[21]], and for estimating K_o as described above. The undrained strength s_u of NC clay normally falls within a limited fraction of the effective overburden stress, usually ranging from $s_u/\bar{p}_o$ = 0.16 to 0.4 (*see Art. 3.3.3*). Schmertmann (1977)[17] has presented a normalized ratio of $s_u/\bar{p}_o$, as measured by the cone penetrometer test, in terms of the OCR as given in Fig. 3.21. It provides

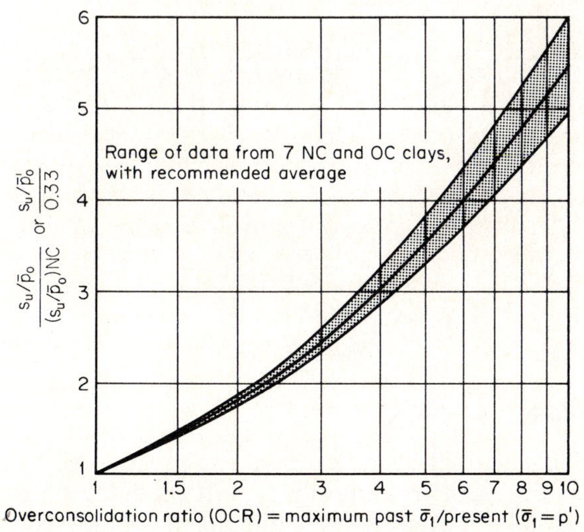

FIG. 3.21 Normalized $s_u/\bar{p}_o$ ratio versus OCR for use in estimating OCR from cone penetrometer value q_c in clays. [From Schmertmann (1977).[17] Reprinted with permission of the Federal Highway Administration.]

a method to approximate the OCR, but variations with clay types occur; for example, the OCR for freshwater montmorillonite clays will be overestimated and the OCR for "quick" clay will be underestimated. The OCR is normally determined from the laboratory consolidation test (*see Art. 4.3.4*)

Passive and Active Stresses

Passive and active stresses occur when shear strength is fully mobilized as a soil mass moves from the at-rest condition. The relationships between ϕ and the principal stresses (σ_v, σ_h) are illustrated for a cohesionless material in Fig. 3.10 and expressed by Eq. 3.12, which defines N_ϕ, often used to designate $\tan^2(45° + \phi/2)$. The Mohr-Coulomb equation can also be expressed in terms of N_ϕ and the principal stresses

$$\sigma_h = \sigma_v N_\phi + 2c\sqrt{N_\phi} \tag{3.25}$$

The *passive state* represents a state of *maximum* resistance mobilized when a force pushes against a soil mass and the soil exerts a resistance to the force, as illustrated by the retaining wall in Fig. 3.22*a*. As the soil element is compressed, σ_v remains unchanged but σ_h increases ($\sigma_h > \sigma_v$). As movement continues, τ decreases from the at-rest condition (1) until $\sigma_v = \sigma_h$ (2) and then increases until failure occurs at (4) as *slip lines form* in the mass and a state of *plastic equilibrium* is reached. The ratio σ_v/σ_h, termed the *coefficient of passive stress* K_p, is expressed as in Eq. 3.12 as

$$K_p = \tan^2(45° + \phi/2) \tag{3.26}$$

The *active state* is a state of minimum resistance occurring when a soil is permitted to stretch as would occur when the retaining wall tilts as shown in Fig. 3.22*b*. As the soil is stretched, σ_v remains unchanged and σ_h decreases (2) until the induced shear stress is sufficient to cause failure (3) and a state of plastic equilbrium is reached. The ratio σ_v/σ_h, termed the *coefficient of active stress* K_a, is expressed as

$$K_a = \tan^2(45° - \phi/2) \tag{3.27}$$

and

$$K_p = 1/K_a \tag{3.28}$$

The *Rankine states of stress* exist when the ratio $K = \sigma_h/\sigma_v$ reach K_p or K_a, and the stresses (conjugate stresses) defined by $K_p\sigma_v$ and $K_a\sigma_v$ are limiting values for σ_h, at which failure occurs. K_p and

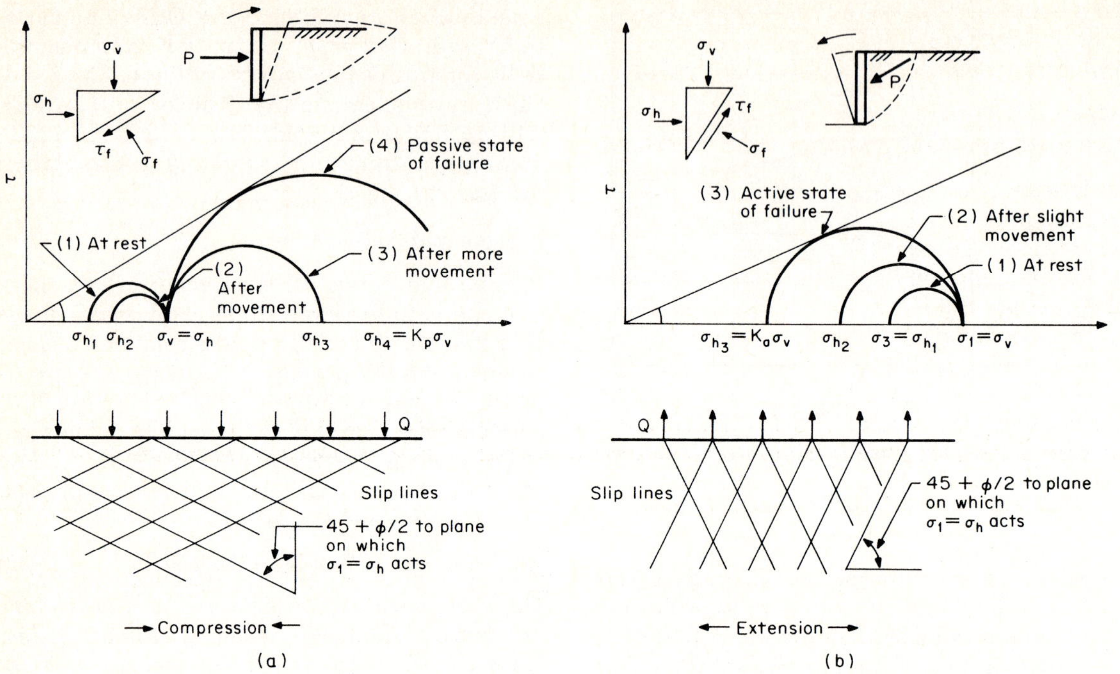

FIG. 3.22 Mohr diagrams for the passive (*a*) and active (*b*) states of stress as illustrated by a retaining wall.

K_a, often termed the *coefficients of lateral earth pressure*, are also found expressed as

$$K_p = \frac{1 + \sin\phi}{1 - \sin\phi} \tag{3.29}$$

$$K_a = \frac{1 - \sin\phi}{1 + \sin\phi} \tag{3.30}$$

The *passive resistance* increases with increasing ϕ and the *active force* decreases with increasing ϕ. Typical values are

ϕ, degrees	K_p	K_a
20	2.04	0.490
30	3.00	0.333
40	4.60	0.217

It is to be noted that K_o ranges between K_p and K_a, and for cohesionless soils beneath a horizontal surface $K_o = 1 - \sin\phi$. For *cohesive soils*, the stress coefficients are expressed as

$$K_p = \tan^2(45° + \phi/2) + \frac{2c}{\sigma_v}\tan(45° + \phi/2) \tag{3.31}$$

$$K_a = \tan^2(45° - \phi/2) - \frac{2c}{\sigma_v}\tan(45° - \phi/2) \tag{3.32}$$

and for $\phi = 0$

$$K_p = 1 + \frac{2c}{\sigma_v} \tag{3.33}$$

$$K_a = 1 - \frac{2c}{\sigma_v} \tag{3.34}$$

Earth Pressures

The significance of the Rankine stress states lies in the earth pressures that result when the active or passive states exist. Fundamental to the concept is that rupture has occurred along slip lines which then define the boundary of an unstable mass of earth, referred to as the *plastic zone*. The magnitude of the pressure exerted in the *lateral* direction may be expressed as the product of the weight of a unit mass times the depth times the lateral stress coefficient as follows

Passive earth pressure: $p_p = K_p \gamma z$ (3.35)

Active earth pressure: $p_a = K_a \gamma z$ (3.36)

To calculate p_p or p_a for practical problems requires the determination of the shape or boundary conditions of the plastic zone often requiring empirical assumptions. For the simple case of a wall retaining a dry, clean sand backfill, the plastic zone is defined by slip lines rising from the base of the wall at angles of $45° - \phi/2$ for the passive state, and $45° + \phi/2$ for the active state, assuming there is no friction between the wall and the soil (see *Art. 9.2.3*). In practice, the resulting states of stress when equilibrium is reached and the shape of the plastic zone depend very much on the type of deformation and the degree of roughness of the contact surface between the soil and its support.

The effect of induced and residual stresses on earth pressures has been mentioned in Art. 3.1.2.

Field Examples

Some field examples of passive and active earth pressures are illustrated in Fig. 3.23.

SLOPES (FIG. 3.23*a*) A small movement along the slope has caused the upper portion of the slope to stretch, mobilizing the active state and caus-

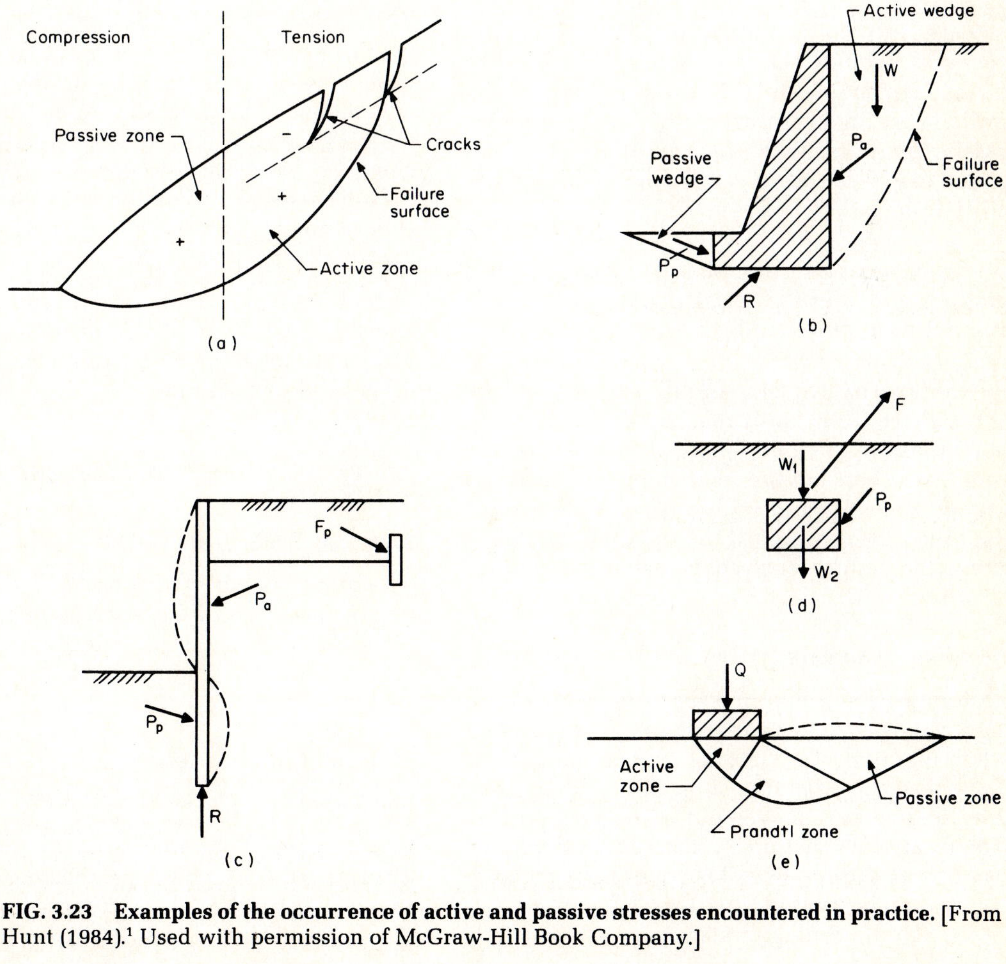

FIG. 3.23 Examples of the occurrence of active and passive stresses encountered in practice. [From Hunt (1984).[1] Used with permission of McGraw-Hill Book Company.]

ing tension cracks to form. The lower portion is resisting movement and entering the passive state. Failure will occur when the passive state is reached fully and shear strength along the failure surface can no longer support the weight of material above it.

RETAINING WALL (FIG. 3.23*b*) As the wall tilts, even a slight amount, the soil behind stretches until it reaches the active state; a failure surface forms the boundary of a soil wedge which exerts an "active" thrust against the wall through its weight. It pushes the wall at the toe into the soil causing passive resistance. Failure of the wall will occur unless the total weight of the wall, the passive resistance at the toe, and the strength of the soil are adequate to resist the active pressure thrust.

FLEXIBLE SHEET-PILE BULKHEAD (FIG. 3.23*c*) Small deflections mobilize active pressures which cause further deflections in the lower portion of the wall. These in turn push into the soil causing passive pressures. The upper portion of the wall is tied to a deadman anchor which provides stability from passive resistance.

EARTH ANCHOR BLOCK (FIG. 3.23*d*) Passive earth pressure provides the resistance against force F attempting to pull the block from the ground.

FOUNDATION LOADING (FIG. 3.23*e*) When the state of plastic equilibrium is reached during application of load to the foundation, an active Rankine zone forms beneath the footing, pushing a radial Prandtl zone (Fig. 6.28) sideways and the passive Rankine zone upwards. If the shearing resistance along the failure surface is exceeded, failure occurs.

Arching Phenomena

Arching occurs when one part of a soil or rock mass yields and the remainder stays in place, or when one part of a yielding support moves out more than the adjoining supports. As the soil in the yielding part moves out of its position between adjacent stationary masses, the relative movement within the soil is opposed by a shearing resistance within the zone of contact between the yielding and stationary masses. The shearing resistance tends to keep the yielding mass in its original position, reduces the pressure on the yielding mass, and increases the pressure on the unyielding mass. The *arching effect* signifies the transfer of pressure from the yielding mass to the unyielding mass and the soil is said to "arch over" the yielding part of the support.

True arching refers to the arch that forms a stable dome in moderately compact or denser sands—as, for example, when they flow or "run" through an opening such as can exist in a sheet-pile wall and then flow stops, or when a looser, lower zone of granular soil in a trench backfill settles away from a compact upper zone. The wedging required to form the dome will not occur in sands that are too loose as they will tend to run through an opening or collapse into a trench.

In *clays*, the term *active arching condition* is used to refer to the transfer of pressure that occurs along a flexible sheet-pile wall from the yielding parts to the unyielding parts at the anchor and below the dredge line. The condition occurs in any soil type when yielding mobilizes shear strength and there is a change in the state of stress behind the wall.

In *rock masses*, arching occurs in jointed zones and the effect is of particular importance in the design of roof supports for tunnels and other underground excavations.

3.3 FACTORS AFFECTING SOIL SHEAR STRENGTH

3.3.1 FACTORS SUMMARIZED

The many variables affecting soil shear strength can be placed in three general categories: (1) material characteristics, (2) external factors, and (3) test method employed to measure the parameters.

Material Characteristics

Soil Types

In *cohesionless granular soils* the friction angle ϕ, in general, can be expected to increase with increasing gradation coarseness, roughness of the grain particles, and relative density. Grain mineralogy can be important: quartz grains are essentially indestructable, but grains of calcareous sands are relatively soft and crushable yielding much lower values for ϕ than those composed of quartz.

In *clay soils*, the strength depends primarily on the degree of prestress as indicated by the consistency and increases as consistency changes from soft to firm to stiff to hard. In normally consolidated clays, the drained strength ϕ generally decreases as the plasticity index increases. As the amount of granular material in a specimen increases its strength normally can be expected to increase.

Other Characteristics

Other significant characteristics affecting strength are the degree of saturation, drainage characteristics, material anisotropy, and metastable conditions. As the saturation increases, in general, strength decreases, but if a soil can drain under applied load its strength normally will be higher than that attained if drainage cannot occur. Material *anisotropy* will cause strength to be highest under compressive forces applied normal to stratification and lowest under tensile forces applied normal to stratification. Strength will be relatively low when compressive forces are applied parallel to stratification as compared to application in the perpendicular direction. *Metastable conditions* (i.e., the soil structure is stable only because of the existence of some supplementary stabilizing influence) can lead to strength loss due to some transient factor as described below.

External Factors

External factors affecting strength can be considered to include various forms of transient natural processes which can result in a decrease or increase in strength, depending upon the factor, and the conditions of the applied stress acting to cause rupture.

Transient Factors Acting to Reduce Strength

Strength can be reduced by chemical weathering, particularly in residual soils; by the leaching of cementing agents; or by lateral strains, such as may occur in slopes, which cause a strength reduction in heavily overconsolidated clays and clay shales. Partially saturated fine-grained cohesionless soils may exhibit apparent cohesion, and other metastable soils such as loess may have grains lightly cemented or bonded by clay films. Upon saturation, the cohesion bonds may be destroyed, resulting in substantial strength loss. Sensitivity of marine clays can increase significantly upon the leaching of salts which increases their susceptibility to liquefaction, particularly under dynamic loads.

Stress History

Stress history refers to the conditions leading to normal consolidation, overconsolidation, or underconsolidation. *Normally consolidated* (NC) soils have never been subjected to a load greater than existing overburden pressure. Normally consolidated clays usually have a consistency ranging from very soft to soft, depending upon their age. Offshore, however, where substantial thicknesses can be deposited, "normally consolidated" clays can be stiff to hard in consistency. NC sands may not necessarily be in a loose condition because their relative density depends very much on their mode of deposition (dunes, beaches, river alluvium, etc.).

Overconsolidated (OC) soils have been subjected to loads exceeding overburden pressures from such occurrences as the weight of glacial ice, desiccation, erosion of overburden, lowering of the groundwater table, or surcharging, and such soils have retained a degree of prestress (preconsolidation in clays, precompression in sands). Depending upon the amount and duration of prestress, clay soils vary in consistency from firm to stiff to hard.

Underconsolidated (UC) soils reside in the ground under a vertical effective stress less than existing effective overburden pressures. This condition exists commonly when a soil stratum is subjected to excess hydrostatic pressures that do not permit the soil to consolidate or compress under overburden pressures. Such conditions may be found in valley formations where groundwater is present as an artisian aquifer.

The term *consolidation* (*see Art. 4.3.4*) refers to clay soils that undergo volume change from the slow dissipation of pore-water pressure under applied stress. Granular soils undergo immediate *compression* upon load application as the grains become rearranged. They can be said to be precompressed or prestressed, but the degree cannot be measured in the laboratory, as it can with clay soils (*see Art. 4.3.4*). *Residual soils* evidence prestress in laboratory consolidation tests, but as these soils are continuously decomposing and weakening with time rather than strengthening under most natural situations, the indications are not preconsolidation in the true sense.

Applied Stress

Strength depends upon the magnitude, time rate of application, and direction of applied stress. Three characteristic levels of strength can be related to applied stress and resulting deformations:

1. Consolidation or compression occurs at relatively low stress levels applied slowly.
2. Creep occurs at relatively low stress levels held for relatively long time intervals.
3. Rupture occurs at either relatively high stress levels or under low stress levels applied rapidly. The occurrence is related to soil drainage, i.e., drained versus undrained loading conditions.

The significance of the direction of the applied stress relates to soil anisotropy as described previously under *Material Characteristics* and to the orientation of the principal stresses along the failure surface as discussed in Art. 3.1.3. Failure by rupture in the ground reduces the strength from its peak value to some residual value by remolding until it approaches or equals its ultimate strength. If there is a *preexisting failure surface*, therefore, strength will be much lower than peak values.

Dynamic or *cyclic loading* such as induced by earthquakes or sea-wave forces produces cyclic strains that tend to reduce the strength of clay soils; the amount of reduction is very much dependent upon cyclic strain amplitude. In free-draining granular soils the effect may be the compaction of loose deposits increasing the ϕ value. *Liquefaction* may occur in fine-grained cohesionless soils when they are saturated and under restricted drainage conditions.

Test Method to Measure Parameters

The measurements of strength parameters can be placed in four categories: (1) drained peak strength, (2) undrained peak strength, (3) ultimate and residual strength, and (4) cyclic strength. Values for strength parameters can be obtained by test directly, or indirectly through correlations made in the laboratory on in the field. A summary of soil laboratory static strength tests is given in Table 3.2, and one of soil in situ static strength tests in Table 3.3. The various methods of performing the triaxial test, the most widely used laboratory test, are summarized in Table 3.4. Tests to measure the undrained strength s_u are summarized in Table 3.6, and laboratory tests to measure dynamic soil properties are summarized in Table 3.9.

3.3.2 DRAINED PEAK STRENGTH ($\bar{\phi}$ or $\bar{\phi}$, $\bar{c}$)

Field Occurrence

The *drained peak strength* prevails in the field under relatively slow loading conditions during which pore pressures can dissipate. Such conditions generally exist in

- Most foundations, except for cases involving rapid load application
- Natural slopes, except for the sudden drawdown case
- Cut slopes, embankments, and retaining structures some time after the completion of construction

Parameter Measurements in Triaxial Compression

Consolidated-Drained Test (CD, S, or Slow Test for $\bar{\phi}$, $\bar{c}$)

The CD test is considered to yield the most reliable parameters for drained strength, but it has the major disadvantage of requiring substantial laboratory time for performance.

A saturated specimen is placed in the triaxial apparatus, subjected to a confining pressure ($\sigma_1 = \sigma_3$) and permitted to drain and consolidate until the pore pressure $u = 0$ (isotropic consolidation). (In recent years the trend has been toward anisotropic consolidation, applying a stress equal to effective overburden pressure in the vertical di-

TABLE 3.2
SUMMARY OF SOIL LABORATORY STATIC STRENGTH TESTS*

Test (F = also field test)	Parameters measured	Comments
Triaxial compression		
CD	$\bar{\phi}$, $\bar{c}$	Most reliable method for effective stresses.
CU	ϕ, c, $\bar{\phi}$, $\bar{c}$ s_u	Strength values higher than reality because disturbance causes lower $w\%$ upon reconsolidation. (Table 3.4, footnote ‡)
UU	s_u	Most representative laboratory value for undrained shear strength in compression.
Triaxial extension	$\bar{\phi}$, $\bar{c}$, s_u	Normally consolidated clays yield values approximately one-third those of compression tests because of soil anisotropy.
Plain strain compression or extension	ϕ	Values are a few degrees higher than those of normal triaxial test except for loose sands; more closely approach reality for retaining structure.
Direct shear box	$\bar{\phi}$, $\bar{c}$, $\bar{\phi}_r$	Values most applicable where test failure surface has same orientation with field failure surface. Values generally lower than triaxial compression values for a given soil, but higher than triaxial extension. Most suitable test for determination of residual strength ϕ_r, from UD samples.
Simple shear	s_u, $\bar{\phi}$, $\bar{c}$	Horizontal plane becomes plane of maximum shear strain at failure.
Unconfined compression	$s_u = \frac{1}{2}U_c$	Strength values generally lower than reality.
Vane shear (F)	s_u, s_r	Applies shear stress on vertical planes.
Torvane (F)	s_u, s_r	Shear occurs in a plane perpendicular to the axis of rotation.
Pocket penetrometer (F)	$s_u = \frac{1}{2}U_c$	Yields approximate values in clays. Used primarily for soil classification by consistency.
California bearing ratio (F)	CBR value	Used for pavement design. Empirical strength correlates roughly with U_c.

*After Hunt (1984).[1] Reprinted with permission of McGraw-Hill Book Company.

TABLE 3.3
SUMMARY OF SOIL IN SITU STATIC STRENGTH TESTS*

Test	Parameters measured	Comments
Vane shear	s_{ur}, s_r (direct test)	Measures undrained strength by shearing two circular horizontal surfaces and a cylindrical vertical surface; therefore, affected by soil anisotropy.
SPT	$\bar{\phi}$, s_u (indirect test)	D_r is estimated from N and correlated with soil gradation to obtain estimates of ϕ (Fig. 3.29, Table 3.5). Consistency is determined from N and correlated with plasticity to obtain estimates of U_c (Fig. 3.35, Table 3.7).
CPT	$\bar{\phi}$, s_u (indirect test)	Various theoretical and empirical relationships have been developed relating q_c to $\bar{\phi}$ [summarized by Mitchell and Lunne (1978)[22]]. See Fig. 3.27. s_u is expressed as $s_u = \dfrac{q_c - \Sigma\gamma z}{N_c}$ (3.42) where N_c is the cone-bearing capacity factor (deep foundation depth correction factor). Pore-water pressure u is measured by some cones (piezocones).
Pressuremeters	s_u (indirect test)	Affected by material anisotropy, s_u is expressed as $s_u = \dfrac{P_L - P_o}{2K_b}$ (3.43)
California bearing ratio	CBR value	Field values generally less than lab values because of rigid confinement in the lab mold.

*After Hunt (1984).[1] Reprinted with permission of McGraw-Hill Book Company.

rection ($\bar{\sigma}_v$) and a stress equal to $K_o\bar{\sigma}_v$ in the lateral direction to better simulate in situ conditions.) After consolidation the deviator stress σ_d is applied slowly until failure occurs while drainage is permitted to continue during deformation. The strain rate is set to approximate field conditions but sufficiently slow to *avoid pore-pressure buildup*, and during the test measurements are made and recorded of strain, deviator stress, and volume change. The CD test may require 40 hours or more under applied stress, whereas performance of the CU test (see following section) may require only 4 to 6 hours on clay soils. Bishop and Henkel (1962)[6] describe the calculation of the approximate time required for CD tests on the basis of the theory of consolidation.

A refinement of the test procedure, the use of *back pressure* either to ensure saturation or to saturate a partially saturated specimen, is described in detail in Fig. 3.24. In any case, the procedure is repeated with specimens of equal index properties with the confining pressures set to bracket the in situ overburden stresses. The results are plotted as Mohr's circles, and a Mohr's envelope is drawn to permit evaluating

TABLE 3.4
TRIAXIAL TEST METHODS*

Test	To measure	Procedure
Consolidated-drained (CD) or (S) compression test	Effective stress parameters $\bar{\phi}$, $\bar{c}$	Specimen permitted to drain and consolidate under confining pressure until u = 0. Deviator stress applied slowly to failure while specimen drains during deformation.†
Consolidated-undrained (CU) or (R) compression test	Total stress parameters ϕ, c Effective stress parameters $\bar{\phi}$, $\bar{c}$	Specimen permitted to drain and consolidate under confining pressure until u = 0. Deviator stress applied slowly to failure, but specimen drainage not permitted. Pore pressures are measured during test (see Pore-Pressure Parameters).
CK_oU test	$\bar{\phi}$, $\bar{c}$	See Note‡.
Unconsolidated-undrained (UU) or (Q) compression test	Undrained strength s_u	Confining pressure applied but no drainage or consolidation permitted to reduce test time. Deviator stress applied slowly to failure with no drainage permitted.
Extension tests as CD, CU, UU	Lateral shear strength	Maintain confining pressure constant and reduce axial stress, or maintain axial stress constant and increase confining pressure until failure.
Plane strain compression or extension test	Parameter ϕ in cohesionless granular soils	Modified triaxial apparatus in which specimen can strain only in axial direction and one lateral direction while its dimension remains fixed in the other lateral direction.

*After Hunt (1984).[1] Reprinted with permission of McGraw-Hill Book Company.

†Back pressure is applied to the pore water to stimulate in situ pore-water pressures, or to saturate partially saturated specimens.

‡CK_oU test: Specimen anisotropically consolidated with lateral pressure at K_op_o and vertical pressure at p_o.

the effective stress parameters as shown in Fig. 3.25 for both an NC and an OC clay.

Consolidated-Undrained Test (CU or R Test for ϕ_u, c_u, $\bar{\phi}$, $\bar{c}$)

The CU test is performed in a manner similar to the CD test except that during loading no drainage is permitted. Total stresses are measured, but since pore pressures commonly are measured, effective stresses also are determined. Strain rates must be sufficiently low, usually on the order of 0.1%/min for the average clay, to permit readings of u. Pore-pressure parameter A ($\Delta u/\Delta\sigma_1$) for evaluating $\bar{\phi}$ and $\bar{c}$ is calculated from measurements made at failure.

As noted under the CD test, a specimen may be anisotropically consolidated to account for soil anisotropy (the CK_oU test). The normal isotropically consolidated CD or CU test is considered to give strength values higher than under real conditions because disturbance during sampling and laboratory handling leads to a smaller water content upon reconsolidation, and $K_o > 1$.

CU test results for an NC clay and an OC clay are given in Fig. 3.26 in terms of both total and effective stresses. The pore pressure decreased during failure of the OC specimen, which was fairly heavily overconsolidated. Mohr's envelopes for a loose and dense sand in terms of total and effective stresses are given in Figs. 3.12 and 3.13.

1. Select values for $\bar{\sigma}_3$. Usually $\bar{\sigma}_3$ = 0.5, 1.0, 2.0 tsf except for high earth dam embankments.
2. Saturate specimen under back pressure. Increase σ_3 and u_o in increments simultaneously until saturation occurs (the specimen takes no more water) and $B = \Delta u/\Delta\sigma_3 \approx 1.0$

$$\sigma_3 = 0.5 \text{ to } (\sigma_3)_2 = 1.0 \text{ to } (\sigma_3)_3 = 1.5 \text{ tsf, etc.}$$

$$u_o = 0.45 \text{ to } (u_o)_2 = 0.95 \text{ to } (u_o)_3 = 1.45 \text{ tsf, etc.}$$

3. Consolidation (after $\Delta v = 0$ during saturation)
   a. Increase σ_3 for back pressure by $\sigma_3' = 0.5$ tsf. For example, $(\sigma_3)_2 = 1.5 + 0.5 = 2.0$ tsf and u remains at $u_o = 1.45$ tsf.
   b. Open drainage and measure Δv for $\epsilon_v = \Delta v/V$ to find new volume and dimensions.
   c. When no more drainage occurs, close valves.

4. Rupture specimen at $\sigma_3' = 0.5$ tsf and measure $\sigma_d(\bar{\sigma}_1 = \sigma_3' + \sigma_d)$. σ_3 and u_o from consolidation are held constant during failure. In the CU test u is measured during test with transducer.

5. Repeat test for $\sigma_3' = 1.0$ and 2.0 tsf.

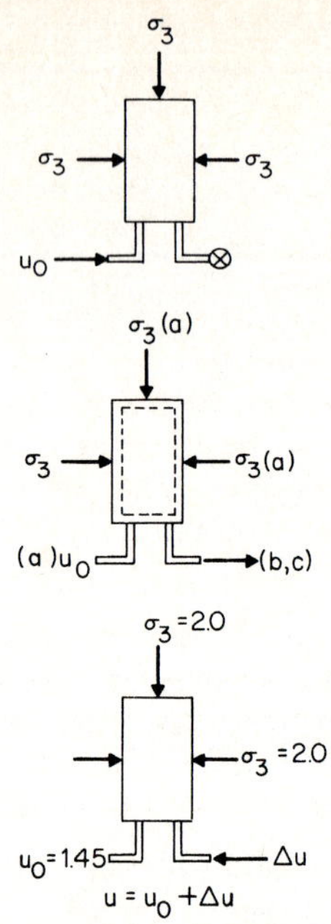

FIG. 3.24 Procedure for performing CD and CU triaxial tests using back pressure to obtain specimen saturation. (Verbal communication from Tiago Franco Vilela, 1982.)

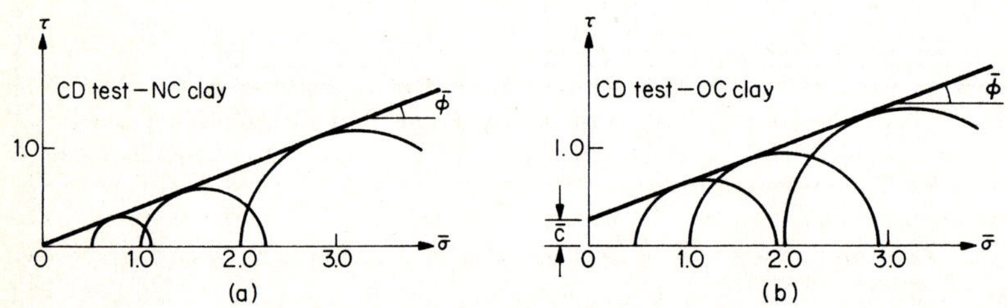

FIG. 3.25 Mohr's stress circles for drained tests on clay: (*a*) normally consolidated sample and (*b*) overconsolidated sample.

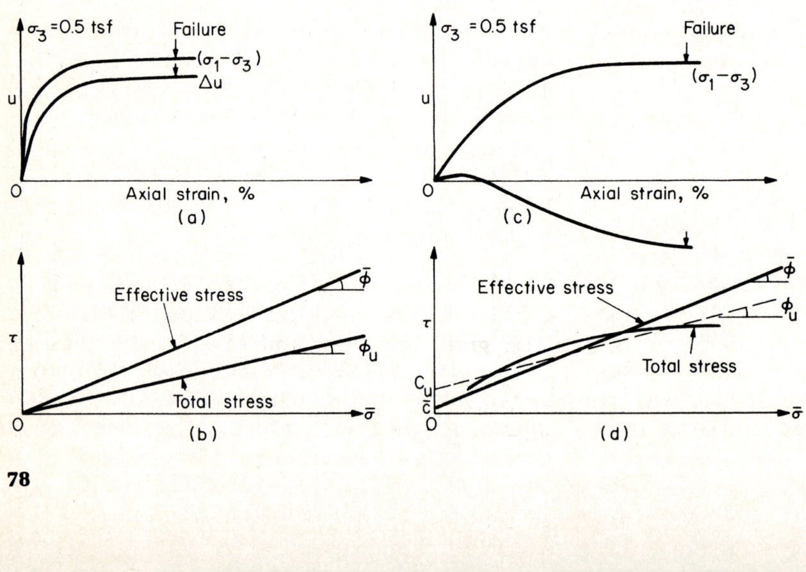

FIG. 3.26 Consolidated-undrained tests on clay samples: (*a*) NC clay, deviator stress versus pore pressure; (*b*) NC clay, Mohr's envelopes for total and effective stresses; (*c*) OC clay, deviator stress versus pore pressure; and (*d*) OC clay, Mohr's envelopes for total and effective stresses.

Triaxial Test Variations

Extension tests measure shear strength under conditions of axial lengthening in the lateral direction and may be performed as CD or CU tests. Failure is induced by either maintaining the confining pressure constant and reducing the axial stress, or maintaining the axial stress constant and increasing the confining pressure. Because of soil anisotropy, values in extension may be only one-half to one-third those obtained in compression; this difference is important to consider in stability analysis (*see Art. 3.5.3*).

Stress control, rather than the normal procedure of strain control, is performed to simulate certain field conditions. For this test the stress level is kept constant under each increment of loading as the specimen strains to failure.

Plain strain tests are a form of triaxial test in which a specimen can strain only in the axial direction and one lateral direction while its dimension remains fixed in the other lateral direction. The test better simulates many retaining-wall problems than does the normal triaxial test. Values for ϕ are a few degrees higher than those obtained from the usual test, except for sands of loose density where the difference is insignificant [Lambe and Whitman (1969)[3]].

Other Tests

Direct Shear Test

The direct shear test usually is performed as a CD test, although it has been used as a CU test when load application rates are high during shear. In any event some drainage always occurs during testing. A normal load is applied to a specimen contained in a horizontally split ring and anisotropic consolidation is permitted, followed by application of a shearing force that slides one-half of the ring with respect to the other until failure occurs.

Because of the form of the test only the normal and shear stress on a single plane are known. If it is assumed that the horizontal plane through the shear box is identical to the theoretical failure plane, usually an inexact assumption, it can be assumed that the stresses at failure are in the ratio of $\tau/\sigma = \tan\phi$ and provide a point on the Mohr envelope. Plotting the test results for a sequence of normal and deviator stresses provides the envelope and the values for $\bar{\phi}$ and $\bar{c}$ in a form similar to Fig. 3.25.

Strain rate should be such that the total time to reach peak strength should exceed $6t_{100}$, where t_{100} is determined from a plot of displacement versus time under the applied preconsolidation load [ISRM (1981)[23]]. The determination of t_{100} is described in Art. 4.3.4. Porous end platens on the specimen are required to permit drainage.

Simple Shear Test

In the simple shear or simple direct shear test, specimen confinement is such that shear deformation is considered to be uniformly distributed as the shearing stress is applied. The principal stresses are oriented initially in the vertical and horizontal directions, but at failure the horizontal plane becomes the plane of maximum shear strain (the lower point on the failure circle in Fig. 3.56*b*).

Indirect Field Tests for $\bar{\phi}$

STANDARD PENETRATION TEST (SPT) Various relationships can be found in the literature between the penetration resistance N and ϕ. Table 3.5 contains conservative values for $\bar{\phi}$ in terms of soil gradation and relative density. Grain angularity and the normal stress also influence the ϕ value.

CONE PENETROMETER TEST (CPT) Various theoretical and empirical correlations have been developed relating cone-tip resistance q_c to $\bar{\phi}$, such as the one given in Fig. 3.27. Schmertmann (1977)[17] suggests determining D_R from q_c in terms of $\bar{\sigma}_v$ (Fig. 3.28) and then entering Fig. 3.29*a* to find $\bar{\phi}$ for a particular gradation. The SPT and CPT are described in detail in Art. 3.4.5 of Hunt (1984).[1]

Strength Parameter—Soil Type Relationships

Granular Soils $\bar{\phi}$

In *cohesionless free-draining sands*, strength is expressed by $s = \sigma_n \tan\bar{\phi}$ (Eq. 3.14) in which $\bar{\phi}$ is affected by gradation, grain shape, relative density, grain minerals, and effective overburden pressure. As shown in Table 3.5 and Fig. 3.29, ϕ increases with gradation coarseness and generally with increasing relative density. Dense

TABLE 3.5
COMMON PROPERTIES OF COHESIONLESS SOILS**

Material	Compactness	D_R, %	N*	γ dry,† g/cm³	Void ratio e	Strength‡ ϕ
GW: well-graded gravels, gravel-sand mixtures	Dense	75	90	2.21	0.22	40
	Medium dense	50	55	2.08	0.28	36
	Loose	25	<28	1.97	0.36	32
GP: poorly graded gravels, gravel-sand mixtures	Dense	75	70	2.04	0.33	38
	Medium dense	50	50	1.92	0.39	35
	Loose	25	<20	1.83	0.47	32
SW: well-graded sands, gravelly sands	Dense	75	65	1.89	0.43	37
	Medium dense	50	35	1.79	0.49	34
	Loose	25	<15	1.70	0.57	30
SP: poorly graded sands, gravelly sands	Dense	75	50	1.76	0.52	36
	Medium dense	50	30	1.67	0.60	33
	Loose	25	<10	1.59	0.65	29
SM: silty sands	Dense	75	45	1.65	0.62	35
	Medium dense	50	25	1.55	0.74	32
	Loose	25	<8	1.49	0.80	29
ML: inorganic silts, very fine sands	Dense	75	35	1.49	0.80	33
	Medium dense	50	20	1.41	0.90	31
	Loose	25	<4	1.35	1.0	27

*N is blows per foot of penetration in the SPT. Adjustments for gradation are after Burmister (1962).[24] See Table 6.4 for general relationships of D_R vs. N.

†Density given is for $G_s = 2.68$ (quartz grains).

‡Friction angle ϕ depends on mineral type, normal stress, and grain angularity as well as D_R and gradation (see Fig. 3.29).

**From Hunt (1984).[1] Reprinted with permission of McGraw-Hill Book Company.

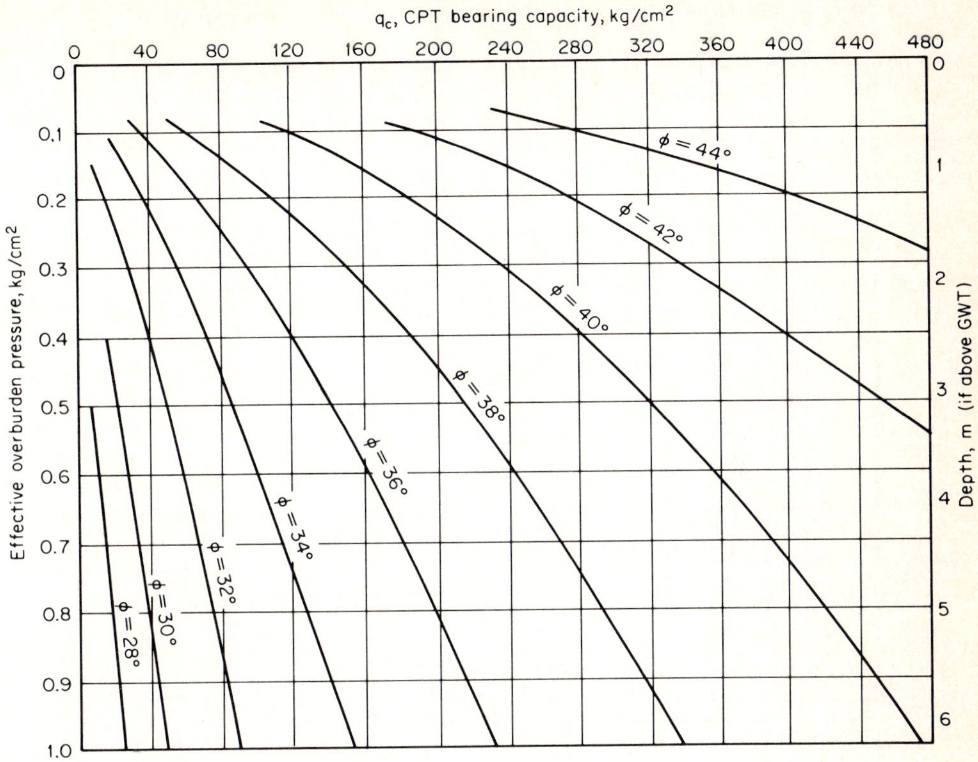

FIG. 3.27 Correlation between effective overburden pressure q_c and ϕ. [From Trofimenkov (1974).[25]]

sands, however, experience a stress drop-off after peak $\bar{\phi}$ is obtained, as shown in Fig. 3.29*b*, and $\bar{\phi}$ drops to an "ultimate" value, which is attributed to grain "interlocking" being overcome, as discussed in Art. 3.2.1. At the ultimate strength, after shearing movement, two samples of the same material, one initially loose and one initially dense, will have a similar void ratio, regardless of the initial void ratio, which is termed the *critical void ratio*. In practice, therefore, the peak value is used for cases where soil deformation is limited to small values, and for cases where the sand may strain past the peak value, $\bar{\phi}$ values are reduced on the order of 75 to 90% depending upon the initial density.

Grain shape and mineral content are important. Angular quartz grains can *increase* $\bar{\phi}$ by about 15% in the loose state and 30% in the dense state over rounded grains. Quartz grains predominate in nature, and most tables and charts give $\bar{\phi}$ values for these relatively indestructable materials. Sands can also be encountered, however, composed of shells, gypsum, and other calcareous material which have low crushing strengths. In the middle latitudes offshore, calcareous sands are abundant. When the calcareous material in the sand exceeds 30%, values for $\bar{\phi}$ are reduced substantially when evaluating pile support for offshore structures (*see Art. 7.3.2*).

Fine Sands and Silts ($\bar{\phi}$, c_a)

Partially saturated silts and fine sands may have an apparent cohesion intercept c_a, due to capillary forces, which is destroyed upon saturation. Loess and other "collapsible soils" obtain a substantial portion of their strength from particle cementation or from clay "braces," the destruction of which causes a significant reduction in strength.

Clays and Mixtures

Mohr's envelope for *normally to slightly overconsolidated* soils under drained testing usually goes through the origin and $c = 0$ (Fig. 3.25), although

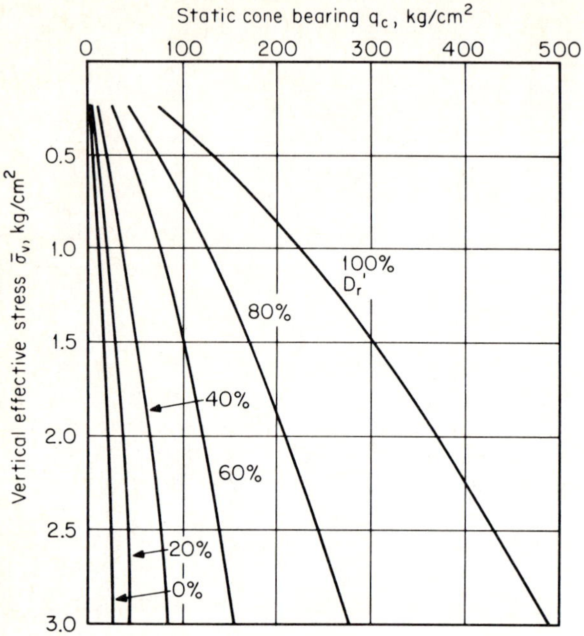

FIG. 3.28 Static cone resistance q_c vs. D_R. [From Schmertmann (1977).[17] Reprinted with permission of the Federal Highway Administration.]

due to very low permeability the strength may not be developed for a very long time. Strength is expressed as $s = \sigma_n \tan \bar{\phi}$, and $\bar{\phi}$ has been found to be related approximately to the PI as given in Fig. 3.30.

Intact, overconsolidated clays may be found during testing to have a small cohesion intercept, which depends on σ_n, but a small error occurs on the safe side if strength is expressed only in terms of $\bar{\phi}$ [Terzaghi and Peck (1967)[21]].

Fissured, overconsolidated clays normally provide a cohesion intercept during drained loading, and strength is expressed as $s = \bar{c} + \sigma_n \tan \bar{\phi}$ (Eq. 3.9). The magnitude of the peak strength varies with the magnitude of the normal stress N, and the strain at which the peak stress occurs also depends on the normal stress as shown in Fig. 3.31. In slopes, therefore, the peak strength cannot be mobilized simultaneously along the failure surface, because of the variation in normal stress. In actuality, slope designs are usually based on the residual strength, and in some circumstances on the ultimate strength (*see Art. 8.2.1*).

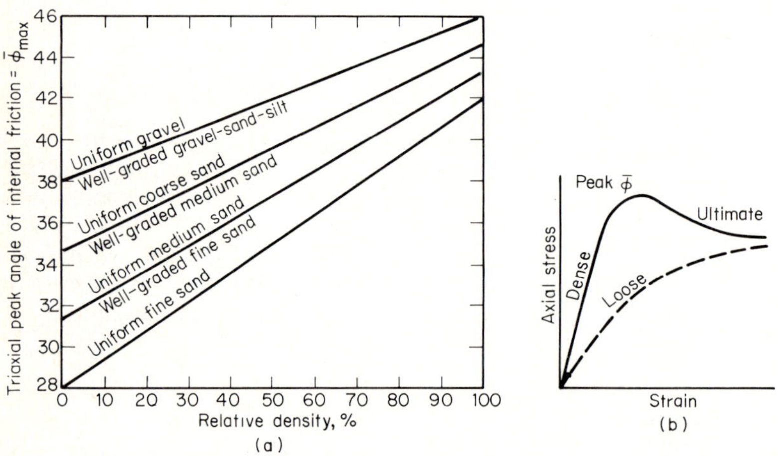

FIG. 3.29 Friction angle and relative density relationships for granular soils. (*a*) Chart for the approximate evaluation of the peak angle of internal friction from relative density. Schmertmann modification of Burmister (1948).[26] NOTES: 1. For quartz sands. 2. Angular grains can increase ϕ by about 15% in the loose state and 30% in the dense state over rounded grains. (*b*) Peak versus ultimate strength. In problems where the sand may strain past the peak strength before a general failure occurs, then a reduced value of ϕ must be used (particularly in the denser cohesionless soils). [After Schmertmann (1977).[17] Reprinted with permission of the Federal Highway Administration.]

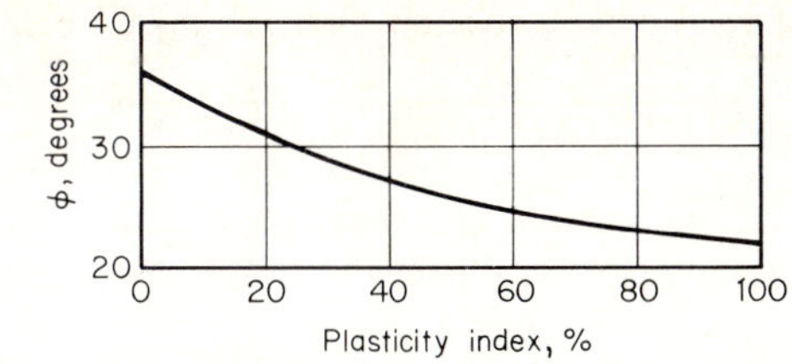

FIG. 3.30 Approximate relationship between ϕ and PI for clays of moderate to low sensitivity under drained conditions. [From Terzaghi and Peck (1967).[21] Reprinted with permission of John Wiley & Sons, Inc.]

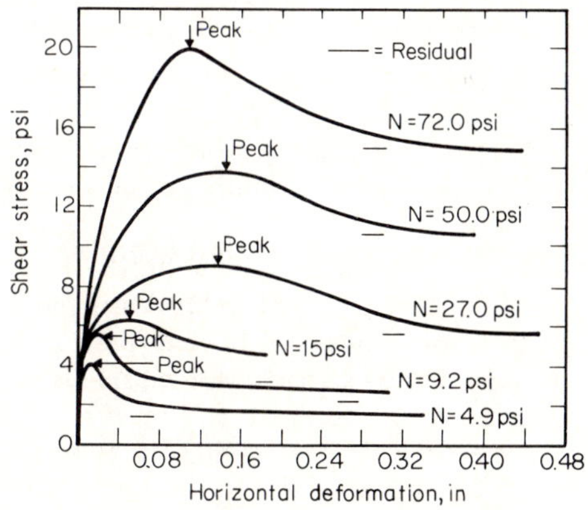

FIG. 3.31 Shear stress versus displacement for undisturbed samples of Lake Agassiz clay tested in direct shear at various normal pressures. [From Peck (1969).[27]]

Mixtures, including most residual soils, especially those from igneous and metamorphic rocks, and some compacted fills, exhibit both $\bar{\phi}$ and $\bar{c}$, and strength is expressed as $s = \bar{c} + \sigma_n \tan \bar{\phi}$. Loess and other collapsible soils with lightly cemented structure should be tested under low confining pressures so as to not destroy the clay braces and other structural supports.

Partially Saturated Soils

Negative Pore Pressures and Soil Suction

To express strength in terms of effective stresses, values of pore pressures must be known, but these values are difficult to measure during laboratory testing of partially saturated, cohesive soils. At low normal loads, negative pore pressures develop during shear, as shown in Fig. 3.32. These negative pore pressures, termed *soil suction*, increase the effective stresses, providing strengths higher than those obtained for saturated specimens, and are of particular significance in the stability analysis of soil slopes and earth dam embankments during the construction phase. Whether or not they persist in the field during failure is a factor not fully investigated.

In an unsaturated soil, menisci exist at the air-water interfaces, and the water is under capillary tension resulting in pore-water pressures with a negative value. The difference between the pore-air pressure u_a and the pore-water pressure u_w, termed the *matrix suction* $(u_a - u_w)$, is equal to the capillary tension. In a system with both air and water under pressure, the effective stress may be written [Bishop et al. (1960)[29]] as

$$\bar{\sigma} = \sigma - u_a + \beta\,(u_a - u_w) \qquad (3.37)$$

and the shear strength expressed as

$$s = \bar{c} + [\sigma - u_a + \beta(u_a - u_w)] \tan \bar{\phi} \qquad (3.38)$$

in which β is a coefficient that depends on the saturation degree.

Laboratory Testing

Coefficient β may be found from the triaxial test; $\bar{c}$ and $\bar{\phi}$ are found by the normal procedure for

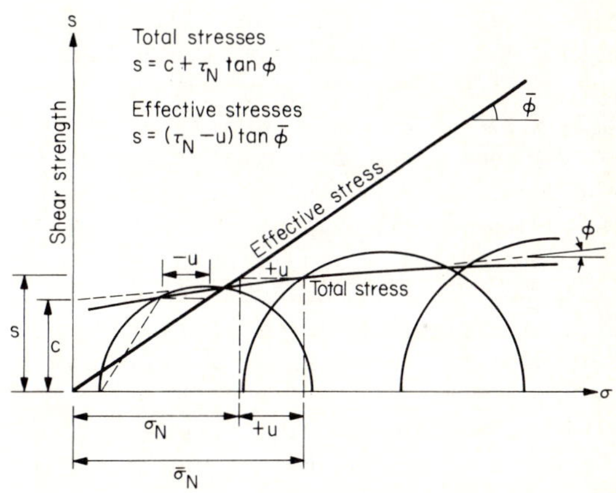

FIG. 3.32 Unsaturated, unconsolidated, undrained shearing strength. [From Lowe (1969).[28]]

saturated specimens; and s, σ, u_w, and u_a are found for a particular specimen. A relationship can be established between β and the saturation degree. For compacted clays a value of $u_a = 0$ is commonly used [Wu and Sangrey (1978)[30]].

Brand (1982)[4] considers that the correct mechanism of failure by wetting in the field can be modeled in the laboratory only by maintaining σ_1 and σ_3 constant and increasing the pore pressure from its initial negative value until failure occurs, a procedure which involves many difficulties and has yet to be standardized.

Fredlund et al. (1978)[31] proposed the following relationship between strength and soil suction:

$$s = \bar{c} + (\sigma - u_a) \tan \bar{\phi} + (u_a - u_w) \tan \phi_b \quad (3.39)$$

where the friction angle ϕ_b is equal to the slope of the plot of matrix suction versus shear strength when $(\sigma - u_a)$ is held constant. Ho and Fredlund (1982)[32] suggest that the increase in strength due to suction can be evaluated from a single, multistage triaxial test, performed in a cell modified to permit measurements of pore-air pressure. They suggest that the strength increase may be expressed as an increase in cohesion to a "total cohesion" C:

$$C = \bar{c} + (u_a - u_w) \tan \phi_b \quad (3.40)$$

In Situ Measurements of Soil Suction

Monitoring of in situ soil suction changes through wet seasons in the residual soils of Hong Kong has been undertaken by several methods [Brand (1982)[4]]. Psychrometric techniques (*see Art. 4.3.6*) for long-term measurements were not successful because of the insensitivity of the instruments in the critical suction range of 0 to 1 atm (0 to 101 kPa). The type tensiometer, used in agriculture, consisting of a small porous ceramic tip connected to a perspex tube and a vacuum gage, performs satisfactorily, but until recently required manual reading. New instruments, called the *Scanivalve* system, provide automatic, continuously recorded measurements [NCE (1983)[33]]. Measurements indicate that during the wet season soil suction decreases to zero along shallow potential failure surfaces, but not necessarily in deeper zones.

3.3.3 UNDRAINED PEAK STRENGTH (s_u, THE $\phi = 0$ CASE)

Field Occurrence

The undrained peak strength prevails when stress is applied rapidly and pore pressures are not permitted to dissipate. This condition, the total stress case (*see Art. 3.2.2*), occurs in the field during

- Rapid placement of embankments or loading of storage tanks over soft soil
- Initial phases of slope excavations or retaining-wall construction in slowly draining soils where time is inadequate to permit drainage
- Sudden drawdown conditions in the upstream slope of a dam embankment, or a period when flood waters recede rapidly from stream banks and the soils remain saturated.

Parameter Measurements in the Laboratory

The CU triaxial test in which total stress parameters c_u and u are measured is described in the previous article. In this article the emphasis is on measurements of the undrained strength s_u, the special case in which $\phi = 0$, generally applicable to NC clays. Laboratory tests to obtain values of s_u are summarized in Table 3.6.

Unconsolidated-Undrained Triaxial Test (UU or Q, Quick Test)

A confining pressure equal to $\bar{\sigma}_v$ is applied, but consolidation is not permitted, and the specimen is loaded to failure without permitting drainage. The deviator stress is carried partially by the pore water and partially by the soil skeleton. The vertical principal stress at failure $\sigma_1 = u + \sigma_d$. Increasing the confining pressure does not increase the diameter of Mohr's circle because the pore pressure increases as much as the confining pressure. The plot of Mohr's circles for the total and the effective stress condition are given in Fig. 3.33, although pore pressures are not normally measured during a test. The *undrained strength* s_u is independent of an increase in normal stress ($\phi = 0$) and is given by

$$s_u = \tfrac{1}{2} \sigma_d \quad (3.41)$$

The UU test can be performed as a compression or an extension test as described in the previous article.

TABLE 3.6
LABORATORY TESTS FOR UNDRAINED SHEAR STRENGTH*

Test	Parameters	Procedure	Comments
Triaxial compression	s_u	See Table 3.4.	UU test.
Unconfined compression†	$U_c = 2s_u$	Unconfined specimen (σ_3 = 0) strained to failure by axially applied load.	Test yields stress-strain curve and U_c. Slight drainage may occur during shear. Values usually conservative because of nonrecovered decrease in effective stress occurring from disturbance during extraction.
Vane shear (miniature vane)	s_u, s_r	Small two-bladed vane is pushed about 2 in into a UD-tube specimen and then rotated while the torque indicated on a gage is recorded to measure peak and ultimate strength.	Test limited to clay soils of very soft to firm consistency (Table 3.7).
Torvane†	s_u, s_r	One-inch-diameter disk containing eight small vanes is pushed about ¼-in penetration in UD sample and rotated while torque is recorded.	Test limited to clay soils of very soft to firm consistency. Tested zone may be disturbed by sample trimming and cutting.
Pocket penetrometer†	$U_c = 2s_u$	Calibrated spring-loaded rod (¼ in diameter) is pushed into soil to a penetration of 6 mm and the gage read for U_c.	Penetration is limited to soils with $U_c < 4.5$ tsf. Data are representative for soils with PI > 12. Below this value the ϕ of granular particles increases strength to above the measured values of s_u.
Simple shear	s_u	See Fig. 3.43.	

*From Hunt (1984).[1] Reprinted with permission of McGraw-Hill Book Company.

†These tests are considered generally as classification tests to provide an index of soil consistency (Table 3.7).

Unconfined Compression Test ($U_c = 2s_u$)

In the unconfined compression test, $\sigma_3 = 0$ and the specimen is loaded to failure, at which point $\sigma_d = U_c$, the *unconfined compressive strength* (Fig. 3.33). Undrained strength is measured, but disturbance during shear of the unconfined specimen decreases the effective stress, and the resulting strength values may be lower than reality. Some drainage may occur during testing, however, increasing the effective stress and the resulting strength values. The test is usually used as an index test for soil consistency classification.

SHANSEP CU Triaxial Test

This test procedure [Ladd and Foott (1974)[34]] attempts to minimize the effects of sample disturbance and assumes normalized behavior of clay. Specimens are consolidated to preconsolidation stresses (p_c) higher than overburden ($\overline{p}_o$), re-

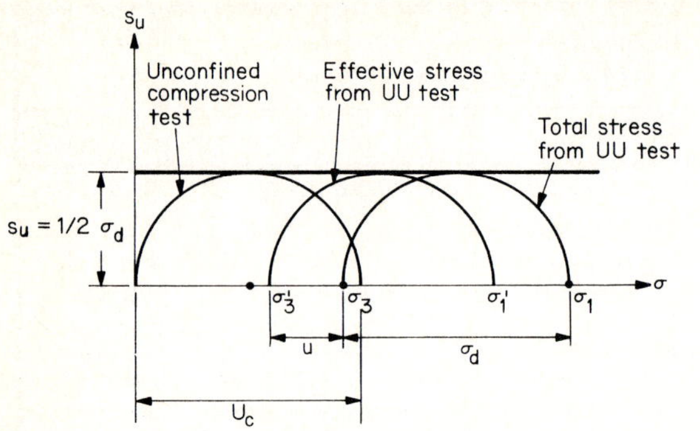

FIG. 3.33 Undrained shear strength in cohesive soils.

bounded to select values of OCR, and then tested in undrained shear to establish the relationship between $s_u/\overline{p}_o$ and OCR for different modes of failure. Normalized behavior requires this relationship to be independent of p_c. The s_u profile is then calculated from the values of $\overline{p}_o$ and OCR versus depth [Baligh et al. (1980)[35]].

Other Tests

In the *simple shear apparatus*, zero volume change during shear is maintained continuously by adjusting the vertical stress during testing. The *miniature vane* and *torvane* provide measurements of s_u and the *pocket penetrometer* provides measurement of U_c. These data are mainly for soil consistency classification.

Parameter Measurements in the Field

Vane Shear Test

The vane shear test is generally considered to obtain the most reliable measurements of s_u. The field vane test data, as applicable to practical problems, are strongly affected by soil anisotropy and loading time-rate effects. *Soil anisotropy* is evaluated by using vanes of varying length-to-width ratios [Aas (1965)[36]], or by comparing the results of UU triaxial compression and extension tests. The anisotropic correction factor is then applied directly to the measured in situ strength. The importance of the correction factor depends on the location of the measured strength along the failure surface (see *Art. 3.5.3*), i.e., loading direction.

Studies of a number of embankments constructed on soft clay have revealed that s_u as determined by field vane tests is generally greater than the actual field strength and that the difference increases with the plasticity of the clay [Casagrande (1960),[37] Bjerrum (1969),[38] 1972,[39] and 1973,[40] and Ladd (1973)[41]]. The cause is considered to be the difference in *loading time rate;* laboratory tests have demonstrated that the more rapidly the load is applied, the greater the strength is. Laboratory tests and the in situ vane test are run in a very short time interval compared with an embankment placement which may require weeks or even months to achieve full load, as is also the case with braced excavation construction. Bjerrum et al. (1972)[43] analyzed a number of embankment failures founded on soft clay and developed a *rate reduction factor* μ_r to be applied to the field vane value. In the cases analyzed the failure occurred during or immediately after construction completion and the construction period varied from less than one to a few months. The proposed relationship between μ_r and PI is given in Fig. 3.34.

Indirect Field Tests

STANDARD PENETRATION TEST Approximate correlations are made between N and U_c as given in Table 3.7 and Fig. 3.35.

CONE PENETROMETER TEST Correlations have been made between cone resistance q_c and s_u. San-

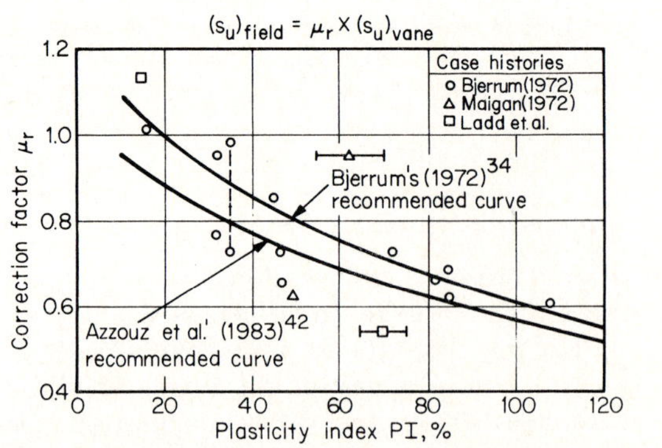

FIG. 3.34 Field vane time rate of loading correction factor versus plasticity index derived from embankment failure. [From Ladd (1973),[41] modified after Azzouz et al. (1983).[42]]

TABLE 3.7
COMMON PROPERTIES OF CLAY SOILS‡

Consistency	N	Hand test	γ_{sat},* g/cm³	Strength† U_c, kg/cm²
Hard	>30	Difficult to indent	>2.0	>4.0
Very stiff	15–30	Indented by thumbnail	2.08–2.24	2.0–4.0
Stiff	8–15	Indented by thumb	1.92–2.08	1.0–2.0
Medium (firm)	4–8	Molded by strong pressure	1.76–1.92	0.5–1.0
Soft	2–4	Molded by slight pressure	1.60–1.76	0.25–0.5
Very soft	<2	Extrudes between fingers	1.44–1.60	0–0.25

$^*\gamma_{sat} = \gamma_{dry} + \gamma_w \left(\frac{e}{1+e}\right)$

†Unconfined compressive strength U_c is usually taken as equal to twice the cohesion c or the undrained shear strength s_u. For the drained strength condition, most clays also have the additional strength parameter ϕ, although for most normally consolidated clays $c = 0$ [Lambe and Whitman (1969)[3]]. Typical values for s_u and drained strength parameters are given in Table 3.30, Hunt (1984).[1]

‡From Hunt (1984).[1] Reprinted with permission of McGraw-Hill Book Company.

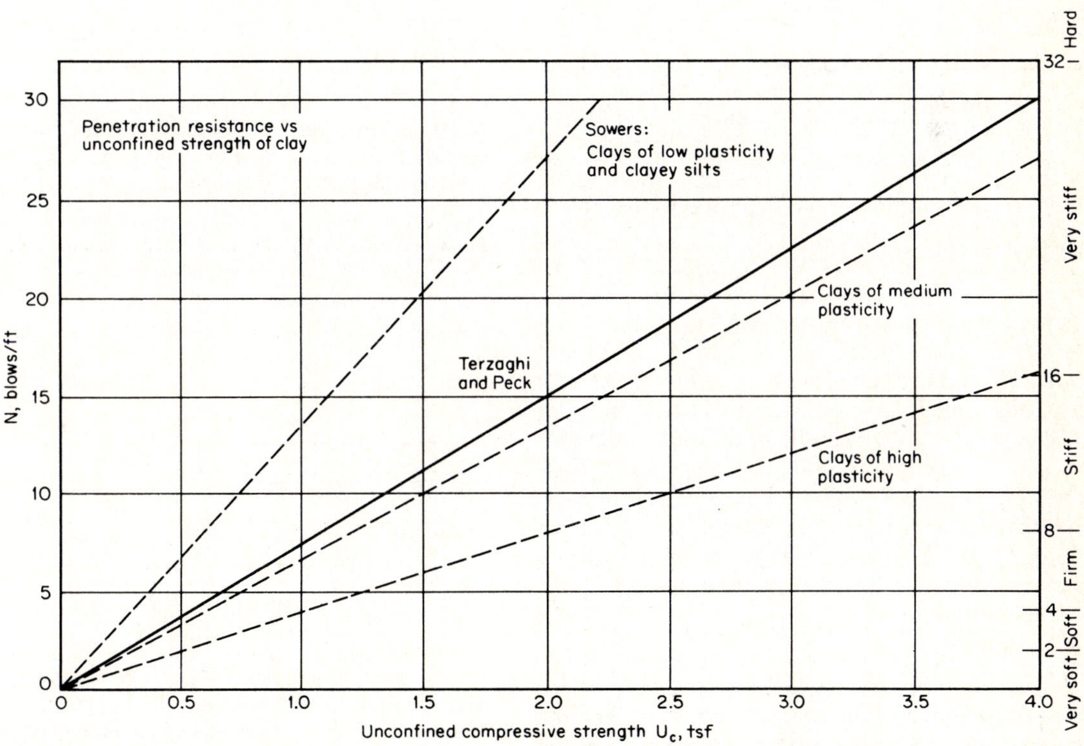

FIG. 3.35 Correlations of SPT *N* values with U_c for cohesive soils of varying plasticities. [From NAVFAC (1971).[44]]

glerat (1972)[45] gives $s_u \approx q_c/15$ for soft to stiff clays and $s_u \approx q_c/30$ for stiff fissured clays. Schmertmann (1977)[17] gives the relationship

$$s_u = \frac{q_c - \Sigma\gamma z}{N_c} \tag{3.42}$$

where $\Sigma\gamma z$ = total overburden pressure at depth z.

N_c = deep foundation correction factor, variable but 10 for stiff clays and 16 for soft clays with the Fugro friction cone. Values should be confirmed by other strength tests for each new location.

PRESSUREMETER TEST Menard (1965)[46] gives the relationship

$$s_u = \frac{p_L - p_{oh}}{2\ K_b} \tag{3.43}$$

where P_L = limiting pressure (see Tables 4.5 and 4.10)

p_{oh} = at-rest horizontal stress (from pressuremeter test)

K_b = a coefficient varying with E_c/P_L (E_c is the pressuremeter modulus)

$2K_b$ = typically 5.5 [Lukas and de Bussy (1976)[47]]

Strength Parameter—Soil Type Relationships

Granular Soils

In saturated, cohesionless, free-draining sands, drained conditions almost always prevail. In silts and fine sands, undrained conditions will exist during sudden drawdown or under dynamic loadings, and analysis will be based on total stress parameters. Partially saturated sands and mixtures, commonly used for earth dam construction, also are evaluated on the basis of total stresses [see discussion in previous article and Lowe (1969)[28]], or effective stresses if pore pressures are measured during testing and construction.

Clays

NORMALLY TO SLIGHTLY OVERCONSOLIDATED s_u prevails, usually falling within a limited fraction of the effective overburden pressures, or $s_u/\bar{p}_o$ = 0.16 to 0.6, based on field vane and K_o triaxial tests. Relationships between $s_u/\bar{p}_o$ and $p_c/\bar{p}_o$ for "aged" and "young" clays in terms of PI are given in Fig. 3.36. For estimating purposes s_u may be taken as an average of

$$(s_u/\bar{p}_o)_{NC} = 0.33 \tag{3.44}$$

or, in terms of PI [Terzaghi and Peck (1967)[21]]

$$s_u/\bar{p}_o = 0.11 + 0.0037\ PI \tag{3.45}$$

If $s_u > 0.6\ \bar{p}_o$ the clay can be considered to be overconsolidated.

Because of layering during deposition, clays, and in particular soft clays, can have substantially higher strengths in the vertical direction than in the horizontal. Bjerrum et al. (1972)[43] found that

- Strength values for NC clay from compression tests were almost 3 times as high as those obtained from extension tests.
- The undrained strength in compression is about 1.5 times the vane strength.
- The ratio between the extension and vane strength was found to be 0.5 in low to medium plasticity clays, increasing toward 1.0 in highly plastic clays.

If values for s_u are accurately represented in analysis, results are usually conservative because of the strength increase in NC clays during the consolidation that will occur during loading, as long as some drainage is permitted, such as may occur during embankment construction.

OVERCONSOLIDATED In OC clays the undrained strength may be larger or smaller than the

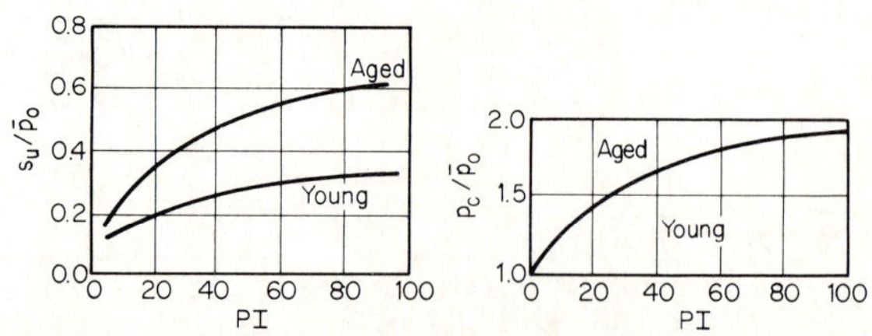

FIG. 3.36 Typical values of $(s_u/\bar{p}_o)$ vane and $p_c/\bar{p}_o$ observed in normally consolidated late glacial and postglacial clays. "Young" clays have been recently deposited and come to equilibrium under $\bar{p}_o$ but have not undergone significant secondary or delayed consolidation. [From Bjerrum(1973).[40]]

drained strength, depending on the OCR (*see Art. 3.2.3*). For OCR ranges between 1.0 and about 4 to 8, the volume tends to decrease during shear and s_u, as for NC clays, is less than the drained strength.

For OCR values greater than about 4 to 8, the volume tends to increase, pore-water pressure decreases, and the undrained strength exceeds the drained value. For high values for OCR the excess may be very large. In heavily overconsolidated clays strong negative pore-water pressures tend to draw water into the soil causing it to swell, thereby reducing its strength. The undrained strength, therefore, often cannot be relied upon, and the $\phi = 0$ concept should not be applied to clays with OCR values in excess of 2 to 4 [Terzaghi and Peck (1967)[21]].

3.3.4 ULTIMATE AND RESIDUAL STRENGTH ($\bar{\phi}_r$ or S_r)

Concepts and Field Occurrence

The ultimate strength of sands has been discussed in Art. 3.3.2.In many cohesive materials, if strain continues past the peak strength, the strength will decrease until some minimum value is reached. Thereafter strength remains constant with increasing strain as shown in Fig. 3.37. This constant strength value has been referred to variously as either the ultimate or the residual strength.

Concepts of ultimate and residual strength have been evolving. In the past, investigators have tended to equate these strengths whereas, in actuality, they represent two different conditions as related to material structure (personal communication from Dr. Ralph Peck). The *ultimate strength* is the lowest strength that a cohesive material can attain without radical change in structure in the confined state, substantially below the peak strength, and often approaches the *remolded strength*. The *residual strength* is a strength lower than the peak strength that may be reached if large strains develop along a discrete failure surface to the extent that the clay particles assume a preferred orientation, which is often disclosed by slickensides.

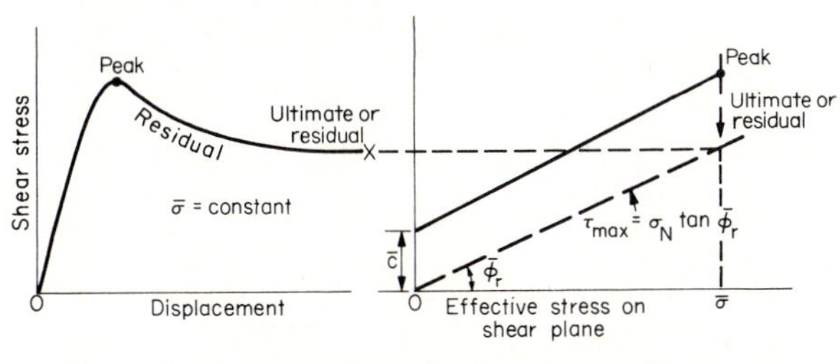

FIG. 3.37 Relationship between peak, residual, and ultimate strength in an overconsolidated material.

In a remolded soil, which has a partly dispersed structure, a triaxial test is likely to fail by bulging and there is no single failure plane. Even at advanced strains not all of the clay plates are parallel to each other, which results in some interlocking. There will be a peak followed by a drop-off of the stress-strain curve to what is now referred to as the "ultimate" value. If the sample is forced to slip on one plane, however, and if the strain is very large or is reversed several times, the particles along the plane become aligned in parallel array, there are no interlocks, and the strength is the minimum possible. This strength is referred to as the "residual friction" value.

In the field, the residual strength may prevail in any mass where rupture and large strains have occurred, such as in slopes that have already experienced failure by sliding, along bedding planes of clay or shale between stiffer layers if bedding-plane shears have occurred, or along recent fault displacements.

Parameter Measurements

Drained Residual Strength $\bar{\phi}_r$

The *direct shear box* is the best method of testing undisturbed samples to obtain measures of the drained residual strength $\bar{\phi}_r$. If the sample contains the preexisting failure surface and the same surface is subjected to shear in the laboratory, the stress-strain curve will exhibit little or no peak. If a new failure surface is formed in the test, the curve will exhibit a peak. In either case the sample is subjected to a number of cycles of sequential back-and-forth movement until a constant strength value is reached.

Mohr's envelope for the residual drained strength passes through the origin (no cohesion intercept) and strength is expressed as $s = \bar{\sigma}_n \tan \bar{\phi}_r$. The post-peak drop-off in a previously unsheared clay, as shown in Fig. 3.37, becomes more pronounced as the degree of overconsolidation increases. In heavily overconsolidated clays and clay shales $\bar{\phi}_r$ is often in the range of only 8 to 15°, and for highly plastic clay shales it can be as low as 3°.

Undrained Residual Strength

The in situ vane shear apparatus is the most common means of determining the undrained residual strength s_r. During vane shear testing torque is applied to the vane continuously beyond the point of peak strength until a constant value is reached. After several rotations, a well-defined cylindrical shear surface is established and the strength is at the residual value.

Thixotropy is a physiochemical phenomenon which causes a remolded soil to gain strength with time. Strength gain also results from consolidation. For these reasons there is a strength gain with time along a failure surface, often resulting is a strength higher than the residual.

Sensitivity

Sensitivity S_t is defined as the ratio of the natural peak strength to the strength when completely remolded at unaltered water content,

$$S_t = s_u/s_r \tag{3.46}$$

Clays are classified by sensitivity as given in Table 3.8. Soils termed "quick" have high sensitivities and are extremely susceptible to vibrations and other forms of disturbance. They can quickly lose all strength, change to a fluid, and flow, even on very shallow slopes.

TABLE 3.8
CLAY CLASSIFICATION BY SENSITIVITY S_t*

Sensitivity s_u/s_r	Classification
<2	Insensitive
2–4	Moderately insensitive
4-8	Sensitive
8–16	Very sensitive
16–32	Slightly quick
32–64	Medium quick
>64	Quick

*From Skempton (1953),[48] Bjerrum (1954).[49]

3.3.5 DYNAMIC OR CYCLIC LOADING (G_d, D; s_u, $\bar{\phi}$, $\bar{c}$)

Field Occurrence

Dynamic or cyclic loading in the field results from:

- Seismic waves from earthquakes or vibrating machinery imposing low- to high-frequency forces on soil or rock materials
- Wave forces imposing low-frequency cyclic forces on offshore structures which are transferred to the supporting soils

It is only in the last two decades or so that substantial interest in the effects of dynamic forces on structures has arisen among geotechnical practitioners. This interest coincided with the growth of nuclear power plant construction and the consequent concern with earthquake forces, and with the growth in offshore petroleum platform construction and the consequent concern with wave forces. The state of the art in terms of in situ and laboratory testing and analytical procedures is slowly, but continuously, being advanced.

Cyclic Stress-Strain Relationships

The Phenomenon

Excitation emanating as a stress wave from an underlying bedrock surface during an earthquake, or from vibrating machinery, or from an offshore structure responding to ocean waves, applies a cyclic shearing stress to soils. Under initial dynamic loading, a soil specimen subject to shear undergoes deformations that are partially irreversible irrespective of the strain amplitude; hence, stress-strain curves in loading and unloading do not coincide.

Small Strain Amplitudes

Strain amplitudes in the order of 0.1 to 0.001% normally result from earthquake loadings. The difference between successive reloading curves tends to disappear after a few loading cycles of a similar amplitude and the stress-strain curve becomes a closed loop, as shown on Fig. 3.38, that can be defined by two parameters: dynamic shear modulus and the damping ratio:

- *Dynamic shear modulus* (G_d or G) (*see also Art. 4.1.3*) is expressed as the equivalent secant

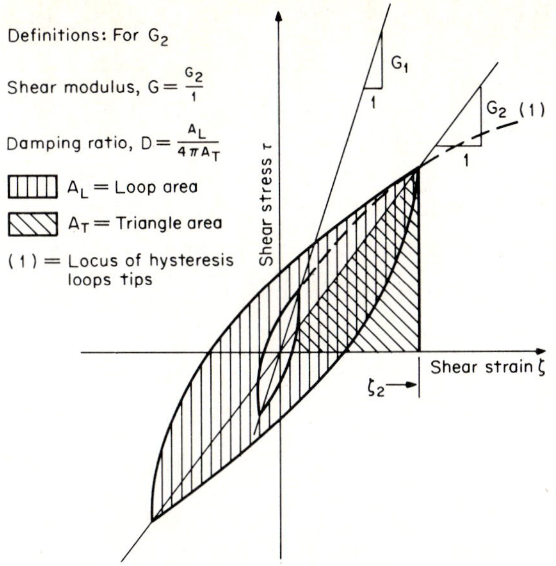

FIG. 3.38 Hysteretic stress-strain relationships from cyclic shear tests at different strain amplitudes. [From USAEC (1972).[50]]

modulus, determined by the slope of a line passing through the ends of the hysteresis loop at the peak stress and strain after each cycle of loading. It has been found to decrease markedly with increasing strain amplitude.

- *Internal damping ratio* (D or λ) pertains to the dissipation of energy during cyclic loading and is defined by the enclosed areas in Fig. 3.38.
- *Poisson's ratio* (ν), required for description of dynamic soil response, varies within relatively close limits and affects seismic response only slightly. General ranges are for cohesionless soils, $\nu = 0.25$ to 0.35, and for cohesive soils, $\nu = 0.4$ to 0.5.

The main factors affecting values for shear modulus and damping ratio in all soils are shear strain amplitude, initial effective mean principal stress, void ratio, shear stress level, and number of loading cycles. In cohesive soils values are affected also by stress history (OCR), saturation degree, effective stress parameters, thixotropy, and temperature.

These relationships apply generally to "stable" soils; i.e., those that undergo elastic and plastic deformations but serve to dampen seismic motion and still maintain some strength level. "Unstable" soils are subject to sudden compaction, or a complete loss of strength by *cyclic liquefaction.*

Effects on Pore Pressure

Wylie and Streeter (1976)[51] hypothesized that the shearing motion of the soil causes slippage of the soil grains which weakens the soil skeleton temporarily. At the time of shear reversal, the particles do not slide. Thus the soil skeleton recovers much of its strength, but in a slightly consolidated state. The consolidation increases pore pressure and reduces effective stress.

In free-draining granular soils, pore pressures reduce rapidly; in loose deposits, substantial compression occurs. In loose, fine-grained soils, such as silts and silty sands, however, the horizontal shaking causes a trend toward zero effective stress since pore pressures cannot dissipate quickly enough. The soils liquefy, losing all strength and becoming fluid: this process is called *cyclic liquefaction*. In saturated clay soils, the seismically induced pore pressures reduce the shear strength [Zeevaert (1972)[52]].

The susceptibility for liquefaction depends on soil gradation; groundwater table position; relative density; boundary drainage conditions and soil stratigraphy; initial effective overburden pressure; and the duration, amplitude, and period of the induced vibrations.

Large Strain Amplitudes

Under cyclic loadings with large strain amplitudes, such as occur from sea wave forces acting on a structure, stress-strain relationships and the effect of cyclic loads on the strength parameters s_u, and $\bar{\phi}$ and $\bar{c}$, are most important. In the present state of the art, soil parameters usually are defined in terms of *p-y* curves which define soil resistance versus deflection (*see Art. 4.4.3*). These curves are developed from empirical relationships which include a number of soil parameters as well as characteristics of the pile foundation.

Parameter Measurements

Soil testing methods fall into two general categories: tests to define the parameter's dynamic

TABLE 3.9
LABORATORY METHODS FOR DETERMINING DYNAMIC SOIL PROPERTIES*

Test conditions	Test and reference	Properties measured†	Stress or strain conditions: Initial	Stress or strain conditions: Dynamic	Strain amplitude
Low frequency	Cyclic triaxial (Seed and Chan, 1966; Castro, 1969)	E, D; stress vs. strain; strength	Axisymmetric consolidation	Pulsating axial or confining stress; constant-amplitude stress	10^{-4} to 10^{-1}
	Cyclic torsion (Zeevaert, 1967; Hardin and Drnevich, 1972a and b)	G, D; stress vs. strain	Axisymmetric consolidation	Pulsating shear stress; constant-amplitude stress or free vibration	10^{-4} to 10^{-2}
	Cyclic simple shear (Seed and Wilson, 1967)	G, D; stress vs. strain; strength	K_0 consolidation	Pulsating shear stress; constant-amplitude stress (strain) or free vibration	10^{-4} to 3×10^{-2}
High frequency	Ultrasonic (Lawrence, 1965; Nacci and Taylor, 1967)	c_p or c_s	Axisymmetric consolidation	Dilation or shear; single pulse wave	10^{-6}
	Resonant column (Afifi, 1970)	c_p or c_s; E or G; D	Uniform or axisymmetric consolidation	Pulsating axial or shear (torsional) stress; constant-amplitude strain	10^{-6} to 10^{-2}

*From Faccioli and Reséndiz (1976).[53] Reprinted with permission of Elsevier Scientific Publishing Company.

†E = dynamic Young's modulus, G = dynamic shear modulus, D = damping ratio, c_p = compression-wave velocity, c_s = shear-wave velocity.

shear modulus and damping ratio for earthquake studies, and tests to determine the effects on soil strength of large-amplitude sea wave forces applied to a structure. There is no single test that can impose the wide range of amplitudes encountered in earthquakes; therefore a series of tests are performed to develop relationships between *G* and *D* in terms of amplitude as illustrated in Figs. 3.43 and 3.44. The NRC (Nuclear Regulatory Commission), *Regulatory Guide* 1.138 (April 1978) recommends performing cross-hole seismic tests in situ, and stress- and strain-controlled cyclic triaxial tests and resonant column tests in the laboratory. The effect of wave forces imposed on soils is usually evaluated with cyclic triaxial or cyclic simple shear tests. Dynamic laboratory tests are summarized in Table 3.9.

Cyclic Triaxial Tests (E_d, D, s_u, $\bar{\phi}$)

The initial stages of cyclic triaxial tests are similar to those of static tests. Sand specimens are fabricated at some specified relative density and clay specimens are selected from undisturbed samples. The specimens are saturated under back pressure and tested under a range of confining pressures. Clay specimens are either anisotropically or isotropically consolidated prior to load application. Sophisticated equipment permits programing variations in the frequency of the applied load as well as control of the wave form (sinusoidal or square, ramp or random). In random loadings actual earthquake histories are imposed. Strain amplitudes vary from intermediate to large: 0.01 to 5% for earthquake loadings and as high as 15% for sea wave forces.

Cyclic triaxial tests only approximate duplication of field stress conditions during earthquakes [see USACE (1972),[50] p. 91] or under sea wave forces, mainly because laboratory specimens undergo deformation in each of the three principal stress directions, whereas in the field soils are deformed primarily in simple shear, or unidirectionally. There are two general types of cyclic triaxial tests: stress-controlled and strain-controlled.

Stress-controlled tests are favored for liquefaction studies [Silver et al. (1976)[54]] but are also used for measurements of *G* and *D*. A confining pressure is imposed on the specimen and a con-

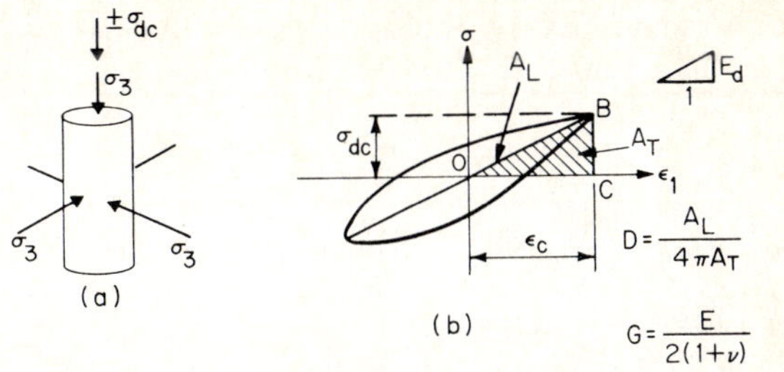

FIG. 3.39 Stress-controlled cyclic triaxial test and definition of Young's modulus E_d, shear modulus G, and damping ratio D.

stant-amplitude cyclic axial load is applied (Fig. 3.39*a*) while no drainage is permitted, and the pore-water pressure, axial load, and axial deformation are recorded. The specimen is subjected to a number of cycles (in the order of 20 to 30) of loading at a given cyclic stress (σ_{dc}), mean effective stress (Eq. 3.49), and frequency of loading. Initial values of σ_{dc} are very low to obtain the lowest possible strains, and then progressively increased to account for large cyclic strains. Specimens are tested under a range of frequencies: for example, 0.5, 2, 5, and 10 Hz. Typical test results for one loading cycle are expressed by the loop of the stress-strain relationship given in Fig. 3.39*b*. The slope of the line through the end points of the loop is considered to represent Young's modulus *E*, and half of the strain produced by one loop is termed the cyclic axial strain ϵ_c. The *damping ratio D* is determined from the areas enclosed in the loop from

$$D = \frac{A_L}{4\pi A_T} \tag{3.47}$$

and since the hysteretic stress-strain curve is determined in compression rather than in shear, the compression modulus obtained directly must be converted to the shear modulus using the relationship

$$G = E_d/(1 + \nu) \quad \text{kg/cm}^2, \text{MPa} \tag{4.10}$$

and cyclic shear strain is given as

$$\zeta = \epsilon_c\,(1 + \nu) \tag{3.48}$$

For a given frequency *f* and number of cycles *N* the data for the shear moduli are plotted versus the shear strain to obtain a family of curves, each curve representing a specimen at its void ratio and mean effective stress, as given in Fig. 3.40. The *mean effective stress* is defined as

$$\bar{\sigma}_o = \frac{\bar{\sigma}_v + 2\bar{\sigma}_h}{3} \tag{3.49}$$

where $\bar{\sigma}_h = K_o \bar{\sigma}_v$ in an anisotropically consolidated test. A similar procedure is followed for the damping ratio (Fig. 3.41). For many soils, such as fine-grained granular materials, after a number of stress cycles, the pore pressure increases to a value approximately equal to the cell pressure. At this point the soil begins to deform

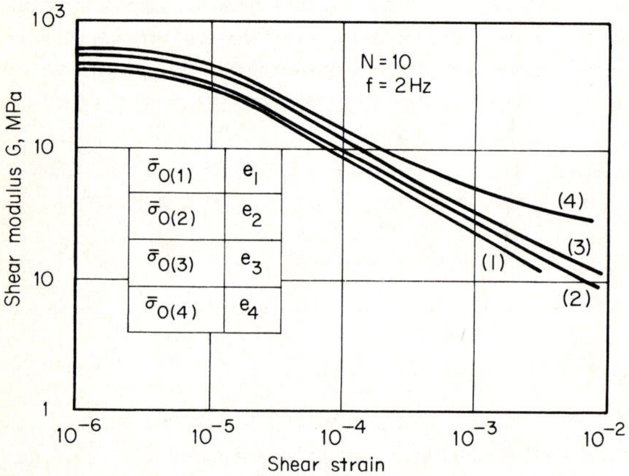

FIG. 3.40 Shear modulus versus shear strain amplitude for an OC clay for a given number of cycles and a frequency, and various values of mean effective stress and void ratio.

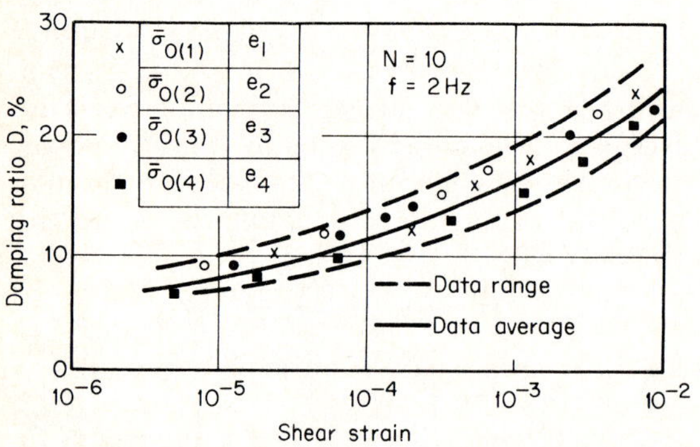

FIG. 3.41 Damping ratio versus shear strain amplitude for a clay for a given number of cycles and a frequency, and various values of mean effective stress and void ratio.

excessively, thereafter exhibiting little load-carrying ability, and a state of *cyclic liquefaction* exists. Testing procedures are described by Silver et al. (1976).[54] To evaluate the liquefaction potential of a given material, tests are usually performed on specimens fabricated at three different confining pressures.

Failure by liquefaction is said to occur when $u = \sigma_3$ which requires a certain number of cycles under a given axial load as shown in Fig. 3.42*a*. A plot of number of cycles versus axial strain is given in Fig. 3.42*b*. Data are normally presented as a family of curves for various relative densities relating the number of cycles to reach liquefaction to the *normalized cyclic stress ratio* SR, as shown in Fig. 3.42*c*, where

$$SR = \Delta\sigma_v / 2\sigma_3 \tag{3.50}$$

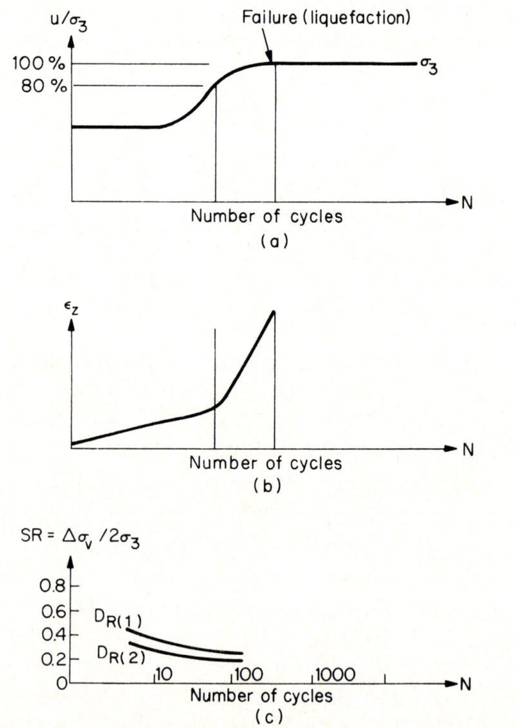

FIG. 3.42 Stress-controlled triaxial test for defining liquefaction characteristics: (*a*) number of cycles versus u/σ_3 and definition of failure; (*b*) number of cycles versus vertical strain; and (*c*) number of cycles to reach liquefaction versus cyclic stress ratio SR for specimens at various values of D_R.

$\Delta\sigma_v$ = the single amplitude cyclic axial stress, and σ_3 = the initial effective confining pressure.

Strain-controlled cyclic triaxial tests have been widely used to evaluate dynamic stress-strain characteristics of soil. In a strain-controlled test the sample is subjected to a cyclic strain amplitude that is kept constant during the test, and the cyclic stress required to maintain this cyclic strain is measured.

One popular approach is the *stage testing technique* [Silver and Park (1975)[55]] which involves:

1. Cycling a specimen at the lowest possible value of vertical strain for a specified number of cycles in the undrained state.
2. Stopping the test to permit the specimen to drain and relieve any excess pore-water pressure.
3. Closing the drainage and cycling the specimen at a slightly higher strain.
4. Repeating the process 5 or 6 times at successively higher strain amplitudes to define modulus and damping values in the shear-strain range of 0.01 to 0.5%.

To evaluate the characteristics of a natural sand formation, a series of tests is performed with specimens fabricated at three different void ratios, then subjected to a minimum of three different levels of shear strain. Clays are tested as undisturbed samples at a range of σ_3.

CK_oU cyclic tests are used to evaluate the effects of wave-induced forces on soil parameters (*see Art. 4.4.3, p-y* curves). In the performance of a typical study, undisturbed samples representing a range in depths and strength are selected and each is consolidated isotropically under a confining pressure equal to $\bar{\sigma}_h$, then saturated under back pressure in the ranges of total stresses followed by anisotropic consolidation under effective overburden stresses. It is to be noted that in offshore projects where one is working in 60 to 100 m of water or more, total stresses are proportionately much higher than effective stresses, and samples tend to expand after sampling.

Cyclic loads are applied in stages to simulate storm wave forces which ideally should be based on statistical data. Periods of 10 sec (0.1 Hz) are common. As many as 12 stages may be run on each specimen for a total number of cycles in the order of 1200, with the initial stages containing a much greater proportion of the total cycles. At the initial stage a very low deviator stress is imposed, then increased during each subsequent stage until at the final stage a maximum value is attained, which is often set to equal the maximum horizontal earth pressure on the deepest specimen sampled. After cycling, the pore pressure is permitted to dissipate and the specimen is loaded statically to failure. During the test the axial stress, vertical deformation, and pore pressures as a functon of the number of cycles are recorded. The test provides data on strength versus deformation as a function of cyclic forces, as well as on the degradation of the undrained shear strength when compared with data obtained from static tests.

Cyclic Simple Shear Tests (G_d, D, s_u, $\bar{\phi}$)

A specimen is placed in the apparatus and subjected to confining pressure. A vertical stress is applied and then cyclic shear imposed laterally on the specimen while no drainage is permitted. Loading frequency and waveform are controlled and strains may be varied from intermediate to large. This test most nearly duplicates field conditions for soils subject to strong earthquake or sea wave motion (large strains) although test boundary conditions will cause values to be somewhat lower than in reality.

The results of cyclic shear tests performed on *dense sands* is given in Fig. 3.43. The pore pressure increases almost linearly with the number of cycles until failure occurs. By performing a number of tests at different stress levels ($\tau_h/\bar{\sigma}_d$), a relationship can be developed between the stress level and the pore-pressure increase per cycle. In *clay soils* the cyclic undrained shear strength is very much influenced by the amplitude of cyclic strain. During testing the specimen is first consolidated under actual field stresses, then cyclically loaded at a particular stress level with the strain recorded as a function of the number of cycles. The test is repeated at various stress levels to obtain a relationship between number of cycles, cyclic shear stress, and cyclic shear strain. For reference see Finn et al. (1969),[57] USAEC (1972),[50] and Hardin and Drnevich (1972).[58,59]

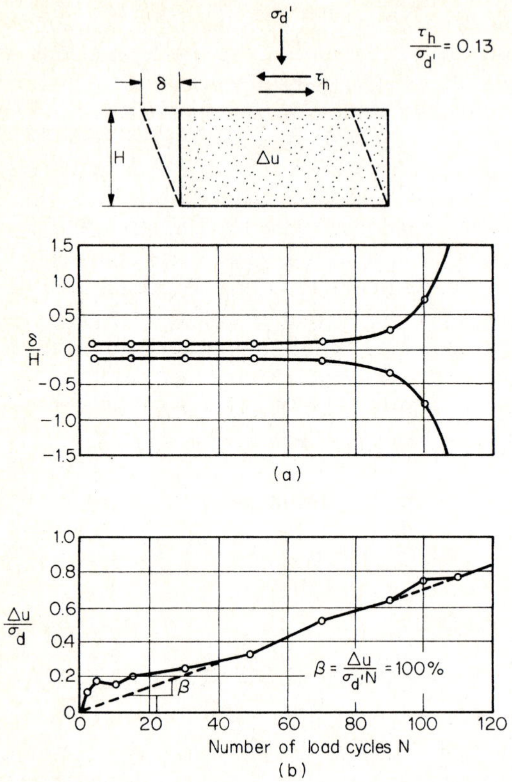

FIG. 3.43 Results of cyclic shear tests on sands in simple shear apparatus: (*a*) deformation versus number of load cycles and (*b*) rate of increase of the ratio of the change in pore pressure normalized with respect to the applied stress versus the number of cycles. [From Eide (1974).[56]]

Resonant Column Test

In the resonant column test [Hardin (1970)[60]], a specimen is subjected to a steady-state forced *torsional vibration* until resonance is obtained. The specimen is first consolidated under a confining pressure and then vibrated by a frequency generator to control the amplitude and frequency of the forced vibration. In the chamber the specimen rests between two end caps, with the base cap fixed and the top cap serving to excite the specimen with a system of magnets and coils. Specimen response is measured with accelerometers attached to the top cap. Resonance is obtained by varying the frequency of the input motion until the specimen response (acceleration) reaches a maximum.

Shear modulus G is a function of the mean effective stress $\bar{\sigma}_o$ (Eq. 3.49), void ratio *e*, time under cyclic loading *t*, shear strain $\zeta_{\theta z}$ (see Fig. 3.44*a*), and period *T*. Shear strain amplitudes are less than in the cyclic triaxial test and are generally specified as less than 0.0001% ranging to as high as the machine's capability permits. Test duration under cyclic stress is in the order of 1000 min for sands and 10,000 min for clays (Fig. 3.44*b*). A series of tests is performed on specimens at different values of D_R or OCR, under several confining pressures. The final data are presented as a plot of normalized shear modulus versus shear strain as given in Fig. 3.44*c*. Tests are run on both solid soil columns and hollow columns; the latter tests give values closer to pure shear, but are much more difficult tests to

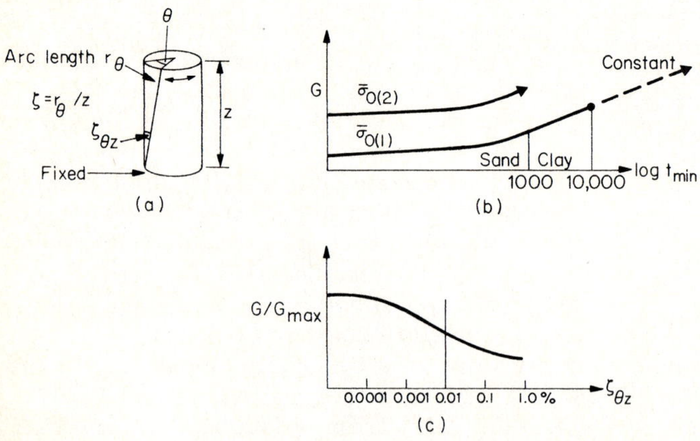

FIG. 3.44 The resonant column test where $G = f(\bar{\sigma}_o, e, t, \zeta_{\theta z}, T)$: (*a*) shear strain ζ defined; (*b*) relationship between *G* and the time of test for sand and clay—specify $\zeta_{\theta z} < 10^{-4}$; and (*c*) plot of shear strain versus normalized shear modulus. (Personal communication, R. D. Woods, 1982.)

set up than are solid column tests. The test appears to be the most practical laboratory technique for measuring G_d, but it has the disadvantage of applying only strain levels lower than those of interest in either earthquake or sea wave studies

In Situ Measurements of G_d

Measurements of G_d are made in situ by *seismic direct methods*, such as the *crosshole test*, or by the *steady-state vibration method* [Hunt (1984)[1]]. The moduli, determined from recorded shear-wave velocities, are likely to be higher than in reality since they are obtained at lower amplitudes than those imposed by earthquakes (see Fig. 3.46). Values obtained are scaled down by comparing the results with those obtained for the same soils by laboratory testing.

The *crosshole test* is illustrated schematically in Fig. 3.45*a*. Boring verticality is checked by the inclinometer. A mechanical impulse source acting first downward and then upward along casing walls provides a strong polarized energy output which generates predominantly shear waves. The waves are recorded and oscilloscope traces are photographed for analysis as shown in Fig. 3.45*b*. Arrival time between the two geophones as a function of their distance provides the shear-wave velocity from which G_d is computed.

A crude estimate of G may be obtained from the SPT *N* value given by Ohsaki and Iwasaki (1973)[62]:

$$G \approx 120\ N^{0.8} \tag{3.51}$$

Practical Aspects

Shear Modulus and Damping Design Values

Values for the shear modulus and damping ratio for earthquake design studies are required over a potentially wide range of ground-motion amplitudes. There is no single test method to date which provides this information; therefore, a testing program involves several different types of tests. Seismic direct (crosshole) tests or the steady-state vibration method provides the parameters for low amplitudes, the resonant column for somewhat larger amplitudes, and the strain-controlled triaxial test for the largest magnitudes of induced shaking.

The test data are used to plot curves for the full range of shear-strain amplitude for the normalized shear modulus G/G_{max} (Fig. 3.46) and the damping ratio *D* (Fig. 3.47). Typical curves for sands and clays are given in the figures. In Fig. 3.46 are given the common ranges in shear-strain amplitudes for the various tests for earthquakes in general, and for those with strong ground motion, with the latter representing the range of in-

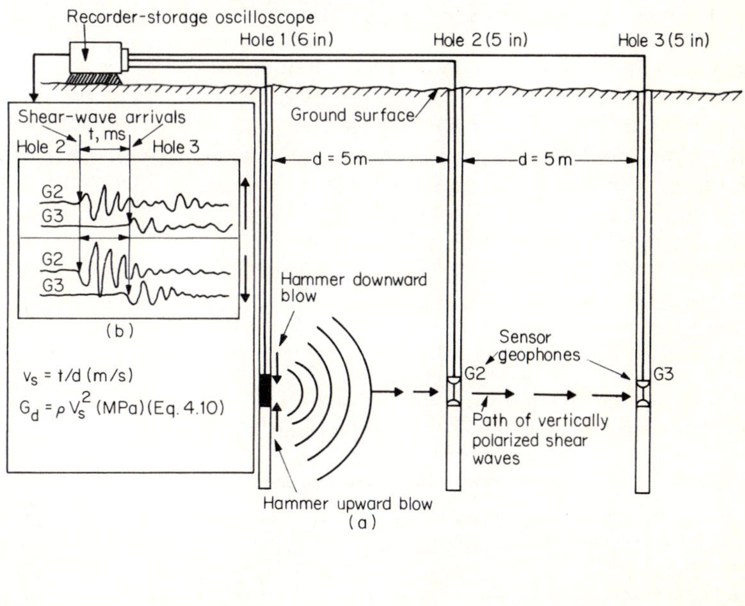

FIG. 3.45 Crosshole direct seismic test to measure shear modulus in situ. (*a*) Borehole configuration. Boreholes are absolutely vertical, cased, and grouted. (*b*) Oscilloscope traces of shear waves. Upper traces are from one blow upward of hammer; lower traces are from one blow downward of hammer; time *t* is given in milliseconds. [After Statton et al. (1978).[61]]

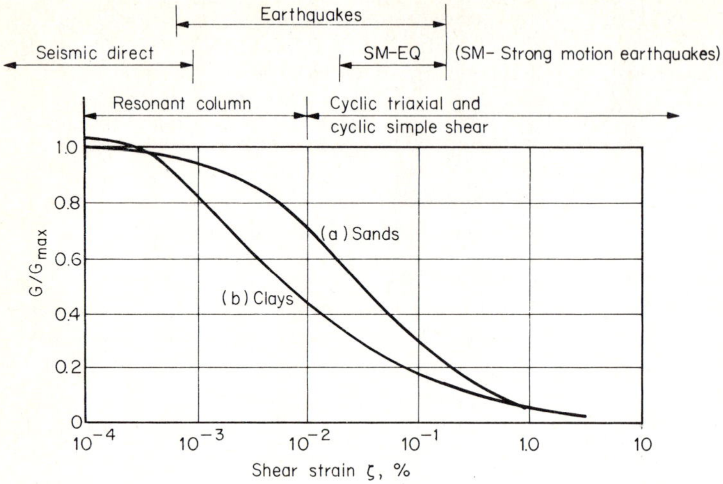

FIG. 3.46 Variation of normalized shear moduli with shear strain for: (*a*) sands; (*b*) saturated clays, and the approximate strain ranges for various test procedures and earthquakes. [After USAEC (1972).[50]]

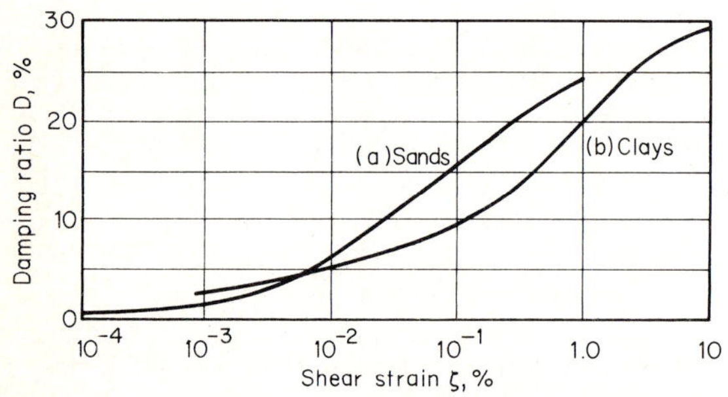

FIG. 3.47 Damping ratio versus shear strain for: (*a*) sands and (*b*) saturated clays. [From USAEC (1972).[50]]

terest for most problems. In selecting values for design analysis, however, consideration is given not only to the range of shear strain as represented by the data but, perhaps more important, to the possibility that the data are not truly representative of real conditions because of the factors inherent in the test method itself, as discussed in this section.

Liquefaction

Field experience has demonstrated that there are a number of conditions under which a soil becomes susceptible to cyclic liquefaction. From the aspects of *gradation*, fine sands and silty sands are most susceptible, especially when they are poorly graded, as shown in Fig. 3.48. Permeability is relatively low and drainage is slow, and to be susceptible the stratum must be below the *groundwater level* and saturated, or nearly so, without the capacity to drain freely. Boundary drainage conditions and soil stratification, therefore, become important.

Relative density is another important factor. Loose soils are most susceptible; D'Appolonia (1970)[63] suggested that liquefaction might occur where D_R values were as high as 50% during ground accelerations in the order of 0.1g, but for sands with D_R in the range of 75% or greater, liquefaction was unlikely. Other influencing factors are the effective overburden pressure and the duration, amplitude, and period of the induced ground motion.

Existing field evidence of liquefaction below level ground is almost exclusively associated with data from the standard penetration test from which estimates of relative density are made based on *N* values corrected for depth (*see Art. 6.2.2*). Although correlations have been developed based directly on corrected *N* values [Seed (1976)[64]], the susceptibility for liquefaction is evaluated by many practitioners through correlations of relative density (from *N* values) and gradation and consideration of the possible groundwater level in relation to the liquefiable stratum at the time of an earthquake, with the objective of establishing the degree of hazard.

Over the past two decades extensive research has been directed toward the study of cyclic liq-

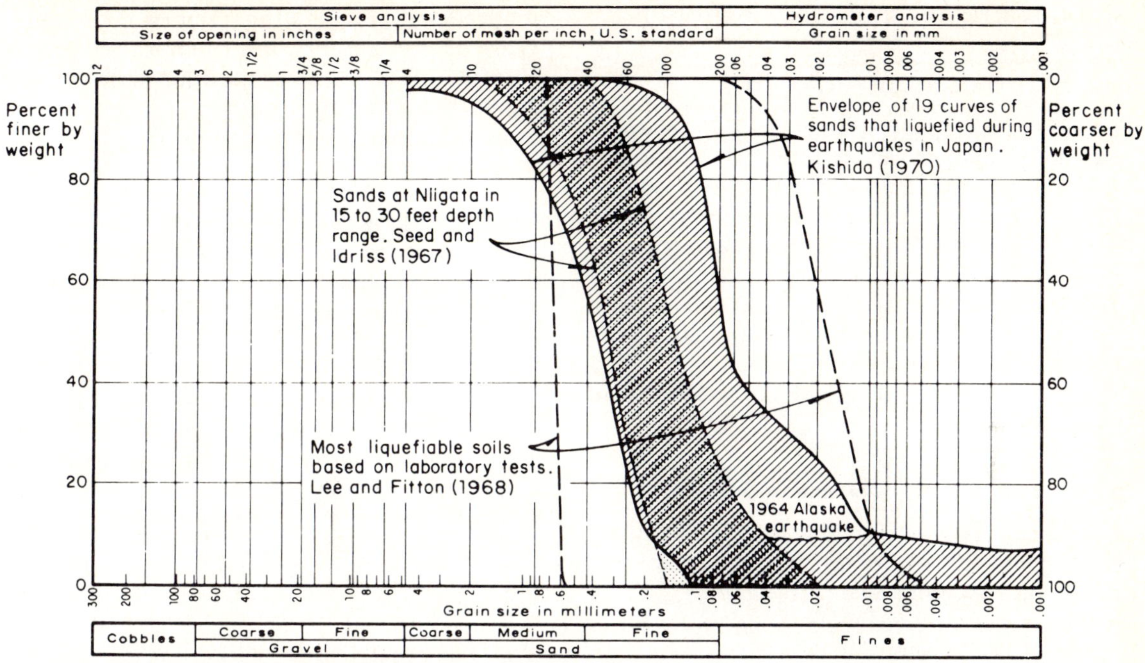

FIG. 3.48 Effect of grain size distribution on cyclic liquefaction susceptibility. [After Shannon and Wilson, Inc. and Agbabian-Jacobsen Associates (1971)[65]; from Finn (1972).[66]]

uefaction in the laboratory employing cyclic triaxial and cyclic shear tests, the shaking table, and other methods. The U.S. NRC has recommended that, for nuclear power plant studies, the liquefaction potential be evaluated on the basis of stress-controlled cyclic triaxial tests. Although these tests in research programs have added substantially to the knowledge concerning the phenomenon, Peck (1979)[67] points out that "there are as yet no case histories that include both thorough laboratory testing under cyclic conditions of the materials in their undisturbed state before the earthquake and good records of the earthquake motions at places where liquefaction has been observed." Laboratory data alone, therefore, are not sufficient evidence for evaluating liquefaction potential, but must be weighted heavily by environmental factors and data from field tests such as the SPT.

Wave-Induced Cyclic Loading

It is only in recent years that extensive laboratory research has been done to evaluate the effect of large-amplitude cyclic strains on clay soils. Andersen et al. (1980)[68] report on a study in which clays at various OCR values were tested for static and cyclic loads in the triaxial and simple shear devices. They found that pore-pressure response measurements and stress concentrations in the samples are testing problems that may be more severe for cyclic than for static loading and that the measured undrained static shear strength depends on the test mode. Triaxial compression, triaxial extension, and simple shear tests follow different effective stress paths and have different undrained static strengths.

It was found that the number of cycles to failure depends on the OCR, and at a certain cyclic shear stress level, a normally consolidated clay is more resistant to undrained cyclic loading than an OC clay. Undrained cyclic loading causes a reduction in the undrained static strength, which seems to be independent of the OCR. For overconsolidated samples, the effective stress parameters, $\bar{\phi}$ and $\bar{c}$, did not appear to be influenced by cyclic loading.

3.4 STRENGTH OF INTACT ROCK AND ROCK MASSES

3.4.1 FACTORS AFFECTING STRENGTH

Factors affecting rock strength include material and "mass" characteristics, external natural factors, applied stresses, and the test method employed to measure the parameters. From the aspects of engineering problems, rock can be placed in two general categories: *intact rock* or rock blocks consisting of fresh to decomposed materials and the *rock mass* consisting of blocks separated by discontinuities. Rock mass characteristics are described in detail in Chap. 6 of Hunt (1984).[1]

Material and "Mass" Characteristics

Mineral composition provides the characteristic rock hardness, which for the more common minerals ranges on Mohs scale [Hunt (1984),[1] Table 5.4] from quartz (7) and feldspar (6) to the softer calcite (3) and gypsum (2). *Cementation* in sedimentary rocks is an important factor. *Texture* refers to the grain sizes, and *fabric* refers to grain orientation, which is generally described as isotropic or anisotropic. In an *isotropic fabric* the mineral grains have a random orientation and mechanical properties are essentially similar in all directions, whereas in an *anisotropic fabric* the grains have planar orientations from mineral cleavage (shales, slate), foliations (gneiss) or schistosity (schists), and the mechanical properties in a direction perpendicular to the planar orientations are much different from those in the parallel direction. *Decomposition* from chemical weathering changes the mineral composition and results in a substantially weaker material. Major *discontinuities*, representing weakness planes, divide the rock mass into blocks, as described in Art. 3.4.3.

External Factors

Transient Conditions

Strength is reduced by weathering processes, by leaching of cementing agents, and in clay shales by lateral strains resulting from erosion of slopes. Mechanical weathering results in the opening of discontinuities, the formation of new discontinuities by fracturing, the opening of grain boundaries, and the fracture or cleavage of individual mineral grains. Chemical weathering results in mineral decomposition with the most significant change occurring in the silicates which eventually become clays. Weathering acts much more strongly along the discontinuities than within a block, thereby providing a "wall strength" along the discontinuities much lower than that of the interior, fresher rock. Although decomposition proceeds at minuscule rates from natural processes in terms of the life of a structure, it can be greatly accelerated by human activities. Excavations, for example, result in rapid disintegration in marine shales even when they are exposed to the elements for a brief period, and schist supporting structures in New York City has been found to soften substantially over a 20-year period from contact with liquids from broken sewer lines (*see Art. 6.3.1*).

Water

Water results in the dissolution of cementing materials, and as the degree of saturation increases the strength generally decreases (see note to Table 3.12). In most cases, however, it is pore- or fissure-water pressure that exerts the greatest influence on intact rock strength.

In a test on dry rock there is no difference between normal stresses and effective normal stresses, but in a saturated specimen the normal stress is reduced by the pore- or fissure-water pressure thereby resulting in a reduction in the peak strength. A sufficiently high increase in pore pressure can result in failure in intact rock stressed initially to near its limit. (Cleft-water pressures on joints act on joint strength, but are not considered an element of strength.)

Residual Stresses

Residual stresses are generally horizontal, compressive stresses stored in a rock mass. Their origin is in folding deformation, metamorphism, slow cooling of magma at great depths, and the erosion of overburden causing stress relief. Relatively common and often high, they can be in the range of 3 to over 10 times overburden stresses. Their effect on strength is a reduction due to the tensile straining which occurs upon reduction of confining stresses.

Marine shales have characteristically been prestressed by high overburden pressures and when uplifted and subjected to erosion, or to excavation, stress release and the resulting strains cause intense fracturing and disintegration in the mass. Significant deflections can occur in river valleys in flat-lying sedimentary rocks, resulting in differential movements along bedding planes and the formation of mylonite. Foliation joints in gneiss can open from tensile straining during excavation and the opening widths can increase from creep over some time interval.

Applied Stress

Strength of fresh to slightly decomposed rock in nature depends on the magnitude, time rate of application, direction of applied stress (as related to anisotropy), and on the magnitude and distribution of confining stresses. Several characteristic levels of strength can be related to applied and confining stresses and resulting deformations.

1. Compression occurs at relatively low stress levels applied slowly.
2. Creep occurs at relatively low stress levels held for relatively long time intervals. Soft rocks, such as halite, will exhibit ductile behavior and undergo plastic shear and creep under low to moderate confining pressures.
3. Rupture seldom occurs in fresh rock under the usual range of applied stresses unless there is no or little confinement in the direction opposed to the applied stress, as in the case of a mine roof. Other exceptions are foliated rocks, such as schist or gneiss, which may undergo plastic shear along the foliation planes under low to moderate confining pressures or coal columns supporting a mine roof.
4. Under very high confining pressures some rocks behave in a ductile manner and plastic shear occurs. Murphy (1970)[69] reports the development of kink structure and plastic shear in calcite under confining pressures in the order of 45,000 psi (3000 kg/cm^2).

With increasing decomposition, intact rock strength approaches the characteristics of a soil, as described under Art. 3.3.1. In rock masses it is generally the orientation of the applied stress and the confining stresses with the direction of the weakness planes that have the greatest influence on strength.

Test Method to Measure Parameters

Intact Rock

Fresh to slightly decomposed specimens of intact rock are tested in the laboratory to measure strength in compression, direct shear, tension, or flexure. Cores of moderately to highly decomposed specimens are tested in uniaxial or triaxial compression or in direct shear. Direct shear is used also to measure the strength of asperities along a joint in a fresh to slightly decomposed specimen.

Rock Masses

In situ tests in rock masses are generally limited to the direct shear test to measure strength along weakness planes or in decomposed rock. A summary of tests to measure rock strength is given in Table 3.10.

3.4.2 INTACT ROCK STRENGTH

Failure Forms

Many rocks are isotropic materials that undergo elastic deformation under stress until a brittle failure occurs, but many other rocks are either anisotropic or soft, or both, and undergo combinations of elastic-plastic deformations under stress as shown in Table 3.11. The isotropic, elastic rocks approach and finally become plastic materials as their degree of decomposition increases. Because of the wide range in failure modes in intact rock it appears that the application of a single failure law to all rock types is not feasible.

Brittle Failure

Griffith Criterion of Tensile Failure

Griffith (1924)[73] published an analysis of the mechanics of brittle failure in glass which has been applied by various investigators to fresh, intact, isotropic rock. The theory postulates that materials contain elliptically shaped flaws or cracks, and in compression large tensile stresses exist near the ends of critically oriented cracks. Frac-

TABLE 3.10
TESTS TO MEASURE ROCK STRENGTH

Material	Test	Strength measured
	LABORATORY TESTS	
Intact rock specimens	Schmidt rebound hardness*	Correlations with uniaxial compressive strength.
	Uniaxial compression*	Uniaxial compressive strength.
	Triaxial compression*	Peak drained or undrained strength, soft to hard rock.
	Direct shear*	Peak or residual drained strength, soft rocks and along joints.
	Point-load test†	A tensile strength for correlations with uniaxial compressive strength, soft to hard rock.
	Pull-apart axially*	Direct tensile strength.
	Brazilian*	Indirect tensile strength.
	Beam or flexure	Flexural or bending strength.
	IN SITU TESTS	
Rock masses	Direct shear box*	Peak or residual drained strength, soft rocks and along discontinuities.
	Triaxial compression box‡	Peak undrained shear strength in soft rock.
	Dilatometer or Goodman jack§	Limiting pressure p_L (borehole test) in soft or weak rock.

*From ISRM (1981).[23]

†From Broch and Franklin (1972).[70]

‡From Hibino et al. (1977).[71]

§From Hunt (1984).[7]

ture occurs when the most severely stressed crack propagates from stresses locally higher than the elemental cohesion. The relationship between shear stress τ_{xy} and the normal stress σ_y at which fracture begins on the boundary of an elliptical flaw, and the uniaxial tensile strength σ_t, has been epxressed as

$$\tau_{xy}^2 = 4\sigma_t(\sigma_t - \sigma_y) \tag{3.52}$$

which is the equation for a parabola in the $\tau_{xy} - \sigma_y$ plane.

General Rupture Criterion: Coulomb-Navier Equation

A complete rupture criterion for brittle rock is shown in Fig. 3.49. The simplest type of rupture occurs when the minor principal stress equals the uniaxial *tensile* strength ($\sigma_3 = \sigma_t$). Rupture is caused by the propagation of a crack that begins at the tip of the flaw that is perpendicular to the minor principal stress direction (Eq. 3.52). The radius of curvature of the Mohr envelope at the point of its intersection with the normal stress axis is equal to twice the uniaxial tensile strength ($2\sigma_t$). Consequently, all Mohr stress circles which fall in this circle of radius $2\sigma_t$ can touch the rupture envelope at only one point, which is defined by τ_{xy} and $\sigma_3 = \sigma_t$. When the normal stress acting on the potential shear surface is *compressive*, rupture occurs when the shear resistance of the plane is overcome; i.e., when the *Coulomb-Navier equation*, which is similar to the Mohr-Coulomb expression (Eq. 3.2), is satisfied:

$$S_s = \tau_{max} = \tau_o + f\sigma_n \tag{3.53}$$

where τ_o = intrinsic shear resistance of the material, or the shear resistance when

TABLE 3.11
DEFORMATION CHARACTERISTICS OF INTACT ROCK IN UNIAXIAL COMPRESSION TO FAILURE*

Deformation stress vs. strain	Rock type	Failure form
I. Elastic 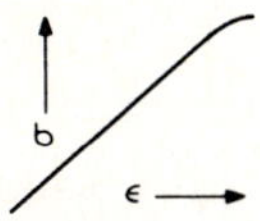	Fine-grained, massive. Basalt, quartzite, diabase dolomite, some strong limestones.	Sudden, explosive, brittle.
II. Elastic-plastic 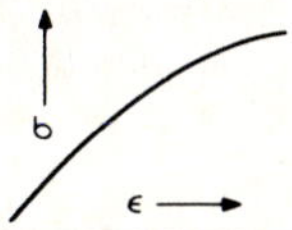	Fine-grained sedimentary. Softer limestones, siltstone, tuff.	Plastic yielding.
III. Plastic-elastic 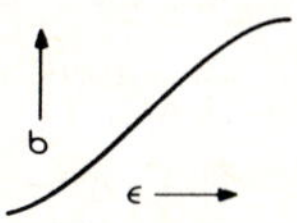	Sandstone, granite, some diabases, schist cored parallel to foliation.	Brittle fracture—stiffness increases as microfissures close.
IV. Plastic-elastic-plastic 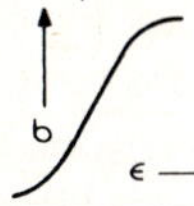	Metamorphic—marble, gneiss.	Inelastic yielding—stiffness increases as microfissures close.
V. Plastic-elastic-plastic 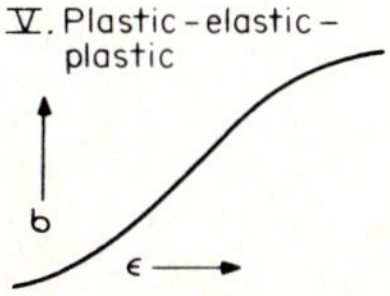	Schist cored perpendicular to foliation.	Inelastic yielding—stiffness increases as microfissures close.
VI. Elastic-plastic creep 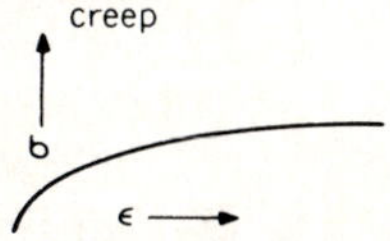	Rock salt.	Inelastic yielding and continuous creep.

*After Miller (1965).[72]

the normal stress $\sigma_n = 0$, which is due to cohesion, cementation, etc.

f = the frictional coefficient; the ratio between the shear stress S_s required to sustain movement, once the intrinsic shear resistance has been overcome, and σ_n.

When the principal stresses (σ_1 and σ_3) apply to a specimen, the crack initiated at the boundary of the flaw will propagate only a short distance before it stops and becomes stable. Propagation occurs when the shear resistance of the flaw zone is overcome and the shear movement continues.

Various other approaches to rock strength have been made [see, for example, Goodman (1980)[75]]. Investigators have demonstrated that the failure envelopes for most rocks lie between a straight line and a parabola which has led to the development of an empirical criterion of failure defined by the envelope to a series of Mohr's circles

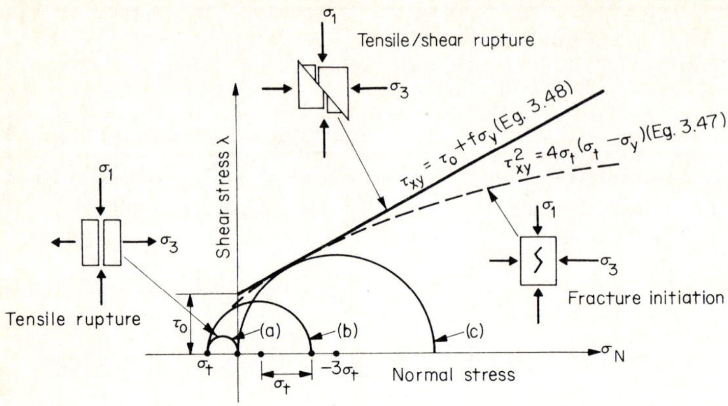

FIG. 3.49 A rupture criterion for brittle rock: (*a*) direct tension; (*b*) Brazilian tension test; and (*c*) uniaxial compression test. [After Hoek (1969).[74]]

as determined from direct tensile, Brazilian, and unconfined compression tests (as shown in Fig. 3.49), and triaxial compression tests.

Plastic Shear

Under the confining pressures encountered in practice (low to moderate), foliated rocks, such as schist and gneiss, will undergo plastic shear along the foliation planes, and soft rocks, such as halite, will behave in a ductile manner and undergo plastic shear and even creep. As rock decomposes and reaches the completely weathered state (Table 3.14) it is treated as a Mohr-Coulomb material. There is very little information in the literature regarding the strength of rock in the transition zone between the fresh, unweathered and the totally decomposed states.

Parameter Measurements with Laboratory Testing

In the reporting of laboratory test results it is essential that a complete description of the specimen be provided, including rock type, weathering degree, mineral composition, grain size, fabric (isotropic or anisotropic), and the orientation of the fabric to the deviator stress in anisotropic specimens. A sketch or photograph of the specimen before and after failure is important. In metamorphic rocks in particular, the fabric in a given core run can vary from nearly isotropic to extremely anisotropic, and to identify the material simply as a "gneiss" is far from adequate for the professional who must work with the data.

Laboratory tests can be divided into four general categories: compressive strength, direct shear strength, tensile strength, and flexural strength, as illustrated in Fig. 3.50.

Uniaxial Compressive Strength (U_c, q_c)

Strength in unconfined compression is measured in the same manner as for soils, except that the apparatus has a far greater capacity [for procedures see ASTM D2938-71a or ISRM (1981)[23]]. Test specimens should be NX cores or larger, with a height-to-diameter ratio of 2.5 to 3.0. Typical values of U_c for fresh specimens of various rock types, together with values for density and Young's modulus, are given in Table 3.12. Ranges of values for various rocks, including a hardness and strength classification, are given in Table 3.13. A useful correlation between the Schmidt hardness, obtained by simple test [see ISRM (1981)[23]], and U_c is given in Fig. 3.51.

Triaxial Compressive Strength (ϕ, or ϕ and τ_o or c, ϕ_r)

Except for capacity, the triaxial shear apparatus is similar to that used for soils. Procedures are given in ASTM D2664-74 and ISRM (1981).[23] Some apparatuses have the capability of testing specimens under high temperatures as well as high confining pressures.

Direct Shear Test (ϕ_a, ϕ_b, ϕ_r, c', c)

If the specimen is decomposed to the extent that it may be trimmed into the direct shear ring (Fig.

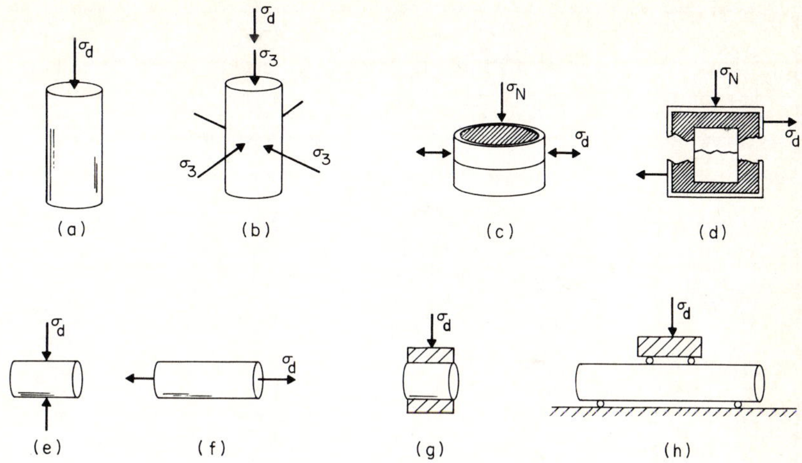

FIG. 3.50 Common laboratory tests to measure strength of rock cores: (*a*) uniaxial compression; (*b*) triaxial compression; (*c*) direct shear for soft specimens; (*d*) direct shear for joints; (*e*) point load; (*f*) direct tension; (*g*) splitting tension (Brazilian); and (*h*) four-point flexural.

3.50*c*), the test is performed like the soil test. If possible, the shearing plane should coincide with the weakness planes in the specimen. Tests to measure the characteristics of joints in fresh to moderately weathered rock are performed by encapsulating the specimen in some strong material within a shear box as shown in Fig. 3.50*d* [ISRM (1981)[23]]. The specimen is permitted to consolidate under a normal force and then sheared to obtain measures of the peak and residual strength as described in Art. 3.3.2. The normal stress is increased, consolidation permitted, and the specimen sheared again. The process is repeated until five values of shear stress versus normal stress are obtained, from which a graph for peak and residual strength is constructed as shown in Fig. 3.52.

Shear strength parameters ϕ_a, ϕ_b, ϕ_r, c', and c are abstracted from the graph, which is often a composite of a number of tests, as shown in Fig. 3.52, where

ϕ_r = residual friction angle.

ϕ_a = apparent friction angle below stress σ_a; point *A* is a break in the peak shear strength curve resulting from the shearing off of major irregularities (asperities) on the shear surface. Between points *O* and *A*, ϕ_a will vary somewhat and is measured at the stress level of interest. [$\phi_a = \phi_u + j$, where ϕ_u is the friction angle obtained for smooth surfaces of rock on rock and angle *j* is the inclination of surface asperities (Fig. 3.53).]

ϕ_b = apparent friction angle above stress level σ_a; it is usually equal to or slightly greater than ϕ_r and varies somewhat with the stress level. It is measured at the level of interest.

c' = cohesion intercept of peak shear strength, which may be zero.

c = apparent cohesion at a stress level corresponding to ϕ_b.

Tensile Strength

Tensile strength is measured in several forms and is generally applied to specimens of fresh to slightly weathered rock.

DIRECT TENSILE TEST Caps are attached to the ends of a cylindrical specimen by resins and pulled apart by cables or other means (Fig. 3.50*f*) [ISRM (1981)[23]]. This method yields the lowest tensile strength value, which is given as

$$S_t = P/\pi r^2 \qquad (3.54)$$

BRAZILIAN TENSILE TEST A core diameter *D* of not less than NX, with length *L* approximately equal to the diameter, is split by the force of steel jaws in a loading frame imposed on the sides of the

TABLE 3.12
COMMON ROCK TYPES AND TYPICAL ENGINEERING PROPERTIES*

Rock type†	Texture	Fabric structure	γ_d, g/cm²	U_c, tsf, kg/cm²	E_r 10⁴ tsf, kg/cm²
		IGNEOUS			
Granite	Coarse to medium	Massive relatively	2.69	700–1750	28–49
Diorite	Coarse to medium	tight, and widely	2.82	700–1750	35–56
Gabbro	Coarse to medium	spaced joints	2.88	1050–2100	49–84
Rhyolite	Fine	Massive, extensive	2.59	700–1750	35–56
Andesite	Fine	jointing, often	2.66	700–1750	42–63
Basalt	Fine	vesicular	2.85	1050–2100	49–90
Obsidian	Glassy	Massive,	2.20	140–560	7–28
Tuff	Coarse	continuous	1.60	14–70	1–7
		Cemented ash, porous			
		METAMORPHIC			
Gneiss	Coarse to medium	Banded to foliated	2.70	700 1400	28–56
Schist	Fine	Foliated	2.67	350–1050	14–35
Slate	Fine	Platy	2.69	700–1400	35–56
Quartzite	Fine	Massive, fine and widely	2.66	1050–2450	42–56
Marble	Fine to very fine	spaced joints	2.69	840–2100	49–70
Serpentine	Various	Massive, often soft	2.53	70– 700	7–35
		SEDIMENTARY			
Conglomerate	Coarse, rounded	Layered, cemented	2.48	350–1050	7–35
Breccia	Course, angular	Layered, cemented	2.53	350–1050	7–35
Sandstone	Medium	Layered, cemented	2.35	280– 840	7–21
Siltstone	Fine	Layered, cemented	1.8–2.4	7– 350	3–14
Shale‡	Very fine	Laminated, compaction shales unstable, cemented shales stable	1.6–2.2	7– 350	3–14
Limestone	Fine	Massive, stratified, soluble, cavities form	2.64	350–1050	14–42
Dolomite	Fine	Massive, some recrystallization	2.67	490–1400	28–56

*After NAVFAC (1971).[44] Properties are for sound, unweathered specimens without voids or fractures, tested dry in the laboratory. Elasticity and strength depend on porosity, cementation, and in foliated, platy, or laminated rocks, on loading direction. Saturated values for U_c and E_r are usually 80 to 90% of the dry values given.

†For detailed descriptions of rock types, composition, textures, fabrics, and structure, see Hunt (1984).[1]

‡See also Table 3.30 of Hunt (1984).[1]

TABLE 3.13
TYPICAL RANGES OF UNCONFINED COMPRESSIVE STRENGTH U_c FOR VARIOUS ROCK TYPES WHEN FRESH*

Hardness	Strength class	U_c, tsf
Extremely hard	Very high	
Very hard to hard	High	2000
	Medium	1000
Moderate	Low	500
Soft	Very low	250
Very soft		100

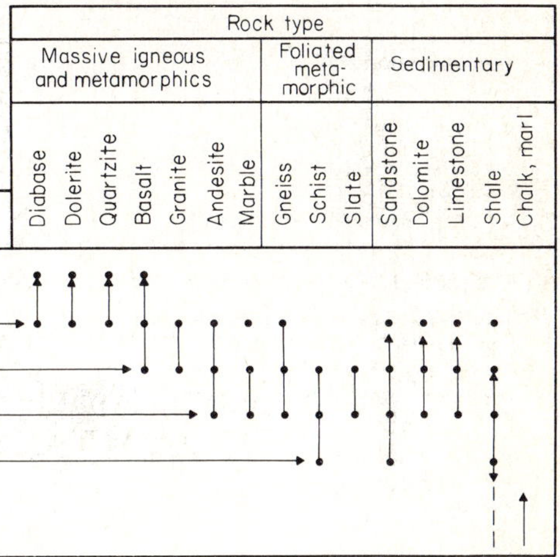

*After Deere (1968)[77] and other sources.

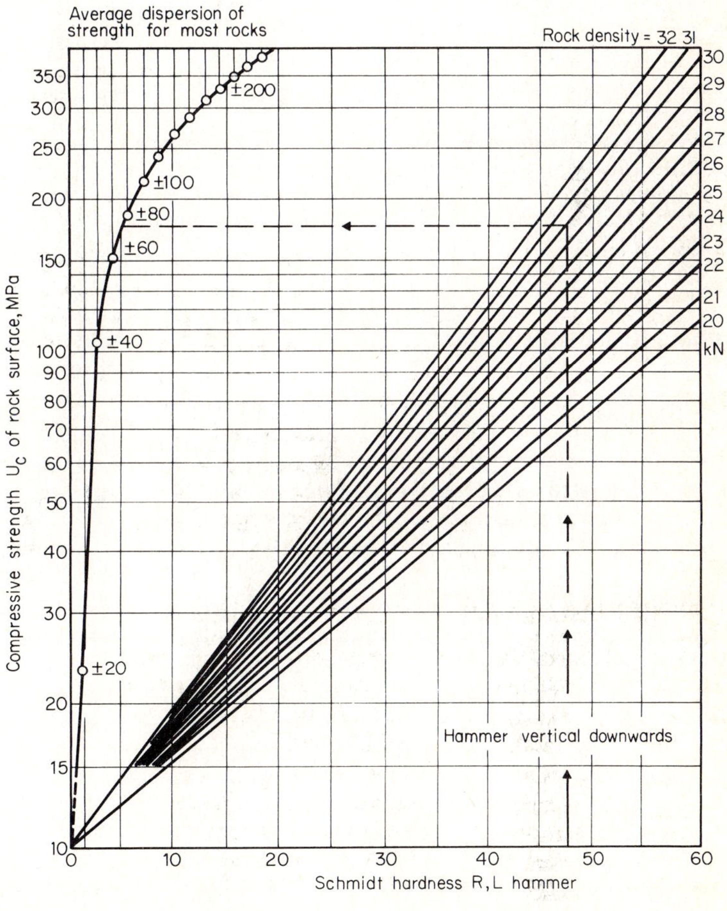

FIG. 3.51 Correlation chart for Schmidt L hammer, relating rock density, uniaxial compressive strength, and rebound number *R* (Schmidt hardness). Hammer vertical downward; dispersion limits defined for 75% confidence. (NOTE: 100 MPa = 14.5 × 10^3 psi = 1021 tsf; 1 kN/m² = 6.3 pcf.) [From ISRM (1981)[23]; after Deere and Miller (1966).[76]]

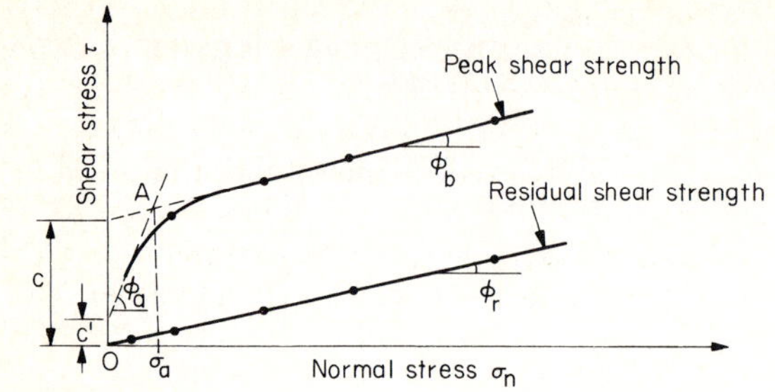

FIG. 3.52 Shear strength–normalized stress graph for direct shear test on rock specimen.

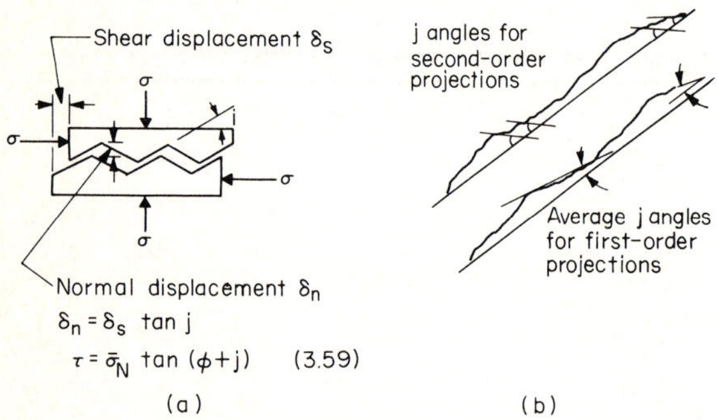

FIG. 3.53 The joint roughness angle *j*: (*a*) experiments on shearing regular projection and (*b*) measurements of *j* angles for first- and second-order projections on rough rock surface. [From Patton (1966),[80] Hoek and Bray (1977).[81]]

specimen (Fig. 3.50g). The tensile strength is then calculated at failure from

$$S_t = \frac{2P}{\pi DL} = 0.636\, P/DL \qquad (3.55)$$

The Brazilian test has been found to give values for tensile strength higher than those of the direct tension test, probably because of the effect of fissures [Goodman (1980)[75]].

POINT-LOAD TEST Compressive loads are applied to failure through hardened steel conical points to the sides of a core specimen with its axis horizontal (Fig. 3.50e). Failure is caused by the development of tensile cracks parallel to the loading axis. Specimen length should be at least 1.4 times the core diameter. The strength factor, the *point-load index*, is given by the empirical expression for a 50-mm core as

$$I_s = P/D^2 \qquad (3.56)$$

Because the point-load strength has been found to vary with core diameter, a correction factor is required to obtain the point-load index for a 50-mm core when other core diameters are tested [Broch and Franklin (1972)[70]]. On the basis of many correlations, the point-load index is used to estimate the uniaxial compressive strength for strong rocks from the relationship

$$U_c = 24\, I_{s(50)} \qquad (3.57)$$

In weak rocks, the correlations should be checked, since Eq. 3.57 can be highly inaccurate when applied to these materials. Light, portable equipment enables field testing of cores.

Flexural Strength

Flexural strength is measured by causing a rock beam to fail in bending under a four-point flexural loading as shown in Fig. 3.50*h*. The flexural strength (modulus of rupture) is the maximum tensile stress on the bottom of the beam corresponding to peak load, and is calculated from simple beam theory assuming elastic conditions. For four-point loading of cylindrical specimens, with loads applied at $L/3$ from each end and reactions at the ends, modulus of rupture (MR) is given [Goodman (1980)[75]] as:

$$T_{MR} = \frac{16 P_{max} L}{3\pi d^3} \qquad (3.58)$$

The flexural strength has been found to be 2 to 3 times as great as the direct tensile strength.

3.4.3 ROCK-MASS STRENGTH

Controlling Factors

A rock mass may be described in general terms as containing blocks of intact rock separated by discontinuities or fractures. *Rock-mass quality* refers to the block condition itself, given in terms of hardness or degree of decomposition, and the discontinuities, representing weakness planes, given in terms of their characteristics.

Rock-mass strength may be controlled by the strength of intact blocks or by the mechanical interaction of the blocks and the discontinuities. Seepage forces, acting along the discontinuities, significantly affect rock-mass strength.

Rock Quality

Elements: Blocks and Discontinuities

Rock blocks, often referred to as intact rock, are generally described in terms of hardness, which relates to rock type (mineral composition) and degree of decomposition. A general classification of intact rock based on hardness, weathering grade, diagnostic features, and attendant uniaxial compressive strength ranges is given in Table 3.14. The relative hardness of fresh specimens of various rock types and typical ranges for U_c are given in Table 3.13. It is noted that the weathering grades apply primarily to the crystalline rocks such as igneous (nonglasses) and metamorphics (quartzite and gneiss). Typical values for ϕ

TABLE 3.14
INTACT ROCK CLASSIFICATION BASED ON HARDNESS AND WEATHERING*

Class	Hardness†	Diagnostic features ( ) = Weathering effects	Weathering grade‡	Symbol	Strength,§ tsf
I	Extremely hard (or strong)	Rings under hammer impact; many blows required to break specimen. (No visible signs of decomposition or discoloration.)	Fresh	F	>2500
II	Very hard to hard (or very strong)	Hand-held specimen breaks with hammer under more than one blow. (Slight discoloration inwards from open fractures, otherwise similar to F.)	Slightly weathered	WS	2500–1000
III	Moderate (or medium strong)	Cannot be scraped or peeled with knife. Hand-held specimen can be broken with single moderate hammer blow. (Discoloration throughout; weaker minerals, such as feldspar, decomposed. Texture preserved.)	Moderately weathered	WM	500–250
IV	Soft (or weak)	Can just be scraped or peeled with knife. Indentations 1 to 3 mm show in specimen with moderate blow with pick end; lower strength specimens can be broken by hand with effort. (Most minerals somewhat decomposed; texture becoming indistinct but fabric preserved.)	Highly weathered	WH	250–50
V	Very soft (or very weak)	Material crumbles under moderate blow with pick and can be peeled with knife, but is hard to hand-trim for test specimen. (Minerals decomposed to soil but fabric and structure preserved; i.e., saprolite.)	Completely weathered	WC	50–10
	Extremely soft or weak	Advanced state of decomposition resulting in plastic soils. (Fabric and structure completely destroyed. Large volume change.)	Residual soil	RS	<10

* After ISRM (1982).[23]

†Hardness depends on rock type (see Table 3.13) as well as weathering grade.

‡Weathering grade applies primarily to crystalline rocks.

§Relationships to be considered only as a general guide, from U_c test.

TABLE 3.15
ROCK-MASS DISCONTINUITIES*

Discontinuity	Definition	Characteristics
Fracture	A separation in the rock mass, a break.	Signifies joints, faults, slickensides, foliations, and cleavage.
Joint	A fracture along which essentially no displacement has occurred.	Most common defect encountered. Present in most formations in some geometric pattern related to rock type and stress field. Open joints allow free movement of water, increasing decomposition rate of mass. Tight joints resist weathering and the mass decomposes uniformly.
Faults	A fracture along which displacement has occurred due to tectonic activity.	Fault zone usually consists of crushed and sheared rock through which water can move relatively freely, increasing weathering. Waterlogged zones of crushed rock are a cause of *running ground* in tunnels.
Slickensides	Preexisting failure surface; from faulting, landslides, expansion.	Shiny, polished surfaces with striations. Often the weakest elements in a mass, since strength is often near residual.
Foliation planes	Continuous foliation surface results from orientation of mineral grains during metamorphism.	Can be present as open joints or merely orientations without openings. Strength and deformation relate to the orientation of applied stress to the foliations.
Foliation shear	Shear zone resulting from folding or stress relief.	Thin zones of gouge and crushed rock occur along the weaker layers in metamorphic rocks.
Cleavage	Stress fractures from folding.	Found primarily in shales and slates; usually very closely spaced.
Bedding planes	Contacts between sedimentary rocks.	Often are zones containing weak materials such as lignite or montmorillonite clays.
Mylonite	Intensely sheared zone.	Strong laminations; original mineral constituents and fabric crushed and pulverized.
Cavities	Openings in soluble rocks resulting from groundwater movement, or in igneous rocks from gas pockets.	In limestone range from caverns to tubes. In rhyolite and other igneous rocks range from voids of various sizes to tubes.

*From Hunt (1984).[1] Reprinted with permission of McGraw-Hill Book Company.

and ϕ_r for weathered specimens are given in Table 3.27 of Hunt (1984).[7]

Discontinuities include joints, faults, slickensides, foliation planes, foliation shear zones, cleavage, bedding planes, and cavities, as described in Table 3.15. All can significantly influence rock quality, especially in Classes I and II (Table 3.14) and to a lesser extent in Class III, where block characteristics begin to control rock strength and deformation characteristics. Relict structures, such as slickensides, even in completely weathered rock or residual soil, however, quite commonly control strength, particularly in slopes, and their presence should always be suspected.

Joints are the most common discontinuity, encountered in some degree in almost all rock types [see Art. 6.4 of Hunt (1984)[1]]. A classification of joints and rock masses based on joint spacing is given in Table 3.16. Joints have the characteristics of orientation, spacing, aperture, types of fillings, roughness, and wall strength. *Roughness* relates to surface characteristics which can, in general form, be stepped, undulating, or planar, each in turn providing profiles that are rough, smooth, or slickensided (previous shear displacement).

Nomenclature

Standard nomenclature to describe rock quality has yet to be developed, and various terms are found in the literature. In general, rock quality is described as ranging from competent or excellent rock to incompetent rock, with the latter ranging from good to very poor quality. The approaches of various investigators are summarized in Art. 5.2.7 of Hunt (1984).[1] A classification based on RQD (rock quality designation from core recovery and core joint frequency) proposed by Coon and Merritt (1970)[78] is as follows: excellent (100 to 90%), good (90 to 75%), fair (75 to 50%), poor (50 to 25%), and very poor (< 25%). Core-drilling procedures and equipment can severely influence RQD [see Art. 2.4.5 of Hunt (1984).[1]]

Competent rock refers generally to solid and massive (joint spacing >1 m), extremely hard to very hard rock (Table 3.14). Experience has shown that in underground openings, the properties of a hard rock mass will be governed by the block properties if the spacing of tight joints is in the order of 1 m or more, and the mass contains no other defects [Hartmann (1966)[79]]. Such rock masses underground are usually self-supporting, although in any given situation consideration must be given to scale effects: i.e., the relationship between block dimensions and the size of the excavation opening, the orientation of the joints bounding the blocks, and joint characteristics. An important factor in deep underground excavations in competent rock is the potential for rock bursts from high residual stresses.

Incompetent rock-mass behavior normally is controlled by the mechanical interaction of the blocks and weakness planes and other geologic factors such as groundwater and residual stresses, as long as the blocks are fresh to only moderately weathered and hard to moderately hard. As rock hardness decreases and the degree of weathering increases, behavior may become controlled by block characteristics, although re-

TABLE 3.16
CLASSIFICATION OF JOINTS BASED ON SPACING*

Description	Joint spacing	Rock-mass designation
Very wide	≥3 m	Solid
Wide	1 to 3 m	Massive
Moderately close	30 cm to 1 m	Blocky/seamy
Close	5 cm to 30 cm	Fractured
Very close	<5 cm	Crushed

*From Deere (1963)[92] and Bieniawski (1974).[93]

lict fractures may still provide potential failure surfaces.

Block Parameters

Fresh or decomposed block parameters include dimensions, unit weight, Young's modulus, Poisson's ratio, cohesion, friction, and total strength at failure. Small-scale weakness planes, such as foliation, schistosity, and cleavage are normally accounted for in block parameters. The orientation of these features to the loading direction affects strength parameters and deformation and always requires careful consideration. In most practical situations it is the block strength of the weaker rocks that is of concern, and strength is expressed in terms of the Mohr-Coulomb or Terzaghi-Coulomb equations.

Joint Strength Characteristics

Parameters

Joint strength parameters ϕ_a, ϕ_b, ϕ_r, c', and c have been described in Art. 3.4.2, under *Direct Shear Test.*

Joint Condition Effects

As the joints close under applied load, resistance to deformation is provided as the strength along the joint is mobilized. Joints filled with clay, mylonite, or some other material will realize their strength characteristics from those of the fillings. Clean joints will realize their strength either from friction along a smooth rock-to-rock surface or from the irregularities of the joint surfaces (asperities). In a sense, joints respond to the effective stress principle in that water pressure directly counteracts the strengthening effect of the applied normal stress, a major cause of failure in slopes.

Asperity strength depends very much on the magnitude of the normal stress. Small asperities of very rough surfaces may shear easily under high stresses, whereas under low normal stresses, such as in slopes, the shearing of asperities of fresh rock is considered uncommon. Weathering along joint walls, however, often leaves conditions less than fresh. In most relationships developed to describe joint surface shear strength, the average angle *j*, the *roughness angle*, of the asperities as related to the joint surface, is considered, as shown in Fig. 3.53.

Estimating Joint Shear Strength

Concepts of joint shear strength have undergone periodic revision in recent years since their initial development only a decade or so ago. Expressions for *high normal stresses* are relatively complex and provide for both shear strength of the asperities and the joint friction angle and the joint roughness angle in terms of normal stress [Ladanyi and Archambault (1970),[82] Hoek and Bray (1977)[81]].

Relationships for *low normal stresses*, such as exist in slopes, are of more practical use. Roughly speaking, joint shear strength will consist of a maximum (peak) or minimum (residual) friction angle or some intermediate angle (depending upon the degree of previous shear displacement), plus a contribution (*j*) due to surface roughness, expressed as

$$\tau = \bar{\sigma}_n \tan(\phi + j) \tag{3.59}$$

Which strength parameter will prevail depends very much on displacement as well as joint roughness. Peak strength is mobilized after relatively small displacements; therefore, it may not be realistic to add a value for joint roughness *j*. As displacement increases, ϕ_r becomes mobilized and the addition of a value for *j* is realistic [ISRM (1981)[23]]. For large displacements, completely planar surfaces, or completely sheared asperities, ϕ_r, without a value for *j*, represents the *minimum* shear strength of the discontinuity. Strength, therefore, can be expressed as Eq. 3.59, or as

$$\tau = \bar{\sigma}_n \tan(\phi_r + j) \tag{3.60}$$

or as

$$\tau = \bar{\sigma}_n \tan \phi_r \tag{3.61}$$

The roughness angle *j* is expressed by Barton (1973)[83] as

$$j = \text{JRC} \log_{10}\left(\frac{U_j}{\bar{\sigma}_n}\right) \tag{3.62}$$

where $j = 0$ on a smooth, planar, rock-to-rock surface

U_j = uniaxial compressive strength of asperities (joint wall)

JRC = *joint roughness coefficient*, varying from 0 to 20. JRC values result from tests and observations, and account for the average value of j which affects the shear and normal stresses acting on the failure surface. A scale of JRC values versus roughness profiles is given in ISRM (1981),[23] p. 29. Some general values are:

- JRC = 20 for rough undulating joints (tension joints, rough sheeting, rough bedding)
- JRC = 10 for smooth undulating joints (smooth sheeting, nonplanar foliation, and bedding)
- JRC = 5 for smooth nearly planar surfaces (planar shear joints, planar foliation, bedding)

Hoek and Bray (1977)[81] suggest that Eq. 3.60 is probably most valid in the range of $\bar{\sigma}_n/U_j = 0.01$ to 0.03. ISRM (1981)[23] suggests using the *Schmidt hammer* to estimate both U_j and ϕ_r, as described in Barton and Choubey (1977).[84] Schmidt hardness correlations with U_c are given in Fig. 3.51, and U_j is taken as U_c. Approximations of U_c based on diagnostic features are made from Table 3.14, which represents a large amount of correlative data. ϕ_r is estimated on the basis of the ratio of the Schmidt hammer rebound value R obtained on the weathered joint wall and the rebound value obtained on unweathered, fresh rock [Barton and Choubey (1977)[84]]. Weathering may reduce U_j to as low as 25% of the fresh rock value. Corrections for reducing the Schmidt hammer rebound value if the hammer is not used vertically downward are given in ISRM (1981),[23] p. 34.

Ranges for friction values for various joint types are given in Table 3.17. If movement has occurred along the joint, the residual strength will govern. Peak ϕ values of fresh rock usually range from about 30 to 75°, commonly averaging about 45°. Where joints have vertical or very steep steps, there will also be a cohesion value. ϕ_r values depend on the degree of wall weathering and rock type, usually ranging from about 25 to 35°, most commonly about 30° in unweathered conditions and falling to around 15° for strong weathering, even in the absence of clay filling [ISRM (1981)[23]].

Parameter Measurements

Laboratory Tests

Block parameters are measured in the laboratory on a small scale, employing the testing procedures for intact rock described in the previous article. The direct shear test is used to test joint strength parameters on a small scale.

In Situ Tests

Direct shear tests are the most practical and reliable method for measuring strength parameters of weak rocks and joints in situ. The procedure is generally similar to the laboratory test described in the previous article, except for scale. A rock block with usual dimensions 70 cm × 70 cm × 35 cm is trimmed carefully to avoid disturbance since it remains attached to the rock mass. The weakness plane to be tested should be located as near to the direction of applied shearing force as possible. The block is confined in a grout-filled steel box and a vertical load applied by a hydraulic jack reacting against rock anchors. Consolidation is permitted and then a shear force is applied with another hydraulic jack until failure. The test is repeated for five different normal stress values to provide the data for the construction of Mohr's envelopes as given in Fig. 3.52 [see ISRM (1981)[23] for details of testing procedures.]

Schmidt hardness tests are performed in situ to obtain values for correlations with U_c as mentioned previously.

Triaxial shear box is similar to the direct shear method except that flat jacks are used to impose lateral confining pressures, as described by Hibino et al. (1977).[64]

Pressuremeter, dilatometer, and Goodman jack are borehole tests to measure strength in relatively weak rock masses. Results are strongly affected by anisotropic conditions (*see Arts. 3.3.3 and 4.1.4*).

Residual stresses are measured with flat-jack tests made on a rock surface, or with borehole

TABLE 3.17
TYPICAL STRENGTH VALUES FOR ROCK-MASS DISCONTINUITIES FOR USE IN PRELIMINARY STABILITY ANALYSIS[a,b]

Defect	Condition	Strength parameter[c]
Joint surface: smooth, unweathered rock[d,e]	Hard massive, well-interlocked or cemented (granite, basalt, gneiss, sandstone, limestone)	ϕ_u = 30–35°
	Hard shaley or schistose (hard shale, slate, phyllite, mica-schist)	ϕ_u = 25–30°
	Soft, laminated or schistose (clay shale, talc schist, chlorite schist)	ϕ_u = 20–25°
Asperity angle j (add to ϕ_u)[e]	Very irregular surface	j = 30–40°
	Average surfaces	j = 5–15°
	Smooth, planar surface	j = 0–2°
Gouge or mylonite from fault or shear zones	Sandy material	ϕ_d = 24–32°
	Intermediate material	ϕ_d = 16–23°
	Clayey material	ϕ_d = 8–15°

[a]After Deere (1976).[85]

[b]For FS = 1.5 to 2.0, ϕ_u = basic friction angle, ϕ_d = drained test.

[c]Generally assume $c = 0$ for weak zones (weathered, altered, sheared), and in no case > 1 tsf even for hard rock surfaces when tested.

[d]To ϕ_u add asperity angle j, or for analysis, $\bar{\phi} = \phi_u + j$.

[e]Reduce $\bar{\phi} = \phi_u + j$ by 5 to 10° for weathered surfaces (Eq. 3.59).

[f]Values for ϕ_r are generally 5 to 10° or more lower than values for $\bar{\phi}$.

tests such as those employing a deformation gage or strain gage [see Art. 4.4.4 of Hunt (1984)[1]].

3.4.4 PRACTICAL ASPECTS

Intact Rock Strength

There is a large variety of practical problems in rock mechanics where the intact strength has application, some of which are illustrated in Fig. 3.54. They are described more fully in subsequent chapters which pertain to particular problem categories.

Compression Tests

Uniaxial compressive strength U_c is used as an index value in classification systems for rock masses and in analysis. In the latter case it is used to evaluate the shearing resistance of asperities along joint walls (Fig. 3.54*a*), the resistance to crushing of mine pillars (Fig. 3.54*c*), and the bond strength or shaft resistance f_s of a rock-socketed concrete pile (Fig. 3.54*f*) or of a rock anchor (Fig. 3.54*g*). Values for U_c are estimated from correlations with data from the point-load test or the Schmidt rebound hardness test, with the latter having the important capability of obtaining measures in situ of joint wall strength. The selection of U_c values for analysis usually is made through empirical correlations based on load test data, back-analysis of failures, etc.

Triaxial compression and *direct shear strength parameters* are used to evaluate the stability of weak or soft rock in slopes and beneath foundations or the shearing resistance along microfissures such as foliations in hard rock masses. The latter case is particularly important in the stability analysis of the heavily loaded foundations of concrete arch dams (Fig. 3.54*d*).

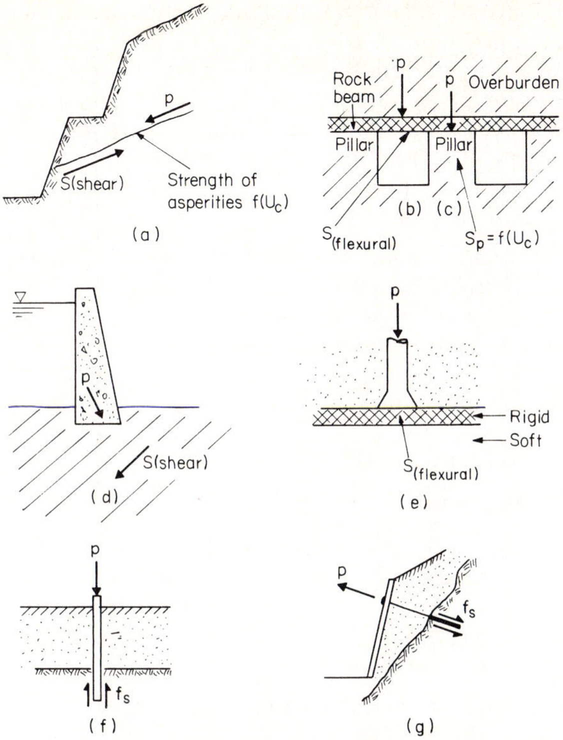

FIG. 3.54 Some applications of rock strength: (*a*) slopes; (*b*) mine or cavern roof; (*c*) mine pillar; (*d*) concrete dam foundation; (*e*) pier foundation; (*f*) pile socket; and (*g*) anchor.

Other Tests

Tensile strength provides the bond strength or shearing resistance along rock anchor and pile shafts, but as described above, analysis usually is based on empirical correlations with U_c.

Flexural strength is of interest in connection with a rock beam, free of defects, supporting a mine roof or overlying a limestone cavern (Fig. 3.54*b*), or in the case of a heavily loaded foundation supported on a beam of sound rock which in turn overlies a substantially weaker rock (Fig. 3.54*e*).

Rock-Mass Behavior

Strength Parameter Applications

Under applied stress a rock mass initially undergoes deformation from compression as the fractures close; eventually, as deformation continues, strength is mobilized along the fractures. Seepage forces may provide additional stresses in the fractures. Rock slopes provide the most common engineering problem where stability against rupture requires analysis (Fig. 3.54*a*). Foundation problems in concrete arch or gravity dam structures may occur less frequently, but when they do stability analysis is extremely important, especially for the abutments. Triaxial compression and direct shear tests provide the strength parameters. Direct shear tests have the advantage that they can readily be performed in situ, and they provide measures of the residual strength, particularly important in slope stability problems. Laboratory triaxial tests provide the advantage of pore-pressure measurements, particularly important in rocks with microfissures (see discussion of the Malpasset Dam failure in Art. 6.3.3).

Analytical Approaches

Intuitive evaluation of the overall rock quality (including the geometry and characteristics of the discontinuities and the characteristics of the material composing the intact blocks), based on experience and judgment, resolves many rock mechanics problems.

Simple analytical approaches based on limit equilibrium relationships also are commonly applied (see the following article).

Finite element methods (FEM) are being used in many current problems, including slopes, dam foundations, caverns, and mines. The finite element model requires detailed representative information concerning the geometry and properties of the weakness planes and rock blocks to produce reliable results. That information, however, is usually difficult to obtain. In all problems, the successful solution must be based on substantial experience and qualified judgment.

Behavior Models

Investigators have developed various models to describe rock-mass behavior from the creation of a discontinuity to its behavior before, during, and after sliding. These models, together with a new "comprehensive" model, are described by Roberds and Einstein (1978).[86]

An *empirical strength criterion for rock masses* has been developed by Hoek and Brown (1980).[87] It provides strength relationships in terms of principal stresses, uniaxial compressive strength, and a normalized shear strength criterion for five basic rock groups (grouped by types) and six categories of rock quality ranging from intact to very poor. The relationships incorporate parameters to account for anisotropy from discontinuities and other factors, and are intended for use in estimating overall rock-mass strength where the discontinuity spacing is small on the scale of the problem and in which complex modes of failure might be expected to occur. They are not to be used for problems in which failure is likely to occur along one or two discontinuities, such as single planar or wedge-type failures.

3.5 STABILITY ANALYSIS

3.5.1 EARLY THEORETICAL CONCEPTS

Plastic Equilibrium

Induced stresses cause movements in the ground which mobilize shearing resistance as the mass enters either the passive (pushed into) state resisting the force, or the active state occurring as the soil is stretched. At the point of maximum shear strength, the material is on the verge of failure and is in a *state of plastic equilibrium*. In the 1850s, Rankine developed his concepts of the *active and passive Rankine states* as applied to earth pressures against retaining walls *(see Art. 3.2.3)*. In practice, in addition to the rigid retaining wall problem, more complicated states of plastic equilibrium are encountered in slopes, in braced excavations, and beneath foundations.

Slip-Line Analysis

When a sufficiently large mass of soil is stressed to its yield condition, plastic flow occurs and slip lines develop in the mass *(see Art. 3.2.3)*. The stresses that develop in the soil at the instant of impending "flow" can be investigated by a set of differential equations.

The mathematical solution to the problem, termed the *slip-line method*, is considerably complicated because of the number and shape of the slip lines and necessary inclusion of soil properties. In 1903 Kotter derived slip-line equations for the case of plane deformations, and in 1920 Prandtl obtained an analytical closed-form solution to Kotter's equations for a footing on a weightless soil. The inclusion of soil weight complicates the problem considerably and, consequently, numerous approximate methods have been developed.

The *limiting equilibrium method* is an approximate approach to the slip-line field method, and is commonly used by practitioners in stability analysis. These methods neglect the stress-strain relationship of soil and consider only the equilibrium and yield conditions.

3.5.2 LIMIT EQUILIBRIUM ANALYSIS

Assumptions

Limit equilibrium analysis considers a limiting value that can be reached when the forces acting to cause failure are in balance with the forces resisting failure. Basic elements common to limit equilibrium analysis are:

1. Failure surface of simple shape (planar, circular, or log-spiral) is assumed.
2. Distribution of the stresses acting along the failure surface to cause failure is also assumed.
3. Mobilized shear strength is estimated and assumed to act simultaneously along the entire failure surface.

Progressive failure cannot be considered directly, although a range of shear-strength values may be applied. Conclusions cannot be drawn about either the distribution of stress in the soil or rock or the deformations that occur.

Analysis

When the basic elements are established, an overall equation of equilibrium is developed in terms of stress resultants, and the problem is solved by simple statics. Equilibrium requires that the sum of the forces in the normal direction must equal zero, as must the sum of the forces in the tangential direction and the sum of the moments about any given point. In practice, most problems are statically indeterminate and assumptions must be made regarding the force systems and the directions of their applications.

Validity

Conditions of Reality

A valid solution to a problem involving the mechanics of a deformable solid requires, in reality, three basic conditions to be satisfied:

1. Stress equilibrium equations
2. Stress-strain relationships
3. Compatibility equations relating strain and displacement

Ideally, in an elastic-plastic material there is a three-stage development as the deforming load is applied: the material undergoes an initial elastic response, then an immediate contained plastic flow, and finally, an unrestricted plastic flow as collapse occurs.

Collapse Condition Evaluated

The solution of most engineering problems, such as the stability of a slope or foundation, requires analysis of the conditions that would exist at the *collapse condition* and the application of safety factors to prevent collapse. The latter requirement is satisfied by limiting equilibrium analysis. The approach is know as *limit design*. Recent developments in soil mechanics are approaching the problem by *limit analysis* techniques *(see Art. 3.5.4)*.

Applications

A number of simplified examples illustrating the applications of the strength parameters in limiting equilibrium analysis are given in Fig. 3.55. Various definitions are used for the factor of safety, FS. In slopes (*a*, *b*) FS is now commonly considered as the ratio of the shearing strength available along the sliding surface to the shearing stresses tending to produce failure along the surface. In some cases of embankments over soft clay or retaining walls subject to overturning (*c*, *d*), FS is expressed as the ratio of the resisting moment to the overturning moment. In other retaining structures (*e*, *f*) or foundations (*g*, *h*), FS may be applied directly to the measured strength parameters; or, for foundations, it may be expressed as the ratio of the ultimate load capacity (failure condition) to a selected value for the design load.

In slopes and embankments over soft ground (*a*, *b*, *c*), the weight of the mass causes the shearing stresses that tend to produce failure. Additional shearing stresses may result from a surcharge load or, in the case of slopes, water pressures in tension cracks or joints. Retaining structures (*d*, *e*, *f*) require the determination of the distribution of active and passive earth pressures as a function of soil weight and strength parameters to enable determination of the forces acting on the structure. In foundation analysis it is the proposed load tending to cause failure (*g*, *h*), resisted by the strength of the soil or rock, which must be evaluated.

Applications are described in detail in subsequent chapters.

3.5.3 SELECTING STRENGTH PARAMETERS

Basic Factors to Consider

Soils

Peak strength prevails in most practical problems, although *residual strength* along preexisting failure surfaces in slopes and the *remolded strength* after driving piles in clays may govern stability. Parameter selection is based on the *degree of saturation* (dry, partially saturated, saturated) and the *drainage conditions* that will pertain in the field during loading. Depending primarily on the ability to assess pore pressures quantitatively, analysis is based on effective or total stresses. In certain cases, *dynamic loadings* may be imposed and the possibility of liquefaction in fine-grained granular soils or of strain softening in clay soils may require evaluation.

In addition to the foregoing factors, the selection of the type of test is made in consideration of the direction of loading as related to *anisotropic conditions* and to the *variation of strength* along the potential failure surface.

Rock Masses

Rock masses in the fresh to slightly decomposed condition may fail in *shear*, *tension*, or *flexure* depending upon the orientation of the confining stresses to the direction of applied load. Slopes normally are analyzed on the basis of effective stresses, and in decomposed masses the strength both of the blocks themselves and along the dis-

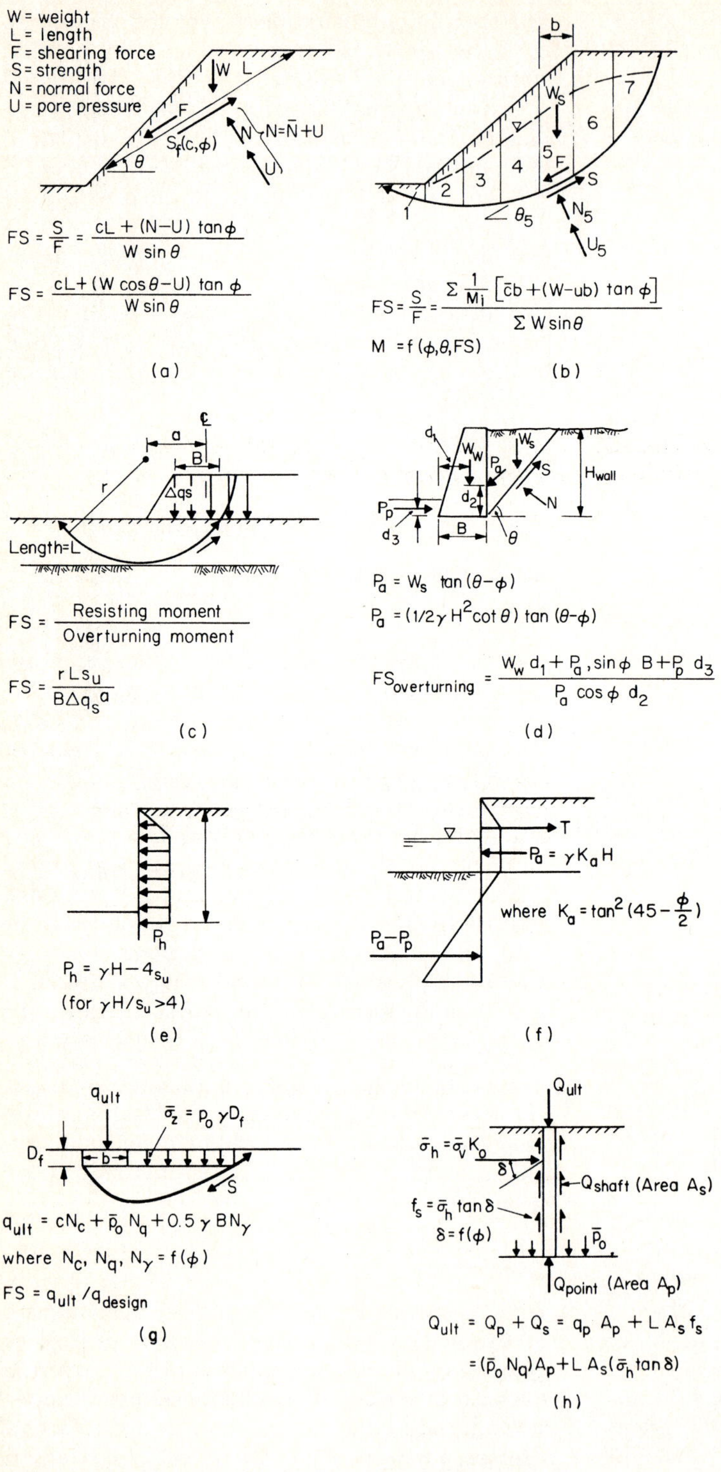

FIG. 3.55 Some applications of strength parameters in stability analysis by limit equilibrium methods: (*a*) rock slopes—simple wedge; (*b*) soil slope—forces considered in the ordinary method of slices; (*c*) embankment on soft clay; (*d*) gravity retaining wall with Coulomb forces and factor of safety FS solved for overturning; (*e*) earth pressures against braced excavation in clay; (*f*) earth pressures against anchored sheet-pile bulkhead in sand; (*g*) ultimate bearing capacity q_d beneath a footing on soils with cohesion; and (*h*) ultimate bearing capacity of a pile in sand.

continuities must be evaluated. Strength along discontinuities may range from *peak* to *residual* depending upon previous straining of the mass. Consideration is given to *anisotropic* conditions in intact rock such as those caused by cleavage, foliations, or schistocity, and to strength variation along the failure surface.

Effective Versus Total Stress Analysis

Effective Stress Analysis

Effective stress analysis is appropriate when the pore pressures are known or can be readily evaluated. These conditions apply if the material is free-draining, or loads are applied slowly enough that drainage occurs. In analysis, shear strength is expressed by the Terzaghi-Coulomb equation (Eq. 3.9) and estimates made of the pore pressures which exist, or may exist at some future date, on the potential failure surface. Parameters $\bar{c}$ and $\bar{\phi}$ are measured in the laboratory with CD and CU triaxial tests or with the direct shear test, in which the residual strength ϕ_r is also measured.

Effective stress analysis is applied to most problems involving natural slopes (Figs. 3.55*a* and *b*), cut slopes and earth dams some time after completion (except for sudden drawdown conditions), gravity retaining structures some time after construction (Fig. 3.55*d*), and shallow foundations (Fig. 3.55g), except where structures such as storage tanks or silos are erected over soft clays and loaded rapidly.

Groundwater conditions require careful appraisal, particularly for slope and foundation stability analysis. Conditions encountered during investigation often are not those that will prevail during failure. Water tables fluctuate seasonally, as well as over a period of several years, as a result of natural or human activities. Because pore- or cleft-water pressures serve to reduce the total normal stress and thereby the frictional component of shearing resistance (see Fig. 3.55*a* and *b*), judgment is required during stability analysis in the selection of the highest water table that might be anticipated during the life of the structure or slope. Cleft-water pressures in vertical or near vertical joints and tension cracks impose an additional force in slopes tending to produce failure.

Total Stress Analysis

Total stress analysis is appropriate when the pore-pressure response in the field is difficult to predict quantitatively or where the effects of pore pressures can be incorporated into the strength parameters by conducting the tests in such a way as to simulate the field conditions. These conditions may apply, for example, if loading is rapid in comparison with the ability of a saturated or partially saturated material to drain and relieve pore pressures. Laboratory tests, such as CU or UU triaxial tests, are performed employing procedures designed to simulate field conditions, or strengths are measured in the field in situ, and the shear strength is determined in terms of total stresses without pore-pressure measurements. The strength parameters are given as c_u and ϕ_u, or s_u, depending on the type of test. It is assumed that the pore pressures which develop in the test specimen will be equal to those which will develop in the field at failure.

In drained tests the maximum strength is measured at maximum obliquity, whereas in the $\phi = 0$ case, the maximum strength s_u is measured by the maximum difference in principal stress. In the CU tests the envelope is often severely curved and the strength relationship cannot be conveniently expressed in terms of c and ϕ except as an approximation for the range of normal stress of interest in the practical problem under study.

Cases where the parameters c_u and ϕ_u are often employed occur in the stability analysis of earth dams during the construction phase where cohesive materials do not have time to drain, or during the sudden drawdown phase when there is not sufficient time for the effective stresses to change.

The *undrained strength parameter* s_u applies to saturated normally consolidated clays of low to moderate sensitivity undergoing loading with no drainage occurring (the $\phi = 0$ concept). Pore pressure is ignored along any failure surface where s_u is used, not because pore pressure does not exist, but because the undrained strength can be expressed independently of the effective stress at failure.

Parameter s_u is applied in the analysis of embankments or foundations over soft clay (Fig.

3.55c and g), and braced excavations (Fig. 3.55e) and pile foundations in soft clay. In the expression given in Fig. 3.55h for Q_{ult}, the unit end bearing q_p and the unit shaft friction f_s are given in terms of s_u, and in the case of f_s consideration is given to the strength reduction to s_r from remolding during pile driving. Strictly speaking, the capacity of a pile foundation in soft clay is determined by the undrained strength only before consolidation of the clay occurs under the loading. Since consolidation increases the strength, the undrained calculation is conservative.

Testing Mode and Failure Surface Orientation

Strength Test Orientation

In *soil formations* strength parameters usually are measured by testing samples obtained either from vertical boreholes or in situ in vertical boreholes. Strength values may be obtained in the axial direction in triaxial compression or extension, in the lateral direction from direct or simple direct shear tests, or by the vane shear test which measures strength mobilized simultaneously on vertical and horizontal planes.

In *rock formations* boreholes can be vertical or inclined for either in situ testing or the retrieval of cores for laboratory testing. The important factor during loading is the load orientation with any weakness planes such as foliations and joints. Strength values for loads applied perpendicular to weakness planes can be substantially higher than values obtained from loads applied inclined or parallel to the planes.

Strength Variation along the Failure Surface

As the principal stresses, and consequently the zones of active and passive stresses, vary along the failure surface, as shown in Fig. 3.8, the strength mode mobilized will vary from compression to direct shear to extension, as shown in Fig. 3.56. A rigorous analysis, therefore, would be based on strength values obtained from various types of tests.

For natural slopes the solution may consider the formation of a tension crack along which no strength is mobilized, at the location of the upper "triaxial extension" in Fig. 3.56a. In slopes in residual soils stability is often evaluated on the basis of direct shear tests performed on saturated specimens, although partially saturated strength values will pertain along the failure surface above the water-table level (*see Art. 3.3.2*). In rock slopes, where the potential failure surface is normally selected on the basis of structural features of the rock mass, analysis by some practitioners has been based on assuming an upper tensile zone, a middle zone of residual strength resulting from deformations during initial straining, and a peak value for strength along the lower portion. In slope studies it is usual to employ relatively low safety factors.

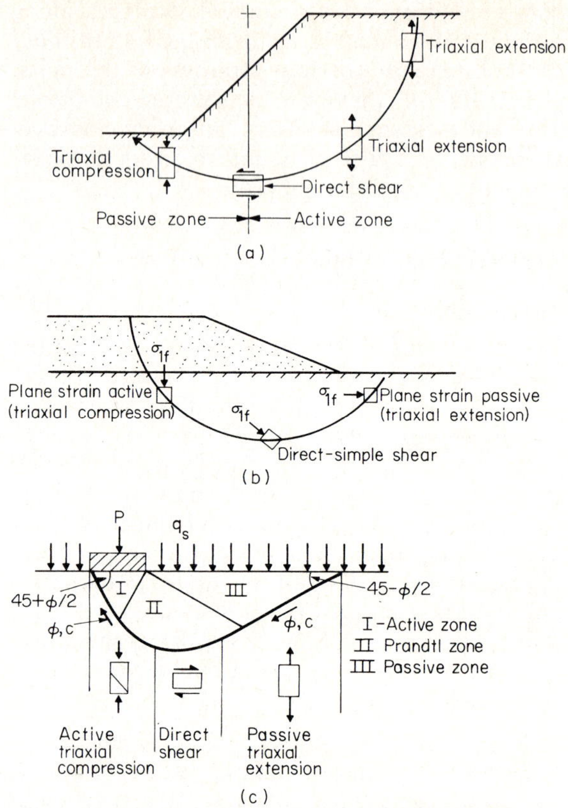

FIG. 3.56 Variations in strength along failure surfaces. (*a*) Slope failure. [After Burmister (1953).[2]] **(*b*) Embankment over soft, nonlayered clay.** [From Ladd (1973).[88]] **(*c*) Shallow foundation in general shear with soil parameters *c* and ϕ acting. (For $\phi = 0$, the rupture surface is always a circle.)**

Embankments on soft ground are illustrated in Fig. 3.56b. Besides the variation in stress along the failure surface, soft clays may be anisotropic in their strength behavior. The simple direct shear test has been found to give good to slightly conservative values of average s_u along a circular arc failure in nonlayered soft clays [Ladd (1973)[80]].

Foundation stability analysis (Fig. 3.56c) usually is based on the results of a single type of test to measure either $\bar{c}$ and $\bar{\phi}$, or s_u, and a generous safety factor is applied because of somewhat broad assumptions made for the development of the bearing-capacity equation *(see Art. 3.5.1, Slip Line Analysis).*

3.5.4 LIMIT ANALYSIS METHOD

Concept

The limit analysis method to evaluate stability is based on a yield criterion and its associated flow rule which considers the stress-strain relationship. To obtain the collapse load, the method employs two theorems for any body of elastic, perfectly plastic material [Chen (1975),[89] Fang (1975),[90] Wroth (1976)[91]].

Theorems

The lower-bound theorem holds that the soil mass will not fail, or will be just at the point of collapse, at an equilibrium of stress balancing the applied load and nowhere violating the yield criterion which includes cohesion c and friction angle ϕ. Only equilibrium and yield are considered and no consideration is given to soil kinematics.

Upper-bound theorem holds that the soil mass will collapse if there is any comparable pattern of plastic deformation for which the rate of *work* of the external loads exceeds the rate of internal *energy* dissipation. To obtain an upper-bound solution, it is necessary to find a compatible failure mechanism (velocity field). The method is based on the theory of metal plasticity. The deformation mode that occurs is expressed as a *velocity field* and the associated dissipation of energy can be computed from the idealized stress-strain relationship (the *flow rule*). By suitable choice of stress (lower-bound theorem) and velocity fields (upper-bound theorem) the collapse load can be closely bracketed.

Applications

In the present state of the art the application of limit analysis to practical problems is considered uncertain by many investigators because of difficulties in obtaining a proper stress-strain relationship for the soil mass. In this book stability analysis is based on limiting equilibrium methods.

REFERENCES

1. Hunt, R. E. (1984) *Geotechnical Engineering Investigation Manual*, McGraw-Hill Book Co., New York.
2. Burmister, D. M. (1953) "The Place of the Direct Shear Test in Soil Mechanics," *Symp. on Direct Shear Testing of Soils, STP 131*, Amer. Soc. for Test. and Mat., Philadelphia, Pa., pp. 3–18.
3. Lambe, T. W. and Whitman, R. V. (1969) *Soil Mechanics*, John Wiley & Sons, New York.
4. Brand, E. W. (1982) "Analysis and Design in Residual Soils," *Conf. on Engineering and Construction in Tropical and Residual Soils*, Honolulu, *Proc. ASCE*, pp. 89–143.
5. Leonards, G. A. (1962) *Foundation Engineering*, McGraw-Hill Book Co., New York, Chap. 2.
6. Bishop, A. W. and Henkel, D. J. (1962) *The Triaxial Test*, Edward Arnold Ltd., London.
7. Lambe, T. W. (1967) "Stress Path Method," *Proc. ASCE, J. Soil Mech. and Found. Engrg. Div.*, Vol. 93, No. SM6, November, pp. 309–331.
8. Lambe, T. W. and Marr, W. A. (1979) "Stress Path Method: Second Edition," *Proc. ASCE, J. Geotech. Engrg. Div.*, Vol. 105, No. GT6, June, pp. 727–738.
9. Kassif, G. and Baker, R. (1969) "Swell Pressure Measured by Uni- and Triaxial Techniques," *Proc. 7th Intl. Conf. Soil Mech. and Found. Engrg.*, Mexico City, Vol. 1, pp. 215–218.
10. Lambe, T. W. (1964) "Methods of Estimating Settlements," *Proc. ASCE, J. Soil Mech. and Found. Engrg. Div.*, Vol. 90, No SM5, September, pp. 47–71.
11. Lo, K. Y. and Lee, C. F. (1973) "Analysis of Progressive Failure in Clay Slopes," *Proc. 8th Intl. Conf. Soil Mech. and Found. Engrg.*, Moscow, Vol. 1, pp. 251–258.
12. Pells, P. J. N. (1973) "Stress Ratio Effects on Construction Pore Pressure," *Proc. 8th Intl. Conf. Soil Mech. and Found. Engrg*, Moscow, Vol. 1, pp. 327–332.

13. Adbelhamid, M. S. and Krizek, R. J. (1976) "At-Rest Lateral Earth Pressure of Consolidating Clay," *Proc. ASCE, J. Geotech. Engrg. Div.*, Vol. 102, No. GT7, July, pp. 721–738.

14. Zeevaert, L. (1953) Discussion of "Evaluation of Transverse Stresses During Consolidation" by P. Habib and A. Puys, *Proc. 3rd Intl. Conf. Soil Mech. and Found. Engrg.*, Zurich, Vol. 1, pp. 32–35.

15. Poulos, H. G. and Davis, E. H. (1972) "Laboratory Determination of In Situ Horizontal Stress in Soil Masses," *Geotechnique*, London, Vol. 22, No. 1, pp. 177–182.

16. Massarsch, K. R. and Broms, B. B. (1976) "Lateral Earth Pressure At-Rest in a Soft Clay," *Proc. ASCE, J. Geotech. Engrg. Div.*, Vol. 102, No. GT10, October, pp. 1041–1048.

17. Schmertmann, J. H. (1977) "Guidelines for CPT Performance and Design," Pub. No. FHWA-TS-78-209, Federal Highway Administration, Washington, D.C.

18. Brooker, E. W. and Ireland, H. O. (1965) "Earth Pressures at Rest Related to Stress History," *Canadian Geotechical J.*, Vol. II, No. 1, February.

19. Mayne, P. W. and Kulhawy, F. H. (1982) "K_o-OCR Relationships in Soil," *Proc. ASCE, J. Geotech. Engrg. Div.*, Vol. 108, No. GT6, June, pp. 851–872.

20. Clemence, S. P. and Finbarr, A. O. (1981) "Design Considerations for Collapsible Soils," *Proc. ASCE, J. Geotech. Engrg. Div.*, Vol. 107, No GT3, March, pp. 305–318.

21. Terzaghi, K. and Peck, R. B. (1967) *Soil Mechanics in Engineering Practice*, John Wiley & Sons, New York.

22. Mitchell, J. K. and Lunne, T. A. (1978) "Cone Resistance as a Measure of Sand Strength," *Proc. ASCE, J. Geotech. Engrg. Div.*, Vol. 104, No. GT7, July, pp. 995–1012.

23. ISRM (1981), *Rock Characterization Testing and Monitoring*, E. T. Brown, ed., published for the *Comm. on Test. Meth. Intl. Soc. for Rock Mech.*, Pergamon Press, Oxford.

24. Burmister, D. M. (1962) "Physical, Stress-Strain, and Strength Responses of Granular Soils," *ASTM STP No. 322*, Amer. Soc. for Test. and Mat., Philadelphia, Pa., pp. 67–97.

25. Trofimenkov, J. G. (1974) "Penetration Testing in the USSR," State of the Art Report, *European Symp. Penetration Testing*, Stockholm, Vol. 1.

26. Burmister, D. M. (1948) "The Importance and Practical Use of Relative Density in Soil Mechanics," *Proc.* Amer. Soc. for Test. and Mat., Philadelphia, Pa., Vol. 48.

27. Peck, R. B. (1969) "Stability of Natural Slopes," *Proc. ASCE, Stability and Performance of Slopes and Embankments*, Berkeley, Calif., August, pp. 437–451.

28. Lowe, J. III (1969) "Stability Analysis of Embankments," *Proc. ASCE, Stability and Performance of Slopes and Embankments*, Berkeley, Calif., August, pp. 1–36.

29. Bishop, A. W., Alpan, I., Blight, G. E., and Donald, I. B. (1960) "Factors Controlling the Strength of Partly Saturated Cohesive Soils," *Proc. ASCE, Res. Conf. Shear Strength of Cohesive Soils*, Boulder, Colo. pp. 503–532.

30. Wu, T. H. and Sangrey, D. A. (1978) "Strength Properties and Their Measurement," *Landslides: Analysis and Control*, SR176, National Academy of Sciences (TRB), Washington, D.C., Chap. 6.

31. Fredlund, D.G., Morgenstern, N.R., and Widger, A. (1978) "Shear Strength of Unsaturated Soils," *Canadian Geotech. J.*, Vol. 15, No. 3, pp. 313–321.

32. Ho, D. Y. F. and Fredlund, D. G. (1982) "Increase in Strength Due to Suction for Two Hong Kong Soils," *Proc. ASCE, Spec. Conf. on Engrg. and Const. in Tropical and Residual Soils*, Honolulu, pp. 263–295.

33. NCE (1983) "Electronic Gadget Gains Ground," *New Civil Engineer International*, Soc. of Civil Engrs., London, p. 47.

34. Ladd, C. C. and Foott, R. (1974) "New Design Procedure for Stability of Soft Clays," *Proc. ASCE, J. Geotech. Engrg. Div.*, Vol. 100, No. GT7, July, pp. 447–461.

35. Baligh, M. M., Vivatrat, V., and Ladd, C. C. (1980) "Cone Penetration in Soil Profiling," *Proc. ASCE, J. Geotech. Engrg. Div.*, Vol. 106, No. GT4, April, pp. 447–461.

36. Aas, G. (1965) "A Study of the Effect of Vane Shape and Rate of Strain on the Measured Values of In Situ Shear Strength of Clays," *Proc. 6th Intl. Conf. Soil Mech. and Found. Engrg.*, Montreal, Vol. 1, pp. 141–145.

37. Casagrande, A. (1960) "An Unsolved Problem in Embankment Stability on Soft Ground," *Pan Amer. Conf. on Soil Mech. and Found. Engrg.*, Mexico, Vol. 2, pp. 721–746.

38. Bjerrum, L. (1969) Contribution to panel discussion, Session 5, "Stability of Slopes and Embankment Foundations," *Proc. 7th Intl. Conf. Soil Mech. and Found. Engrg.*, Mexico City, Vol. 3, pp. 412–413.

39. Bjerrum, L. (1972) "Embankments on Soft Ground: State of the Art Paper," *ASCE Specialty Conference on Performance of Earth and Earth Supported Structures*, Purdue University, Lafayette, Ind., Vol. II.

40. Bjerrum, L. (1973) "Problems of Soil Mechanics and Construction on Soft Clays; State-of-the-Art Report to Session I," *Proc. 8th Intl. Conf. Soil Mech. and Found. Engrg.*, Moscow, Vol. 3, pp. 111–159.

41. Ladd, C. C. (1973) Discussion of "Application of the Field Vane Test," *Proc. 8th Intl. Conf. Soil Mech. and Found. Engrg.*, Moscow, Vol. 3, pp. 108–110.

42. Azzouz, A. S., Baligh, M. M., and Ladd, C. C. (1983) "Corrected Field Vane Strength for Embankment Design," *Proc. ASCE, J. Geotech. Engrg. Div.*, Vol. 109, No. GT5, May, pp. 730–734.

43. Bjerrun, L., Clausen, C. J. F., and Duncan, J. M. (1972) "Earth Pressures on Flexible Structures: A State-of-the-Art Report," *Proc. 5th European Conf. Soil Mech. and Found. Engrg*, Madrid, Vol. 1, pp. 169–196.

44. NAVFAC (1971) "Soil Mechanics, Foundations and Earth Structures," *Design Manual DM-7*, Naval Facilities Engineering Command, Alexandria, Va.

45. Sanglerat, G. (1972) *The Penetrometer and Soil Exploration*, Elsevier Publishing Company, Amsterdam.

46. Menard, L. F. (1965) "Rules for the Computation of Bearing Capacity and Foundation Settlement Based on Pressuremeter Results," *Proc. 6th Intl. Conf. Soil Mech. and Found. Engrg.*, Montreal, Vol. 2, pp. 295–299.

47. Lukes, R. G. and de Bussy, B. L. (1976) "Pressuremeter and Laboratory Test Correlations for Clays," *Proc. ASCE, J. Geotech. Engrg. Div.*, Vol. 102, No. GT9, September, pp. 945–962.

48. Skempton, A. W. (1953) "The Colloidal Activity of Clays," *Proc. 3rd Intl. Conf. Soil Mech. and Found. Engrg.*, Zurich, Vol. 1, pp. 57–61.

49. Bjerrum, L. (1954) "Geotechnical Properties of Norwegian Marine Clays," *Geotechnique*, London, Vol. 4, No. 2, p. 49.

50. USAEC (1972) "Soil Behavior Under Earthquake Loading Conditions," *National Tech. Info. Serv. Pub. TID-25953*, U.S. Department of Commerce, Oak Ridge National Laboratory, Oak Ridge, Tenn., January.

51. Wylie, E. B. and Streeter, V. L. (1976) "Characteristics Method for Liquefaction of Soils," *Proc. ASCE, Numerical Methods in Geo-Mechanics*, New York, Vol. II, pp. 938–954.

52. Zeevaert, L. (1972) *Foundation Engineering for Difficult Soil Conditions*, Van Nostrand-Reinhold Book Company, New York.

53. Faccioli, E. and Resendiz, D. (1976) "Soil Dynamics: Behavior Including Liquefaction," *Seismic Risk and Engineering Decisions*, Lomnitz and Rosenblueth, eds., Elsevier Scientific Publishing Co., New York, pp. 71–140.

54. Silver, M. L., Chan, C. K., Ladd, R. S., Lee, K. L., Tiedemann, D. A., Townsend, F. C., Valera, J. E., and Wilson, J. H. (1976) "Cyclic Triaxial Strength of Standard Test Sand," *Proc. ASCE, J. Geotech. Engrg. Div.*, Vol. 102, No. GT5, May, pp. 511–524.

55. Silver, M. L. and Park, T. K. (1975) "Testing Procedure Effects on Dynamic Soil Behavior," *Proc. ASCE, J. Geotech. Engrg. Div.*, Vol. 101, No. GT10, October, pp. 1061–1083.

56. Eide, O. (1974) "Marine Soil Mechanics—Applications to North Sea Offshore Structures," *Pub. 103, Norwegian Geotech. Inst.*, Oslo, p. 19.

57. Finn, W. D. L., Pickering, D. J., and Bransby, P. L. (1969) "Sand Liquefication in Triaxial and Simple Shear Tests," *Soil Mechanics Research Report Series No. 10*, University of British Columbia, Vancouver.

58. Hardin, B. O. and Drnevich, V. P. (1972) "Shear Modulus and Damping in Soils: Measurement and Parameter Effects," *Proc. ASCE, J. Soil Mech. and Found. Engrg. Div.*, Vol. 98, No. SM6, June, pp. 603–624.

59. Hardin, B. O. and Drnevich, V. P. (1972) "Shear Modulus and Damping in Soils: Design Equations and Curves," *Proc. ASCE, J. Soil Mech. and Found. Div.*, Vol. 98, No. SM7, July, pp. 667–692.

60. Hardin, B. O. (1970) "Suggested Methods of Test for Shear Modulus and Damping of Soils by Resonant Column," *ASTM STP479, Special Procedures for Testing Soil and Rock for Engineering Purposes*, Amer. Soc. for Test. and Mat., Philadelphia, Pa., pp. 516–529.

61. Statton, C. T., Auld, B., and Fritz, A. (1978) "In Situ Seismic Shear-Wave Velocity Measurements and Proposed Procedures," *ASTM STP654, Dynamic Geotechnical Testing*, Amer. Soc. for Test. and Mat., Philadelphia, Pa., pp. 56–65.

62. Ohsaki, Y. and Iwasaki, R. (1973) "On Dynamic Shear Moduli and Poisson's Ratio in Soil Deposits," *Soils and Foundations*, Vol. 13, No. 4.

63. D'Appolonia, E. (1970) "Dynamic Loadings," *Proc. ASCE, J. Soil Mech. and Found. Engrg. Div.*, Vol. 95, No. SM1, January, p. 49.

64. Seed, H. B. (1976) "Evaluation of Soil Liquefaction Effects on Level Ground During Earthquakes," State of the Art Paper, *Liquefaction Problems in Geotechnical Engineering*, ASCE Preprint 2752, New York, pp. 1–104.

65. Shannon and Wilson, Inc. and Agbabian-Jacobsen Assoc. (1971) "Soil Behavior under Earthquake

Loading Conditions," Report prepared for USACE, *Contract W-7405-eng.-26* (Reference No. 44).

66. Finn, U. D. (1972) "Soil Dynamics-Liquefaction of Sands," *Proc. Intl. Conf. on Microzonation*, Seattle, November, Vol. 1, pp. 87–112.
67. Peck, R. B. (1979) "Liquefaction Potential: Science Versus Practice," *Proc. ASCE, J. Geotech. Engrg. Div.*, Vol. 105, No. GT3, March, pp. 393.
68. Andersen, K. H., Pool, J. H., Brown, S. F., and Rosenbrand, W. F. (1980) "Cyclic and Static Laboratory Tests on Drammen Clay," *Proc. ASCE, J. Geotech. Engrg. Div.*, Vol. 106, No. GT5, May, pp. 499–530.
69. Murphy, D. J. (1970) "Soils and Rocks: Composition, Confining Level and Strength," *doctoral thesis*, Department of Civil Engineering, Duke University, West Lafayette, Ind.
70. Broch, E. and Franklin, J. A. (1972) "The Point Load Strength Test," *Intl. J. Rock Mech.*, Min. Sci., Vol. 8.
71. Hibino, S., Hayashi, M., Kanagawa, T., and Motojima, M. (1977) "Forecast and Measurement of the Behavior of Rock Masses During Underground Excavation Works," *Proc. Intl. Symp. on Field Measurements in Rock Mech.*, Zurich, Vol. 2, pp. 935–948.
72. Miller, R. P. (1965) "Engineering Classification and Index Properties for Intact Rock," Ph.D. thesis, University of Illinois, Urbana, Ill.
73. Griffith, A. A. (1924) "Theory of Rupture," *First Intl. Cong. Appl. Mech.*, Delft, The Netherlands, pp. 55–63.
74. Hoek, E. (1969) "Brittle Failure of Rock," *Rock Mechanics in Engineering Practice*, Stagg and Zienkiewicz, eds., John Wiley & Sons, New York, Chap. 4.
75. Goodman, R. E. (1980), *Introduction to Rock Mechanics*, John Wiley & Sons, New York.
76. Deere, D. U. and Miller, R. P. (1966) "Classification and Index Properties for Intact Rock," *Tech. Rep. AFWL-TR-65-116*, AF Special Weapons Center, Kirtland Air Force Base, Albuquerque, New Mexico.
77. Deere, D. U. (1968) "Geological Considerations," *Rock Mechanics in Engineering Practice*, Stagg and Zienkiewicz, eds., John Wiley & Sons, New York, Chap. 1.
78. Coon, J. H. and Merritt, A. H. (1970) "Predicting In Situ Modulus of Deformation Using Rock Quality Indexes," *Determination of the In Situ Modulus of Deformation of Rock, ASTM STP477*, Amer. Soc. for Test. and Mat., Philadelphia, Pa., pp. 154–173.
79. Hartmann, B. E. (1966) *Rock Mechanics Instrumentation for Tunnel Construction*, Terrametrics, Wheatbridge, Colo.
80. Patton, F. D. (1966) "Multiple Modes of Shear Failure in Rock," *Proc. 1st Intl. Cong. on Rock Mech.*, Lisbon, Vol. 1, pp. 509–513.
81. Hoek, E. and Bray, J. W. (1977) *Rock Slope Engineering*, 2d ed., Inst. of Min. and Met., London.
82. Ladanyi, B. and Archambault, G. (1970) "Simulation of Shear Behavior of a Jointed Rock Mass," *Proc. 11th Symp. on Rock Mech.*, AIME, New York, pp. 105–125.
83. Barton, N. R. (1973) "Review of a New Shear Strength Criterion for Rock Joints," *Engineering Geology*, Elsevier Publishing Co., Amsterdam, Vol. 7, pp. 287–332.
84. Barton, N. R. and Choubey, V. (1977) "The Shear Strength of Rock Joints in Theory and Practice," *Rock Mechanics*, Springer-Verlag, Berlin, Vol. 10.
85. Deere, D. U. (1976) "Dams on Rock Foundations—Some Design Questions," *Proc. ASCE, Rock Engineering for Foundations and Slopes*, New York, Vol. 2, pp. 55–85.
86. Roberds, W. J. and Einstein, H. H. (1978) "Comprehensive Model for Rock Discontinuities," *Proc. ASCE, J. Geotech. Engrg. Div.*, Vol. 104, No. GT5, May, pp. 553–569.
87. Hoek, E. and Brown, E. T. (1980) "Empirical Strength Criterion for Rock Masses," *Proc. ASCE, J. Geotech. Engrg. Div.*, Vol. 106, No. GT9, September, pp. 1013–1035.
88. Ladd, C. C. (1973) Discussion of "Major Test Variables for Consolidated-Undrained Laboratory Strength Tests," *Proc. 8th Intl. Conf. Soil Mech. and Found. Engrg.*, Moscow, pp. 110–115.
89. Chen, W. F. (1975) *Limit Analysis and Soil Plasticity*, Elsevier Scientific Publishing Co., New York.
90. Fang, H. Y. (1975) "Stability of Earth Slopes," *Foundation Engineering Handbook*, Winterkorn and Fang, eds., Van Nostrand-Reinhold Co., New York, Chap. 10.
91. Wroth, C. P. (1976) "Review of Earth Pressure Theory and Earth Behavior," *Geotechnical Lecture Series on Lateral Earth Pressures*, Boston Soc. of Civ. Eng., pp. 1–51.
92. Deere, D. U. (1963) "Technical Description of Rock Cores for Engineering Purposes," *Rock Mech. Engrg. Geol. 1*, pp. 18–22.
93. Bieniawski, Z. T. (1974) "Geomechanics Classification of Rock Masses and Its Application to Tunneling," *Proc. 3d Intl. Cong. Rock Mechs.*, Denver, Colo. Vol. IIA, pp. 27–32.

CHAPTER FOUR

DEFORMATION WITHOUT RUPTURE

4.1 DEFORMATION IN GEOLOGIC MATERIALS

4.1.1. PRACTICAL SIGNIFICANCE

Occurrence

Deformation in geologic materials can be elastic or plastic, or a combination of both, as well as occasionally viscous. During shear failure, deformation occurs during displacement (shear strain) along the failure surface as discussed in the previous chapter.

Prior to failure, deformation, which is the subject of this chapter, occurs in the form of:

Compression as a result of elastic deformation; plastic deformation from the closure of voids, fissures, or fractures; or combinations of elastic-plastic deformation; or

Expansion during swelling or elastic-plastic straining.

Parameters of deformation are given as moduli or from the consolidation test. The more common applications occur in the evaluation of foundation settlements, deflections for the various cases of soil-structure interaction problems, and deflections around underground excavations such as tunnels and caverns. Some examples are given in Fig. 4.1.

Foundations

Excessive settlements from compression or heave from swelling can result in distortion of buildings, pavements, or other structures, affecting their usefulness and even causing structural failure and collapse. The tolerance of a structure to resist settlement or heave deformation is related directly to its design and the properties of the structural elements. Foundation design is often based on an *allowable bearing value*, defined as the maximum load that can be applied to a given geologic formation without causing a settlement exceeding the tolerances for a given structure.

Settlement magnitudes and distributions depend on material properties, stratigraphy, size of loaded area, structural rigidity, and the rapidity of load application. In analysis, the magnitude and distribution of the stresses imposed on the geologic strata are determined, and settlement magnitudes, distributions and time rates are computed by employing deformation moduli or consolidation test data for the new stress conditions.

Soil-Structure Interaction

Soil-structure interaction problems usually make use of beam- or plate-on-an-elastic subgrade solutions and include long footings, mat foundations, retaining structures, and laterally loaded piles. These problems involve evaluation of the reaction between the soil and the structure with which it is in continuous, direct contact, as well as the subsequent deflections.

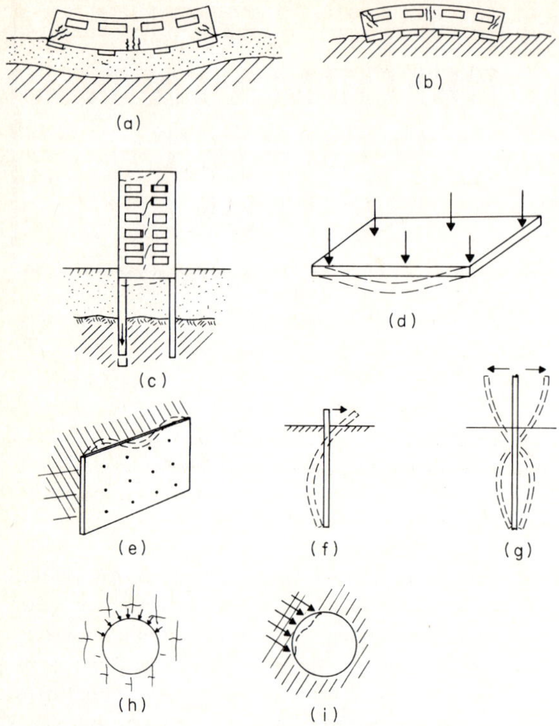

FIG. 4.1 Some practical field problems requiring study of deformations: (*a*) building settlement, shallow foundations; (*b*) building heave; (*c*) building settlement, deep foundations; (*d*) mat foundation deflection; (*e*) anchored retaining wall deflection; (*f*) laterally loaded pile deflection; (*g*) cyclically loaded pile deflection; (*h*) small deflections around tunnel cause arching; and (*i*) large deflections around tunnel can cause lining distortions.

Excessive deflections may result in high bending moments and structural failure, and in the case of retaining structures may result in backslope subsidence which can cause settlement of adjacent structures. Expressions for soil load-deformation characteristics are given as moduli or as *p-y* curves and structural properties.

Underground Openings

Small deflections in rock masses can be beneficial because they result in arching (*see Art. 3.2.3*), but large deflections result in excessive earth pressures on linings and other support systems. Evaluations usually are based on empirical relationships, but deformation moduli are employed in sophisticiated analyses, such as the FEM.

4.1.2 MODES OF DEFORMATION

Elastic Deformation

Ideal Elastic Media

In an elastic medium, the strain ϵ is instantly and totally recoverable. Stress σ is directly proportional to strain (Fig. 4.2*a*) according to Hooke's law. The constant of proportionality is expressed by Young's modulus E (also modulus of elasticity) as:

$$E = \sigma/\epsilon \qquad \text{tsf, kg/cm}^2\text{, kPa, etc.} \qquad (4.1)$$

Poisson's ratio ν is the inverse ratio between strain in the direction of applied stress and the induced strain ϵ_L in a perpendicular direction:

$$\nu = \frac{\epsilon_L}{\epsilon} = \frac{\epsilon_L}{E\sigma} \qquad (4.2)$$

Young's modulus and Poisson's ratio are referred to as *elastic constants.*

For a geologic body to exhibit elastic characteristics defined by only one value of E and of ν it must be isotropic, homogeneous, and continuous. For it to be *isotropic,* the particles must be oriented so that the ratio of stress to strain is the same regardless of the direction of the applied stress; i.e., the elastic properties in every direction through any point are identical. For it to be *homogeneous,* the body must have identical properties at every point in identical directions. *Continuity* refers to structure; if a mass is continuous it is free of planes of weakness or breaks.

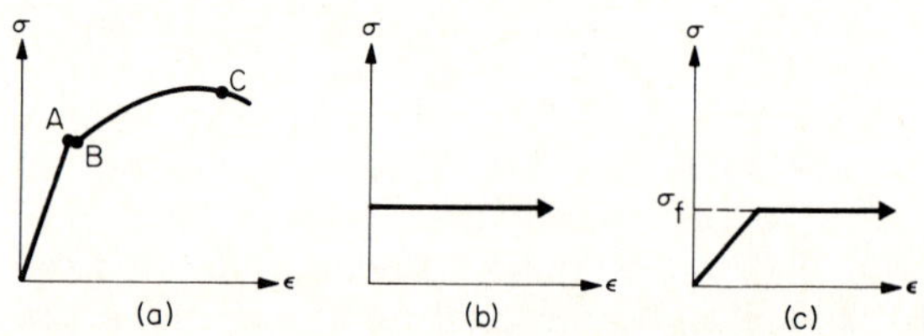

FIG. 4.2 Stress-strain relationships of elastic and plastic deformation. (*a*) Elastic deformation in medium steel to point *A*, the proportional limit; *B* is the yield point; and rupture occurs at *C*. (*b*) Rigid-plastic deformation. (*c*) Elastic-plastic deformation.

Validity for Geologic Materials

Hard, massive, intact rock approaches an elastic material under most stress levels prior to rupture. Most rocks and rock masses, however, are to some extent anisotropic, nonhomogeneous, and discontinuous, and are termed quasi-elastic, semi-elastic, or inelastic. In fresh intact specimens, deformation varies with rock type as related to mineral hardness, grain bonding, and fabric, as shown in Table 3.11.

Soils are essentially inelastic, but demonstrate pseudoelastic properties under low stress levels as evidenced by initial stress-strain linearity. "Elastic" deformation, however, is immediate, and in many soil types it does not account for the total deformation occurring over long time intervals because of the process of consolidation.

Soils and rock masses basically are inelastic materials; therefore, their load-deformation relationships are seldom linear and their stress-strain characteristics are not normally considered to yield truly elastic moduli. Concepts and formulas from elastic theory, however, are used in analysis for lack of more suitable concepts, so that the nonlinear curves obtained from testing must be replaced by straight lines or approximated by a series of straight lines. For most soil and rock formations Young's modulus and Poisson's ratio are not constants, but rather are quantities describing material behavior for a particular stress condition.

Plastic Deformation

Material exhibiting plastic behavior undergoes permanent and continuous deformation when the applied stress reaches a characteristic level (Fig. 4.2*b* and *c*) and in geologic materials can occur in several modes including "pure" compression, consolidation, expansion, and shear.

Pure Compression

Under an applied stress, pure compression occurs when sand particles are packed more closely together decreasing the void space, when fissures close in intact rock, or when joints close in a rock mass. Because rock masses are often discontinuous, they may undergo an initial plastic deformation as fractures close, then an elastic deformation of intact blocks, followed by additional plastic deformation.

Consolidation

Consolidation is the slow process of compression under applied stress that occurs as water is extruded from the voids of clay soils. The one-dimensional theory of consolidation of Terzaghi (1943)[1] is valid for the assumption that strains are small, the soil is saturated, and flow is laminar. Most soils exhibit "primary" and "secondary" consolidation, and in some soils the latter phenomenon, which is not well understood, can be of very significant magnitude. A substantial time delay in compression, which increases generally as the plasticity of the clay increases, occurs in clay soils under a given applied stress.

Expansion

Expansion occurs as an increase in volume from swelling or from plastic extension strain. Soils and rock masses containing active clay minerals have an affinity for absorbing water and swelling; heavily overconsolidated soils and rock masses containing residual stresses undergo plastic extension straining upon a decrease in confining stresses.

Shear

Plastic deformation during shear is described in Art. 3.2.3 for the cases of the active and passive states of stress, in Art. 6.2.4 for the case of the bearing capacity of foundations, and in Art. 8.1.3 for the case of slopes.

Viscous Deformation and Creep

In a viscous material the rate of deformation is roughly proportional to the applied stress. *Creep* is a time-dependent deformation wherein strains occur beyond elastic compression, consolidation, or shear at a constant stress level below failure. It is not thoroughly understood but occurs in varying degrees in all types of geologic materials. Secondary compression of soils is sometimes considered a form of creep.

Most rocks exhibit both immediate and delayed deformation under applied stress and are there-

TABLE 4.1
DEFORMATION MODULI

Parameter*	Definition	Expression		Normal application
		ELASTIC MODULI (STATIC)		
Young's modulus	Ratio of unit axial stress to unit strain.	$E = \sigma/\epsilon$	(4.1)	Rock masses, sands, strong granular cohesive soils.
Shear modulus	Ratio of unit shear stress to unit shear strain (modulus of rigidity).	$G = \dfrac{E}{2(1 + \nu)}$	(4.3)	Not commonly used statically.
Bulk modulus	Ratio of unit stress to unit strain applied in all directions simultaneously (modulus of incompressibility).	$B = \dfrac{E}{3(1 - 2\nu)}$	(4.4)	Not commonly used statically.
Constrained modulus	Ratio of unit stress to unit strain applied axially to a specimen confined laterally.	$D = \dfrac{E\,(1 - \nu)}{(1 + \nu)(1 - 2\nu)}$	(4.5)	Structure of large areal extent underlain by relatively thin compressible soil deposit.
		DYNAMIC ELASTIC PARAMETERS—SEE TABLE 4.2		
		MODULI FROM THE STRESS-STRAIN CURVE		
Initial tangent modulus (Fig. 4.4)	Initial portion of curve.	$E_i = \sigma/\epsilon$	(4.12)	Usually taken as E for geologic materials.
Tangent modulus (Fig. 4.4)	Modulus at specific point. Often taken at point of maximum curvature before rupture.	$E_t = \dfrac{d\sigma}{d\epsilon}$	(4.14)	The lowest value for E usually reported.
Secant modulus (Fig. 4.4)	σ and ϵ taken between two particular points.	$E_{se} = \dfrac{\Delta\sigma}{\Delta\epsilon}$	(4.13)	Define E for a particular stress limit.
		OTHER MODULI		
Compression modulus (Fig. 4.5)	Lateral deformation caused by applied stress from pressuremeter.	$E_c = E\alpha$	(4.15)	In situ measures of E for materials difficult to sample.
Modulus of subgrade reaction	A unit of pressure to produce a unit of deflection.	$k_s = p/y$ kg/cm^3	(4.16)	Beam or plate on an elastic subgrade problems.

*Unless noted, units are tsf, kg/cm^2, kN/m^2.

fore termed *viscoelastic* [Goodman (1980)[2]]. Relatively soft rocks such as halite, tar sands, and some shales undergo a mass flow form of creep under low deviator stress due to intracrystalline gliding. Hard rocks exhibit creep as the network of cracks increases in length and intensity under relatively high deviator stresses (*see Art. 3.4.2, Brittle Failure*). Creep may occur also in foliated rock masses containing residual stresses, when stress-relieved by excavation.

4.1.3 DEFORMATION PARAMETERS

Deformation Moduli

Applications

Load-deformation relationships can be expressed in terms of deformation moduli, which can take a number of forms. *Moduli* are quantities expressing the amount of change in the form or dimensions of a body occurring as a result of applied stresses, given as the ratio of unit stress to unit strain, without regard to time.

Moduli are used to calculate the strain for a particular applied load, or to determine the limiting load for a desired deformation or strain. Tabulated values of moduli are useful for materials from which the retrievel of undisturbed samples is difficult or impossible, such as rock masses, decomposed rock, residual soils, gravelly glacial till, and cohesionless sands. Moduli are used also for the estimation of deflections in soil-structure interaction problems.

The Various Forms

Deformation moduli are expressed in a number of ways which relate to either the form of deformation or the stress-strain relationships obtained from a particular test method as summarized in Table 4.1.

Elastic moduli include Young's modulus E and the shear modulus G which relate primarily to changes in shape, and the bulk modulus B and the constrained modulus D which relate primarily to changes in volume as defined in Table 4.1.

Dynamic elastic moduli include the dynamic Young's modulus E_d, the dynamic shear modulus G_d, and the dynamic bulk modulus K, as defined in Table 4.2. They are obtained from the compression and shear-wave velocities of test-generated sonic waves which are functions of

TABLE 4.2
DYNAMIC ELASTIC PARAMETERS

Parameter	Expression		
Compression-wave velocity	$V_p = \{[K + (4/3)G]/\rho\}^{1/2}$ m/s		(4.6)
Shear-wave velocity	$V_s = (G/\rho)^{1/2}$ m/s		(4.7)
Mass density of materials	$\rho = \gamma/g$ kg/m³		(4.8)
Dynamic Poisson's ratio	$\nu = (V_p^2/2V_s^2 - 1)/(V_p^2/V_s^2 - 1)$	*Appropriate Values* Igneous rocks—0.25 Sedimentary rocks—0.33 Sands: 0.25–0.35; Clays: 0.4–0.5	
Dynamic Young's modulus*	$E_d = \rho(3V_p^2 - 4V_s^2)/(V_p^2/V_s^2 - 1)$	or $E_d = 2pV_s^2(1 + \nu)$	(4.9)
Dynamic shear modulus*	$G_d = \rho V_s^2 = E_d/2(1 + \nu)$		(4.10)
Dynamic bulk modulus*	$K = \rho(V_p^2 - 4V_s^2/3) = E_d/3(1 - 2\nu)$		(4.11)

*Units are pascals or tsf or kg/cm².

the elastic properties and mass densities of the transmitting medium (Table 4.2).

Stress-strain curve moduli, obtained from axial loading of soil or rock, include the initial tangent modulus E_i, the tangent modulus E_t, and the secant modulus E_{se}, as defined in Table 4.1 and illustrated in Fig. 4.3. In a truly linear elastic material all three moduli are equal.

The *initial tangent modulus* closely approximates Young's modulus and is the modulus given most frequently for rock E_r. *Tangent modulus* values vary with the point selected on the stress-strain curve and are selected to define the relationship at a particular stress or strain, depending upon the problem under study.

The *secant modulus* is the slope of a straight line connecting two separate points on the stress-strain curve. The values vary with the location of the points. As the points come closer together, E_{se} eventually becomes equal to E_t. E_{se} often is used as a "conservative" modulus to define a limiting stress or limiting deformation and is commonly the modulus given for soils E_s when used for calculating deformations. E_s is often taken as the secant modulus between deviator

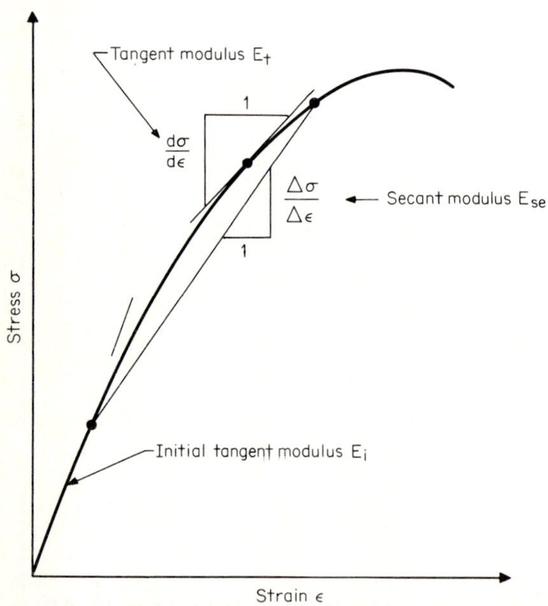

FIG. 4.3 Forms of moduli from the stress-strain curve.

stresses of one-third and one-half peak deviator stress in the triaxial test [Lambe and Whitman (1969)[3]]. For clays it is also often taken as the slope of the consolidation curve when plotted linearly as $\Delta h/h$ vs. p [CGS (1978)[4]].

The *compression modulus* E_c is obtained from the in situ pressuremeter test, and describes deformation in a lateral direction (see following article). In some soil formations E_c approaches Young's modulus, and in other cases correction factors are used for settlement analysis (see Table 4.10). In any event, although values are very much affected by material anisotropy since the test normally is performed in vertical boreholes, it is a favored method for computing settlements resulting from the compression of sands.

The *modulus of subgrade reaction* or *spring constant* (k_s) is obtained by the in situ plate-load test *(see Art. 4.4.1)* or laterally loaded pile test *(see Art. 4.4.3)*. k_s is defined as the pressure per unit area required to produce a unit of deflection, and is given in units of kg/cm³ or t/ft³. It is applied to soil-structure interaction problems.

Consolidation Parameters

Several parameters are obtained from the laboratory consolidation test performed on undisturbed specimens of clays, including:

Coefficient of compressibility a_v, which represents the ratio of a change in strain to a change in pressure expressed as

$$a_v\ (\text{cm}^2/\text{gm}) = \frac{e_0 - e_1}{\Delta p\ (\text{gm}/\text{cm}^2)} \qquad (4.17)$$

Coefficient of volume compressibility m_v, which represents the compression per unit of original thickness, due to a unit increase in pressure, expressed as

$$m_v(\text{cm}^2/\text{gm}) = \frac{a_v(\text{cm}^2/\text{gm})}{1 + e_0} \qquad (4.18)$$

Compression index C_c, which is equal to the tangent of the slope angle of the virgin compression curve when pressure is plotted to a logarithmic scale, expressed as

$$C_c = \frac{e_1 - e_2}{\log (p_1/p_2)} \qquad (4.43)$$

Coefficient of consolidation c_v, which expresses the coefficient of volume compressibility m_v in terms of the coefficient of permeability k and is applied to compute the time required to reach some particular percent consolidation, given as

$$c_v(\text{cm}^2/\text{s}) = \frac{k}{\gamma_w m_v} \tag{4.19}$$

or as

$$c_v = \frac{T_v h^2}{t} \tag{4.48}$$

Evaluations of consolidation test parameters and applications in settlement analysis are discussed in Art. 4.3.4. Parameter c_v is used for estimating the rate of settlement, and C_c, a_v, and m_v for estimating the settlement at 100% consolidation. C_c is particularly useful for correlations and estimating settlements in normally consolidated clays. It is interesting to note that for estimating long-term settlement, E_s is taken as equal to $1/m_v$, or the slope of the consolidation curve when plotted on a linear $\Delta h/h$ versus load graph [CGS (1978)[4]].

4.1.4 MEASURING DEFORMATION MODULI

Tests Summarized

In Situ Tests

Measurements of moduli can be made by a variety of methods, some applicable only to soils, others only to rock masses, and some to both soils and rocks. Most methods are direct, some are indirect, and for most materials, measurements are best made in situ. The methods are summarized in Table 4.3 and described briefly in this article. More detailed descriptions are given in Chap. 3 of Hunt (1984).[5] Typical values for Young's modulus and Poisson's ratio for soil, rock, and other materials are given in Table 4.4. Young's modulus for a large variety of intact rock specimens is given in Table 3.12 and for a variety of rock masses in Fig. 4.9 and Table 5.8.

Requirements to be met in order to measure the in situ modulus by static testing include

1. The rock or soil deformation and the stress producing it must be measurable
2. An analytical method of describing the geometry of the stress-deformation relationship must be developed

The methods of calculating moduli are related to testing procedures. Because a modulus is the ratio of unit stress to unit strain, and strain is the change in length per unit of length, in an in situ load test the *deflection* that is measured must be related in some manner to the depth of the stressed zone to determine strain. Strain is determined by volumetric measurements, by the use of instrumentation, or by the application of the Boussinesq equations which describe the stresses in a semi-infinite solid under applied load.

Laboratory Tests

Static soil moduli are measured in the triaxial test (E_s, B), simple direct shear test (G), and consolidometer (D). Dynamic moduli are measured in the cyclic triaxial test (E_d), cyclic simple shear test (G_d), or resonant column test (G_d) as described in Art. 3.3.5.

Rock static moduli of intact cores are measured in the triaxial test, or more commonly in unconfined compression. Dynamic moduli are measured in the ultrasonic column test (ASTM D2845).

Soil Formations

General

Young's modulus E_s is measured in the laboratory by triaxial tests. In sands it is measured in situ with the pressuremeter, by plate- or full-scale load tests and by seismic techniques, or it may be obtained approximately from correlations with the cone penetrometer test (CPT) or the standard penetration test (SPT).

Poisson's ratio for soils is evaluated from the ratio of lateral strain to axial strain during a triaxial test with axial loading. Its value varies with the strain level and becomes constant only at large strains in the failure range [Lambe and Whitman (1969)[3]]. It is generally more constant under cyclic loading. For cohesionless soils it ranges from 0.25 to 0.35 and for clays from 0.4 to 0.5.

Shear modulus may be obtained by using the laboratory value for Young's modulus in Eq. 4.3, or by direct test as mentioned in the previous section.

TABLE 4.3
IN SITU TESTS TO MEASURE DEFORMATION MODULI

SOIL FORMATIONS

Category—test or method	Applications	Limitations
	DEFORMATION MODULI (STATIC)	
Pressuremeter	Measures E in materials difficult to sample undisturbed such as sands, residual soils, glacial till, and soft rock in a test boring.	Modulus values only valid for linear portion of soil behavior; invalid in layered formations; not used in weak soils.
Plate-load test	Measures modulus of subgrade reaction used in beam-on-elastic-subgrade problems. Surface test.	Stressed zone limited to about 2 plate diameters. Performed in sands and overconsolidated clays.
Lateral pile-load test	Used to determine horizontal modulus of subgrade reaction.	Stressed zone limited to about 2 pile diameters. Time deformation in clays not considered.
Full-scale foundation-load tests	Obtain E in sands and design parameters for piles.	Costly and time-consuming.
	DYNAMIC PROPERTIES	
Seismic direct methods	Borehole measurements of S wave velocity to compute E_d, G_d, and K.	Very low strain levels yield values higher than static moduli.
Steady-state vibration method	Surface measurement of shear wave velocities to obtain E_d, G_d, and K.	Small oscillators provide data only to about 3 m. Rotating mass oscillator provides greater penetration.
	PENETRATION RESISTANCE	
Standard penetration test (SPT)	Correlations provide measures of granular soil D_r, ϕ, E, allowable bearing value, and clay soil consistency. Samples recovered.	Correlations empirical. Not usually reliable in clay soils. Sensitive to sampling procedures.
Static cone penetrometer test (CPT)	Continuous penetration resistance can provide measures of end bearing and shaft friction. Correlations provide data similar to SPT.	Samples not recovered, material identification requires borings or previous area experience.

TABLE 4.3
IN SITU TESTS TO MEASURE DEFORMATION MODULI

ROCK MASSES		
Category—tool or method	**Applications**	**Limitations**
	DEFORMATION MODULI (STATIC)	
Dilatometer or Goodman jack	Measures *E* in lateral direction.	Modulus values valid for linear portion of load-deformation curve. Results affected by borehole roughness and layering.
Large-scale foundation-load test	Measure *E* under footings or bored piles. Measure shaft friction of bored piles.	Costly and time-consuming.
Plate-jack test	Measure *E*; primarily used for tunnels and heavy structures.	Requires excavation and heavy reaction or adit. Stressed zone limited by plate diameter and disturbed by test preparation.
Borehole plate-jack test	Measure *E* in confined condition.	Relatively small stressed zone may not be representative.
Flat-jack test	Measure *E* or residual stresses in a slot cut into the rock.	Stressed zone limited by plate diameter. Test area disturbed in preparation. Requires orientation in same direction as applied construction stresses.
Radial jacking tests (pressure tunnels)	Measure *E* for tunnels. Data most representative of in situ rock tests and usually yields the highest values for *E*.	Very costly and time-consuming and data difficult to interpret. Preparation disturbs rock mass.
Triaxial compression test	Measure *E* of rock block.	Costly and difficult to set up. Disturbs rock mass during preparation.
	DYNAMIC PROPERTIES	
Seismic direct methods	Obtain dynamic elastic moduli, *E*, *G*, *K*, and ν in boreholes.	Very low strain levels yield values higher than static moduli.
3-D velocity logger	Measure velocity of shear and compression waves (V_s, V_p) from which moduli are computed. Borehole test.	Penetrates to shallow depth in borehole.

TABLE 4.4
TYPICAL RANGES FOR ELASTIC CONSTANTS OF VARIOUS MATERIALS*

Material	Young's modulus E_s, tsf, kg/cm²	Poisson's ratio ν
	SOILS	
Clay:		
Soft sensitive	20–40 ($500s_u$)	
Firm to stiff	40–80 ($1000s_u$)	0.4–0.5
Very stiff	80–200 ($1500s_u$)	(undrained)
Loess	150–600	0.1–0.3
Silt	20–200	0.3–0.35
Fine sand:		
Loose	80–120	
Medium dense	120–200	0.25
Dense	200–300	
Sand:		
Loose	100–300	0.2–0.35
Medium dense	300–500	
Dense	500–800	0.3–0.4
Gravel:		
Loose	300–800	
Medium dense	800–1000	
Dense	1000–2000	
	ROCKS	
Sound, intact igneous		0.25–0.33
and metamorphics	6–10 × 10⁵	0.25–0.33
Sound, intact sandstone		0.25–0.30
and limestone	4–8 × 10⁵	
Sound, intact shale	1–4 × 10⁵	
Coal	1–2 × 10⁵	
	OTHER MATERIALS	
Wood	1.2–1.5 × 10⁵	
Concrete	2–3 × 10⁵	0.15–0.25
Ice	7 × 10⁵	0.36
Steel	21 × 10⁵	0.28–0.29

Material	E_s
ESTIMATING E_s FROM N (SPT)†	
Soil type: Silts, sandy silts, slightly cohesive mixtures	$4N_1$
Clean fine to medium sands and slightly silty sands	$7N_1$
Coarse sands and sands with little gravel	$10N_1$
Sandy gravel and gravels	$12N_1$

*After CGS (1978)[4] and Lambe and Whitman (1969).[3] †After NAVFAC (1982).[23] NOTE: Use N values corrected for depth N_1.

Pressuremeter

Tests are performed in boreholes with the pressuremeter, a cylindrical probe that is inflated by pressure to cause a thrust in the horizontal radial direction. Apparati of various capacities are available for testing soils and rock. Results may be strongly affected by material anisotropy and borehole disturbance. The latter condition is reduced by using a self-boring pressuremeter, such as the Cankometer, which is designed primarily to test soft to firm clays [for example, see Clough and Denby (1980)[6]].

Curves of volume change in the probe versus applied pressure are obtained from the test, such as the example given in Fig. 4.4. The *compression modulus* E_c is obtained from the slope of the compression curve between the seating or initial pressure P_i and the "creep" pressure P_f from the expression

$$E_c = K(\Delta p/\Delta v) \tag{4.20}$$

where K is a constant for the apparatus accounting for borehole diameter, probe size, and Poisson's ratio. The limiting pressure P_L is used to es-

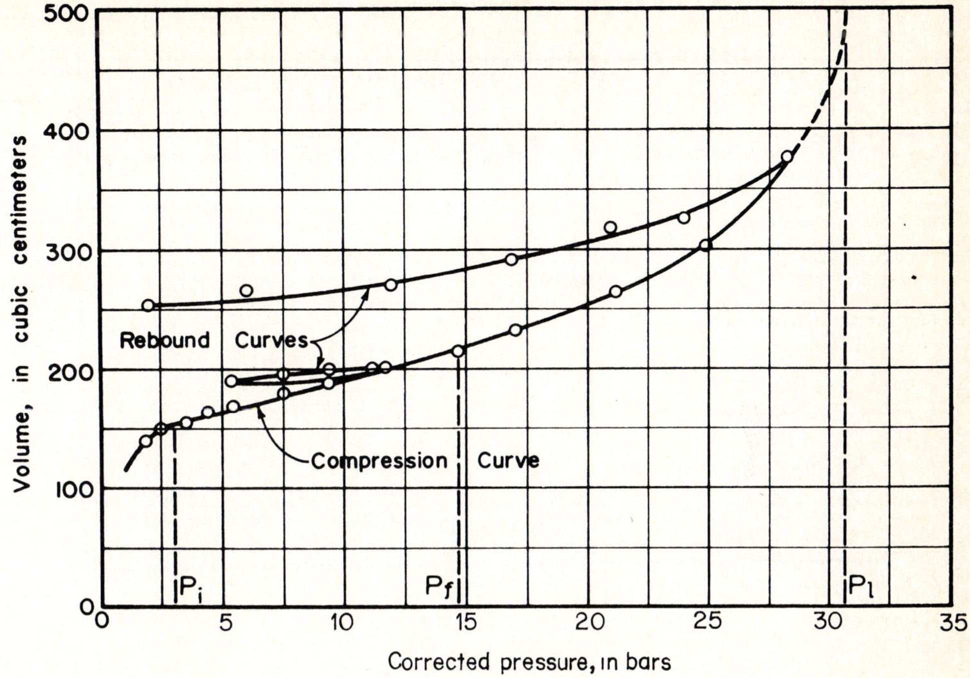

FIG. 4.4 Typical pressuremeter results.

timate the undrained strength s_u as expressed in Eq. 3.42 (Table 3.3). In settlement analysis, a correction factor α is applied to E_c as described in Art. 4.3.3. Typical values for E_c and P_L for various soil types are given in Table 4.5.

Load Tests

E_s for sands may be obtained from full-scale load tests on footings, or from plate-load tests. Plate-load tests *(see Art. 4.4.1)* are used to obtain values of the vertical modulus of subgrade reaction for sands and overconsolidated clays, and laterally loaded pile tests are used to obtain values for the horizontal modulus of subgrade reaction *(see Art. 4.4.3)*.

Seismic Direct Techniques

Moduli for strain levels on the order of 10^{-6} cm/cm can be determined from shear-wave velocities obtained from crosshole seismic surveys *(see Art. 3.3.5,* Figs. 3.45 and 3.46). E_d and G_d are computed from V_s as expressed in Eqs. 4.9 and 4.10 given in Table 4.2. The method has been used to evaluate compression in medium-dense to dense sands.

Under loads in the order of 2 to 3 tsf, strains in dense sands are small, approximately 10^{-3}, but higher than the strains occurring during dynamic testing. In granular soils E and G have been found to decrease in magnitude with increasing strain levels *(see Art. 3.3.5)*. A relationship to reduce the shear and compression moduli obtained under low strain levels of field testing to those moduli for use in analysis for higher strains is given in Fig. 4.5. In general, the value of E for a strain in the order of 10^{-3} is about $3E$ for a strain in the order of 10^{-6}. Seismic direct methods, therefore, may be used to estimate values of E_s in medium dense to dense sands for settlement computations where small strains are critical [Swiger (1974)[8]].

Correlations with Field Penetration Test Data

CPT correlations with E have been commonly taken as

$$E_s = 2q_c \tag{4.21}$$

where q_c is the cone penetration resistance. To adjust for soil compressibility, Vesić (1967)[10] suggests a higher E/q_c ratio and a modification of Eq.

TABLE 4.5
VALUES OF COMPRESSION MODULUS E_c AND ULTIMATE PRESSURE P_L FROM MENARD PRESSUREMETER TESTING

Soil type	E_c, kg/cm², tsf	P_L, kg/cm², tsf
Peat and very soft clays	2–15	0.2–1.5
Soft clays	5–30	0.5–3.0
Firm clays	30–80	3.0–8.0
Stiff clays	80–400	6–25
Loose silty sands	5–20	1– 5
Silts	20–100	2–15
Sands and gravels	80–400	12–50
Till	75–400	10–50
Recent fill	5–50	0.5–3
Ancient fill	40–150	4.0–10
Glacial till (Ohio)	600–1500	Ward (1972)[47]
Residual soil (schist)	20–160	Ward (1972)[47]
Decomposed schist	200–500	Ward (1972)[47]
Residual soils (schist and gneiss; SM, ML, saprolite)	50–2000	Martin (1977)[48]

*From SOLS SOILS (1975).[7]

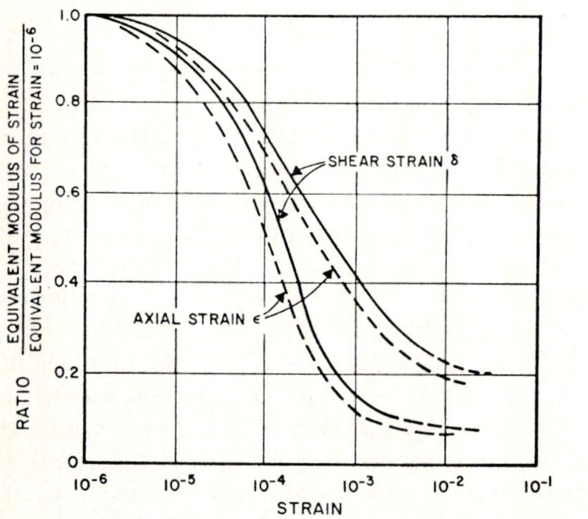

FIG. 4.5 Strain modulus for sands. [After Seed (1969)[9]; from Swiger (1974).[8]]

4.21 to include relative density for *deep foundations* as follows:

$$E = 2(1 + D_R^2)q_c \qquad (4.22)$$

Schmertmann (1977)[11] gives relationships for estimating E_c (compression modulus) from q_c obtained using Delft mechanical tips as follows:

- Dense sand, $E/q_c = 1$
- Loose sand, $E/q_c = 1.5$
- Silt, $E/q_c = 2$
- Insensitive clay, $E/q_c = 3$
- Very sensitive clay, $E/q_c = 20$

The error range, however, could easily be ±25% possibly ±100%. For *shallow square footings* on sand he recommends $E_s = 2.5q_c$ and for *long footings* he recommends $E_s = 3.5\ q_c$ for electric, friction-jacket cones.

SPT correlates roughly with q_c (Fig. 4.6). In granular soils gross estimates for E_s may be made by determining D_R for a particular grain size distribution and N value (Table 3.5), and then selecting a value for E_s from Table 4.4. A general correlation appears to be $E_s = 10$ N (tsf, kg/cm²).

Rock Masses

General

Because of mass defects from discontinuities, values for E_r in rock masses usually are obtained from in situ tests. A wide range in values of E can be expected from performing various types of tests in similar rock-mass conditions or performing the same type of test at various locations in the rock mass. More representative values are

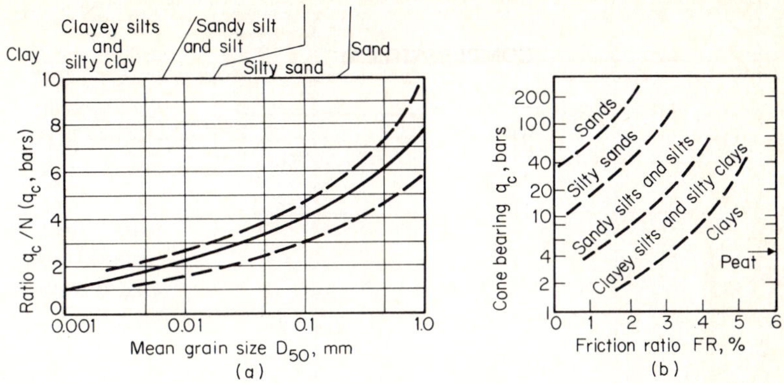

FIG. 4.6 SPT-CPT correlations. (*a*) Variation of q_c/N ratio with mean grain size. (*b*) Friction ratio FR used in test to estimate soil type. FR = f_s/q_c where f_s = sleeve friction and q_c = cone bearing. For discussion, see Hunt (1984).[5] (NOTE: 1 bar = 100 kPa = 1.02 kg/cm².) [After Robertson et al. (1983)[11a]; part (*b*) adapted from Douglas and Olsen (1981).[11b]]

obtained for design by relating in situ field static measurements to in situ dynamic measurements and laboratory dynamic measurements.

Static Tests In Situ

Pressuremeters, including the dilatometer [Rocha (1970)[12]] and the Goodman jack [Goodman et al. (1972)[13]], have been designed to obtain a compression modulus in rock in the same manner as in soils. Since it is affected by anisotropic conditions, whenever possible, the test should be performed with the instrument oriented to stress the mass in the same direction as the projected imposed loads. The modulus in the radial direction is expressed by

$$E_r = (1 - \nu)\,\Delta p\,\frac{r}{\Delta d} \tag{4.23}$$

where p = applied pressure
r = radius
d = radial deformation

The *borehole plate-jack test* is performed by applying successively higher pressures, in loading and unloading cycles, to the flattened end of a borehole and measuring the resulting deflections of the rock mass [ISRM (1981)[15]]. A modulus is calculated from the slope of the load-deflection curve. Although the loaded area may be small compared to the size of the usual foundation, the test has the advantages that it can be performed at a number of elevations in the same borehole, and the stressed rock is relatively undisturbed.

The *plate-jack test,* usually performed in adits, applies load only to a limited area and may not effectively stress an adequately large mass to be representative. In addition, it is a surface test, and stress relief during test preparation may affect mass properties. It is described in ISRM (1981)[15] and by Goodman (1980).[2]

The *flat-jack test* is performed in a shallow slot cut into the rock mass from its surface, and also stresses a limited volume. It is described in Goodman (1980)[2] and Rocha (1970).[12]

Radial-jacking tests, including both water pressure and mechanical chambers, are performed in tunnels [ISRM (1981)[15]]. Although costly compared to other methods, they stress relatively large volumes and permit anisotropic conditions to be measured when instrumentation is installed to monitor strains.

Full-scale pile-load tests are described in Art. 7.5.3.

The *in situ triaxial compression test,* which employs flat jacks installed to confine a rock block cut from the surrounding mass but attached at its base, is described by Hibino et al. (1977).[15]

Seismic Techniques

Seismic velocities, obtained in situ by direct methods (*see Art. 3.3.5*), reflect the conditions of a larger rock mass than in situ static testing. Because of the short pulse duration and low stress level of seismic testing, however, E_{seis} is always lower than the static modulus E_r, although the ratio approaches unity as the rock mass quality approaches that of an elastic material.

Ultrasonic tests are performed in the laboratory on intact cores to obtain measures of V_p and V_s

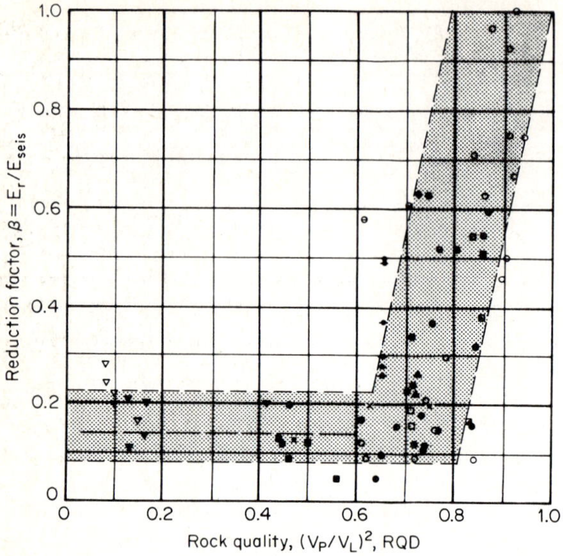

FIG. 4.7 Variation in modulus reduction factor β with rock quality. (E_r = static rock modulus from load tests; E_{seis} = modulus from seismic velocity; V_p = field stress velocity, compressional; V_L = laboratory sonic velocity of sound cores.) [From Deere et al. (1967).[16]]

One approach to selecting a design modulus is based on the *modulus reduction factor* β, Deere et al. (1967),[16] which relates the ratio E_r/E_{seis} to *rock quality* as determined from the RQD or from the velocity index, $(V_p/V_L)^2$. V_p is the field compression wave velocity and V_L is laboratory sonic velocity. A relationship between the modulus reduction factor and RQD, or the velocity index, based on data from a number of test sites, is given in Fig. 4.7. It is seen that the value for β becomes relatively uniform below a rock quality rating of "good" (*see Art. 3.4.3*). A similar graph should be developed for each site of a specific major project.

A more recent approach cited by Goodman (1980)[2] is based on the concept that measurements of shear-wave frequency f or wavelength received at a standard distance from a constant type of seismic source tend to correlate with E_{seis}/E_r. Studies by Schneider (1967)[17] and Bieniawski (1978)[18] yield a linear relationship between E_r and the shear-wave frequency obtained by a hammer energy source used with the single-channel engineering seismograph, expressed as

$$E_r = 0.054f - 9.2 \qquad (4.24)$$

[ASTM D2845-69 (72)]. To perform the test, specimen ends are smoothed and attached by resins to transducers. A sonic pulse is generated and the wave forms displayed on an oscilloscope from which V_p and V_s are interpreted. Tests should be performed under several levels of applied stress, and in anisotropic materials, specimens should be oriented to measure the effects of anisotropy. An evaluation of rock quality is obtained from a comparison of the field and laboratory velocities.

The Design Modulus

It is seldom satisfactory to select a modulus for design of a major project based on only one type of test; at the very least both static and dynamic moduli should be measured. There is relatively little information in the literature regarding design moduli for rock masses, and foundations on rock commonly are designed on the basis of allowable bearing values developed empirically from experience.

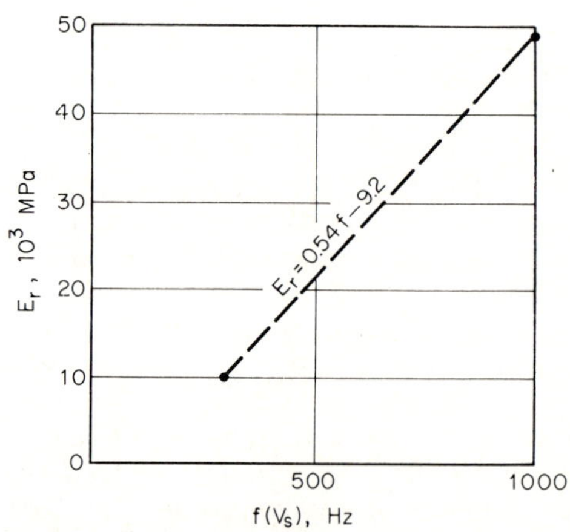

FIG. 4.8 Relationship between transmitted vibration frequency f, as recorded for the shear-wave velocity V_s with a single-channel seismograph, and the in situ deformation modulus E_r. [After Bieniawski (1978).[18]]

as shown in Fig. 4.8, in which E_r is in MPa and f in Hz.

Correlations with Uniaxial Compressive Strength U_c

E_r may be estimated from correlations with the uniaxial compressive strength U_c. The correlation, given in Table 3.12, also includes bulk density γ, and Fig. 4.9 includes a large range of rock types. The Schmidt hardness rebound value R may be used to estimate values for E_r through correlations with U_c and γ such as given in Fig. 3.51. Modulus reductions for weathering effects may be estimated on the basis of the ratio of the rebound value R obtained from weathered rock to the rebound value obtained from fresh, unweathered specimens *(see Art. 3.4.3)*.

4.2. INDUCED STRESSES

4.2.1 CONCEPTS

Significance

When a load is applied to the surface of a mass, or within a mass, additional stresses are induced in the mass that is under "at rest" or geostatic stresses *(see Art. 3.1.2)*. To evaluate the compressive deformation that will occur it is necessary to compute the magnitude of stress increase in the material beneath or around the loaded area.

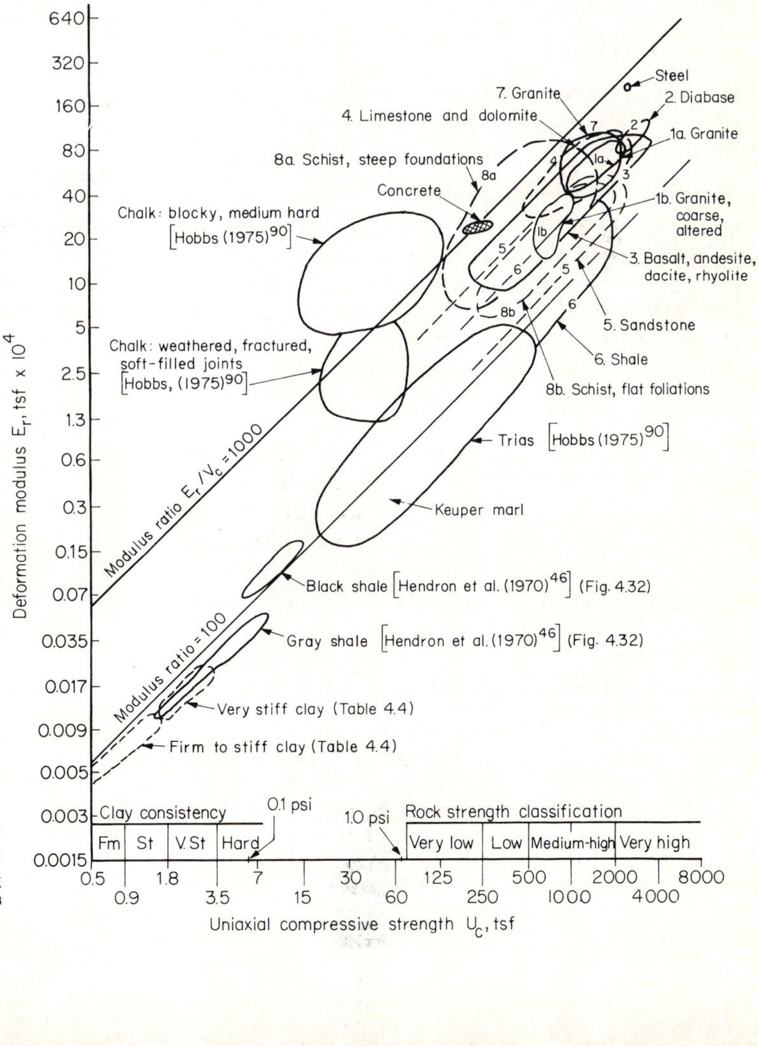

FIG. 4.9 Relationships between uniaxial compressive strength and deformation modulus for various rock types and clays. [After Peck (1976)[88] and Deere (1968).[89]]

Theory

Basis

Computations of imposed stresses are based on elastic theory, considered applicable to soil and rock formations as long as there are constant ratios between stresses and corresponding strains. This condition exists in geologic materials along the lower portion of the stress-strain curve (Fig. 4.3) where the curve is essentially linear; the ratio changes as rupture is approached.

Validity

Theory validity is influenced by geologic and foundation loading conditions. With regard to geologic conditions, the stress-strain relationships of soils and rock masses are generally disproportional. In soils, elastic theory is most applicable to strata of saturated or partially saturated cohesive materials subjected to moderate increments of stress. It is generally not applicable to cohesionless soils for which the elastic modulus depends markedly on confinement [Perloff (1975)[19]].

Most theoretical solutions are based on surface loading, but most practical applications are for foundations bearing at some depth. Theoretically, settlement under a flexible, continuously loaded foundation takes the form of a "dish" due to the shape of the pressure bulb distribution (Fig. 4.11). In sands, however, deflections will be uniform under a rigid foundation, but under a flexible foundation, settlement will occur in an irregular pattern with the larger settlement near the edge of the footing (*see Art. 6.2.3*). In clays, settlement under a rigid foundation uniformly loaded will be uniform, but under a flexible foundation it will be dish-shaped in accordance with theory.

Elastic theory is used in deformation analysis for lack of better techniques. Computed stress values may be in error by as much as ±25%; therefore, the results of analysis can seldom be considered precise and must be applied with judgment and an understanding of the theoretical limitations. This means that when the "allowable bearing value" is selected, a relatively generous factor of safety is employed.

4.2.2 BOUSSINESQ STRESSES

Point Loads

In 1885 Boussinesq presented equations that express stress components caused by a perpendicular point surface load Q at points within an elastic medium extending infinitely in all directions below a horizontal surface, as shown in Fig. 4.10*a*. The stresses are directly proportional to the applied load and decrease markedly with depth. In geomechanics problems the vertical stress component is used much more frequently than the other stresses.

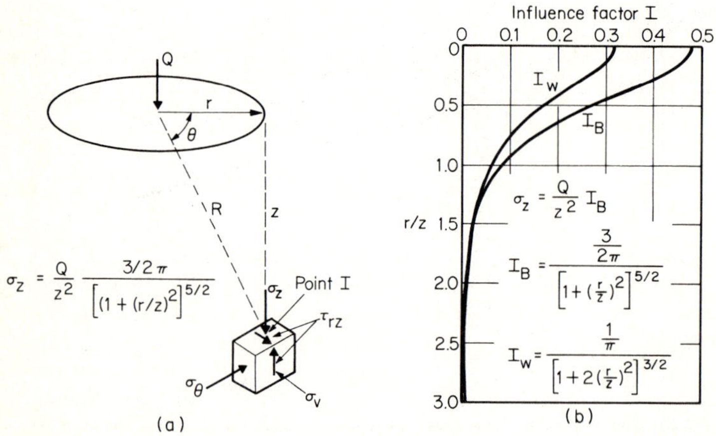

FIG. 4.10 Stresses in an elastic, semi-infinite medium caused by a surface, vertical point load: (*a*) stresses at a point in the mass and (*b*) values for influence factors I_B and I_W for calculating vertical stress from surface load Q. [After Taylor (1948).[20] Reprinted with permission of John Wiley & Sons, Inc.]

Vertical stress is expressed as

$$\sigma_z = \frac{Q}{z^2} \frac{3/2\,\pi}{[1 + (r/z)^2]^{5/2}} \tag{4.25}$$

or

$$\sigma_z = \frac{Q}{z^2} I_B \tag{4.26}$$

where I_B, termed the *influence factor*, equals the expression which forms the last part of Eq. 4.25. From Eq. 4.25, values of I_B can be computed for different r/Z ratios and presented as shown in Fig. 4.10*b*. Also shown is the stress distribution for the Westergaard influence factor I_W derived for thinly layered systems (*see Art. 4.2.3*). In analysis, the influence factor is used directly as a percentage of the applied load for a given r/Z ratio as shown in the example given on Table 4.6. *These stresses are added to the geostatic stresses when determining the magnitude of deformation or settlement.*

Loaded Surface Areas

The "Pressure Bulb" beneath Square, Strip, and Circular Loadings

Subsurface stresses resulting from loading acting over an area can be determined by integration of Eq. 4.25, where the loadings on infinitely small increments of the foundation area are assumed to be point loads. Such integrations have been performed for a wide variety of foundation shapes and forms of uniform loadings. Elastic solutions for many problems in soil and rock me-

TABLE 4.6
BOUSSINESQ (I_B) VERSUS WESTERGAARD (I_W) VERTICAL STRESSES AS PERCENT OF q_s FOR VARIOUS FOOTINGS WITH UNIFORM LOADINGS

	Square				Circular				Strip	
	Edge		Center		Edge		Center		Center	
Z/B	I_B	I_W	I_B	I_W	I_B	I_W	I_B	I_W	I_B	I_W
0	50	50	100	100	50	50	100	100	100	100
0.25	48	38	92	70	44	32	90	62	95	80
0.50	43	28	70	46	35	23	64	41	82	60
0.75	35	21	48	32	26	18	44	28	67	48
1.00	27	16	34	22	20	14	28	19	55	39
1.25	20	12	25	16	15	10	20	14	47	33
1.50	16	9	19	12	12	7	14	10	39	29
1.75	12	7	14	9	9	5	12	8	33	25
2.00	10	5	12	7	7	4	8	5	30	22
2.50	5	3	7	5	5	3	6	3	24	18
3.0	4	2	5	3	3	2	4	2	20	15
4.0	3	2	3	2	2	2	2	2	16	12
5.0	3	1	2	1	1	1	1	1	12	9
10.0	—	—	—	—	—	—	—	—	6	5

←——— $\%q_s$ ———→

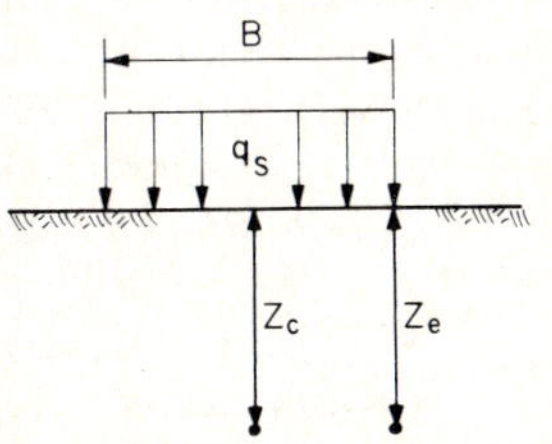

Example: Given square footing width $b = 5$ ft, footing pressure $q_s = 2$ tsf. Find $\Delta\sigma_v$ at depth $z = 5$ ft under center and edge of footing for Boussinesq stresses.

Solution: $Z/B = 5/5 = 1.0$

$\Delta\sigma_v(\text{center}) = 34\%\ (2\text{ tsf}) = 0.68$ tsf

$\Delta\sigma_v(\text{edge}) = 27\%\ (2\text{ tsf}) = 0.54$ tsf

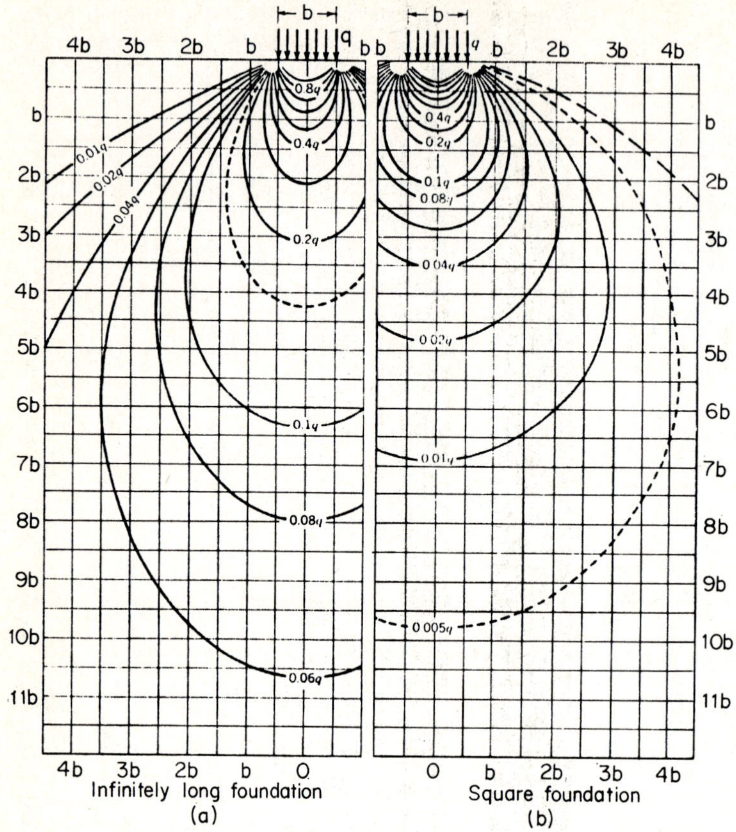

FIG. 4.11 Contours of equal vertical Boussinesq stress beneath: (*a*) infinitely long footing and (*b*) square footing. [From Sowers and Sowers (1951).[22] Reprinted with permission of The Macmillan Company.]

chanics can be found in Poulos and Davis (1974).[12]

Contours of equal vertical stress beneath a loaded area form a "bulb" of pressure as shown in Fig. 4.11. The stress distribution beneath an infinitely long *strip footing* is shown in Fig. 4.11*a* and that for a *square footing* is shown in Fig. 4.11*b*. It is seen that for square footings, stresses at a depth equal to twice the footing width are about 10% of the applied stress, and below this depth for most problems imposed stresses are considered insignificant.

Graphs for *strip footing with rectangular load* (similar to Fig. 4.11*a* but more detailed for shallow depths) and *triangular strip load* are given in Figs. 4.12 and 4.13, and for a *circular loaded area* in Fig. 4.14 which may be used to estimate stresses beneath square areas. For rapid estimation of the influence factor for Boussinesq and Westergaard stresses for square, circular, and strip loads, Table 4.6 is useful. *Stress relief from excavation* is also evaluated from these graphs and tables.

Rectangular Areas

Another form of influence graph is given as Fig. 4.15, which applies to the corner of a uniformly loaded rectangular area. It is used to find the stresses beneath a corner, beneath any point within an area, or beneath some point outside the area of loading. The influence value I is plotted versus m and n, which are interchangeable, defined as $m = x/z$, $n = y/z$, where x and y represent loaded area dimensions. Stress beneath

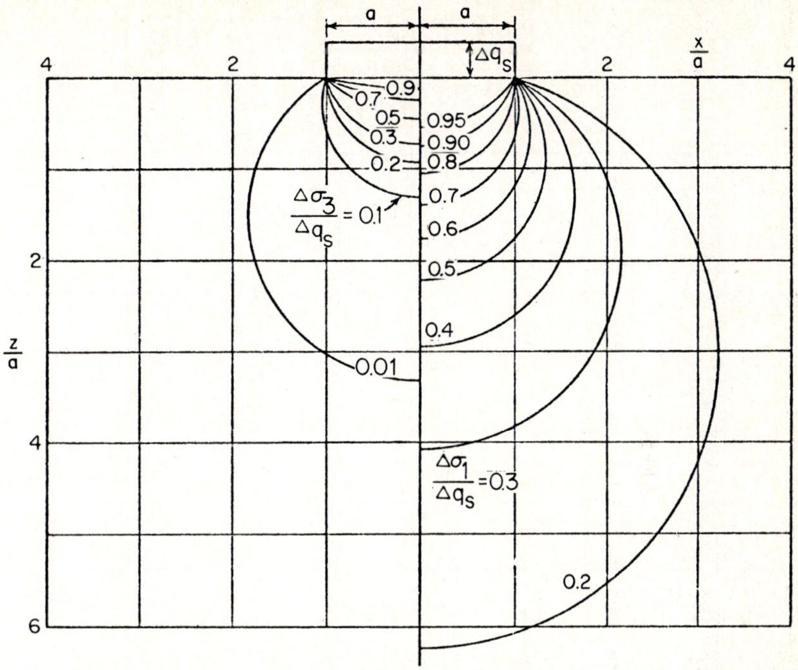

FIG. 4.12 Contours of principal Boussinesq stresses under strip load. Stresses caused by a strip load are independent of Poisson's ratio. [From Lambe and Whitman (1969).[3] Reprinted with permission of John Wiley & Sons, Inc.]

Example: **Find σ_v and σ_h at z = 10 ft.**

1. Given: γ = 0.105 kcf, K_o = 0.5, Δq_s = 3 ksf, $a = b/2$ = 10 ft, z/a = 1.0

2. Solution:

	$\sigma_v = \sigma_1$	ksf	$\sigma_h = \sigma_3$	ksf
a. Initial stresses:	γ_z =	**1.05**	$K_o\gamma z$ =	**0.525**
b. Stress increments:	**(0.82)(3) =**	**2.46**	**(0.18)(3) =**	**0.540**
c. Final stresses:		**3.51**		**1.065**

The stress beneath the center of the loaded area is found by superposition. The area is divided into small rectangles and the *I* values found beneath their common corner and summed to find total *I*. To determine stresses outside the loaded area, a rectangle is subdivided with a common point over the point where *I* is desired, and the *I* values for the unloaded areas are subtracted from those for the loaded area. When evaluating stresses overlapping from adjacent structures, the same procedure is followed, and the stresses from the two adjacent structures are summed. Figure 4.16 illustrates the general approach.

The *influence area chart* provides a simple means of determining stresses beneath *irregularly loaded areas*. The chart, developed by Newmark (1942)[25] and given as Fig. 4.17*a*, is suitable for all shapes under uniform load. Each subdivision on the chart is an "influence area" with a value computed to be 0.005 for the stress σ_z beneath some particular point.

To find the stress beneath point *D* in Fig. 4.17*b*, for example, the following procedure is used:

1. Prepare the building plan to a scale such that length *AB* is equal to depth z.

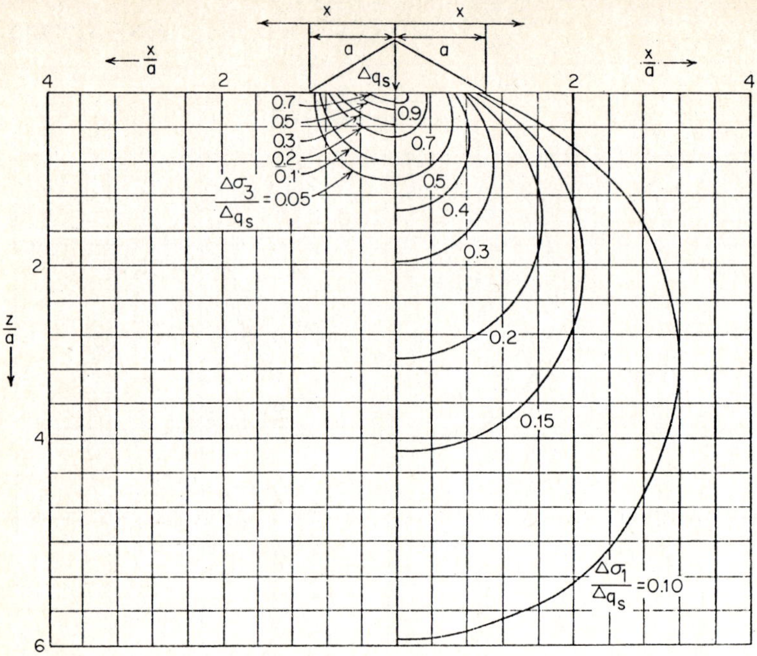

FIG. 4.13 Contours of principal stresses under triangular strip load. [From Lambe and Whitman (1969).[3] Reprinted with permission of John Wiley & Sons, Inc.]

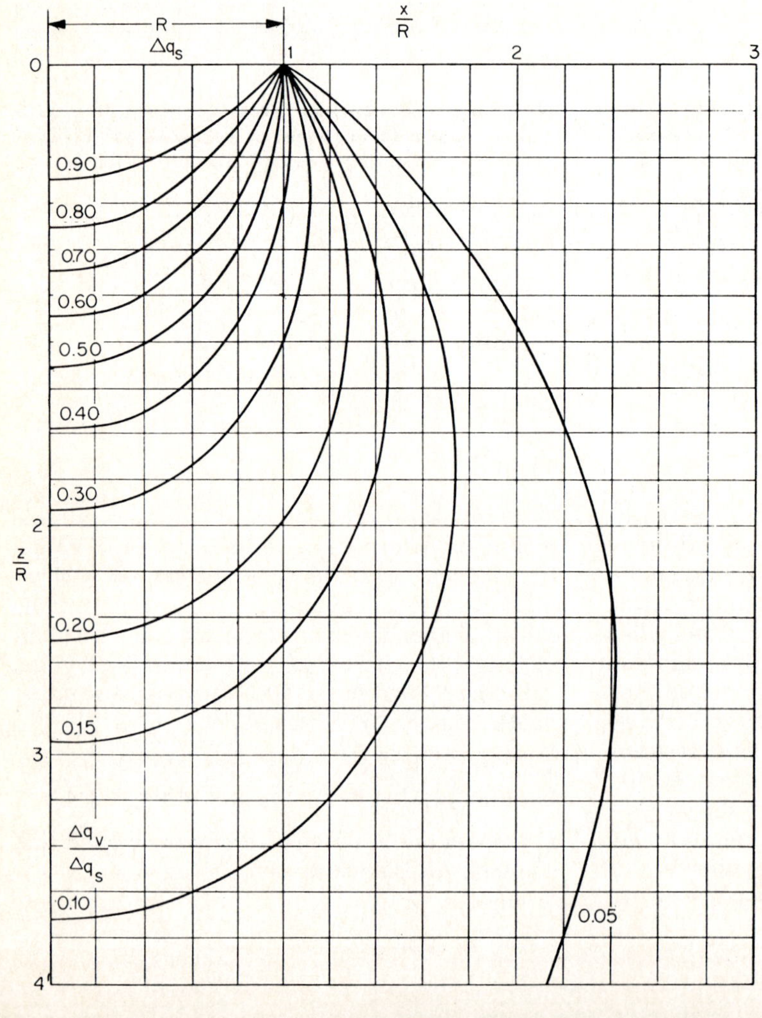

FIG. 4.14 Vertical stresses induced by a uniform load on a circular area. [From Lambe and Whitman (1969).[3] Reprinted with permission of John Wiley & Sons, Inc.]

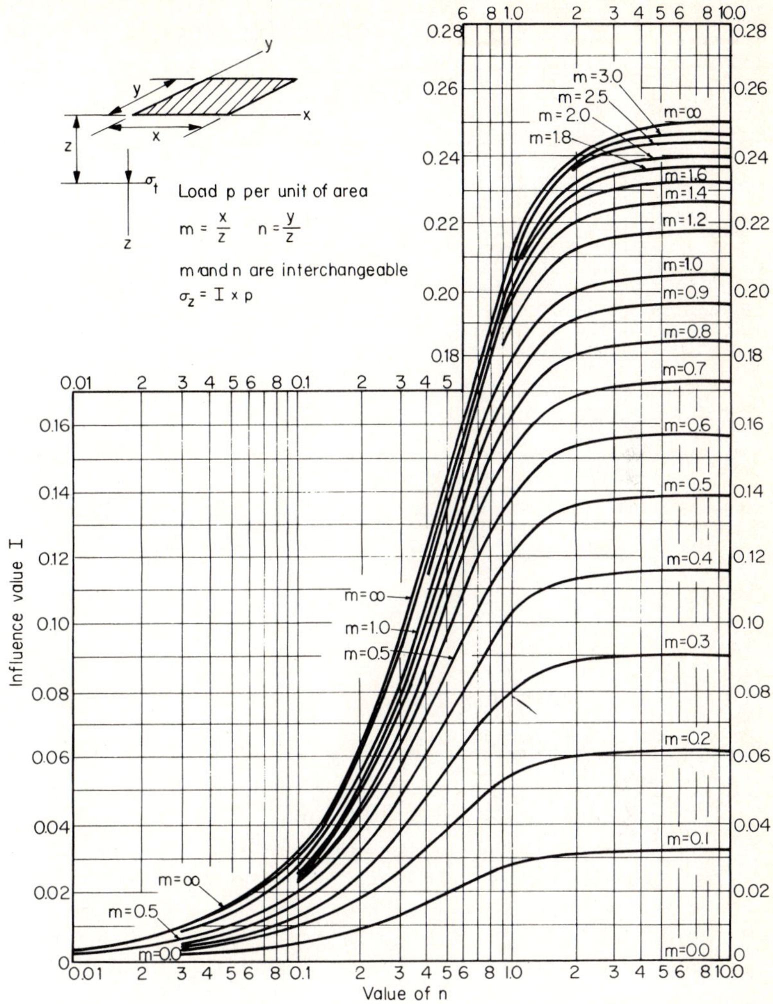

FIG. 4.15 Influence value *I* for vertical stress under corner of uniformly loaded rectangular area. [From NAVFAC (1982)[23]; from Fadum (1948).[24]]

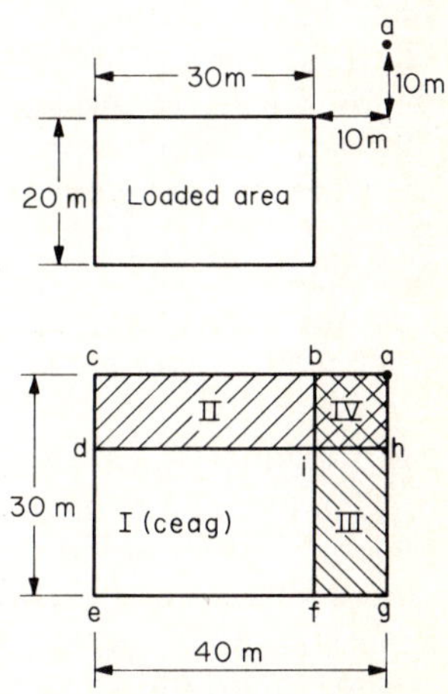

FIG. 4.16 Solution to find stress changes at depth outside of a loaded rectangular area.
Problem: Beneath point *a* at depth z = 20 m find $\Delta\sigma_v$ resulting from loaded mat foundation.
Solution:

Area	*m*(*x*/*z*)	*n*(*y*/*z*)	*I*
I (*ceag*)	30/20 = 1.5	40/20 = 2.0	0.223
II (*cdah*)	10/20 = 0.5	40/20 = 2.0	−0.135
III (*bafg*)	30/20 = 1.5	10/20 = 0.5	−0.131
IV (*baih*)	10/20 = 0.5	10/20 = 0.5	0.085
			$\Delta\sigma_v$ = 0.042 t/m^2

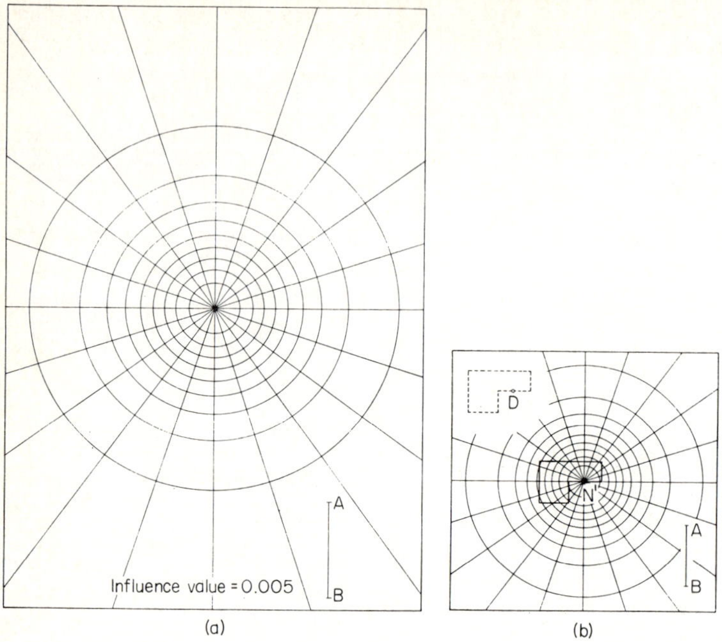

FIG. 4.17 Influence chart for evaluating vertical pressure beneath irregularly shaped areas. (*a*) Influence chart. [After Newmark (1942)[25]; from Terzaghi and Peck (1967).[26]] **(*b*) Application of influence chart.** [From Terzaghi and Peck (1967).[26] Reprinted with permission of John Wiley & Sons, Inc.]

2. Overlay the building plan on the chart with point D over point N' and count the number of influence areas within the building area.
3. The stress at depth z equals the number of influence areas times the average foundation pressure times 0.005.

Approximation Method

Stresses can be approximated beneath any uniformly loaded area by distributing the pressure on the loaded area downward at $2V$ to $1H$ to the depth where the stress is desired, as shown in Fig. 4.18. The stress is assumed to be uniform over the stressed area and is found by dividing the foundation load by the stressed area as follows

$$\Delta\sigma_v = \frac{Q}{(B + Z)(L + Z)} \quad (4.27)$$

This method most closely depicts stresses in deep layers, at depths in the order of $4B$ or greater. At shallower depths the pressure-bulb effect results in computed stresses less than Boussinesq stresses under the footing center and greater under the footing edge.

4.2.3 LAYERING EFFECTS

Thinly Layered Formations (Westergaard Analysis)

Westergaard (1938)[27] developed an expression for an influence factor for thinly layered soil systems, in which he assumed that thin layers of homogeneous and anisotropic material alternated between closely spaced, infinitely thin sheets of inextensible material which would permit compression but no lateral deformation. For the

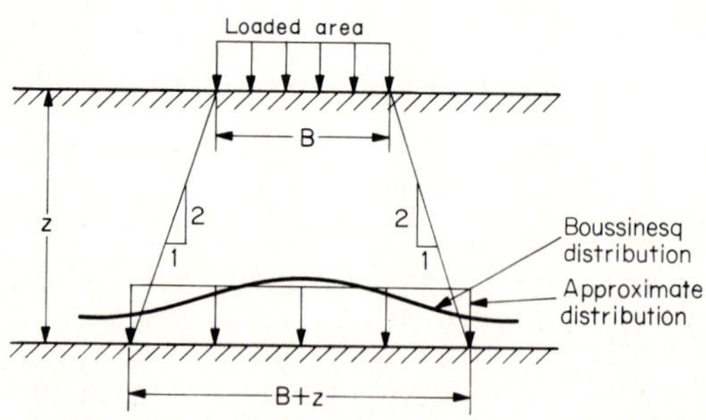

FIG. 4.18 Comparison of approximate stress distribution with the Boussinesq distribution beneath a loaded area.

case of $\sigma = 0$, the Boussinesq equation, Eq. 4.25, reduces to that given for I_W in Fig. 4.10*b*. Formulas based on Westergaard's equation give maximum values for vertical stresses approximately two-thirds those of the Boussinesq stresses at shallower depths, as shown in Table 4.6.

As maximum settlement computed according to the Boussinesq distribution usually is higher than actually occurs, the Westergaard relationships are believed by some practitioners to be closer to reality because of the reinforcing effect of thin, relatively strong layers common to many soil formations, such as varved clays, for example.

Two-Layered Systems

Rigid Base Layer

Burmister (1956)[28] addressed the problem of a weak upper layer overlying a relatively rigid layer in which the stresses have a tendency to concentrate at the interface of the rigid base due

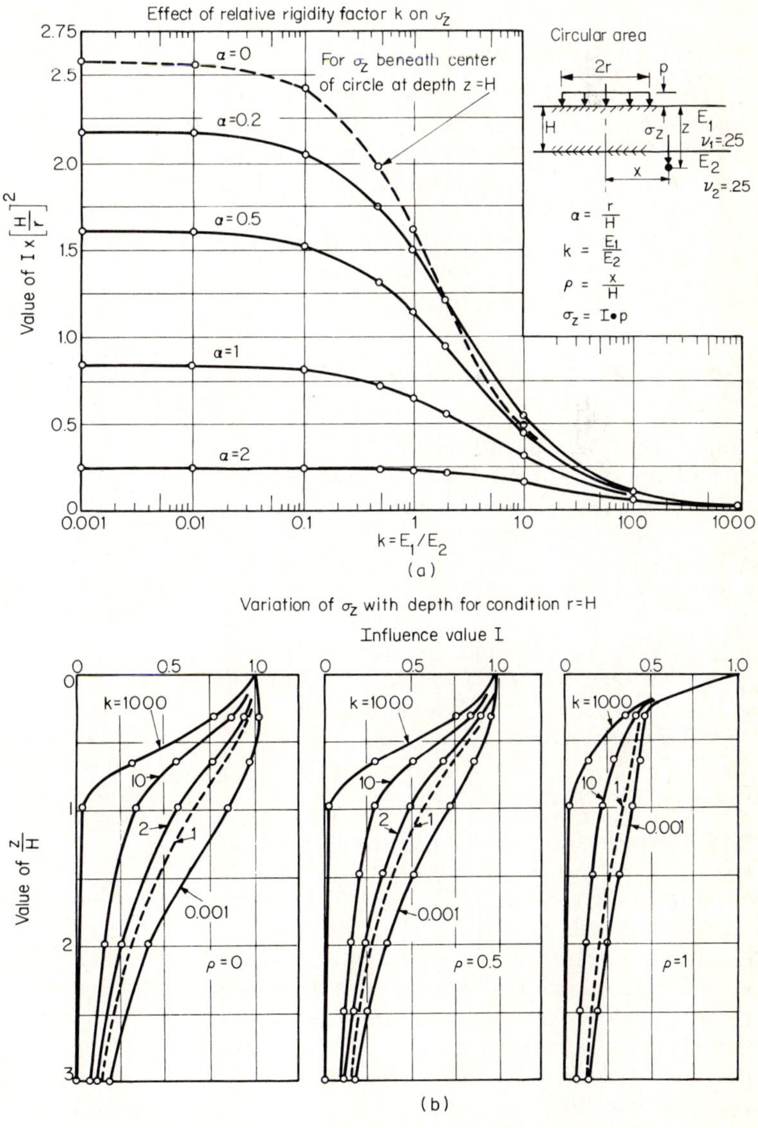

FIG. 4.19 Influence values for vertical stresses beneath uniformly loaded circular area on two-layered foundation: (*a*) beneath center and (*b*) beneath center, with $\rho = x/H = 0$, $x/H = 0.5$, and $x/H = 1.0$. [For example, if $\alpha = 1$, $k = 10$, $\rho = L$, $z/H = 1$, then $I = 0.21$, from right panel of part (*b*). [From NAVFAC (1982)[23]; after Mehta and Veletsos (1962).[30]]

to restrictions of the shear deformations at the interface. In Fig. 4.19, this condition is represented by the case where the rigidity factor $k = E_1/E_2$ is less than 1.

Reinforcing Layer

Burmister (1943)[29] addressed the problem of a relatively rigid layer overlying a weaker layer, such as the case for pavements, or a compacted fill over soft soils. The upper relatively rigid layer tends to produce a bridging action over underlying softer soils, resulting in stresses lower than the Boussinesq, as shown by the ratios of the moduli k in Fig. 4.19*b*.

The size of the loaded area in relation to the stratum thickness is most significant. A large building with heavy floor loads supported on a relatively thin layer of compacted fill, for example, will not benefit significantly from the fill rigidity; on the other hand, narrow wheel loads on relatively thick pavements will impose relatively small stresses on the supporting soils.

4.2.4 ROCK MASSES

Influencing Factors

Stratification and discontinuities affect the stress distribution system in rock masses, and the Boussinesq stresses do not truly apply unless the rock is sound and massive. Factors affecting stress distribution include the orientation of joint systems, shapes of blocks and their spatial arrangement in the system, joint characteristics, deformability and strength of intact rock composing the blocks, rock-structure interaction (the ratio of corresponding rigidities), and the number of blocks in contact with the foundation.

Empirical Distributions

Model tests performed by Gaziev and Erlikman (1971)[31] show that the penetration of maximum stress varies according to the angle made by the applied load with the direction of bedding and major discontinuities. Stress distributions for various bedding and joint orientations are shown in Fig. 4.20. It is seen that in all cases the depth of stress penetration in a jointed mass is greater than the Boussinesq penetration. This is caused by the deformations that occur along the joints.

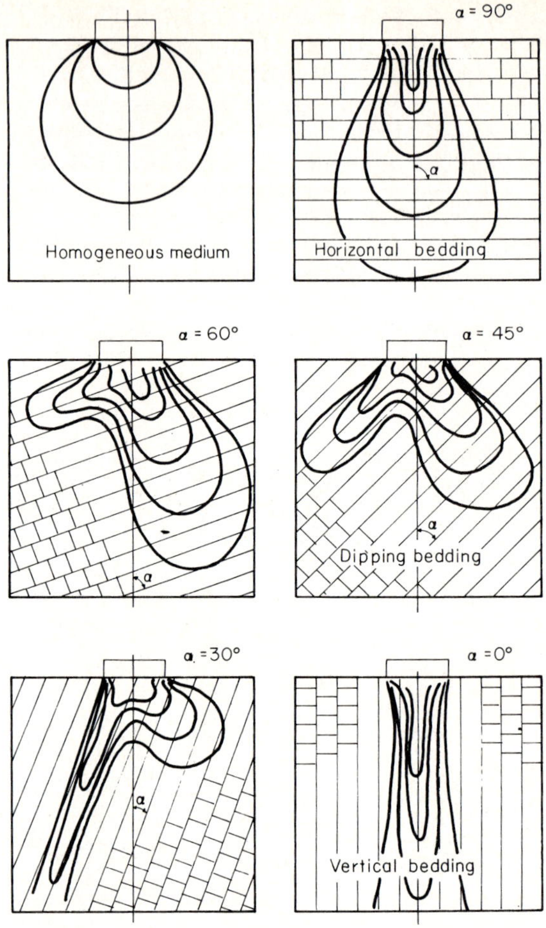

FIG. 4.20 Stress distribution in regularly jointed anisotropic rock. [From Gaziev and Erlikman (1971).[31]]

In horizontally bedded strata, deformation is at a maximum, but the depth of penetration is at a minimum; in vertically dipping strata, deformation is at a minimum, but the depth of penetration is at a maximum.

Theoretical Approach

In practical rock mechanics, stresses and deflections are of greatest concern in foundations for massive and thin-arch concrete dams, in large caverns, and in deep, open excavations. These problems are being approached by numerical modeling techniques, most commonly the finite element method. As described by Zienkiewicz

(1968),[32] (1971),[33] the region of influence of the foundation or excavation is subdivided into elements with the model designed to pictorialize rock-mass conditions including discontinuities. Each element is assigned a set of elastic properties including the joint properties of "normal stiffness" and "shear stiffness" [see Goodman (1980),[2] pp. 188 and 307]. For dams, pressure distribution and shear on the foundation are input, and the stresses in each element and the displacement of a set of points throughout the medium are obtained, as well as the stress-deformation characteristics of the dam.

4.2.5 PILE FOUNDATIONS

Mindlin's Stress Coefficients (I_s, I_p) for Single Piles

Stress distribution around single, deep, embedded foundations is complex, as the stresses can be imposed by the shaft alone (friction pile), the base alone (end-bearing pile), or by shaft-base combinations, as discussed in Art. 7.3.2. The Boussinesq formula cannot be applied because of the assumption that a load acts only on the boundary of a halfspace, whereas a pile transfers its loads to the interior of the halfspace.

Using elastic theory, Mindlin (1936)[34] and Antunes Martins (1948)[35] developed approximate influence values for a friction pile and for an end-bearing pile as given in Table 4.7. These values have been plotted as isobars of vertical stress in Fig. 4.21. A uniform distribution of shaft resistance has been assumed for Fig. 4.21*a* and can be used for friction piles in clay. The change in vertical stress is computed from Eqs. 4.28 and 4.29 as given in Table 4.7.

In practice, evaluation of single piles is based more commonly on limit design methods employing safety factors to computed shaft and point resistances, as described in Art. 7.4.1, than on the theory of elasticity.

Pile Groups

Overlapping stresses between adjacent piles complicate the stress distributions about a pile group. Analysis normally is based on assuming a mat foundation at a depth equal to two-thirds the pile length (equivalent pier method) for friction piles, or at the pile tips for end-bearing piles, and applying Boussinesq stresses (*see Art. 7.4.2*). The resulting computed settlements are considered approximations.

4.3 ANALYSIS OF COMPRESSION AND EXPANSION

4.3.1 APPROACHES TO ANALYSIS

Compression

Settlements from the compression of rock masses, residual soils, glacial till, and sands are estimated from deformation moduli and elastic-theory stress distributions. Compression of sands is estimated also by elastic strain concepts. Settlement estimates of the compression of clays are based on stress-strain relations from tests.

The various methods and relationships for computing settlements are presented in this article; the overall aspects of settlement analysis for structures supported on shallow foundations are discussed in Art. 6.2.3 (soils) and Art. 6.3.2 (rock), and for deep foundations in Art. 7.4.

Expansion

Theories are not available to define the characteristics of most cases of expansion in soil and rock due to swelling, frost heave, or extension straining. Volume changes, swelling pressures, and time-delay aspects, for the most part, are phenomena that only can be approached empirically, and then only in a general manner. Practical solutions require recognizing that the problem exists and then designing conservatively to mitigate or avoid expansion effects, as described in Arts. 6.2.6 and 7.2.3. The characteristics and some aspects of analysis of expansive materials are reviewed in Art. 4.3.6.

4.3.2 USING DEFORMATION MODULI

Applications and Validity

As a practical matter, deformation moduli are used to evaluate settlements in most geologic materials. Results of analysis will be approximations since soil and in situ rock are seldom truly isotropic, homogeneous, and continuous materials. Their inelastic properties influence the distribution of stresses under load and the resulting strains.

TABLE 4.7
MINDLIN'S STRESS COEFFICIENTS FOR PILES*

	r/L									
Z/L	**0**	**0.1**	**0.2**	**0.3**	**0.5**	**0.7**	**1.0**	**1.5**	**2**	**3**
FRICTION PILE										
0.25	0.698	0.568	0.355	0.209	0.079	0.035	0.009	0.001	0.000	0.000
0.50	0.654	0.522	0.442	0.345	0.184	0.091	0.031	0.005	0.001	0.000
1.0	∞	1.550	0.750	0.478	0.249	0.145	0.067	0.024	0.000	0.001
1.25	0.902	0.790	0.593	0.432	0.246	0.152	0.078	0.027	0.010	0.001
1.5	0.419	0.404	0.364	0.312	0.216	0.147	0.085	0.033	0.013	0.003
2.0	0.180	0.178	0.172	0.163	0.139	0.112	0.077	0.039	0.019	0.005
3.0	0.068	0.068	0.067	0.066	0.062	0.057	0.048	0.034	0.022	0.009
POINT LOAD										
0.25	0.174†	0.166†	0.142†	0.090†	0.021†	0.009	0.012	0.004	0.001	0.000
0.50	0.755†	0.678†	0.476†	0.268†	0.034†	0.032	0.024	0.062	0.029	0.000
1.0	∞	0.120	0.119	0.110	0.096	0.078	0.051	0.021	0.008	0.001
1.25	3.89	2.59	1.221	0.494	0.145	0.082	0.051	0.023	0.011	0.001
1.5	1.04	0.948	0.736	0.513	0.237	0.118	0.061	0.026	0.012	0.002
2.0	0.289	0.281	0.263	0.239	0.183	0.130	0.076	0.035	0.012	0.004
3.0	0.086	0.086	0.084	0.082	0.076	0.068	0.055	0.036	0.026	0.008

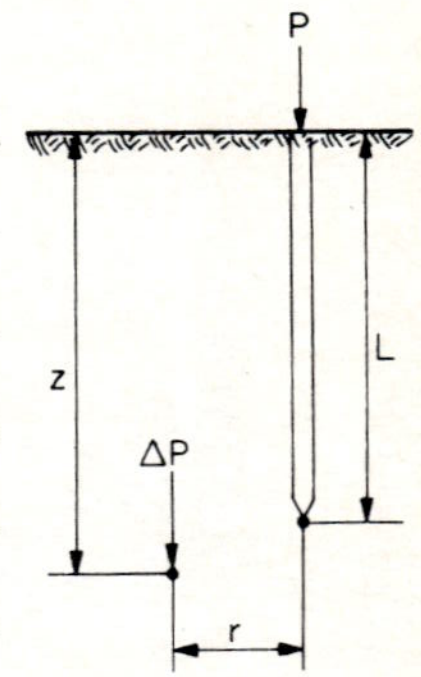

$$\Delta P_s = \frac{P_s I_s}{L^2} \quad (4.28)$$

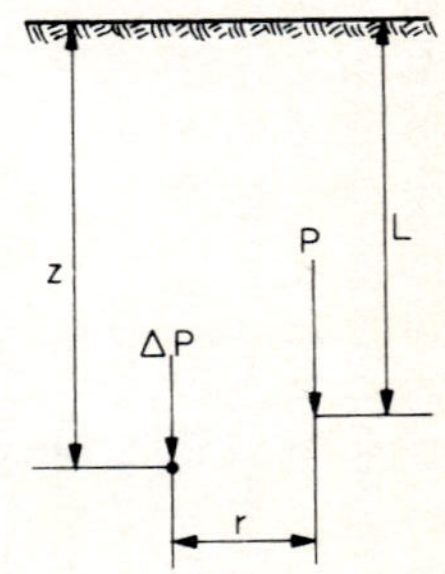

$$\Delta P_p = \frac{P_p I_p}{L^2} \quad (4.29)$$

*After Antunes Martins (1948).[35]

†Tension.

Soil Moduli Values

Typical moduli values for soils and other materials, which have been obtained from the early portion of the stress-strain curve at low stress levels, are given in Table 4.4. In soft, cohesive soils they will yield settlement estimates much lower than reality; in all cohesive or slow-draining materials, they will not account for the settlement time delay caused by the process of consolidation.

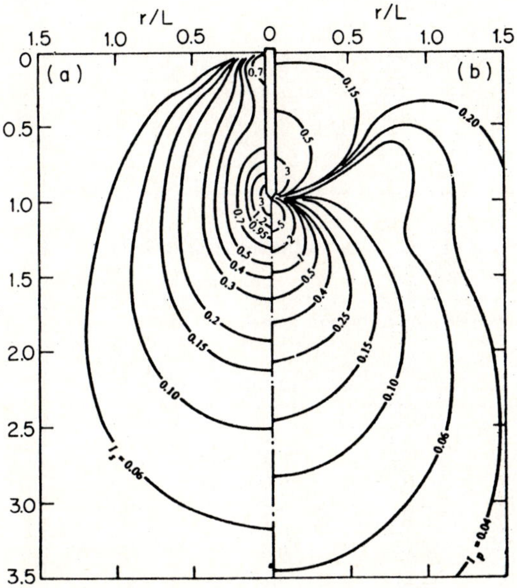

FIG. 4.21 Mindlin's vertical stress coefficients plotted as isobars around a pile supported by (*a*) shaft friction and (*b*) end bearing. [After Grillo (1948).[35a]]

Rock Moduli Values

Moduli values for rock are given in Table 3.12 and Fig. 4.9. The higher value for a given rock type is for fresh, intact specimens. As a rough approximation, discontinuities could decrease these values to the following amounts:

- Very wide joint spacing, 40% E_r
- Wide joint spacing (1 to 3 m), 25% E_r
- Moderately close spacing (30 cm to 1 m), 10% E_r

These values generally are applicable to joint apertures as large as 2.5 cm, filled with soil or rock debris [CGS (1978)[4]]. Decomposition also will reduce the values for fresh, intact rock until moduli are attained in the ranges for the stronger soils.

Analyses based on deformation moduli result in what is often referred to as the "immediate elastic settlement" and do not account for creep or any other time-delay phenomena.

TABLE 4.8
INFLUENCE FACTOR *I* FOR LOADED AREAS ON AN ELASTIC HALFSPACE OF INFINITE DEPTH*

Shape	Rigidity	Center	Corner	Edge/middle (long side)	Average
Circle	Flexible	1.00	—	0.64	0.85
	Rigid	0.79	—	0.79	0.79
Square	Flexible	1.12	0.56	0.76	0.95
	Rigid	0.82	0.82	0.82	0.82
Rectangle	Flexible				
	$L/B = 2$	1.53	0.76	1.12	1.30
	$L/B = 5$	2.10	1.05	1.68	1.82
	$L/B = 10$	2.56	1.28	2.10	2.24
	Rigid				
	$L/B = 2$	1.12	1.12	1.12	1.12
	$L/B = 5$	1.60	1.60	1.60	1.60
	$L/B = 10$	2.0	2.0	2.0	2.0

*From NAVFAC (1982)[23]; after University of California (1965).[91]

Settlement Computations

Elastic Theory Basis

Based on elastic theory [Terzaghi (1943)[7]], the immediate settlement of a corner of a flexible, rectangular surface can be expressed in terms of Young's modulus E as follows:

$$\rho = q_s B \frac{1 - \nu^2}{E} I \tag{4.30}$$

where ρ = settlement
q_s = applied uniform surface stress
B = least lateral footing dimension ($2R$ for circular footing)
ν = Poisson's ratio (since the range for ν is not great for a given soil, and the effect on the computation is not large, ν is usually selected from tables such as Table 4.4)
E = elastic modulus (E_s for soil, E_r for rock)
I = influence factor

Values for the influence factor I for various shapes, either flexible or rigid, for calculating

TABLE 4.9
INFLUENCE FACTOR *I* FOR LOADED AREAS ON AN ELASTIC HALFSPACE OF LIMITED DEPTH OVER A RIGID BASE*

H/B	Center of rigid circular area diameter = *B*	Corner of flexible rectangular area				
		L/B = 1	*L/B* = 2	*L/B* = 5	*L/B* = 10	(Strip) *L/B* = ∞
		FOR ν = 0.50				
0	0.00	0.00	0.00	0.00	0.00	0.00
0.5	0.14	0.05	0.04	0.04	0.04	0.04
1.0	0.35	0.15	0.12	0.10	0.10	0.10
1.5	0.48	0.23	0.22	0.18	0.18	0.18
2.0	0.54	0.29	0.29	0.27	0.26	0.26
3.0	0.62	0.36	0.40	0.39	0.38	0.37
5.0	0.69	0.44	0.52	0.55	0.54	0.52
10.0	0.74	0.48	0.64	0.76	0.77	0.73
		FOR ν = 0.33				
0	0.00	0.00	0.00	0.00	0.00	0.00
0.5	0.20	0.09	0.08	0.08	0.08	0.08
1.0	0.40	0.19	0.18	0.16	0.16	0.16
1.5	0.51	0.27	0.28	0.25	0.25	0.25
2.0	0.57	0.32	0.34	0.34	0.34	0.34
3.0	0.64	0.38	0.44	0.46	0.45	0.45
5.0	0.70	0.46	0.56	0.60	0.61	0.61
10.0	0.74	0.49	0.66	0.80	0.82	0.81

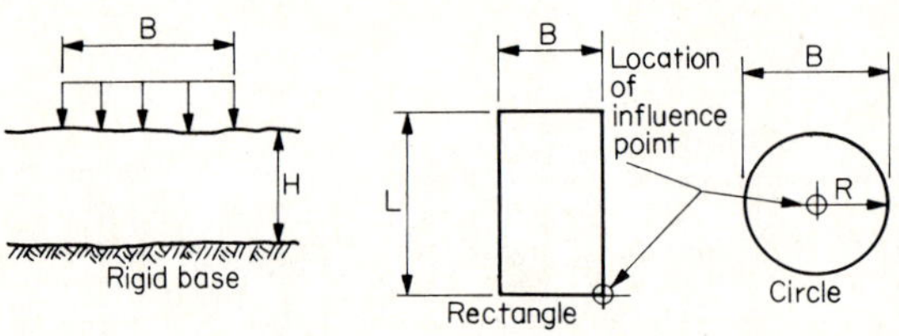

Notation for loaded areas, shown in plan view

*From NAVFAC (1982)[23]; after University of California.[91]

settlements of points on loaded areas at the surface of an elastic halfspace of infinite depth are given in Table 4.8 and at the surface of an elastic halfspace of limited depth over a rigid base on Table 4.9.

Using Eq. 4.30 for the case illustrated in Fig. 4.27, but assuming $\nu = 0.33$, $E_s = 100\ \text{kg/cm}^2$, and $I = 0.82$, the computed settlement is 2.2 cm.

Rectangular flexible foundations were analyzed by Terzaghi (1943)[1] to develop values for I for various *ratios of footing width to length and width to depth of embedment*. These relationships have been expanded by Janbu et al. (1956)[36] into the family of curves given in Fig. 4.22. The expression for settlement computations given by Bowles (1974)[37] is similar to Eq. 4.30

$$\rho = q_s B \frac{1 - \nu^2}{E} I_1 I_2 \tag{4.31}$$

where I_1 = influence factor to account for depth to a rigid boundary
I_2 = influence factor to account for footing depth

Critical Point Method

Beneath the critical point of a foundation the settlement is independent of the footing rigidity *(see Art. 6.2.3)*. The settlement under the point as computed from the Boussinesq solution for flexible footings, therefore, will be equal to the settlement of a rigid footing of the same area supporting the same load. Assuming a constant value for E_s, the settlement can be calculated for the critical point [Kany (1959),[38] CGS (1978)[4]] from the relationship

$$\rho = \frac{B\, q_{net}}{E_s} f_c \tag{4.32}$$

where q_{net} = net bearing pressure ($q_s - \gamma_t D_f$)
f_c = settlement coefficient, given in Fig. 4.23.

Using Eq. 4.32 for the case illustrated in Fig. 4.27, but assuming $E_s = 100\ \text{kg/cm}^2$, and the depth of the compressible layer z = 4 m ($f_c = 0.65$), the computed settlement is 1.74 cm. This method is considered useful for preliminary analysis but not for final design [CGS (1978)[4]].

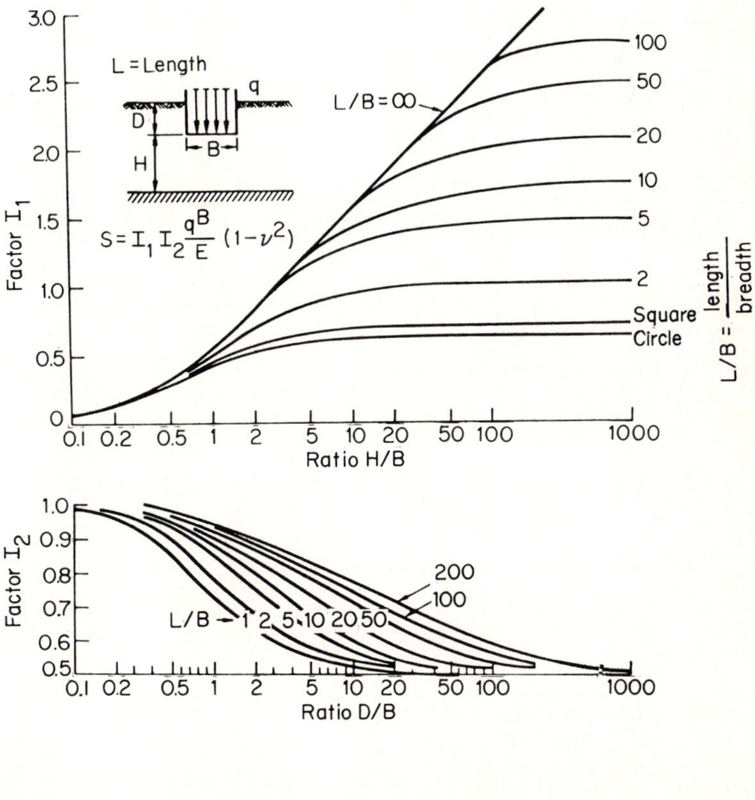

FIG. 4.22 Influence factors for use in analysis of elastic settlements, Eq. 4.31. [From Bowles (1974)[37]; after Janbu et al. (1956).[36]]

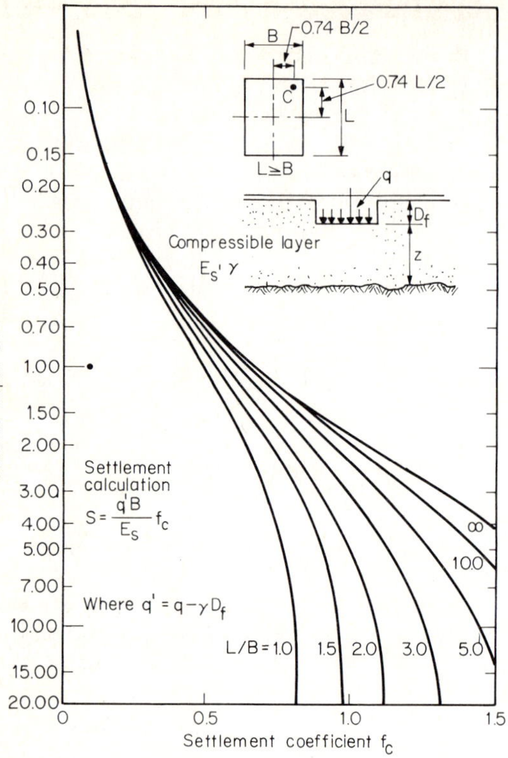

FIG. 4.23 Curves for settlement coefficient f_c in terms of relative depth z/B for calculation of settlement at the critical point C. [From CGS (1978)[4]; after Kany (1959).[38]]

The settlements of *stratified formations* with different moduli, occurring within a depth of $3B$, are evaluated from the expression

$$\rho = B\, q_{net} \left\{ \frac{f_1}{E_{s1}} + \frac{f_2 - f_1}{E_{s2}} - \cdots \right\}$$

$$= B\, q_{net} \sum \frac{f_n - f_{n-1}}{E_{sn}} \qquad (4.33)$$

where $f_n = f_1, f_2, f_3$, etc.; the settlement coefficients obtained assuming depths z of the bottoms of the compressible strata are equal to z_1, z_2, z_3, etc., for strata 1, 2, 3, etc.; and E_{s1}, E_{s2}, etc., are the moduli.

Pile Foundations

Settlements of pile foundations for offshore structures are computed on the basis of load-displacement curves (t-z and Q-z) developed from deformation moduli as discussed in Art. 7.4.1.

4.3.3 COMPRESSION IN SANDS

Characteristics

Compression of sands composed of quartz particles, or other rigid material, is essentially immediate under application of load, due to a rearrangement of particles and a closing of voids. Long-term compression or creep may occur in sands under some conditions which are not well-defined. In addition to densification from void closure, silica sands can undergo small amounts of elastic compression under relatively high loads, whereas "soft" sands composed of particles of calcareous materials may suffer crushing of grains as well as void closure.

Under a rigid foundation over silica sands, deflections will be uniform, but under a uniformly loaded flexible foundation settlements will occur in an irregular pattern, with the larger settlements occurring near the footing edge (*see Art. 6.2.3*). In sands the maximum differential settlement is close to the maximum total settlement.

Estimating Settlements: Various Methods

Settlements in sands are estimated on the basis of elastic theory concepts (previous article), load tests, elastic strain concepts, or the pressuremeter modulus. Foundations are designed on the basis of an allowable bearing value or allowable soil pressure.

Allowable Bearing Value

Shallow foundations in sands are most commonly designed for an allowable bearing value or allowable soil pressure obtained from the empirical penetration test (SPT). Charts have been developed correlating N values from the SPT for various footing sizes that will limit settlement to 1 in (2.54 cm), as described in Art. 6.2.2. N values should be adjusted for the effect of depth and overburden pressure (*see Art. 6.2.2*).

Load Tests

Full-scale load tests on footings or piles to obtain actual relationships among bearing pressure, foundation dimensions, and settlement may be economically feasible on projects where settlement is critical or a very large number of individual foundations are required.

The *simple plate-load test (see Art. 4.4.1)* formerly was used to estimate settlement but now is used infrequently by most practitioners because of the very limited depth of stress influence when compared to a normal full-size footing. This limitation can be overcome by performing a series of tests at various depths to obtain a profile of moduli. In uniform cohesionless soils, soil rigidity increases in depth along with the overburden and confining pressures. For the conditions where the increase in rigidity is uniform with depth, the depth of the foundation below ground surface is less than the foundation width, and the width is not greater than 20 ft, the following expression [McCarthy (1982)[39]] may be used to estimate settlement:

$$\rho = \frac{4qB^2}{K_v(B+1)^2} \tag{4.34}$$

where K_v = vertical modulus of subgrade reaction for a one-foot-square plate bearing on the ground surface (Table 4.14).

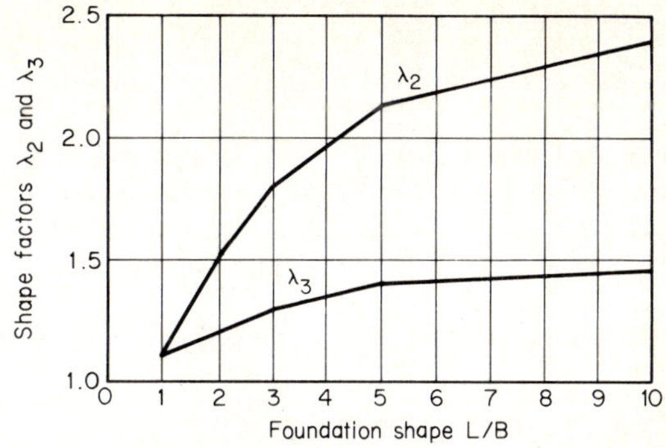

FIG. 4.24 Shape factors for settlement calculations using the pressuremeter modulus. [After SOLS SOILS (1975).[7]]

Using Eq. 4.34 for the case illustrated in Fig. 4.27, and assuming K_v = 150 kips/ft², the settlement is calculated to be

$$\rho = \frac{(4)(3.0)(6.7 \times 6.7)}{(150)(6.7+1)(6.7+1)} = 0.06 \text{ ft} = 1.8 \text{ cm}$$

TABLE 4.10
STRUCTURAL FACTOR α FOR VARIOUS SOIL AND ROCK CONDITIONS*

	Peat		Clay		Silt		Sand		Sand and Gravel	
	E_c/P_L	α	E_c/P_L	α	E_c/P_L	α	E_c/P_L	α	E_c/P_L	α
SOIL MATERIAL										
Overconsolidated or very dense			16	1	14	0.67	12	0.5	10	0.33
Normally consolidated or dense		1	9–16	0.67	8–14	0.5	7–12	0.33	6–10	0.25
Underconsolidated or loose			7–9	0.5	5–8	0.5	5–7	0.33		

ROCK CONDITIONS†			
Wide joint spacing	Moderately close joint spacing	Close spacing	Very close spacing, low strength
α = 0.67	α = 0.5	α = 0.33	α = 0.67

*After CGS (1978),[4] SOLS SOILS (1975).[7]

†Reduction based on joint spacing (Table 3.16) and rock strength (Table 3.13).

Pressuremeter Method

The *Canadian Foundation Engineering Manual* [CGS (1978)[4]] considers that settlement predictions based on pressuremeter tests are presently the most reliable for cohesionless granular materials *(see also Art. 6.2.3)*. Footing settlement is estimated as follows [after SOLS SOILS (1975)[7]]

$$\rho = \frac{p}{4.5E_c}\left[2B_o\left(\lambda_2\frac{B}{B_o}\right)^{\alpha} + (\alpha\lambda_3 B)\right] \qquad (4.35)$$

where E_c = pressuremeter modulus
ρ = mean footing contact stress added to soil
B_o = reference width, equal to 30 cm
B = foundation width
λ_2, λ_3 = shape factors (Fig. 4.24)
α = soil structure factor (Table 4.10)

The correction factor α, originally termed the rheological factor, was developed by Menard (1965)[40] to account for the difference between the horizontal earth pressure during test and the vertical earth pressure that will apply under foundation loads. Many practitioners correct the pressuremeter modulus to arrive at an equivalent soil modulus, as follows:

$$E_s = E_c/\alpha \qquad (4.36)$$

As shown in Table 4.10, in some cases E_c is very nearly the same as E_s, with the greatest differences occurring in sands and gravels. Using Eq. 4.35 for the case illustrated in Fig. 4.27, and assuming a value for E_c = 30, the settlement is calculated to be 2.1 cm. Calculated values are extremely sensitive to the values for E_c selected for analysis.

Elastic Strain Method

Concept

The elastic strain concept proposed by Schmertmann (1970)[41] is currently favored by many prac-

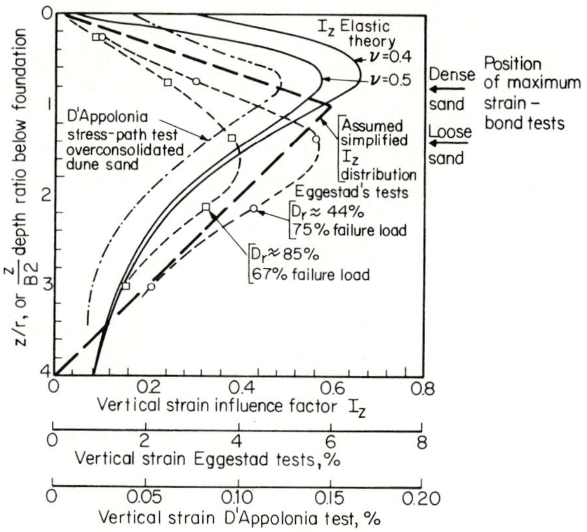

FIG. 4.25 Theoretical and experimental distributions of vertical strain below center of loaded area. [After Schmertmann (1970).[41]]

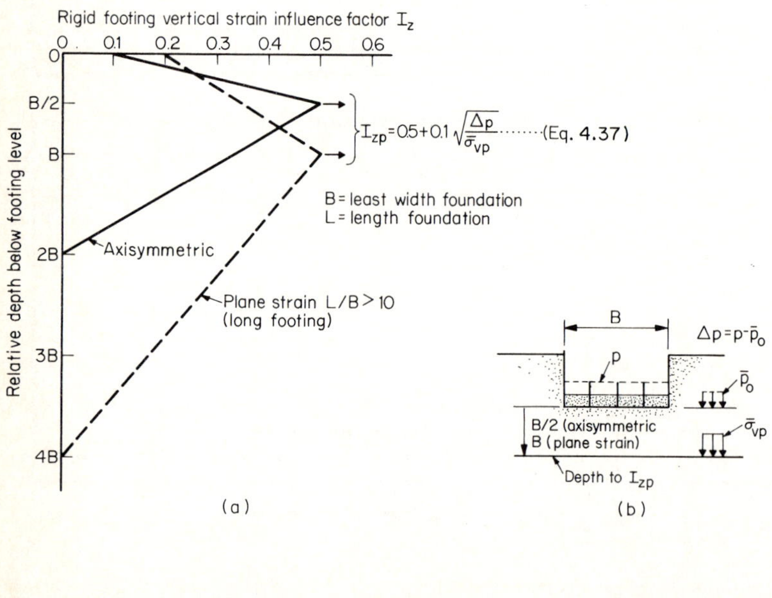

FIG. 4.26 Modified strain influence factor diagrams for use in Schmertmann method for estimating settlement over sand: (*a*) modified strain influence factor distributions and (*b*) explanation of pressure terms in Eq. 4.37. In part (*a*) note that the peak value for the influence factor I_{zp} is found from Eq. 4.37. [From Schmertmann et al. (1978).[43]]

titioners. Theoretical model studies and other investigations have shown that the vertical strain under shallow footings bearing on homogeneous free-draining soils varies from a low value directly beneath the footing to a maximum value at a depth in the range of $z = B/2$, diminishing thereafter with depth to $I_z = 0$ at a depth of $z/B = 2$. This is significantly different from the distribution based on the increase in vertical stress from elastic theory as shown in Fig. 4.25. *Strain influence factor* I_z reaches a maximum value of 0.6 and extends to a maximum depth of 2B as shown in the figure. The method is often referred to as the "$2B - 0.6I_z$" strain distribution.

Modified strain influence factor diagrams have been presented by Schmertmann (1977),[42] (1978)[43] which differentiate axisymmetric footings from long footings (plane strain) as shown in Fig. 4.26. The *peak value* for the influence factor I_{zp} is found from Eq. 4.37, given in the figure.

Settlement Calculations

Schmertmann (1970)[41] gives the following relationship which summarizes by integration the strains occurring in a number of layers beneath a footing or pier

$$\rho = C_1 C_2 \, \Delta p \sum_{0}^{2B} \left(\frac{I_z}{E_s} \right) \Delta z \qquad (4.38)$$

where p = settlement

C_1 = a factor to account for strain relief due to embedment, defined as $C_1 = 1 - 0.5\,(\overline{p}_o/\Delta p)$, where $\overline{p}_o$ = effective overburden pressure and Δp = pressure increase

C_2 = a factor to account for creep, $C_2 = 1 + 0.2 \log (t_{yr}/0.1)$, where t_{yr} = settlement in some time period after application of $\overline{p}_o + \Delta p$; the creep factor can be ignored if the lateral pressure is less than the creep pressure determined from pressuremeter tests

Δp = pressure increase at footing depth

B = footing width

I_z = strain influence factor from Fig. 4.26; in analysis, I_{zp} is determined and the strain influence diagram constructed for either a square or a long footing; in a homogeneous stratum use I_{zp} at $B/2$ or B; in layered strata take the value at the midpoint of each layer as illustrated in Fig. 4.27.

E_s = deformation modulus for each stratum in the strain influence zone; to find the equivalent modulus for sands from q_c values obtained by CPT, Schmertmann (1978)[43] suggests the following: $E_s = 2.5q_c$ for square footings and $E_s = 3.5q_c$ for long footings

Δz = layer thickness

An example of the elastic strain method is given in Fig. 4.27; other examples of multilayered conditions, for which it is most useful, can be found in Schmertmann (1977)[42] and Perloff (1975).[19] In the example in Fig. 4.27, if $E_s = 100$ kg/cm² is used for the full depth, the immediate settlement (no creep) is 2.4 cm.

Simplified Settlement Computations

Using Eq. 4.38 and substituting for E_s in terms of the SPT value N, a simple expression for estimating immediate settlement in sands can be formulated. Assuming $E_s = 10N$ (tsf) (*see Art. 4.1.4*), and considering a uniform soil with E_s constant with depth to at least $2B$, a footing at the surface ($C_1 = 0$), and Δp, B, and ρ in feet, then for $I_{zp} = 0.6$

$$\rho = \frac{\Delta p}{E_s} \left[\frac{1}{2} (2B)(0.6) \right] = \frac{\Delta p}{10N} (0.6B) \quad \text{ft} \qquad (4.39)$$

or

$$\rho = \frac{(0.6)(\Delta pB)\,12}{10N} = 0.7\, \Delta pB/N \quad \text{in} \qquad (4.40)$$

This simple expression usually overpredicts settlement in overconsolidated sands. In the example in Fig. 4.27, for $N_1 = 9$, settlement = 1.78 cm using Eq. 4.40 (N_1 is corrected N value; *see Art. 6.2.3*).

Summary

A summary of the settlement predictions for a square footing of $B = 2$ m and $d_f = 1$ m by the various methods supported on a loose-to-medium compact sand above the water table is given in Table 4.11. It is seen that the results from all of the methods were quite close. The allowable bearing value, $q_a = 1.5$ kg/cm² (1.5 tsf),

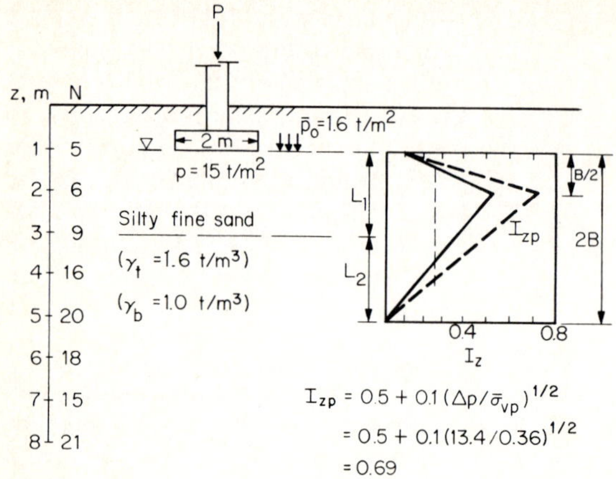

SUMMATION OF STRAIN INFLUENCE

Layer	Δz, cm	N_1	E_{avg}, kg/cm²	L_z, m*	I_z	$(I_z/E_s)\ \Delta z$, cm/kg/cm²
1	200	9	100	1	0.73	1.46
2	200	14	140	3	0.24	0.34
						$1.80 = \Sigma(I_z/E_s)\ \Delta z$

*L_z = depth to midpoint of layer below foundation level.

Find settlements from

$$\rho = C_1 C_2\ \Delta p (I_z/E)\ \Delta z \tag{4.38}$$

$$C_1 = 1 - 0.5(\bar{p}_o/\Delta p);\ \bar{p}_o = 1.6 \times 1.0 = 1.6\ \text{t/m}^2;\ p = 15\ \text{t/m}^2$$

$$\Delta p = 15 - 1.6 = 13.4\ \text{t/m}^2 = 1.34\ \text{kg/cm}^2$$

$$C_1 = 1 - 0.5(1.6/13.4) = 0.94$$

$$C_2 = 1 + 0.2 \log t_{yr}/0.1,\ \text{for 10 yr} = 1.4$$

$$\rho = (0.94)(1.4)(1.34)(1.80) = 3.2\ \text{cm; or for } C_2 = 0,\ \rho = 2.27\ \text{cm}$$

FIG. 4.27 Computation of settlements in a layered soil deposit using elastic strain concepts for a square footing.

was selected on the basis of N values from design charts given in Art. 6.2.2 which refer to shallow footings on sands above GWL and total settlements less than 1 in.

When selecting a method for analysis the important factors to consider are the sensitivity of the structure to settlement, the footing width and the L/B ratio, the footing depth, the soil stratification, and the validity of the moduli data. The elastic strain method is particularly useful for large footings over a stratified deposit with various values for E_s where a more accurate estimate of settlement is required than would be obtained by other methods.

4.3.4 CONSOLIDATION IN CLAYS

Characteristics

Consolidation

Consolidation is the form of compression occurring in clays that results from the slow dissipation of pore-water pressures. Two aspects are evaluated analytically:

1. Total amount of compression occurring under a given applied load, and
2. Relationship between time and compression for a given load increment. Analysis is based on Terzaghi's one-dimensional theory of con-

TABLE 4.11
SUMMARY OF EXAMPLES FOR COMPUTING IMMEDIATE SETTLEMENTS IN SANDS*

Method	Relationship	Equation	Modulus value	Settlement cm	Settlement in
Elastic theory	$\rho = \dfrac{0.89 q_s BI}{E_s}$ for $\nu = 0.33$	4.30	$E_s = 100$	2.2	0.8
Critical point	$\rho = (B\, q_{net}/E_s) f_c$	4.33	$E_s = 100$	1.7	0.7
Subgrade reaction	$\rho = 4qB^2/k_v(B+1)^2$	4.35	$k_v = 150$	1.8	0.7
Pressuremeter	$\rho = \dfrac{p}{4.5E_c}\left[2B_o\left(\lambda 2 \dfrac{B}{B^o}\right)^{\alpha} + (\alpha\lambda_3 B\right]$	4.36	$E_c = 30$	2.1	0.8
Elastic strain	$\rho = C_1 C_2\, \Delta p \sum_0^{2B} \left(\dfrac{I_z}{E_s}\right) \Delta z$	4.38	$E_{s1} = 100$, $E_{s2} = 140$	2.3	0.9
Simplified	$\rho = 0.7\, \Delta\, pB/N$	4.41	$N = 9$	1.8	0.7

*For conditions in Fig. 4.25 ($q_s = 1.5$ kg/cm^2; $B = 2$ m, $d_f = 1.0$ m).

solidation which is valid for the assumptions that strains are small, the soil is saturated, and flow is laminar.

Lateral Strains

Lateral strains are not permitted in the one-dimensional consolidation test, but there are cases where they contribute significantly to settlement, such as during the relatively rapid application of load to soft soils. Analysis of this case, as well as approaches to settlement analysis based on triaxial test data, are found in Lambe and Whitman (1969),[3] Lambe (1973),[44] and Kavazanjian and Mitchell (1980).[45]

Stresses and Deflections

In clays, settlements under a uniformly loaded rigid foundation will be uniform, but under a flexible foundation will be dish-shaped in accordance with theory. Settlement under the center of a rigid circular footing is approximately 21% less than under the center of a flexible circular footing *(see Art. 6.2.3).*

Computing Compression

Compression Versus Pressure Curve

The laboratory consolidation test [Art. 3.5.4 in Hunt (1984)[5]] provides the data from which is plotted a curve representing either strain versus pressure or void ratio versus pressure, where pressure is plotted on a log scale as shown in Fig. 4.28 (e-log-p curve). Unload-reload cycles, usually made soon after the preconsolidation pressure p_c has been identified, provide data for correction of the compression curve to reduce conservatism in analysis caused by the slight expansion of the sample occurring during sampling and laboratory handling. The recompression portion of the curve is extended back from the initial curve at p_c to form a new curve with a flatter initial slope. This flat portion of the curve is referred to as having a slope C_r.

Settlement Analysis

For settlement analysis, the compression of an in situ clay stratum is assumed to be directly related to the compression of the test specimen, expressed by

$$\rho = \Delta H = \frac{\Delta e}{1 + e_o} H \tag{4.41}$$

where H = stratum thickness
Δe = the change in void ratio resulting from an increase in pressure from the effective overburden pressure ($\bar{p}_o$ or $\bar{\sigma}_{vo}$, Art. 3.1.2) to the pressure imposed on the stratum by the foun-

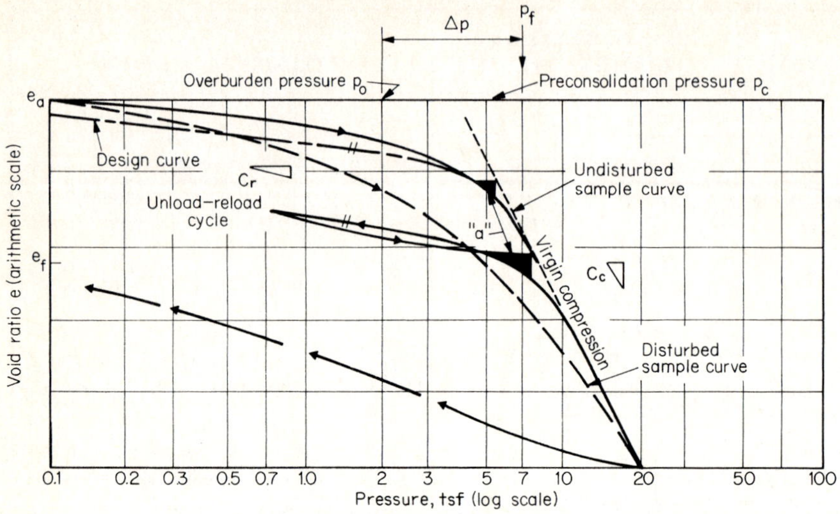

FIG. 4.28 Pressure versus void ratio curve (*e*-log-*p* curve) from the consolidation test.

dation load (p_f or $\Delta\sigma_v$; p_f is determined by procedures given in Art. 4.2.2)

e_o = the initial void ratio corrected for the slight expansion occurring during sampling and extrusion

Note: For estimating settlements from CPT data see Art. 6.2.2, Clays.

Compression Index C_c

The compression index C_c is the slope of the virgin compression curve on a semilog plot of *e* vs. *p*. It is used to compute settlements for NC clays from data from a single test or from an average value from a number of tests, with the expression

$$\rho = \frac{C_c H}{1 + e_o} \left(\log \frac{\bar{p}_o + p_f}{\bar{p}_o}\right) \tag{4.42}$$

where $\bar{p}_o$ = effective overburden pressure at midthickness of consolidating layer

p_f = added pressure at midthickness of consolidating layer due to foundation

$$C_c = \frac{e_1 - e_2}{\log (p_1/p_2)} \tag{4.43}$$

Some references suggest adding the settlement along the recompression line (C_r) into Eq. 4.42; however, a significant slope along the recompression portion normally indicates sample disturbance and its inclusion tends to overestimate settlements unless the curve has been corrected.

C_c is estimated from the following expressions:

For NC clays of moderate to low sensitivity [Terzaghi and Peck (1967)[26]]:

$$C_c \approx 0.009\,(\text{LL} - 10\%) \tag{4.44}$$

For silts and clays [McCarthy (1982)[39]]:

$$C_c \approx 0.54(e_o - 0.35) \tag{4.45}$$

$$C_c \approx 0.0054(2.6w - 35) \tag{4.46}$$

where LL = liquid limit

w = natural water content

Collapsible Soils

Certain soils, such as loess, "porous" clays, or recent valley alluvium in an arid climate, have a lightly cemented structure which is destroyed during saturation, resulting in its collapse. The *double consolidation test* is used to provide data for estimates of collapse magnitude *(see Art. 4.3.6)*.

Consolidation Time Rate

The Phenomenon

Stress conditions existing in the pore water at the center of a clay stratum in the field are illustrated in Fig. 4.29. When the foundation load is

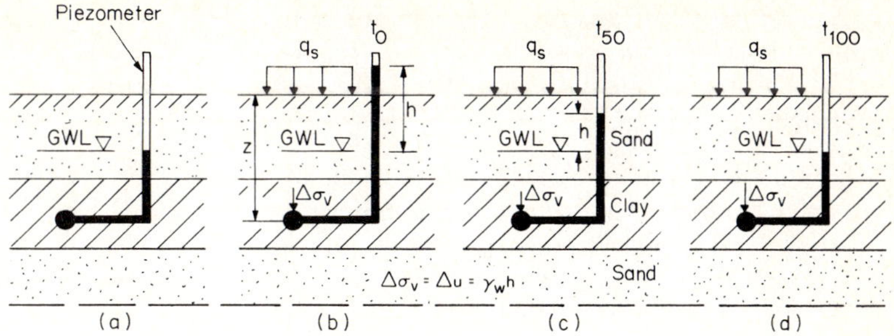

FIG. 4.29 Pore-water pressure changes during clay consolidation: (*a*) initial condition—piezometer reads GWL; (*b*) full foundation load applied—water under pressure rises to h = $\Delta\sigma_v/\gamma_w$ above GWL; (*c*) piezometer reading drops during consolidation, until eventually full consolidation has occurred; and at (*d*) piezometer level equals GWL.

applied, initially it is carried totally by the pore water causing the excess hydrostatic pressure (Fig. 4.29*b*), which is highest beneath the center of the foundation and decreases toward the edges. With time, the stratum drains and the pore-pressure levels accordingly decrease (Fig. 4.29*c*). The rate of decrease is lowest in the center of the clay layer and highest at the clay-sand interfaces, which would be reflected in piezometer levels. Finally, at 100% consolidation, pore pressure returns to equilibrium (Fig. 4.29*d*) and the effective stress on the clay stratum has increased.

Settlements computed from the stress-strain curve or *e*-log-*p* curve will be total amounts as the curve is constructed from the void ratios for fully dissipated pore pressure, or full consolidation, at each load. As consolidation is a time-related phenomenon, a thick clay stratum may require 50 years to reach full consolidation; therefore, the *time rate of consolidation* must be considered in terms of the life of the structure, rate-of-load application, or other factors.

A compression versus time curve for one load cycle is given in Fig. 4.30. It is divided into two parts:

1. *Primary compression*, or consolidation which occurs while the excess pore pressures dissipate.
2. *Secondary compression*, which is a slow process of compression continuing beyond primary consolidation. The phenomenon is not clearly understood but is considered a form of creep.

Calculating Primary Consolidation Time Rates

Time *t* to reach a given percent consolidation *U* is expressed as

$$t = \frac{T_v}{c_v} H^2 \tag{4.47}$$

where T_v = the *theoretical time factor*, a pure number that has been determined for all conditions of importance and is given in terms of *U* as described below

H = one-half the thickness of the stratum if it has drainage at both interfaces, or the full thickness of the stratum if it has a drainable layer at only one interface

c_v = coefficient of consolidation

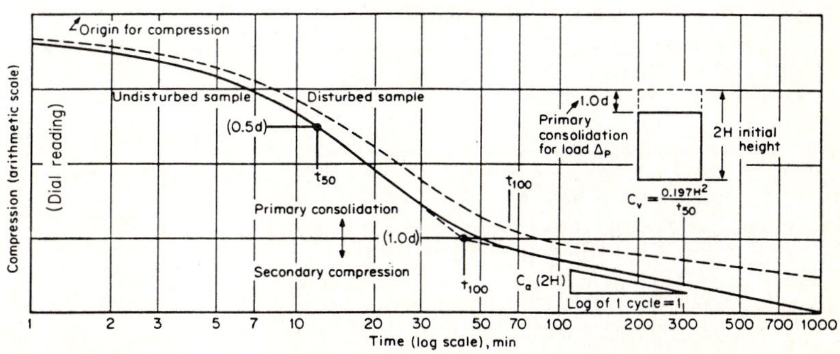

FIG. 4.30 Compression versus time for one load cycle of the consolidation test.

The *theoretical time factors* T_v for various conditions of loading and drainage are given in Fig. 4.31. The three curves represent the following conditions:

C_1— applies to the common cases of a consolidating stratum free to drain through both its upper and lower surfaces, regardless of the distribution of the consolidation pressure, or the case of a uniform distribution of consolidation pressure throughout a layer free to drain through only one surface.

C_2— applies to a stratum with an impervious base subjected to a consolidation pressure decreasing from some value Δp_t at the top to zero at the bottom, such as the case where the consolidating stratum is very thick compared to the size of the loaded area.

C_3— applies to a stratum with an impervious base acted upon by no force other than its own weight, such as the case of a recently placed hydraulic fill.

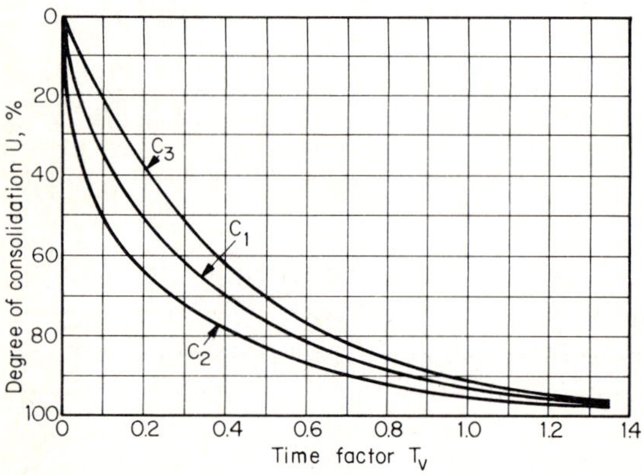

VALUES FOR CURVE C_1	
U	T_v
10	**0.0077**
20	**0.0314**
30	**0.0707**
40	**0.126**
50	**0.197**
60	**0.286**
70	**0.403**
80	**0.567**
90	**0.848**
100	**Infinity**

FIG. 4.31 Relationship between time factor T_v and percent consolidation U for three cases of field drainage and loading. C_1 is the common case. [From Terzaghi and Peck (1967).[26] Reprinted with permission of John Wiley & Sons, Inc.]

The *coefficient of consolidation* c_v is determined from the time curves in the range of loading of practical interest. Tangents are drawn to the primary section of the curve at its point of inflection and to the secondary section to locate d_{100} (the dial reading at 100% primary consolidation). The corrected dial reading for t_o is located by laying off, above a point corresponding to a time of about 0.1 min, a distance equal to the vertical distance between this point and the point for which the dial reading is four times greater. The dial reading for 50% compression is found as the midpoint between d_0 and d_{100}, and the corresponding value for t_{50} is determined in seconds. The theoretical time factor for $U = 50\%$ is 0.197 (Fig. 4.31, curve C_1), and c_v is computed from

$$c_v = \frac{T_v h^2}{t} = \frac{0.197 h^2}{t_{50}} \qquad \text{cm}^2/\text{sec, in}^2/\text{min, ft}^2/\text{d} \tag{4.48}$$

where h = one-half the thickness of the sample as it has double drainage.

The *square root of time method* is an alternate procedure for determining c_v and can be found described in Lambe and Whitman (1967).[3]

The percent of primary consolidation, in terms T_v, that will be attained at some particular time in the field is then computed from the c_v value using Eq. 4.47, or nomographs such as given in Fig. 4.32. Results are usually only approximations because field time rates are normally much higher than laboratory rates primarily because stratification permits lateral drainage. In studies where time rates are critical, piezometers are employed to measure in situ pore pressures.

Calculating Secondary Compression

In soft clays and organic soils, secondary compression can be very significant and may equal primary consolidation in magnitude. It can be estimated from the time curve for the desired stress since its rate is essentially independent of

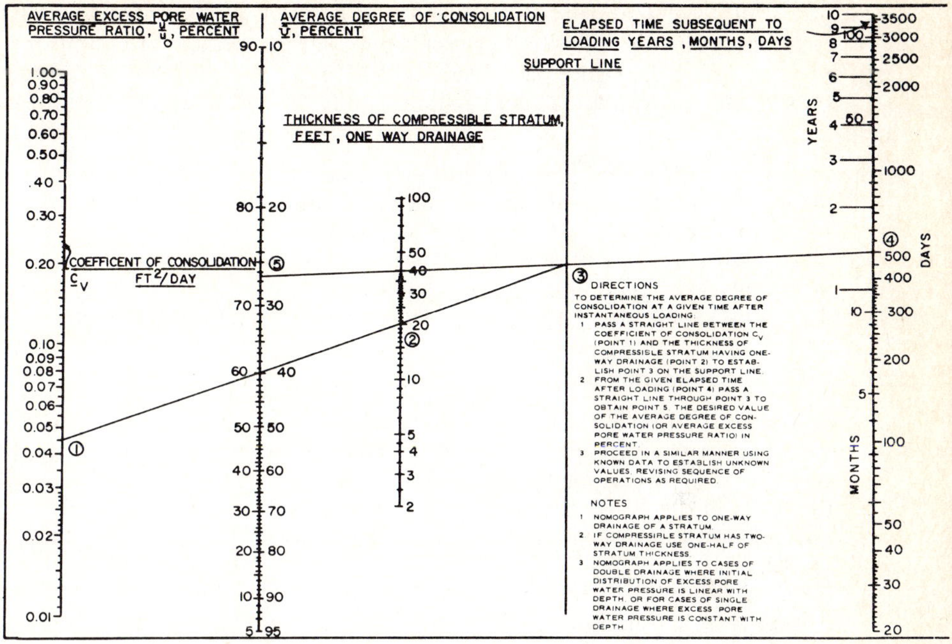

FIG. 4.32 Nomograph for consolidation degree *U* with vertical drainage. [From NAVFAC (1982).[23]]

the stratum thickness in the field. As the relationship between magnitude and time on the semilog plot is essentially a straight line (Fig. 4.30), it can be expressed [Perloff (1975)[19]] as

$$\Delta e = -C\alpha \log t_1/t_2 \tag{4.49}$$

where C_α = the *coefficient of secondary compression* represented by the slope of the time curve
t_1 = time at the start of secondary compression
t_2 = time at which secondary compression is desired

Organic Soils

Physical characteristics of many organic deposits, including their highly compressible nature and the presence of gas and air pockets and fibrous materials, cause them to be only generally susceptible to analysis using consolidation theory for calculating time rates of settlement. In addition, secondary compression is usually very large.

In addition to theoretical limitations, the recovery of undisturbed samples is difficult, especially in highly fibrous material. Compression amounts and time-rate relationships are best determined by full-scale load tests, fully instrumented to measure settlements and pore-water pressures [Art. 4.5.2 in Hunt (1984)[5]].

4.3.5 COMPRESSION IN ROCK MASSES

Characteristics

Magnitude

Compression in rock masses depends on *rock quality (see Art. 3.4.3)*. Competent rock will undergo elastic compression, but only under extremely high loads will magnitudes be significant, since many rocks are stronger than concrete. In moderately hard to hard masses of incompetent rock compression occurs as fractures close or displacement occurs along the fracture surfaces. Additional compression is attributed to intact block deformation which increases with increasing rock softness.

Time Rate

The harder the rock mass, the more immediate will be the deformation as fractures close or displace, although long-term creep may occur along fractures. As rock softness increases as a result of weathering, clay minerals form, and the long-term compression effects may be significant.

Stresses and Deflections

Stress distributions and deflections in sound competent rock occur in accordance with theory, but in fractured rock masses there are large variations which depend on fracture orientations, as discussed in Art. 4.2.4.

Estimating Settlements

Settlements in rock masses are based on elastic theory concepts (*see Art. 4.3.2*), full-scale load tests, or the pressuremeter method; or foundations are designed on the basis of an allowable bearing value. Poor quality rock is difficult to assess.

Allowable Bearing Value

Foundations in rock are most commonly designed on the basis of an allowable bearing value obtained from codes and charts developed from experience for particular geographic areas and rock conditions (*see Art. 6.3.2*). Values indicated usually are a function of rock type and rock quality which is roughly defined by simple field tests, such as the Schmidt hammer (*see Art. 3.4.3*).

Elastic Theory

In most cases E_r, as measured by in situ testing (*see Art. 4.1.4*), and the use of Eq. 4.30, provides the basis for analytical evaluation of deformation. In softer rocks the pressuremeter method may be used. For rock, the initial tangent modulus usually is reported as the deformation modulus, although a tangent modulus may be used to define a limiting stress or deformation. In any event, moduli apply to immediate deformation and do not consider creep and other long-term effects prevalent in weaker rocks.

Obtaining representative values for moduli is a difficult task. In general, the moduli near the rock surface are significantly less than those deeper in the mass because of weathering effects, but rock quality often varies, both with depth and laterally, from poor to good and back to poor because of differential weathering and other factors. In situ tests should cover the full range of conditions as revealed by core borings in terms of core hardness, RQD, and recovery values. General relationships between E_r and U_c for a wide range of rock types are given in Fig. 4.9.

Poor Quality Rock and Residual Soils

Pressuremeter Method

In soft to poor quality rocks, such as shales or those in the highly to completely weathered grades, measurements of moduli are made by full-scale load tests or by the pressuremeter. The pressuremeter is favored by many practitioners because of its relative simplicity and economy. Analysis is performed in the manner described under Art. 4.4.3; the structural correction factor α for various rock conditions is given in Table 4.8.

Shales

In a study made for bridge foundations to be supported on shales, Hendron et al. (1970)[46] evaluated the compressibilty characteristics of the shales from laboratory strain-controlled triaxial and consolidation tests and in situ pressuremeter tests. The in situ tests gave values 5 to 13 times the laboratory values for the drained moduli; Hendron et al. attribute the difference to disturbance in the laboratory samples and the direction of stress imposed by the pressuremeter in the field against the highly layered materials.

Two materials were tested: a gray shale and a stronger black shale. A comparison of the field and laboratory moduli versus water content for each is given in Fig. 4.33. The undrained E_r was found to be $150U_c$ for the black shale and $60U_c$ for the gray shale.

In this case, using the pressuremeter values in computing strains would result in values substantially lower than reality because of the rock anisotropy due to bedding. For estimating settlement Hendron et al. recommend using values equal to one-third the pressuremeter moduli or 3 times the laboratory drained moduli, to account for sample orientation and disturbance. The undrained moduli to be used for design, therefore, would range from 250 to 400 kg/cm^2

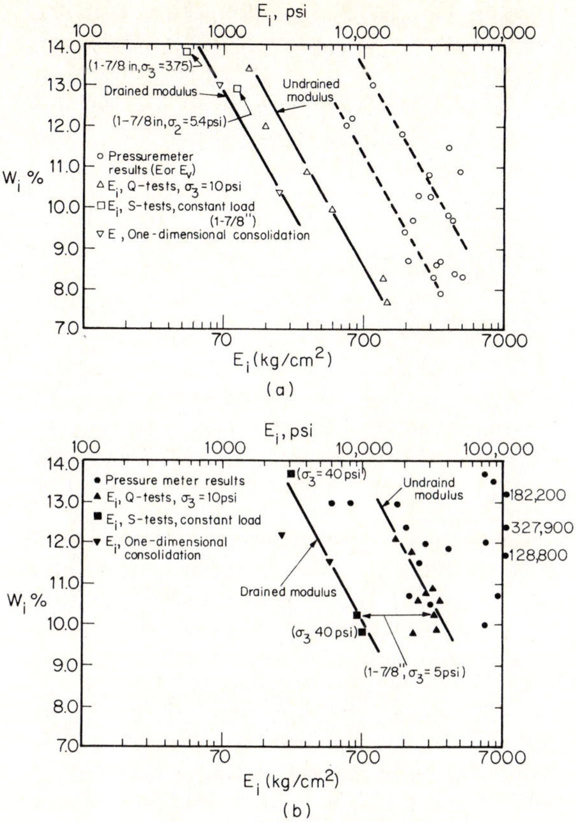

FIG. 4.33 Comparison of field and laboratory moduli for: (*a*) gray shale and (*b*) black shale from Rock Island, Ill. [From Hendron et al. (1970).[46]]

for the gray shale and from 500 to 2500 kg/cm^2 for the black shale.

Decomposed Schist

Ward (1972)[47] reports on an investigation for a settlement-sensitive structure which was to be supported by a mat bearing on a residual soil, derived from schist, which graded to an unweathered schist. The rock surface, generally about 48 ft below the ground surface, was somewhat irregular.

Deformation properties of the residual soil and the decomposed schist were measured by pressuremeter tests, laboratory consolidation tests, and consolidated-undrained triaxial tests. The undrained modulus in the residual soils (*N*-8 to 30) ranged from 19 to 158 tsf, generally increasing with depth, and in the decomposed schist (*N* = 50 to 100/3 in), from 218 to 500 psf, with the higher values obtained at a maximum test depth of 75 ft. The values obtained by the pressuremeter generally were lower than the laboratory values. Design values selected for the residual schist were 100 and 200 tsf for the two depth ranges, and for the decomposed schist, 200 and 275 tsf for the two depth ranges. Martin (1977)[48] also reports data from pressuremeter testing in residual soils derived from gneiss and schist.

4.3.6 EXPANSION AND COLLAPSE IN SOILS AND ROCKS

Expansion in soil formations and rock masses occurs in several forms including swelling, frost heaving, and extension straining. Collapse in soils can be considered as a deformation problem, whereas in rocks it is more closely associated with rupture.

Swelling

Occurrence and Characteristics

Clay soils and certain minerals readily undergo volume change by swelling when wetted or shrinking when dried. When in the dry state, or when less than fully saturated, some clays have a tremendous affinity for moisture, and in some cases may swell to increase their volume by 30% or more. Pressures in excess of 8 tsf can be generated by some swelling materials when they are confined. Once a material is permitted to swell, however, the pressures reduce. There is a time delay in the phenomenon. Noticeable swell may not occur for over a year after completion of construction, depending upon the soil's access to moisture, but may continue for 5 years or longer. In rock masses, swelling occurs in marine and clay shales, which commonly contain montmorillonite, the highly expansive clay mineral; in pyritic shales; and in gneiss and other metamorphic rocks containing seams of montmorillonite. Heave from swelling soils and rocks is covered in some detail in Art. 10.6 of Hunt (1984).[5]

Predicting Expansion Potential

The phenomenon of absorption and swell is complex and not well understood, but it appears to be basically physicochemical in origin. Swell potential is related to the percentage of the material in the clay fraction (less than 2 μm), the fineness of the clay fraction, the clay structure, and the type of clay mineral. Montmorillonite has the highest potential for swell, followed by illite, with kaolinite being the least active. Thus, mineral identification is one means of investigating swell potential.

Swell potential has been related by Skempton (1953)[49] to *activity*, defined as the ratio of the plasticity index to the percent finer by weight than 2 μm. The activity of various types of clay minerals as a function of the plasticity index and the clay fraction is given in Fig. 4.34. As a result of correlations between index tests and swell tests performed in the laboratory on compacted samples of various types of clays, relationships have been developed between index tests and potential expansiveness such as given in Fig. 4.34. A similar relationship developed by the USBR (1974)[53] is given in Table 4.12, in which the estimates of probable expansion are for very light loads, such as those exerted by canal linings.

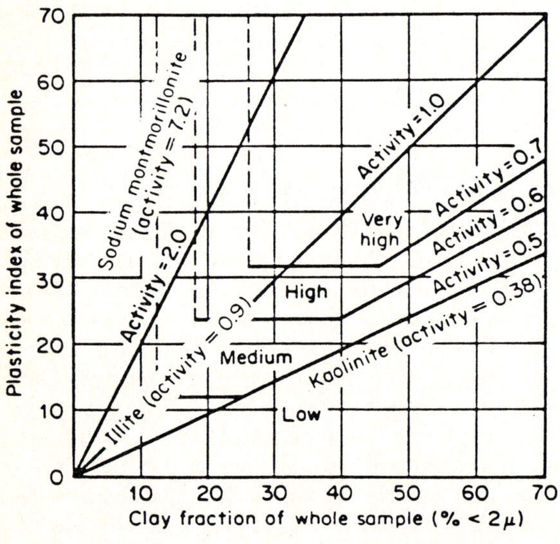

Potential expansiveness	Inch per foot of soil
Very high	1.0
High	0.5
Medium	0.25
Low	0

FIG. 4.34 Proposed modified chart for determining expansiveness of soils and activity of typical clay minerals. [After Williams and Donaldson (1980)[51]; van der Merwe (1975).[52]]

Index tests have limitations, and interpretations should be used only as guidelines. The tests are made on remolded specimens; remolding destroys the natural structure and modifies swelling characteristics.

Laboratory consolidation tests on undisturbed specimens provide an indication of potential heave and the swelling pressures that may develop. These tests are performed with three different procedures:

1. *Percent swell measurements:* Specimen is placed in the consolidometer at natural moisture content and protected against moisture loss. Loads are added in the same manner as in the normal consolidation test, although initial loads may be higher, and compression is measured and recorded until the final foundation design load is reached. At design load the specimen is immersed in water, saturation is permitted, and the volume increase is measured as a function of time until movement ceases.
2. *Maximum swell pressure:* Specimen is placed in the consolidometer under an initial seating load and immersed in water; then, loads are added as necessary to prevent swelling and to determine the maximum swelling pressure.
3. *Swell versus pressure:* Specimen is placed in the consolidometer at natural moisture content, loaded in the normal manner to some preselected load, and allowed to come to equilibrium. The sample is immersed and the increase in height observed until equilibrium is reached. The vertical pressure is then reduced by a factor of 2, the swell again observed, and the procedure repeated until the vertical pressure is reduced to zero. A curve is plotted of swelling pressure versus percent expansion.

The reliability of predictions based on consolidation-swell tests depends on the degree to which the initial moisture content of the test specimen agrees with the actual moisture con-

TABLE 4.12
RELATION OF SOIL INDEX PROPERTIES TO PROBABLE VOLUME CHANGES FOR HIGHLY PLASTIC SOILS*

Data from index tests†			Estimation of probable expansion,‡ percent probable total volume change (dry to saturated condition)	Degree of expansion
Colloid content, % < 0.001 mm	Plasticity index	Shrinkage limit, %		
>28	>35	<11	>30	Very high
20–31	25–41	7–12	20–30	High
13–23	15–28	10–16	10–20	Medium
<15	<18	>15	<10	Low

*From USBR (1974).[53]

†All three index tests should be considered in estimating expansive properties.

‡Based on a vertical loading of 1.0 psi as for concrete canal lining. For higher loadings the amount of expansion is reduced, depending on the load and clay characteristics.

tent at the start of construction, and the extent to which the final moisture content in the field beneath the structure approaches that attained by the specimen during testing. In general, it can be expected that the laboratory tests represent maximum conditions.

Estimating Magnitudes and Time Rates

Magnitudes of maximum swelling pressures and amount of heave beneath a structure may be estimated from tests performed in the consolidometer on representative, undisturbed samples. Whether or not the pressures and volume changes actually will develop depends upon many factors which cannot be well defined. Time rates cannot be evaluated analytically, and evaluations must be based on experience. Estimates of swell magnitudes are also made with the soil suction approach. The practical approach, in general, is to recognize that the problem exists and to develop conservative designs to mitigate or avoid the problem.

The *soil suction* approach to estimating expansion magnitude has been receiving substantial attention in recent years [Snethen (1980),[54] Krazynski et al. (1980)[55]]. Soil moisture suction is a macroscopic property of soil which indicates the intensity with which a soil attracts water. Total soil suction refers to the sum of what are termed the "matrix" and "osmotic" components. Matrix suction results from the surface attractive forces for water and cations and the surface tension effects of water in soil, and is dependent on both water content and surcharge pressure. Osmotic suction arises from the presence of soluble salts in the pore water and is identical with osmotic attractive forces. Mechanisms that cause soil volume to change by shrinking or swelling use energy, and changes is soil suction reflect the amount of energy in the pore water used by the mechanisms during volume change.

The change in free energy due to movement of free pure water from cracks and fissures into the pore water has been expressed in terms of an equivalent total soil suction as related to relative humidity. *Thermocouple psychometers* are used to measure the relative humidity in the soil in situ and in undisturbed samples in the laboratory. *Relative humidity* is the ratio between the vapor pressure of the pore water in the soil (atm) and the vapor pressure of free pure water (atm). A semilog plot (Fig. 4.35) of soil suction versus natural water content for a given sample represents its matrix soil suction. The osmotic soil suction is the value at which soil suction becomes stable with increasing water content. From Fig. 4.35 the matrix soil suction without surcharge pressure τ_m^o is expressed as

$$\log \tau_m^o = A - Bw \qquad (4.50)$$

where τ_m^o = matrix soil suction at atmospheric pressure

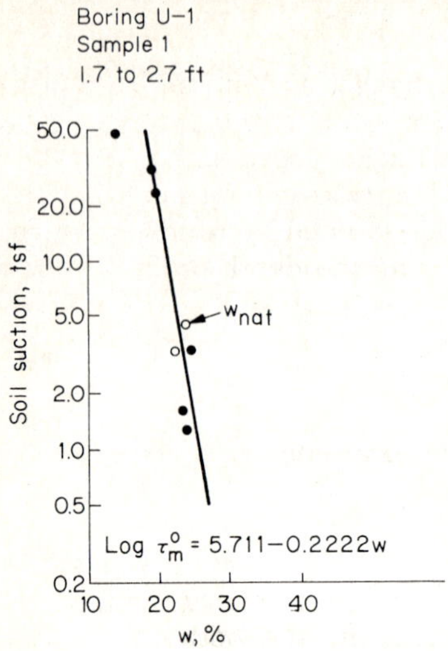

FIG. 4.35 Soil suction versus water content. [From Snethen (1980).[54]]

A = y intercept (a constant), calculated from Eq. 4.50 for soil suction = 1 tsf (95.8 kN/m²)

B = slope of soil suction versus w curve, calculated as the inverse change in w over one cycle of the soil suction scale

w = water content, %

Heave is predicted for the expansive clay stratum from the expression:

$$\frac{\Delta H}{H} = \frac{C_\tau}{1 + e_0}[(A - B\,w_0) - \log(\tau_{mf} - \alpha\sigma_f)] \qquad (4.51)$$

where H = stratum thickness, ft

C_τ = suction index, $\alpha G_s/100B$

α = compressibility factor (PI < 5, $\alpha = 0$; PI = 5 to 40, $\alpha = 0.0275$ PI − 0.125; PI > 40, $\alpha = 1$)

e_0 = initial void ratio

w_0 = initial moisture content, %

τ_{mf} = final matrix suction, tsf

σ_f = final applied pressure ($p'_o - \Delta p$)

Note: The final matrix soil pressure and the final applied pressure are functions of the assumed depth of the active zone.

The *depth of active zone*, the depth within which a soil moisture deficiency exists, is a transient value influenced by soil type, soil structure, topography, and climate. No universally applicable rules exist for establishing it, but it is usually greater than the depth of seasonal moisture variation because of periodic long dry spells. Snethen (1980)[54] suggests setting the depth of the active zone at either the upper boundary of constant levels of soil suction (Fig. 4.36*a* and *b*), or below the first major change in soil suction (high to low, or low to high) in S- or Z-shaped profiles (Fig. 4.36*c* and *d*). A number of assumptions can be made in estimating the final matrix soil suction (future equilibrium moisture conditions). Snethen (1980)[54] suggests that the most reasonable values are obtained from using the saturated water content in Eq. 4.51, then using the results in Eq. 4.50. In either case experience is still limited and caution is required when applying the results.

The use of thermocouple psychometers reportedly is a relatively simple, inexpensive, and ac-

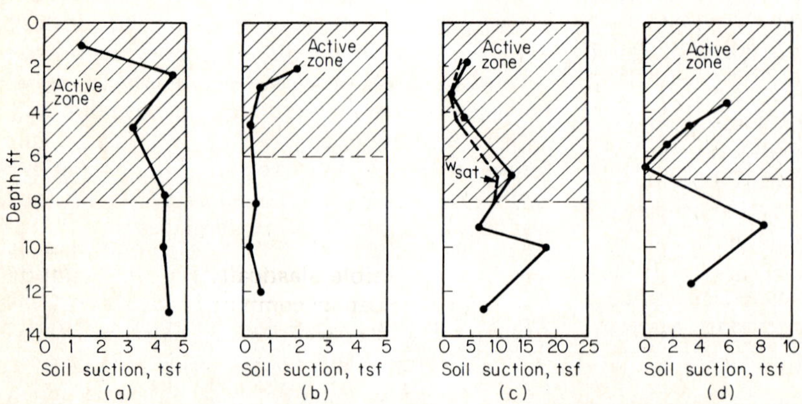

FIG. 4.36 Interpretation of active zone from plots of soil suction versus depth: (*a*) Jackson, Miss.; (*b*) Lake Charles, La.; (*c*) Hayes, Kan.; and (*d*) Mewcastle, Wyo. [After Snethen (1980).[54]]

curate method of measuring soil suction and providing a basis for evaluating soil expansivity. In some geographical areas, they are being installed to measure long-term changes in soil suction and thereby obtain information to establish the active zone limits.

Frost Heave

Occurrence and Characteristics

In climates with temperatures that remain below freezing for some period of time, the moisture in the soil interstices freezes and thereby increases its volume, resulting in surface heave. Upon thawing there is an attendant volume decrease and subsequent subsidence. Freezing depth in the ground is a function of the freezing time interval, the mean seasonal or annual temperature, the depth to the water table, and soil characteristics [see Art. 7.7.3 of Hunt (1984)[5]].

As water crystallizes upon freezing, its volume expands approximately 9%. This expansion has little effect on clean sands and gravels, even when saturated, and results in little ground heave. It is the development of *ice lenses* in fine-grained soils that results in significant ground heave which is extremely irregular in magnitude. Ice lens growth, related to soil capillarity, requires that a source of water, such as the groundwater table, be sufficiently close to the freezing zone that it can be drawn into it. All factors causing moisture migration through frozen soil are not fully understood, but in general it is attributed to a thermal gradient where capillary flow is from higher to lower temperatures, and to osmotic flow which is related to ion concentrations.

Soil susceptibility to frost heave and ice lenses depends on its gradation. In uniform soils ice lenses do not develop unless the grains are smaller than 0.01 mm; fairly uniform soils must contain at least 10% of grains smaller than 0.02 mm. Ice lenses are not formed in the field in soils with less than 1% of the grains smaller than 0.02 mm [Terzaghi and Peck (1967)[26]]. Silty soils, therefore, with high rates and distances of capillary rise, have the highest potential for frost heave, as shown in Table 2.6. In pure clays the rate of capillary rise is much less and the susceptibility, therefore, is lower.

Prediction

Volume changes and pressures from frost heave are seldom evaluated in practice. Predictions of the potential are based primarily on climatic factors and secondarily on geologic conditions. The problem is avoided by prudent design based on charts giving the depth of frost penetration for given locales, such as Fig. 6.15 for the United States.

Extension Straining

Occurrence and Characteristics

Extension straining occurs below the bottoms of excavations in soils upon removal of overburden, and below the bottoms and outside the walls of excavations in rock masses containing high residual stresses. In clays in particular, the resulting heave can be substantial. In rock masses, bottom heave or wall deflections depend on the degree of residual stress, rock type, and the orientation of the joints with respect to the excavated wall or bottom face. There is a time delay associated with the phenomenon, particularly in rocks, and expansion can continue over a period of many months or longer. Residual stresses are discussed in Art. 6.6 of Hunt (1984).[5]

Predicting Potential

The potential for heave from tensile straining in soils may be evaluated by laboratory tests. Prediction is based on soil types anticipated in excavation; heave is negligible in sands (except where below the water table and subjected to uplift pressures), but can be significant in soft clays and highly overconsolidated clays.

In rock masses, residual stresses are measured in situ with the flat-jack test, or with borehole overcoring techniques employing strain gages, deformation meters, or other techniques [Art. 4.4.4 in Hunt (1984)[5]].

Estimating Magnitudes

Soft clays deform at constant water content like an incompressible elastically isotropic material; hence, heave can be computed by means of elastic theory and use of a soil initial tangent modulus [Terzaghi and Peck (1967)[26]]. To obtain a representative value for E_i, samples are tested under undrained conditions in the triaxial device under a cell pressure approximately equal to the effective overburden pressure. The axial load is alternately increased moderately and then reduced to zero several times, and the value of E_i is taken as the initial slope of the stress-strain curve under the last load increase. These procedures are followed to avoid grossly overestimating the heave.

Heave in *overconsolidated clays* may be evaluated from the consolidation test by analyzing the expansion that occurs during unloading cycles. The unload curve is reconstructed between the existing overburden pressure and the pressure from excavation unloading and the corresponding void ratios obtained. Heave is calculated from Eq. 4.41.

Extension straining in rock masses may be evaluated with the finite element method as discussed in Art. 9.35.

Collapsing Soils

Susceptible Soils

Soils susceptible to collapse are characterized by a temporary internal support system consisting of cementing or other bonding agents including iron oxide, calcium carbonate, clay-welding of grains, clay bonds, clay bridges, or silt bonds. Collapse, or a sudden decrease in the volume of the voids, occurs as a result of wetting which leaches out cementing agents or softens clay bonds and bridges. *Increasing water content* under an applied load can result in gradual settlement, or in a sudden collapse as the soil bonds are weakened. An applied load of *critical magnitude* can cause a sudden collapse of the soil structure when the bonds break in a brittle type of failure, even at natural moisture content.

Susceptible soils are loess, valley alluvium in a low-rainfall climate, and certain residual soils such as "porous" clays. The collapse phenomenon and the characteristics of susceptible soils are described in detail in Arts. 7.5.3 and 10.5 of Hunt (1984).[5]

Predicting Collapse and Settlement Potential

Collapsing soils include the classes SM, SC, ML, and CL, and characteristically have low natural densities, in the order of 1.4 g/cm^3 (87 pcf) or less, and high strengths and low compressibility at low natural moisture contents. Natural density measurements, therefore, can be used to judge settlement potential from relationships such as given in Table 4.13.

Collapse potential is evaluated from field and laboratory test data for four possible conditions: either gradual compression or collapse at some

TABLE 4.13
SETTLEMENT UPON SATURATION VS. NATURAL DENSITY: LOESSIAL SOILS FROM KANSAS AND NEBRASKA*

	Density			
D_R	pcf	g/cm^3	Settlement potential	Surface loading
Loose	<80	<1.28	Highly susceptible	Little or none
Medium dense	80–90	1.28–1.44	Moderately susceptible	Loaded
Dense	>90	>1.44	Slight, provides capable support	Ordinary structures

*From USBR (1974).[53]

NOTES 1. For earth dams and high canal embankments, σ = 85 pcf (1.36 g/cm^3) has been used as the division between high-density loess requiring no foundation treatment and low-density loess requiring treatment.

2. Moisture contents above 20% will generally result in full settlement under load.

critical load under natural moisture conditions, and either gradual compression or collapse under "wetted" conditions. The field plate-load test is useful for evaluating the compression and the *critical load* at which collapse occurs at natural moisture, and also under wetted conditions for which additional tests are necessary. In the normal laboratory consolidation test, loads are applied to the specimen maintained at its natural water content, and at the proposed foundation stress the specimen is wetted to determine the collapse potential. The effects of saturation on the *e*-log-*p* curve of a porous clay from Brazil are given in Fig. 4.37. Typical consolidation curves for Missouri River Basin loess, tested at natural water content and after prewetting, are given in Fig. 4.38.

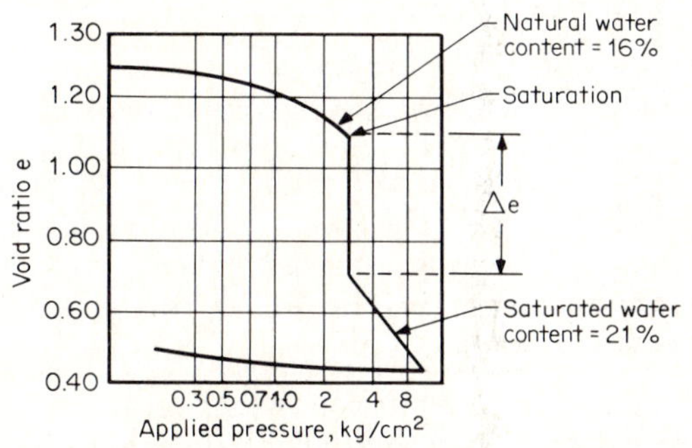

FIG. 4.37 Effect of saturation on the pressure versus void ratio curve of a porous clay from Brazil. [From Vargas (1972).[56]]

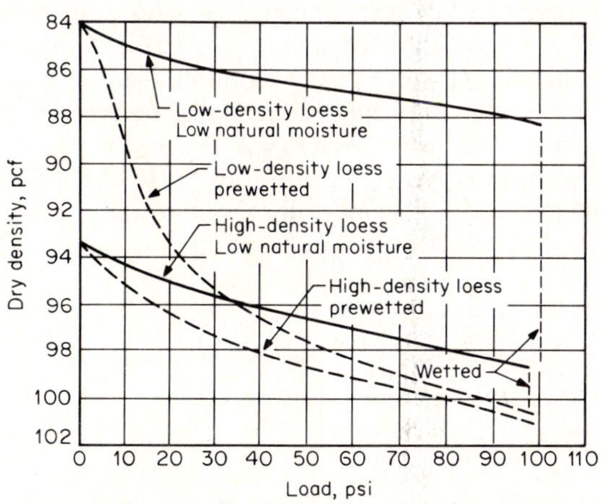

FIG. 4.38 Typical consolidation curves for Missouri River basin loess. [From Clevenger (1958).[57]]

The *double consolidation test* has been proposed to provide data for the calculation of estimates of collapse magnitude [Jennings and Knight (1975),[58] Clemence and Finbarr (1981)[59]]. Two specimens of similar materials (preferably from block samples) are trimmed into consolidometer rings and placed in the apparatus under light [1 kPa (0.01 tsf)] seating loads for 24 hr, after which one specimen is submerged and the other kept at its natural water content for an additional 24 hr. Loads are then applied on each specimen in the normal manner.

The *e*-log-*p* curve for each test is plotted on the same graph, along with the overburden pressure p_o and the preconsolidation pressure p_c for the saturated specimen. The curves for a normally consolidated soil ($p_c/\overline{p}_o$ = 0.8 to 1.5, Art. 3.2.3) are given in Fig. 4.39, and for an overconsolidated soil in Fig. 4.40 ($p_o > 1.5$). The curve for the natural water content condition is relocated to the point (e_0, p_0) given in the figures. For an increase in foundation stress Δp on wetted soil, the settlement ρ is estimated from the expression

$$\rho = \frac{\Delta e_s}{1 + e_0} + \frac{\Delta e_c}{1 + e_0} \tag{4.52}$$

Collapse in Rock Masses

In rock masses, collapse occurs as a result of roof-arch rupture in cavernous, soluble rock, or as a result of the rupture of pillars or roof spans in underground mines, as discussed in Art. 6.3.3. The effect on the surface can range from subsidence to the opening of a large void. Evaluations are based on strength considerations.

4.4 SOIL-STRUCTURE INTERACTION

4.4.1 BASIC CONCEPTS

Applications

Evaluation of soil-structure interaction problems is based on the concept that the behavior of a structure in full contact with soil and the behavior of the soil are interdependent.

In soil mechanics this concept is most important where the contact pressure between a relatively flexible structure and the supporting soils must be evaluated in order to determine the deflection

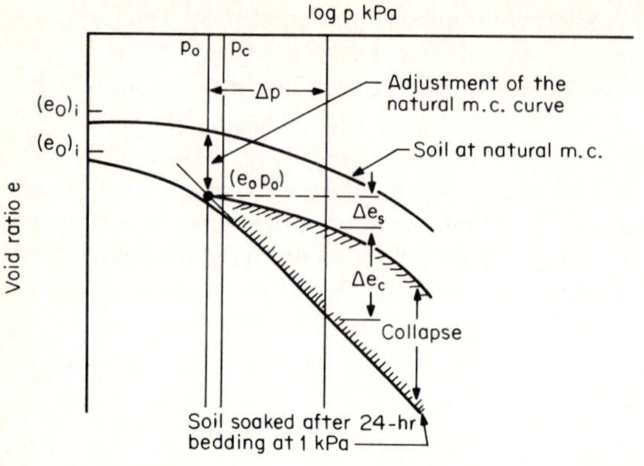

FIG. 4.39 Double consolidation test *e*-log-*p* curves and adjustments for a normally consolidated soil. [From Clemence and Finbarr (1981).[59]]

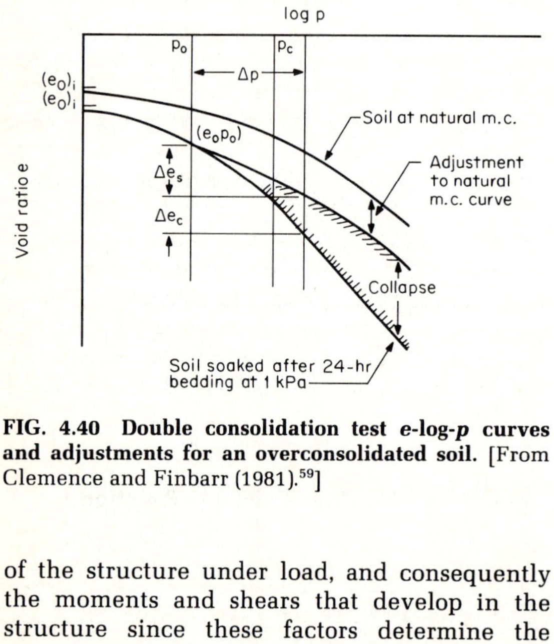

FIG. 4.40 Double consolidation test *e*-log-*p* curves and adjustments for an overconsolidated soil. [From Clemence and Finbarr (1981).[59]]

of the structure under load, and consequently the moments and shears that develop in the structure since these factors determine the strength required in terms of steel and concrete areas. The major types of problems requiring consideration of soil-structure interaction are

- Semi-rigid mat foundations and long continuous footings
- Deep foundations subjected to lateral loads
- Flexible retaining structures such as thin concrete walls
- Pavements
- Foundations subjected to earthquake loadings

Analytical methods used to solve soil-structure interaction problems have been undergoing substantial changes in recent years with the increasing applications of computer technology. One general computer program is called SOIL-STRUCT.

This article is provided as a brief introduction to the soil-structure interaction problem to give a general idea of the basic concepts for the solutions of various types of problems, as well as the soil parameters required for the solutions.

Beam- or Plate-on-an-Elastic-Subgrade Problem

Approaches

Methods for determining shears and bending moments in flexible to semirigid structures are based on consideration of the flexibility of the structure and the deformation characteristics of the adjoining soil. The basis for solutions to these problems is the *beam- or plate-on-an-elastic-subgrade theory*. Historically, there are two approaches to the problem:

1. The *Winkler foundation*, which treats the soil mass supporting the foundation as a series of springs, and
2. Treatments of the foundation bed as an elastic solid.

Mathematical solutions can be obtained in either instance. Hetenyi (1946)[60] devoted an entire text to solving the differential equations and considered various boundary conditions for the Winkler foundation. In recent years, the more common approaches include the *finite difference method* using the Winkler foundation concept and the *finite element method*.

Structural Relationships

The relationship between soil reaction, and moment, shear, and deflection of a beam, obtained from mechanics of materials, is illustrated in Fig. 4.41.

The beam, subjected to load p (Fig. 4.41*a*), undergoes deflection *y* (Fig. 4.41*b*), which results in

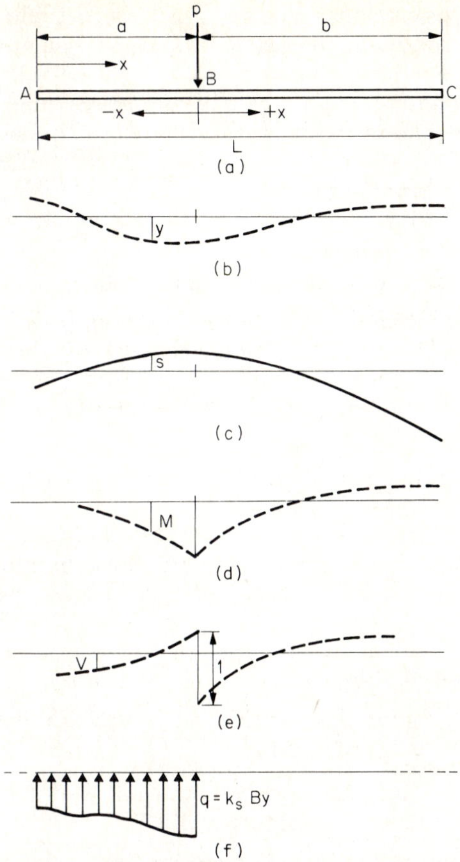

FIG. 4.41 Theoretical solution of the beam-on-an-elastic foundation: (*a*) infinite or finite beam and alternate origins; (*b*) deflection; (*c*) slope; (*d*) moment; (*e*) shear; and (*f*) probably pressure assuming transverse deflection (across width *B*). [From Bowles (1974).[37] Used with permission of McGraw-Hill Book Company.]

slope S (Fig. 4.41c), moment M (Fig. 4.45*d*), and shear V (Fig. 4.41e), defined as follows:

$$\text{Slope,} \qquad S = EI\frac{dy}{dx} \qquad (4.53)$$

$$\text{Moment,} \qquad M = EI\frac{d^2y}{dx^2} \qquad (4.54)$$

$$\text{Shear,} \qquad V = EI\frac{d^3y}{dx^3} \qquad (4.55)$$

where x = coordinates along the beam
y = deflection at x
E = modulus of elasticity of beam
I = moment of inertia of beam

Fig. 4.41*f* illustrates the probable pressure assuming constant transverse deflection (across width *B*). *Soil reaction q* at any given point on the elastic foundation is given by

$$q = \frac{d}{dx}V = EI\frac{d^4y}{dx^4} \qquad (4.56)$$

$$\text{or} \qquad q = k_sB\,y \qquad \text{or} \qquad y = q/k_sB \qquad (4.57)$$

where k_s = modulus of subgrade reaction.

Soil Reaction

The basic assumption of the Winkler foundation is that the soil acts as a medium consisting of "springs," and the deflection at a point is influenced only by the soil reaction at the point. The ratio between pressure and deflection at the point can be expressed as the *modulus or coefficient of subgrade reaction* or *soil spring constant* k_s. In the application of this hypothesis to continuous footings, mat foundations, laterally loaded deep foundations, or pavements, it is assumed that the soil behaves like an infinite number of coil springs, each of which is not affected by the others, as shown in Fig. 4.42. If a mat is pile-supported, each pile has a spring constant of *EA/L* (Fig. 4.43).

Coefficient of Vertical Subgrade Reaction k_s

Definition

The modulus of subgrade reaction k_s is defined as the unit pressure required to produce a unit of deflection, given as

$$k_s = p/y \text{ force/length}^3, \text{ or lb/in}^3, \text{kg/cm}^3, \text{etc.} \qquad (4.58)$$

where p = pressure per unit area between the contact surface of the loaded area and the supporting subgrade (unit soil reaction of force per unit length)
y = deflection produced by the load

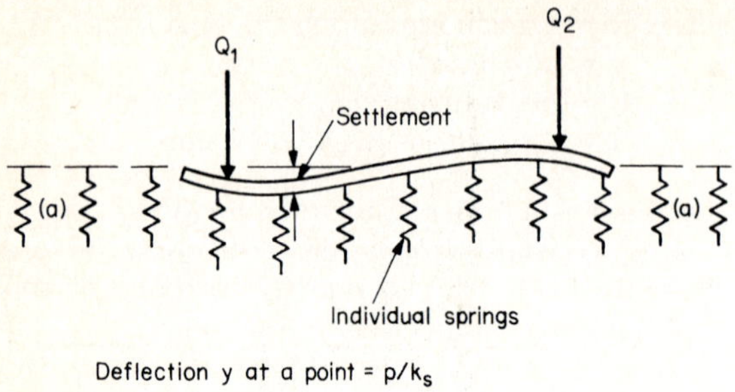

FIG. 4.42 Simplified expression of a beam-on-an-elastic foundation (the Winkler hypothesis). Ground outside footing at (*a*) is not affected.

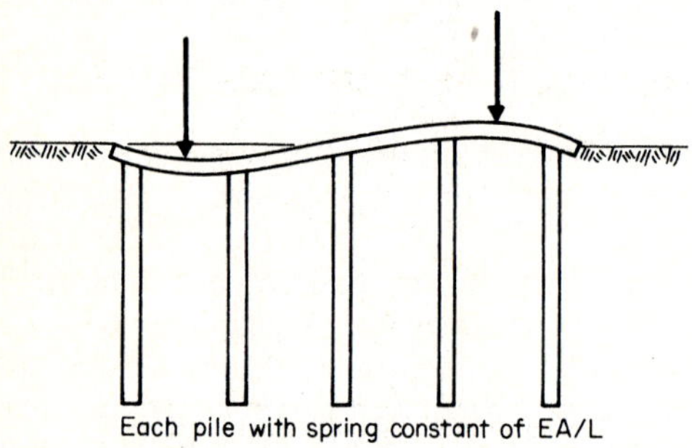

FIG. 4.43 Pile-supported mat.

The value of k_s depends on the elastic properties of the subgrade and the dimensions of the loaded area; it is obtained from k_t, the modulus from testing, by applying various relationships. The horizontal modulus of subgrade reaction is described in Art. 4.4.3.

Measurements of k_t

Values are obtained from plate-load tests, or extrapolated from laboratory triaxial, consolidation, or CBR tests [see Bowles (1975)[61]]. The plate-load test, the most common method, usually is performed on either 12- or 30-in-dia. plates, with smaller plates stacked on top to provide rigidity. Load is applied by jacking hydraulically against a reaction and a deformation curve is obtained for a series of loads. The test modulus k_t is computed from the slope of the load-deformation curve prior to the yield point [see Art. 3.4.4 in Hunt (1984)[5]].

Limiting values for sands are given in Table 4.14 and for overconsolidated clays in Table 4.15. These values, obtained by Terzaghi (1955),[62] are for one-foot-square plates, or beams one foot wide, and must be modified by various relationships to account for footing shape and dimensions, since larger areas stress greater depths. Values for k_t are also given in Fig. 4.44, which includes values for the constant of horizontal subgrade reaction, n_h *(see Art. 4.4.3)*.

Other relationships have been developed for clays. Broms (1964)[63] analyzed the results of a plate-load test on clay and presented k_t in terms of the *secant modulus* at 50% of the failure load (E_{50}) and $\nu = 0.5$ as

$$k_t \approx 1.67\ E_{50} \qquad (4.59)$$

Skempton gives an approximate relationship for clays at E_{50} as $E_{50} = 25 - 100U_c$, and therefore

$$k_t \approx 40 - 160U_c \qquad (4.60)$$

Design Modulus k_s

Many factors influence the value of the design modulus k_s, including

- Size of footing—For the same soil, k_s decreases with increasing footing width *B*
- Shape of footing—For footings with the same width *B* and same uniform load *p* and supported on the same soil, the value k_s decreases with increasing footing length
- Depth of footing—The modulus for sand increases with depth, whereas that for clays is more or less independent of depth
- Location below the water table substantially reduces k_s values
- Stressed zone—For validity, the material in the tested zone must have the same deformation characteristics as the material in the foundation stressed zone, which is only likely under very uniform soil conditions.

Because of the influencing factors, a number of relationships have been developed from which values for k_s are obtained. As a general approach, Vesić (1961)[64] analyzed an infinite horizontal beam on an elastic foundation and showed that

TABLE 4.14
VALUES OF VERTICAL MODULUS OF SUBGRADE REACTION (k_t) FOR PLATES ON SAND*

Relative density of sand	Loose†	Medium†	Dense†
Dry or moist sand, limiting values for k_t	20–60	60–300	300–1000
Dry or moist sand, proposed values for k_t	40	130	500
Submerged sand, proposed values for k_t	25	80	300

*From Terzaghi (1955).[62]

†k_t in ton/ft³ for square plates, 1 × 1 ft, or beams 1 ft wide (1 ton/ft³ = 0.032 kg/cm³ = 32 tonne/m³).

TABLE 4.15
VALUES OF VERTICAL MODULUS OF SUBGRADE REACTION (k_t) FOR PLATES ON PRECONSOLIDATED CLAY*

Consistency of clay	Stiff†	Very stiff†	Hard†
Values of q_u, ton/ft²	1–2	2–4	>4
Ranges for k_t, square plates	50–100	100–200	>200
Proposed values, square plates	75	150	300

*From Terzaghi (1955).[62]

†k_t in ton/ft³ for square plates 1 × 1 ft, or long strips 1 ft wide. (1 t/ft³ = 1.15 lbs/in³).

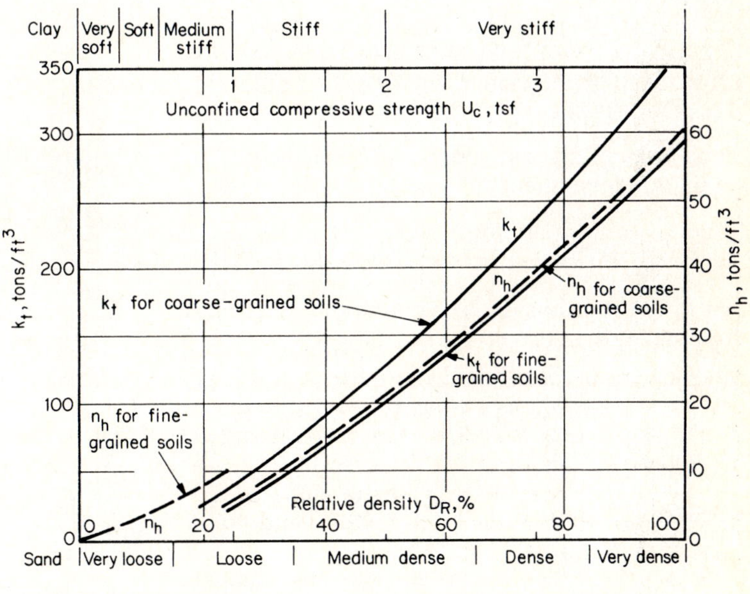

FIG. 4.44 Approximate relationship between moduli of subgrade reaction k_t and n_h as a function of relative density and consistency. [From NAVFAC (1982).[23]]

k_s could be obtained from laboratory triaxial test values for E_s as follows:

$$k_s = \frac{0.65}{B} \sqrt[12]{\left(\frac{E_s B^4}{E_b I}\right)} \left(\frac{E_s}{1 - \nu_s} 2\right) \quad (4.61)$$

where $E_b I$ = beam stiffness
E_s = soil modulus
γ_s = Poisson's ratio, soil
B = beam width

Sands

Terzaghi (1955)[62] proposed the following expression for spread or continuous footings with width B or a mat acted upon by line loads with line spacings at B:

$$k_s = k_t \left(\frac{B + 1}{2B}\right)^2 \quad (4.62)$$

valid for $p < q_u/2$.

He considered the expression valid for spread and continuous footings, for beams with contact pressures smaller than one-half the ultimate bearing capacity q_u of the subgrade per unit of area of the base of the supporting structures, and for slabs and mats supporting concentrated loads, if none of the concentrated loads is greater than one-half of the ultimate bearing capacity of an equivalent circular footing of radius R (*see Art. 6.2.4*). For spread footings in the usual range of 2 to 3 m, however, Eq. 4.62 may give values as much as 100% too small for dense to medium dense sands [Bowles (1975)[62]].

Teng (1962)[65] gives a general expression to account for embedment and footing size as follows:

$$k_s = k_t \left(\frac{B + 1}{2B}\right)^2 \left(1 + \frac{2D}{B}\right) \quad (4.63)$$

but k_s is not to exceed *twice* the value obtained from Eq. 4.62.

For immediate settlement ρ, with the modulus of elasticity increasing linearly with depth, NAVFAC (1982)[23] gives the following relationships:

- Shallow footings, $D \leq B$, for $B \leq 20$ ft,

$$\rho = \frac{4q\,B^2}{k_t(B + 1)^2} \quad \text{see also Eq. 4.34} \quad (4.64)$$

- Shallow footings, $D \leq B$, for $B \geq 40$ ft,

$$\rho = \frac{2q\,B^2}{k_t(B + 1)^2} \quad (4.65)$$

- Deep foundation, $D \geq 4B$, for $B \geq 20$ ft, use Eq. 4.65.

Interpolate for intermediate values of B for shallow footings, and for continuous footings multiply the settlement computed for width B by 2. Equations 4.64 and 4.65 are for GWL at least 1.5 B below footing base; if GWL is at footing base use $k_t/2$ in computing settlement.

Stiff, Overconsolidated Clay

For contact pressures less than one-half the ultimate bearing value, Terzaghi (1955)[62] proposed for *square footings:*

$$k_s = k_t/B \quad (4.66)$$

or for long beams, *continuous footings*, or *mats* (B is the distance between line loads represented by column points), where L can be assumed equal to infinity:

$$k_s = 0.67\,k_t/B \quad (4.67)$$

For *rectangular areas* with width B and length mB, Terzaghi proposed a decreasing value for k_s given by:

$$k_r = k_s \left(\frac{m + 0.5}{1.5m}\right) \quad (4.68)$$

4.4.2 CONTINUOUS FOOTINGS AND MAT FOUNDATIONS

Validity of Subgrade Theory

The general assumptions upon which the subgrade theory is based are:

1. Value for k_s is independent of the pressure, and
2. k_s has the same value for every point of the surface acted on by the contact pressure.

Errors due to assumption 1 are relatively small, providing that p is less than about one-half the ultimate bearing value. Far greater inaccuracies result from assumption 2 which implies that the deformation characteristics of the materials beneath the rigid plate will be those of the stressed zone beneath the foundation (*see Art. 4.2.2*). If the material within the zone stressed by the foundation is uniform or becomes less deformable with increasing depth, then the approach can be considered valid. If weaker, more compressible soils than those tested are present beneath the foundation, however, then design based on a subgrade modulus is not appropriate unless a settlement analysis indicates the foundation deflections will not be excessive. A major error can also be introduced by the use of the relations between k_s and k_t, which often are quite inaccurate.

Continuous Footings

Elastic Line Method

Soil reaction q at a given point along a beam is expressed by Eq. 4.57, and from the basic assumption of the Winkler foundation it may also be expressed by Eq. 4.56; therefore

$$EI\frac{d^4y}{dx^4} = -yBk_s \tag{4.69}$$

Equation 4.69 is used to analyze continuous or combined footings. The general solution is given by Hetenyi (1946).[54] A large number of mathematical manipulations are required to arrive at deflections along the length of the foundation, as well as the moments and shears. Discussions are given in Teng (1962),[65] (1975).[61]

Finite Difference Method

The finite difference method basically replaces the differential equations given by Eq. 4.53 (slope), Eq. 4.54 (moment), Eq. 4.55 (shear), and Eq. 4.69 (soil reaction). The footing is treated as a flexible member consisting of sections, usually of equal length h. Instead of being supported by a continuous soil contact, each section is supported by equivalent concentrated reactions at panel points, in the form of the Winkler foundation.

The solution is obtained by successive differentiation of the average slope $\Delta y/\Delta x$ at each panel point, which assumes a *finite value* as opposed to dy/dx, which approaches a limit as $dx \to 0$ in the elastic line method. The deflection y is established in terms of derivatives $\Delta x/\Delta y$, and moments and shears are determined from the relationships of mechanics of materials [see Bowles (1975)[61]]. The method requires the solution of 10 or more simultaneous equations, thus necessitating the use of a computer. The finite element method is somewhat similar to the finite difference method.

Mat Foundations

Approaches

When less than rigid, mats can be analyzed by

1. Theory of plates on elastic foundations [Hentenyi (1946)[60]]
2. Finite difference method
3. Finite element method

Elastic Plate Method

The mat is evaluated as a plate and the effect of a column load on the surrounding area is considered. The effect of concentrated loads on a mat has been found to dampen out rapidly, so by superimposing all of the column loads within the area of influence, the total effect of the column loads at any point is determined. Since the effect of each load is transmitted through the mat in a radial direction, the use of polar coordinates is convenient.

The moments and deflections at any point are calculated from expressions for the radial moment M_r, the tangential moment M_t, and the deflection w [see Teng (1975)[66] for expressions]. These expressions relate column load P, distance of the point under investigation from the column load along the radius r, and functions for shear, moment, and deflections in terms of r/L, where L is the *radius of effective stiffness* given by

$$L = \sqrt[4]{\frac{D}{k_r}} \tag{4.70}$$

where k_r = coefficient of subgrade reaction adjusted for mat size, and

D = flexural rigidity of the mat expressed in terms of its properties E and ν, and thickness

Finite Difference Method

The differential equation for the deflection of a mat is

$$\Delta^4 w = \frac{q - k_s w}{D} = \frac{\delta^4 w}{\delta_x^4} - 2\frac{\delta^4 w}{\delta_x^2 \delta_y^2} - \frac{\delta^4 w}{\delta^4 y} \quad (4.71)$$

where w = deflection
q = subgrade reaction per unit area of mat
k_s = modulus of subgrade reaction
D = mat rigidity

To solve the differential equation by finite differences, the mat is divided into square areas ($h \times h$) and an equation is written for each point of intersection in terms of deflection. When these simultaneous equations are solved by computer, the deflections at all points are determined rapidly. The method is based on the assumption that the Winkler foundation (i.e., a bed of uniformly distributed elastic springs) can be substituted for the subgrade.

Finite Element Method

The problem of plates on an elastic foundation has been transformed into a problem of matrix structural analysis by the finite element method. In the procedure developed by Haddadin (1971)[67] the mat is idealized as a mesh of finite elements interconnected only at the corners of the elements (nodes) and the soil may be modeled as a set of isolated springs, or as an elastic isotropic halfspace. The same mesh applies to the soil, and its stiffness (represented by elastic springs) is concentrated at the nodal points as shown on Fig. 4.45*a*.

Three degrees of freedom exist at each node (Fig. 4.45*b*): a vertical deflection, a rotation about the x axis, and a rotation about the y axis. An external load can exist in the direction of each of these deformations. The internal stress resultants (moments and shears), as shown in Fig. 4.45*c*, are related to the degrees of freedom of the element by a stress matrix derived from the finite element deflection function.

Mat stiffness is evaluated by adding the stiffness of the elements adjacent to each node. The soil is divided into a mesh identical with that of the mat, and it is assumed that the rectangular areas surrounding each node, defined by the centerlines of adjacent elements, undergo uniform deflection. The soil, therefore, is idealized as a set of isolated springs capable of resisting only compression. Mat or plate stiffness is given by

$$K_{ij} = \Sigma k_{ij} \quad (4.72)$$

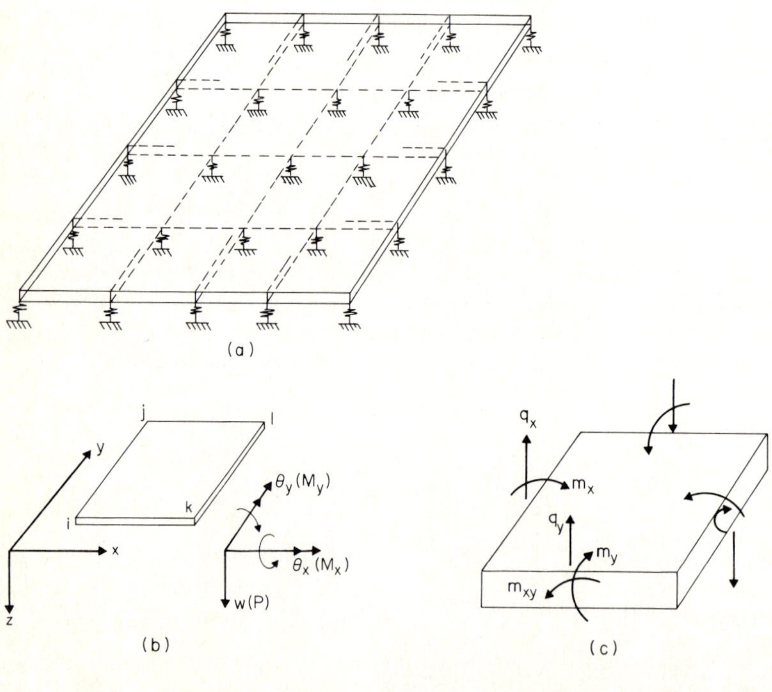

FIG. 4.45 Mat foundation—finite element approach to analysis: (*a*) structural idealization of mat and supporting soil by finite elements; (*b*) three degrees of freedom at the nodal point of a rectangular plate element, forces and corresponding displacements; and (*c*) internal forces of an element. [From Haddadin (1971).[67]]

where K_{ij} = the stiffness matrix of the plate and k_{ij} = stiffness matrix of the element. The *soil stiffness* is given by

$$K_{fi} = \sum_{e} \frac{A_e}{4} k_{fe} \tag{4.73}$$

where K_{fi} = soil stiffness at node i
A_s = area of element surrounding node i
k_{fe} = coefficient of subgrade reaction for the element under consideration

The *contact pressure* at a node is given by:

$$Q_i = k_{fi} w_i \tag{4.74}$$

where w_i = deflection at node i.

4.4.3 LATERALLY LOADED DEEP FOUNDATIONS

Design Concepts

Beam-on-an-Elastic Foundation

As a pile deflects under a lateral load applied at its top, moments and shears develop and the pile reacts as a cantilever as shown in Fig. 4.46. Such lateral loads may be static or cyclic, the latter case common in off-shore structures. A design objective is to determine the maximum deflections to which the pile may be subjected in order to provide the pile section necessary to resist the resulting moments.

The problem has been treated by various forms of solutions. Older solutions treated the problem as one of earth pressure, assuming the development of active and passive resistances, which in reality do not develop except at complete failure. In addition, these theories furnish only two-dimensional solutions, considering the pile as a plane section of a cantilever bulkhead, as illustrated in Figs. 9.85 through 9.87, in *Art. 9.4.4.* These solutions are still used to analyze short, rigid piles (*see Art. 7.6.3*). A detailed discussion of the laterally loaded pile problem is presented in Poulos and Davis (1980).[69]

The *long, flexible pile,* which applies to many practical situations, is now treated as a *beam-on-an-elastic foundation* by most practitioners. Solutions vary from simple to complex; in general, the pile is assumed divided into spring segments, each of which experiences lateral deflection y and soil resistance p in units of force per unit pile length, as shown in Fig. 4.47. Hetenyi (1946)[60] and others analyzed the problem using a

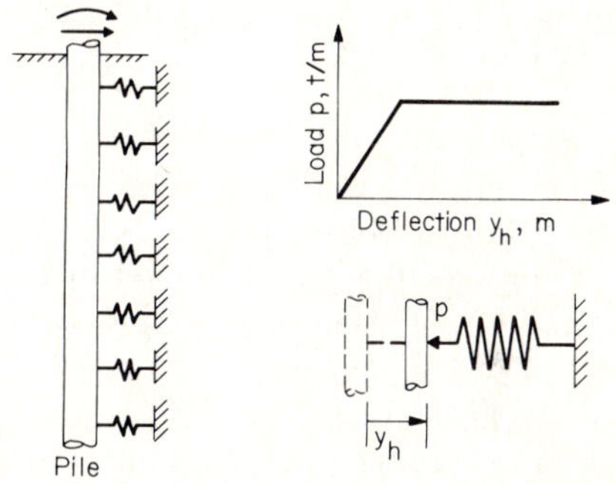

FIG. 4.47 Design principles for laterally loaded piles. Boundary conditions are: horizontal force, bending moment, angle of rotation, and degree of restraining. [From Eide (1974).[68]]

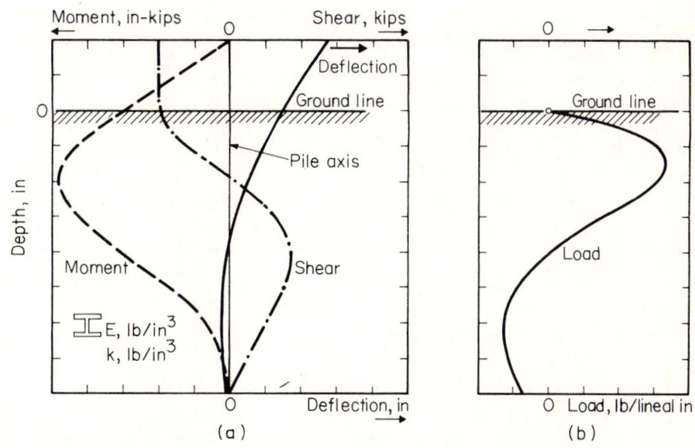

FIG. 4.46 Vertical pile loaded by horizontal forces: (*a*) deflection, shear, and moment and (*b*) load.

fourth-order differential equation derived on the basis that a long pile is a linearly elastic beam and that the soil reaction may be represented as a line load (Eq. 4.69) expressed as

$$EI\frac{d^4y}{dx^4} + P_x\frac{d^2y}{dx^2} + p = 0 \quad (4.75)$$

where P_x = axial load on pile
y = lateral deflection of pile at point x along its length
p = soil reaction per unit length

$$p = -k_h x \quad (4.76)$$

or

$$p = -E_s y \quad (4.77)$$

EI = flexural rigidity of pile

Soil Reaction

The soil reaction per unit length of pile is expressed by one of several parameters as follows:

- Horizontal modulus of subgrade reaction K_h, n_h
- Secant modulus E_s, which is used for Young's modulus, since it better defines load deformation in the range of approaching failure
- *p-y* curves, which relate soil resistance to pile deflection at various depths below the surface. In general, these curves are *nonlinear* and depend upon several parameters including depth, soil shear strength, and for dynamic loads, the number of load cycles.

Solutions

A computer solution, developed by Matlock and Reese (1960)[70] as a general approach to the problem, is in common use following the development of charts expressing dimensionless coefficients for deflection, moment, and shear in terms of relative pile stiffness (*see Art. 7.6.4*). A relatively simple solution, it is used primarily for the case of k_h increasing linearly with depth. Soil modulus n_h is determined by test (*see the following section*) and the relative stiffness factor T is calculated from:

$$T = \left(\frac{EI}{n_h}\right)^{1/5} \quad (4.78)$$

The linear relationship of soil modulus with depth provides only a rough approximation, as it does for beams, even if moduli are used that are functions of depth, since soil stratification, time effects, the effect of load intensity, and structural factors cannot be considered.

More complete solutions for static and dynamic loads have been developed based on finite differences and *p-y* curves for problems where it is necessary to consider the complexity of the soil system, variable pile structural stiffness, variable head fixity, and the stiffness of the supported structure. This approach was suggested by McClelland and Focht (1958).[71] Reese (1977)[72] describes a computer program available to provide the complete solution. The differential equation expressing the pile as a linearly elastic beam and the soil reaction as a line load (Eq. 4.75) is given in finite difference form. This is accomplished by discretizing the pile length using a constant increment h as shown in Fig. 4.48. Substituting Eq. 4.58 into Eq. 4.75 results in

$$\frac{d^2M}{dx^2} + P_x\frac{d^2y}{dx^2} + ky = 0 \quad (4.79)$$

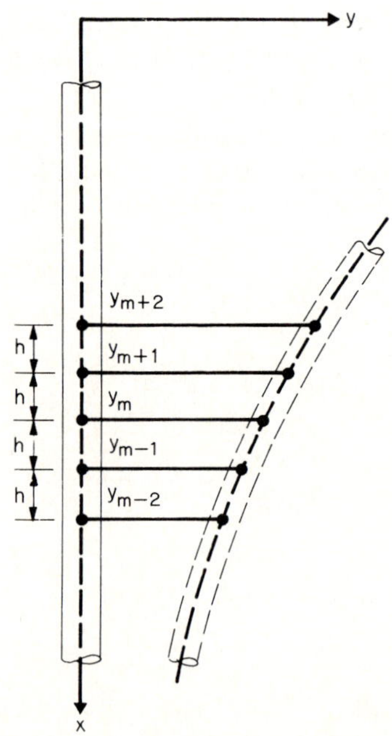

FIG. 4.48 Representation of deflected pile—length discretized using constant increment *h*. [From Reese (1977).[72]]

where $ky = p$. The elements of this expression containing M and P_x are then expressed in terms of deflection and structural rigidity to obtain the difference equations.

Computer program COM 622 (Analysis of Laterally Loaded Piles by Computer) was developed by the Computing Center, University of Colorado, to solve the problem. The soil-load deformation relationship used in COM 622 is expressed as follows:

- Clay soils: $E_s = -p/y$ (E_s = secant modulus)
- Cohesionless soils: $E_s = Kz$, where K is a constant since E_s is assumed to vary linearly with depth, as described in the following section.

Horizontal Modulus of Subgrade Reaction k_h

Distribution along Pile Shaft

The concept of a modulus of subgrade reaction is valid for vertical as well as horizontal beams and slabs. Terzaghi (1955)[62] suggested that the probable distributions of moduli along a pile shaft was generally uniform with depth in *stiff clays* and proportional to depth in sands as illustrated in Fig. 4.49.

Measurements of k_h and n_h

The *constant of horizontal subgrade reaction* n_h is the value measured in *pile-load tests* from which are computed values for k_h, the horizontal modulus of subgrade reaction, as described below. Tests are performed on a pile of 1-ft width to obtain a load-deflection curve for n_h or are performed on the design pile to obtain design values for k_h. The use of the inclinometer to measure deflection versus depth permits the determination of k_h versus depth (*see Art. 7.6.4, Lateral Load Tests*).

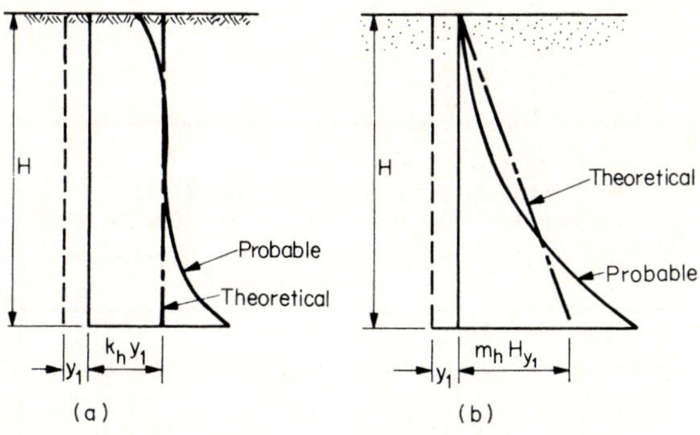

FIG. 4.49 Theoretical and probable horizontal pressures on piles: (*a*) clay and (*b*) sand. [From Terzaghi (1955).[62]]

Pressuremeter testing [CGS (1978)[4]] is used to obtain values for k_h as related directly to E_c or P_L as follows

Cohesionless soil:

$$k_h = 3.3\frac{E_c}{B} \approx 25\frac{P_L}{B} \qquad \text{kN/m}^3 \qquad (4.80)$$

Cohesive soil:

$$k_h = 1.6\frac{E_c}{B} \approx 16\frac{P_L}{B} \qquad \text{kN/m}^3 \qquad (4.81)$$

where pressuremeter modulus E_c and limiting pressure P_L are in kPa units and pile diameter B is in meters. Typical values for E_c and P_L are given in Table 4.5.

Sands and Normally Consolidated Clays

It is generally assumed that the modulus increases with depth; therefore

$$k_h = n_h z \qquad \text{tsf} \qquad (4.82)$$

or in terms of beam width B,

$$k_h = n_h z/B \qquad \text{tons/ft}^3\text{, kN/m}^3 \qquad (4.83)$$

where n_h is the constant of horizontal subgrade reaction for piles embedded in sands. For *sands*, on the assumption that the elastic modulus depends only on the effective overburden pressures and relative density, Terzaghi (1955)[62] showed that

$$n_h = A\gamma/1.35 \qquad (4.84)$$

where A is a coefficient depending upon density. Values for factors n_h and A are given in Table 4.16.

Loose to medium dense sands are particularly susceptible to densification under repetitive loading and thus allow a pile more freedom of movement under cyclic loading. Broms (1964)[62a] suggests reducing n_h for *cyclic loading effects* by

TABLE 4.16
VALUES OF FACTORS A AND n^* FOR ESTIMATING HORIZONTAL MODULUS k_h FOR A PILE 1 FT WIDE EMBEDDED IN MOIST SUBMERGED SAND AND FOR A HORIZONTAL STRIP 1 FT WIDE†

Relative density of sand	Values for A		Values for n_h*	
	Range	Adopted	Dry or moist sand	Submerged sand
Loose	100–300	200	7 (2,300)	4 (1,300)
Medium	300–1000	600	21 (6,800)	14 (4,500)
Dense	1000–2000	1500	56 (18,000)	34 (11,000)

*n_h is given in tons/ft³ or (kN/m³); $k_h = n_h \cdot z/B$; $r_h = A\gamma/1.35$.
†After Terzaghi (1955).[62]

¼ n_h for low D_R values to ½ n_h for soils with high D_R values. In *soft, normally consolidated clays* values for n_h of 1 to 2 tons/ft³ (0.03 to 0.06 kg/cm³) have been observed. In *soft organic silts* values as low as 0.5 tons/ft³ have been reported. Approximate values for n_h in terms of shear strength and relative density are given in Fig. 4.44.

Overconsolidated Clays

Terzaghi (1955)[62] gives the following relationship, considered conservative, for estimating k_h in overconsolidated clays

$$k_h = \frac{1}{B} k_{ht} = \frac{1}{B} k_s$$
$$= 0.67\ k_t/B \qquad \text{tons/ft}^3 \qquad (4.85)$$

where k_h = coefficient for pile of width B
k_{ht} = coefficient for pile of unit width (1 ft)
k_s = coefficient for horizontal beam of unit width
k_t = coefficient of horizontal square plate, 1 ft wide (Table 4.15)

The factor 0.67 occurs in going from an infinitely long beam to a square plate (Eq. 4.67). Vesić (1961)[64] found that the factor 0.67 is not a constant, but varies with beam rigidity.

Broms (1964)[63] expressed the subgrade modulus (k_∞, Eq. 4.61, from Vesić) in terms of the shear strength of the clay and the pile material, and has written Eq. 4.85 as:

$$k_\infty = \alpha\ (k_t/B) \qquad (4.86)$$

where $\alpha = n_1 \cdot n_2$ and n_1 and n_2 are functions of soil strength and pile properties as given in Table 4.17. These values may be used for long piles where the dimensionless length $\beta L > 2.25$ and:

$$\beta = \left(\frac{k_h B}{4E_p I}\right)^{1/4} \qquad (4.87)$$

For large-diameter piers, Francis (1964)[73] recommends that values for k_h obtained from the equation of Vesić (Eq. 4.61) be doubled to account for complete embedment. Considering that the term $\sqrt[12]{B^4/I}$ is 1.28 for a solid circular section, he ex-

TABLE 4.17
FACTORS n_1 AND n_2 FOR COMPUTING SUBGRADE MODULUS k_h* FOR CLAYS†

q_u, tsf	n_1	Pile material	n_2
0.5	0.32	Steel	1.00
0.5–2.0	0.36	Concrete	1.15
2.0	0.40	Timber	1.30

*$k_h = n_1 \cdot n_2\ (k_t/B)$.
†From Broms (1964).[63]

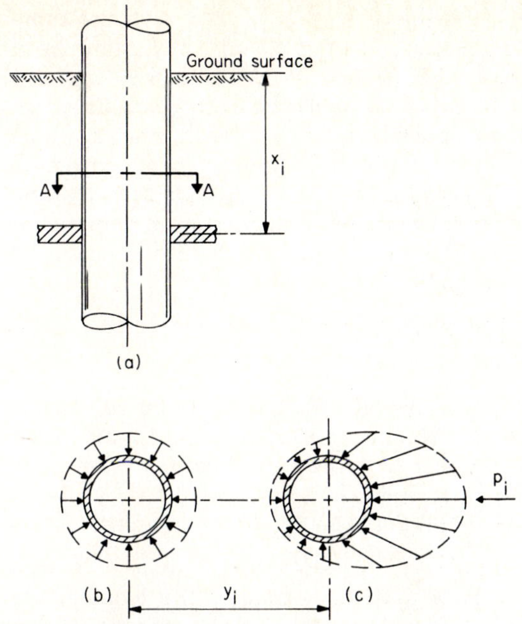

FIG. 4.50 Graphical definition of *p* and *y*: (*a*) side view of pile; (*b*) vertical view, section *A-A*, earth pressure distribution prior to lateral loading; and (*c*) section *A-A*, earth pressure distribution after lateral loading and deflection through *y*. [From Reese (1977).[72]]

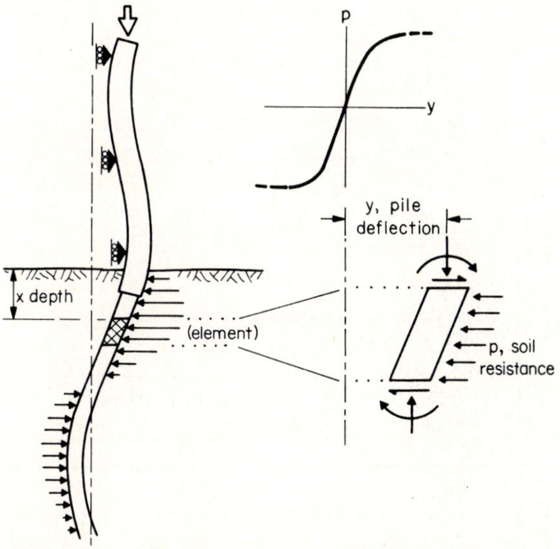

FIG. 4.51 Elements in analysis of laterally loaded piles by difference equation method. [From McClelland (1972).[77]]

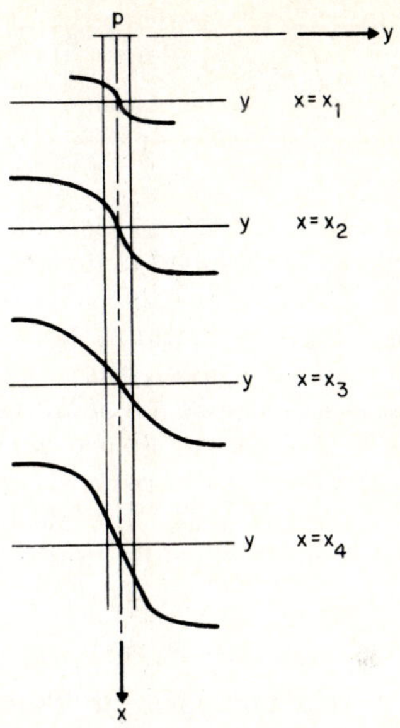

FIG. 4.52 Set of *p-y* curves. [From Reese (1977).[72]]

pressed the *pier modulus K* ($K = k_h d$, where d = pier diameter) as

$$K = \frac{1.60E_s^{13/12}}{E^{1/12}(1 - \nu^2)} \quad \text{tsf} \tag{4.88}$$

p-y Curves

Concept

As a pile deflects under a laterally applied load *p*, the resulting soil reaction varies with the pile deflection *y*. Figure 4.50 illustrates the concept showing the distribution of earth pressures prior to lateral loading (*b*) and after lateral loading (*c*) which caused displacement (*y*). The deflection of the pile through distance *y* (Fig. 4.51) generates unbalanced stresses; their integration yields an unbalanced force (*p*) per unit length of pile.

To solve the pile deflection problem, the prediction of a set of *p-y* curves is developed (Fig. 4.52). Potential deflections are particularly severe offshore where piles may stand freely in 100 m of water or more. Normally, however, a depth below the mudline only in the order of 10 pile diameters need be considered. The construction of *p-y* curves is described in API (1982),[78e] de

Ruiter and Beringen (1978),[75] Bogard and Matlock (1980),[76] and DNV (1977).[78b]

Sands

Former method of API (1971)[78] is illustrated in Fig. 4.53, which shows a load-deformation curve for laterally loaded piles in sands considering the soil as an elastic-plastic medium. For the elastic phase, the ratio of p to y is given as a density coefficient A (Table 4.14) times the effective vertical pressure, divided by 1.35d in accordance with Terzaghi's analysis of the horizontal modulus of subgrade reaction (Eq. 4.84). The plastic part of the p-y curve is established by limiting the soil resistance p_u to

$$p_u = 3\bar{\sigma}_v K_p p \qquad (4.89)$$

Most practitioners familiar with the problem, however, consider that the actual sand response will be stiffer than indicated by these criteria, resulting in calculated deformations and moments somewhat on the high side [McClelland (1972)[77]]. The method of Reese, Cox, and Koop is now recommended [API (1982)[78a]].

The *method of Reese, Cox, and Koop* (1974)[79] is based on static and cyclic load tests on 24-in-dia. piles in clean fine to silty fine sand. Relationships for the construction of the p-y curves based on their approach are given in Fig. 4.55c and Fig. 4.56. The expressions for the ultimate resistance near the ground surface p_{3s} (Eq. 4.96), ultimate resistance well below the ground surface p_{3d} (Eq. 4.97), and the deflection (y_1) (Eq. 4.98) are somewhat complex (see Fig. 4.55c). They relate the effective soil friction angle, effective unit soil weight, pile penetration below the mudline, coefficients of active and at-rest earth pressures, pile diameter, initial soil modulus, and correction coefficients for cyclic or static loadings as functions of the ratio of pile penetration to diameter. The CPT is the favored method to measure the soil parameters in situ.

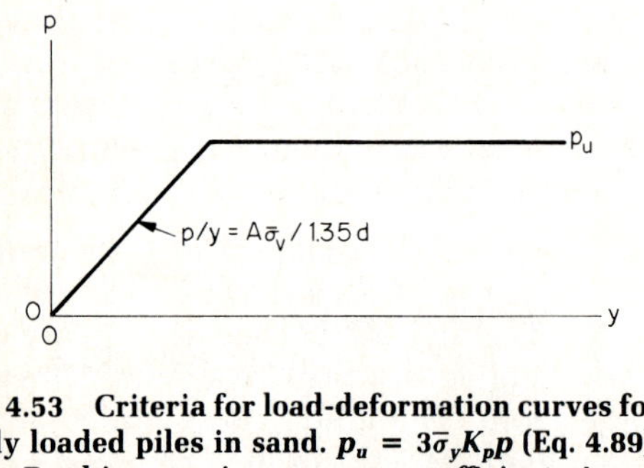

FIG. 4.53 Criteria for load-deformation curves for laterally loaded piles in sand. $p_u = 3\bar{\sigma}_y K_p p$ (Eq. 4.89) and K_p = Rankine passive pressure coefficient. A = sand density coefficient, ranging from 200 for loose sand to 2000 for dense sand. For other values of A, see Table 4.16. [From API (1971)[78]; McClelland (1972).[77]]

Soft to Firm Clays

The p-y curve criteria developed for soft to firm clays by Matlock (1970)[80] for static and cyclic loads, based on large-scale load tests, are given in Fig. 4.54. Figure 4.55a shows p-y curve development and Fig. 4.56 gives an example.

As shown in Fig. 4.54, the ultimate soil resistance p_u is used to normalize the loads on the vertical axis. The deflection occurring at 50% of the ultimate stress y_c is used to normalize the deflection on the horizontal axis. If soft clay is confined so that plastic flow around the pile occurs only in horizontal planes, the ultimate static resistance per unit length of pile p_u may be expressed as

$$p_u = N_p s_u D \qquad (4.90)$$

where s_u = undrained shear strength
D = pile diameter
N_p = a nondimensional ultimate resistance coefficient varying from about 3 for a cylindrical pile near the soil surface to a maximum of about 9 at depth x_r, the critical depth

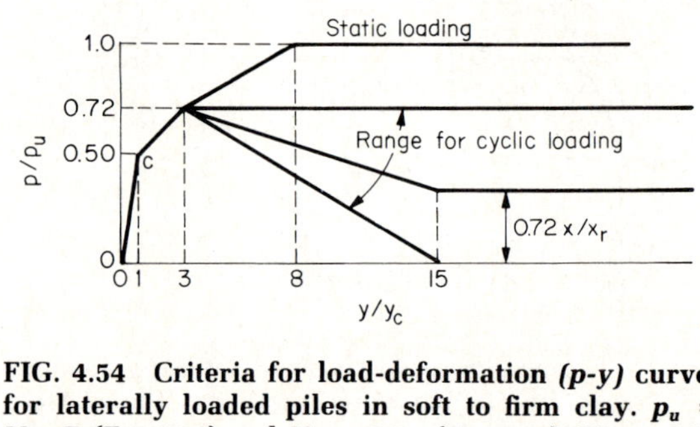

FIG. 4.54 Criteria for load-deformation (p-y) curves for laterally loaded piles in soft to firm clay. $p_u = N_p s_u D$ (Eq. 4.90) and $Y_c = 2.5\epsilon_o$ (Eq. 4.93). [From Matlock (1970).[80]]

N_p is estimated from

$$N_p = 3 + \frac{\gamma' x}{s_u} + J\frac{x}{D} \tag{4.91}$$

where γ' = effective soil unit weight
x = depth
J = coefficient of soil restraint related geometrically to the failing wedge or soil prism around the pile. J appears to range from about 0.25 for soft clays to 0.5 for firm clays.

The critical depth x_r ($N_p = 9$ when $x > x_r$) is estimated from

$$x_r = \frac{6D}{(\gamma' d/s_u) + J} \tag{4.92}$$

ESTIMATING PILE DEFLECTION Relationships from Skempton's (1951)[81] expression for the immediate deflection of a long strip footing at one-half the ultimate load supported by clay soils is used for estimating deflection.

$$y_c = 2.5\epsilon_c D \tag{4.93}$$

where ϵ_c = strain occurring at one-half the maximum stress on a laboratory triaxial stress-strain curve. Therefore,

$$y_{50} = \epsilon_{50} D \tag{4.94}$$

CYCLIC LOADING Test results reported by Matlock (1970)[80] show that cyclic loads, common in offshore structures, produce a softening of the soil resistance with the peak resistance after cycling being only about 0.7 of the resistance to a single static load. Resistance to large strains decreases significantly in the soil near the mudline. The *p-y* curve for depth $x = 0$ follows the lower boundary of the shaded area in Fig. 4.54, with *p* becoming zero at large deflections. Below critical depth x_r, the applicable *p-y* curve follows the upper boundary of the shaded area. Complete loss at the mudline is assumed to occur when deflections there reach 15*y*. For deflections greater than 15*y*, the pseudoplastic resistance is established by

$$\frac{p}{p_u} = 0.72\frac{x}{x_r} \tag{4.95}$$

Laboratory CK_oU cyclic triaxial tests, performed to measure clay strength degradation, are described in Art. 3.3.5.

Overconsolidated Clays

Criteria for predicting *p-y* curves for laterally loaded piles in stiff clays, described by Reese and Welch (1975),[82] are presented in Fig. 4.55*b* and Fig. 4.56.

4.4.4 TIED-BACK RETAINING WALLS

Concepts

Retaining structures in braced excavations normally are analyzed by the use of earth pressure theories and empirical concepts to arrive at designs that will provide support against failure. Forces in anchors or braces and bending moments in the wall are computed on the basis of the distribution of lateral earth pressures *(see Art. 9.3.3).*

Wall deflection and backslope subsidence, often a serious problem because of possible detrimental effects on adjacent structures, cannot be mathematically analyzed by the classical theories or empirical methods. In recent years the finite element method has been employed to evaluate wall performance in terms of deflections, and for continuous walls such as concrete diaphragm walls, the problem has been approached by modeling the wall as slabs-on-elastic subgrades.

Approach

The problem of a tied-back retaining wall is illustrated in Fig. 4.57*a*. In the analysis of the problem presented by Tsui and Clough (1974),[83] it is assumed that plane strain loading conditions (Fig. 4.57*b*) reasonably represent the actual three-dimensional loading of the anchors. (Plane strain conditions assume that the wall is continuous and that the discontinuous tiebacks can be simulated by planar supports at the same elevations as the real supports.)

Wall behavior was described by the slab-on-an-elastic-subgrade theory (Fig. 4.57*c*). Bending moments were expressed as the radial moment M_r

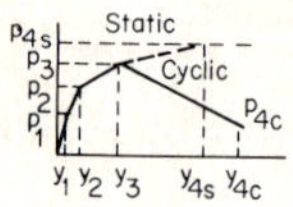

$$p_u = N_p s_u D \tag{4.90}$$

$$0 < x < x_r: \quad N_p = 3 + \frac{\gamma x}{s_u} + 0.25\frac{x}{D} \tag{4.91}$$

$$x > x_r: \; N_p = 9$$

$$x_r = \frac{6D}{\dfrac{\gamma D}{s_u} + 0.25} \tag{4.92}$$

$$y_{50} = \epsilon_{50} D \tag{4.94}$$

$$p_1 = 0.3\,p_u \qquad y_1 = 0.55\,y_{50}$$

$$p_2 = 0.5\,p_u \qquad y_2 = 2.5\,y_{50}$$

$$p_3 = 0.72\,p_u \qquad y_3 = 7.5\,y_{50}$$

$$p_{4s} = p_u \qquad y_{4s} = 20\,y_{50}$$

$$p_{4c} = \frac{x}{x_r}\,p_3 \qquad y_{4c} = 37.5\,y_{50}$$

(a)

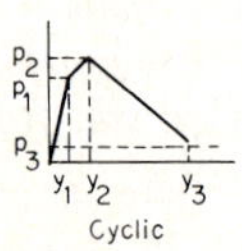

$$p_u = N_p s_u D$$

$$0 < x < x_r: \quad N_p = 2 + \frac{\gamma x}{s_u} + 2.83\frac{x}{D}$$

$$x \geq x_r: \; N_p = 11$$

$$x_r = \frac{9D}{\dfrac{\gamma D}{s_u} + 2.83}$$

$$y_{50} = \epsilon_{50} D$$

Cyclic

$$y_1 = 0.92\,A_1\,y_{50} \qquad p_1 = 0.82\,B_1\,p_u$$

$$y_2 = 1.84\,A_1\,y_{50} \qquad p_2 = B_1\,p_u$$

$$y_3 = 7.4\,A_1\,y_{50} \qquad p_3 = (0.936\,B_1 - 0.42\,A_1)\,p_u$$

(b)

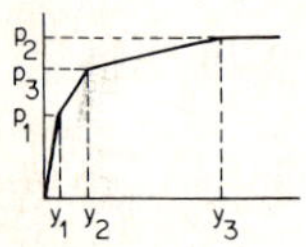

$$p_{3s} = A_2 \gamma x \left[\frac{K_0 \cdot x \cdot \tan\phi' \sin\beta}{\tan(\beta - \phi')\cdot\cos\alpha} + \frac{\tan\beta}{\tan(\beta - \phi')}(D + x\cdot\tan\beta\cdot\tan\alpha) + K_0\cdot x\cdot\tan\beta\,(\tan\phi\cdot\sin\beta - \tan\alpha) - K_A D \right] \tag{4.96}$$

$$p_{3d} = A_2 D \left[K_a \cdot \gamma' \cdot x(\tan^8\beta - 1) + K_0 \cdot \gamma' \cdot \tan\phi \cdot \tan^4\beta \right] \tag{4.97}$$

$$y_3 = \frac{3}{80} D$$

$$p_2 = \frac{B_2}{A_2} p_3$$

$$y_2 = \frac{1}{60} D$$

$$p_1 = K_1 y_1 x \qquad y_1 = \left(\frac{p_2}{K_1 x \; y_2^{1/n}} \right)^{\frac{n}{n-1}} \tag{4.98}$$

(c)

Legend

S_u = Undrained shear strength
γ' = Effective unit weight of soil
ϵ_{50} = Strain at S_u level
D = Pile diameter
x = Penetration
A,B = Correction coefficients
ϕ' = Effective angle of friction
$\beta = 45° + \dfrac{\phi'}{2}$
$\alpha = \dfrac{\phi'}{2}$
K_A = Active soil pressure
K_0 = Coefficient of lateral soil pressure
K_1 = Initial soil modulus
$n = \dfrac{1.25}{\dfrac{A_2}{B_2} - 1}$

Correction coefficients: A,B

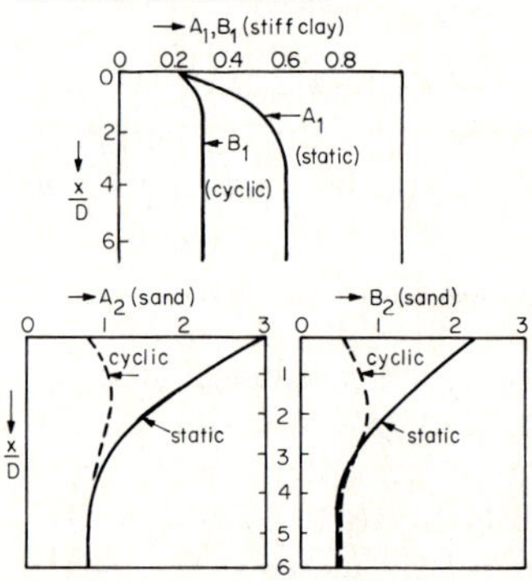

FIG. 4.55 Summary of *p-y* curve techniques: (*a*) soft clay; (*b*) stiff clay; and (*c*) sands. [From de Ruiter and Beringen (1978).[75]]

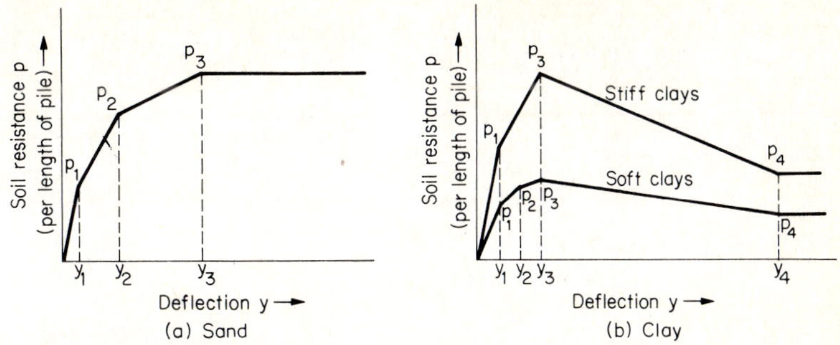

OVERCONSOLIDATED SOIL PROFILE

Depth, m	Soil type	Soil resistance *P*, MN/m, for given deflection *Y*, mm								$\epsilon_{50}/\bar{\varphi}$	s_u, kN/m²
		P_1	Y_1	P_2	Y_2	P_3	Y_3	P_4	Y_4		
0	Clay	0.08	2.8	0.09	5.6	0.05	22.5				150
2	Clay	0.35	7.0	0.43	14.0	0.10	55.9				160
5	Clay	0.72	8.4	0.88	16.8	0.08	67.1			1%	175
30	Clay	1.23	8.4	1.51	16.8	0.14	67.1				300
											k1 (MN/m³)
30	Sand	7.3	7	13.6	25.3	21.6	57.0				
40	Sand	9.7	7	18.1	25.3	28.8	57.0			35°	35
50	Sand	12.2	7	22.7	25.3	36.0	57.0				

FIG. 4.56 Examples of *p-y* curves for 1.52-m-dia. piles in (*a*) sand and (*b*) clay under cyclic loading. [From de Ruiter and Beringen (1978).[75]]

and the tangential moment M_t which depend on distance from load P, slab deflection w, Poisson's ratio ν, and slab flexural rigidity D. *Slab deflection* was defined in terms of load P, the slab characteristics, and a parameter L_o, termed the characteristic length of the soil-slab system, defined as

$$L_o = \sqrt[3]{\frac{2D(1-\nu_s^2)}{E_s}} \tag{4.99}$$

where E_s and ν_s are the elastic soil parameters. The characteristic length L_o is particularly useful as it accounts for the relative effects of both soil and slab stiffness and may be compared to the spacing of prestress loads as described in Art. 9.3.4. It is also used to estimate the error or the applicability of plane strain simulation for any combination of wall rigidity, soil rigidity, and horizontal brace or tie spacing *(see Art. 9.3.4)*.

4.4.5 EARTHQUAKE LOADINGS (SSI ANALYSIS)

Ground Motion

Dynamic Forces

The large amounts of energy released during earthquakes travel through the earth as various types of seismic or elastic waves with varying oscillation frequencies and amplitudes or displacements. The oscillating particles in the wave possess velocity and exert a force due to their acceleration *(see discussion in Art. 5.6.2)*. In rock and other nearly elastic geologic materials, these dynamic forces result in transient deformations which are recovered under the low strains of seismic waves. Interest in the properties of these materials lies primarily in their ability to transmit the seismic waves.

In weaker deposits, such as alluvium, colluvium, and eolian soils, however, the base rock excita-

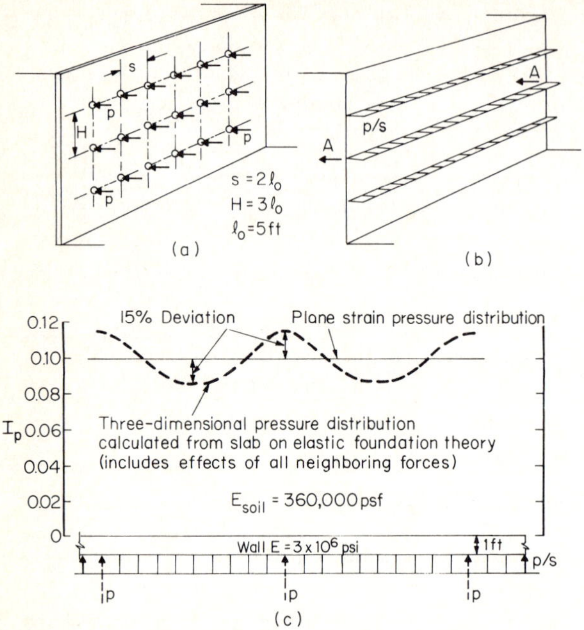

FIG. 4.57 Comparison of plane strain and actual pressure distribution behind 1-ft-thick concrete wall: *(a)* actual problem; *(b)* plane strain simulation; and *(c)* wall pressure along section *A-A*. [From Tsui and Clough (1974).[83]]

tion transmitted to the soils is usually amplified. In addition, some materials respond to cyclic shear forces by densifying, liquefying, or reducing in shear strength (unstable soils) as discussed in Art. 3.3.5.

Significance

The effect of the dynamic forces, therefore, can be considered in two broad categories:

1. The effect on structures subjected to the forces transmitted through the ground which result in *ground shaking*
2. The effect on the geologic material itself, primarily in the form of *response to cylic shear forces.*

Structural Response

Reaction to Strong Ground Motion

Dynamic forces are imposed on structures by strong ground motion. Structural response is related to the interaction between the characteristics of the structure as described by its mass, stiffness, and damping capability, and the characteristics of the ground motion as described by the combined influence of the amplitude of ground accelerations, their frequency components, and the duration of shaking. Base excitation of the structure from ground shaking results in horizontal and vertical displacements for some time interval and imposes strains, stresses, and internal forces on the structural elements.

Forces on Structural Members

Static loads result in stresses and deflections, but *dynamic loads* result in time-varying deflections which involve accelerations. These engender inertial forces which resist the motion and which must be determined for the solution of structural dynamics problems. The complete system of inertial forces acting on a structure is determined by evaluating the accelerations, and therefore the displacements, acting at every point on the structure.

The deflected shape of a structure may be described in terms of a "lumped mass" idealization, in which it is assumed that the entire mass of the structure is concentrated at a number of discrete points at which accelerations are evaluated to define the internal forces developing in the system. The simple *single-degree-of-freedom systems* illustrated in Fig. 4.58 serve to demonstrate the forces acting on a structure. In both cases the system consists of a single rigid (lumped) mass M so constrained that it can move with only one component of simple translation. The dynamic forces acting on a simple building frame founded on the surface may be represented by a simple mass-spring-dashpot system as shown in Fig. 4.58*a*, where the dashpot is an energy absorber. Translational motion is resisted by weightless elements having a total spring constant K (stiffness) and a damping device which absorbs energy from the system. The damping force F_D is proportional to the velocity of the system.

During earthquakes the motion is excited by an external load $P(t)$ which is resisted by an inertial force F_1, a damping force F_D, and an elastic force F_s. These resisting forces are proportional re-

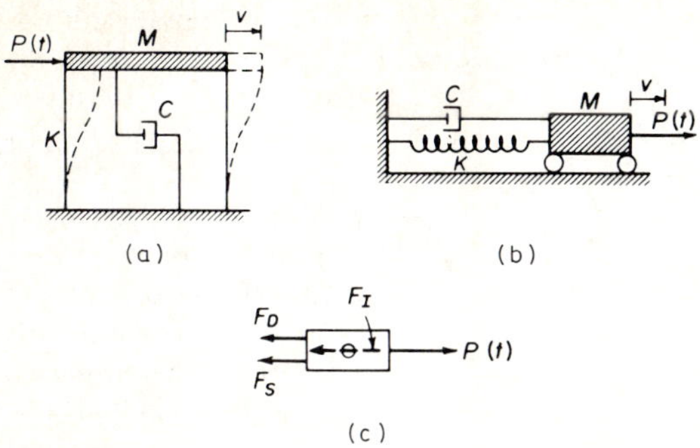

Inertia force $F_1 = M\ddot{v}$
Damping force $F_D = C\dot{v}$ **$P(t) = F_1 + F_D + F_S$**
Elastic force $F_S = Kv$

where M = mass; C = damping coefficient; K = total spring constant; v = displacement; and $P(t)$ = external load.

FIG. 4.58 Single-degree-of-freedom systems: (*a*) simple frame; (*b*) spring-mass system; and (*c*) forces acting on mass.

spectively to the acceleration, velocity, and displacement of the mass as follows: inertial force $F_1 = M\ddot{v}$, damping force $F_D = C\dot{v}$, and elastic force $F_s = Kv$, where v is the displacement. Equilibrium requires that

$$P(t) = F_1 + F_D + F_s \qquad (4.100)$$

where the force $P(t)$ is the *ground acceleration input*, equal to the product of the mass and the acceleration, or

$$P(t) = M\, a_{max} \qquad (4.101)$$

Dynamic Analysis

Applications

Methods of analysis for design vary from the relatively simple, conventional approach based on g forces, to comprehensive analytical procedures that consider the combined influence of ground acceleration, frequency, and to some extent, ground-shaking duration in relation to the period and damping characteristics of the structure. The selection is based on the degree of earthquake hazard for a given locale and the risk factor which pertains to the consequences of structural failure. For example, structures such as nuclear power plants, 50-story buildings, long suspension bridges, and large dams have a high risk in terms of public safety and today are evaluated with respect to earthquake loadings by comprehensive dynamic analysis.

Soil-structure interaction (SSI) analysis is performed for those high-risk cases where the structure will be supported on a soil foundation rather than the usually more desirable rock foundation. The purpose of the SSI analysis is to evaluate a coupled bedrock-soil-structure system including resonance and feedback effects, for foundations on or below the surface.

Feedback from Structural Oscillations

For relatively light structures founded on strong rock, the influence of oscillation feedback from the structure is minimal and the structural model of excitation is essentially that of the ground motion. For massive structures on strong soils or conventional structures on deep, relatively weak deposits, however, feedback of structural oscillations to the underlying soils may significantly affect the motion at the soil-structure interface, which in turn may result in amplification or reduction of the structural response.

Deformation of the soil formation also may be caused by the feedback from horizontal, vertical, or rotational oscillatory motion of the structure. The problem is complicated by founding below ground level, the usual procedure for most heavy structures. *Stable soils* undergo plastic and elastic deformations under seismic loadings, but dampen seismic motion, maintain some characteristic strength level, and are amenable to SSI analysis. Both dynamic input to the soil from the excitation of the underlying rock (which basically applies a shearing stress to the soils) and the feedback from structural oscillation should be considered in dynamic analysis, for which dynamic soil properties should be used.

Unstable soils, such as loose silty fine sands below the water table, are subject to sudden and essentially complete loss in strength by liquefaction or sudden compression resulting in subsidence and are not readily considered in SSI analysis.

Halfspace Analysis

There are two general approaches to SSI analysis. In the simpler approach, the soil effects on structural response are presented by a series of springs and dashpots in a theoretical halfspace surrounding the structure, as shown in Fig. 4.59. This approach has limitations, particularly when applied to buried structures, and is best used to analyze surface structures. The method is described in Seed et al. (1975).[84]

For the problem illustrated, the horizontal earthquake motion was specified at the ground surface in the free field (a location where interaction between soil and structure is not occurring). It was assumed that the ground surface motions for the relatively thin soil deposit were the result of vertically propagating shear waves, and the maximum accelerations and corresponding time histories were found for other depths down to the bedrock surface. A maximum acceleration at the foundation base level was selected. The procedure is described in Schnabel et al. (1971).[85]

Soil characteristics were represented by *dynamic spring constants* and an analysis of structural response was made for damping ratios of 7 to 15%, which led to values of maximum acceleration at the base of the structure of 0.38 and 0.32g respectively. As of 1974, the approach does not consider soil damping, can only be applied to one- or two-layer soil systems, and provides no means for determining strains induced in the soils. Induced strains very much affect the soil deformation modulus G used in the determination of the spring constant.

As described by Whitman and Richart (1967),[86] values for the dynamic soil spring constant may be established by several methods.

1. Use formulas for spring constants derived from elastic theory and evaluate the elastic constants either from in situ shear-wave-velocity measurements or from laboratory tests.
2. Determine spring constant from small-scale plate bearing tests using static repeated loadings.
3. Deduce spring constant from the results of small-scale vibrator tests.
4. Use the concept of an elastic subgrade modulus together with tables or charts correlating subgrade modulus to soil type.

Finite Element Method of Analysis

The finite element model idealizes the soil continuum as a system of finite elements interconnected at a finite number of nodal points as shown in Fig. 4.60. In most cases the soils are considered to be equivalent linear-elastic materials. Soil resistance is described by formulating matrices for stiffness and mass, and a nodal solution or a time-marching integration is effected, depending upon the capability of the particular computer program employed. Response-time histories of displacement, velocity, and acceleration can be computed for each nodal point. Soil

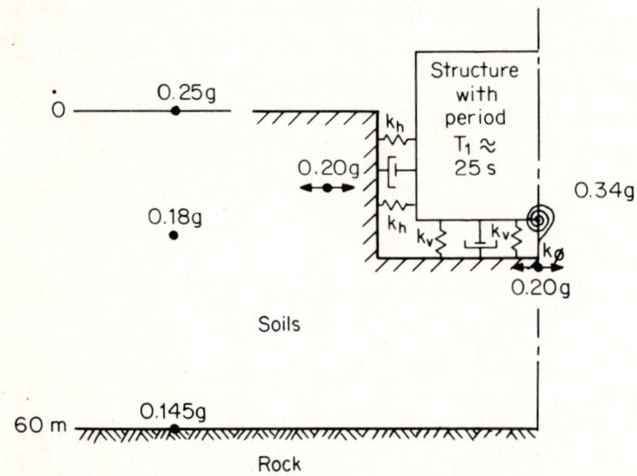

FIG. 4.59 Soil-structure-interaction model for halfspace analysis. [From Seed et al. (1975).[84]]

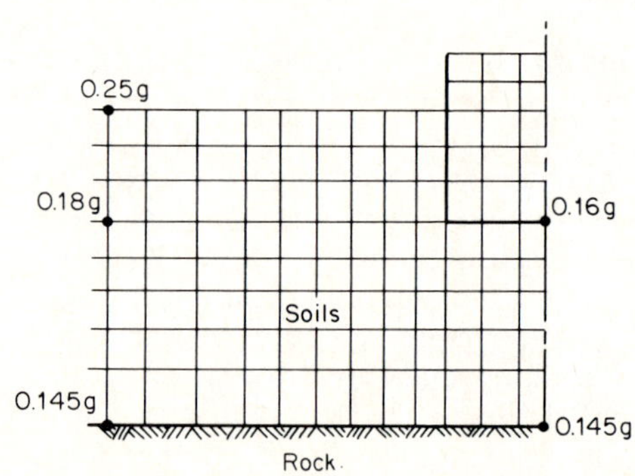

FIG. 4.60 Finite element model of soil-structure system of Fig. 4.58. [From Seed et al. (1975).[84]]

characteristics required include shear modulus, Poisson's ratio, soil unit weight, and damping coefficients. (Dynamic properties are described in Arts. 3.3.5 and 4.1.4.) The variations of the shear moduli and damping coefficients with strain are considered.

Input or control motion is based on actual earthquake data, or on synthetic records, to obtain vertical and horizontal excitation in terms of g, and can be specified as located in the free field at the surface or at foundation level. A deconvolution analysis is necessary to determine compatible base-rock formation to produce the specified motions at the control point. This is accomplished with the computer program SHAKE. The base rock motion is then used for a two-dimensional analysis (plain strain) of the soil-structure system, leading to an evaluation of the motions at any selected points such as the base of the structure, the operating floor, etc. LUSH is a computer program used for the two-dimensional analysis. In comparing an FEM analysis of the problem evaluated by the halfspace method (Fig. 4.59), Seed et al. (1975)[84] found that the maximum acceleration at the base of the structure would be only 0.16g, or roughly half that found by the halfspace analysis. The FEM analysis is discussed also by Idriss and Sadigh (1976).[87]

For a detailed description of earthquakes and their geotechnical aspects, see Chap. 11 of Hunt (1984).[5]

REFERENCES

1. Terzaghi, K. (1943) *Theoretical Soil Mechanics*, John Wiley & Sons, New York.
2. Goodman, R. E. (1980) *Introduction to Rock Mechanics*, John Wiley & Sons, New York.
3. Lambe, T. W. and Whitman, R. V. (1969) *Soil Mechanics*, John Wiley & Sons, New York.
4. CGS (1978) *Canadian Foundation Engineering Manual*, Canadian Geotech. Soc., Montreal.
5. Hunt, R. E. (1984) *Geotechnical Engineering Investigation Manual*, McGraw-Hill Book Co., New York.
6. Clough, G. W. and Denby, G. M. (1980) "Self-Boring Pressuremeter Study of San Francisco Bay Mud," *Proc. ASCE, J. Geotech. Engrg. Div.*, Vol. 106, No. GT1, January, pp. 45–63.
7. SOLS SOILS (1975) "The Menard Pressuremeter—Interpretation and Application of Pressuremeter Test Results to Foundation Design," *Sols Soils*, No. 26, Paris.
8. Swiger, W. F. (1974) "Evaluation of Soil Moduli," *Proc. ASCE, Conf. on Analysis and Design in Geotech. Engrg.*, Austin, Vol. II, pp. 79–92.
9. Seed, M. B. (1969) "Influence of Local Soil Conditions in Earthquake Damage," *Proc., 7th Intl. Conf. Soil Mech. and Found. Engrg.*, Spec. Sess. 2, Soil Dynamics, Mexico City.
10. Vesić, A. B. (1967) "A Study of Bearing Capacity for Deep Foundations," *Final Rept., Project B–189*, Georgia Inst. Tech., Atlanta, March.

11a. Robertson, P. K., Campanella, R. G., and Wightman, A. (1983) "SPT-CPT Correlations," *Proc. ASCE, J. Geotech. Engrg. Div.*, Vol. 109, No. GT11, November, pp. 1449–1459.

11b. Douglas, B. J. and Olsen, R. S. (1981) "Soil Classification Using Electric Cone Penetrometer," *Proc. ASCE, Symp. on Cone Penetration Testing and Experience*, St. Louis, Mo., October.

12. Rocha, M. (1970) "New Techniques in Deformability Testing of In Situ Rock Masses," *Determination of the In Situ Modulus of Deformation of Rock*, ASTM STP 477, Amer. Soc. for Test. and Mat., Philadelphia, Pa., pp. 39–57.
13. Goodman, R. E., Van, T. K., and Heuze, F. E. (1972) "The Measurement of Rock Deformability in Boreholes," *Proc. 10th Symp. on Rock Mechs.*, AIME, pp. 523–555.
14. ISRM (1981) *Rock Characterization and Monitoring*, E. T. Brown, ed., Pub. for Comm. on Testing Methods, Intl. Soc. for Rock Mechs., Pergamon Press, Oxford, England.
15. Hibino, S., Mayashi, M., Kanagawa, T., and Motojima, M. (1977) "Forecast and Measurement of the Behavior of Rock Masses During Underground Excavation Works," *Intl. Symp. Field Measurements in Rock Masses*, Zurich, Vol. 2, pp. 935–948.
16. Deere, D. U., Hendron, A. J., Patton, F. D., and Cording, E. J. (1967) "Design of Surface and Near-Surface Construction in Rock," *Proc. 8th Symp. Rock Mechs.*, Univ. of Minnesota, Amer. Inst. Min. and Metal. Engrs., Chap. 11.
17. Schneider, B. (1967) "Moyens nouveaux de reconaissance des massifs rocheux" *Supp. to Annales de L'Inst. Tech. de Batiment et des Travaux Publics*, Vol. 20, No. 235–236, pp. 1055–1093.

18. Bieniawski, Z. T. (1978) "Determining Rock Mass Deformability—Experience from Case Histories," *Intl. J. Rock Mechs. Min. Sci.* Vol. 15, No. 5, pp. 237–248.

19. Perloff, W. H. (1975) "Pressure Distribution and Settlement," *Foundation Engineering Handbook*, Winterkorn and Fang, eds., Van Nostrand-Reinhold, New York, Chap. 4.

20. Taylor, D. W. (1948) *Fundamentals of Soil Mechanics*, John Wiley & Sons, New York.

21. Poulos, H. G. and Davis, E. H. (1974) *Elastic Solutions for Soil and Rock Mechanics*, John Wiley & Sons, New York.

22. Sowers, G. B. and Sowers, G. F. (1951) *Introductory Soil Mechanics and Foundations*, 2d ed., The Macmillan Co., New York.

23. NAVFACS (1982), *Design Manual 7.1, Soil Mechanics*, Dept. of the Navy, Naval Facilities Engrg. Command, Alexandria, Va.

24. Fadum, R. E. (1948) "Influence Values for Estimating Stresses in Foundations," *Proc. 2d Intl. Conf. Soil Mechs. and Found. Engrg.*, Rotterdam.

25. Newmark, N. M. (1942) "Influence Charts for Computation of Stresses in Elastic Foundations," *Engrg. Exp. Stat. Bull.*, Series No. 338, Univ. of Illinois, Urbana, Ill.

26. Terzaghi, K. and Peck, R. B. (1967) *Soil Mechanics in Engineering Practice*, John Wiley & Sons., New York.

27. Westergaard, H. M. (1938) "A Problem in Elasticity Suggested by a Problem in Soil Mechanics: Soft Material Reinforced by Numerous Strong Horizontal Sheets," *Contributions to the Mechanics of Solids, Stephen Timoshenko 60th Anniversary Volume*, The Macmillan Co., New York.

28. Burmister, D. M. (1956) "Stresses and Displacement Characteristics of a Two-Layer Rigid Base Soil System: Influence Diagrams and Practical Applications," *Proc. HRB*, Highway Research Board, Washington, D. C., Vol. 35, pp. 773–814.

29. Burmister, D. M. (1943) "Theory of Stresses and Displacements in Layered Systems," *Proc. HRB*, Highway Research Board, Washington, D.C., Vol. 23, pp. 127–148.

30. Mehta, M. R. and Veletos, A. S. (1962) "Stresses and Displacement in Layered Systems," *Structural Research Series No. 178*, Univ. of Illinois, Urbana, Ill.

31. Gaziev, E. G. and Erlikman S. A. (1971) "Stresses and Strains in Anisotropic Rock (Model Studies)," *Proc. Intl. Symp. Rock Mechs.* (Rock Fracture), Nancy, France, Sec. 11–1.

32. Zienkiewicz, O. C. (1968) "Continuum Mechanics as an Approach to Rock Mass Problems," *Rock Mechanics in Engineering Practice*, Stagg and Zienkiewicz, eds., John Wiley & Sons, New York, Chap. 8.

33. Zienkiewicz, O. C. (1971), *The Finite Element Method in Engineering Practice*, McGraw-Hill Book Co., New York.

34. Mindlin, R. D. (1936) "Force at a Point in the Interior of a Semi-Infinite Solid," *Physics* 7, pp. 195–202.

35. Antunes Martins (1948) "Influence Scale and Influence Chart for the Computation of Stresses due Respectively to Point Load and Pile Load," *Proc. 2d Intl. Conf. Soil Mechs. and Found. Engrg.*, Rotterdam, Vol. 6, pp. 70–73.

35a. Grillo, O. (1948) "Influence Scale and Influence Chart for the Computation of Stresses Due, Respectively, to Surface Point Load and Pile Load," *Proc. 2d Intl. Conf. Soil Mechs. and Found, Engrg.*, Rotterdam, Vol. 6, pp. 70–73.

36. Janbu, N., Bjerrum, L., and Kjaernsli, B. (1956) "Veiledning ved løsning av fundamenteringsoppgaver" ("Soil Mechanics Applied to Some Engineering Problems"), in Norwegian with English summary, *Norwegian Geot. Inst., Pub. 16.*

37. Bowles, J. B. (1974) *Analytical and Computer Methods in Foundation Engineering*, McGraw-Hill Book Co., New York.

38. Kany, M. (1959) *Beitrag zur Berechnung von Flachengrundungen*, Wilh. Ernst & Sohn, Berlin.

39. McCarthy, D. F. (1982) *Essentials of Soil Mechanics and Foundations*, 2d ed., Reston Publishing Co., Inc., Reston, Va.

40. Menard, L. F. (1965) "Rules for the Computation of Bearing Capacity and Foundation Settlement Based on Pressuremeter Results," *Proc. 6th Intl. Conf. Soil Mechs. and Found. Engrg.*, Montreal, Vol. 2, pp. 295–299.

41. Schmertmann, J. H. (1970) "Static Cone to Compute Static Settlement Over Sand," *Proc. ASCE, J. Soil Mechs. and Found. Engrg. Div.*, Vol. 96, No. SM3, May.

42. Schmertmann, J. H. (1977) *Guidelines for CPT Performance and Design, Pub. No. FHWA-TS-78-209*, Federal Highway Administration, Washington, D.C.

43. Schmertmann, J. H., Hartman, J. P., and Brown, P. R. (1978) "Improved Strain Influence Diagrams," *Proc. ASCE, J. Geotech. Engrg. Div.*, Vol. 104, No. GT8, August, Technical Note, pp. 1131–1135.

44. Lambe, T. W. (1973) "Predictions in Soil Engineering," *Geotechnique*, Vol. 23, No. 2, London.

45. Kavazanjian, E. and Mitchell, J. K. (1980) "Time-Dependent Deformation of Clays," *Proc. ASCE, J. Geotech. Engrg. Div.*, Vol. 106, No. GT6, June, pp. 593–610.

46. Hendron, A. J., Jr., Mesri, G., Gamble, J. C., and Way, G. (1970) "Compressibility Characteristics of Shales Measured by Laboratory and In Situ Tests," *Determination of the In Situ Modulus of Deformation of Rock*, STP 477, Amer. Soc. for Test. and Mat., Philadelphia, Pa., pp. 137–153.

47. Ward, J. S. (1972) "Insitu Pressuremeter Testing on Two Recent Projects," *Soils*, J. S. Ward & Assocs., Caldwell, N.J., November, pp. 5–7.

48. Martin. R. E. (1977) "Estimating Settlements in Residual Soils," *Proc. ASCE, J. Geotech. Engrg. Div.*, Vol. 103, No. GT3, March, pp. 197–212.

49. Skempton, A. W. (1953) "The Colloidal Activity of Clays," *Proc. 3d Intl. Conf. Soil Mechs. and Found. Engrg.*, Zurich, Vol. 1, pp. 57–61.

50. Basu, R. and Arulanandan, K. (1973) "A New Approach to the Identification of Swell Potential for Soils," *Proc. 3d Intl. Conf. on Expansive Soils*, Haifa, Israel.

51. Williams, A. A. B. and Donaldson, G. W. (1980) "Building on Expansive Soils in South Africa: 1973–1980," *Proc. 4th Intl. Conf. Expansive Soils*, Denver, Colo., Vol. II, pp. 834–844.

52. Van der Merwe, D. H. (1975) Contribution to Speciality Session B, "Current Theory and Practice for Building on Expansive Clays," *Proc. 6th Reg. Conf. for Africa on Soil Mechs. and Found. Engrg.*, Durban, South Africa, Vol. 2, pp. 166–167.

53. USBR (1974) *Earth Manual*, U.S. Bureau of Reclamation, Federal Center, Denver, Colo.

54. Snethen, D. R. (1980) "Characterization of Expansive Soils Using Soil Suction Data," *Proc. ASCE, 4th Intl. Conf. on Expansive Soils*, Denver, Colo. Vol. 1, pp. 54–75.

55. Krazynski, L., Escario, V., Fredlind, D. G., Lytton, R. L., Holland, J., McKeen, R. G., Mitchell, P., and Williams, A. A. B. (1980) "Panel discussion, Session 2," *Proc. ASCE, 4th Intl. Conf. on Expansion Soils*, Denver, Colo., Vol. 2, pp. 683–715.

56. Vargas, M. (1972) "Fundação de Barragems de Terra Sobre Solos Porosos," *VIII Seminario Nacional de Grandes Barregems*, São Paulo, November.

57. Clevenger, W. A. (1958) "Experience with Loess as a Foundation Material," *Proc. ASCE, J. Soil Mech. and Found. Engrg. Div.*, Vol. 82, No. SM3.

58. Jennings, J. E. and Knight, K. (1975) "A Guide to Construction on or with Materials Exhibiting Additional Settlement Due to 'Collapse' of Grain Structure," *6th Regional Conf. for Africa on Soil Mechs. and Found. Engrg.*, September, pp. 99–105.

59. Clemence, S. P. and Finbarr, A. O. (1981) "Design Considerations for Collapsible Soils," *Proc. ASCE J. Geotech. Engrg. Div.*, Vol. 107, No. GT3, March, pp. 305–317.

60. Hetenyi, M. (1946) *Beams on Elastic Foundation*, Univ. of Michigan Press, Ann Arbor, Mich.

61. Bowles, J. E. (1975) "Combined and Special Footings," *Foundation Engineering Handbook*, Winterkorn and Fang, eds., Van Nostrand-Reinhold Co., New York, Chap. 16.

62. Terzaghi, K. (1955) "Evaluation of the Modulus of Subgrade Reaction," *Geotechnique*, Vol. 4, London, pp. 297–326.

62a. Broms, B. B. (1964) "Lateral Resistance of Piles in Cohesionless Soils," *Proc. ASCE, J. Soil Mech. and Found. Engrg. Div.*, Vol 90, No. SM3, pp. 123–156.

63. Broms, B. B. (1964) "Lateral Resistance of Piles in Cohesive Soils," *Proc. ASCE, J. Soil Mech. and Found. Engrg. Div.*, Vol. 90, No. SM2.

64. Vesić, A. B. (1961) "Beams on an Elastic Subgrade and Winkler's Hypothesis," *Proc. 5th Intl. Conf. Soil Mechs. and Found. Engrg.*, Paris, Vol. 1, p. 845.

65. Teng, W. C. (1962) *Foundation Design*, Prentice-Hall, Englewood Cliffs, N.J.

66. Teng, W. C. (1975) "Mat Foundations," *Foundation Engineering Handbook*, Winterkorn and Fang, eds., Van Nostrand-Reinhold Co., New York, Chap. 17.

67. Haddadin, M. (1971) "Mats and Combined Footings—Analysis by Finite Element Method," *ACI J.*, American Concrete Institute, December.

68. Eide, O. (1974) "Marine Soil Mechanics—Applications to North Sea Offshore Structures," *Offshore North Sea Technology Conf.*, Norwegian Geotechnical Institute Pub. No. 103, Oslo.

69. Poulos, H. G. and Davis, E. H. (1980) *Pile Foundation Analysis and Design*, John Wiley & Sons, New York.

70. Matlock, H. and Reese, L. C. (1960) "Generalized Solutions for Laterally Loaded Piles," *Proc. ASCE, J. Soil Mech. and Found. Engrg. Div.*, Vol. 86, No. SM5, October.

71. McClelland, B. and Focht, J. A., Jr. (1958) "Soil Modulus for Laterally Loaded Piles," *Trans. ASCE*, Vol. 123, Part VII, Paper No. 2954, pp. 1049–1086.

72. Reese, L. C. (1977) "Laterally Loaded Piles: Program Documentation," *Proc. ASCE, J. Geotech. Engrg. Div.*, Vol. 103, No. GT4, April, pp. 287–304.

73. Francis, A. J. (1964) "Analysis of Pile Groups with Flexural Resistance," *Proc. ASCE, J. Soil Mech. and Found. Engrg. Div.*, Vol. 90, No. SM3.

74. API (1979) *Recommended Practice for Planning, Designing and Constructing Fixed Offshore Platforms*, 10th ed., *Amer. Petroleum Inst.*, Washington, D.C.

75. de Ruiter, J. and Beringen, F. L. (1978) "Pile Foundations for Large North Sea Structures," *Marine Technology*, Vol. 3, No. 2.

76. Bogard, D. and Matlock, H. (1980) "Simplified Calculations of *p-y* Curves for Laterally Loaded Piles in Sand," *Offshore Tech. Conf.*, Houston, Tex.

77. McClelland, B. (1972) "Design of Deep Penetration Piles for Ocean Structures," *Terzaghi Lectures: 1963–1972*, ASCE, New York (1974), pp. 383–421.

78. API (1971) "Recommended Practice for Planning, Designing and Constructing Fixed Offshore Platforms," Amer. Petroleum Inst., *RP 2A*, January.

78a. API (1982) "Recommended Practice for Planning, Designing and Constructing Fixed Offshore Platforms," Amer. Petroleum Inst., *RP 2A*, 13th ed., January.

78b. DNV (1977), *Rules for the Design, Construction and Inspection of Offshore Structures*, Det Norske Veritas, Oslo, App. F, Foundations.

79. Reese, L. C., Cox, W. R., and Koop, F. D. (1974) "Analysis of Laterally Loaded Piles in Sand," *Offshore Technology Conf. Paper OTC 2080*, Houston, Tex.

80. Matlock, H. (1970) "Correlations for the Design of Laterally Loaded Piles in Soft Clay," *Preprint, 2d Annual Offshore Technology Conf.*, Houston, Tex., Vol. 1, pp. 577–594.

81. Skempton, A. W. (1951) "The Bearing Capacity of Clays," *Proc. Building Research Cong.*, Div. 1, London.

82. Reese, L. C. and Welch, R. C. (1975) "Lateral Loading of Deep Foundations in Stiff Clay," *Proc. ASCE, J. Geotech. Engrg. Div.*, Vol. 101, No. GT7, July, pp. 633–649.

83. Tsui, Y. and Clough, G. W. (1974) "Plane Strain Approximations in Finite Element Analysis of Temporary Walls," *Proc. ASCE, Conf. on Analysis and Design in Geotechnical Engrg.*, Univ. of Texas, Vol. 1, pp. 137–197.

84. Seed, H. B., Lysmer, J., and Hwang, R. (1975) "Soil-Structure Interaction Analysis for Seismic Response," *Proc. ASCE, J. Geotech. Engrg. Div.*, Vol. 101, No. GT5, May, pp. 439–458.

85. Schnabel, P., Seed, H. B., and Lysmer, J. (1971) "Modifications of Seismograph Records for Effects of Local Soil Conditions," *Report No. EERC 71-8*, Earthquake Engineering Research Center, Univ. of California, Berkeley, December.

86. Whitman, R. V. and Richart, F. E., Jr. (1967) "Design Procedures for Dynamically Loaded Foundations," *Proc. ASCE, J. Soil Mech. and Found. Engrg. Div.*, Vol. 93, No. SM6, November, pp. 169–191.

87. Idriss, I. M. and Sadigh, K. (1976) "Seismic SSI of Nuclear Power Plant Structures," *Proc. ASCE, J. Geotech. Engrg. Div.*, Vol. 102, No. GT7, July, pp. 663–682.

88. Peck, R. B. (1976) "Rock Foundations for Structures," *Rock Engineering for Foundations and Slopes, Proc. ASCE*, New York, Vol. 2, pp. 1–21.

89. Deere, D. U. (1968) "Geological Considerations," *Rock Mechanics in Engineering Practice*, Stagg and Zienkiewicz, eds., John Wiley & Sons, New York, Chap. 1.

90. Hobbs, N. B. (1975) "Factors Affecting the Prediction of Settlement of Structures on Rock: With Particular Reference to Chalk and Trias," *Proc. Conf. on Settlement of Structures*, Pentech Press, London, pp. 579–610.

91. University of California (1965) "Stresses and Deflections in Foundations and Pavements," Dept. of Civ. Engrg., Univ. of California, Berkeley, Fall.

PART II

SITE PLANNING AND PREPARATION, GROUND IMPROVEMENT, EARTH STRUCTURES, AND FOUNDATIONS

CHAPTERS

PURPOSE AND SCOPE

Part II covers the geotechnical elements of site selection, planning, and preparing a site for construction; methods of improving the natural ground; the fabrication of earth structures, including load-bearing fills, earth- and rock-fill dams, and pavements; and shallow and deep foundations, including the use of soil and rock anchors.

Site selection and planning require a thorough investigation and a comprehensive assessment of geologic conditions to identify potentially hazardous conditions which should be avoided, to permit preparation of site plans optimizing land use from the viewpoint of geologic conditions, and to permit preparation of estimates of site development costs.

Site preparation for building, which always includes earthwork and the installation of foundations, also may require consideration of the impact of construction on adjacent structures from ground vibrations, dewatering excavations, and other factors. Earthwork includes stripping, grading, excavation, and filling, performed with due attention to surface and subsurface water control.

Natural ground conditions may be improved during site preparation by compaction, reinforcement, treatment of pore water, or stabilization with admixtures, and such techniques should be considered in conjunction with foundation selection and design criteria development. Techniques of *fabricating earth structures* in layers by mechanical compaction must be applied properly to obtain a load-bearing fill suitable for building or pavement support or a properly functioning earth dam.

From the viewpoint of the geotechnical engineer, the satisfactory performance of a structure depends mainly on the adequacy of its foundation, which, however, should not be more costly than required for safety. *Shallow foundations*, including spread and continuous footings and mats, usually are designed on the basis of an allowable soil- or rock-bearing value selected to restrict settlements to within tolerable limits, but may require considerations of the potential for failure by ground rupture or of heave from ground swelling.

Deep foundations include piles, piers, and caissons, and for piles in particular, there is a large assortment to chose from. Design normally is based on selecting a pile capacity by static analysis, dynamic methods, or load tests, or all three in combination, and then applying safety factors and considering group effects. Settlement analysis is performed after pile group dimensions are established. In some cases, in particular for offshore structures, lateral forces on piles are a most important consideration. Soil and rock anchors are designed from principles similar to those used to evaluate piles.

CHAPTER FIVE

SITE PLANNING AND PREPARATION, GROUND IMPROVEMENT, AND EARTH STRUCTURES

5.1 SITE SELECTION AND DEVELOPMENT PLANNING

5.1.1 SITE SELECTION FACTORS

Geologic Assessment

With respect to geologic conditions, the selection of a site for a particular project requires assessments of, firstly, the potential major geologic hazards and, secondly, the more routine conditions. The latter are classed as poor to favorable *(see Art. 5.1.2)* as regards to their possible impact on foundations, site preparation, and other elements of development. Potential owners often consider the land cost of development to be merely the cost of the raw land, whereas in reality the *true land cost* equals the cost of the raw land plus the cost of preparing the site for development plus the cost of foundations. Such analysis is particularly significant when several alternate sites are being considered for purchase.

Evaluations are based on comprehensive geotechnical investigations, which in many cases should be performed prior to purchase of land or the initiation of development planning. A site may be abandoned if ground conditions are so hazardous or poor as to require costly, or even impractical, solutions to provide suitable support for structures.

Site Preparation Factors

Site development planning and cost analysis require evaluation of a number of important factors included under the general category of site preparation, or the preliminary phases of construction, some of which are illustrated in Fig. 5.1, as follows:

- *Earthwork,* which includes stripping, grading, excavating, and filling
- *Surface water control* from the aspects of flooding and erosion
- *Ground improvement* by the removal or strengthening of weak geologic materials
- *Slope retention structures* to reduce the extent of earthwork grading
- *Foundation requirements* including types of foundations, excavation of material, retention of excavation walls, and dewatering to permit work in the dry
- *Impact on adjacent structures* from dewatering, excavation, blasting, pile driving, etc.

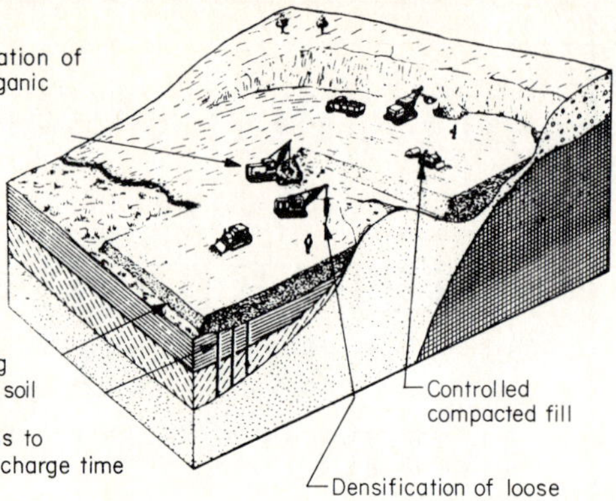

FIG. 5.1 Some typical procedures for site preparation and ground improvement for a marginal site for building construction. Surface drainage and erosion also require control. (Sketch courtesy of Converse Consultants.)

Ground Improvement Potential

Many ground conditions which may appear to be unfavorable for economical development can be improved. Many techniques are available to decrease permeability or to increase the strength and decrease the compressibility of soil formations and rock masses, and these should be considered during the early site planning stages. Applications vary with the type of soil formation or rock-mass condition, the depth to which improvement is desired, and the proposed construction. Some techniques are applicable also to postconstruction corrective measures.

5.1.2 GEOLOGIC FACTORS

Rating Categories

Thorough geologic information must be obtained not only for the site area, but also regionally, especially if the designer has no prior experience in the area. Geologic conditions, as they may impact on surface construction, can be placed in three general categories:

- *Hazardous:* A regional or local phenomenon that can result in disaster, and for which treatment may not be economically feasible or even practical. Such locations should be avoided in many cases.
- *Poor:* A local condition, occasionally regional, requiring costly, special treatments for development.
- *Favorable:* Local or regional conditions for which normal foundation treatments are suitable. The treatments are a function of the structure's characteristics as well as of the ground conditions.

Hazardous Ground Conditions

Sites located over or in close proximity to *faults* in seismically active regions may be subjected to ground rupture and severe ground shaking [Art. 11.3.1 in Hunt (1984)[1]] and should be avoided.

Liquefaction may occur in loose to medium dense saturated fine sands in seismically active regions, resulting in complete loss of ground support or large lateral movements in areas adjacent to rivers and bays [Art. 11.3.3 in Hunt (1984)[1]]. These sites should be avoided if possible since treatment is costly and the potential risk may remain high.

Ground collapse may occur where sites are underlain by dormant or active mines, particularly coal mines, or by cavernous limestone [Arts. 10.3 and 10.4 in Hunt (1984)[1]]. Collapse in the latter case often is triggered by groundwater withdrawal from adjacent sites over which the designer has no control. Treatments are costly and if the risk will remain high, then such sites are best avoided.

Slope failures, such as landslides, avalanches, or flows, are particularly prone to occur in moist climates in glacial lacustrine clays, clay shales, colluvium, thick deposits of residual soils on steep slopes, dipping stratified rock masses, or any slope cut too steeply for the natural conditions and not properly stabilized. Some failure forms occur suddenly, without warning, contain enormous masses which attain great velocities, and are essentially unpreventable. A site with a high hazard potential should be avoided [Chap. 9 in Hunt (1984)[1]].

Floodplains and other relatively low ground require provisions for flood protection with adequately high grades, diking, and surface drainage control. Flood potential is increased by regional subsidence and floodplain development over which the designer has no control. Sites must be selected carefully [Art. 8.2.3 in Hunt (1984)[1]].

Poor Ground Conditions

Collapse and subsidence occur on saturation in loess, "porous" clays, and the lightly cemented, low-density recent alluvium of arid-climate valleys. Partial saturation may also result in subsidence. Saturation must be prevented; the collapsible zone must be removed by excavation, precollapsed, or bypassed; or the structure must be designed with large potential settlement in view [Art. 10.5 in Hunt (1984)[1]].

Swelling clays and rocks, generally highly overconsolidated and associated with climates with long dry periods, undergo large volume changes with changes in moisture content. Even small moisture changes can result in volume increase, high uplift pressure, and building heave brought about merely by a structure covering the ground. The most suitable foundation treatment is to bypass the "active" zone which is usually the zone above the deepest water table as affected by climatic changes [Art. 10.6 in Hunt (1984)[1]].

Soft to firm clays rupture readily if their supporting capacity is exceeded, and even under low applied loads they undergo long-term consolidation of significant magnitude. Except for very light loads, deep foundations normally are required where the depth is too great for removal and replacement of the compressible soils.

Organic soils vary widely from organic silts and clays to rootmat, are the most compressible of natural materials, and seldom provide suitable support for structures, except for preloaded fills. Where beneath structures, they usually are either removed and replaced with strong soils or bypassed with deep foundations. In the latter case, downdrag forces from consolidation of the organic soils under fill loads can be substantial and may even exceed the structural load. Preloading reduces the downdrag forces.

Groundwater, as indicated by the position of the water table (or positions as indicated by perched water tables or artesian conditions) impacts in a number of ways. It must be considered not to be a static geologic feature but rather to undergo large seasonal and periodic variations. Excavations below the water table in granular soils require dewatering; an unforeseen water table rise can cause uplift of basement floors, or of a structure itself; water table changes affect the active zone in swelling soils; and groundwater pumping can result in surface subsidence by causing ground impression or even ground collapse, such as occurs when overburden weight over the arch of a sinkhole or limestone cavern is increased.

Shallow rock may require costly blasting for removal for site grading and deep excavations.

Favorable Ground Conditions

Cohesive granular soils, such as sand-clay mixtures, are relatively strong and with suitable allowable bearing values often provide support for moderately to heavily loaded foundations.

Cohesionless granular soils, including medium dense to dense sands and sand-gravel mixtures, usually provide good support for most loading conditions. Loose sands and silts, however, may require deep foundations or ground improvement if a suitable bearing value is not available for shallow foundations. Sands are subject to scour from leaking water mains or beneath bridge foundations.

Shallow rock usually provides the best foundation support if it exists. For building foundations on level ground, support capability depends on rock quality, primarily in terms of the degree of decomposition, although the characteristics of discontinuities, particularly cavernous conditions, may govern design. For structures on slopes or for concrete dams, in addition to rock quality, capacity depends upon the potential seepage forces which may develop.

5.1.3 INVESTIGATIONS

Investigation Planning

The importance of a completely thorough investigation to define the geologic conditions on which the evaluations necessary to achieve safe and economical construction are based cannot be over emphasized. The practitioner should not accept the tendency of many clients to reduce investigation costs if it is believed that such reductions will jeopardize *confidence* in having adequate knowledge regarding site geotechnology.

The basic considerations in planning an investigation are the lateral and vertical extent to which the proposed construction or development will significantly influence the geologic environment or be influenced by the environment.

Lateral Extent of Influence

The lateral extent of influence can vary from the immediate area of the project, to many meters beyond if dewatering of excavations is anticipated, to an entire valley or region where deep mineral extraction is planned or is occurring, and even to several hundred kilometers where area seismicity is of concern.

Vertical Depth

For structures on or near the surface, the vertical influence is a function of several variables, including the size and loading of the structure which imposes changes on the natural stress conditions to some significant depth *(see Art. 4.2)*; unfavorable geologic conditions which require explorations to depths adequate to define suitable support for deep foundations; and excavations which require explorations to some distance below their maximum depth to determine groundwater, soil, and rock conditions both within and below the excavation zone. Deep excavations for tunnels and mines and the deep extraction of fluids require knowledge of geologic conditions between the point of excavation or extraction from the surface, and at times between even deeper points and the surface.

Investigations can be divided into a number of *phases* based on purpose, such as feasibility, preliminary, etc., with each phase containing various *stages*, such as exploration, testing, etc. Economies are eventually realized by performing investigations in steps, since the scope of each step depends upon the information gathered in the previous ones.

Stages

Each stage has a different objective, and each requires interpretation, analysis, and evaluation.

1. *General geologic conditions*, locally and regionally, are identified through the retrieval of existing data, interpretation of remote-sensing imagery (terrain analysis), and field reconnaissance. Engineering geology maps are prepared [see Art. 2.2 in Hunt (1984)[1]].
2. *Subsurface sections* are prepared from exploration data obtained by test borings and pits, geophysical methods, etc. [see Art. 2.3 in Hunt (1984)[1]].
3. *Samples* are recovered of the soils and rocks for identification and laboratory testing [see Art 2.4 in Hunt (1984)[1]].
4. *Measurements of engineering properties* are made in the laboratory and in situ [see Chap. 3 in Hunt (1984)[1] and Chaps. 2, 3, and 4 in this volume].
5. *Instrumentation* is installed to monitor ground and structural response to changing field conditions [see Chap. 4 in Hunt (1984)[1]].

Feasibility and Preliminary Site Planning Phases

Feasibility

This phase is imperative when the investigator has no prior experience in the project area, or the site is located in virgin or rural country, or for projects covering large land areas. The purpose is to obtain a general knowledge of regional and site conditions, with emphasis on those which are potentially so hazardous or unfavorable that they would cause construction to be either risky with respect to safety or extraordinarily costly. Investigation includes Stage 1, and often parts of 2 and 3. If the site is high-hazard, the decision may be made to abandon it and to search for one with more favorable conditions.

If the site is marginal, preliminary planning can consider optimum utilization of favorable areas and avoidance of unfavorable areas to minimize construction costs. On sites with no obviously severe constraints, preliminary designs and cost estimates may be prepared. In addition, the data gathered provide the basis for intelligent planning of more detailed investigations, as the selection of proper methods and tools is very much dependent upon geologic conditions.

Preliminary Site Planning

Using the preliminary information as a basis, the designer can lay out the structure or structures on the site, maximizing use of favorable conditions, and determine structure dimensions, loading conditions, and other features such as final grades and basement levels. If soil conditions are poor, consideration may be given to providing a relatively light structure that is flexible to increase its settlement tolerance for support on shallow foundations, or to ground improvement. If the groundwater level is high or hard rock is shallow, consideration may be given to avoiding deep excavations.

Among the most common causes of extra construction costs are groundwater and rock at unanticipated elevations and sites covered with old miscellaneous fills. Poor building performance often results from buried stream channels or limited thicknesses of buried weak soils, not accounted for in design, as well as misinterpretation of the supporting capabilities of the geologic materials. Adequate feasibility studies should disclose the possibility that such conditions exist or identify their existence in a general manner to be confirmed by more comprehensive study.

Preliminary or Final Design Phase

Purpose and Scope

The preliminary design phase usually proceeds when locations, dimensions, and loadings of the proposed construction have been defined. Its purposes are the comprehensive determination of the distribution of the geologic materials and their engineering characteristics and the establishment of criteria for foundation design and other aspects of site development. Investigation includes Stages 2, 3, and 4, and on many projects may be combined with the feasibility phase. On routine projects with favorable subsurface conditions this phase may be final.

The data are used for the selection and dimensioning of foundations and retaining structures, for groundwater control and slope stabilization, and for choosing tunnel and pavement support systems, in conjunction with ground improvement techniques. Where subsurface conditions are complex and imply potential difficulties, their presentation in three-dimensional diagrams, as illustrated in Fig. 5.2, is much more informative than the standard geologic sections.

Information Required from Designer

For this phase of investigation it is necessary that the designer provide information on the dimensions of the structure, final grades in reference to existing grade (both above grade required for site filling and below grade required for basements, pits, etc.), and loading conditions. Loading data should include total area loads and wall and column loads, divided into dead and live loads.

Dead loads normally consist of the weight of the structure itself, including permanent installations supported by the structure. Permanent installations placed toward the end of construction completion, such as turbines or other heavy machinery, should be considered separately from the weight of the structure itself.

Live loads are the loads imposed on a structure by snow, wind, and seismic forces, as well as by transient loads placed on floors and bulk storage materials placed in tanks, silos, etc. Floor loads in warehouses and bulk storage in tanks normally are much more significant than the weight of the structure in producing settlements, and in the cases of silos and tanks in producing both settlement and possibly ground rupture. *Dynamic live loads* are imposed by rotating or reciprocating machinery, blasting and other construction activities, and earthquakes. Determination of earthquake loadings may be part of the geotechnical investigation [see Art. 11.5 in Hunt (1984)[1]].

Supplemental Final Design Phase

Sites with difficult conditions requiring the designer to change building locations or dimensions or to confirm or modify designs after completion of the preliminary phase are subject to a supplemental final investigation. The purpose is

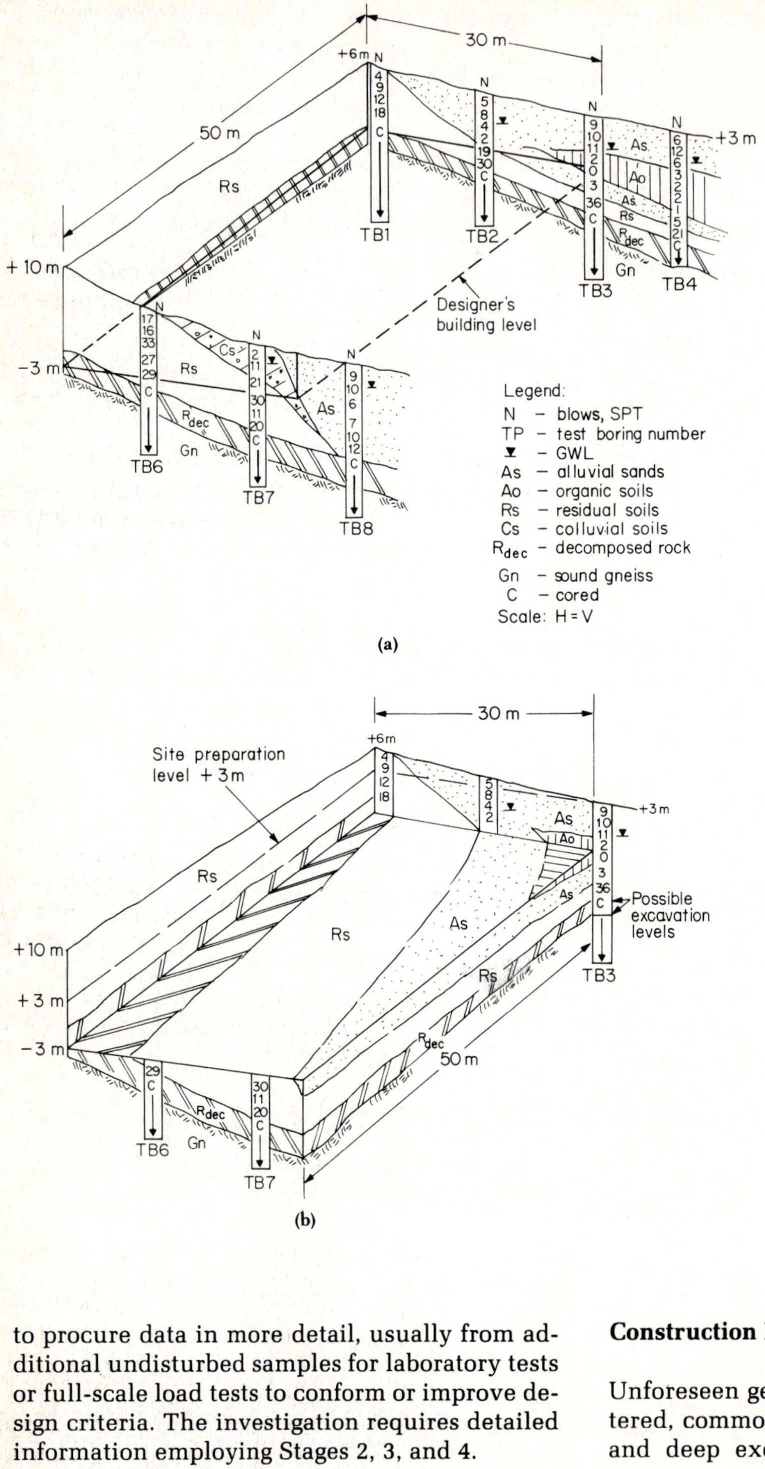

FIG. 5.2 Geologic diagrams illustrate conditions at and below the building level proposed by the designer for a Class I structure in a nuclear power plant complex where support on material of only one type is required. The diagrams show clearly the conditions for foundations and for excavation support. To reach decomposed rock will require a maximum excavation of 13 m and to reach sound rock, 15 m. Backfill would be controlled fill or rollcrete depending on requirements of the licensing agency. (*a*) Geologic conditions. (*b*) Geologic conditions at designer's building level.

to procure data in more detail, usually from additional undisturbed samples for laboratory tests or full-scale load tests to conform or improve design criteria. The investigation requires detailed information employing Stages 2, 3, and 4.

Construction Phase

Unforeseen geologic conditions may be encountered, commonly on sites with complex geology and deep excavations, or unconventional designs may be developed along with contingency plans. In either event, additional investigation may be required during almost any stage of construction—Stages 2, 3, 4, and frequently 5. Instrumentation is installed to monitor changing groundwater levels; stability and movements of slopes and retaining structures; settlements of fills, embankments, and foundations; ground subsidence and its effect on adjacent structures; and deformations and stresses in retaining structures and underground support systems.

Instrumentation provides an early warning system and the data base for decisions to invoke contingency plans when deformations and stresses exceed those anticipated during design.

Postconstruction Phase

Use of instrumentation (Stage 5) is good practice for heavy foundation loadings or other conditions in which settlements are anticipated; for retaining structures in critical conditions; and for slopes, dams, and tunnels. The purpose is to provide early warning of potentially troublesome conditions and to provide data to advance the state of the art and reduce the degree of conservatism on similar projects.

5.1.4 SITE PREPARATION

Factors

This article covers components of the site preparation including earthwork (site grading), runoff and seepage control, impact on adjacent structures, and ground improvement techniques, all of which impact the cost. In the remainder of the chapter, although the emphasis appears to be on ground improvement, it is seen that site preparation and ground improvement are closely related. Retaining structures and foundations are also elements under the general subject of site preparation that impact the cost and are described in subsequent chapters.

Earthwork (Site Grading)

The amount and nature of earthwork depends upon topographic and geologic conditions and includes stripping, grading, compacting fill (structural fill), and disposal of excavated waste materials.

Stripping of an upper topsoil layer is necessary for buildings, pavements, and other structures and can be a major cost factor where large paved areas are planned such as for parking, roadways, and airfields. The thickness of stripping should be well defined by investigation, for which numerous hand-auger holes and test pits are useful *(see also Art. 5.2.4)*.

Grading, which consists of excavation and filling and is necessary for the preparation of level building sites, can be costly where topography is highly irregular, and in some conditions retaining structures are required. Excavation grading costs are affected primarily by the presence of rock and its rippability by mechanical equipment versus the need for blasting [see Art. 3.2.2 of Hunt (1984)[1]]. A chart relating rock rippability to seismic *P*-wave velocities as measured by field tests is given as Table 5.1. It is often economical to remove soft soils where they extend to depths of 3 to 4 m below the surface. Depressions on which structures, pavements, or other settlement-sensitive structures will be located should be brought to grade with controlled, compacted fill.

Compacted fill is also used for the construction of site grades for flood protection, to support ground floors of buildings, and for dam and roadway embankments. In a site grading plan, the suitability of excavated material for use as compacted fill requires assessment. Where large areas are involved, as for industrial and commercial sites, roadways, railroads, and airports, the attempt is made to balance the cut and fill volumes in the selection of final grades. In other cases, such as filling lowlands or constructing large roadway embankments and earth dams, *borrow* sources for fill must be located. This is an important element of investigation.

Disposal of waste materials from excavations is a problem to be considered for such works as open-pit mines, underground mines, and tunnels.

Runoff and Seepage Control

Surface Water Runoff and Erosion

In all site preparation plans for earthwork, consideration is given to proper site grading to con-

TABLE 5.1
ROCK RIPPABILITY AS RELATED TO SEISMIC *P*-WAVE VELOCITIES *(COURTESY CATERPILLAR TRACTOR CO.)*

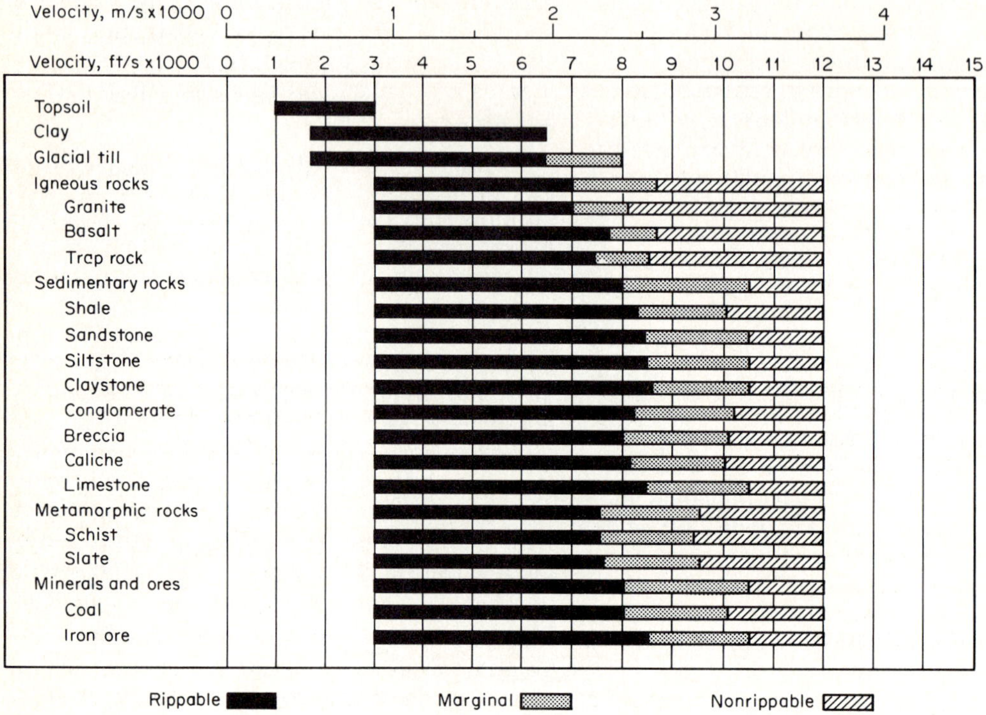

*Rippability based on Caterpillar D9 tractor with mounted hydraulic No. 9 ripper.

trol runoff and erosion and prevent flooding. Proper grading avoids ponding water, controls flow velocities, and directs runoff to ditches for offsite discharge. On sites covering large areas, erosion and the subsequent siltation of adjacent bodies of water can be a severe problem unless provided for and controlled. Erosion control measures including linings for channels and ditches and drainage control, vegetation, and other methods for slopes are described in Art. 8.2.2 of Hunt (1984).[1]

Groundwater Control

Surface and subsurface excavations below the water table require control of seepage which is accomplished either by dewatering or by the installation of barriers, summarized in Art. 2.3.3 and described in detail in Art. 8.4 of Hunt (1984).[1] Suitable dewatering methods depend on geologic conditions and can include sump pumps, wellpoints, deep gravity or pumped wells, interceptor drains, or electroosmosis, all of which are applicable also to slope stability treatment. *Barriers* are provided by concrete or clay cutoff walls, slurry walls, grouting, or freezing.

Impact on Adjacent Structures

Site preparation and construction can affect adjacent structures in a number of ways as follows:

Ground subsidence causing building settlements can result from dewatering excavations [see Art. 10.2.1 of Hunt (1984)[1]], from deflection of excavation retaining structures *(see Art. 9.3.4)*, from tunneling operations *(see Arts. 10.2.4 and 10.3.5)*, from driving piles in free-draining soils *(see Art. 7.2.2)*, or from the superposition of stresses *(see Art. 6.2.3)*. In areas underlain by cavernous limestone, dewatering can cause ground collapse [see Art. 10.4.2 of Hunt (1984)[1]].

Ground heave can be caused by driving piles in clays (see Art. 7.2.2).

Vibrations result from blasting rock, driving piles, or other construction activity (see Art. 5.6).

Ground Improvement

The removal of undesirable materials and the control of groundwater are two general methods of improving ground conditions. Another large group of methods improves ground conditions by strengthening the geologic materials. Most methods are applicable during the site preparation phases of development, while others can be used to correct poor ground conditions which are causing settlements of existing structures or other undesirable events. The *objective* of ground improvement during site preparation may be, for example, to change unfavorable conditions into those more suitable for the support of structures on shallow foundations, or to reduce downdrag forces on deep foundations, or to reduce pavement thickness requirements, or to make use of geologic materials for the construction of dams and embankments.

Technique Classes

Compaction of soils to increase their density is achieved by a number of methods. Fills and embankments are constructed by compacting borrow materials in layers with mechanical equipment. Surface compaction in situ also is achieved with mechanical equipment. Surcharges are used to strengthen weak soil formations generally by the application of a surface load and at times by dewatering to increase overburden pressures. Hydrocompaction is used for collapsible soils. Vibroflotation, the Terra-probe, and dynamic compaction are methods used to increase soil density in situ. Compaction grouting is used in both soil and fractured rock masses.

Ground reinforcing is accomplished by the inclusion of some foreign material into the soil formation. Injections with grout are suitable in any soil or rock formation where flow is possible. Geotextiles are installed in excavations or embankments. Columns of stone, sand, or lime are installed in weak soil formations; the column material is selected after consideration of the objectives and the type of geologic material. Bolts and cable anchors are installed in fractured rock masses.

Pore-water treatments to increase strength in saturated soils include vacuum wellpoints in silt and silty sand layers, electroosmosis in silts, and freezing. The latter may be particularly useful in deep deposits of silty sands and free-draining materials.

Soil stabilization is also accomplished with admixtures which may include gravel, cement, bitumen, lime, or calcium and sodium chlorides.

Geologic Conditions and Some Improvement Techniques

Surficial soils can be stripped when they are undesirable, subjected to surface compaction in situ if they are free-draining, or stabilized with admixtures if they are nonorganic.

Shallow weak soil formations, such as a few meters of soft organic or clay soil, or collapsible soils, often are treated economically by excavation and replacement with a compacted fill of good soil material (the excavation-backfill method). Geotextiles are used to support embankments constructed over weak soils where some settlement is tolerable. Surface compaction of loose sands to maximum depths of about 2 m can be achieved with some types of mechanical equipment.

Deep deposits of weak organic soils or clays are often strengthened by preloading or surcharging for some interval of time. Sand or wick drains reduce the time required for the preload operation. Stone columns or sand piles penetrating through organic soils or soft clays, or lime columns in soft clays (nonsaline), may serve to reduce settlement of embankments or lightly loaded structures to within tolerable limits. In cases where the weak stratum is buried, a surcharge may be applied on the surface or by dewatering to increase overburden weight. Buried organic strata have been strengthened by dynamic compaction or, in a case where an existing building is undergoing settlement, by compaction grouting.

Deep formations of loose sands are strengthened by vibroflotation or dynamic compaction or in some cases by surface compaction if relatively small spread-footing foundations are suitable for support of light to moderate loads. Buried deposits of loose sands are strengthened by grouting, even for postconstruction remedial treatments.

Deep formations of silty sands and silts are strengthened by vacuum wellpoints, grouting, or freezing, usually prior to excavation. Electroosmosis is applied to silts for slope stabilization or prior to excavation.

Expansive soils are generally not suitable for treatment for the preparation of structure foundations by methods such as presaturation or the use of lime mixtures [see Art. 10.6.4 of Hunt (1984)[1]], but stabilization with lime is a suitable treatment to improve the strength of clay soils and to reduce their activity where the treated soil will support pavements.

Collapsible soils of substantial depth have been treated by hydrocompaction, or ground saturation, to cause collapse of their lightly cemented structure. Dynamic compaction should also improve their supporting capability. Lime stabilization has been used on occasion to arrest settlements occurring in light structures [see Art. 10.5.4 of Hunt (1984)[1]].

Rock masses are strengthened by grouting or the use of rock bolts or cable anchors.

Slopes in soil or rock undergo an improvement in stability with the control of surface and subsurface drainage as discussed in Chap. 7.

A general summary of ground-strengthening techniques for various geologic conditions is given in Table 5.2.

5.2 CONTROLLED COMPACTED FILL FOR EARTH STRUCTURES

5.2.1 BASIC FACTORS

Description and Applications

Controlled compacted fill is the placing of soil or other nonorganic earth materials in layers of predetermined thickness and moisture content and compacting it by suitable mechanical equipment to achieve a specified density, in the presence of experienced technical surveillance and inspection. *Rollcrete* is a compacted aggregate-cement mixture.

Applications are numerous and under this article are considered to include

- Fills used to support structures and pavements where the objective is to obtain a high supporting capacity for the imposed loads.
- Construction of earth- and rock-fill dams where the objectives are to obtain adequate strength to assure stability without undergoing significant deformations and to provide a stable barrier so as to contain water with a controlled amount of seepage through, around, and beneath the embankment.
- Compaction of pavement layers including base, subbase, and subgrade.

Material Selection Factors

Soils have various levels of quality for use as fills, and the selection of borrow material for fill requires consideration of the purpose of the fill, the material available, and climatic conditions during placement. In general, organic soils are never used for load-bearing fills, and soft, saturated clays are used only under very unusual circumstances such as the impervious core for an earth dam in a wet climate.

Structure foundations and floors are best supported on free-draining, well-graded granular soils, with less than 10 to 15% passing the no. 200 sieve. Moisture buildup from capillary forces resulting in building dampness and possible frost heaving is minimized, and when such soils are compacted properly allowable soil pressures in excess of 4 tsf can be obtained. Silts and clays are avoided where possible because of the difficulties of obtaining adequate support capacity by compaction, as well as construction difficulties arising from their response to increased moisture during rainy periods. Other materials are suitable, such as cohesive mixtures, but capillary cutoffs are required immediately beneath the structure floor, and allowable bearing values are somewhat lower than for clean coarse-grained granular materials. *Hydraulic fills* of silt-sand mixtures, uncompacted except for surface compaction at final grade, have been used to support lightly loaded structures such as houses.

Earth- and rock-fill dams are often constructed of a wide range of materials. Since the major objec-

TABLE 5.2
GROUND-STRENGTHENING TECHNIQUES SUMMARIZED

Conditions	Technique	Application
Low grades	Compacted sand fill	Minimize structure settlements*
Miscellaneous fill		
Shallow	Excavate-backfill	Minimize structure settlement
Deep	Dynamic compaction	Reduce structure settlement†
	Sand columns	Reduce structure settlement
Organics		
Shallow	Excavate-backfill	Minimize structure settlement
	Geotextiles	Support low embankments
Deep	Surcharge	Reduce structure settlement
	Geotextiles	Support low embankments
	Sand columns	Reduce structure settlement
Buried	Surcharge	Reduce structure settlement
	Dynamic compaction	Reduce structure settlement
	Compaction grouting	Arrest existing structure settlement
	Sand columns	Reduce structure settlement
Soft clays		
Shallow	Excavate-backfill	Minimize structure settlement
	Geotextiles	Support low embankments
Deep	Surcharge	Reduce structure settlement
	Geotextiles	Support low embankments
	Sand columns	Reduce structure settlement
	Lime columns	Reduce structure settlement
Buried	Surcharge	Reduce structure settlement
	Dynamic compaction	Reduce structure settlement
	Compaction grouting	Arrest existing structure settlement
	Sand columns	Reduce structure settlement
	Lime columns	Reduce structure settlement
Clays, surface	Gravel admixture	Base, subbase, low-quality pavement
	Lime admixture	Stabilize roadway base and subbase
	Freezing	Temporary arrest of settlement
Loose silts		
Shallow	Excavate-backfill	Minimize structure settlement
	Salts admixture	Dust palliative
	Surface compaction	Increase support capacity‡
Deep	Surcharge	Reduce structure settlement
	Stone columns	Increase support capacity
	Electroosmosis	Increase slope strength temporarily
Buried	Vacuum wellpoints	Improve excavation bottom stability

(*continued*)

TABLE 5.2
GROUND-STRENGTHENING TECHNIQUES SUMMARIZED

Conditions	Technique	Application
Loose sands		
Shallow	Surface compaction	Increase support capacity
	Cement admixture	Base, subbase, low-quality pavement
	Bitumen admixture	Base, subbase, low-quality pavement
Deep	Vibroflotation/Terra-probe	Increase support capacity
	Dynamic compaction	Increase support capacity
	Stone columns	Increase support capacity
	Wellpoints	Increase stable cut-slope inclination
	Freezing	Temporary stability for excavation
Buried	Penetration grouting	Arrest existing structure settlement
	Freezing	Temporary stability for excavation
Collapsible soils		
Shallow	Excavate-backfill	Minimize structure settlement
Deep	Hydrocompaction	Reduce structure settlement
	Dynamic compaction	Increase support capacity
	Lime stabilization	Arrest building settlement
Liquefiable soils	Dynamic compaction	Increase density
	Stone columns	Pore-pressure relief
Expansive soils	Lime admixtures	Reduce activity in compacted fill
Rock masses		
Fractured	Compaction grouting	Increase strength
	Penetration grouting	Increase strength
	Bolts and cable anchors	Stabilize slopes and concrete dam foundations
	Shotcrete or gunite	Reinforce slopes
	Subhorizontal drains	Stabilize slopes

*Minimize structure settlement signifies that settlement will be negligible under moderate foundation loads if the technique is applied properly.

†Reduce structural settlement signifies that after application of the technique significant settlement which must be anticipated in the design of the structure may still occur.

‡Increase support capacity signifies that proper application of the technique will result in an increase in bearing capacity and a decrease in compressibility on an overall basis.

tive of a dam is to act as a water barrier to control seepage flows, dams are either constructed totally of material with low to moderate permeability (homogeneous section) or divided into zones containing very low permeability areas for seepage control (core section) and relatively high permeability material to provide embankment strength (shell sections), as illustrated in Fig. 5.3. Filter layers are provided between the core and shells and at other locations, and riprap is placed on the upstream slope for protection against wave action.

Roadway embankments are constructed of a wide range of materials since they are required to support only their own weight and the weight of traffic. Organic soils, soft clays, and silts are avoided, but clay-silt-sand mixtures are used frequently, as are hydraulic fills. Fills and subgrades beneath the pavement are desired to

FIG. 5.3 Zoned embankment of an earth dam under construction, Round Valley, N.J. From left to right: impervious clay core (at location of small tractor), filter layer of crushed stone, shell material of slightly clayey sand, and facing of large stone fragments on outer slope.

have free-draining properties to prevent moisture buildup from capillary forces resulting in frost heaving in cold climates, expansion of clay soils, or pumping of softened fine-grained soils under wheel loads. *Pavements* are described in Art. 5.2.3.

Design and Installation

Design of fills for support of structures, for earth dams, and for roadway embankments is based on optimum moisture–maximum density relationships of the various fill materials as established by laboratory test. Design of pavement base, subbase, and subgrade layers often is based on values from the California bearing ratio (CBR) test, as well as on moisture-density relationships.

Construction of a controlled fill should be performed to standards established by a set of specifications which are designed in consideration of the purpose of the fill and the characteristics of the various types of materials to be used. Performance specifications normally state the densities required and the thickness of each lift as compacted. In procedure specifications, the maximum lift thickness in a loose-dumped condition, the range in moisture content permitted in the borrow material, the acceptable range of compacted moisture content, and the type, minimum weight, and number of passes per lift of the compaction equipment may also be specified. Where large volumes of fill are required, equipment specifications are often based on the construction of test fills.

Borrow Sources

An important element of the site investigation is the location of suitable borrow areas. Evaluations are made of the on-site material available from cuts and excavations and from off-site borrow sources. Borrow exploration requires primarily ground reconnaissance and airphoto interpretation, supplemented by exploratory pits and auger or machine borings [see Chap. 2 of Hunt (1984)[1]]. Large samples, in the order of 70 to 100 lb each, are required for laboratory testing.

Fill requirements vary with project type. For a building, controlled fill may require only one type of "select" material, whereas borrow for dams normally includes a wide range of materials: clays for core material, granular soils or clay-sand-gravel mixtures for shell material, a rock quarry for a source of riprap for upstream facing, and coarse-grained granular soils for filters. Hydraulically excavated soils can make excellent fill when they are nonorganic and primarily composed of sands, but for load-bearing fills, stock piling and drying are usually necessary prior to placement and compaction.

5.2.2 DESIGN CRITERIA: STRUCTURAL FILLS

Moisture-Density Relationships

Soil Compaction

The mechanical and hydraulic properties of a given soil type are related directly to its density: strength increases and compressibility and permeability decrease with increasing density for most soil types. Density is increased in fills by *compaction* with various types of mechanical equipment. Moisture-density relationships are established by laboratory testing.

Optimum Moisture Content and Maximum Density

If completely dry soil is placed in a mold in layers and each layer compacted under the weight of a falling hammer with a tamping foot, a particular dry density for the material and specific test is obtained. In general, if a small percentage of water is added and the test repeated, a higher dry density γ_d is obtained because the water lubricates the soil grains, permitting greater compaction. By repeating the test on samples mixed with increasing percentages of water a curve is developed relating dry density to moisture content, and a peak value is found beyond which the addition of moisture causes a decrease in dry density as shown in Fig. 5.4. The peak value on the curve is referred to as the *maximum density* and the corresponding water content as the *optimum moisture or optimum water content.* As shown on the figure, the condition of 100% saturation is never reached as some air always remains trapped in the specimen.

Typical Curve Shapes

Moisture-density curve shapes vary for different material types and gradations. Uniformly graded sands may undergo a *decrease* in density at low moisture content as the moisture increases because capillary forces cause a resistance to compaction, or rearrangement of soil grains. This phenomenon is referred to as *bulking*. As moisture is added, a relatively gentle curve with a poorly defined peak is obtained, as shown in Fig. 5.5. Curves for well-graded granular soils and clay-sand mixtures usually have a well-defined peak, whereas curves for clay soils usually have more gently rounded peaks similar to uniformly graded sands, as shown in Fig. 5.6.

The moisture-density relationship varies also with the compactive energy applied. Several variations of the laboratory compaction test have been developed.

Standard Compaction Test

In the early 1930s, while working on earth dams, R. R. Proctor developed the *standard compaction test* as a means of establishing design and control criteria (ASTM D698-78 or AASHTO T99-74). The standard test, also known as the Proctor test, employs an energy of 12,400 ft·lb/ft^3. Three sequential layers of soil are compacted with a 5½-

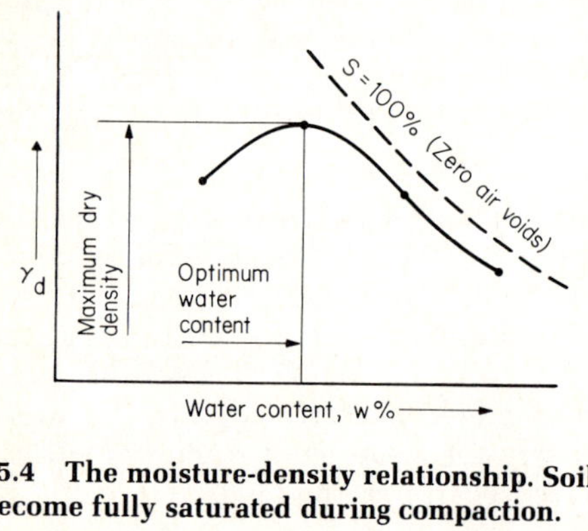

FIG. 5.4 The moisture-density relationship. Soil does not become fully saturated during compaction.

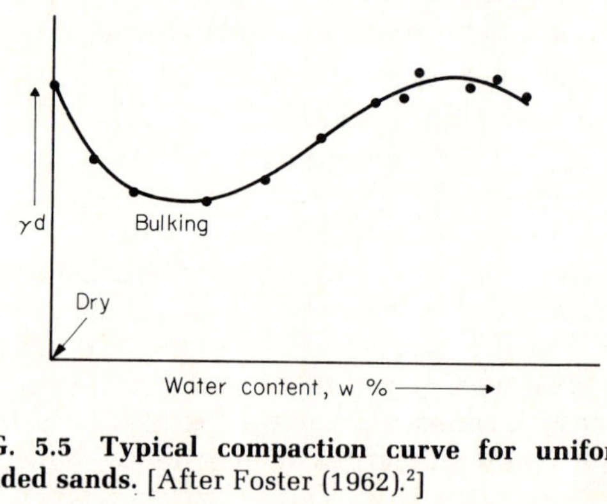

FIG. 5.5 Typical compaction curve for uniformly graded sands. [After Foster (1962).[2]]

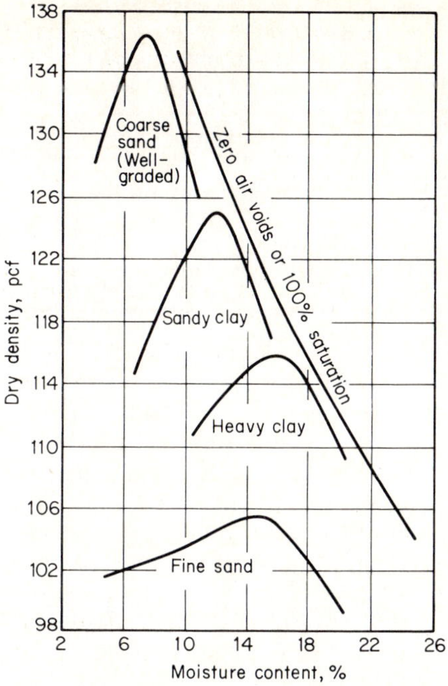

FIG. 5.6 **Typical standard Proctor curves for various materials.**

lb hammer dropped 25 times from a height of 12 in, in a 4-in-dia. mold with a volume of 1/30 ft^3. The energy compared favorably with the energy of field compaction equipment at the time of Proctor's work.

Modified Compaction Test

It was found subsequently that high fills, constructed in accordance with specifications based on the standard density, suffered from compression under their own weight. This fact, combined with the development of aircraft and truck traffic with heavier wheel loads, led to the development of the *modified Proctor compaction test*, in which a substantially higher energy of 56,250 ft·lb/ft^3 is used. This is obtained by compacting the soils in the same 1/30-ft^3 mold, but in five layers with a 10-lb hammer dropped 25 times from a height of 18 in (ASTM D1557-78 or AASHTO T180-74).

Materials containing significant amounts of gravel are compacted in a 6-in-dia. mold having a volume of 0.075 ft^3, with 56 blows for each of the five layers. Curves for the same soil obtained for three different compaction energies are given in Fig. 5.7. As the maximum density increases under increasing energy, the optimum moisture content decreases.

Soil type can considerably influence the difference between the maximum densities obtained in the standard and modified tests. For fine uniform sands the difference is only 2 or 3 pcf; for well-graded silty or clayey sands or gravels the difference is commonly 8 to 10 pcf, and it may be as high as 15 pcf for some clays [Sherard et al. (1963)[4]].

Other compaction tests include the British standard 1377:1948, in which the energy applied is similar to that in the AASHTO standard test, and tests developed by the state of California for earth dams, in which the energies are similar to those in the modified test.

Typical compacted properties for a wide range of soil types, including ranges of maximum unit dry weight (standard compaction) and optimum

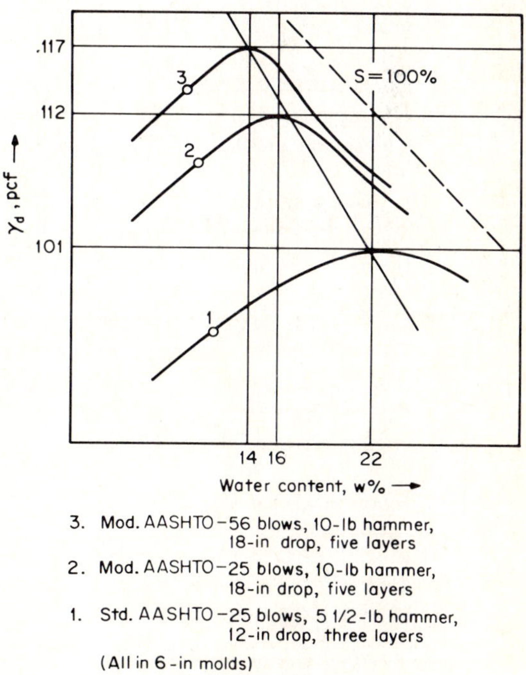

FIG. 5.7 Effect of different compactive energies on a silty clay. [After Turnbull (1950).[3]]

TABLE 5.3
TYPICAL PROPERTIES OF COMPACTED SOILS*

Group symbol	Soil type	Range of maximum dry unit weight, pcf	Range of optimum moisture, %	Typical value of compression: Percent of original height, At 1.4 tsf (20 psi)	Typical value of compression: Percent of original height, At 3.6 tsf (50 psi)	Typical strength characteristics: Cohesion (as compacted), psf	Typical strength characteristics: Cohesion (saturated), psf	Typical strength characteristics: Effective stress envelope φ, degrees	Typical strength characteristics: tan φ
GW	Well-graded clean gravels, gravel-sand mixtures	125–135	11–8	0.3	0.6	0	0	>38	>0.79
GP	Poorly graded clean gravels, gravel-sand mix	115–125	14–11	0.4	0.9	0	0	>37	>0.74
GM	Silty gravels, poorly graded gravel-sand silt	120–135	12–8	0.5	1.1	. . .	. . .	>34	>0.67
GC	Clayey gravels, poorly graded gravel-sand-clay	115–130	14–9	0.7	1.6	. . .	. . .	>31	>0.60
SW	Well-graded clean sands, gravelly sands	110–130	16–9	0.6	1.2	0	0	38	0.79
SP	Poorly-graded clean sands, sand-gravel mix	100–120	21–12	0.8	1.4	0	0	37	0.74
SM	Silty sands, poorly graded sand-silt mix	110–125	16–11	0.8	1.6	1050	420	34	0.67
SM-SC	Sand-silt clay mix with slightly plastic fines	110–130	15–11	0.8	1.4	1050	300	33	0.66
SC	Clayey sands, poorly graded sand-clay mix	105–125	19–11	1.1	2.2	1550	230	31	0.60
ML	Inorganic silts and clayey silts	95–120	24–12	0.9	1.7	1400	190	32	0.62
ML-CL	Mixture of inorganic silt and clay	100–120	22–12	1.0	2.2	1350	460	32	0.62
CL	Inorganic clays of low to medium plasticity	95–120	24–12	1.3	2.5	1800	270	28	0.54
OL	Organic silts and silt-clays, low plasticity	80–100	33–21	. . .	. . .	. . .	. . .	. . .	. . .
MH	Inorganic clayey silts, elastic silts	70–95	40–24	2.0	3.8	1500	420	25	0.47
CH	Inorganic clays of high plasticity	75–105	36–19	2.6	3.9	2150	230	19	0.35
OH	Organic clays and silty clays	65–100	45–21	. . .	. . .	. . .	. . .	. . .	. . .

*From NAVFAC Manual DM 7 (1982).[5] All properties are for condition of "standard Proctor" maximum density, except values of k and CBR which are for "modified Proctor" maximum density. Typical strength characteristics are for effective strength envelopes and are obtained from USBR data. Compression values are for vertical loading with complete lateral confinement. (. . .) indicates insufficient data available for an estimate.

Typical coefficient of permeability, ft/min	Range of CBR values	Range of subgrade modulus k_s, lb/in³
5×10^{-2}	40–80	300–500
10^{-1}	30–60	250–400
$>10^{-6}$	20–60	100–400
$>10^{-7}$	20–40	100–300
$>10^{-3}$	20–40	200–300
$>10^{-3}$	10–40	200–300
5×10^{-5}	10–40	100–300
2×10^{-6}	...	
5×10^{-7}	5–20	100–300
10^{-5}	15 or less	100–200
5×10^{-7}	...	
10^{-7}	15 or less	50–200
...	5 or less	50–100
5×10^{-7}	10 or less	50–100
10^{-7}	15 or less	50–150
...	5 or less	25–100

moisture, typical strength, and permeability characteristics, and values for CBR and subrade reaction modulus are given in Table 5.3.

Specifying Moisture Density

General Criteria

Specifications for compaction of a given soil type define a percent of *maximum dry density* to be attained, which falls in a certain moisture range as shown in Fig. 5.8. The specifications for clean granular soils often are based on a percent relative density. For clay soils the range of moisture content often is specified.

To economize on construction compactive effort, the lowest acceptable densities are specified depending upon the purpose of the compacted fill. Some typical values are as follows:

- Non-load-bearing fills used for general site grading: 90 to 92% standard
- Low- to medium-height dams and roadway embankments: 95 to 98% standard
- Fills to support foundation loads and high dam embankments: 95 to 98% modified
- Pavement layers: see *PAVEMENTS*

The foregoing specification guides apply generally to materials which develop a well-defined peak such as sandy clays and other mixtures with a wide range of grain sizes.

Sands and Gravels

Clean sands and gravels in the completely dry condition exhibit a maximum density more or less the same as their peak density at optimum

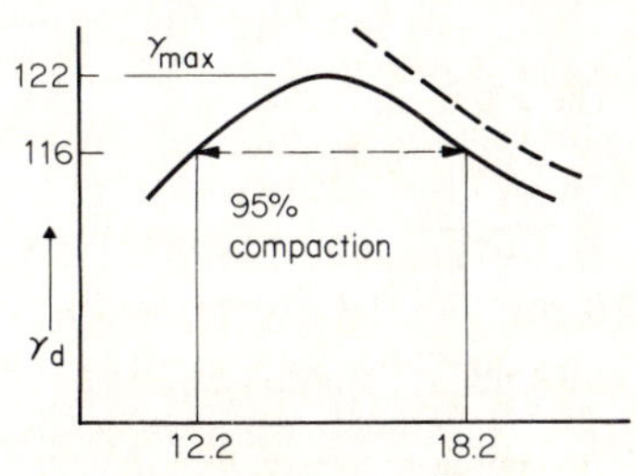

FIG. 5.8 Dry density and range in moisture content for a specified 95% compaction.

moisture content, which is not well-defined by testing, as shown in Fig. 5.5. In other words, the density of free-draining material is not greatly affected by the moisture content except at lower values where bulking occurs. In addition, Burmister (1948)[6] pointed out that clean granular soils compacted to 90% maximum dry density are usually in the loose state of relative density ($D_R < 40\%$) and that at 95% maximum density D_R is only about 70%. As discussed in Art. 3.2.3 of Hunt (1984),[1] a relative density of only 40% represents an appreciably compressible material. There has been a trend, therefore, to specify for compacted fills of clean granular soils a *relative density* rather than a percent compaction, with some typical values as follows:

- To limit objectionable settlement under heavy vibrating machinery and heavy aircraft wheel loads: $D_{R(min)} = 90\%$ [Burmister (1948)[6]]
- Load-bearing fills for settlement-sensitive structures: $D_R = 70$ to 75%

The relative density approach generally is not applicable to material with more than about 35% silt since reliable values for minimum density are impossible to obtain in the laboratory [see Art. 3.2.2 of Hunt (1984)[1]].

Clay Soils

Clay soils present several problems in achieving in the field the maximum density, which in a general way is associated with the highest strength and lowest compressibility. One problem is the correlation of the laboratory density obtained by dynamic compaction with a hammer and that obtained in the field under the kneading action of the commonly used sheepsfoot roller. Kneading remolds the clay to a greater degree than the standard laboratory test, thereby producing lower densities. Nevertheless, the standard laboratory compaction tests are used normally to determine the required field densities. To obtain field densities for control, correlations are made with densities measured in test fills.

Another problem is specifying the acceptable field moisture to obtain the desired density, since it is not practical to attempt to achieve peak density as measured by laboratory test. If the clay is compacted at a moisture content too far above optimum, high pore pressures will develop and a relatively weak fill will result. If the clay is compacted at a moisture content too far below optimum, individual clay fragments may be subjected to expansion when the moisture content increases at some future time and, as water causes softening at points of contact, collapse may occur, resulting in volume decrease. The net volume change in this case will be the sum of the two effects. Clays and clay silts are usually compacted slightly on the *wet side of optimum* to prevent cracking in an earth dam and the development of seepage channels.

Since the properties of some clays, particularly halloysites, are changed by hydration, laboratory compaction tests performed on specimens that are either dried or wetted from the natural moisture content to obtain the moisture-density curve are more representative of field compaction densities than are tests performed starting with a completely air-dried or oven-dried specimen [Hilf (1975)[7]].

Rollcrete

Rollcrete is a recent innovation being used to construct dams as an alternate to rock fill and mass concrete and to construct massive foundations for heavy structures. It is a cement-aggregate mixture with a low cement content, in the order of 80 lb/yd^3 and aggregate sizes to a maximum of 3 in. Depending upon plasticity, up to 7% of the total weight of all aggregate may be fines (conventional concrete normally contains about 1%). Water content of the mix is critical, and moisture must be sufficient to dampen fines so they coat and adhere uniformly to the rock particles. At Willow Creek Dam in Oregon, the first major project application in the United States [ENR (1982)[8]], water-cement ratios ranged from about 0.40:1 to 1.60:1 because of ambient temperatures, but it was found that the wetter the mix, the better the compaction.

The material is hauled from the batch plant to the site by earth-moving equipment such as scrapers, spread by dozer, and compacted by vibratory rollers. At Willow Creek, after four passes, the mix normally compacted about 1 in in a 12-in lift, but lifts were permitted to range from 9½ to 13½ in. To avoid a potentially weak plane between lifts, a rich bedding mix is placed down when the surface is exposed for more than a day.* Test data show that, at any age and cement content, strengths for rolled concrete are higher than typically would be achieved for comparable mass concrete placed conventionally. Temperatures are kept low by using cooled aggregate and low-heat cement.

5.2.3 PAVEMENTS

Basic Factors

The objective of a pavement is to provide support to repetitive wheel loads with a structure that will not unduly wear, deform, or break up. This is accomplished with a layered system consisting of a *surface* of asphalt or concrete overlying a *base course*, and at times a *subbase*, placed on a soil foundation or *subgrade* as shown in Fig. 5.9. Stabilized soils *(see Art. 5.4.4)* provide a pavement surface of soil materials suitable for secondary and tertiary roads with light to moderate traffic and wheel loads. Stabilized soils are used also for bases and subbases.

Design Parameters

The parameters to be considered in pavement design include

- Loadings: Loads and tire pressures, and the uses, purposes, and anticipated traffic as they affect the design life.
- Pavement thickness: Each layer composing the pavement and the supporting subgrade must have adequate strength to resist the shear stresses imposed by traffic loads, normally evaluated in terms of CBR values or, for subgrades, the modulus of subgrade reaction.
- CBR values: A strength index based on a 100% standard to which other materials are compared.

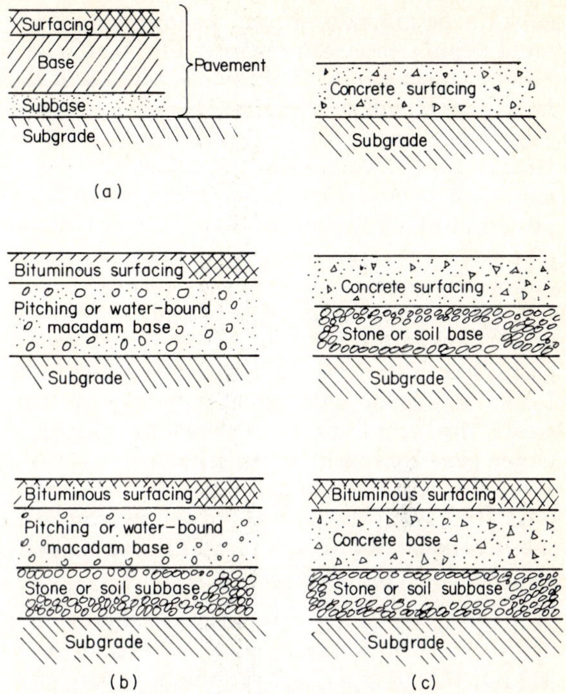

FIG. 5.9 Basic structural elements of some typical pavement sections: (*a*) basic pavement structure; (*b*) typical bituminous pavements; and (*c*) typical concrete pavements. [After TRRL (1974).[9] Reproduced with permission of the TRRL. Crown copyright 1974.]

- Amount of traffic: The amount of traffic will affect wearing surface and pavement thickness.
- Other subgrade factors to consider besides strength are swelling, "pumping," susceptibility to frost heaving. Provisions for adequate drainage are a most important aspect of pavement design. For *swelling clay* treatment *see Art. 6.2.6.*

California Bearing Ratio (CBR)

The CBR value is the result of a field or laboratory test providing a measure of the static penetration resistance of soils. It is used primarily in the design of highway and airport pavements, and resulted from a series of experiments performed by the California Department of Highways in 1929.

In the *laboratory* the test is conducted in accordance with ASTM D1883-67. A specimen of base,

*At Willow Creek Dam, specifications for determining how long a lift could be exposed before a special bedding mix was required permitted 1600°F·hr. Subsequent to dam completion and reservoir filling, leakage occurred through the joints, prompting a comment from the designer that if the specifications require 400 to 500°F·hr leakage would not be expected, but over 500°F·hr some joint treatment will be required [ENR (May 19, 1983, p. 13)].

subbase, or subgrade is compacted in a 6-in-dia. mold using a procedure similar to the standard Proctor density test. Slight modifications to the procedure, developed by the U.S. Army Corps of Engineers, can be found described in the *Soils Manual* of the Asphalt Institute (1969),[10] for example. After compaction a surcharge weight is placed on the soil surface and the mold is immersed in a tank for 4 days to permit saturation, and measurements of swell are recorded. Following soaking, the mold is placed in a testing machine [see Fig. 3.61 of Hunt (1984)[1]], and a plunger with an area 3 in^2 (diameter of 1.91 in) is forced to a penetration of 0.1 in into the soil while the penetration resistance is recorded. Other penetrations may be made, but the value at 0.1 in is the one used normally. The CBR value is determined as the ratio of the test resistance to the resistance of crushed rock at the same penetration [1000 psi (70 kg/cm^2)].

In the *field* the CBR test is performed to obtain a measure of the natural subgrade resistance by jacking the plunger against the reaction of a truck. Field testing should never be performed during the dry season when the surface soils are dry and have their highest apparent strength.

Aggregate

Aggregate for asphalt and concrete pavements is selected on the basis of its durability, but for concrete pavements the possibility of reaction with cement must be considered. Durability testing and the reactivity between cement and aggregate are discussed in Art. 3.2.1 of the companion volume.

Design Criteria

Basic criteria for the design of flexible pavement layers and the supporting subgrade are based on CBR values, although subgrades have been designed and evaluated on the basis of the subgrade reaction modulus *(see Art. 4.4.2)*. An important factor to consider is stress drop-off with depth beneath wheel loads which affects the density requirements for the various layers and the subgrade.

A CBR value of 100 is used as the standard reference, as it indicates excellent material with resistance and support properties similar to those of crushed rock. An approximate correlation between CBR values, various soil classes, and the rating of soils for subgrade, subbase, or base for light-traffic pavements is given in Fig. 5.10. It is seen that CBR values less than 5 represent a poor subgrade for light-traffic pavements, values less than 20 are unacceptable for light-traffic subbases, and values less than 80 are unacceptable for base courses.

Pavement design is primarily empirical. Pavements that are nonrigid are classed as *flexible* and are composed of a bitumen surfacing and various other layers as shown in Fig. 5.9*b*. Various organizations have performed field tests on pavements with differing designs and thicknesses, and through correlations between field performance and laboratory data have developed design charts for various wheel loads, traffic intensities, and pavement designs. The design curves given as Fig. 5.11 were developed by the U.S. Army Engineers for flexible pavements car-

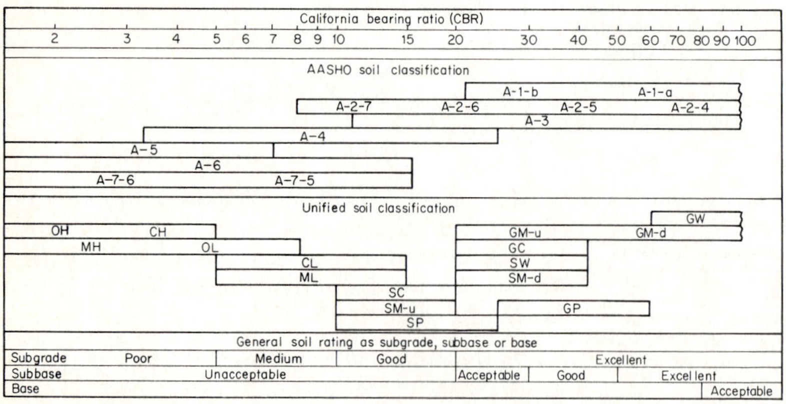

FIG. 5.10 Approximate correlation of soil ratings for use in design of light traffic pavements. [From the Asphalt Institute (1970).[11]]

rying wheel loads to 70,000 lb traveling continuously. Wheel load, total pavement thickness, and CBR value of the subgrade are related. Charts are available for heavier wheel loads. The total design thickness includes the surface, base, and subbase (or surfacing and base, depending upon the materials available).

The general acceptability of materials to be used as base, subbase, or subgrade is given in Table 5.4, along with modified compaction-density ranges for a heavy-duty pavement. Because of the stress drop-off with depth occurring beneath wheel loads, specified density requirements decrease with depth, as illustrated in Fig. 5.12.

Concrete pavement thicknesses often are estimated on the basis of CBR values, but thicknesses are reduced somewhat because of the rigidity of the concrete surface.

Drainage control is imperative for successful pavement performance since pavement failures are usually related in some manner to groundwater or surface infiltration. Failure causes and control measures are discussed in Art. 8.4.7 of Hunt (1984)[1] and illustrated in the sketches given in Fig. 5.13.

Frost protection is necessary in cold climates. As the depth of frost penetration increases, pavement thickness is increased accordingly. Full protection against frost heaving is considered to

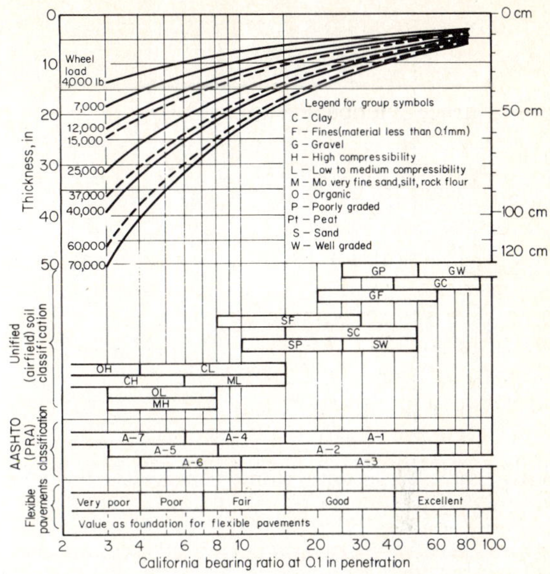

FIG. 5.11 Chart for the design of flexible pavements, giving combined thickness of surface and base related to CBR values of the underlying materials. [From Tschebotarioff (1973).[12] Reprinted with permission of the McGraw-Hill Book Company.]

be achieved with a pavement thickness equal to the maximum depth of frost penetration; a thickness of one-half the depth of frost penetration generally is considered as the minimum required for protection, as long as some risk of

TABLE 5.4
GENERAL EVALUATION CRITERIA FOR PAVEMENTS*

Section	Material	CBR value, %	Density, % modified (heavy-duty pavement)†
Base	Good quality crushed rock‡	80	103 to 101% decreasing with depth
	Good quality gravel‡	50 to 80	
Subbase	Good quality soil‡	30 to 50	101 to 98% decreasing with depth
Subgrade	Very good	20 to 30	98 to 83% decreasing with depth
	Good to fair	10 to 20	
	Questionable to fair	5 to 10	
	Poor	<5	

*After Turnbull [1968].[13]

†See Fig. 5.11.

‡Specifications for base and subbase limit the PI and the percent passing the no. 200 sieve because of strength decrease that occurs upon soaking or saturation if the limits are exceeded, as follows: High quality base: PI <5, fines <10%. High quality subbase: PI <12, fines <10%.

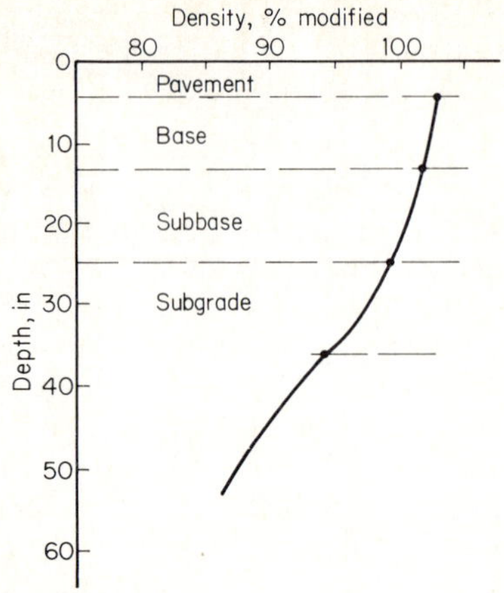

FIG. 5.12 Compaction requirements with depth for heavy-duty pavements. [from Turnbull (1968).[13]]

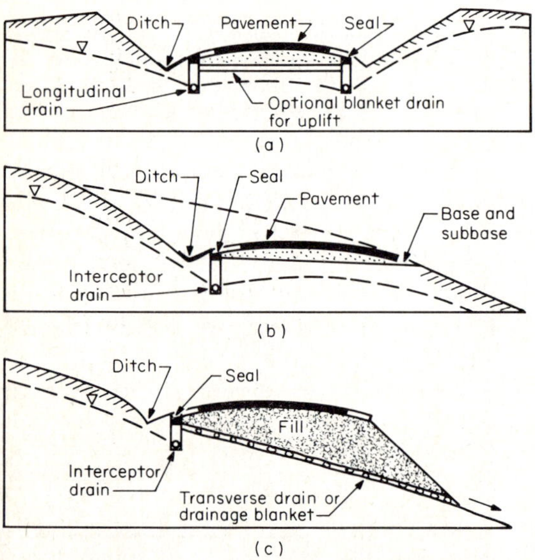

FIG. 5.13 Some methods of pavement drainage control: (*a*) shallow cut in level ground; (*b*) side-hill cut; and (*c*) side-hill fill. [From Hunt (1984).[1] Reprinted with permission of the McGraw-Hill Book Company.]

pavement deflection can be tolerated, such as in parking lots, for example. The approximate maximum depth of frost penetration in the United States is given in Fig. 6.15. The actual depth of frost penetration depends upon climate, soil type, and water-table location as discussed in Art. 7.7.3 of Hunt (1984).[1] The frost susceptibility of soils is summarized in Table 2.6, including the relationship to capillary rise.

Construction

General Requirements

Specifications are prepared covering subgrade preparation, pavement thicknesses and composition, compacted densities, and other factors. Subgrade treatment varies with the quality of the material. Poor quality material may be excavated and replaced with good quality soils, may be excavated and reused as compacted material, or may be compacted in situ from the surface, as discussed in the following article. Each successive pavement layer is then placed and compacted to its specified thickness and density.

Turnbull (1968)[14] notes that by far the majority of heavy-duty pavement deficiencies are the result of improper specifications or construction practices rather than design deficiencies.

Lightweight Fills over Weak Soils

Roadway construction over weak soils requires a fill for support, and it is mostly the fill weight that causes objectionable settlements which, in many cases, are reduced by surcharging *(see Art. 5.3)*. Various approaches have also been used to reduce the fill weight. The Louisiana State Highway Department is experimenting with the use of *clam shells* to construct a 4.4-mile section of U.S. 90 near Raceland, where the marshland soils have almost no penetration resistance for depths over 12 ft. The Michigan State Highway Department is using "foamed" *concrete* to reduce loads over soft ground on bridge approach fills [ENR (1976)[15]]. A combination of cement, water, and a foaming agent, the concrete weighs as little as 20 pcf (0.32 t/m^3).

The California Department of Transportation has used *sawdust* to lighten fill loads, and *reinforcing fabric* at the fill base, to provide a uniform load distribution for a test fill constructed over the weak San Francisco Bay muds [Seim, Walsh, and Hannon (1981)[16]]. The sawdust was incorporated in the fill in the lower elevations where saturation would prevent oxidation and rotting.

5.2.4 FIELD COMPACTION

Specifications

Controlled compacted fill is placed in the field in accordance with a set of specifications prepared in consideration of the materials to be used and the purpose of the fill. Specification content will vary with the engineer's local experience relative to that of local contractors, particularly with regard to equipment type, loose lift thicknesses, and number of passes required to achieve the desired results. The two basic elements of the specifications are site preparation and fill compaction.

Site Preparation

Ordinarily, all brush and other vegetation, the organic topsoil layer, and limited thicknesses of weak organic or clay soils are excavated prior to fill placement. Several examples are given in Art. 5.2.5.

When stripping near or below the water table in fine-grained soils, pumping or boiling (Fig. 5.14) may result from temporary high pore pressures caused by the weight of the excavating equipment. Small local areas are best treated by a cover of sand or gravel and not otherwise disturbed. Large areas that become softened may require excavation and replacement with stone or other coarse material or treated as an excavation-backfill problem, as will be discussed.

Fills which are relatively thick, such as for buildings, earth dams, or other embankments, are usually placed directly on the subgrade prepared in the foregoing manner. Subgrades for heavy-duty pavements are either excavated and the material reused as compacted fill, or are *compacted in situ* to the specified densities as described in the preceding article. In many cases for buildings and pavements, the specifications call for *proof-rolling* the subgrade surface with a heavy roller to disclose soft or weak areas requiring treatment.

Fill Compaction

As a minimum, specifications normally cover those materials that are acceptable, the required minimum dry densities, the maximum permissible thickness of each lift after compaction, and the maximum particle size, which must be less than lift thickness. Some typical compacted density criteria are given in Art. 5.2.2. Typical lift thicknesses after compaction are 6 in for load-bearing fills for structures and 6 to 9 in for embankments and dams, depending upon the type of equipment used.

FIG. 5.14 Dozer operation during stripping caused pumping of groundwater in clayey silts. A small pad of sand will alleviate the problem.

Permissible ranges in moisture content also may be specified. Moisture content ranges usually run on the order of ±2% of optimum for most cases, but for dams over 15 m high with a clay core, specifications may call for a range of −1 to +2% of optimum. A range in the type of equipment to be used for compaction also may be specified, and frequently the minimum weight and number of passes are given.

Quality control is provided by observation of filling operations and in situ testing.

Compaction Equipment

There is no single piece of equipment totally suitable for the compaction of all soil types, and the inappropriate use of equipment for a given soil type will result not only in a poor quality fill but also in construction delays. Depending upon soil type, compaction is achieved by rolling, vibrating, kneading, or tamping, and equipment can be classed generally as smooth-drum rollers (static or vibratory), tamping-foot rollers (sheepsfoot), rubber-tired rollers, and rammers. Where equipment movement is not confined, cohesionless soils are most effectively compacted by rolling, especially with a vibrating roller, and cohesive soils (clays to clay-sand-gravel mixtures) by the kneading action of either a sheepsfoot or penumatic-tired roller. Vibrating pads and rammers are used in confined spaces such as trenches.

Static Smooth-Drum Rollers

Static smooth-drum rollers impose a limited pressure on the soils because of their relatively large contact area and are most effective in compacting relatively thin lifts, on the order of 7 to 9 in loose thickness, of crushed rock, gravel, or clean medium or coarse sand. They are used generally to compact base and subbase layers in pavements. In areas of large cohesive fills compacted by equipment such as the sheepsfoot roller, which leaves an irregular surface, smooth drum rollers are effective in smoothing and sealing the surface at the end of each day or some other period of operation. Rainfall on a surface compacted by sheepsfoot will not drain, and further placing of lifts and compaction is not possible until the surface is scraped off or dried. Sealing also aids in fill erosion protection. After rolling, however, prior to placing the next lift, the sealed surface must be scarified, especially in an earth dam, otherwise a weakness plane will remain in the embankment. When of necessity rollers are used to compact clays, lifts must be kept thin, in the order of 6 in loose thickness.

Vibrating Smooth-Drum Rollers

Vibrating smooth-drum rollers (Fig. 5.15) are the most effective equipment for compacting clean granular soils in fills or subgrades and are being used also for rock fills, soil cement, asphalt, and rollcrete. In most models a separate motor drives a system of eccentric weights to cause a high-frequency, low-amplitude oscillation of the vibrating drum or drums. Compaction effect depends primarily on static weight (linear load), frequency and amplitude of vibrations, and roller speed. The depth effect is approximately proportional to the drum weight, and weights from 10 to 15 tons are used to roll thick layers. Moorhouse and Baker (1968)[17] reported on the compaction of a clean, poorly graded sand where eight coverages achieved relative densities in excess of 70% to a depth of 5.5 ft (1.7 m), and increases in density have been reported to depths as great as 3 to 5 m [CII (1981)[18]].

Frequency curves for rollers are normally flat, and as a rule maximum compactive effort in soils is achieved at frequencies between 25 and 50 Hz (1500 and 3000 vibrations per minute). An in-

FIG. 5.15 Vibrating smooth-drum roller is most efficient for compacting clean sands.

crease in *amplitude* gives an increase in compactive effort in the entire frequency range, but amplitudes are limited by equipment design considerations. Vibratory rollers designed to compact large volumes of soil and rock fill in thick layers should have an amplitude in the range of 1.5 to 2.0 mm with a corresponding suitable frequency of 25 to 30 Hz (to limit excessively high bearing stresses developing in the drum). For asphalt and stabilized base compaction, the optimum amplitude is 0.4 to 0.8 mm and the suitable frequency range is 33 to 50 Hz [CII (1981)[18]]. Modern machines have variable frequencies and dual amplitudes. *Optimum roller speeds* normally range between 3 and 6 km/hr for soil and rock fill, although for high-density requirements, soils difficult to compact, and thick layers, speeds of 3 to 4 km/hr are recommended. Although vibratory compaction is now a standard method for all rock fills, specially designed rollers with a thick drum of high-grade steel are necessary.

Some rollers are equipped with "compaction meters," an electronic device which receives continuous signals from an accelerometer mounted on the drum. It provides a measure of a relative value for the deformation modulus and, through correlations, the degree of compaction obtained. It is particularly useful for rock fills; in fine-grained soils it is necessary to determine water content of the soil.

Tamping-Foot Rollers (Sheepsfoot Rollers)

Tamping-foot rollers are covered by projecting feet or studs (Fig. 5.16) and compact by a combination of tamping and kneading action, making them most effective in cohesive soils. They come in a large variety of sizes and weights and the tamping feet may be clubbed or tapered, the former being more suitable in heavy clays. Contact pressures on the feet range in general from 100 psi (7 kg/cm^2) to over 600 psi (42 kg/cm^2). Feet length vary, ranging up to 10 in (25 cm). Ballasting is achieved by adding water or wet sand to the drum.

For a 6-in compacted layer, satisfactory densities can be obtained with 6 to 12 passes if the water content is right. For most soils little extra density is obtained after 6 passes except where the water content is on the dry side of optimum [Sherard et al. (1963)[4]]. Many engineers specify a minimum of 10 to 12 passes, however, to improve uniformity of the embankment, since extra rolling is relatively inexpensive. Large units with long feet will compact lifts of 12 in loose thick-

FIG. 5.16 Cleaning the feet of a small tapered-foot sheepsfoot roller pulled by crawler-tractor for the compaction of an earth dam core section (see photograph in Fig. 5.3).

ness in 3 to 5 passes. During compaction, with each successive lift the thickness of the compacted layer increases until finally the sheep's feet "walk out" onto a compacted surface. For most efficient compaction the roller should walk out enough to lift the drum off the construction surface at least 1 or 2 in during the last few passes. Failure to walk out indicates an excessively high water content, permitting shear failure of the soil, but this does not necessarily indicate that adequate compaction is not being attained. Lightening the roller load may result in walkout. When the device is pulled by crawler-tractors, maximum speeds are in the order of 4 mph (6 km/hr); self-propelled rollers can attain speeds in the order of 6 mph (9 km/hr), in either case slower than rubber-tired rollers.

Advantages of the tamping-foot roller over other types for cohesive soils result from the kneading action of the feet, which causes successive layers to be fused and thus increase dam stability, from an increase in the blending of soils and simplification of adding water or drying as needed, and from soil compaction over a wider range of moisture content. It is also effective in breaking down large pieces of soft rock. Disadvantages result from its relatively slow operation; the rough surface of finished lifts which collect rainwater requiring drying; lower compacted densities; and a relatively large amount of pore air compacted into the embankment. In addition, because it is normally lighter than a rubber-tired roller, it does not readily reveal soft zones.

Vibrating tamping-foot rollers, such as the self-propelled unit shown in Fig. 5.17, are useful in silty and clayey sands. In the photo the machine is compacting a silt-sand-gravel mixture in which the fines range from 20 to 35%. The borrow soils are totally dry since the environment is desert, and all moisture must be added and thoroughly mixed with the soil. For the mixing the tamping feet are helpful.

FIG. 5.17 Self-propelled vibrating tamping-foot roller compacting a very silty, gravelly sand in a desert environment.

Pneumatic-Tired Rollers

Pneumatic-tired rollers are the most versatile equipment for general compaction as they are effective in both cohesive and cohesionless soils. Each consists of a ballast box mounted between two axles, supported on a number of closely spaced pneumatic tires arranged to make overlapping tracks so as not to create ruts and leave uncompacted areas. Many specifications require that each wheel be supported independently. Compaction takes place by kneading action and some designs provide a "wobble-wheel" effect to increase kneading. Wheel loads, tire sizes and inflation pressures, number of passes, and loose lift thicknesses are varied depending upon the soil type and the equipment available.

Earth dam embankments generally are compacted with the 50-ton rubber-tired roller to final lift thicknesses of 9 in [Sherard et al. (1963)[4]]. Pressure applied by the tires is roughly equal to the tire air pressure for any roller of a given weight. In the field, as tire pressures and roller passes are increased, the maximum density increases and the optimum water content decreases. Field tests are required with a particular piece of equipment to determine the number of passes and tire pressures required to achieve the specified density. Roller speeds are generally in the range of 4 to 8 mph (6 to 12 km/hr) with the higher speed a limit, since excessive speeds tend to increase the number of passes required.

Advantages over the tamping-foot roller are provided by the heavier weight, which requires fewer passes, permits thicker lifts, and readily reveals poorly compacted zones. In addition, the smooth final surface acts as a seal, providing protection against rainfall, and when the cohesive soils are composed of large-size material, such as coarse gravel, better compaction and less equipment wear result. The major disadvantage when used to compact cohesive materials in earth dam construction is the relatively weak bond between lifts when compared to the sheepsfoot roller. For this reason many practitioners favor the latter equipment. The bond is improved by scarifying the surface before placing the subsequent lift, but construction costs are increased. In addition, there is less soil blending than with the sheepsfoot, and the moisture content should be just right since adding water or drying slows the operation and requires scarifying.

Construction Traffic

Some compaction is achieved by construction traffic of trucks, scrapers, and bulldozers and is most effective primarily in coarse granular soils. Traffic patterns require control to avoid rutting of the fill.

Confined Areas

Small *vibrating pads* such as the "nervous turtle" are used to compact clean granular backfill in trenches, around footings, and in other areas with space limitations. Pneumatic and internal combustion *rammers*, available in various weights, are necessary for the compaction of cohesive soils in confined areas and useful in granular soils as well.

Material Moisture Treatment

Dry Materials

In arid regions in particular, the natural water content of soils normally is too low. The most efficient procedure for increasing water content in very dry soils is by sprinkling or irrigating borrow areas for some period of time. The water content of dry soils is seldom raised by irrigation to values too high for compaction.

Where the natural water content is only a few percent below optimum, the soil spread loosely on the construction surface may be sprinkled from trucks prior to compaction. In cases such as the recompaction of loose subgrade soils for airfield pavements in an arid environment it is necessary to add substantial quantities of water by sprinkling, then to scarify with plow equipment (Fig. 5.18), repeating the sequence several times

FIG. 5.18 Scarifying very silty gravelly sands of an airfield subgrade in a desert environment as moisture is added to raise the moisture content to suitable levels for compaction.

to achieve uniform distribution of the proper moisture content.

Wet Materials

Processing overly wet soils usually is more troublesome than processing overly dry soils, especially cohesive types. In general, compaction is avoided during rainfalls, and the surface of a cohesive soil fill is maintained sealed and sloped to avoid ponding. Prior to placing the next loose lift, it may be necessary to scrape off the top few inches before scarifying the fill to permit drying (*see Art. 5.4.4*, Construction). If the loose materials are too wet, they also are scarified to permit drying prior to compaction. Construction schedules often are arranged to avoid compaction of cohesive fills during rainy seasons because of construction delays and added costs. In Sweden and Norway, where the summer construction season is predominantly rainy, impervious cores for dams have been placed in a wet condition and compacted with dozers or tractors with little or no control of density. Permeabilities are satisfactory although postconstruction settlement of the core occurs, the amount varying with the core material [Sherard et al. (1963)[4]].

Freezing

Work often continues in temperatures below freezing as long as the desired densities are attained. Adding about 1% salt or $CaCl_2$ to borrow materials permits compaction in very low air temperatures. When the construction season is ended because of cold weather, completed earth dam fills are covered with a layer of several feet of loose soil to prevent freezing. The thickness depends on the frost depth for a particular region. Care is also required to protect subfloor fills and shallow interior footings in uncompleted, unheated buildings during the winter months. Thawing of a frozen fill can result in substantial settlement or voids in an embankment.

Compaction Quality Control

Observational Procedures

There are two phases to the control of compacted fill quality: observation and testing. Observation is required to assure that the entire filling and compacting procedure is proper, and testing is required to prove and document the densities obtained. Compacted fills should be placed under the full-time observation of an experienced engineer or technician who observes the type of material being placed; notes the loose and compacted lift thicknesses; estimates the field moisture content; records the equipment type, weight, and number of passes; and observes the reaction of the fill to moving construction equipment. The latter condition is most important.

In fine-grained cohesionless soils with excessive moisture, for example, high pore pressures will develop, making roller compaction impossible and resulting in weaving and cracking of the fill surface. The material should be scarified and permitted to dry before attempting compaction. Fill delivery equipment may cause rutting and destroy a newly compacted layer unless the traffic pattern is such as to avoid successive vehicles in the same wheel tracks. Construction traffic also will show weak or soft zones in the fill where the material should be removed, then replaced and compacted.

In cohesive soils, excessively high water content is indicated when after several passes a heavy rubber-tired roller ruts excessively or a sheepsfoot roller continues churning the construction surface and shows no sign of walking out. Excessively low water content will result in a sheepsfoot roller walking out several inches after only 3 or 4 passes, and a rubber-tired roller will leave a smooth, hard and stiff surface, in either case leaving the underlying soil improperly compacted. At proper water content sheepsfoot rollers will walk out after about 6 to 8 passes, as described previously, and heavily loaded rubber tires will produce some small penetration of the surface.

In Situ Density Testing: Volumetric Methods

In situ measurements of compacted densities frequently are made on a grid basis, requiring one test per so many square meters, depending upon the fill purpose and project dimensions. The USBR requires one test per 2000 yd^3 for embankment construction as a minimum [USBR (1974)[19]], which for 8-in lifts amounts to one test per every 9000 yd^2 or an area of 95 yd × 95 yd. Industrial buildings supported on load-bearing fills should have a minimum of one test every 1000 yd^2. In addition, tests are made in areas considered to have inadequate compaction as determined by the engineer. Test depths range from the fill surface, to 6 in, to as deep as 12 to 18 in.

In situ measurement of compacted wet density of most soils is made with the sand cone density apparatus (Fig. 5.19), the water balloon apparatus, and nuclear moisture-density devices. In the first two procedures a hole about 6 in deep and 6 in in diameter is excavated, the material stored in a sealed container, and the wet weight recorded. Hole volume is measured with either a "calibrated" sand, or with the rubber balloon filled with water (ASTM D1556-64 and D2167-66). An important factor to consider during performance of these tests when material is on the wet side is the possibility of hole squeezing and reducing in volume if the technician stands too close. The result will be a large weight of material for a small volume and abnormally high densities. In fine-grained soils test samples may be obtained by driving short sections of UD tubes, and in very coarse materials the excavated hole may be lined with plastic and filled with a measured volume of water.

Moisture Content Measurements

The test specimen recovered by the methods requiring a hole is weighed and a portion placed in

FIG. 5.19 Sand-cone density apparatus used to measure wet density in situ. In the photograph, a test is being made to check compaction in the second layer from the surface.

an oven for drying and the determination of the field moisture content. Because of the delay from oven drying, the dry density results are not available until the following day. This can result in construction delays if it it is necessary to remove and replace or recompact unsatisfactory fill not disclosed by observational methods.

Several methods are available for rapidly measuring the water content, including heating the soil in an open pan over a hot plate or gas flame, the alcohol burning method [HRB (1952)[20]], the alcohol method with a hydrometer [Bouyoucos (1931)[21]], and a *moisture meter* using calcium carbide to generate acetylene in a closed container connected to a pressure gage [Reinhold (1955)[22]]. These methods are suitable for cohesionless materials as they yield results similiar to oven drying, but give only approximate results for many types of clay soils. Checking should always be done with oven-dried specimens.

Nuclear Moisture-Density Gauge

Measurement of wet density is made from the surface either by direct transmission or backscatter of gamma rays, and moisture content by transmission of neutron rays when using the nuclear moisture-density gauge. Direct transmission measurements are made from a probe driven into the ground to a depth of 8 in. Procedures are described in ASTM D2922-81. In use since 1949, for many years the device had been found to give a large standard error, as much as ±2.60 pcf [Dunn and McDougal (1970)[23]]. In their instruction manual of 1979, Troxler Electronic Laboratories Inc. claims a precision of ±1 pcf for correctly calibrated direct transmission measurements.

The method is rapid and, with modern equipment frequently calibrated, yields satisfactory results, especially on large projects where soil types used as fills do not vary greatly. Results should be checked occasionally by volumetric test.

Rock Fills

During exploration and design phases the rock quality at potential quarry sites is evaluated in terms of rock hardness and the shapes and amount of fine particles resulting from quarrying operations. The latter consideration is most critical in hard rock since a large percentage of fines causes construction problems. In most hard rock formations it is possible to vary the average rock size and fine content by varying quarry operations.

In the United States, pervious rockfill sections of dams are almost always constructed by placing layers 1.5 to 3 ft thick and compacting with large vibrating rollers. In the past dumping in high, sluiced lifts was often practiced [Sherard et al. (1963)[4]]. In Europe rock is often placed in thin lifts (on the order of 1 to 2 times the maximum rock size) and compacted with heavy vibrators moved from location to location, by vibrating rollers, or by the impact of falling heavy steel tamping plates.

5.2.5 APPLICATIONS TO BUILDING SITES

Preparing sites for buildings where controlled fill is required can fall into several major categories:

1. Fills placed on a stripped, compacted subgrade (*see Art. 5.2.4*) to raise general site grade for such purposes as flood protection or to provide dock-high loading in an industrial building

2. Fills placed for structures after excavating weak soils from lowlands (the excavation-backfill method)
3. Fills combined with benches that are constructed for hillside developments

Excavation-Backfill Method

In most cases the purpose of the excavation-backfill method is to improve ground conditions to preclude the necessity for deep foundations where they would be more costly or more risky, or both.

Case 1: Miscellaneous Fill and Organic Soils Overlying Thick Sand Stratum

For the construction of the proposed school building illustrated in Fig. 5.20, a few feet of new fill were required to raise the site grade. Evaluations of subsurface conditions determined that the excavation of the miscellaneous fill and peat layer and replacement of them with a controlled compacted fill would be the most economical and safest solution. Alternate schemes considered were deep end-bearing piles to rock (very costly for a light building), short piles in the sand stratum, and compacted concrete piles bearing in the sand at shallow depth so as not to overstress the underlying clays. All deep foundations would require design for downdrag resulting from consolidation of the peat stratum under the new fill load.

As shown in the sketch, excavation proceeded with a dragline because of the high water table. After the excavation was dewatered by pumping, water inflow was relatively low, permitting control by sump pumping. A dozer entered the excavation to clean the remaining muck and expose the underlying sands, which provided support for the controlled fill. The first few lifts con-

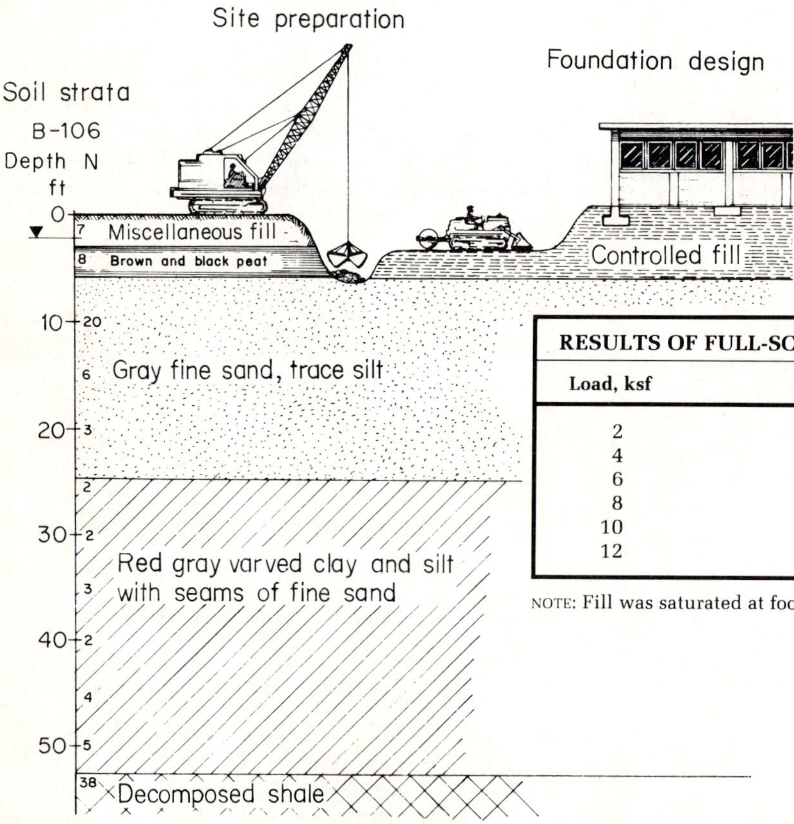

RESULTS OF FULL-SCALE LOAD TEST ON FOOTING	
Load, ksf	**Deflection, in**
2	0.031
4	0.062
6	0.089
8	0.130
10	0.198
12	0.256

NOTE: Fill was saturated at footing elevation by heavy rains.

FIG. 5.20 High-school building supported on spread footings designed for an allowable bearing value of 4 ksf on compacted fill, Ridgefield Park, N.J. Site prepared by excavating miscellaneous fill and peat. (Courtesy of Converse Consultants.)

sisted of a select borrow with a minimum of fines (less than 10 to 15%) to preclude the development of pore pressures and make compaction with the vibrating roller more efficient.

Upon completion of the fill operation a full-scale load test was performed on a footing to demonstrate to the school board the satisfactory nature of the operation, since 20 years ago, when the project was constructed, compacted fills for buildings had not yet received full acceptance in the eastern United States. As shown in the figure, the net deflection under 12 kips/ft^2 (6 kg/cm^2) was 0.256 in (0.65 cm) at a time when the fill was saturated at footing elevation by heavy rains.

Case 2: Organic Soils over Thick Varved Clays in Flood-Prone Area

A problem similar to Case 1 is illustrated in Fig. 5.21, although for several reasons conditions are more unfavorable for construction. A larger thickness of fill (9 ft) is required to raise the site from the surface of the sand stratum to final grade, for flood protection and to provide a dock-high floor for truck loading (thereby imposing relatively heavy loads on the natural soils over a large area); the sand layer is only 10 ft thick at most; and the underlying firm to very soft varved clays extend to depths in the order of 110 ft. Friction piles in the clay stratum, the only feasible alternate to the excavation-backfill method, would require design to support downdrag from about 6 ft of new fill in addition to the building and floor loads.

The sand stratum shown in Fig. 5.21 is not present in many locations of the New Jersey Meadowlands, the general locale of Case 2, and immediately below the meadowmat excavation encounters the varved clay surface shown in Fig. 5.22, which is prepared as shown in the figure.

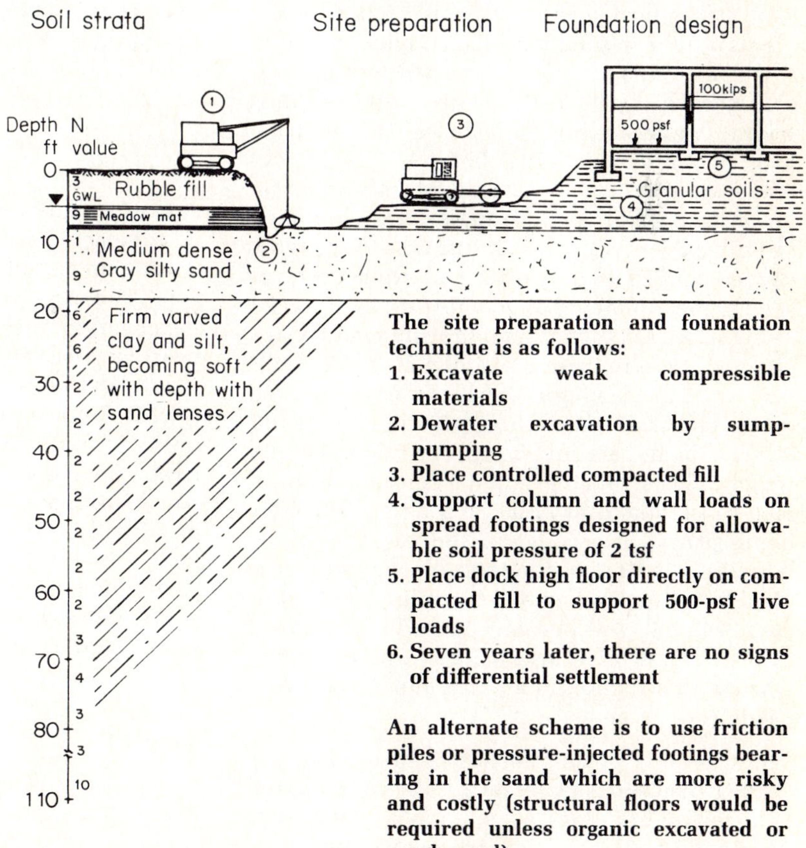

The site preparation and foundation technique is as follows:

1. **Excavate weak compressible materials**
2. **Dewater excavation by sump-pumping**
3. **Place controlled compacted fill**
4. **Support column and wall loads on spread footings designed for allowable soil pressure of 2 tsf**
5. **Place dock high floor directly on compacted fill to support 500-psf live loads**
6. **Seven years later, there are no signs of differential settlement**

An alternate scheme is to use friction piles or pressure-injected footings bearing in the sand which are more risky and costly (structural floors would be required unless organic excavated or surcharged).

FIG. 5.21 Excavation-backfill method used to support warehouse with 0.25-tsf floor loads overlying more than 100 ft of soft varved clays in East Rutherford, N.J. (Courtesy of Converse Consultants.)

FIG. 5.22 Exposed varved clays after stripping organic layer (see Fig. 5.21). First lift of clean medium to coarse sand is being placed to provide a stable working platform. Note settlement plate in left center of photograph.

Analysis of the consolidation of the varved clays under the loads imposed by the controlled fill, building weight, and significant design floor load (500 psf) indicated that postconstruction settlements would be tolerable. The major concern in this situation is the possible occurrence of large dish-shaped settlements across the building width, which are not detrimental to the structure in limited amounts, but the loss of roof grade results in the ponding of rainwater on buildings of large areas. In some cases, therefore, buildings are designed to provide a convex-shaped roof to permit drainage when postconstruction settlements of significant amounts are anticipated; as an alternate, the site may be surcharged to eliminate detrimental settlements *(see Art. 5.3)*. A natural benefit from the varved clays [see Art. 7.6.4 of Hunt (1984)[1]] is their relatively high rate of horizontal permeability which allows relatively rapid consolidation. In fact, substantial consolidation occurs as the fill is being placed, especially in the case described here, where most of the total fill load of 9 ft is placed in an "unbuoyed" condition, and after completion it will become partially submerged as the water level returns to its natural elevation. During construction, therefore, the fill acts as a partial surcharge.

The foregoing procedure has been used in the Meadowlands of New Jersey for the construction of dozens of buildings, some up to four stories in height, without the occurrence of detrimental settlements where proper practices are used. At times, settlement in the order of 4 to 6 in occurs differentially across the width of a large building, but the effect is measurable only in terms of the building elevations, without causing structural distress [Lobdell (1970)[24]].

Case 3: Fill-Supported High-Rise over Deep Clay Formation

A 12-story apartment building supported on 15 ft of compacted fill placed after the removal of a weak clayey silt stratum is illustrated on Fig. 5.23. The structure is supported on spread footings bearing in the fill designed for an allowable bearing pressure of 4 tsf. The predicted settlement from consolidation of the supporting firm to stiff Cretaceous clays was 1.5 in beneath the building center with a maximum differential between the center and the building edge of 0.5 in. Newly imposed loads included the average building load of 0.65 tsf plus a portion of the fill load. The alternate solutions considered as more costly were a mat foundation and basement bearing on the firm to stiff clay and a friction pile foundation.

Case 4: Variable-Thickness Organic Soils over Sands

Several classes of fill material were used to prepare a large site for a shopping center as illustrated in Fig. 5.24. The original site area was

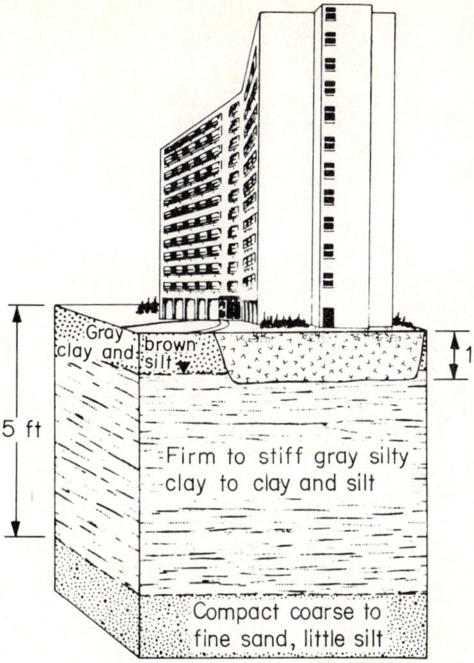

FIG. 5.23 Twelve-story apartment building in Keyport, N.J. supported on spread footings bearing in compacted fill for an allowable bearing pressure of 4 tsf. (Courtesy of Converse Consultants.)

swampy with an upper stratum of soft organic soils extremely variable in thickness, ranging to a maximum of over 10 ft over a sandy stratum in the order of 30-ft-thick overlying rock. Heavy column loads for the proposed department-store structure precluded their support on spread footings placed in compacted fill, since in some areas fill thickness would be inadequate to avoid overstressing the underlying loose to medium dense sands and differential settlement would occur in the large settlement-sensitive structure. After consideration of vibroflotation *(see Art. 5.4.1)*, friction piles and end-bearing piles to rock, pressure-injected footings (compacted concrete piles) with 100-ton capacity were selected as the most economical foundation, shown in Fig. 5.24. Rather than being pile-supported slabs, floors were supported on compacted fill.

Within the building areas, the site was prepared by excavation of the soft organic soils. Since these were found frequently in deep pockets, complete dewatering to provide a dry excavation was impractical and the pockets were filled with pea gravel (Class I material) to achieve acceptable densities under high water conditions. Above the water level, sand and gravel (Class II material) was placed and compacted to provide support for the floors. The free-draining soil was specified to preclude frost heaving during construction and damp floors from capillary action after construction. In the areas outside the buildings, most of which were covered by parking, the organic soils were not excavated. An extra thickness of fill, in the order of 12 in, was placed over the general site fill (Class III material—common granular soil) and permitted to settle for 1 year, after which the pavement was constructed. The performance of the buildings and pavement, which have been in place for over 15 years, has been entirely satisfactory.

Hillside Developments

Sites for housing and other hillside developments are frequently prepared by the construction of a number of platforms and benches. In many cases structures will be partially in cut and partially in fill and the construction of a high-quality fill is imperative to avoid differential settlement. Fill stability against sliding is a major problem to be evaluated and stability normally is achieved by adequate surface drainage and drainage between the fill and the natural soils *(see Art. 8.4.3)*.

In areas where creep of residuum is a problem, as in expansive clay shales, fills at times are stabilized by key benches and a buttress key, together with adequate site drainage, as illustrated in Fig. 5.25. A discussion of the problems caused by the dipping alternating beds of nonexpansive sandstone and expansive clay shales is given in Art. 10.6.3 of Hunt (1984).[1] Drilled piers installed below the deepest GWL were found to be the only solution for avoiding differential building heave under these geologic conditions.

5.3 SURCHARGING WEAK SOILS

5.3.1 APPLICATIONS

Surcharging, or preloading, is a technique applied primarily to strengthen weak soils such as

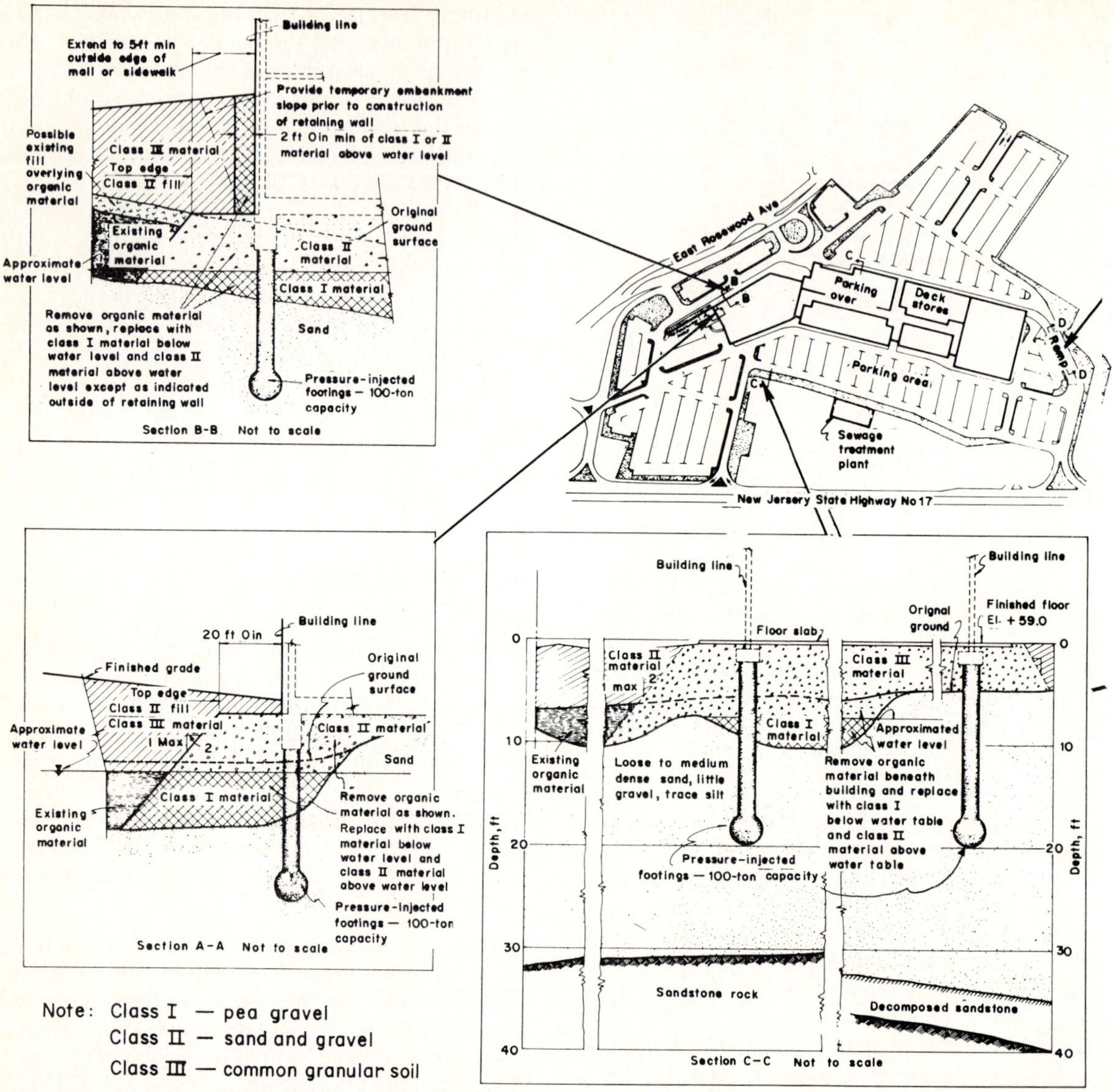

FIG. 5.24 Site preparation and foundation installation for department store complex in Paramus, N.J. Original site was swampy and covered with up to 10 ft of organic material. (Courtesy of Converse Consultants.)

organics and soft clays, although it is also applicable to other materials such as collapsing soils and loose sands. Weak soils are preloaded where the required depth of excavation, usually greater than 3 m, makes the excavation-backfill method uneconomical, and the proposed structure can tolerate some postconstruction settlement.

Preloading commonly is used to improve soil conditions for support of roadway or other pavement embankments, for support of light buildings and storage tanks, and for reducing downdrag loads on piles. The procedure usually is limited to structures that impose a relatively small average load, or can suffer moderate

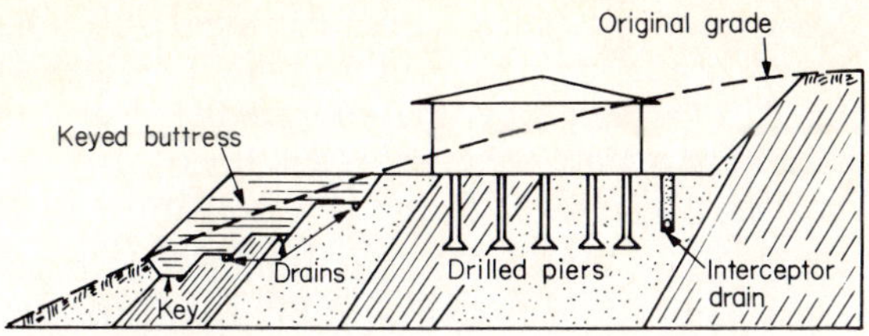

FIG. 5.25 Site grading and foundation solutions for housing development on hillside of dipping sandstones and expansive shales in Menlo Park, Calif. [From Meehan et al. (1975).[25]]

amounts of postconstruction settlement, since surcharging frequently does not eliminate all of the significant settlements.

Strength increase in noncollapsing and cohesive soils is the result of consolidation under preload which ranges generally from 1.5 to 2.0 times design load. The preload is left in place for some time interval, generally ranging from several months to a year or more. Preload is obtained from several sources: a lightly compacted earth fill; water retained within a diked area with an impervious bottom; the use of the structure itself, such as a flexible steel storage tank loaded with water; or the lowering of the groundwater level. Collapsible soils are preloaded by saturation which destroys the weak bonds between grains by a process termed "hydrocompaction."

5.3.2 SURCHARGE METHODS

Surface Loadings

Fill that will remain in place to support the final structure is compacted by the normal procedures. The excess or surcharge load placed above final construction grade can be loose fill, ponded water, or the structure itself, if it can be loaded with water and tolerate large settlement.

Most surcharges are constructed of earth, but the Port Authority of New York and New Jersey has used water placed within a large diked area with an impervious bottom to preload deep deposits of organic soils as part of a large-scale land development for warehousing in Port Elizabeth, New Jersey. Areas as large as 43 acres (17.5 hectares) were preloaded with 22 ft (6.7 m) of water. Each pond bottom was lined with a polyvinyl membrane, and a system of drains installed beneath the membrane prevented uplift pressures which would result from localized leakage through the liner.

Earth Fills

Most surcharging is accomplished by constructing over the weak soils an earth fill (Fig. 5.26) which extends to some predetermined distance beyond the proposed structure. Surcharge fill design requires analysis executed as carefully as for any structure, and a number of factors require consideration including the shear-strength parameters and time-rate consolidation characteristics of the weak soils and the time available for preload. The objective is to preload the soils to the maximum preconsolidation possible within some time schedule, normally controlled by project requirements.

Settlement Analysis

The first factor to consider is the total and differential settlement to be anticipated beneath the proposed embankment or structure load, and the time-settlement relationships if there were no surcharge (*see Art. 4.3.4*). Analysis is then performed to determine the surcharge load and the time required to reduce these settlements to de-

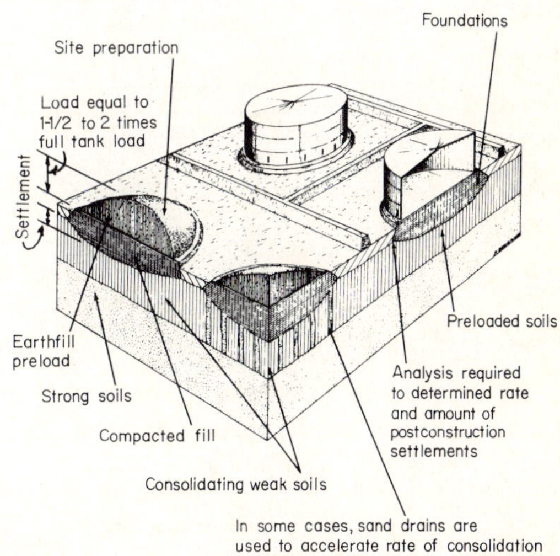

FIG. 5.26 Surcharging storage-tank foundations with earth, with and without sand drains [From Hunt (1967).[26]]

sirable limits. Surcharge loads ranging from 1.5 to 2.0 times the proposed structure loading generally are used, and analysis is performed to determine the time required to achieve consolidation and prestress. If the time is excessive, then sand or wick drains are considered to accelerate consolidation (see the following article).

Stability

A very important factor is the lift thickness that can be placed on the weak soils without causing ground rupture (*see Art. 8.2.4*). Failure not only causes loss of surcharge material, but also results in a further weakening of the underlying soils and additional delays to the operation. To avoid failure, initial lift thicknesses generally are limited to the range of 1 to 2 m. A time interval passes to permit some gain in strength from consolidation before the next lift is placed, and the procedure is repeated until the full surcharge load is attained.

Pore-water pressures in undrained loading, which occur under an embankment surcharge placed over soft clay, may be predicted on the basis of parameter A measured in the CU triaxial test (*see Art. 3.3.2*) with the following procedure:

1. Calculate initial conditions prior to placing the preload to find the vertical effective stress $\bar{\sigma}_v$ and the static pore pressure u_s for the clay middepth at z (Fig. 4.29).
2. Find the stress increments at depth z from the preload ($\Delta\sigma_3$ and $\Delta\sigma_1$) from Fig. 4.12.
3. Employing parameter A measured in the laboratory, calculate the excess pore-water pressure from

$$\Delta u = \Delta\sigma_3 + A(\Delta\sigma_1 - \Delta\sigma_3) \qquad (3.15)$$

4. Measure field pore pressures with piezometers to check computed values.
5. Using effective stress paths (Fig. 3.16), determine the values of pore pressure and parameter A at any stage of loading.

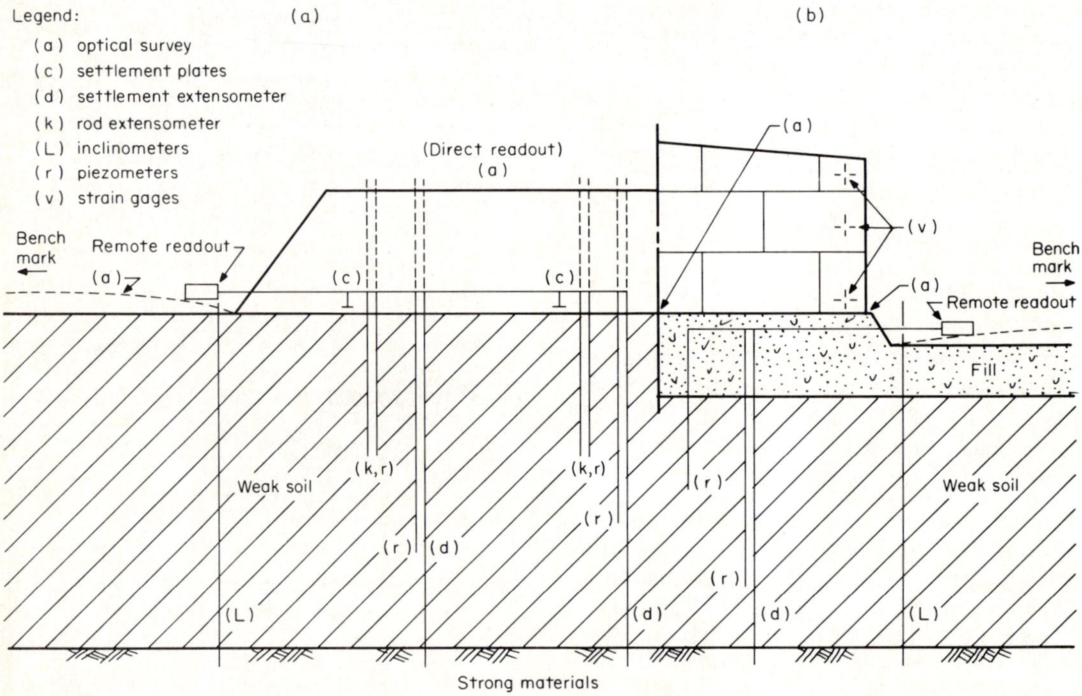

FIG. 5.27 Instrumentation for embankment or steel storage tank over soft ground: (*a*) embankment and (*b*) steel storage tank. [From Hunt (1984).[1] Reprinted with permission of McGraw-Hill Book Company.]

The rate of loading relative to the soil rate of consolidation should be such that excessive pore pressures do not develop; that is, drained loading and consolidation occur.

At times, fills are placed to purposely cause rupture and weak soil displacement, or explosives are used to displace deep deposits of organic soils from beneath fills. These methods frequently result in trapping weak materials beneath the fill, and postconstruction settlement may be both large and differential.

Monitoring

In the field, ground behavior is monitored carefully by instrumentation (Fig. 5.27) to determine when additional loads may be placed and when the surcharge operation is complete. Settlements are measured with settlement plates which measure deflections at the fill base, or with extensometers which can measure settlement occurring at various depths within the compressible stratum. Piezometers are installed to measure pore pressures which indicate the degree of consolidation. Inclinometers are used to measure horizontal deformations within the weak soil stratum to give additional warning of impending failure. Instrumentation is described in detail in Chap. 4 of Hunt (1984).[1]

Fill-thickness, settlement, and pore-pressure data are plotted together to illustrate the time rate and percent of consolidation occurring as shown in Fig. 5.28. Percent consolidation is determined from the computed total consolidation settlement.

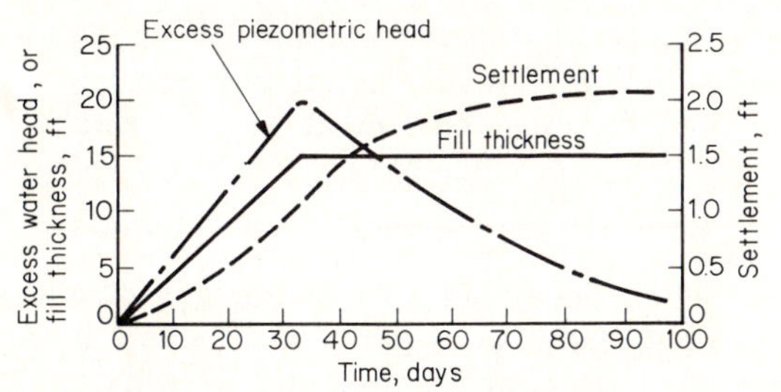

FIG. 5.28 Typical results of measurements during a surcharging operation.

Flexible Steel Storage Tanks

Economic Analysis

In the evaluation of foundations for flexible steel storage tanks, such as are used in the petroleum industry, the major factors are the foundation costs and the time available for site preparation when constructing over weak ground. The results of a particular study comparing the foundation preparation cost per tank with the depth of weak soil and the construction time required is given in Fig. 5.29. Because piled foundations can exceed the cost of the tank itself, the surcharge method is always given strong consideration. In Fig. 5.29*a*, there is a savings in the earth preload method if scheduling permits reusing the earth for subsequent tank sites.

Using the Tank for Preload

Steel storage tanks are quite flexible and can withstand a substantial amount of settlement

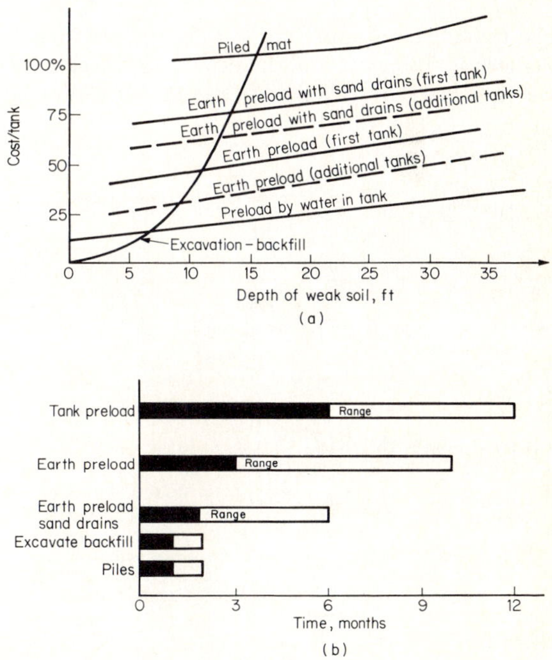

FIG. 5.29 Economic comparison of various site-preparation and foundation techniques for steel storage tanks: (*a*) construction cost comparison and (*b*) construction time comparison. [From Hunt (1967).[27]]

without undergoing rupture. Where suitable fill is not available or is very costly, or where early tank delivery is scheduled and time is not available to mount an earth surcharge, the tank itself may be used for preloading (Fig. 5.30). The major limitation to this method is the amount of preload that can be attained, unless the tank is ordered with excess capacity above anticipated storage requirements. Surcharging is most feasible for petroleum products lighter than water such as crude oil, with a specific gravity of about 86%, and gasoline, with a specific gravity of about 74%; the corresponding surcharge loads are about 1.2 and 1.4 times the design load, respectively. The surcharge is more effective than might be indicated by the ratio of surcharge to design load since in many cases tanks are full only for relatively short times because products are withdrawn; however, gasoline and oils may be stockpiled for 3 to 6 months or longer prior to delivery.

Differential Settlement

When using steel tanks for surcharge, the major concern is the control of deflections occurring in the tank *(see Art. 6.1.2 and Fig. 6.8)*. If uniform

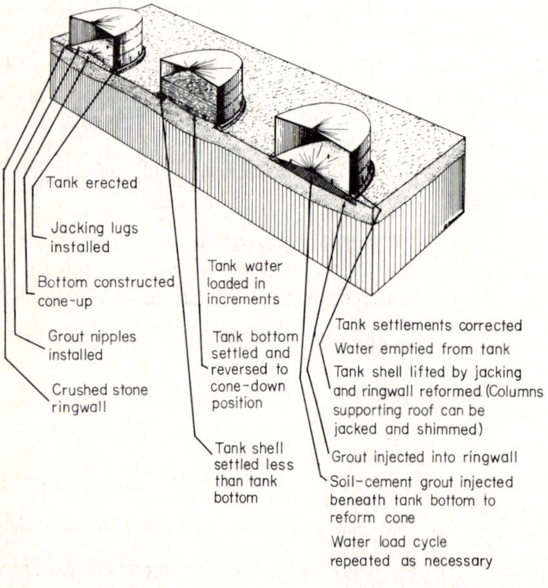

FIG. 5.30 Preloading steel storage tanks with water prior to service. [From Hunt (1967).[28]]

weak soil conditions prevail, the main interest lies in the differential settlement between the shell and the bottom, since the bottom will settle at a much higher rate than the shell and significant overstress may then develop in the weld connections between the shell and the bottom. With irregular soil formations, differential displacement along the perimeter as well as the tilting of the tank require consideration. Measurements of deformations around the shell and in the bottom during surcharging, as well as pore-pressure measurements, are imperative. When differential warping is a concern, strain gages are installed to monitor shell stresses. Tank monitoring is illustrated in Fig. 5.27*b*.

Because the bottom settles about twice as much as the shell where the tank is underlain by relatively uniform strata, the measurement of the bottom settlement is used to control the preload schedule as long as tilting and perimeter deformations are tolerable. Tanks used as a means for preload should never be constructed with bottoms formed cone-downward, since with this configuration the amount of differential movement tolerable between the bottom and the shell is extremely limited.

Tank bottoms should always be formed cone-upward, as shown in Fig. 5.30, and the bottom will deflect to the cone-downward position during preloading. The amount of total deflection permissible is determined for a given tank diameter prior to preloading. Experience has shown, for example, that if a tank 42 m in diameter is constructed with a bottom coned upward 100 cm, that bottom can move safely 200 cm through a complete cone reversal. If the bottom is reformed cone-upward by mud-jacking, it can go through several cycles of cone reversal without its structural stability being affected if weld connections are well made. The entire tank is supported on a cushion of compacted sand at least 3 ft thick.

Correcting Tank Settlement

Settlement occurring during preloading can be corrected readily prior to placing the tank in service. If the thickness and compressibility of the weak stratum yield small to moderate settlement, it may be practical to construct the tank on a raised foundation pad which settles during preloading to the final desired grade, and no further correction to the tank itself is required.

Where large settlement is anticipated, shell and bottom distortions are corrected either by floating the empty tank off the foundation pad, reconstructing the pad to a new grade, and then refloating the tank onto the new pad; or by releveling the tank in place (Fig. 5.30). This is achieved by jacking the perimeter of the empty tank to the desired level, raising the crushed stone ringwall under the perimeter, and then "mud-jacking" the bottom to reform it to the cone-upward position.

Staged Loading

Where large settlement is anticipated it may be necessary to repeat the preload-releveling program in a number of stages. Figure 5.31 presents the actual loading schedule and settlement diagrams for a tank 52 ft (15.6 m) in diameter and 40 ft (12 m) in height that underwent a total of almost 5 ft (1.4 m) of bottom settlement during five successive stages of preloading. Between each loading stage, as shown in the figure, the tank was releveled and the bottom was reformed. The initial water load was placed too rapidly, and the 10 m of water in the tank caused some soil remolding as well as consolidation; bottom settlement was excessive, approaching failure. The second loading stage (October-November) proceeded more slowly, and consolidation was achieved. During the beginning of the third stage the load was applied very rapidly and soil failure, indicated by the high settlement rate, began. The load was immediately reduced and then reapplied at a much slower rate. The much reduced settlement during the fourth and fifth stages indicates that preloading was effective.

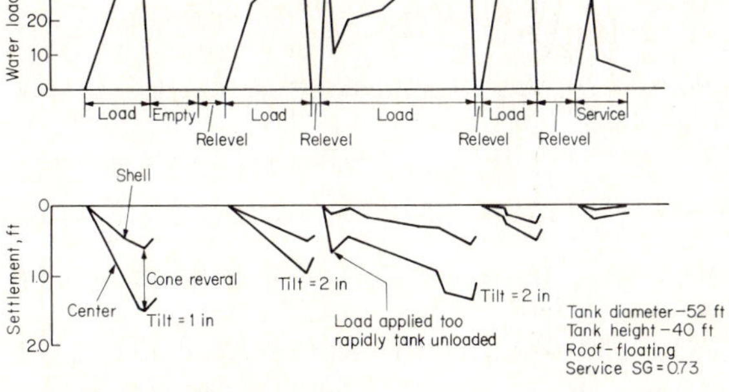

FIG. 5.31 Storage-tank preloading: load and time versus settlement. [From Hunt (1967).[28]]

This particular project involved 56 storage tanks surcharged by the tank-loading method. Soil conditions consisted of a few meters of hydraulic fill overlying up to 15 m of soft organic alluvium. Pile support for these tanks would have cost several millions of dollars in 1956, at the time of the project. Due to deficient early studies, time was not available to achieve the degree of prestress desirable, and large postconstruction settlement was anticipated. It was necessary to monitor tank settlement for many years of service and to remove a tank from service occasionally for releveling. Tall asphalt tanks with a loaded specific gravity of about 1.03 were particularly troublesome because of their tendency to tilt. With proper program scheduling, however, including early ordering and erection of the tanks, this method can be effective and economical.

Other Surcharging Methods

Dewatering

Lowering of the water table by wellpoints or deep wells is an effective surcharge method, particularly where a thick granular stratum overlies a weaker, very compressible stratum. For every foot that the water table is lowered the overburden pressure is increased by an amount equal to the bouyant force of water, or 62.4 pcf (1 t/m^3). If the water level is lowered 33 ft (10 m), a surcharge pressure of 1 tsf results, equivalent to about 18 ft (5.4 m) of earth surcharge. Ground subsidence from groundwater withdrawal is a well-documented phenomenon as described in Chap. 10 of Hunt (1984).[1]

Hydrocompaction

Thick formations of collapsible soils have been treated by "hydrocompaction." The area to be treated is surrounded by a dike and then filled with water which seeps downward saturating the soils, destroying the lightly cemented structure, and causing collapse [see Art. 10.5.4 of Hunt (1984)[2]].

5.3.3 VERTICAL DRAINS TO ACCELERATE PRELOADING

Applications and Types

When the time delay under a normal preloading scheme is excessive, the installation of sand or wick drains can reduce substantially the time required for consolidation and preload.

Sand Drains

A system of sand drains, or sand columns, customarily varying from about 2.5 to 5 m in spacing, is installed through the weak stratum and connected to a free-draining blanket of granular soil, as illustrated in Fig. 5.32. As the fill is placed in the normal surcharging method, pore pressures increase but, because the drainage paths are very much shorter than they are without the drains, consolidation occurs much more rapidly. Sand drains may reduce the time required for surcharging to as little as 1/100 or less, but in scheduling, the extra time for their installation must be considered. A large-diameter storage tank, for example, may require up to 6 weeks for installation of sand drains on 2.5 m centers, with two rigs working, where the weak soils are thick. Wellpoints can be placed in the drains to accelerate drainage as an alternate procedure to reduce the amount of earth surcharge.

Wick Drains

Wick drains have been used for over 20 years in Europe and Japan in lieu of the generally more costly sand drains, and in the United Stated installations have been increasing rapidly since 1978 [Morrison (1982)[29]]. The wicks, composed of polyethylene or PVC with either paper or polyester filters forming a sandwich, are in the order of 100 mm wide by 4 mm thick and are contained on rolls. Threaded into the ground by various methods, spacing is commonly in the order of one and a half to two wick drains per one sand drain, although a 1:1 ratio has been used [Morrison (1982)[29]].

In comparing the less expensive paper filters with the polyester filters, it appears that paper filters have a lower permeability and may be more susceptible to damage, but the polyester filters with a higher permeability may allow fines to pass into the core and may stretch while in place. Therefore, one type may be more suitable for a particular soil condition than the other, and on large projects installation in a test surcharge will indicate which filter type is more efficient. As of 1984, industry standards have yet to be developed, but in time selection should be possible with less uncertainty. Installation to depths in excess of 30 m is possible with polyester filters. For additional references, see Hansbo (1979),[30] Seim et al. (1981),[16] and Fellenius (1981).[31]

Analysis

Radial Drainage Problem

Analysis is performed to estimate the sand-drain spacing required to achieve the desired degree of

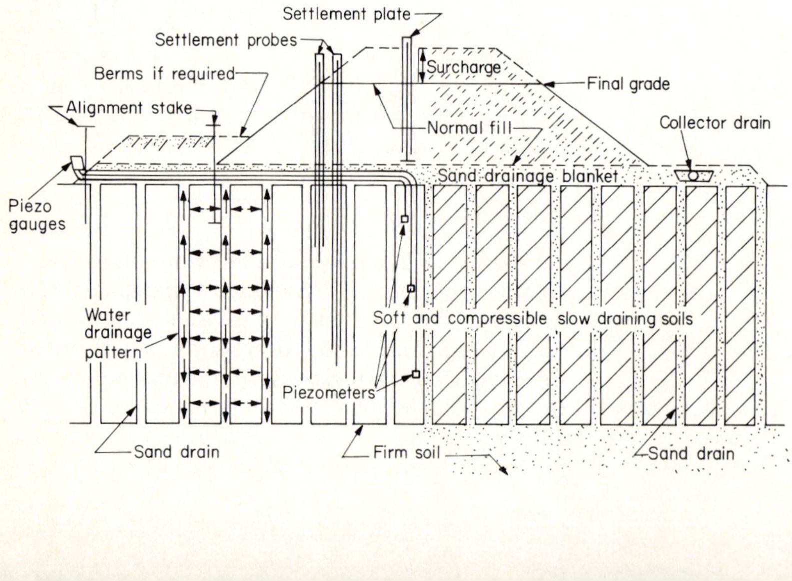

FIG. 5.32 Typical sand drain installation with surcharge.

consolidation within the time available. The drain diameter, usually ranging from 12 to 18 in, has much less effect on the time rate of consolidation than does the drain well spacing. Consolidation analysis for vertical drains is a problem in radial drainage in the horizontal direction. An expression for the time to obtain a given percent consolidation U is obtained from Eq. 4.47 by substitution which yields

$$t = \frac{T_h}{c_h} d_d^2 \tag{5.1}$$

where t = elapsed time
T_h = time factor for horizontal drainage
c_h = coefficient of horizontal consolidation
d_d = diameter of zone draining to the well (Fig. 5.33)

The value of T_h varies with the ratio of the draining area d_d to the well diameter d_w, given as n. The variation of the time factor T_h with the percent consolidation U for various values of n is given in Fig. 5.34.

Evaluating c_h

The coefficient of horizontal consolidation c_h is measured in the laboratory consolidation test on a specimen trimmed into the consolidometer ring with orientation perpendicular to the usual direction to simulate drainage in the horizontal direction, or by placing a normally oriented specimen in a porous ring and placing plates on the sample top and bottom to prevent vertical drainage.

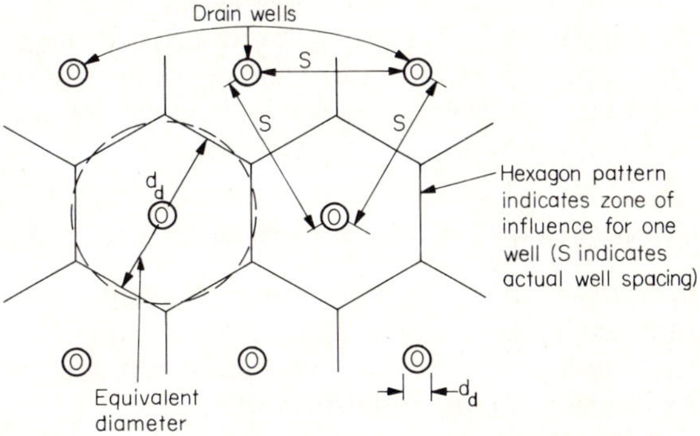

FIG. 5.33 Pattern of sand-drain installation and equivalent well diameter. [After Barron (1948).[32]]

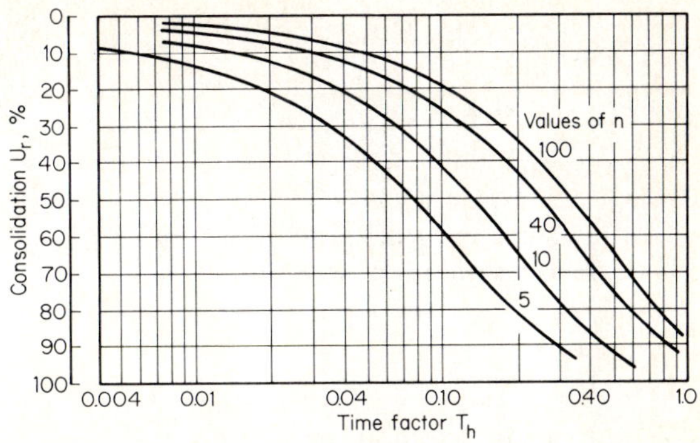

FIG. 5.34 Variation of time factor T_h with percent consolidation U for various values of n, the ratio of draining area of diameter d_d to the well diameter d_w. [From Barron (1948).[32]]

The driving of the mandrel during installation causes remolding or smearing of the adjacent soils, which distorts stratification. The remolded annular space about a sand drain can extend for a distance of one-half to one-third the drain diameter. As smear has the effect of decreasing permeability, it is accounted for in analysis by increasing the diameter of the zone draining to the well d_d or by using values for c_h obtained from laboratory consolidation tests performed on remolded specimens. In any event, computed time rate of consolidation relationships can be considered only as approximations, as discussed in Art. 4.3.4.

A nomograph for estimating the average degree of percent consolidation for various values of c_h, sand drain diameter and spacings, and time, is given as Fig. 5.35.

Installation

Drainage Blanket

The first step of the installation procedure is to place a working platform for the equipment. The platform, consisting of free-draining soils, also serves as a drainage blanket when connected to the drains (Fig. 5.32). Typical specifications for the blanket and for the drain soils require a material with a grain size ranging between the no. 4 and no. 100 sieves, with a maximum of 3% passing the no. 200 sieve and 7% passing the no. 100 sieve. The blanket thickness must be ade-

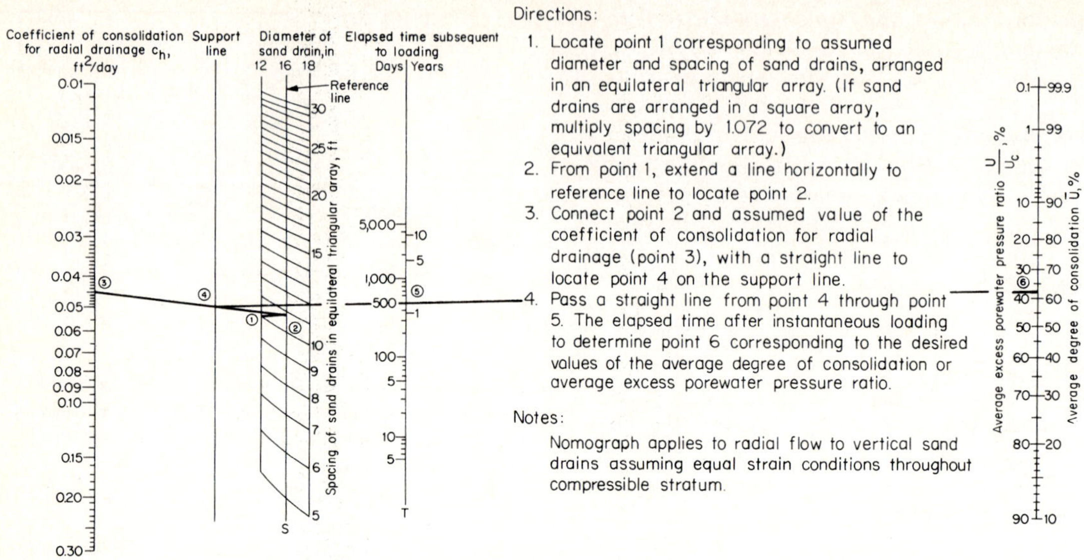

FIG. 5.35 Nomograph for consolidation with radial drainage to vertical sand drains. [From NAVFAC (1982).[5]]

quate to allow for loss of grade during consolidation of the weak soils, and it is connected to perimeter or collector drains.

Sand Drains

Holes for sand drains normally are formed with a closed-end mandrel, although augered drains and other types have been used. The metal plug is either pulled or left in the ground. Prior to pulling, the mandrel is filled with the specified sand and sometimes subjected to air pressure to force the sand into the mandrel and assure continuity of the sand column. The mandrel is then pulled from the ground while air pressure is applied beneath the top lid, and the sand column is completed.

Wick Drains

Installation of wick drains is accomplished with a conventional vibratory pile hammer in conjunction with a specially fabricated wick-feeding apparatus, or by a static method in which the wick is pushed into the ground, or in some cases by jetting. The first two "dry" methods aid in making wick drains less costly than sand drains, which may require not only substantial quantities of water but also sand of a narrow gradation, which may not be readily available. In sensitive soils, whose strength and permeability are severely affected by remolding, installation by pushing or jetting a mandrel of the smallest possible size is considered advisable to minimize disturbance.

Monitoring

During fill placement it is imperative that lift thicknesses and time intervals between placement be controlled carefully to avoid rupture or any lateral displacement of the weak soils. Rupture or excessive displacement will cause shearing of sand drains and kinking of wick drains and result in the loss of their effectiveness.

5.4 OTHER SOIL-STRENGTHENING TECHNIQUES

In addition to the use of moving machines to compact soils in layers for pavements and to build fills for embankments, earth dams, and structure support and in addition to the surcharging of weak soil strata, as discussed in the preceding articles, a number of techniques are available for strengthening soil formations. These can be placed in four major categories:

1. Compaction of weak strata in situ
2. Reinforcing the ground by the insertion of some foreign material
3. Treatment of pore water
4. Soil stabilization with admixtures

5.4.1 COMPACTION IN SITU

Vibroflotation (Vibrocompaction)

Description of Process

Loose sands above or below the water table can be effectively densified by the patented process termed *vibroflotation* (also called *vibrocompaction*). A cylindrical vibrator, with lengths ranging from about 3 to 5 m, a diameter of 40 cm, and weighing about 3 to 6 tons (depending upon length), suspended from a crane, is jetted into the sand deposit to the depth where compaction is to begin by a pressured stream of water discharged from the cylinder tip. The limiting depth has been about 12 m. At depth the vibrator is activated, which causes the soil to compact. Vibration is continued as the cylinder is raised slowly to the surface. The void around the cylinder caused by the vibrations is filled continuously from the surface as the cylinder is raised, as shown in Fig. 5.36. Water jets at the top of the cylinder create flow to carry the sand fill to the hole bottom.

Results and Limitations

The process provides relative densities up to about 70% to a distance of about 1 m from the cylinder; therefore, spacings of 2 m or less are used to achieve an average relative density of 70% or more. Spacing is often increased up to 2.4 m in coarser sands. Allowable bearing values in deposits compacted by vibroflotation are in the range of 2 to 3 tsf. Application is generally limited to granular soils with not more than about 20% of the material passing the no. 200 sieve, and preferably less than 10%. The greater the percentage of fines, the slower the removal of excess water required for densification [Mitchell (1970)[33]].

Terra-Probe

The Terra-probe is also a patented process used to densify loose sands to depths below the surface in the order of 12 m. A 30-in-dia. open-ended tubular probe is vibrated into the ground by a unit similar to those used in vibratory pile driving to the planned penetration depth and then withdrawn slowly while vibrations continue. Spacings are closer than with the vibrofloation process, in the order of 5 ft (1.5 m). The operation is most efficient where groundwater is within 2 to 3 m of the surface, although water jets can be incorporated into the probe to assist in penetration and densification. Ponding the surface also helps the procedure. Although the Terra-probe method is more rapid than vibroflo-

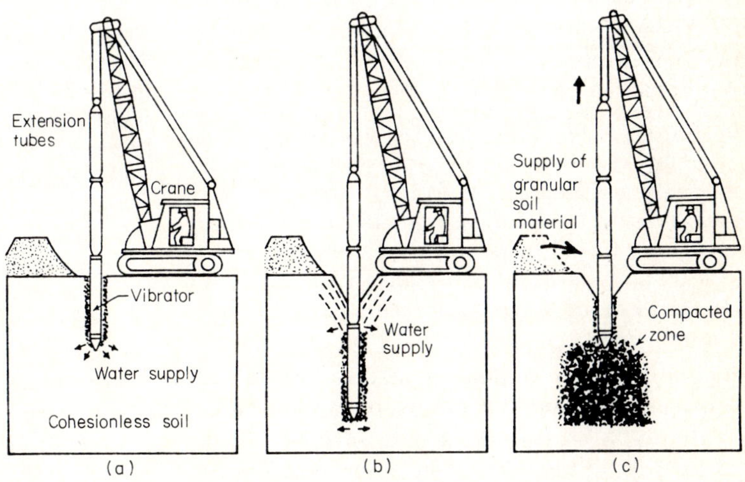

FIG. 5.36 Sketch of vibroflotation or vibrocompaction process: (*a*) vibrator penetrates to desired depth, (*b*) vibrocompaction begins at depth, and (*c*) granular soil is added as vibrator is raised.

tation, the achieved relative densities are somewhat lower [McCarthy (1982)[34]]. Since the procedure does not require a sand fill, it has the benefit of being applicable to offshore sites.

Dynamic Compaction

Description of Process

A pounder in the form of a concrete block or steel plates, weighing in the range of 10 to 40 tons, is dropped from a standard crane from a height varying from 15 to 45 m to impact on the ground surface. The high-impact-imparted energy results in the formation of craters. Little or no soil heave occurs around the crater after a number of drops in the same location, usually between 5 to 10. Pounder weight, drop weight, number of drops per pass, drop spacing, and number of passes needed to achieve required compaction levels depend upon the thickness of the ground to be compacted, the characteristics of the materials, the groundwater level, and the consistency of the material underlying the soil to be compacted.

Field trials with test borings or CPT before and after compaction are required for any given site to determine the necessary procedures, especially since experience with the process is relatively recent. It was developed commercially by Louis Menard.

Applications

As reported in the literature, dynamic compaction has been used to improve a variety of unfavorable soil conditions so as to achieve satisfactory performance of shallow foundations or other loadings as follows:

Soft silty clay fill, 5 to 8 m thick overlying a stiff silty clay, was compacted to provide an allowable bearing value of 2 tsf for footings for an air cargo terminal. *Saturated dredged sand fill,* 6 m thick, was raised in relative density from 40 to 70% to provide support for an airport runway and taxiways [Ramaswamy, Lee, and Daulah (1981)[35]].

Soft silty clay from 3 to 6 m thick *overlying peaty clay* averaging 5 m in thickness was improved to an allowable bearing value of 1 tsf for footings for light buildings. The compaction resulted in mechanical mixing of the silty clay with the peaty clay [Ramaswamy, et al. (1979)[36]]. According to Menard, where *sand fill overlies soft organic soils,* mechanical mixing is possible if the underlying weak soil of high void ratio is less than about 6 m thick and if the overlying sand does not exceed about 5 m in thickness. Compaction results in closely spaced, very large sand columns. as well as mixing [Ramaswamy, Lee, and Daulah (1981)[35]].

Random fill, 3 to 15 m thick, was compacted by a 12-ton concrete block dropped from a height of 9 m six times at each footing location, causing approximately 0.76 m of compaction at each location. Subsequent investigation revealed that compaction influence was in excess of 6 m in depth [Ramaswamy et al. (1981)[35]].

Compaction Grouting

Description of Process

Compaction or displacement grouting is performed by injecting into a relatively weak soil mass a grout of very thick consistency. The effectiveness depends on soil compressibility, rather than permeability, and the effect of the grouting operation is to compact the surrounding soil as well as to leave an irregularly shaped column or bulb of grout. Its diameter reflects the relative compressibility of the material.

To assure maximum restriction of the compacted zone, injection may be performed in vertical stages, usually from the top down [Warner (1978)[37]]. A hole is drilled from the surface to the top of the zone to be compacted, steel casing is inserted in the hole and cemented in place. The hole is then extended by drilling through and beyond the casing for a depth usually of 4 to 8 ft (1.2 to 2.4 m) for the first grout stage. Stiff-consistency grout is then injected until slight movement at the surface is noted or an excessive pressure is reached [500 to 600 psi (35 to 42 kg/cm^2)]. After hardening of the first grout stage, the hole is redrilled through the grout and advanced another 4 to 8 ft for a second stage, and grouting is repeated. The process is repeated until the desired competent bearing stratum is reached. The grout material used most often is a fine sand mixed with about 12% cement and water to form a stiff, mortar-like material with a maximum slump of 1 in (25 mm) [Brown and Warner (1973)[38]].

Applications

The process appears most applicable where existing structures are suffering detrimental settlement due to a buried weak stratum of limited thickness. The technique generally is not suitable where overburden pressures are low, as for example beneath shallow foundations or close to hillside slopes, unless mud-jacking is desired. In one case a court building supported on friction piles was settling detrimentally because of consolidation of several strata of organic soils underlying the pile tips. The settlement was arrested by compaction grouting of the organic layers [Warner (1978)[37]]. The compressible strata were at depths of 37 and 47 ft. The grout patterns selected were triangular layouts around individual piles and box layouts around pile-supported continuous footings.

Applications for the control of subsidence during soil tunneling are discussed in Art. 10.2.4.

Mud-jacking is a form of displacement grouting used to raise highway pavements, railway tracks, building floors, and storage tanks *(see Art. 5.3.2)* which have suffered detrimental settlement. Injection is done at relatively low pressures and to shallow depths with a mixture of sand, portland cement, and water (Table 5.5).

Surface Compaction

As described in Art. 5.2.4, smooth-drum vibrating rollers have achieved relative densities in excess of 70% to a depth of 1.7 m and are effective to depths as great as 5 m, when heavy enough and when used to compact a stratum of clean, poorly graded sand from the surface. Surface compaction usually is necessary to "top off" densification by vibroflotation, dynamic compaction, etc.

5.4.2 GROUND REINFORCING

Injections or Penetration Grouting

Applications

Grout is a material in liquid form injected under pressure into the ground where it fills the voids and hardens. Grouts are commonly used to control groundwater and seepage [see Art. 8.4.2 of Hunt (1984)[1]], but some types are used also to strengthen soil formations. In either case, the procedure is applicable where treatment must be carried out in place, where soil removal is uneconomical or impractical, or adjacent to or beneath existing structures which cannot be disturbed.

Various types of grouts have various viscosities, and since penetration is dependent upon the size of the interstices of the soil deposit, their application varies with the grain size distribution of the soil. In general, sand-cement mixtures may be used to strengthen free-draining soils including types GW and GP; cement to strengthen GW, GP, and SW; and clay-cement to strengthen GW, GP, SW, and SP. These are the grouts that cost the least. The more expensive sodium silicates can penetrate materials with a permeability coefficient down to 10^{-4}, and resins to a k value of 10^{-6}, or into the range of gravelly or sandy clays. These grouts have the advantage that the incorporation of additives permits very short set times. As silica gels have a viscosity approaching water and contain no solid particles, penetration is usually excellent. Grout characteristics are described in Art. 8.4.2 of Hunt (1984)[1] and summarized in Table 5.5. To determine the suitability of a particular grout type, in situ test sections are injected and then sampled with equipment such as the Denison core barrel to retrieve specimens for laboratory strength and consolidation testing.

Installation

Because grout pipes are normally installed by jetting, it is necessary to first install cased holes when improving conditions next to or beneath existing foundations. The spacing depends on the grout penetration and the degree of strengthening desired. Since columns of grouted areas will reinforce a compressible stratum, it is not necessary to have grouted zones in contact as it is when an impervious barrier is required. Grout pressure control is imperative to avoid displacement of soil and the adjacent structure or foundations. Treatment of existing foundations must be performed carefully and requires substantial experience, as well as the monitoring of structural movements. In *penetration grouting* the materials treated usually are nonorganic silts or

TABLE 5.5
COMMON CEMENT AND CHEMICAL GROUTS*

Grout type	Composition	Application
Sand-cement	Loose volume sand-cement ratio varies from about 2:1 to 10:1. Bentonite or fly ash additives reduce segregation and increase pumpability. Water-cement ratios from about 2:1 to 5:1 by volume.	Grouting large foundation voids, mud-jacking, and contact grouting around structure periphery. Slush grouting without pressure to fill surface irregularities in rock foundation. Depending upon sand gradation in mix, will penetrate gravels with D_{10} about ¾ in. For usual mix, strengths vary with water content from 100 to 700 psi.
Clay-cement	Loose volume clay-cement ratio varies from about 3:1 to 8:1. Water-clay ratios from about ¾:1 to 2:1 by volume. Water-cement ratios from about 3:1 to 10:1.	Comparatively large voids and fractures where clay is added for economy. Penetrates coarse sand with D_{10} as small as 1 mm; fissures in the range of 0.06 to 0.01 mm depending upon pressures, water-cement ratio, and cement types. Not appropriate for large voids with vigorous groundwater movement. Strength depends on water-cement ratio and averages about 100 psi for typical mix.
Portland cement	Water-cement ratios generally between 1:1 and 1:4. To reduce bleeding and segregation, add bentonite, silica gels, pozzolans. To increase pumpability, add ligno-sulphonates. To accelerate set, add calcium chloride.	Penetrates coarse sand with D_{10} as small as 1 mm; fissures in the range of 0.06 to 0.01 mm depending upon pressures, water-cement ratio, and cement type. Not appropriate for grouting large voids with vigorous groundwater movement. Most common material for decreasing permeability and increasing strength. Strength depends on water-cement ratio and averages about 100 psi for average mix.
Bentonite gel	Dispersed clay slurry with flocculating agent such as aluminum sulfate to cause suspension to coagulate after injection.	Materials down to fine sand with D_{10} between 0.2 and 1.0 mm, depending on grain size of clay. Relatively low cost, but slurry may be removed by vigorous flow. No significant strength increase.
Sodium silicate (single stage)	Sodium silicate with a setting agent such as sodium aluminate in water solution. May be combined in cement or soil-cement mixes.	Sands with D_{10} as small as about 0.08 mm. Setting agent used to obtain set time ranging from few minutes to several hours. Compressive strength about 100 psi. (*Two stage:* successive injection of sodium silicate and calcium chloride produces strength ranges from 500 to 1000 psi.)
Epoxy resins	See Hunt (1984),[1] Art. 8.4.2.	

*After NAVFAC (1971).[37]

sands, or mixtures thereof, substantially stronger than the organic soils or soft clays treated by *compaction grouting*, and more susceptible to imposing pressures on existing foundations.

Columns Formed in Situ

Columns formed of gravel, sand, lime, or admixtures have been formed in situ in weak soil formations to reduce settlement to tolerable magnitudes as well as for other purposes.

Stone Columns (Vibroreplacement)

Columns of coarse gravel-size fill, well-graded between ½ and 3 in, are installed to penetrate through compressible layers of soft cohesive and organic soils, of inorganic silts, and of silty sands to reduce settlement to within tolerable limits and to achieve higher safe bearing capacity. In fine sands and silty fine sands below the water table and susceptible to liquefaction, Seed and Booker (1977)[39] propose gravel drains to relieve pore pressures during cyclic loadings [see Art. 11.3.3 of Hunt (1984)[1]].

Installation is performed with equipment similar to that used for vibroflotation. As illustrated on Fig. 5.37, the vibrator is sunk to a predetermined depth and then withdrawn, leaving a hole of slightly greater diameter, which is subsequently filled with coarse gravel in a series of charges. With each charge the vibrator repenetrates and displaces the backfill horizontally into the natural soils and simultaneously compacts the material beneath the tip, finally forming an irregular but basically cylindrical stone column. In saturated fine sand or silt, the water level in the hole must be maintained at least level with the GWL during hole advancement and gravel installation. Stone column diameter usually varies between 0.6 and 1.0 m and depends mainly on the strength of the surrounding soil.

An application of the stone column technique is illustrated in Fig. 5.38. A description of design approaches is given in CGS (1978).[40]

Sand Columns

Sand piles or sand columns were installed to reinforce a loose mixture of incinerated rubbish fill about 10 m thick to permit shallow foundation support for a new terminal building at LaGuardia Airport, New York City. The rubbish fill overlies about 15 m of very compressible bay muds, and all other structures in the airport are supported on deep pile foundations [ENR (1979)[42]].

During the 40 years the airport has been in operation the consolidation of the bay mud under the rubbish fill load has resulted in about 2 m of settlement of the site, large differentials between structure and pavement grades, breaking of utility lines, and downdrag forces on piles. The latter condition has imposed actual loads twice those of design on the piles, and many have

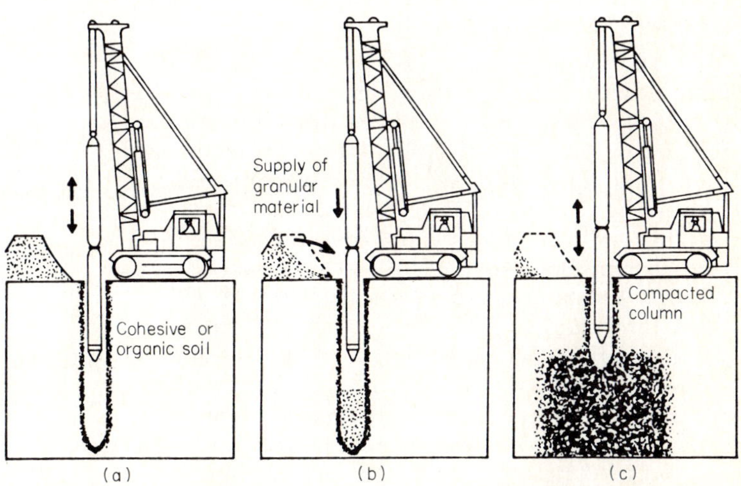

FIG. 5.37 Sketch of vibroreplacement method forming a stone column: (*a*) vibrator penetrates to predetermined depth, (*b*) hole charged with gravel, and (*c*) vibrator compacts gravel column. [After Munoz and Mattox (1977).[41]]

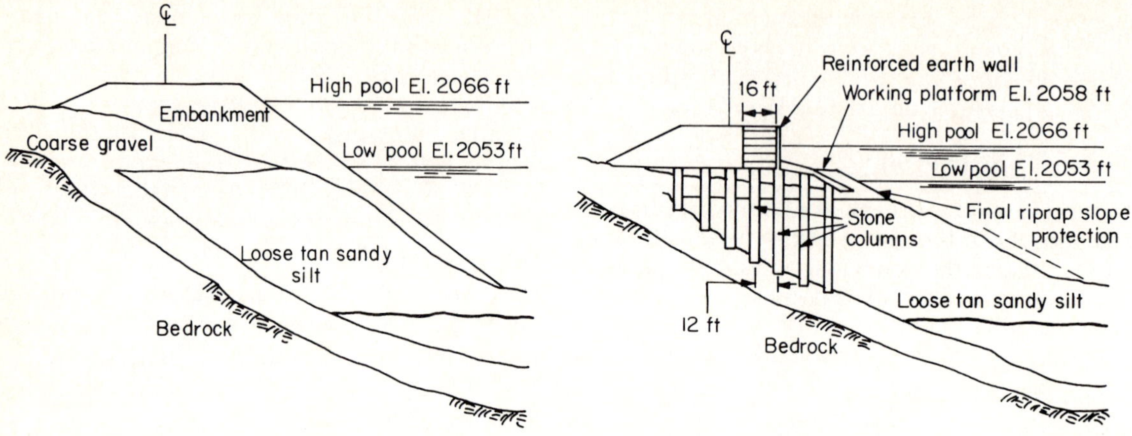

FIG. 5.38 An example of the application of the stone column technique combined with a reinforced earth wall to support a railroad embankment along a lakefront. (*a*) Embankment proposed for location underlain by loose sandy silt poses a stability problem. (*b*) Solution: embankment constructed partially as reinforced earth wall, supported on stone columns. [From Munoz and Mattox (1977).[41]]

failed, especially single piles supporting hanger floors.

The sand piles are clustered around footing locations with the objective of reducing differential settlement and permitting the new building to settle at about the same rate as the general site area. Provisions for releveling the structure by jacking and shimming columns are provided to enable correction of differential settlement should it become too severe for structural integrity.

Lime Columns

Lime columns are being used in Sweden beneath light structures and roadway embankments to reduce settlement resulting from consolidation of thick deposits of soft glacial clays. Unslaked lime or quicklime is mixed with the clay in situ by a specially made auger-type machine to form columns up to 10 m long and 50 cm in diameter [Broms and Bowman (1979)[43]]. The lime-clay mixture substantially increases the strength of the natural clays, and the columns act to transfer surface loads to deeper, stronger strata. The increase in shear strength from lime mixing was found by Broms and Bowman to be most significant in fresh and brackish water-deposited glacial clays, but in soft organic clays and saltwater-deposited clays the strength increase was found to be low.

The technique has been applied also to decrease the negative skin friction on pile-supported buildings and bridge abutments, to prevent the lateral displacement of soil around pile foundations from creep, to increase the stability of clay slopes, to reduce lateral earth pressures on sheet-pile and other retaining structures, and as a replacement for sheet piles in deep excavations to prevent bottom heave.

Deep Chemical Mixing Method (DCM)

Since 1976, the Japanese have been treating very soft seabed alluvial soils for support of waterfront retaining structures [ENR (1983)[43a]]. For about \$34/yd^3, material with an unconfined compressive strength of 0.02 to 0.04 tsf has been increased in strength to 20 to 40 tsf. Operating from a barge, a plant produces a slurry consisting of conventional portland cement, a solidifying material, and seawater. The chemical is a pozzolith material containing lignin and an air-entrained water-reducing admixture, which is mixed in situ with the soft seabed soils.

To initiate the operation, eight 10-in-dia., 171-ft-long shafts are inserted into a stable stratum by rotating excavating blades mounted near their pointed tips. The shafts are then raised at more than 3 ft/sec as slurry is pumped from the barge and injected into the soft soils through 1½-in-dia. pipes. Blades at the shaft ends are rotated to blend the discharged slurry with the soil. A pozzolan reaction is produced, and 28 days later the treated stratum is rigid and able to support construction. Since the shafts are relatively closely spaced and positioned in two rows of four each, the strengthened mass is approximately in the shape of a block.

Geotextiles

Description

Large rolls of fabrics of synthetic polymers, originally called *filter cloth*, are a modern innovation for use in drainage systems, erosion control, and earth reinforcement. Various fabrics have been developed, including a polymer mesh reportedly having the strength of mild steel [NCE (1980)[44]].

Applications

Various applications of geotextiles for earth reinforcement are illustrated in Fig. 5.39. In *road construction* they reduce the pavement thickness required over weak subgrades and permit embankment construction over soft soils. In pavements they are used only for low-standard roads because some rutting is necessary to strain the fabric and enable it to develop the required tensile stress. Most road applications have been for low-standard, aggregate-surface roads on very soft subgrades. It has been suggested that for the reinforcing effect to be significant, the subgrade CBR value should be less than 3 [*Civil Engineering* (1980)[45]]. In one reported case a bituminous concrete surface placed over at least 0.8 m of crushed stone base was constructed on woven filter fabric placed over a variable thickness of peat and organic silt ranging up to 3.6 m thick. Old miscellaneous fill left in place or crushed stone placed in soft pockets provided a working platform for placing the fabric [Belongia (1981)[46]]. The pavement performance after 2 years of service is reported to be good.

In *embankment construction* the fabrics reinforce the edge of the fill permitting compaction to the very edge, thereby achieving higher densities than normally obtained and improving stability and erosion resistance.

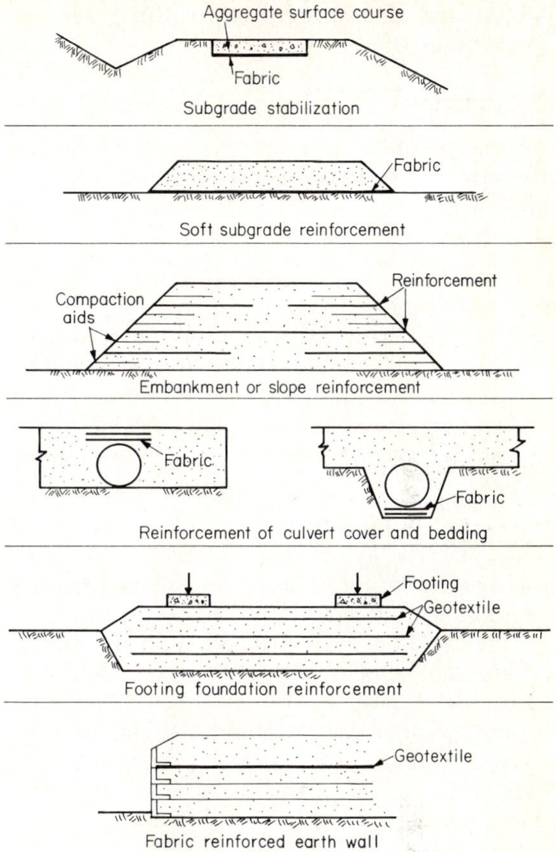

FIG. 5.39 Geotextile applications for earth reinforcement.

5.4.3 PORE-WATER TREATMENTS

Wellpoints

When sandy soils are dewatered with wellpoints, capillary forces result in an apparent cohesion enabling cut slopes to remain stable even when cut to relatively steep inclinations.

Vacuum wellpoints can produce a small change in pore pressure that can cause silt in an excavation bottom to change from a soft "quick" condition to a strong, firm condition [Art. 8.4.3 of Hunt (1984)[1]]. In a sewer trench, for example, a stable foundation is provided for the support of the bedding stone and pipe.

Electroosmosis

Electroosmosis may be used to increase the strength of saturated silts when encountered as thick deposits in open excavations, slopes, or tunnels.

Two electrodes are installed to the depth to be dewatered and a direct current applied. The induced current causes the pore water to flow from the anode to the cathode where it is removed by pumps. Soil strength is increased by the seepage pressure of flowing water which adds to the intergranular pressure [for a discussion, see Lambe (1962)[47]].

Freezing

Freezing, or the formation of ice walls, provides a temporary expedient during construction of open excavations, tunnels, and shafts to control seepage and temporarily increase strength. The technique is most useful in thick deposits of "running sands" and saturated silts encountered at depths that may exceed 300 m, or where grout materials may contaminate water supply. The procedure for deep shaft construction is described in Art. 8.4.2 of Hunt (1984).[1]

To temporarily arrest rapidly occurring and dangerous differential settlement of a 26-story building in São Paulo, Brazil, a stratum of soft clay beneath the deep foundations was frozen while a new foundation of deep piers bypassing the soft soils was installed [Dumont-Villares (1956)[48]].

5.4.4 SOIL STABILIZATION WITH ADMIXTURES

Soil stabilization has been described in its broadest sense as the alteration of any soil property to improve its engineering performance [Lambe (1962)[47]]. This article covers stabilization by the mixing of surficial soils with additives to form new materials resistant to erosion and suitable for use as pavements.

Clay-Gravel (Clay-Concrete)

The oldest method of soil stabilization is the mixing of granular soils with clay which provides a binder, improves strength and compactability, and decreases erodability. A pavement suitable for light to moderate traffic on secondary roads is thus provided in many climates.

A satisfactory pavement, often referred to as *clay-concrete*, will include the following mixtures: about 22 to 38% silt and clay, about 26% sand, and about 36 to 50% gravel [TRRL (1974)[9]]. Too much clay and silt (>50%) and too little gravel (<20%) results in a slippery pavement; too little silt, clay, and sand (< about 40%) results in corrugated surfaces. The gravel particles can be laterites as well as quartz or other similarily hard material. (In normal pavement design, the clay-silt binder is replaced with either cement or asphalt.)

Soil-Cement

Description

Soil-cement, used for pavement bases and surfaces and erosion protection, is formed by mixing pulverized soil with measured amounts of cement and water and then compacting the material placed over a prepared subgrade. As the cement hydrates, a hard, durable structural material is formed. When made properly, it does not soften under cycles of wetting and drying or of freezing and thawing, although when it is used as a pavement surface abrasion resistance to traffic is low.

In general, the coarser the soil, the less cement is required, as given in Table 5.6. The best results are obtained with well-graded soils having less than 50% particles finer than 0.074 mm (no. 200 sieve) and a PI less than 20%. The use of admixtures of certain chemicals reduces the amount of cement required and accelerates set [Lambe (1962)[47]].

Developing Specifications for Mixtures

Laboratory control tests are performed on mixtures compacted in standard Proctor molds. The specimens are subjected to the deteriorating effects of 12 cycles of wetting and drying (ASTM D559-57) and 12 cycles of freezing and thawing (ASTM D560-57), performed on two sets of specimens after 7 days of moist-curing. After each cycle the specimens are brushed with a steel-wire brush and the amount of loose material re-

TABLE 5.6
CEMENT REQUIREMENTS OF AASHTO SOIL GROUPS*

AASHTO soil group	Usual range in cement requirement, %		Estimated cement content, also used in moisture-density test, % wt	Cement contents for wet-dry and freeze-thaw tests, % wt
	By volume	By weight		
A-1-a	5– 7	3– 5	5	3- 5- 7
A-1-b	7– 9	5– 8	6	4- 6- 8
A-2	7–10	5– 9	7	5- 7- 9
A-3	8–12	7–11	9	7- 9-11
A-4	8–12	7–12	10	8-10-12
A-5	8–12	8–13	10	8-10-12
A-6	10–14	9–15	12	10-12-14
A-7	10–14	10–16	13	11-13-15

*From PCA (1959).[49]

corded. Specifications provide for the amount of material loss permitted by brushing, varying from 7% for cohesive soils to 14% for granular soils.

Alternate specifications, much simpler to use, are based on the unconfined compressive strength after 7 days of curing and saturation: clean sands should have a $U_c \geq 230$ psi. The required compressive strength varies with clay-silt content, ranging from 300 psi at 25% but decreasing to about 200 psi at about 50% [PCA (1959)[49]].

Pavement design is described in PCA (1959).[49] Base courses in the United States are usually 6 to 8 in thick, and design is based on CBR values in accordance with normal practice.

Erosion Control

Soil-cement has been used as erosion-resistant linings for earth dams, highway and railroad embankments adjacent to waterways, in drainage and irrigation channels, and along shorelines, as well as in many other applications. When used as protection against erosion caused by flowing water, it is usually constructed as a 2-ft (60-cm)-thick facing measured normal to the slope along which it is compacted in horizontal lifts. The slope must consist of stable material.

Rollcrete is a mixture of sand and gravel to which cement is added at the rate of about 120 kg/m^3, as well as water, to make a stiff, zero-slump mix. The material is then spread on the surface, graded, and compacted with vibratory rollers in 45-cm lifts to form a cap very resistant to erosion caused by flowing water.

Soil-Bitumen (Soil-Asphalt)

Sands and clays are stabilized also by mixing with bitumen or asphalt. In *sands* the bitumen acts as a binder to increase strength; in *clays* the result is a waterproofed material with little or no strength increase. The amount of bituminous binder added to sands depends upon the void ratio and ranges from about 4 to 10%, whereas for clays, waterproofing can be achieved by the addition of as little as 1 to 2% bitumen [Winterkorn (1975)[50]].

When it is used as a base course, the best results are obtained with soils of the following properties: greater than 50% finer than no. 4 sieve, 35 to 100% finer than no. 40 sieve, greater than 10% but less than 50% finer than no. 200 sieve, LL less than 40%, and PI less than 18% [Lambe (1962)[47]]. Pavement design is described in a reference publication from The Asphalt Institute (1970).[11]

Soil-Lime

Mixtures

Hydrated lime is used to stabilize clay soils and clay-sand mixtures. Experience shows that 2 to 4% of $Ca(OH)_2$ by weight of soil will stabilize clay gravel materials to serve as bases for pavements; for heavy clay soils, 5 to 10% of lime is appropriate where the material is to be used as a base and 1 to 3% where it is to be used as a subbase

[Winterkorn (1975)[50]]. Lime is also used with admixtures including lime-portland cement, lime-bitumen, and lime-fly ash (hydrated lime, 4 to 8% of soil weight, plus fly ash, 8 to 20% of soil weight) [Lambe (1962)[47]].

As most knowledge of lime stabilization is empirical, trial mixtures should always be prepared and tested prior to construction.

Roadway Stabilizaton

Stabilization of roadways constructed of clay soils is particularly important in wet climates where free-draining materials are not readily available. With the addition of lime, PL increases, LL decreases, and in some instances the net result is a decrease in PI by a factor of 3. This causes the clay to decrease markedly in its normal ability to shrink and swell with changes in moisture content. In addition, maximum density decreases, optimum moisture increases, and the unconfined compressive strength can increase as much as 60 times thereby substantially increasing supporting capability.

Lime-stabilized bases and subbases form effective moisture barriers to downward percolation or capillary action. Lime stabilization of clay soils, therefore, is used to provide base courses beneath all types of pavements and to provide surfaces for secondary and tertiary roads.

Calcium and Sodium Chlorides (Salts)

Salts are commonly used as a *dust palliative*, although they have been used as additives in the construction of granular-stabilized pavement surface and base courses. They have the effect also of reducing or preventing frost heave in soil by lowering the freezing point of water. Where leaching by groundwater movement can occur, the treatment is temporary.

When used as a dust palliative, salt is added in amounts equal to about 0.5% of the soil weight and is particularly useful for stabilizing silty soils in dry climates.

Construction

Pulverizing

The construction of soil-stabilized mixtures must be carried out with substantial care. After removal of the topsoil layer, the soil to be treated is broken up into small particles by rotary comminution machines. In clay soils pulverizing is most easily accomplished by farm plows, if the soil is slightly on the dry side of optimum and friable. Wet clays are very difficult to work with and, to make the material more friable for proper mixing, 3% of lime is plowed into the clay and allowed to act for a day or two.

Mixing

Mixing with cement, lime, or bitumen is performed by machine which varies depending upon the type of stabilizer being added. Stationary plants provide the most thorough but most costly mixing. Wet sands and gravels usually do not present problems and are treated directly. Wet clays must be friable, and soils that are too dry are wetted to increase their moisture content to near optimum. Cement-sand mixtures should be compacted slightly on the dry side of optimum and cement-fine-grained mixtures slightly on the wet side of optimum.

Compaction

Following mixing, the stabilized soil is compacted with rollers to a specified percent of standard or modified maximum density; overcompaction should be avoided. The pavement is then sealed or otherwise protected against drying to permit the proper curing required for hydration and hardening of cement. A protective cover can be provided by a bituminous material, a layer of earth, waterproof paper, or moist straw. Bituminous material permits immediate use after curing but should be blotted with sand.

5.5 IMPROVING ROCK-MASS QUALITY

5.5.1 TREATMENTS AND PURPOSES

Treatment Categories

There are five main categories of techniques for treating rock masses:

1. *Excavation* and the removal of soft, weathered, or highly fractured or otherwise undesirable materials to provide for support on rock of adequate quality, to provide for a stable slope, or to remove permeable material from beneath dams.

2. *Injections with grout* to increase rock strength and improve support characteristics for foundations and tunnels and to decrease permeability and cleft-water pressures, thereby reducing seepage and improving the stability of dam foundations, tunnels, and other underground openings.
3. *Rock bolts or cable anchors* to reinforce slopes and underground excavations and to improve uplift resistance for concrete dams.
4. *Surface treatments with shotcrete or gunite* to reinforce loose masses of small rock blocks, to deter rock decomposition in slopes, or to reinforce the exposed surfaces in tunnels and other underground excavations.
5. *Drains* to relieve cleft-water pressures, including subhorizontal drains, galleries, and deep-pumped or gravity wells for slopes or simple drain holes for dam foundations.

Project Type and Treatment

Foundations

The emphasis in this article is on foundations, and in particular on dam foundations. Rock-mass quality beneath a proposed dam can be improved to reduce seepage losses and pressures, increase strength, and decrease deformability. Common methods used are grouting to reduce seepage quantities and forces and to increase strength and grouting and anchors to increase strength. The capability of some rocks to lose strength with time from deterioration must also be considered.

Other Project Types

Rock-slope stabilization is described in Art. 8.4.4, open excavations for structures in Art. 9.3.5, underground openings in Art. 10.3.5, and some aspects of foundations on rock in Art. 6.3.4.

5.5.2 UPLIFT PRESSURE AND SEEPAGE CONTROL

Necessity for Control

Earth Dams

Foundation seepage has two major aspects: first, flow quantity and reservoir loss and second, uplift pressures. In earth dams the major concern usually is with flow and reservoir loss through the abutment and foundation rock since uplift pressures are controlled by embankment drainage design. The simplest procedure is to excavate fractured rock until tight rock is encountered and construct either a core trench of impervious material or a concrete cutoff wall; or, if the fractured rock is too deep for economical excavation, treatment consists of a grout curtain connected to a grout blanket, impervious core trench, or cutoff wall (Art. 8.4.5).

Grouting necessity depends primarily on dam height, rock permeability, and the value of the water loss. Historically, dams with a height of 50 ft or less founded on rock not excessively fractured generally have not been grouted, whereas foundations for dams retaining 100 ft or more of water head usually have been grouted in some way [Sherard et al. (1963)[4]]. In general, for grouting to be feasible the permeability of the foundation rock must decrease at some depth that can be economically reached by a grout curtain, since permeable rock below the curtain will cause the curtain to have negligible influence, even at depths of several hundred feet. The same conditions generally are true for the abutment rock; impervious rock must be reachable within some economical lateral distance from the embankment for the grout curtain to be effective.

In borderline cases, as for example if the rock is intensely fractured with very fine cracks or clay-filled joints into which cement grout will not penetrate, and the amount of seepage loss is not objectionable, or if it is desired to construct the dam as quickly as possible, grouting can be omitted and relief wells or drains installed which should safely control foundation seepage.

Concrete Dams

The need for control is assessed in a manner similar to that for earth dams with respect to rock quality, except for the additional, very important factor of uplift forces (*see Art. 6.3.3*) which are controlled with a grout curtain and relief wells. In either case, pressure tests are performed during exploration to evaluate the necessity for grouting as described in Art. 2.3.4 of Hunt (1984).[1] Careful assessment of rock characteristics is required, particularly for thin-arch concrete dams. Failure of Malpasset Dam (*see Art. 6.3.3*) has been judged to have been caused by excessive uplift pressures developing in the tightly jointed, finely fissured gneiss foundation and abutment rock.

Injections or Penetration Grouting

Grout Types and Mixes

Grouting of finer fractures employs a mixture of cement and water, although chemicals are being used with increasing frequency where permeabilities are low, especially where an increase in strength is desired. To reduce the amount of cement where large openings are to be grouted, sand, clay, rock flour, and other inert materials are added. Grout types and their applications are described in Art. 8.4.2 of Hunt (1984)[1] and summarized in Table 5.5.

Grout mixes or *consistencies* are given in terms of the water-cement ratio on the basis of volume measure. To enable penetration into fine fissures, a water content higher than required for cement hydration is used. For most rocks grouting is started with a consistency between 3:1 and 5:1. If the pressure increases rapidly and the hole takes little grout, the mix is thinned. If building up grout pressure with the maximum pump speed is difficult, then grout take is excessively high and the mixture is thickened or some inert material added as mentioned above. Experimentation is required on any given project to develop optimum mixes, and changes in the program often are required as rock conditions are found to vary as indicated by grout takes.

Installation

To construct a grout curtain a series of percussion holes, 2½ to 3 in diameter and inclined 15 to 20° upstream, are drilled and grouted to some depth and spacing preselected based on evaluation of core borings, geophysical surveys, and pressure tests. A second series of holes is drilled and grouted in between those of the first series, and a third series may be added in line if takes remain large (Table 5.7). The pattern can be extended laterally to increase the width of the grouted zone. Spacing depends on joint concentrations and the effectiveness of the curtain desired. Typical *spacing* starts at about 6 to 10 m, but must be much closer if a positive cutoff is required in heavily fractured rock, as beneath an earth dam with erodible core material [see discussion of Teton Dam failure in Art. 8.3.4 of Hunt (1984)[1]]. Final spacing on most projects is between 2 and 3 m. The *dip and bearing* of the holes should be chosen to be as normal to the major joint sets as possible to provide for maximum intersection of the joints; otherwise, many will remain ungrouted.

TABLE 5.7
DEGREE OF GROUT TAKE VERSUS ROCK TIGHTNESS*†

Degree of tightness (grout absorption)	Grout take, kg/m‡
Very low	0–12.5
Low	12.5–25
Moderately low	25–50
Moderate	50–100
Moderately high	100–200
High	200–400
Very high	400

From Deere, 1976.[50a]

†For high dams final grout holes should arrive at low grout absorptions (12.5 to 25 kg/m).

‡Multiply values by 0.0067 for approximate quantities in sacks per foot.

Grouting may proceed downward from the top of the hole after drilling and flushing *(stage grouting)* or upward from the bottom of the hole with the use of packers *(packer grouting)*, similar in purpose and nature to water pressure tests [Art. 2.3.4 of Hunt (1984)[1]].

Blanket grouting is used to make the upper portions of heavily jointed rock stronger and less pervious, and is commonly located in the upstream foundation area under concrete dams or under the thin impervious core of high rock-fill dams as illustrated in Fig. 5.40. A number of closely spaced holes are drilled and grouted on a grid pattern over an area where the high permeability of the rock would not be treated successfully by a grout curtain.

Checking grout-curtain efficiency is necessary and is performed with NX or larger core borings, and with water pressure tests (WPT) before and after grouting as described in Arts. 2.3.4 and 8.4.2 of Hunt (1984).[1] The recommended satisfactory upper limit of water loss in water pressure test-

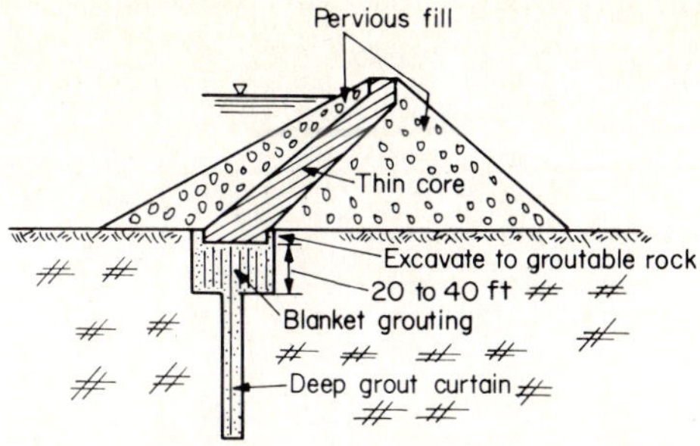

FIG. 5.40 Blanket grouting used under thin core of rock-filled dams where upper rock surface is heavily fractured. [From Sherard et al. (1963).[4] Reprinted with permission of John Wiley and Sons, Inc.]

ing for ground curtain effectiveness is between 5 to 7 L/min/m at 10 kg/cm^2 pressure [Fecker (1980)[51]]. Evaluations of curtain effectiveness are made from plots of depth versus geology, water loss, and grout takes.

After grouting completion and before placing the earth dam core section the rock surface requires treatment to assure that no seepage occurs along the core-rock contact. Procedures are numerous and depend on rock characteristics [see Sherard et al. (1963),[4] for example].

Stress release and expansion may be caused by a deep excavation and require consideration in the selection of the grout curtain location. Recompression from the weight of a heavy concrete dam will fracture the curtain and destroy its purpose. In such cases postconstruction grouting from galleries may be more desirable.

Grouting Pressures

Grout is injected under pressure to obtain penetration and to expel the excess water so as to ensure adequate setup strength. The highest safe pressures should be used, particularly where a reservoir will develop hydrostatic pressures which may be large enough to displace the rock mass and cause external uplift or internal fracturing. Maximum safe pressures are determined by in situ hydraulic fracturing tests.

There is a difference of opinion among grouting specialists regarding the benefits of increasing rock fracturing by grouting. In the past, practice in the United States has related grouting pressures to overburden thickness: 1 psi per foot (1 kg/cm^2 per 4.2 m) of depth, to prevent ground heave and further rock fracturing. European practice accepts much higher pressures, as high as 1 kg/cm^2 per meter of overburden. Pressures as high as 100 kg/cm^2 have been used to compress rock masses [see Art. 8.4.7, Tunnels, in Hunt (1984)[1]]. Advantages of higher pressures are [Rigny (1974)[52]]:

- Fractures open under higher pressures, and when they close a tight bond between the rock and cement is achieved.
- Grout will penetrate further, decreasing the number of holes required.
- Water is expelled from the grout, resulting in material of higher strength.

Disadvantages of excessively high pressures are:

- Fracturing can be increased and the rock mass significantly weakened, especially in the abutment rock.
- Surface displacement along with increased fracturing can occur, particularly in horizontally bedded sedimentary rocks, when grout pressures are much higher than the weight of the overlying rock.

A chart giving permissible grout pressures for various rock conditions as developed by the Water Resources Commission of New South Wales, Australia, is given as Fig. 5.41. Also shown are the U.S. and European criteria.

Drain Holes

Installation

Grout curtains can reduce uplift pressures beneath the dam and reduce pore pressures in the abutments, but it is always good practice to install drain holes to ensure that pore pressures are relieved, since complete cutoff is seldom achieved with a grout curtain. The suggested location of a grout curtain and relief well system beneath a *concrete arch dam* is shown in Fig. 5.42. Drain spacing depends on rock conditions but often ranges from 3 to 10 m on centers, extending to perhaps 75% of the depth of the grout curtain, with holes of 2 to 3 in diameter.

In *earth dams*, drainage holes may be drilled beneath the dams at locations where the inter-

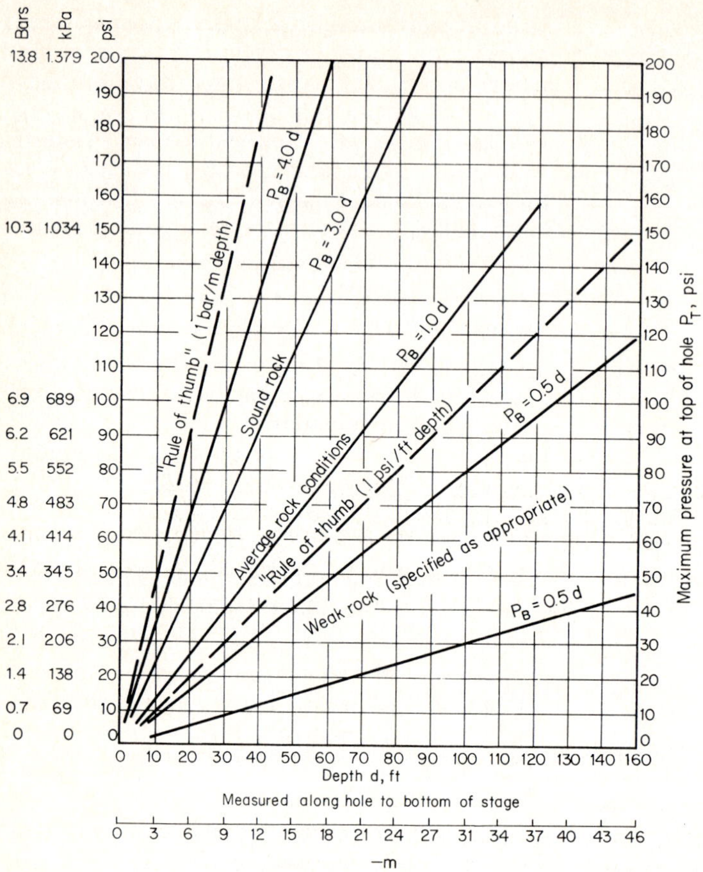

FIG. 5.41 Typical permissible grout pressures for various rock-foundation conditions. Pressures are shown at the surface; P_B denotes pressure at bottom of stage and d is depth to there. [From Houlsby (1977).[43]]

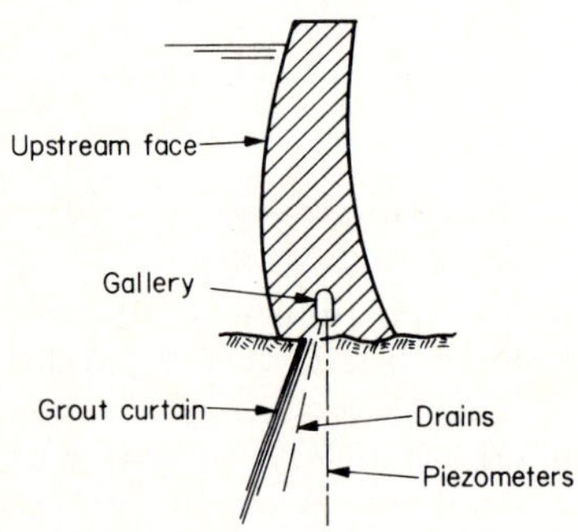

FIG. 5.42 Schematic of grout curtain and drains beneath thin-arch concrete dam.

cepted water can be discharged into the internal drainage system, in which case stability is improved but drain flows cannot be observed and monitored. As an alternative they can be installed in the abutments and the foundation downstream, a less effective method from the viewpoint of stability but one which permits observation and analysis of the grout curtain leakage. In some cases, inclined holes can be installed from the downstream toe both to improve stability and to permit observation of the leakage.

Field Control

Piezometers should be installed to permit measurement of pore pressures. If pore pressures develop that are higher than those considered permissible by the design criteria, additional drains are installed for their reduction. Drains should be checked periodically to ensure that they are functioning properly and are not plugged. Christensen (1974)[54] reports plugging of drain holes within 3 years of completion of a 190-m-high concrete arch dam in California. The plugging material, calcite, was encountered in the drains in zones 1 to 3 m thick at depths 15 m below the gallery floor and 6 m into the hard, massive, moderately jointed, metavolcanic rock. The calcite was thought to have been leached from the grout curtain.

5.5.3 ROCK REINFORCEMENT

Rock Bolts and Cable Anchors

Applications

Rock-mass strength can be increased with rock bolts and cable anchors or by grouting. Rock bolts are used in slope stabilization *(see Art. 8.4.4)*; in open excavations *(see Art. 9.3.5)*; in tunnels, caverns, and mines *(see Arts. 10.3.4 and 6.3.3)*; and in concrete dam foundations to provide resistance to uplift and sliding. The latter condition can exist where weak foundations are interbedded with strong layers. In open or closed excavations they are particularly useful to retain high residual stresses causing floor uplift or wall deflection. They also have application where there is a concern that high grouting pressures will cause displacement of shallow rock, which can first be strengthened by anchors and then grouted to create a much stronger mass. Cable anchors are used primarily in slope stabilization where a large mass requires reinforcing.

Rock-Bolt Types

Rock bolts are tensile units used to place or keep the mass in compression and should be installed as nearly perpendicular to joints as practical. The *ordinary types* consist of rods installed in drill holes by driving and wedging, by driving and expanding, or by grouting with mortar or resins as illustrated in Fig. 5.43. Bolt heads are then attached to the rod and torqued against a metal plate to impose the compressive force on

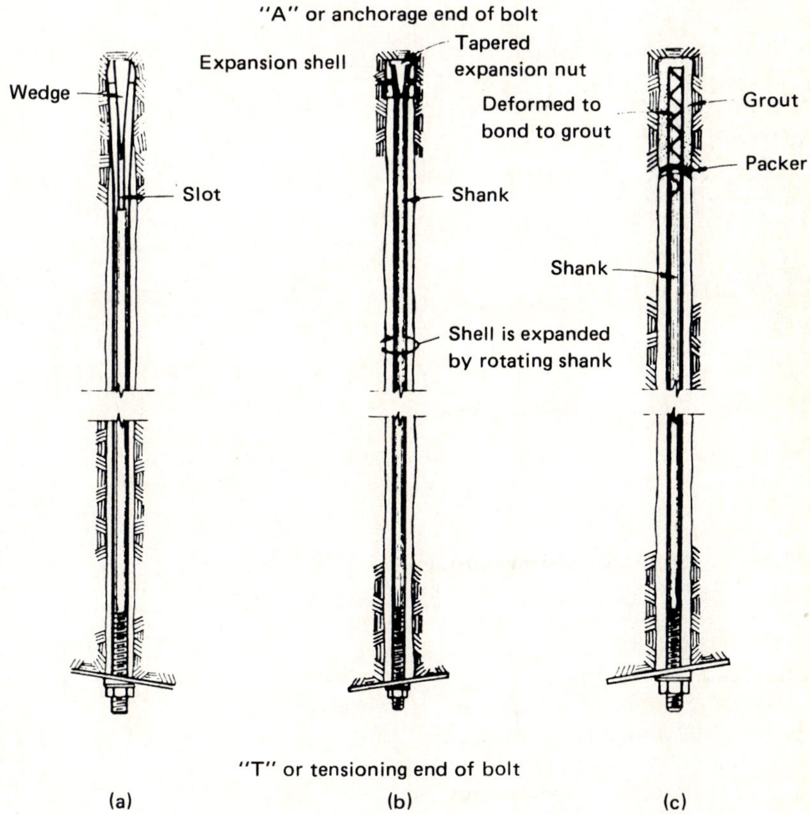

FIG. 5.43 Types of ordinary rock bolts: *(a)* drive-set or slot and wedge bolt; *(b)* torque-set or expansion bolt; and *(c)* grouted bolt. [From Lang (1972).[54a]]

the mass. Weathering of rock around the bolt head may cause a loss in tension; therefore, heads are usually protected with concrete or other means or used in conjunction with concrete straps in high-risk or poor rock conditions. The ordinary bolt is subjected to loss in tension with time from several other sources including corrosion from attack by aggressive water, anchorage slip or rock spalling around and under the bearing plate, and block movement along joints, pinching the shaft.

Fully grouted rock bolts (Fig. 5.44) provide more permanent bolts than the ordinary types. Grouting with resins is becoming more and more common because of easy installation and the rapid attainment of capacity within minutes of installation. Care is required during grouting to minimize grout spread, which results in decreasing mass drainage, especially where bolts are closely spaced. Drain holes may be required in slopes or other cases where hydrostatic pressures are a concern.

Shotcrete

Shotcrete is used to reduce weathering susceptibility in weak rocks such as shales and to reinforce fractured rock masses in slopes and tunnels. It usually consists of a wet-mix mortar with aggregate as large as 2 cm (¾ in) which is projected by air jet directly onto the rock face. The force of the jet compacts the mortar in place, bonding it to the rock, which first must be cleaned of loose particles and blocks. Application is in 8- to 10-cm (3- to 4-in) layers, each of which is permitted to set before application of subsequent layers. Weep holes are installed in slopes to relieve seepage pressures behind the face. Since shotcrete acts as reinforcement and not as support, it is often used in conjunction with rock bolts. The tensile strength can be increased significantly by adding 25-mm-long wire fibers to the concrete mix.

Wet-mix steel fiber *microsilica shotcrete* has been used extensively in Norway in recent years for tunnel support, slope protection and rein-

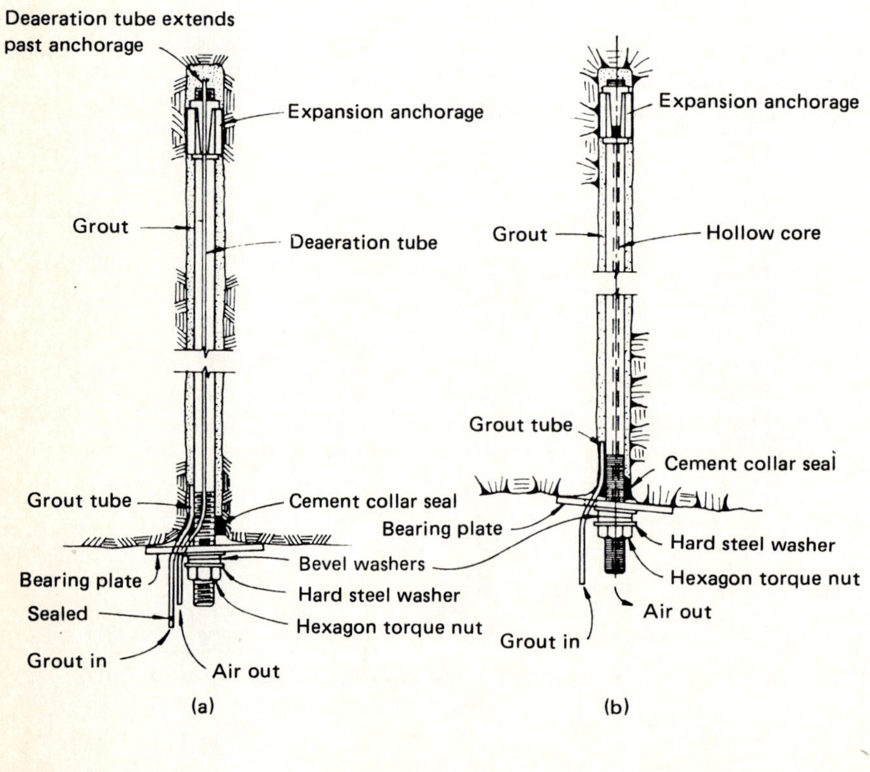

FIG. 5.44 Fully grouted rock bolts: (*a*) grouted solid expansion anchorage bolt and (*b*) hollow-core grouted rock bolt. [From Lang (1972).[54a]]

forcement, repair of concrete structures, and surface protection for steel structures (*Civil Engineering*, ASCE, October 1983). Microsilica consists of ultrafine, super pozzolanic, amorphous silica (SiO_2) particles in the 0.1- to 0.2-μm size. The water/cement ratio is reduced to about 0.20 instead of the average 0.30 to 0.40. The shotcrete has minimum 28-day strengths of 9000 psi compressive and 1500 psi flexural and an average permeability of less than 10^{-12} m/s.

Compaction Grouting

Installation

Foundations for dams and other heavy structures often are improved with compaction or "consolidation" grouting to increase the elastic or deformation moduli. Compaction grouting normally is carried out by treating a grid of holes spaced 3 to 12 m on centers to depths ranging from 10 to 30 m, depending on rock quality [Lancaster-Jones (1969)[55]]. If grout takes are large it is normal to treat a superimposed grid that reduces the spacing to 2 m or less. Clay-filled fractures are difficult to treat. If the compressibility of the rock mass is of concern, the clay is removed by high-pressure jetting to the degree possible. A series of boreholes is drilled and water or air injected alternately as the clay is washed from the adjacent holes. The operation is repeated in sequence in the adjacent holes. Dispersants may be added to soften the clay and facilitate its removal.

Rock Improvement

Lancaster-Jones (1969)[55] reviewed and summarized the literature regarding the improvement that can be obtained with compaction grouting. The results are given in Table 5.8 in terms of either the static or dynamic modulus before or after grouting and in Table 5.9 in terms of the increase in seismic velocity. The data indicate that sound, tightly jointed rock will not be significantly improved (and seldom needs to be), but that in sound to moderately sound rock with open joints the best results are obtained, and even poor quality rock can be improved significantly.

Limestone Cavities

Grouting of cavities and fractures in limestones is common practice in site preparation for foundations, but large quantities of grout can cause the operation to be expensive. Other problems associated with grouting limestone formations are the difficulties in locating the caverns before construction and the removal of clay fillings. Washing out often is not successful.

During the installation of pier foundations for a large industrial complex in Brazil, numerous caverns large enough for a person to stand in were encountered. As many foundations and parts of some structures were already constructed, it was decided to undertake an extensive grouting program as the only reasonable solution to provide adequate support for the heavy structures. To obtain penetration into the many fractures leading from the caverns, high pressures were used, even at shallow depths. At pressures of 5 kg/cm^2 and at depths less than 10 m, one structure on a mat foundation was lifted 18 cm and consequently broke in half. Pressures were subsequently reduced to 4 kg/cm^2 without further damage to structures.

Where successful, grouting not only strengthens the limestone mass, but has the important advantage of impeding groundwater movement and therefore cavity growth, even where pumping is anticipated.

5.5.4 ROCK DETERIORATION PREVENTION

Dam Foundations

Soluble Rocks

Possible deterioration of rock with time beneath a dam foundation requires consideration. Rocks that are highly soluble (such as limestone; sandstone that is cemented with gypsum, calcite, or other carbonates; or even dikes of calcite in metamorphic or igneous rocks) are susceptible to attack by moving water.

The failure of the St. Francis concrete gravity-arch dam in 1928, which resulted in the loss of more than 500 lives, is attributed to the rapid softening of conglomerate foundation rock during filling of the reservoir [Christensen (1974)[54]].

TABLE 5.8
IMPROVEMENT IN STATIC/DYNAMIC MODULUS BY INJECTION*

Rock type	Mean E_s or E_d, tsf, kg/cm²		Increase, %	Modulus
	Before injection	After injection		
Granite, sound	340,000	355,000	4.5	Dynamic
	188,000	224,000	19	Static
	157,500	210,000	33	Static
	145,000	160,000	10	Dynamic
Granite, slightly weathered	125,000	195,000	56	Static
	118,500	185,000	56	Static
	71,000	104,500	47	Static
	67,500	102,500	52	Dynamic
	43,500	75,500	80	Static
Granite altered	21,000	25,500	22	Static
	20,000	34,500	72	Static
	10,900	16,500	51	Static
Granite, porphyritic	230,000	330,000	43	Dynamic
Quartz phyllite	430,000	655,000	52	Dynamic
	402,000	651,000	62	Dynamic
	300,000	662,000	121	Dynamic
	250,000	594,000	138	Dynamic
Schistose quartzite	345,000	350,000	1.5	Dynamic
	135,000	190,000	41	Dynamic
	125,000	155,000	24	Dynamic
	100,000	140,000	40	Dynamic
	60,000	125,000	104	Dynamic
	55,000	150,000	175	Dynamic
	41,250	147,500	257	Dynamic
Augen gneiss	110,000	126,000	15	Static
	49,000	66,250	35	Static
Fissured gneiss	8,000	30,000	275	Static
	800	1,600	100	Static
Mica schist	33,600	49,000	46	Static
Sandstone and conglomerate	234,000	272,500	17	Dynamic
	117,000	188,000	61	Static
	91,000	250,000	175	Dynamic
Conglomerate	24,000	75,000	212	Static
Sound sandstone	234,000	363,000	55	Static
Broken sandstone	98,500	176,000	78	Static
Limestone	158,000	482,000	205	Static
	121,500	154,000	27	Static
	64,000	106,000	66	Static
	42,650	202,000	375	Static
	16,400	36,700	124	Static

*After Lancaster-Jones (1969).[55]

TABLE 5.9
IMPROVEMENT IN SEISMIC VELOCITY V_s BY INJECTION*

Rock type	Mean V_s, m/s		Increase, %
	Before injection	After injection	
Quartz phyllite	4350	5380	24
	4225	5330	26
	3600	5400	50
	3325	5095	53
Paleozoic sediments, primarily schists	3500	4090	17
Mica schist	2919	3620	24
Schist	2750	3800	38
Hard sandstone	2300	2620	14

*After Lancaster-Jones (1969).[55]

The conglomerate was cemented with clay and gypsum. Collapse could have been precipitated by settlement or piping or caused by foundation failure or even uplift, as the dam had no underdrain system, although the failure did occur in the area of the dam overlying the conglomerate.

Grout curtains are used to prevent or impede deterioration of soluble materials.

Shales

Many shales weather rapidly upon exposure [Art. 6.7.3 of Hunt (1984)[1]] and if the foundation excavation remains open for some time the rock quality can deteriorate substantially. Exposure to weathering should be minimized by making excavations in sections as small as practical, preventing water infiltration, and immediately covering the bottom with foundation concrete, cyclopic concrete, or compacted earth. In slopes, deterioration is retarded with shotcrete or gunite.

Building Foundations

Soluble Rocks

Cavity growth and roof collapse as a result of lowering the water table by groundwater withdrawal is a major concern [Art. 10.4.2 of Hunt (1984)[1]] which is prevented or impeded by grouting as discussed above.

Shales should be treated as described above.

Foliated metamorphic rocks (gneisses and schists) may undergo severe deterioration from contaminated groundwater resulting from leaking sewage pipe *(see Art. 6.3.4)* indicating the necessity for inspecting and maintaining such systems beneath cities.

5.6 GROUND VIBRATIONS FROM CONSTRUCTION ACTIVITY

5.6.1 SOURCES

Vibrations of structures are caused by elastic waves originating either from earthquakes [see Chap. 11 of Hunt (1984)[1]], water waves, or sources induced by human activities. In an existing structure they can be human-induced by vibrating machinery within the structure or by construction activity, the more significant forms of which are pile driving and blasting. Under certain conditions structural damage can result.

Rock Blasting

This article is devoted primarily to the effect on structures of blasting rock for nearby excavations, although the principles are applicable to other forms of construction activity.

Whether for open excavations or for tunnels, rock blasting criteria are developed for projects in terms of weight of charge per delay, distance between the structure to be protected and the blast area for various types of geologic materials, and an acceptable level of ground motion at the structure to prevent damage.

Blasting nomenclature and procedures for open excavations are described in Art. 6.3.4.

5.6.2 INDUCED VIBRATORY MOTION

Elastic Waves

Source

The energy released by detonation of an explosive first crushes the rock in the immediate vicinity, then moves through the next few meters as a shock front or strain pulse which becomes an oscillatory wave in which the particles along the travel path move in orbits that repeat cyclically. In simple two-dimensional illustration, the oscillation is in the shape of a wave with a crest and a trough, as shown in Fig. 5.45. The wave energy produces *displacement* in the rock or other materials within their elastic limit, and after the energy wave has passed they return to their original volume and shape. The waves in this region are called *elastic waves*. There are two general forms: body waves and surface waves.

Body Waves

The initial shock outside the zone of rock shattering applies a compressive force to the rock which causes wave motion in the elastic zone. Particles in the wave path move back and forth, and the wave is termed variously as longitudinal, compressional, push-pull, or primary. The term *primary (P wave)* is derived from the fact that the compressional waves travel the fastest of all elastic wave types and are the first to arrive at a distant point.

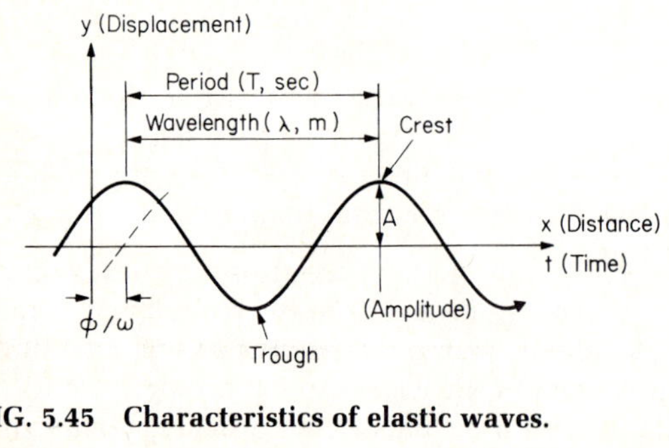

FIG. 5.45 Characteristics of elastic waves.

Where the initial pressure pulse, or the P wave which it generates, strikes a free surface or a change in material in a direction other than normal, the shape of the transmitting material is changed in a form other than compressional. The change is in the form of shear, and new waves are generated termed *shear* or *secondary (S waves)*.

P waves can travel in any matter that resists compression or volume change: a solid, liquid, or gas. S waves, however, can travel only in a solid, because their existence depends on the ability of the transmitting material to resist changes in shape (the shear modulus, Art. 3.3.5). As both P waves and S waves travel into and through the bodies which transmit them, they are referred to as *body waves*.

Surface Waves

There are also waves, termed *surface waves*, that travel along the free surface of a solid elastic material bounded by air or water. There are several types, each defined by the motion through which a particle in its path moves as the wave passes. *Rayleigh waves* (*R*) cause a particle to move in an elliptical orbit, or "push-up, pull-down." *Love waves* (*Q*) cause the particles to vibrate transverse to the direction of wave advance, with no vertical displacement. Atomic bomb testing revealed two other wave forms: C waves, where the particles oscillate diagonally rather than vertically or horizontally, and H waves, which move elliptically in the vertical direction as do Rayleigh waves, but first push down, then pull up.

In any given explosion, however, only one or all or combinations of waves might be generated. The detection of P, S, and R waves from seismograms is described in Art. 2.3.2 of Hunt (1984).[1]

Wave Characteristics

Propagation Velocity

Elastic waves have the properties of velocity V, wavelength λ, and frequency f, related by the expression:

$$V = \lambda f \tag{5.2}$$

All elastic waves in the ground rely for their existence and velocity on the elastic properties of the transmitting medium. Travel or *propagation velocity* V of P waves and S waves can be expressed in terms of elastic moduli and material density as

$$V_p = \sqrt{\frac{K + (4/3)G}{\rho}} \qquad (4.6)$$

$$V_s = \sqrt{\frac{G}{\rho}} \qquad (4.7)$$

where K = bulk modulus
G = shear modulus
ρ = material density

as discussed in Art. 4.1.3.

As mentioned above, P waves travel with the highest velocities. Among the surface waves, Rayleigh waves, with a velocity of about 0.9 V_s, arrive last at a distant point; earlier arrivals are generally the C waves, followed by the H waves, then the Q waves. Not all waves, however, are generated by every explosion.

Other Quantities

Seismic waves (Fig. 5.45) can be described by their quantities of vibratory motion; i.e., amplitude, wavelength, period, and frequency. *Amplitude A* is the displacement from the mean position, or one-half the maximum displacement. It differs along the wavelength and differential displacement may cause rupture of structures. *Wavelength* λ is the distance between two crests, and *period T* is the time of a complete vibration, or the time a wave travels distance λ. *Frequency f* is the number of vibrations per second or oscillation in cycles per unit of time, given normally in *hertz* (Hz), or cycles per second. *Circular frequency* ω defines the rate of oscillation in terms of radians per unit of time; 2π rad are equal to one complete cycle of oscillations. Period, frequency, and circular frequency are related as follows:

$$T = \frac{1}{f} = \frac{2\pi}{\omega} \qquad (5.3)$$

Waves cause the shaking of a structure as their crests pass beneath and the magnitude of the acceleration depends on the frequency.

Surface Response

Ground Motion

As the seismic waves reach the surface, ground motion occurs which can be described in terms of displacement, velocity, and acceleration, which are in turn derived from the quantities of the seismic waves. *Displacement* y at a given time t is a function of position x and time t in Fig. 5.45, expressed as

$$y = A \sin \frac{2\pi}{\lambda}(x + vt) \qquad (5.4)$$

Velocity v, termed the *particle* or *vibrational velocity*, is expressed as

$$v = \lambda/T = f\lambda = dy/dt = \dot{y} \qquad (5.5)$$

Acceleration a is expressed as

$$a = d^2y/dt^2 = \ddot{y} \qquad (5.6)$$

and is given in terms of the acceleration of gravity g where 1g = 32 ft/s^2 = 980 cm/s^2 or 1g = 980 gal.

Structural Response

Vibrational or particle velocity is a key element in evaluating the energy transmitted to a structure by seismic waves. The *maximum vibrational velocity* imposed by the elastic wave as it passes beneath a structure in terms of circular frequency ($\omega = 2\pi f$) and amplitude is

$$v_{max} = 2\pi f A \qquad (5.7)$$

The *greatest acceleration* to which the structure is subjected is

$$a_{max} = 4\pi^2 f^2 A \qquad (5.8)$$

The *greatest force* applied to the structure is the product of the mass and the maximum acceleration, expressed as follows:

$$F_{max} = ma_{max} = \frac{W}{g}a_{max} = \frac{W}{g}(4\pi^2 f^2 A) \qquad (5.9)$$

where W = weight. It is seen that the force varies as the square of the frequency.

Kinetic energy is possessed by a body by virtue of its velocity of movement; the energy of a body is the amount of work it can do against the force applied to it, and *work* is the product of the force required to displace a mass a unit of distance at the standard acceleration of gravity (hence the significance of g in earthquake engineering). Kinetic energy KE is expressed by

$$KE_{max} = \frac{1}{2} mv^2_{max} = \frac{1}{2}\frac{W}{g} v^2_{max} \qquad (5.10)$$

where v = the velocity with which a structure moves back and forth as the seismic wave passes [Leet (1960)[56]].

Monitoring Ground Motion

Ground motion is monitored by small seismographs called *vibration monitors* (see Art. 4.2.5 of Hunt (1984)[1]]. They can be designed to measure and record vibration velocity, amplitude and frequency, or acceleration. A nomograph giving the relationship between frequency, acceleration, vibration velocity, and amplitude is given as Fig. 5.46. It is of interest for a single, strong pulse such as that caused by an explosion. Earthquakes are in the low frequency range, usually between 0.2 and 5 Hz.

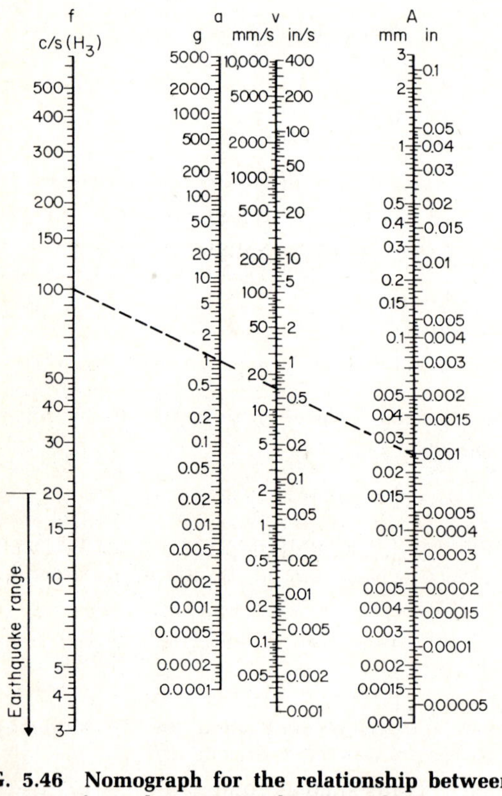

FIG. 5.46 Nomograph for the relationship between frequency *f*, acceleration *a*, vibration velocity *v*, and amplitude *A*. [From Langefors and Kihlstrom (1963).[57] Reprinted with permission of John Wiley and Sons, Inc.]

Ground motion is measured in three orthogonal components: two horizontal directions and the vertical.

5.6.3 DAMAGE TO STRUCTURES

General

Basic Elements

Frequency and amplitude are the basic elements of harmonic motion wave form from which can be determined acceleration, particle velocity, force, and kinetic energy. Human response, limit for machines and machine foundations, and damage to structures in terms of frequency and amplitude are given in Fig. 5.47.

Acceleration of Gravity

Initially the acceleration of gravity was used as a criterion of blasting damage, as it is of earthquake loadings, but acceleration from blasting usually is much lower than for earthquakes yet contains sufficient energy to cause damage. The acceleration of gravity is 386 in/s^2. A quarry blast might have an acceleration of 36 in/s^2 (0.09g) causing no damage, and an earthquake might have an acceleration of 101 in/s^2 (0.38g), only four times as great, and yet cause substantial damage, because the energy level is 450 times greater [example from Leet (1960)[56]]. Consequently, various investigators have pointed out that the best guide for determining the damage potential from blasting vibrations is an evaluation of the energy they possess.

Energy Ratio Approach

Damage criteria for structures subjected to blast-induced vibrations were advanced by Crandell (1949)[59] based on measured vibration levels in the ground near structures during a number of tests. He developed a relationship termed the *energy ratio*, expressed as

$$ER = \frac{a^2}{f^2} \qquad (5.11)$$

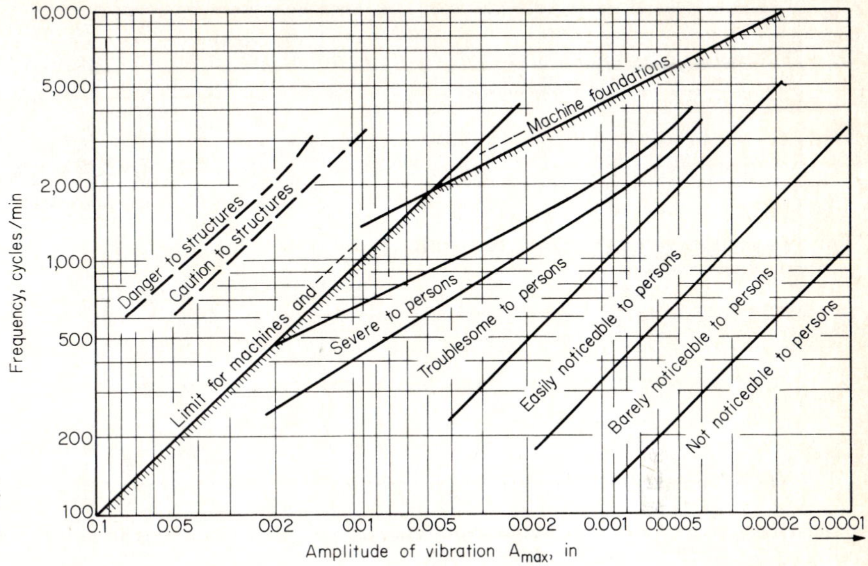

FIG. 5.47 Scale of vibration damage to structures and limits of human perception [From Richart (1960)[57]].

Crandell's tests showed that when the energy ratio was less than 3.0, nearby structures were safe; 3.0 to 6.0 is a caution zone; and 6.0 or greater, a danger zone. The criterion was the cracking of plaster, considered to be the most sensitive element of a structure to vibration damage. In the United States, local building codes have used the energy ratio criterion.

Peak-Particle-Velocity Approach

Acceptable Limits

Work by various investigators [Edwards and Northwood (1960),[60] Langefors et al. (1958)[61]] indicated that damage from blasting is more closely related to particle velocity than to either displacement or acceleration. The energy ratio is proportional to the particle velocity squared, as can be found from the relationship between velocity and kinetic energy. From this and the work of Crandell and others has been developed the generally accepted level of *peak particle velocity* of 2.0 in/s (50 mm/s) to prevent cracking of plaster and masonry in residential structures [Nichols et al. (1971),[62] Neff (1972[63]]. The U.S. Bureau of Mines [Nichols et al. (1971)[62]] recommends the safe vibrational criterion of 2.0 in/s particle velocity as a probability type criterion: "If the observed particle velocity exceeds 2.0 in/sec in any of the three orthogonal components, there is a reasonable probability that damage will occur to residential type structures." The studies by the U.S. Bureau of Mines were based on quarry blasting.

Geologic and Structural Factors

Both geologic conditions and the type and condition of the structure influence the energy level that will cause damage. Gustafsson (1973)[64] presents a relationship between damage risk, vibrational velocity v, and the propagation velocity V for various types of geologic materials based on over 100,000 observations in Sweden; the information is summarized in Table 5.10.

A value of $v = 70$ mm/s is given as the limit for no noticeable cracking in well-constructed houses for rock with $V = 4500$ to 6000 m/s. In older buildings of poorer quality v is decreased to 50 mm/s. Gustafsson reports values as high as $v = 100$ mm/s without causing any damage in many buildings, and states that "robust" concrete structures can stand values of more than 200 mm/s ($v \approx 8.0$ in/s). *Accelerations,* however, become important when blasting is carried out close to structures with sensitive equipment or instruments such as computer installations.

Table 5.10 illustrates also that the softer geologic materials have lower acceptable velocities than do sounder rocks such as granite. If a site is hor-

TABLE 5.10
DAMAGE RISK OF RESIDENTIAL HOUSING RELATED TO VIBRATION VELOCITY *v* WITH RESPECT TO WAVE VELOCITY *V* AND FOUNDATION GEOLOGY*

Wave velocity V, m/s (propagation velocity)	1000–1500 (Sand, gravel, clay under groundwater level)	2000–3000 (Moraine, slate, soft limestone)	4500–6000 (Granite, gneiss, hard limestone, quartzite, sandstone, diabase)	Result in normal housing units
Vibration velocity v, mm/s (particle velocity)	18	35	70†	No noticeable cracking
	30	55	100	Fine cracks and fall of plaster (threshold value)
	40	80	150	Cracking
	60	115	225	Severe cracking
			300	Fall of rock in tunnels‡
			600	Cracks appear in rock‡

*From Gustafsson (1973).[64]

†In older buildings of poorer quality, reduce to 50 mm/s. Robust concrete structures can stand values over 200 mm/s.

‡From Langefors and Kilhstrom (1963).[57]

izontally stratified or of massive rock with horizontal isotropy, little difference in wave propagation would be expected. If there is a structural dip, geologic complexity, anisotropy, or lineations such as foliations or joint systems, propagation might differ with direction. The effect of blasting on a particular structure founded on particular geologic conditions must be determined by field-testing to measure velocities in situ.

For *tunnel damage* in rock, Langefors and Kilhstrom (1963)[57] give the following criteria: v = 300 mm/s (12 in/s) for the "fall of rock in unlined tunnels," and v = 600 mm/s (24 in/s) is correlated with the formation of new cracks in the rock.

Blasting Variables

Explosives: Weight and Distance Relationships

As the seismic waves travel through the ground they lose energy through attenuation. Variables influencing particle velocity *v* at the structure are the radial distance *R* from the blast, the total weight *W* of the explosives per delay, and the nature of the transmitting medium. Various empirical relationships have been developed for these variables, for which the U.S. Bureau of Mines (1971)[62] gives

$$v = k(R/W^{1/2})^{\beta} \tag{5.12}$$

where *k* and *β* are constants for local geologic conditions to be determined at each blasting site and $R/W^{1/2}$ is a function defining the scaled distance.

To obtain values for *k* and *β*, field tests are made at the site to measure the variation in particle ve-

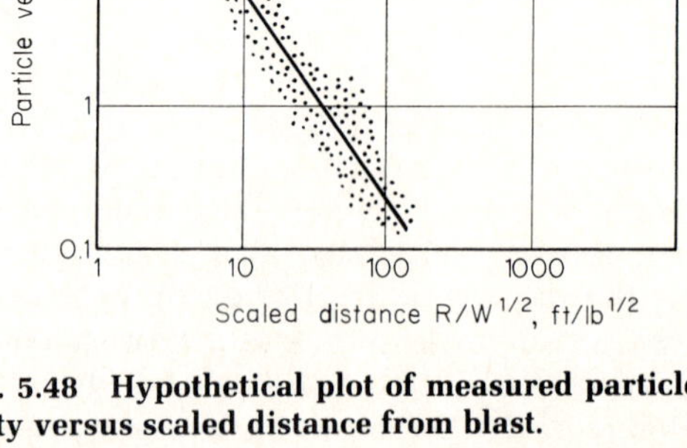

FIG. 5.48 Hypothetical plot of measured particle velocity versus scaled distance from blast.

locity for various ratios of R and W. A plot is made on a log-log scale of v versus the scaled distance $R/W^{1/2}$ as shown in Fig. 5.48. As Eq. 5.12 plots as a straight line, the best fit is made from the data and k is given by the v intercept at a scaled distance of unity and β by the slope of the line. A plot is then made for the site relating the damage-threshold value of interest to charge per delay and blast distance such as given in Fig. 5.49. For this figure, values of $k = 200$ and $\beta = -1.5$ were used in solving Eq. 5.12.

Charge Delay

The charge weight W used in Fig. 5.49 is the maximum weight per delay. To avoid the influence of sequential delays too closely spaced, Ambraseys and Hendron (1968)[66] recommend using a delay interval of approximately one-fourth of the transit time. For example, if the structure is 500 ft from the proposed blast and a criterion of 2 in/s is adopted, Fig. 5.49 gives a maximum allowable charge of 800 lb per delay. If the seismic velocity of the rock is 10,000 ft/s

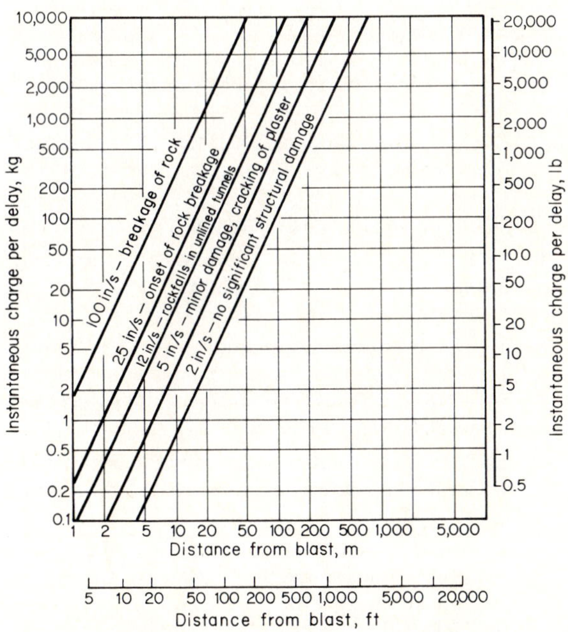

FIG. 5.49 Variation of maximum particle velocity with distance from blast and weight of charge per delay for the case $k = 200$ and $\beta = -1.5$. [From Hoek and Bray (1977)[65]].

the delay time should be greater than 12.5 ms to ensure that the ground motions are governed by the maximum charge in a single delay.

REFERENCES

1. Hunt, R. E. (1983) *Geotechnical Engineering Investigation Manual*, McGraw-Hill Book Co., New York.
2. Foster, C. R. (1962) "Field Problems: Compaction," *Foundation Engineering*, G. A. Leonards, ed., McGraw-Hill Book Co., New York, pp. 1000–1024.
3. Turnbull, W. J. (1950) "Compaction and Strength Tests on Soil," presented at *ASCE Annual Meeting*, January.
4. Sherard, J. L., Woodward, R. J., Gizienski, S. F., and Clevenger, W. A. (1963) *Earth and Earth-Rock Dams*, John Wiley & Sons, New York.
5. NAVFAC (1982) *Design Manual DM 7.1: Soil Mechanics*, Dept. of the Navy, Naval Fac. Engrg. Comm., Alexandria, Va.
6. Burmister, D. M. (1948) "The Importance and Practical Use of Relative Density in Soil Mechanics," *Proc. Amer. Soc. for Test. and Mat.*, Philadelphia, Pa., Vol. 48.
7. Hilf, J. W. (1975) "Compacted Fill," *Foundation Engineering Handbook*, Winterkorn and Fang, eds., Van Nostrand-Reinhold, New York, Chap. 7.
8. ENR (1982) "Rolled Concrete Triumphs," *Engineering News-Record*, October 21, pp. 35–37.
9. TRRL (1974) *Soil Mechanics for Road Engineers, Dept. of the Environment, Transport and Road Research Lab., Her Majesty's Printing Office*, London, 10th Impression.
10. The Asphalt Inst. (1969) *Soils Manual for Design of Asphalt Pavement Structures*, The Asphalt Inst., College Park, Md.
11. The Asphalt Inst. (1970) *Thickness Design, Manual Series No. 1*, The Asphalt Inst., College Park, Md.
12. Tschebotarioff, G. P. (1973) *Foundations, Retaining, and Earth Structures*, 2d ed., McGraw-Hill Book Company, New York.
13. Turnbull, W. J. (1968) "Problem Areas in Design of Flexible Pavements," Preprint of lecture presented to Brazilian highway engineers sponsored by Instituto de Pesquisas Rodoviarios do Brasil and USAID, April.
14. Turnbull, W. J. (1968) "Problem Areas in Specifi-

cations for and Construction of Flexible Pavements," Preprint of lecture presented to Brazilian highway engineers sponsored by Instituto de Pesquisas Rodoviarios do Brasil and USAID.

15. ENR (1976) "Foamed Concrete Is Lightweight Fill," *Engineering News-Record*, July 22, p. 11.

16. Seim, C., Walsh, T. J., and Hannon, J. B. (1981) "Wicks, Fabrics and Sawdust Overcome Thick Mud," *Civil Engineering*, ASCE, July, pp. 53–56.

17. Moorhouse, D. C. and Baker, G. L. (1969) "Sand Densification by Heavy Vibratory Compactor," Proc. ASCE, J. Soil Mech. and Found. Engrg. Div., Vol. 95, No. SM4, July, p. 985.

18. CII (1981) "Vibratory Rollers: Development Trends," *Construction Industry International*, Vol. 7, No. 10, October, pp. 90–111.

19. USBR (1974) *Earth Manual*, 2d ed., U.S. Bureau of Reclamation, U.S. Govt. Printing Office, Washington, D. C.

20. HRB (1952) "Compaction of Embankments, Subgrades, and Bases," *Bull. No. 58*, Highway Research Board, Washington, D. C., p. 40.

21. Bouyoucos, G. J. (1931) "The Alcohol Method for Determining the Water Content of Soils," *Soil Scientist*, Vol. 32, pp. 173–179.

22. Reinhold, F. (1955) "Soil-Cement Methods for Residential Streets in Western Germany," *World Construction*, March–April p. 41.

23. Dunn, W. L. and McDougal, F. H. (1970) "Minimizing Errors in Gamma-Ray Surface-Type Density Gages: Existing Gages and New Design Concepts," *Bull. No. 301*, Highway Research Board, Washington, D. C.

24. Lobdell, H. L. (1970) "Settlements of Buildings Constructed in the Hackensack Meadows," *Proc. ASCE, J. Soil Mech. and Found. Engrg. Div.*, Vol. 96, No. SM4, July, p. 1235.

25. Meehan, R. L., Dukes, M. T., and Shires, P. O. (1975) "A Case History of Expansive Claystone Damage," *Proc. ASCE, J. Geotech. Engrg. Div.*, Vol. 101, No. GT9, September.

26. Hunt, R. E. (1967) "Engineered Tank Pads Can Lower Costs," *Oil and Gas Journal*, November.

27. Hunt, R. E. (1967) "Preloading: an Effective Tank Site Preparation," *Oil and Gas Journal*, December.

28. Hunt, R. E. (1967) "Foundations for Cylindrical Steel Storage Tanks," Pub. of Joseph S. Ward & Assocs., Caldwell, N.J.

29. Morrison, A. (1982) "The Booming Business in Wick Drains," *Civil Engineering*, ASCE, March, pp. 48–51.

30. Hansbo, S. (1979) "Consolidation of Clay by Band-Shaped Prefabricated Drains," *Ground Engineering*, July.

31. Fellenius, B. H. (1981) "Discussion on 'Consolidation of Clay' by Band-Shaped Prefabricated Drains," *Ground Engineering*, October.

32. Barron, R. (1948) "Consolidation of Fine-Grained Soils by Drain Wells," *Tran. ASCE*, Vol. 113.

33. Mitchell, J. K. (1970) "In-Place Treatment of Foundation Soils," *Proc. ASCE, J. Soil Mech. and Found. Engrg. Div.*, Vol. 96, No. SM1, January p. 73.

34. McCarthy, D. F. (1982) *Essentials of Soil Mechanics and Foundations*, 2d ed., Reston Pub. Co., Reston, Va., Chap. 12.

35. Ramaswamy, S. D., Lee, S. L., and Daulah, I. U. (1981) "Dynamic Consolidation—Dramatic Way to Strengthen Soil," *Civil Engineering*, ASCE, April, pp. 70–73.

36. Ramaswamy, S. D., Aziz, M. A., Subrahamanyam, R. V., Khader, M. H. A., and Lee, S. L. (1979) "Treatment of Peaty Clay by High Energy Impact," *Proc. ASCE, J. Geotech. Engrg. Div.*, Vol. 105, No. GT8, August, pp. 957–967.

37. Warner, J. (1978) "Compaction Grouting—A Significant Case History," *Proc. ASCE, J. Geotech. Engrg. Div.*, Vol. 104, No. GT7, July, pp. 837–847.

38. Brown, D. R. and Warner, J. (1973) "Compaction Grouting," *Proc. ASCE, J. Soil Mech. and Found. Engrg. Div.*, Vol. 99, No. SM8, August, pp. 589–601.

39. Seed, H. B. and Booker, J. R. (1977) "Stabilization of Potentially Liquefiable Sand Deposits Using Gravel Drains," *Proc. ASCE, J. Geotech. Engrg. Div.*, Vol. 103, No. GT7, July, pp. 757–768.

40. CGS (1978) *Canadian Foundation Engineering Manual*, Part 3, Deep Foundations, *Canadian Geotechnical Society*, Montreal.

41. Munoz, A., Jr. and Mattox, R. M. (1977) "Vibroreplacement and Reinforced Earth Unite to Strengthen a Weak Foundation," *Civil Engineering*, ASCE, May, pp. 58–61.

42. ENR (1979) "Terminal Is Designed to Sink Along with Site," *Engineering News-Record*, September 6, pp. 18–20.

43. Broms, B. B. and Boman, P. (1979) "Lime Columns—A New Foundation Method," *Proc. ASCE, J. Geotech. Engrg. Div.*, Vol. 105, No. GT4, April, pp. 539–556.

43a. ENR (1983) "Weak Seabed Strata Yields Strong Support," *Engineering News-Record*, March 10, pp. 30–31.

44. NCE (1980) "Tensar Set for 'Major Assault' on

Geotextiles," *NCE International, Institute of Civil Engineers*, London, September, pp. 46–47.

45. Civil Engineering (1980) "Geotextiles for Soil Improvement," *Civil Engineering*, ASCE, July, p. 83.

46. Belongia, D. L. (1981) "Highway Applications of Geotextiles," *Civil Engineering*, ASCE, May, pp. 60–63.

47. Lambe, T. W. (1962) "Soil Stabilization," *Foundation Engineering*, G. A. Leonards, ed. McGraw-Hill Book Co., New York, Chap. 4.

48. Dumont-Villares, A. (1956) "The Underpinning of the 26-Story 'Campanhia Paulista de Seguros' Building, São Paulo, Brazil," *Geotechnique*, London, Vol. 6.

49. PCA (1959) *Soil Cement Construction Handbook, Portland Cement Assoc.*, Chicago, Ill.

50. Winterkorn, H. F. (1975) "Soil Stabilization," *Foundation Engineering Handbook*, Winterkorn and Fang, eds., Van Nostrand-Reinhold Co., New York, Chap. 8.

50*a*. Deere, D. U. (1976) "Dams on Rock Foundations—Some Design Questions," *Proc. Rock Engineering for Foundations and Slopes*, ASCE, Vol. 2, pp. 55–85.

51. Fecker, E. (1980) "The Influence of Jointing on the Failure of Teton Dam: A Review and Commentary," *Bull. Intl. Assoc. Engrg. Geol.*, No. 21, pp. 232–238.

52. Rigny, P. (1974) "European vs. U.S. Grouting Practices," *Foundations for Dams*, ASCE, New York, pp. 37–46.

53. Houlsby, A. C. (1977) "Engineering of Grout Curtains to Standards," *Proc. ASCE, J. Geotech. Engrg. Div.*, Vol. 103, No. GT9, September, pp. 953–970.

54. Christensen, D. L. (1974) "Unusual Foundation Developments and Corrective Action Taken," *Foundations for Dams*, ASCE, New York, pp. 343–370.

54*a*. Lang, T. A. (1972) "Rock Reinforcement," *Bull. Assoc. Engrg. Geol.*, Vol. IX, No. 3, Summer, pp. 215–239.

55. Lancaster-Jones, P. F. F. (1969) "Methods of Improving the Properties of Rock Masses," *Rock Mechanics in Engineering Practice*, Stagg and Zienkiewicz, eds., John Wiley & Sons, New York, Chap. 12.

56. Leet, L. D. (1960) *Vibrations from Blasting Rock*, Harvard Univ. Press, Cambridge, Mass.

57. Langefors, U. and Kilhstrom, B. (1963) *The Modern Technique of Rock Blasting*, John Wiley & Sons, New York.

58. Richart, F. E., Jr. (1960) "Foundation Vibrations," *Proc. ASCE, J. Soil Mech. and Found. Engrg. Div.*, Vol. 86, No. SM8, August.

59. Crandell, F. J. (1949) "Ground Vibration Due to Blasting and Its Affects on Structures," *J. Boston Soc. of Civ. Engrs.*, April, pp. 222–245.

60. Edwards, A. T. and Northwood, T. D. (1960) "Experimental Studies of the Effects of Blasting on Structures," *The Engineer*, Vol. 210, No. 30, September, pp. 538–546.

61. Langefors, U., Kilhstrom, B., and Westerberg, H. (1958) "Ground Vibrations in Blasting," *Water Power*, February, pp. 335–338, 390–395, 421–424.

62. Nicholls, H. R., Johnson, C. F., and Duvall, W. J. (1971) *Blasting Vibrations and Their Effects on Structures*; Bull. 656, U.S. Bureau of Mines.

63. Neff. T. L. (1972) "Qualitative Factors for Evaluating Stability," *Proc. 14th Symp. on Rock Mech., New Horizons in Rock Mechanics*, ASCE, New York (1973), pp. 311–328.

64. Gustafsson, R. (1973) *Swedish Blasting Techniques*, SPI, Gothenburg, Sweden.

65. Hoek, E. and Bray, J. W. (1977) *Rock Slope Engineering*, 2d ed., Institution of Mining and Metallurgy, London.

66. Ambraseys, N. N. and Hendron, A. J. Jr., (1968) "Dynamic Behavior of Rock Masses," *Rock Mechanics in Engineering Practice*, Stagg and Zienkiewicz, eds., John Wiley & Sons, New York, Chap. 7.

CHAPTER SIX

SHALLOW FOUNDATIONS

6.1 INTRODUCTION TO FOUNDATIONS

6.1.1 BASIC FACTORS

Purpose

The purpose of a foundation is to transfer the loads of a structure to the ground in a manner that provides adequate support for the structure and results in its satisfactory performance. Adequate foundation support requires that there be no danger of ground rupture (the bearing capacity) which would result in collapse or overturning of the structure and that ground compression (the allowable bearing value) or heave, which would result in detrimental deflections in the structure, not occur.

Types

Foundation types can be placed in two general categories:

Shallow foundations derive their support entirely from their base, and their depth is less than or in the order of their least width. Included are spread or spot footings, combined and continuous footings, and mats or rafts as illustrated in Fig. 6.1.

Deep foundations derive their support in most cases from their base and from friction acting along their sides, or friction or base support alone, and the depth is substantially greater than the least width. They are grouped as piles, including bored piles often referred to as piers, and caissons, as illustrated in Fig. 6.1.

Another type which can fall into either group is termed floating or compensated foundations (Fig. 6.1).

Failures

Major causes of foundation "failure" are summarized as follows:

Ground rupture and collapse conditions include

- General shear failure in soils such as soft to stiff clays and organic materials
- Erosion and scour in granular soils from broken water pipes or flowing natural waters
- Liquefaction of fine-grained granular soils during earthquakes
- Ground collapse caused by tunneling, old mines, and limestone cavities
- High seepage forces in rock masses supporting concrete dams

Vertical ground displacements and detrimental structural deflection are caused by

- Excessive compression or consolidation of soils or rock masses

- Expansion and heave from swelling soils or rock and frost penetration in fine-grained soils
- Ground subsidence from lowering of the groundwater table, or excavations for tunnels, retention of adjacent structures, or construction of adjacent structures
- Ground collapse caused by tunneling, old mines, limestone cavities, and saturation of "collapsible" soils
- Deterioration of concrete or other foundation materials, due to groundwater or other environmental conditions

Investigations

The selection of the optimum foundation for a given site and the evaluation of design criteria required for safety and economy must be based on a thorough investigation as described in Art. 5.1.3. There are many factors to be considered besides surface and subsurface conditions, including site development plans and the characteristics of the proposed structures as discussed. The presentation of subsurface conditions on three-dimensional sections (Fig. 5.2) is extremely useful for the interpretation and evaluation of complex geology.

6.1.2 CRITERIA FOR SATISFACTORY PERFORMANCE

General

The satisfactory performance of a foundation requires adequate depth, safety against failure by ground rupture, and limited deformation. These are the normal factors to consider after it has been determined that the site is not subjected to the major hazards of faulting, liquefaction, ground collapse, slope failures, and flooding *(see*

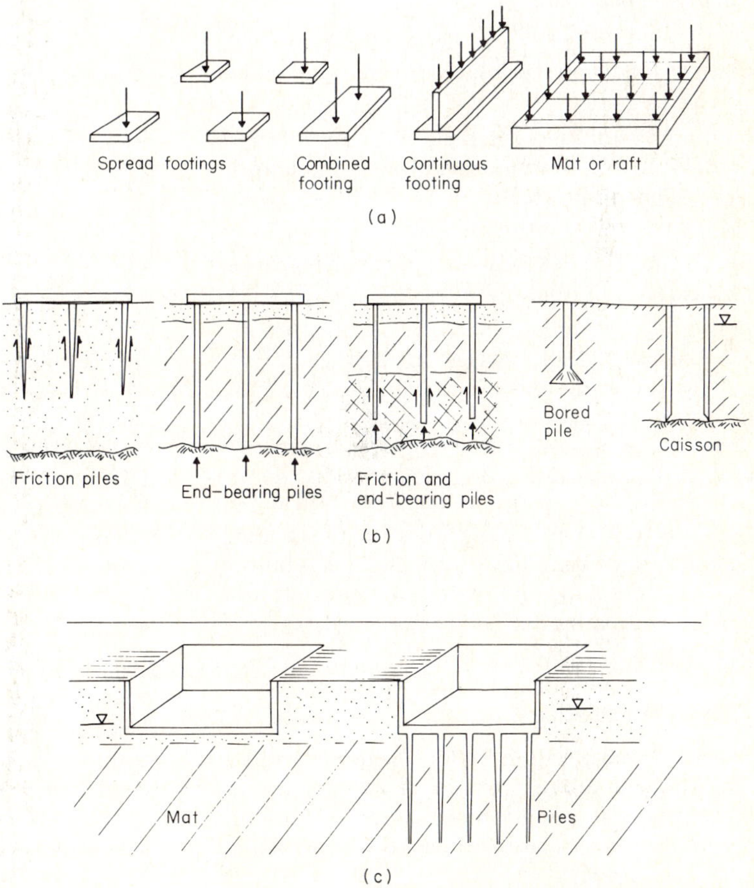

FIG. 6.1 Various foundation types: (*a*) shallow foundations; (*b*) deep foundations; and (*c*) compensated or floating foundations.

Art. 5.1.2). Satisfactory performance can usually be provided for a given site by several foundation types or foundations in concert with ground improvement methods (*see Chap. 5*). The choice is based on economic comparisons and since shallow foundations normally are the least costly, their feasibility is examined first, often together with ground-improvement methods. If they prove to be impractical, then deep foundations are evaluated.

Evaluation Factors

The major factors requiring consideration in the evaluation of foundations include the nature of the structure itself, foundation depth requirements, ground capacity in terms of deformation and rupture, and in some cases earthquake forces.

Aspects of Structure

Loaded area dimensions: These dimensions influence the depth to which the ground will be stressed.

Loading conditions: Intensity and rate of application affect both ground rupture and compression. High time rates in weak soils can lead to rupture before compression occurs.

Flexibility: The flexibility of a structure provides its tolerance to differential deflections.

Foundation elevation: Placing the structure on a fill to raise grade increases stresses on the supporting soils, whereas excavation for a basement decreases stresses. In some conditions the inclusion of a basement where one was not considered previously can change foundation requirements from deep to shallow. A basement, however, will require consideration of the added costs of excavation dewatering, slope retention, and protection against postconstruction groundwater compared to the costs of a deep foundation.

Foundation Depth

The first consideration in the selection of a foundation depth is the provision for adequate protection against ground freezing, or swelling soils, or uplift or overturning forces, or scour, whichever case prevails.

The second consideration is the depth required to attain a suitable foundation support, which will be equal to or greater than the depth determined from the first consideration. In either case, foundations for adjacent structures must be considered to avoid disturbing them.

Ground Deformations

In the case where the foundation will be supported directly by the underlying soil or rock stratum, the *stratum* is assigned an "allowable bearing value", i.e., the load per unit area that the material can support without undergoing objectionable deformations. If a large loaded area, such as a mat or storage tank, placed directly on grade will stress more than one soil stratum, the *structure* is assigned an allowable load per unit area. The criterion for selecting the foundation and the bearing value or allowable load per unit area is the *settlement tolerance* of the structure. Swelling soils normally are not assigned an allowable bearing value since the problem is one of avoiding ground heave, not containing it.

Ground Rupture

Although the allowable bearing value, based on settlement tolerances, is often far below the capacity of a soil or rock to resist failure by rupture (about one-third the rupture capacity), the *bearing capacity* against rupture failure must be considered. Cases where the bearing capacity is most critical include structures designed to tolerate large settlements such as steel storage tanks or embankment fills over soft ground where loads are applied rapidly; or cases where high live loads relative to dead loads will be imposed, such as chemical plant towers, grain silos, storage tanks, or transmission-line towers; or under certain rock conditions.

Earthquake Forces

Ground shaking during earthquakes imposes dynamic forces on the structure transmitted through the foundations. These forces have the components of acceleration of gravity, frequency of vibration, and duration, as well as other less significant elements, and are resisted by the stiffness and damping characteristics of the structure. Ground response usually takes the form of transmission of cyclic forces to the foundations, but may be characterized also by liquefaction and subsidence of granular soils and the reduction in strength of weak cohesive soils, as discussed in Art. 4.4.5.

Settlement of Structures

Importance

The most important criterion for the design of foundations for most structures is the restriction of settlement to within tolerable limits. Settlement results from ground compression (*see Art. 4.3*) under foundation loadings which can take several forms, as illustrated in Fig. 6.2: uniform settlement, tilt, or nonuniform settlement. The latter two cases represent *differential settlement.*

Nonuniform settlement resulting in distortions of the structure normally are more critical to its performance than is total settlement or even tilt, unless the structure is connected to utilities or another structure. Excessive tilt may lead to a bearing-capacity failure.

Causes of Differential Settlement

Differential settlement can result from a variation in stresses imposed by structural loadings as related to the configurations of the foundation element or from a variation in subsurface conditions. For the average light to moderately loaded building, the more common geologic causes are inadequate foundations bearing over thick deposits of soft clays or organic soils covering large areas and swelling clays. Except for heavily loaded foundations or sensitive structures, settlement is seldom a problem in granular, residual, or glacial-till soils, although confirmation of the allowable value is always required.

Sands are subject to scour from leaking water or sewer mains, and the possibility of such events should be guarded against. A 20-year-old, 12-story building in Rio de Janeiro suddenly tilted almost 1 m and leaned against an adjacent building because of scour from a leaking sewer pipe [ENR (1977)[1]]. Collapse was prevented by emergency shoring of the structure with heavy steel beams, and the structure was eventually demolished.

Effects of Differential Settlement

In *buildings*, structural distress caused by differential settlement is manifested by diagonal cracks in walls, jamming of doors, breaking of windows, overloading and rupture of columns, and dislocation of beams. Collapse can occur in extreme cases. Where the structure is pile-supported and the floor is not, the floor may settle away from the walls causing utilities to rupture and imposing additional loads on the piles, possibly leading to overloading. Tilting may affect the alignment of sensitive machinery such as turbines.

Various examples of the effects of differential settlement on structures are illustrated in Figs. 6.3 through 6.7 (also see Figs. 1.1 and 1.2). All of the structures illustrated are pile-supported and located in the northeastern United States. The incidence of poor performance of pile-supported structures is high in that region because of the relatively large number of light industrial buildings constructured over areas characterized by old miscellaneous fill overlying deep deposits of

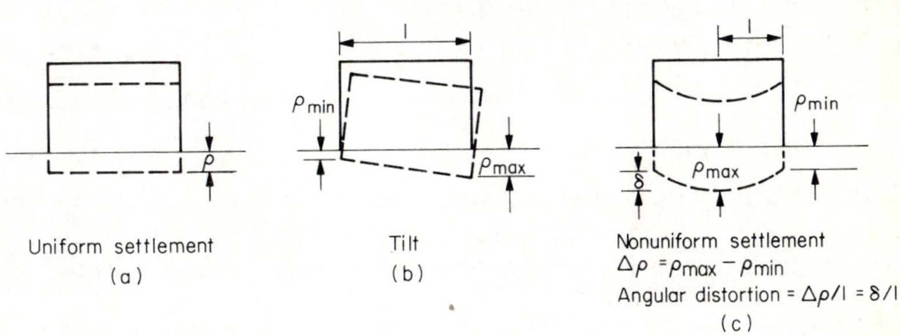

FIG. 6.2 Forms of settlement: (*a*) uniform; (*b*) tilt; and (*c*) nonuniform.

FIG. 6.3 Differential settlement apparent along wall of warehouse, Queens, New York, 1960 (see also Figs. 6.4 and 1.1). Walls and columns are pile-supported, lightly loaded floors are fill-supported. Soil conditions of miscellaneous fill over thick organic soils are subjecting piles to both downdrag and high lateral forces. Overhead doors are jammed.

FIG. 6.4 Settlement cracks and corner rupture in building shown in Figs. 6.3 and 1.1.

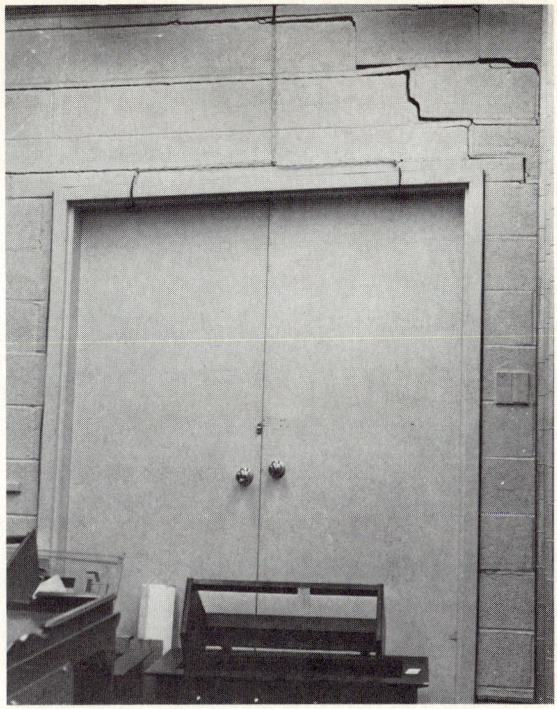

FIG. 6.5 Effects of differential settlement on interior walls. Note different elevations of door knobs. Building columns were pile-supported, but floors and some walls were supported on fill placed over organic soils. Newark, N.J., 1955.

FIG. 6.6 Separation of floor from pile-supported portion of wall in the building in Fig. 6.5. Note the electric cables under stress and the shimming under the boiler. Many utility lines were ruptured.

FIG. 6.7 Fill-supported warehouse floor overlying organic soils underwent over 1 m of settlement in locations subjected to floor loads in the order of 500 psf (0.25 kg/cm²). Pile-supported columns and walls suffered tilting from lateral earth pressures and movements. Floor separation from wall and some wall cracking are apparent in the photo.

organic soils where deep foundations normally are required. Other regions have their own characteristic soil conditions resulting in structural distress, such as the expansive clays of Houston and Denver or the bay muds of San Francisco, unless proper engineering procedures are followed.

Cylindrical steel storage tanks are simple, flexible structures, but when supported on a pad of engineered, compacted fill placed over compressible soils, they can be subjected to several forms of distress from settlement as illustrated in Fig. 6.8. *Total settlement* can overstress interconnecting piping and result in loss of the pad elevation which can subject the tank bottom to corrosion from water trapped within the fire walls surrounding the tank. *Tilting* can effect gaging of the tank contents, jam floating roofs, and overstress the shell. *Differential floor settlement* can overstress floor welds or a column-supported roof. *Differential movement between the bottom and the shell* can overstress weld connections, possibly resulting in lower shell bulging and eventually tank collapse. *Differential perimeter*

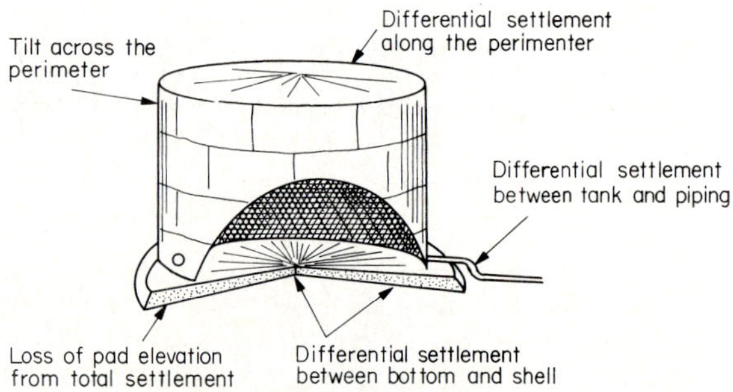

FIG. 6.8 Effects of settlement on a steel storage tank.

settlement causes shell warping and can lead to tank rupture. This condition occurs where soil conditions vary significantly beneath the tank.

Using flexible steel storage tanks for preload, some aspects of differential settlement and correcting tank settlement are discussed in Art. 5.3.2. Criteria for evaluating the performance of large steel tanks from the various aspects of settlement effects are given by Marr et al. (1982).[2]

Permissible Settlement for Buildings

Total settlement affects interconnecting utility lines, connections to adjacent structures, adjacent pavements, etc. Limiting total settlement of a structure frequently is used as an indirect means of controlling the amount of differential settlement as they are often related. The following values are suggested [CGS (1978)[3]] for maximum total settlement when other factors do not govern as described above:

Structures on clay: 8 cm (3 in)

Structure on sand: 4 cm (1½ in)

Differential settlement, however, is often the controlling factor in structural performance. The major reference has been Skempton and MacDonald (1956),[4] who used the maximum net slope of the deflection curve (Fig. 6.2, δ/l) to correlate settlement with damage in a study of 98 case histories. The evaluation of the parameter requires the removal of rigid body settlement and tilt as shown in Fig. 6.2. Bjerrum (1963)[5] published the limiting angles of distortion for various conditions as given in Fig. 6.9. Grant et al. (1974)[6] presented the results of 95 additional case studies and updated the findings of Bjerrum. The differences are minor.

It has been concluded that cracking of panels, in frame buildings, or walls in wall-bearing structures, is likely to occur if $\delta/l > 1/300$, and structural damage to columns and beams is likely to occur if $\delta/l > 1/150$. For $\delta/l = 300$, the largest value of the associated differential settlement between columns is typically 45 mm (1¾ in) for foundations on clay and 30 mm (1¼ in) for foundations on sand. For a given differential settlement, the allowable settlement is greater for a raft foundation than for a footing foundation, in the order of 25%. Since the rigidity and width of rafts vary considerably, however, the settlement limits for rafts may deviate widely from the average values.

Monitoring Deflections

Building settlement is monitored with optical survey, the water level device, settlement points, inclinometers, and piezometers as illustrated in Fig. 6.10 and discussed in Art. 4.5.2 of Hunt (1984).[7] Instrumentation for monitoring deflections of flexible steel storage tanks is illustrated in Fig. 5.27. Monitoring provides data on the magnitudes and time rates of total and differential settlement and facilitates estimating the ultimate settlements. Depending upon the performance of the structure and the predicted magnitudes, corrective measures may or may not be taken. Corrective measures are described in Chap. 5 under Ground Improvement.

Foundation Heave

Causes

Foundation heave can result from frost action *(see Art. 6.2.1)*, uplift from hydrostatic forces in an excavation bottom *(see Art. 9.3.3)*, stress relief in soils or rock in excavations *(see Arts. 4.3.6 and 9.3.3)*, and swelling of soil or rock (this article). Frost action (except for permafrost), hydrostatic uplift, and stress relief normally are phenomena affecting shallow foundations, whereas swelling materials can affect either shallow or deep foundations.

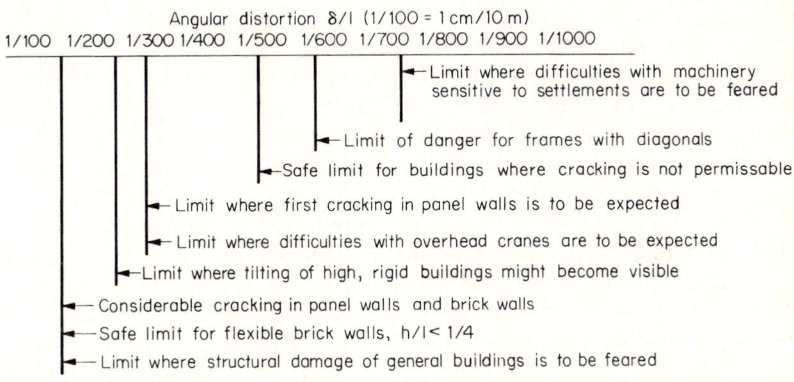

FIG. 6.9 Limiting angular distortions for structures. [From Bjerrum (1963).[5]]

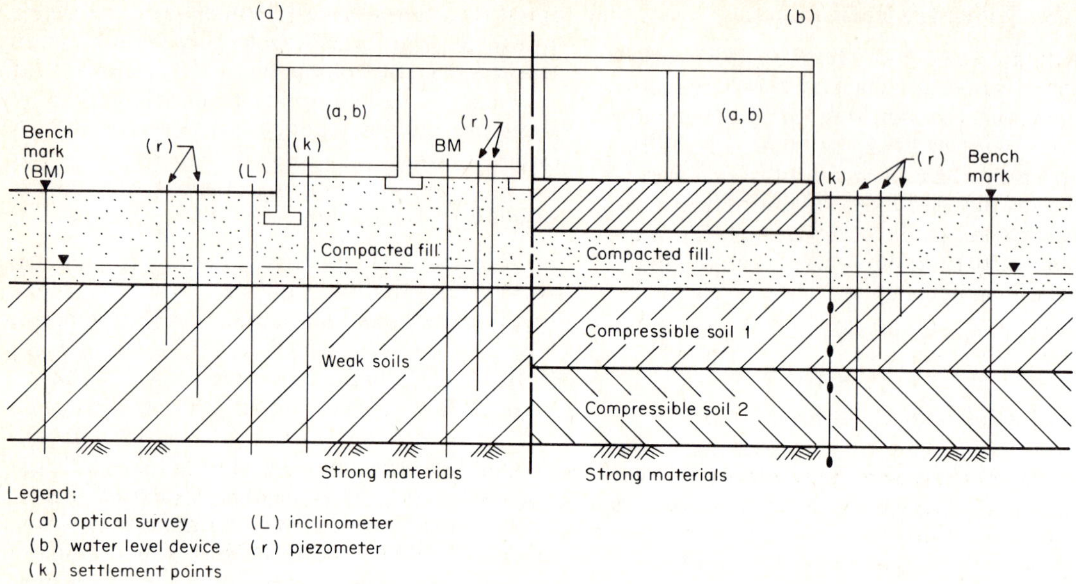

FIG. 6.10 Instrumentation for building settlements: (*a*) spread-footing-supported light structure and (*b*) mat-supported heavy structure. [From Hunt (1984).[7] Reprinted with permission of McGraw-Hill Book Company.]

Swelling of Soils

Damage to structures in the United States each year from swelling soils alone is reported to amount to $700 million [ENR (1976)[8]] indicating the severity of the potential problem. In most soils and rocks, expansion, which can result in severe structural distortions, is most prominant in materials containing montmorillonite clay. The problem can be especially serious for light structures because of the large volume increases, but heave can occur in heavy structures as well since swell pressures as high as several tons per square foot or more can develop.

Characteristics of swelling soils are unusual. The higher the foundation load and the more rigid the pressure confining the clay, the higher will be the swell pressure. Once the soil is permitted to expand, however, the pressures reduce. An important aspect is the time delay for occurrence where it is thought that the soil is protected from free water. Often a year or more passes after completion of a structure before significant volume changes occur in the soils and result in structural distress. Case studies have shown also that swelling stops after 4 to 5 years in many instances. Time delay results from a natural slow increase in moisture content, although if the soil is exposed to free water, swell occurs immediately. Swelling in soils and rocks is described in detail in Art. 10.6 of Hunt (1984).[7]

Swell potential is measured in the laboratory by the determination of the type of clay mineral present, simple index tests, and swell tests performed in the consolidometer, as described in Art. 6.3.6. In general, clay soils are highly expansive if the colloidal content (percent less than 0.002 mm) is greater than 20% and the PI is greater than 25% (Fig. 4.34).

Environmental factors affect swell potential for a given clay type. Highly expansive soils are most active in climates with long periods of little or no rainfall, during which the soil dries and shrinks, large cracks open on the surface, and fissures develop throughout the soil mass and increase its permeability. When the ground surface is covered with a structure, ground moisture increases with time and swelling occurs. High volume changes occur usually in the zone above the lowest permanent water table.

Foundation Solutions for Swelling Soils

In general, swelling is difficult to prevent, and normally it is only reduced by the various treatment methods available. The most satisfactory solution usually is deep foundations penetrating to below the zone of seasonal volume changes, or "active zone" *(see Art. 4.3.6)*, and protected from the development of uplift pressures along the shaft. Light structures have been supported satisfactorily on shallow "rigid" mats or continuous foundations where some movement can be tolerated and differential deflections are not too severe.

Other solutions have been attempted but results often are not satisfactory. Areas have been preflooded to allow expansion, after which the foundation has been designed to contend with settlement of the softened material and the attempt has been made to maintain a balance between swell and consolidation. Injections of lime have met with some success, but in highly active clays it may aggravate swelling as the lime is normally added with water. Excavation of the upper portion of the swelling clays, mixing with lime, and then replacing the soil-lime mixture as a compacted fill is a costly procedure and not always satisfactory.

6.2 SHALLOW FOUNDATIONS ON SOILS

6.2.1 SELECTION: FOUNDATION TYPE AND DEPTH

Foundation Types (See Fig. 6.1)

Spread or spot footings support single columns and are the simplest and least costly of foundations. In soil, regardless of the allowable bearing pressure, they are never less than 12 in (30 cm) in width to avoid the possibility of punching into the soil by shear failure, and their sides are of equal, or nearly equal, dimensions.

Combined or strap footings are spread footings extended to support two columns close together, and are used where a footing will extend over a property line, in which case the dimension of the outer footing is reduced by sharing the load with an adjacent inner footing.

Continuous or strip footings are combined footings extended to support a column row or a wall. If they are designed to be adequately stiff they can bridge local zones of soft ground. When spread footings occupy up to about 30% of the foundation area, or when allowable differential settlement criteria cannot be satisfied, continuous footings may be more economical. They are made circular to support chimneys, towers, or storage tanks. Design may be based on the beam-on-an-elastic-subgrade concept *(see Art. 4.4.2)*.

A *mat or raft* beneath the entire structure may be more economical when the footings occupy about 50% of the building area. In addition to the advantages of load distribution and bridging local soft zones, the excavation and foundation formwork are simplified. Mats and continuous footings are considered as less than rigid, whereas spread footings are practically rigid. The moments and shears in reinforced concrete mats usually are based on the slab-on-an-elastic-subgrade concept *(see Art. 4.4.2)*. The analysis requires a modulus of subgrade reaction k_s which often is difficult to establish with accuracy. Conservative values are selected for k_s and a generous amount of steel is added to the foundation [Teng (1975)[9]].

Compacted fills often are constructed as a soil mat *(see Art. 5.2.2)* to support cylindrical steel storage tanks without the necessity for concrete foundations. Free-draining material is used for the fill to aid in reducing bottom corrosion.

Floating or compensated foundations are used where weak, compressible soils extend to substantial depths. In a floating foundation, either the building weight is approximately equal to the full weight (including water) of the soil removed from the building site, or the building weight is balanced against hydrostatic uplift or buoyancy. The use of compensated foundations in Mexico City is described in Art. 7.4.2.

Selection

Bases

Selection of the foundation type normally is based on consideration of the column spacing, necessity for wall support, and the allowable soil pressure. Foundation dimensions are increased to restrict the contact pressure to values within the allowable soil pressure. If strong soils overlie relatively weak soils at shallow depths, the result can be greater settlement than that anticipated for the allowable soil pressure assigned to the bearing stratum.

If shallow foundations are found to be unsuitable from the aspect of bearing capacity or settlement for the existing site conditions, then the feasibility of ground improvement (Chap. 5) or a deep foundation (Chap. 7) is evaluated in terms of relative economics.

Examples of Shallow Foundations

Several examples of shallow-foundation case histories are given in Figs. 6.11 through 6.14, which illustrate the evaluation of such selection factors as structure life, strong supporting soils over relatively weak soils, moderately weak supporting soils and high loads, and the effect of a high water table on basement construction. Examples of shallow foundations supported on controlled fill are given in Figs. 5.20, 5.21, and 5.23.

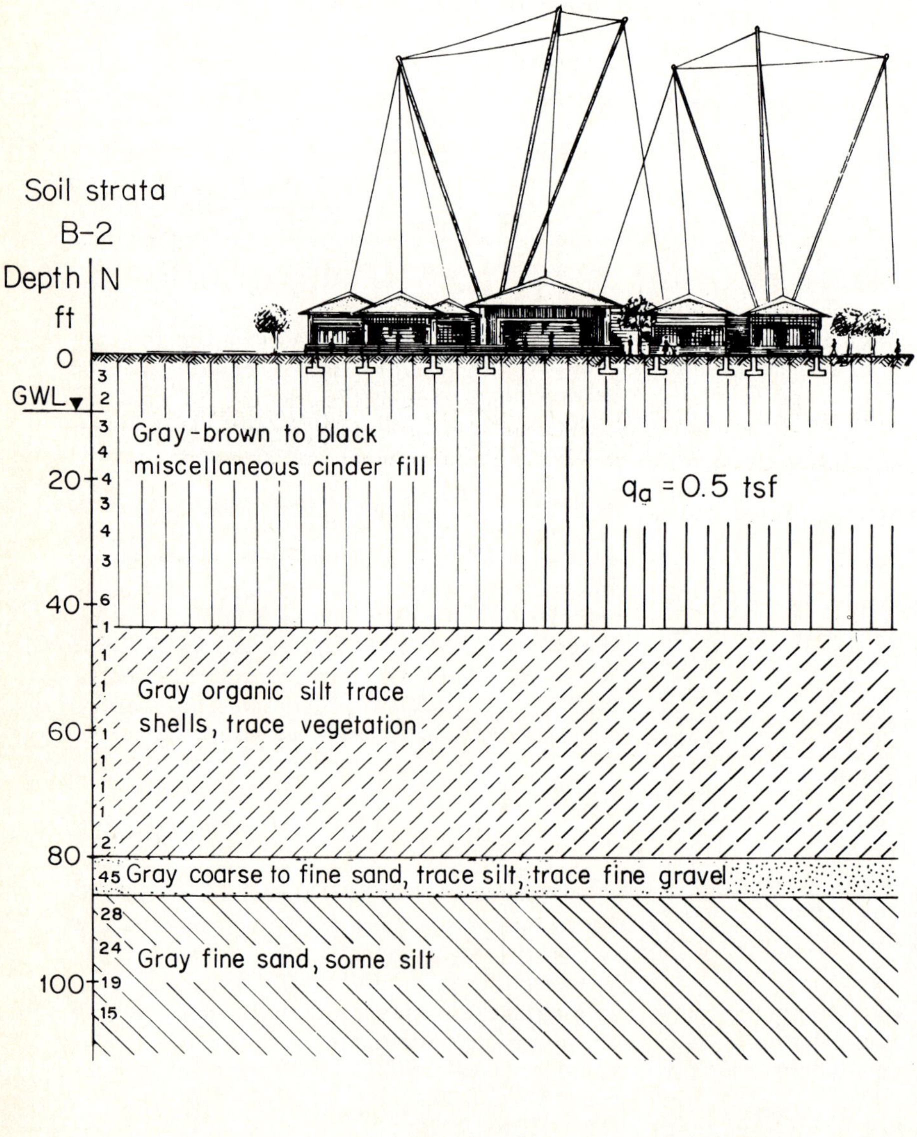

FIG. 6.11 Example of light structure with short life supported on spread footings overlying weak soils (The New Jersey Pavilion, New York World's Fair). (Courtesy of Converse Consultants.)

It is interesting to compare the solution in Fig. 6.11, a light structure with a short life, to that shown in Fig. 7.3, a heavy structure with a desired long life, both founded above similar soil conditions. The light structure, with column loads to 250 kips, was supported on spread footings bearing in the cinder fill with a $q_a = 0.5$ tsf. The cinder fill had been in place for 23 years. Building settlement was nominal and rapid and was negligible after construction.

The 25-story building in Fig. 6.12 was supported on spread footings designed for a $q_a = 5$ tsf bearing in compact granular soils, whereas the 33-story building in Fig. 6.13 was supported on a mat designed for a $q_a = 4$ tsf bearing in granular soils underlain by a firm silty clay stratum. In both cases GWL is substantially below founding level.

Thick deposits of glacial lakebed clays decreasing in consistency with depth provided suitable support for the rigid mat beneath the 15-story building in Fig. 6.14. Designed for a $q_a = 2$ tsf, the mat was stiffened by shear walls, by the exterior basement walls, and by beams on the underside in some locations. The excavation for a two-level garage substantially reduced the net building load added to the soil. A permanent underdrain system was installed to control the high

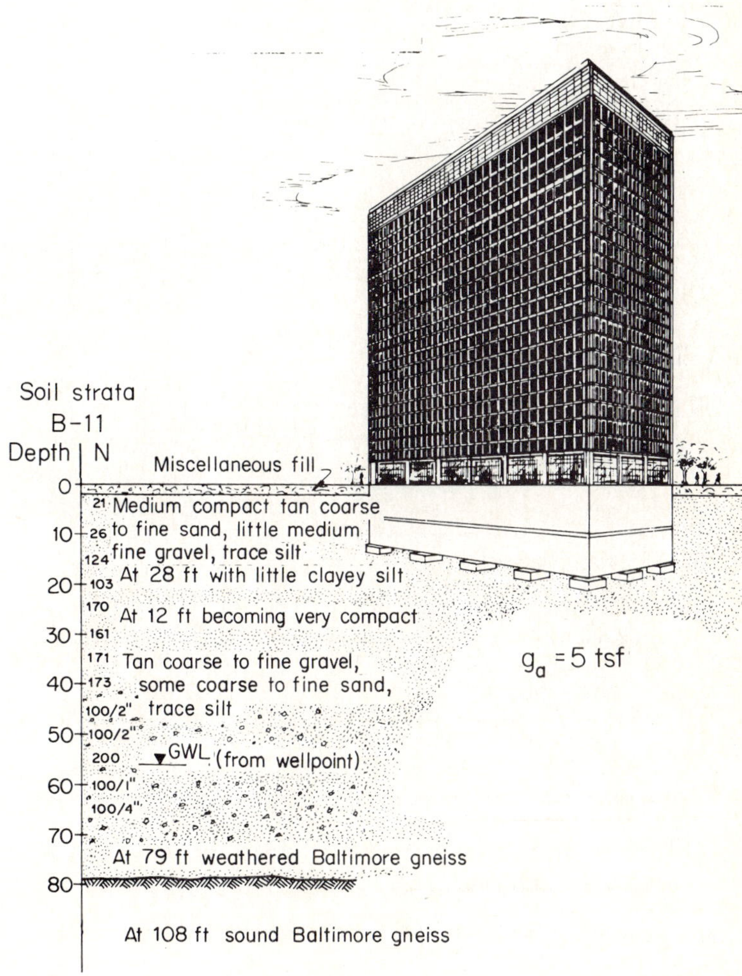

FIG. 6.12 Twenty-five story building in Baltimore, Md., supported on spread footings bearing in compact granular soils. (Courtesy of Converse Consultants.)

groundwater level and to provide dry garage levels and eliminate hydrostatic uplift. Subsequent building settlement was insignificant.

Foundation Depth

Selection Factors

The first step in foundation analysis is the determination of the minimum or adequate foundation depth, since it is independent of foundation type. Unless the structure has a basement, adequate depth depends on soil conditions, climate, the presence of adjacent structures, and, in special cases, on the presence of uplift or overturning forces or the necessity for scour protection. Footings or mats normally are placed at a minimum depth below the ground surface unless they can be influenced by frost or ground collapse or require embedment to resist uplift or overturning forces.

Freezing Ground

In cold climates exterior footings must be placed below the maximum depth of frost penetration in all soils except free-draining granular soils above the water table, such as those classed as GW or SW. In these soils a depth of three-fourths of the frost penetration depth is considered adequate. The depth of frost penetration in the United States is given in Fig. 6.15.

Interior footings and floors supported on other than free-draining material of adequate thick-

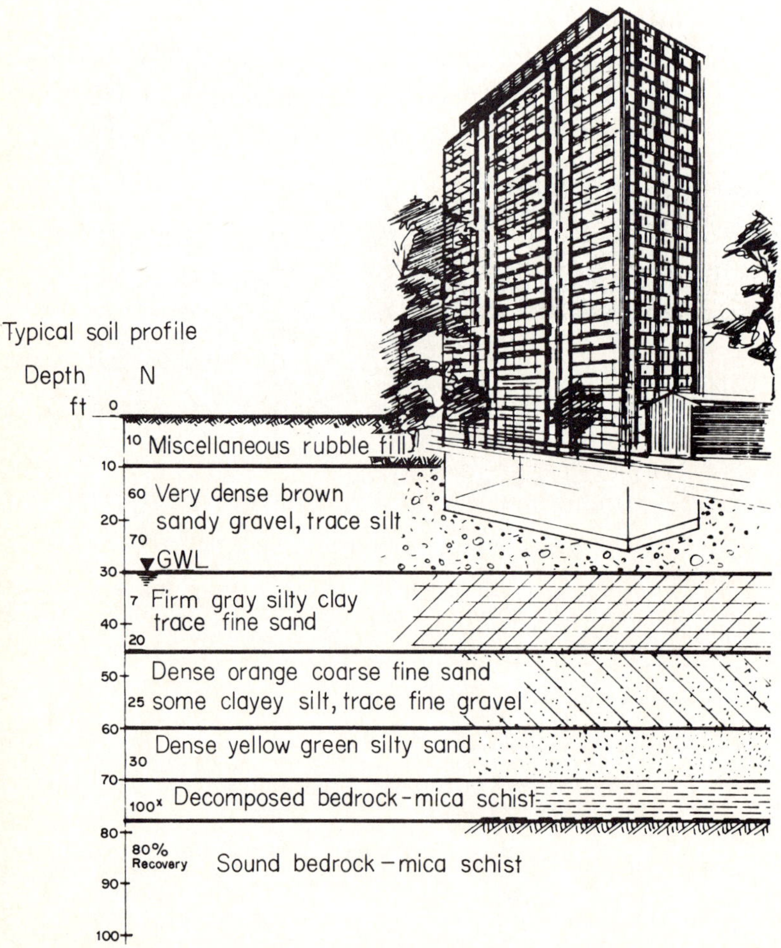

FIG. 6.13 Thirty-three story building in Philadelphia, Pa., supported on a mat foundation. Maximum column loads are 520 tons and allowable bearing pressure is 4 tsf. (Courtesy of Converse Consultants.)

ness must be protected in unheated buildings, or during the winter months of construction in cold climates when the structure is unheated. Protection is provided by insulation with earth, straw, or some other material. Frost heave and subsequent thaw beneath unprotected footings will cause differential movements and leave the supporting soils in a weakened condition. Insulation or other forms of protection also are required to prevent ground freezing beneath cold-storage or refrigerated buildings.

In *permafrost*, foundations often are placed *below the depth of thaw*, and insulation is required to prevent heat from the structure causing thaw [see Art. 7.7.3 of Hunt (1984)[7]].

Other Soil Conditions

In *expansive clays*, the depth of a shallow foundation normally is not a significant factor from the aspects of swelling and heave. Light structures are often founded directly on the clays at shallow depths, as discussed in Art. 6.2.6. In certain *collapsible soils*, such as porous clays, foundations are placed below the collapsible zone (*see Art. 6.2.7*).

Adjacent Structures

A foundation cannot be placed much deeper than that of an existing adjacent structure, unless it is underpinned or otherwise given temporary support. If the new footings are lower than and too close to the existing footings, the existing footings may suffer support loss. A rule of thumb regarding the acceptable orientation is given in Fig. 6.16. If the new footing is placed higher than or at the same level as those of an adjacent structure, consideration must be given to the imposition of additional stresses on the existing foundation or its subsoil, which may result in differential settlement.

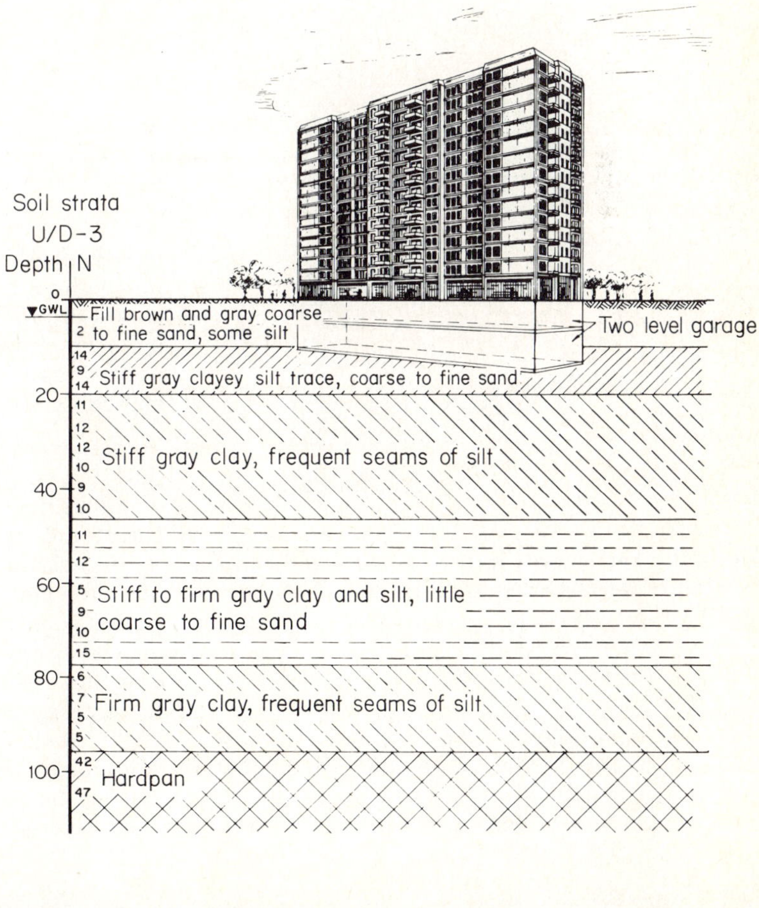

FIG. 6.14 Fifteen-story apartment building in Cleveland, Ohio, supported on a mat foundation bearing in stiff sandy clayey silt and founded below the water table. Allowable soil pressure is 2 tsf. (Courtesy of Converse Consultants.)

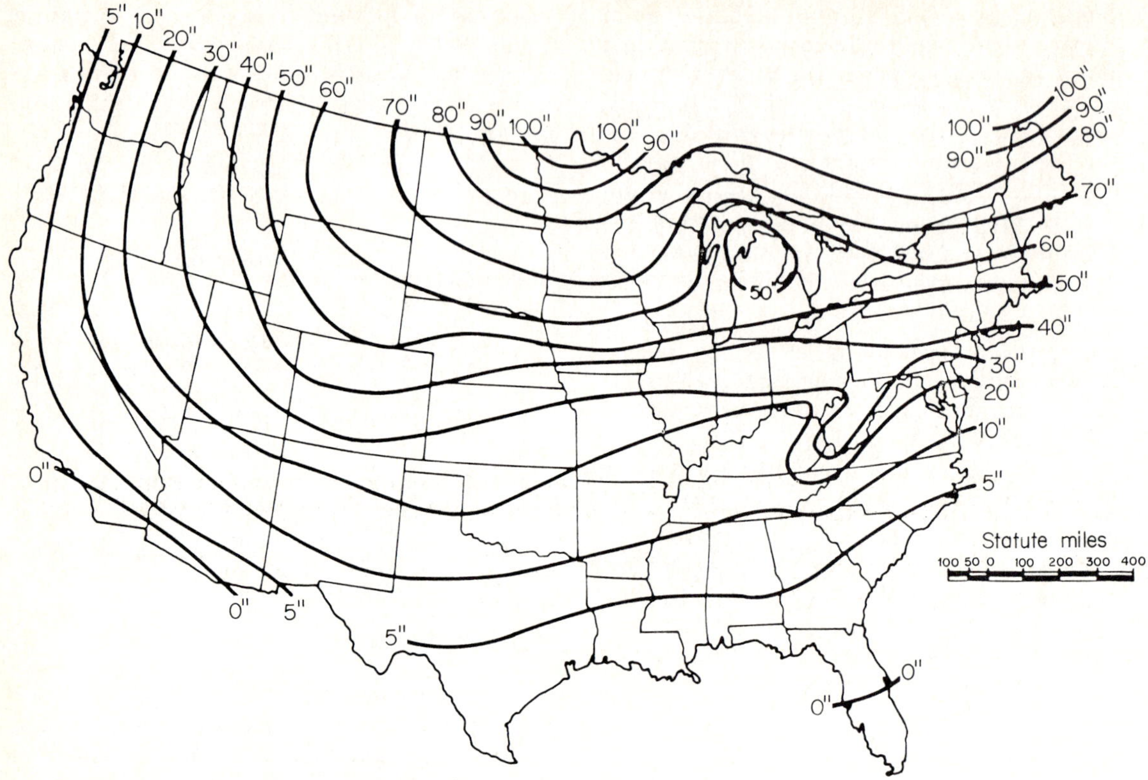

FIG. 6.15 Extreme frost penetration, in inches, based on state average. [From NAVFAC (1982)[13] and National Oceanic and Atmospheric Administration, Environmental Data and Information Service, Asheville, N.C.]

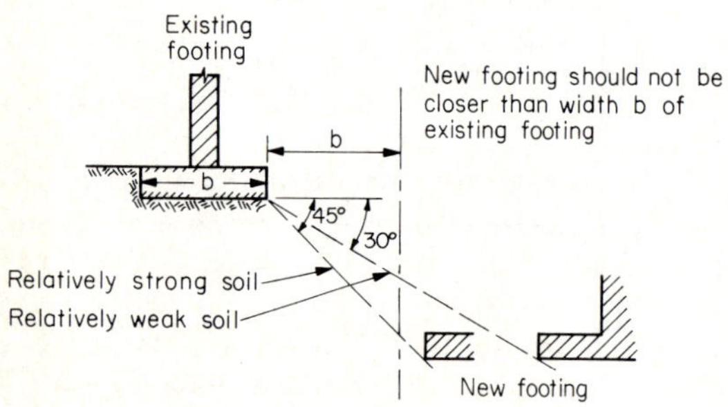

FIG. 6.16 Approximate location of adjacent footings between an existing and a new structure. [After Sowers (1962).[11]]

Basements

For basement design, careful consideration must be given to the maximum possible height that can be reached by a rising water table which, even for a short period, can result in a "wet" basement or even floor slab upheaval. Basements below the water table should be made impermeable and protected by *underdrain systems* (Fig. 6.17) which when working properly will relieve uplift pressures on slabs. Underdrains are discussed in Art. 8.4.7 of Hunt (1984).[7]

In cases where an underdrain system is not practical (as, for example, if excessive demands would be made on the drainage systems), or temporary failure of the system would have disastrous consequences, the floor slab is designed to resist the maximum anticipated uplift pressures. This is achieved by thickening the slab, by adding walls or beams, or by using anchors. In an interesting case in New York City, excess hydrostatic pressure from a water table perched on a "tight" rock formation was relieved by drilling holes into underlying, highly fractured rock, permitting the perched water to drain and relieve the pressures.

Scour

Waterfront or river structures require protection against scour erosion from current or wave action. Soil susceptibility depends on particle size, cohesion, and water velocity and ranges from high in silts and fine sands in average water velocities to low for cohesive materials unless velocities are high for long times. Scour from leaking sewer pipes has been mentioned in the previous article.

"The depth at which a river bed may be scoured by a flood is difficult to measure and not easy to believe. In the absence of definite information relevant to a particular bridge site, a well-known rule is to carry the foundation to a depth of 4 times the difference between the flood and dry weather levels, unless rock is found nearer to the surface" [Smith (1977)[12]]. In a case study of bridge failure causes, Smith (1977)[12] found that out of 143 failures studied, 46% were the result of scour. It was by far the largest single cause of bridge failure, although only rarely are large bridges casualties.

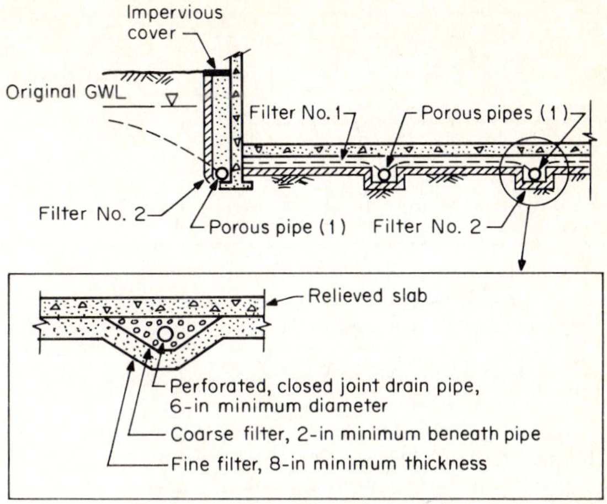

FIG. 6.17 Typical underdrain system for structures. Pipes (1) lead either to gravity discharge or to a storage basin for removal by pumping. [From Hunt (1984).[7] Reprinted with permission of McGraw-Hill Book Company.]

6.2.2 THE ALLOWABLE BEARING VALUE q_a

Evaluations

Foundation dimensions depend on the *allowable soil bearing value* q_a (also *allowable soil pressure*), which is a function of the settlement tolerances of a structure. A number of empirical and analytical procedures are in use for evaluating q_a.

Empirical procedures, described in this article, commonly are based on penetration test data (SPT or CPT). They are suitable generally for predominantly granular soils where conservatism in selecting an allowable bearing value will not result in overly costly design. In cohesive soils, especially those that are relatively weak and compressible, they provide only a guide for preliminary estimates. In the use of empirical procedures, settlement generally is considered to be restricted to within 1 in (2.5 cm). Analytical methods are used for more critical situations (*see Art. 6.2.3*).

In cohesive soils, for nonuniform conditions where material deformability increases with depth, and for cases where a refined value of q_a is required, a settlement analysis is performed as discussed in Art. 6.2.3. When there is concern about ground rupture, analysis is performed to determine the bearing capacity q_b as discussed in Art. 6.2.4.

Building Codes and Other "Experience" Charts

Presumptive Bearing Value

Preliminary estimates of q_a are obtained from building codes and charts developed by various agencies and investigators which give values for q_a in terms of density for granular soils and consistency for cohesive soils, applicable to spread foundations. In building codes these values are referred to as "presumptive" bearing values and are based on local experience from satisfactory and unsatisfactory performance, usually giving q_a as the pressure that will limit settlement to 1 in (2.5 cm) total or differential.

Nominal values for q_a for spread foundations in the United States, as given by *U.S. Navy Design Manual DM7.2* (1982),[13] are presented in Table 6.1, and the values as given by the Building Code of the City of New York are presented in Table 6.2. Values for clay soils vary widely depending on the location; a summary of values from a number of U.S. cities is given in Table 6.4.

TABLE 6.1
NOMINAL VALUES FOR ALLOWABLE BEARING PRESSURE FOR SPREAD FOUNDATIONS*

Type of bearing material	Consistency in place	Allowable bearing pressure, tsf: Ordinary range	Allowable bearing pressure, tsf: Recommended value for use
Massive crystalline igneous and metamorphic rock: granite, diorite, basalt, gneiss, thoroughly cemented conglomerate (sound condition allows minor cracks)	Hard, sound rock	60 to 100	80
Foliated metamorphic rock: slate, schist (sound condition allows minor cracks)	Medium hard sound rock	30 to 40	35
Sedimentary rock: hard cemented shales, siltstone, sandstone, limestone without cavities	Medium hard sound rock	15 to 25	20
Weathered or broken bedrock of any kind except highly argillaceous rock (shale)	Soft rock	8 to 12	10
Compaction shale or other highly argillaceous rock in sound condition	Soft rock	8 to 12	10
Well-graded mixture of fine and coarse-grained soil: glacial till, hardspan, boulder clay (GW-GC, GC, SC)	Very compact	8 to 12	10
Gravel, gravel-sand mixtures, boulder-gravel mixtures (GW, GP, SW, SP)	Very compact	6 to 10	7
	Medium to compact	4 to 7	5
	Loose	2 to 6	3
Coarse to medium sand, sand with little gravel (SW, SP)	Very compact	4 to 6	4
	Medium to compact	2 to 4	3
	Loose	1 to 3	1.5
Fine to medium sand, silty or clayey medium to coarse sand (SW, SM, SC)	Very compact	3 to 5	3
	Medium to compact	2 to 4	2.5
	Loose	1 to 2	1.5
Fine sand, silty or clayey medium to fine sand (SP, SM, SC)	Very compact	3 to 5	3
	Medium to compact	2 to 4	2.5
	Loose	1 to 2	1.5
Homogeneous inorganic clay, sandy or silty clay (CL, CH)	Very stiff to hard	3 to 6	4
	Medium to stiff	1 to 3	2
	Soft	0.5 to 1	0.5
Inorganic silt, sandy or clayey silt, varved silt-clay-fine sand (ML, MH)	Very stiff to hard	2 to 4	3
	Medium to stiff	1 to 3	1.5
	Soft	0.5 to 1	0.5

*From NAVFAC (1982).[13]

NOTES:

1. Variations of allowable bearing pressure for size, depth, and arrangement of footings are given in the text.
2. Compacted fill, placed with control of moisture, density, and lift thickness, has allowable bearing pressure of equivalent natural soil.
3. Allowable bearing pressure on compressible fine grained soils is generally limited by considerations of overall settlement of structure (Fig. 6.9).
4. Allowable bearing pressure on organic soils or uncompacted fills is determined by investigation of individual case.
5. Allowable bearing pressure for rock is not to exceed the unconfined compressive strength.

Correlation Basis: Penetration Tests

The "density" criterion given in U.S. codes is the relative density D_R as estimated from the N values of the SPT as given in Table 6.3. "Consistency" values for cohesive soils are estimated from the SPT or from unconfined compression tests as given in Table 6.4 (see also Table 3.7 for simple hand test).

The SPT is popular in the United States and South America for evaluating q_a, whereas the

TABLE 6.2
ALLOWABLE BEARING PRESSURES FOR ROCK AND SOIL*

Material	Q_{all}, tsf	Description
Hard, sound rock	60	See Table 6.13.
Medium hard rock	40	See Table 6.13.
Intermediate rock	20	See Table 6.13.
Soft rock	8	See Table 6.13.
Hardpan	12	Groups of GM, GC, and SW, well-cemented and free of lenses of fines and soft rock.
Hardpan	8	Groups GM, GC, and SW, poorly cemented with fine-grained matrix or lenses of fines.
Gravelly soils	10	Groups GW, GP, GM, GC; compact, well-graded.
	6	Groups GW, GP, GM, GC; loose, poorly graded.
	8	Groups SW, SP, SM; compact, well-graded.
	4	Groups SW, SP, SM; loose, poorly graded.
Sands, coarse-medium	N × 0.1 6 max. 3 min.	Groups SW, SP, SM; with less than 10% of material retained on no. 4 sieve.
Sands, fine	N × 0.1 4 max. 2 min. (except for vibratory loads which require study)	
Hard clay	5	Groups SC, CL, CH; clay requires picking for removal, fresh sample cannot be remolded by finger pressure.
Medium clay	2	Can be removed by spading, can be remolded by substantial finger pressure.
Soft clay	—	Requires soil testing and analysis. Can be remolded with slight finger pressure.
Silts, dense	3	Groups ML and MH; requires picking for removal.
Silts, medium	1.5	Can be removed by spading.
Silts, loose	—	Requires soil testing and analysis.
Varved silts	2 max.	Higher values permitted when preconsolidated.
Organic soils	1 max.	Untreated.
Organic soils	2 max.	Treated by preloading.

*After Building Code, City of New York (1968).[14]

CPT is more popular in Europe. [For a discussion of penetration tests see Art. 3.4.4 of Hunt (1984).[7]]

Limitations

Building codes and charts are useful for obtaining limiting values for q_a, which are usually conservative, but for clays and settlement-sensitive structures on sands and silts more refined analytical procedures should be used. Codes have several important limitations. Soil descriptions are brief and often vague and do not account for such properties as expansion and collapse. In addition, the following are assumed to have no ef-

TABLE 6.3
CORRELATIONS BETWEEN COMPACTNESS, RELATIVE DENSITY D_R, AND N VALUE FROM THE SPT FOR SANDS*

Compactness	D_R†	N (SPT)
Very loose	<0.15	<4
Loose	0.15–0.35	4–10
Medium dense	0.35–0.65	10–30
Dense (compact)	0.65–0.85	30–50
Very dense	0.85–1.0	>50

*For correlations with various gradations see Table 3.5.
†From Gibbs and Holtz (1957).[15]

TABLE 6.4
CORRELATIONS BETWEEN CONSISTENCY, N, U_c, and q_a

Consistency	N (SPT)	U_c, tsf	q_a, tsf*
Very soft	2	0.25	—
Soft	2–4	0.25–0.5	0.25–1.0
Medium (firm)	4–8	0.5–1.0	1.0–2.0
Stiff	8–15	1.0–2.0	2.0–3.0
Very stiff	15–30	2.0–4.0	3.0–6.0
Hard†	30	4.0	6.0–12.0

*From a sampling of building codes from a number of U.S. cities.
†Including hardpan or glacial till.

fect on bearing pressure: underlying strata; water-table location; structure type and design-load conditions; and foundation size, shape, and depth.

Cohesionless Granular Soils

Soil Type Factors

A number of investigators have developed relationships for estimating q_a in *sands* that take into consideration footing dimensions and water-table depth. The procedures are suitable for studies where conservative estimates of q_a are adequate, such as light to moderately loaded structures which are relatively insensitive to small deflections, supported on a relatively uniform deposit of sandy soil underlain by stronger soils. More refined estimates of settlement result from analytical procedures described in Art. 6.2.3.

Gravels are relatively incompressible, and q_a is normally estimated from codes and charts. Undisturbed sampling is not feasible and SPT and other penetration test values are not very significant, especially where large particle sizes are present.

Silts commonly are found thinly layered and thus considered as an integral part of a sand stratum. SPT values in saturated thick deposits may not be reliable because temporary pore-pressure changes influence the penetration resistance. Settlement estimates are most reliable when based on properties measured in the laboratory or in situ.

SPT Basis for Sands

The most commonly used relationship in the United States for estimating q_a was developed by Terzaghi and Peck (1948),[16] based on correlations between plate load tests, experience with foundations, and the SPT. Found with time to be overly conservative, the relationships have been updated. A more recent correlation between SPT values and q_a to restrict settlement to 1 in for foundations of various widths and width-to-depth ratios, bearing on sands above the water table, is given in Fig. 6.18 (*see discussion in Art. 6.2.4,* Bearing Capacity Analysis).

To use the chart, footing width B is taken as the side of a square, the smaller dimension of a rectangular footing, the width of a long continuous footing, or the diameter of a circular footing, where footing depth $D_f \leq B$. N is corrected for the depth factor C_N (Fig. 6.20), and average values from the boring logs are used for a depth equal to about the width of the largest footing. Then q_a is determined for the footing subjected to the greatest load, i.e., the largest footing. Smaller footings are dimensioned based on q_a and then Fig. 6.18 is used with B and D_f/B to check that q_a is not governed by bearing capacity (*see Art. 6.2.4*), i.e., the left-hand rising branch of the design curve, which represents $q_b/2$ (or $q_{ult}/2$). If so, then B is reproportioned for a smaller q_a. If the water table is located at or may rise to ground surface, the chart values are multiplied by the correction factor C_w = 0.5. For depth of GWL =

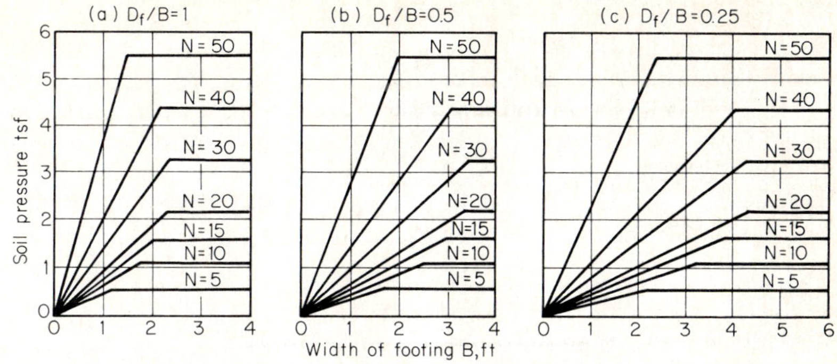

FIG. 6.18 Design chart for proportioning shallow footings on sand (γ = 100 pcf) for total settlement of less than 1 in and deep GWL. [From Peck et al. (1974).[17] Reprinted with permission of John Wiley & Sons, Inc.]

D_w, measured from the surface of the surcharge surrounding the footing (Fig. 6.28), use

$$C_w = 0.5 + 0.5 \frac{D_w}{D_f + B} \tag{6.1}$$

Based on investigations by Schmertmann (1970)[19] and others, the Terzaghi and Peck curves [also Terzaghi and Peck (1967)[18]] are considered conservative for footing widths $B < 10$ ft (3 m) and unconservative for larger footing widths. Curves for estimating q_a as a function of N and B as developed by several investigators are given in Fig. 6.19. A wide disagreement is apparent. Oweiss (1979)[20] provides a summation of methods of estimating settlement in sands based on SPT data.

Water-Table Effects

Complete submergence with water table at footing depth approximately doubles settlement, but if the water level is below the base of a footing, at a depth equal to or greater than one-half the footing width, it will have an insignificant effect. According to Sanglerat (1972),[22] for an intermediate case, the settlement from N values should be adjusted by the ratio of $\bar{\sigma}_t/\bar{\sigma}_s$ where

$\bar{\sigma}_t$ = effective overburden pressure for the dry condition at a depth of $B/2$ beneath the footing base; and

$\bar{\sigma}_s$ = effective overburden pressure for the submerged state at a depth of $B/2$ for the highest anticipated water level.

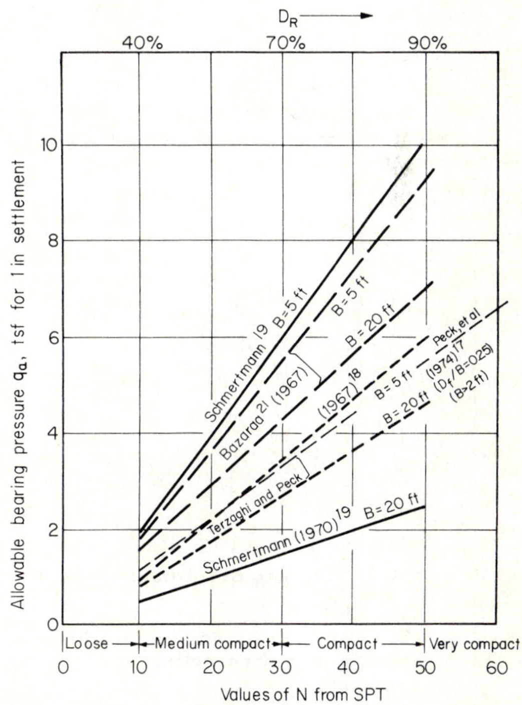

FIG. 6.19 A comparison of q_a vs. N for 1 in of settlement as determined by methods proposed by several investigators for footings with width B (ft) on dry sands.

The correction factor proposed by Terzaghi and Peck is given in Eq. 6.1. Meyerhof (1965)[23] and Schmertmann (1970)[19] believe that no correction is required as it is assumed that the N values reflect the presence of water, *unless* there is a possibility that the water table will rise, as is often the case. For cases where there is such a possibility, a correction factor is required.

SPT Depth Correction Factor C_N

The effect of depth on N values due to overburden pressures has been studied by a number of investigators, and correction factors C_N have been proposed to obtain a design value N_1 [for a discussion see Art. 3.4.4 of Hunt (1984)[7]].

$$N_1 = C_N N \tag{6.2}$$

In Eq. 6.2, N is the field SPT value. Seed (1979)[24] presents recommended curves for determining C_N in terms of effective overburden pressure as given in Fig. 6.20. The relationships represent an averaging of the work by several previous investigators.

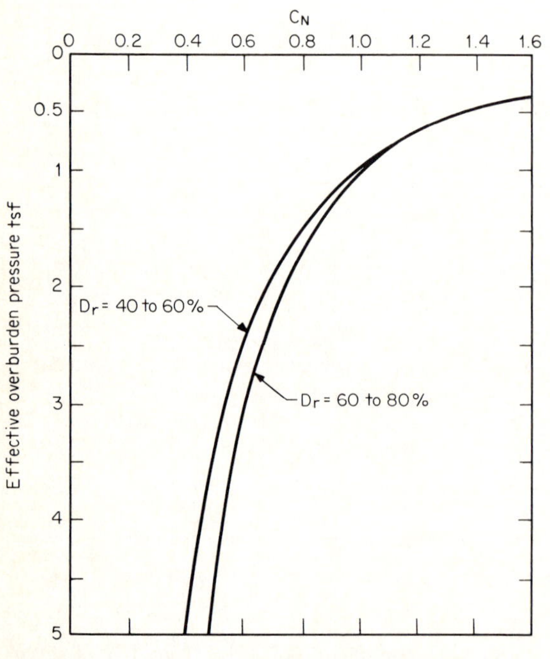

FIG. 6.20 Recommended curves for determination of C_N based on averages from various investigators. [From Seed (1979).[24]]

The results of tests by Marcuson and Bieganousky (1977)[25,26] and others indicate that correlations between SPT, D_R, and q_a should be used with caution and only qualitatively. The results are very sensitive to drilling and sampling equipment conditions as well as the operating techniques and small variations in soil-mass conditions. They concluded that even in controlled laboratory conditions a single value for N can represent a spread of $\pm 15\%$ in D_R in sands.

Static Cone Penetrometer Test CPT

The CPT has been widely used in Europe for many years and is being used in the United States with increasing frequency [for a discussion see Arts. 2.3.4 and 3.4.5 of Hunt (1984)[7]]. European experience [Sanglerat (1972)[22]] indicates that the relationship between the cone-tip resistance q_c and the allowable bearing value q_a for spread footings at depths in the order of 1 m can be expressed as

$$q_a = q_c/10 \tag{6.3}$$

The relationship has been found applicable to silty sands, clayey sands, and stiff clays. Correlations between q_c and N are given in Table 4.6. In general, in sands $q_c \approx 4N$.

Meyerhof (1956)[27] developed a correlation between q_a/q_c, footing width B, and footing depth-to-width ratios D/B as given in Fig. 6.21, which is expressed by the relationship

$$q_a = q_c B(1 + D/B) \cdot 1/40 \tag{6.4}$$

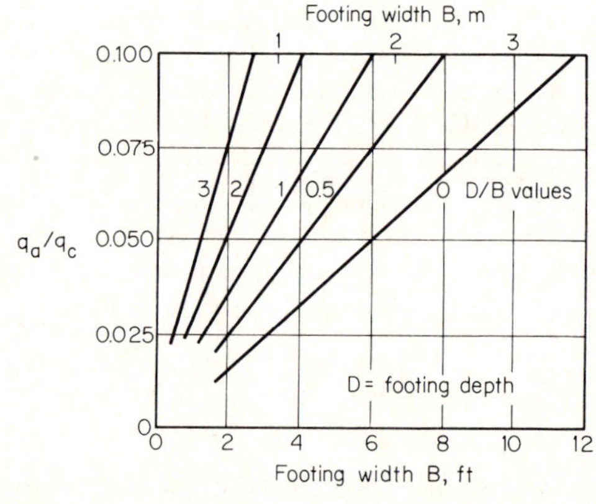

FIG. 6.21 Estimating q_a from cone penetrometer test point resistance q_c in sands. [From Meyerhof (1956).[27]]

where B and D are in meters. For the case of B and D in feet,

$$q_a = q_c B(1 + D/B) \cdot 1/120 \qquad (6.5)$$

Schmertmann (1970)[19] considers that settlement of small foundations appears greatly overestimated and that of large foundations underestimated when using the approach of Meyerhof.

Clays and Organic Soils

Determining q_a

Determination of design values for q_a for normally to slightly overconsolidated clays (soft to firm) should be based on settlement analysis using laboratory or field data. In most cases the same procedures should be used with saturated stiff intact clays *(see Art. 6.2.3)*. Organic soils are most reliably evaluated on the basis of large-scale field-load tests.

In overconsolidated clays (stiff to hard) q_a may be approximated from correlations with the unconfined compressive strength U_c, as measured roughly with the pocket penetrometer in the field. Values for q_a are estimated by applying a safety factor, FS = 3, to the following expression from Terzaghi and Peck (1967)[18] for the net ultimate bearing capacity q_d *(see Art. 6.2.4)*:

$$q_a = q_d/3$$
$$= 2.5U_c(1 + 0.2D_f/B)(1 + 0.2B/L) \qquad (6.6)$$

where footing width and length are B and L, and foundation depth D_f does not exceed 2.5B.

SPT Reliability

SPT values in clayey silts, clays, and organic soils are extremely unreliable and serve only as a preliminary guide to compressibility or for correlations where the investigator has previous experience in an area. The author has experienced N values in stiff clayey silts, with a precompression range of 2 tsf and higher, ranging as low as 3 to 5.

CPT Relationships

Based on literally thousands of tests in limited geographical areas in Europe, various correlations have been made between consolidation characteristics of clays and organic soils and the cone resistance q_c (Eq. 6.3). Other correlations have been developed for initial approximations of q_a based on settlement estimates. For normally consolidated clays, a relationship between q_c, water content w, and the compressibility index C_c is given in Table 6.5. Settlement is estimated from the expression

$$\rho = HC_c[\log(\bar{p}_o - p_f) - \log \bar{p}_o] \qquad (6.7)$$

which is the expression for settlement from a plot of change in percent compression versus load change on a log scale from the laboratory consolidation test, in which H = stratum thickness, $\bar{p}_o$ = effective overburden pressure, and p_f = stress imposed by foundation *(see Art. 4.3.4)*.

Another approach for estimating settlements for NC and OC clays and silty and organic soils employs a coefficient α. Rewriting Eq. 4.18 for m_v (coefficient of volume compressibility) to include Eq. 4.17 for a_v (coefficient of compressibility) and combining with Eq. 4.4.2 results in

$$\rho = H(p_f)(m_v) \qquad (6.8)$$

Relating m_v to soil modulus and to q_c gives [Sanglerat (1972)[22]]

$$m_v = 1/E = 1/\alpha q_c \qquad (6.9)$$

Values for α are given in Table 6.6.

6.2.3 SETTLEMENT ANALYSIS ρ

Reliability

With respect to geologic conditions, the accuracy of settlement estimates is influenced mainly by soil type and layering. The most important factor in analysis is the application of truly representative soil parameters, and data procured for some soil types is much more representative than data procured for others. Of somewhat less importance is the actual distribution of stresses and resulting displacements of foundations as related to elastic theory which vary with soil type and layering, as well as foundation rigidity.

Analytical procedures, discussed in Art. 4.3, are based generally on either laboratory consolidation tests of undisturbed samples or elastic moduli measured or estimated by various procedures.

TABLE 6.5
CORRELATION OF q_c, C_c, AND w*

Point resistance q_c, tsf, kg/cm^2	Water content w	Compressibility index C_c
>12	<30	<0.2
<12	<25	<0.2
	25–40	0.2–0.3
	40–100	0.3–0.7
<7	100–130	0.7–1
	>130	>1

*From Sanglerat (1973).[22]

TABLE 6.6
VALUES OF THE COEFFICIENT α FOR CLAYEY, SILTY, AND ORGANIC SOILS*

Soil type	Point resistance q_c, tsf, kg/cm^2	Coefficient α
CL—low plasticity clay	<7	3–8
	7–20	2–5
	>20	1–2.5
ML—low plasticity loam	<20	3–6
	>20	1–2
OH—very plastic clay	<20	2–6
MH—very plastic loam	>20	1–2
OL—very organic loam	<12	2–8
T-OH—peat and very organic clay	<7	
	$50 < w < 100$	1.5–4
	$100 < w < 200$	1–1.5
	$w > 300$	<0.4

*From Sanglerat (1972).[22]

Soil Formation Factors

Free-draining granular soils do not yield undisturbed samples suitable for laboratory compression testing in the consolidometer. The most reliable settlement data are obtained from full-scale load tests. Methods have been developed for estimating settlement empirically on the basis of penetration test data *(see Art. 6.2.2)* or from elastic theory on the basis of soil moduli.

Coarse-grained cohesive soils, such as residuum or glacial till, are often difficult to sample undisturbed, in which case settlement calculations are based on elastic moduli concepts. These materials often exhibit substantial prestress, which in the case of residual soils is "apparent," or not based on previous loadings (see also NC to OC clays below).

Heavily overconsolidated clays, especially those that are fissured and located above the water table, often have expansive tendencies that are more significant than their compressibility under most foundation loadings.

Collapsible soils such as loess, "porous" clays, and fine-grained valley alluvium in arid climates may yield low SPT values and yet appear strong, standing on near vertical slopes to substantial heights in open cuts. Collapse of their structure occurs at the natural water content at some critical pressure, or upon or approaching full saturation. Gradual compression may occur under applied load as the percent of saturation is increased. Collapsible soils are discussed in Art. 6.2.6.

Normally consolidated to somewhat overconsolidated intact clays are investigated by laboratory consolidation tests and classical settlement analysis. *Residual soils* containing undecomposed mica often give compressibilty parameters from consolidation testing that approximate field conditions.

Organic soils, when evaluated on the basis of consolidation test data, yield settlement estimates that can be considered only as approximations because of the influence of root fibers and gas in the soil. A large amount of secondary compression, often equal to primary compression, is to be expected. The often open structure from root fibers results in very rapid primary compression, but secondary compression continues for many years. Rootmat and other highly fibrous materials can undergo as much as 50% compression under light loads within a few weeks time and can even shrink this amount when dewatered in an open excavation. Compressibility characteristics often are best evaluated by large-scale load tests.

Stresses and Deflections

Validity of Elastic Theory

Settlement analysis by some methods available requires the determination of the stresses imposed on the supporting soils by the structure, but the concepts available do not permit accurate determinations in many cases. Imposed stresses are usually determined from isotropic elastic theory *(see Art. 4.2.1),* which is valid if the load distribution remains uniform at the foundation-soil contact and the modulus of the supporting material remains constant within the full depth beneath the foundation influenced by the load. Uniform load at the soil-foundation contact exists for a completely flexible foundation such as a steel storage tank, but does not exist for rigid or semirigid loaded areas.

Stress distribution is also influenced by soil type, thin stratification, and layering. Completely rigorous solutions are not available for these conditions, and the stress variation from the theoretical can only be approximated.

Effects of Foundation Rigidity and Soil Type on Displacements

FLEXIBLE FOUNDATIONS The theoretical Boussinesq stress beneath the center of a uniformly loaded circular flexible area is twice that beneath the perimeter, except at the surface. For a uniformly loaded rectangular flexible area the stresses beneath a corner are one-fourth the center stress, and beneath the midpoint of the longer side are one-half the center stress. The result is a dish-shaped settlement pattern.

Under a flexible loaded area, clays settle in a nearly theoretical dish-shaped pattern (Fig. 6.22*a*), whereas sands settle in an irregular pattern with the larger settlements occurring near the footing edge (Fig. 6.22*c*). In clays the maximum differential settlement usually is much less than the maximum settlement, but in sands the maximum differential settlement is close to the maximum settlement.

RIGID FOUNDATIONS The settlement is more or less *uniform* over the footing-soil contact area in sands or clays (Fig. 6.22*a* and *b*). In reality this requires that the contact pressures must decrease on the inside of the loaded area and increase on the outside. The settlement under the center of a rigid circular footing is approximately 21% less than under the center of a circular uniform flexible load [Lambe and Whitman (1969)[28]]. Load-settlement relationships for rigid rectangular footings are given in Whitman and Richart (1967).[29]

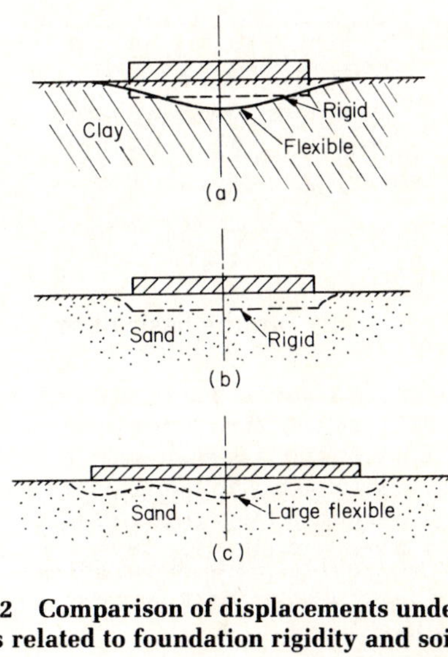

FIG. 6.22 Comparison of displacements under loaded areas as related to foundation rigidity and soil type.

Some Stress Distribution Factors

SUPERPOSITON OF STRESSES Stresses at the interface of a compressible soil layer or at a particular depth can be approximated by taking a 2:1 distribution from the horizontal as shown in Fig. 6.23*a*. The discrepancy between the elastic (Boussinesq) solution and the approximate solution decreases as the ratio of the depth considered to the size of the loaded area increases; i.e., the greatest discrepancy exists at the shallower depths. Stresses from adjacent foundations may be *superimposed* (Fig. 6.23*b*), resulting in larger settlements along the inner faces of the two footings and tilting toward each other. In critical cases, analysis with Boussinesq stresses should be performed.

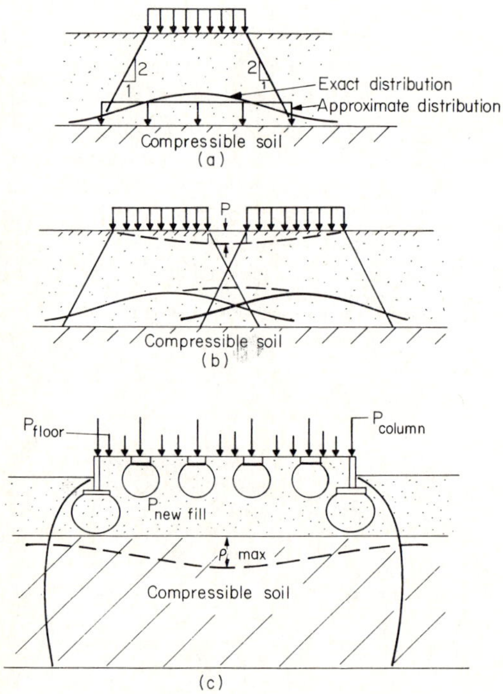

FIG. 6.23 Distribution of stresses under some typical conditions: (*a*) relationship between vertical stress below square uniformly loaded area as determined by elastic and approximate solutions; (*b*) superposition of stresses from adjacent footings causes them to tilt toward each other; and (*c*) stress distribution beneath footings and fill-supported structure. The overall load of the structure, the new fill, and the floor loads will govern settlement.

AREA EFFECTS: In Fig. 6.23*c*, settlement is governed by the area load imposed by the structure, its floor loads, and the new fill weight, rather than by footing stresses which primarily affect the compacted fill.

SOIL LAYERING EFFECTS A rigid layer overlying a less rigid layer (Fig. 4.19), such as a compact sand overlying soft clay, results in stresses at the interface beneath the loaded area significantly less than those given by the Boussinesq solution. The decrease depends on the elastic moduli ratios. Thinly layered systems, such as varved clays, are analyzed by the Westergaard solution (Fig. 4.10*b*). The relatively stronger layers serve to reinforce the system and reduce the Boussinesq stresses directly beneath the loaded area. A compressible stratum over a rigid base, such as rock, results in a significant stress increase at the interface beneath a loaded area of limited extent (Fig. 4.19) and the stresses may be as much as 40% over the Boussinesq. The foregoing discussion applies primarily to loaded areas relatively small in extent and becomes less correct as the loaded area increases. Under a uniform load of infinite extent, the vertical stress everywhere is the applied stress and there is no distribution.

Sands

Concepts

Most of the significant settlement resulting from the compression of sands occurs immediately upon load application, although long-term creep has been observed. The problem of estimating settlement in sands has been approached by numerous investigators who have developed various procedures, all of which can be considered to yield only estimates. The most reliable data for sands is obtained from full-scale load tests which should be considered on large projects with many footings.

The long-used plate-load test (*see Art. 4.4.1*) is of limited value as it only significantly stresses the soil to a depth of about 1.5 plate diameters, or a little over a meter for a 30-in-dia. plate. Soils below that depth will not be stressed as they would be with a larger footing. It is feasible, however, to perform tests at a series of depths to obtain a profile of subgrade modulus values.

Analytical Approaches

Compression in sands may be evaluated by elastic theory *(see Art. 4.3.2)*. Some approaches are based on elastic stress concepts and require values for the soil modulus and Poisson's ratio for solution. The elastic strain concept also requires a value for E_s. Values for E_s do not appear to be well established for soils as yet, but they can be estimated from charts using correlations with penetration test results (SPT, CPT) or from formulas relating penetration test data directly. Values also can be obtained from pressuremeter testing or, for low strain levels, from crosshole seismic testing. All of these methods provide in situ measurements *(see Art. 4.1.4)*. Estimating settlements in sands is discussed in Art. 4.3.3, including the plate-load test, pressuremeter, and elastic strain methods. A summary of various methods is given in Table 4.11.

As a typical example, the value of q_a for a spread footing bearing at a depth of 3 ft in a loose to medium dense silty sand is shown in Fig. 6.24, for 1 in (2.5) of settlement.

Eight methods were applied to evaluate the problem. If the stratum were buried, then p_f would be used for q_a in the various expressions given to compute settlement. In general, methods using tables and charts (nos. 1, 2, and 3) yield the more conservative values for q_a, which confirms experience indicating that allowable soil pressures in sands are usually underestimated by these methods. Their application is discussed in Art. 6.2.2. Methods 4, 5, and 6 are used infrequently and are included for comparison. Methods 7 and 8 appear to be the methods currently having the greatest merit, and method 8 has the advantage that creep effects can be evaluated. The most critical element in analysis is employing a representative value for the soil modulus, measurements of which are most commonly made in situ with the pressuremeter or obtained through correlations with the CPT.

It should be noted that interior footings are usually placed at nominal depths below grade, and their settlement should be carefully evaluated because of the likelihood of differential deflections between exterior and interior footings.

Clays

Concepts

Data from laboratory consolidation tests provide the basis for estimating settlement in saturated clays and cohesive organic soils with low organic contents. Analytical procedures permit estimating settlement magnitudes *(see Art. 4.3.4)* and consolidation theory permits estimating the time rates. Other saturated cohesive soils from which undisturbed samples can be retrieved often yield reasonably accurate results.

Analysis: Water Table below Mat Foundation Depth

An example of the procedures followed to calculate the settlement beneath the corner and center of a mat foundation due to consolidation of an underlying clay stratum is given in Fig. 6.25. The mat is bearing above the water table in sand which has been determined to contribute negligible settlement. Where there are several compressible layers, similar procedures are followed for each and their contributory compression summed. If settlement time rates are of interest, they are calculated using procedures given in Art. 4.3.4.

Analysis: Water Table above Mat Foundation Depth

Factors to be considered for the case of a structure placed in an excavation below the water table in a relatively thin sand stratum overlying a preconsolidated clay stratum are illustrated in Fig. 6.26. The initial conditions are illustrated in (1). Lowering the water table prior to excavation (2) increases overburden pressures and starts consolidation of the clay if p_c is approached or exceeded. Excavation to foundation level (3) reduces p_o by unloading and may result in foundation bottom heave. Construction of the foundation and the structure (4) causes recompression of the heaved soil plus consolidation. The net effect of recompression and consol-

Given: $P = 60T$ and N values

Find: q_a for $\rho = 1$ in

Preliminary solution to find B: From Table 6.1 select $q_a = 1.5$ tsf

$$B = (60/1.5)^{1/2} = 40^{1/2} = 6.6 \text{ ft}$$

$$D_f/B = 3/6.6 \approx 0.5$$

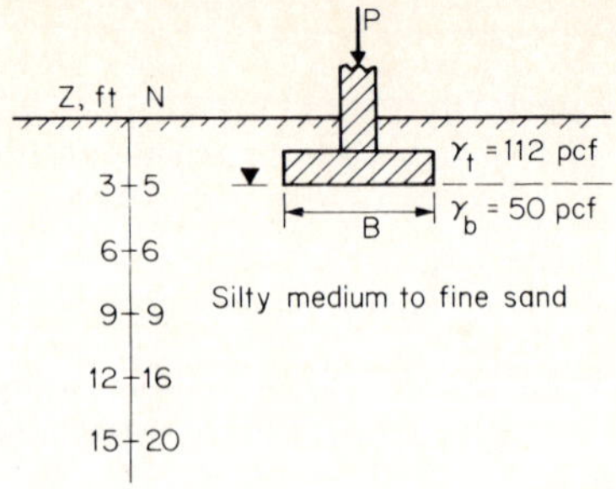

Correct N values:

$$N_1 = C_N N \text{ (Fig. 6.20)}$$

$$p'_o \text{ at 6 ft} = 3 \times 112 = 336$$

$$3 \times 50 = \underline{150}$$

$$486 \text{ psf} = 0.25 \text{ tsf}$$

$$N_1 \text{ at 6 ft} = 1.6 \times 6 = 9$$

$$\text{at 9 ft} = 14$$

Find: q_a for $B = 6.6$ ft by various methods as follows (see also Table 4.9):

Method	Source	q_a, tsf
1. Table (does not consider B)	NAVFAC, DM 7, Table 6.1	1.5
2. Code (does not consider B)	New York City, Table 6.2	1.0
3. Chart [Peck et al. (1974)[17]]	Fig. 6.18, use correction C_w	$1.0 \times 0.65 = 0.7$
4. Chart [Bazaraa (1967)[21]]	Fig. 6.19	1.8
5. Chart [Meyerhof (1956)[27]]	Fig. 6.21, $q_c = 4N = 24$, $D/B = 0.5$, $q_a/q_c = 0.08$	1.9
6. Elastic theory *(see Art. 4.3.2)*	$\rho = qaB \dfrac{1-\nu^2}{E_s} I_w$ (Eq. 4.30)	1.7 (see *a*)
7. Pressuremeter *(see Art. 4.3.3 and Eq. 4.3.5)*	$\rho = \dfrac{q_a}{4.5E_c}\left[2B_o\left(\lambda_2 \dfrac{B}{B_o}\right)\alpha + \alpha\lambda_3 B\right]$	1.8 (see *b*)
8. Elastic strain concept [Schmertmann (1970)[19]]	$\rho = C_1C_2q_a \sum_0^{2B} \left(\dfrac{I_z}{E_z}\right)\Delta z$ (Eq. 4.38)	1.8 (see *c*)

a. Estimate E_s for methods 6 and 8. There are several procedures besides actual testing:

(1) From Table 4.4 for silty sand,

$$E_s = 200 \text{ tsf for dry soil}$$

$$E_s = 200/2 = 100 \text{ tsf, saturated soil}$$

(2) From Art. 4.1.4,

$$E_s = 2.5q_c$$

where $q_c = 4N$ (Fig. 4.6)

$$E_s = 2.5 \times 4 \times \frac{9 + 14}{2} = 115 \text{ tsf}$$

(3) From compression modulus E_c (Table 4.5), use

$$E_c = 30 \text{ tsf for loose to medium dense silty sands}$$

From Eq. 4.36,

$$E_s = E_c/\alpha = 30/0.33 = 91 \text{ tsf}$$

Therefore use $E_s = 100$ tsf, and $\nu = 0.33$, $I_w = 0.82$ (rigid, square footing)

Method 6 is solved:

$$\rho = q_aB\frac{1-\nu^2}{E_s}I_w \qquad 0.08 = q_a(6.6)\frac{0.89}{100}0.82 \qquad q_a = 1.7 \text{ tsf}$$

FIG. 6.24 Settlement analysis of spread footings supported on sand.

b. Method 7 (pressuremeter):

$$\rho = \frac{q_a}{4.5E_c}\left[2B_o\left(\lambda_2\frac{B}{B_o}\right)\alpha + \alpha\lambda_3 B\right]$$

$$2.5 = \frac{q_a}{4.5(30)}(2 \times 30)\left(1.1 \times \frac{200}{30}\right)^{0.33} + (0.33 \times 1.1 \times 200) \quad \text{cm}$$

$$q_a = (2.5)(135)/188.4 = 1.8\ \text{kg/cm}^2;\ \text{for}\ \rho = 1.0\ \text{in},\ q_a = 1.8\ \text{tsf}$$

c. Method 8 (elastic strain theory) (see Fig. 4.26). Check settlement for $q_a = 1.8\ \text{kg/cm}^2$

$$\overline{p}_o = 1.6\ \text{t/m}^2;\ p = 18\ \text{t/m}^2 - 1.6\ \text{t/m}^2 = 16.4\ \text{t/m}^2$$

$$\rho = C_1C_2\,p\sum_0^{2B}\left(\frac{I_z}{E_s}\right)\Delta z = (0.95)(1.62)(1.80) = 2.77\ \text{cm} = 1.1\ \text{in}$$

$$C_1 = 1 - 0.5(\overline{p}_o/\Delta p) = 1\ \text{ms}\ 0.5(1.6/16.4) = 0.95$$

$$C_2 = 0\ \text{for immediate settlement}$$

FIG. 6.24 (*Continued*) Settlement analysis of spread footings supported on sand.

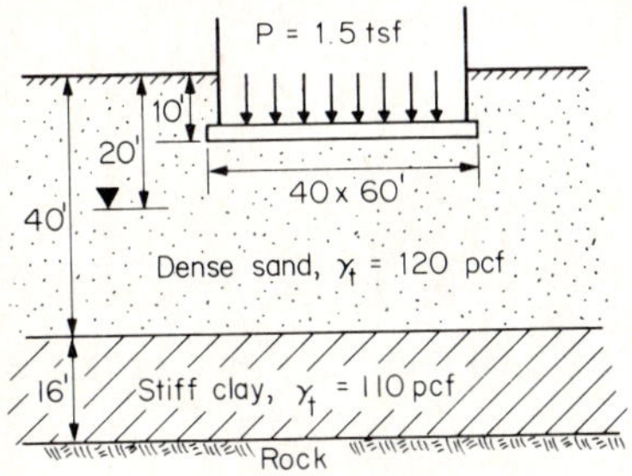

Clay Properties: w = 42%, LL = 54%, PI = 28%, U_c = 1.2 tsf, and G_s = 2.7

Find: C_c and e_o from laboratory consolidation tests or estimate from correlations.

$$C_c = 0.009(\text{LL} - 10\%) = 0.396 \qquad \text{(Eq. 4.44)}$$

$$e = wG_s/100 = 1.13\ \text{(for saturated soil)}$$

1. *Find:* $\overline{p}_o$ at mid-depth of clay

$$20 \times 0.120 = 2.40$$

$$20 \times (0.120 - 0.0624) = 1.15$$

$$8 \times (0.110 - 0.0624) = \underline{0.38}$$

$$p'_o = 3.93\ \text{ksf}$$

$$= 1.97\ \text{tsf}$$

2. *Find:* net load at base of mat

DL + permanent LL = 1.5 tsf

Soil removed:

$$10 \times 0.120 = 1.20\ \text{ksf} = \underline{0.6\ \text{tsf}}$$

$$p_n = \text{net pressure} = 0.9\ \text{tsf}$$

3. *Find* p_f at mid-depth of clay. From Fig. 5.13

a. Stress under corner:

$$m = x/z = 40/48 = 0.83$$

$$n = y/z = 60/48 = 1.25$$

$$\sigma_z = I \times p_n = 0.173 \times 0.9 = 0.156\ \text{tsf}$$

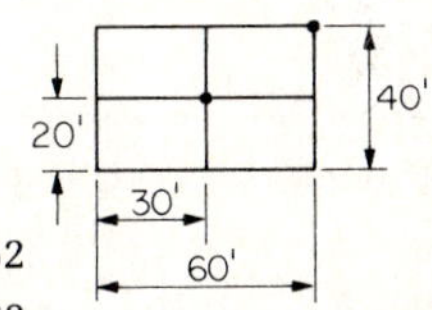

b. Stress under center:

$$m = y/z = 20/48 = 0.52$$

$$n = x/z = 30/48 = 0.63$$

$$\sigma_z = 4I \times p_n = 4 \times 0.085 \times 0.9 = 0.306\ \text{tsf}$$

4. *Calculate:* total settlement

$$\rho = \frac{H}{1 + e_o} C_c \log \frac{\overline{p}_o + p_f}{\overline{p}_o} \qquad (4.42)$$

a. At corner:

$$\rho = \frac{16 \times 12}{1 + 1.13} \times 0.396 \times \log \frac{1.97 + 0.156}{1.97}$$

$$\rho = 1.2\ \text{in}$$

b. At center:

$$\rho = \frac{16 \times 12}{1 + 1.13} \times 0.396 \times \log \frac{1.97 + 0.31}{1.97}$$

$$\rho = 2.3\ \text{in}$$

5. For bearing capacity analysis see Fig. 6.33.

FIG. 6.25 Settlement analysis for mat foundation over clay stratum.

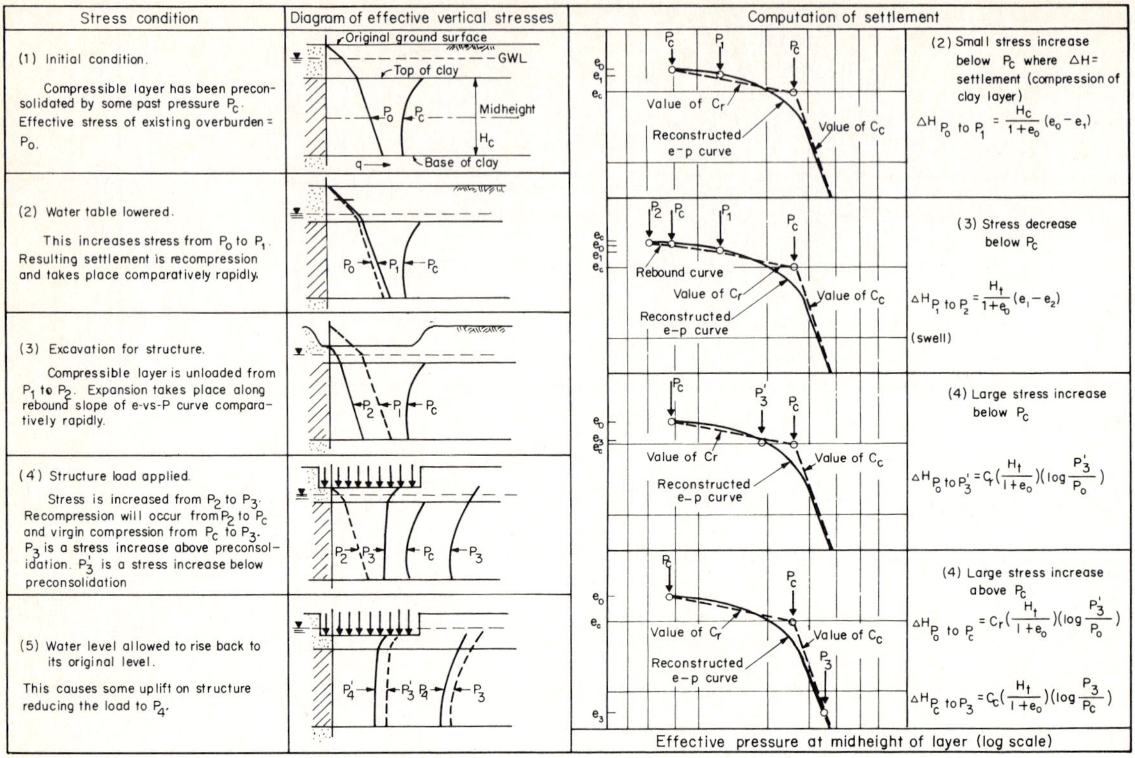

FIG. 6.26 Computation of total settlement for various loading conditions. [From NAVFAC (1982).[13]]

idation will be greater than that from consolidation alone. The applied building load, however, is reduced by the weight of the overburden removed, and if it should equal the building load there is essentially no load increase on the clay stratum over the in situ load and the result is a floating foundation.

During construction (5) the water level is allowed to return to its natural level and the imposed stresses are reduced by the buoyant force of water. Postconstruction building settlement is computed for this condition, and the possibility of building uplift or floor heave must be evaluated. For this analysis to be valid, it must be *ensured* that the water level will be permitted to return by the end of construction or objectionable settlement may result. In complexes involving a number of structures, dewatering may continue for several years, an event which may be beyond the control of the investigator who should take it into account during analysis.

Coarse-Grained Cohesive Soils

Total settlement in other materials besides sands, which are difficult to sample undisturbed, such as residual soils and glacial till with large-size particles, can be evaluated on the basis of elastic theory from pressuremeter data, with the limitation that time-delay effects cannot be accounted for. As previously mentioned, some saturated residual soils can be analyzed by consolidation theory.

6.2.4 ULTIMATE BEARING CAPACITY (q_b or q_{ult})

Basic Factors

Definition

As an increasing load is applied to a soil mass by a foundation, deformation occurs and, if the load becomes large enough, slip lines eventually begin to form, the strength along some critical surface is exceeded, and rupture and collapse occur.

Failure occurs at the ultimate foundation load Q which in terms of the foundation area is referred to as the *ultimate bearing pressure* ($q_b = Q/A$) or *ultimate bearing capacity*. q_b (or q_{ult}) is not a constant number applicable to a given soil, but rather is extremely variable depending upon such factors as the foundation depth and dimensions, the direction and rate of applied loads, soil homogeneity and properties, and the water-table depth. The problem usually is solved by limit equilibrium concepts *(see Art. 3.5.2)*.

Foundation Design Criteria

Shallow foundations normally are designed to ensure that settlement occurring from soil compression will be tolerable and the allowable bearing value q_a is selected by applying procedures described in the preceding articles. There are many structures, however, such as grain storage silos, storage tanks, or embankments, that may be subjected to rapidly applied loads that do not permit consolidation and the stability against shear failure, and soil rupture must be evaluated. The allowable bearing value normally is substantially less than the ultimate value, on the order of one-half to one-third of the ultimate, but it is prudent in foundation problems to examine the possibility of a bearing-capacity failure.

Deep foundations usually are designed by first determining the capacity of single units, applying a generous value for FS, and then evaluating single-unit or group settlement (and group capacity), depending upon the spacing between units (Chap. 7).

Shear Failure, Three General Modes

General Shear Failure

General shear failure is characterized by a well-defined failure surface extending to the ground surface with a failure zone consisting of a series of wedges and slip surfaces as shown in Fig. 6.27. A bulge will appear on the ground surface adjacent to the foundation, and unless the footing is tied rigidly to the structure it will rotate and tilt.

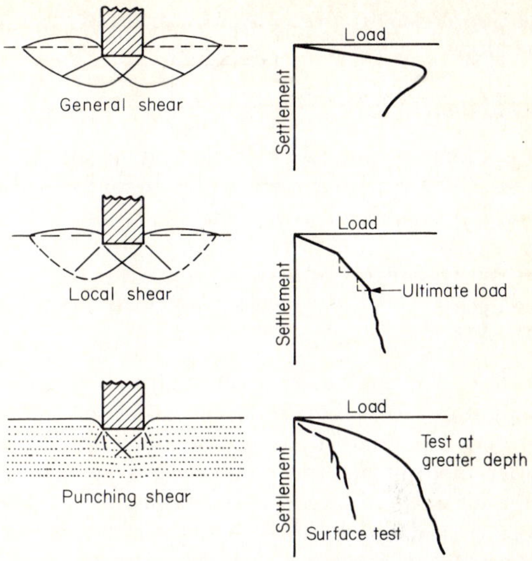

FIG. 6.27 Bearing-capacity failure modes for shallow foundations. [From Vesić (1963).[30]]

Failure is sudden. It occurs in soils that are relatively incompressible with a definite shear strength and in saturated, normally consolidated clays if the load is applied rapidly enough to cause undrained conditions to prevail.

Local Shear Failure

In local shear a failure pattern of wedges and slip surfaces starts at the edge of the footing, as in the general shear case, and there is bulging of the ground surface and significant footing displacement. The slip surfaces do not reach the ground surface, however, ending somewhere in the soil mass. The footing embeds itself deeper and deeper into the soil mass (Fig. 6.27) in a condition transitional between general and punching shear.

Punching Shear Failure

Punching shear involves rapid vertical footing displacement as load is applied but does not result in adjacent ground heave (Fig. 6.27). It can occur under a footing placed on the surface of very loose sand, a footing placed on dense sand at a depth imposing high loads, a footing bearing in a sand stratum underlain by a weak, compressible stratum, or in weak soils under slow load application permitting the drained state to develop.

Prediction

The shear failure mode is related to soil compressibility in that relatively incompressible soils fail in general shear and compressible soils usually fail in local or punching shear. There are still, however, no general numerical criteria for prediction of which mode will occur [Vesić (1975)[31]].

Vesić (1973)[32] proposed a rational parameter for evaluation of the relative compressibility, termed the *rigidity index*, which relates the shear modulus to the soil-strength parameters and the vertical stress at the depth of interest. Recent studies by Ismael and Vesić (1981)[33] indicate that the analysis proposed by Vesić can successfully determine the compressibility factors for surface foundations on compressible soils, but for footings with substantial overburden pressure the compressibility factors are underestimated (overly conservative). In addition, they concluded that, with increasing overburden pressure, the failure mode changes to local or punching shear, regardless of soil compressibility. They also concluded that Terzaghi's approach, given under Bearing-Capacity Analysis, is suitable for cases with overburden pressure and fairly conservative for surface or near-surface footings.

Bearing-Capacity Theory

Limitation

Failure of a loaded foundation is defined clearly only for the case of *general shear* where a peak ultimate load in relatively incompressible soil is reached simultaneously with the appearance of slip lines, foundation collapse, and surface bulging (Fig. 6.27). In contrast, the failure load for local and punching shear is less clearly defined and difficult to establish as the failure curves have no well-defined peaks (Fig. 6.27). When the theory is used for evaluations of relatively compressible soils, reduction factors are required.

Theoretical Basis

Bearing-capacity theory is based on solutions developed for the rigid-plastic solid of the classical theory of plasticity in which the solid is assumed to exhibit no deformation prior to shear failure and plastic flow under constant stress after failure. The problem is illustrated in Fig. 6.28.

At ultimate load Q_b, imposed on a continuous footing of infinite length L, a state of plastic equilibrium develops in the soil. The active Rankine zone (I) pushes the Prandtl zone (II) radially, which in turn pushes the passive Rankine zone (III) upwards. The distribution of the compressive and shearing stresses that develop is shown in Fig. 6.29. The shape of the Zone II boundary (CD) depends on ϕ and the ratio $\gamma_2 B/q_s$ where q_s is a surcharge (usually $q_s = \gamma_1 D_f$). An important aspect of the geometry is the angle ψ since it significantly affects the solution for q_b (unit ultimate bearing value). Terzaghi (1943)[34] considered that $\psi = \phi$, but it has been shown that ψ actually is somewhat greater than $45 + \phi/2$, representing slip lines in the active state (Fig. 3.22*b*). When $\gamma B/q_s \rightarrow 0$, for the condition of a hypothetical weightless soil as solved by Prandtl (see *Art. 3.5.1*), the failure surface is a log spiral. For the $\phi = 0$ case, the curve is a circular arc.

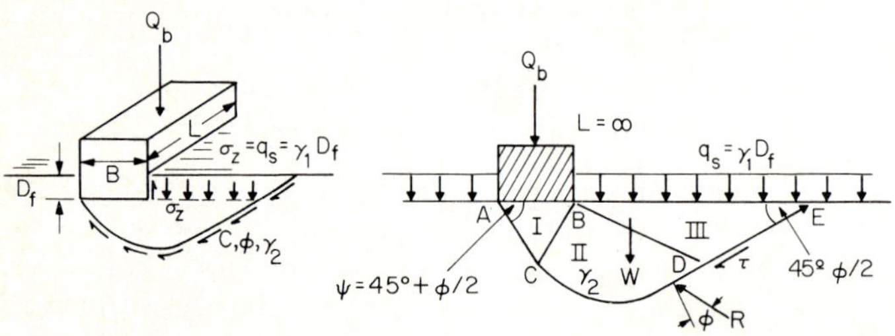

FIG. 6.28 The problem of the bearing capacity of shallow foundations failing in general shear with parameters *c* and ϕ. Boundaries are simplified. I = active Rankine zone; II = Prandtl zone; III = passive Rankine zone.

The bearing-capacity problem is really one of passive earth pressure. The shearing stresses at the contact face BC of Zone I are expressed by Coulomb's equation (Eq. 3.8) in which shear stress s is replaced by the total passive resistance and the normal stress σ_n is replaced by the normal component of the passive earth pressure per unit of area of the contact face. At the failure instant, the pressure on BC is equal to the resultant of the passive resistance plus cohesion along the failure surface BC. The total passive resistance includes the weight of the mass in Zones II and III, a cohesive resistance along the failure surface AE, and a surcharge q_s from the overburden soil.

Terzaghi's Expression

A closed form for this problem in plasticity has been found only for the case of weightless soil [Prandtl (1921)[37]]. The real nature of soil with weight can be considered approximately (with an error <20% for ϕ between 30 and 40°) by Terzaghi's (1943)[34] expression for drained loading conditions for a long rectangular footing (the Buisman-Terzaghi equation):

$$q_b = cN_c + qN_q + 0.5\gamma BN_\gamma \qquad (6.10)$$

where q_b = ultimate bearing capacity = Q_b/B
B = width of smaller footing side
c = cohesion
γ = effective soil unit weight
q = overburden stress at footing base = γD_f
N_c, N_q, N_γ = the *bearing-capacity factors*, dimensionless functions of ϕ

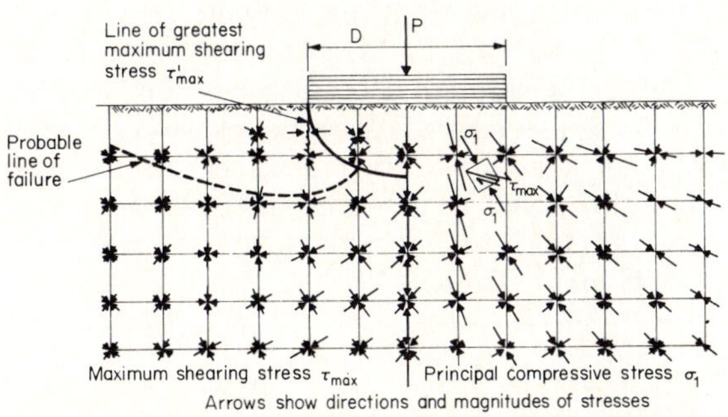

FIG. 6.29 Distribution of compressive and shearing stresses beneath a circular loaded area during general shear. [After Converse (1939)[35]; from Tschebotarioff (1973).[36] Reprinted with permission of McGraw-Hill Book Company.]

The three parts of Eq. 6.10 provide for soil cohesion, foundation depth (overburden pressure), and foundation width, with ϕ a factor in all three parts. Even a small amount of cohesion substantially increases q_b. The first term is a function of parameters ϕ and c, the second term is a function of the vertical confining effective stress ($\bar{\sigma}_v = \gamma D_f$) at the elevation of the footing base, and the third term is a function of the footing width on the average weight of the soil mass under the footing. All three terms also are a function of the assumed shape of the rupture surface and of ϕ. [For *deep foundations (see Art. 7.3.3)*, the third term is very small in comparison to the second and is neglected.]

A number of values are available for the *bearing-capacity factors*, which are exclusively a function of ϕ and the failure surface geometry. The current trend of many practitioners is to use values developed by Prandtl (1921)[37] and Reissner (1924)[38] for N_c and N_q and those developed by Caquot and Kerisel (1953)[39] for N_γ, as given in Table 6.7.

Bearing-Capacity Analysis

Factors to Consider

- Drained versus undrained conditions and, for the drained case, general versus local shear
- Foundation shape and depth and base inclination
- Loading eccentricity and inclination
- Ground-slope condition
- Groundwater-table condition

Drained Loading Conditions

Drained loading conditions occur normally where loads are applied gradually as the structure is erected and parameters c and ϕ are mobilized. *General shear* prevails for relatively strong materials, and bearing capacity is commonly found for continuous footings by Eq. 6.10. For square and circular footings Terzaghi (1943)[34] proposed

$$q_b = \frac{Q}{A} = 1.2\, cN_c + qN_q + s_\gamma BN_\gamma \qquad (6.11)$$

where the shape factor s_γ = 0.4 for square footings and 0.6 for circular footings.

TABLE 6.7
BEARING-CAPACITY FACTORS

ϕ	N_c	N_q	N_γ	N_q/N_c	tan ϕ
0	5.14	1.00	0.00	0.20	0.00
1	5.38	1.09	0.07	0.20	0.02
2	5.63	1.20	0.15	0.21	0.03
3	5.90	1.31	0.24	0.22	0.05
4	6.19	1.43	0.34	0.23	0.07
5	6.49	1.57	0.45	0.24	0.09
6	6.81	1.72	0.57	0.25	0.11
7	7.16	1.88	0.71	0.26	0.12
8	7.53	2.06	0.86	0.27	0.14
9	7.92	2.25	1.03	0.28	0.16
10	8.35	2.47	1.22	0.30	0.18
11	8.80	2.71	1.44	0.31	0.19
12	9.28	2.97	1.69	0.32	0.21
13	9.81	3.26	1.97	0.33	0.23
14	10.37	3.59	2.29	0.35	0.25
15	10.98	3.94	2.65	0.36	0.27
16	11.63	4.34	3.06	0.37	0.29
17	12.34	4.77	3.53	0.39	0.31
18	13.10	5.26	4.07	0.40	0.32
19	13.93	5.80	4.68	0.42	0.34
20	14.83	6.40	5.39	0.43	0.36
21	15.82	7.07	6.20	0.45	0.38
22	16.88	7.82	7.13	0.46	0.40
23	18.05	8.66	8.20	0.48	0.42
24	19.32	9.60	9.44	0.50	0.45
25	20.72	10.66	10.88	0.51	0.47
26	22.25	11.85	12.54	0.53	0.49
27	23.94	13.20	14.47	0.55	0.51
28	25.80	14.72	16.72	0.57	0.53
29	27.86	16.44	19.34	0.59	0.55
30	30.14	18.40	22.40	0.61	0.58
31	32.67	20.63	25.99	0.63	0.60
32	35.49	23.18	30.22	0.65	0.62
33	38.64	26.09	35.19	0.68	0.65
34	42.16	29.44	41.06	0.70	0.67
35	46.12	33.30	48.03	0.72	0.70
36	50.59	37.75	56.31	0.75	0.73
37	55.63	42.92	66.19	0.77	0.75
38	61.35	48.93	78.03	0.80	0.78
39	67.87	55.96	92.25	0.82	0.81
40	75.31	64.20	109.41	0.85	0.84
41	83.86	73.90	130.22	0.88	0.87
42	93.71	85.38	155.55	0.91	0.90
43	105.11	99.02	186.54	0.94	0.93
44	118.37	115.31	224.64	0.97	0.97
45	133.88	134.88	271.76	1.01	1.00
46	152.10	158.51	330.35	1.04	1.04
47	173.64	187.21	403.67	1.08	1.07
48	199.26	222.31	496.01	1.12	1.11
49	229.93	265.51	613.16	1.15	1.15
50	266.89	319.97	762.89	1.20	1.19

Local shear applies to weaker materials and Terzaghi accounted for soil compressibility by reducing the ultimate strength parameters c_g for ϕ_g (general shear) as follows:

- Soft to firm clays:

$$c_{local} = \tfrac{2}{3} c_g \qquad (6.12)$$

- Loose sands:

$$\tan \phi_{local} = \tfrac{2}{3} \tan \phi_g \qquad (6.13)$$

This results in an abrupt transition from general to local shear but is considered to be suitable for cases with overburden pressure although conservative for surface to near-surface loadings [Ismael and Vesić (1981)[33]]. See, for example, case 2 in Fig. 6.33.

Sands (c = 0)

In sands, where $c = 0$ (see for example, case 3 in Fig. 6.33), Eq. 6.10 for the gross ultimate bearing capacity becomes

$$q_b = \gamma D_f N_q + 0.5\gamma B N_\gamma \qquad (6.14)$$

The *net* ultimate bearing capacity $q_b' = q_b - \gamma D_f$, or the gross capacity minus overburden pressure at footing level. In terms of q_b', Eq. 6.14 can be expressed in the form [Peck et al. (1974)[17]] for the allowable bearing value as

$$q_a = \frac{q_b'}{FS} = \left[\gamma (N_q - 1) \frac{D_f}{B} + 0.5\gamma N_\gamma \right] \frac{B}{FS} \qquad (6.15)$$

where FS = safety factor. For a particular value of D_f/B and a given sand deposit, the expression within the brackets is a constant. Thus, the relation between footing width B and the net allowable soil pressure q_a for a given value of FS can be plotted as in Fig. 6.18 as a family of straight lines radiating from the origin. In Fig. 6.18, FS = 2. Each line corresponds to a different value for ϕ as determined roughly from the N value of the SPT as given in Fig. 6.30. Note that the values for N_γ given in Fig. 6.30 are lower than those given in Table 6.7.

Bearing-Capacity Equation: General Form

In recent years Eq. 6.10 has been expanded by various investigators to provide for the various factors that influence q_b as follows:

$$q_b = cN_c s_c b_c i_c g_c + qN_q s_q b_q i_q g_q + 0.5\gamma B N_\gamma s_\gamma b_\gamma i_\gamma g_\gamma \qquad (6.16)$$

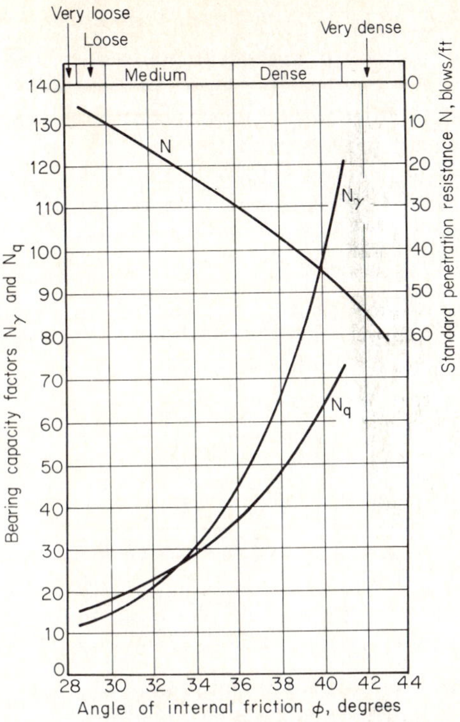

FIG. 6.30 Curves showing the relationship between bearing-capacity factors and ϕ, as determined by theory, and rough empirical relationships between bearing-capacity factors or ϕ and values from the SPT (*N*). [From Peck et al. (1974).[17] Reprinted with permission of John Wiley & Sons, Inc.]

where the factors are *s*, shape; *b*, base inclination; *i*, inclination and eccentricity of load; and *g*, ground slope.

Undrained Loading (ϕ = 0)

Undrained conditions (see, for example, case 1 in Fig. 6.33) occur in saturated intact clays when loads are applied rapidly, such as the cases with high live loads when storage tanks or silos are filled quickly with a product or during hydrostatic testing of storage tanks. Since for $\phi = 0$, $N_q = 1.0$ and $N_\gamma = 0$ (Table 6.7), the bearing capacity from Eq. 6.10 is expressed as

$$q_b = N_c s_c s_u + q \tag{6.17}$$

where s_u = undrained shear strength (*see Art. 3.3.3*)
$N_c s_c$ = 5.14 for continuous footings
$N_c s_c$ = 6.2 for round or square footings (Table 6.8)
$N_c s_c = 5.14(1 + 0.2B/L)$ for footings of width *B* and length *L* (Table 6.8)

When a fill is placed over a uniform deposit of soft soils, embankment stability (*see Art. 8.2.4*) can be estimated from Eq. 6.17, and expressed as

$$q_{ult} = 5.14 s_u/FS \tag{6.18}$$

Shape Factors (s_c, s_q, s_γ)

Equation 6.10 was developed for an infinitely long strip footing. Semiempirical factors have been developed for rectangular, square, and round footings for use in Eq. 6.16, as given in Table 6.8.

Footing Base Inclination Factors (b_c, b_q, b_γ)

Footing-base inclination results in a reduction in bearing capacity. The reduction may be estimated by applying the following factors from Vesić (1970),[41] (1975)[31]:

$$b_c = b_\gamma - \frac{1 - b_\gamma}{N_c \tan \phi} \tag{6.19}$$

and

$$b_q = b_\gamma = (1 - \alpha \tan \phi)^2 \tag{6.20}$$

TABLE 6.8
SHAPE FACTORS AND BEARING-CAPACITY EQUATION*

Base shape	s_c	s_q	s_γ
Strip	1.0	1.0	1.0
Rectangle	$1 + (B/L)(N_q/N_c)$	$1 + (B/L) \tan \phi$	$1 - 0.4B/L$
Square and circle	$1 + (N_q/N_c)$	$1 + \tan \phi$	0.60

*After DeBeer (1967)[40] as modified by Vesić (1970).[41]

where α, base inclination (Fig. 6.31), is given in radians and is less than 45°. The limiting value for the $\phi = 0$ case is

$$b_c = 1 - [2\alpha/(\pi + 2)] \tag{6.21}$$

Ground Slope Factors (g_c, g_q, g_γ)

If a footing is adjacent to a slope there will be a reduction in bearing capacity [Vesić (1970)[41]]. The use of the equations given below, however, does not consider the effect of shearing stresses in the slope which may be significant where the slope angle $i > \phi/2$ (Fig. 6.31). For these cases the general slope instability is examined in Art. 6.2.5.

$$g_c = g_\gamma - \frac{1 - g_\gamma}{N_c \tan \phi} \tag{6.22}$$

$$g_q = g_\gamma = (1 - \tan i)^2 \tag{6.23}$$

where i, the slope angle, is given in radians and is less than 45° and less than ϕ. The limiting value for g_c for the $\phi = 0$ case is

$$g_c = 1 - [2i/(\pi + 2)] \tag{6.24}$$

The existence of ground slope for the $\phi = 0$ case requires the addition of the weight term in the bearing-capacity equation, which becomes

$$q_b = N_c s_c s_u g_c + q \cos i + 0.5\gamma B N_\gamma s_\gamma \tag{6.25}$$

where

$$N_\gamma = -2 \sin i \tag{6.26}$$

Equations 6.23 through 6.26 are theoretically valid for $i < 45°$ and $i < \phi$. See, for example, case 4 in Fig. 6.33.

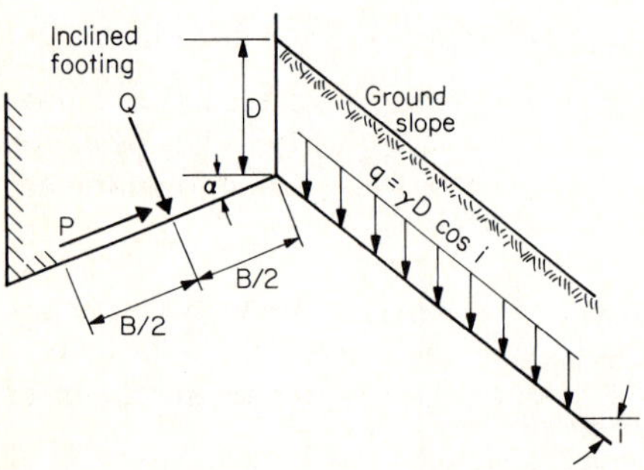

FIG. 6.31 Conditions of an inclined footing and sloping ground. [From Vesić (1970).[41]]

Inclination and Eccentricity Factors (i_c, i_q, i_γ)

Load inclination and eccentricity e have the effect of reducing the bearing capacity [Vesić (1970)[41]]. In general, footings should be designed so eccentricity e from the footing center (Fig. 6.32) does not exceed $B/6$ or there will be a danger of lifting of the opposite footing edge and loss of contact. The inclination correction factors are expressed as follows:

$$i_c = i_q - \frac{1 - i_q}{N_c \tan \phi} \tag{6.27}$$

and for $\phi = 0$,

$$i_c = 1 - \frac{mP}{B'L'cN_c} \tag{6.28}$$

$$i_q = \left[1 - \frac{P}{Q + B'L'c \cot \phi}\right]^m \tag{6.29}$$

$$i_\gamma = \left[1 - \frac{P}{Q + B'L'c \cot \phi}\right]^{m+1} \tag{6.30}$$

where P = horizontal component of load
Q = vertical component of load
B' = effective width $= B - 2e$ (load on short side, L unchanged)
L' = effective length $= L - 2e$ (load on long side, B unchanged)
m = exponential factor relating B/L or L/B ratios
For load on short side: $m_b = (2 + B/L)/(1 + B/L)$
For load on long side: $m_L = (2 + L/B)/(1 + L/B)$

In a retaining-wall problem (*see Art. 9.2.3 and Fig. 9.26*) loads P and Q are found. In other cases Vesić (1975)[31] suggests using $P = 0.5P_{max}$ where

$$P_{max} \leq Q \tan \phi + A'c_a \tag{6.31}$$

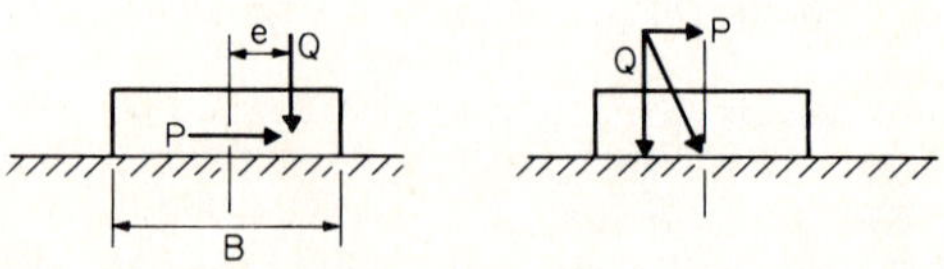

FIG. 6.32 Eccentric and inclined footing loads.

where A' = effective area such as $B'L$
c_a = footing-soil adhesion
$c_a = s_u$, undrained conditions
$c_a = 0$, undrained conditions

and P_{max} represents a limit for Q.

Note that if $e > B/4$, uplift begins to occur on the side with less load, and for design, e normally should not exceed $B/6$ (see, for example, Fig. 9.26).

The foregoing relationships proposed by Vesić (1970),[41] (1975)[31] yield values of i factors slightly higher than those proposed by Brinch Hansen (1970).[41a]

Depth Factors

The foregoing relationships ignore the shearing resistance of the overburden, which is normally justified by the fact that the overburden soil is weaker than the bearing stratum. Depth correction factors developed by various investigators are based primarily on data from driven piles and are not considered applicable to shallow foundations. Therefore, Vesić (1975)[31] advises that depth factors not be introduced into the bearing-capacity analysis.

Groundwater Table

Bearing capacity is reduced by a high water table, which reduces overburden pressures and destroys apparent cohesion. It is accounted for as follows:

$$z_w > B \qquad \text{no influence} \tag{6.32}$$

$$z_w \leq B \qquad \text{use } \gamma = \gamma_b + \left(\frac{z_w}{B}\right)(\gamma_t - \gamma_b) \tag{6.33}$$

$$z_w \leq 0 \qquad \text{use } \gamma = \gamma_b$$

where z_w = highest groundwater depth from footing base of width B.

Two-Layered Soil Profiles

When foundations are to be supported on distinctly different soil layers, the most critical case is a footing bearing in a relatively rigid layer underlain by one substantially weaker, such as a stiff clay or a dense sand over a soft-clay stratum. In this case, if the strong stratum is not adequately thick, failure is likely to occur by punching. Vesić (1970),[41] (1975)[31] describes several approaches to the problem, some of which are relatively complex.

The bearing capacity for clays is given by the general expression

$$q_b = s_{u1}N_m + q \tag{6.34}$$

where s_{u1} = undrained shear strength of the upper layer
N_m = a modified bearing-capacity factor which depends on the ratio of the shear strengths of the two layers $k = s_{u2}/s_{u1}$, the relative thickness of the upper layer H/B, and foundation shape. (H = depth between footing base and surface of second layer.)

For the case of a *stiff clay layer over a soft clay layer* (see case 5, Fig. 6.33), assuming simple punching shear around the footing perimeter, N_m is expressed as

$$N_m = 1/\beta + ks_cN_c \text{ (where } N_m \leq s_cN_c) \tag{6.35}$$

where

$$\beta = BL/[2(B + L)H] \tag{6.36}$$

which is termed the *punching index* and $k = s_{u2}/s_{u1}$, which is termed the *shear strength ratio*.

For the case of a *cohesionless sand* as the upper layer (see case 6, Fig. 6.33), assuming that slip lines will be vertical in the sand, with ϕ = 25 to 50°, q_b is expressed

$$q_b = q_b'' \exp\{0.67\,[1 + (B/L)](H/B)\} \tag{6.37}$$

where q_b'' = bearing capacity of underlying clay layer for a fictitious footing of the same size and shape as the actual footing resting on its surface

The critical depth of the upper layer, beyond which the bearing capacity will be little affected by the presence of the lower layer, can be given by:

$$(H/B)_{crit} = \frac{3B \ln (q_b'/q_b'')}{2[1 + (B/L)]} \tag{6.38}$$

Case 1: Deep firm clay stratum, undrained conditions

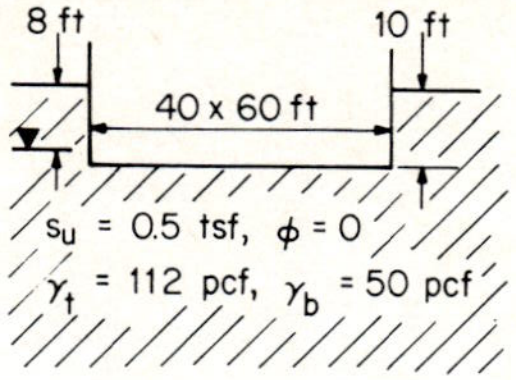

$q_b = N_c s_c s_u + q$ (Eq. 6.17) B/L = 0.67

$q = [(8 \times 12) + (2 \times 50)]/2000 = 0.50$ tsf

$s_c = 1 + (40/60)(0.2) = 1.13$ (Table 6.8)

$q_b = (5.14 \times 1.13 \times 0.5) + 0.5 = 3.4$ tsf ←

Case 2: Deep firm clay stratum, drained conditions (use local shear)

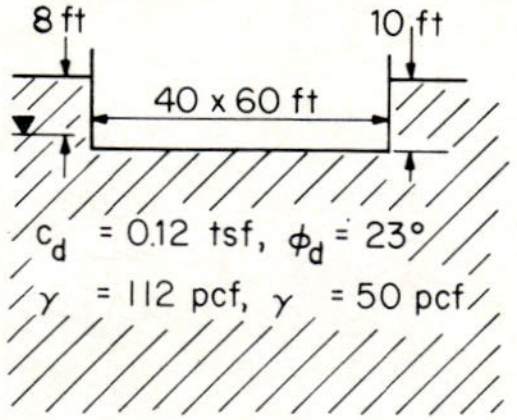

$q_b = cN_c s_c + qN_q s_q + 0.5\,\gamma B N_\gamma s_\gamma$ (Eq. 6.16)

$2/3\,c_d = 0.08$ tsf, $2/3\,\phi = \tan^{-1}(0.67)(0.42) = 16°$

$N_c = 11.63$, $N_q = 4.34$, $N_\gamma = 3.06$, $N_q/N_c = 0.37$

$s_c = 1 + (0.67)(0.33) = 1.22$ (Table 6.8)

$s_q = 1 + (0.67)(0.23) = 1.15$, $s_\gamma = 1 - 0.4\,(0.67) = 0.73$

$q_b = (0.08 \times 11.63 \times 1.22) + (0.5 \times 4.34 \times 1.15)$
$+ (0.5 \times 50/2000 \times 40 \times 3.06 \times 0.73 = 4.75$ tsf ←

Case 3: Deep sand stratum

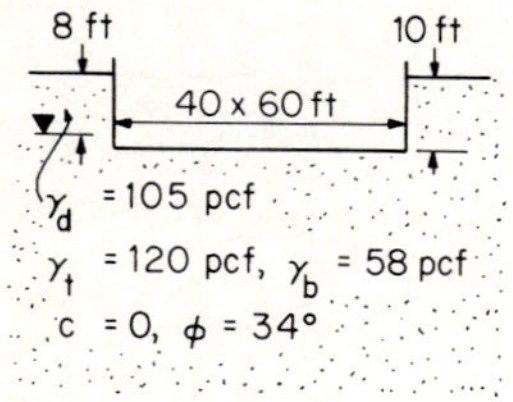

$q_b = qN_q s_q + 0.5\gamma\, BN_\gamma s_\gamma$ (Eq. 6.16 for c = 0)

$q = [(8 \times 105) + (2 \times 58)]/2000 = 0.48$ tsf

$N_q = 29.44$, $N_\gamma = 41.06$,

$s_q = 1 + (0.67)(0.67) = 1.45$, $s_\gamma = 0.73$

$q_b = (0.48 \times 29.44 \times 1.45) + (0.5 \times 58/2000 \times 40 \times 41.06 \times 0.73) = 38.0$ tsf ←

Case 4: Ground slope for Case 1 (firm clay with undrained conditions)

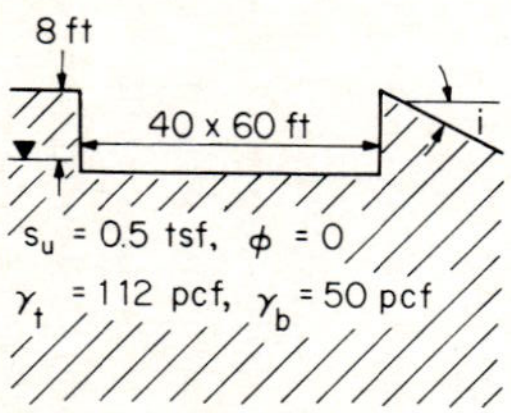

$q_b = N_c s_c q_c s_u + q \cos i + 0.5\gamma\, BN_\gamma s_\gamma$ (Eq. 6.25)

Slope angle i = 20° = 0.349 rad

$q_c = 1 - 2i/(\pi + 2) = 1 - (2 \times 0.349/3.14 + 2) = 0.86$

$N = -2 \sin i = -2\,(0.349) = -0.70$ (Eq. 6.26)

$q_b = (5.14 \times 1.13 \times 0.86 \times 0.5) + (0.5 \times 0.94)$
$-(0.5 \times 50/2000 \times 40 \times 0.7 \times 0.73) = 2.7$ tsf ←

Case 5: Two-layered profile, stiff clay over soft clay, undrained conditions

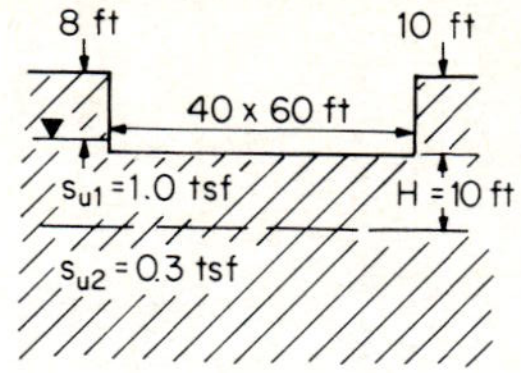

$q_b = s_{u1} N_m + q$ (Eq 6.34)

$N_m = 1/\beta + k\, s_c N_c$ (Eq. 6.35)

$\beta = BL/[2(B + L)H] = (40 \times 60)/(2 \times 100 \times 10) = 1.2$

$k = s_{u2}/s_{u1} = 0.3$, q = 0.5 from Case 1

$N_m = 1/1.2 + (0.3 \times 1.13 \times 5.14) = 2.57$

$q_b = (1.0 \times 2.57) + 0.5 = 3.07$ tsf ←

Case 6: Two-layered profile, medium dense sand over stiff clay (Fig. 6.25)

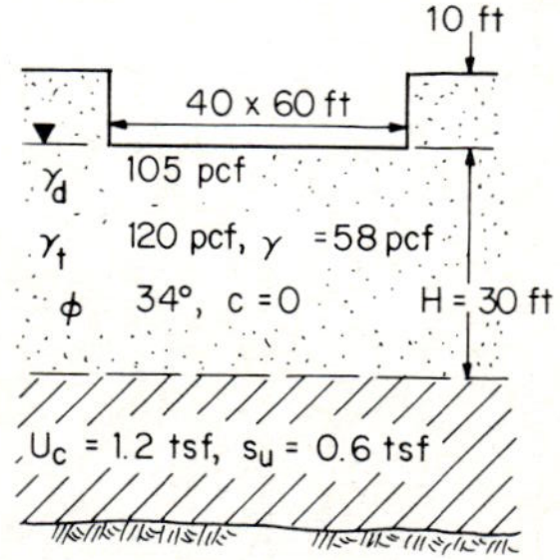

1. Find q for sand from Eq. 6.16 for c = 0
   Use bearing capacity and shape factors from Case 3

   $q = 10 \times 105/2000 = 0.53$ tsf

   $q_b = (0.53 \times 29.44 \times 1.45) + (0.5 \times 0.029 \times 40 \times 41.06 \times 0.73) = 41$ tsf

2. For the clay

   $q_b = q_b'' \exp\{0.67\,[1 + (B/L)](H/B)\}$ (Eq. 6.37)

   $q_b'' = N_c s_c s_u + q$ (Eq. 6.17)

   $= (5.14 \times 1.13 \times 0.6) + [(10 \times 105) + (30 \times 58)]/2000 = 4.9$ tsf

3. Find critical depth (Eq. 6.38)

   $H_{crit} = \dfrac{3 \times 40 \ln(41/4.9)}{2[1 + (0.67)]} = \dfrac{255}{3.34} = 76 \text{ ft} > 30 \text{ ft}$

   Therefore, q_b will be affected by the presence of the clay layer.

4. Find q (From Eq. 6.37)

   $q_b = 4.9 \exp\{0.67\,[1 - 0.67]\,(30/40)$

   $= 4.9 \exp 0.84$

   $= 4.9\, e^{0.84} = 11.35$ tsf ←

FIG. 6.33 Examples of bearing-capacity analysis for various conditions for foundation with B = 40 ft, L = 60 ft, and D_f = 10 ft.

Factors of Safety (FS)

The allowable bearing pressure q_a is defined as

$$q_a = q_b/\text{FS} \tag{6.39}$$

In selecting values for FS, a number of factors need to be considered including the risk involved with failure (disastrous to serious), the degree of confidence in the adequacy of the data (accurate measurements of soil parameters, stratigraphy well-defined), and the likelihood of the total plus live loads really occurring. Some general conditions are as follows:

FS = 2 applies generally to cases where soil conditions are well-defined, the hazard is not great, and the full design loads are unlikely.

FS = 3 applies when soil conditions are well-defined, failure risk is high, and the design loads are likely to occur.

FS = 4 applies when soil conditions are not well-defined, failure risk is high, and the full design loads are likely.

A settlement analysis is required to assure that q_a provides for settlement within tolerable limits.

Pressuremeter Analysis

In using pressuremeter data to determine the ultimate bearing capacity, it is assumed that q_b is a function of the limit pressure P_L [see Art. 3.4.4 of Hunt (1984)[7]], q_b is expressed by Menard (1965)[42] as follows:

$$q_b = p_o + k_g(P_L - p_{oh}) \tag{6.40}$$

where p_o = total overburden pressure
k_g = bearing-capacity factor, a function of soil type and foundation geometry (Fig. 6.34)
P_L = limit pressure (within a zone extending 1.5 times the foundation width above and below foundation level)
p_{oh} = total horizontal earth pressure at foundation level

FS = 3 is applied to the term $k_g(P_L - p_{oh})$. The term $(P_L - p_{oh})$ is referred to as the *net limit pressure* (P_{Ln}); typical values for various soil types are given in Table 6.9. In obtaining values for k_g from Fig. 6.34, the *equivalent depth* D_{fe} is first determined. In a uniform material, D_{fe} = foundation depth D_f. Additional discussion can be found in CGS (1978),[3] Baguelin et al. (1978),[43] and SOL SOILS (1975).[44]

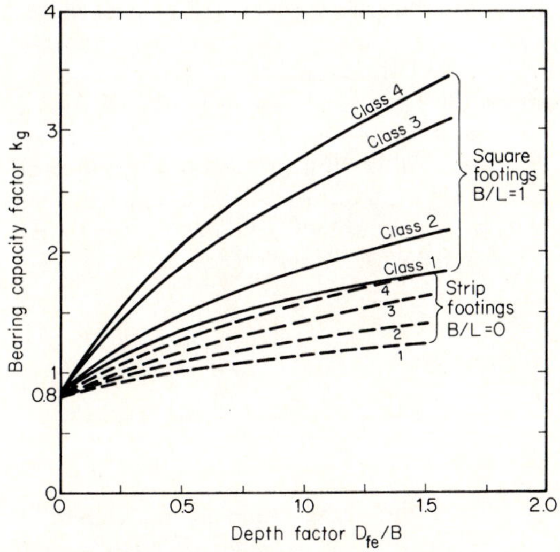

FIG. 6.34 **Factor k_g for various soil types versus depth factor D_{fe}/B for evaluating bearing capacity from pressuremeter data.** [After SOLS SOILS (1975).[44]]

Equivalent depth D_{fe}:

$$D_{fe} = \frac{1}{P_{Le}} \int_0^{Df} P_{Ln}(z) \tag{6.41}$$

where P_{Le} = equivalent limit pressure
$= \sqrt[3]{(P_{L1})(P_{L2})(P_{L3})}$
P_{Ln} = net limit pressure at depth z

P_{L1}, P_{L2}, and P_{L3} are also net limit pressure $(P_L - p_{oh})$

$P_{L1} = P_L$ at elevation 1 width B above foundation level

$P_{L2} = P_L$ at foundation level

$P_{L3} = P_L$ at depth of 1 width B below foundation level

TABLE 6.9
SOIL CATEGORIES AND TYPICAL VALUES OF NET LIMIT PRESSURE P_{LN}*

Material	Class	Net limit pressure P_{Ln},† kg/cm²
Clays: soft to firm	1	0–12
Silts: loose		0–7
Sands: loose	2	4–8
Silts: dense		12–30
Clays: stiff		18–40
Rock: soft, decomposed, friable		10–30
Sands and gravels: medium dense	3	10–20
Rock: moderately soft, decomposed		40–100
Sands and gravels: very dense	4	30–60
Rock: medium to high strength		60–100+

*After Menard (1965)[42] and CGS (1978).[3]
†$P_{Ln} = P_L - P_{oh}$.

Pressuremeter testing has several limitations [see Art. 3.3.4 of Hunt (1984)[7]]. The severest effects on measured values for P_L are from material anisotropy and from hole boundary conditions, and values may not be representative.

6.2.5 FOUNDATIONS ON SLOPES

Analysis

Assessment of Overall Slope Stability

A footing bearing on a slope or behind it, as shown in Fig. 6.35*a*, presents a two-phase analytical problem: (1) assessment of the stability of the overall slope due to the imposed concentrated loads and (2) evaluation of the bearing capacity of the footing (Fig. 6.35*b*).

Slope stability is assessed by using procedures described in Chap. 8, such as the modified Bishop method in which the footing imposing the *allowable bearing pressure* is placed over a slice. If the slope is found to be potentially unstable then stabilization measures are required as discussed in Chap. 8 and 9. *Creep*, the slow downslope movement of a shallow zone, must

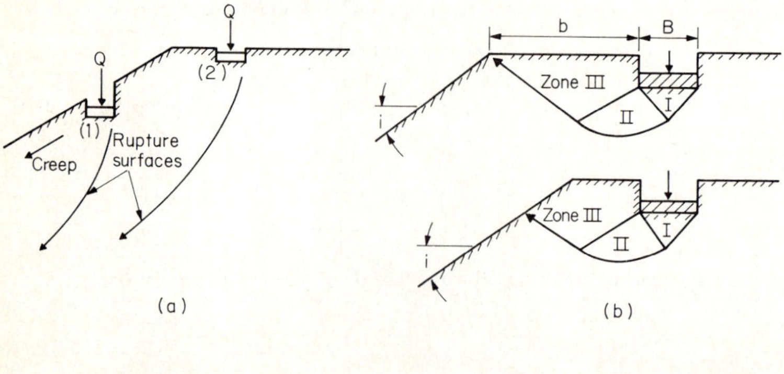

FIG. 6.35 Footings on slopes. A problem in slope stability (*a*) and foundation bearing capacity (*b*). In (*b*), distance *b* (critical) generally is between 2 to 3*B*.

also be considered since high lateral forces may be imposed on foundations.

Foundation Stability

Placing a foundation on a slope, or near its crest, results in a reduction in bearing capacity because the passive Rankine zone (Zone III, Fig. 6.35*b*) may be greatly reduced or eliminated entirely. In general, if a footing is placed at least a distance of 2 to 3*B* from the edge of the slope, there will be no theoretical reduction in bearing capacity. Where the footing is in the slope, and the slope angle *i* and the friction angle ϕ are less than 45°, the bearing-capacity reduction factors may be evaluated using procedures given in Art. 6.2.4, *Ground Slope Factors*.

Meyerhof's method (1957)[4] is useful for the case of a continuous footing bearing near the slope crest. He modified the bearing-capacity equation (Eq. 6.10), but without the depth factor:

$$q_b = cN_{cq} + 0.5\gamma BN_{\gamma q} \qquad (6.42)$$

Values for the factors N_{cq} and $N_{\gamma q}$ for continuous footings are given in Fig. 6.36. The slope stability factor N_s, defined in the figure, provides a means for relating the footing width *B* to the slope height *H*. The factors are shown to vary with the ratio of footing depth to width *D*/*B*, the ratio of distance of footing from edge of slope to width *b*/*B*, slope inclination *i*, and friction angle ϕ.

EXAMPLE Find width *B* for FS = 3 of footing resting behind slope crest.

Slope conditions: H = 60 ft, i = 60°, no GWL. Stiff clay, γ = 100 pcf, s_u = 0.5 tsf, ϕ = 0°.

Footing design criteria: Continuous-footing supporting wall. Q = 2 tsf per running foot, D = 4 ft, b = 4 ft.

Solution: Assume B = 4 ft, then D/B = 1 and b/B = 1. Since $B < H$, N_s = 0, from Fig. 6.36, for i = 60°, N_{cq} = 4.9

$$q_b = cN_{cq} = 0.5 \times 4.9 = 2.45 \text{ tsf}$$

$$q_a = 2.45/3 = 0.8 \text{ tsf}$$

$$B_{\text{required}} = Q/q_a = 2.0/0.8 = 2.5 \text{ ft, so use 3 ft}$$

Troublesome Practical Cases

It is often necessary to construct transmission towers or roadway viaducts on steep hillsides. In humid climates, soil creep can impose substantial forces on supports for structures which must be considered along with the rupture stability problem.

Transmission Towers

Large transmission towers supported on four legs often have one or two legs that impose an inclined load downslope as shown in Fig. 6.37 with the remaining legs placed in a platform cut into the slope or possibly bearing upslope. Small amounts of movement are not critical to the function of the tower as it will tend to move as a unit and is relatively flexible, but if the downslope legs undergo substantial movement, severe tower distortion will occur.

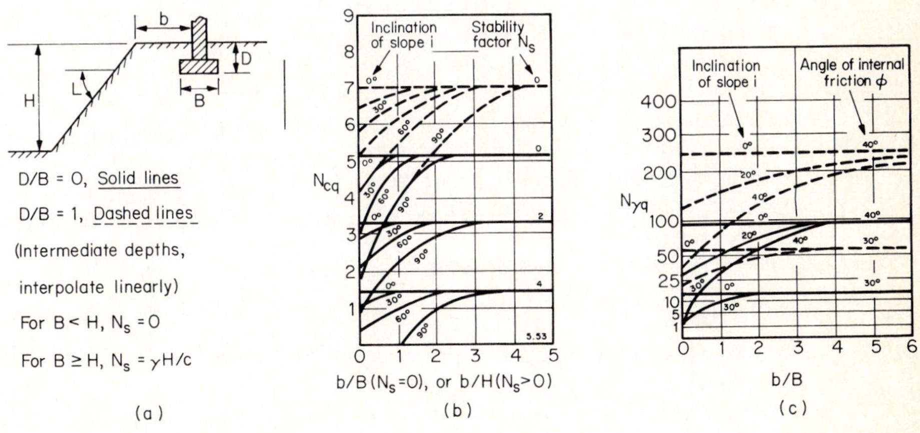

FIG. 6.36 Bearing-capacity factors for continuous footings behind slopes: (*a*) general geometry; (*b*) cohesive soil; and (*c*) cohesionless soil. [After Meyerhof (1957).[45]]

FIG. 6.37 Foundation preparation to lower a transmission-tower footing on a steep slope subject to creep. The other three legs are founded on a platform cut into the slope and the tower is guyed for temporary support. Apparent is the standard foundation for strong soils of a metal grillage buried in earth.

Tower foundations on relatively strong soils normally are metal grillages buried in earth. On slopes subject to creep, foundations can be supported by small anchored walls or by tiebacks installed through a reinforced concrete footing to resist lateral forces or buried at depths below the potentially active creep zone.

Viaduct Foundations

Viaduct foundations should extend through the active creep zone and be either protected by an anchored wall or tied back with anchors through the foundation as described in Art. 7.6.5. In some cases, piers have been enclosed by concrete cylinders designed to move with the creep, but it must be recognized that estimating total creep movement over the life of the structure and providing adequate "slack" between pier and cylinder may not provide adequate protection because of the problems inherent in estimating slope movements. The cylinders can be replaced to provide adequate space for movement if inspection is performed periodically to permit estimating when excessive closure is occurring.

6.2.6 SWELLING AND COLLAPSING SOILS

Considerations

In certain geographical regions, soils with a tendency either to swell or to collapse are found over large areas and when their characteristics are not properly accounted for in design, detrimental distortions of structures occur. They have certain characteristics in common. In general, they are both found primarily in regions of low rainfall with periodic relatively wet seasons, and they both can react unfavorably to increases in moisture content.

Because of the characteristics of these materials, foundation evaluations generally are not based on rigorous mathematical analysis, but rather on empirical considerations with the objectives to either avoid or minimize the problem. Most important in the selection of suitable foundation treatments is the recognition that the potential problem exists and the evaluation of its magnitude.

Swelling Soils

General

The normal problem to be resolved with swelling soils is the volume increase occurring during swelling and the volume decrease occurring during drying. Swelling soils are discussed in detail in Art. 10.6 of Hunt (1984),[7] and their prediction and estimations of swell magnitudes and time rates are discussed in Art. 4.3.6 above.

The safest solution often is provided by extending foundations through the "active" zone with the sides protected from soil contact which could impose adhesive forces resulting in uplift. Floor areas should not be in contact with the ground surface. Other solutions anticipate ground movements. If swell occurs, however, the addition of loads to the foundations, such as live loads, may result in settlement. If low swelling pressures are anticipated, it may be economical to support the structure on a clean granular fill of adequate thickness to provide the weight necessary to resist heaving. Bearing capacity is seldom a consideration except on slopes where fissures in the often desiccated materials provide weakness planes.

Shallow Foundations

Spread and continuous footings may be supported on compacted free-draining materials placed after partially excavating the active clays where low swelling pressures are anticipated. Various degrees of foundation movement may occur, depending on clay activity, depth of removal, depth of water table, and climate. In highly active clays, soil removal may improve conditions but usually will not totally resolve the problem. In addition, the granular backfill will provide the conditions for ponding of water and thereby aggravate the problem. Stiffening continuous footings with heavy reinforcing reduces structural distortion; therefore, estimates of the potential amounts of heave are important.

"Waffle" mats, or stiffened slabs, have been used with success in Texas to support light structures such as houses [Meyer and Lytton (1966),[46] Wray (1980)[47]]. The waffles are shallow beams that give the mat rigidity, and although the mat will rise as the soils swell, it will move as a unit and the structure will not be subjected to distortion. Differential movement will occur between the structure and utilities, which should be provided with adequate flexibility, and between the structure and adjacent paved areas, where it cannot be cheaply prevented (see *Pavements*).

Floors

Swelling is reduced by partially excavating the clay soils from beneath the floor area and replacing them with crushed stone or gravel. The procedure allows the soil to "breathe," which reduces moisture increase, and when the soils expand they squeeze into the gravel voids and their swelling pressures are reduced. Protection should be provided to prevent free water from entering the gravel and saturating the soils. Paving the area around the structure prevents rainfall infiltration and helps to reduce the problem. Even with these procedures some swelling is to be expected, the amount varying with clay activity, water-table depth, and climate. The safest solution is to support floors structurally, providing a space between the floor and the ground surface to permit expansion.

Basement and Other Retaining Walls

Lateral expansion may impose high pressures on basement or other retaining walls. Preventive measures are excavation and backfill with inert material, good surface and subsurface drainage, and an impervious blanket on the surface extending up to 20 ft (6 m) out from the building [Turnbull (1969)[48]]. Anchors may be required to resist high lateral forces.

Pavements

A number of treatments have been used which generally serve to reduce, and not eliminate, pavement movement. Success depends, as always, on environmental conditions. Turnbull (1969)[48] has summarized the common procedures as follows:

- Excavate a portion of the active soils and replace them with inert material as described for floors, or replace with the excavated soils compacted on the wet side of optimum moisture (usually economical where coarse-grained materials are not available), or replace with the excavated soils mixed with lime or some other stabilizing agent *(see Art. 5.4.4)*.
- Eliminate surface water infiltration at pavement edges and install adequate underdrainage.
- Provide special treatment for buried pipe and culvert locations.

Corrective Treatment

Heaved structures, floors, and pavements may be releveled by mud-jacking *(see Art. 5.4.2)*, although wet mixes may aggravate the problem.

Deep Foundations

Deep foundations such as bored piles (drilled piers) are often the most satisfactory solution to avoid the swell problem, as discussed in Art. 7.2.3.

Collapsing Soils

General

Collapsing soils can undergo compression, resulting in foundation settlement in three different manners: a sudden collapse of their lightly cemented structure at natural moisture content when the *critical load* is applied, a gradual compression under increasing degree of saturation, and a sudden collapse of the structure when saturation is approached or reached. The origin, occurrence, and general characteristics of these soils are described in Art. 10.5 of Hunt (1984)[7] and the prediction of collapse potential and settlement estimates are discussed in Art. 4.3.6 of this volume.

Treatment and Support of Structures

Ground treatment and structure support procedures depend upon the collapse potential and magnitude and the tolerance of the structure to differential deflections.

Preventing ground wetting in the area of the structure is the first objective where collapse potential is high. Adequate site drainage should be provided to prevent ponding, and all runoff should be collected and directed away from the structure. Locating septic tanks and leaching fields near the structure should be avoided, and all utilities and storm drains should be constructed carefully to ensure tightness. In some cases, underground water lines near structures have been double-piped or encased in concrete to ensure protection against exfiltration. In a desert environment, watering lawns, commonly accomplished by flooding, has been a major cause of housing settlement and should be avoided. It is best to landscape areas with natural desert vegetation.

Shallow foundations, designed for a q_a sufficiently below the *critical pressure* to avoid collapse at natural water content, are generally suitable for support of light to moderately heavy structures where the risk of wetting is low. The critical pressure is best determined by in situ plate-load tests, and q_a is based on FS = 2 to 3, depending upon the settlement tolerances of the structure. In some cases, such as large grain elevators, where the load and required size of a mat foundation impose bearing pressures on the order of the critical pressure, the structures have been permitted to settle as much as 1 ft (30 cm), provided that tilting is avoided. This solution has been applied to foundations on relatively uniform deposits of loess with natural water contents in the order of 13%. In collapsible soils derived from residual soils, however, such solutions may not be applicable because of the likelihood that variation in properties will result in large differential settlements. In all cases, it must be considered that settlement will occur as soil moisture content rises above the natural value.

Ground treatment methods have met with varying degrees of success. *Hydrocompaction,* or preflooding, has been used successfully to precollapse soils beneath dams and canals, but its use results in the possibility of long-term consolidation of clay soils under the increased overburden pressures, and it is generally not considered a risk-free solution for important buildings. *Dynamic compaction (see Art. 5.4.1)* should be a suitable method for increasing natural densities. *Lime stabilization* has been used where collapsible soils were causing detrimental settlement in a housing development [Sultan (1969)[49]]. A water-lime mixture was pumped under high pressure into 2-in-dia. holes to depths of about 5 ft, and significant movements were arrested. *Excavation-compacted backfill* procedures are suitable where the depth of the collapsible zone is limited, as is normally the case with residual soils.

Deep foundations, installed to depths beyond the zone of potential collapse, are used for settlement-sensitive structures. If the possibility for wetting exists, designs should provide for downdrag imposed by the collapsing soils.

Floors for large industrial buildings often are supported on the collapsible soils or on compacted fills and protection against ground wetting provided. The risk of subsidence must be accepted with the understanding that some relevelling by "mud-jacking" may be required in the future.

6.2.7 FOUNDATION VIBRATIONS

Aspects of Vibration Phenomena

Sources and Forms

Elastic waves traveling through the earth cause ground motion, which is transmitted through foundations to a structure as vibrations. This form of foundation vibration results from earthquakes, as discussed in Art. 11.4, *Earthquake-Resistant Design: An Overview* in Hunt (1984),[7] or from construction activity such as rock blasting, pile driving, etc., as discussed in Art. 5.6 of this volume.

Vibrations also result from rotating machinery (compressors, turbines, etc.) and reciprocating machines (internal combustion engines, punch presses, etc.) and are transmitted through the foundations to the soil as described briefly in this article. This form of vibration is termed *steady-state vibration.*

Characteristics of Elastic Waves and Soil Response

In their simplest form, elastic waves are described by the quantities of harmonic motion (*see Art. 5.6.2*) including wavelength, period or frequency, and amplitude, which induces motion with the characteristics of displacement, particle or vibrational velocity, and acceleration. Elastic waves travel through a medium at a characteristic propagation velocity.

Excitation emanating from some source as a stress wave applies a cyclic shearing stress to soils. The resistance to shear deformation of soils under cyclic strain is described by the *shear modulus G (see Art. 3.3.5)*. Internal dissipation of energy occurs during cyclic loading, which is described by the *internal damping ratio D (see Art. 3.3.5)*. Shear modulus, damping ratio, and mass density are the most important soil characteristics for most problems involving small amplitudes resulting from the sources described above. (Sea wave forces, for example, involve large-amplitude cyclic loadings as described in Art. 7.6.4.)

Analysis of foundation vibrations is a problem in soil-structure interaction because of the nature of the elastic response of the soil to the harmonic motion and its damping effects (*see Art. 4.4.5*). Machine-induced vibrations are most significant in loose to medium-dense cohesionless soils.

Monitoring Vibrations

Earthquake-induced seismic waves are monitored with *seismographs, accelerographs* (strong-motion seismographs), or *seismoscopes* [see Art. 11.2.3 of Hunt (1984)[7]].

Vibrations induced by blasting, pile driving, and rotating or reciprocating machinery are monitored by small portable seismographs or *particle-velocity seismographs* which measure frequency, acceleration, and displacement of surface particles in three orthogonal directions of vibration and in some cases are designed to measure particle velocity directly [Art. 4.2.5 of Hunt (1984)[7]].

Measuring Dynamic Deformation Moduli of Soils

As discussed in Art. 3.3.5, there are a number of methods of measuring the dynamic deformation moduli of soils (E_d, G_d) in the laboratory and in situ. The damping ratio is measured in the laboratory. In situ methods include *seismic direct techniques* to generate compression, shear, and Rayleigh waves and *steady-state vibration methods* to generate Rayleigh waves.

Effect of Vibrations on Foundations and Structures

In the case of vibrations induced from some exterior source, such as earthquakes or blasting, the primary concern is with damage to the structure which occurs when the motion exceeds some limiting value usually expressed in terms of velocity or acceleration of the foundation.

For steady-state vibrations, the concern is with the induced vibrations causing harmful effects to the machine itself or to structures or people in the immediate vicinity. Limiting values are usually expressed in terms of an *allowable displacement at specified frequencies.*

General relationships between vibration damage to structures and the limits of human perception and motion frequency and amplitude are given in Fig. 5.47.

Steady-State Vibrations

Vibratory Motion

In the design of foundations for rotating or reciprocating machines, criteria usually provide for a *maximum amplitude of vertical motion.* Machine characteristics provided by the manufacturer establish the maximum periodic force produced at a rated velocity.

The fundamentals of vibratory or harmonic motion are described briefly in Art. 5.6.2. In Fig. 5.45, displacement y can be described at any time t by the expression

$$y = A \sin (\omega t - \phi) \qquad (6.43)$$

where A = amplitude
ω = circular frequency or rate of oscillation, radians per unit t
ϕ = phase angle representing the shift of the sinusoidal curve along the time axis from the zero position

Sinusoidal loadings Q may be expressed by:

$$Q = Q_o \sin \omega t \qquad (6.44)$$

where Q_o may be either a constant or a function of the rotating frequency ω (rad/sec). Expressions for force can then be developed in a manner similiar to that of Eq. 5.9, relating foundation mass to vibration frequency and amplitude.

Foundation Analysis

A current approach to analysis [Richart (1975)[50]] is to represent the vertical response of a rigid foundation on an elastic halfspace by a lumped mass-spring–dashpot system with a single degree of freedom, as illustrated in Fig. 6.38. [See also Art. 11.4.2 of Hunt (1984).[7]]

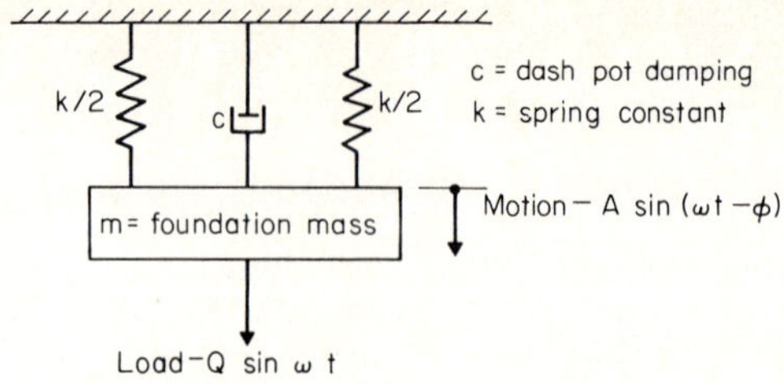

FIG. 6.38 Representation of forced vibrations of a single-degree-of-freedom system with damping.

The effective mass m of the lumped system is the foundation mass; the spring constant k *(see Art. 4.4.5, Dynamic Analysis)*, as established by the theory of elasticity for the soil, is given in an expression relating shear modulus, Poisson's ratio, and the foundation "equivalent" radius; and a viscous damping coefficient c is given in terms of the parameters of the foundation and the soil which represents energy dissipation in the system. The mass moves through displacement y with load Q at some operating frequency.

STATIC DISPLACEMENT In the analysis for a single-cylinder compressor the initial step is to compute the approximate base area required to limit the static displacement y_s to some preselected value, such as 0.002 in, caused by the periodic vertical force. This is accomplished with an expression relating displacement, force, and base area to two soil properties, Poisson's ration ν and shear modulus G, as follows:

$$y_s = \frac{(1-\nu)Q_o}{4Gr_o} = 0.002 \qquad (6.45)$$

where r_o = required equivalent foundation radius,
Q_o = maximum vertical compressor force, and
ν = Poisson's ratio (usually taken as 0.33 for sands, Art. 3.3.5)

From the r_o value computed, a value is assumed for design and a corrected static displacement value is then computed.

DYNAMIC DISPLACEMENT Using r_o for width, foundation dimensions then are selected to establish the total block and machine weight W from which the *mass* (or inertia) *ratio* B_y is computed from an expression relating foundation mass and equivalent diameter, soil density, and Poisson's ratio as follows:

$$B_y = \frac{1-\nu}{4}\frac{W}{\gamma r_o^3} \qquad (6.46)$$

The *geometric damping ratio* D_y is computed from an expression relating the mass ratio to a constant

$$D_y = \frac{0.425}{\sqrt{B_y}} \tag{6.47}$$

and the *maximum dynamic vertical displacement* y at the *resonant frequency* is found from

$$y = y_s\left(\frac{1}{2D_y\sqrt{1 - D_y^2}}\right) \tag{6.48}$$

Note: Equations 6.46 and 6.47 are from Richart, Hall, and Woods (1970).[51] Soil damping values of *internal damping* may be important in establishing response to rotational modes of vibration (rocking and torsional modes). Geometric damping usually governs for translational modes. The *dynamic magnification factor* relates y to y_s. If it is small the amplitude limitation factor is satisifed even if resonance occurs.

RESONANCE Soil-foundation system resonance occurs when the operational speed (rpm) of a machine coincides with the natural frequency (cpm) of the machine-soil system (*not* the natural frequency of *only* the soil formation). At resonance there may be a large increase in the amplitude of oscillations and in the dynamic vibratory forces; therefore, in machine-foundation design, an objective is to avoid resonance, or at least to evaluate its effect on the dynamic vertical displacement as described above. The natural frequency f_n of the soil-foundation system may be computed from a relationship between the spring constant k_y and the mass (block plus machine weight) as follows:

$$f_n = \frac{1}{2\pi}\sqrt{\frac{k_y}{m}} \tag{6.49}$$

where k_y is found from

$$k_y = \frac{4Gr_o}{1 - \nu} \tag{6.50}$$

Other Conditions

The foregoing analysis pertains to vertical vibrations in harmonic motion such as are developed by a single-cylinder compressor. Other conditions that may require analysis are sliding, rocking or torsional vibrations, and nonharmonic modes such as periodic, random, or transient motions, as described by Richart (1975).[50]

Soil Factors

Vibrations usually have a negligible effect on strong clays or other cohesive soils, but always require careful evaluation where foundation support is provided by sands, particularly those that are well-graded and dry or submerged, at a relative density less than about 70 to 75%. The continued vibrations may lead to large cumulative settlement.

Solutions

If deflections are unacceptable or if resonance is undesirable, then analyses are made for foundations of larger dimensions and consideration is given to increasing the embedment depth.

Loose sand should be compacted to achieve high values of D_R by methods such as described in Art 5.4.1. An existing foundation system reacting unfavorably to vibrations can be treated by grouting (*see Art. 5.4.2*). Machinery should be separated from the buildings and floors and supported on its own foundation of adequate mass to eliminate excessive vibrations. Small foundations have been supported on springs, and large foundations have been supported on cork or asbestos fiber to screen vibrations from the ground and reduce the influence on the adjacent structural elements.

ADDITIONAL REFERENCES Refer to Tschebotarioff (1973)[36] and Converse (1962).[52]

6.3 SHALLOW FOUNDATIONS ON ROCK

6.3.1 GENERAL CONSIDERATIONS

Support Conditions

Routine

The support of major structures on rock traditionally is the objective of the civil engineer, as only a relatively few rock formations cannot adequately support foundation loads of 4 tsf, the maximum pressure imposed by the majority of structures, even nuclear power plants. Foundations on rock usually are an economical and troublefree solution as long as the rock is reasonably shallow, relatively unweathered, and sound.

Nonroutine

Structures imposing high loads such as tall buildings, towers, chimneys, power plants, bridges, and concrete dams require conservatism and careful evaluation in all cases, and even routine structures of any importance require careful evaluation where rock conditions are unfavorable. If rock is shallow, or exposed by basement excavation, support will be provided by spread or continuous footings or a mat; if rock is deep, support will be provided by piles or piers. Where rock is exposed for examination, the engineer or geologist will have much more confidence in judging rock quality than he will for the case of deep foundations where examination is either impossible, not practical, or limited in area.

Critical problems involve buildings with very heavy column loads (1000 tons or more) founded in less than excellent quality rock, bridge foundations constructed in canyon walls, and foundations for concrete gravity, buttress, or arch dams where, in addition to rock quality, seepage pressures must be considered. In these and similar problems, rock conditions must be evaluated carefully with respect to both rupture and deformation as affected by rock quality, rock surface irregularity, and groundwater conditions.

Design Criteria

Allowable Bearing Value q_a

Most structures founded on rock are designed for an allowable bearing value to restrict settlement to within tolerable limits. q_a is selected from building codes, from empirical correlations with rock strength and joint conditions, or from evaluations of deformation moduli measured by in situ testing.

Stability against Rupture

Conditions where rock-mass stability against rupture can be the most important design criteria exist with concrete dams, construction over old mines, foundations on slopes, bridge foundations installed in canyon walls, or any situation where tensile forces can develop. Orientation of rock-mass discontinuities and their strengths are important parameters requiring evaluation. Foundation failure by rupture also is possible under a variety of unfavorable rock conditions, even for normal building loads.

Ground Factors

Rock Quality

The major factor influencing rock-mass performance is rock quality (*see Arts. 3.4.3 and 4.3.5*) described in terms of the rock blocks themselves (specifically their hardness, which depends on rock type and mineral composition, and their degree of decomposition or weathering) and in terms of the discontinuities separating the blocks (their spacing, opening width and filling, and fracture-surface characteristics). Except for weak rocks, such as clay shales, chalk, marl, and those that are highly weathered and decomposed, the major factors affecting support capacity are mass defects such as closely spaced joints, fault and shear zones, and cavities. In addition, some rock masses tend to heave from swelling or relief of high residual stresses. Rock-mass quality can be improved, as discussed in Arts. 5.5 and 6.3.4.

Rock Conditions

Whenever there is the possibility that rock will be encountered in excavation or be used to support structures, thorough subsurface investigation is required to adequately define its surface conditions and mass characteristics. A variety of rock-mass conditions is illustrated in Fig. 6.39, a number of which can be troublesome when design criteria are being established or during construction.

In Fig. 6.39*a* a residual soil containing corestones and boulders has formed over a granite mass with an irregular surface resulting from weathering along joints. Heavy loads placed on a boulder mistaken for bedrock may result in objectionable foundation deflections. The irregular rock surface will require treatment. Foliated metamorphic rocks, such as gneiss and schist, in humid climates (Fig. 6.39*b*) characteristically have profiles gradually changing with depth through four zones: residual soil without relict

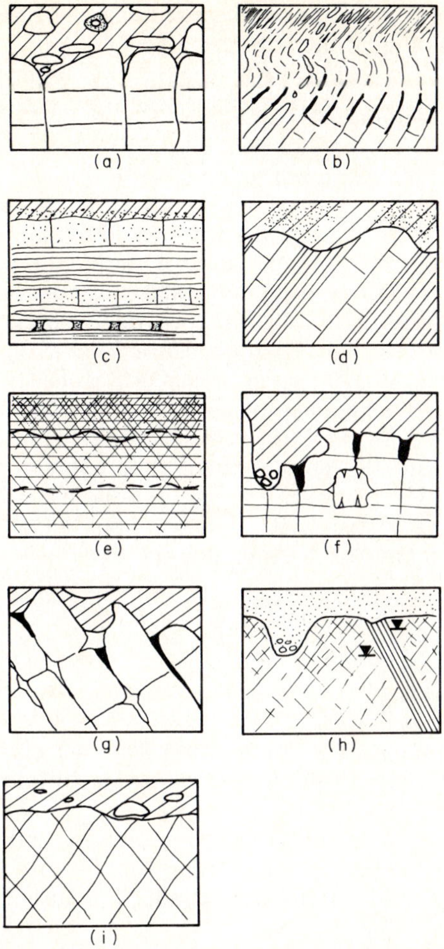

FIG. 6.39 Rock surface conditions: (*a*) decomposed granite with corestones and boulders; (*b*) decomposed gneiss; (*c*) residual soil over sandstones and shales with a mined coal seam; (*d*) swelling and nonswelling residual soils over dipping clay shales and sandstones; (*e*) fragmentation in marine shales; (*f*) horizontally bedded pure limestone with soil cover; (*g*) dipping beds of pure limestone with soil cover; (*h*) buried crystalline rock with ancient channel and fault zone; and (*i*) glaciated crystalline rock with bouldery till cover.

structure, saprolite, weathered rock, and finally fresh rock. Transitions are gradual and usually quite irregular, and sound rock can be at depths of 30 m or more.

Horizontally bedded sandstones and inactive shales (Fig. 6.39*c*) are found with a relatively thin residual cover of uniform thickness. The presence of active or abandoned coal mines can be hazardous. Dipping beds of sandstone and clay shales (Fig. 6.39*d*) may decompose, leaving an irregular surface and a residuum that varies in swelling tendencies from high to low activity. q_a will vary across the site. Massive beds of marine and clay shales (Fig. 6.39*e*) are characterized by a highly fragmented and active upper zone. Fragmentation decreases with depth, but when exposed in open excavation the rock disintegrates in a relatively short time and rock quality decreases. These rocks have high swelling tendencies.

Argillaceous and other impure limestones react to weathering in a manner more or less similar to sandstones and shales (Fig. 6.39*c*), but pure limestones and other soluble rocks, such as gypsum, can be found with extremely irregular surfaces and cavities in the rock mass, as illustrated for horizontally bedded rock in Fig. 6.39*f* and dipping strata in Fig. 6.39g. Residual soil cover is clayey and generally of limited thickness. Conditions are often very hazardous, particularly where groundwater is being or will be withdrawn, since this activity results in increased overburden pressures resulting in roof collapse over cavities. Excavations of large areas in horizontally bedded sedimentary rock may be subject to floor heave from extension straining (*see Art. 4.3.6*).

Three other important rock conditions are illustrated in Fig. 6.39*h*. Where bedrock is covered by alluvium, much or all of the residual soil may have been removed by erosion, leaving an irregular rock surface and even perhaps a deep-buried ancient channel, particularly troublesome to dam and bridge foundations. Fault zones in any rock mass are a source of relative weakness and if there is a potential for possible movement during earthquakes, such zones should not be used for support of any structure in seismically active areas or for the support of critical structures such as nuclear power plants or very tall buildings in any area. A fault may also cause a variation in groundwater level. Also indicated in Fig. 6.39*h*, as well as in other figures, is the condition that joint and weathering intensity in a rock mass normally decrease with depth; therefore, rock quality normally increases. Some aspects of discontinuities are described in Art. 6.3.3.

A large portion of the northern United States has had the bedrock surface scoured of residual soil and weathered rock by glaciation. The remaining rock usually is relatively sound and fresh, overlain by glacial tills and outwash or stratified drift. In Fig. 6.39*i* a clayey till with occasional boulders overlies a fairly uniform rock surface. It may not be possible to define the location of all boulders during explorations but their presence must be suspected in certain till formations and accounted for in foundation design.

Rock identification and classification, mass characteristics, and attendant hazards are described in detail throughout Hunt (1984),[7] in particular Chaps. 5, 6, and 10. Rock decomposition and residual soils are described in Arts. 6.7 and 7.2.

Other Factors

Excavation in rock requires consideration of rock hardness and fracture intensity as it impacts on the degree of difficulty of rock removal, and blasting procedures must be carefully planned to avoid decreasing rock quality by increasing fracture intensity. Depending upon rock type and fracture intensity, groundwater control may be troublesome. A not uncommon problem in deep excavations in rock is high residual stresses which cause heaving of excavation floors and significant wall deflections as discussed in Art 9.3.5. After the completion of large concrete dams, high seepage forces in the rock mass can substantially reduce its stability.

Rock deterioration with time ordinarily would not be considered a significant factor during the life of a structure. Feld (1966)[53] however, reported a number of instances in New York City where foundation rock was "hard, ringing" material when the foundations were installed but in 10 to 20 years was found to be altered and disintegrated to the point where it was easily removed by pick and shovel when exposed in an adjacent excavation. The rock, Manhattan mica schist, was found altered to depths as great as 25 ft. Bearing pressures as high as 25 tsf were reported, but no indication of distress in the buildings was observed. The cause of deterioration appeared to be downward seepage of sewage from broken sewer lines, or perhaps polluted rainwater. In one case, a broken steam line 50 ft distant from a tunnel excavation was attributed to softening of the schist.

Chemical reactivity from certain rocks can cause the deterioration of concrete. Pyritic or "alum" shales can cause foundation or floor heave during sulfide alteration of the pyrite; in addition, when the pyrite oxidizes, sulfuric acid is produced which has reportedly turned foundations and floors to "mush" over a period of years [Grattan-Bellew and Eden (1970)[54]]. Under floors the process is inhibited by spraying the rock surface immediately upon exposure with an impermeablizing agent such as asphalt, since oxidation does not appear to occur under conditions of complete saturation [Peck (1976)[55]]. The phenomenon does not appear to affect more deeply embedded footings. Aggressive salts also attack concrete, causing its deterioration below the water table (Art. 8.3.4 of Hunt (1984).[7]

6.3.2 ALLOWABLE BEARING PRESSURE

General

Selection of q_a

An abundance of information does not exist in the literature regarding highly stressed foundations on rock, especially where there is a need to truly evaluate rock quality in terms of jointing and decomposition and to select a value for q_a which will limit settlement to acceptable amounts. Values for q_a can be selected from building codes, estimated from laboratory tests in conjunction with empirical correlations, or computed on the basis of in situ measurements of E_r. The rock pressures for routine structures founded on favorable rock conditions are normally based on values selected from building codes. Poor quality rock can present troublesome conditions even for normal structures.

Influence of Rock Quality

Rock-mass quality can be classified on the basis of two separate and distinct groupings of characteristics, aside from the major defects of caverns and large faults: hardness and joint conditions. *Rock hardness* is related to mineral composition, and a distinction is made between the fresh, unweathered state and the weathered or decomposed state. A classification by hardness including weathering grade and unconfined compressive strength is given in Table 3.14, and for various types of rock when fresh in Table 3.13. A classification based on joint frequency as measured by the RQD is given in Table 6.10 for fresh rock. It includes values of allowable bearing pressure.

The values given in Table 6.10 for q_a are based on the fact that most fresh rocks, especially those of better quality, have U_c strengths higher than concrete (Fig. 4.9) and are substantially higher than permitted in building codes or used by most engineers. More conservative values are assigned to the better quality classes because the high loads requiring high q_a values pertain to very important structures such as high-rise buildings, and beneath any given site rock quality can vary through various degrees which may not be disclosed by investigation.

A general classification of rock quality in terms of strength U_c of fresh specimens, degree of weathering, and joint spacing, as related to typical values for q_a found in building codes and used by most practitioners, is given in Table 6.11. Also included is the basis for selecting q_a by testing.

Excellent quality rock has a bearing capacity greater than concrete and is characteristic of fresh, massive crystalline rocks (such as diabase, dolerite, granite, some gneisses, and quartzite with joints tightly closed and spaced greater than 1 m). Confirmation requires good recovery (RQD > 90%) of cores of high to very high strength.

Good quality rock has medium strength when fresh and is characteristic of some igneous rocks, gneiss, some schists, sandstones, dolomite, and limestone, as well as some igneous rocks when slightly weathered, with joints tightly closed and spaced greater than 1 m. Confirmation requires good core recovery (RQD > 90%) and correlations with U_c values. Quality decreases with decreasing joint spacing and with other characteristics, and can be accounted for by empirical correlations or in situ load testing.

Fair quality rock has low strength when fresh and is characteristic of some schists, sandstones, and shales and some slightly weathered rocks with joints tightly closed and spaced greater than 1 m. Joint characteristics can be accounted for by empirical correlations and overall rock quality measured in terms of E_r with in situ testing in critical cases.

Poor to very poor quality rock has very low strengths, as is typical of marine and clay shales, chalk, and moderately to completely weathered igneous and metamorphic rocks; the mass is blocky, fractured, or crushed (Table 3.16). Bearing values are best determined by in situ and laboratory testing. The potential for swell of shales must be evaluated.

Empirical Estimates

Building Codes

Usual practice is to select a "presumptive bearing value" from a building code. Codes relate to

TABLE 6.10
ALLOWABLE CONTACT PRESSURE ON FRESH JOINTED ROCK*

		Allowable pressure†	
Rock quality	**RQD**	**tsf**	**kN/m²**
Excellent	90–100	200–300	20,000–30,000
Good	75–90	120–200	12,000–20,000
Fair	50–75	65–120	6,500–12,000
Poor	25–50	30–65	3,000–6,500
Very poor	0–25	10–30	1,000–3,000
	0	<10	<1,000

*After Peck et al. (1974).[17]

†If tabulated value exceeds unconfined compressive strength of intact samples of the rock, allowable pressure equals U_c.

TABLE 6.11
ROCK ALLOWABLE BEARING VALUE q_a AS RELATED TO ROCK QUALITY AND SELECTION METHOD

Rock quality	Strength fresh*	Weathering grade†	Jointing‡	q_a, tsf, kg/cm²§	Selection method
Excellent	Very high to high	Fresh	Very wide to wide	≥100	U_c, rock description
Good	Medium	Slightly weathered	Very wide to wide	100–50	U_c, in situ load test
Fair	Low	Slightly weathered	Very wide to wide	50–25	U_c, in situ load test, pressuremeter
Poor	Very low	Moderately weathered	Blocky	25–10	In situ load tests, pressuremeter, U_c
Very poor	Very low	—	Fractured to crushed	≤10	In situ load tests, pressuremeter
		Highly to completely weathered	—	≤10	Pressuremeter, soil lab testing

*Fresh strength ranges for various rock types from Table 3.13. Strength should be measured in unconfined compression at natural moisture content.

†Weathering grade from Table 3.14; mainly affects crystalline rocks because of gradual transition in grades and weathered zone thickness.

‡Jointing from Table 3.16; blocky, fractured, or crushed applies to fresh rock or any weathering grade in reducing q_a.

§Values for q_a are general ranges found in building codes and the literature. They vary from location to location for a given rock type as its characteristics vary. See Table 6.12 for code ranges. For q_a vs. RQD for fresh, jointed rock, see Table 6.10.

local experience and "presumptive" usually signifies an approximate value for q_a to limit settlement to tolerable amounts [normally in the order of 1 in (2.5 cm)], "to be confirmed by the engineer's experience and judgment." Codes often are lacking in detailed descriptions of rock characteristics or rock quality, but generally are very conservative.

A summary of q_a values from a number of codes from the major cities of the United States and the Institution of Civil Engineers in London is given in Table 6.12. The wide ranges among values for a given rock type reflect local rock quality and are found especially for foliated and sedimentary rocks. The New York City Building Code, which provides relatively detailed descriptions of rock quality and its determination, is given in modified form in Table 6.13. The NAVFAC (1971)[13] code is given in Table 6.1.

Core Strength Basis U_c

Results from laboratory uniaxial compression tests (or point-load test or Schmidt hardness test correlated with U_c, Art. 3.4.2) provide a basis for estimating q_a, with reduction factors to account for joint conditions. Peck (1976)[55] considers that for fresh, intact rock, $q_{ult} \geq 3U_c$, and U_c may be used for q_a (see Table 6.10). To obtain reliable strength data, rock specimens should be tested at natural moisture content and not be permitted to

TABLE 6.12
PRESUMPTIVE BEARING VALUES FOR ROCK FROM VARIOUS BUILDING CODES q_a

Rock type	Presumptive bearing value, tsf*	
	U.S. Codes	London, U.K.†
Massive crystalline	50–100	100
Sound foliated	20–60	100
Schists‡	20–40	30
Sound sedimentary	15–100	20–40
Soft, broken (excluding shales)	8–12	20 (soft sandstone)
Shales, soft	1–8	10 (clay shales)
Shales, broken	1.5	–
Hard, solid chalk		6

*Some codes provide for a maximum value equal to 20% U_c.

†From Institution of Civil Engineers, London [Thornburn (1966)[56]].

‡Steeply dipping beds often require reduction.

air-dry before testing, as the strength of many rocks is affected by changes in moisture content. Immersion of air-dried specimens in water can reduce U_c to 75 to 90% of the air-dried value, except for expansive shales, for which the reduction will be much greater.

When used to estimate q_a for rock masses, U_c is reduced to account for joint conditions which increase compressibility. Some codes account for rock fractures in the manner described in Table 6.13, others decrease q_a to 20% of U_c. Practitioners, in general, use ⅕ to ⅛ of U_c for q_a [Teng (1962)[57]]. A more refined procedure is presented by the Canadian Geotechnical Society [CGS (1978)[3]] which accounts for joint spacing and other factors. It is applicable to "sound" rock defined as having a value for $U_c > 850$ kPa (8.5 tsf), with the rock surface perpendicular to the foundation, a load with no tangential component, and rock with no open discontinuities.

$$q_a = K_{sp} U_c \tag{6.51}$$

where q_a = allowable bearing pressure
U_c = average unconfined compressive strength of cores, as determined by ASTM D2938-71
K_{sp} = empirical coefficient depending on spacing of discontinuities and including a value for FS = 3, as given in Table 6.14

Other rock-mass defects such as fault zones, slickensides, foliation planes and shear zones, cleavage, bedding planes, and cavities (Table 3.15) all require special consideration when the allowable bearing value is evaluated.

Settlement Analysis Based on Values of E_r

Applications

Major structures to be supported on rock with quality less than excellent, or average structures to be supported on rock of poor to very poor quality, can be designed on the basis of q_a calculated from elastic theory and deformation moduli *(see Art. 4.3.5)*. In settlement analysis consideration must be given to the potential variation in the stress distribution in the rock mass from the Boussinesq *(see Art. 4.2.4)* caused by the orientation of the discontinuities.

Deformation Moduli

Values for E_r are best measured by in situ testing (plate jack, flat jack, radial jacking and pressure tunnels, or the pressuremeter) as described in Art. 4.1.4. Full-scale load tests employing anchors for reactions permit measurement of deflections under loads of 2000 tons or more *(see Art. 7.5.3)*. In situ testing normally stresses a smaller zone than does a foundation. Mass ef-

TABLE 6.13
ALLOWABLE BEARING PRESSURES FOR ROCK*

Characteristic	Rock material class: Hard, sound	Medium hard	Intermediate	Soft
Q_{all}, tsf†‡	60	40	20	8
Rock type	Crystalline: gneiss, diabase, schist, marble, serpentinite	Same as hard rock	Same as hard to medium-hard rock, and cemented sandstones and shales	All rocks and uncemented sandstones
Struck with pick or bar	Rings	Rings	Dull sound	Penetrates
Exposure to air or water	Does not disintegrate	Does not disintegrate	Does not disintegrate	May soften
Fractures: Appearance	Sharp, fresh breaks	Cracks slightly weathered	Show weathered surfaces	Contain weathered zones
Width	<3 mm	<6 mm	Weathered zone to 25 mm width	Weathered zone to 30 cm width
Spacing	<1 m	<60 cm	Weathered zone spaced as close as 30 cm	Weathered zone filled with stiff soil
Core recovery§	>85%	>50%	>35%	>35%
SPT				>50 blows/ft

*After New York City Building Code (1968).[14]

†Allowable bearing value applies only to massive crystalline rocks or to sedimentary or foliated rocks where strata are level or nearly level, and if area has ample lateral support. Tilted strata and their relation to adjacent slopes require special consideration.

‡Allowable bearing for hard to intermediate rock applies to foundations bearing on sound rock. Values can be increased by 10% for each 12 in (30 cm) of penetration into rock of equal or better quality, but shall not exceed twice the basic values.

§Rock cored with double-tube, diamond core barrel, in 5-ft (1.5-m) run.

fects, primarily as influenced by discontinuities, are accounted for by in situ and laboratory seismic testing and the application of a modulus reduction factor to obtain a *design modulus* as described in Art. 4.1.4.

Values for E_r vs. U_c for intact specimens from a wide range of rock types are given in Fig. 4.9. Steel, concrete, and clays are also given for comparison. Values for γ, E_r, and U_c for sound weathered specimens are given in Table 3.12 and for E_r and ν in Table 4.4. Table 5.8 summarizes E_r for many rock types of varying quality both before and after being treated with grout injections which can improve quality substantially, especially in poorer quality rocks.

TABLE 6.14
VALUES FOR K_{sp} IN TERMS OF JOINT SPACING*†

Description	Spacing	K_{sp}
Very wide	>3 m	0.4
Wide	1 to 3 m	0.25
Moderately close	30 cm to 1 m	0.1

*After CGS (1978).[3]

†Validity: Discontinuity thickness, <5 cm, or <2.5 cm if filled with soil or rock debris. Foundation width > 30 cm. Sedimentary or foliated rocks must be level or nearly so.

The Schmidt rebound hammer hardness test *(see Art. 3.4.2)* is useful for estimating E_r through correlations with U_c (Fig. 3.51) as a large number of tests can be made quickly and economically. Rebound hardness value R is entered in Fig. 3.51 for the specimen bulk density γ to find U_c, and U_c is entered into Fig. 4.9 or Table 3.12 to find a value for E_r. A modulus value reduced for weathering effects may be estimated on the basis of the ratio of the rebound value R obtained on weathered rock to the rebound value obtained on a fresh, unweathered specimen *(see Art. 3.4.3)*.

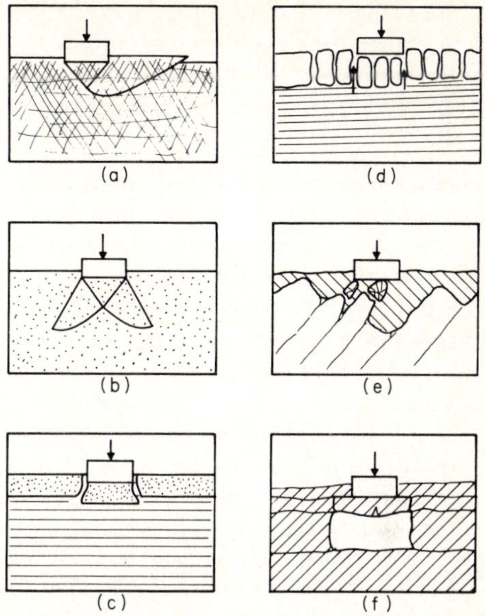

FIG. 6.40 Some failure forms by rupture in rock masses: (*a*) general shear in soft or decomposed rocks; (*b*) local or punching shear in brittle, porous rock; (*c*) punching shear in rigid over compressible layer; (*d*) unconfined compression along vertical joints in hard rock over weaker rock; (*e*) local crushing of pinnacles in limestone; and (*f*) cavern roof failure in beam action. [Parts *a* through *e* after Sowers (1976).[62]]

6.3.3 BEARING CAPACITY AND STABILITY

Important Factors

Failure Occurrence

Failures of shallow building foundations by rock rupture are seldom heard of, although the possibility is very real in poor quality rock if strength is not properly evaluated and eccentric or high loads occur. Under some conditions rock can have relatively low bearing values, as illustrated in Fig. 6.40. Shallow foundations also may be affected by the collapse of roof arches over caverns or the rupture of mine pillars (Fig. 3.54*b* and *c*).

There are also situations where rock-mass conditions require assessment of stability for all categories of rock quality such as foundations and abutments of concrete dams (Fig. 6.41), bridge piers bearing in steep canyon walls, and anchor ties for towers or other structures installed to resist tensile forces.

Rock Quality

Rupture resistance is related to rock quality, primarily as identified by decomposition degree and discontinuities *(see Art. 3.4.3)*. For shallow foundations for buildings, rock hardness, joint intensity, and surface irregularity are the normal controlling factors. Large structures such as concrete dams will be founded on hard rock, and it is their stability against sliding that is of major concern. The controlling factors are the orientations and characteristics of the major weakness planes such as master joints, faults, foliation and bedding-plane shear zones, continuous clay seams, and foliated structure.

Rock Strength Factors

In softer rocks, the strength of the intact mass or intact blocks normally governs resistance to rupture, although relict failure surfaces such as slickensides or other weakness planes may control mass strength. Strength parameter measurements of intact rock are described in Art. 3.4.2. In rock of low to very low strength the ultimate capacity may be evaluated on the basis of pressuremeter data *(see Art. 6.2.4)*.

In relatively hard rock, strength is provided by the blocks of intact rock and the interaction between the blocks acting along discontinuities. The identification of the extent, width, and material properties of the weakness planes by comprehensive investigation is the most critical element in the stability analysis for a major project. Some typical strength values for various rock-mass discontinuities that may be used in preliminary stability analysis are given in Table 3.17. Deere (1976)[61] suggests that in most cases it is prudent to assume that cohesion will be zero, because it is very likely that at least some small displacement has occurred along the discontinuity in past geologic history and the residual strength prevails (Fig. 3.52). Strength measurement for rock masses is described in Art. 3.4.3. It is to be noted that, in general, in the literature all discontinuities (except cavities) are often referred to as "joints."

Seepage Forces

High seepage forces under concrete dams cause uplift pressures on the dam base or in the rock mass and can be significantly higher in rock under tension than in rock under compression.

Analytical Approaches

Failure by rupture of rock beneath shallow foundations may be investigated by a number of methods, the selection of which depends upon conditions of the applied load and rock-mass characteristics. In some cases, building foundations are analyzed by the use of classical bearing-capacity theory, in other cases by theories from mechanics of materials. The stability against sliding of large concrete dams commonly is evaluated on the basis of such general forms of limit-equilibrium analysis as are employed for slopes.

Building Foundations: Shallow Conditions

Some forms of possible bearing-capacity failures in rock masses supporting building foundations are shown in Fig. 6.40. If the rock is homogeneous in its weakness, as may be the case with plastic rocks such as shales and those that are highly decomposed (Fig. 6.40*a*), the condition of general shear may be applicable *(see Art. 6.2.4)*. In more compressible, brittle rock, such as badly weathered sandstone, granite and porous limestone in which a loose, lightly cemented structure prevails, failure may be in the form of local or punching shear (Fig. 6.40*b*) and capacity is evaluated as described in Art. 6.2.4 [Sowers (1976)[62]].

Layered systems are more complex, as for example in the case of a footing supported on a thin, rigid rock slab underlain by much weaker rock (Fig. 6.40*c*). Bearing capacity is calculated by considering the deformation of the lower layer and analyzing the upper layer as a beam. In vertically jointed rocks underlain by a weaker stratum, failure may occur in unconfined compression (Fig. 6.40*d*); or, where rock pinnacles are surrounded by soil, failure may occur by local crushing (Fig. 6.40*e*). For a footing underlain by a cavity (Fig. 6.40*f*), the roof slab is analyzed as a beam, as described in Art. 10.3.3.

Construction over Caverns and Mines

Collapse Phenomena

Subsurface investigation must be performed with extreme care where construction is contemplated over cavernous ground or over old mines as described in Arts. 10.3 and 10.4 of Hunt (1984).[7] The size of the openings and the conditions of the roof arches must be defined in both cases; and, for old mines, the condition of the pillars and of their roof and foundation must be known.

In *cavernous soluble rock* the potential for cavern growth or increased load on the roof arch due to groundwater withdrawal is a critical factor in stability and collapse, even if evaluation reveals existing conditions to be stable. Where it has been determined that the collapse potential for large caverns is real, with a high risk to overlying structures, it is best that the site be abandoned. Relatively small caverns and cavities can be treated by grouting, as discussed in Art. 6.3.4. Analysis of the roof arch is described in Art. 10.3.3.

Underground coal mines, excavated by the room and pillar method, can present substantial problems for analysis and evaluation. In old mines, the potential for ground subsidence, which can result in structural distress, is a function of whether or not collapse has occurred. If it has, then the problem remains to evaluate the amount of void remaining, if any. If collapse has not occurred, then the conditions of the pillars and the roof requires evaluation. In the experience of Gray and Meyers (1970),[63] if there are at least 30 m of relatively competent rock overlying the mine, if the percent extraction does not exceed 50%, and the pillars and their foundation support are sound, there is a low probability of collapse as long as roof spans are not excessive.

Old mines usually collapse as a result of the punching of a pillar into either the roof or floor stratum, commonly formed of clay shales which soften with time under the humid conditions of a closed mine. As pillars successively collapse under the additional load, the mine closure results in surface subsidence. An FEM analysis of ground subsidence is described in Art. 10.3.5 of Hunt (1984).[7] Analysis of pillar support and roof stability is described in Art. 10.3.3.

Foundation Support Solutions

New mines should be excavated on the basis either of total extraction permitting collapse to occur during mining operations (if not detrimental to overlying structures) or partial extraction leaving sufficient pillar sections and adequate roof spans to prevent collapse and resulting subsidence at some future date.

Old mines existing beneath proposed construction require predictions of distortions from subsidence and their probability of occurrence. Assessment of the probability of collapse is based on the mine depth (or depths); percent extraction; pillar, roof, and floor conditions; and the amount of closure and surface subsidence that has occurred previously. Where no or small surface distortions are anticipated, foundations may include mats, doubly reinforced continuous footings, or articulated or flexible design to provide compensation for some differential movement of the structure. Such conditions exist where pillar support is adequate, in the order of 50% or greater; where complete or substantial collapse of the mine has occurred; or where the mined coal seam is at a substantial depth, overlain by competent rock.

Where collapse has not occurred and the mine is at relatively shallow depths, and where large distortions are anticipated, small distortions cannot be tolerated, or pillar support is questionable, solutions may include:

- Relocating the project to a trouble-free area
- Providing mine-roof support with construction of piers in the mine or installation of grout columns, or complete grouting of all mine openings from the surface within the confines of a grout curtain installed about the site periphery
- Installing drilled piers from the surface to strong bearing below the mine floor

Concrete Dams: Stability Analysis

Stability Factors

Water stored behind a concrete dam will tend to produce an overturning moment as well as impose a horizontal thrust against the dam. Overall stability analysis against sliding is required where rupture in fractured rock is a possibility, especially in abutment rock, which is often of lower quality than the rock in the valley bottom. Especially critical are weakness planes, such as foliation or bedding-plane shear zones, parallel or nearly so to the foundation base, which daylight downstream. The condition is worsened by a downstream-sloping rock surface or an excavation for a downstream powerhouse.

In high dams, particularly arch types, there is the possibility that the overturning moment can cause the development of high tensile stresses and fissures in the rock at the dam heel, thereby increasing rock permeability and seepage. Overturning forces for a gravity dam are shown in Fig. 6.41*a* *(see also discussion in Art. 9.2.4)*.

Sliding Stability Analysis: Normal Approach

The normal analysis of sliding is illustrated in Fig. 6.41*b*, which is similar in concept to procedures used to analyze slopes *(see Art. 8.3)*. The factor of safety FS is taken as the ratio of the resisting forces to the driving forces, expressed by Deere (1976)[61] as

$$FS = \frac{(V - U)\tan\phi + cL}{P_w} \tag{6.52}$$

where for a horizontal potential failure surface, per unit length of dam,

V = total vertical force, due to dam weight plus water weight above sloping upstream face plus weight of rock above failure surface

U = total vertical uplift force acting on the failure surface

ϕ, c = strength parameters along failure surface (see discussion under Rock Strength Parameters; $c = 0$ for a weakness plane usually is a wise assumption)

L = length of potential failure surface

P_w = horizontal thrust of reservoir water, plus silt, ice, or earthquake load

The case of a passive wedge providing additional resistance against sliding is illustrated in Fig. 6.41*c*; however, the passive wedge resistance should be added with caution, and larger values for FS than normal should be employed. Since the stress-deformation characteristics of the weak seam and the passive wedge are often different, the peak strengths will not be reached simultaneously and therefore will not be additive. One consideration is to reduce the passive resistance by some factor. Deere (1976)[61] suggests that where realistic values are used for strength parameters in stability analysis, FS = 1.5 to 2.0 should be acceptable in Eq. 6.52.

Uplift pressure U is distributed along the base of the dam. The common assumption is that the pressure varies linearly from full reservoir head at the upstream face to full tailwater head at the downstream face, as shown in Fig. 6.42. For most modern concrete dams, especially since the collapse of the thin-arch Malpasset Dam (see dis-

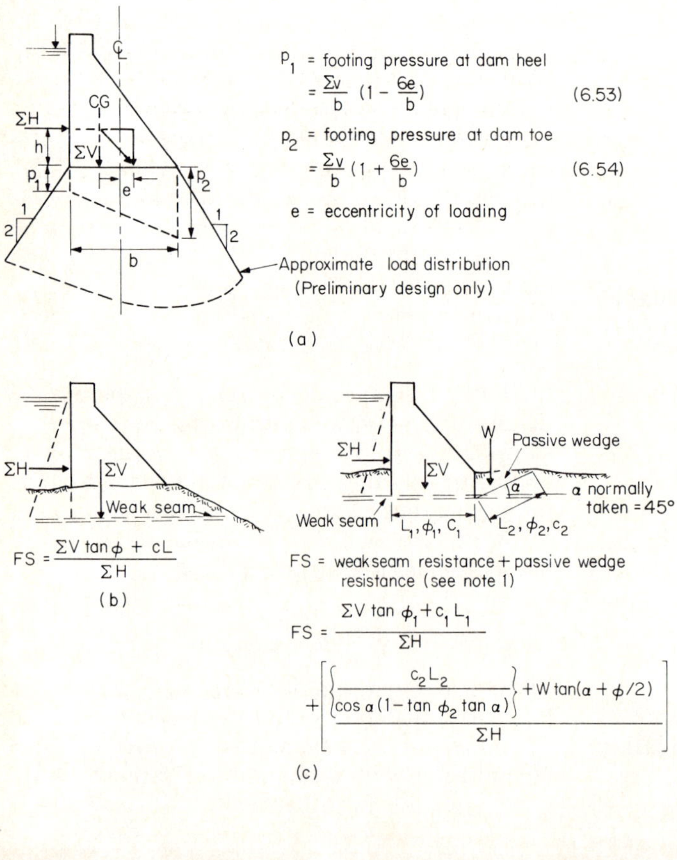

FIG. 6.41 Stability analysis of concrete gravity dam: (*a*) safety against overturning requires resultant kept within middle third of base; (*b*) sliding resistance of 1-ft-wide strip; (*c*) sliding resistance with passive earth pressure. (Note: peak strengths from weak seam and passive wedge may not be additive on account of progressive failure.) [From Underwood and Dixon (1976).[64]]

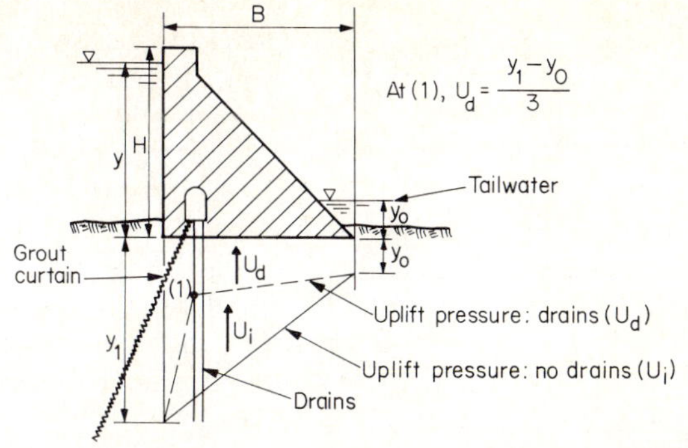

FIG. 6.42 The effect of drains on uplift pressures beneath a concrete gravity dam. [After Deere (1976).[61]]

cussion below), a grout curtain and drains are installed near the upstream portion of the base from a gallery within the dam, as shown in the figure. The drains are considered to be effective in reducing the uplift pressure to a value equal to one-fourth to one-third of the difference between the reservoir head and the tailwater head, and the pressure distribution then becomes as shown in Fig. 6.42. Deere (1976)[61] suggests that a value of one-third is adequate, although some design firms use three-fourths.

Drain spacing depends on rock quality but often ranges from 3 to 10 m on centers. The depth extends to perhaps 75% of that of the grout curtain. Drains should be inspected regularly to ensure that clogging is not occurring, either from the deposition of minerals carried in the groundwater or from the grout curtain. For the case of possibly inoperable drains, Deere (1976)[61] suggests that FS against sliding should be at least 1.0. Grout curtains and drains are discussed in Art. 5.5.2.

Deformation Analysis: FEM Approach

The FEM is becoming the common approach to analysis of rock-structure interaction for major dam foundations and abutments. The usual formulation of the FEM, however, is based on linear behavior of the rock mass under stress and assumes that no tensions develop in the mass, which often is not the case.

The initial elastic analysis will reveal the elements where tension exists. Zienkiewicz (1969)[65] describes a "no-tension" analysis approach to handle nonlinearity, an important concept since a jointed rock can have little resistance to tensile forces. The most critical tensile stresses usually develop at the heel of the dam, and the "no-tension" analysis will determine if the tensile or fissured region will remain stable. FEM analysis for dams also is described by Londe and Tardieu (1975).[66]

Malpasset Dam Failure

Seepage pressure buildup is considered to be the cause of the rupture of the thin-arch Malpasset Dam, near Fréjus, France, on December 2, 1959, during which over 400 people lost their lives [Jeager (1972)[67] and Arts. 8.3.2 and 8.3.4 of Hunt (1984)[7]]. The tightly jointed and finely fissured foundation gneiss was considered to have more than adequate strength to support the compressive stresses and the arch thrust from the 60-m-high dam. Failure studies have concluded that as the water level rose behind the dam, the tensile strains opened joints. This reduced the rock-rock contact area, thus increasing the area against which the water pressure could act, permitting full reservoir hydrostatic pressure to be exerted against the wedge illustrated in Fig. 6.43*b*. Progressive displacement of the dam continued over a period of about a year as the water level rose until rupture suddenly occurred along a crushed zone in the rock, and the dam collapsed. Since the Malpasset failure it has become standard practice to drain the foundation rock of concrete dams, regardless of rock quality.

6.3.4 CONSTRUCTION ASPECTS

Specifications and Inspection

Purpose and Scope

When a structure is to be supported on shallow foundations bearing on rock, specifications are required to define design depth, desired rock type and quality, rock treatment, and excavation blasting. Documents with different purposes and scopes are required of the geotechnical engineer for the project designer, the contractor(s), and the field engineers.

Design Depth, Rock Type, and Quality

Design depth should be related to the rock quality desired, with an estimated range in eleva-

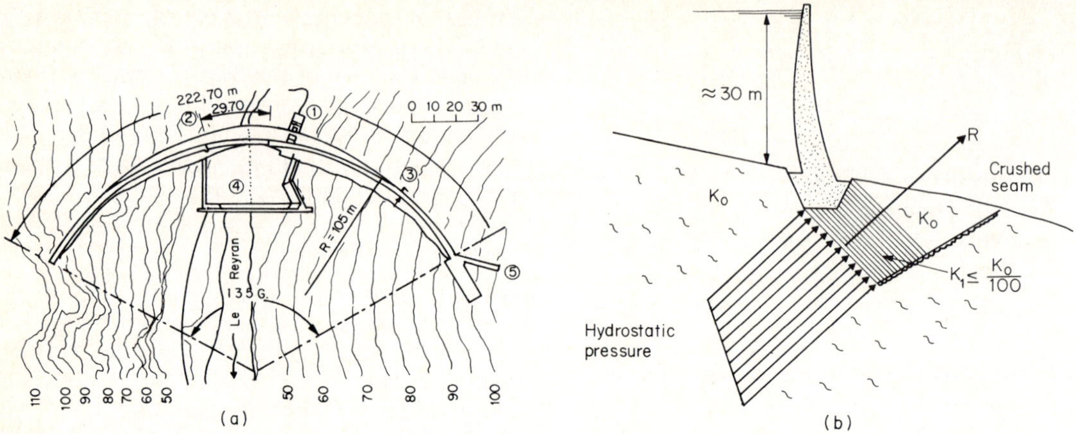

FIG. 6.43 Malpasset Dam (thin-arch) failure. (*a*) General plan: 1. outlet, 2. spillway, 3. location of failure initiation, 4. spillway apron, 5. buttress of left abutment. (*b*) Typical section near left abutment showing suggested failure mode. [From Londe (1966).[68]]

tions at which it will be encountered on the basis of interpretations of the exploration data. For the field engineer carrying out inspection, *rock quality* may be defined in terms of rock type, hardness, and joint or fracture conditions. Hardness criteria may be qualitative, as given in Table 3.14 under "Diagnostic Features," or quantitative, based on the Schmidt hardness test (*see Art. 6.3.2, Deformation Moduli*). The acceptability of fracture conditions is related to spacing and width of opening, as given in Table 6.14. Fracture variations from these criteria are brought to the attention of the design engineers who will prescribe special treatments.

Rock Treatment

Rock treatment is placed in two categories: surface and subsurface. On routine projects with favorable rock conditions, only surface treatment is necessary. Surface treatment involves stripping undesirable rock and cleaning the surface prior to concreting. Unfavorable conditions require special treatment. Earth dams also require special considerations to ensure that no seepage will occur along the embankment or core-rock contact (*see Art. 5.5.2*). Subsurface treatment or ground improvement is required to strengthen the rock mass, reduce hazardous conditions, or control groundwater flow and seepage forces.

Excavation and Blasting

Exploration data should be adequate to enable contractors to judge the depth to which rock is rippable (*see Art. 5.1.4*) and where blasting will be required. As an alternative to blasting, wedging methods, which are relatively expensive, have been used. Holes are drilled at about 2-ft centers, and cracks are produced between the holes by expanding hydraulic jacks in the holes or by driving feather wedges into the holes with a jackhammer. Pneumatic rockbreakers mounted on tractors are an alternate in some conditions.

Excavation blasting must be carefully planned, performed, and controlled to avoid a decrease in the quality of the rock outlining the excavation limits. Specifications are based on performance and are particularly stringent for the zones approaching the final excavation depths and walls.

Rock Treatment

Unfavorable Surface Conditions

Once the final proposed design grade is reached and all boulders and rock of undesirable quality have been removed, unfavorable conditions may still remain in the form of open or soil-filled fractures or cavities of excessive width, or the excavation floor may be subject to heave from swelling clay shales or from high residual stresses. Some cases and solutions are illustrated in Fig. 6.44.

Fractures or cavities are cleaned of soil or other relatively weak material, usually to depths of 1.5 to 2.0 times the opening width, and filled with lean or "dental" concrete (Fig. 6.44*a*). In the case of a wide zone of poor quality rock, such as a deeply weathered zone or seam of schist, it may be decided that redesigning the foundation to span the weak zone (Fig. 6.44*b*) is a more economical approach than excavation and backfill. In heavily jointed rock, blanket grouting is used beneath concrete and earth dams to both strengthen the rock and decrease its permeability *(see Art. 5.5.2)*. Rock bolts are used to restrain floor heave (Fig. 6.44*c*) and wall deflections resulting from relief of residual stresses *(see Arts. 5.5.3 and 9.3.5)*.

Excavations in expansive shales should be made in sections as small as practical as final depth is

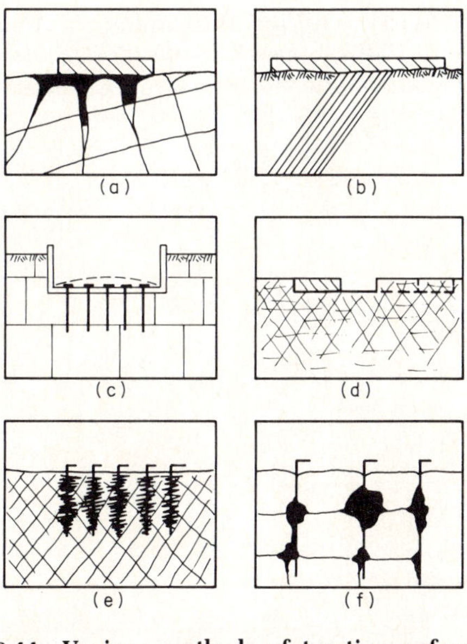

FIG. 6.44 Various methods of treating unfavorable rock conditions: (*a*) clean surface cavities and fill with lean concrete; (*b*) span over wide, deep weak zone; (*c*) anchor slab against residual stress-relief heave; (*d*) in active shales, excavate and pour concrete in sections after placing asphalt coating if pyritic; (*e*) grout to strengthen heavily jointed rock or fill cavities in cavernous rock (*f*).

approached (Fig. 6.44*d*), water infiltration prevented, and the bottom covered immediately with concrete. As mentioned in Art. 6.3.1, the surface of pyritic or "alum" shales may be sprayed with asphalt (Fig. 6.44*d*) immediately upon exposure at final grade to inhibit the process of pyrite oxidation and the formation of the sulfuric acid which will attack the concrete.

Subrock-Surface Improvement

Low quality rock, particularly when the poor quality is due to high porosity or intense fracturing, can be strengthened substantially by grouting, as discussed in Art. 5.5.3 (Fig. 6.44*e*). The grouting of limestone caverns (Fig. 6.44*f*), also discussed in Art. 5.5.3, meets with various degrees of success, depending upon cavern extent beneath the site and how well their locations and sizes have been identified by exploration. Grouting usually proceeds upward along vertical joints from some predetermined depth to encounter the cavities. Such treatments can be quite costly, but in limestone they provide some protection against cavern growth due to dewatering. As mentioned in Art. 6.3.3, construction should be avoided over caverns too large to treat economically where the risk of collapse remains high. Construction over old mines also is discussed in Art. 6.3.3.

Grout curtains are used beneath concrete and earth dams to control seepage flow and uplift, which has the added benefit of inhibiting rock deterioration *(see Arts. 5.5.2 and 5.5.4)*. Caution is required, however, since complete cutoff is seldom achieved.

Open-Excavation Blasting Control

Objectives

The objective in excavation blasting control is rock-quality preservation, including maintaining the compression modulus and joint tightness without increasing the joint intensity. In addition, for concrete dams, control maintains low permeability and high concrete-rock-base friction. Stepped excavations, often required for nuclear power plants and concrete dam complexes, must preserve the rock walls by not introducing new tensile cracks or opening existing joints, and should have minimal backbreak along the top of cut.

In very large excavations where the rock becomes a resource, it is also an objective to obtain sizes suitable for crushing for aggregate with a minimum of fines or sizes suitable for construction of a rock-fill dam, slope protection riprap, or a jetty or breakwater.

Blasting Fundamentals and Nomenclature

Explosive *strength* is a measure of the work done by a certain weight or volume of explosive in breaking up a rock mass. Volume or bulk strength is related to weight strength by specific gravity. A higher specific gravity means a smaller blast-hole volume for a given explosive energy, but the explosive cost is higher. Explosives are purchased on the basis of weight, but the energy they release is evaluated on the basis of bulk strength.

The *powder factor* p_f, or specific charge, equals the explosive weight required to break a unit volume of rock, given in lb/yd^3 or kg/m^3. It is related to the explosive type, and rock type and quality. For example, for sound granite, the powder factor will be in the order of 1.0 lb/yd^3 (0.5 kg/m^3) for gravity loaded ANFO (ammonium nitrate–fuel-oil mixture), a common explosive; $p_f = 0.7$ lb/yd^3 for gelatine (nitric acid–cotton gel with nitroglycerine); and $p_f = 0.5$ lb/yd^3 for ANFO in relatively weak rock. For large projects, tests are run in situ for various explosives and rock conditions to correlate powder factor with seismic compression-wave velocities to obtain optimum values.

The *optimum charge* equals the powder factor times the burden volume, where the *burden volume* equals the distance of the line of holes from the free face *(burden) b* times the hole *spacing s* times the bench height *h*, as illustrated in Fig. 6.45. The charge may vary with depth in a given hole, as when a high energy bottom load is required to remove a heavy toe burden, or in vertical holes where an inclined break is desired. Experience has shown that an effective utilization of the explosive charge for *production blasting* is given by using a *blast-hole diameter* equal to about 1/40 of the burden *b* [Hoek and Bray (1977)[69]]. Large blast holes are more likely to result in flyrock and airblast problems, however, than are smaller holes. The *effective burden* b_e is the distance between a row of holes and the

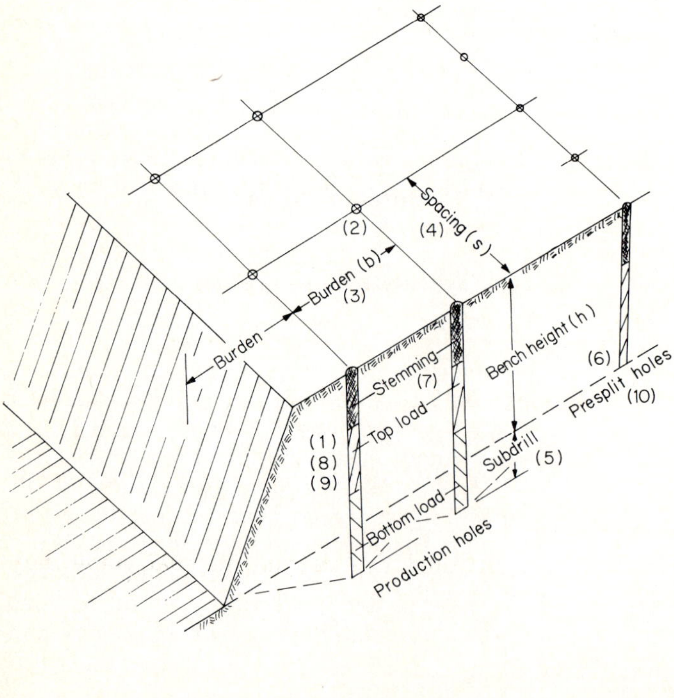

FIG. 6.45 Bench-blasting nomenclature and factors: (1) type, weight, and distribution of explosives; (2) blasthole diameter; (3) burden; (4) spacing; (5) subdrill depth; (6) hole inclination; (7) stemming; (8) initiation sequence; (9) delays between successive holes; and (10) presplitting. [After Hoek and Bray (1977).[69]]

nearest free face, and it must be chosen correctly to avoid flyrock and still permit the blast to efficiently fragment the rock. It depends on the blasthole pattern and firing sequence. Experience suggests that an *effective spacing* s equal to 1.25 times burden gives good results in most rocks, but in good quality rock, improved fragmentation may be obtained with $s = 4$ to $8b$. Joints reduce blast effectiveness and require closer spacing.

Holes are *subdrilled* (Fig. 6.45) to a depth of 0.2 to 0.3 times the spacing during production blasting, to provide for rock breakage to bench level for machine removal. Care is required to avoid overfragmentation of the bench, and where bench quality is critical subdrilling may be reduced or even eliminated. If an inclined free face is required, all blast holes are drilled inclined. *Stemming*, or filling the upper portion of the holes with drill cuttings or finely crushed rock, is necessary to avoid upward flyrock and to direct the explosive effort into the rock mass. Optimum stemming depends on rock quality, but usually varies from 0.7 to $2b$.

Following drilling and charging, the blast is tied up in a pattern with detonating cord. To maintain rock quality and improve blast effectiveness, the blast is divided into a number of sequential detonations by means of *delays*. The front row is detonated and the burden moves away from the mass to create a new free face, followed by the second row, and so on. In production blasting, delay intervals of 1 to 2 ms/ft of burden (3 to 6 ms/m) are typical [Hoek and Bray (1977)[69]].

Controlled Blasting Techniques

In *cushion blasting* the main excavation is carried out by production blasting up to the last row of blastholes along the limit line. The perimeter holes are drilled at a reduced spacing, in the order of $0.8b$, and loaded lightly with well-distributed charges with sand in between the charges. The charges in all holes are fired simultaneously [Peck et al. (1974)[17]].

In *presplitting blasting*, a row of closely spaced (1½- to 3-ft centers), small-diameter (2- to 4-in) holes is drilled along the line of the final face and lightly charged. The charge is decoupled from the rock by leaving an air space between the charge and the blasthole walls by placing the charge in a plastic tube which is inserted into the hole with spacers. The row is fired *before* the main charge and the close spacing of the holes results in the formation of a clean fracture between holes. Presplitting is not very effective in obtaining a smooth face in hard rocks, well-jointed with open joints inclined to the presplit line, nor in soft or foliated rocks.

Smooth-wall blasting is similar to presplit blasting except that the final row is fired *after* the main blast. The technique is used for cleanup to minimize the danger of rockfalls where jointing has created loose, blocky conditions and the face has been heavily blasted, and is favored in tunneling.

In *line drilling* the main excavation is carried up to two to three rows from the limit line, and the remaining rock is slabbed off a row at a time by the use of delays. Hole spacing is reduced toward the boundary and the holes lightly charged. Perimeter holes are spaced at two to three diameters and not charged.

Operations for Foundations

At all times the major concern is with preservation of rock quality at the excavation limits, but during the initial phases of a large excavation, operation planning should provide for use of the rock material as a resource. For example, a low-velocity blasting agent helps to reduce the percentage of fines and maintain natural block size, whereas a high-velocity blasting agent will result in high fragmentation. If there are existing nearby structures or slopes, charge weights and delays are selected to control particle velocity as discussed in Art 5.6.3. In general, no blasting should be performed within 100 ft (30 m) of concrete or grout with less than 7 days strength or within 50 ft (15 m) of any concrete or grout regardless of age [Lutton (1977)[70]], conditions which are hardly practical in many field situations.

Production blasting in less-sensitive areas may begin with a layout of 4- to 8-in (10- to 20-cm) holes on a 10-ft (3-m) spacing or greater. Depths of individual lifts depend on final grade, but often are initially between 10 and 15 ft (3 to 5 m), with shallower depths considered more desirable for control. (In quarries and large road cuts holes may be 6 to 8 in dia. at 20-ft spacings, although depths are usually limited to about 40 ft because of difficulty in maintaining alignment. In foundation excavation, small holes and closer spacing result in better fragmentation, which facilitates removal.) The powder factor for ANFO will usually range from 0.5 to 1.0 lb/yd^3 for relatively weak and relatively strong rock, respectively.

As the excavation approaches final lines and grades, hole depths, spacing and powder factors are reduced. A line of presplit holes often is drilled ahead of the main blasting area both in time and spatially about half the length of the adjacent production blast. Less commonly, delays of 100 to 200 ms are the only separation. Typical blasthole arrays in confined structure excavations are in the order of 5-ft (1.5-m) squares with 3-in (7.5-cm) holes. Upon approaching within 15 ft (5 m) of final grade, the holes are not loaded below two-thirds of the distance to grade which, in effect, reduces the last two lifts to 10 and 5 ft (3 and 1.5 m) or less and necessitates a reduction in hole spacing [Lutton (1977)[70]].

Special care is required along the top and at the outside corners of stepped-structure excavations; this is provided by line drilling in place of presplitting. Final trimming to grade is accomplished with light drills and very light charges and by jackhammers.

REFERENCES

1. ENR (1977) "12-Story Building Tilts in Rio," *Engineering News-Record*, November 17.
2. Marr, W. A., Ramos, J. A., and Lambe, T. W. (1982) "Criteria for Settlement of Tanks," *Proc. ASCE, J. Geotech. Engrg. Div.*, Vol. 108, No. GT8, August, pp. 1017–1039.
3. CGS (1978) *Canadian Foundation Engineering Manual*, Part 2, "Shallow Foundations," *Canadian Geotechnical Society*, Montreal, March.
4. Skempton, A. W. and MacDonald, D. H. (1956) "The Allowable Settlement of Buildings," *Proc., Inst. of Civ. Engrs.*, London, Vol. 5, Part III, pp. 727–784.
5. Bjerrum, L. (1963) "Discussion," *European Conf. on Soil Mechs. and Found. Engrg.*, Vol. II, Wiesbaden, West Germany, p. 135.
6. Grant, R., Christian, J. T., and Vanmarke, E. H. (1974) "Differential Settlement of Buildings," *Proc. ASCE, J. Geotech. Engrg. Div.*, Vol. 100, No. GT9, Sept., pp. 973–991.
7. Hunt, R. E. (1984) *Geotechnical Engineering Investigation Manual*, McGraw-Hill Book Co., New York.
8. ENR (1976) "Expansive Soils Culprit in Cracking," *Engineering News-Record*, November 4, p. 21.
9. Teng, W. C. (1975) "Mat Foundations," *Foundation Engineering Handbook*, Van Nostrand-Reinhold Co., New York, Chap. 17.
10. HRB (1952) "Frost Action in Roads and Airfields," *Highway Research Board Spec. Report No. 1*, Pub. 211, National Academy of Sciences-National Research Council, Washington, D.C.
11. Sowers, G. F. (1962) "Shallow Foundations," *Foundation Engineering*, G. A. Leonards, ed., McGraw-Hill Book Co., New York, Chap. 6.
12. Smith, D. W. (1977) "Why Do Bridges Fail?" *Civil Engineering*, ASCE, November, pp. 59–62.
13. NAVFAC (1982) *Design Manual DM 7.2: Foundations and Earth Structures*, Department of the Navy, Naval Facilities Engineering Command, Alexandria, Va.
14. Building Code, City of New York (1968) "Building Code—Local Law No. 76 of the City of New York" effective December 6, 1968, amended August 22, 1969, *The City Record*, New York.
15. Gibbs, H. J. and Holtz, W. G. (1957) "Research on Determining the Density of Sands by Spoon Penetration Testing," *Proc. 4th Intl. Conf. on Soil Mechs. and Found. Engrg.*, London, Vol. 1, pp. 35–39.
16. Terzaghi, K. and Peck, R. B. (1948) *Soil Mechanics in Engineering Practice*, John Wiley & Sons, New York.
17. Peck, R. B., Hanson, W. E., and Thornburn, T. H. (1974) *Foundation Engineering*, 2d ed., John Wiley & Sons, New York.
18. Terzaghi, K. and Peck, R. B. (1967) *Soil Mechanics in Engineering Practice*, 2d ed., John Wiley & Sons, New York.
19. Schmertmann, J. H. (1970) "Static Cone to Compute Static Settlement Over Sand," *Proc. ASCE, J. Soil Mech. and Found. Engrg. Div.*, Vol. 96, No. SM3, May, p. 1045.
20. Oweiss, I. S. (1979) "Equivalent Linear Model for Predicting Settlements of Sand Bases," *Proc. ASCE, J. Geotech. Engrg. Div.*, Vol. 105, No. GT12, December, p. 1525.

21. Bazaraa, A. R. S. (1967) "Use of S. P. T. for Estimating Settlements of Shallow Foundations on Sand," Thesis, University of Illinois, Urbana, Ill.

22. Sanglerat, G. (1972) *The Penetrometer and Soil Exploration*, Elsevier Scientific Pub. Co., Amsterdam, The Netherlands.

23. Meyerhof, G. G. (1965) "Shallow Foundations," *Proc. ASCE, J. Soil Mech. and Found. Div.*, Vol. 91, No. SM2.

24. Seed, H. B. (1979) "Soil Liquefaction and Cyclic Mobility Evaluation for Level Ground During Earthquakes," *Proc. ASCE, J. Geotech. Engrg. Div.*, Vol. 105, No. GT2, February, pp. 201–255.

25. Marcuson, W. F. and Bieganousky, W. A. (1977) "Laboratory Standard Penetration Tests on Fine Sands," *Proc. ASCE, J. Geotech. Engrg. Div.*, Vol. 103, No. GT6, June, pp. 565–588.

26. Marcuson, W. F. and Bieganousky, W. A. (1977) "SPT and Relative Density in Coarse Sands," *Proc. ASCE, J. Geotech. Engrg. Div.*, Vol. 103, No. GT11, November, pp. 1255–1305.

27. Meyerhof, G. C. (1956) "Penetration Tests and Bearing Capacity of Cohesionless Soils," *Proc. ASCE, J. Soil Mech. and Found. Engrg. Div.*, Vol. 82, No. SM1, p. 866.

28. Lambe, T. W. and Whitman, R. V. (1969) *Soil Mechanics*, John Wiley & Sons, New York.

29. Whitman, R. V. and Richart, F. E., Jr. (1967) "Design Procedures for Dynamically Loaded Foundations," *Proc. ASCE, J. Soil Mech. and Found. Engrg. Div.*, Vol. 93, No. SM6, November, pp. 169–193.

30. Vesić, A. S. (1963) "Bearing Capacity of Deep Foundations in Sands," *HRB* No. 39, Highway Research Board, National Academy of Sciences, National Research Council, pp. 2–153.

31. Vesić, A. S. (1975) "Bearing Capacity of Shallow Foundations," *Foundation Engineering Handbook*, Winterkorn and Fang, eds., Van Nostrand-Reinhold, New York, Chap. 3.

32. Vesić, A. S. (1973) "Analysis of Ultimate Loads of Shallow Foundations," *Proc. ASCE, J. Soil Mech. and Found. Engrg. Div.*, Vol. 99, No. SM1, January, pp. 45–73.

33. Ismael, N. F. and Vesić, A. S. (1981) "Compressibility and Bearing Capacity," *Proc. ASCE, J. Geotech. Engrg. Div.*, Vol. 107, No. GT12, December, pp. 1677–1691.

34. Terzaghi, K. (1943) *Theoretical Soil Mechanics*, John Wiley & Sons, New York.

35. Converse, F. J. (1939) "The Practical Use of Shear Test Data," paper presented to *36th Ann. Con. American Road Builders Assoc.*, San Francisco.

36. Tschebotarioff, G. P. (1973) *Foundations, Retaining, and Earth Structures*, McGraw-Hill Book Co., New York.

37. Prandtl, L. (1921) "Ueber die Eindringungsfestigkeit Plastisher Baustoffe und die Festigkeit von Schneiden," *Zeitschrift fur Angewandte Mathematik und Mechanik*, Vol. 1., No. 1, pp. 15–20.

38. Reissner, H. (1924) "Zum Erddruckproblem," *Proc. 1st Natl. Conf. on Appl. Mech.*, Delft, The Netherlands, pp. 295–311.

39. Caquot, A. and Kérisel, J. (1953) "Sur le terme de surface dans le calcul des fondations en milieu pulvérulent," *Proc. 3rd Intl. Conf. on Soil Mechs. and Found. Engrg.*, Zurich, Vol. 1, pp. 336–337.

40. de Beer, E. E. (1967) "Proefondervindelijke bildrage tot de studie van het gransdraagvermogen van zand oder funderingen opstall: Bepaling von der vormfactor, s_b," *Annales des Travaux Publics de Belgique*, Vol. 68, No. 6, pp. 481–506; Vol. 69, No. 1, pp. 41–88; Vol. 69, No. 4, pp. 321–360; Vol. 69, No. 5, 395–442; Vol. 69, No. 6, 495–522.

41. Vesić, A. S. (1970) "Research on Bearing Capacity of Soils," unpublished.

41a. Brinch Hansen (1970) "A Revised and Extended Formula for Bearing Capacity," *Bull. 28*, Danish Geotech. Inst., Copenhagen, pp. 5–11.

42. Menard, L. F. (1965) "Rules for the Computation of Bearing Capacity and Foundation Settlement Based on Pressuremeter Tests," *Proc. 6th Intl. Conf. Soil Mechs. and Found. Engrg.*, Montreal, Vol. 2, pp. 295–299.

43. Baguelin, F., Jezequel, J. F., and Shields, D. H. (1978) *The Pressuremeter and Foundation Engineering*, Trans. Tech. Pub., Germany.

44. SOL SOILS (1975) "The Menard Pressuremeter: Interpretation and Application of Pressuremeter Test Results to Foundation Design," *SOL SOILS* No. 26, Paris.

45. Meyerhof, G. G. (1957) "The Ultimate Capacity of Slopes," *Proc. 4th Intl. Conf. Soil Mechs. and Found. Engrg.*, London, Vol. 1, pp. 384–386.

46. Meyer, K. T. and Lytton, R. M. (1966) "Foundation Design in Swelling Clays," paper presented to Texas Section, ASCE, October.

47. Wray, W. K. (1980) "Analysis of Slabs-on-Ground Over Expansive Soils," *Proc. ASCE, 4th Intl. Conf. on Expansive Soils*, Denver, Co., Vol. 1, pp. 558–581.

48. Turnbull, W. J. (1969) "Expansive Soils—Are We Meeting the Challenge?" *Proc. 2d Intl. Research*

and Engrg. Conf. on Expansive Clay Soils, Texas A&M University, College Station, Tex. pp. 6–8.

49. Sultan, H. A. (1969) "Foundation Failures on Collapsing Soils in the Tucson, Arizona Area," *Proc. 2d Intl. Conf. on Expansive Clay Soils,* Texas A&M University, College Station, Tex.

50. Richart, F. E., Jr. (1975) "Foundation Vibrations," *Foundation Engineering Handbook,* Winterkorn and Fang, eds., Van Nostrand-Reinhold Co., New York, Chap. 24.

51. Richart, F. E., Jr., Hall, J. R., and Woods, R. D. (1970) *Vibrations of Soils and Foundations,* Prentice-Hall, Englewood Cliffs, N.J.

52. Converse, F. J. (1962) "Foundations Subjected to Dynamic Forces," *Foundation Engineering,* G. A. Leonards, ed., McGraw-Hill Book Co., New York, Chap. 8.

53. Feld, J. (1966) "Age Change in the Bearing Capacity of Mica Schist," *Proc. 1st Intl. Cong. Intl. Soc. for Rock Mechs.,* Vol. 2, pp. 523–524.

54. Grattan-Bellew, P. E. and Eden, W. J. (1975) "Concrete Deterioration and Floor Heave Due to Biogeochemical Weathering of Underlying Shale," *Canadian Geotechnical J.,* National Research Council of Canada, Ottawa, Vol. 12, No. 3, August, pp. 372–378.

55. Peck, R. B. (1976), "Rock Foundations for Structures," *Rock Engineering for Foundations and Slopes, Proc. ASCE,* New York, Vol. 2, pp. 1–21.

56. Thornburn, S. (1966) "Large Diameter Bored Piles Founded on Bedrock," *Proc. Symp. on Large Bored Piles,* Inst. of Civ. Engrs., London.

57. Teng, W. C. (1962) *Foundation Design,* Prentice-Hall, Englewood Cliffs, N.J.

61. Deere, D. U. (1976) "Dams on Rock Foundations—Some Design Questions," *Rock Engineering for Foundations and Slopes, Proc. ASCE,* New York, Vol. 2, pp. 55–62.

62. Sowers, G. F. (1976) "Foundation Bearing in Weathered Rock," *Rock Engineering for Foundations and Slopes, Proc. ASCE,* New York, Vol. 2, pp. 32–42.

63. Gray, R. E. and Meyers, J. F. (1970) "Mine Subsidence and Support Methods in the Pittsburgh Area," *Proc. ASCE, J. Soil Mech. amd Found. Engrg. Div.,* Vol. 96, No. SM4, July.

64. Underwood, F. B. and Dixon, N. A. (1976) "Dams on Rock Foudations," *Rock Engineering for Foundations and Slopes, Proc. ASCE,* Vol. 2, pp. 125–146.

65. Zienkiewicz, O. C. (1969) "Continuum Mechanics as an Approach to Rock Mass Problems," *Rock Mechanics in Engineering Practice,* Stagg and Zienkiewicz, eds., John Wiley & Sons, New York, Chap. 8.

66. Londe, P. and Tardieu, B. (1975) "Practical Foundation Design for Dams," *Design Methods in Rock Mechanics, 16th Symp. on Rock Mechs.,* University of Minnesota, Minneapolis, Minn., September, *Proc. ASCE* (1977), pp. 115–138.

67. Jeager, C. (1972) *Rock Mechanics and Engineering,* Cambridge University Press, Cambridge, England.

68. Londe, P. (1966) Panel discussion, *1st Intl. Cong. Intl. Soc. Rock Mechs.,* Lisbon, Vol. 3, pp. 449–453.

69. Hoek, E. and Bray, J. W. (1977) *Rock Slope Engineering,* Inst. of Min. and Metall. London.

70. Lutton, R. J. (1977) "Constraints on Blasting Design for Construction," *Design Methods in Rock Mechanics, Proc. ASCE, 16th Symp. on Rock Mech.,* University of Minnesota, Minneapolis, Minn. (1975), pp. 365–369.

CHAPTER SEVEN

DEEP FOUNDATIONS

7.1 INTRODUCTION

7.1.1 DEFINITIONS

A *deep foundation* is defined as one with a depth greater than its least width and includes foundations with structural members referred to as piles, piers or bored piles, and caissons. Its purpose is to transfer loads through materials too weak or compressible to support shallow foundations to more suitable material (*see Art. 6.1* for an introduction to foundations). In many cases support is derived from friction along the sides as well as from end bearing, although some deep foundations are supported only by side friction and others only by end bearing. Soil and rock anchors are also included in this chapter because of the analytical similarities of the more common deep anchors with deep foundations.

Piles are structural members with cross-sectional areas that are small compared to their length, constructed of timber, concrete or steel, or composites of these materials, normally installed by driving by hammer or vibrating, and occasionally by augering. Predrilling or other procedures may be necessary to permit penetration through strong soils or obstacles to the desired depth or tip elevation.

Piers are structural members of relatively large diameter constructed of concrete placed in a preexcavated hole, either hand-dug or drilled, and at times are referred to variously as *bored piles*, *large-diameter piles*, *drilled piers*, or *drilled caissons*. The shaft can be enlarged at the base in certain soil conditions, resulting in a belled or underreamed pier. These are used normally to obtain higher capacities than are possible with piles. *Caissons* refer to very large concrete boxes or tubes that are sunk into place. When caissons are sunk to substantial depths below ground and water level, excavation may be performed under air pressure, hence the term *pneumatic caisson*.

Anchors usually are long, very slender structural members installed in soil or rock to provide resistance against lateral earth pressures or uplift.

7.1.2 CLASSIFICATION OF PILES AND PIERS

Piles can be found classed variously by material composition, by installation method, by ground effect, and by function as a foundation. There is a tendency in Europe and South America not to use the term pier, but rather to refer to large-diameter or bored piles.

Material Classes

These include timber, steel, concrete, or composite piles.

Installation Classes

Driven piles are installed by hammer impact or by a vibrating machine. They tend to densify looser granular materials, thereby increasing soil resistance, and to remold cohesive soils, thereby temporarily reducing soil strength along the shaft.

Bored piles are installed normally in a hole pre-excavated by machine or hand-dug. A variation is the *auger pile*. Bored piles tend to relieve the lateral earth pressure, thereby reducing soil strength along the shaft to values below those of driven piles. Most of their support frequently is considered in analysis to be derived from end bearing, although shaft friction is often substantial.

Ground-Effect Classes

Displacement piles displace a substantial volume of soil during installation and generally include all driven piles, although steel H-section piles often are referred to as "nondisplacement" piles because of their relatively small volume. In cohesionless granular soils there is a tendency for compaction, whereas in clays heaving of the ground surface and nearby piles often results. Heave volume as a percent of the pile volume may vary in the foundation from 50% in clay to 30% in soft silty clay. To reduce displacement in clays, driven piles are installed in predrilled holes.

Nondisplacement piles include bored piles, although H piles can, as noted above, be found referred to as nondisplacement piles. They usually tend to derive their support from end bearing, but in many cases support is derived from both end bearing and shaft friction.

Function Classes

Friction piles (also termed *floating piles*) provide support for compressive loads by distribution of most of the load to the soils along the shaft through friction and adhesion. Displacement piles are used as friction piles and are most suitable in granular soils. Of necessity, they are used at times in clays, but careful analysis is required to evaluate the effects of remolding and thixotropy, and adjacent ground movements.

End-bearing piles transfer compressive loads through water or weak soil to strong material and include both displacement and bored piles. *Compaction piles* improve the support characteristics of the sands into and below which the piles are driven. In many cases both displacement and nondisplacement piles derive support from both shaft friction and end bearing with the distribution depending upon pile type and geologic conditions.

Tension piles provide anchorage against uplift by mobilization of frictional forces along the shaft.

Batter piles provide resistance against horizontal or inclined loads and can be either end-bearing or friction piles. They are used commonly to support retaining walls, bridge piers, abutments, trestle bents, and waterfront structures where ship-docking forces are imposed.

7.1.3 PERFORMANCE FACTORS

General

The subject of deep foundations can be one of the more complex areas of foundation engineering not only because there are many types to choose from but also because their design requires the consideration of many factors, some of which are not well defined. In a foundation the various types can act as single units or as groups, and can be subjected to axial loads in excess of structural loads (from "downdrag") as well as uplift and lateral forces. Their performance is very much dependent upon proper installation practices, which can be even more important than the analytical phase of a project.

Evaluation and Design Procedures

Evaluation and design of a deep foundation proceeds stepwise through the following elements:

1. *Perform subsurface investigation* to accurately determine the stratigraphy and to obtain measures of strength and compressibility properties both in the laboratory and in situ. In many instances piles supported in sands are designed on the basis of standard penetration test (SPT) *N* values. Useful data also are obtained with the cone penetrometer test (CPT) and the pressuremeter. Data from the latter test are particularly useful for laterally loaded deep

foundations and for materials difficult to sample or test with the SPT or CPT, such as strong gravelly soils or soft rock.

2. *Select deep foundation types* suitable for the given conditions. Selection is based on economic considerations which are influenced by the types and magnitudes of the loads to be supported, the depth to suitable supporting material (length of unit required), the nature of the material to be penetrated, and pile deterioration and required life.
3. *Evaluate analytically* the *axial capacity* of single units of the alternate feasible types statically and dynamically employing measured soil properties. The objective is to estimate the length required to develop the capacities of the various types. Axial compressive forces are considered in terms of the resisting strength of the soil or rock, and tensile forces (uplift) in terms of strength and swelling characteristics. Consolidation of weak soil strata can impose axial loads from downdrag on end-bearing piles or piers. Bending moments and deflections from lateral loads are considered to evaluate the adequacy of the section and the necessity for batter foundations. Safety factors are applied to the computed ultimate capacities to achieve allowable or design capacities.
4. *Perform load tests* on test piles, and sometimes on test piers, to substantiate the results of the static and dynamic analyses and to acquire additional data on the soil and rock characteristics. Axial compression, tension, or lateral load tests may be performed. (On many routine pile projects, load tests are performed usually after the piles are installed as permanent foundations.)
5. *Compute pile or pier spacing* on the basis of imposed loads, and evaluate the *group* axial and lateral capacity. Economic comparison of several types may be made.
6. *Perform settlement analysis* for single elements or groups, as the case may be. Settlement is normally a concern with friction piles or piers in clays, in all cases where the pile tips or piers are underlain by material weaker and more compressible than those in which they are supported, and in heavily loaded end-bearing units supported in soils or on rock of less than good to excellent quality.
7. *Make the final selection* based on an economic comparison of the unit costs including the cost of material, which is a function of length, diameter, composition, and load-carrying capacity (number of units required), as well as on consideration of installation problems which relate to geologic and general physical conditions (installation equipment access, adjacent buildings, obstacles, etc.) of the site. It may be apparent at the end of Step 3 that a particular type is most practical, or it may be necessary to perform load tests on several types in Step 4. Normally, however, the final selection is made during Step 5 when the number of units required to support the imposed loads is determined.
8. *Inspect the installation* continuously and *perform load tests* on selected completed units. *Dynamic testing* of steel and concrete piles is performed during driving installation. Large-diameter concrete piles may be *proof-tested* for concrete quality by various methods.

Causes of Failure

Performance failure of deep foundations may result from inadequate analysis, improper installation procedures, and postconstruction deterioration.

Inadequate Analysis

Inadequate analysis may result in excessive deflection due to selection of nonrepresentative strength parameters, inadequate penetration, neglected consideration of downdrag forces, imposed stresses on underlying strata, and heave from inadequate protection in swelling clays. Excessive lateral movement and moments may develop from inadequate consideration of lateral forces imposed at the head by wind, wave, and ship loads or along the shaft imposed by lateral earth pressures induced by nearby excavations or fills on soft soils or hillside slope movements.

Installation Procedures

Installation procedures may weaken deep foundations as structural members when overdriving results in breakage of timber and concrete and in dogleg steel H piles; when shell piles contain voids in the concrete which may result if mix or placement is improper; when compacted or injected concrete has suffered a decrease in shaft diameter from pinching if it has been installed uncased through soft ground; when deep piers contain voids, decreased diameters, or poor quality concrete conditions that occur particularly below the GWL when casing is withdrawn; and when proper concreting procedures are not followed. Groundwater infilling or heave of the excavation bottom from excess hydrostatic pressure also can weaken bored piles.

Postconstruction Deterioration

Timber piles may be subjected to rot in the fluctuating water zone, marine borers below water levels, and termites above GWL. *Steel,* if unprotected, will suffer corrosion in salt water, when exposed either in the shaft or in tension cracks in precast, prestressed, or bored piles. *Concrete* and *steel,* if unprotected, can suffer corrosion from chemical reaction, especially where driven through fills with highly acidic conditions, always a problem around chemical plants.

7.2 TYPES, SELECTION, AND INSTALLATION

7.2.1 GENERAL

Deep foundation selection is based on consideration of the types and magnitudes of the loads to be supported, the depth required to achieve capacity, the nature of the materials to be penetrated, and the desired life of the foundation.

Driven piles are most suitable for penetrating relatively weak alluvial deposits to provide support for light to heavy loads. Depending upon pile type, lengths usually range up to about 175 ft (53 m) and capacities up to 200 tons onshore, and greater than 500 ft (150 m) and capacities to 4000 tons offshore where penetration through water depths in excess of 333 ft (100 m) often is required for oil production platforms.

Bored piles, or *drilled piers,* are most suitable in strong clay soils where compressive high capacity and/or uplift resistance is required. Bored piles have been used also to achieve high capacities where supported in rock after penetrating alluvium with boulders or other obstacles which would cause difficult driving conditions. They have been installed to lengths in excess of 166 ft (50 m) and for capacities in excess of 1000 tons.

Caissons are used in all types of soil conditions where very heavy loads require support, but nowadays they are most commonly used for subaqueous foundations such as for bridge support. They have been used in water depths as great as 260 ft (78 m).

Selection is based on economic analysis. On a given project there are often several choices of driven piles, and under certain conditions bored piles or caissons should also be considered. The costs of alternate types are compared based on the costs per unit area of support for the given loading conditions. Cost analysis elements include pile length required, load capacity achieved per unit, cost per unit length of installation, and cost of pile caps (the latter may approximate the cost of a shallow foundation).

7.2.2 DRIVEN PILES

Types Based on Composition

Some of the more common types of driven piles are illustrated in Fig. 7.1, and a summary of optimum lengths and capacities, usual applications, and some advantages and disadvantages of each are given in Table 7.1. The customary ranges of working loads on driven piles of various dimensions are provided in Table 7.2.

Timber piles usually derive their support from shaft friction and are driven as displacement piles in sands or soft to firm clays (Fig. 7.2). Capacities usually range from 30 to 40 tons and available lengths are normally limited to 30 to 60 ft, with lengths up to about 80 ft obtained by special order. Capacity depends on type of timber, pile circumference, and length of penetration in the bearing strata. Timber is not suitable for hard driving conditions caused by rubble, boulders, or compacted fill unless special procedures are used (see *Installation*). Protection against termites, marine borers, or rot in the zone of fluctuating water level requires consideration *(see Art. 7.3.2).*

Steel piles include *H sections* and *pipe,* the latter driven as *open-* or *closed-end.* Onshore optimum lengths range from 40 to 100 ft and optimum capacities range from 40 to 200 tons for H piles and 40 to 120 tons for pipe piles. H or open-end pipes are commonly used as end-bearing piles on rock

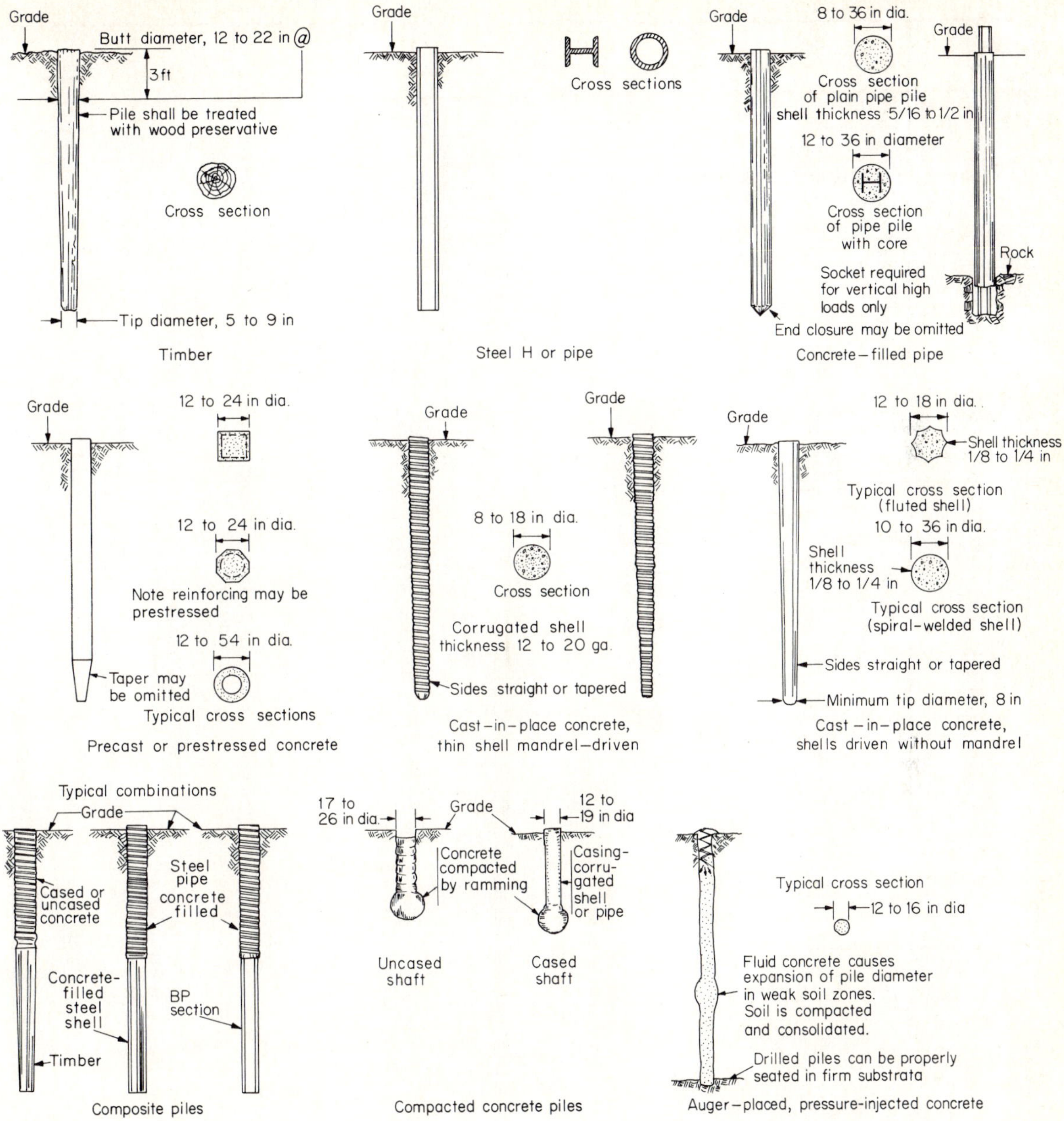

FIG. 7.1 Types of piles. All are driven except the augered pile, at lower right. [After NAVFAC (1982).[1]]

(Fig. 7.2). They can be fitted with hardened tips to permit penetration through soft to hard rock and can be grouted into predrilled rock sockets to provide uplift resistance or batter-pile stability on sloping rock surfaces. Cutoff or splicing for length adjustment, a normal requirement for irregular rock surfaces, is relatively easy. H piles are also used as end-bearing units in strong soils and weathered rock, and occasionally as friction piles in thick clay formations. Pipe piles are also

TABLE 7.1
TYPICAL CHARACTERISTICS AND APPLICATIONS OF DRIVEN PILES*

Type	Capacity, T	Lengths, ft	Applications	Advantages	Disadvantages
Timber piles	10–50	30–60	Friction pile in sands, silts, clays.	Relatively low cost. Decay-resistant when permanently submerged. Easy to handle.	Difficult to splice. Vulnerable to damage during hard driving. Vulnerable to decay where intermittently submerged, or to attack by marine borers, unless treated.
Steel H piles	40–200	40–100†	End bearing on rock or in strong soil. Friction pile in clays.	Easy to splice. Small displacement results in easier driving through dense granular soils or stiff clays. Able to penetrate small obstructions.	Vulnerable to corrosion where exposed. Vulvernable to damage or deflection by major obstructions.
Steel pipe piles Onshore: D_{max} = 24 in	40–120	40–100†	End bearing on rock or in strong soil. Friction and end bearing in deep soils.	Easy to splice. When open-ended can be drilled through obstructions and drilled into rock. Inspection possible with closed-end.	Vulnerable to corrosion where exposed. Vulnerable to damage or deflection by major obstruction.
Offshore: D_{max} = 84 in	To 4000	>500 ft	Same as for onshore.	Same as for onshore.	Same as for onshore.
Concrete-filled steel pipe	40–200 (without cores) 500–1500 (with cores)	40–175†	Same as for steel pipe onshore. Provides high bending resistance.	Easy to splice. Small displacements when open-ended and can drill through obstructions.	High initial cost. Displacement with closed-end pipe.

Concrete: precast	20–200	40–50	Friction pile in sand, gravel, clay. End bearing on rock or strong soil.	Corrosion resistance. Hard driving possible. Cylindrical shapes provide bending resistance. Splicing feasible.	Vulnerable to handling. Considerable displacement. Splicing difficult unless sectional designs used. Cutoff of excessive length difficult and costly.
Concrete: prestressed	20–200	60–100	Same as for precast.	Same as for precast except for splicing. Handling facilitated.	Considerable displacement. Splicing difficult. Cutoff difficult and costly.
Concrete: cast-in-place mandrel-driven thin shell	30–70	30–100	Friction piles in sand and silt.	Initial economy. Tapered sections provide relatively high bearing resistance.	Difficult to splice after concreting. Redriving not recommended. Thin shells vulnerable during driving. Considerable displacement.
Concrete: cast-in-place fluted shell driven without mandrel	30–70	30–80	Friction piles in sand and silt.	Can be redriven. Shell not easily damaged.	Hard to splice after concreting. Considerable displacement.
Composite piles	30–80	60–120	Friction or end bearing, depending on composition.	Provides considerable length at reduced cost.	Difficult to obtain good joint between two materials. Capacity governed by weakest material used.
Compacted concrete	60–150	10–60	End bearing in coarse-grained granular soils.	Can achieve capacity on relatively thin sand stratum. Depth requirements readily predicted. High blow energy can overcome obstructions. High uplift resistance when reinforced.	Base compaction in cohesive soils not feasible. Shafts through clays require protection when in groups.

*See text for additional comments and conditions.

†Length readily increased by splicing even exceeding maximum length given.

TABLE 7.2
CUSTOMARY RANGE OF WORKING LOADS ON DRIVEN PILES*,†

Type	Load, t
Timber (8-in tip diameter)	10–30
Concrete, precast or prestressed	
10-in dia.	20–60
18-in square	70–200
Steel pipe or shell, concrete-filled, not mandrel-driven	
10¾ × 0.188 pipe	30–50
10¾ × 0.250 pipe	40–70
12¾ × 0.250 pipe	50–80
14 × 0.312 pipe	60–90
16 × 0.375 pipe	100–120
Monotube, 7-gage	30–50
Steel pipe or shell, concrete-filled, mandrel-driven‡	
Raymond Step-taper with 10¼-in point	30–50
Raymond Step-taper with 12⅛-in point	40–70
12-in corrugated, 16-gage	30–60
10-in × 0.125 pipe	30–50
Steel H section§	
HP 10 × 42	50–75

*From Peck, Hanson, and Thornburn (1972).[2] Used with permission of John Wiley & Sons.

†Indicated maximum loads can be exceeded if freeze or setup occurs after pile has been driven to resistance corresponding to tabulated value.

‡Use of the mandrel permits driving these piles to a resistance great enough to warrant working loads based on the full structural capacity of the pile.

§When driven with adequate hammer to resistance indicated by wave equation, H piles may be stressed to as much as 12,600 psi under working loads; in soils likely to deform the tips, the same stress may be allowed if the piles are equipped with drive points.

driven as end-bearing units in strong soils or weathered rock and as a combination of friction and end-bearing piles in thick, granular soil formations. Offshore oil production platforms commonly are supported by large-diameter [to 84 in (213 cm)], thick-walled pipe piles driven to capacities as great as 4000 tons, with lengths ranging from 250 ft (75 m) in shallow water to 500 ft (150 m) or greater in deep water. Steel piles are vulnerable to damage or deflection by major obstructions and to corrosion where exposed *(see Art. 7.3.2)*.

Cast-in-place concrete piles can be cased or uncased. *Concrete-filled pipe piles* have optimum capacities of 40 to 200 tons and optimum lengths of 40 to 175 ft. If piles are installed with a steel H section as a core, capacities of 500 to 1500 tons are feasible. Significant advantages are the relatively high bending resistance provided, ease of cutoff or splicing, and ability to be drilled through obstructions. They are used for end bearing on rock or in strong soils, or for friction and end bearing in deep soil formations (Fig. 7.3).

Cast-in-place tapered thin-shell piles are driven by a mandrel and are most efficient as displacement, friction piles in sands where lengths and capacities higher than those possible with timber piles are desired, as for example on the order of up to 100 ft and 70 tons. A common trade name is Raymond Step-Tapered Piles. Concrete must be placed with care to avoid creating cavities in the shaft. An application is illustrated in Fig. 7.4. *Cast-in-place fluted shell piles* (a common trade name being Monotube) are driven without a mandrel because the shell is adequately heavy.

Compacted concrete piles or *pressure-injected footings* (a common trade name being Franki piles), cast-in-place, are useful to achieve high

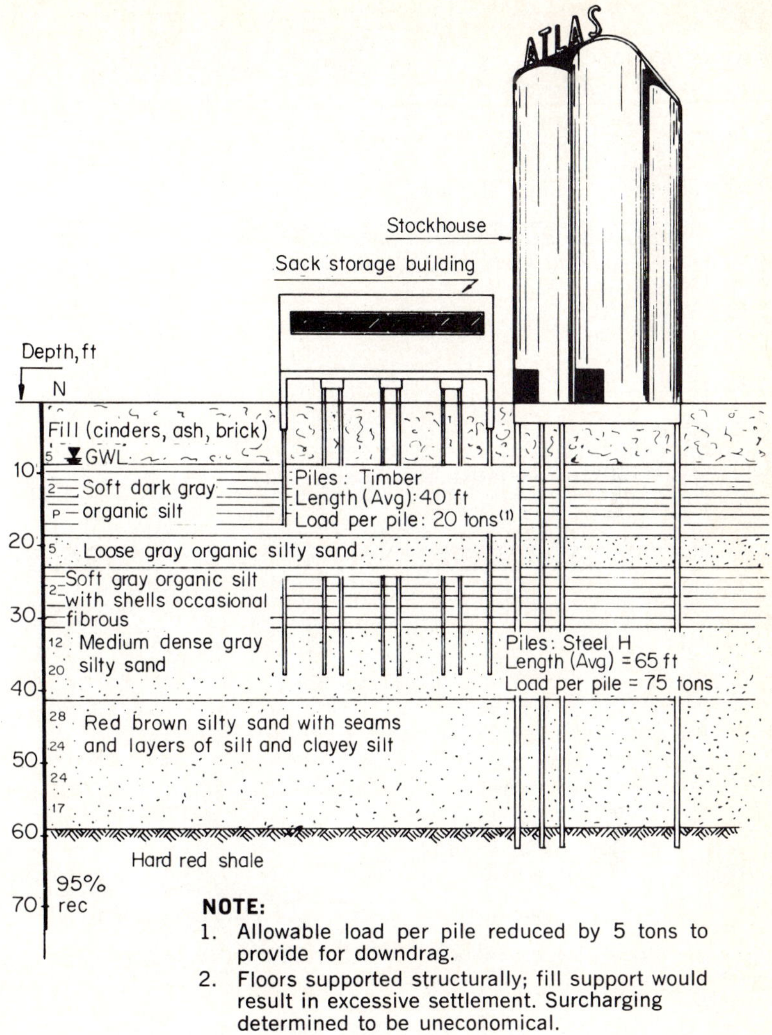

FIG. 7.2 Two pile types used on the same project, Newark, N.J. Short timber friction piles support a sack storage building and steel H piles end bearing on rock support cement storage silos. (Courtesy of Converse Consultants, Inc.)

capacities (normally to 150 tons) in relatively short lengths in sand strata (Fig. 5.24). They can develop substantial capacities even in a relatively thin sand stratum as illustrated in Fig. 7.5. In the case shown, however, consideration must be given to downdrag loads imposed by the weight of the new fill overlying the organic soils and to the potential compressibility of the underlying varved clay stratum, as will be discussed in subsequent articles.

Compacted concrete piles are constructed by forming a bulb of "no slump" concrete in the bearing stratum by compaction under up to 140,000 ft·lb (19 T·m) of force resulting from a 7000-lb hammer dropped 20 ft. In the photo (Fig. 7.6) the dry concrete in the foreground is being placed into a casing where it will form a plug. The falling hammer drives the plugged casing to the desired pre-estimated depth, at which point the casing is secured and the concrete driven from the casing. As more concrete is added, a bulb is formed below the casing which densifies the sand for a distance of several diameters, achieving high pile capacities. (The pore pressures that develop in silts and the remolding of clays cause compacted concrete piles to be a poor choice in these materials.) Allowable pile capac-

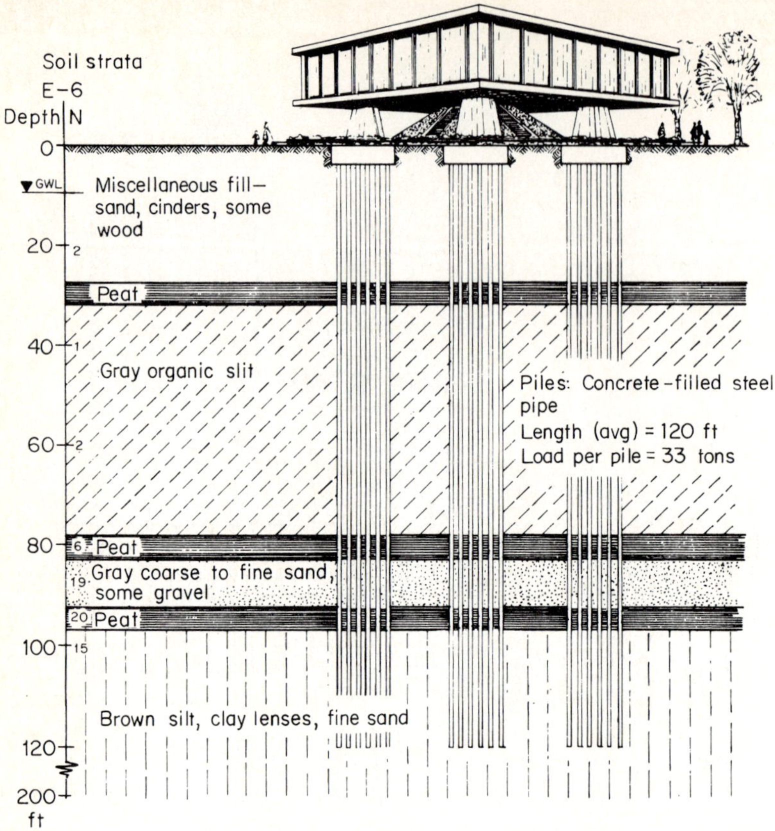

FIG. 7.3 Concrete-filled steel pipe piles, 36 m in length, penetrated through almost 30 m of weak soils to support the U.S. Pavilion at the New York World's Fair. (Courtesy of Converse Consultants, Inc.)

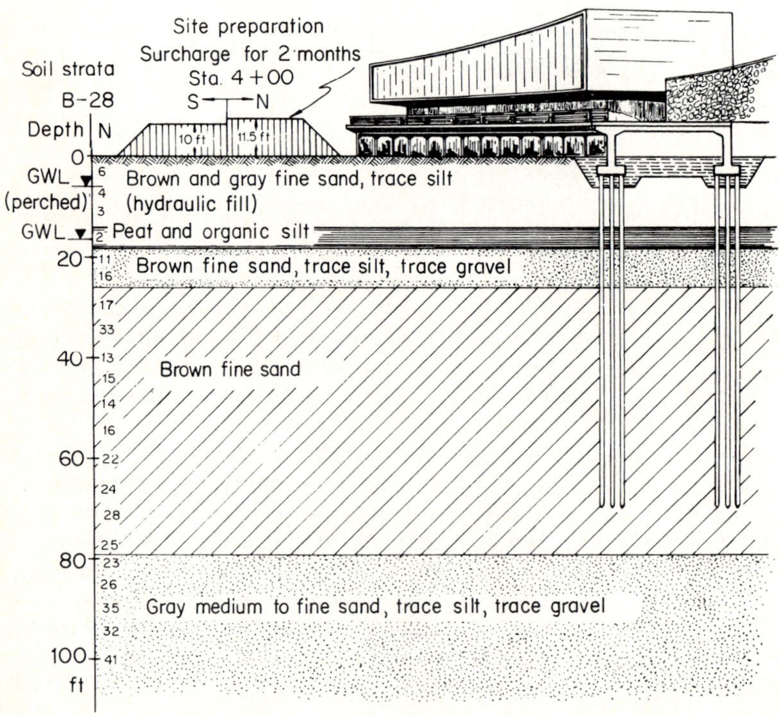

FIG. 7.4 Airlines terminal at John F. Kennedy International Airport, New York, supported on 50-ton-capacity cast-in-place concrete-tapered-shell piles. Surcharge was placed first to preload fill and organic soils to minimize pavement settlement. (Courtesy of Converse Consultants, Inc.)

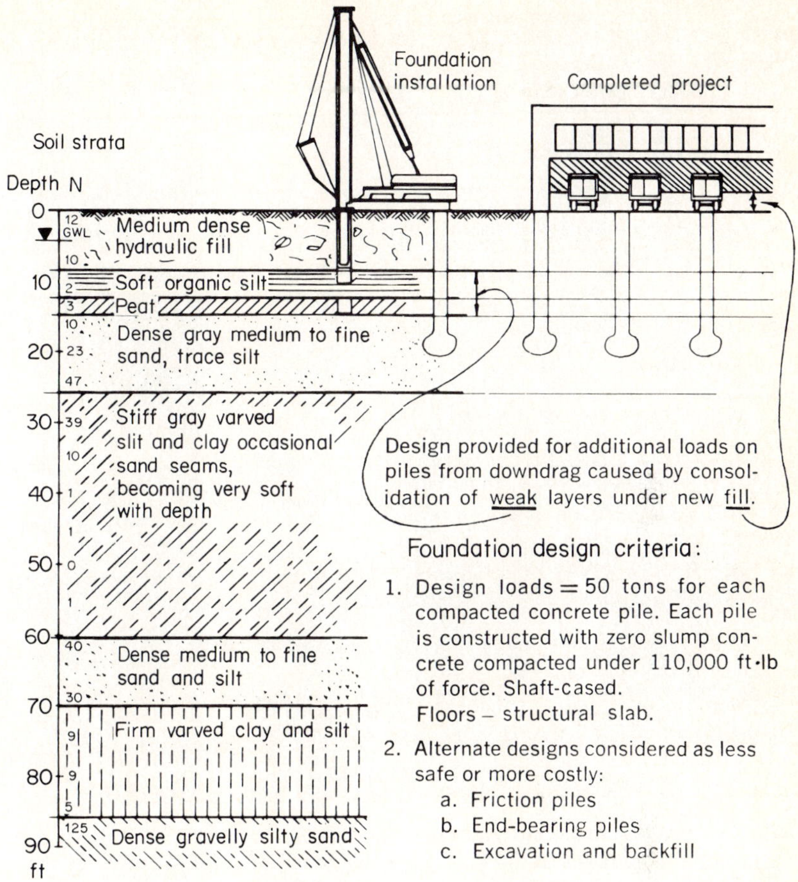

FIG. 7.5 Compacted concrete piles, bearing in sand stratum overlying very soft varved clays, support trucking terminal in South Kearny, N.J. (Courtesy of Converse Consultants, Inc.)

ity is a function of the total volume of compacted base and the number of hammer blows. Upon completion of the bulb the shaft is formed of compacted concrete as the casing is extracted. In soft soils, where there is a danger of shaft pinching and diameter reduction, or where there is concern with pile bending, a *cased variant* is used. When the bulb is formed, a corrugated steel shell is inserted in the drive tube, a plug of concrete is placed in the shell bottom on the bulb, and the plug is driven to drag the shell into the upper portion of the still plastic bulb. The drive pipe is then removed and the remainder of the shell filled with concrete. In some cases reinforcing is installed to increase strength in bending or tension. The irregularly shaped tops of a group of completed, uncased, reinforced piles is shown in Fig. 7.7 and the completed foundation in Fig. 7.8.

Precast concrete piles are fabricated without and with *prestressing*. Used as friction piles in sands, gravels and clays, and end-bearing piles in strong soils or on rock, capacities range from 20 to 200 tons and optimum lengths from 40 to 50 ft for precast piles and 60 to 100 ft for prestressed piles. Advantages over other pile types include resistance to corrosion, suitability for hard driving conditions, and bending resistance (for cylindrical shapes in particular). Handling and installation must be performed with care, however, to avoid cracking which would expose the reinforcing steel to corrosion, and cutoff and splicing is difficult and costly. Hollow cylindrical pre-

FIG. 7.6 Placing no-slump concrete into casing in preparation to form the bulb of a compacted concrete pile in Minas Gerais, Brazil.

FIG. 7.7 Exposed upper portion of compacted concrete piles showing reinforcing. Piles will be cut to excavation bottom.

stressed spun-concrete piles with diameters up to 5 ft or more and wall thicknesses of 4 to 6 in have been developed for high capacities, such as are required for bridge piles.

Composite piles (Fig. 7.1) may be constructed of cased or uncased concrete over timber, concrete-filled steel shell over steel pipe of H section, or other combinations. Optimum lengths are 60 to 120 ft for capacities of 30 to 80 T from friction or end bearing, depending on composition. They provide considerable length at a reduced cost but because of the difficulties of joining two dissimilar materials they are used with less frequency than other types. Pipe steps are easy to join and widely used, however.

Interpile is a trade name for a *cast-in-situ concrete pile*. A metal plow point is positioned in a hopper at the pile location. A pusher is inserted in the plow point and the hopper is filled with grout. A hammer drives the pusher, advancing the plow point as the grout surrounds the pusher and fills the hole created by the plow point. At the desired depth the pusher is removed from the pile shaft leaving the plow point and the concrete pile. Most suitable for sands, the plow point densifies the surrounding materials and the concrete shaft is formed in direct contact providing for high shaft resistance. Ultimate capacities of 120 tons have been reported for only 21 ft of embedment in loose fine sands of a 14-in-dia. pile.

Insert or grouted piles have been developed for offshore oil platform support and are particularly useful in calcareous sands to reduce lengths nec-

FIG. 7.8 Finished foundation of compacted concrete piles.

essary to achieve high capacities, as discussed in Art. 7.2.5.

Installation of Standard Driven Piles

The System

Driven piles are forced into the ground by the impact of a hammer or by a vibratory driver. A standard lifting crane is used to suspend *leads* which incorporate rails to guide operation of the *hammer*. The *spotter* is a horizontal member connecting the lower end of the leads to the crane base. It is extended or retracted for plumbing the leads over the location of a vertical pile, or to permit driving piles on a batter. *Vibratory drivers* rest directly on the pile and are moved from location to location by the crane.

Hammers and Drivers

Drop hammers, falling freely from the top of the leads, are the simplest driving system. Their use is limited generally to underdeveloped regions or to special piles such as compacted concrete, as described in the preceding section, because of the cumbersomeness of equipment for heavy hammers and their slowness of operation.

Single-acting hammers use steam or compressed air acting against a piston to raise a ram, which drops freely to strike the pile (Fig. 7.9). A cap, called the anvil or helmet, is mated to the ram bottom and contains a cushion block to blunt the sharp impact force on the pile and avoid pile damage.

Double-acting hammers use steam or compressed air both to raise the ram and to force it downward.

Differential-acting hammers are similar to double-acting hammers except that the steam or air pressure remains constant.

Diesel pile hammers use an explosion of atomized diesel oil to raise the ram, which is permitted to fall freely when single-acting or open-ended (Fig. 7.9), or is bounced off compressed air when double-acting (closed-ended) (Fig. 7.9), which increases the striking frequency. Driving

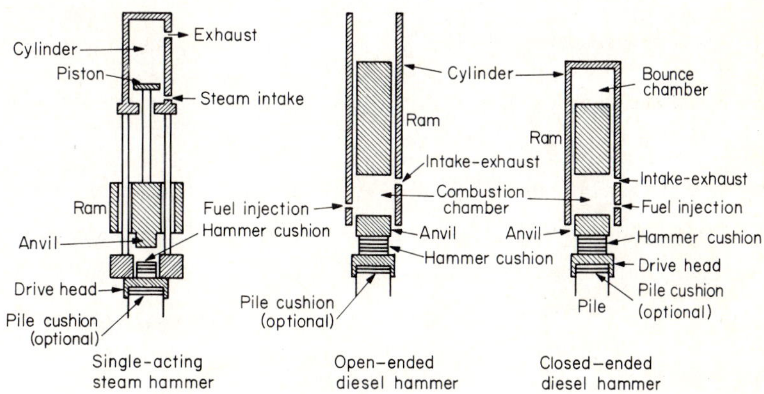

FIG. 7.9 Types of pile-driving impact hammers.

characteristics differ appreciably from those of drop or steam hammers. As the hammer descends, fuel is injected into the combustion chamber and ignited at approximately the instant of contact. The pressure of the burning gas also acts on the anvil, increasing the magnitude and duration of the driving force.

Vibratory drivers use counterrotating eccentric weights (Fig. 7.10) phased so that the lateral force components cancel and the vertical force components add. The pile thus is vibrated into the ground. Driving rates are up to 8 times faster than those of impact hammers in noncohesive soils with nondisplacement piles, the case for which vibratory drivers are most efficient. As there is no impact, pile damage is almost eliminated and the installation is much quieter.

Cushions

Pile-head protection is provided by a *hammer cushion (cap block)* between the ram and drive head or anvil and drive head, and a *pile cushion* may be inserted between the drive head and the pile. Not only do these elements protect the pile top from overstress, but they also have a significant effect on the stress waves developed in the pile during driving and, as a consequence, on the pile's driving characteristics, penetration depth, and, to some extent, capacity.

Pile cushion blocks are generally classed as soft or hard. *Soft cushions*, commonly made of wood and asbestos, result in a relatively long stress wave with a lower peak stress. The effect is longer hammer life and less potential pile damage, but the peak force generated may not be adequate to produce the desired pile capacity. *Hard cushions*, commonly made of alternating disks of aluminum and micarta, result in relatively short stress waves and higher peak stresses. The selection of the most efficient cushion depends on such factors as the desired pile capacity, the hammer's driving force, and pile impedance. Close inspection during pile driving is required. By improper manipulation of cushion blocks on the part of an operator, small penetrations indicating high bearing capacity may be obtained resulting in an inadequate pile appearing acceptable when evaluated by a pile-driving formula.

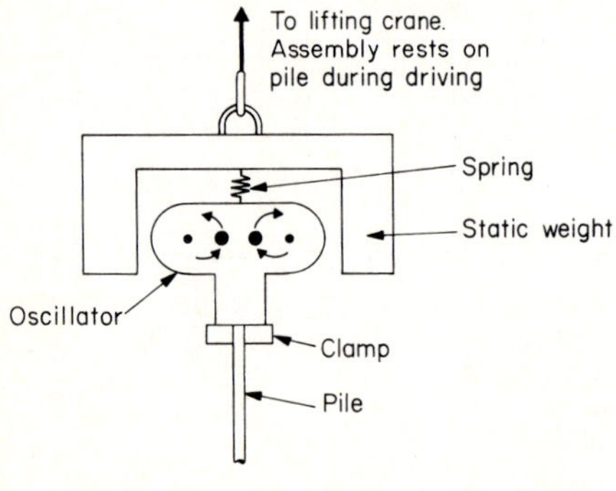

FIG. 7.10 Elements of a vibratory pile driver.

Energy Ratings

Hammers are rated by energy and drivers by force. Selection is based on the pile capacity required, and in general, for economy of installation the pile hammer should have as large an energy input as can be used without damaging the pile. In many cities minimum hammer energy and minimum driving resistance are controlled by building codes, such as given in Table 7.3. Normal maximum rated energy is in the order of 50,000 ft·lb, but for driving piles in offshore conditions hammers with rated energies as high as 1.450×10^6 ft·lb have been developed in recent years. Tables describing the current characteristics of pile hammers and drivers can be obtained from manufacturers such as MKT Corp., Vulcan Iron Works, Raymond Builders, Link-Belt Speeder Corp., or Delmag.

Hammer rating is based on gross energy per blow (ram weight times fall height for the simple drop hammer). *Hammer efficiency* is the ratio of the energy delivered at impact to the gross rated energy. Depending upon equipment maintenance and operation, efficiency may be much lower than rated by the manufacturer. (See Table 7.20 for typical values.) Dynamic testing methods provide a means for evaluating hammer efficiency *(see Art. 7.5.4)*, a necessary input to many resistance formulas *(see Art. 7.5.1)*.

Wave equation analysis (see Art. 7.5.2) is an important procedure for evaluating driving stresses in piles and selecting the hammer-cap–block-pile cushion combination most suitable for the geologic conditions and for developing relationships between driven pile capacity and the num-

TABLE 7.3
MINIMUM REQUIRED HAMMER ENERGY AND MINIMUM DRIVING RESISTANCE FOR VARIOUS PILE TYPES*

Pile capacity, t	Hammer energy, ft/lb	Friction piles, blows/ft	Piles bearing on hardpan, blows/ft	Piles bearing on decomposed rock: Nondisplacement, blows/ft	Piles bearing on decomposed rock: Displacement, blows/ft	Piles bearing on sound to medium rock
STEEL H, PIPE, PRECAST AND CAST-IN-PLACE CONCRETE, AND COMPOSITE PILES (EXCEPT TIMBER)†						
Up to 20	15,000	19	19	48	48	
	19,000	15	15	27	27	
	24,000	11	11	26	16	
30	15,000	30	30	72	72	
	19,000	23	23	40	40	
	24,000	18	18	26	26	
40	15,000	44	50	96	96	
	19,000	32	36	53	53	
	24,000	24	30	34	34	
50	15,000	72	96	120	120	Five blows per ¼ in (minimum hammer energy of 15,000 ft/lb)
	19,000	49	54	80	80	
	24,000	35	37	60	60	
	32,000	24	25	40	40	
60	15,000	96		240	240	
	19,000	63		150	150	
	24,000	44		100	100	
	32,000	30		50	50	
70 and 80	19,000 24,000 32,000		Five blows per ¼ in (minimum hammer energy of 15,000 ft/lb)	Five blows per ¼ in (minimum hammer energy of 19,000 ft/lb)		
TIMBER PILES‡						
Up to 20	7,500–12,000					
Over 20–25	9,000–12,000 single acting 14,000–16,000 double acting					
Over 25–30	12,000–16,000 single acting 15,000–20,000 double acting					

*After New York City Building Code (1968).[3]

†*Note:* table does not take into account such variables as hammer and cushion characteristics or the weight, length, and stiffness of the pile. The judgment of a competent foundation engineer is essential.

‡The minimum driving resistance for timber piles may be determined from the *Engineering News* formula, with the proviso that the weight of the striking part of the hammer shall not exceed 3 times the weight of the pile.

ber of blows per unit of penetration required for a particular hammer in particular geologic conditions. It is also useful for evaluating the hammer blow count at which a pile could refuse for a given pile-hammer-cushion combination.

Driving Resistance

To attain an allowable capacity, piles installed by impact hammers are driven to a particular *resistance* measured by the number of blows required for the last inch of penetration, with a maximum criterion established for various pile types to reduce damage potential. Penetration versus allowable capacity is often computed from resistance formulas *(see Art. 7.5.1)*.

Timber piles should not be driven to more than 3 to 4 blows/in of final penetration by steam hammers delivering 15,000 ft·lb of energy, to reduce damage potential.

Concrete and steel piles usually are driven to a specified resistance of 6 to 8 blows/in, with resistance greater than 10 blows/in constituting excessively severe requirements. Heavy steel pipe piles used offshore often are driven to blow counts in the order of 25 to 30 blows/in.

Relief of Troublesome Driving Condition

Jetting, by discharging a stream of water at a location near the pile point or along its sides through a small diameter pipe, is used to loosen dense layers of sand and gravel or compacted fill where penetration through underlying softer soils is required. Friction piles should not be jetted through resistant strata to aid in reaching final penetration since jetting substantially reduces soil strength along the shaft by factors from one-half to one-eighth below those of unjetted soils [McClelland (1972)[4]].

Rotary drilling also is used to ease penetration through dense sand and gravel layers or to increase penetration into strong soils and weathered rock.

Predrilling by auger or rotary methods is used in saturated plastic clays where pile driving may cause large displacements, resulting in ground heave lifting adjacent piles, or undesirable lateral displacements affecting adjacent structures. The purpose of the predrilling is to remove part of the soil from the location of the pile to be driven. In OC clays, however, excessively high fluid pressures during rotary drilling may cause hydraulic fracturing of the soil and uplift of adjacent installed piles. A wet auger method with lower fluid pressures may be preferable.

Spudding is performed by hard metal points (spuds), driven through stiff clays or soft rock at shallow depths and then withdrawn to ease driving penetration.

Electroosmosis has been used to reduce the driving time of steel piles in stiff clays both onshore and offshore [NCE (1979)[5]]. Electric current is induced to pass between two electrodes, causing water to be drawn from the soil to the pile wall; this lubricates the soil/pile interface. Anode/cathode configurations vary with physical conditions. Decreased driving resistance also results in decreased hammer energy requirements, decreased pile damage potential, and a lower noise level.

Verticality

It is desirable to install piles as near to the vertical as possible, but except for closed-end pipe piles, measurements are either not possible or not practical. Deviation from verticality by as much as 5 to 10% of the pile length, however, appears to be acceptable [Peck et al. (1974)[2]].

Pile Type Factors of Installation

Timber Piles

Potential problems are splitting or brooming of the tip or butt, splitting or bowing of the body, and pile breaking during driving.

HARD DRIVING CONDITIONS Protect pile butt with steel ring and pile tip with steel boot and do not overdrive.

DRIVING EQUIPMENT With drop and single-acting hammers, use relatively low ram drops. With double-acting and diesel hammers use low-velocity blows and incorporate a soft cushion in the cap block.

HAMMER SIZE GUIDELINE Maximum rated energy/blow = 3000 ft·lb × pile diameter in inches [CGS (1978)[6]].

Steel Piles

Potential problems are less than with other types and installation is relatively easy.

HARD DRIVING CONDITIONS H pile tips may be damaged and separation of web and flanges may occur when overdriven through very dense gravel or materials with boulders or other hard obstructions or into rock, unless hardened points are provided. Closed-end pipes may deflect and open-end pipes may be deformed when obstructions are encountered. Closed-end pipes are protected with conical tips, or provided with hardened points for penetration of soft rock. Open ends can be provided with a special driving shoe, and the inside must be checked periodically for the formation of a soil plug.

DRIVING EQUIPMENT All hammer types are satisfactory, but with rated energy limited to 3000 $ft \cdot lb/in^2$ of cross-sectional area [CGS (1978)[6]].

DRIVING CAP Use hard cushion for cap block. Pile cushions are not required.

Precast and Prestressed Concrete

Potential problems are excessive tensile stresses developing during low driving resistance, which result in horizontal cracks, and crushing of pile tip or butt in compression during hard driving.

HARD DRIVING CONDITIONS Protect butt with a steel plate and collar cast with the pile and protect the tip with cast steel shoes or steel points for end-bearing penetration into rock.

DRIVING EQUIPMENT Wave equation analysis is recommended for hammer selection because hammer-pile compatibility is necessary for satisfactory performance. Vibratory drivers are not recommended because they develop high tensile stresses.

SUGGESTED HAMMER GUIDELINES Ram mass to pile mass should not exceed 2:1 for steam/air hammers and 4:1 for diesel hammers, with a minimum hammer energy to pile mass ratio of 2. For drop hammers, use a maximum height of 30 in. Reduce ram velocity or drop height for low soil resistance such as during early driving or driving through soft soils [(CGS (1978)[6]].

DRIVING CAP Use slightly loose fit to avoid torsion or bending forces, and insert hard hammer cushion inside driving helmet and soft pile cushion on pile head to avoid damage.

Applications and Geologic Conditions

Rock

It is common practice when heavy structures are involved to first consider the feasibility of providing support through end-bearing piles driven to rock. On hard, sound rock the pile capacity is usually its capacity as a structural member. Steel H or pipe piles which can be fitted with hardened tips or grouted into predrilled sockets normally are used. An important advantage over concrete piles is the relative ease of splicing to add length and cutoff of stickup after driving, conditions common above irregular rock surfaces.

Strong Materials

Materials such as *glacial till, residual soils,* and *weathered rock* also often provide adequate support for high capacities in combination of end bearing and shaft friction to steel piles, concrete-filled pipe, or precast or prestressed concrete piles.

In *OC clays* (very stiff to hard), nondisplacement bored piles may be the more practical solution for high capacities. If the formation contains water-bearing sand strata, however, displacement piles driven into predrilled holes may be more economical.

SANDS If hard rock is at depths not economically reachable or if structural loads do not require high-capacity piles, then displacement piles obtaining support from either shaft resistance or both shaft resistance and end bearing are considered. Where an adequate thickness of sand exists within a depth of about 20 m and structural loads are light to moderate, timber piles are suitable. In thick deposits of sandy soils, where capacities higher than can be achieved by timber piles are required, cast-in-place concrete, tapered shell piles, or similar displacement-type piles are considered for friction bearing.

Compacted concrete piles have the advantage of achieving high capacities from end bearing in relatively short penetrations and can provide adequate support even in a relatively thin stratum in the order of 3 m thick.

Where soil conditions require very long piles, such as in interbedded alluvial strata, or high capacities are required and rock or very strong material is at depths not economically reachable, displacement piles such as precast or prestressed concrete or concrete-filled pipe, driven closed-end into granular strata, will achieve support from end bearing and friction as illustrated in Fig. 7.3. In stratified soils where support will be obtained from penetration in a sand stratum underlain by soft clay, care is required that the piles are not overdriven or the clay will be overstressed and settlement will occur.

NC CLAYS If a strong bearing stratum is very deep and the supporting material is of necessity normally to slightly overconsolidated clays (soft to firm consistency), various solutions are possible. H piles, although relatively expensive when compared to some other types, have the advantage of developing significant side friction with a minimum of disturbance and heave. Where displacement piles are used, such as timber or step-taper concrete, various installation procedures must be considered to control heave, as discussed under *Adjacent Structures*. Friction piles in clays are particularly efficient where the shear strength increases with depth and the compressibility decreases because the stresses are transferred to the deeper zone as pile movement occurs. The use of a basement in combination with friction piles (*compensated friction pile foundation*), where the basement reduces the building load to be carried by the piles, is a useful solution where soft soils extend to great depths, as in Mexico City (*see Art. 7.4.2*).

Obstructions

Obstructions cause timber and concrete piles to break and steel piles to bend and "dogleg" under heavy driving. Dense compacted fill may require preboring prior to driving piles, and rubble fill with concrete blocks and other impenetrable materials requires excavation. Boulders are particularly troublesome and may be encountered in the lower portions of some alluvial deposits, in some glacial tills, and in colluvial deposits in tropical climates with mountainous terrain. Large boulders are most frequently removed by drilling and blasting, but extreme care is required so as not to damage existing piles. An alternate procedure, which requires a redesign of the foundation, is to drive additional piles around the boulder.

In alluvial soils and glacial deposits, boulders often are located near pile founding level, except for an occasional glacial erratic. When originating from colluvium, however, they can be found deposited in weak soils many meters above founding level. This condition is common along the coastline of Brazil, where the mountains of crystalline rock fall to the sea. Ancient slope failures carried granite boulders as large as 10 m across and frequently at least several meters in size onto the coast and into the coastal valleys. A rise in sea level washed away the smaller particles, then redeposited sands and organic soils to bury the boulders in some locations as much as 8 to 20 m below the surface, often in large concentrations and occasionally within weak strata. Such conditions are extremely hazardous to deep foundations and their presence can be confirmed reliably only with a large number of subsurface explorations by using the CPT or rotary probes as the most economical approach.

Adjacent Structures

The impact of installing displacement piles on adjacent structures must be considered. Driving piles in sands will impose dynamic loads on the adjacent structure, which may cause damage (*see Art. 5.6*). If the adjacent structure is supported on shallow foundations, the vibrations from pile driving can cause densification of the sand, which results in settlement. Horn (1966)[7] describes several case histories including one where pile driving in cohesionless soil caused settlement as large as 15 cm within the driving area and ground settlement as distant from the site as 23 m.

In clay soils, significant movements in adjacent structures can be caused by soil displacement and the development of large pore pressures. The longer the piles are, the more serious the effect can be, even though the piles are driven through the clay stratum to bear on stronger material. Zeevaert (1972)[8] reports on a project in Mexico City where driving piles to point bearing at depths of 32.5 m through soft silty clay caused heave in the order of 35 cm in the center of the foundation area and on the order of 26 cm at the perimeter. Closely spaced (1.5-m) timber piles were used. The ground surface returned to normal two months after completion of driving but the possible effect on close-proximity adjacent structures is apparent.

Pile installation effects on adjacent structures can be minimized by using nondisplacement piles, jetting end-bearing piles, preboring prior to installation of driven piles as already discussed, or by using bored or auger piles as discussed in the following article.

Corrosive Environments require consideration as discussed in Art. 7.3.2, *Structural Capacity*.

7.2.3 BORED PILES (DRILLED PIERS)

Description

Bored piles, also termed *drilled piers*, *drilled caissons*, and *large-diameter piles*, are nondisplacement piles installed by excavating a shaft by machine (or by hand) which may or may not be cased, and then filling the shaft with concrete. Shaft diameters are usually at least 60 cm and have been excavated to diameters as great as 3 m [Woodward, Gardner, and Greer (1972)[9]]. In favorable soil conditions, the bottoms can be belled by underreaming, greatly increasing end bearing and uplift capacity. Underreams of 6 m diameter or larger have been installed. Types of bored piles are illustrated in Fig. 7.11.

Auger piles are a small-diameter form of bored piles (Fig. 7.1).

Applications

Most Favorable Conditions

Bored piles are a favorable foundation solution in strong, cohesive soils where an excavated

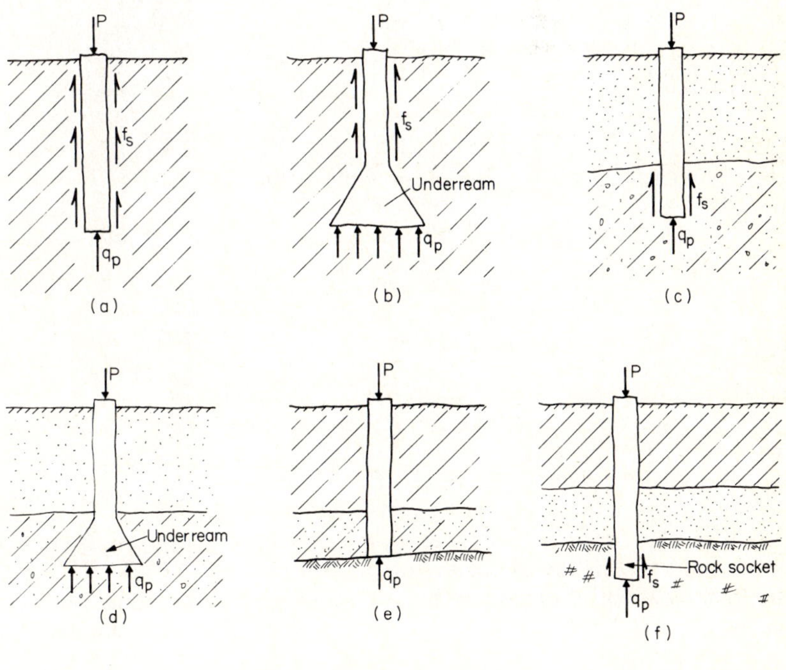

FIG. 7.11 Principal types of bored piles (drilled piers): (*a*) and (*b*) in strong cohesive soils; (*c*) and (*d*) penetrating through relatively weak soil into strong soil, and (*e*) and (*f*) rock-supported.

hole will remain open and high capacities are desired. If the soils are adequately strong, the pile capacity is increased substantially by underreaming or belling. Other advantages are the elimination of ground heave and displacement, pile driving noise, and vibrations; the piles can be drilled through small boulders and can be protected against uplift from swelling clays or downdrag from consolidating soft soils by double-casing or other methods; in most cases, their bottoms can be inspected to ensure that adequate bearing material has been reached and that the bottom is clean.

Difficult Conditions

They also find application where geologic conditions may be potentially troublesome, but where high-capacity piles bearing in strong, cohesive soils or rock are desired. Saturated sands and soft soils below GWL, water-bearing lenses or layers in strong cohesive soils, dry cohesionless soils above GWL, and boulders all present difficult conditions for bored-pile installation. Special treatment utilizing metal casing or mud slurry is required and substantially increases construction costs.

Large depths are possible with bored piles, however, even in difficult conditions. As of 1973, the deepest drilled pier in the United States was 48.5 m, installed for the Courts Tower in Cleveland (ENR, August 16, 1963); at the site of Unit 2 of Brazil's nuclear power plant at Angra dos Reis, State of Rio de Janeiro, a few piers were drilled to depths greater than 50 m and socketed into diorite rock. Even greater depths have been attained elsewhere.

Augered Piles

Augered piles (Fig. 7.1) have been used in sands and stiff clays where a minimum of disturbance is required. The hole is augered and, as the auger is removed, concrete is injected into the hole through the stem of the auger. Pile capacities are usually in the order of 30 tons, and the diameter is about 40 cm. Care is required during concreting to ensure that pinching of the shaft does not occur.

Excavation for Bored Piles

Favorable Conditions

In strong, cohesive soils (OC clays, glacial till, residual soils) or soft rock, excavation is made by a helix auger or a bucket auger (Fig. 7.12), or on occasion by hand when mechanical equipment is not available. The latter method causes bored piles to be a favored foundation solution in countries where labor is relatively low-cost. Underreaming for bells is achieved by extendable blades mounted in the bucket. Casing is not required unless the sides tend to squeeze or water-bearing layers are encountered. The bottom is inspected to ensure that all loose soil has been removed, the steel reinforcement is placed, and the concrete is poured. Installation under favorable conditions is relatively straightforward and simple.

FIG. 7.12 Bucket auger mounted on Calwell machine excavating a drilled pier hole.

Difficult Conditions

Casing is installed to prevent hole collapse where loose or soft saturated or otherwise unstable soils are to be penetrated; it is used with or without a mud slurry. Without a mud slurry, and for small diameter piles in the order of 70 cm, the casing is twisted down under pressure by a machine such as the Benoto (Fig. 7.13) until penetration into strong cohesive soils is accomplished and a seal is formed. In larger-diameter piles, in the order of 1 m or more, the casing is twisted by reciprocating rotation under force ranging from 15 to 25 tons from a larger, more powerful machine such as a Wirth or a Bade (German manufacture), as shown in Fig. 7.14. Double-wall casing for extra strength with carbolloy cutting teeth on the end (Fig. 7.15) is used for penetrating into strong soils or into weathered rock to form a seal and prevent caving where rock sockets are desired. The relatively weak soils down to the strong soils are removed from within the casing by a "hammergrab," or small clamshell (Fig. 7.16).

FIG. 7.13 Benoto machine designed to rotate casing into the ground and excavate the materials with a "hammergrab" as the casing penetrates under force.

Mud slurry may be used alone or in combination with casing. A bucket auger (Fig. 7.12) excavates the soft alluvial soils while a bentonite slurry is pumped into the hole to prevent collapse [see Art. 8.4.2 of Hunt (1984)[12a]]. When the caving strata have been passed, casing with an inside diameter slightly larger than the auger is inserted and seated in the strong soils by rotation with the kelly bar on the auger machine under vertical force. The slurry may then either be bailed out and the remainder of the hole advanced in the dry, or the hole may be advanced without removal of the slurry and concrete placed by tremie methods.

There has been discussion for some time as to whether or not bentonite slurry affects the bond between the concrete and the reinforcing steel or between the concrete and the soil (shaft friction). Experiments have confirmed that the use of slurry does not affect bonding or shaft friction if the slurry is not permitted to remain for more than a few days [Woodward, Gardner, Greer (1972)[9]]. In one case familiar to the author, holes drilled with bentonite for test piles were permitted to remain open for over a month before the piles were made. Strain-gage data from load testing showed a minimum of shaft friction for the mud-cased holes, whereas for cased holes drilled without mud, the shaft friction carried most of the load on the pile in the same geologic conditions. It is apparent that the mud forms a thick cake along the hole sides with the passing of time, and proper construction techniques are necessary to maintain the desired shaft friction after completion of the pile.

Boulders encountered during hole-making operations prevent penetration of the auger or casing and require removal. When large boulders are encountered below GWL, an expedient method for their removal requires pumping the water or mud slurry from within the casing, placing an airlock over the casing (where substantial depths

FIG. 7.14 The Wirth machine (left) is being prepared for drilling and the Bade machine (right) has rotated the casing into the ground.

FIG. 7.15 Carbolloy cutting teeth on double-wall casing permit penetration into strong residual soils to effect a seal for drilling rock.

FIG. 7.16 Excavating soft alluvial soils from within casing with small clamshell or "hammergrab" as Wirth machine advances casing by twisting under force.

are involved), hand-excavating the soil around the boulder to expose it, drilling a line of 1-in holes with a jack-hammer at 30 to 50 cm spacing around the boulder and in its center, and then breaking the boulder with light dynamite charges to permit removal in pieces.

Penetrating Strong Soils and Rock

Excavation into strong soils proceeds with the bucket auger or by rotary drilling methods. For deep holes where penetration to rock is desired, roller bits (Fig. 7.17) are used for residual and other strong soils and soft rock, and special roller bits with hardened steel "buttons" (Fig. 7.18) are used to penetrate hard rock such as thick pegmatite dikes and moderately weathered to fresh granite gneiss. Prior to drilling into rock, the mud is pumped from the hole and drilling proceeds with use of water. As the bits advance, they are followed by the casing, generally about a meter above the bit tip, until the casing is seated in weathered rock if rock sockets are to be drilled.

Rock-socket collapse may occur in unfavorable conditions such as steeply inclined layers alternating between hard and soft rock or highly decomposed rock or clay zones when the socket is being drilled below the casing level. This problem may be resolved by removing the drill bit, filling the bottom 2 to 3 m of the hole with concrete, waiting a few days for some set, and then redrilling.

FIG. 7.18 Roller bit with hardened steel "buttons" for penetrating hard, sound rock.

Pile-tip conditions in rock can be critical, especially for small socket lengths. Pile depths should be predetermined from test borings as closely as possible and the rock cuttings during

FIG. 7.17 Saw-tooth roller bits for cutting strong soils and soft rock used with Wirth or Bade machine.

pile drilling examined by geologists to confirm that the desired rock material is penetrated. At final tip elevation, the bottom is cleaned by "air lift" to remove all cuttings and sediment. The bottom depth is measured carefully after cleaning and again just prior to placing the reinforcing steel, similar to the cage shown in Fig. 7.65. With deep piers below GWL the hole remains flooded; inspection of bottoms is then costly and difficult and requires divers, assuming that the hole diameter is adequate for entry.

Verticality

Piles should be installed as near to the vertical as possible, and specifications often call for a maximum inclination of 3%. It generally is not practical to measure verticality of piers ranging from 30 to 50 m in length under water although the inclinometer could be used for this purpose. The inclination of very long piles, however, usually is not a problem as long as the casing is installed vertically, since the heavy drilling element (Fig. 7.17), under an additional force of 15 to 25 tons from the drilling machine, tends to drill vertically even in steeply dipping rock beds of varying hardness.

Concreting

Dry Hole

Concrete is normally permitted to fall freely from the surface in a dry hole but it should be poured through a centering chute to avoid contact with the hole sides, which would result in segregation. Adequate compaction below a depth of 1.5 m results from the free fall, but above 1.5 m the concrete is vibrated to produce a uniform strength. A slump of 6 in is suitable under most conditions, but higher slumps have been used in heavily reinforced piers and in small-diameter piers from which casing will be removed. It is important that the hole bottom be dry to avoid concrete segregation. It may be necessary to install a small pump in the bottom which is removed just prior to concreting, or if the quantity of water is small, to lay bags of cement on the bottom to absorb the excess water.

Tremie Concrete

Placing concrete under water by the tremie method can be the most critical phase of pier construction. Slump ranges from 17 to 22 cm and a 6-hr retarder is advisable to provide for delivery or other delays. The operation begins with the tremie pipe lowered to within 30 cm of the hole bottom. It must be well cleaned inside to permit unimpeded flow of the concrete and contain a valve, either mechanical or a rubber ball, at the bottom to keep the pipe closed. The pipe is filled with concrete and the valve opened to permit flow into the hole. As flow occurs, the tremie pipe is raised slowly and the depth to the top of the concrete and the depth to the bottom of the tremie pipe are monitored constantly to ensure immersion. Tremie operation requires careful timing and close coordination among the workers, as the tremie must at all times remain submerged in the concrete to a depth of at least ½ m, but not greater than 1 m or the pipe may become stuck (Fig. 7.19). If immersion is lost, reimmersion generally is not possible and the operation is stopped and the concrete surface treated as a cold joint. If the tremie is lost in the rock socket it is best to stop concreting, immediately pull the steel from the still-unset concrete, and redrill the hole.

Pulling Casing

If the casing is left in the hole it may be considered as part of the pier section in bearing, but because of cost it is removed in most cases. Where

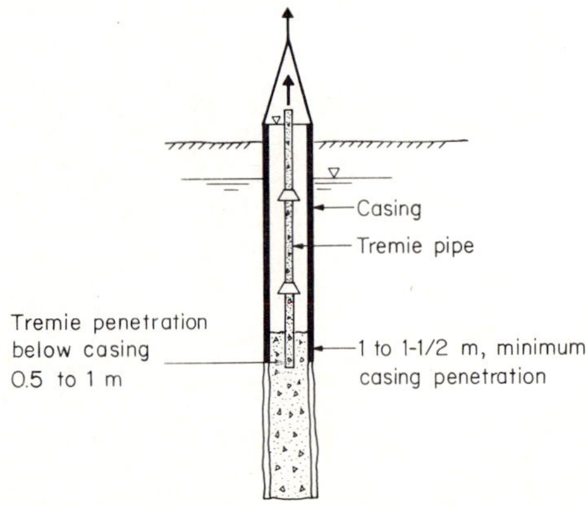

FIG. 7.19 Proper placing of tremie concrete requires penetration control as concrete surface rises and the tremie pipe and casing are raised.

removal is anticipated, the reinforcing cage is fitted with roller spacers (Fig. 7.65), permitting casing withdrawal without withdrawal of the cage. In a tremie hole, removal is extremely critical to prevent pier damage. When the tip of the casing is reached during concreting, it must be raised slowly and kept about 1 m or so below the level of concrete (Fig. 7.19). If it is raised above the concrete, the concrete will become contaminated from the surrounding soils and the hole may collapse; if kept too far below the concrete surface, it will either not be possible to withdraw, or worse, it will pull the concrete and reinforcing with it, if it is possible for the reinforcing to part at a splice. If the steel is pulled and separates at a splice, it is necessary to stop concreting, send down a worker to cut the steel at the concrete surface, drill holes, grout a new splice connection, treat the surface as a cold joint, and recommence concreting, all a very costly and time-consuming operation, especially at large depths below water level. If the casing remains, it can be pumped dry to permit work; if it has been pulled, it must be reinstalled in an oversize to cover the completed concrete section.

Causes of Pier Defects

The major causes of concrete pier defects that can occur during casing removal are illustrated in Fig. 7.20. Defects that can occur in general and the conditions leading to these defects are summarized as follows:

- Inadequate concrete strength, caused by segregation during placement, by improper placement at cold joints, or by the use of poor concrete. Figure 7.21*a* illustrates a lightly cemented sand zone resulting from segregation.
- Displacement or contamination of concrete by caving or by collapse of the hole walls or the underream.
- Voids or discontinuities (Fig. 7.21*b*) in the shaft or underream, caused by the use of concrete that is too stiff, by collapse of the casing or liner, by squeezing of a soft formation, or by hang-up of the concrete in the casing as it is being pulled.
- Separation of cement from the aggregates, as well as voids, caused by even a few inches of water left in the bottom of the hole before concreting.
- Inadequate bearing, resulting from an inadequate foundation investigation and evaluation, from error in identification of the proper bearing material during drilling, or even occasionally from an unauthorized change in the dimensions of the bearing area or an omission of the underream, or failure to properly clean the hole bottom. The photo in Fig. 7.21*c* shows the contact between high-quality concrete and hard, sound gneiss in a core sample from a well-executed pier.

Proof-Testing Integrity

Upon completion, piles are selected for proof-testing either at random or because there is a question of their quality. Core boring or other methods that investigate only the center portion of the pier will not provide information on the outer portions where the steel is located and

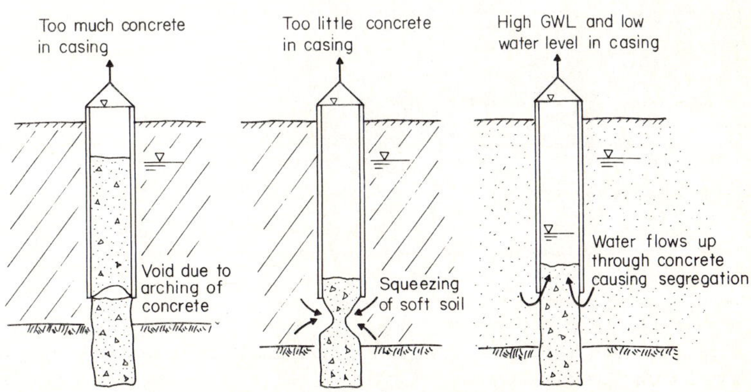

FIG. 7.20 Potential problems leading to inadequate shaft concrete during casing removal.

(a)

(b)

(c)

FIG. 7.21 Test boring cores from completed bored piles: (*a*) lightly cemented sand zone about 30 cm thick, resulting from segregation; (*b*) voids and cracks may permit entry of water and steel corrosion; and (*c*) interface between high-quality concrete (lower core) and hard, fresh granite gneiss.

where defects are most likely to occur. Seismic velocity measurements, which involve sending a vibratory wave down the pile which is reflected and refracted back up the pile, have been used to deduce breaks and other deficiencies from the nature of the returned signal [Hearne et al. (1981)[10]].

A more satisfactory method, particularly for large-diameter piles, is to cast parallel tubes into the concrete during installation. Quality is tested by passing a seismic wave or radioactive source down one tube and a pick-up down another. The tubes should be as far apart as possible and are generally mounted on the reinforcing. Seismic testing and nuclear probes are discussed in Hunt (1984).[12a]

Expansive Soils

The Active Zone

Bored piles extending through the "active zone" into the inactive zone for support are the most successful foundation solutions in expansive materials under many conditions. As discussed in Art. 4.3.6, however, the active zone often is difficult to identify with certainty. It is defined as the depth in which a soil moisture deficiency exists which is a transient value influenced by soil type, soil structure, topography, and climate and because of periodic long dry spells is usually greater than the depth of seasonal moisture variation. An increase in soil moisture in the active zone can result in the development of substantial uplift forces along the pile shaft from the surrounding soil unless adequate protection is provided.

Perched Water-Table Condition

Underreamed piers providing foundations for a medical complex near San Antonio, Texas, have heaved as much as 6 in since about 1962 because of somewhat unusual soil conditions [ENR (1979)[11]]. The piers were drilled into a dry, expansive clay overlain by 22 ft of clay which included a sandy gravel containing a perched water table. Pier penetration through the sandy gravel permitted water to seep to the expansive clays, resulting in swelling and foundation heave. Recent construction in the area has resorted to mats and continuous footings. This case provides an example of the necessity of remaining alert for unexpected subsurface conditions and of always questioning the suitability of a foundation type which may be considered as standard practice for a given geographic location, as drilled underreamed piers are for the Texas coastal plain.

Protective Measures

Protection against swelling uplift has been provided by placing sufficient steel in the pile to resist tension, by painting casing left in place with a graphite-based lubricant, or by isolating the pier from the clays. One method of isolation is to drill an oversized hole through the expansive zone, use a cardboard or light metal liner as a form for the shaft, and fill the annular space around the liner with sand or fine gravel. The annular space must be protected to prevent infiltration of water from the surface, which could otherwise reach the unprotected zone and cause swelling. Another procedure is to fill the annular space with bentonite slurry.

In some cases a pipe has been used inside the shaft in lieu of casing. The pipe, of smaller diameter than the drilled hole, is coated with a bituminous mastic for its length in the expansive zone and is encased in concrete. When the pier is gripped and lifted by the expanding clay, the annulus of external concrete breaks in tension near the bottom of the expansive layer and is free to move along the mastic thereby relieving the uplift forces on the bearing portion of the shaft.

Metal pipe piles placed in prebored holes have been protected against uplift by painting their sides with a graphite-based lubricant.

7.2.4 CAISSONS

Applications

Modern caissons are large, heavy, box-shaped structures that are excavated or sunk to a predetermined depth. A very costly foundation type, they are used usually for the construction of bridge piers or other heavy waterfront structures. In waterfront construction their cost is compared to the cost of drilled piers *(see Art. 7.2.3)* or cofferdams *(see Art. 9.4.5)*, and they often become advantageous where water depths exceed 10 to 12 m, structure loads are high, the foundation is subjected to large lateral forces, and the subsurface conditions include boulder formations that require removal.

Types

Open Caissons

Open caissons, sunk in place with the top and bottom open during construction, are of two general types: (1) constructed from the ground surface or from a sand island; or, (2) a concrete box floated over the pier location.

When sunk from land, the concrete box is constructed with a number of openings through which the soil is removed by grab buckets as the caisson sinks under its own weight. The weight must be adequate to overcome the shaft friction along the sides and as the caisson sinks, additional sections are added. Loads may be added or jetting methods used to assist penetration. Upon reaching founding level a concrete seal, usually about 2 to 5 m thick, is deposited under water by tremie procedures. After the concrete has matured, the caisson is pumped dry and filled with concrete. The open-caisson method does not permit thorough cleaning and inspection of the bottom, but piers can be extended to great depths at relatively low construction costs.

Floating caissons are cast on land with wall spaces open for buoyancy. Filling the wall spaces with concrete after the caisson has been floated over the site location causes it to sink to the bottom from where excavation and sinking proceeds as described for the land caisson. A variation is a caisson with a closed bottom, launched and floated to the site, and sunk into position by filling with sand, gravel, or concrete to bear on a level prepared surface. Protection against scour must be provided as the penetration depth for this variation usually is low.

Pneumatic Caissons

Pneumatic caissons are closed at the top and open at the bottom, enabling excavation to proceed from within as the caisson sinks (Fig. 7.22). They may be constructed from land or from sand islands or may be prefabricated and floated into place. Overburden materials are excavated by hand or machine from a working chamber while compressed air is used to keep water from enter-

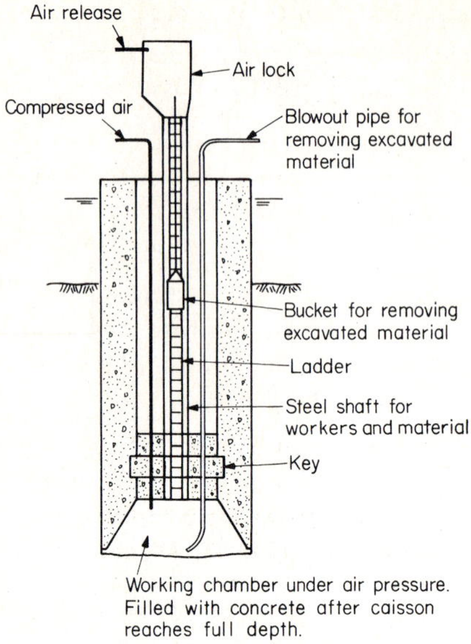

FIG. 7.22 Section through typical pneumatic caisson. [From Peck et al. (1974).[2] Used with permission of John Wiley and Sons, Inc.]

ing the chamber. Because both excavation and concreting are done in the dry, the result is assurance that the proper bearing stratum has been reached and that the concrete will be of adequate quality. Rock and boulders can be readily removed. Penetration depth below water, however, is limited to about 36 m (50 psi or 3.4 atm) as higher pressures are beyond human endurance. In addition, pneumatic caissons are costly because of the time limitations for workers under air pressure (Table 7.4). Economies are realized by first constructing the caisson as an open type until it is sunk to near founding level and then converting to a pneumatic caisson for final foundation preparation.

7.2.5 OFFSHORE CONDITIONS

Geologic Features

Water Depths and Sea Conditions

Expansion since the early 1950s in offshore exploration and development of petroleum resources has moved the structures involved from the relatively shallow gulf waters of the middle latitudes to deeper gulf waters as well as into the deeper waters and more severe climates of the northern latitudes. In the middle latitudes severe storms are infrequent and water depths in general range from 30 to 100 m along continental shelves, although in the constant search for oil deeper and deeper waters are being penetrated. As of August 1981, the deepest water penetrated by offshore works is 1025 ft (308 m), at the site of the Cognac platform in the Gulf of Mexico.

In the North Sea, where storms are frequent and sea conditions can be very severe, platforms were installed during the 1970s to water depths of 70 to 150 m.

Soil Conditions

Soil conditions in offshore water [Art. 7.4.6 of Hunt (1984)[12a]] frequently consist of sands of varying densities and soft clays and may include hard layers or zones or reef structures. Stiff to hard clays are encountered at increasing depths and are common to the North Sea region.

In the middle latitudes where warm climates prevail, the sands can be lightly cemented and calcareous providing an extremely low capacity for driven piles in comparison to the capacity obtained in silica sands *(see Art. 7.3.2).*

TABLE 7.4
ALLOWABLE CONTINUOUS WORKING HOURS IN COMPRESSED AIR AND DECOMPRESSION TIME*

Air pressure, psi	Up to 12	12–16	16–20	20–24	24–28	28–32
Maximum hours per shift	4	3	2	2	2	1
Maximum hours work under pressure per 24-hr day	8	6	4	4	3	2
Total decompression time, min	3	33	43	92	98	85

*From Proctor and White (1977).[12]

Hazards

In addition to wind, wave, and current forces, there are geologic hazards to be considered such as earthquake loadings and seafloor instability. Underwater slides have been known to be caused by earthquakes, as for example off the Grand Banks in 1929 [Art. 11.3.4 of Hunt (1984)[12a]], but Hurricane Camille in August 1969 demonstrated that ocean waves can produce pressures on the seafloor that can trigger slides [Focht and Kraft (1977)[13]]. During Hurricane Camille two structures were destroyed by massive seafloor movements and a third was displaced about 1.2 m. The soil movements extended to as much as 46 m below the seafloor even though the bottom slope was less than 0.5%. Seafloor instability is discussed in Art. 9.2.12 of Hunt (1984).[12a]

Foundations

Structure-Foundation Systems

Fixed production platforms are usually supported on pile groups beneath the platform legs (Fig. 7.23*a* and *b*) although pile-supported mats placed on the seafloor have been used in waters 30 m deep in hurricane zones where soil conditions are poor. Platforms for very deep water have been fabricated on land and then sunk and attached to a pile-supported template structure. Gravity structures, prefabricated on land and sunk into place on the seafloor have been supported on mats (Fig. 7.23*c*) or similar foundations where seafloor soils are relatively strong.

Anchored floating structures include tension-leg platforms for deep water conditions for which the platform is attached by cables to tension piles installed in the seafloor (Fig. 7.23*d*) and single buoy moorings (SBM) tied to cables which may extend laterally for several hundred meters to where they are attached to tension piles installed in the seafloor (Fig. 7.23*c*). In either case there are no compressive forces, and in the case of an SBM lateral forces are usually in the order of several hundred tons.

Jack-up rigs for exploration are supported on spread footings (Fig. 7.23*f*) or mats. Where supported on spread footings the primary concern is

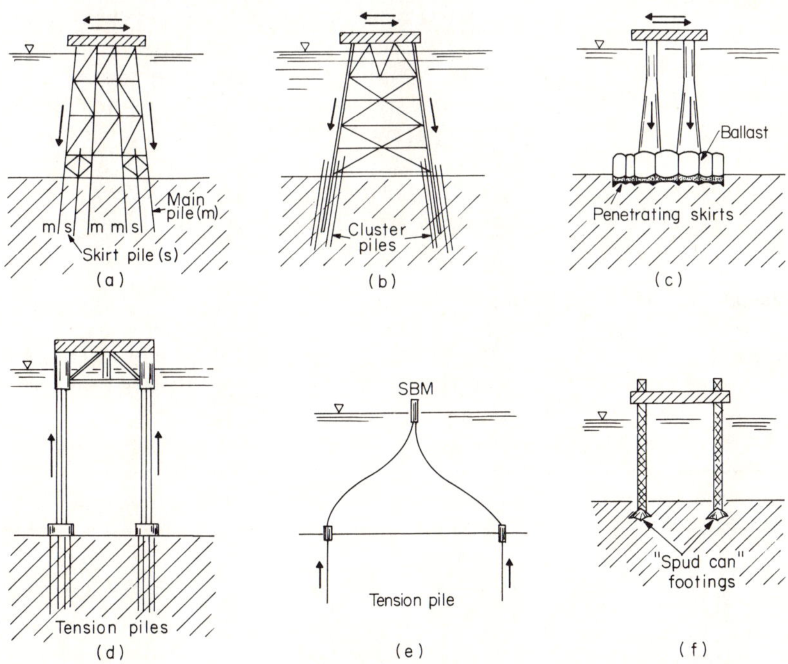

FIG. 7.23 Various types of petroleum industry offshore structures: (*a*) and (*b*) pile-supported production platforms; (*c*) gravity-structure production platform; (*d*) tension-leg platform for deep water; (*e*) single-buoy mooring system (SBM); and (*f*) jack-up exploration drilling platform.

with bearing-capacity failure (punch-through) as the rig is being jacked up and its surcharge tanks filled.

Driven Piles

Offshore drilling and production platforms often are supported on steel pipe piles which may reach 84 in (2.13 m) in diameter with wall thicknesses to 2.5 in. Axial capacities ranging from 1000 to 4000 tons are typically attained with design pile penetrations of 200 to 400 ft (60 to 120 m). Special design considerations of pile-soil interaction in axial and lateral directions, as well as static and cyclic loads, are required.

In relatively weak soils (NC sands and clays) and average sea conditions (100-year wave amplitude less than 15 m), large-dimension platforms are typically supported on 4 to 8 main piles 1.0 to 1.2 m in diameter penetrating 80 to 100 below the seafloor and a small number of secondary "skirt-piles" (Fig. 7.23*a*) which generally penetrate to lesser depths than the main piles. The "skirt-piles" are installed to ensure better distribution of horizontal and vertical loads, particularly in the case of soft soils. In overconsolidated soils and severe sea conditions (100-year wave of about 30 m) foundations for structures with four main legs usually consist of a group of 4 to 12 piles for each leg (Fig. 7.23*b*) [Le Tirant (1979)[14]].

Insert or Grouted Piles

In difficult soil conditions such as dense sands, hard clay, or reefs, the required penetration cannot be readily attained with conventional hammers, and in calcareous sands driving disturbance substantially reduces support capacity [Kraft and Lyons (1974)[15]]. Insert, or grouted, piles (Fig. 7.24*a*) were developed for these conditions, enabling the design capacities to be achieved with shorter penetrations *(see Art. 7.33).*

To construct an insert pile, a main pile is driven to refusal and a steel pipe is inserted in a larger-diameter hole predrilled to some predetermined depth, usually in the order of 30 m below the mudline. The bentonite used to drill the hole is displaced with cement grout by one of several methods [Kraft and Lyons (1974)[15]].

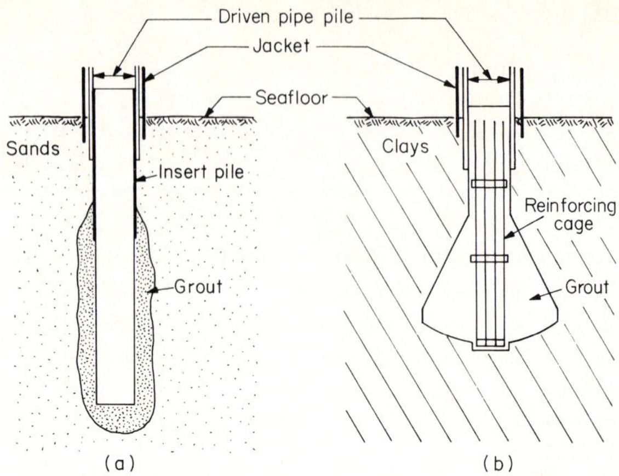

FIG. 7.24 Two methods of offshore pile installation to reduce penetration requirements in hard driving conditions: (*a*) grouted insert pile and (*b*) underreamed pile. [After Kraft and Lyons (1974).[15]]

In the North Sea grouted piles have yielded unfavorable experiences, especially as related to installation time, and the development of hammers with rated energies as high as 1.45×10^6 ft·lb (2 MN·m) has permitted penetration through dense sands and hard clays to the desired capacities [de Ruiter and Beringen (1978)[16]]. Grouted piles, however, still are a useful solution in calcareous sands where driving substantially reduces soil-support characteristics *(see Art. 7.3.2).*

Drilled and Underreamed Piles

For conditions in the North Sea where conventional hammers could not provide adequate penetration in hard clays, drilled and underreamed piles were developed to reduce the required penetration depth (Fig. 7.24*b*). Installation is time-consuming and difficult in deep and rough water and the technique has fallen into disuse with the advent of very-high-capacity drive hammers as discussed above [de Ruiter and Beringen (1978)[16]].

Design Considerations Summarized

Focht and Kraft (1977)[13] summarize the major categories for consideration in the design of marine structure foundations as follows:

1. Site evaluation to select optimum foundation is based on deep penetration borings, geophysical surveys, and in situ testing.
2. Axial pile performance is evaluated for single piles and pile groups (driven, grouted, or underreamed piles) and the time-load environment.
3. Small footing performance considers mobile rig (jack-up) footings and template mud mats.
4. Installation considerations include seafloor conditions, template stability, skirt penetration, pile drivability, breakout resistance, and predrilling and grouting operations.
5. Seismic evaluation is made for ground response characteristics and soil-structure interaction.
6. Laterally loaded pile performance is evaluated for single piles and pile groups for static and cyclic loads.
7. Large footing performance is evaluated for bearing capacity, liquefaction, cyclic degradation, sliding, and settlement.
8. Seafloor stability is evaluated from the aspects of mudslides, creep flow, contemporaneous faults, and liquefaction.

Design guidelines covering foundations for offshore structures are published by API (1982)[22] and DNV (1977).[13a]

7.3 AXIAL CAPACITY: STATIC ANALYSIS

7.3.1 GENERAL CONSIDERATIONS

Forces and Capacities

Forces imposed on deep foundations can be variable, including axial compressive forces from structure and downdrag loads, axial tensile forces from overturning moments or uplift from swelling clays, and horizontal forces from a wide variety of sources.

The *capacity* of deep foundations has two elements: the capacity of the unit as a structural member in compression, tension, bending, and buckling and the supporting capacity of the surrounding geologic materials. The limiting design load for end-bearing piles driven to sound rock usually is based on structural capacity; otherwise, the design of deep foundations is based on the support capacity of the geologic medium in which they are founded.

Geotechnical analysis of the support capacity can be approached by using the theory of plasticity (bearing capacity, conventional, or static analysis), by dynamic analysis with the wave equation (for driven piles), or from load test data. The last is the most reliable approach, but in many situations it is necessary to first estimate the penetration depth required for the test pile, which may be accomplished by examination of the geologic sections and the application of experience or by static and dynamic analysis. Additionally, load tests for design are ordinarily performed only on major or critical projects. On the average project they are performed after or during the pile installation. Load tests are performed for offshore projects only infrequently, and design relies primarily on static analysis checked by dynamic testing during installation of driven piles.

Dynamic pile-driving formulas are used frequently as a basis for evaluating the capacities of small-diameter driven piles. This approach can be dangerous, leading to poor foundations, unless the engineer has prior experience in the site locale or unless the results are correlated with the analysis of load test data or static or dynamic analysis

Elements of static analysis of the axial capacity of deep foundations include the general expression for the ultimate capacity of a single unit *(see Art. 7.3.2)* and its components: end bearing *(see Art. 7.3.3)*, positive shaft resistance *(see Art. 7.3.4)*, negative shaft friction or downdrag *(see Art. 7.3.5)*, and group effects *(see Art. 7.3.6)*. Settlement analysis is covered in Art. 7.4 and analysis by pile-resistance formula, the wave equation, and load testing in Art. 7.5. Lateral capacity is covered in Art. 7.6.

Design Procedure

Design procedures have the basic purpose of resulting in the selection of the most economical foundation unit that will provide a suitable deep foundation. For any given type of unit, the most significant cost factor to be evaluated is the length required to obtain the optimum capacity, which is based on the effective penetration. Other factors subsequently evaluated include material availability, installation conditions, and unit deterioration and life.

Examination is made of the geologic sections obtained from explorations, the type or types of deep foundations are selected that appear most suitable for the subsurface conditions, and the necessity for geotechnical analysis is judged.

- Units bearing on sound (competent) rock, including end-bearing piles, bored piles, and caissons, generally will not require rigorous analysis, nor will many cases of high-capacity piles driven to some penetration into strong materials such as residual or glacial till soils where support will be derived from end bearing and shaft friction. Compacted concrete piles are not suitable for capacity analysis because of the nature of installation, but they should be evaluated for potential settlement where underlain by clay strata.
- Geotechnical analysis should be performed for friction piles in all soils, piles deriving support from friction and end bearing in relatively weak materials such as alluvial sands and clays, and bored piles and caissons supported in clays, residual soils, or less-than-competent rock.

After a pile type is selected and its static vertical capacity evaluated, the number of units is calculated from the structural loads and the spacing determined. Closely spaced piles act as a group and widely spaced piles act singly; the spacing will influence the additional analysis required to evaluate the effect of horizontal and dynamic loads and to evaluate anticipated settlement. Economic study of groups formed of several types of units may be warranted.

Capacity Evaluation

Sands

Capacity of piles or piers in granular soils may be evaluated using the bearing-capacity theory *(see Art. 7.3.2 through 7.3.4)*, correlations with the SPT or CPT results *(see Arts. 7.3.3 and 7.3.4)*, dynamic analysis with the wave equation *(see Art. 7.5.2)*, or load tests *(see Arts. 7.5.3)*. Load tests are the most reliable method, and for compacted concrete piles, the only reliable method.

In *loose sands*, driven piles with a heavy taper or consisting of compacted concrete are most efficient. Depth-versus-capacity relationships for driven piles are evaluated first by static analysis and then by dynamic analysis before and during driving. (*Note:* Piles of 20- to 30-T capacity with taper of 1 in for every 2.5 ft seldom penetrate more than 25 ft even in loose sands.)

In *soft materials overlying dense sand*, penetration into the dense sand is estimated from SPT or CPT data and the penetration record obtained during driving is evaluated. The best procedure is to run a series of load tests at varying depths.

In *sands overlying clays*, a settlement analysis is always required *(see Art. 7.4)*.

Very Fine Sand and Silt below GWL

Fine-grained granular soils below the groundwater table are difficult to analyze due to pore-pressure changes during driving.

In *loose conditions* pore pressures accumulate, reducing effective stresses and shearing resistance, and may even result in liquefaction. Reducing the progress rate of pile installation permits pore pressures to dissipate, thereby somewhat relieving the problem.

In *dense conditions* negative pore pressures develop, giving a falsely high resistance since *relaxation* may occur. Driving causes the soil below the pile tip to dilate, thereby generating the negative pore pressures and a temporary strength increase. When the pressures dissipate, the resistance reduces (relaxation). Dynamic analysis of redriving records provides an evaluation of the capacity obtained.

Clays

Driving resistance of *friction piles* in clay is likely to be low because of disturbance to clay structure, but after driving the strength may increase markedly over a period of time. Strength gain *(freeze)* may be caused partly by thixotropic processes and partly by consolidation of the highly stressed clay immediately surrounding the piles [Peck et al. (1974)[2]].

All analytical methods should be used with caution, especially in medium- to high-shear strength clays supporting *friction piles*. As freeze is likely to constitute the greatest part of pile capacity and is not related to stress transmission in the pile during driving, use of dynamic driving formulas or analysis based on the wave equation is likely to give an erroneous conception of capacity. Capacity versus length estimates in all cases are made based on static calculations for both friction piles and end-bearing units.

For major projects it is advisable to perform full-scale load tests as part of final design or at the beginning of construction, including pullout tests for evaluation of shaft friction along driven piles. Routine projects are designed on the basis of an adequate safety factor applied to static analysis.

In very stiff to hard clay, point resistance is based on static analysis and guided by dynamic analysis. In the case of stiffness increasing with depth, dynamic analysis provides for an estimate of the capacity that can be developed for a given penetration per blow. Load tests are necessary to confirm and evaluate freeze.

Rock

In sound rock, other than soft stratified rock, point and shaft resistance may be estimated on the basis of unconfined compression-strength values from tests on rock cores.

Pressuremeter tests or load tests are used for weathered, closely jointed, or soft rock to evaluate end bearing. Load tests are used to evaluate shaft friction along heavily loaded bored piles.

7.3.2 ULTIMATE CAPACITY: SINGLE UNIT

Support Criteria

Elements: End Bearing and Shaft Resistance

A deep foundation can be supported only by tip or point resistance (end bearing Q_p), only by frictional resistance along the shaft (friction bearing Q_s), or by a combination of both as illustrated in Fig. 7.25.

Point capacity may be evaluated from a modification of the bearing-capacity equation,

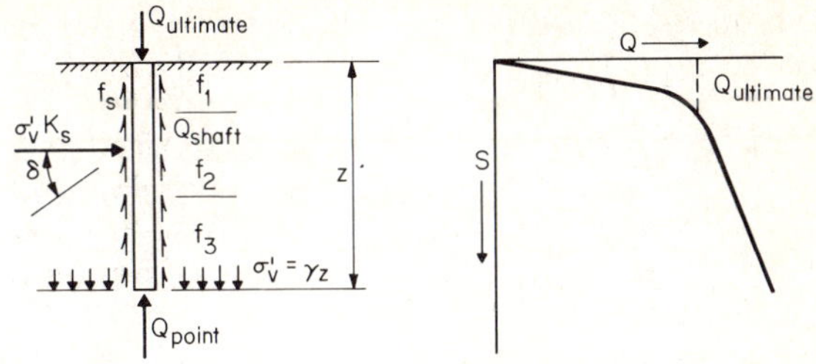

$$Q_{\text{ultimate}} = Q_{\text{point}} + Q_{\text{shaft}} = q_p A_p + \Sigma f_s A_s$$

where q_p = ultimate unit resistance at tip
f_s = frictional resistance of each stratum
A_p = tip area
A_s = shaft area

FIG. 7.25 Elements of axially loaded static pile capacity.

$$q_u = cN_c + qN_q + 0.5\gamma BN_\gamma \quad (6.16)$$

and shaft friction may be evaluated from the Mohr-Coulomb expression for shear strength,

$$s = c + \sigma_n \tan \phi \quad (3.8)$$

Ultimate Capacity Q_u

The ultimate capacity for a single deep foundation under compressive (downward) loads, supported by both end bearing and shaft friction, in soils with parameters c and ϕ, may be expressed as follows:

$$Q_u = \text{end bearing} + \text{shaft resistance} = q_p A_p + \Sigma f_s A_s - W \quad (7.1)$$

where q_p = ultimate unit tip resistance = $qN_q + cN_c$
A_p = pile tip area = $\pi/4B^2$, where B = pile tip diameter (circular pile)
f_s = frictional resistance for each stratum, the values for which are summed, for sands = $K_s\bar{p}_o \tan \delta$, for clays = $c_a = \alpha c$
A_s = shaft area in a particular stratum per unit length = πB (circular pile)
W = pile weight, usually ignored as it is nearly equal to the weight of material displaced or removed

Therefore,

$$Q_u = (qN_q + cN_c)A_p + \sum_0^L (K_s\bar{p}_o \tan \delta + c_a)A_s\, dz \quad (7.2)$$

where q = mean normal stress at the tip = $[\bar{p}_o(1 + 2K_o)]/3$, often taken as γz

$\bar{p}_o$ = effective vertical stress at the tip or at the midpoint of a cohesive stratum ($\bar{p}_o = \gamma z$

K_o = coefficient of earth pressure at rest

c = cohesion

N_q and N_c = bearing-capacity factors for deep foundations

K_s = lateral earth pressure factor acting on the shaft ($\bar{\sigma}_n = \bar{p}_o K_s$)

δ = friction mobilized between pile and soil (related to ϕ)

c_a = adhesion between pile and soil = αc, where α is a strength factor varying with the OCR and installation method

dz = penetration depth in a particular stratum

L = stratum thickness

Other notations found in various expressions include:

z_b = bearing depth

z_c = critical depth

q_c = static cone-point resistance

Note: Large-diameter-pipe piles such as those used offshore, with diameters greater than 1 m or so, may not plug when driven. For this case an additional shaft friction value is taken for the inside of the pipe, and the end bearing is based on the contact area of the pipe. The value obtained is compared with that obtained with a closed-end pipe (plugged) and the lesser value used for design [de Ruiter and Beringen (1978)[16]].

Allowable or Design Capacity Q_a

The factor of safety used to obtain the allowable pile capacity varies with the method of installation and the reliability of the measurements of soil properties. Q_a for driven piles onshore normally is taken as $Q_u/3$ to $Q_u/2$ and for bored piles $Q_u/2$, as long as the soil below the base is less compressible than the soil around the shaft [CGS (1978)[6]]. In certain cases settlement analysis is required (see *Art. 7.4*). For piles offshore $Q_a = Q_u/1.5$.

End Bearing versus Shaft Resistance (Q_p vs. Q_s)

Point and shaft resistance are not mobilized simultaneously because the mobilization of Q_s and Q_p require different amounts of displacement. In general, Q_s is mobilized by displacements in the order of ½ to 1 cm. Once Q_s has reached its ultimate value, however, subsequent pile deformation may result in a decrease in frictional capacity, thereby resulting in Q_u less than the sum of Q_p and Q_s (see also the discussion in Art. 7.4.1, *Load Transfer Mechanisms*).

SANDS Piles in granular soils derive their load-carrying capacity from both Q_p and Q_s. The relative contributions of the two elements to total capacity depend essentially on relative density and soil shear strength and on pile displacement characteristics. In the usual practical situation of piles driven into a sand stratum with D_R increasing with depth, Q_s provides less than one-half the total capacity.

CLAYS In soft to firm clays ($c_u < 1$ tsf), support is derived almost totally from shaft resistance. In stiff or stronger clays ($c_u > 1$ tsf), Q_p and Q_s are both mobilized to provide support, and in very stiff clays or strong cohesive glacial tills, essentially total capacity may be derived from Q_p for large-diameter bored piles.

ROCK MASSES If the stratum at the pile tip is rigid, no movement of the base normally will occur and little or no shaft friction will be mobilized. Elastic compression of a pile, however, will mobilize some shaft friction, but only a small percentage of the total value [Kezdi (1975)[17]]. In rock masses the ratio of Q_p to Q_s depends upon pile diameter, depth of socket, and rock quality and can vary from a total capacity of only Q_p or only Q_s, or $Q_p + Q_s$. (In deep sockets in sound rock, stresses may never reach the pile tip, as will be discussed.)

Factors Affecting Reliability of Analysis

General

The soil-pile interaction problem is complex, and rigorous computation of pile capacity is not possible in the present state of the art because of many factors, some not well defined, that influence the various elements expressed in Eq. 7.2.

Substantially affecting the values of soil-strength parameters and the lateral earth pressure along the shaft are the pile type (displacement or nondisplacement), pile material (steel, concrete, wood), method of installation (driven, bored, or pushed), and penetration depth. Time delay between installation and building construction is important for piles obtaining support in cohesive materials because strength may increase with time (freeze), and in dense fine-grained granular soils because strength may decrease with time (relaxation).

Critical Depth Concept for Sands

Conventional analysis assumes that the vertical stresses represented by q and $\overline{p}_o$ in Eq. 7.2 are the effective vertical stresses caused by overburden and increase linearly with depth. Large-scale experiments and field load tests, however, have shown that point resistance and average shaft friction in a homogeneous sand deposit *do not increase* indefinitely with penetration depth [Vesić (1964),[18] Kerisel (1964)[19]]. Below a certain critical depth z_c there is little or no increase because of soil compressibility, arching, and other factors [Meyerhof (1976)[20]], and the effective stress adjacent to the pile is not necessarily equal to the effective overburden pressures, but reaches a limiting value at some depth.

There have been various approaches to the problem of evaluating capacities in terms of the critical depth concept, as will be discussed in following articles. One approach commonly used is to establish limiting values for f_s and q_p. Such approaches as yet have not been refined to account for varying relative densities and may not accurately account for the variation of capacity with penetration, as the limiting resistances become operative only at relatively large penetrations, in the order of 25 m or more.

API Design Criteria

The American Petroleum Institute (1982)[22] has published recommended limiting criteria for various granular soil gradations for the design of very-high-capacity pipe piles in quartz sands, when used to support offshore structures. These values, based on investigations and experience, are given in Table 7.5, where f_{max} and q_{max} are the

TABLE 7.5
DESIGN CRITERIA FOR AXIAL CAPACITY OF DRIVEN PIPE PILES IN MEDIUM DENSE QUARTZ SANDS*

Soil type	ϕ,† degrees	δ,† degrees	f_{max}, tsf	N_q	q_{max}, tsf
Clean sands	35	30	1.0	40	100
Silty sand	30	25	0.85	20	50
Sandy silt	25	20	0.7	12	30
Silt	20	15	0.5	8	20

Calcareous soils: reduction in axial capacity‡

Carbonate content, %	Percent of axial capacity for quartz sands	
Up to 30	$f_{max} = 100$	$q_{max} = 100$
30 to 45	$f_{max} = 32$	$q_{max} = 80$
Above 45	$f_{max} = 28$	$q_{max} = 70$

*After API (1982)[22] and deRuiter and Beringen (1978).[16]

†For calcareous sands use $\phi = 19°$ and $\delta = 14°$.

‡After Agarwal et al. (1977).[21]

limiting values for shaft friction and end bearing, respectively. These values usually are reached at depths in the order of 24 m below the seafloor [McClelland (1972)[4]] and are for silica sands. They require reduction when soft, carbonate sands are encountered, as discussed below.

Sand Minerals and ϕ

In the following discussions regarding end bearing and shaft friction, it must be kept in mind that the capacities determined from tabulated ϕ values apply only to silica sands or sands composed of materials of similiar hardness. *Calcareous sands* are soft-grained and their contribution to the capacity of driven piles is extremely low in comparison to that provided by hard-grained silica sands. Skin-friction values for piles driven into calcareous sands may be one-fifth or less of values for medium-dense silica sand, and their end-bearing values may be only one-half of values for silica sand [Focht and Kraft (1977)[13]]. The reduction of skin friction and end bearing has been found to relate to the silica-carbonate sand percentage in a given stratum [Agarwal et al. (1977)[21]]. Suggested design values, based on an evaluation of load test data, are given in Table 7.5. Penetration tests, however, should reflect the difference in grain hardness, but as yet no correlations have been developed. [Calcareous sands are described in Art. 7.4.6 of Hunt (1984).[12a]]

Clay Soils

In clays the most critical factor is the accurate measurement of shear strength. Two general approaches are used in analysis: total stresses (undrained strength) and effective stresses (drained strength). The shaft area in contact with the soil depends on clay consistency and pile type, and in certain situations the actual contact area may be in doubt.

Rock Masses

In rock masses the major factor influencing capacity is rock quality. In excellent to good quality (hard, sound rock) the strength of concrete may govern bond strength in a rock socket. In less than good quality rock, where high capacities are desired, assessment of both end bearing and shaft resistance can be made reliably only with load tests.

Structural Capacity

Stresses in end-bearing piles under axial working (static) loads may be taken as those appropriate for the pile materials under direct compression. There is normally no danger of buckling of piles of conventional dimensions because of inadequate lateral support, even in very soft soils (*see Art. 7.6.4*). For piles decreasing in size with depth, the critical section is at the tip.

Timber Piles

Maximum static stresses are 1200 psi for southern pine and Douglas fir; the *USDA Wood Handbook No. 72* gives stress values for other species. To maintain the cross section after installation, consideration must be given to creosote impregnation for protection against termites, marine borers, or rot, the latter prevalent in the zone of fluctuating water levels. In temperate zones protection against marine borers by creosoting usually exceeds 40 years, but in tropical climates it provides protection for only 15 years or so. It is not necessary in cold northern waters.

Steel Piles

Maximum static stresses normally are in the order of one-half the yield stress (or, for 36-ksi steel, 18 ksi), although during driving stresses may approach the yield stress. In cases where hard driving is likely to deform the tips, piles should be equipped with hardened drive points. Loss of cross section from corrosion requires consideration. Corrosive environments normally do not occur in the zone of soil penetration, except in miscellaneous fills of trash, rubbish, etc., but do exist in the zones just above and below the ground line and between tide levels in seawater, or in the salt spray zone. Protection is provided by extra steel thickness, factory-applied epoxy coatings, concrete encasement of the most vulnerable zones, or cathodic methods.

Concrete: Precast or Cast-in-Place

In general, working stresses should not exceed 33% of the 28-day unconfined compressive strength of the concrete in test cylinders. For piles in which concrete placement is locally difficult, as in small-diameter tips, irregularities in shell side walls, reinforcing cages, or batter piles, only 25% of the 28-day strength is recommended

[Peck et al. (1974)[2]]. Slump, in general, should be controlled between 3 to 6 in or more. For a straight, smooth, vertical pipe, a 3½ in slump is suitable, whereas in a vertical pile with a corrugated shell or reinforcement, or in a batter pile, the slump should be at least 6 in. Compacted concrete piles are installed with "no-slump" concrete. [Dry concrete: 0 in of slump with about 3.5 gal of water per cement bag; the 28-day strength is taken equal to 3000 psi. CGS (1978).[6]]

Concrete piles ordinarily are not subject to *deterioration* even above the water table, although high concentrations of magnesium or sodium sulfate salts (over 1000 ppm of SO_3 in the pore water) may cause damage. Reinforcement is attacked by salts in seawater and spray which enters through cracks in the concrete, resulting in rusting which causes the concrete cover to spall. Dense, high-quality concrete is the best protection against attack.

Prestressed Concrete

A reduction in the compressive capacity is required due to the application of the prestress. For most practical cases the working load may be taken as [NAVFAC (1982)[1]]:

$$f_c = 0.33f'_c - 0.27f_{pe} \tag{7.3}$$

where f_c = working load or maximum allowable stress
f'_c = 28-day strength of the concrete
f_{pe} = effective prestress on the gross section

Prestressed concrete piles are less susceptible to corrosion because tension cracks are minimized if the piles are properly handled and installed.

7.3.3 END BEARING (POINT CAPACITY)

Bearing-Capacity Equation

Calculations may be made from a modification of the bearing-capacity equation, usually expressed for deep foundations as:

$$Q_p = (qN_q + cN_c)A_p \tag{7.4}$$

Deep foundations deriving support at their tip fail in *punching shear* which has not been well defined mathematically. Several modes of failure have been suggested by various investigators as illustrated on Fig. 7.26.

Bearing-capacity factors for deep foundations, especially N_q, are agreed generally to be significantly higher than those for shallow foundations. Values of N_q in terms of ϕ, as suggested by various investigators, are given in Fig. 7.27. The values given by Vesić (most conservative) seem to be supported best by full-scale evidence [Murphy (1973)[24]], although those of Meyerhof (Fig. 7.28) consider the critical depth. The values given by Berezantsev are favored by many investigators [Poulos and Davis (1980),[25] CGS (1978)[6]]. The value of N_c, for soils with cohesion, increases from about 6 to 10 with increasing undrained strength and increasing pile diameter, but for piles of conventional diameter it is usually taken as 9. Values of N_c as a function of ϕ are given in Fig. 7.28. The soil weight factor N_γ, a function of foundation width, is ignored as it is very small compared with the depth factor.

The *installation* method impacts significantly on end bearing. For a given initial ϕ, bored piles have a point resistance of only about one-third to one-half of that of driven piles, and compacted

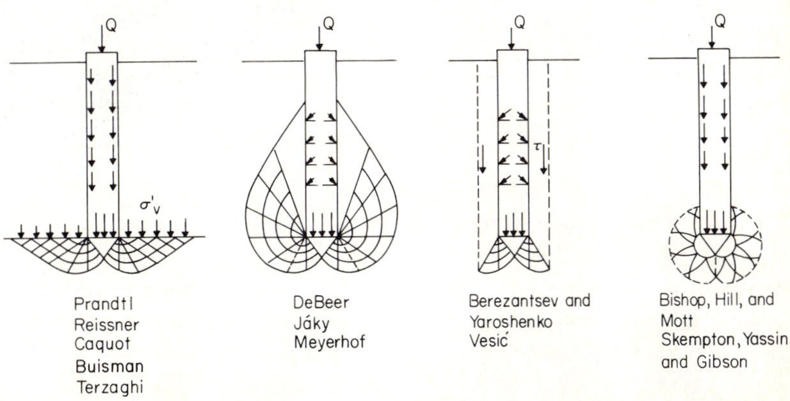

FIG. 7.26 Assumed failure patterns under deep foundations. [From Vesić (1967).[23]]

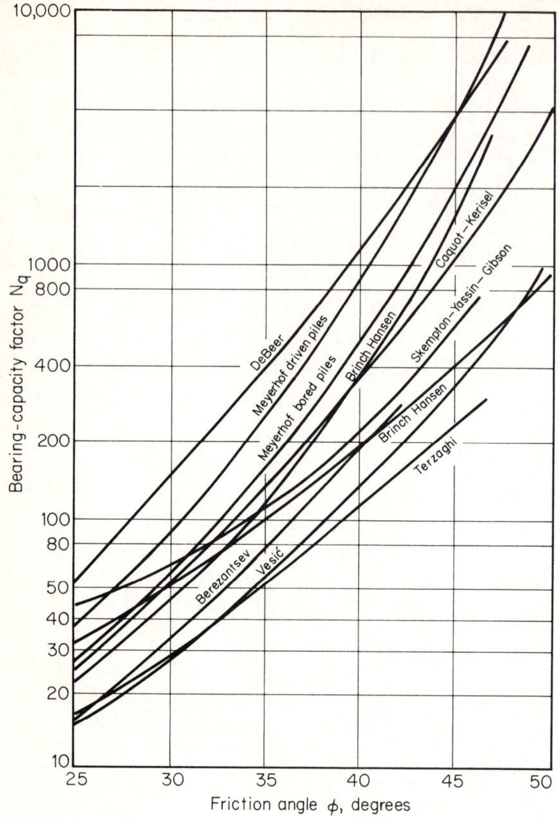

FIG. 7.27 Bearing-capacity factors for deep circular foundations. Berezantsev's curve appears favored by practitioners. [From Vesić (1967).[23]]

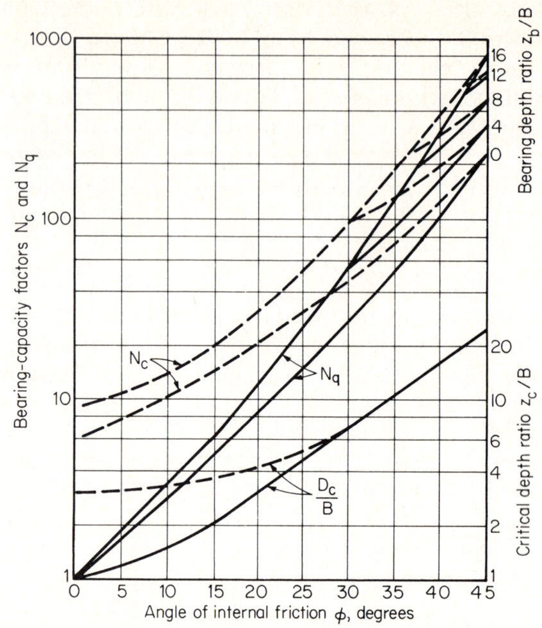

FIG. 7.28 Bearing-capacity factors and critical-depth ratios for driven piles. [From Meyerhof (1976).[20]]

concrete piles driven with great impact energy have up to about twice the unit point resistance of driven piles of constant section.

Sands

Bearing-Capacity Equation

The unit end-bearing resistance q_p in cohesionless soils is given by

$$q_p = \bar{p}_o N_q \qquad (7.5)$$

where $\bar{p}_o$ = effective overburden pressure
N_q = bearing capacity factor in terms of ϕ (Fig. 7.27)

There have been various approaches to the critical-depth problem [Poulos and Davis (1980)[25]]. In the approach proposed by Meyerhof (1976),[20] until z_c is reached, N_q increases roughly linearly with the ratio z_b/B (z_b = bearing depth, B = pile diameter) as shown in Fig. 7.28. N_q reaches its maximum value at approximately one-half of the critical depth ratio z_c/B, beyond which the conventional bearing-capacity theory no longer applies; i.e., at depths greater than about 7 (ϕ = 30°) to 22 (ϕ = 45°) pile diameters into the bearing stratum. In Eq. 7.5, therefore, $\bar{p}_o = L\gamma$ above z_c, where L = pile length, and $\bar{p}_o = z_c\gamma$ below z_c.

Another approach is to establish *limiting values* for q_p and N_q for various gradations such as are given in Table 7.5 for medium-dense silica and calcareous sands.

Various methods are available for *estimating* ϕ *(see Art 3.3.2)*. Most commonly used are the SPT and the CPT; as an alternative, q_p is evaluated directly from CPT and SPT results as described below. Empirical relationships have been developed to consider the relative density of the soil and pile type. To evaluate *loose sands* densified by a *driven pile*, ϕ values beneath the pile tip may be taken as the final values subsequent to driving, as given by Kishida (1967)[26]:

$$\phi = \frac{\phi_1 + 40}{2} \qquad (7.6)$$

where ϕ_1 = internal friction angle prior to pile installation. In *bored piles* where installation may loosen the soils at the base, it is suggested that ϕ be taken as $\phi = \phi_1 - 3$ [Poulos and Davis (1980)[25]].

For *insert (grouted) piles* offshore, Eide (1974)[26a] suggests $N_q(\phi)$ = 40 to 500, and for dense sands, q_{max} = 60 kg/cm².

Static Cone Penetration Test (CPT) Method

The CPT acts as a model pile test and provides a relatively reliable method for evaluating pile capacity in sands. The cone tip resistance q_c may be used directly to estimate end bearing, or it may be used to estimate ϕ from which N_q is found for insertion in Eq. 7.5. Finding ϕ is described in Art. 3.4.5 of Hunt (1984).[12a]

From many load test–CPT correlations, engineers in the Netherlands have found that the pile end bearing is governed by the cone resistance over a zone of 0.7 to 4.0 pile diameters below the pile tip and 8 diameters above the pile tip. q_p is found from the plot of q_c results as shown in Fig. 7.29 and given by the expression

$$q_p = \frac{q_{c1} + q_{c2}}{2} \tag{7.7}$$

where q_{c1} = Average q_c over distance y below the pile tip (path *a-b-c*). Sum q_c values in both the downward (path *a-b*) and upward (path *b-c*) directions using actual q_c values along path *a-b* and the minimum-path rule along path *b-c*. Compute q_{c1} for y values from 0.7 to 4.0*B* and use the minimum value obtained.

q_{c2} = Average q_c over a distance of 8*B* above the pile tip (path *c-e*). Use the minimum-path rule as for path *b-c* in the q_{c1} computations. Ignore any minor x peak depressions if in sand, but include any in the minimum path if in clay.

Limiting values are applied to the CPT analysis results because of various influencing factors such as uncertain size effects, difference in mobilization rate of Q_p and Q_s, grain crushing, creep effects, and negative pore pressures during CPT penetration in very dense fine sand or silty sands

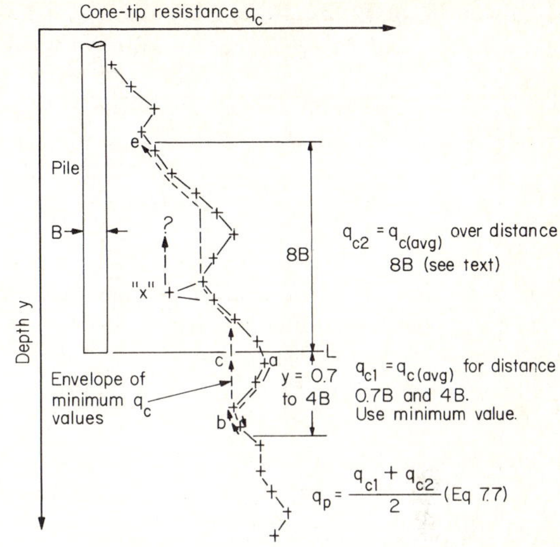

FIG. 7.29 Dutch procedure for predicting pile-tip capacity from CPT data. [From Schmertmann (1978).[27]]

[Schmertmann (1978),[27] de Ruiter and Beringen (1978)[16]] as follows:

q_c cutoff: 300 kg/cm² (maximum test value considered valid for any soil)

Sands: $q_{p(max)}$ = 150 kg/cm²

Very silty sands: $q_{p(max)}$ = 100 kg/cm²

Gravelly coarse sand and sand (OCR = 2 to 4): $q_{p(max)} = 0.67\, q_p$

Fine gravel and dense sand (OCR = 6 to 10): $q_{p(max)} = 0.5\, q_p$

q_a: Use FS = 2.25 with electric tip data, and 3.0 with mechanical tip data [Schmertmann (1978)[27]]

Acceptable pile penetration generally is considered as a maximum for studies offshore if the API criteria are used and a minimum if the CPT method is used for determining limiting values because the CPT results are considered as more reliable than the API procedure (Table 7.5) [de Ruiter and Beringen (1978)[16]].

Standard Penetration Test (SPT) Method

N values may be used to estimate ϕ (Table 3.5 and Fig. 3.29) or from the empirical relationship suggested for piles by Kishida (1967)[26]:

$$\phi = \sqrt{20N} - 15° \tag{7.8}$$

for use in Eq. 7.5.

From empirical correlations between the CPT and the SPT, Meyerhof (1976)[20] suggests that N from immediately below the pile-tip elevation can be used to estimate q_p for a pile driven into a sand stratum to a depth z_b from the expression:

$$q_p = \frac{0.4Nz_b}{B} \leqq 4N \quad \text{kg/cm}^2 \qquad (7.9)$$

where $q_p = 4N$ is a limiting value. It appears that load test data for piles driven into sand and gravel support the validity of Eq. 7.9 for q_p, whereas for piles driven into *nonplastic silt* better agreement is obtained by using an upper limit of approximately $q_p = 3N$.

N values should be corrected for depth as discussed in Art. 6.2.2 and shown in Fig. 6.20. A safety factor of 4 should be applied and $q_a = q_p/4$.

Clays

Bearing-Capacity Equation

In clays, the ultimate unit point resistance is expressed as

$$q_p = cN_c \qquad (7.10)$$

and the total point resistance or end bearing as

$$Q_p = cN_cA_p \qquad (7.11)$$

where c = the undrained shear strength (c_u or s_u) as measured in undrained triaxial tests, vane shear tests, or with the CPT (*see Art. 3.3.3*).

A range of values for N_c for driven circular or square piles with various depth ratios z_c/B is given in Fig. 7.28. In saturated, homogeneous clay under undrained conditions, theory and observation have shown that the value of N_c below z_c (generally considered as four base diameters) varies with sensitivity and deformation characteristics of the clay from about 5 for very sensitive, brittle, normally consolidated clay, to about 10 for insensitive, stiff, overconsolidated clay [Meyerhof (1976)[20]]. A value of 9 normally is used for bearing-capacity estimates for driven and bored piles in nonfissured (intact) clays. Values on the order of 7 have been used in the OC fissured London clays.

Disturbance of the soil by pile installation affects the initial point resistance, but subsequent consolidation and thrixotropic effects (freeze) can increase the value to exceed the initial value by the end of construction.

Clay-Sand Mixtures

In clay-sand mixtures with ϕ and c acting, the ultimate unit point resistance is expressed by Meyerhof (1976)[20] as

$$q_p = cN_c + \bar{p}_oN_q \leqq q_m \qquad (7.12)$$

in which $\bar{p}_o$ is the effective overburden pressure and q_m is the limiting value of unit point resistance below z_c.

Pressuremeter Method

Application

The pressuremeter (*see Arts. 3.3.3 and 4.1.4*) is particularly useful to obtain values of shear strength and evaluate the bearing capacity of deep foundations in materials difficult to sample or otherwise test in situ, such as sands, soft rock, residual soils, or glacial till.

Ultimate Bearing Capacity

The relationship for q_p as given by Menard (1975)[28] is

$$q_p = \bar{p}_{oh} + k_g(P_L - \bar{p}_{oh}) \qquad (7.13)$$

where $\bar{p}_{oh}$ = at-rest horizontal stress (from test data)

P_L = limiting or ultimate pressure (from test data)

k_g = a bearing factor varying from 0.8 to 9 for deep foundations depending upon embedment and soil conditions

Values for k_g for various soil classes are given in Table 7.6. A factor of safety of 3 is applied to the term $k_g(P_L - \bar{p}_{oh})$ to find q_a.

Vesić (1972)[29] has presented a modification of Menard's approach based on his work regarding the expansion of cylindrical and spherical cavities in a soil mass, which appears to yield results less conservative than Menard's [Murphy (1973)[24]].

TABLE 7.6
PRESSUREMETER BEARING CAPACITY FACTOR k_q FOR DEEP FOUNDATIONS*

Material	Class	P_{Ln},† tsf	Bearing capacity factor k_g,‡ Driven piles	Bored piles
Clays: soft to firm	1	0–12	2.0	1.8
Silts: loose		0–7		
Sands: loose§	2	4–8	3.6	3.2
Silts: dense		12–30		
Clays: stiff		18–40		
Rock: soft, decomposed, friable		10–30		
Sands and gravels: medium dense	3	10–20	5.8	5.2
Rock: moderately soft, decomposed		40–100		
Sands and gravels: very dense	4	30–60	9.0	7.1
Rock: medium to high strength		60–100+		

*After CGS (1978),[6] Menard (1975).[28]

†Net limiting pressure, $P_{Ln} = P_L - P_{oh}$.

‡For penetrations of $z_{tip}/B = 10$ or more; k_q decreases for shallower depths for a given B.

§Calcareous sands: <2 [From Tirant (1979)].

Limitations

In using the results from pressuremeter testing, the limitations of the method and the equipment must be considered, as discussed in Art. 3.5.4 of Hunt (1984).[12a]

Rock Masses

General

Reference should be made to the discussion regarding shallow foundations on rock *(see Art. 6.3)*, which also applies to deep foundations.

Rock-Mass Quality

In the evaluation of end bearing (and shaft resistance) for deep foundations supported on rock masses, the consideration of rock quality in terms of discontinuities and decomposition is the most important factor for a given rock type. The elements of rock-mass quality and its measurement are described in Hunt (1984).[12a] Article 6.3.2 in this volume includes descriptions of rock-quality classes (excellent, good, fair, poor, and very poor) and a summary of q_a as related to rock quality and the method of selecting q_a is given in Table 6.11.

Deep foundations bearing on rock normally impose high loads, and an adequate subsurface investigation is imperative to ensure the continuity of quality with depth, once a tip elevation is selected. Cavernous limestone, expansive shales, and formations that weather differentially, such as often occurs in metamorphic rocks with igneous intrusions (Fig. 7.30) may create hazardous foundation conditions. A number of possible failure forms involving rupture in rock masses are illustrated in Fig. 6.40. Thorough geologic study will reveal where these conditions may exist.

Cavernous or pinnacled limestones (Figs. 6.40*e* and *f*) are particularly hazardous and the only certain procedure to investigate for suitable support is to test-bore every pier before construction or to "proof-test" every pier following its excavation. Proof-testing is accomplished by either core-drilling or air-drilling to confirm that an adequate thickness of sound rock exists beneath each pier, usually in the order of 3 diameters, as long as caverns are not large. The analysis of structures built over large caverns and mines can be approached using methods discussed under Art. 6.3.3, but the possibility of limestone cavern collapse in the future from groundwater pumping must be considered.

FIG. 7.30 Joint conditions and a gouge zone in this granite gneiss will provide poor foundation support. Other nearby areas are characterized by continuously sound, widely jointed rock, indicating the possible wide variation at a given site and the need for thorough investigation. [From Hunt (1984).[12a]]

Assessment of Allowable Bearing Value q_a

Building codes provide an empirical basis for selecting q_a, usually given in terms of the "presumptive bearing value" *(see Art. 6.3.2)*. A summary of such values from a number of building codes is given in Table 6.12, and allowable bearing pressures for rock from the New York City Code, which provides some detail regarding rock quality, is given in Table 6.13. Table 7.7 provides values for q_a in terms of weathering and cementation and in terms of joint spacing. Values from codes are considered generally to be conservative, providing for FS = 3 or more. Caution is required in their use, however, since not only may they not consider rock of less than good quality, but they usually apply to rock-mass characteristics for a given geographic location and often are not applicable to the same rock type in an other location where its characteristics are different.

Core-strength analysis based on measured values of U_c is used to assess intact rock, and the results are modified empirically to account for discontinuities *(see Art. 6.3.2)*.

Rock quality less than good should raise concern regarding its capacity to provide suitable support, because of its softness, degree of decomposition, or intensity of fracturing. Consideration is given to performing full-scale load tests to measure load-deflection relationships, or other in situ tests performed to obtain moduli values for correlations with RQD and seismic velocity values as discussed in Art. 6.3.2. The *pressuremeter* is useful for estimating the ultimate bearing pressure as discussed in the previous section, using Eq. 7.13 and applying FS = 3 to the right-hand portion of the equation to find q_a [CGS (1978)[6]]. The bearing factor k_g for piers in rock is given by CGS in terms of the ratio of socket depth to pier diameter as follows:

Ratio of socket depth to pile diameter	0	1	2	3	5	7
Bearing factor k_g	0.8	2.8	3.6	4.2	4.9	5.2

Note that 0.8 is the minimum value for surface foundations from Fig. 6.34, and 5.2 is the value for bored piles in soft rock (Class 3) from Table 7.6.

Foundation Types

Driven slender piles supported on rock are installed to refusal in accordance with one of the dynamic driving formulas and their capacity, considered for end-bearing support only, is often the strength of the member in compression. They are installed in accordance with local codes which usually require load tests for capacity confirmation. Care is required to assure that the piles are not damaged by overdriving, and hardened steel tips are used to obtain rock penetration. Pile tips are *socketed* and *grouted* into rock with steeply sloping surfaces where overlain by weak soils to obtain resistance to move-

TABLE 7.7
ALLOWABLE END-BEARING PRESSURES, TSF, FOR PILES FOUNDED IN ROCK AND ROCKLIKE MATERIAL*

Degree of weathering or effect on cementation	Joint spacing		
	Wide	Narrow	Very narrow
Unweathered, strongly cemented	>80	>80	>70
Slightly weathered, still well cemented	50	40	30
More severely weathered, weakly cemented	20	15	15
Disintegrated, noncemented	Soil criteria apply		

* After the German Standard DIN 4014, Part 2.

NOTES:

1. For crystalline and sedimentary rocks.
2. Skin friction values are added based on an amount equal to 10% of the allowable end bearing values for the various rock qualities penetrated.
3. Skin friction in weathered rock should not exceed 1.2 tsf.

ment by sliding, especially where batter piles incline in the direction of the rock surface. Socketing also provides uplift resistance. Where penetration is achieved, shaft resistance is usually ignored for driven piles.

Bored piles (or *drilled piers*) founded on, or socketed into, rock are common and effective foundations for supporting the large loads of high-rise buildings, industrial structures, and bridges. The rock surface must be clean of all cuttings and other loose, soft, or otherwise compressible materials. Where rock penetration is shallow, shaft resistance is ignored, but with increasing depth in rock masses it becomes very significant. There is not, as yet, a clearly described basis for designing piers on rock [D'Appolonia et al. (1975)[30]]. The bearing capacity of less-than-good-quality rock and the bond between concrete and the socket perimeter are poorly defined parameters. Because of the high loads normally applied to piers, load tests are extremely expensive and seldom carried to failure.

7.3.4 POSITIVE SHAFT RESISTANCE

General Expression

Total shaft resistance (or *total side friction*) is a function of the pile surface area and the shearing resistance mobilized along the shaft, which may be expressed as

$$Q_s = \Sigma(\Delta L)(A_s)(f_s) \qquad (7.14)$$

where L = embedment length in a given stratum

A_s = circumferential shaft area = $\pi B/$ unit length

f_s = shaft resistance = $c_a + \bar{p}_o K_s \tan \delta$ (from Coulomb's expression)

Stress conditions acting along a pile shaft are shown in Fig. 7.31. A modification of Coulomb's expression is used to calculate the shearing resistance which becomes mobilized along the

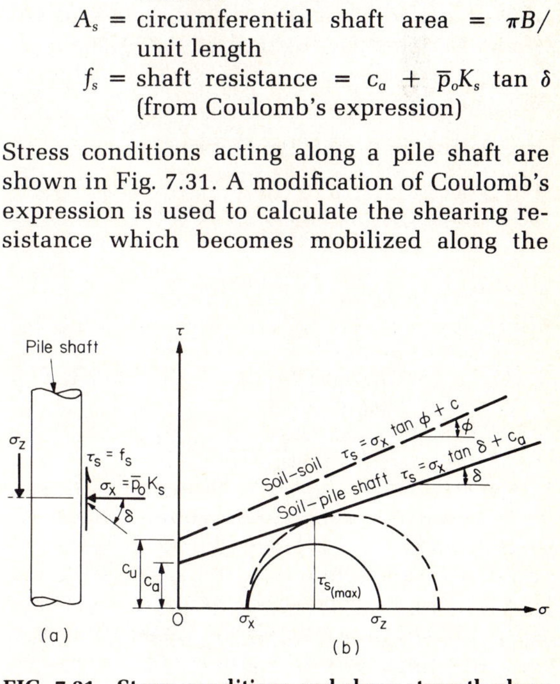

FIG. 7.31 Stress conditions and shear strength along pile shaft.

shaft by displacements in the order of ½ to 1 cm (3⁄16 to ⅜ in). In all materials f_s is very much affected by pile type and installation method. In sands f_s depends on the lateral stress from overburden pressure and soil-pile friction; in rock or clays f_s depends on rock or soil-pile adhesion. Approximate values for f_s used for *preliminary capacity estimates* in soils are given in Table 7.8.

Criteria for evaluating the various elements of the expression for f_s are not well defined and the state of the art is very much in transition. Analysis relies heavily on empirical relationships.

Sands

Shaft Friction Resistance f_s

In a free-draining soil the unit shaft resistance f_s is the product of the soil-pile friction angle δ, which varies with the roughness of the pile material in contact with the soil (timber, concrete, steel), and the lateral earth pressure acting along the pile:

$$f_s = \bar{\sigma}_{ho} \tan \delta \qquad (7.15)$$

$\bar{\sigma}_{ho}$ is the initial effective horizontal stress, a function of the *lateral earth pressure coefficient* K_s, and the effective vertical stress $\bar{\sigma}_{vo}$ taken at the stratum midpoint ($K_s = \bar{\sigma}_{ho}/\bar{\sigma}_{vo}$). Therefore, f_s may be expressed as:

For straight shaft piles in compression

$$f_s = K_s\bar{\sigma}_{vo} \tan \delta \quad \text{or} \quad K_s p_o \tan \delta \qquad (7.16)$$

For straight shaft piles in tension or pullout

$$f_s = K_s\bar{p}_o \tan \delta \, \rho \qquad (7.17)$$

where $\bar{p}_o = \gamma z$ until z_c and limiting value γz_c below z_c

ρ = reduction coefficient for pullout (usually 0.7 to account for a reduction in confining pressure)

f_s depends on many factors including the original horizontal stress in the ground represented by K_o (at-rest coefficient); pile size, shape, and penetration depth; soil compressibility, and installation method. Driving may densify some soils and cause relaxation in others, whereas bored piles may loosen granular soils. Tension load tests on piles installed by driving and jetting have shown that jetting very substantially reduces shaft friction, and therefore should be avoided wherever f_s is counted on for support. Because of the various factors, the most reliable method to establish values for f_s, or (K_s tan δ), is with full-scale load tests.

Usually, Q_{as} is taken as $Q_s/3$.

Critical Depth Factor

Factor K_s, and therefore f_s, are not uniformly distributed along the shaft but rather decrease from maximum values near the top of a pile driven into sand to minimum values near the pile tip. The actual value reached by lateral friction in sand is influenced by the arching effect that modifies the stress state around the pile. Research and experience indicate that lateral friction increases approximately linearly to a *critical depth* of about 10*B* in very loose sands and to about 20*B* for dense sand. For greater depths than z_c, f_s remains constant.

Most investigators have taken the approach of establishing *limiting values* for K_s and f_s. An alternate approach, considered more realistic by Poulos and Davis (1980),[25] assumes $\bar{p}_o = \bar{p}_o$ to some critical depth z_c, beyond which the value for $\bar{p}_o$ remains constant, which is an idealized distribution in accordance with the behavior observed by Vesić (1967)[23] from full-scale load test data.

Approaches Based on Evaluating K_s and δ

Selection of values for K_s and δ is not well defined in the present state of the art. Particularly difficult is the selection of a value for K_s. As f_s decreases along the shaft, the corresponding coefficient of lateral earth pressure K_s also decreases from a maximum, near the pile top, which may approach the passive earth pressure coefficient K_p, to a minimum near the tip, where K_s may be less than the initial earth pressure at rest K_o. Load tests on short piles above z_c in homogeneous, loose sand have shown that the value of K_s for a given ϕ value can scatter considerably from a lower limit of roughly K_o for open-end pipe and bored or jacked piles to about $4K_o$ or more for displacement piles driven into dense sands. Driving displacement piles in sand can have the effect of densifying the sands for distances as great as 8 diameters from the pile.

TABLE 7.8
FRICTION BETWEEN PILES AND SOIL

Material*	Identification*	Ordinary range of values per square foot of bounding area of pile*	*N* number of blows per foot on 2-in OD sampling spoon from 140-lb weight falling 30 in†
Fine-grained soils:			
Very soft clay	Fist penetrates several inches easily	125 ± 125	0–2
Mud		250 ± 200	
Silt		300 ± 200	
Soft clay	Thumb penetrates several inches easily	400 ± 200	2–4
Silty clay		600 ± 200	
Sandy clay		600 ± 200	
Medium clay	Thumb dents several inches—moderate effort	700 ± 200	4–8
Sandy silt		800 ± 200	
Firm clay	Thumb dents several inches—great effort	900 ± 200	
Dense silty clay		1200 ± 300	
Stiff clay	Thumb dents readily—great effort to penetrate	1500 ± 400	8–15
Very stiff clay	Thumbnail indents readily	3000 ± 1000	15–30
Hard clay	Thumbnail indents with difficulty	Over 4000	Over 30
Coarse-grained soils:			
Very loose sand and silt or clay		100 ± 50‡	0–6‡
Medium sand and silt or clay		500 ± 100‡	6–30‡
Dense sand and silt or clay		700 ± 100‡	30–50‡
Very dense sand and silt or clay		900 ± 100‡	Over 50‡
Sand:		1200 ± 500§	
Very loose			0–4
Loose			4–10
Medium			10–30
Dense			30–50
Very dense			Over 50
Sand and gravel		2000 ± 1000	
Gravel		2500 ± 1000	

*From Chellis (1948).[108]

†From Tables 3.5 and 3.7.

‡From Liver and Jennings (1952).[109]

§If not micaceous, muddy, or under hydrostatic pressure or vibration.

Therefore, K_s depends on sand density, earth pressure at rest K_o, volume changes during pile insertion, penetration depth, and pile loading direction (compression or tension). K_s values generally are assumed to range between K_a and K_o (where $K_o = \bar{p}_h/\bar{p}_o$). For NC sands, $K_o \approx 1 - \sin \phi$ (generally around 0.5), increasing with OCR and in soils subject to cyclic action [Kraft and Lyons (1974)[15]]. Typical values for K_s are shown in Table 7.9.

In general, tan δ ranges between 0.67 and 1.0ϕ [Kraft and Lyons (1974)[15]]. Various practitioners have presented a range in values for δ as a function of soil gradation and pile type [for example, Kezdi (1975)[17]]. Current practice for large-diameter piles appears to favor the values recommended by the American Petroleum Institute (Table 7.5), where ϕ ranges from 20 to 35° depending upon gradation and $\delta = \phi - 5$ between steel and sand. Woodward et al. (1972)[9] give $\delta = 0.7\phi$ for smooth-sided piles and $\delta = 0.85\phi$ for corrugated piles. Typical values are summarized in Table 7.10.

TABLE 7.9
TYPICAL VALUES FOR LATERAL EARTH PRESSURE FACTOR K_s*

Pile type	K_s†
Driven	0.7 to 3.0
Driven with pilot hole	0.4 to 0.7
Drilled	0.1 to 0.4
Insert	0.7 (compression) 0.5 (tension)

*From Le Tirant (1979).[14]
†Increases with OCR or D_R.

Limiting values recommended for driven or grouted piles offshore are

$$K_s \tan \delta = 0.7 \tan (\phi - 5) \text{ in compression} \tag{7.18}$$

$$K_s \tan \delta = 0.5 \tan (\phi - 5) \text{ in tension} \tag{7.19}$$

f_{max} = 1.0 tsf for medium-dense clean sands (see Table 7.5 for other gradations)

Some practitioners are adapting these relationships for evaluating long piles onshore.

EXAMPLE Determine the capacity of a steel pipe pile, B = 0.32 m, driven to 24.4 m penetration in sand grading from fine to coarse with depth.

Data: $N_{avg} = 16$, GWL = 2.4 m, $\gamma_t = 1.73$ t/m³, $\gamma_b = 0.73$ t/m³, $K_{s(assumed)} = 1.5$

Note: For computing shaft friction it is convenient to set up a diagram such as given in Table 7.11 indicating the variation of $\bar{p}_o$ with depth. The product of the total area from the diagram ($K_s \tan \delta$) and the pile circumference provides the total shaft friction acting along an increment of embedded pile length.

Solution: ϕ is determined from soil gradation and N, and Table 7.11 is prepared with data input. Assuming limiting depth $D = 10B = 3.2$ m,

TABLE 7.10
TYPICAL VALUES FOR PILE-SOIL FRICTION FACTOR*

Pile material	δ, degrees	tan δ	Alternate for δ
Concrete (rough, cast-in-place)	33	0.65	$\delta = 0.85\phi$
Concrete (smooth)	30	0.58	$\delta = 0.70\phi$
Steel (corrugated)	33	0.65	$\delta = \phi$
Steel (smooth) (see Table 7.5)	—	—	$\delta = \phi - 5°$
Wood	22	0.40	—

*Compiled from various sources: see text.

TABLE 7.11
DATA INPUT FOR ANALYSIS OF FRICTION PILE IN SANDS USING STANDARD METHOD

Depth, m	ϕ, degrees	$\delta(\phi - 5)$, degrees	tan δ	K_s	ΔL, m	$\bar{p}_o$, t/m³	$\bar{p}_o$, area
0–2.4	25	20	0.36	1.5	2.4	4.15	4.98
2.4–3.2	25	20	0.36	1.5	0.8	4.74	3.55
3.2–18.3	32	27	0.51	1.5	15.1	4.74	71.57
18.3–20.8	30	25	0.47	1.5	2.5	4.74	11.85
20.8–24.4	33	28	0.53	1.5	3.6	4.74	17.06

$$
\begin{aligned}
Q_s &= \Sigma\ (\Delta L)(A_s)(f_s) \\
&= \Sigma\ [(\text{pile circumference})(K_s) \\
&\quad \times (\text{area } \bar{p}_o \text{ diagram})(\tan\delta)] \\
&= (1.01)(1.5)(4.98)(0.36) \\
&\quad + (1.01)(1.5)(3.55)(0.36) \\
&\quad + (1.01)(1.5)(71.57)(0.51) \\
&\quad + (1.01)(1.5)(11.85)(0.47) \\
&\quad + (1.01)(1.5)(17.06)(0.53) \\
&= 82.09 \text{ tons}
\end{aligned}
$$

$$Q_p = \bar{p}_o N_q A_p \text{ for } N_q,\ \phi = (\phi_1 + 40)/2 = 36.5 \qquad (7.6)$$

$$N_q = 98 \text{ (Fig. 7.27)}$$

$$Q_p = (4.74)(98)\left[\frac{(3.14)(0.32)^2}{4}\right] = 37.36 \text{ tons}$$

Total ultimate capacity $= Q_s + Q_p = 119$ tons; $Q_a = Q_u/2 = 59$ tons

Alternate Approach: K_s tan δ as a Function of $\bar{\phi}$

Poulos and Davis (1980)[25] suggest using an idealized distribution of $\bar{p}_o$ assuming that $\bar{p}_o$ increases linearly as a function of γz, until some critical depth z_c is reached, beyond which $\bar{p}_o$ remains constant. Based on the investigations of Vesić (1967)[23] and Meyerhof (1976),[20] Poulos and Davis provide relationships between K_s tan δ and $\bar{\phi}$ and between z_c/B and $\bar{\phi}$ to provide for z_c without resorting to limiting values which only become operative at substantial penetrations in the order of 25 m or more. They advise that the relationships given in Fig. 7.32 may be subject to amendment with future investigations since, for example, the dependence of K_s tan δ on pile materials is not defined.

They suggest using curve *b* for driven piles and curves *c* for jacked and bored piles and to find ϕ for use in *b* for *driven piles*, the following: $\phi = 0.75\ \bar{\phi}_1 + 10$, where $\bar{\phi}_1$ is prior to driving. For *bored or jacked piles* $\phi = \bar{\phi}_1$ for use in Fig. 7.32*c*. For large-diameter piles with capacities greater than 300 t, this approach will yield average values for f_s considerably above those normally adopted for design, such as those given in Table 7.5. For these conditions, Poulos and Davis suggest using values for K_s tan δ given in Fig. 7.32*c* as derived by Meyerhof for driven piles.

EXAMPLE Pile and other data are as given in the previous example for the standard approach. $\bar{\phi}_1$ prior to driving is estimated from *N* corrected for depth, and $\phi = 0.75\ \bar{\phi}_1 + 10$. K_s tan δ is from Fig. 7.32*b* as a function of ϕ. These are summarized with other data in Table 7.12.

Critical depth: Assume $z_c = 8B = 2.4$ m. For $\phi = 28.75$, $z_c/B = 5.0$ (Fig. 7.32*a*), or $z_c = 5.0 \times 0.32 = 1.56$, indicating the assumption is justified. At z_c, $\bar{p}_o = 1.56 \times 1.73\ (\text{t/m}^2) = 2.7\ \text{t/m}^2$

Compute:

$$
\begin{aligned}
Q_s &= A_s \Sigma \bar{p}_o K_s\, dz \tan\delta \\
&= \pi 0.32 \Bigg\{ \Bigg[\left(\frac{0 + 2.7}{2}\right) 1.56 \\
&\quad + 2.7\ (2.4 - 1.56) \Bigg] 1.0 \\
&\quad + [2.7(18.3 - 2.4)1.30] \\
&\quad + [2.7(20.8 - 18.3)1.18] \\
&\quad + [2.7(24.4 - 20.8)1.35] \Bigg\} \\
&= 81.7 \text{ tons}
\end{aligned}
$$

$$Q_p = (2.7)(98)(0.08) = 21.28$$

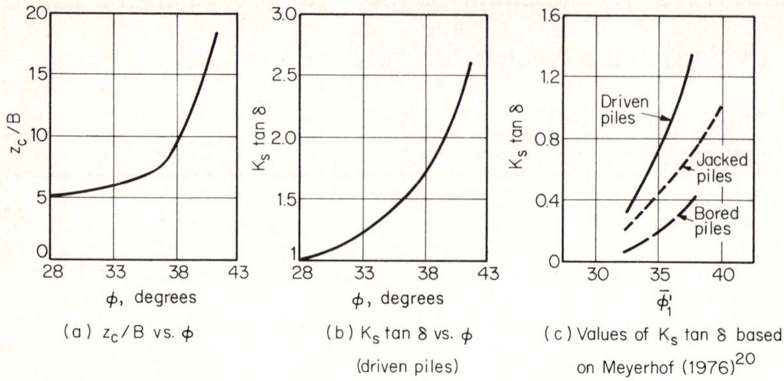

FIG. 7.32 Determining f_s in sands. Values of z_c/B and K_s tan δ vs. φ and $\bar{\phi}_1$. [From Poulos and Davis (1980).[25] Used with permission of John Wiley & Sons, Inc.]

For driven piles: $\phi = 0.75\bar{\phi}_1 + 10$ **(*a* and *b*)**

For bored piles: $\phi = \bar{\phi}_1 - 3$ **(*a*)**

$\phi = \bar{\phi}_1$ **(*c*)**

where $\bar{\phi}_1 = \phi$ prior to installation.

TABLE 7.12
DATA INPUT FOR ANALYSIS OF FRICTION PILE IN SANDS USING METHOD OF POULOS AND DAVIS (1980)[75]

Depth, m	$\bar{\phi}_1$, degrees	ϕ (degrees) = $0.75\,\bar{\phi}_1 + 10$	K_s tan δ
0–2.4	25	28.75	1.00
2.4–18.3	32	34	1.30
18.3–20.8	30	32.6	1.18
20.8–24.4	33	34.75	1.35

Total capacity = $Q_s + Q_p$ = 103 tons; Q_a = 51.5 tons

The difference between the two examples lies mainly in the selection for the critical depth and its influence on $\bar{p}_o$, especially on end bearing. A comparison of the various methods of determining Q_s in sands is summarized in Table 7.13.

Taper Correction Factor F_ω

To provide for the installation effects of tapered piles, Poulos and Davis (1980)[25] express Q_s as

$$Q_s = \Sigma(\Delta L)(A_s)(f_s)(F_\omega) \qquad (7.20)$$

where F_ω = the taper correction factor given in terms of the taper angle ω and φ on Fig. 7.33.

CPT Analysis

Estimates of f_s can be obtained from the CPT method using semiempirical and conservative relationships, with limiting values provided to account for z_c. Various approaches have been proposed. Meyerhof (1976)[20] suggests that preliminary estimates be obtained by assuming that the ultimate shaft friction of a driven cylindrical pile is given approximately by the local sleeve friction f_L of the CPT, whereas for bored piles f_s

TABLE 7.13
COMPARISON OF METHODS TO EVALUATE Q_s IN SANDS*,†

Method	$\bar{p}_o$	K_s	tan δ	φ	K_s tan δ	Limits
API	Compute Compute	0.4 compression 0.7 tension	tan (φ − 5) tan (φ − 5)	φ from SPT Corrected $\bar{p}_o$	Compute Compute	$f_{s(max)}$ $f_{s(max)}$‡
Conventional	Compute	0.7–3.0	tan (φ − 5) (smooth steel)	Corrected $\bar{p}_o$	Compute	Z_c (critical depth)
	Compute	Table 7.9	tan φ (corrugated steel)	Corrected $\bar{p}_o$	Compute	Z_c (critical depth)
	Compute	Table 7.9	tan 0.7φ (smooth concrete)	Corrected $\bar{p}_o$	Compute	Z_c (critical depth)
	Compute	Table 7.9	tan 0.85φ (CIP concrete)	Corrected $\bar{p}_o$	Compute	Z_c (critical depth)
	Compute	Table 7.9	0.4 (wood)	Corrected $\bar{p}_o$	Compute	Z_c (critical depth)
Poulos and Davis (1980)[25]	Compute	—	—	$\bar{\phi} = 0.75\phi + 10$§	§	Z_c (critical depth)
				$\bar{\phi} = \phi - 3$¶	¶	Z_c (critical depth)
				$\bar{\phi} = \phi$¶		

*To solve: $Q_s = \Sigma\Delta Q_s = \Sigma\bar{p}_o(K_s \tan\delta)A_s\,\Delta z$, in which $f_s = \bar{p}_o(K_s \tan\delta)$.

†Tapered piles: $Q_s = \Sigma f_s A_s F\omega$, where $F\omega$ = taper factor. See Fig. 7.33. Use B_{mean} and Z_{mean}.

‡Limit values, Table 7.5, including calcareous sands.

§Driven piles, Fig. 7.32*b*; critical depth from Fig. 7.32*a*.

¶Bored or jacked piles, Fig. 7.32*c*.

= 0.33 to 0.5 f_L. For design use, de Ruiter and Beringen (1978)[16] suggest the *lowest value* of f_s, f_L, or f_1 where

$f_s = \beta q_c$, where q_c = tip resistance, and β = empirical coefficient, 1/300 for compression, 1/400 for tension

f_L = local sleeve friction [CGS (1978)[6] suggest $Q_s = 2f_L A_s$, and FS = 3.0]

f_1 = limiting values = 0.12 MN/m² = 1.2 kg/cm² = 1.2 tsf

Schmertmann (1978)[27] suggests evaluating Q_s as follows:

$$Q_s = C_s, C_c\left[\sum_{z=0}^{8B}\left(\frac{z}{8B}\right)f_L A_s + \sum_{8B}^{L} f_L A_s\right] \qquad (7.21)$$

where Q_s = total ultimate side-friction resistance

C_s, C_c = correction factors: C_s for sands, C_c for clays (Fig. 7.34)

z = depth to f_L value considered

B = pile width or diameter (8*B* is critical depth for CPT)

f_L = local sleeve friction resistance

A_s = pile-soil contact area per f_L depth interval

L = total embedded length of pile

Limiting values: f_L = 1.2 tsf for sands and 1.0 tsf for silty sands

SPT Analysis

Expressions for f_s have also been given in terms of relative density D_R since it reflects soil compressibility. For driven piles, f_s does not exceed 1 tsf; for cast-in-place piles it is even less. To es-

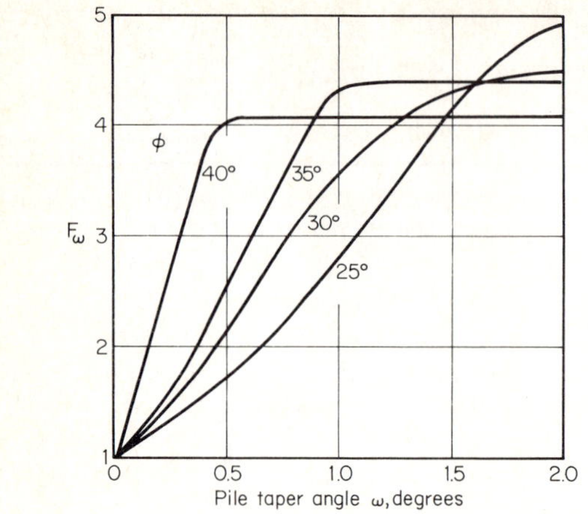

FIG. 7.33 Pile-taper factor F_ω in terms of ϕ and pile taper ω_o. [After Nordland (1963).[32] From Poulos and Davis (1980).[25] Used with permission of John Wiley & Sons, Inc.]

timate f_s, Vesić (1970)[34] suggests the following empirical formulas in which D_R is evaluated from the CPT or SPT:

For driven piles:

$$f_s = (0.08)10^{1.5}D_R^4 \tag{7.22}$$

For bored piles and piers:

$$f_s = (0.025)10^{1.5}D_R^4 \tag{7.23}$$

Meyerhof (1959),[35] (1960)[36] has related $Q_{s(\max)}$ to the SPT N value as follows:

For driven displacement piles:

$$Q_s = 0.02\overline{N}A_s \leqq 1 \text{ tsf} \tag{7.24}$$

For bored piles and foundation piers:

$$Q_s = 0.006\overline{N}A_s \leqq 0.4 \text{ tsf} \tag{7.25}$$

where $\overline{N}$ is the average N value within the embedded pile length, corrected for depth. For piles of small displacement, such as H piles, $f_s/2$ can be used conservatively, and $Q_{\max} = 0.5$ tsf.

In the usual practical situation of piles driven into sand strata with D_R increasing with depth, f_s provides less than one-half of the total capacity. For practical purposes it can be assumed that the lateral friction increases linearly to a depth of $15D$, where it reaches a value of f_s given by the above relationships.

Pressuremeter Analysis

f_s acting at any depth along the surface of a deep foundation may be estimated from the limit pressure P_L at that depth, as given in Fig. 7.35.

Granular materials:

- Use curve A for nondisplacement concrete piles and displacement steel piles. For nondisplacement steel piles apply a 50% reduction.

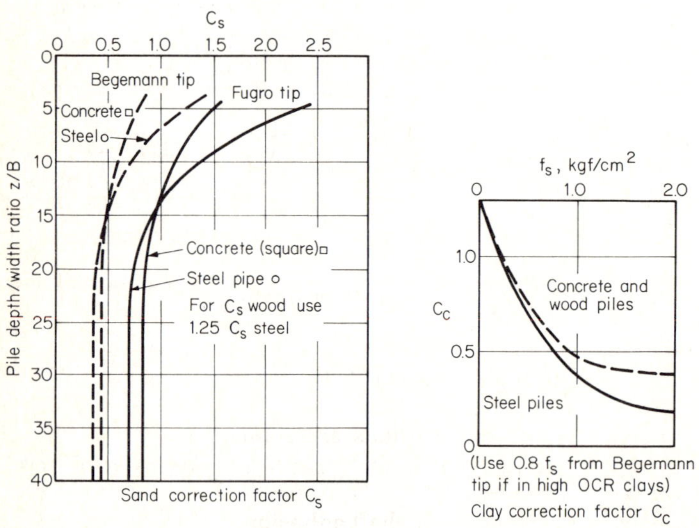

FIG. 7.34 Nottingham's (1975)[33] correction factors for use in Eq. 7.21 to compute Q_s from CPT data. [From Schmertmann (1978).[27]]

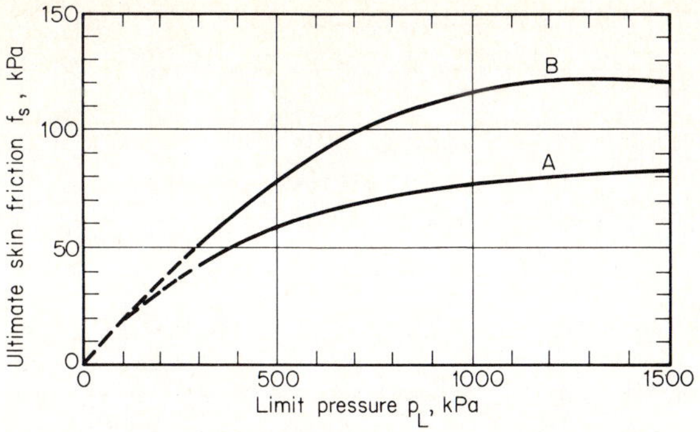

FIG. 7.35 Ultimate shaft friction f_s in terms of limiting pressure P_L from pressuremeter tests. [From CGS (1978),[6] after Menard (1975).[28]]

- Use curve *B* for displacement concrete piles.

Cohesive soils:

- Use curve *A* directly for concrete and timber piles; apply a 25% reduction for steel piles.

Upper limits for ultimate skin friction [From Baguelin et al. (1978)[37]]:

- Concrete displacement piles in granular soil, 1.22 tsf
- Concrete displacement piles in cohesive soil, or steel displacement piles in granular soil, 0.82 tsf
- Steel displacement piles in cohesive soil, 0.62 tsf
- Nondisplacement piles in any soil, 0.40 tsf

A factor of safety of 2 is applied [CGS (1978)[6]] to the ultimate skin friction.

Tension Loads (Pullout Capacity)

In sands, f_s will vary from 0.5 to the value obtained in compression, depending upon relative pile stiffness. Short, stiff piles will have the lowest coefficient, and long, slender piles will have the highest. For long, heavily loaded pipe piles offshore, the API (1983) recommends taking $0.5f_s$ for pullout resistance in sands, plus pile weight.

Clays

Support Factors

Shaft area in Eq. 7.14 is normally taken as the shaft circumference which often is not an exact assumption. In *NC clays* (soft to firm), when piles are extracted they are usually coated with a thin clay layer, indicating that failure occurred in the soil just outside the perimeter, not between the pile and soil. For driven cylindrical or square piles, therefore, using the shaft area is conservative. In steel H piles, pull tests have shown that clay soil usually remains between the flanges; therefore, the perimeter area is taken as the perimeter of a rectangle enclosing the flanges, a conservative assumption. In *OC (stiff) clays* driven piles may possess an imperfect soil-pile contact, and bored piles may display an irregular soil surface which can reduce the pile contact area upon which soil-pile adhesion depends.

Shaft resistance in clays may be analyzed by two general approaches: the undrained strength (total stress) analysis or the drained strength (effective stress) analysis. Pile driving causes remolding, resulting in a reduction in shear strength and an increase in pore pressure in NC clays. In time, some or all of the strength will recover as consolidation occurs and excess pore pressures dissipate. Indeed, as the horizontal stress after pile driving is greater than the initial stress, and as consolidation leads to a decreased void ratio, the strength may be higher after consolidation than before pile driving. In stiff, OC clays, however, negative pore pressures may be produced during driving which dissipate with time, weakening the clay to values below the initial strength. For offshore piles the trend appears to be to use effective stress analysis which, under the assumptions commonly made, leads to less conservatism in design.

End bearing and *shaft resistance* are both mobilized to provide support in stiff or stronger clays ($s_u > 1$ tsf), whereas in soft to firm clays ($s_u < 1$ tsf) support is derived almost totally from shaft resistance [CGS (1978)[6]].

Undrained Strength (Total Stress) Analysis

Three general methods are in use employing the undrained strength: the α coefficient method, the λ coefficient method, and the CPT method. The first two methods are semiempirical, with the coefficients based on evaluations of load test data; the CPT method is based on in situ measurements of shaft adhesion.

COEFFICIENT α METHOD In homogeneous, saturated clays, the ultimate side friction f_s in Eq. 7.14 has, in the past, usually been taken as equivalent to the ultimate shaft adhesion c_a equal to the average undrained shear strength c_u or s_u along the embedded pile length, reduced by an empirical coefficient α termed the *friction ratio:*

$$f_s = c_a = \alpha s_u \qquad \text{or} \qquad \alpha = c_a/s_u \tag{7.26}$$

The coefficient α depends on the nature and strength of the clay; pile type, dimensions, and installation method; and time effects. Values vary widely and decrease rapidly with increasing shear strength.

- *Driven Piles:* α values range from roughly unity for soft clay to 0.4 or lower for stiff clay as shown in Fig. 7.36, from which the curve of Woodward is recommended [Schmertmann (1978)[27]] for relatively short piles onshore. The API curve [API (1982)[22]] is used for long piles offshore. The rationale of the approach for driven piles is that the contact between the pile and the clay becomes more imperfect with increasing soil stiffness throughout the pile length, and therefore a reduction factor is necessary. In addition, in stiff clays, negative pore pressures induced by driving dissipate with time, causing the clay to weaken. Tomlinson (1970)[38] found that α may be markedly influenced also by the soil strata overlying the clay, as well as by s_u, as shown in Fig. 7.37.
- *Bored Piles:* α values range from 0 to 0.5 depending upon the method of installation, pier geometry, and site soil conditions [Reese et al. (1976)[39]]. Design parameters for drilled piers in clay are summarized in Table 7.14.
- *Grouted Piles:* Kraft and Lyons (1974)[15] suggest using α = 0.5 to 1.0 for NC clays, α = 0.3 to 0.5 for OC clays, and f_{max} = 1.0 to 2.0 tsf.

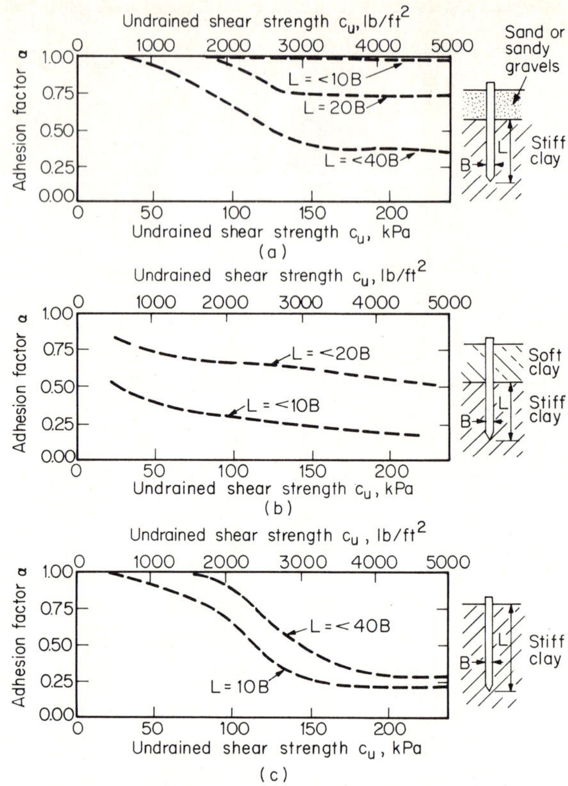

NOTE: (1) Curves not applicable to H- or cruciform-section or to bored or driven and cast-in-place piles.
(2) Safety factor should not be less than 2.5 except for design based on adequate loading test data.

FIG. 7.37 Adhesion (friction) factor α versus undrained strength for various strata conditions: (*a*) piles driven through overlying sands or sandy gravel: (*b*) piles driven through overlying soft clay; and (*c*) piles without different overlying strata. [After Tomlinson (1970),[38] from CGS (1978).[6]]

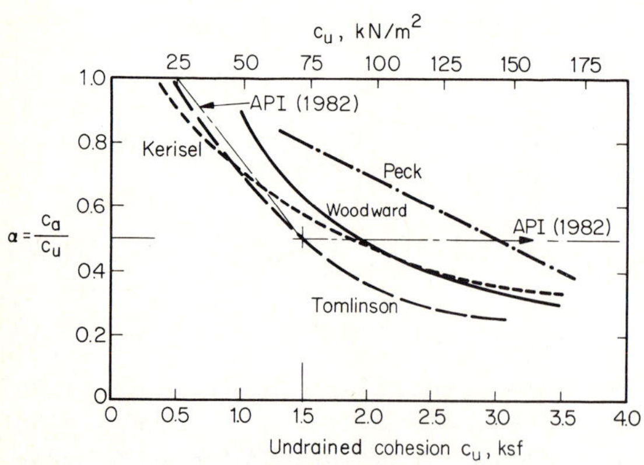

FIG. 7.36 Friction ratio α versus undrained cohesion c_u as determined by various investigators for driven piles. [After McClelland (1972).[4]]

TABLE 7.14
DESIGN PARAMETERS FOR DRILLED PIERS IN CLAY*

Parameter	Straight-sided shafts			Belled bottoms		
	A	B	C	D	B	C
α_{avg}	0.5	0.3 (see **E** below)	0	0.3	0.15 (see **F** below)	0
Limiting side shear, tsf	0.9	0.4 (see **E** below)	0	0.4	0.25 (see **F** below)	0
N_c	9	9	9	9	9	9

*After Reese, Touma, and O'Neill (1976).[39]

A. Pier installed dry or with lightweight slurry and displacing slurry directly with fluid concrete.

B. Installed with slurry along some portion of the hole such that entrapment of slurry between sides and natural soil is possible.

C. Base resting on soil significantly stiffer than the soil around the stem.

D. Installed dry.

E. Increase values to Category A for segment installed dry.

F. Increase values to Category D for segment installed dry.

▪ *Example:* For a driven pile of 14-in-square prestressed concrete:

Site Conditions:

0 to 20 ft, NC clay:

$$\gamma_t = 0.1 \text{ ksf}$$
$$s_u = 0.6 \text{ ksf}$$
$$\alpha = 1.0$$
$$f_s = 0.6 \text{ ksf}$$

20 to 60 ft, OC clay:

$$\gamma_t = 0.125 \text{ ksf}$$
$$s_u = 2.0 \text{ ksf}$$
$$\alpha = 0.5$$
$$f_s = 1.0 \text{ ksf}$$

Assume: Pile length $L = 40$ ft and $A_s = (14/12 \times 4) = 4.67 \text{ ft}^2$

Capacity:

$$Q_s = \Sigma(\Delta L)(A_s)(f_s) \qquad (7.14)$$
$$Q_s = (20)(4.67)(0.6) + (20)(4.67)(1.0)$$
$$= 150 \text{ kips}$$
$$Q_p = cN_cA_p = (2)(9)(14/12^2) = 24.5 \text{ kips}$$
$$Q_s + Q_p = 174 \text{ kips}$$

COEFFICIENT λ METHOD An alternate approach employes a coefficient λ to estimate the shaft capacity of heavily loaded pipe piles penetrating clays supporting offshore structures where penetration lengths are greater than about 50 ft. The rationale for this approach is that friction along the pile must depend on lateral earth pressure against the pile, which depends in turn on the Rankine passive pressure, $2s_u$ *(see Art. 9.2.3)*, for undrained conditions during driving. From evaluating load test data, Vijayvergiya and Focht (1972)[40] expressed the shaft capacity as

$$Q_s = \lambda(\bar{\sigma}_m + 2s_{um})A_s \qquad (7.27)$$

where: λ = a dimensionless coefficient
σ_m = mean effective vertical stress between ground surface and pile tip (Fig. 7.38*a*)
s_{um} = average undrained shear strength along pile
A_s = pile surface area

Lambda was found to be a function of pile penetration as shown in Fig. 7.38*b*, which is a modification of the original curve of Vijayvergiya and Focht based on additional load test data analyzed by Kraft et al. (1981)[41] employing regression analysis. They suggest that a minimum value of λ =

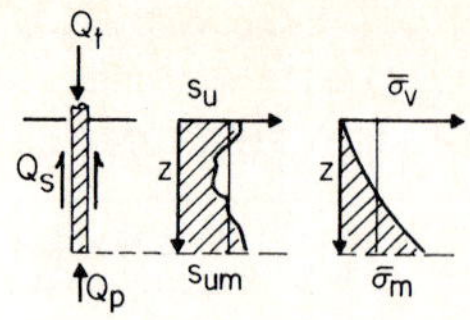

$f_s = \lambda(\bar{\sigma}_v + 2s_u)A_s$

$Q_s = \lambda(\bar{\sigma}_m + 2s_{um})A_s$

(a)

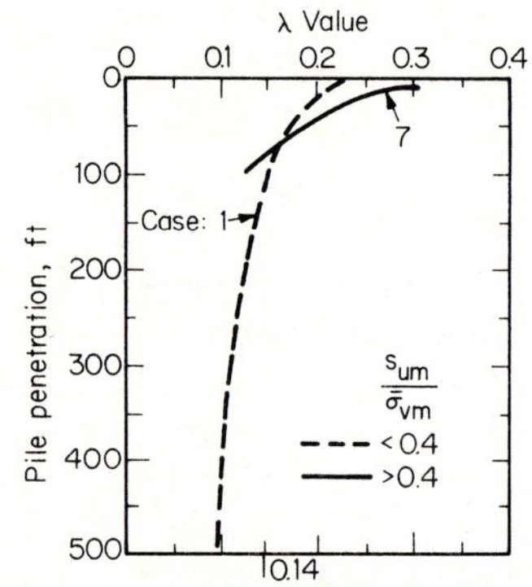

For $s_u/\bar{\sigma}_{vm} < 0.4$: $\lambda = 0.296 - 0.032 \ln(L)$

For $s_u/\bar{\sigma}_{vm} > 0.4$: $\lambda = 0.488 - 0.078 \ln(L)$

(b)

FIG. 7.38 Coefficient lambda: (*a*) definition of s_{um} and $\bar{\sigma}_{vm}$ and (*b*) λ value versus pile penetration. Suggested value for λ_{min} = 0.14. [From Kraft et al. (1981).[41]]

0.14 be considered in analysis, and state that the lambda concept provides the most consistent and reliable evaluations of long piles in NC or OC clays in the present state of the art.

EXAMPLE Determine the shaft capacity of a 24-in-dia. pipe pile driven to 200 ft of penetration below the seafloor in clay, for the soil parameters given in Table 7.15. (The α method yields Q_s = 1184 kips.)

$$\begin{aligned} Q_s = \Sigma\Delta Q_s &= 0.14(0.52 + 1.2)(6.28)(20) \\ &+ 0.14(1.87 + 2.0)(6.28)(30) \\ &+ 0.14(3.85 + 2.4)(6.28)(40) \\ &+ 0.14(6.75 + 3.2)(6.28)(60) \\ &+ 0.14(9.99 + 6.0)(6.28)(50) \\ &= 1580.6 \text{ kips} \end{aligned}$$

In an API analysis for layered soils, a common condition encountered offshore, Vijayvergiya and Focht (1972)[40] suggest calculating ΔQ_s for sand layers using clay parameters from adjacent layers, then subtracting the values for the sand layers to find $Q_{s(clay)}$. Values for the sand layers are then calculated in the normal manner to find $Q_{s(sand)}$, which is added to $Q_{s(clay)}$ to find the total shaft resistance. In conventional analysis of layered or mixed soils Eq. 7.2 is used.

Drained Strength (Effective Stress) Analysis

LONG-TERM EFFECTS In a study on many load tests for steel piles in clay, McClelland (1972)[4] found that capacities predicted based on the undrained strength and alpha correlations were in the order of 0.4 times the observed capacities to depths of about 20 m and ranged from about 0.75 to 1.4 times the observed capacities between 20 and 40 m. He concluded that estimating f_s through correlations of α and s_u may be unsatisfactory, and that there is "a growing conviction that pile support in clay is frictional in character—that load transfer is dependent upon the effective lateral pressure acting against the side of a pile after it is driven." (The λ approach provides for this condition and has been referred to as a "quasieffec-

TABLE 7.15
DATA INPUT: ANALYSIS OF FRICTION PILE DRIVEN IN CLAY, λ METHOD

Layer	Depth below seafloor, ft	ΔZ, ft	γ_b, kcf	σ_v, ksf	σ_v (mean), ksf	s_{um}, ksf
1	0–20	20	0.052	0 + (0.052 × 20) = 1.04	(0 + 1.04)/2 = 0.52	0.6
2	20–50	30	0.055	1.04 + (0.055 × 30) = 2.69	(1.04 + 2.69)/2 = 1.865	1.0
3	50–90	40	0.058	2.69 + (0.058 × 40) = 5.01	(2.69 + 5.01)/2 = 3.85	1.2
4	90–150	60	0.058	5.01 + (0.058 × 60) = 8.49	(5.01 + 8.49)/2 = 6.75	1.6
5	150–200	50	0.060	8.49 + (0.060 × 50) = 11.49	(8.49 + 11.49)/2 = 9.99	3.0

tive stress" method.) Meyerhof (1976)[20] also suggested that consideration be given to effective stress analysis based on lateral earth pressures K_s around driven and bored piles together with drained strength parameters.

Immediately after pile driving, shaft adhesion is closely given by the undrained shear strength of remolded clay, but because consolidation occurs during construction, adhesion for support after construction usually can be taken as the effective drained strength parameters $\bar{c}$ and $\bar{\phi}$ of remolded clay failing close to the shaft. These parameters are obtained from CD triaxial tests on a remolded specimen, reconsolidated to the vertical overburden pressure.

Effective unit side friction f_s in homogeneous clay is then taken as

$$f_s = c + K_s\bar{p}_o \tan \delta \leqq s_u \tag{7.28}$$

In Eq. 7.28 it is assumed that excess pore pressure sometime after pile installation and loading is negligible compared with the effective overburden pressure. Because of the relatively small disturbance area, the excess pore pressures usually dissipate after about one or more months. Meyerhof concluded that "the ultimate skin friction of a pile in saturated clay can be approximately estimated from the drained shear strength of remolded soil for which the cohesion may usually be taken as zero." Equation 7.28 is then written

$$f_s = \bar{p}_o K_s \tan \delta \tag{7.16}$$

Expressing $K_s \tan \delta$ as β, the *skin friction factor,* results in

$$f_s = \bar{p}_o\beta \tag{7.29}$$

NC CLAYS For bored and driven piles in saturated soft clay K_s may be expected to be close to K_o (at-rest earth pressure) as in the case for loose sand, and $K_o = 1 - \sin \phi$. Burland (1973)[42] expressed f_s as

$$f_s = \bar{p}_o\beta = \bar{p}_o(1 - \sin \phi) \tan \phi \tag{7.30}$$

where $\bar{\phi}$ = effective friction angle for the clay, and

$$\beta = (1 - \sin \bar{\phi}) \tan \bar{\phi} \tag{7.31}$$

β provides a lower limit and varies from 0.24 ($\bar{\phi}$ = 20°) to 0.29 ($\bar{\phi}$ = 30°). For preliminary estimates Meyerhof (1976)[20] suggests the following limiting values for β:

- *Driven piles* into soft to medium clays, $\beta = 0.3$, providing the depth of penetration is not greater than about 10 m and

$$f_s = 0.3\bar{p}_o \tag{7.32}$$

- Piles exceeding penetration depth of about 60 m, $\beta = 0.15$

(*Note:* These values can be used for cylindrical, H, or tapered piles, as shaft adhesion is not significantly affected by pile taper. For H piles the perimeter area is taken as the area of the rectangle enclosing the flanges and the web.)

K_s also may be estimated from semiempirical relationships based on either the average s_u and $\bar{p}_o$ along the shaft, or more simply from s_u. Meyerhof states that using a correlation between the PI and the values of s_u/p_o and ϕ of NC clay indicates that K_o varies roughly between $1.5s_u/p_o$ for highly plastic clay and over $2s_u/p_o$ for low-plasticity clay. The unit skin friction may then be expressed as

$$f_s = 1.5s_u \tan \phi \tag{7.33}$$

in which ϕ would be expected to decrease from the peak value for short piles to the residual value for long piles. For *bored piles,* f_s may be taken approximately as

$$f_s = s_u \tan \phi \tag{7.34}$$

OC CLAYS In stiff, saturated clay, the ultimate skin friction is also estimated from Eq. 7.16 on the basis of the drained strength of remolded soil, provided that K_s is known from previous pile load tests or the value K_o can be estimated from field or laboratory tests. Roughly,

$$K_o = (1 - \sin \phi) \sqrt{\text{OCR}} \tag{7.35}$$

where OCR is the overconsolidation ratio (*see Art. 3.2.3*) and ϕ is taken as $\bar{\phi}_r$.

For piles in stiff clays, Burland (1973)[42] suggests taking $K_s = K_o$ and $\bar{\phi}_r$ to provide an *upper limit* for f_s in bored piles and a *lower limit* for driven piles. In the stiff, fissured, OC London clays with K_o ranging from about 3 at shallow depths to unity at large depths, analysis of pile load tests show that for *driven piles* $K_{s(avg)}$ varies from roughly K_o to more than $2K_o$, corresponding to a range of β from roughly unity to over 2. For *bored piles* K_s varies from about 0.7 to $1.2K_o$ corresponding to a range of β roughly from 0.7 to 1.4 [Meyerhof (1976)[20]].

Allowable Capacity

Where load tests confirm computations using either total or effective stress analysis, $Q_a = Q_s/2.5$; where no load tests are performed, $Q_a = Q_s/3$ [CGS (1978)[6]].

In Situ Strength Tests

CPT methods all require estimating s_u from cone data as described in Art. 3.4.5 of Hunt (1984).[12a] There are three general methods, given by Schmertmann (1978)[27]:

1. Use of the Woodward curve, Fig. 7.36, to find α vs. s_u.
2. Use of the value for λ in Eq. 7.38 ($s_u = s_{um}$).
3. Direct use of f_L (local sleeve friction, see Sands) but reduce $f_L = f_s$ in accordance with the Woodward curve before entering Fig. 7.34 to find C_c.

Note: Use all three methods and select results most reasonable for local experience, and apply FS = 2.25 (electric tip) and FS = 3.0 (mechanical tip).

The *pressuremeter test* is described under Sands.

Tension Loads

Tension pile capacity, where the piles penetrate into clays and cohesive silts, can be approximated from the expression given for f_s, plus pile weight.

Belled piers are analyzed usually by two methods and the lower value of pullout resistance selected for design. A simple method is to use resistance from shaft friction along the diameter of a cylinder equal to the bell diameter, plus the weight of the pier and soil mass (the friction cylinder method, Fig. 7.85*a*). The second method involves bearing-capacity theory as described in Art. 7.7, *Soil and Rock Anchors*.

Rock Sockets

Significant Factors

When piles are socketed into rock, pile capacity often will be derived from both end-bearing and shaft resistance (*see Art. 7.3.3*). The latter, resulting from pile-rock adhesion, also referred to as the *bond strength*, can be a function of the concrete strength or of the rock strength.

Capacity proportions shared by end-bearing and shaft resistance are related to rock quality and to the ratio of socket length to diameter. In poor-quality rock shaft adhesion may be relatively low, especially where the tip bears on good quality rock. It will also be negligible with large diameters and shallow penetration, but it can be substantial with relatively smaller diameters and deep penetration, and indeed often provides total support. Some codes, such as those of Chicago and the Institution of Civil Engineers of the United Kingdom, permit an increase in bearing value for each 1 ft (30 m) of embedment with a limit equal to twice the code-specified value.

Pile or pier design in rock masses is not yet clearly defined in geotechnical engineering and there is little information in the literature regarding shaft friction which becomes a very significant factor when high-capacity units penetrate into poor quality rock and fail to reach that of good to excellent quality. Shaft resistance can be reliably determined only by load tests, which are particularly important where erroneous assumptions will lead to unconservative design, or to socket penetrations substantially in excess of those actually required.

Concrete Strength Factor

In rock of good to excellent quality (*see Art. 6.3.2*) bond strength may be controlled by concrete strength. Normally, an allowable adhesion equal to 0.03 to $0.05f'_c$ (ultimate concrete strength) is used, depending on the rock quality [D'Appolonia et al. (1975)[30]]. The New York City Building Code, for example, gives an allowable "presumptive" adhesion between sound, hard rock and concrete of 200 psi (0.05 × 4000 psi concrete) (14.4 tsf) which is to be verified by tests.

Assessment of Rock-Pile Adhesion f_s

Good to excellent quality rock approaches an elastic material and both load tests and theory demonstrate that f_s is not distributed uniformly along the socket wall, but rather peaks at the top of the socket, then diminishes rapidly with penetration depth. Mattes and Poulos (1969)[43] present a theoretical solution of stress distribution and load transfer as shown in Fig. 7.39 as a function of normalized shear stress and normalized depth within a rock socket.

The mobilized shear stresses τ are normalized by dividing them by the theoretical average shear stress τ_{av} for the socket, or

$$\tau_{av} = \frac{P_s}{\pi D_s L_s} \qquad (7.36)$$

where P_s = applied load at top of socket
D_s = socket diameter
L_s = socket length

τ, computed from load distribution curves from load tests, is given as

$$\tau = \frac{\Delta p}{D_s \Delta L} \qquad (7.37)$$

where Δp = load change between top and bottom of each element (measured by instrumentation bounding the element)
ΔL = length of each element

Normalized shear-stress distribution is a function of the stiffness ratio K, defined as the ratio of the pile modulus to the rock modulus. Koutsoftas (1981)[44] suggests that apparently a precise knowledge of K for using elastic solutions in rock-socket analysis is not required, and a range for K between 10 and 50 should provide good results.

SOUND MICA SCHIST In a *socket load test* reported on by Koutsoftas (1981),[44] 24-in-dia. casing was driven to the top of sound rock at z = 70 ft and a rock socket drilled to L_s = 30 ft. The pile consisted of a core beam within 6000 psi concrete, instrumented with Carlson-type strain meters at seven locations in the rock socket. The rock consisted of sound mica schist with foliations 45 to 60° from the horizontal and tight joints moderately spaced (2 to 3 ft). Property ranges from core tests included low to medium strength (U_c = 300 to 700 tsf) and E_r = 21 to 42 × 10^4 tsf (from the initial tangent modulus). Under the maximum applied butt load of 920 t, the pile settlement was 0.4 in of which 0.05 in was socket compression. Socket load was about 750 t, with the remaining 18% carried by skin resistance along the casing penetrating the upper 70 ft of alluvium and decomposed schist (Fig. 7.40). Load distribution along the socket (Fig. 7.41) shows that 80% of the load was carried in the first 10 ft (5B) of the socket with practically no load transfer to the last 7 to 10 ft, and that the load-penetration depth increased as the load increased. The test results are plotted also in Fig. 7.39. The range in τ_{av} (f_s) obtained from the data analysis is given in Table 7.16. The measured maximum bond stress was 219 psi or $0.05U_c$ (U_c = 4380 psi), in the lower range for the mica schist (f'_c = 0.05 gives 300 psi maximum for 6000 psi concrete).

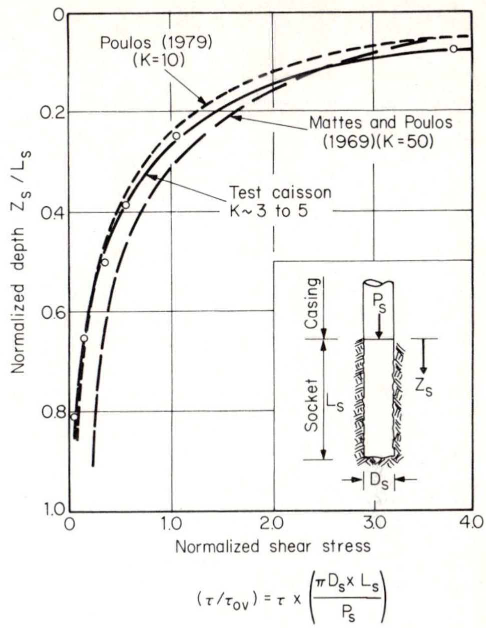

FIG. 7.39 Normalized shear-stress distribution along a rock socket. [From Koutsoftas (1981).[44]]

GRANITE GNEISS Socketed pile load test in granite gneiss is reported by Costa Nunes and Fernandes

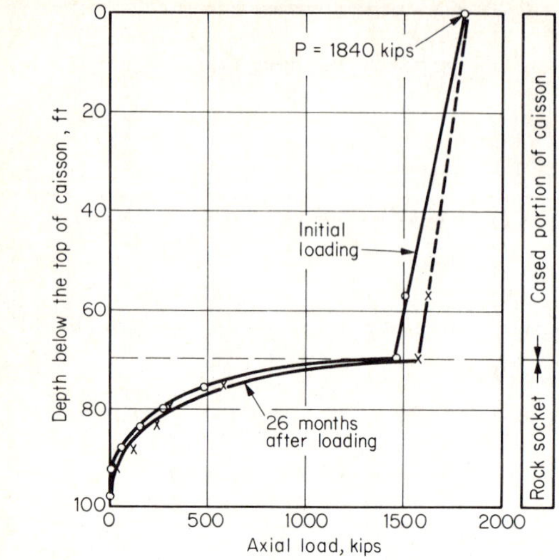

FIG. 7.40 Load distribution for test caisson at maximum applied load. [From Koutsoftas (1981).[44]]

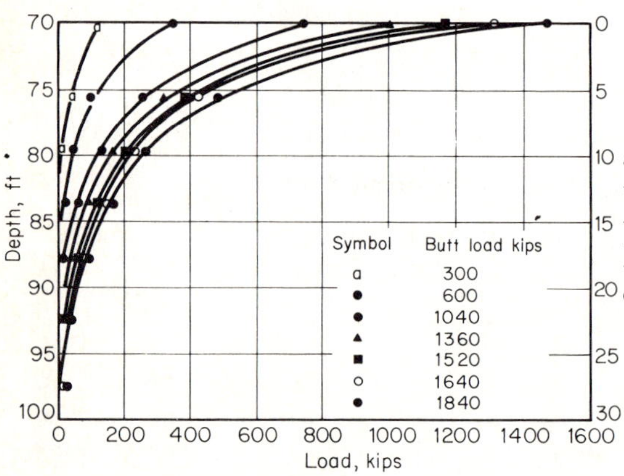

FIG. 7.41 Load distribution within test socket at various stages of loading. [From Koutsoftas (1981).[44]]

TABLE 7.16
MEASURED VALUES FOR τ_{av} FROM LOAD TEST IN MICA SCHIST*

Socket depth, ft	τ_{av}, tsf	Vertical movement, in
0–5.5	15	0.028†
0–5.5	9	0.015
5.5–9.5	4	0.012†
9.5–13.5	2	0.005†

*From Koutsoftas (1981).[44]

†Under maximum butt load of 750 t.

(1981).[45] A 110-cm-dia. asphalt-coated casing was driven to the top of moderately to slightly weathered granite gneiss at a depth of 26.8 m after excavating with a bentonite slurry through marine, colluvial, and residual soils and weathered rock. The rock socket was drilled to $L = 3.5$ m and the pile tip placed at $L_s = 3.3$ m over a flexible metal pad to eliminate end bearing. Instrumentation included strain gages and telltales. Rock RQD averaged 50% and recovery 90%. Maximum butt load was 1100 t and socket load 774 t, with the remaining 30% carried by skin resistance along the 26.8 m of overlying casing. Total settlement was about 1 cm and socket compression about 2 mm. Shaft friction distributions for various test loads are summarized in Table 7.17, where it is seen that load transfer drops off rapidly with depth, dissipating to very low values in about 3.2 m or 3*B*.

CLAY SHALES Full-scale load tests on drilled piers installed by the dry method in cretaceous clay shales in Oklahoma indicated that average shaft resistance in the socket was approximately 70% of the in situ shear strength measured by pressuremeter tests ($S_{max} = 5.4$ tsf) and was fully mobilized at much smaller movements (< 6.4 mm) than required to fully mobilize end-bearing resistance [Goeke and Hustad (1980)[46]]. It appeared that most of the pier support was derived from shaft friction within about 10 ft of penetration into the shale. Pier diameter was 76 cm, and in one test failure occurred at an applied load of 560 tons.

Summary

Some values for f_s from the literature are given in Table 7.18. Important factors to consider in the selection for design are tolerable structural settlement, concrete strength, socket depth and load transfer mechanism, rock type, and rock-mass quality. It appears from the three load tests described above, and others, that in good to excellent quality rock, socket depths greater than about 3 to 5*B* may not provide a significant increase in shaft resistance. Load-transfer analysis is best applicable to rock masses approaching elastic materials. New design procedures using elastic solutions (*see Art. 7.4.1*) are proposed by Rosenberg and Journeaux (1976)[47] and Pells and Turner (1979).[48]

TABLE 7.17
MEASURED VALUES FOR τ_{aw} FROM LOAD TEST IN GRANITE GNEISS*

Socket depth, m	Socket load,† t				
	P = 774	P = 670	P = 525	P = 482	P = 477
0–1.1	12.7	11.0	8.6	7.9	7.8
1.1–2.0	6.5	5.6	4.4	4.1	4.0
2.0–3.24	1.3	1.2	0.84	0.82	0.81

*From Costa Nunes and Fernandes (1981).[45]

†τ at top of socket = 17 kg/cm² at P = 774 t.

TABLE 7.18
BOND STRENGTH VALUES GIVEN FOR VARIOUS ROCK TYPES*,†

Rock type	Bond strength f_s, tsf	Condition	$\%U_c$	Reference
Granite gneiss	6.0_{avg}, 17_{max}	Unfailed‡	—	Costa Nunes and Fernandes (1981)[45]
Mica schist	15(max)	Unfailed ‡	0.05	Koutsoftas (1981)[44]
Sandstone	26	Unfailed	0.2	Thorne (1977)[49]
Sandstone	30	Failed	0.1	Thorne (1977)[49]
Shales	10–25	Failed	≈0.07	Thorne (1977)[49]
Shale, fragmented	2.5–7	Failed	—	Thorburn (1966),[50] others
Shale, expansive	0.75 (wetted)	Failed	—	Jasper and Shtneko (1969)[51]
Highly fractured	0.8–1.5	Allowable	—	Poulos and Davis (1980)[25]
Weathered	1.2	Ultimate	—	German Standard

*Suggested range of f_s = 0.05 to $0.10U_c$, depending upon magnitude of tolerable settlement.

†In sound rock, concrete strength might govern: $f'_{c(all)} = 0.05U_c$.

‡Socket compression was 2 mm or less.

ESTIMATING f_s FOR HIGH-QUALITY ROCK Values ranging from $f_{s(all)}$ = 0.05 to $0.10U_c$, as suggested by Horvath and Kenny (1979),[52] appear reasonable for high-quality rock, with the selection based on acceptable settlements (*see Art. 7.4.1*). For very small deflections, for example, Koutsoftas (1981)[44] found $f_s = 0.05U_c$ for the lower-strength cores tested. The German Standard (Table 7.7) suggests using $f_s = 0.01q_a$ in all conditions with $f_{s(max)}$ = 1.2 tsf in weathered rock. Poulos and Davis (1980)[25] suggest using $f_{s(all)} = 0.05U_c$ or $0.05f'_c$, whichever value is lower.

ESTIMATING f_s FOR POOR QUALITY ROCK The data are meager for these conditions. As mentioned above, the German Standard gives $f_s = 0.10q_a$ with a maximum value of 1.2 kg/cm² for weathered rock. In highly fractured rocks, Poulos and Davis (1980)[25] suggest, $f_{s(all)}$ = 0.8 to 1.5 tsf. For a fragmented shale, Thorburn (1966)[50] reported $f_{s(ult)}$ = 2.5 tsf, but in expansive shales caution is required for wetted conditions for which Jasper and Shtneko (1969)[51] reported $f_{s(ult)}$ = 0.75 tsf. Goodman (1980)[53] suggests for soft, clay-rich rocks such as weathered clay shale,

$$f_s = \alpha S_u \tag{7.38}$$

where S_u = undrained shear strength, or

$$f_s = U_c\left[\frac{\alpha}{2\tan(45 + \phi/2)}\right] \tag{7.39}$$

where α = a factor ranging from 0.3 to 0.9, but if the socket surface is artificially roughened may be considerably higher [Kenny (1977)[53]].

7.3.5 NEGATIVE SHAFT FRICTION (DOWNDRAG)

Description

Piles that penetrate through soft clays or organic soils can be subjected to loads that may even exceed structural loads. Such forces result from downdrag or negative shaft friction (Fig. 7.42).

Floating Piles

In the less serious case, a friction pile driven to bear in soft to medium clay can greatly increase the soil compressibility and cause high pore pressures because of disturbance. As the soft soil around the pile consolidates, it transfers its weight to the pile through shaft friction, causing additional downward force. The *short-term* negative shaft friction can be estimated roughly from the undrained strength of clay or silt as given in $c_a = \alpha c_u$ (Eq. 7.26). The long-term negative friction can be estimated from $f_s = \overline{p}_o K_s \tan \delta$ (Eq. 7.16), or $f_s = \overline{p}_o \beta$ where β, the skin friction factor, equals $K_s \tan \delta$. For preliminary estimates, Meyerhof (1976)[20] gives values for β for uncoated piles ranging from 0.3 for penetration to a depth of about 15 m to about 0.2 for piles longer than about 60 m.

Downdrag is usually of little importance for friction (floating) piles, but may become extremely significant for end-bearing piles, especially when recent fill is placed over compressible soils. The case of fill placed over floating piles is treated as a *settlement problem (see Art. 7.4).*

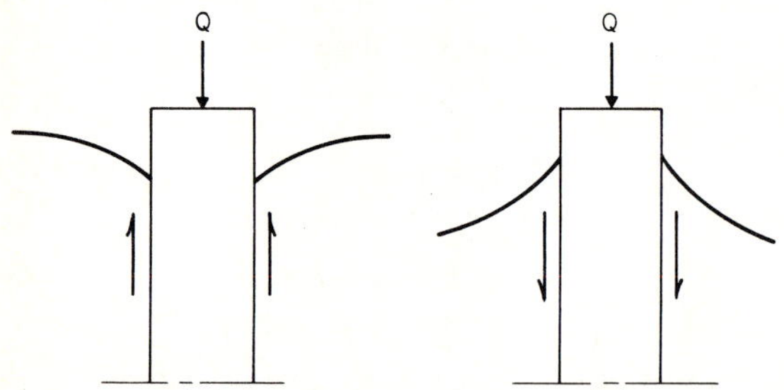

FIG. 7.42 Positive and negative skin friction: (*a*) support from shaft friction on pile moves down relative to soil and (*b*) negative shaft friction as soil moves down relative to pile.

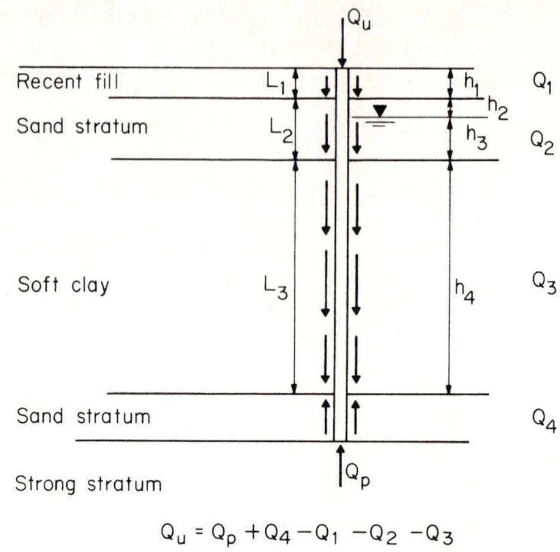

FIG. 7.43 Forces on a pile from downdrag, or negative shaft friction.

End-Bearing Piles

A far more serious situation exists when end-bearing piles are driven through fill, or fill is placed after pile installation, and the fill is underlain by soft compressible soil. As the weak soil consolidates, it moves downward relative to the pile and the fill load is transferred to the pile as shown in Fig. 7.43. Lowering of the groundwater table will impose additional loads. If the forces are too high, failure at the tip may result.

Poulos and Davis (1975)[55] report on several case studies involving downdrag loads caused by fill over soft soils. In one case, 8 m of fill in place for 2 to 4 years prior to pile installation, overlying up to 30 m of soft, silty clay, imposed a downdrag load on end-bearing steel pipe piles of 120 t. In another case, 10 to 15 m of fill overlay 30 to 40 m of soft clay, and the downdrag loads on the steel-pipe piles were measured to range from 200 to about 270 t.

Analysis

Significant Factors

The computation of downdrag forces under fill loads is somewhat complicated, but its necessity is supported by the fact that these conditions cause a large number of pile-foundation failures. Complications arise because the effects of consolidation and the increase in shear strength with time must be considered, as well as the effects of how the pile achieves its capacity. Conditions can range from full slip between the pile and the soil for friction piles in soft clay to soil slip and elastic compression of the pile for end-bearing piles.

Total Downdrag Force

A common situation is illustrated in Fig. 7.43. Strata below the soft clay will provide support essentially from end bearing; if the tip is bearing in a thick sand stratum, however, support will be derived from end bearing and shaft friction. Poulos and Davis (1975)[55] present a series of parametric solutions for the magnitude and rate of development of maximum downdrag force and axial movement of an end-bearing pile in a consolidating soil layer subjected to surcharge loading. Their results compare favorably with a number of full-scale load tests.

In Fig. 7.43 the total downdrag force P_N on the pile due to the consolidating clay stratum from the fill and the sand overlying the clay can be expressed as

$$P_N = (P_{NFS}N_RN_T)_{clay} + (P_{NFS})_{fill} + (P_{NFS})_{sand} \quad (7.40)$$

where P_{NFS} = downdrag force (relationship is similar to that for shaft friction, Eq. 7.14)

N_R = a reduction factor for cases where full slip does not occur (surcharge, end-bearing piles, stiff clays)

N_T = a reduction factor for the effects of consolidation and the delayed installation of the pile

and

$$P_{NFS} = \pi BL\left[\bar{c}_a + K_s \tan\delta\left(\frac{\gamma L}{2} + q_s\right)\right] \quad (7.41)$$

where $\bar{c}_a$ = effective stress value of soil-pile adhesion for the cases of no overburden or surcharge layer; taken as zero in NC clays

$\bar{c}_{ae}$ = an equivalent soil-pile adhesion that develops from an overburden layer

$= \bar{c}_a + \bar{p}_oK_s \tan\delta$ or for NC clays

$= \bar{p}_o\beta$, where $\beta = 0.3$ for moderate depths in NC clays, or $\beta = c_u/\bar{\sigma}_v$ [Johannessen and Bjerrum (1965)[56]]

$\bar{p}_o$ = initial effective stress from overlying soil stratum $= \gamma h_2 + \gamma h_3$

δ = effective stress value of soil-pile friction

K_s = lateral stress coefficient

γ = soil unit weight (use γ_b below GWL)

L = pile length in compressible stratum

q_s = applied surcharge stress, or overlying stratum weight γh

B = pile diameter

Therefore, in an NC clay:

$$P_{NSF(clay)} = \pi BL\{\bar{c}_{ae} + \beta[(\gamma_{b(clay)} \times L/2) + q_s]\} \quad (7.42)$$

where

$$\bar{c}_{ae} = \bar{p}_o\beta = (\gamma_t h_2 - \gamma_w h_3)\beta \quad (7.43)$$

$$q_s = \gamma_1 h_1 \quad (7.44)$$

Pile-slip effects are accounted for with factor N_R applied to $P_{NSF(clay)}$. Poulos and Davis developed the curves given in Fig. 7.44 for N_R for a pile stiff-

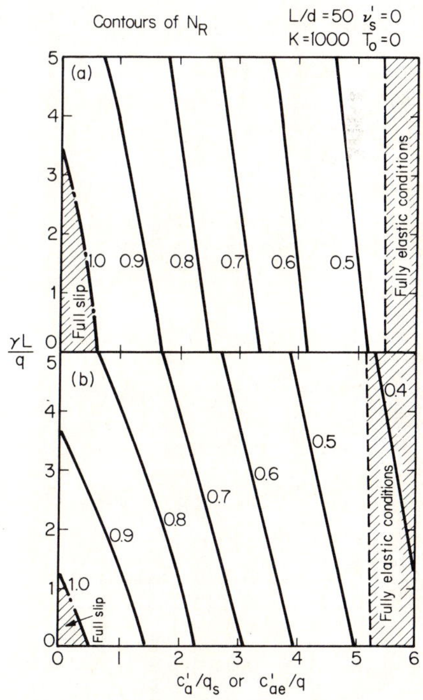

FIG. 7.44 Downdrag reduction factor N_R: (*a*) $K_s \tan\delta = 0.2$ and (*b*) $K_s \tan\delta = 0.4$. [From Poulos and Davis (1975).[55]]

ness $K = 1000$ and a ratio of $L/B = 50$. K, the *pile stiffness factor*, represents the soil-pile interaction and is expressed by

$$K = (E_p/E_s)R_A \tag{7.45}$$

where E_p and E_s are Young's modulus for pile and soil and R_A is the area ratio ($4A_p/\pi B^2 = 1$, if the pile is solid). For NC clays the ratio of $\bar{c}_{ae}/q_s$ is used to find N_R; in OC clays $\bar{c}_a$ is used. The curves show, for example, that if the soil-pile adhesion in the clay is 6 times the surcharge load q_s, the reduction factor would be about 60% and fully elastic conditions would exist.

Time effects are accounted for with factor N_T, for which the curves given in Figs. 7.45 (one-way drainage) and 7.46 (two-way drainage) represent N_T in terms of the expression for time rate of consolidation:

$$T_o = \frac{c_v t_o}{L^2} \tag{7.46}$$

which is a rearrangement of Eq. 4.47, where T_o is the theoretical time factor, t_o is the time delay before pile installation, c_v is the coefficient of consolidation, and L is the stratum thickness. The curves are not greatly affected by L/B, ν_s (Poisson's ratio for soil), or K, unless K is less than 100.

Fill and sand layer downdrag effects are added to the value $P_{NSF}N_RN_T$:

$$P_{fill} = \pi B(\bar{p}_{of}/2)(h_f)(K_s \tan\delta) \tag{7.47}$$

$$P_{sand} = \pi B(\bar{p}_{os}/2)(h_s)(K_s \tan\delta) \tag{7.48}$$

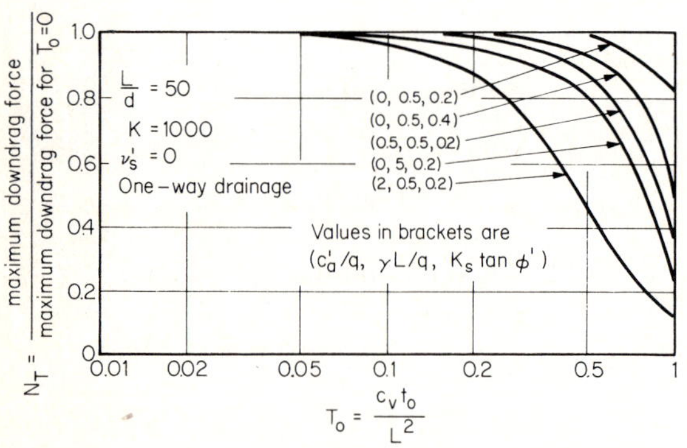

FIG. 7.45 Downdrag reduction factor N_T for one-way drainage. [From Poulos and Davis (1975).[55]]

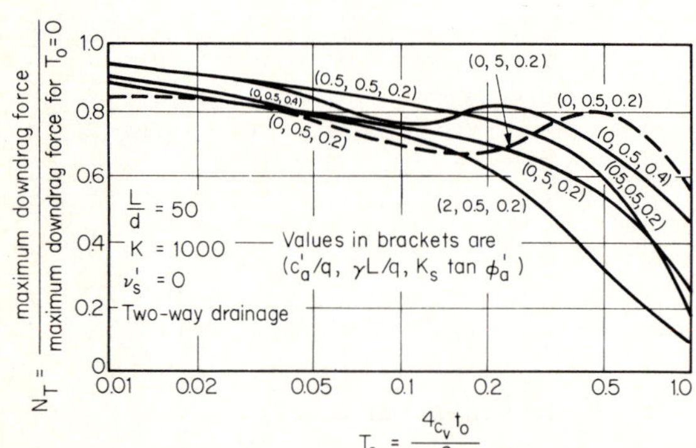

FIG. 7.46 Downdrag reduction factor N_T for two-way drainage. [From Poulos and Davis (1975).[55]]

Example

Data: In Fig. 7.43, fill thickness is 10 ft, sand stratum is 10 ft, soft clay stratum is 50 ft. GWL is 3 ft below the sand surface. Fill has been in place for a year and 16-in-dia. pipe piles are to be installed to bear in the strong stratum to support axial structural loads of 50 tons.

Fill and sand: $\gamma_t = 125$ pcf $= 0.125$ kcf, $\phi = 35°$

Soft clay: $\gamma_t = 119$ pcf $= 0.119$ kcf, $\beta = 0.3$, $C_v = 205$ ft^2/year

Analysis to find maximum load applied to pile, including downdrag:

1. From Eq. 7.43:

$$\bar{c}_{ae} = \bar{p}_o\beta = [(10 \times 0.125) - (7 \times 0.062)](0.3) = 0.25 \text{ ksf}$$

2. Applied stress from fill:

$$q_s = 10 \times 0.125 = 1.25 \text{ ksf}$$

$$\bar{c}_{ae}/q_s = 0.25/1.25 = 0.20$$

$$\gamma L/q_s = (0.125 - 0.057)(50)/1.25 = 2.28$$

3. From Eq. 7.42 downdrag from clay layer:

$$P_{NSF} = (\pi)(1.33)(50)\left[0.20 + 0.3\left(\frac{0.057 \times 50}{2} + 1.25\right)\right] = 218 \text{ kips}$$

4. Evaluate effect of 1-year delay after fill placement:

From Eq. 7.46, $T_o = (205)(1)/(50^2) = (0.082)(4) = 0.328$

From Figs. 7.44*a* and *b*, interpolating for $K_s \tan \delta = 0.3$ and values calculated in No. 2 above, find $NR = 0.9$

From Fig. 7.46, assuming two-way drainage and interpolating, for $T_o = 0.328$ find $N_T = 0.55$

From Eq. 7.40, $P_N = (218)(0.55)(0.90) = 108$ kips

Total axial load on pile = 108 + 100 = 208 kips

5. Add downdrag from sand layer: assume $\bar{c}_a = 0$, $\delta = 0.71\phi = 25\%$, $K_s = 0.5$, $K_s \tan \delta = 0.23$, and full slip occurs.

At GWL: $\bar{p}_o = (13)(0.125) = 1.625$ ksf

At top of clay: $\bar{p}_o = (7)(0.125 - 0.062) = 0.438 + 1.625 = 2.063$ ksf

From Eqs. 7.47 and 7.48,

$$P_{\text{NFS(fill)}} + P_{\text{NFS(sand)}} = (\pi)(1.33)\left[\left(\frac{0 + 1.625}{2}\right)(10) + \left(\frac{1.625 + 2.062}{2}\right)(10)\right](0.23) = 25.5 \text{ ksf}$$

6. Final total axial load = 208 + 25 = 233 kips. Reevaluation of the pile as a structural member and of its tip support is required.

Reducing Downdrag

Downdrag can be reduced or minimized by:

1. Surcharging the compressible stratum with an overload before the piles are installed (*see Art. 5.3*).
2. Applying electroosmosis.
3. Applying special coatings to the pile shaft, such as bitumen, which yield plastically. Unless adequately thick, however, driving through fill and sands may remove some of the coating, decreasing its effectiveness.
4. Enlarging the tip or lower pile portion to exceed the shaft diameter, which also reduces shaft friction.
5. Predrilling a hole larger than the pile shaft to the bearing stratum, filling it with bentonite, and driving a coated pile. This procedure may reduce downdrag by 90% or more and is the most effective method.

7.3.6 PILE GROUP EFFECTS

Pile Spacing

After a pile type has been selected and its capacity evaluated, the imposed structural loads are then distributed over a number of piles and various configurations result. Depending upon loading conditions, the piles can be evenly distributed beneath the entire structure, or can be distributed in clusters or groups beneath columns. A light building, such as a warehouse with large open areas, usually will have the columns and walls supported on groups of piles and floors supported on widely spaced piles, evenly distributed.

Because driven piles often are not installed in the exact position required and may not be perfectly vertical and because some bending moment is transmitted to the pile cap by the column, many building codes require that a minimum of three piles be used to support a major column and two piles be used to support a foundation wall at each support point. Bored piles with high capacities often are used as single units.

Group Behavior

Group Efficiency

The behavior of a pile acting as a single unit is not the same as that of a pile acting in a group. Although relatively rigorous procedures have been presented for analyzing single piles, the evaluation of group behavior is primarily empirical. There is much load test data for single piles, but very little is available for pile groups besides some laboratory model studies.

Group efficiency refers to the ratio of the ultimate capacity of a pile group to the ultimate capacity of a single pile times the number of piles in the group. It is related primarily to soil conditions, pile diameter and spacing, number of piles in the group (to a lesser extent), and whether the pile cap is freestanding or in contact with the ground surface. The following discussion pertains to freestanding groups.

Sands

Driving of adjacent piles in sands causes a densification of the material supporting the cluster resulting in an *increase* in efficiency, or a group capacity per pile that may be higher than the capacity of a single pile, depending upon the spacing. Model tests show [Kezdi (1975)[17]] that in a sand with a ϕ of 35°, the efficiency can range from 2.0 for piles with a center spacing of about two diameters to about 1.3 for a center spacing of four diameters. Some practitioners use these values in design. Efficiency drops to one at a spacing of about seven diameters and the piles begin to act as single units. It has been found to increase as ϕ decreases to as high as three for closely spaced piles in loose sands. In design, *minimum spacing* should equal 2.5B (average) [CGS (1978)[6]].

Driving piles in sands affects only side friction with point resistance practically unaffected, even at small pile spacings. Meyerhof (1976)[20] recommends that the ultimate group capacity in sand not underlain by a weak deposit should generally be taken as the sum of the single pile capacities.

Bored pile groups in sands, however, have ultimate load capacities less than the sum of single bored capacities due to overlap of the individual shear zones near the pile and the absence of soil compaction. The ultimate group capacity, where the sands are not underlain by a weak deposit, should be taken as about two-thirds the sum of the single pile capacities at the customary spacing of a few diameters [Meyerhof (1976)[20]]. This suggestion is based on theoretical considerations and data from some load tests.

Clays

Group action in clay soils causes a *reduction* in efficiency, if the piles are *freestanding*, because of the overlapping zone of shear deformation in the surrounding soil. Full-scale load tests on friction piles *driven* into soft and medium clays, and similar model tests, have shown that at customary pile spacings of three to four pile diameters the ultimate group capacity may be only about two-thirds of the sum of the capacities of single piles using the drained, remolded shear strength for side friction [Meyerhof (1976)[20]]. Point resistance for driven piles bearing in NC clays should also be reduced by two-thirds.

A *pile cap in contact* with ground surface tends to force block behavior to prevail in NC clays, even at large pile spacings. Group effects for piles in OC clays ($c_u > 1$ tsf) are usually not considered in determining group capacity.

Block Behavior

CLAY SOILS WITH PILE CAP IN CONTACT WITH GROUND SURFACE In the solution given in Terzaghi and Peck (1967),[57] it is assumed that the group of piles with the enclosed soil (Fig. 7.47) forms a rigid block which transfers the load to the soil below as a unit and fails as a unit. The ultimate bearing capacity of the block (termed *equivalent pier*) is computed as the sum of the shaft resistance available around the block perimeter plus the base capacity, or

$$P_{\text{group}} = 2L(a + b)f_s + (qN_q + cN_c)a \times b \qquad (7.49)$$

$$Q_a = P_{\text{group}}/3.$$

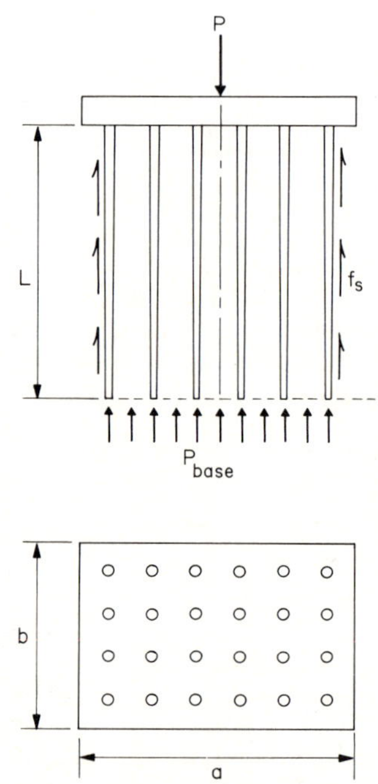

FIG. 7.47 The equivalent pier for pile-group capacity. [From Terzaghi and Peck (1967).[2] Used with permission of John Wiley & Sons, Inc.]

PILE GROUP UNDERLAIN BY A WEAK DEPOSIT This case is also analyzed by the equivalent pier method.

NC CLAYS Here, if there is no negative side friction, the ultimate group capacity can be estimated either from block failure of the equivalent pier or from the ultimate capacities of individual piles, using the smaller value for the ultimate capacity. This results in an abrupt transition between the two conditions. Poulos and Davis (1980)[25] provide an empirical relationship for intermediate conditions based on pile spacing. Since the pore pressures in the clay mass surrounding the group can be expected to dissipate only partially by the end of construction, Terzaghi and Peck suggest that the ultimate group capacity for block failure be based on the initial undrained strength for both side and base resistance.

DOWNDRAG The total negative side friction on a point bearing group passing through consolidating soft clay can be taken as the smaller value of either the sum of the negative side friction forces on the individual piles using the drained remolded shear strength of the clay or the weight of the soil within the pile group plus the undrained shear strength of the clay on the perimeter surface area of the equivalent pier.

SANDS OVERLAIN BY SOFT SOILS Estimates of the ultimate base resistance of the group with points resting above the critical depth in a sand stratum should be based on the effective overburden pressure reduced by negative side friction on the piles. Below the critical depth the base resistance can be estimated directly from the limiting value for Q_p (Eq. 7.5) without reduction for negative side friction [Meyerhof (1976)[20]].

PILE CAP CONTRIBUTION IN SANDS Poulos and Davis (1980)[25] suggest taking the bearing capacity of a strip footing of half width equal to the distance from the edge of the cap to the outside of the pile to evaluate the contribution of the pile cap to group load capacity.

Efficiency Formulas

Various efficiency formulas have been in use for some time to compute the reduction in capacity for freestanding groups, but none are rational and few consider soil properties. The basis of the Converse-Labarre method, the most commonly used, is that adjacent piles reduce the area available for the development of shearing resistance. The basis for the Converse-Labarre method for clays is given in Fig. 7.48.

Efficiency Factors Summarized

A summary of efficiency factors for various conditions is given in Table 7.19.

7.4 SETTLEMENT ANALYSIS

7.4.1 SINGLE PILES

General

There is a relatively large amount of literature describing various approaches to the problem of analyzing settlement of single piles. In practice, for the greater majority of real situations, settlement is provided for by selecting Q_a based on an adequate value for FS applied to Q_u. Analysis is more commonly required for large-diameter units such as drilled piers and for pile groups in or overlying compressible clays.

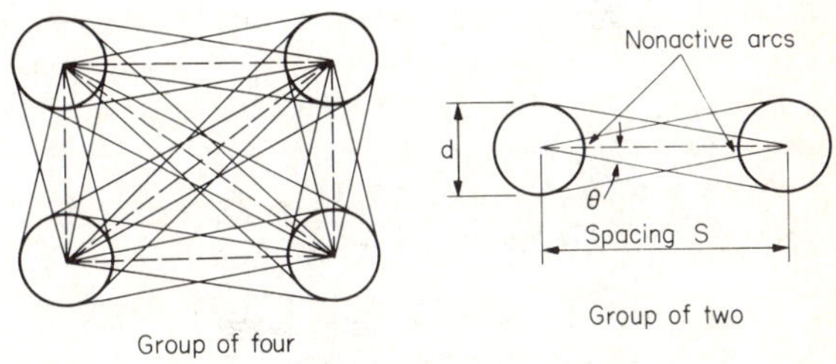

PREMISE: Adjacent piles reduce the area available for developing shear. The reduction occurs at the nonactive arcs as shown.

$$E = 1\,[(n-1)m + (m-1)n + \sqrt{2}(m-1)(n-1)\arctan(d/\pi\,Smn)$$

where E = efficiency
S = pile spacing, ft
d = pile diameter, ft
m = number of rows
n = number of piles per row

FIG. 7.48 Basis of Converse-LaBarre method to reduce pile efficiency due to group action in clays. [From Murphy (1973).[24]]

TABLE 7.19
SUMMARY OF PILE GROUP EFFICIENCY FACTORS

Material	Pile type	Efficiency	Comments	Reference
Sands	Driven	$E = 2$	S_f only, spacing 2 dia.	Kezdi (1975)[17]
		$E = 1.3$	S_f only. spacing 4 dia.	Kezdi (1975)[17]
		$E = 1.0$	S_f only, spacing 6 dia.	Kazdi (1975)[17]
		$E = 1.0$	Ultimate group capacity	Meyerhof (1976)[20]
	Bored	$E = 0.67$	Ultimate group capacity Up to few diameters spacing	Meyerhof (1976)[20]
		$E = 1.0$	Over few diameters	Meyerhof (1976)[20]
Underlying weak soils		Varies	Analyze by equivalent pier	Terzaghi and Peck (1967)[54]
Clay	Friction	$E = 0.67$		Meyerhof (1976)[20]
soft to medium			S_f and Q_p for 3 to 4 dia. Cap not bearing on surface	
stiff		$E = 0.67$	Q_p for 3 to 4 dia. Cap not bearing on surface	Meyerhof (1976)[20]
Clays		Varies	Cap bearing on surface, analyze block failure	Meyerhof (1976)[20]
Clays (negative S_f)		Use either $E = 0.67$ of sum of S_f on individual piles, or weight of soil within groups plus s_u along perimeter surface of equivalent pier.		

The *total settlement* of a single deep foundation is the sum of the immediate settlement due to point load, the immediate settlement due to shaft load, the elastic shortening of the unit, and long-term settlement from consolidation and creep. Current analytical methods, all relatively complex, can be classed into three broad categories [Poulos and Davis (1980)[25]]:

1. Theory of elasticity methods which employ Mindlin's equations (1936)[57] for subsurface loading within a semiinfinite mass (Table 4.7 and Fig. 4.21).
2. Load-transfer methods, which use measured relationships between pile resistance and pile movement at various points along the pile.
3. Finite element method [see Desai and Christian (1977)[59] for description].

At present, the simplest approach to settlement analysis for single units appears generally to be based on limit-design methods predicated exclusively on elastic behavior and the determination of immediate and total settlements. The *t-z* and *Q-z* load-transfer approach is finding increasing favor for evaluating piles for offshore structures.

Limit Analysis

Before analyzing settlement by limit analysis, the relationship between shaft resistance and base support must be considered. This is accomplished best by a study of instrumented load tests. As discussed in Art. 7.3.2, shaft and base resistance are not mobilized simultaneously.

In general, under small loads shaft movements and those of the immediately adjacent soil are equal. As applied load increases, shaft resistance becomes fully mobilized at some points and eventually slip occurs. With units acting in end bearing, full shaft mobilization usually occurs at much smaller movements than are required to fully mobilize base resistance. Finally, load increases are resisted almost fully by base resistance until bearing capacity is reached and failure occurs.

Sands

For piles in sands, Vesić (1970)[34] indicates that the effect of lateral friction on settlement can be neglected and that settlement is a function of only the force acting on the pile point as given in the expression

$$S_b = \frac{\beta P_b}{(1 + D_R)^2 P_b} \tag{7.50}$$

where S_b = base settlement
D_R = relative density
P_b = ultimate base resistance
β = installation method coefficient: driven piles = 0.04, jacked piles = 0.05, and bored piles = 0.18

S_b can also be expressed in terms of the soil modulus E_s as follows

$$S_b = \frac{P_b}{d_b E_s} \tag{7.51}$$

where d_b = base diameter
F_s = soil modulus after pile installation
= $25\,(1 + D_R^2)\overline{p}_o$ for driven piles and
= $5(1 + D_R^2)\overline{p}_o$ for bored piles
$\overline{p}_o$ = mean normal stress at pile tip = γz

Clays

Primarily because of time-delay effects, the calculation of settlement for single piles in clay is not well formulated, and if a pile is supported by both shaft resistance and end bearing, the distribution of the total load between these two elements of capacity must be known. Because of these factors Q_a for single piles in clay usually is based on analysis of bearing capacity (*see Art. 7.3.3*) with FS normally taken as 3.

Elastic settlement of a *friction pile* in clay may be computed from

$$S_f = \frac{P_s I_s}{L E_s} \tag{7.52}$$

where L = length
I_s = settlement influence factor, a function of L/B

Poulos and Davis (1968)[60] developed influence charts for rigid, cylindrical piles embedded in a uniform, saturated clay resting on a rigid base, as given in Fig. 7.49.

End Bearing and Shaft Friction Mobilized

A method found to yield results approximating observed pile behavior is described by Woodward, Gardner, and Greer (1972).[9] Based on limit analysis, it permits calculation of ultimate settlement for the limiting loads based on elastic concepts and is particularly useful for analysis of drilled piers.

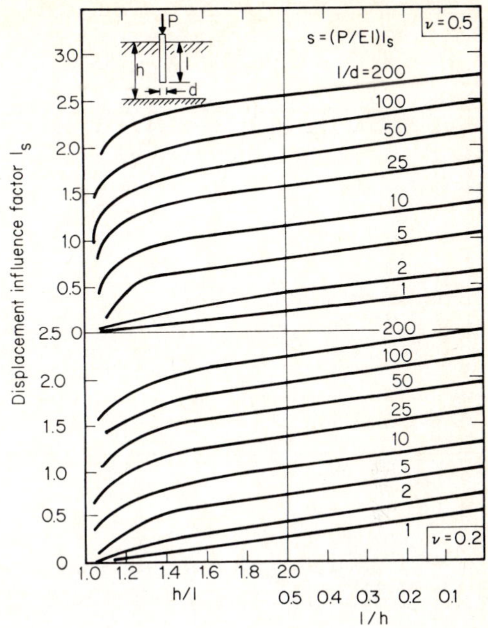

FIG. 7.49 Elastic displacement influence factor for single friction pile in clay. [From Poulos and Davis (1968).[60]]

The *total elastic settlement* can be expressed as

$$S = S_b + S_s + S_{es} \tag{7.53}$$

where

$$\text{Base settlement } S_b = \frac{p_b d_b I_{bb}}{E_b} \tag{7.54}$$

$$\text{Shaft settlement } S_s = \sum \frac{\Delta P_s I_{bs}}{E_s L} \tag{7.55}$$

$$\text{Shaft shortening } S_{es} = \frac{(P_b + \lambda P_s)L}{a_s E_p} \tag{7.56}$$

where p_b = ultimate unit base resistance
d_b = base diameter
E_b, E_s, E_p = Young's moduli for material at base, along shaft, and in pile
P_s, P_b = ultimate shaft and base resistance
a_s = shaft cross-sectional area
L = pile length
I_{bb}, I_{bs} = base deflection influence factors as

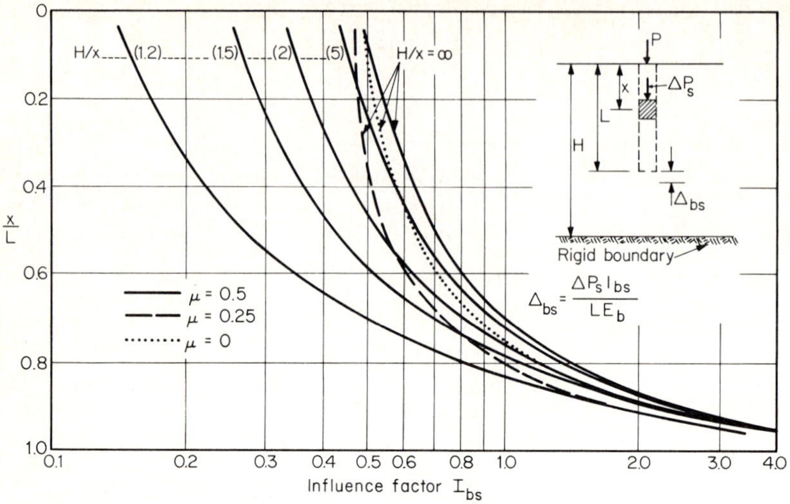

FIG. 7.50 Influence factor I_{bs} for base elastic displacement for shaft load increments. [From Woodward, Gardner, and Greer (1972).[9] Used with permission of McGraw-Hill Book Company.]

derived from the integration of the Mindlin solution for a point load in the interior of an elastic halfspace (Figs. 7.50 and 7.51)*

λ = ratio of average to total shaft load. It is a function of the shape of the load-distribution curve and is influenced by long-term creep and consolidation. Best determined by load tests, it ranges generally from between one-third to two-thirds, with 0.6 reported for sands and 0.5 for the Beaumont clays.

Theoretical considerations indicate that the greater part of single-pier settlement occurs as an immediate distortion. Combined shear and consolidation (volumetric) settlement can be estimated, however, from Eqs. 7.54 through 7.56 by deriving E_b and E_s in terms of effective stress from laboratory triaxial shear tests, ideally following stress paths simulating field loading conditions. Immediate settlement can be estimated from undrained shear test results for representative "undisturbed" samples [Woodward, Gardner, Greer (1972)[9]].

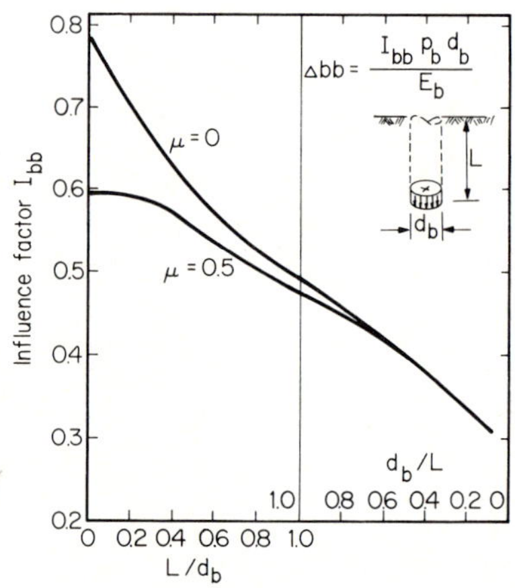

FIG. 7.51 Influence factor I_{bb} for elastic displacement of rigid base with uniform bearing pressure. [From Woodward, Gardner and Greer (1972).[9] Used with permission of McGraw-Hill Book Company.]

Rock Sockets

Some data and discussion on settlement of piles socketed in rock are given in Art 7.3.4, *Rock Masses* (see also discussion in Art. 6.3.2, *Settlement Analysis Based on Values of E_r*). Pells and

*The influence of shaft resistance on base deflection may be considered by dividing the shaft load into a series of point loads (ΔP_s) acting at the center of each of the associated pile segments. Poulos and Davis (1980)[25] provide more rigorous solutions, simulating shaft load increments by incremental line or ring loads. As a general guide, five segments are used for short, stiff piles, 10 for slender piles with L/B to 50, and 15 to 20 segments for $L/B > 50$.

Turner (1979)[61] suggest that elastic settlement of a pile socketed in rock can be computed from

$$\rho = \frac{P}{E_r a} I_p \tag{7.57}$$

where ρ = settlement
P = applied load
E_r = rock modulus
a = socket radius
I_p = influence factor, based on socket geometry and the ratio of pier modulus to rock modulus as given in Fig. 7.52

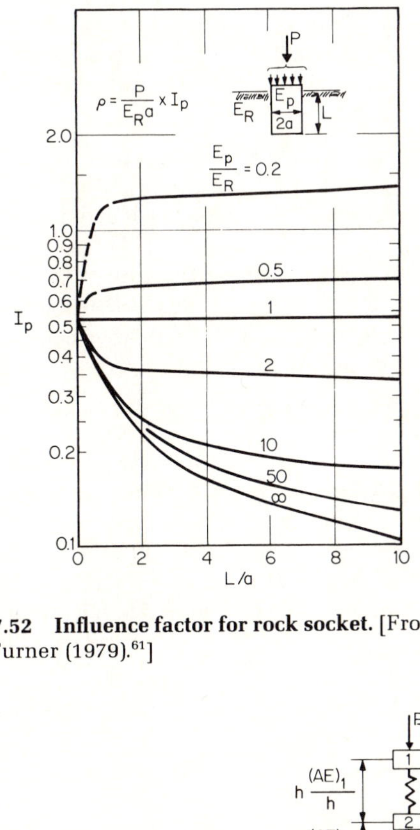

FIG. 7.52 Influence factor for rock socket. [From Pells and Turner (1979).[61]]

Eq. 7.57 is used also to calculate E_r when deflections are measured during load test.

Load Transfer Analysis: *t-z* and *Q-z* Curves

The application of *t*-z and *Q*-z curves to analysis of single-pile and group settlement is being used with increasing frequency, particularly for evaluating piles for offshore structures where static load tests are seldom practical. Until recently the most commonly used procedures were empirical, based on data from tests on relatively short, small-diameter piles performed on land. These data, however, do not adequately cover the variety of factors influencing pile performance, including pile diameter and length, axial pile stiffness, and the distribution of soil strength and stiffness along the pile shaft. The theoretical approach proposed by Kraft et al. (1981)[63] appears to provide a rational basis for evaluating pile settlement. Their concept correlates well with data obtained from pile load tests.

Basic Considerations

The shaft is considered to be divided into segments with shaft friction represented by a series of elastic springs and the base represented by a single spring as shown in Fig. 7.53*a*. Supporting resistance of the shaft and base are each equal to the product of spring displacement and spring constant, given in terms of the elastic moduli of the supporting materials.

A family of curves is developed from an expression including the shear modulus to describe

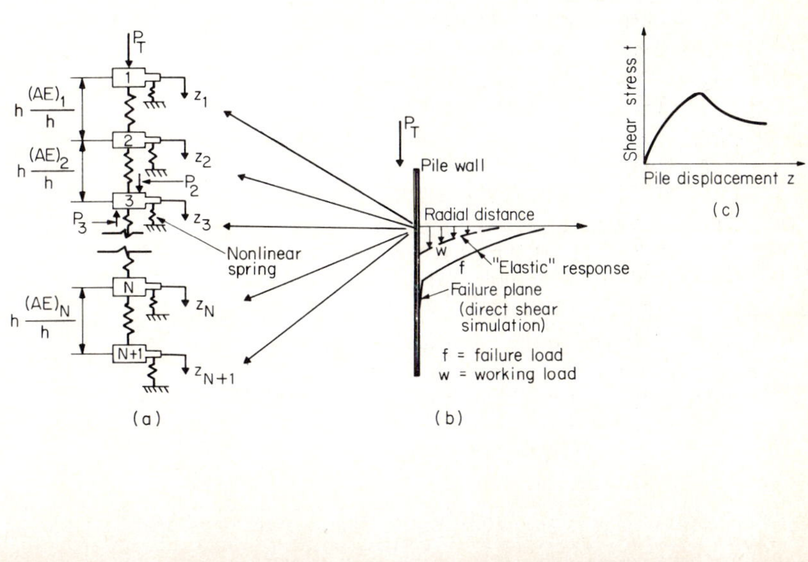

FIG. 7.53 Analysis of *t-z* curve: (*a*) finite difference model of axially loaded pile; (*b*) exaggerated displacement profile; (*c*) *t-z* curve prepared for pile elements. [After Kraft et al. (1981).[63]]

load displacement along the pile shaft (*t-z* curves), and a load-displacement curve (*Q-z*) is computed for the pile tip from values for Young's modulus (secant modulus). Methods based on elastic moduli, however, provide only for immediate and final total settlement and not for intermediate time-dependent deformations.

t-z Curves

The relationship between shear stress at the pile shaft (load transfer *t*) and pile displacement z for some increment of length is represented by a *t-z* curve. Prefailure *t-z* relationships are based on a theoretical elastic model, whereas postfailure relationships are modeled considering the residual stress-deformation behavior at the soil-pile interface. *Prefailure t-z curves* are based on a concentric cylinder and radial soil-displacement model. Soil load displacements may be described by the shear modulus which varies radially and with depth. From a load-displacement equation and a hyperbolic expression for nonlinear stress strain (Eq. 7.58), Kraft et al. (1981)[63] developed Eq. 7.59 for calculating pile displacement or "*t-z* response."

$$G = G_i \left[1 - \left(\frac{\tau_o R_f}{\tau_{max}} \right) \right] \tag{7.58}$$

$$z_s = \frac{\lambda_o r_o}{G_i} \ln \left(\frac{\dfrac{r_m}{r_o} - \psi}{1 - \psi} \right) \tag{7.59}$$

where G = secant shear modulus
G_i = initial shear modulus at small strains (measured from crosshole or resonant column tests)
τ_o = applied shear stress on pile-soil interface (pile load)
τ_{max} = shear stress at failure
r_o = pile radius
r_m = radial distance of zone of influence (limit within which deformation occurs)
$r_m = 2.5\,L\rho\,(1 - \nu)$, where L = pile length, ρ = ratio of soil shear moduli at depths $L/2$ and the pile tip, and ν = Poisson's ratio
$\psi = \tau_o R_f/\tau_{max}$
R_f = stress-strain curve fitting constant taken as 0.9

Postfailure portion of the *t-z* curve requires consideration of the maximum shaft friction and the displacement at which it is mobilized, the residual shaft friction at large displacements, and displacement behavior between maximum and residual stresses. The maximum friction or maximum load transfer value t_{max} is computed by conventional means such as Eq. 7.26. The direct shear test is used to model postfailure behavior and measure the residual shear strength although it is recognized as an imperfect simulation due to boundary conditions. In addition, the direct shear test stress-strain curve represents conditions along a failure surface near the pile wall (Fig. 7.53*b*), but does not account for elastic strains occurring in the soil some distance from the pile.

t-z curve development is illustrated in Fig. 7.54. Up to the point of failure the pile displacement–shear stress relationship is described by Eq. 7.59, with t_{max} determined by conventional means as described above. Thereafter the curve is governed by the direct shear test results along the failure surface and Eq. 7.59 for soil away from the failure surfce.

The *t-z* displacement Δz that occurs after failure may be taken as equal to the displacement δz that occurs in moving from the maximum to the residual stress, although in some cases a correction factor for δz may be required for elastic rebound occurring in the soil beyond the failure surface. Elastic rebound is represented by Δz_e, which is computed from Eq. 7.59, taking the difference for $\tau_o = t_{max}$ and $\tau_o = t_{res}$. The value of

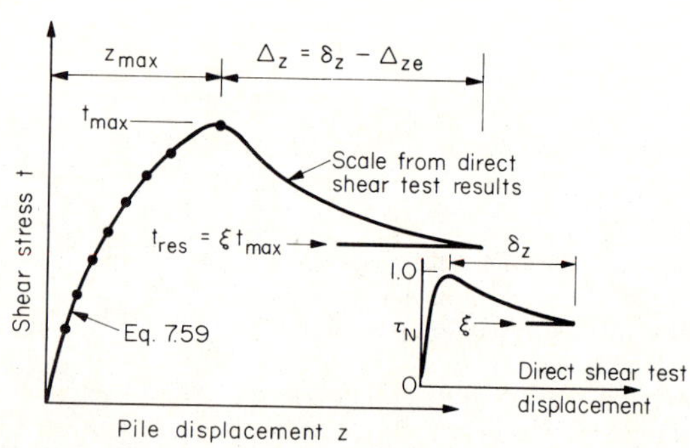

FIG. 7.54 Development of *t-z* curve (τ_n = normalized shear-stress τ/τ peak). [From Kraft et al. (1981).[63]]

Δz_e increases with an increase in pile diameter, but δz is probably insensitive to pile diameter. Kraft et al. (1981)[63] give the maximum difference between Δz and δz as less than 10% for a 12-in (0.3-m)-dia. pile and typical soil conditions. Evaluation is made to determine if a correction of the direct shear test curve is required; if not, the value obtained for δz from the test is used directly.

Examples of *t*-z construction for sand, stiff clay, and soft clay, each for a different pile diameter, are given in Fig. 7.55. Comparisons are shown for *t*-z curves developed from empirical approaches. Pile displacement is plotted versus normalized stress t/t_{max}, and the postpeak response is approximated by a straight line from t_{max}−90% of the maximum pile movement greater than z_{max}.

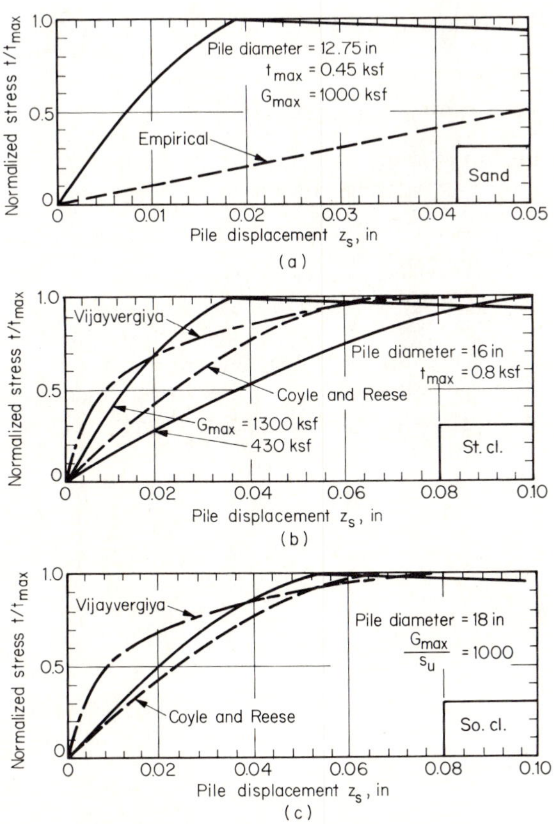

FIG. 7.55 Comparison of theoretical *t-z* curves with empirically developed curves for: (*a*) sand, (*b*) stiff clay, and (*c*) soft clay. [From Kraft et al. (1981).[63]]

Q-z Curve Development

Load-displacement response at the pile tip is modeled by a *Q*-z curve developed from Eq. 7.60, which describes the elastic solution for a rigid punch.

$$z = \frac{q_b D(1 - \nu^2)}{E_s} I_b \qquad (7.60)$$

where q_b = stress at pile tip
D = tip diameter
E_s = Young's modulus taken as the secant modulus
I_b = influence coefficient which varies with pile depth as shown in Fig. 7.51, with values typically between 0.78 and 0.5

Results

From the *t*-z and *Q*-z curves the deformations occurring along the pile segments plus the tip deformation are summed to find the pile movement versus axial load, and a curve is prepared of axial load versus pile movement. The computer program AXPILE is useful for single pile analysis.

Limitations

Although the method appears to give good correlations with available load test data, Kraft et al. (1981)[63] conclude that several steps are needed for improvement:

1. Define the distribution and magnitude of the ultimate load transfer along the pile shaft and at the tip.
2. Predict the soil movements (and thus negative shaft friction) caused by consolidation as excess pore pressure induced by pile installation dissipates (especially in soft clays).
3. Incorporate the effects of soil creep.
4. Evaluate changes in soil properties when piles are installed in sands.
5. Evaluate group efficiencies, especially if load-displacement predictions are needed at loads near failure.

7.4.2 GROUP SETTLEMENT

General

Stress Distribution

Group settlement usually is of more concern than group capacity. If the piles are within or underlain by compressible strata, settlement will be larger in magnitude than it would be if measured by load test on a single pile, because of the larger zone of stress influence as shown in Fig. 7.56. Total group settlement will be the sum of the elastic shortening of the piles, the immediate point settlement, and the long-term or consolidation settlement. Elastic shortening is group independent, but the other two elements vary with group size since it controls the volume of material participating in the development of pile resistance.

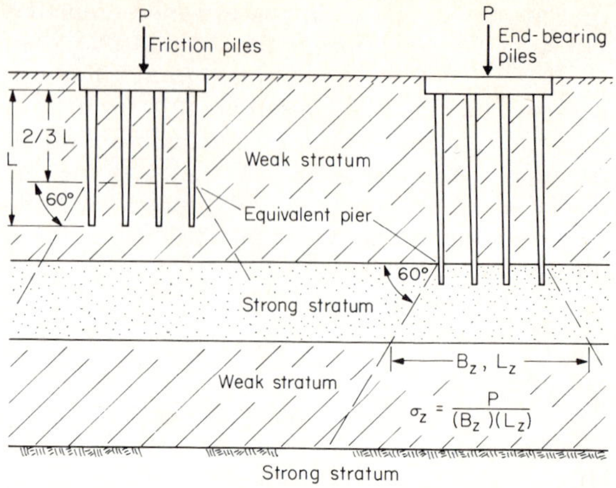

FIG. 7.57 Approximate stress distribution beneath two examples of the equivalent pier. [After NAVFAC (1982).[1]]

Equivalent Pier Approach

Group settlement usually is evaluated on the basis of the equivalent pier approach, as illustrated on Fig. 7.57. For friction piles the load is assumed to be acting at an effective depth of two-thirds of the pile embedment in the bearing stratum, the presence of piles below that depth is ignored, and for group end bearing, the equivalent loaded area is taken as the pile-tip elevation, or the depth where support begins in the strong stratum as shown in the figure.

Pile Interaction Analysis

Sophisticated settlement analysis based on evaluating the contributions of adjacent piles to soil-pile interaction is described by Poulos and Davis (1980).[25]

Pile Cap Effects and Load Distribution

Pile cap location and rigidity influence not only group capacity, as discussed in Art. 7.3.6, but also influence the load distribution within the group. A rigid cap will result in equal pile settlement under concentric loads. By analogy to a rigid footing, the outside piles, therefore, will carry the highest percentage of load. A relatively flexible cap will cause the load to concentrate toward the central piles. Soil type also affects distribution. Research has indicated that in clays, where the cap is less rigid than the soil, the cen-

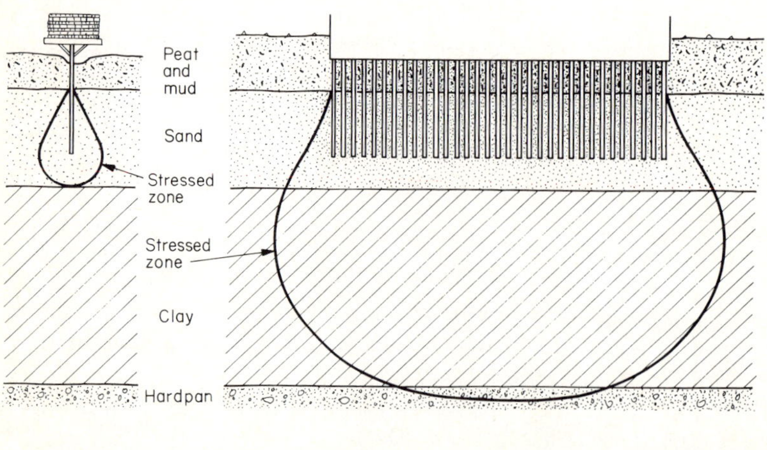

FIG. 7.56 Difference in stressed zones beneath single-pile load test and pile group of friction piles.

tral piles carry the most load. Analytical approaches to pile cap effects and the design of pile-supported rafts are given in Poulos and Davis (1980).[25]

Settlement Tolerances

For pile foundations in sands not underlain by more compressible soil, a maximum settlement of 25 mm frequently is permitted for buildings on small pile groups and 50 mm for piled rafts. For clay soils, values about twice those given for sands usually are acceptable [Skempton and MacDonald (1956)[64]].

Settlement Analysis

Property Measurements

Settlement analysis may be based on laboratory test data as described in Art. 4.3.4, or on field load tests *(see Art. 7.5.3)*, penetration tests, or pressuremeter tests *(see Art. 4.3.3)*. As is the case in all geotechnical analysis, the most significant factors are representative measurements of properties which rely on the quality of undisturbed samples for laboratory testing or the validity of in situ tests, and the questionable accuracy of such measurements usually invalidates the necessity for highly sophisticated analytical approaches.

Sand

Settlement of pile groups in sands, not underlain by a more compressible material, can be estimated by means of the equivalent pier concept from the *SPT N value* using empirical correlations, with the following expression from Meyerhof (1976)[20]:

$$S = \frac{2p\sqrt{B}(I)}{N} \tag{7.61}$$

where B = width of pile group
p = net foundation pressure
N = average corrected SPT result within the seat of settlement (depth $\approx B$ in homogeneous soil)
I = influence factor of effective group embedment given by

$$I = 1 - \frac{D'}{8B} \geqq 0.5 \tag{7.62}$$

where D' = the effective depth of the group

In *silty sand* the settlement estimate obtained for sand should be multiplied by 2.

CPT data are used to estimate maximum settlement from the expression

$$S = \frac{pBI}{2q_c} \tag{7.63}$$

where q_c = average static cone resistance within the settlement zone.

Comparisons of settlement estimates based on the SPT or CPT with observed settlement show that the observed maximum settlement is somewhat smaller than estimated, as has been found with shallow foundations. The approach, therefore, is considered generally to be conservative, especially in view of the densification that occurs during pile driving which is not present during the in situ penetration testing.

Negative skin friction resulting from an upper consolidating layer *(see Art. 7.3.6)* causes a corresponding downdrag force which must be included in the net foundation pressure for settlement estimates for pile groups end bearing in sands. For the case of a pile group passing through a clay stratum into a sand stratum, the effective depth of the piles in the sand stratum governs any reduction of the estimated settlement due to the depth of group embedment.

Clays

Settlement of a pile group bearing in clay, or above a clay stratum, is estimated in the same manner as for shallow foundations, but by treating the group as an equivalent pier. Allowance is made for the effective embedment and the compressible stratum thickness. In NC or sensitive clays, if large settlements are to be avoided, the net foundation pressure should be kept substantially below the difference between quasi-preconsolidation and effective overburden pressures [Meyerhof (1976)[20]]. Settlement can be decreased further if a *compensated friction pile foundation* is used, for which piles are installed before basement excavation to reduce heave and recompression of the clay as discussed in the following section.

The *time rate* of settlement in clays is relatively impossible to predict with certainty because of the large variation in permeability that can be expected in the soils along the shaft. Local experience is the only sound basis for time-rate estimates.

Creep

Creep is a contribution to settlement, in addition to immediate compression and consolidation, that occurs most significantly with heavily loaded piles or piers in OC clays. The phenomenon is a time-dependent plastic deformation occurring under conditions of sustained shear stress and can result in important long-term settlement. A method of analyzing the time-dependent behavior in a soil exhibiting creep properties is presented by Booker and Poulos (1976).[65] Interpretation of long-term pile load tests is required to obtain the necessary soil parameters.

Load-Transfer Analysis

In recent years there has been a trend to analyze the settlement of pile groups with the computer program PILGP1 which uses *t-z* curves to model pile response, adjusting the input single pile *t-z* curves for the interaction effects to account for group behavior [Kraft et al. (1981)[63]].

Differential Settlement

Occurrence

Within a structure, differential settlement can occur between pile groups, or in the case of large groups or pile-supported rafts, it will occur as a result of differential stress distribution caused by soil type and raft rigidity as discussed in the previous article.

In very deep deposits of soft clays, where structures of varying load conditions are erected in close proximity as in Mexico City, various forms of differential movement occur. Buildings on piles can be particularly troublesome. In Mexico City, end-bearing piles usually are driven into a sand stratum at a depth of about 33 m. As the ground surface tends to settle away from the building because of the *regional subsidence phenomenon* [Art 10.2.4 of Hunt (1984)[12a]], the load of the overburden soils is transferred to the piles from negative side friction. If the piles do not have sufficient capacity to support both the building and downdrag loads, additional settlement results. If pile capacities are adequate, however, the structure will settle less than the surrounding ground and the subsiding adjacent ground settles away from the building. Utilities are installed with flexible connections and allowances are made in the first-floor design for sidewalks and roadways moving downward from the building when adjacent ground settlement is anticipated.

Compensated Friction-Pile Foundation

This approach (Fig. 6.1), developed by Zeevaert (1972),[8] requires a high degree of analytical sophistication and a large amount of experience for application. Analysis is aimed at causing the building to settle at the same rate due to downdrag and consolidation as the ground is subsiding; therefore, the building does not sink below the surface or rise above it by significant amounts. The solution has been used successfully for medium to large structures. Tower Latino Americana in Mexico City, 43 stories in height, is supported on a combination of end-bearing piles and a compensated raft foundation [Zeevaert (1957)[66]]. The piles were driven into the sand stratum at 33 m [Fig. 7.60 of Hunt (1984)[12a]], and a 13-m-deep excavation was made afterward for the raft. It was completed in 1951, and as of 1957 settlement from consolidation of the underlying clay strata had been as predicted, or about 10 cm annually. Building settlement was, however, almost equal to that of the surrounding ground surface.

7.5 CAPACITY EVALUATION: DYNAMIC METHODS AND LOAD TESTS

In addition to the static methods described in Art. 7.4, the capacity of driven piles is evaluated by dynamic analysis involving resistance formulas and the wave equation. Driven and bored piles are evaluated by static load tests; driven piles are also evaluated by dynamic testing.

7.5.1 RESISTANCE FORMULAS FOR DRIVEN PILES

General

Formulas have been in use for over 100 years that relate the dynamic energy used to force the pile into the ground by hammer impact to the penetration achieved, which result in an empirical evaluation of the static capacity of a single pile. The *dynamic pile formulas*, or *resistance formulas*, relate energy input to energy used plus energy lost during driving. Energy used equals the driving resistance times the pile movement; energy is lost through plastic and elastic compression of the pile cap, the pile, and the soil. Thus, driving resistance, related to *allowable pile load*, is determined from a relationship between energy input and energy lost to pile penetration during a hammer blow. The various formulas reflect the relationship: pile resistance × pile penetration = energy applied minus losses.

Application

Pile driving involves numerous variables, for many of which the measurement of precise values is difficult. It is generally accepted among geotechnical practitioners, therefore, that dynamic formulas cannot be used to predict static capacity with any degree of accuracy and large safety factors are employed. Even so, the evaluation of static capacity by dynamic formulas is still in wide use.

In the practical application of one of the dynamic driving formulas, a desired capacity for the pile selected is assigned, and then for a typical hammer energy the penetration or "set" into the bearing stratum is specified, taking the average penetration in inches per blow for the final 6 in of driving.

Common Formulas

A large number of dynamic formulas have been developed; the most commonly used in the United States probably is the *Engineering News Formula*, dating from 1888. Developed initially for timber piles in sand, presumably the equation yields a safety factor of 6, but experience has shown it to vary from 0.5 to as high as 16. It is now considered the least reliable of the commonly used formulas [Poulos and Davis (1980)[25]]. It is expressed in a modified version as follows:

$$Q_a = \frac{2E}{S + C} \tag{7.64}$$

where Q_a = "supposed" design static resistance, lb (also given as R)

E = delivered energy (hammer weight $W \times$ drop height H for single-acting hammers), ft·lb

S = set, in (penetration into soil per blow)

C = an empirical constant = 1.0 for drop hammers, 0.1 for steam hammers when driven or pile weight W_p is smaller than striking weight W_s, or

C = 1.0 W_p/W_r, or 0.1 W_p/W_r when driven weight is larger than striking weight

In terms of the final "set":

$$S = \left(\frac{2E}{Q_a}\right) - C \tag{7.65}$$

Other commonly used formulas, all having about the same reliability, and all substantially superior to the Engineering News Formula, are the Hiley, Janbu, and Danish formulas. The *Hiley formula*, much used in Great Britain, includes factors measured in the field during driving to account for movement of the pile top due to temporary elastic compression in the cap, pile, and soil, as well as factors to account for the pile weight. The *Danish formula*, relatively simple, is expressed as

$$Q_{ult} = \frac{W_r H e_f}{S + 0.5S_o} \tag{7.66}$$

where Q_{ult} = pile axial capacity

W_r = weight of hammer or ram

H = height of hammer or ram drop

e_f = pile hammer efficiency (Table 7.20)

E = rated hammer energy

E = $W_r H e_f$

S = set

S_o = elastic compression of the pile = $[(2EL)/AE_p]^{1/2}$

L = pile length

A = pile cross-sectional area

E_p = pile material modulus of elasticity

In terms of final "set":

$$S = \left(\frac{E}{3Q_a}\right) - 0.5S_o \tag{7.67}$$

Note: If set is to be in inches per blow, all lengths should be expressed in inches; the units of ram weight and pile capacity must be the same.

TABLE 7.20
VALUES OF HAMMER EFFICIENCY e_f*

Hammer type	e_f
Drop hammer released by trigger	1.00
Drop hammer actuated by rope and friction winch	0.75
McKiernan-Terry (MKT) single-acting hammers	0.85
Warrington-Vulcan single acting hammers	0.75
Differential-acting hammers	0.75
McKiernan-Terry (MKT), Industrial Brownhoist, National & Union double-acting hammers	0.85
Diesel hammers	1.00

*From Chellis, (1962).[31] Used with permission of McGraw-Hill Book Co.

Reliability

Difficulties in evaluating the many sources of energy loss during driving as well as the temporary effects of driving on soil properties reduce the reliability of dynamic formulas. Energy loss is affected by equipment efficiency, ram-pile contact, and transmission through the pile itself. Driving through granular fills or intermediate strata of granular soils will falsely influence penetration resistance possibly resulting in the predetermined or specified "set" even though the tip is in weak soils. The shear resistance of fine-grained soils (clays and silts) during pile installation is not the shear resistance that will support the pile after construction as "freeze" occurs. In dense silty sands, or in cohesionless silts, temporary high pore pressures during driving may yield a falsely high penetration resistance. When pore pressures dissipate (relaxation), support is lost, and in rare cases piles settle overnight under their own weight, even to be lost from sight.

Vibratory or *"sonic" driver* penetration resistance cannot be analyzed by pile-driving formulas and static analysis and load testing are required for evaluating the necessary pile penetration.

7.5.2 WAVE-EQUATION ANALYSIS

Theoretical Concepts

Wave Propagation

The impact of the driving ram or hammer induces longitudinal stress waves to travel along the pile. Initially these waves are compressive, travel downward and result in pile penetration, but various factors can cause their modification. For example, in a fixed-end pile under hard driving conditions, the wave is reflected as a compression wave and stresses could double, possibly causing pile damage; in a free-end pile under easy driving conditions, the compression wave may be reflected as a tension wave and stresses could cancel, resulting in little penetration per blow, falsely indicating hard driving.

Wave-equation analysis has permitted a better understanding of the nature of stresses created in piles of different types driven by hammers of differing weight, as well as a rational basis for understanding pile dynamics.

Pile Reaction to Driving Forces

Pile capability to transmit longitudinal force is measured by its *impedance*, given as the product of mass density, cross-sectional area, and stress wave velocity, the latter a function of pile modulus of elasticity. The force developed at the pile tip depends on pile impedance, furnished hammer energy, the nature of the pulse delivered by the hammer, stress-transmission characteristics of the cushions and pile-head assembly, the distribution pattern of resistance along the pile exerted by the soil, and the proportion of total resistance divided between the shaft and the tip.

The problem has been treated by computerized solution of the one-dimensional wave equation first applied by Glanville et al. (1938)[67] [Cummings (1940),[67a] Smith (1960)[67b]] and later refined by others. Various parameters are used to define the pile penetration per hammer blow as a function of soil resistance, modeled as illustrated in Fig. 7.58. The pile is assumed to consist of a series of elements, each possessing a weight W_n, connected by springs with stiffnesses K_n, associated with the elastic properties of the pile material. Also provided are ram velocities at impact, values for weights and stiffnesses of the ram and drive head, and cushion properties. Soil properties are represented by ultimate soil resistance R_u and by soil-damping J and "quake" Q values.

Soil Parameters

Ultimate soil resistance R_u is proportioned along the shaft and a percentage is assumed mobilized at the base as point resistance.

Damping values J are a measure of the increase in soil resistance with increasing pile penetration velocity, and *quake values* Q represent the elastic displacement required to mobilize maximum soil resistance. The selection of values for R_u, J, and Q usually is based on empirical correlations with soil type or evaluated by dynamic test methods *(see Art. 7.5.4)*.

Typical values for f_s as a percentage of R_u, and of J_p and Q, are given in Table 7.21. Bowles (1974)[69] suggests that quake and damping are not very critical, and as long as reasonable values are selected (Q = 0.1 to 0.15 and J_p = 0.15 to 0.30), the computed results will be satisfactory for the most part. The side damping constant J_s is usually taken as equal to $J_p/3$, where J_p is point damping.

Analysis

Detailed description of the wave-equation concepts and the selection of the various factors, including the soil parameters, is found in Poulos and Davis (1980).[25] A computer program and a description of its derivation are found in Bowles (1974).[69] Results of wave-equation analysis are shown typically as curves of the resistance during driving R_u versus the hammer-blow count, as shown in Fig. 7.58.

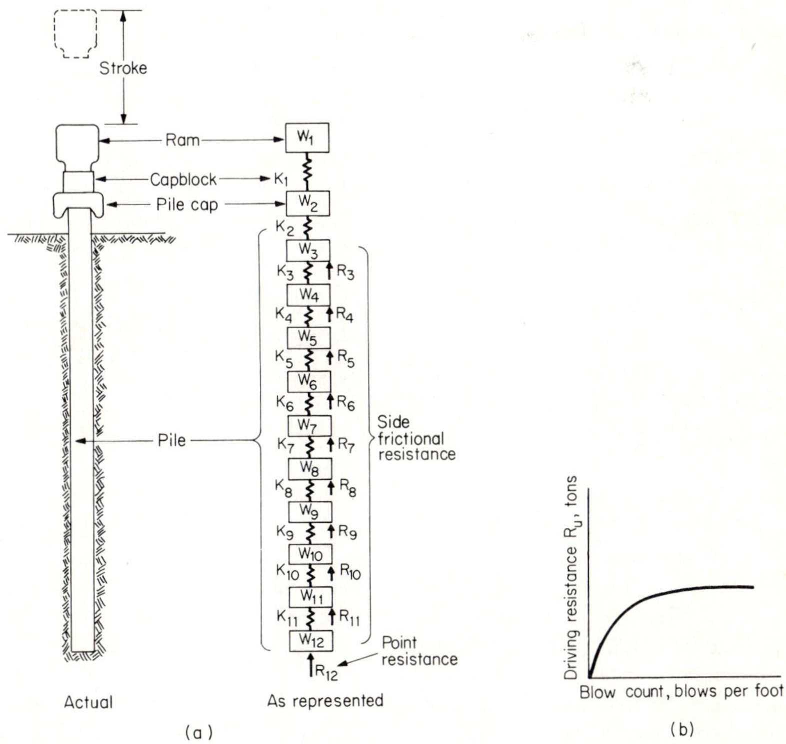

FIG. 7.58 Representation of pile for wave-equation analysis and typical results. (*a*) Model of system. (*b*) Typical results. [From Smith (1960).[67]]

TABLE 7.21
EMPIRICAL VALUES FOR QUAKE (Q), POINT DAMPING (J_p) AND PERCENT SIDE ADHESION ($\%R_u$)*,†

Soil type	Q, in	J_p, sec/ft	Side adhesion, $\%R_u$
Coarse sand	0.10	0.15	35
Sand-gravel mixtures	0.10	0.15	75–100
Fine sand	0.15	0.15	100
Sand and clay, or loam, at least 50% of pile in sand	0.20	0.20	25
Silt and fine sand underlain by hard strata	0.20	0.20	40
Sand and gravel underlain by hard strata	0.15	0.15	25

*After Forehand and Reese (1964).[68]

†J_s typically taken as $J_p/3$.

The wave equation attempts to model the pile, the driving system, and the soil. The problem still remains that although the pile can be modeled accurately, the modeling of the driving system depends on its performance in an impact situation which is affected by equipment condition, as well as other factors. Soil modeling is even more difficult as assumptions must be made regarding the various soil parameters. The method is particularly useful, however, when combined with the information obtained by the Case method of dynamic pile testing *(see Art. 7.5.4)*.

Applications

Selection of Driving Equipment

Dynamic analysis is used to select the hammer-cap block-pile cushion combination most appropriate for the geologic conditions, especially where neither the design engineer nor the contractor has prior experience under these conditions.

Driving Stresses in Piles

Maximum stresses developed in a pile are those experienced momentarily during driving. Dynamic analysis is the most reliable method available for estimating driving stresses in steel, concrete, or timber piles, especially when combined with field measurements of strain and acceleration *(see Arts. 7.5.3 and 7.5.4)*.

Pile Bearing Capacity

Relationships between driven pile capacity and the number of blows per unit of penetration required for a particular hammer in particular geologic conditions are established by dynamic analysis, particularly in free-draining soils. In cohesive soils, increasing in stiffness with depth, the analysis provides an estimate of depth and capacity that can be developed for a given penetration per blow, or in stiff or hard clay it can provide a guide to point resistance. Analysis of redriving records in dense, very fine sands and silts provides an evaluation of the capacity obtained after relaxation occurs.

Limitations

Dynamic analysis is most suitable in free-draining granular soils, and is, in general, unreliable for friction piles in clays because of freeze effects after driving. Where practical, results should be correlated with load test data and static analysis, particularly in fine-grained saturated soils.

7.5.3 STATIC AXIAL LOAD TESTS

General

Allowable pile axial load is confirmed by load tests on almost all projects involving driven piles, but only occasionally on high-capacity bored piles because of cost factors. Loads normally are applied statically, but new dynamic methods for driven piles are increasing in use (see Art. 7.5.4). Lateral load tests are described in Art. 7.6.4.

Load testing is the most reliable method of evaluating single pile or group capacity in free-draining material. In clays, load tests are most valid when performed after the clay has had time to consolidate and freeze; otherwise, results are likely to be overly conservative. In all cases test installations should duplicate the proposed construction methods and the same installation equipment should be used. Pile groups supported by shaft friction in clays, or underlain by compressible clay strata, must be evaluated by settlement analysis in addition to load testing.

Test Arrangements

Load Application

Basic elements of a static load test include the load application system, a deflection measurement system, and the pile. Several methods of applying loads to a test pile are illustrated in Fig. 7.59, which can include jacking against two uplift anchor piles as in Fig. 7.59*a*, or against a heavy weight (reaction block) as in Fig. 7.59*b*, or for very heavy loads, against a relatively light cone or block that can be tied down with anchors as in Fig. 7.59*c* and in the photograph in Fig. 7.60. When reaction piles are used they must be separated by an adequate distance (usually at least five diameters) to avoid stress influence on the test pile and a reduction in its capacity. Pile load test details are often specified by local building codes and also can be found in ASTM D1143-69.

Four setups for high-capacity load tests using anchors for a project with over 800 bored piles are shown in Fig. 7.60. The tests are designed to apply 600 tons to 70-cm-dia. piles and 2200 tons to 130-cm-dia. piles bearing in residual soil or rock. As the posttest removal of the heavily reinforced blocks proved difficult and time consuming, reaction blocks were prefabricated in conical shapes with locations for anchor inserts for subsequent tests. The piles were set relatively close together to permit horizontal load testing by using adjacent piles in reaction.

Measurements

Measurements of pile deflections are made as loads are applied, and a number of instruments

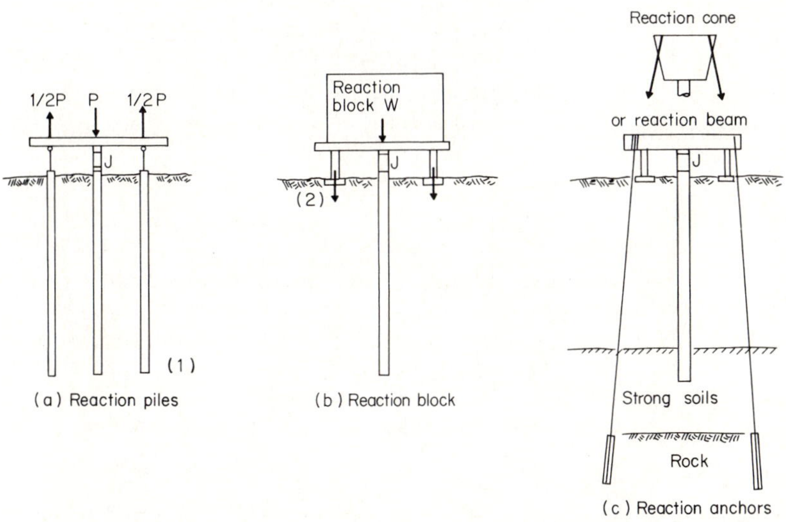

FIG. 7.59 Various types of load tests. In (*a*) reaction piles (1) must be spaced so as not to influence stresses of test pile. In (*b*) the footings must provide adequate support for the reaction block at (2). In (*c*) anchors provide very high reactions for little weight.

FIG. 7.60 Pile-load tests using anchored reaction blocks. Two blocks on left: 130-cm-dia. test pile under 2200-ton test load. Two blocks on right: 70-cm-dia. piles under 600-ton test load. (Photo by A.J. Sabatino.)

and procedures are possible [Art. 4.5.2 of Hunt (1984)[12a]]. Instrumentation can be installed to measure deflection of the pile head, deflections in the pile for various lengths along the shaft, and deflections at the pile tip (except for timber piles where deflection measurements are possible only at the head).

Head deflection measurements are made with both dial gage extensometers and optical instruments (levels) using a vernier gage, or simply by a wire held under constant tension between two supports and passing across a scale attached to the pile. A setup for a large-diameter bored pile is shown in Fig. 7.61 and the sketch in Fig. 7.62. The dial gages and the vernier are attached to beams supported independently so as to be completely free of the loading arrangement. The entire loading and measurement system should be covered to minimize the effects of temperature changes, and records of temperatures in the test area should be kept.

As the net deflection of the pile is desired for given loads, the total measured deflection must be reduced by the pile's elastic compression. In a friction pile, shortening (elastic compression) is measured by instrumentation; in a purely end-bearing pile it can be computed from the expression

$$\Delta L = PL/AE \tag{7.68}$$

FIG. 7.61 Applying load to 130-cm-dia. pile with high-capacity electric hydraulic jack (not shown): (1) four 600-ton jacks, (2) pile head, (3) optical measurements, (4) dial gages, and (5) jack for low loads.

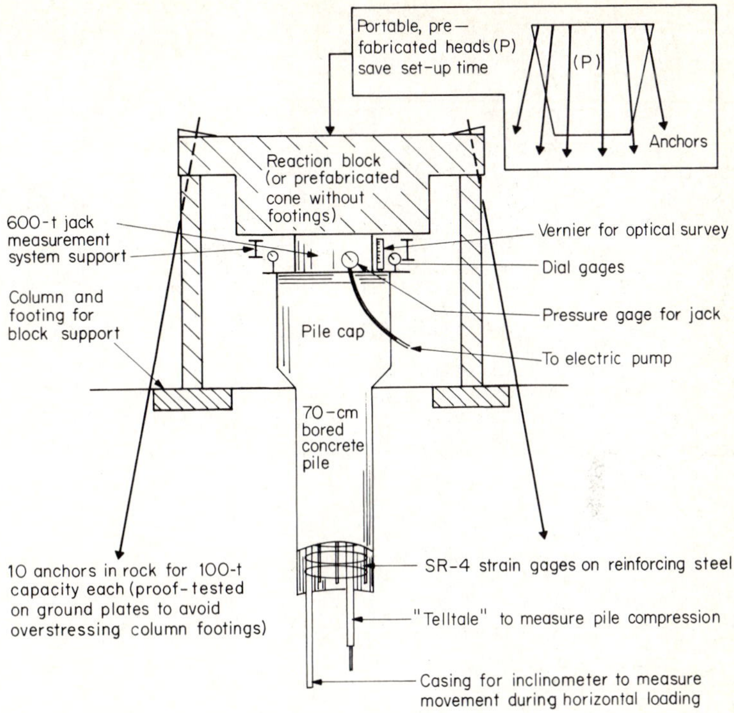

FIG. 7.62 Setup for 600-t-load test on bored, cast-in-place reinforced concrete pile of 70-cm-dia. bearing in residual soils.

where L = pile length, in
P = load, lb
A = pile area, in²
E = modulus of elasticity; common values are
steel, $E_s = 30 \times 10^6$ psi
concrete, $E_c = 4 \times 10^4$ psi
timber, $E_t = 1.4 \times 10^6$ psi (average value)

For two materials, such as concrete-filled pipe, the steel area is converted to an equivalent concrete area, proportional to the elastic modulus of the materials, or $A_s \times E_s/E_c = A_c$, and

$$\Delta L = \frac{PL}{A_cE_c \times A_s(E_s/E_c)E_c} \qquad (7.69)$$

Shaft deflections are measured with strain-gages and/or telltales installed at various depths (Fig. 7.63). The strain gages are mounted directly on the pile steel as shown in Fig. 7.64, and the tubes for the telltales are tied to the steel (Fig. 7.65) or otherwise inserted in the pile during concreting. The telltale consists of a closed-end, small-diameter pipe, oil-filled, in which a steel rod is free to move downward with the pile. Dial gages at the pile head measure the rod movement. A rod extending to the pile tip, which because of its freedom from load does not shorten but moves downward with the pile, provides the measure of the net pile movement under load. Side friction along the shaft can be computed from the strains measured along the shaft with the gages and telltales, since the strain measurements indicate the proportion of the applied load reaching the depth of measurement.

On small-diameter piles the installation of too much instrumentation may affect the elastic properties of the pile; although, at the least, a single telltale should be installed to the tip, along with strain gages. Also shown in the photograph in Fig. 7.65 is the special casing in which an inclinometer was lowered to measure lateral movements during horizontal load testing *(see Art. 7.6.4)*.

Testing Procedures

Slow maintained load test (slow ML test), recommended by the ASTM (1969),[71] is the most

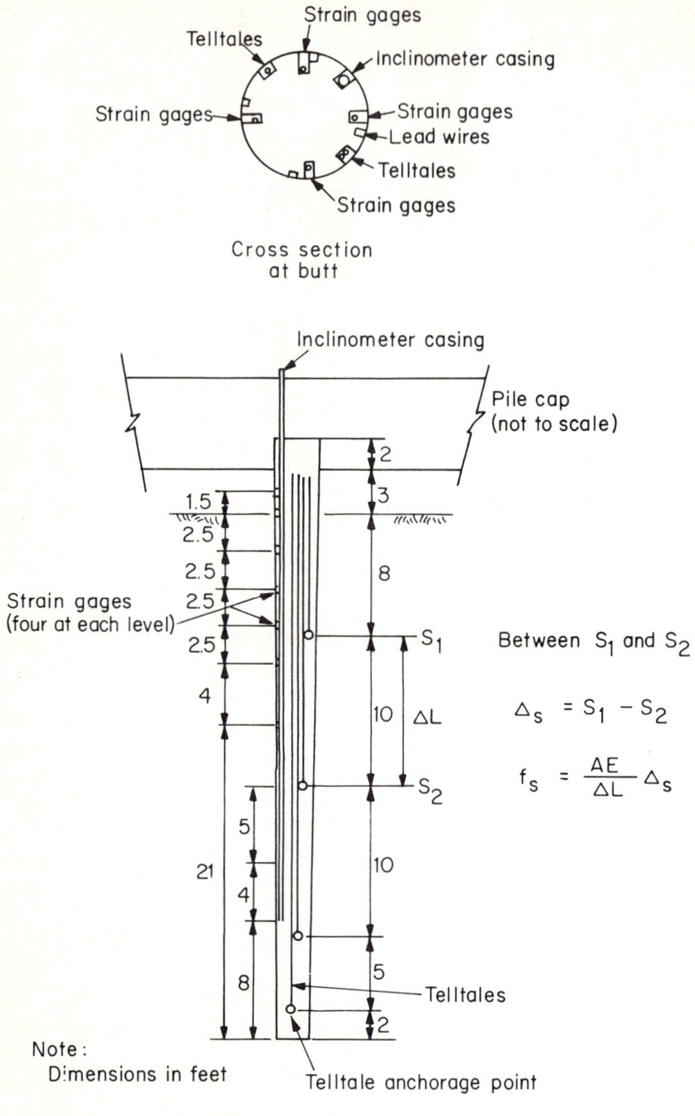

FIG. 7.63 Instrumentation to monitor load distribution in a concrete or bored pile. [After Stevens et al. (1979).[70]]

common test method in North America. The pile is loaded in eight equal increments to 200% of the anticipated pile working load, then unloaded in four equal decrements. Each load is maintained until the settlement rate has decreased to 0.001 ft/hr (0.3 mm/hr), i.e., 0.002 in/10 min (0.05 mm/10 min), or for 2 hr, whichever comes first. The 200% load is maintained for 24 h, and normally the total test will require about 70 h or more to perform, depending upon geologic conditions. The settlement criterion of 0.05 mm/10 min is often called the "zero" settlement, which is misleading as it represents significant settlement on a long-term basis (1.8 m/yr). Fellenius (1975)[72] recommends locking the jack piston once the load increment is reached and the load (jack pressure) is permitted to drop to an equilibrium value. This modification reduces testing time without affecting test results.

Constant rate of penetration test (CRP test) is the reverse of the slow ML method. The pile head is

FIG. 7.64 Mounted SR-4 strain gage for 600-t load test. (Photo by A.J. Sabatino.)

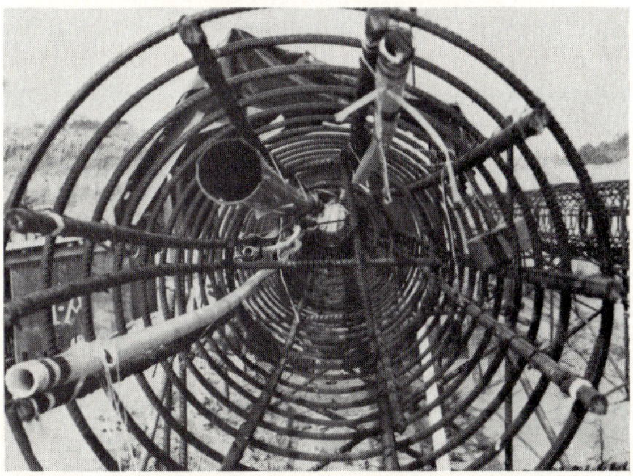

FIG. 7.65 Reinforcing cage for 600-t-load test pile with inclinometer casing and telltale casing installed. Note roller spacer on right. (Photo by A.J. Sabatino.)

forced to deflect at a predetermined rate, usually 0.02 in/min (0.5 mm/min), and the force required to achieve the penetration is recorded. The test is carried to a total penetration of 50 to 75 mm or the maximum capacity of the reaction arrangement, and is usually completed in about 2 to 3 hr. In addition to the short time for performance, the test results in distinctive load-deflection curves for various soil types which provide information on pile behavior. It is recommended generally for testing piles in clay, and for all tests when only the ultimate capacity is to be measured.

Cyclic testing involves unloading the pile from a few or all load levels in the slow ML test and the pile is repeatedly loaded and unloaded. The test objective is to separate shaft resistance from end resistance, but the test yield value is normally smaller than the ultimate capacity since the shaft resistance is reduced by the cycling. Van Weele (1957)[73] describes the use of the cyclic test to evaluate the shaft and base resistance relationship.

Quick maintained load test (quick ML test), proposed by Fellenius (1975),[72] is comparable to the CRP test as it eliminates the influence of time-dependent pile movements which are measured in the slow ML test. The pile usually is loaded to 300% of anticipated allowable load in 200 small increments, each equal to 15% of the allowable load. Each load is maintained for a period of 15 min with readings taken every 3 min. Total test duration is 3 to 5 h.

A comparison of times required to perform the four types of load test is given in Fig. 7.66, and a comparison of the load-movement behavior for each for a friction pile in clay is given in Fig. 7.67.

Test Interpretations

General

The slow ML test and the cyclic test may show only that failure has occurred, and for these tests interpretation of a value for the failure load is difficult. Ultimate or peak failure value is obtained

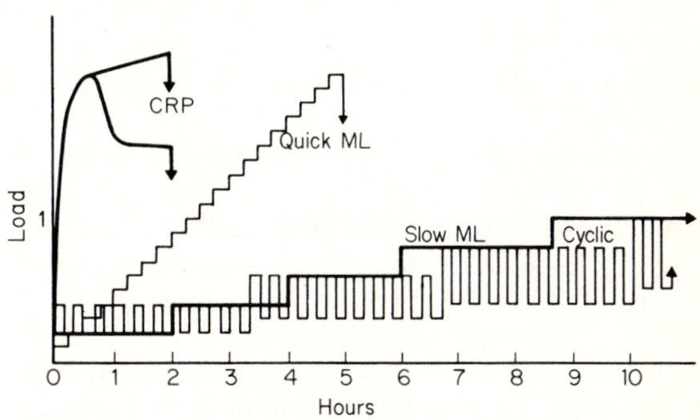

FIG. 7.66 Comparison or required time for four load-test procedures, friction piles in clay. [From Fellenius (1975).[72]]

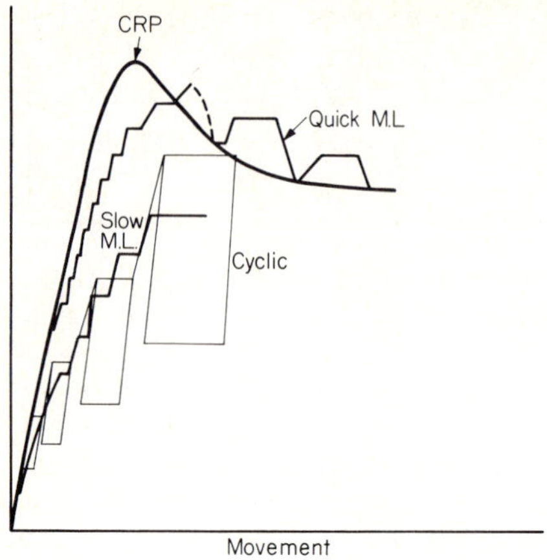

FIG. 7.67 Comparison of typical load-movement behavior for four test procedures for friction piles in clay. [From Fellenius (1975).[72]]

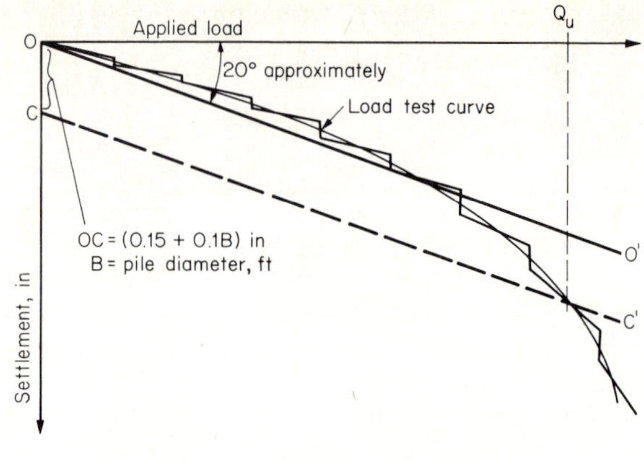

FIG. 7.68 Procedure to find ultimate pile capacity from load test. [After Davisson (1973).[74]]

in the CRP test, and Q_a is taken as $Q_u/2.5$. Peak failure is normally clearly defined, however, only in soft or loose soils. The quick ML test also defines approximately the failure load.

Slow ML Test

Interpretations of the slow ML test have been approached by numerous arbitrary or empirical rules and even building codes can be found to differ substantially. One simple approach used by many practitioners is to take Q_a as not exceeding ½*P* causing a net settlement of not more than ½ in (12.5 mm), or a gross settlement of 1 in (25.4 mm) taking into account the elastic shortening of the pile, whichever value is the smaller.

Peck et al. (1974)[2] suggest the approach of Davisson (1973),[74] in which tip settlement at failure is defined in terms of pile diameter *B*, as shown in Fig. 7.68. A similar approach is described in the CGS Manual (1978).[6] Elastic pile deflection is computed from PL/AE and plotted on the load-settlement diagram from the test data as line OO'; for best interpretation, the diagram scales should be selected to provide the slope of OO' at about 20°. Dashed line CC' is drawn parallel to OO' with an intercept on the settlement axis at OC equal to $(0.15 + 0.1B)$ in, where B = pile diameter in feet, to provide a measure of tip settlement required to develop the capacity. Ultimate load Q_u is defined as the load at which line CC' intersects the load settlement curve. Because of the effects of time deformation, the procedure is recommended only for tests where load increments are applied for periods not exceeding about 1 hr.

Large-Diameter Bored Piles

Fully instrumented pier-load tests provide measurements of tip pressure, shaft friction, total deflection, and tip deflection. Shaft friction is computed from Eq. 7.70 from the strains measured along the shaft with gages and telltales (S_1, S_2, etc. in Fig. 7.63) since the strains indicate the proportion of the applied load reaching the measurement depth.

$$f_s = A_s E_c \frac{S_1 - S_2}{\Delta L} \qquad (7.70)$$

Deformation moduli also are computed from the test results from stress-strain data, as described in Art. 7.4.1. In rock masses, moduli values are related to RQD and recovery values from core samples and U_c test data. Several load tests run for differing and bracketing conditions of rock quality will provide data for correlations across a given site.

Limitations

As normally performed, load tests on single piles provide limited data. They do not account for long-term settlement and downdrag in cohesive soils, nor for the effect of group action.

7.5.4 DYNAMIC AXIAL LOAD TESTING

Method

A dynamic testing method to evaluate pile capacity was developed at Case Western Reserve University in Ohio. Known as the *Case method* of dynamic pile analysis [Gobel and Rausche (1970),[75] La Fond (1977),[76] Rausch (1978)[77]], it utilizes transducers attached either to the top of the pile or along the shaft to measure force and acceleration caused by hammer impact. Instrumentation removes the unknown factors of energy loss in the driving system and permits indirect elimination of the dynamic pile characteristics, thus providing a more accurate measurement of driving resistance. It is being used extensively offshore where full-scale load tests are usually not practical.

Data Recording and Processing

The instrumentation, data recording, and processing system of the Case method are illustrated in Fig. 7.69. Installation is relatively simple and the equipment is readily portable, fitting in a container about the size of a large suitcase.

Instrumentation consists of accelerometers and strain transducers. Two piezometric accelerometers (Fig. 7.69*a*. a) are mounted near the pile top on opposite sides (to cancel gross bending effects) to pick up the acceleration signal of hammer impact. Strain transducers (Fig. 7.69*a*. b) are also attached on opposite sides of the pile (Fig. 7.69*a*. c) near its top. The signals from the two accelerometers and two strain transducers are led through a combination box (Fig. 7.69*a*. d) and cable (Fig. 7.69*a*. e) to an analyzing unit (Fig.

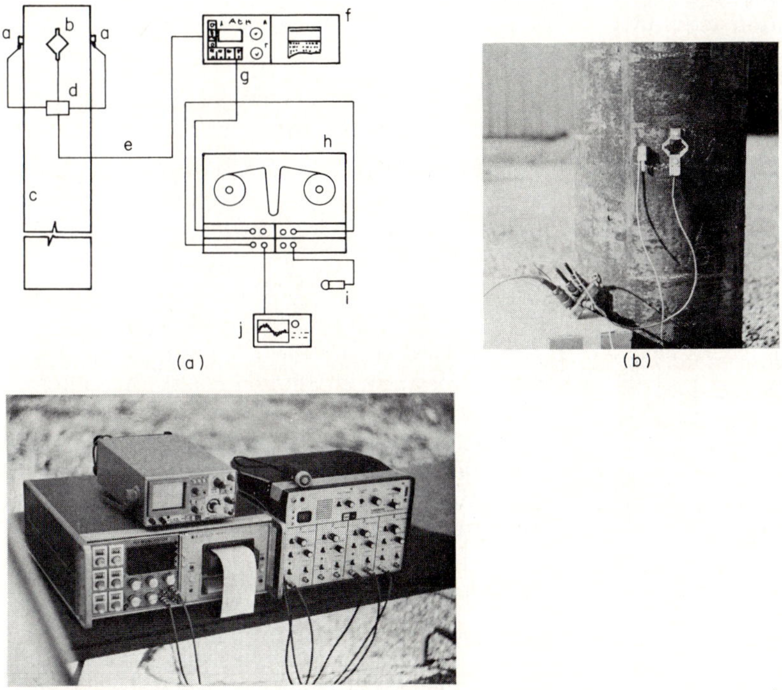

FIG. 7.69 Instrumentation for dynamic pile testing using the Case method. (*a*) Instrumentation elements— a. piezometric accelerometers, b. strain transducers, c. pile, d. combination box, e. cable, f. analyzing unit, g. signals, h. analog magnetic tape, i. microphone for voice records, j. oscilloscope. (*b*) Photo of items a through e. (*c*) Photo of items f, h, i and j. [Part (*a*) From Rausche (1978).[77] Photos courtesy of Petro-Dynamics, Inc.]

7.69a. f) which conditions, amplifies, calibrates, and integrates the signals to provide the following in analog form onsite:

- Maxima of force, velocity, and acceleration
- Peak value of energy transferred to the pile
- Pile bearing capacity according to the Case method of analysis

Signals (Fig. 7.69a.g) are also monitored on an oscilloscope (Fig. 7.69a.j) and stored on an analog magnetic tape (Fig. 7.69a.h) together with voice records (Fig. 7.69a.i). Field data stored on the magnetic tape are later processed to give a simulated load-deflection curve and a force distribution between point bearing and side friction.

Analysis and Applications

Pile Capacity Evaluation

Pile capacity is evaluated dynamically for every hammer blow as the pile penetrates through various soil strata. The instrumentation measures the force and acceleration of the pile providing the dynamic factors necessary for computing resistance, i.e., pile velocity at its top, wave velocity in the pile material, and the force at the pile top.

Driving Stresses and Pile Damage

The forces delivered to the pile are computed automatically by the analyzing unit from the strain readings and from these forces the stresses in the pile from every hammer blow are calculated. Control of hammer throttle settings or steam pressures, or a change in cap block or cushions, or even the hammer, therefore, is possible if damage appears likely. Force and velocity records appearing on the oscilloscope screen provide the skilled observer with information on structural pile damage. Piles already driven to penetration can be tested for structural damage or for long-term setup by restriking with a few blows.

Hammer Performance

Hammer efficiency is accurately evaluated since the energy delivered to the pile is calculated from the force and velocity records, thus providing more reliable input to wave-equation analysis.

Pile-Soil Interaction

The CAPWAP method (Case Pile Wave Analysis Program) uses the velocity records as a pile-top boundary condition and performs a wave equation form of analysis. A soil-resistance distribution is assumed, and if it is correct then good agreement between measured and computed pile-top force is obtained. By iteratively improving the pile-top force match, the most likely set of soil-strength parameters producing the measured data is found. It is also possible to estimate side-friction and end-bearing distribution and the location along the pile where side friction is greatest. The soil constants are then input into the standard wave-equation analysis from which optimum hammer size and blow count to be expected during driving can be obtained.

7.6 LATERAL CAPACITY AND DEFLECTION

7.6.1 FORCES AND RESISTANCE

Lateral Forces

Deep foundations can be subjected to two general forms of lateral forces:

1. *Forces applied at the pile top* result from wind and earthquake loads on tall buildings, transmission towers, or overhead cranes; ship breasting, mooring, or docking forces on waterfront structures; and wind, wave, and current forces on offshore structures.
2. *Lateral loads along the pile shaft* result from earth pressures, a common occurrence where piles support a bridge abutment with a backfill overlying weak soils through which the piles penetrate, failure and lateral movement of an inadequately supported excavation into fill overlying soft soils (such as for a utility trench), or pile-supported viaducts along hillsides subjected to slope movements.

Pile Resistance

Vertical piles resist horizontal forces by deflecting to mobilize their strength and that of the surrounding soil. The horizontal load capacity may be limited by:

- *Ultimate soil capacity* which if exceeded results in very large horizontal movements and foundation failure,

- *Bending moments* which may generate excessive bending stresses in the piles resulting in their structural failure, or
- *Pile-head deflections*, which may be too large to be compatible with the superstructure.

Analysis

Analytical methods presently available for evaluating horizontally loaded deep foundations must be considered as highly empirical, not only from the aspect of how the pile reacts to the imposed forces, but also because of the uncertainties of measuring the soil response.

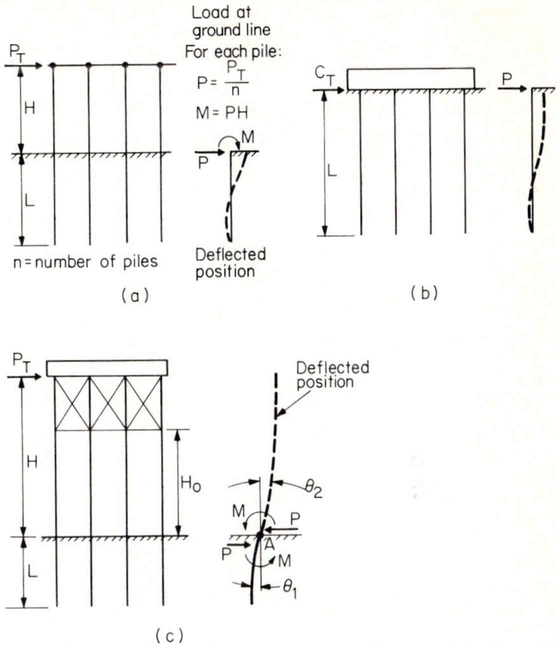

FIG. 7.70 Head conditions for laterally loaded piles: (*a*) free-head or hinged-end condition; (*b*) fixed-head or rigid-cap condition; and (*c*) superstructure elevated above ground line. [From NAVFAC (1982).[1]]

7.6.2 PILE BEHAVIOR

Controlling Factors

Fixity Conditions and Soil-Pile Stiffness

Pile behavior under lateral loads applied at the head is controlled by the degree of fixity imposed at the pile head by the structure, the flexural stiffness of the pile relative to the stiffness of the soil surrounding its upper portion, and by the actual stiffness of the soil itself. As shown in Fig. 7.70, a pile can react either singly with a free-head or hinged-end condition or singly or in a group with a fixed-head or rigid-cap condition. Based on the load-deformation characteristics of the soil and pile, a pile is classed in a general manner as either a rigid (short) pile or a flexible (long) pile.

Short, rigid piles are sufficiently stiff to rotate under lateral load without appreciable distortion from their axis, and the lateral rotation at depth may be described with reference to a center of rotation as shown in Fig. 7.71. For this case to prevail the base must be *free to rotate.* The maximum horizontal load that can be applied to the pile is limited by the maximum horizontal reaction that can be mobilized in front of the pile. Analysis is by semiempirical methods or by use of pressuremeter data.

Long, flexible pile behavior, shown on Fig. 7.72, is considered to be closely related to the behavior of a beam on an elastic foundation *(see Art. 4.4.1).* Capacity is not usually considered to be limited by soil capacity, but rather by the magnitude of deflection and resulting bending moments in a pile. The principal problem is to compute the amount of deflection that the pile or pile group can be expected to undergo, from which are computed the moments and shears on which to base the selection of pile cross section necessary to resist the bending forces. When these forces become excessive, selected piles usually are designed inclined to the vertical (battered) to provide axial reaction to the lateral load component.

Soil Restraint

For a pile cap resting on the surface, it is reasonable to allow a horizontal restraint at the top of each pile of as much as 1500 lb if the soils have an N value greater than about 5, for silt or clay, and for sands irrespective of D_R [Peck et al. (1974)[2]].

In denser or stiffer soils greater loads can be resisted, but the use of higher values requires careful study of soil conditions and structural requirements. Soil reaction may be presented by a

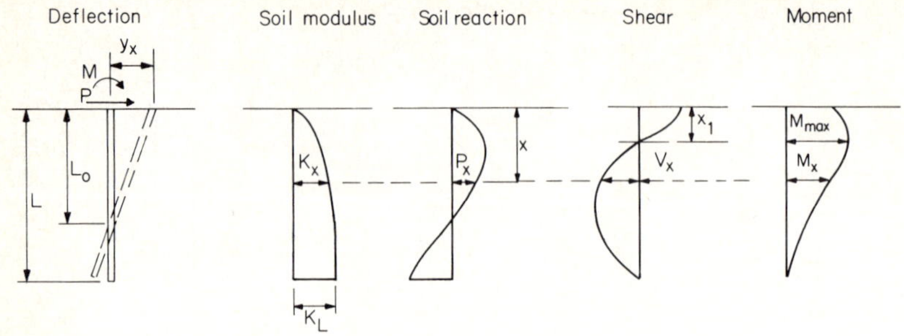

FIG. 7.71 Rigid pier behavior. Distinction between elastic and rigid behavior occurs at point where $M_{max} = M_{yield}$ of pile section. M_{max} is located at x_1. [From Woodward et al. (1972).[9] Used with permission of McGraw-Hill Book Company.]

For cohesionless soils: $x_1 = 0.82(P/dK_p\,\alpha)^{1/2}$

For cohesive soils: $x_1 = (p/9s_u d) + 1.5d$

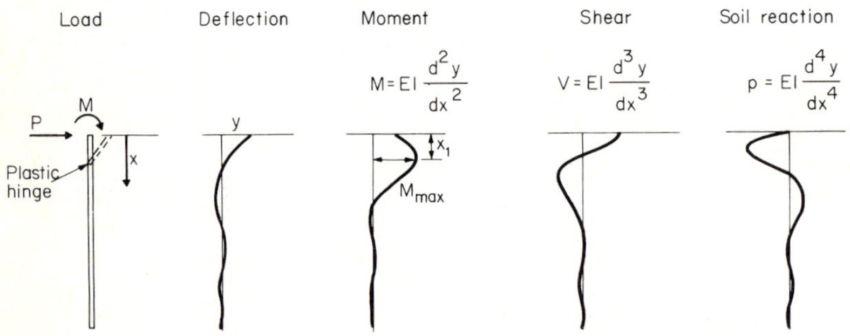

FIG. 7.72 Elastic pier behavior. [After Matlock and Reese (1960).[78]]

coefficient of subgrade reaction or elastic modulus.

Batter Piles

When horizontal forces exceed those that can be withstood by vertical piles alone, selected piles are designed inclined to the vertical (battered) to provide axial reaction to the lateral load component. The horizontal component of the axial load is then considered to be the horizontal load per batter pile. Common batters range from 1*H*:12*V* to 5*H*:12*V*; batters in excess of 3*H*:12*V* require special driving equipment.

Elastic Versus Rigid Behavior

Soil-Pile Interaction

From the theory of a beam-on-an-elastic subgrade (*see Art. 4.4.3*), the soil reaction to a lateral load may be expressed as

$$EI\frac{d^4y}{dx^4} + P_x\frac{d^2y}{dx^2} + p = 0 \qquad (4.75)$$

and if the buckling load is ignored, as is usually the case, Eq. 4.75 is expressed as

$$EI\frac{d^4y}{dx^4} + p = 0 \qquad (7.71)$$

where p = the soil reaction per unit length ($p = k_s = k_h x$, or $p = -E_s y$)
k_h = horizontal modulus of subgrade reaction (*see Art. 4.4.3*)

From Eq. 7.71, the demarcation between rigid and elastic pile behavior can be estimated in terms of a *relative stiffness factor* (R or T) which expresses a relation between pile flexural stiffness and soil stiffness and is dependent upon the soil modulus of subgrade reaction. For a soil with a modulus *increasing linearly with depth* (cohesionless soils and NC clays) the stiffness factor T is expressed as

$$T = \left(\frac{EI}{n_h}\right)^{1/5} \tag{7.72}$$

where n_h = the constant of horizontal subgrade reaction (Fig. 4.44), and for soil with a *constant modulus* (OC clays), the stiffness factor is expressed as

$$R = \left(\frac{EI}{k_s}\right)^{1/4} \tag{7.73}$$

Short Versus Long Piles

Kezdi (1975)[17] defines short piles as those with an L/B ratio of less than 10. Morgan and Poulos (1968)[79] state that the distinction between long and short piles occurs at the point where the maximum moment (M_m) becomes equal to the yield moment (M_y) of the pile cross section. In other words, for short piles $M_y > M_m$. They give the depth at which the maximum moment occurs (Figs. 7.71 and 7.72) as follows:

Cohesionless soils:

$$x_1 = 0.82\left(\frac{P_u}{BK_p\gamma}\right)^{1/2} \tag{7.74}$$

Cohesive soils:

$$x_1 = \frac{P_u}{9c_uB} + 1.5B \tag{7.75}$$

where P_u = ultimate capacity
c_u = undrained strength
B = pile diameter
K_p = passive earth-pressure coefficient

Both theoretical analysis and load test results indicate that the soil properties from the surface down to a depth equivalent to about $2.5R$ for free-head piles and $1.5R$ for fixed-head piles essentially control behavior. For large-diameter piles or piers, Woodward et al. (1972)[9] define short versus long members in terms of stiffness factors R and T as follows:

Constant modulus, free-head:

Short (rigid) pier: $L < 2R$

Long (elastic) pier: $L \geq 3.5R$

Linearly increasing modulus, free-head:

Short pier: $L \leqq 2T$

Long pier: $L \geqq 4T$

7.6.3 ULTIMATE LATERAL CAPACITY

Short, Rigid Piles

In most practical situations the primary concern is with lateral deflections of long, flexible piles, but on occasion it is necessary to analyze the capacity of a short, rigid pile, such as a straight-shaft, nonsocketed bored pile.

Lateral Soil Bearing Capacity q_h

The lateral soil bearing capacity q_h at a given depth along a pile in cohesionless soil can be expressed as a function of the overburden pressure $\bar{p}_o$ and passive earth pressure K_p, and in a cohesive soil as a function of the undrained shear strength c_u. Broms (1964)[80,81] proposed the following relationships for q_h:

Cohesionless soil:

$$q_h = 3K_p\bar{p}_o \quad (\text{for } K_p \text{ see Art. 3.2.3}) \tag{7.76}$$

Cohesive soil:

$$q_h = 9c_u \quad (\text{see Art. 7.3.3}) \tag{7.77}$$

These relationships apply to short-term loadings, as would be the case, for example, of an overhead crane supported on a single pier, subjected to crane and wind loads. Equation 7.76 applies to both long-term and short-term loadings since pore pressures dissipate rapidly in a free-draining soil. For long-term conditions in cohesive soils, the parameters $\bar{c}$ and $\bar{\phi}$ apply.

Ultimate Lateral Capacity P_u: Pile Free to Rotate

The ultimate lateral capacity of short, rigid piles, free to rotate (Fig. 7.73) can be found from static equilibrium methods as follows.

1. *Free-head condition:*
   - Cohesionless soil:

$$P_u = \frac{0.5BL^3K_p\gamma'}{H + L} \tag{7.78}$$

   - Cohesive soil:

$$P_u = \left(\frac{L_o'^2 - 2L'L_o' + 0.5L'^2}{L' + H + 1.5B}\right)(9c_uB) \tag{7.79}$$

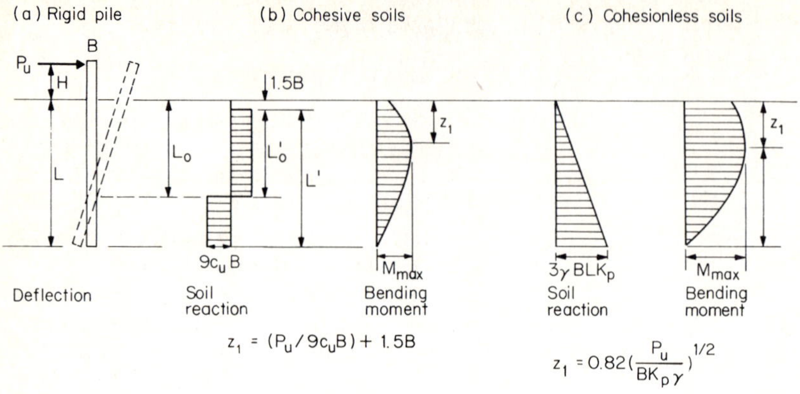

FIG. 7.73 Free-head, short, rigid piles—ultimate load conditions. [After Broms (1964).[80,81]]

where L = embedded length of pile
H = distance of resultant lateral load above ground surface
B = pile diameter
L' = embedded pile length measured from a depth of 1.5B below ground surface
L_o = depth to center of rotation

$$= (H + 2/3L)/(2H + L) \qquad (7.80)$$

L_o' = depth to center of rotation measured from a depth 1.5B below the surface

2. *Fixed-head conditions (Fig. 7.74):*
- Cohesionless soils:

$$P_u = 1.5\gamma' BL^2K_p \qquad (7.81)$$

- Cohesive soils:

$$P_u = 9c_uB(L - 1.5B) \qquad (7.82)$$

Note: For a pile group, B is interpreted as the width of the group perpendicular to the direction of force P [Hansen (1961)[82]].

Pressuremeter Analysis

Menard (1962)[83] considered that there is a close analogy between soil behavior around a short rigid laterally loaded pile and the pressuremeter test results. He provides the following relationship, which has been confirmed by full-scale load tests, for computing P_u:

$$P_u = P_LB(L - B) \qquad (7.83)$$

where P_L = limit pressure from pressuremeter tests

Long Flexible Piles

Failure Conditions

For long flexible piles (Fig. 7.72) the ultimate lateral resistance is a function of the flexural resistance of the shaft. Failure occurs when the maximum resisting moment or yield moment M_y of the pile is exceeded by the maximum imposed moment M_m and a plastic hinge develops at the location of the maximum positive moment.

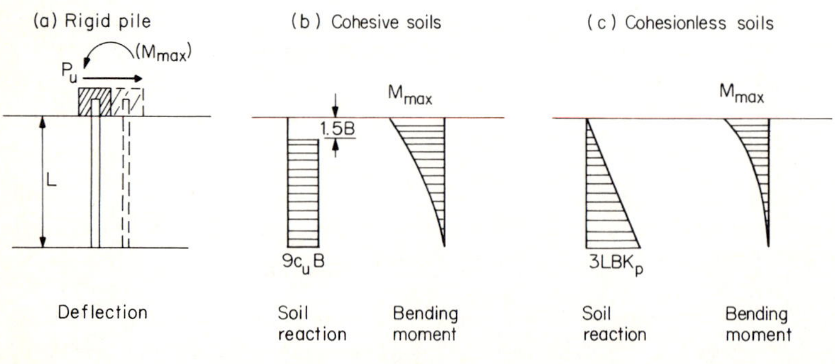

FIG. 7.74 Fixed-head, short, rigid pile—ultimate load conditions. [After Broms (1964).[80,81]]

Ultimate Lateral Capacity

In analysis it is assumed that the ultimate soil resistance will develop on the portion of the pile *above* the hinge and can be expressed as follows.

1. *Free-head conditions:*
   - Cohesionless soils:

$$P_u = \frac{M_y^+}{H + 0.54(P_u/\gamma' B K_p)^{1/2}} \qquad (7.84)$$

   where M_y occurs at approximately 0.82 $(p_u/\gamma' B K_p)^{12}$ (Eq. 7.75)
   - Cohesive soils:

$$P_u = \frac{M_y^+}{H + 1.5B + f} \qquad (7.85)$$

   where M_y occurs at a depth below the surface of $(f + 1.5B)$, and

$$f = P_u/9c_uB \qquad (7.86)$$

2. *Fixed-head conditions:*
   - Cohesionless soils:

$$P_u = \frac{2M_y^+}{H + 0.54(P_u/\gamma' B K_p)^{1/2}} \qquad (7.87)$$

   where M_y occurs approximately at 0.82 $(P_u/\gamma' B K_p)^{12}$
   - Cohesive soils:

$$P_u = \frac{2M_y^+}{1.5B + 0.5f} \qquad (7.88)$$

   where M_y occurs at depth f (Eq. 7.86)

Note: In practice, long, elastic pile capacity is normally governed by the magnitude of deflection and bending moments in the pile, rather than by soil capacity.

Other Conditions

Solutions for piles "intermediate" between short and long, for socketed piles, and for piles subjected to inclined loadings can be found in Poulos and Davis (1980).[25]

References

Discussions concerning the ultimate capacity of long and short piles can be found in Hansen (1961),[82] Broms (1964),[80,81] Morgan and Poulos (1968),[79] Woodward, Gardner, Greer (1972),[9] Kezdi (1975),[17] and Poulos and Davis (1980).[25]

7.6.4 PILE DEFLECTION ANALYSIS

Approaches to the Problem

The problem of the deflection of long, flexible piles has been approached by many methods (see *Art. 4.4.3*), but the more common current solutions are based on the beam-on-an-elastic-subgrade theory as applied by Reese and Matlock (1956)[84] and Matlock and Reese (1960).[78] Simple solutions involve hand computations and nondimensional charts with soil load-deformation relationships given in terms of a soil modulus of subgrade reaction (*see Art. 4.4.3*). Complex solutions are based on nonlinear relationships and require computer analysis. Soil load-deformation relationships are modeled in terms of p-y curves (*see Art. 4.4.3*), and analysis can provide for complex layered soil systems, variable pile structural stiffness, variable head fixity, and the stiffness of the supported structure. Other methods, such as those that assume a constant moment with depth, or treat the pile as a simple cantilever beam, are considered as irrational and lead to ultraconservative designs.

Soil Moduli

Selecting Values

In the determination of deflections, moments, and shears of a pile employing the beam-on-an-elastic-subgrade theory, the value selected for the soil modulus has a very significant effect on results, and it is the variable most difficult to obtain with reasonable accuracy. As discussed in Art. 4.4.3, the contact pressure along piles in sands and NC clays is generally considered to increase linearly with depth, and in stiff OC clays it is generally considered to be uniform with depth.

Soil moduli are commonly given in terms of the horizontal modulus of subgrade reaction k_h or k_s as follows:

Cohesionless soils:

$$k_s = k_h B \qquad (7.89)$$

where $k_h = n_h z/B$ where n_h = the constant of horizontal subgrade reaction.

Cohesive soils: Various relationships are given; the following is suggested by Francis (1964)[85] for

OC clays in terms of the elastic parameters:

$$K_s = k_h B = \frac{1.60E_s^{13/12}}{E^{1/12}(1 - \nu)^2} \quad (7.90)$$

or as expressed simply by Davisson (1970)[86] for conservative values for clays in general:

$$k_h = 67c_u/B \quad (7.91)$$

Expressions and tables and charts for the estimation of k_h and n_h are given in Art. 4.4.3. Pressuremeter testing is useful for evaluating k_h, but the most reliable values are obtained from lateral load tests.

Lateral Load Tests

Lateral load tests may be run on two adjacent piles by either pulling the piles together or jacking them apart. The latter procedure will minimize the effects of overlapping stresses if the piles are closely spaced. Such tests were performed on the piles under the reaction blocks in the photograph in Fig. 7.60.

In most cases it is not sufficient to measure only the displacement of the pile head versus the applied horizontal load, which is done by theodolite and precision dial gages. Bending strains are measured by strain gages placed on reinforcing steel or the body of a steel pile, and deflections as a function of depth may be measured by an inclinometer run through previously installed casing (Fig. 7.63). The deflection with depth and the soil profile for a 130-cm-dia. bored pile are given in Fig. 7.75. By extrapolation from the curves, values were obtained for a mean coefficient of horizontal subgrade reaction at a lateral load of 10 t at a depth of 10 m for tests run on 70- and 130-m piles as follows: 70-cm pile, $k_{10} = 5.5$ kg/cm²; 130-cm pile, $k_{10} = 2.5$ kg/cm². The value for the 70-cm pile represents a 50% reduction to account for a low water table during the test which will rise to pile-cap level after construction. From the pile head to a depth of 10 m the coefficient k_d for design was expressed as

$$k_d = k_{10}\left(\frac{x}{10}\right)^{1/2}$$

where x = desired depth. Below a depth of 10 m, $k_d = k_{10}$ in a uniform distribution. A check using Terzaghi's value of $n_h = 14$ for medium-compact saturated sands (Table 4.16) to compute k_h at a

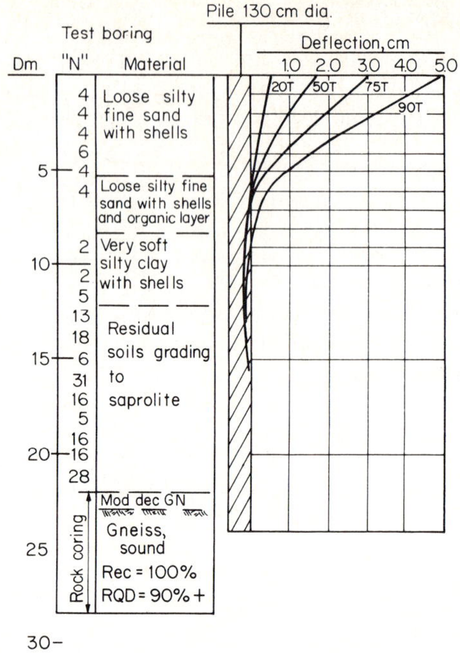

FIG. 7.75 **Deflection under various loads—horizontal load test on 130-cm-dia. pile.**

depth of 10 m for a pile with a diameter of 130 cm yields

$$k_h = 14 \text{ t/ft}^3 \times 0.032 \times 1000 \text{ cm}/130\text{cm} = 3.5 \text{ kg/cm}^2$$

For studies of dynamic horizontal forces, cyclic loads are applied.

Simple Analytical Methods

Cohesionless Soils and NC Clays

Most problems involving laterally loaded long piles may be analyzed by the relatively simple method presented by Reese and Matlock (1956).[84] Used for the case of k_h increasing linearly with depth (cohesionless soils and NC clays), it requires as the initial steps obtaining values of the soil modulus constant n_h and the calculation of the *relative stiffness factor* T given as Eq. 7.72. T is also referred to as the *characteristic length.*

Dimensionless coefficients have been developed for deflection, moment, and shear for various cases of the ratio of pile length L to the effective length L/T and these are plotted as a function of a depth coefficient Z ($Z = z/T$) both for the applied moment M_t and the applied lateral force P_t.

Free-head conditions for $L \leqq 4T$ are provided for in the charts given as Fig. 7.76 which present curves relating L/T to Z and the dimensionless coefficients for deflection y and moment m. In the analysis the appropriate curve is selected for L/T and the coefficients B_y, B_m and A_y, A_m for the applied moment and applied lateral force are obtained for various depth increments Z. (On some versions of the chart L/T is given as Z_{max}.) The moments and deflections for various depths are then computed using the coefficients and the following expressions:

Moment:

$$m = A_m P_t T + B_m M_t \tag{7.92}$$

Deflection:

$$y = A_y \frac{P_t T^3}{EI} + B_y \frac{M_t T^2}{EI} \tag{7.93}$$

Fixed-head condition for $L \leqq 4T$ for the case of 100% fixed is included in the charts given as Fig. 7.77. The procedure is similiar to the free-end case except that there is no applied moment and the maximum shear occurs at the pile top and is equal to the ratio of the total lateral load divided by the number of piles, or $p = p_t/n$ and

Moment:

$$M_p = F_m(PT) \tag{7.94}$$

Deflection:

$$y = F_y \left(\frac{PT^3}{EI}\right) \tag{7.95}$$

where F_m and F_y are the moment and deflection coefficients at depth z.

The case of the structure elevated above the ground surface (Fig. 7.70) can be found in NAVFAC (1982).[1]

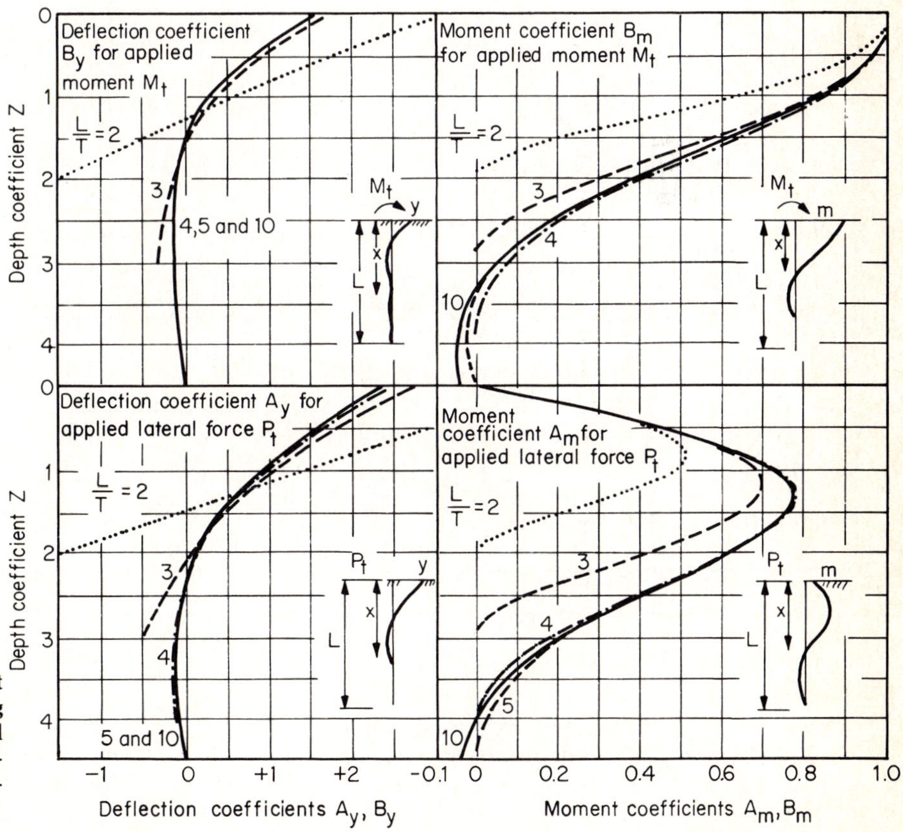

FIG. 7.76 Deflection and moment coefficients for modulus increasing linearly with depth—free-head conditions. [After Reese and Matlock (1956),[84] as given in Woodward, Gardner, and Greer (1972).[9]]

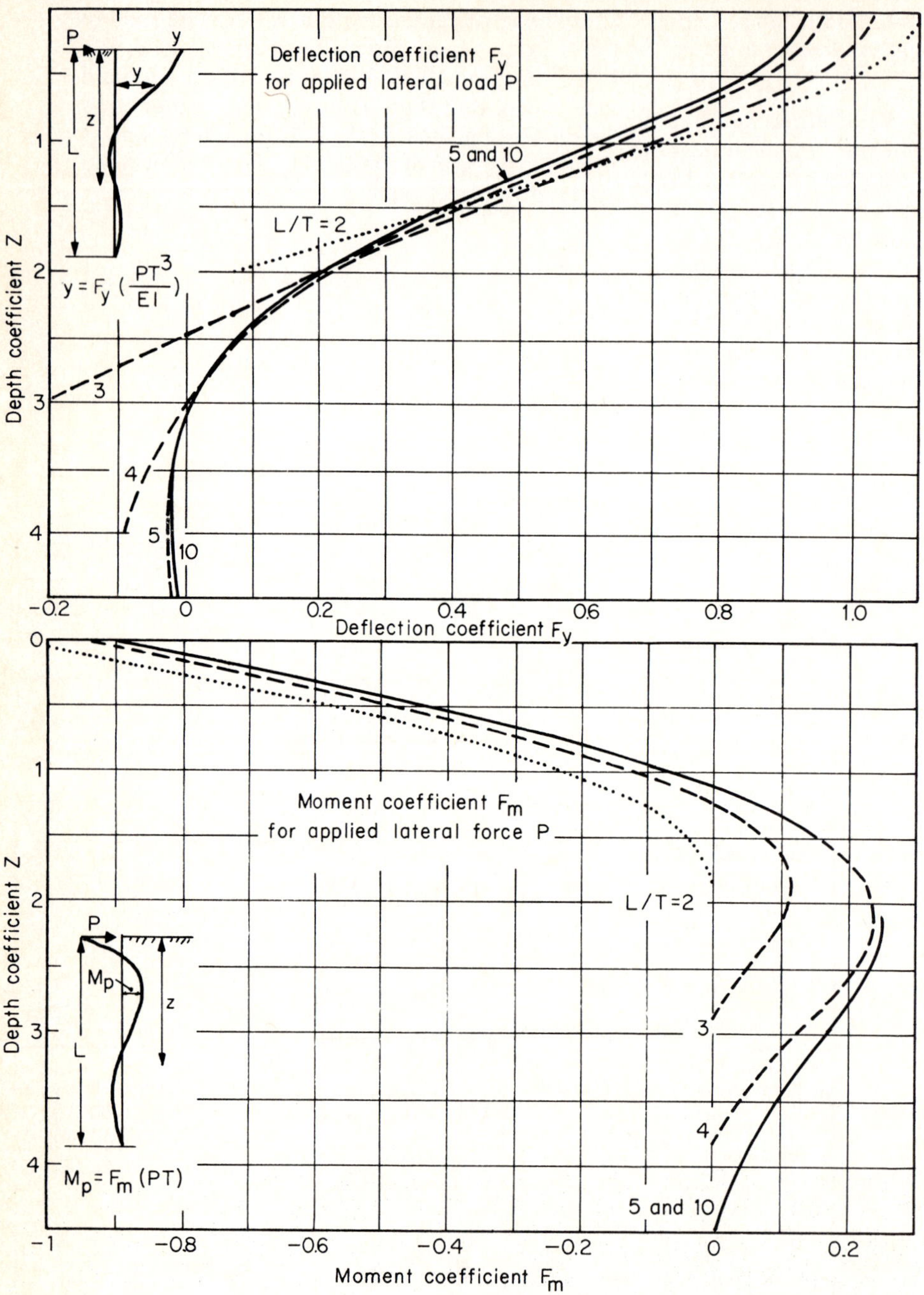

FIG. 7.77 Influence values for laterally loaded pile fixed against rotation at the ground surface. [From Reese and Matlock (1956).[84]]

Very long piles ($L \geq 4T$) have been addressed by Matlock and Reese (1961),[87] who express deflection and moment as follows:

Moment:

$$M = C_m(PT) \tag{7.96}$$

Deflection:

$$y = C_y\left(\frac{PT^3}{EI}\right) \tag{7.97}$$

where C_m and C_y are moment and deflection coefficients (Figs. 7.78 and 7.79) plotted as functions Z and M/PT.

Depending upon the angular restraint provided by the pile cap, values of M/PT will range from zero for the free-head case to -0.93 for the fixed head case [Poulos and Davis (1980)[25]]. Davisson (1970)[86] suggests that the degree of fixity that can usually be developed is $M/PT = -0.4$ to -0.5, for which the positive and negative moments in the pile are approximately equal.

Soil Modulus Constant with Depth (OC Clays)

For the case of soil modulus constant with depth (stiff to hard clays), the NAVFAC Manual (1982)[31] adjusts n_h by the relationship given in Eq. 4.68, and then uses the adjusted values to find T and subsequently the coefficients given on the charts in Fig. 7.76 and 7.77.

Davisson and Gill (1963)[88] have performed complete solutions to develop similiar charts presenting deflection and moment coefficients. In the use of these charts, also found in Woodward, Gardner, Greer (1972),[9] the soil modulus is used to determine the pier stiffness factor R as given in Eq. 7.73.

Limitations

The linear relationship of soil moduli provides only a rough approximation of deflection and moment since, even if moduli are used that are functions of depth, it is not possible to consider

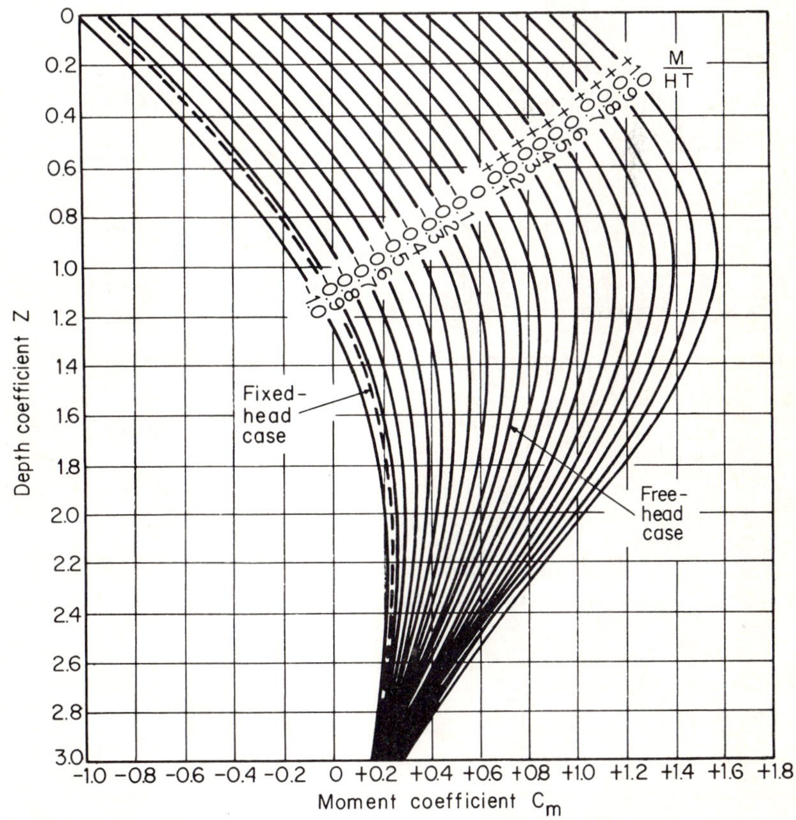

FIG. 7.78 Curves of deflection coefficient C_m for very long piles. [After Matlock and Reese (1961),[87] as given in Poulos and Davis (1980).[25] Used with permission of John Wiley & Sons, Inc.]

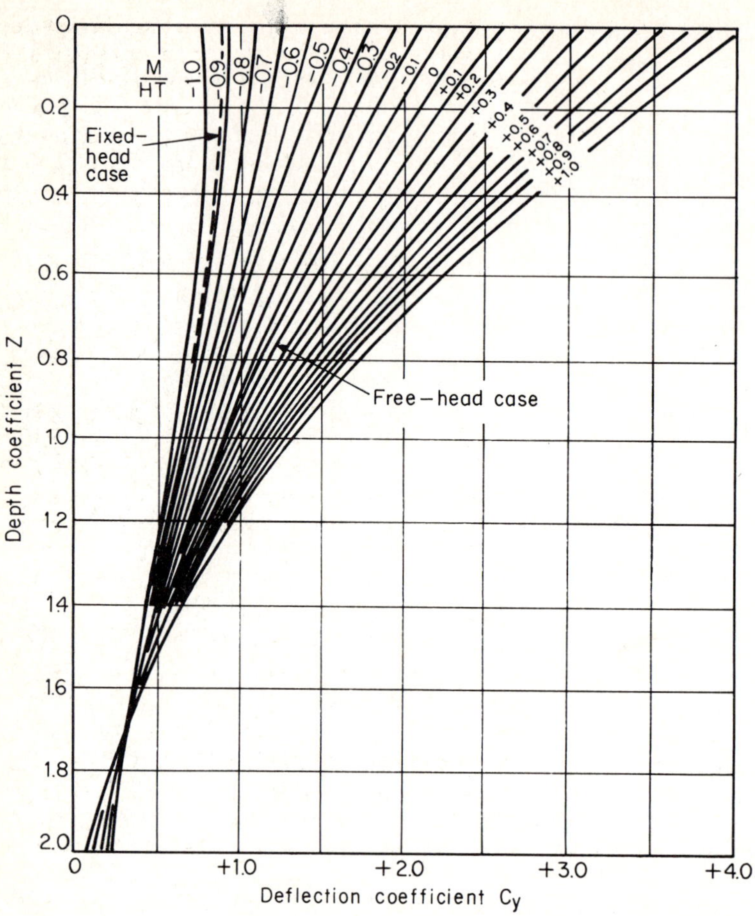

FIG. 7.79 Curves of moment coefficient C_y for very long piles. [After Matlock and Reese (1961),[87] as given in Poulos and Davis (1980).[25] Used with permission of John Wiley & Sons, Inc.]

soil stratification, time effects, the effect of load intensity, and structural factors.

Finite Difference Solution and p-y Curves (Nonlinear Analysis)

General

The more complete solution to the laterally loaded pile problem is based on finite difference equations (*see Art. 4.4.3*). As it can provide for complex layered soil systems, variable pile structural stiffness, variable head fixity, and the stiffness of the supported structure, it is a strong tool for evaluating offshore structures which are usually subjected to large lateral forces which often govern the foundation design.

The problem has been generalized for computer solution in beam-column type programs by Matlock and Reese (1960).[78] A recent program is coded COM 622 [Reese (1977)].[89] (*See Art. 4.4.3* for source.)

Solution

To solve the problem requires prediction of a set of *p*-*y* curves as described in Art. 4.4.3. With the curves, the finite difference equations can be solved to yield deflection, pile rotation, bending moment, shear, and soil reaction for any load capable of being sustained by the deep foundation.

In the solution the pile is divided into spring segments, each of which experiences lateral deflection *y* and soil resistance *p* in units of force per unit pile length, as shown in Fig. 4.47. The differential equation representing the pile as a linearly elastic beam and the soil reaction as a line load is given as Eq. 4.79. In solving the equation for clays a secant modulus of soil reaction is often used, i.e., $E_s = -p/y$. For cohesionless soils, E_s often is assumed to vary linearly with depth; i.e., $E_s = Kz$, where K is a constant. The concept of these *p-y* curves is presented in Fig. 4.50: part *a* is a section through a deep foundation below the ground surface, part *b* shows the stress distribution around the pile before loading at some depth x_i, and part *c* shows the stress distribution around the pile after lateral loading. The deflection of the pile through distance *y* (Fig. 4.51) generates unbalanced soil stresses; integration of the soil stresses yields an unbalanced force *p* per unit length of pile. (It should be noted that repetitive or cyclic loads in NC clays produce a softening of the soil resistance with a peak resistance after cycling being only about 0.7 of the resistance under a single static load.)

Once the *p-y* data are known, the bending moments and deflections are computed using beam-column type computer programs. Generally, piles are assumed to be partly fixed-headed in an offshore superstructure, but lower degrees of fixity can be considered. The computer program follows an iteration procedure to calculate pile moments and deflections for any given set of *p-y* data, proceeding until compatibility exists between soil modulus tried and soil modulus obtained.

Comparison with Simpler Methods

In the analysis of step-tapered piles bearing in a layered soil system, Sanders (1976)[90] found that for a given set of conditions the results from analysis with the simple approach required about 52% more steel in the pile to resist bending moments than the results from using the COM 622 computer program.

Buckling

Failure by pile buckling is an uncommon occurrence, since the lateral support provided by practically any soil is adequate for its prevention. Long piles extending through water or very soft materials are susceptible, however, and require analysis. Bjerrum et al. (1957)[91] suggested that buckling needs to be considered only if

$$\frac{I}{A^3} < \frac{P_{\max}^2}{4cE} \tag{7.98}$$

where c = coefficient of lateral displacement equal to the horizontal subgrade reaction times B, the pile width.

Buckling analysis is described in Poulos and Davis (1980).[25]

7.6.5 LATERAL LOADS ALONG THE PILE SHAFT

Some Field Conditions

There are cases where lateral forces resulting from earth pressures can result in excessive bending and even failure in shear. Such conditions are particularly troublesome where the piles penetrate soft clay or other weak materials, or unstable slopes.

Excavations

One form of excavation problem occurs when piles are driven through fill overlying soft clays and an excavation is made adjacent to the piles or within the group. On one occasion a trench was excavated adjacent to a group of precast concrete piles which resulted in a lateral movement of fill over soft organic soils causing the piles to fail in bending. In another case an excavation was made into a timber pile group in preparation for cutting the piles and preparing the conical form for the bottom of a concrete tank in a sewage treatment plant. Deformation of the soft soil under the fill caused the piles to incline toward the center of the excavation as it proceeded downward.

Bridge Abutments over Soft Ground

Another case where consideration of bending stresses in piles is required is where a bridge abutment is supported on piles penetrating soft soils and retaining a backfill load.

Viaducts on Unstable Hillsides

Viaducts supported on piers passing through unstable soils on slopes may be subjected to very high lateral forces, the most common cause of which is creep.

Bridge Abutment over Soft Soils

Analysis

The stability conditions of a pile-supported bridge abutment retaining fill over a weak soil stratum are illustrated on Fig. 7.80*a*. Tschebotarioff (1973)[92] provides a criterion to be used to

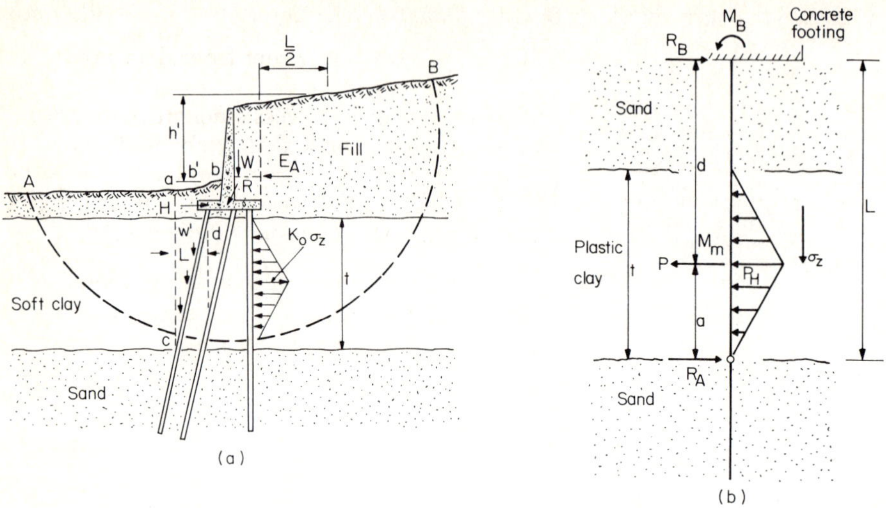

FIG. 7.80 The stability problem of a pile-supported retaining wall over soft soils: (*a*) general stability conditions and (*b*) bending stresses in piles under the heel of the abutment. [From Tschebotarioff (1973).[92] Used with permission of McGraw-Hill Book Company.]

evaluate when special attention to excessive bending stresses is required; that is, when the weight of the embankment fill behind the wall exceeds the yield point p_y of the clay, taken as $p_y = 3c_u$. To analyze the problem a triangular distribution of lateral earth pressure is assumed (Fig. 7.80*b*) in which the maximum value of p_H occurs at the center of the clay layer; it is expressed as

$$p_H = K_o\sigma_z B \tag{7.99}$$

where K_o = 0.4, the coefficient of at-rest earth pressure

σ_z = the increment of vertical stress at the center of the clay layer under the heel of the embankment due to the backfill weight

B = pile width

The bending moment in the pile can then be estimated from the following expressions pertaining to structures:

$$M_B = -\frac{Pa(L^2 - a^2)}{2L^2} \tag{7.100}$$

$$M_M = \left(P\frac{a}{2}\right)\left(2 - \frac{3a}{L} + \frac{a^3}{L^3}\right) \tag{7.101}$$

where
$$P = 0.9\,\frac{p_H t}{2} \tag{7.102}$$

The resulting bending stresses in the piles account for the observed tilting of pile-supported abutments toward their backfill. The abutment heel will settle more than the toe, thus inducing a rotational movement of the abutment and a horizontal displacement at the level of the bridge seat.

Solutions

Tschebotarioff (1973)[92] recommends that for these cases, nondisplacement piles, such as H piles, should be used to minimize soil remolding, that they should be oriented to provide resistance to bending with their maximum section, and that piles of reverse batter should be installed to ensure stability in all directions.

The problem can also be relieved by preloading the weak soils (*see Art. 5.3*) to increase their strength, or by constructing spill-through abutments which reduce the lateral pressures in the soft soils.

Hillside Viaducts

Conditions

Viaducts supported on long columns and deep piles require protection against creep and other forms of slope instability. The Immigrantes Highway between Santos and São Paulo, Brazil, climbs the coastal mountain range where slope conditions are notoriously unstable. The viaducts, as long as 1900 m, are supported on piers as high as 100 m and extending as deep as 20 m to bear on rock [ENR (1977)].[93]

Solution

To relieve the creep pressure from the piers, each was surrounded by an off-center reinforced concrete cylinder of up to almost 5 m ID that extends through the anticipated creep zone. The ring is offset to provide about 1.2 m of space between the pier and the uphill slope to permit slope movements, but without earth pressure being exerted on the piers. Considering a 70-year structural life, 1.2 m permits an average movement of only 1.7 cm per year. It may be necessary, therefore, to periodically reconstruct the ring to provide space for slope movements, the success of which depends on periodic observations of slope movements.

7.6.6 GROUP ACTION

General

Group action under laterally applied loads is not well defined. Although theoretical concepts have been developed, very little instrumented performance of pile groups under either load test or actual structures has been published in the literature to confirm theoretical analysis. It would be convenient for analysis to establish single-pile versus group efficiency values, but such values are difficult to obtain, even during load tests, because of the problem of equating single-pile head fixity with fixity in the group. Analytical methods usually assume a single-pile deflection equal to the deflection in the group to evaluate bending moments.

Ultimate Capacity

Analytical Approaches

Poulos and Davis (1980)[25] suggest taking the lateral load capacity of a single pile times the total number of piles in the group or the lateral load capacity of an equivalent single block containing the piles in the group and the soil between them together with the block width perpendicular to the loading direction. In the latter case, for example in Eqs. 7.81 and 7.82, B is interpreted as the width of the group perpendicular to the direction of force P [Hansen (1961)[82]].

Group Efficiency

Not much is known except from data provided by some model pile tests, but groups under lateral loads appear to be less efficient than groups under axial loads. Model tests show efficiencies for four to nine pile groups ranging from 0.4 to 0.7 varying with soil type and the L/B ratio. Poulos and Davis (1980)[25] define efficiency E as

$$E = \frac{\text{ultimate lateral load capacity of group}}{P_u \text{ for single pile} \times \text{number of piles in group}}$$

Deflection

Simple Approach

It has been considered that group action produces a reduction in the coefficient of subgrade reaction k_s which is a function of pile spacing in the direction of loading [CGS (1978)[6]] as given in Table 7.22. Pile spacing normal to the direction of loading is said to have no influence provided it is greater than $2.5B$. Poulos and Davis (1980),[25] however, consider that, because the subgrade-reaction model is not continuous, the effects of interaction between piles cannot be properly considered.

Pile Interaction Analysis

Analytically the problem of pile-group deflection under lateral loads has been approached by considering the interaction between piles [Poulos (1971)[94]]. It is an extension of the elastic analysis of single piles based on equations derived from the Mindlin (1936)[58] solution for stress distributions. Discussions can be found in Poulos and

TABLE 7.22
SUBGRADE REACTION OF PILE GROUPS RELATED TO PILE SPACING*

Spacing	K_{group}
$8B$	$1.00K_s$
$6B$	$0.70K_s$
$4B$	$0.40K_s$
$3B$	$0.25K_s$

*From CGS (1978).[6]

Davis (1980)[25] and analysis for offshore platforms is described by de Ruiter and Beringen (1978).[16]

A computer program recently developed by the Norwegian Geotechnical Institute, termed SPLICE, is coupled to the structural analysis program FASTRUDL to model cyclic loads on offshore platforms. The analysis can consider group effects, horizontal loads from mudflows, the influence of adjacent loads imposed by nearby jack-up rig foundations, and other conditions [Clausen et al. (1981)[95]]. Lateral analysis is based on the Poulos approach and *p-y* curves.

Based on a study of full-scale load tests on circular pile groups in soft clay, Bogard and Matlock (1983)[95a] present procedures for the development of nonlinear resistance-deflection curves which reproduce both the magnitude and the distribution of lateral resistance with depth reasonably well over a wide range of pile deflections, for both static and cyclic loading of the pile groups tested. The approach is based on modifying values for soil resistance *p* and pile deflection *y* to include effects of group interaction on both the lateral soil resistance and the pile-group deflections.

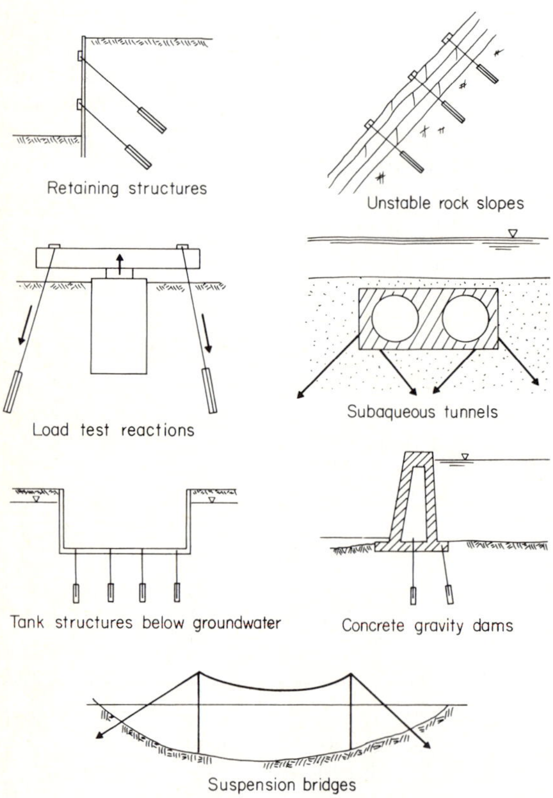

FIG. 7.81 Some applications of soil and rock anchors.

7.7 SOIL AND ROCK ANCHORS

7.7.1 CLASSIFICATIONS

Function

On the basis of function, soil and rock anchors may be divided into two broad categories as follows:

1. To resist *lateral earth pressures:* tiebacks for permanent or temporary retaining structures, or for stabilizing rock slopes or tunnel linings.
2. To resist *uplift,* or *other tensile forces* from wind, hydrostatic pressure, or structural reaction. Examples include reactions for pile load tests, guyed structures, tension roofs, subaqueous tunnels, gravity and other concrete dams, and concrete structures below the water table such as basements and tanks.

Some applications are shown in Fig. 7.81.

Load-Transfer Mechanisms

The purpose of an anchor is to transfer load from a reaction to some existing element in the ground. Two broad categories include

1. *Grouted anchors:* Load is transferred from a tendon to grout and then from the grout to soil or rock by friction along a straight shaft, or in soil bearing against an underream, or both by friction and bearing. Three common types of *tendons* are
   - High-tensile bar, generally used for lengths under 40 ft (12 m) and capacities to 60 t working load
   - Strands or cables of 0.5 to 0.6 in dia. assembled in circular bundles and used for long, high-capacity anchors
   - Button-head wire of 0.25 in dia. assembled in circular bundles (Fig. 7.82) used for intermediate loads and lengths where the exact length of anchor is known because the button heads are factory applied

2. *Mechanical anchors:* Load is transferred directly to the soil by helical auger, expanding bit, or buried plate or footing. Mechanical anchors generally are shallow.

Depth and Failure Mechanism

For analysis, anchors are divided into two broad categories based on depth and the failure mechanism as follows:

1. *Shallow anchors:* Failure is accompanied by a distinct surficial heave or by "breakout." The depth-to-diameter ratio is generally less than 5 to 10.
2. *Deep anchors* show no surficial manifestation of failure and the depth-to-diameter ratio is generally greater than 5 to 10.

Anchor Life

Anchors are considered temporary or permanent depending upon their purpose. *Temporary anchors* commonly are used in sheeted excavations and their life considered as that of the construction period. Any loss in diameter, and therefore in capacity, due to corrosion is provided for by adequate tendon thickness.

Permanent anchors, such as are used for anchored concrete curtain walls *(see Art. 9.2.5)*, however, must have adequate corrosion protection. In Brazil such protection is provided by coatings and wrappings with several materials as shown in Fig. 7.83. Anchor bars are first bathed in a solution to remove rust and other impurities, then coated with a metallic paint followed by an epoxy coating. A steel tube is placed over the bulb or fixed portion of the tendon, and a plastic (usually PVC) tube over the "free" portion. The space between the tubes and the tendon is then grouted with cement. The ends of several anchors ready for installation are shown in Fig. 7.82. To date, after over 25 years of service, anchor corrosion has not been found to be a problem in Brazil.

FIG. 7.82 Cable anchors ready for installation in wall shown in Fig. 9.30.

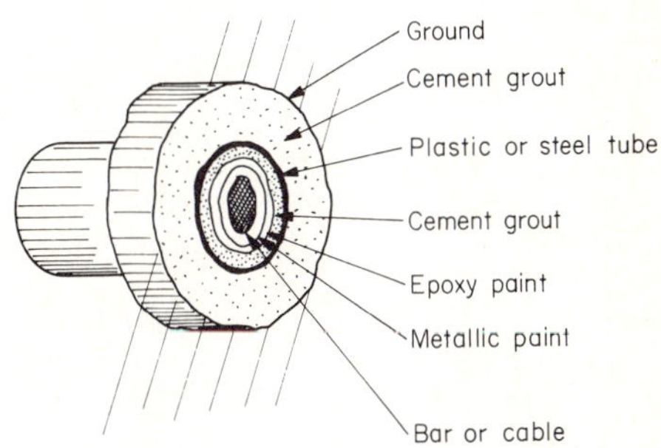

FIG. 7.83 Corrosion protection for anchors in Brazil.

7.7.2 INSTALLATION

Conditions

The major condition for the use of anchors is that a suitable stratum for anchorage exist within economical reach. Weathered or competent rock, or strong soils such as residual soils or glacial till, are the most common anchorage materials. Very stiff to hard cohesive soils and medium-dense to dense sands also provide good anchorages. Stiff cohesive soils and loose granular soils can provide adequate anchorages, but at much reduced capacities.

A severe condition that must be evaluated carefully prior to installation is the presence of underground utilities or other obstacles that will present substantial hazards to anchor installation. In addition, when anchors are to be installed beneath adjacent structures, permission is required of the owner.

Grouted Anchors

Straight Shaft

Grouted, straight shaft anchors can be installed in drilled or augered holes, or by driving.

Rotary drilled hole procedures consist of boring a hole (Fig. 9.30) (rock cores may be recovered in the process), installing casing (not necessary in strong materials such as residuum or rock), inserting the tendon, and injecting neat portland cement grout (without additives) around the lower or fixed portion as the casing is withdrawn. The remaining portion of the tendon (free section) is backfilled with sand or very lean grout, or left open where there is no danger of hole closure. It also can be protected with a sleeve, but in any case the objective is to ensure that there is no load transfer from the fixed portion directly to the anchored structure. A plate is placed on the end of the tendon against the wall and the fixed portion is then posttensioned with a jack to a proof-test load and then the load is locked at the design value. After the wall is completed, the anchor loads are retested and the anchor head covered with concrete for corrosion protection.

Grout pressures in sands range usually between 60 and 200 psi, and high grouting pressures can result in a substantial increase in anchor capacity in some geologic formations, as shown in Fig. 7.88. Care must be taken, however, that pressures are not so high as to cause heaving of the overlying ground surface, especially if structures or pavements are resting there and, as a general rule, should not exceed overburden pressure.

Regroutable anchors can be installed if there is a question as to the capacity that will be attained under the initial grouting pressures. These contain a perforated pipe and the initial grouting is done under low pressure. If the desired capacity is not attained, grout can be reinjected under higher pressures [140 to 420 psi (10 to 30 kg/cm^2)] which cracks the initial grout and permits additional grout penetration. In general, caution is required for pressures over 20 kg/cm^2 (278 psi) to avoid ground heave.

Grouted rock bolts are described in Art. 5.5.3.

Auger-drilled holes are made by hollow-stem continuous-flight augers normally 6 to 10 in dia. The tendon is placed through the hollow stem of the auger before or after drilling, and in cohesionless soils grout is then pumped under pressure through the hollow stem. As the auger is withdrawn the grout fills the hole. In clays the auger is withdrawn before grouting and pressure grouting is not used.

Driven or Bauer injection anchors are illustrated in Fig. 7.84. A 3- to 6-in casing fitted with a slip-fit conical point is driven to the desired penetration. A threaded anchor rod is inserted into the casing and screwed into the point; a grout line is connected and grout injected as the casing is withdrawn by hydraulic jacks. Plastic sleeves over the rod prevent bonding in the free section.

Underreammed Anchors

Belled or underreammed anchors (Fig. 7.87) are drilled by machine auger with an underream device to form a series of bells in stiff to hard clays or soft rock where the hole will not collapse prior to grout injection and where maximum soil-contact area is required to utilize soil strength. The bar is installed manually and the grout pumped or gravity placed into the bond length of the anchor.

Mechanical Anchors

Reaction footing or plate is placed in the bottom of an excavation, a tendon attached, and the hole backfilled with compacted soil when in earth or with concrete when in rock. Such anchors normally are of the shallow type (Fig. 7.85).

Rock bolts, of the mechanical type, are installed in drilled holes by driving or wedging, or driving or expanding *(see Art. 5.5.3)*.

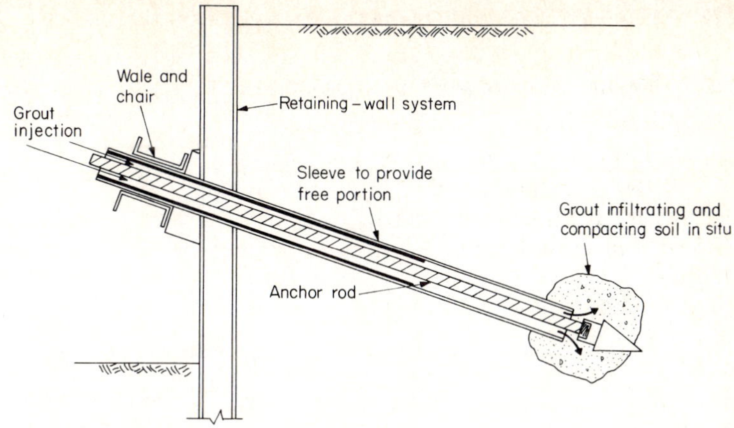

Installation procedure:

1. Drive casing with slip-fit point
2. Place tread bar into casing and screw into point
3. Withdraw casing slightly
4. Inject grout and withdraw casing completely

FIG. 7.84 Installation of the Bauer injection anchor. [From Larson et al. (1972).[96]]

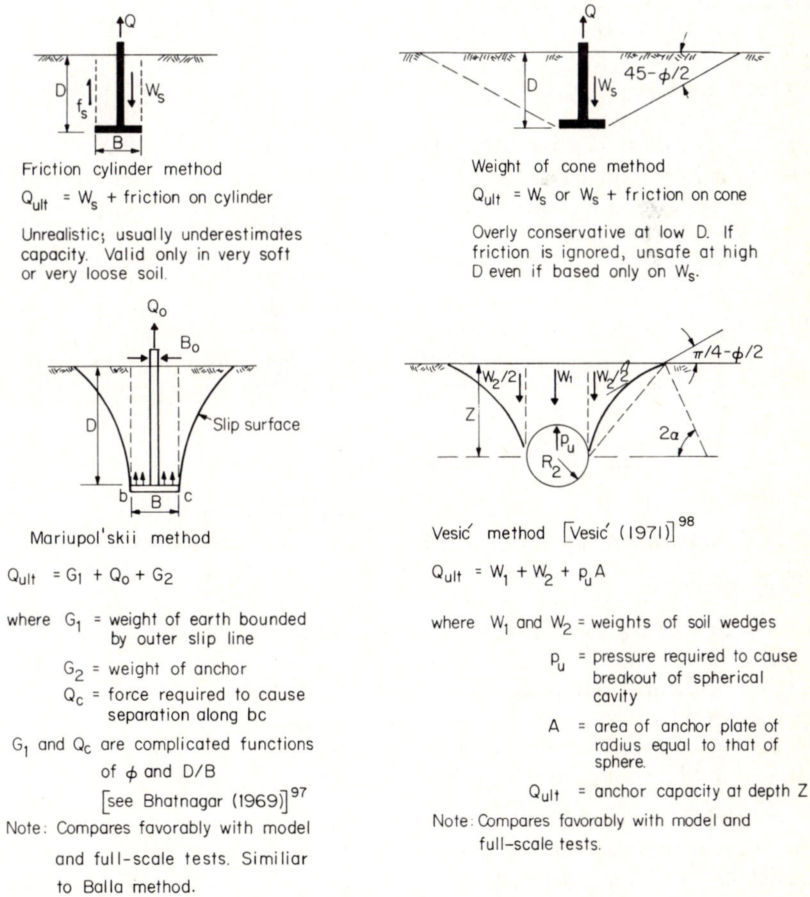

FIG. 7.85 Methods of evaluating the capacity of shallow anchors in soil. [From Murphy (1974).[99]]

7.7.3 FAILURE MECHANISMS AND CAUSES

Shallow Anchors

Shallow anchors loaded vertically break out of the ground by rupture, occurring generally where the depth-to-diameter ratio is in the range of less than 5 to 10. As illustrated in Fig. 7.85, resistance to pullout in soil comes from the weight of the overburden imposed on the anchor in addition to soil strength mobilized along some failure surface. As shown in the figure, there are a number of approaches to defining the failure surface and the soil volume in the breakout zone. Factors for the evaluation of breakout can be found in Bhatnagar (1969)[97] and Vesić (1971).[98]

For vertical anchors *in sound rock* the friction cylinder method (Fig. 7.85*a*) is used to evaluate resistance and

$$Q_{ult} = W + f_s + Q_o \tag{7.103}$$

where W = anchor weight
f_s = shaft friction
Q_o = force required to cause separation along the base B

Note that in deep sockets the upper zone could carry all of the load, as discussed in Art. 7.3.4.

In fractured rock, the NAVFAC (1971)[112] method assumes a breakout zone extending out 30° from the base of the socket of depth D (for inclined anchors, D parallels the base, not the ground surface). Allowable anchor force Q_a is given as

$$Q_a = 2.1\, D^2\, C_{all} \tag{7.104}$$

where C_{all} = allowable rock shear stress, taken as 0.15 tsf, unless pullout tests are performed
D_{min} = 1.25 $(Q_a)^{1/2}$, or minimum depth of socket

Deep Anchors

Plates or footings buried deeply in soil (depth/diameter > 5 to 10) fail in the form of an expanding cylindrical cavity. Two general analytical approaches are used: Mariupolśkii's method [Bhatnagar (1969)[97]] and Vesić's method (1971).[98]

Grouted, straight shaft anchor behavior factors are illustrated on Fig. 7.86. Load is transferred from tendon to grout to soil by friction. Anchor capacity *increases* in proportion to an increase in

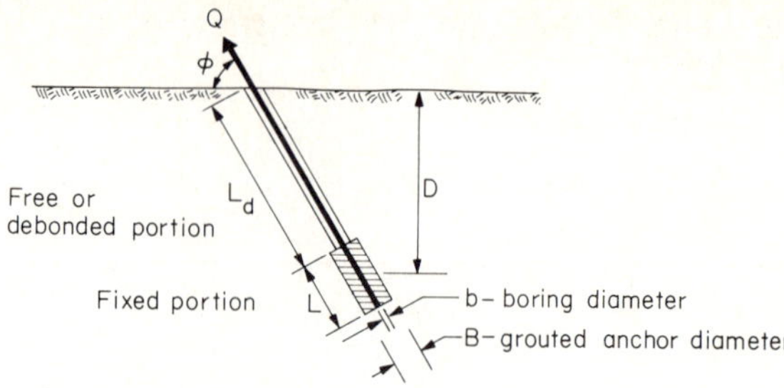

FIG. 7.86 Properties of straight-shaft grouted anchors. [After Murphy (1974).[99]]

anchor diameter, bulb length, overburden pressure, and soil properties, as discussed in the following article. Higher permeability, for example, results in higher grout take, which effectively increases the diameter. Higher grouting pressures can also result in a substantial increase in anchor capacity in some geologic formations as already mentioned. Capacity *decreases* as a result of soil remolding during injection, or consolidation or creep deformation of the soil around the bulb. Release of prestressing due to soil deformation is discussed by Costa Nunes (1966).[100] Corrosion can result in destruction of the tendons as mentioned, or even of the grout. Capacity decrease is discussed also in Art. 9.2.5.

Multiple underreammed anchors (Fig. 7.87) fail when end-bearing resistance of the bells and adhesion or cohesion resistance along the shaft and underream is exceeded. Capacity is discussed in the following article.

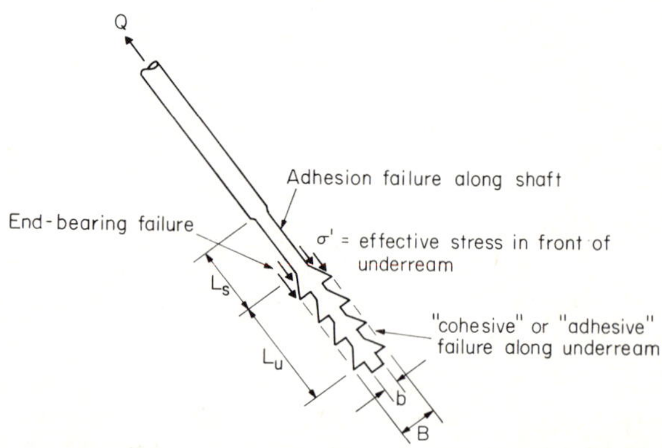

FIG. 7.87 Factors influencing behavior of multiple underream anchors. [After Murphy (1974).[99]]

7.7.4 CAPACITY EVALUATION: DEEP GROUTED ANCHORS

Estimating Capacities

Predicting anchor capacities requires substantial experience and judgment. Difficulties result from uncertain anchor diameters in grouted anchors, the effect of the installation procedure, the amount of developable friction between soil or rock and anchor, and the effective stress acting on the anchor.

Capacities may be estimated from empirical data, or computed on the basis of theoretical relationships. When computed, results should be checked against experience values for similar geologic conditions, and in any case a number of sample anchors should be load-tested as standard procedure prior to production installation. It is generally desirable to make conservative anchor capacity predictions because substitute anchors cost 3 to 4 times as much as production anchors, and in addition result in project delays. About 10% of production anchors fail on difficult projects and about 1% on simpler or routine projects. Regroutable anchors can reduce the failure rate since progressively higher grout pressures are used.

Simple relationships for estimating anchor lengths are given by Dismuke (1975)[102] as follows:

- Pressure-grouted sands and gravels: 6 T/ft
- Pumped or gravity-grouted sands and gravels: 3 T/ft
- Pumped or gravity-grouted clays: 2 T/ft
- Short, belled anchors in clay: 50 T per anchor
- Grouted rock anchor: 4 T/ft
- Typical ultimate capacities for soils: see Table 7.23.

Relationships between ultimate anchor capacity and grout pressures for hard limestone, sand and gravel, silt, and chalk are given in Fig. 7.88.

In the *Rock and Soil Anchor Manual* [Nicholson Anchorage Division of Nicholson Construction Co.)[105] the following high test loads are cited as being held by single anchors: rock, 750 tons; clayey glacial till, 240 tons; shale, 200 tons; sand and gravel, 186 tons; and fine sand, 150 tons. Maximum practical total lengths are given as ranging from 150 to 200 ft.

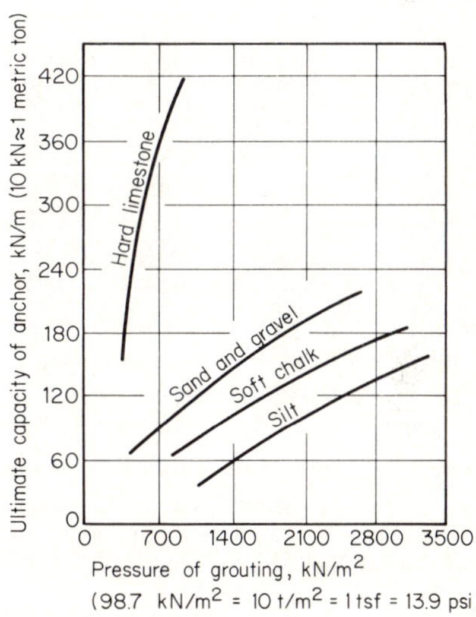

FIG. 7.88 Ultimate anchor capacity versus grout pressure. [From Littlejohn (1970).[101]]

TABLE 7.23
TYPICAL ANCHOR CAPACITIES IN SOILS*,†

Soil type	Ultimate load, T/m	kips/ft
Clean sands/gravels	16–33	10–20
Clean medium to coarse sands	12–25	7–15
Silty sands	8–16	5–10
Very stiff to hard clay	3–10	2–6

*From Goldberg (1976).[103]

†Small-diameter (4-in), high-pressure anchors.

Straight Shaft-Anchor Analysis

There remains a substantial variation among investigators and practitioners regarding the selection of parameters involved in analysis of straight shaft-anchor capacity. Knowledge is particularly incomplete on the distribution of the stresses from the fixed portion of the anchor into the surrounding ground and on the ground disturbance caused by grout injection.

Friction Cylinder Approach

The simplest approach is to ignore the effect of overburden pressures and assume that the bond between the grout and the surrounding material f provides the anchor and resists tie-rod tension T, expressed by

$$T = \pi BLf \qquad (7.105)$$

where B = hole diameter, L = fixed length, and f for design varies from 0.5 tsf (or $0.25U_c$) for stiff clays, to 1.0 tsf for dense sands, to 1.5 tsf for sound rock [Wu (1975)[104]]. Typical bond stresses for rock anchors are given in Table 7.24, to which values for a safety factor ranging from 1.5 to 2.5 are applied. Alternatively, for f in rock, Nicholson[105] uses the lower value of 0.1 grout-crushing strength ($0.1U_c$) versus 0.1 rock-crushing strength ($0.1U_c$) and applies FS = 3.

TABLE 7.24
ULTIMATE BOND STRESSES FOR ROCK ANCHORS: GUIDE VALUES*,†

Rock type	Ultimate bond stresses between sound nondecayed rock and anchor grout plug: psi	kg/cm² (tsf)
Granite and basalt	250–450	18.0–32.3
Dolomitic limestone	200–300	14.4–21.6
Soft limestone‡	150–220	10.8–15.8
Slates and hard shales	120–200	8.6–14.4
Soft shales‡	30–120	2.2–8.6
Sandstone	120–250	8.6–18.0
Concrete	200–400	14.4–28.8

*From PCI (1974).[106]

†FS range to be applied: 1.5 to 2.5.

‡Bond strength must be confirmed by pullout tests which include time creep tests.

Overburden pressures are considered in the following approach where the ultimate capacity Q_u is expressed [Murphy (1974)[99]] as

$$Q_u = \pi BL(c_a + \sigma_n \tan \delta) \qquad (7.106)$$

where c_a = shaft adhesion = s_u for undrained behavior in clays, and 0 in sands

σ_n = normal stress at the midpoint of the fixed portion. Various investigators assume different values. It can be equal to:

- p_o, overburden stress, the most popular assumption
- average of vertical and horizontal stress $[(\sigma_v + K_o\, \sigma_v)/2]$, which is more conservative than using the overburden stress
- normal stress on the plane of the anchor as determined from the Mohr's circle using vertical and lateral stresses as principal stresses which is the most accurate procedure if grout pressure "lock-in" is not counted on
- some fraction of the grout pressure, usually ½

B, b = the diameter of the drilled hole if the grout take is low (b), or if grout take is high assume that the grout fills the voids uniformly along the fixed portion and the diameter increases to B.

δ = ϕ is usually acceptable and it must be equal to or greater than ϕ; it is best determined from tests of the grout-soil interface.

General Formula

Costa Nunes (1974)[107] presented a relationship that considers a large number of factors which he derived from comparing relationships developed by a number of investigators and which he refers to as the general formula, expressed as

$$Q_u = [A_{fr}] + [A_{bt}] + [\pi b\zeta_e L\zeta_L(p\zeta_p \tan \delta + c_a)] \qquad (7.107)$$

where A_{fr} = frontal resistance of anchor

A_{bt} = bottom resistance of anchor

b = boring diameter (generally applicable for low grout pressures, i.e. $<$ 100 psi)
ζ_e = factor of diameter increase from grouting
L = length of fixed portion
ζ_L = factor for length reduction due to distribution of stresses on the anchor
p = average normal pressure on the anchor resulting from earth pressure or from residual grout pressure
ζ_p = factor of normal pressure increase resulting from the residual grout pressure when not included in p
$\tan\delta$ = friction factor
$c_a = c\alpha$ = soil-grout adhesion
c = cohesion
α = a soil-grout adhesion parameter (varies from 0.3 to 0.5; Fig. 7.36)

Costa Nunes observed that most differences in anchor resistance computations result from consideration of the factor ζ_p, normal pressure increase from grout pressure, and that, in general, where grout pressures are low [100 psi (7 kg/cm^2)] there is no significant increase in anchor diameter.

Gravel-Packed Anchors

Gravel-packed anchors are used in cohesive soils primarily to increase the value for the undrained shear strength coefficient α. After filling the original anchor hole with angular gravel, a small closed-end casing is driven into the hole to displace the gravel into the surrounding clay. Grout, injected as the casing is withdrawn, penetrates the gravel, both increasing the effective anchor diameter and improving the strength along the grout-soil interface.

The ultimate capacity, considering both frictional resistance and end bearing, is found from Eq. 7.108 [Littlejohn (1970)[101]]. The coefficient α is assumed to be increased to between 0.6 and 0.75. The anchor dimensions are as defined in Fig. 7.86.

$$Q_u = \frac{\pi}{4}(B^2 - b^2)N_c c_u + \pi b L_s c_u \qquad (7.108)$$

Combined Straight Shaft and Underream

Cohesionless Soils

A modification of Eq. 7.107 has been expressed for a combined straight shaft and underreammed anchor in cohesionless soils [Murphy (1974)[99]]:

$$Q_u = Q_{\text{shaft}} + Q_{\text{underream}} + Q_{\text{end}}$$

or

$$Q_u = [\pi b L_s \bar{\sigma}_{\text{avgshaft}} \tan\delta] + [B L_u \bar{\sigma}_{\text{avgunder}} \tan\delta] + \left[\bar{\sigma}_{\text{end}} N_q (B^2 - b^2)\frac{\pi}{4}\right] \qquad (7.109)$$

where N_q = bearing-capacity factor for deep foundations (Fig. 7.27). In this relationship the assumption of a value for K_o in the computation for the normal stress has a very large effect on the computed anchor capacity and must be selected carefully.

Cohesive Soils

Q_u in purely cohesive soils is governed by adhesion along the fixed portion plus pullout bearing capacity of the underreammed bells (Fig. 7.87). Due to disturbance, adhesion in stiff to hard clays is about 30 to 40% of c_u as measured for the undisturbed state [Goldberg (1976)[103]]. Q_u can be estimated from the expression from Littlejohn (1970)[101]:

$$Q_u = \pi B L_u c_u + \frac{\pi}{4}(B^2 - b^2)N_c c_u + \pi\, b L_s c_a \qquad (7.110)$$

where c_a = adhesion (can use remolded, consolidated, drained shear strength or $c_u\alpha$)
c_u = undrained shear strength (along L_u disturbance may reduce to $0.75c_u$)
N_c = 9, the bearing-capacity factor for a circular pile

Using Grout Take Volume

Relationships have been developed to estimate anchor capacity where the effective diameter B had been determined from the grout take during test anchor installation.

Grout take volume:

$$V = \frac{\pi}{4}(B^2 - b^2)L \qquad (7.111)$$

Cohesionless Soil: Straight Shaft

$$Q_u = BL\bar{\sigma}_{avg} \tan \delta \qquad (7.112)$$

where $\bar{\sigma}_{avg} = (\bar{\sigma}_v + K_o\bar{\sigma}_v)/2$, and $\delta = \phi$.

Cohesive Soil: Straight Shaft Anchor

$$Q_u = \pi BLc_a \qquad (7.113)$$

Pressuremeter testing is a useful method for developing soil-pressure versus volume-change graphs for evaluating *B* and grout take.

Load Testing

Performance

As anchor capacity relies strongly on installation procedures, it is prudent to proof-test each one to 100 to 120% of design load; a few anchors should be tested to 150% of design load. Loads normally are left on for 10 to 20 min; longer periods are used if there is a concern for creep. To perform the test, load is applied in increments by a calibrated hydraulic jack, and movement of the rod end is measured by a dial gage. Each load increment is sustained until movement is equal to or less than 0.001 in/min for a period of 4 min, at which point deflection is recorded. Load is then locked off at between 75 and 100% of design load.

Data Interpretation

For each anchor test the load-deformation curve is plotted and compared with the computed elastic deformation curve for the free section and for the total rod length. Experience and judgment are required to evaluate whether the anchor is acceptable and whether failure has occurred. Evaluation is based on the shape of the load-deformation curve and total deflection. One approach to interpretation is illustrated in Fig. 7.89. The anchor may be considered *acceptable* if:

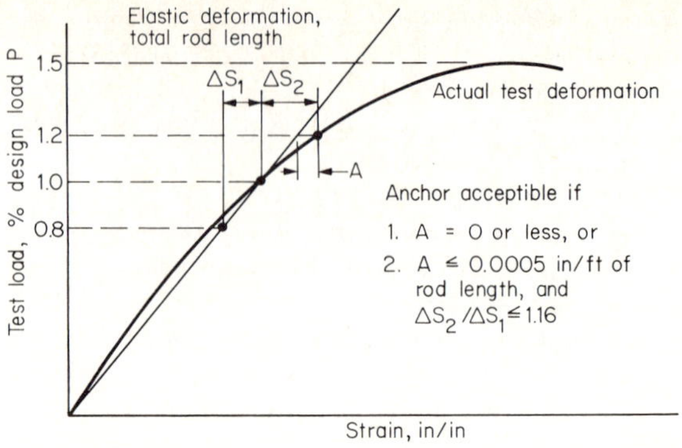

FIG. 7.89 Evaluation of acceptability of a test-loaded anchor. [From Larson et al. (1972).[96]]

1. Total strain at 1.2 design load *P* is less than elastic deformation for the total rod, or
2. If strain at 1.2*P* is greater than total rod elastic strain, the deflection at 0.8*P* is less than the maximum theoretical elongation and the deflection (ΔS_2) from 1.0 to 1.2*P* is less than or equal to 1.16 times the change in length from 0.8 to 1.0*P* ($\Delta S_2/\Delta S_1 = 1.16$).

Failure could be considered as beginning at the point where the actual anchor deformation curve crosses the elastic deformation curve.

Additional References

A useful general reference is available from the Nicholson Anchorage Division of the Nicholson Construction Co., *Rock and Soil Anchor Manual* in which typical applications, design criteria, and guide specifications for high-capacity anchors are described, with comments on installation procedures. Additional references are Schnabel (1982)[110] and Goldberg et al. (1976).[111]

REFERENCES

1. NAVFAC (1982) *Design Manual DM7.2: Foundations and Earth Structures*, Department of the Navy, Naval Facilities Engineering Command, Alexandria, Va., May.
2. Peck, R. B., Hanson, W. E., and Thornburn, T. H. (1974) *Foundation Engineering*, 2d ed., John Wiley & Sons, New York.
3. City of New York (1968) "Building Code—Local Law No. 76 of the City of New York, effective December 6, 1968, amended to August 22, 1969," The City Record, Manhattan, New York.
4. McClellend, B. (1972) "Design of Deep Penetration Piles for Ocean Structures," *Terzaghi Lectures: 1963–1972*, ASCE, (1974) pp. 383–421.

5. NCE (1979) "Electro-osmosis aids US pile drivers," *NCE International*, Inst. of Civ. Engr., London, August, pp. 46–47.
6. CGS (1978) *Canadian Foundation Engineering Manual*, Part 3, "Deep Foundations," Canadian Geotechnical Society, Montreal.
7. Horn, H. M. (1966) "Influence of Pile Driving and Pile Characteristics on Pile Foundation Performance," *Lecture Notes*, New York Metropolitan Section, ASCE, Soil Mech. and Found. Group.
8. Zeevaert, L. (1972) *Foundation Engineering for Difficult Subsoil Conditions*, Van Nostrand-Reinhold Co., New York.
9. Woodward, R. J., Gardner, W. S., and Greer, D. M. (1972) *Drilled Pier Foundations*, McGraw-Hill Book Co., New York.
10. Hearne, T. M., Stokoe, K. H. II, and Reese, L. C. (1981) "Drilled Shaft Integrity by Wave Propagation Methods," *Proc. ASCE, J. Geotech. Engrg. Div.*, Vol. 107, No. GT10, October, pp. 1327–1344.
11. ENR (1979) "Groundwater Bottled Up to Check Moving Piles," *Engineering News-Record*, October 18, p. 26.
12. Proctor, R. V. and White, T. L. (1977) *Earth Tunneling with Steel Supports*, Commercial Shearing Inc., Youngstown, Ohio.

12a. Hunt, R. E. (1984) *Geotechnical Engineering Investigation Manual*, McGraw-Hill Book Co., New York.

13. Focht, J. A., Jr. and Kraft, L. M., Jr. (1977) "Progress in Marine Geotechnical Engineering," *Proc. ASCE, J. Geotech. Engrg. Div.*, Vol. 103, No. GT10, October, pp. 1097–1118.

13a. DNV (1977) *Rules for the Design Construction and Inspection of Offshore Structures*, Appendix F, "Foundations," Det Norske Veritas, Oslo (also Sec. 9).

14. Le Tirant, P. (1979) *Seabed Reconnaissance and Offshore Soil Mechanics for the Installation of Petroleum Structures*, Inst. Français du Pétrole Pub., Editions Technip, Paris, p. 508.
15. Kraft, L. M., Jr. and Lyons, C. G. (1974) "State-of-the-Art: Ultimate Axial Capacity of Grouted Piles," *Offshore Tech. Conf.*, Houston, ASCE, OTC Paper 2081 (preprint).
16. de Ruiter, J. and Beringen, F. L. (1978) "Pile Foundations for Large North Sea Structures," *Marine Geotechnology*, Vol. 3, No. 2.
17. Kezdi, A. (1975) "Pile Foundations," *Foundation Engineering Handbook*, Winterkorn and Fang, eds., Van Nostrand-Reinhold Co., New York, Chap. 19.
18. Vesić, A. S. (1964) "Investigation of Bearing Capacity of Piles in Sand," *Proc. Conf. on Deep Foundations*, Mexican Soc. of Soil Mech., Mexico City, Vol. 1, p. 197.
19. Kerisel, J. (1964) "Deep Foundations, Basic Experimental Facts," *Proc. Conf. on Deep Foundations*, Mexican Soc. of Soil Mech., Mexico City, Vol. 1, p. 5.
20. Meyerhof, G. G. (1976) "Bearing Capacity and Settlement of Pile Foundations," *Proc. ASCE, J. Geotech. Engrg. Div.*, Vol. 102, No. GT3, March, pp. 197–228.
21. Agarwal, S. L., Malhotra, A. K., and Banerjee, R. (1977) "Engineering Properties of Calcareous Soils Affecting the Design of Deep Penetration Piles for Offshore Structures," *9th Annual Offshore Tech. Conf.*, Houston, ASCE, OTC Paper 2792, pp. 503–507.
22. API (1982) "Recommended Practice for Planning, Designing and Constructing Fixed Offshore Platforms," 13th ed. Amer. Petroleum Inst., APT RP 2A, January.
23. Vesić, A. S. (1967) "Ultimate Loads and Settlements of Deep Foundations in Sand," *Proc. Symp. on Bearing Capacity and Settlement of Foundations*, Duke University, Durham, N.C.
24. Murphy, D. J. (1973) "Deep and Shallow Foundations," in-house publication, Joseph S. Ward and Assocs., Caldwell, N.J.
25. Poulos, H. G. and Davis, E. H. (1980) *Pile Foundation Analysis and Design*, John Wiley & Sons, New York, p. 397.
26. Kishida, H. (1967) "Ultimate Bearing Capacity of Piles Driven into Loose Sand," *Soil and Foundations*, Vol. 7, No. 3, pp. 20–29.

26a. Eide, O. (1974) "Marine Soil Mechanics—Applications to North Sea Offshore Structures," Pub. 103, Norwegian Geotech. Inst., Oslo, p. 19.

27. Schmertmann, J. H. (1978) *Guidelines for Cone Penetration Test: Performance and Design*, Pub. FHWA-TS-78-209, Fed. Highway Admin., Office of Research and Development, Washington, D.C.
28. Menard, L. F. (1975) "Interpretation and Application of Pressuremeter Test Results," *Sols-Soils*, Paris, Vol. 26, pp. 1–23.
29. Vesić, A. S. (1972) "Expansion of Cavities in Infinite Soil Mass," *Proc. ASCE, J. Soil Mech. and Found. Engrg. Div.*, Vol. 98, No. SM3, March.
30. D'Appolonia, E., Ellison, R. D., and D'Appolonia, D. J. (1975) "Drilled Piers," *Foundation Engineering Handbook*, Winterkorn and Fang, eds., Van Nostrand-Reinhold Co., New York, Chap. 20.

31. Chellis, R. D. (1962) "Pile Foundations," *Foundation Engineering*, G. A. Leonards, ed., McGraw-Hill Book Co., New York, Chap. 7.

32. Norlund, R. L. (1963) "Bearing Capacity of Piles in Cohesionless Soils," *Proc. ASCE, J. Soil Mech. and Found. Engrg. Div.*, Vol. 89, No. SM3, pp. 1–35.

33. Nottingham, L. (1975) "CPT and Model Pile Study," results given in Schmertmann (1978).[27]

34. Vesić, A. S. (1970) "Load Transfer in Pile-Soil System," *Design and Installation of Pile Foundations and Cellular Structures*, Fang and Dismuke, eds., Envo Pub. Co., Lehigh, Pa., pp. 47–74.

35. Meyerhof, G. G. (1959) "Compaction of Sands and Bearing Capacity of Piles," *Proc. ASCE, J. Soil Mech. and Found. Engrg. Div.*, Vol. 85, No. SM6.

36. Meyerhof, G. G. (1960) "The Design of Franki Piles with Special Reference to Groups in Sands," *Proc. Symp. on Design of Pile Foundations, 6th Intl. Cong. Bridge and Structural Engr.*, Stockholm.

37. Baguelin, F., Jézéquel, J. F., and Shields, D. H. (1978) *The Pressuremeter and Foundation Engineering*, Trans Tech Publications, Clausthal, Germany.

38. Tomlinson, M. J. (1970) "Some Effects of Pile Driving on Skin Friction," *Proc. Conf. on Behavior of Piles*, Inst. of Civ. Engr., London, pp. 59–66.

39. Reese, L. C., Touma, F. T., and O'Neill, M. W. (1976) "Behavior of Drilled Piers Under Axial Loading," *Proc. ASCE, J. Geotech. Engrg. Div.*, Vol. 102, No. GT5, May, pp. 493–510.

40. Vijayvergiya, V. N. and Focht, J. A., Jr. (1972) "A New Way to Predict the Capacity of Piles in Clay," *Proc. OTC, 4th Annual Conf. Offshore Tech. Conf.*, Houston, ASCE, Vol. 2, pp. 865–874.

41. Kraft, L. M., Focht, J. A., Jr., and Amerasinghe, S. F. (1981) "Friction Capacity of Piles Driven into Clay," *Proc. ASCE, J. Geotech. Engrg. Div.*, Vol. 107, No. GT11, November, pp. 1521–1541.

42. Burland, J. B. (1973) "Shaft Friction of Piles in Clay—A Simple Fundamental Approach," *Ground Engineering*, Vol. 6, No. 3, May, pp. 30–42.

43. Mattes, N. S. and Poulos, H. G. (1969) "Settlement of a Single Compressible Pile," *Proc. ASCE, J. Soil Mech. and Found. Engrg. Div.*, Vol. 95, No. SM1, Proc. Paper 6356, January, pp. 189–207.

44. Koutsoftas, D. C. (1981) "Caissons Socketed in Sound Mica Schist," *Proc. ASCE, J. Geotech. Engrg. Div.*, Vol. 107, No. GT6, June, pp. 743–757.

45. Costa Nunes, A. J. da and Fernandes, C. E. de M. (1981) "Computed Friction in a Pile's Rock Socket," *Proc. 10th Intl. Conf. Soil Mech. and Found. Engrg.*, Stockholm, Vol. II, pp. 677–680.

46. Goeke, P. M. and Hustad, P. A. (1980) "Instrumented Drilled Shafts in Clay-Shale," *Proc. ASCE, Symp. on Deep Foundations*, Atlanta, October 1979, pp. 149–165.

47. Rosenberg, P. and Journeaux, N. L. (1976) "Friction and End Bearing Tests on Bedrock for High Capacity Socket Design," *Canadian Geotech. J.*, Vol. 13, August, pp. 324–333.

48. Pells, P. N. J. and Turner, R. M. (1979) "Elastic Solutions for the Design and Analysis of Rock-Socketed Piles," *Canadian Geotech. J.*, Toronto, Vol. 16, No. 3, pp. 481–487.

49. Thorne, C. P. (1977) "The Allowable Loadings of Foundations on Shale and Sandstone in the Sydney Region. Part 3: Field Test Results," paper presented to Sydney Group, Australian Geomech. Soc., Inst. Engr., Australia.

50. Thorburn, S. (1966) "Large Diameter Bored Piles Founded on Bedrock," *Proc. Symp. on Large Bored Piles*, Inst. of Civ. Engr., London.

51. Jasper, J. L. and Shtenko, V. W. (1969) "Foundation Anchor Piles in Clay-Shales," *Canadian Geotech. J.*, Vol. 6, pp. 159–174.

52. Horvath, R. G. and Kenney, T. C. (1979) "Shaft Resistances of Rock-Socketed Drilled Piers," Preprint 3698, paper presented at ASCE Conf., Atlanta, Ca., October.

53. Goodman, R. E. (1980) *Introduction to Rock Mechanics*, John Wiley & Sons, New York.

54. Kenney, T. C. (1977) "Factors to be Considered in the Design of Piers Socketed in Rock," *Proc. Conf. on Design and Construction of Deep Foundations*, Sudbury, Ontario, Canadian Soc. for Civ. Engr.

55. Poulos, H. G. and Davis, E. H. (1975) "Prediction of Downdrag Forces in End Bearing Piles," *Proc. ASCE, J. Geotech. Engrg. Div.*, Vol. 101, No. GT2, February, pp. 189–204.

56. Johannessen, I. J. and Bjerrum, L. (1965) "Measurement of the Compression of a Steel Pile to Rock Due to Settlement of the Surrounding Clay," *Proc. 6th Intl. Conf. Soil Mech. and Found. Engrg.*, Montreal, Vol. 2, pp. 261–264.

57. Terzaghi, K. and Peck, R. B. (1967) *Soil Mechanics in Engineering Practice*, 2d ed., John Wiley & Sons, New York.

58. Mindlin, R. D. (1936) "Force at a Point in the Interior of a Semi-Infinite Solid," *Physics*, Vol. 7, pp. 195–202.

59. Desai, C. S. and Christian, J. T. (1977) *Numerical Methods in Geotechnical Engineering*, McGraw-Hill Book Co., New York.

60. Poulos, H. G. and Davis, E. H. (1968) "The Settlement Behavior of Single Axially Loaded Incompressible Piles and Piers," *Geotechnique*, Vol. 18, No. 3, Inst. of Civ. Engr., London, September, pp. 351–371.

61. Pells, P. J. N. and Turner, R. M. (1979) "Elastic Solutions for the Design and Analysis of Rock-Socketed Piles," *Canadian Geotech. J.*, Vol. 16, No. 3.

62. Glos, G. H. III and Brigs, O. H. Jr., (1983) "Rock Sockets in Soft Rock," *Proc. ASCE, J. Geotech. Engrg. Div.*, Vol. 109, No. GT4, April, pp. 525–535.

63. Kraft, L. M., Ray, R. P., and Kagawa, T. (1981) "Theoretical *t*-z Curves," *Proc. ASCE, J. Geotech. Engrg. Div.*, Vol. 107, No. GT11, November, pp. 1543–1561.

64. Skempton, A. W. and MacDonald, D. H. (1956) "The Allowable Settlement of Buildings," *Proc. Inst. of Civ. Engr.*, London, Vol. 5, Part III, pp. 727–784.

65. Booker, J. R. and Poulos, H. G. (1976) "Analysis of Creep Settlment of Pile Foundations," *Proc. ASCE, J. Geotech. Engrg. Div.*, Vol. 102, No. GT1, pp. 1–14.

66. Zeevaert, L. (1957) "Foundation Design and Behavior of Tower Latino Americano in Mexico City," *Geotechnique*, Vol. VII, No. 3, Inst. of Civ. Engrs., London, September.

67. Glanville, W. H., Grime, G., Fox, E. N., and Davies, W. W. (1938) "An Investigation of the Stresses in Reinforced Concrete Piles During Driving," British Bldg. Research Board, *Tech Paper No. 20*, D.S.I.R.

67*a*. Cummings, A. E. (1940) "Dynamic Pile-driving Formulas," *J. Boston Soc. Civ. Engr.*, Vol. 27, January, pp. 6–27.

67*b*. Smith, E. A. L. (1960) "Pile Driving Analysis by Wave Equation," *Proc. ASCE, J. Soil Mech. and Found. Engrg. Div.*, Vol. 86, No. SM4, August, pp. 35–61.

68. Forehand, P. W. and Reese, J. L. (1964) "Prediction of Pile Capacity by the Wave Equation," *Proc. ASCE, J. Soil Mech. and Found. Engrg. Div.*, Vol. 90, No. SM2, pp. 1–25.

69. Bowles, J. E. (1974) *Analytical and Computer Methods in Foundation Engineering*, McGraw-Hill Book Co., New York.

70. Stevens, J. B., Holloway, D. M., Moriwaki, Y., and Demsky, E. C. (1980) "Pile Group Response to Axial and Lateral Loading," *Proc. ASCE, Symp. on Deep Foundations*, Atlanta, Ga., October, 1979, pp. 396–419.

71. ASTM (1969) "Method of Test for Load-Settlement Relationship for Individual Piles Under Vertical Axial Loads," Part II, Designation 1143, Amer. Soc. for Test. and Mat., Philadelphia, Pa.

72. Fellenius, B. H. (1975) "Test Loading of Piles and New Proof Testing Procedures," *Proc. ASCE, J. Geotech. Engrg. Div.*, Vol. 101, No. GT9, September, pp. 855–869.

73. Van Weele, A. F. (1957) "A Method of Separating the Bearing Capacity of a Test Pile into Skin Friction and Point Resistance," *Proc. 4th Intl. Conf. Soil Mech. and Found. Engrg.*, London, Vol. 2.

74. Davisson, M. T. (1973) "High Capacity Piles," *Innovations in Foundation Construction*, Soil Mech. Div., Illinois Sect. ASCE, Chicago, pp. 81–112.

75. Goble, G. G. and Rausche, F. (1970) "Pile Load Test by Impact Driving," paper presented to Highway Research Board Annual Meeting, Washington, D.C., January.

76. LaFond, K. J. (1977) "Applications of Dynamic Pile Analysis," PILETIPS, Assoc. Pile and Fitting Corp., Clifton, N.J., May–June.

77. Rausche, F. (1978) "Pile Driving Measurements on the Weather Platform Installations," paper presented at the European Offshore Petroleum Conf. and Exhibition, London, October, pp. 423–426.

78. Matlock, H. and Reese, L. C. (1960) "Generalized Solutions for Laterally Loaded Piles," *Proc. ASCE, J. Soil Mech. and Found. Engrg. Div.*, Vol. 86, No. SM5, Proc. Paper 2626, October.

79. Morgan, J. R. and Poulos, H. G. (1968) "Stability and Settlement of Deep Foundations," *Soil Mechanics—Selected Topics*, I. K. Lee, ed., Butterworths, London, Chap. 9, pp. 528–609.

80. Broms, B. B. (1964) "Lateral Resistance of Piles in Cohesive Soils," *Proc. ASCE, J. Soil Mech. and Found. Engrg. Div.*, Vol. 90, No. SM2, March, pp. 27–64.

81. Broms, B. B. (1964) Lateral Resistance of Piles in Cohesionless Soils," *Proc. ASCE, J. Soil Mech. and Found. Engrg. Div.*, Vol. 90, No. SM3, May, pp. 123–156.

82. Brinch Hansen, J. (1961) "The Ultimate Resistance of Rigid Piles Against Transversal Forces," *Danish Geotech. Inst. Bull. 12.*

83. Menard, L. (1962) "Comportement d'une fondation profonde soumise à des efforts de renversement," *Sols-Soils*, No. 3:4, pp. 9–23.

84. Reese, L. C. and Matlock, H. (1956) "Non-dimensional Solutions for Laterally Loaded Piles with Soil Modulus Assumed Proportional with Depth," *Proc. 8th Texas Conf. Soil Mech. and Found. Engrg.*, Spec. Pub. No. 29, Bureau of Engrg. Research, University of Texas, Austin, September.

85. Francis, A. J. (1964) "Analysis of Pile Groups with Flexural Resistance," *Proc. ASCE, J. Soil Mech. and Found. Engrg. Div.*, Vol. 90, No. SM3.

86. Davisson, M. T. (1970) "Lateral Load Capacity of Piles," *HRB No. 133*, Highway Research Board, Washington, D.C., pp. 104–112.

87. Matlock, H. and Reese, L. C. (1961) "Foundation Analysis of Offshore Pile-Supported Structures," *Proc. 5th Intl. Conf. Soil Mech. and Found. Engrg.*, Vol. 2, pp. 91–97.

88. Davisson, M. T. and Gill, H. L. (1963) "Laterally Loaded Piles in a Layered Soil System," *Proc. ASCE, J. Soil Mech. and Found. Engrg. Div.*, Vol. 89, No. SM3, pp. 63–94.

89. Reese, L. C. (1977) "Laterally Loaded Piles: Program Documentation," *Proc. ASCE, J. Geotech. Engrg. Div.*, Vol. 103, No. GT4, April, pp. 287–304.

90. Sanders, R. L. (1976) "Lateral Load Design—Step-Taper Piles," *Foundation Facts*, Vol. XI, No. 1, pp. 16–20. Publication of Raymond Intl. Inc., Houston, Tex.

91. Bjerrum, L., Hansen, J. B., and Sevaldson, R. (1957) "Geotechnical Investigations for a Quay Structure in Horten," *Pub. 28*, Norwegian Geotech. Inst., Oslo.

92. Tschebotarioff, G. P. (1973) *Foundations, Retaining and Earth Structures*, 2d ed., McGraw-Hill Book Co., New York.

93. ENR (1977) "A String of Tunnels and Viaducts Conquers Brazil's Terrain," *Engineering News-Record*, April 21, pp. 16–17.

94. Poulos, H. G. (1971) "Behavior of Laterally Loaded Piles: II Pile Groups," *Proc. ASCE, J. Soil Mech. and Found. Engrg. Div.*, Vol. 97, No. SM5, pp. 733–751.

95. Clausen, C. J. F., Aas, P. M., and Hasle, E. (1981) "SPLICE—A computer program for analyzing structure-soil-pile interaction problems," paper presented in Session 4, *Intl. Symp. on Offshore Engrg.*, Offshore Brazil '81, Rio de Janeiro, September.

95*a*. Bogard, D. and Matlock, H. (1983) "Procedures for Analysis of Laterally Loaded Pile Groups in Soft Clay," *Proc. ASCE, Conf. on Geotech. Prac. in Offshore Engrg.*, University of Texas, Austin, pp. 499–535.

96. Larson, M. L., Willette, W. R., Hall, H. C., and Gnaedinger, J. P. (1972) "A Case Study of a Soil Anchor Tie-back System," *Proc. ASCE, Spec. Conf. Performance on Earth and Earth-Supported Structures*, Purdue University, Vol. 1, Part 2, pp. 1341–1366.

97. Bhatnagar, R. S. (1969) "Pull-Out Resistance of Anchors in Silty Clay," *Soil Mech. Series No. 18*, Duke University Durham, N.C.

98. Vesić, A. S. (1971) "Break-Out Resistance of Objects Embedded in the Ocean Bottom," *Proc. ASCE, J. Soil Mech. and Found. Engrg. Div.*, September.

99. Murphy, D. J. (1974) "Earth and Rock Anchors—A Review," paper presented at *ASCE Spec. Conf. on Analysis and Design in Geotechnical Engineering*, workshop on analysis and design of flexible retaining structures, Austin, Tex.

100. Costa Nunes, A. J. da (1966) "Slope Stabilization—Improvements in the Techniques of Prestressed Anchorages in Soils and Rocks," *Proc. 1st Intl. Cong. Intl. Soc. of Rock Mech.*, Lisbon, Vol. II, Theme 6, pp. 141–146.

101. Littlejohn, G. S. (1970) "Anchorages in Soil—Some Empirical Design Rules," *The Consulting Engineer*, Consulting Engineers Council, Washington, D.C., May.

102. Dismuke, T. D. (1975) "Cellular Structures and Braced Excavations," *Foundation Engineering Handbook*, Winterkorn and Fang, eds., Van Nostrand-Reinhold Co., New York, Chap. 14, pp. 445–480.

103. Goldberg, D. T. (1976) "Lateral Support Systems," *Geotechnical Lecture Series on Lateral Earth Pressure*, Boston Soc. of Civ. Engrs.

104. Wu, T. H. (1975) "Retaining Walls," *Foundation Engineering Handbook*, Winterkorn and Fang, eds., Van Nostrand-Reinhold Co., New York, Chap. 12.

105. Nicholson (undated) *Rock and Soil Anchor Manual*, Nicholson Anchorage Division, Nicholson Construction Co., Bridgeville, Pa. (received by author March 1982).

106. PCI (1974) "Tentative Recommendations for Prestressed Rock and Soil Anchors," Prestressed Concrete Inst., Chicago, September.

107. Costa Nunes, A. J. da (1974) "Tied-Back Walls," *Proc. ASCE, Spec. Conf. on Analysis and Design in Geotech. Engrg.*, Austin, Tex., Vol. II, pp. 146–149.

108. Chellis, R. D. (1948) "A Study of Friction Pile Values," *Proc. 2nd Intl. Conf. Soil Mech. and Found. Engrg.*, Rotterdam, Vol. 5, pp. 142–146.

109. Liver, H. L., Jr. and Jennings, R. C. (1952) *The Prepakt Reporter*, October-November.

110. Schnabel, H., Jr. (1982) *Tiebacks in Foundation Engineering and Construction*, McGraw-Hill Book Co., New York.

111. Goldberg, D. T., Jaworski, W. E., and Gordon, M. D. (1976) "Lateral Support Systems and Underpinning," Vol. III, *Construction Methods*, Report No. FHWA-RD-75-130, Federal Highway Administration, Washington, D.C.

112. NAVFAC (1971) *Design Manual DM7: Soil Mechanics, Foundations and Earth Structures*, Department of the Navy, Naval Facilities Engineering Command, Alexandria, Va., May.

PART III

SLOPES, RETAINING STRUCTURES, AND EXCAVATIONS

CHAPTERS:

PURPOSE AND SCOPE

Part III covers natural and cut slopes in soil and rock, earth-dam embankments and embankments over soft ground; walls and other systems for the support of cut slopes, open excavations for structures, and waterfront structures; and underground excavations in soil and rock for tunnels, caverns, and mines.

Slope stability problems can be considered to include uncut and cut natural slopes in soil or rock and fabricated embankments for earth dams or where constructed over soft ground such as for roadways. There are many analytical methods to choose from, and their application primarily relates to the geometric configuration of the potential failure surface which in turn is a function of geologic conditions or, in the case of an earth-dam embankment, its cross section. The potential failure surface is assumed to be either circular, irregular, or planar and in the latter case the potential failure mass may be divided into two or more blocks or wedges. Stability is provided by proper geometric configurations such as slope inclinations that are not excessive and by external and internal drainage, or at times by retention or reinforcement in conjunction with drainage systems.

Surface retaining structure analysis and design, be it for cut slopes, open excavations, or waterfront structures, require determination of the lateral earth pressures resulting from soil weight and surcharges. Classical theory provides the basis for determining earth pressures for some cases, whereas in others empirical rules are employed. Actual earth-pressure magnitudes and distributions are related closely to wall movements and flexibility, which in some cases vary with construction procedures, and magnitudes and distributions often differ substantially from those determined by classical theory. In addition to earth-pressure retention, walls must be stable against ground rupture, overturning, and sliding.

Cut slopes in soils are retained by gravity walls, of which there are many types, or by anchored concrete walls. The selection of wall type is based strongly on consideration of the height of the cut to be retained. Open excavations, such as for structures, are retained by soldier piles and lagging, sheet piling, or concrete walls, which in most cases are supported by either cross-bracing, rakers and berms, or anchors (tiebacks). Selection of the wall system is based on consideration of the soil types to be retained, groundwater conditions, depth and width of excavation, and wall life (temporary versus permanent). In all cases groundwater control and backslope subsidence, which may result in excessive settlement of adjacent structures, are important considerations. Anchored concrete walls constructed with the slurry trench method are becoming increasingly popular, especially in poor ground conditions such as those caused by loose sands below the water table or soft soils.

Open excavations in rock often are supported by anchors, the design and installation of which can be very much affected by high residual stresses.

Retaining structures for waterfront conditions may be provided by flexible walls such as cantilever or anchored sheet-pile walls, by relieving platforms, or by gravity structures of stone or concrete or cofferdams. Selection is based on consideration of the wall purpose (quays, wharfs, land reclamation, water retention for dry excavation) and the water depth and the loads to be supported by the structure.

Underground excavations refer to tunneling in soil and rock and caverns and mines in rock. Support requirements, divided generally into temporary and permanent conditions, normally are assessed based on empirical relationships, and in either soil or rock the possibility of ground-surface subsidence requires consideration. The most important factor in successful tunneling operations is the construction method for which tunnel boring machines are a major technological advance in both rock and soil. In poor soil conditions slurry-face boring machines are finding increasing application. In large rock excavations, the "New Austrian" method involving shotcrete and wire mesh, and anchors, is an important innovation.

CHAPTER EIGHT

SLOPE AND EMBANKMENT STABILITY

8.1 STABILITY PROBLEMS AND ANALYSIS

8.1.1 SLOPE STABILITY PROBLEMS

Categories

General categories of slope problems for which mathematical analysis of stability is feasible are illustrated in Fig. 8.1. Included are two general categories: (1) slide failure forms in natural uncut slopes and cut slopes in hillsides and open excavations in soil or rock, and (2) embankments for dams, embankments over soft ground, waterfront structures, and sidehill fills.

Slide failure forms refers to conditions where failure by rupture occurs essentially simultaneously at every point along single or multiple failure surfaces and the mass moves simultaneously as a unit or a group of units. *Other forms* of slope failures such as avalanches, flows, failure by lateral spreading, and progressive failure cannot be analyzed in the present state of the art, although all of these forms may be initiated at times by a slide failure at the toe of the mass. The only defense against these other failure forms is recognition that their potential exists. For discussion on slope failure forms, their recognition and treatment, see Chap. 9 of Hunt (1984).[69]

Analytical Objectives

Stability

For *natural slopes* the objective is to determine the factor of safety (FS) against sliding for the existing slope angle in view of the occurrence of changing groundwater conditions and earthquake forces. Other transient conditions, such as increasing slope inclination from erosion or tectonic activity, or decreasing strength due to weathering, are long-term effects that may be accounted for in critical situations. The approach is to evaluate the probability and effects of failure and if they are intolerable, to provide measures to achieve stability, or to avoid construction within the area possibly affected by a slope failure.

Cut slopes and dam-embankment slopes are designed to provide stable slope angles in view of the slope height, materials, and pore-water conditions. Embankments constructed over soft ground and sidehill fills are designed on the basis of evaluations of the shear strength available in the foundation materials.

Deformations

Deformations occurring in slopes and embankments are a concern, not only because they often precede failure, but because of the effects on surface or buried structures. Approximations can be made by the finite element method *(see Art. 1.4.3)*, but the method is seldom used for soil slopes because of the necessity for closely defining a relatively large number of parameters, many of which are difficult to assess accurately. The FEM is finding application for rock slopes and earth-dam embankments, as will be dis-

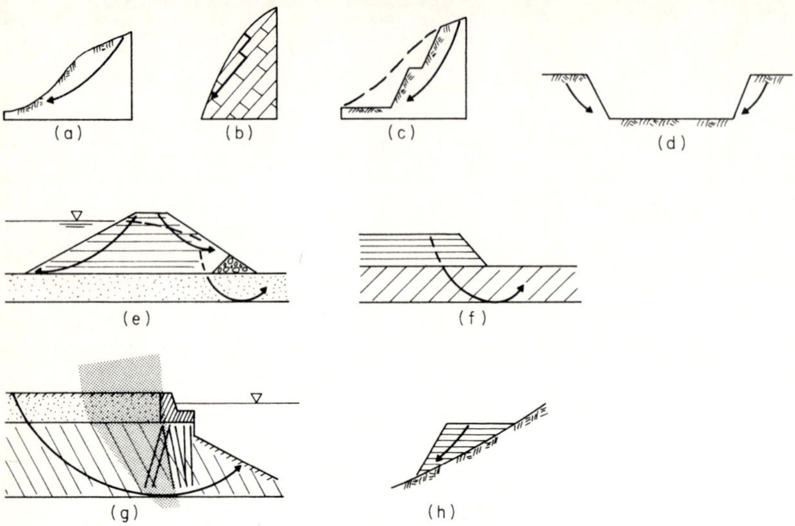

FIG. 8.1 Categories of slope stability problems: (*a*) natural soil slope; (*b*) natural rock slope; (*c*) cut slope; (*d*) open excavation; (*e*) earth dam embankment; (*f*) embankment over soft soils; (*g*) waterfront structure; (*h*) sidehill fill.

cussed, but the emphasis in this chapter is on analysis of stability against failure by rupture.

8.1.2 LIMIT EQUILIBRIUM ANALYSIS

Failure Criteria

Failure, sudden or gradual, is caused by overstress in a slope or its foundation with respect to the available strength. *Limit equilibrium analysis (see Art. 3.5.2)* is the basis for most methods available for slope stability evaluations. It usually considers that the Mohr-Coulomb failure criterion ($s = c + \sigma_n \tan \phi$, Eq. 3.8) is satisfied along an assumed failure surface. A free body is taken from the slope and the known or assumed forces acting on the body and the shearing resistance required for stability are calculated. Most practical problems are statically indeterminate and assumptions must be made regarding the force systems and their directions of applications in order to render the problem determinate.

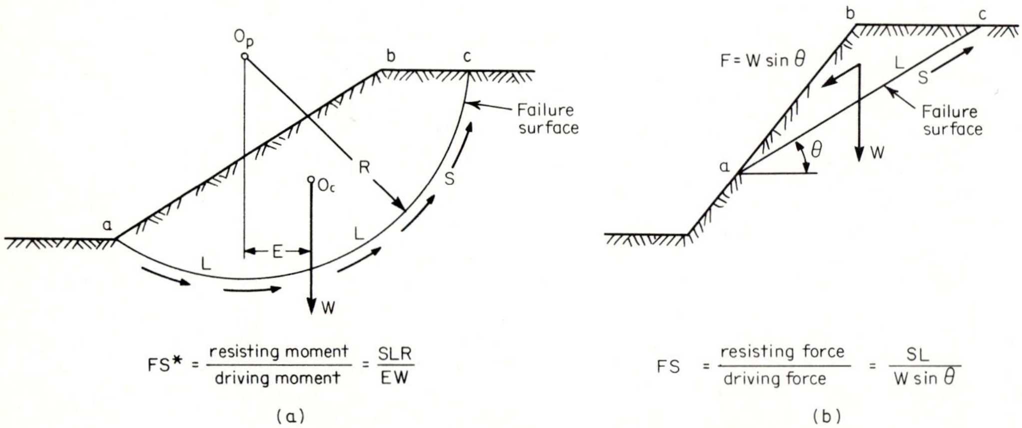

FIG. 8.2 Forces acting on cylindrical failure surfaces. (*a*) rotational cylindrical failure surface with length *L*. Safety factor against sliding FS. (*b*) Simple wedge failure on planar surface with length *L*. *The expression for FS is generally considered unsatisfactory: see text.

Forces Acting on the Whole Free Body

A cylindrical failure form is shown in Fig. 8.2*a*, in which the failing mass is assumed as a whole free body. The mass weight W acting through lever arm E, causing a driving moment, is resisted by the strength S mobilized along the failure surface of length L acting through its lever arm R. In the simple wedge (Fig. 8.2*b*), strength S acting along the planar failure surface of length L resists the driving forces resulting from gravity acting on weight W. These conditions represent early approaches to stability analysis.

Slices as Free Bodies

Other force systems involve dividing the potential sliding mass into "slices" as illustrated in Fig. 8.3 for an "infinite slope," one in which the failure surface is relatively shallow compared to its length, and in Fig. 8.4 for a "finite" slope with a circular failure surface. The geometry and the forces that may act on a slice are shown in Fig. 8.5.

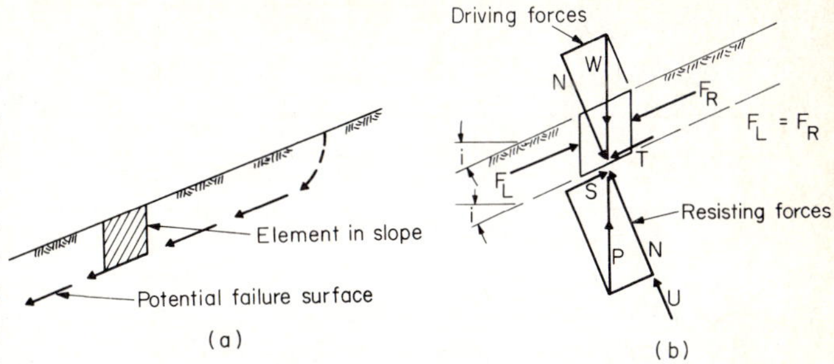

FIG. 8.3 Infinite slope (*a*) and forces on an element (*b*).

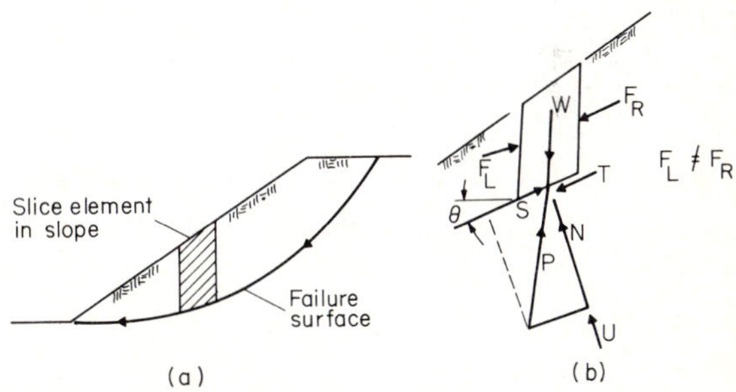

FIG. 8.4 Finite slope with cylindrical failure surface (*a*) and forces on an element (*b*).

In general, where there is *no slope seepage* the forces include material weight W, normal stress N and shearing stress T distributed along an assumed failure surface, and forces acting along the sides of the free body, or slice. Where there is *slope seepage* (the usual condition for failure), pore pressures on the assumed failure surface become an added normal force. The boundary water force U is equal to the sum of the boundary water pressures and is either estimated or found by flow-net construction. The significance of pore-water pressures is seen in the expression $S = (N - U) \tan \phi$ (from Eq. 3.14) in which they directly reduce the frictional component of strength. *Earthquakes* impose dynamic forces in the form of acceleration of gravity which affects the weight component.

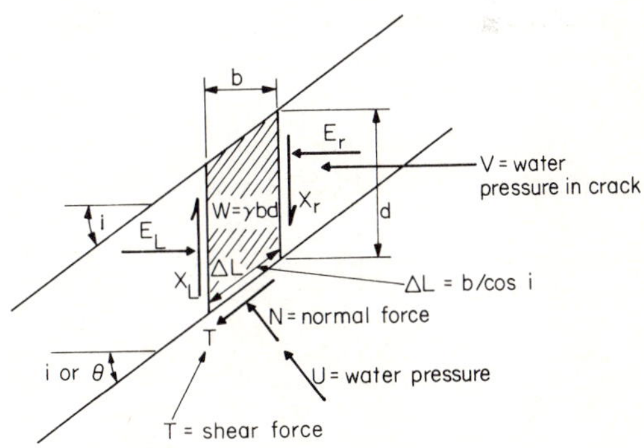

FIG. 8.5 Slice geometry and forces.

These conditions normally apply to soil slopes, and analysis is based on numerical solutions which vary from simple to complex, depending upon slope geology, stress conditions, and the force system assumed.

Blocks and Wedges

Simple planar failures involve a single surface which can define a wedge in soil or rock (Fig. 8.2*b*), a sliding block (Figs. 8.6*a* and *b*), or a wedge in rock with a tension crack (Fig. 8.6*c*). The forces acting are similar to those given in Fig. 8.5 except that there are no forces X_R, X_L, E_R, E_L. Solutions are by simple mathematical relationships.

Complex planar failures involve a number of planes dividing the mass into two or more blocks

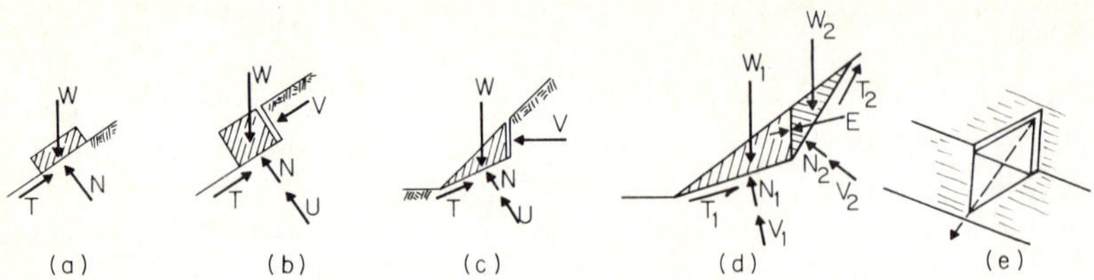

FIG. 8.6 **Some forms of planar failures in finite slopes: (*a*) simple sliding block; (*b*) sliding block with pore- and cleft-water pressures; (*c*) simple wedge; (*d*) multiple wedges with interwedge forces; and (*e*) wedge with intersecting inclined failure surfaces.**

or wedges (Fig. 8.6*d*). Numerical methods, which consider interwedge forces, are relatively complex and the solution may be approached by graphical methods. Multiple wedges are common in rock slopes, some soil slopes, and zoned earth-dam embankments. A variation encountered in rock slopes is a single wedge bounded by inclined planes (Fig. 8.6*e*).

Factor of Safety (FS)

The *factor of safety* can be found in the literature expressed in a number of forms, one of which is given in Fig. 8.2*a*. The form preferred by most practitioners experienced in slope stability analysis is to consider *FS equal to the ratio of the total shearing resistance available along the failure surface to the total shearing resistance required to prevent failure*. This required shearing resistance is equal to the total shearing force along the failure surface. In terms of the unit shearing strength including cohesion, $\tau = \bar{c} + \bar{\sigma} \tan \bar{\phi}$ (Eq. 3.9), FS may be expressed as

$$\text{FS} = \frac{\bar{c} + \bar{\sigma} \tan \bar{\tau}}{\tau} \tag{8.1}$$

or, in terms of total shearing strength and stresses,

$$\text{FS} = \frac{cL + N \tan \phi}{T} \tag{8.2}$$

or, for a failure surface with N and T in terms of the weight force W,

$$\text{FS} = \frac{Cl + (W \cos \theta) \tan \phi}{W \sin \theta} \tag{8.3}$$

which is the basic equation for many forms of stability analysis. These equations are given here to illustrate the basic forms and are described in detail in subsequent articles.

Information Required for Analysis

The following information is required for the analytical assessment of the stability of a slope:

- Location, orientation, and shape of a potential or existing failure surface which is controlled by material type and structural features. Shape can be simple or multiple planar, simple or multiple wedges bounded by surfaces that are planar, cylindrical or log spiral, or irregular
- Distribution of the materials within and beneath the slope, divided generally into homogeneous wherein the properties are more or less similar in all directions and nonhomogeneous wherein soils are stratified and rock masses contain major discontinuities
- Material types and representative shear-strength parameters: granular, cohesionless soils and clean rock joints ($c = 0$); soils or rock joints with cohesion (c, ϕ); fissured clays, soils, and rock masses with preexisting failure surfaces (ϕ_r); and soft clays with undrained conditions ($\phi = 0$)
- Drainage conditions: drained or undrained in view of short-term (construction) or long-term (postconstruction or natural slope) conditions (total versus effective stress analysis)
- Distribution of piezometric levels along the potential failure surface and an estimate of the maximum value that may prevail
- Slope geometry to its full height

Validity of Limit Equilibrium Analysis

Influence of Geologic Conditions

The foregoing factors are relatively well-known for constructed slopes but are difficult to assess with confidence for natural slopes. Planar failure surface locations usually are established by inspection of geologic sections, particularly in nonhomogeneous and anisotropic materials. In relatively homogeneous materials, circular failure surfaces are analyzed by trial and error to find the critical failure surface.

Particularly critical for natural slopes is the selection of representative values for strength and the location of weak zones. In essentially homogeneous clays, for example, such as glacial marine and lacustrine deposits, many slope failures have occurred even though rigorous analysis based on strength measurements obtained from properly executed laboratory triaxial tests and in situ vane tests determined that FS against failure was adequate. The reasons are unclear, but are generally considered to be either thin undetected zones with very low strengths, or perhaps a strength change with time caused by changes in the groundwater regime, especially if the clays are sensitive.

If a natural slope is *undergoing deformation*, shape and location of the failure surface can be reasonably established by exploration and the use of devices such as the inclinometer, and the existing FS can be reasonably well calculated. Important unknown factors are the maximum rainfall that may occur, the pore pressures that could develop (as in any slope), and the reduction in resistance occurring with increasing deformation.

Analytical Weaknesses

In connection with existing stable slopes there are a number of factors not considered in the present state of the art of analysis, including initiation and growth of failure surfaces; stress distribution; energy release; progressive failure; and initiation, development, and rate of movement. Some slopes, such as those in heavily overconsolidated clays, have high residual stresses in excess of overburden pressures, which can substantially affect interslice forces.

The most important weakness in limiting equilibrium analysis, however, is the assumption that FS along the entire failure surface is constant, whereas in reality it is not, since the ratio of shear strength to shear stress along the failure surface is not constant. The shear stresses vary, as does the strength, a function of the normal effective stress which also varies. For most cases, however, limit equilibrium analysis provides a reasonably accurate assessment of slope stability provided that the conditions of geometry, pore pressure, and strength are representative of the real conditions.

Solutions in Practice

Limit equilibrium analysis is performed by solving equations or by graphic methods suited to the geometry and complexity of the problem. Most solutions are based on numerical methods varying from simple hand calculations, to use of charts, to complex calculations by programmable hand calculators, to very complex solutions requiring computers with large storage capacity. The advent of the programmable hand calculator and "personal" (*PC*) or desk-top computer has made possible the solution of many practical problems employing some of the more complex relationships as a standard procedure, whereas previously one resorted to either simple hand calculations, graphics, charts, or complex analysis with large-capacity computers.

One can find in the literature a vast array of expressions for slope stability analysis which vary with the shape of the failure surface, the geometry of the free body or bodies, and the force system acting. A summary of the more common limit equilibrium methods is given in Table 8.1. There are two broad categories as described briefly in a preceding section: infinite slope problems and finite slope problems. The expressions are usually given in terms of total shear strength and shear stresses and a weight term W, but can also be found in the literature in terms of unit shear strength and shear stresses, with the weight term replaced by an expression, often complex, derived in terms of geometry and mass density. In addition some expressions are based on total and submerged weights, whereas others are based on total weights and pore pressures.

TABLE 8.1
SLOPE-STABILITY ANALYSIS BY LIMIT EQUILIBRIUM METHODS*

Failure surface	Method	Assumptions	Reference
Straight line	Infinite slope	Constant slope with unlimited extent.	Taylor (1948)[1]
	Culmann	Planar failure surface intercepting toe of slope.	Culmann (1866)[2]
Straight lines	Wedge or wedges	Sliding block or blocks with lateral forces.	Hoek and Bray (1977)[3] Seed and Sultan (1967)[4] Morgenstern (1968)[5] NAVFAC (1982)[42]
Circular arc	Slices	Lateral forces on each side of slice are equal.	Fellenius (1936)[14]
	Friction circle	Resultant acting on rupture arc tangential to a concentric circle of radius $R \sin \phi$.	Taylor (1948)[1]
	Bishop	Total force system acting on sides of slice considered.	Bishop (1955)[7]
	Modified Bishop	Total force system simplified.	Bishop (1955)[7]
	Spencer	Interslice forces parallel.	Spencer (1967),[8] (1981)[9]
Irregular	Morgenstern and Price	Based on complete Bishop force system.	Morgenstern and Price (1965)[10]
	Janbu	Allows for interslice forces.	Janbu (1954),[11] (1971)[12]

*Methods are those most commonly used. Many have been modified by various investigators to account for pore pressures, tension cracks, etc., which were not provided for in their original derivation. Various other methods exist, including log-spiral-shaped failure surfaces.

Selection of the analytical method is based primarily on consideration of the type of geologic material, degree of isotropy and homogeneity, and location of discontinuities in both natural slopes and earth embankments. Analytical methods vary also in regard to whether total or effective stresses are acting, as will be discussed below. A summary of the general methods of stability analysis and the generally applicable geologic conditions are given in Table 8.2.

Systems of Units

For convenience of solution, the examples in this chapter are executed primarily in the English system of units, particularly, in kips/ft^2 for strength or pressure and kips/ft^3 for unit weights. The metric system, which is used less frequently, is also very convenient. The English to SI conversions for strength, pressure, and unit weight are somewhat awkward, so as an aid to readers familiar with one or the other system, conversions are summarized in Table 8.3. Other conversions are given in Appendix C.

8.1.3 EARLY CONCEPTS

General

Many analytical methods still in use are either direct applications or modifications of methods developed more than 50 years ago. A brief discussion of the early methods and their origins will serve to provide the background for understanding the solutions to the various problems covered in subsequent articles.

Modern slope stability technology has its roots in the studies of numerous slides along the Swedish railways performed in the early 1920s. The slides occurred in cuts made in saturated NC to slightly OC clays and the studies demonstrated that failure involved an approximately cylindrical surface. In 1927, as a result of these studies, Fellenius[6] published solutions for simple slopes

TABLE 8.2
GENERAL METHOD OF STABILITY ANALYSIS FOR SLIDE FAILURES AND APPLICABLE GEOLOGIC CONDITIONS

General method of analysis	Geologic conditions
Infinite slope (Depth small compared to length of failure surface)	Cohesionless sands. Residual or colluvial soils over shallow rock. Stiff fissured clays and marine shales in the highly weathered zone.
Limited slope: simple wedge (Single planar failure surface)	Sliding block. Interbedded dipping rock or soil. Faulted or slickensided materials. Stiff to hard, intact cohesive soil, vertical to very steep slope. Sidehill fills.
Limited slope: wedges-blocks (Multiple planar failure surfaces)	Sliding blocks in rock masses. Closely jointed rock with intersecting sets. Weathered interbedded sedimentary rock. Clay shales and stiff fissured clays. Stratified natural soils and zoned earth dams. Sidehill fills over colluvium.
Limited slope: circular arc (Whole free body analyzed)	Homogeneous, isotropic slopes and embankments.
Limited slope: circular arc (Method of slices)	Thick residual or colluvial soil. Soft marine or clay shales. Soft to firm cohesive soils. Stratified soils: natural or in dams. Embankments over soft ground. Sidehill fills.

TABLE 8.3
CONVERSIONS: ENGLISH TO METRIC SI SYSTEM OF UNITS

Unit	English system	Metric	SI system
Length	ft × 0.3 = m	m × 3.33 = ft	m × 3.33 = ft
Force	1 ksf = 1000 psf		1 kN/m^2 = 1 kPa
	psf × 0.048 = 1 kPa		kPa × 20.8 = psf
	ksf × 48 = 1 kPa		kPa × 0.0208 = ksf
	1 tsf =	1 kg/cm^2 $\simeq$ 10 T/m^2 =	98.07 kPa
Weight	1 kip = 1000 lb		
	pcf × 0.157 = kN/m^3		kN/m^3 × 6.37 = pcf
	kcf × 157 = kN/m^3		kN/m^3 × 0.00637 = kcf
	62.4 pcf =	1 g/cm^3 = 1 T/m^3 =	9.807 kN/m^3

Pore pressure calculations

u (English) = h_w(ft) × 0.0624 = ksf
u (metric) = h_w(m) × 1.0 = t/m^2
u (SI) = h_w(m) × 9.8 = kN/m^2

NOTE: psf = pounds per square foot; ksf = kips per square foot; pcf = pounds per cubic foot; and kcf = kips per cubic foot. kN/m^3 = kilonewtons per cubic meter and kPa = kilopascal. tsf = tons per square foot; kg/cm^2 = kilograms per square centimeter; and T/m^2 = tons per square meter.

without seepage, based on circular failure surfaces, which provided the basis for the friction circle method, the ordinary method of slices, and various chart solutions that are still in use. Prior to that event, slopes usually were evaluated on the basis of earth-pressure theories or the assumption of planar failure surfaces.

Earth Pressures and Slope Failures

A typical slope in homogeneous soil with parameters c and ϕ is illustrated in Fig. 8.7. Small deformations tend to cause a passive resistance to develop downslope and an active state to occur upslope with the formation of tension cracks and shearing displacements to occur along curved failure surfaces.

Slope stability is related to earth-pressure theory through the "second Rankine problem" in which failure conditions are considered to exist within a semi-infinite mass whose surface is inclined. During failure, slip lines form in the mass *(see Art. 3.5.1)* and in stability analysis the conditions required to cause failure along any one of an infinite number of potential slip lines are evaluated. The analytical solution for cohesionless soils, in which the slip lines are straight, was developed by Rankine in 1857; that for cohesive soil, in which the slip lines are curved, was developed by Résal in 1910.

Rankine's solution was based on the assumptions that the slope angle i is less than the friction angle ϕ and that the slightest deformation of the soil is sufficient to mobilize its full frictional resistance and to immediately produce an *active state* if the soil is expanding parallel to the surface and a *passive state* if it is compressing parallel to its surface. Expressions have been developed for the passive and active earth pressures, which are the same expressions given for earth pressures that develop against retaining walls, as discussed in Art. 9.2.3 (Eqs. 9.8 and 9.9).

Conditions are similar for cohesive soils except that both active and passive slip-line orientations change with depth because of the cohesion, and the net result is a curved slip-line field (Fig. 8.7*a*). In addition, the active state does not exist down to a depth of

$$z_o = \frac{2c}{\gamma} \tan (45° + \phi/2) \tag{8.4}$$

since failures to this depth are tensile and not shear and the concept of Rankine stress states is not valid.

Simple Wedges and Vertical Banks

Culmann or Simple Wedge Method

In 1866, Culmann[2] published an analysis of a steeply inclined clay slope in which he assumed that the failure surface would be planar and would pass through the slope toe, as shown in Fig. 8.8. He also assumed that shear strength was constant along the failure surface in a homoge-

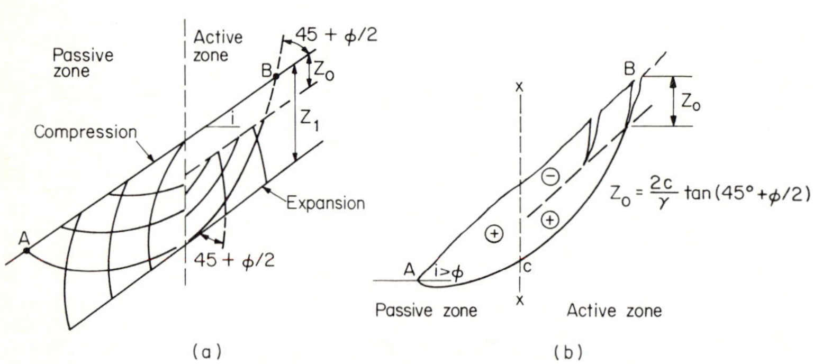

FIG. 8.7 Earth pressures acting in a slope. The passive resistance at the lower portions is most significant. (*a*) Distribution of stresses in slopes with $i > \phi$. (*b*) A case of rupture where (+) indicates zones of passive stress and (−) indicates zones of active stress subject to tension cracks.

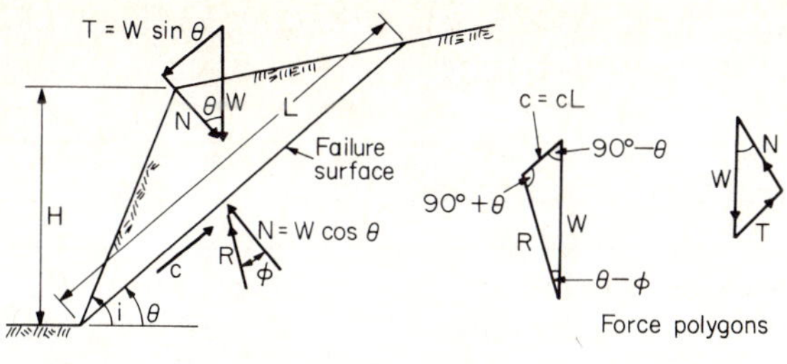

FIG. 8.8 Forces acting on Culmann's simple wedge in a clay slope.

neous section and that there were no seepage forces.

Two force polygons are given in the figure. The one on the left shows the original force system assumed, and the one on the right illustrates the relationship between weight force W, normal force N, and shearing force T from which are determined $N = W \cos \theta$ and $T = \sin \theta$. Substitution of these terms in Eq. 8.2 yields Eq. 8.3:

$$FS = \frac{cL + (W \cos \theta) \tan \phi}{W \sin \theta} \tag{8.3}$$

where

$$W = 1/2\gamma LH \frac{\sin (i - \theta)}{\sin i} \tag{8.5}$$

The *critical failure surface* θ_c is found by using the law of sines for the force polygon shown in the figure and substituting for W and C ($C = cL$), then taking the first derivative of the expression to find

$$\theta_c = \frac{1}{2}(i + \theta) \tag{8.6}$$

Substituting Eq. 8.6 into Eq. 8.5 and rearranging terms yields an expression for a *stability number* N_s in terms of θ_c and the critical height H_c, given in Taylor (1948)[1] as

$$N_s = \left(\frac{c}{\gamma H_c}\right)_\theta = \frac{1 - \cos(i - \phi)}{4 \sin i \cos \phi} \tag{8.7}$$

For a vertical slope ($i = 90°$) and $\phi = 0$, it is seen that the critical height $H_c = 4c/\gamma$, as given in Eq. 8.10 below, for the case of no tension crack.

The Culmann method generally is considered to give reasonably accurate results in slopes that are vertical or nearly so in materials with c and ϕ parameters, but does not give satisfactory results in slopes flatter than about 75°. It is used extensively in Brazil to analyze the stability of anchored curtain walls that are usually constructed nearly vertical to retain slopes in residual or colluvial soils (*see Art. 9.2.5*). Seepage pressures are accounted for in the expression given as Eq. 8.8. The geometry and force system of the Culmann method often are applied to analyze rock masses where the failure surface is assumed.

$$FS = \frac{cL + (W \cos \theta - U) \tan \phi}{W \sin \theta} \tag{8.8}$$

Vertical Banks: Critical Height and Tension Cracks

When the slope angle i is greater than ϕ in Coulomb's equation ($s = c + \sigma_n \tan \phi$), the uppermost part of the soil adjoining the face is in a state of tension [Terzaghi (1943)[13]]. The *critical height* of a vertical slope is the maximum height which it may have before the state of tension is relieved by the formation of tension cracks. To estimate the critical height, the simplest and crudest assumption is that the soil is in an active Rankine state; the sliding surface is planar, rising at an angle of $\theta = 45° + \phi/2$, as shown in Fig. 8.9*a*; and there is no tension crack. It can be shown that when the total horizontal earth pressure on a vertical section is equal to zero, the critical height is

$$H_c = \frac{4c}{\gamma} \tan (45° + \phi/2) \tag{8.9}$$

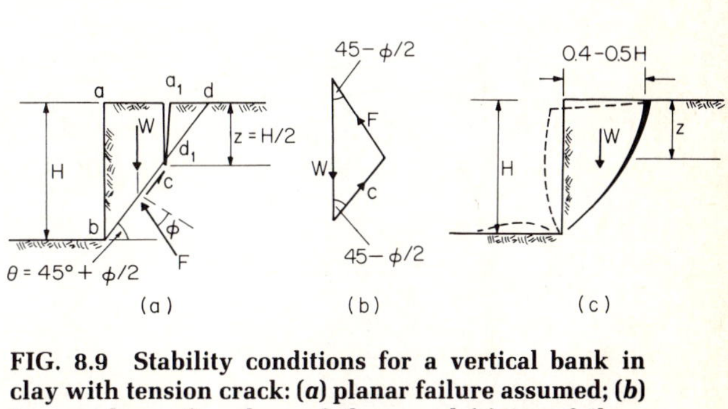

FIG. 8.9 Stability conditions for a vertical bank in clay with tension crack: (*a*) planar failure assumed; (*b*) force polygon for planar failure; and (*c*) true failure surface is curved. [Parts (*a*) and (*b*) from Terzaghi (1943).[13] Used with permission of John Wiley & Sons, Inc.]

and for the $\phi = 0$ case,

$$H_c = \frac{4c}{\gamma} \tag{8.10}$$

which is the value found in the Culmann analysis.

In actuality, the failure surface is not planar, but rather is *curved* (Fig. 8.9c), and Fellenius (1927)[6] showed that

$$H_c = \frac{3.85c}{\gamma} \tag{8.11}$$

Experiments have shown that tension cracks develop to a depth of about $z = H_c/2$. Using this assumption, and considering that wedge a_1dd_1 in Fig. 8.9*a* does not participate in a slope failure, and using a curved failure surface, Terzaghi (1943)[13] found that

$$H_c = \frac{2.67c}{\gamma} \tan (45° + \phi/2) \tag{8.12}$$

and for the $\phi = 0$ case

$$H_c = \frac{2.67c}{\gamma} \tag{8.13}$$

Water in the tension crack will increase the forces tending to cause failure, thereby decreasing the critical height, whereas the tension crack decreases the available shearing resistance along the failure surface and reduces the driving moment. Terzaghi suggested that since the factors of water forces in the crack and the driving moment reduction are practically compensatory, the effect of tension cracks in analysis can be disregarded as long as a reduced value for cohesion along the failure arc is used to compensate for the loss in effective length. In the foregoing discussion the tension cracks are those due to gravity and not to shrinkage. The depth of shrinkage cracks may be estimated from soil-suction techniques (*see Art. 4.3.6*).

Bearing-Capacity Analysis

The vertical slope in Fig. 8.10 was examined by Terzaghi (1943)[13] from the viewpoint of a bearing-capacity problem (*see Art. 6.2.4*). As strains develop in the slope under gravity forces, the upper portion goes into tension and cracks form. The tension crack location *B* defines a block of soil to be supported by the soil foundation. For the case in Fig. 8.10*a*, with *c* and ϕ acting, Terzaghi considered that the support could be expressed by the bearing-capacity equation (Eq. 6.10) without the depth term, since $D_f = 0$, from which he developed an expression for the critical slope height.

For the $\phi = 0$ case in Fig. 8.10*b*, the bearing capacity is expressed from Eq. 6.17 as $q_b = N_c s_c s_u + q = 5.14 s_c$, or $5.14c$, and the critical height for depth factor $d_f = (D + H)/H = \infty$ is

$$H_c = \frac{5.14c}{\gamma} \tag{8.14}$$

Comparing Eq. 8.14 with any of the expressions developed for H_c as a function of slope stability in the preceding section it is seen that slope stability and not bearing capacity governs in vertical slopes. Equation 8.14 is used, however, in stability analysis for embankments (*see Art. 8.2.4*).

By substituting a midpoint circle (Fig. 8.11*a*) for the failure surface in Fig. 8.10*b*, Fellenius found the critical height for $d = \infty$ to be

$$H_c = \frac{5.52c}{\gamma} \tag{8.15}$$

as discussed in the next section.

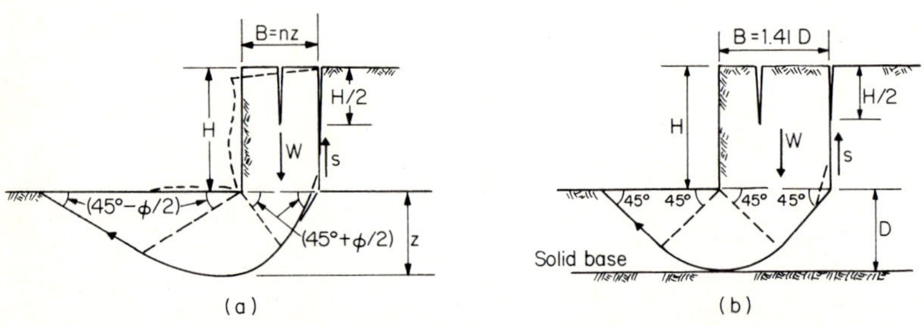

FIG. 8.10 Slope and base failure of a vertical bank: (*a*) $\phi > 0$ and (*b*) $\phi = 0$. [From Terzaghi (1943).[13] Used with permission of John Wiley & Sons, Inc.]

Inclined Slopes with Circular Failure Surfaces

The $\phi = 0$ Case

Fellenius (1927)[6] analyzed the slope illustrated in Fig. 8.11*a*, by assuming that equilibrium requires that the sum of the moments about the center O_p of rotation should be equal to zero:

$$WE - cLr = 0$$

or
$$c = \frac{WE}{rL} \tag{8.16}$$

from which
$$FS = \frac{rcL}{WE} \tag{8.17}$$

A computation of the values W, E, and L from the geometrical data shown in the figure demonstrates that c is a function of i, θ, and ω as in:

$$c = \gamma H \frac{1}{f(i,\theta,\omega)} \tag{8.18}$$

Slope failure occurs along a critical toe circle, or that toe circle for which the required c is a maximum. Fellenius grouped the terms $f(i,\theta,\omega)$ into a single factor called the *stability number* N_s, whose value depends only on the slope angle i, and

$$c = \frac{\gamma H}{N_s} \tag{8.19}$$

or, in terms of the critical height H_c,

$$H_c = \frac{cN_s}{\gamma} \tag{8.20}$$

Fellenius found that for slopes with $i > 53°$ the critical circle passes through the toe (Fig. 8.11*a*), whereas for slopes with $i < 53°$ the critical circle corresponds to a base failure or deep rupture (Fig. 8.11*b*), and in either case the center of the circular arc is situated on a vertical line passing through the midpoint of the slope. The circle depth below its center he defined with a depth factor $d_f = (D + H)/H$ as shown in Fig. 8.11*b*, and found that for $d_f = \infty$

$$H_c = \frac{5.52c}{\gamma} \tag{8.15}$$

for any slope angle up to 53°.

The stability number concept provided the basis for the development of stability charts such as those of Taylor (1948)[7] and Janbu (1954),[11] the latter given as Fig. 8.43. These charts relate the slope angle i to the stability number N_s defined as

$$N_s = \frac{\gamma H}{c} \tag{8.21}$$

Using the value for N_s obtained from the chart for a given value of i, FS for the slope is found from Eq. 8.21 as

$$FS = N_s \frac{c}{\gamma H} \tag{8.22}$$

It is seen on the chart (Fig. 8.43) that the limits for Janbu's stability number $N_c = 3.83$ at $i = 90°$ and 5.53 for $d = \infty$ essentially correspond with those developed for H_c in Eqs. 8.11 and 8.15 respectively, and that the toe circles occur for i above 53°.

The $\phi > 0$ Case

Where $\phi > 0$, shearing resistance is determined by $s = c + \sigma_n \tan \phi$, and failure may occur by either a slope or a base circle, with slope failures occurring along a critical toe circle. Fellenius (1927)[6] expressed the cohesion required to prevent slip in terms of ϕ and Eq. 8.18 as

$$c = \frac{\gamma H}{f(i,\theta,\omega,\phi)} \tag{8.23}$$

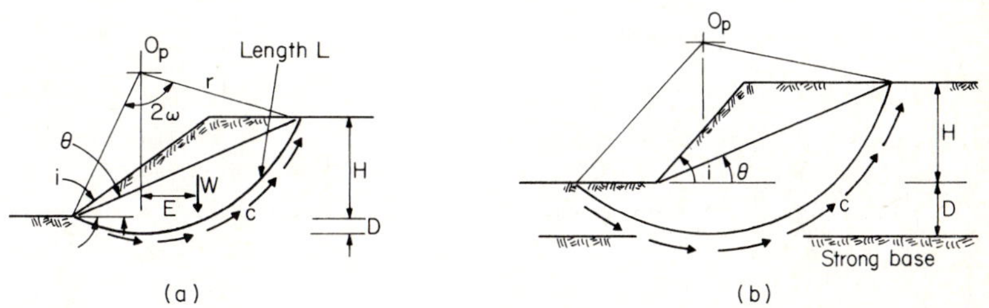

FIG. 8.11 Location of the failure circle for the $\phi = 0$ case: (*a*) toe circle with $i > 53°$ and (*b*) base failure with $i < 53°$.

and as for the $\phi = 0$ case, grouped the terms $f(i,\theta,\omega,\phi)$ into a stability number N_s. The cohesion required to prevent slip along a critical circle is given by Eq. 8.21. In this case, N_s depends on i and ϕ. Charts were also developed for simple analysis of slopes with $\phi > 0$. The charts for both the $\phi = 0$ and $\phi > 0$ cases were developed initially for total stresses, but in recent years charts have been prepared for application to effective stress analysis, such as those given in Art. 8.2.3.

The Friction Circle Method

The friction circle method was developed for analysis of homogeneous slopes exhibiting both c and ϕ [Taylor (1937),[27] (1948)[1]]. An essentially graphical method, it was in use for many years and provided the basis for a number of chart solutions still in use. The general basis for the concept is illustrated in Fig. 8.12. In addition to the above references its application is described in detail in Lambe and Whitman (1969).[16]

Fellenius Method of Slices

The foregoing discussion applies to simple homogeneous slopes in which the assumption of a uniform strength throughout is valid and seepage forces are not accounted for. In 1936, Fellenius (1936)[14] published a solution based on dividing the potential failure mass into slices and determining values for the "actuating force" and "resisting force" using vector polygons. FS is the ratio of the sum of the resisting forces to the actuating force.

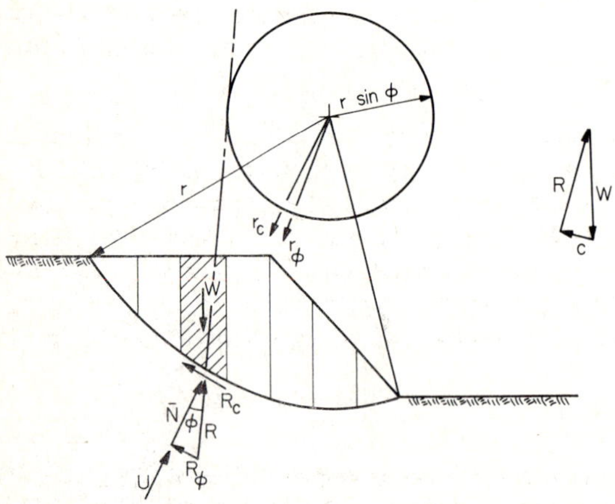

FIG. 8.12 The friction circle concept.

An analytical derivation of the slice method to account for seepage pressures was published by Taylor (1948)[1] and subsequently the method underwent various modifications until it became adaptable to analyzing nonhomogeneous slopes as well as homogeneous slopes for either total or effective stress conditions. In its present form it is known as the *ordinary method of slices (see Art. 8.2.3).*

8.2 STABILITY ANALYSIS: SOIL SLOPES

8.2.1 FACTORS TO CONSIDER

Synopsis

Geometric and time factors form the basis for selection of the proper method for soil slope stability analysis. Geometric factors refer to the project type (natural to cut slopes, earth dam embankments, embankments over soft ground, or sidehill fills) and stratigraphy (isotropic, homogeneous; isotropic, nonhomogeneous; or anisotropic). The time factor of short- or long-term conditions controls whether undrained or drained conditions prevail and thus, whether total-stress or effective-stress analysis is more suitable. A summary of stability analysis methods for soil slopes and applicable conditions covered in this article is given in Table 8.4. Soil-strength parameters, pore-water pressures, and in some cases earthquake forces are input to the analytical process. Depending upon natural conditions, all may be difficult to assess with accuracy.

Strength Parameter Selection

General

Soil type, strain history, and the loading time element are the major factors to consider in the selection of strength parameters *(see Arts. 3.3 and 3.5.3).* Strain history controls whether the peak, ultimate, or residual strength will prevail. The time element refers to short-term or undrained conditions (total stress analysis usually being most convenient), or long-term or drained conditions (effective stress analysis usually being most convenient). A general summary of applicable strength parameters for various geologic conditions is given in Table 8.5.

TABLE 8.4
SUMMARY OF STABILITY ANALYSIS METHODS FOR SOIL SLOPES

Method	Basis	Application	Reference
PLANAR FAILURE: INFINITE SLOPE			
		Homogeneous sands, soil over rock, weathered clay shales	Taylor (1948)[1] Lambe and Whitman (1969)[16]
CIRCULAR FAILURE: EFFECTIVE STRESSES			
Friction circle	Graphical	Simple homogeneous slopes	Taylor (1937),[27] (1948)[1]
Friction circle	Charts	Simple homogeneous slopes	Cousins (1978)[35]
Ordinary slices	Mathematical (hand calculators)	Nonhomogeneous slopes	Fellenius (1936)[14]
Modified Bishop	Mathematical (programmable calculator, PC's)	Nonhomogeneous slopes	Bishop (1955)[7] (preferred method)
CIRCULAR FAILURE: TOTAL STRESSES			
Friction circle	Charts	Simple, homogeneous slopes c_u, ϕ_u; or $\phi = 0$	Taylor (1937),[37] (1948),[1] Janbu (1954)[11]
Ordinary slices	Mathematical	Nonhomogeneous slopes	Fellenius (1936)[14]
Modified Bishop	Mathematical	Nonhomogeneous slopes	Bishop (1955)[7]
PLANAR FAILURE: WEDGES AND BLOCKS			
Simple wedge	Mathematical	Homogeneous clay slopes	Culmann (1866)[2]
Two blocks	Graphical	Sloping core earth dam	Seed and Sultan (1967)[4]
Two or three blocks	Mathematical	Soil or rock slopes	Huang (1983)[47]
Multiple blocks	Graphical/ mathematical	Nonhomogeneous	NAVFAC (1982)[42]
Multiple blocks	Mathematical	Nonhomogeneous slopes	Janbu (1973)[12]
IRREGULAR FAILURE SURFACES			
Slice method	Mathematical (needs large computer)	Nonhomogeneous slopes	Morgenstern and Price (1965)[10]
Slice method	Mathematical (programmable calculator, PC's)	Nonhomogeneous slopes	Janbu (1973)[12]

Strength measurements are obtained from tests performed either in the laboratory on undisturbed specimens or in the field *(see Arts. 3.3.2 through 3.3.4)*. Test method selection is based on soil type and drainage conditions to be evaluated and should also consider the relationship between strength and the failure surface orientation, as illustrated in Fig. 3.56*a* (compression, extension, or direct shear). Back analysis for failed slopes for similar geologic conditions provides data for correlations, and many examples can be found in the literature. A general summary of strength parameters is referenced in App. D of Hunt (1984).[69]

Drained Peak Strength (Effective Stresses)

The drained peak strength *(see Art. 3.3.2)* applies when soil water is permitted to drain during loading, thereby dissipating pore-water pres-

TABLE 8.5
FIELD CONDITIONS AND STRENGTH PARAMETERS ACTING AT FAILURE

Material	Field conditions	Strength parameter*
Cohesionless sands	Dry Submerged slope Slope seepage with top flow line coincident with and parallel to slope surface	$\phi(i = \phi)$ $\bar{\phi}(i = \bar{\phi})$ $\phi(i = \bar{\phi}/2)$
Clays (except stiff fissured clays and clay shales)	Undrained conditions Drained conditions	$s_u(\phi = 0)$ $\bar{c}, \bar{\phi}$†
Stiff fissured clays and clay shales and existing failure surfaces	Without slope seepage With slope seepage	$\bar{\phi}_r(i \approx \bar{\phi}_r)$† $\bar{\phi}_r(i \approx \bar{\phi}_r/2)$†
Cohesive mixtures	Undrained conditions Drained conditions	c_u, ϕ_u $\bar{c}, \bar{\phi}$†
Rock joints	Clean surfaces With fillings Clean but irregular surfaces after failure	ϕ, or $\phi + j$‡ $\bar{c}, \bar{\phi}$, or $\bar{\phi}$ $\phi_r + j$‡

*i = stable slope angle.

†Pore-water pressures: Reduce frictional resistance in accordance with $(N - U) \tan \bar{\phi}$.

‡j = angle of asperities.

sures due to the loading and mobilizing effective stresses. The material may be saturated but seepage is steady. These *long-term conditions* apply in most natural slopes, or some time after a cut is made in a slope, or after the completion of construction of an earth dam (the downstream face or steady-state reservoir level upstream face). Strength is expressed as $s = \bar{c} + (\sigma_n - u) \tan \bar{\phi}$.

Applicable strength parameters are: cohesionless sands and NC clays ($\bar{\phi}$, $c = 0$); OC intact clays ($\bar{\phi}$, c is usually taken as $= 0$); sand-clay mixtures ($\bar{\phi}$, $\bar{c}$); OC fissured clays (usually $\bar{\phi}_r$ pertains); and pre-existing failure surfaces ($\bar{\phi}_r$). If pore pressures do not exist in the slope, then the total stress equals the effective stress, but natural slopes seldom fail without the presence of excess pore pressures. Effective-stress analyses are not directly applicable to *partially saturated* soils because of negative pore pressures (the soil suction effect), as described in Art. 3.3.2 and as will be discussed further.

Undrained Peak Strength (Total Stresses)

The undrained peak strength *(see Art. 3.3.3)* pertains when soils are stressed rapidly compared to their drainage and consolidation rate, thereby not permitting pore-water pressures to dissipate, and total-stress analyses are usually applicable. These short-term conditions apply when a natural slope is excavated quickly, during construction of an earth dam, during rapid drawdown of the reservoir level behind a dam's upstream slope or along a stream bank, or during rapid construction of an embankment over soft ground. The critical time is often at the end of construction. Strength is expressed as $s = c + \sigma_n \tan \phi$, or for the $\phi = 0$ case as s_u.

Applicable strength parameters for total stress analysis are the undrained strengths c_u and ϕ_u, or s_u. The s_u, or $\phi = 0$ case, is applicable basically to soft clays. Short-term conditions also can be analyzed on the basis of effective stresses, but pore-water pressures must be accounted for either by prediction in terms of total stress changes or from field observations using piezometers. The latter is the common procedure.

The peak drained or undrained strength occurs at relatively small strains and may be measured by the maximum obliquity (CD or CU test) or by the maximum difference in principal stresses (UU test) *(see Art. 3.5.3).*

Residual or Ultimate Strength

The *ultimate or minimum strength* occurs after the peak strength is reached under continuing large strains. The *residual strength*, which may or may not be lower than the ultimate strength, but is lower than the peak strength, usually prevails along pre-existing failure surfaces *(see Art. 3.3.4).* Large strains in cohesive materials cause remolding along failure surfaces and in the field often are evidenced by shiny, striated surfaces (slickensides). Such conditions are common in fault zones, in natural slopes in residual soils with relict weakness planes, in stratified deposits containing weak layers along which movement has occurred, and in overconsolidated fissured clays.

Applicable strength parameters are the undrained residual or remolded strength s_r and the drained residual strength $\bar{\phi}_r$. In effective stress analysis the cohesion intercept should seldom be used in analyzing *existing landslides* in soil or rock, as the analysis is likely to indicate strengths unrealistically high and on the unsafe side [Patton and Hendron (1974)[15]]. Analysis is based on $\bar{\phi}_r$, or some value between $\bar{\phi}$ and $\bar{\phi}_r$. Residual strength for clay, sand, and silt mixtures can range from $\bar{\phi}_r = 12$ to 24°, and for clays $\phi_r = 4$ to 14°. Stiff fissured clays and clay shales often contain zones or layers at $\bar{\phi}_r = 9$ to 12°, and in colluvium $\bar{\phi}_r$ may range from 15 to 20°.

Progressive Failure

Progressive failure, which occurs in many natural slopes, results from strains at various rates and magnitudes and can occur in several forms.

- Failure proceeds upslope in the form of a series of slumps after a failure at the slope toe, either on a periodic basis or during a total failure of the slope.
- Strength reduces with time from peak to residual values at some point in the slope as strains occur, or as a result of the variation in normal stress acting along the failure surface.
- Shear strength along the entire failure surface during rupture varies from values ranging from peak to residual strength.

In analysis, progressive failure is accounted for approximately in several ways:

- In residual or colluvial soils, or soils where failure by lateral spreading may occur, such as glacial lacustrine or marine soils, by determining that the stability at the toe is adequate
- Applying strengths obtained in the laboratory at strains between peak and residual strength, or in some cases using the peak strength along part of the failure surface and the residual strength along the remainder

Pore-Water Pressures and Seepage Forces

Significance

In *natural slopes* almost all failures occur during or immediately after periods of heavy rain or during wet seasons, indicating that pore-water and seepage pressures are a most important aspect of stability analysis. These pressures result generally from rising water-table levels. In *constructed embankments*, failures during or after construction are usually the result of high pore pressures induced by embankment loads or rising water levels and in addition, for earth dams, the result of high seepage forces.

Pore-water pressures *(see Art. 2.3.2)* reduce the resistance along the sliding surface in accordance with $S = (N - U) \tan \phi$. Seepage pressures *(see Art. 2.3.2)* reduce the effective soil weight along the lower portions of a potential failure surface, thereby reducing passive resistance, and piping can cause a complete loss of support from passive resistance at the toe. In addition, the material weight in the mass overlying the potential failure surface is increased by saturation or partial saturation, thereby increasing the "driving" forces.

Analysis

In general, for the *no-flow condition* (Fig. 2.8*a*) pore pressure *u* is taken as the height of the phreatic surface above the failure surface times the weight of water or

$$u = \gamma_w z_w \quad (2.7)$$

If *flow is occurring*, two force systems are applicable as illustrated by Fig. 2.8*b*: buoyant weight of material, seepage force, and buoyant effective stresses; or total weight of material, boundary pore pressures, and effective boundary stresses. The latter is usually the more convenient approach.

Pore-pressure ratio r_u is a parameter used to account for the effect of pore-water pressures in computerized methods of stability analysis. It is defined as the ratio between the total upward force due to water pressure and the total downward force due to overburden pressure acting on the failure surface (Fig. 8.32), expressed as

$$r_u = \gamma_w h/\gamma_t d = u/\gamma_t d \quad (8.24)$$

Since the unit weight of soil is approximately equal to twice the unit weight of water, r_u may be expressed approximately as

$$r_u = \frac{\text{cross-sectional area of sliding mass under water}}{2 \times \text{total cross-sectional area of sliding mass}} \quad (8.25)$$

Predicting Pore-Water Pressures

EARTH DAMS Prior to construction, flow nets (*see Art. 2.3.2*) are drawn to locate the phreatic surface from which pore pressures acting along a potential failure surface can be determined and

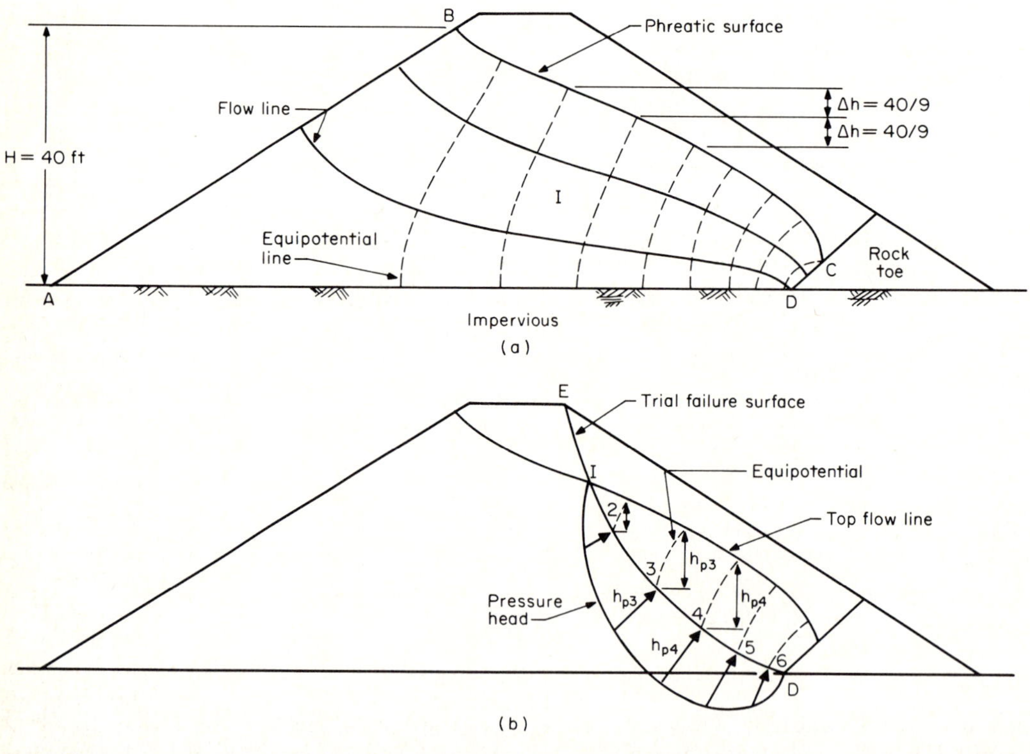

FIG. 8.13 Unconfined flow net through earth dam with rock toe to control toe seepage. (*a*) Use of flow net to find seepage quantity and gradient. Given: k = 0.0005 ft/s, H = 40 ft, as $\mathcal{S} = N_f/N_e$ = 2.65/9 = 0.294. Seepage through dam = $Q/L = kH\mathcal{S}$ = 0.0005 × 40 × 0.294 = 59 × 10^{-4} ft^3/min/ft. Gradient in square I = $i_I = \Delta h/l_1$ = 40/9/11.2 = 0.40. (*b*) Use of flow net to find pore-water pressures on a failure surface: (1) lay off trial failure surface on flow net, (2) measure pressure heads as elevation differences between phreatic line and trial surface at equipotential lines, and (3) pore-water pressure = $h_p\gamma_w$. [From Lambe and Whitman (1969).[16] Adapted with permission of John Wiley & Sons, Inc.]

seepage forces evaluated, as illustrated in Fig. 8.13, for effective stress analyses. Piezometers are used after construction to measure actual pore pressures.

NATURAL SLOPES In an existing natural stable slope the water levels are determined by piezometers and the phreatic surface drawn on a slope section, or in some cases flow nets are drawn if seepage forces are a concern. The validity of the analysis as it pertains to the potential for failure, however, is based on the relationship of the measured phreatic surface to the potential failure surface and whether or not the location is the most critical that might occur during high

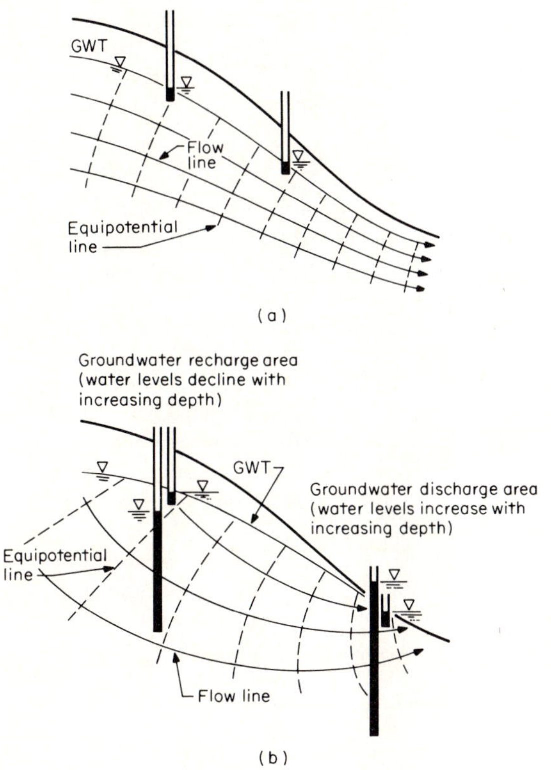

FIG. 8.14 Comparison of the normal concept of groundwater flow in slopes with the more typical occurrence causing instability: (*a*) groundwater flow assumed parallel to the groundwater table (common in geotechnical literature, but seldom found in practice) and (*b*) typical groundwater flow in slopes. [From Patton and Hendron (1974).[15]]

rainfall conditions. If insufficient GWL data are available or if dry-season conditions are present, a flow net might be drawn as in Fig. 8.14*a*, whereas in reality wet season conditions are more commonly as shown in Fig. 8.14*b* with high piezometric levels in the downslope discharge area.

The major uncertainty in evaluating the stability of a natural slope lies in estimating the maximum elevation the phreatic level may feasibly attain since saturation relieves the strengthening effect of negative pore pressures in a partially saturated soil *(see Art. 3.3.2, Soil Suction)* and causes an increase in pore-water pressure on the potential failure surface. An approach proposed by Lumb (1962),[17] (1975)[18] to evaluate the potential rise in the phreatic surface during rainfalls considers the infiltration of water into the horizontal surface of a porous medium of saturated permeability k and porosity n. He showed that the advance of the *wetting front* of thickness h after time t is characterized approximately by

$$h = \frac{kt}{n(S_f - S_o)} \tag{8.26}$$

where t = length of a wet season or given storm
S_o = initial degree of saturation
S_f = final degree of saturation

In Hong Kong, Eq. 8.26 is used frequently for the prediction of design pore pressures in the residual soils derived from granite and rhyolite by superposition of the wetting front depth onto the GWL at the end of the wet season. Hong Kong wetting band thicknesses are usually calculated as being about 6 ft (1.8 m) for each return of the unusually wet seasons which come about once every 10 years [Brand (1982)[19]]. For a discussion of the relationship between slope failures and rainfall see Art. 9.3.4 of Hunt (1984).[69]

Embankments over Soft Ground

Placement of an embankment over soft clays causes an increase in pore pressures. Effective stress analysis is performed by employing a value for u calculated for the imposed stress conditions from parameter A measured in the CU triaxial test *(see Art. 3.2.2)*, as will be discussed. Stability is re-estimated during construction from pore pressures measured with piezometers.

Natural Slopes and Geologic Conditions

Drained versus Undrained Conditions

Pore pressures normally are governed by steady seepage in natural slopes and *effective stresses* prevail. Stability analysis employs drained strength parameters and pore-water pressures. *Undrained* conditions may occur during sudden drawdown, as occur after stream levels have risen, causing ground saturation and an increase in pore-water pressures, and then the levels drop rapidly.

Natural slopes in soils with cohesion can be separated into two zones for stability analysis: one completely saturated and the other only partially saturated. Strength parameters for the saturated zone are based on effective stresses as measured by testing saturated specimens. In the partially saturated zone, effective stresses cannot be applied directly because of negative pore pressures. These negative pore pressures, which cannot be evaluated precisely, influence effective stresses. This may be an explanation for the fact that many stable slopes in residual soils are calculated to have values for FS less than unity. For analysis, strength parameters may be modified as discussed in Art. 3.3.2 concerning *soil suction* effects, or as an approximate alternative, obtained by testing undisturbed specimens at natural water content in the direct shear apparatus. Much more significant in analysis is the correct assumption for the maximum feasible pore-water pressures along the potential failure surface.

The foregoing pertains primarily to sand-clay mixtures such as are common beneath residual soil slopes where, depending upon the degree of saturation, strengths may be as little as 50% of the strength at field moisture. In intact clay slopes in the unsaturated zone, capillary forces are likely to cause tension cracks along which there is no shearing resistance. In a stable slope, the tension crack depth will be a function of the saturation degree and probably will extend to the upper level of constant soil suction *(see Art. 4.3.6)*. Tension cracks are discussed in Art. 8.1.3.

Coarse-Grained Cohesionless Soils ($c = 0$)

Stability in clean sands depends primarily on groundwater conditions. With a low GWL (Fig. 8.15*a*), failure occurs as surface sloughing until the slope flattens to the friction angle ($i = \phi$). The same condition applies to a submerged slope. If seepage occurs parallel to the slope as the water table rises, the stable slope angle becomes approximately $\frac{1}{2}\bar{\phi}$ when the top flow line is coincident with the slope (Fig. 8.15*b*). Analysis is by the finite slope method *(see Art. 8.2.2)*. Consideration must be given to the possible flow forms of mass movement in loose, unsaturated fine sands with apparent cohesion, which become saturated or subject to earthquake forces.

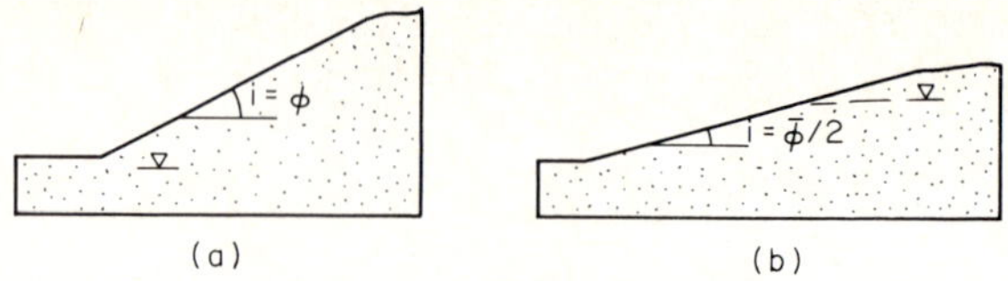

FIG. 8.15 Slope in sands: (*a*) low groundwater level and (*b*) high groundwater level.

NC to Slightly OC Clays ($c = 0$, or $\bar{c}$, $\bar{\phi}$)

Failure may occur as a circular arc to form a slope, toe, or base circle as illustrated in Fig. 8.16 and discussed in Art. 8.1.3. The location of the failure surface is primarily a function of soil strength; a slope circle occurs where strength increases significantly with depth, whereas if strength is constant with depth a base circle occurs. Groundwater influences stability by acting along relatively pervious seams and in tension cracks and by increasing soil weight and changing soil strength during fluctuations. A stiffer, upper desiccated zone is common among many slopes composed predominantly of soft clays. In isotropic, homogeneous conditions, analysis may

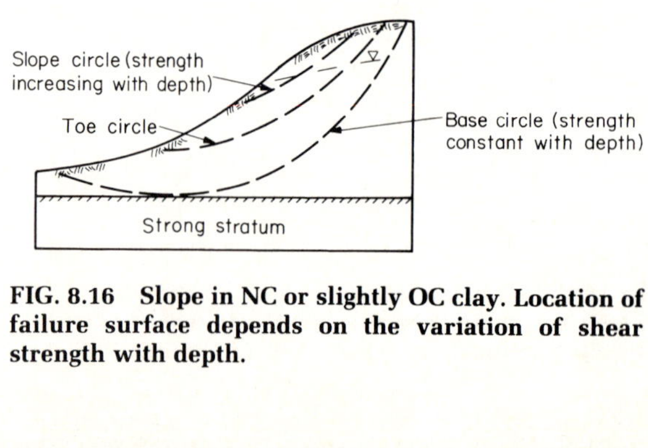

FIG. 8.16 Slope in NC or slightly OC clay. Location of failure surface depends on the variation of shear strength with depth.

be performed with the friction circle method; otherwise, methods of slices are used.

Analysis of intact clay slopes, including normally consolidated to slightly overconsolidated soils in the firm to stiff range, that are not fissured, contains many uncertainties as discussed under *Cut Slopes*.

"Quick" or sensitive clay slopes *(see Art. 3.3.4 and Table 3.8)* require special consideration as they are extremely sensitive to rapid increases in shear strain or pore-water pressure. Many slides occur in these soils from erosion at the toe, sudden drawdown of flood waters, or vibrations such as those induced by earthquakes. Stability of natural slopes in sensitive clay or loose, saturated sand should be analyzed by effective stress procedures using pore pressures associated with the worst condition rather than strength parameters representing the undrained peak laboratory strength [Bjerrum (1973)[20]]. Failure in these soils appears to occur commonly along a thin seam of relatively low shear strength, resulting in a long, planar failure surface bounding a wedge [Leonards (1982)[21]], although such failures may be initiated by a circular toe failure (Fig. 8.17). Continuous soil sampling is necessary during investigation to identify possible weak zones, but due to the possibility of their not being detected, the slope should be considered as stable only if FS $\geq$ 1.5.

Slides in these soils, characteristic of glacial lacustrine and marine deposits, are often examples of retrogressive failing by lateral spreading. In other cases, under increased or sustained shearing stress, the high void ratio structure of "quick" clays collapses. Since there is little opportunity for drainage, pore pressures increase until the effective stress along the potential failure surface reaches zero. At this point the soil liquefies and flows as it has no effective shear strength. Failures are common after stream flood stages drop.

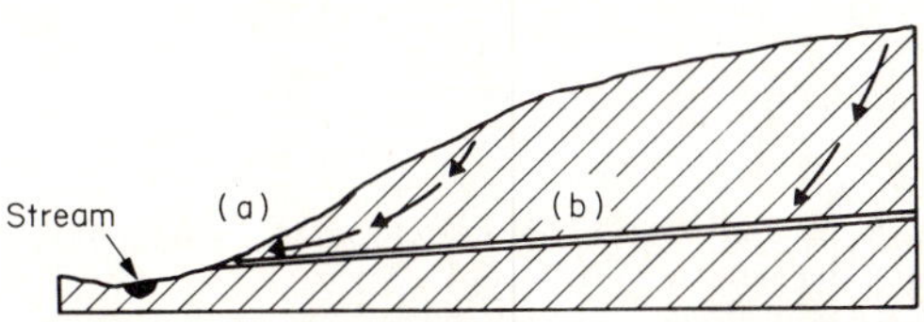

FIG 8.17 Slope failures in sensitive clays may be initiated by a toe circle at (*a*) and usually have a relatively long planar surface (*b*) defined by a very weak zone.

Stiff-Fissured Clays and Clay Shales ($\bar{\phi}_r$)

Deterioration of interparticle bonds from heavy prestress in stiff fissured clays and clay shales causes strength to reduce with time from peak to near-residual values associated with irreversible time-dependent deformations in the slope. Water in the slope accelerates the strain-softening process through the increase in pore pressures and swelling of expansive materials. Seasonal variations in rainfall and water level have a severe effect on stability and progressive failure is common [Bjerrum (1967)[22]]. Natural slopes in these materials usually rest with FS near to unity and analysis is based on $c = 0$ and the residual shear strength $\bar{\phi}_r$.

Failures may occur in several modes: shallow surface slumping, the development of local wedge failures along joints and weakness planes, the development of a major wedge failure along a weakness plane as shown in Fig. 8.18*a* or on a long, shallow weakness plane bounding the weathered and unweathered zones in the infinite slope failure mode (Fig. 8.18*b*). The failure form, therefore, is very much related to the occurrence of major weakness zones in the form either of the weathered zone or of weak seams or fissures *(see also Art. 8.3.2)*.

Sand-Clay Mixtures ($\bar{c}$, $\bar{\phi}$, or $\bar{\phi}_r$)

Sand-clay mixtures in slopes are common in residual and colluvial soils, loess, and occasionally alluvium. In *homogeneous mixtures*, in general,

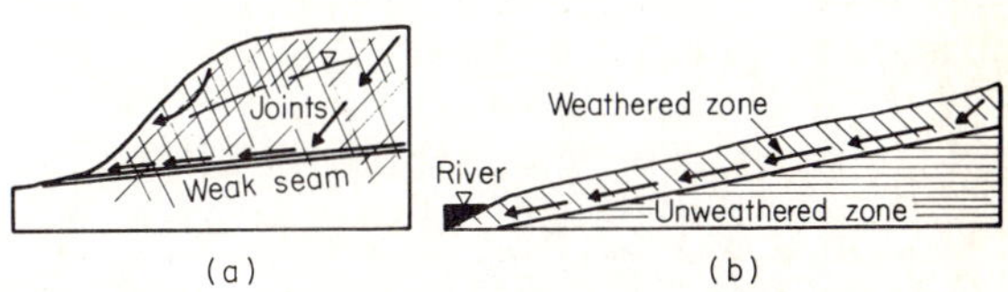

FIG. 8.18 Failures in stiff fissured clays. In (*a*) the existing slope is following the pattern of the fissures, and potential failures may be a shallow circle or a large wedge rupturing along a weakness plane. In (*b*) failure is in the form of the infinite slope following the boundary between the weathered and unweathered zones.

with deep groundwater, failure occurs on a shallow straight or slightly curved surface (Fig. 8.19*a*); with high groundwater, failure occurs on a relatively shallow toe circle whose position is determined primarily by the groundwater elevation (Fig. 8.19*b*). The presence of a tension crack influences the location of the failure surface. Analysis may be performed with the friction circle or similar method using parameters $\bar{c}$ and $\bar{\phi}$ and accounting for pore pressures.

Residual soil profiles are characterized by several transitional zones: mature residual soil to saprolite with relict rock structure to weathered and fractured rock to unweathered rock. Shallow *noncircular* failure surfaces along the base of the residual soil zone are most common, but failure can occur also between any two adjacent zones or along relict joints in the residuum (Fig. 8.20). The effective stress conditions described previously for saturated versus partially saturated soils apply particularly to residual soils. Janbu's method of slices is useful for analysis of the usually noncircular failure surfaces.

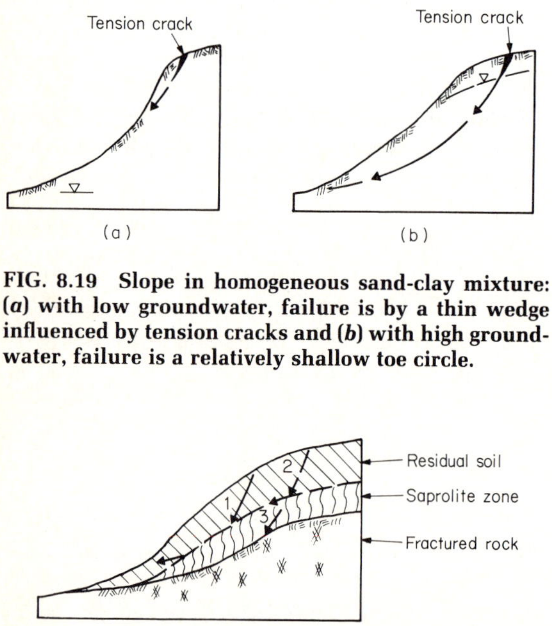

FIG. 8.19 Slope in homogeneous sand-clay mixture: (*a*) with low groundwater, failure is by a thin wedge influenced by tension cracks and (*b*) with high groundwater, failure is a relatively shallow toe circle.

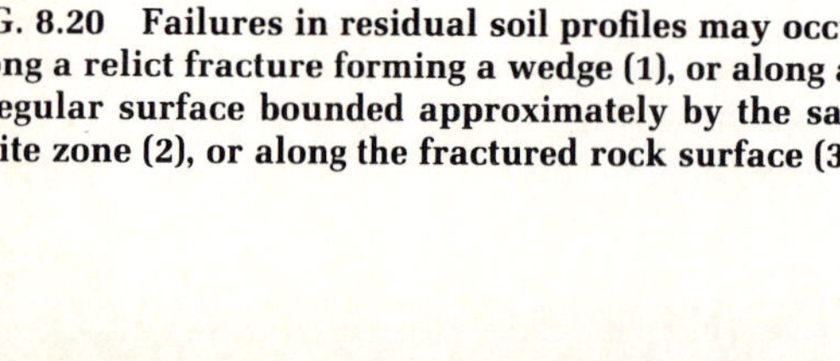

FIG. 8.20 Failures in residual soil profiles may occur along a relict fracture forming a wedge (1), or along an irregular surface bounded approximately by the saprolite zone (2), or along the fractured rock surface (3).

Colluvium, formerly residual soils or other materials, is common in hilly or mountainous terrain in humid climates. In warm, humid climates in residual soils derived from gneiss, it is usual for the normally steep slopes to fail progressively, forming a series of scarps as shown in Fig. 8.21*a*. Deposits can be thick, or thin and underlain by rock. Movement can occur periodically during wet seasons, stabilizing during dry periods, or a very heavy rainfall can cause the entire mass to move downslope in a total failure, often along the rock surface. During progressive failure, as each slump occurs (usually after an initial toe failure), passive resistance of the material upslope decreases, tension cracks are formed (permitting the entry of water which imposes lateral hydrostatic stresses), and another slump occurs. An alternate case of a single old, mature slump failure is shown in Fig. 8.21*b*. The failure surface is located by inclinometer readings and analysis is based on the drained residual strength $\bar{\phi}_r$. Colluvial soil slopes often are resting at an FS very near unity with a slope angle i approximately equal to $\bar{\phi}_r/2$ (see Art. 8.2.2, *The Infinite Slope*).

Loess is usually found naturally on near-vertical slopes of substantial height due to its lightly cemented structure. When it rests on a rock surface, its sliding may take the form of block movement. If the mass in some way becomes saturated, the cementing bonds are destroyed and the soil flows as a slurry. Earthquake forces can cause collapse of the structure, resulting in high pore-air pressures and massive flows. Nor-

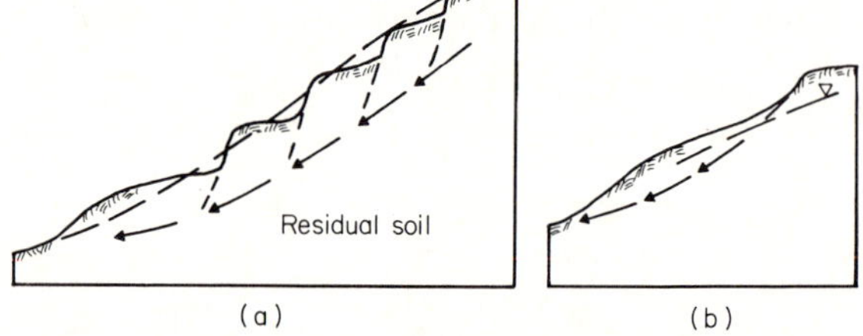

FIG. 8.21 Colluvial soil development in residual soil slopes derived from gneiss in a warm, humid climate. In (*a*) blocks move downhill during rainy periods and stabilize during dry spells, forming a series of rotational slumps. In (*b*) an old, mature slide rests below its head scarp.

mal stability analysis, however, assumes that the slope will remain unsaturated, and effective strength parameters are employed, as measured under low confining pressures so as not to destroy the natural structure.

Stratified Soil Profiles

In stratified soil profiles the location of the failure surface is controlled by the relative strength and orientation of the strata. For analysis, planar failure surfaces (Fig. 8.22) may be assumed which divide the failed mass into one or more wedges. For a failure surface below the toe, the mass is divided into passive and active wedges bounding a central wedge (Fig. 8.55).

Cut Slopes

Effective versus Total Stress Analysis

In the selection of effective versus total stress analysis there are three stages of construction to consider: intermediate stages, the end of construction with saturated or partially saturated soil, and long-term stability.

INTERMEDIATE STAGES If the construction period is sufficiently long to permit drainage in a saturated soil, then stability is based on effective stress parameters with estimated or measured pore pressures. If construction time is short compared to the consolidation time and soil drainage characteristics, then analysis is based on total stresses.

END OF CONSTRUCTION In partially saturated soil, if drainage has been permitted, analysis can be based on either total stresses with ϕ_u and c_u determined by CU triaxial tests, or $\bar{c}$ and $\bar{\phi}$ with estimated or actual measured pore pressures. In

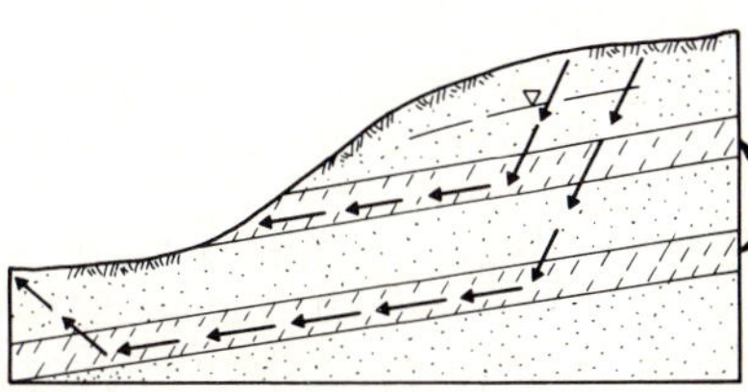

FIG. 8.22 Stratified soil profile. Location of the failure surface depends on strength and orientation of the strata.

saturated soil, total stress analysis and effective stress analysis using measured pore pressures as a check against stability may be performed.

LONG-TERM CONDITIONS For the postconstruction period, analysis is performed based on effective stress parameters with measured pore pressures, with consideration of the possible rise in GWL. The *minimum factors of safety* acceptable for slopes in the residual soils of Hong Kong in terms of risk category (*see Art. 8.4.2*) for a 1 in 10 year storm are: 1.2, low risk; 1.3, significant risk; and 1.4, high risk [Brand (1982)[19]].

Soil Condition

For most geologic conditions, in general, cut-slope situations are similar to those of natural slopes except that short-term conditions during construction must be considered. In all slopes, stability is decreased by cutting, and since in many geologic conditions the natural slope is resting near equilibrium, an increase in slope inclination is quite likely to result in movements or even failure unless precautions are taken. This is particularly true of stiff fissured clays and colluvial soil slopes.

FINE-GRAINED GRANULAR SOILS Silt and sand mixtures may stand essentially stable on near-vertical slopes to substantial heights during construction dewatering operations due to *apparent cohesion* caused by pore-water tensions. This is a temporary condition. Its loss when the water table is permitted to rise may result in failure of the entire slope or, when the bank dries in the sun, may result in local sloughing.

INTACT CLAYS A layered profile, such as shown in Fig. 8.23 is common in intact clays, in which an upper zone strengthened by desiccation often overlies a lower weaker zone, which in turn is underlain by a stronger zone. Stability analysis for the construction phase employs total stresses (the $\phi = 0$ case) in some method of slices. In a state-of-the-art review of cut slopes in *soft clays*, Leonards (1982)[21] concluded that analysis of long-term stability still contains many uncertainties. For example, most failures apparently involved planar slip surfaces developing along thin, weak zones, indicating that slip-circle analysis is not applicable, and cuts induce zones of

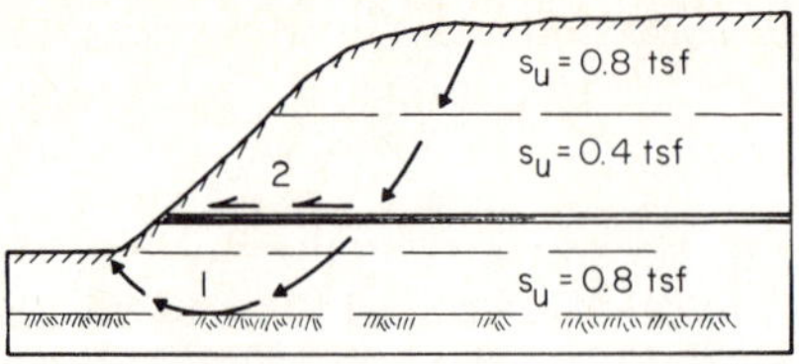

FIG. 8.23 Stability analysis of a cut slope in intact clay with an irregular profile is performed usually employing the undrained strength with a slip-circle (1) and slice method for the construction phase, but planar failure along a weak seam (2) is a strong likelihood.

overstressing of sufficient magnitude to cause creep strains. Creep strains may result in a reduction in the effective strength parameters which can cause a progressive undrained failure. Leonards noted that years ago, prior to modern techniques of investigation and analysis, cut slopes in soft clays were designed for FS = 1.5 and failures seldom occurred, whereas the application of modern techniques and much reduced values of FS, in the order of 1.2, has led to unstable slopes with inclinations cut too steep.

STIFF FISSURED CLAYS AND CLAY SHALES When unloaded by cutting, these clays and shales tend to expand and induce shear stresses, which in the presence of water cause the material to reduce in strength toward residual values with time. Since these materials normally rest with FS near unity, cutting is likely to result in failure.

RESIDUAL OR COLLUVIAL SOILS OVER SHALLOW ROCK When cut as shown in Fig. 8.24, such soils are likely to fail retrogressively as tension cracks separate the mass into blocks. With high enough groundwater and rainfall, total failures are common. Only analysis of the total slope is feasible, unless a tension crack has formed.

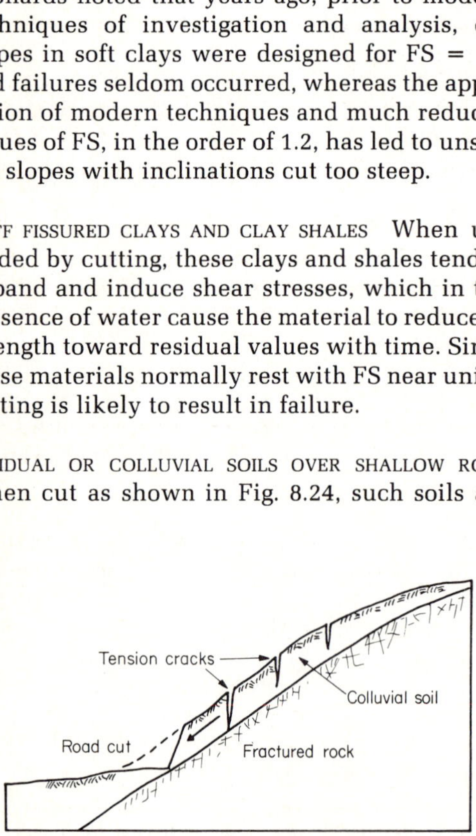

FIG. 8.24 Road cut in colluvium over a shallow rock surface is likely to result in failures as tension cracks form and blocks fail in sequence.

Earth-Dam Embankments

Effective versus Total Stress Analysis

During construction of an earth-dam embankment, shear stresses and pore-water pressures on potential failure surfaces increase as the fill height is raised in successive lifts. The installation of piezometers to monitor pore pressures provides the control for determining when the embankment has gained sufficient strength to permit placement of the next series of lifts. Analysis may be based on total stresses using parameters obtained from unconsolidated, undrained tests performed on unsaturated specimens at compacted densities and checked during construction by effective stress analysis with measured pore pressures.

Unsaturated clay cores may require special consideration because of negative pore pressures which do not apply in effective stress analysis but which are considered to provide a strength increase *(see Art. 3.3.2, Soil Suction)*. Because the time interval during which negative pore pressures in partially saturated soils persist in the field is uncertain, the increase in strength from soil suction should be neglected in conservative analysis [Lowe (1967)[24]]. In the total stress expression, the effective stress parameters $\bar{c}$, $\bar{\phi}$ should be used in the negative pore-pressure range, and $\bar{c}$ usually equals zero as shown in Fig. 3.32. Alternatively, more refined analysis has employed Eq. 3.38 and the procedure described in Art. 3.3.2, assuming a value of $u_a = 0$ for compacted clays, and during construction performing a check analysis based on measured pore pressures. When there is a significant difference between pore-air and pore-water pressure, pore-air pressure may be measured in embankments with either pneumatic or vibrating-wire diaphragm-type piezometers with coarse porous tips (low entry value) [Sherard (1981)[25]].

Following construction the excess pore pressures begin to dissipate and the embankment material gains strength. A check is made based on effective stresses and measured pore pressures.

LONG-TERM STABILITY AFTER RESERVOIR FILLING As the water level rises behind the embankment, it becomes saturated and a phreatic surface is attained as shown in Fig. 8.13. The *downstream face* is designed based on effective stresses and predicted pore pressures, incorporating a design section to prevent the development of excessive seepage pressures. Large clay cores may be analyzed based on total stresses. The *upstream face* is designed for *sudden drawdown conditions* (Fig. 8.25) based on total stresses and the undrained strength, or effective stresses considering excess pore pressures. For the total stress case, soil unit weight during drawdown increases from the submerged to the total unit weight.

Natural slopes in the reservoir area require evaluation of the effects of submergence on stability. Decreased stability is caused by the rise in the water table upslope and the change in soil weight in the submerged portion. The situation is critical where large slope movements are already occurring or relict failure surfaces are present. [See the discussion of the Vaiont Dam in Art. 9.2.3 of Hunt (1984).[69]]

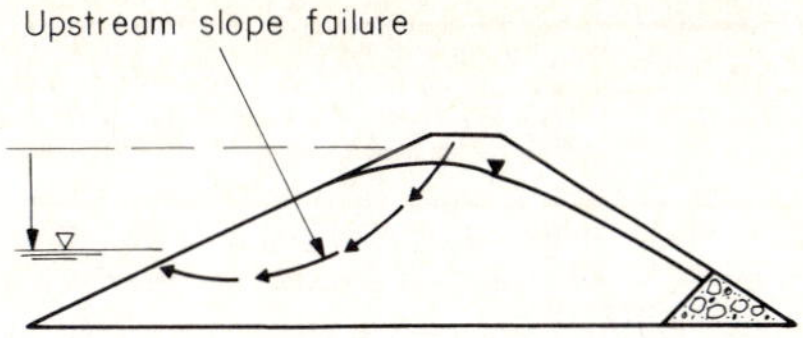

FIG. 8.25 Water level conditions in an earth dam after sudden drawdown in relatively slow-draining materials.

A *summary* of total versus effective analysis is given in Table 8.6, and suggested values for FS for earth and rockfill dams in current usage for various conditions are given in Table 8.7. A brief discussion of design treatments for earth dams is given in Art. 8.4.5, and monitoring systems are illustrated in Fig. 8.84.

Failure Surface Shapes

In homogeneous embankments, analysis is based on a circular arc failure surface. In zoned embankments, slice methods for irregular failure surfaces have been used, although analysis by wedge methods is popular. Examples are given in Figs. 8.26 and 8.53. Figure 8.26*a* shows a

TABLE 8.6
TOTAL VERSUS EFFECTIVE STRESS ANALYSIS*

Condition	Preferred method	Comment
Stability at intermediate times	$\bar{c}$, $\bar{\phi}$ analysis with estimated pore pressures	Actual pore pressures must be field-checked
End of construction; partially saturated soil; construction period short compared to soil consolidation time	Either method: $\bar{c}_u$, ϕ_u from CU tests, or $\bar{c}$, $\bar{\phi}$ plus estimated pore pressures	$\bar{c}$, $\bar{\phi}$ analysis permits check during construction using actual pore pressures
End of construction; saturated soil; construction period short compared to consolidation time	Total stress or s_u analysis with $\phi = 0$ and $c = s_u$	$\bar{c}$, $\bar{\phi}$ analysis permits check during construction using actual pore pressures
Long-term stability	$\bar{c}$, $\bar{\phi}$ analysis with pore pressures given by equilibrium groundwater conditions	Stability depends on amount of water-table rise and pore-pressure increase

*After Lambe and Whitman (1969).[16] Adapted by permission of John Wiley & Sons, Inc.

TABLE 8.7
MINIMUM VALUES FOR FS FOR EARTH AND ROCKFILL DAMS[a,b]

Case	Design conditions	FS_{min}	Shear strength[c]	Remarks
I	End of construction	1.3[d]	Q or S[e]	Upstream and downstream slopes
II	Sudden drawdown from maximum pool	1.0[f]	R, S	Upstream slope only, use composite envelope
III	Sudden drawdown from spillway crest	1.2[f]	R,S	Upstream slope only, use composite envelope
IV	Partial pool with steady seepage	1.5	$\frac{R+S}{2}$ for $R < S$	Upstream slope only, use intermediate envelope
V	Steady seepage with maximum pool	1.5	$\frac{R+S}{2}$ for $R < S$	Downstream slope only, use intermediate envelope
VI	Earthquake (Cases I, IV, and V with seismic loading)	1.0	—[g]	Upstream and downstream slopes

[a]From Wilson and Marsal (1979).[26]

[b]Not applicable to embankments on clay shale foundations; higher FS values should be used for these conditions.

[c]Q = quick (unconsolidated-undrained test), S = slow (consolidated-drained test), and R = intermediate (consolidated-undrained test).

[d]For embankments more than 50 ft high over relatively weak foundation use FS_{min} = 1.4.

[e]In zones where no excess pore-water pressures are anticipated use S strength.

[f]FS should not be less than 1.5 when drawdown rate and pore-water pressures developed from flow nets are used in stability analysis.

[g]Use shear strength for case analyzed without earthquake. (Values for FS are based on pseudostatic approach.)

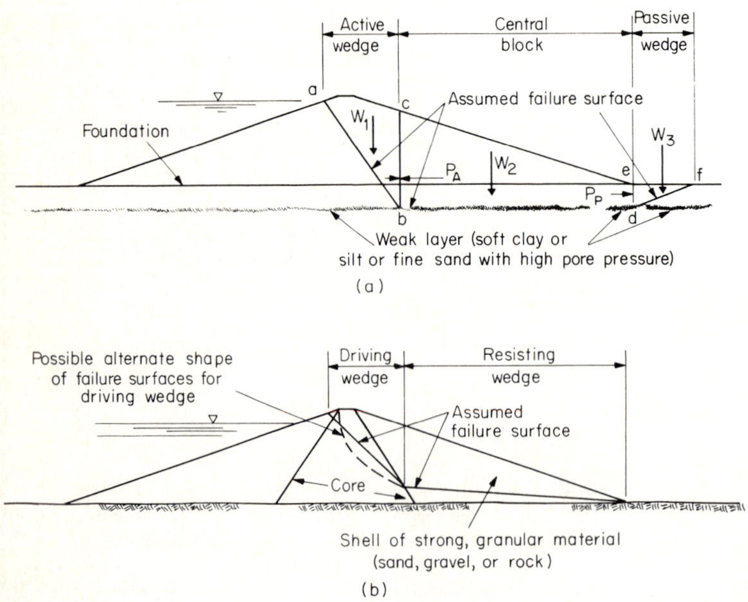

FIG. 8.26 Situations where wedge (sliding block) analyses are applicable: (*a*) dam on foundation containing weak horizontal layer and (*b*) dam on strong foundation with a core and granular shells. [From Sherard et al. (1963).[23] Used with permission of John Wiley & Sons, Inc.]

wedge analysis for the case of an embankment supported on a foundation with a weak layer.

Deformation Analysis

Strains within an embankment from excessive deformations can result in the formation of cracks which become channels for water to exit on the downstream slope, possibly resulting in dangerous piping erosion, and can dislocate or cut off internal drainage layers. The effect also can be very critical in core trenches. The FEM is commonly employed in deformation analysis.

Embankments, Jetties, and Breakwaters over Soft Ground

Embankments

During construction of a roadway or other embankment placed over a soft clay foundation, shear stresses and pore-water stresses increase as the fill is placed in successive lifts. Failure by rupture depends on the fill height and the rate of loading and can be in the form of a rotational slip or a translation along a planar surface. The latter condition occurs where the weak soil is relatively thin compared to the embankment width, or where a continuous very weak seam is present (Fig. 8.27*a* and *b*).

Minimum stability usually occurs at the end of construction, and analysis is based on total stresses (the $\phi = 0$ case). If the strains of the fill and the foundation soils differ greatly, as they do in the case of an engineer-controlled fill, the FS will exceed unity, but the strength of the fill usually is ignored in calculations as it often fails in tension. As a less conservative alternative, the real fill may be replaced by an ideal clay ($\phi = 0$) and assigned a fictitious value for cohesion in terms of *s*, employing Eq. 3.11, or the embankment-foundation system can be analyzed based on total stresses using, for example, the Bishop method. Major fills are designed to be constructed in stages with some time delay between stages to permit consolidation and strength increase. Pore pressures are monitored during construction and analysis is made based on effective stresses using $\bar{c}$ and $\bar{\phi}$ from CU tests. The pore-pressure parameter *A* from the test is compared with measured values for *u* to determine when strength is adequate for subsequent lifts (*see Art. 5.3.2*). Vertical drains (*see Art. 5.3.3*) may be installed to accelerate consolidation times which usually are very long in soft clayey soils.

"Mud waves" (Fig. 1.4) occur with total rupture, which is often sudden, but may be preceded by large plastic displacements and a bulge at the toe as shown in Fig. 8.27*c*. Once failure occurs, the available strength is reduced from s_u to s_r, and conditions are much worse than those existing naturally. In critical situations, embankments

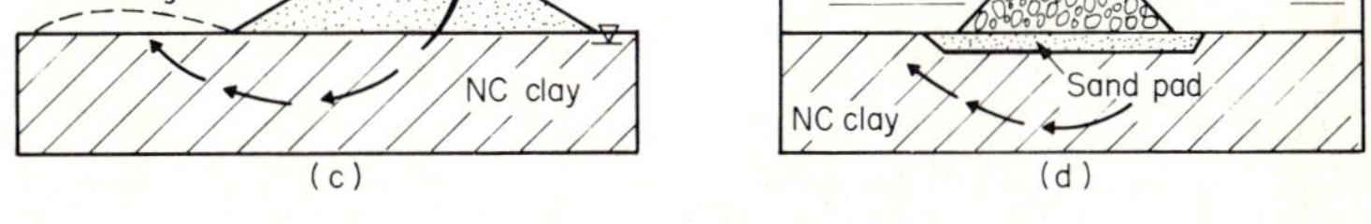

FIG. 8.27 Embankments over soft ground: (*a*) planar failure along shallow rock surface; (*b*) planar failure along weak seam such as very weak clay or fine sand in which pore pressures develop; (*c*) circular failure in deep, uniform soft clay; and (*d*) sand pad under rock-fill breakwater to prevent blocks penetrating the soft clay. NOTE: in (*a*), (*b*), and (*c*) failure can occur left and right simultaneously.

should be monitored with inclinometers as well as piezometers and other instruments as illustrated in Fig. 5.27.

Jetties and Breakwaters

Construction of rock fills for jetties and breakwaters over soft soils should not proceed by dumping blocks directly on the soft soils, since the procedure causes large stress concentrations and soil remolding with subsequent very large fill settlement. This condition is avoided by dredge-removal of 2 to 5 m of the soft soils followed by their replacement with coarse sand to form a foundation pad for the rock fill as illustrated in Fig. 8.27*d*.

Jetties and breakwaters constructed of rock blocks and riprap can be expected to develop large values of intergranular friction as settlement occurs, although the value is difficult to assess. Stability analysis usually is performed considering the entire structure to fail as a unit.

Sidehill Fills

The stability of embankments placed on hillsides depends upon the quality of the compacted fill, the strength of the foundation soils, the pore-water pressures, and the slope inclination. Usually, failure occurs some time after construction when the water table rises and pore pressures increase, normally beginning as a toe failure, as shown in Fig. 8.28, either in the fill itself or in the supporting soils. Usually a circular failure results, but at times the entire fill may slide along a planar surface. Adequate design and construction require the removal of soft foundation soils and provision for the interception of upslope seepage and for free drainage beneath the fill.

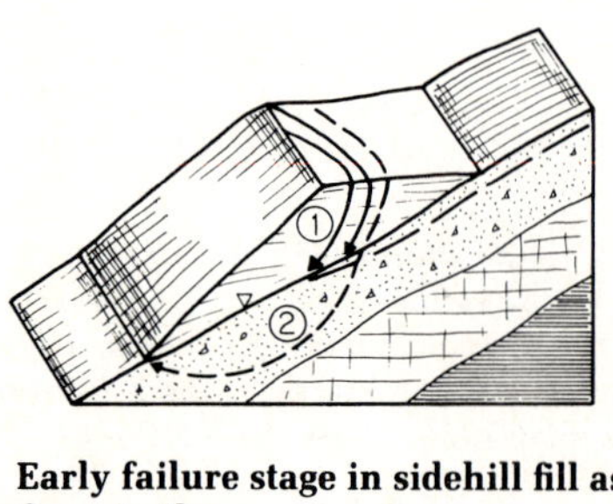

FIG. 8.28 Early failure stage in sidehill fill as concentric cracks form in the pavement.

Analysis is performed using effective stresses and assuming full pore-pressure development beneath the fill base if the fill is constructed of slow-draining cohesive materials. Free-draining materials in the embankment will relieve pore pressures, but failure may still occur in the foundation soils, as in any slope, if they are cohesive and subjected to pore-pressure increase of sufficient magnitude.

8.2.2 INFINITE SLOPE ANALYSIS

Applications

Infinite slope analysis may be applied where the depth to a planar failure surface is small compared to its length, which is considered as unlimited. Such conditions are found in

- Cohesionless materials, such as clean sands
- Cohesive materials, such as residuum or colluvium over a sloping rock surface at shallow depth
- OC fissured clays or clay shales with a uniformly deep, weathered zone such as shown in Fig. 8.18*b*
- Large slabs of sloping rock layers underlain by a weakness plane

The discussion regarding analysis is given in some detail, since the fundamentals apply to most methods of stability analysis regardless of the shape of the failure surface.

Cohesionless Materials

Dry Sands ($i = \phi$)

Stresses acting on a free body from a slope in dry sand are illustrated in Fig. 8.29*b*. It is assumed that the failure zone is very wide along the slope (direction normal to the section) and only stresses acting on the section are considered; i.e., the resistance at the lateral extremities of the failure zone are considered as insignificant compared to the area of the failure surface, the vertical stresses on the sides of the free body are equal, and the entire slope moves as a continuous mass ($E_R = E_L$, $X_R = X_L$).

For equilibrium, the total weight $W = bd\gamma_t$ (Fig. 8.29*c*), which imposes a total normal force N on

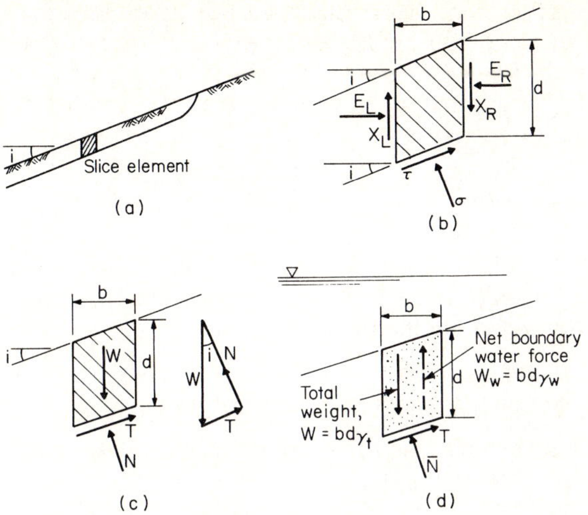

FIG. 8.29 Equilibrium of an infinite slope in sand: (a) the infinite slope; (b) stresses on soil element; (c) forces acting on element of dry sand; (d) totally submerged slope in terms of boundary water forces and submerged weights. [After Lambe and Whitman (1969).[16] Adapted with permission of John Wiley & Sons, Inc.]

the potential failure surface, is balanced by the total shear force T along the failure surface, or

$$W - T\sin i - N\cos i = 0$$
$$T\cos i - N\sin i = 0$$

$N = W\cos i$ and $T = W\sin i$, therefore $T = N\tan i$. When sliding begins, full shearing resistance is mobilized at $T = N\tan\phi$ and $i = \phi$, or in terms of FS as defined in Art. 8.1.2

$$FS = \frac{N\tan\phi}{N\tan i} = \frac{\tan\phi}{\tan i} \tag{8.27}$$

The *angle of repose* is the maximum angle at which a slope in dry sands is stable.

The stresses on the failure plane are:

$$\bar{\sigma}_h = \frac{N}{b/\cos i} = \frac{bd\gamma_t\cos i}{b/\cos i} = d\gamma_t\cos^2 i \tag{8.28}$$

$$\tau = \frac{T}{b/\cos i} = \frac{bd\gamma_t\sin i}{b/\cos i} = d\gamma_t\cos i\sin i \tag{8.29}$$

Submerged Slope ($i = \bar{\phi}$)

Fully submerged in a static body of water, the maximum slope angle i is $\bar{\phi}$. In Fig. 8.29*d*, $\overline{N}$ is the resultant of the effective stresses and T is the resultant of the shear stresses. In analysis the buoyant weight of soil γ_b is used in Eqs. 8.28 and 8.29 which results in unit effective stresses on the failure surface, the resultant of which is given as $\bar{\sigma}_n$. The unit shear stress τ is carried by the soil skeleton and when fully mobilized shear strength is expressed as $s = \bar{\sigma}_n \tan\bar{\phi}$ (Eq. 3.14) and $i = \bar{\phi}$.

Seepage Parallel to the Slope ($i \approx \phi/2$)

Free body forces acting on an element from a slope with seepage parallel to and coincident with the face (Figs. 8.30*a* and *b*) are the weight of the element W, the resulting total effective normal stress $\overline{N}$, a total shear stress T, water pressure on the two vertical sides of the element, and a resultant pore pressure U acting normal to the failure surface, parallel to the normal force.

Boundary pore-water pressures (Fig. 8.30*c*) are found from the flow net (Fig. 8.30*d*) constructed as straight lines with flow lines parallel to the slope. Total head (*see Art. 2.3.2*) is constant along *AB*, and the pressure head varies from zero at *A* to the difference in elevation between *A* and *B* at

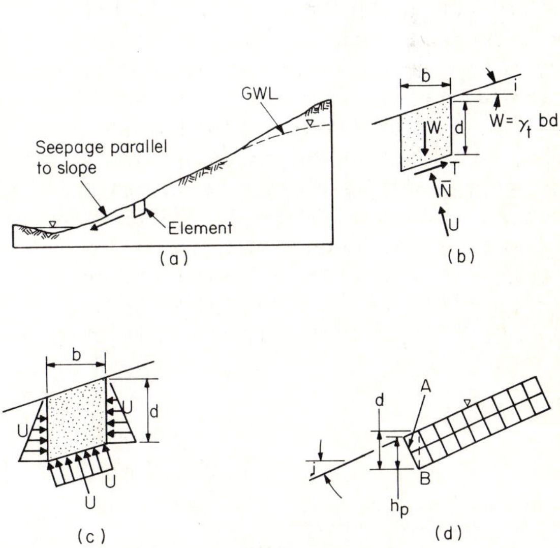

FIG. 8.30 Equilibrium of an infinite slope with seepage parallel to the slope (a); (b) force equilibrium; (c) boundary pore-water pressures; and (d) flow net of seepage parallel to the slope. [After Lambe and Whitman (1969).[16] Adapted with permission of John Wiley & Sons, Inc.]

$B(h_p)$. At depth d,

$$u = h_p\gamma_w = d\gamma_w \cos^2 i \tag{8.30}$$

and

$$U = (d\gamma_w \cos^2 i)\frac{b}{\cos i} = bd\gamma_w \cos i = W_w \cos i \tag{8.31}$$

The *analysis of force equilibrium* is as follows:

$$\overline{N} + U = \gamma_t\, bd \cos i$$

and

$$\overline{N} = \gamma_b\, bd \cos i = W_b \cos i \tag{8.32}$$

$$T = \gamma_t\, bd \sin i = W \sin i \tag{8.33}$$

from which

$$\frac{T}{\overline{N}} = \tan\overline{\phi} = \frac{\gamma_t}{\gamma_b}\tan i = \frac{\tau}{\overline{\sigma}} \tag{8.34}$$

and

$$\tan i = \frac{\gamma_b}{\gamma_t}\tan\overline{\phi} \tag{8.35}$$

It is seen that the effect of the pore pressure is to reduce the total normal stress, and the stable slope angle becomes a function of the ratio of buoyant soil weight to total soil weight as given in Eq. 8.35. Since the ratio of γ_b/γ_t is about ½ for sands, the maximum slope angle with seepage forces is about ½ that for a dry sand, or $i \approx \phi/2$. It is interesting to note that in the field, failed slopes in colluvium and slopes in fissured clays are often found residing at slope angles between $\overline{\phi}_r$ and $\overline{\phi}_r/2$, and the slope angle itself provides an approximation of strength values since these materials commonly rest at FS $\approx$ unity.

In terms of total weight and forces, FS is given as

$$\text{FS} = \frac{(N - U)\tan\overline{\phi}}{T} = \frac{(W\cos i - U)\tan\overline{\phi}}{W\sin i} \tag{8.36}$$

Materials with Cohesion

Seepage at Slope Surface

For the condition of a cohesive material and seepage parallel to and coincident with the slope surface, with the failure surface depth d known, as in the case of colluvium over a rock surface or dipping rock beds underlain by a clay seam, FS in terms of unit shearing stresses is expressed from Eq. 3.9 as

$$\text{FS} = \frac{\overline{c} + \overline{\sigma}\tan\overline{\phi}}{\tau} = \frac{\overline{c} + (\gamma_b d\cos^2 i)\tan\overline{\phi}}{\gamma_t\, d \sin i \cos i} \tag{8.37}$$

where $\overline{\sigma}$ is defined in Eq. 8.28 (but substituting γ_b for γ_t), and τ is defined by Eq. 8.29 (see also Eqs. 8.32 and 8.33).

Seepage Parallel to Surface at Depth d_1

The critical depth d_c for equilibrium may be found from Eq. 8.37 as

$$d_c = \frac{\overline{c}}{\gamma_t}\,\frac{1}{\cos^2 i[\tan i - (\gamma_b/\gamma_t)\tan\overline{\phi}]} \tag{8.38}$$

By rearranging, Eq. 8.38 may be given as

$$\frac{\overline{c}}{\gamma_t\, d_c} = \cos^2 i\,[\tan i - (\gamma_b/\gamma_t)\tan\overline{\phi}] \tag{8.39}$$

where $\overline{c}/\gamma_t\, d_c$ equals a dimensionless number termed the stability number N_s. It reduces the number of variables and enables the preparation of charts for representations of stability relationships [Taylor (1948)[7]].

For seepage parallel to the surface but at depth d_1, from Eq. 8.37,

$$\text{FS} = [\cos^2 i\,[(\{1 - (d_1/d_c)]\,[(\gamma_t - \gamma_1)/\gamma_t]\}\tan i - (\{(\gamma_b/\gamma_t) + (d_1/d_c) \times [(\gamma_1 - \gamma_b)/\gamma_t]\}\tan\phi)]\}/(c/\gamma_t\, d_c)$$

where γ_1 = unit weight of soil within depth d_1. d_c is greater than d_1 as it represents the failure surface

General Relationship

The general relationship for soils with cohesion acting along a surface of length L in terms of total weight and forces acting on an element may be given as

$$\text{FS} = \frac{cL + (W\cos i - U)\tan\overline{\phi}}{W\sin i} \tag{8.41}$$

The following articles all pertain to finite or limited slopes which can have a failure surface with a shape of a circular arc, log-spiral, plane, or it can be irregular. Log-spirals are not discussed since they yield results similar to those obtained for circular arcs.

8.2.3 CIRCULAR FAILURE SURFACE: EFFECTIVE STRESS ANALYSIS

General

Soil slopes that are homogeneous, or approximately so, usually fail along a surface approaching a circular arc of finite length. If there is a relatively very weak or very strong layer within the potential failure surface, however, it will control the location of the failure surface and a major portion is likely to be planar.

For effective stress analysis a number of methods are commonly used, all based on limit equilibrium concepts which are in turn based on slip line theory. Some methods assume the failure surface to be a log-spiral, but the results are similar to those obtained with a circular arc. In most methods the critical failure surface is found by trial and error, but its exact location in analysis is not nearly as significant to the accuracy of the results as is the selection of representative strength parameters and the groundwater conditions that might prevail at failure.

The various analytical methods vary in complexity of solution with the force system assumed. Common to all is a free body acted on by gravity (soil weight) and external forces and normal stress and shear stress distributed along an assumed failure surface. The most important of the external forces on the free body usually is porewater pressure, although some methods consider interslice forces and in some cases earthquake loadings are important.

The more common methods in this article are divided into two general groups: those applicable to simple, homogeneous slopes and those used for nonhomogeneous slopes. Homogeneous slopes normally are analyzed with the use of charts, such as those of Taylor (1948)[1] or more recently Cousins (1978),[35] the latter included herein, and the entire potential failure mass is considered as a free body.

In the second group, the potential failure mass is divided into slices and solutions are mathematical. Included in this group are the *ordinary method of slices* (also known as the *Fellenius method* or the *Swedish method*) and the *complete Bishop* and *modified* or *simplified Bishop methods*. The modified Bishop yields values for FS very near to those of the complete Bishop, which is generally considered to be the most accurate analytical method available. The modified Bishop, therefore, is the most popular method for effective stress analysis of slope stability, especially since the advent of programmable calculators and "personal" or desk-top computers. The ordinary method yields values for FS that are usually, but not always, overconservative. In homogeneous materials the friction circle method *(see Art. 8.1.3)* will yield values for FS very close to those obtained by the modified Bishop method.

Simple Homogeneous Slopes: Stability Charts

Applications

Stability charts are used mainly as a guide for preliminary designs or as a rough check for the more complex solutions. A number of investigators have provided charts for various conditions. All are somewhat limited in application, whether by the geometric configuration of the problem, by soil strength parameters provided for, or by slope inclination, and most were developed for evaluating simple embankment slopes, such as for earth dams, based on the assumption of average soil parameters. Those summarized in Table 8.8 are based on circular failure surfaces and effective stresses.

Cousins' Charts

Charts developed by Cousins (1978)[35] are based on the friction circle method, incorporating some of the work of previous investigators *(see Art. 8.1.3)*. Employing the pore-pressure ratio, they are suitable for slopes up to 45° inclination. Four sets of charts were developed for the analysis of critical toe circles and base circles and for the location of the critical failure circle for toe and base circles. The geometric terms are illustrated in Fig. 8.31.

For stability analysis, the charts relate a minimum stability number N_f to the slope angle i in

TABLE 8.8
SUMMARY OF CHART SOLUTIONS FOR STABILITY ANALYSIS OF SIMPLE SLOPES WITH EFFECTIVE STRESSES

Reference	Basis	Comments
Bishop and Morgenstern (1960)[28]	Modified Bishop	Limited to slope angles of 11 to 27°.
Morgenstern (1963)[29]	Bishop and Morgenstern (1960)	For earth slope during rapid drawdown of reservoir level.
Spencer (1967)[8]	Friction circle and modified Bishop	Limited to slope angles to 34°, for toe circles only, requires iteration to find FS.
Janbu (1967)[30]	Friction circle and modified Bishop	For toe circle only.
Hoek and Bray (1977)[3]	Friction circle	Provides for steady seepage and five idealized cases of phreatic surface location in slope.
Huang (1975)[31]	Modified Bishop	For embankments, homogeneous dams, and soil slopes.
Huang (1977),[32] (1978)[33,34]	Ordinary method of slices	Sidehill fills on soil or rock slopes, particularly mine spoils.
Cousins (1978)[35]	Friction circle	For toe and base circles for slope angles to 45° (see text).
Huang (1979),[36] (1980)[37]	Ordinary method of slices	For nonhomogeneous dams, such as refuse dams on slopes; widely differing soft strength parameters are averaged.

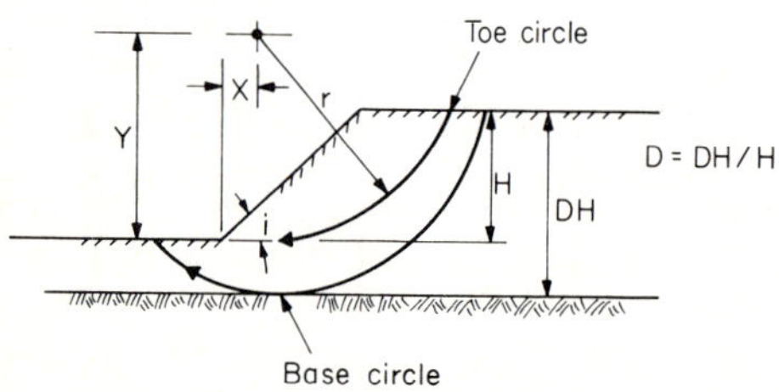

FIG. 8.31 Geometric notation for Cousins' charts.

terms of a factor $\lambda_{c\phi}$ and a depth factor D, with separate charts for varying values of the pore pressure ratio r_u. The stability number (see Eq. 8.21) is given as

$$N_f = \frac{\gamma H\,(\mathrm{FS})}{\bar{c}} \tag{8.42}$$

The factor $\lambda_{c\phi}$ also groups terms relating to soil properties and slope height, given as

$$\lambda_{c\phi} = \frac{\gamma H \tan \phi}{\bar{c}} \tag{8.43}$$

The depth factor D is applied to base failures and defined as the ratio of vertical distance DH from top of slope to the failure arc to the slope height H as shown in Fig. 8.31.

The *pore-pressure ratio* r_u, used in many chart forms of analysis *(see Art. 8.2.1)*, is defined as the ratio of pore-water pressure to total overburden pressure at a particular depth, normally taken as the failure surface, as shown in Fig. 8.32. It is given by

$$r_u = \frac{u}{\sigma_v} = \frac{\gamma_w h}{\gamma_t d} \tag{8.24}$$

Toe circle charts are given in Fig. 8.33 and depth factor charts on Fig. 8.34, each for three different

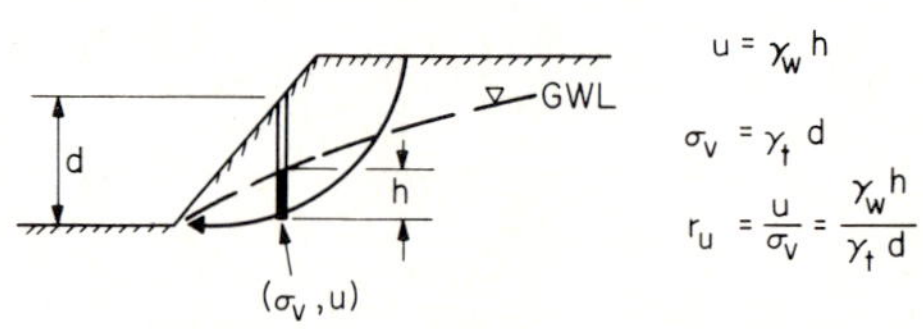

FIG. 8.32 Definition of pore-pressure ratio r_u.

FIG. 8.33 Cousins' charts for stability numbers for toe circles: (*a*) $r_u = 0$; (*b*) $r_u = 0.25$. [From Cousins (1978).[35]]

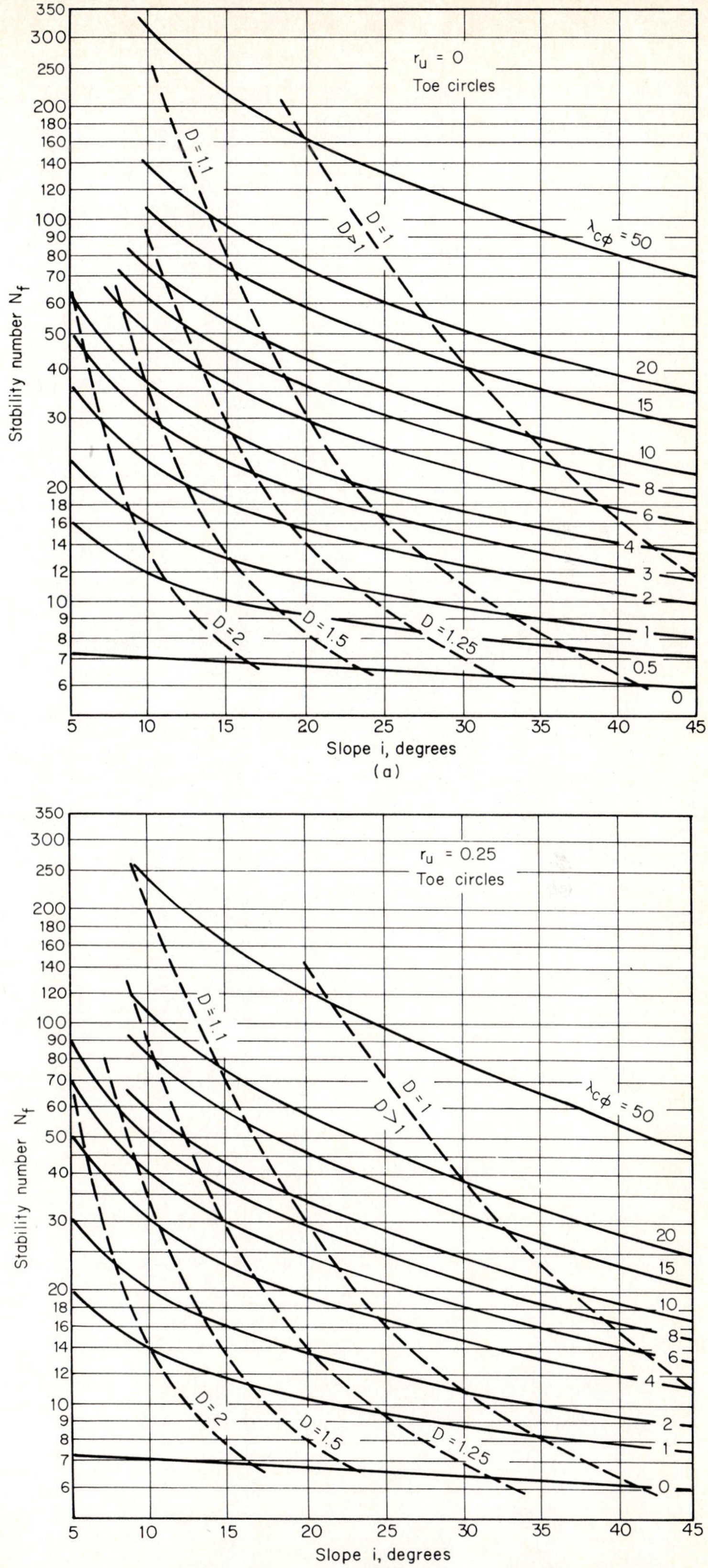

FIG. 8.33 (*Continued*) Cousins' charts for stability numbers for toe circles: (*c*) r_u = 0.5. [From Cousins (1978).[35]]

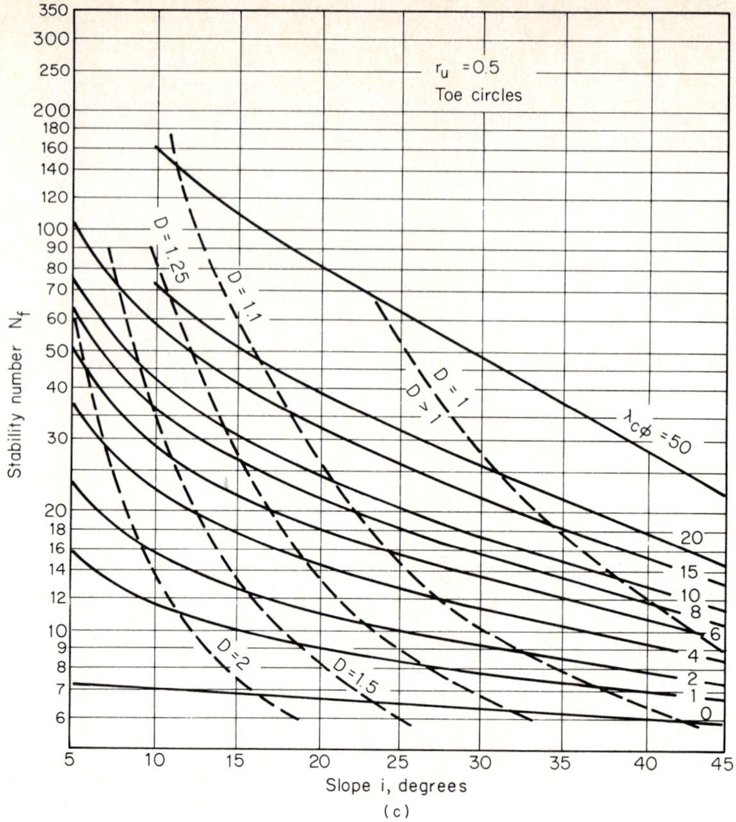

FIG. 8.34 Cousins' charts for stability number for depth factor *D*: (*a*) r_u = 0; (*b*) r_u = 0.25; and (*c*) r_u = 0.5. [From Counsins (1978).[35]]

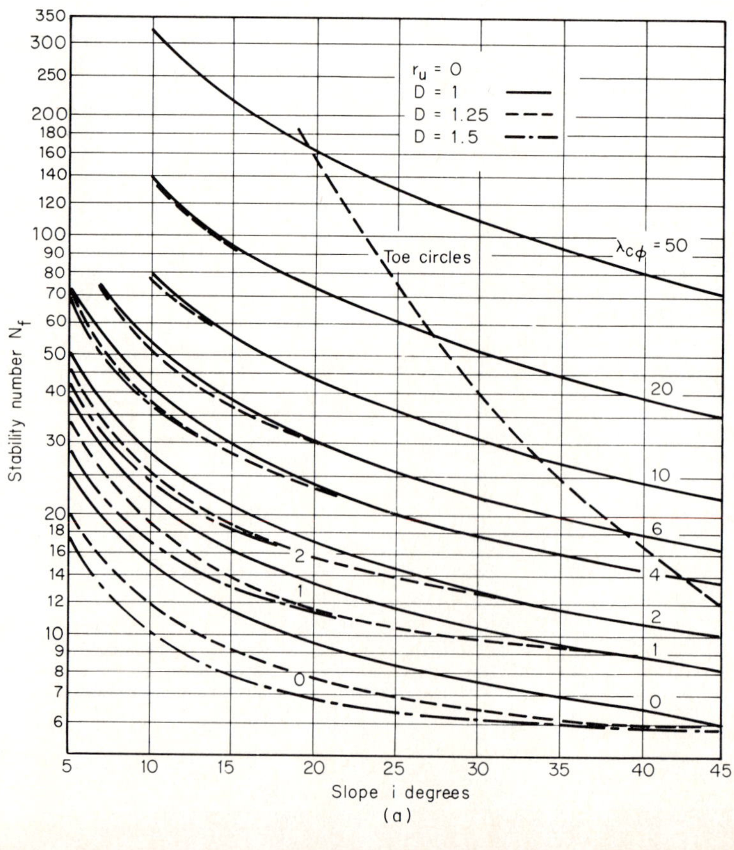

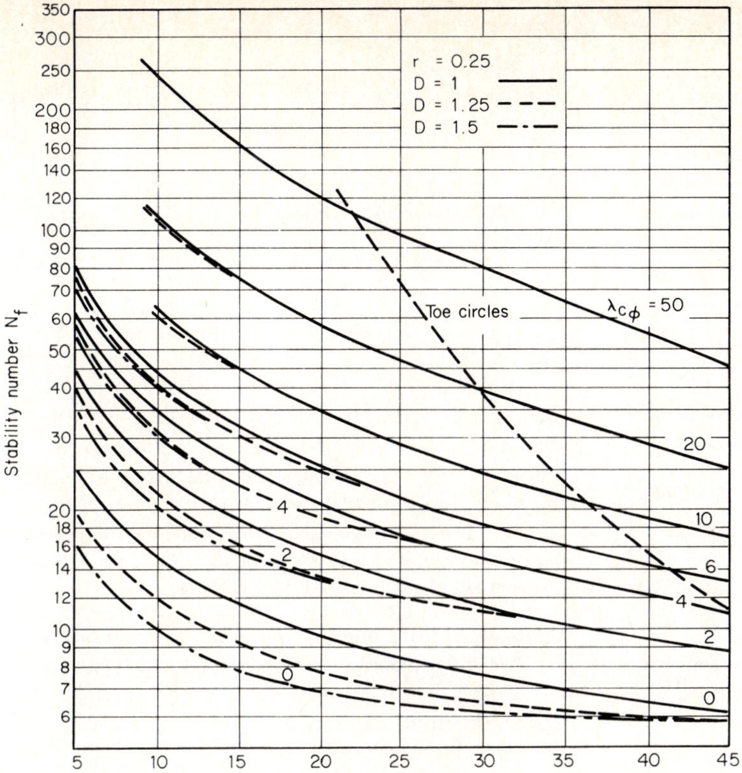
r = 0.25
D = 1
D = 1.25
D = 1.5
Toe circles
$\lambda_{c\phi}$ = 50
20
10
6
4
2
0
Stability number N_f
Slope i degrees
(b)

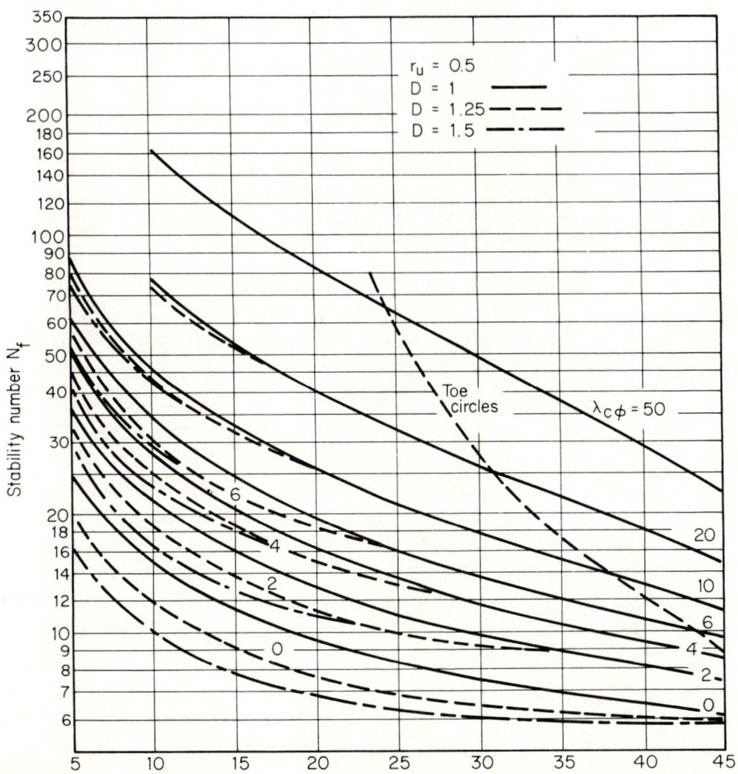
r_u = 0.5
D = 1
D = 1.25
D = 1.5
Toe circles
$\lambda_{c\phi}$ = 50
20
10
6
4
2
0
Stability number N_f
Slope i, degrees
(c)

values of r_u (0, 0.25, 0.5). To use the charts, values for $\lambda_{c\phi}$ and r_u are calculated, and the charts bracketing r_u are entered for the given slope angle i to find the stability number N_f for the bracketing r_u values. Results are very sensitive to r_u and the applicable value for N_f is found by interpolating the values obtained, using the actual r_u value. The maximum value for D is found from the problem geometry. Figure 8.33 is entered first to find N_f for the given parameters and if D is significantly greater than 1 as found from the chart, then Fig. 8.34 is entered to find a lower value for N_f. The factor of safety is found from

$$FS = \frac{N_f \bar{c}}{\gamma H} \tag{8.44}$$

The x and y coordinates for locating a toe circle or a depth factor circle are found in a similar manner using Figs. 8.35 and 8.36.

The charts also are applicable to total stress analysis, as will be discussed in the following article.

Example Using Cousins' Charts

As an example, Cousins' charts are used to analyze the slope given in Fig. 8.37, where $i = 40°$, $H = 43$ ft, $\bar{c} = 200$ psf, $\bar{\phi} = 33°$ and $\gamma = 125$ pcf. The pore pressure ratio $r_u = U/W = 0.25$. From Eq. 8.43

$$\lambda_{c\phi} = \frac{\gamma H \tan \bar{\phi}}{\bar{c}} = \frac{(125)(43)(\tan 33)}{200} = 17.5$$

Entering Fig. 8.33*b* with $\lambda_{c\phi} = 17.5$ and $i = 40°$, find $N_f = 26$ and $D = 1$. Failure is calculated from a toe circle and

$$FS = \frac{N_f \bar{c}}{\gamma H} = \frac{(26)(200)}{(125)(43)} = 0.97$$

for the critical circle. The coordinates of the critical circle are found from Fig. 8.35*b*.

$$(Y/H) \tan i = 145$$
$$(X/H) \tan i = -0.2$$

and

$$Y = 1.45\ H/\tan i = 74 \text{ ft}$$
$$X = -0.2\ H/\tan i = -10 \text{ ft}$$

Solutions for two other circles are shown in the figure for slope angles of 30° which illustrate the increase in FS with a decrease in the slope angle in the same soil materials and the increase in circle depth and decrease in FS with a decrease in ϕ.

Nonhomogeneous Slopes: Ordinary Method of Slices

General

Following his studies of numerous slides along the Swedish railways in the 1920s (*see Art. 8.1.3*), Fellenius (1936)[14] published a method of slices based on circular failure surfaces, which is known as the *Fellenius* or *Swedish method*. It has been modified for effective stress analysis and is now known generally as the *ordinary method of slices*. The solution is by trial and error and locates the critical failure surface, or that circle radius with the lowest value for FS.

Although it has been replaced in general practice by the simplified Bishop method (next section), which generally yields much more accurate results for effective stress analysis, it is described here in detail because it is useful for total stress analysis and is readily solvable by hand calculations. The Bishop method is time-consuming unless a chart or a computer is used. All slice methods are also useful for analysis of simple homogeneous slopes.

Distributed Stresses and Resultant Forces

Slice methods are based on the reasonable assumption that the normal stresses acting on a point of the failure arc should be influenced mainly by the weight of soil above the point. If a potential failure mass is divided into a number of vertical slices, as shown in Fig. 8.38, the equilibrium of each slice may be considered. As illustrated in Fig. 8.39, the force system acting along the potential failure surface for the ordinary method of slices includes the slice weight W, normal force $\bar{N}$, pore-water force U, and shearing resistance T. It is assumed that the forces acting on the *sides* of the slice have *zero resultant* in the direction normal to the failure arc for that slice. (The significance of slice side forces is discussed in the following section on the Bishop method.) Equilibrium is obtained by summing forces in a direction normal to the failure arc.

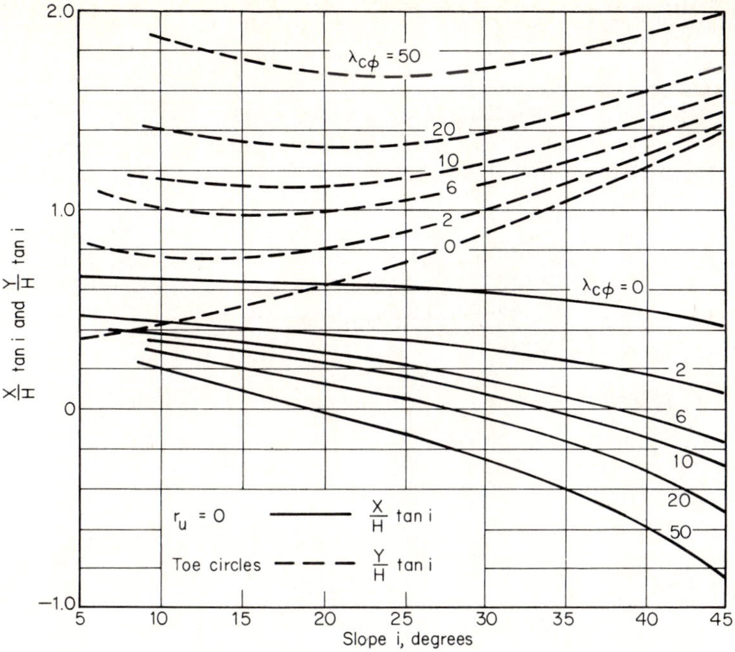

(a)

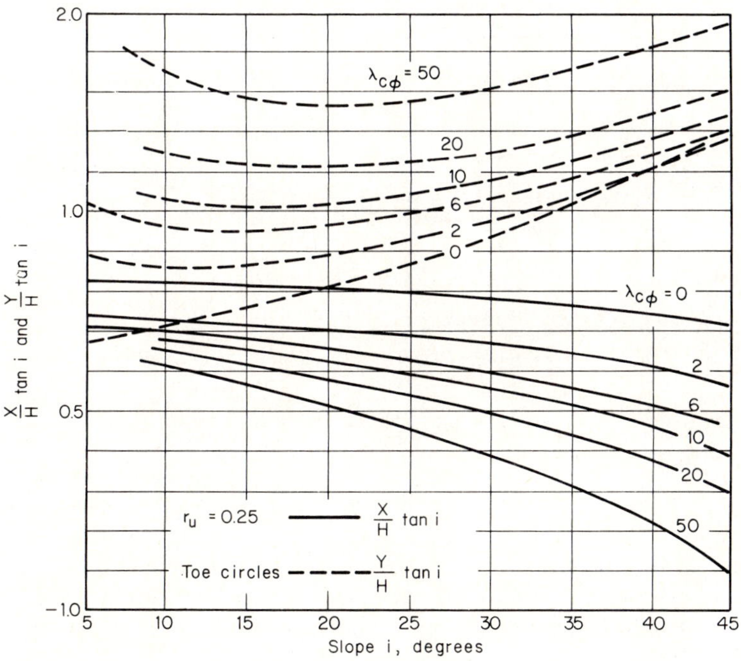

(b)

FIG. 8.35 Cousins' charts for *X* and *Y* coordinates for critical toe circles: (*a*) r_u = 0; (*b*) r_u = 0.25. [From Cousins (1978).[35]]

FIG. 8.35 (*Continued*) Cousins' charts for *X* and *Y* coordinates for critical toe circles: (*c*) r_u = 0.5. [From Cousins (1978).[35]]

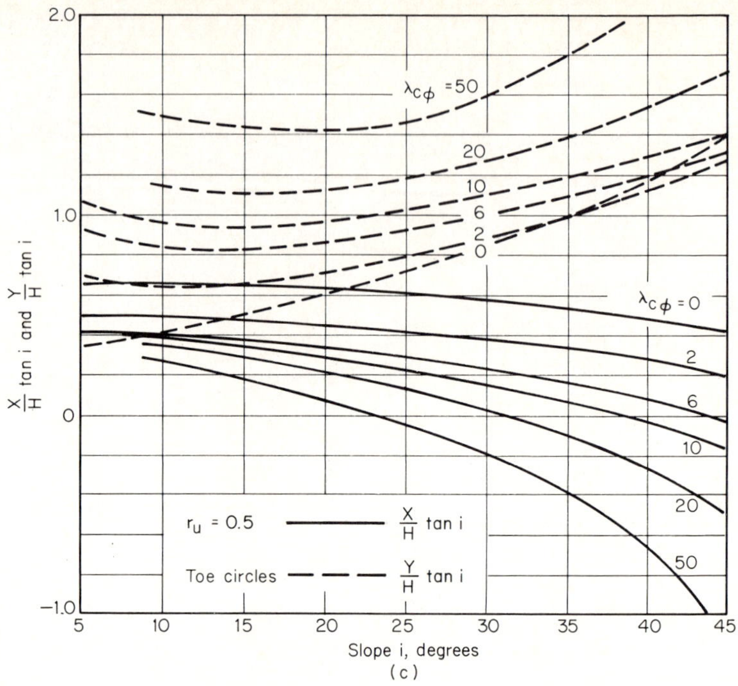

FIG. 8.36 Cousins' charts for coordinates of critical slip circles for depth factor *D* = 1: (*a*) r_u = 0; (*b*) r_u = 0.25; and (*c*) r_u = 0.5. [From Cousins (1978).[35]]

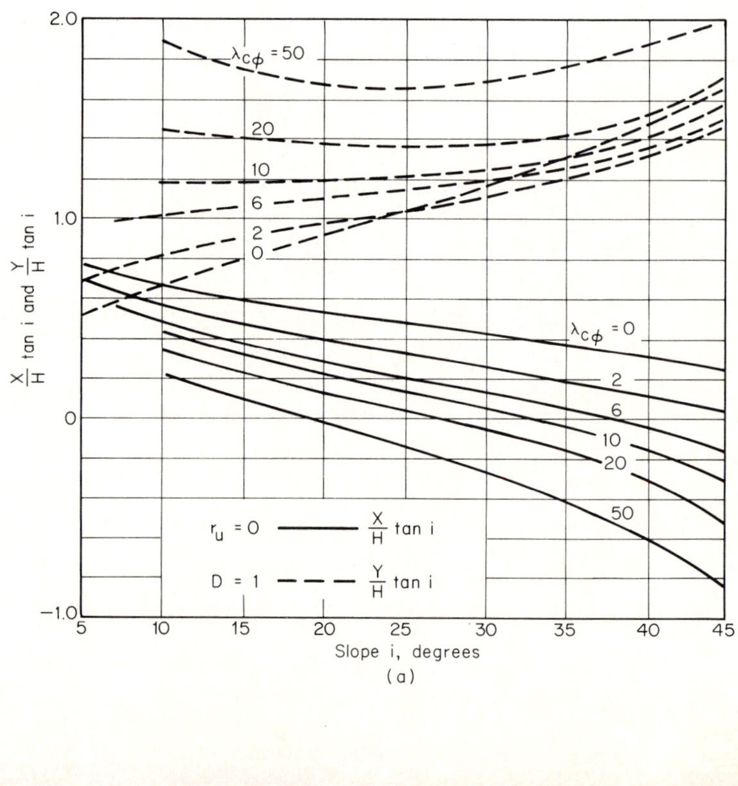

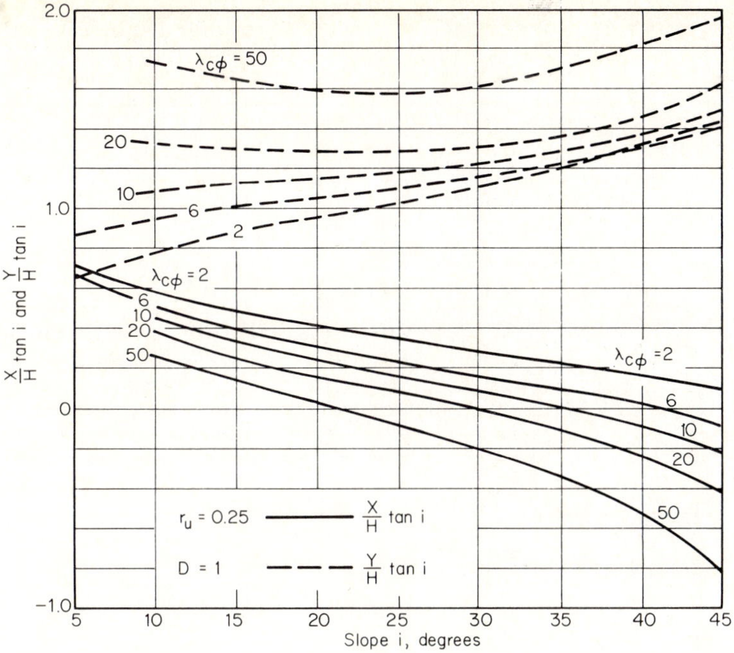
$\lambda_{c\phi}$ = 50
20
10
6
2
$\lambda_{c\phi}$ = 2
6
10
20
50
$\lambda_{c\phi}$ = 2
6
10
20
50
r_u = 0.25 $\frac{X}{H}$ tan i
D = 1 $\frac{Y}{H}$ tan i
$\frac{X}{H}$ tan i and $\frac{Y}{H}$ tan i
Slope i, degrees

(b)

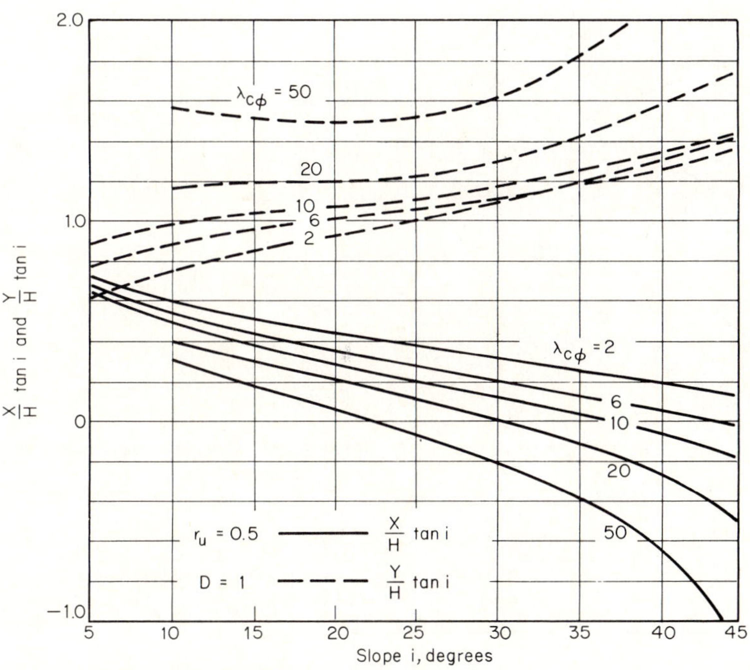
$\lambda_{c\phi}$ = 50
20
10
6
2
$\lambda_{c\phi}$ = 2
6
10
20
50
r_u = 0.5 $\frac{X}{H}$ tan i
D = 1 $\frac{Y}{H}$ tan i
$\frac{X}{H}$ tan i and $\frac{Y}{H}$ tan i
Slope i, degrees

(c)

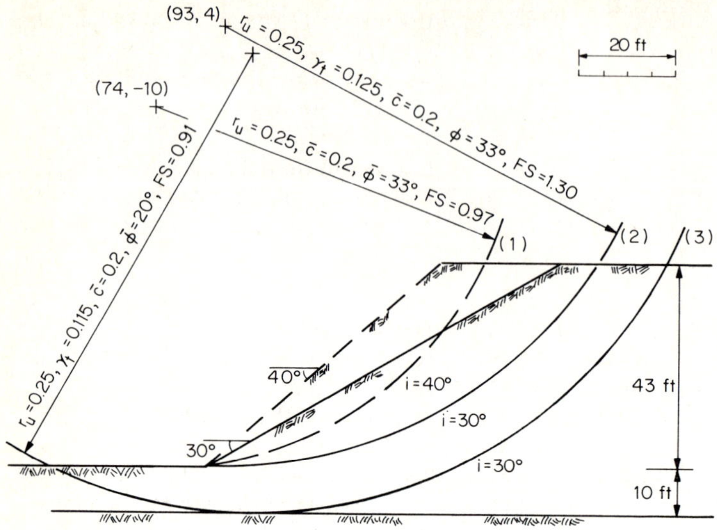

FIG. 8.37 Effective stress-stability analysis using Cousins' charts. Note increase in FS with decrease in slope angle *i* (curves 1 and 2), and increase in circle depth and decrease in FS with decrease in ϕ (circle 3). Circles 1 and 2 represent residual soil; circle 3 represents alluvial clayey silt.

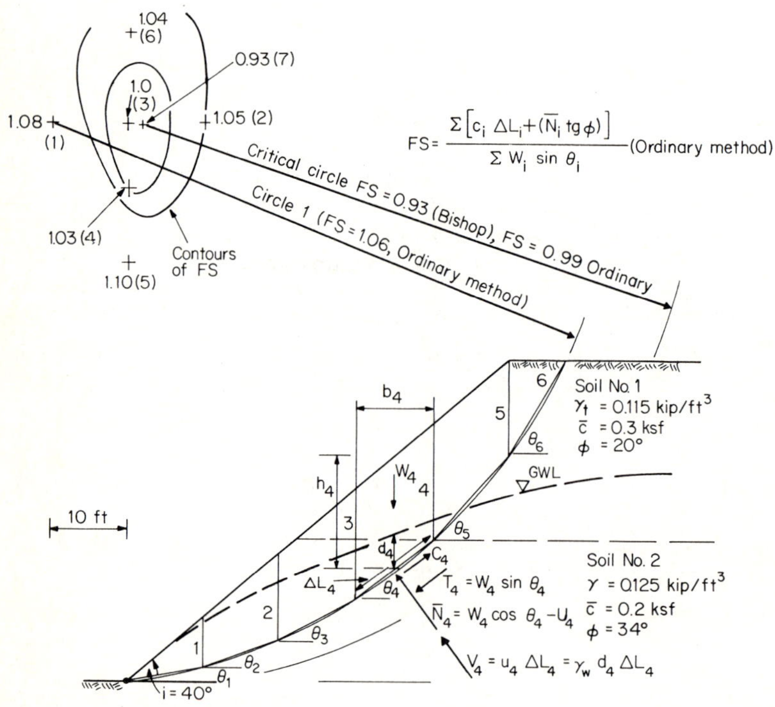

FIG. 8.38 Stability analysis with slice methods.

For this force system,

$$\bar{N}_i + U_i = W_i \cos \theta_i$$

or

$$\bar{N}_i = W_i \cos \theta_i - U_i = W_i \cos \theta_i - u_i \Delta L_i \tag{8.45}$$

The factor of safety FS is found by summing the ratio of shearing resistance to shearing force for all slices along the failure surface, given by

$$FS = \frac{\Sigma(c_i \Delta L_i + \bar{N}_i \tan \bar{\phi}_i)}{\Sigma W_i \sin \theta_i} \tag{8.46}$$

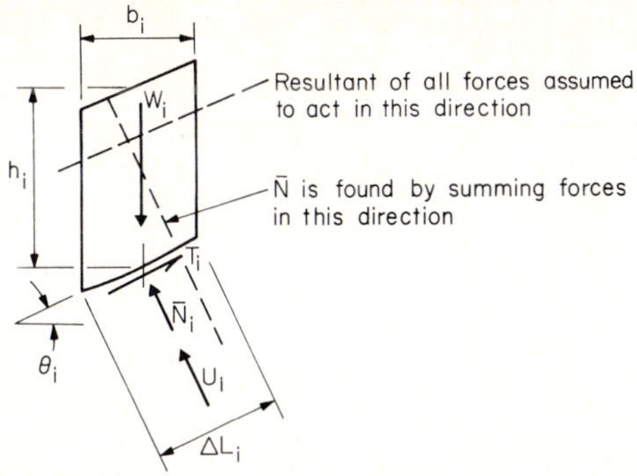

FIG. 8.39 Force system considered in the ordinary method of slices. [From Lambe and Whitman (1969).[16] Used with permission of John Wiley & Sons, Inc.]

or

$$FS = \frac{\Sigma\,[c_i \Delta L_i + (W_i \cos\theta_i - U_i)\tan\bar{\phi}_i]}{\Sigma W_i \sin\theta_i} \tag{8.47}$$

Derivations of Eq. 8.47 can be found in Lambe and Whitman (1969)[16] or McCarthy (1982)[38] (*see also Art. 8.2.2, The Infinite Slope*).

Procedure

The method proceeds as trial and error with solutions obtained for failure arcs with differing radii until a minimum value for FS is found. To begin, the slope with soil strata and estimated phreatic surface (*see discussion in Art. 8.2.1*) is drawn to scale. A failure arc is assumed to act at some radius (charts are useful for estimating critical circle coordinates) and the potential failure mass is divided into vertical slices. Slice boundaries are selected to conform with the slope configuration, and for convenience the slice base should cross only one soil type. Slice-chord lengths should be close to the arc length to obtain representative values for angle θ, otherwise unconservative FS values result. In the order of five to seven slices usually are adequate for most problems.

To aid in analysis, a table is constructed such as given in Table 8.9. Geometric parameters scaled from the drawing (Fig. 8.38) are b_i, h_i (to find W_i from unit soil weight γ_t); d_i (to find u_i); ΔL_i (to find U_i); and angle θ_i (to find $\bar{N}_i$ and the shearing force). It should be noted that angle θ may be negative in the toe area. If there is more than one soil type in a slice, an average value for soil weight is taken. Soil parameters c, $\bar{\phi}$, and γ_t are measured in the laboratory for each stratum. In the example given in Table 8.9, the FS for circles of *other radii* need to be determined to find the lowest value for the slope. Convergence is fastest when 0 is moved horizontally and values are plotted as shown in Fig. 8.38.

A *tension crack* tends to reduce the length of the failure arc as discussed in Art. 8.1.2. Experience has shown that they usually range in depth from ½ to ⅓H [Tschebotarioff (1973)[39]]. Water in the tension crack increases the driving force by $V = \frac{1}{2}\gamma_w z$ (see Fig. 8.48). Spencer (1967)[8] suggests that results are conservative if tension cracks are not considered in analysis as long as the water pressure in the tension crack is ignored. For dynamic analysis, see Art. 8.2.7.

Validity of Method

The ordinary method provides values for FS that fall below the lower bound for most solutions that satisfy statics and is conservative. In cases with low ϕ values and moderate slope angles, FS may fall only 10 to 15% below the range of the more exact solutions, but can be as much as 60% below where ϕ values are high and slopes are steep. In cases where $\phi = 0$ or is small, the method has been found to yield results similar to those of the more refined methods because for these cases, normal stresses on the base of the failure surface do not influence the available shear strength [Johnson (1974)[40]].

With the ordinary method, FS = 1.06 was found for the slope shown in Fig. 8.38, whereas with the simplified Bishop method, FS = 1.08 was found, as presented in the following section.

Nonhomogeneous Slopes: Bishop Slice Methods

Complete Bishop Method

In the complete Bishop method [Bishop (1955)[7]], the complete system of forces acting on a slice is considered. In addition to $\bar{N}_i$, U_i, and T, the complete system includes along each side of the slice, shear stresses (x_i with width b_i), effective normal stresses (E_i), and pore-water pressures (U_L and U_r), as shown in Fig. 8.40. The procedure

TABLE 8.9
SOLUTION TO THE ORDINARY METHOD OF SLICES

No.	Slice												
	ΔL_i, ft	c_i, ksf	$c_i \Delta L_i$, kips/ft	b_i, ft	W_i, h_i, ft	γ_t, kips/ft^3	W_i, kips	θ_i, degrees	$W_i \cos \theta_i$, kips	u_i, $\gamma_w d$ (0.062d), kips/ft	U_i, ($u_i \Delta L_i$), kips	$\overline{N}_i$, $W_i \cos \theta_i - U_i$, kips	$\overline{N}_i \tan \overline{\phi}$, kips/ft (1)	$W_i \sin \theta_i$, kips/ft
1	10.0	0.2	2.00	10.0	3.5	0.125	4.38	11	4.30	0.22	2.20	2.10	1.42	0.84
2	10.5	0.2	2.10	10.0	9.2	0.125	11.50	19	10.87	0.42	4.41	6.46	4.36	3.74
3	11.5	0.2	2.30	10.0	13.3	0.125	16.63	29	14.54	0.45	5.18	9.37	6.32	8.06
4	12.8	0.2	2.56	10.0	15.0	0.125	18.75	36	15.12	0.28	3.58	11.59	7.82	11.02
5	15.5	0.3	4.65	10.0	14.0	0.115	16.1	48	10.77	0	0	10.77	3.92	11.97
6	14.2	0.3	4.26	7.2	6.5	0.115	5.4	60	2.70	0	0	2.70	0.98	4.65
			$\Sigma = 17.9$										$\Sigma = 24.82$	$\Sigma = 40.30$

$$FS = \frac{\Sigma[c\Delta L + \overline{N} \tan \overline{\phi}]}{\Sigma W \sin \theta} = \frac{17.9 + 24.82}{40.30} = \frac{42.72}{40.30} = 1.06$$

Given: $\overline{c}_1 = 0.2$ ksf, $\overline{\phi}_1 = 34°$; $\overline{c}_2 = 0.3$ ksf, $\overline{\phi}_1 = 20°$.

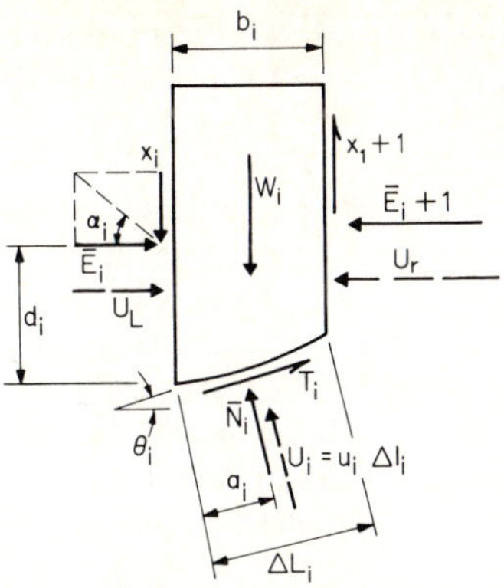

FIG. 8.40 Complete force system acting on a slice. [From Lambe and Whitman (1969).[16] Used with permission of John Wiley & Sons, Inc.]

for solution is analytical in which distributions of $(x_i - x_{i+1})$ are found by successive approximation until a number of equilibrium conditions are satisfied. Computations are considerable and a computer is required for the analysis.

Modified or Simplified Bishop Method

The modified Bishop method [Bishop (1955),[7] Janbu et al. (1956)[41]] is a simplification of the complete Bishop method in which it is assumed that the total influence of the tangential forces on the slice sides is small enough to be neglected; i.e., $\Delta x_i \approx 0$. The normal forces $\bar{N}_i$ are found by considering the equilibrium of forces in the *vertical direction* (Fig. 8.41) and FS is given as

$$\mathrm{FS}_2 = \frac{\Sigma\{[\bar{c}_i b_i + (W_i - u_i b_i)\tan\bar{\phi}_i]\cdot 1/M_i(\theta)\}}{\Sigma W_i \sin\theta_i} \tag{8.48}$$

where $M_i(\theta)$ is a grouping of terms given as

$$M_i(\theta) = \cos\theta_i\left(1 + \frac{\tan\theta_i \tan\bar{\phi}_i}{\mathrm{FS}_1}\right) \tag{8.49}$$

Substitution of Eq. 8.49 into Eq. 8.48 results in FS on both sides of the equation. For solution a value for FS_1 is assumed and FS_2 calculated, and the procedure is repeated by trial and error until $\mathrm{FS}_1 = \mathrm{FS}_2$. Solutions converge rapidly and yield results slightly lower than those obtained from the complete Bishop method and therefore are conservative. As a computational aid, $M_i(\theta)$ in terms of θ and $\tan\bar{\phi}/\mathrm{FS}$ is given in Fig. 8.42. The

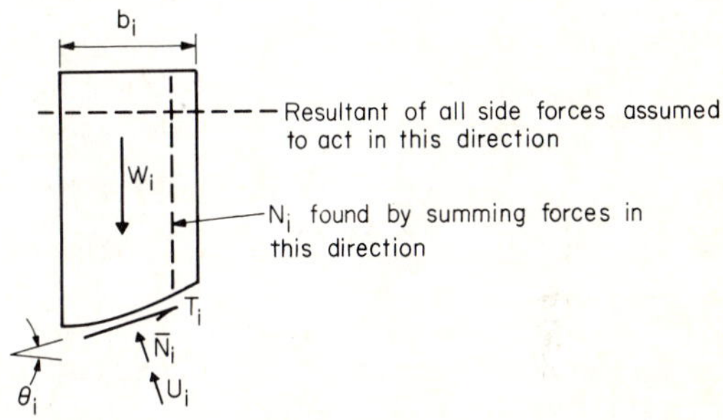

FIG. 8.41 Forces considered in the simplified Bishop method of slices. [From Lambe and Whitman (1969).[16] Used with permission of John Wiley & Sons, Inc.]

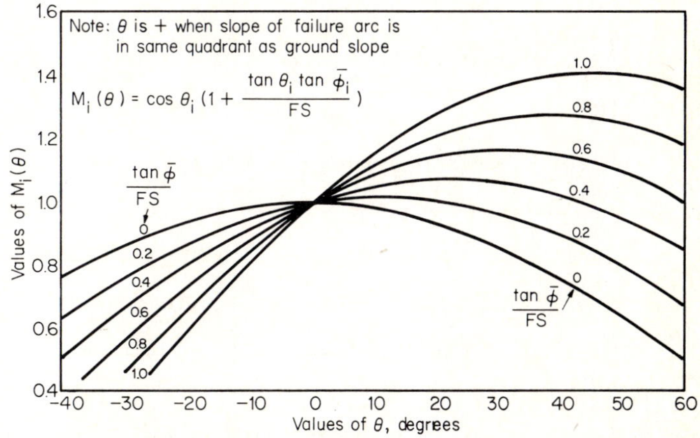

FIG. 8.42 Graph for determination of $m_i(\theta)$ used in modified Bishop method. [After Bishop and Morgenstern (1960).[28]].

simplified Bishop method is the method preferred by most practitioners for analyzing homogeneous and nonhomogeneous slopes where failure will be in the circular arc form and either effective or total stress analysis is desired.

Analysis with the Simplified Bishop Method

The procedure for analysis is similar to the one given previously for the ordinary method of slices. The slope and soil stratigraphy are drawn to scale, the phreatic surface estimated, and a value for FS_1 is assumed. For hand calculations based on Eq. 8.48 a table is set up for data input such as given in Table 8.10, which includes the simplified Bishop analysis for the example given in Fig. 8.38 for the ordinary method. In both solutions the tension crack is ignored. For the circle given in Fig. 8.38, the FS (Bishop) = 1.08 whereas the FS (ordinary) = 1.06. A number of trials for circles of varying radii yield FS = 0.93 for the critical circle shown on the figure. With experience the complete solution for a given slope takes about 1 hour with a programmable calculator, or minutes with a personal computer.

Dynamic analysis with the simplified Bishop method is discussed in Art. 8.2.7.

Limitations

The Bishop methods will yield *unrealistically high* values for FS when failure surface angle θ in the lower slope area approaches about 30°, which can develop for failure circles that are deep or have their center above the vicinity of the crest-of-slope. For these conditions the ordinary method yields overly conservative (low) values [McCarthy (1982)[38]].

8.2.4 CIRCULAR FAILURE SURFACE: TOTAL STRESS ANALYSIS

General

As discussed in Art. 8.2.1, the total stress equals the effective stress plus the neutral stress due to pore-water pressures. Total stress analysis pertains to "end-of-construction" or "short-term" conditions where slope drainage and resulting changes in pore pressures and intergranular stresses do not have time to occur.

Preliminary analysis of simple homogeneous slopes is performed using charts. Homogeneous and nonhomogeneous slopes are analyzed by slice methods. Solutions for the $\phi = 0$ case are relatively simple because the variation of the normal stress along the failure surface is not important. Slopes also are analyzed under undrained conditions employing effective stresses, but pore-pressure parameters must be obtained either by prediction in terms of total stress (*see Art. 8.2.1*) or from field observations. Such analyses are imperative for cuts in NC clay and embankments over soft ground. When pore pressures are governed by steady seepage conditions, stability analysis is based on effective stresses.

Homogeneous Slopes: Chart Solutions

The $\phi = 0$ Case

In his study of slope failures in Sweden (*see Art. 8.1.3*), Fellenius found that in homogeneous NC clay slopes failure occurred along a more or less circular arc, the center of which was situated on a vertical line passing through the center of the slope (Fig. 8.11). He expressed the critical height as

$$H_c = \frac{4c}{\gamma} \frac{1}{f(i, \theta, \omega)} \tag{8.50}$$

and found that for slopes with $i > 53°$ the critical circle passes through the toe, whereas for slopes with $i < 53°$ the critical circle corresponds to a base failure or deep rupture (Fig. 8.11).

The findings of Fellenius provided the basis for the development of chart solutions. The grouping of terms $f(i, \theta, \omega)$ in Eq. 8.50 resulted in the dimensionless factor, the stability number N_s, related to FS by

$$N_s = \frac{c}{(FS)\gamma_t H} \tag{8.51}$$

In homogeneous clay slopes the value of N_s can be given as a function of the slope angle i and a depth factor d_f and presented in simple chart form, as by Taylor (1937),[25] (1948).[1]

The chart given as Fig. 8.43 was developed by Janbu (1954),[11] who defined the *stability number* as

$$N_o = \frac{\gamma_t H(FS)}{c} \tag{8.52}$$

therefore

$$FS = N_o \frac{c}{\gamma_t H} \tag{8.53}$$

TABLE 8.10
SOLUTION FOR THE SIMPLIFIED BISHOP METHOD (EXAMPLE: FIG. 8.38)

Slice no. (1)	$b[(\gamma h - u)\tan\bar{\phi} + c]$: γ, kips/ft^3 (2)	h, ft (3)	u, kips/ft (4)	ϕ, degrees (5)	c, ksf (6)	b, ft (7)	Result, kips (8)	$\frac{\tan\bar{\phi}}{FS_1}$ (9)	$M_i(\theta)$ (10)	Result $M_i(\theta)$ (11)	W, kips (12)	θ, degrees (13)	$W \sin\theta$, kips (14)
1	0.125	3.5	0.22	34	0.2	10.0	3.48	0.613	1.10	3.17	4.38	11	0.84
2	0.125	9.2	0.42	34	0.2	10.0	6.92	0.613	1.15	6.05	11.50	19	3.74
3	0.122	13.3	0.45	34	0.2	10.0	10.18	0.613	1.17	8.67	16.73	29	7.87
4	0.118	15.0	0.28	34	0.2	10.0	12.76	0.613	1.17	10.94	17.70	36	10.40
5	0.115	14.0	0	20	0.3	10.0	8.86	0.331	0.92	9.68	16.10	48	11.97
6	0.115	6.5	0	20	0.3	7.2	4.12	0.331	0.79	5.25	5.85	60	4.66

- Trial 1:

$$FS_1 = 1.10$$

$$FS_2 = \frac{\sum \frac{b[(\gamma h - u)\tan\bar{\phi} + \bar{c}]}{M_i(\theta)}}{\Sigma W \sin\theta} = \frac{\text{Col. 8/Col. 10}}{\text{Col. 14}} = \frac{\text{Col. 11}}{\text{Col. 14}} = \frac{43.76}{39.48} = 1.11$$

- Trial 2:

$$FS_1 = 1.08$$

$FS_2 = 1.08$. Therefore, circle No. 1 is solved for FS = 1.08.

- Prepare new sketch for other circle coordinates and repeat solution to find FS.

Notes: See Fig. 8.42 for values for $M_{i(\theta)}$.
For trial 2, etc.:

$$FS_{(\text{trial})} = \frac{(\text{Col. 8 from trial 1/Col. 10 from new trial})}{\text{Col. 14 from trial 1}} = \frac{\text{New Col. 11}}{\text{Col. 14 (trial 1)}}$$

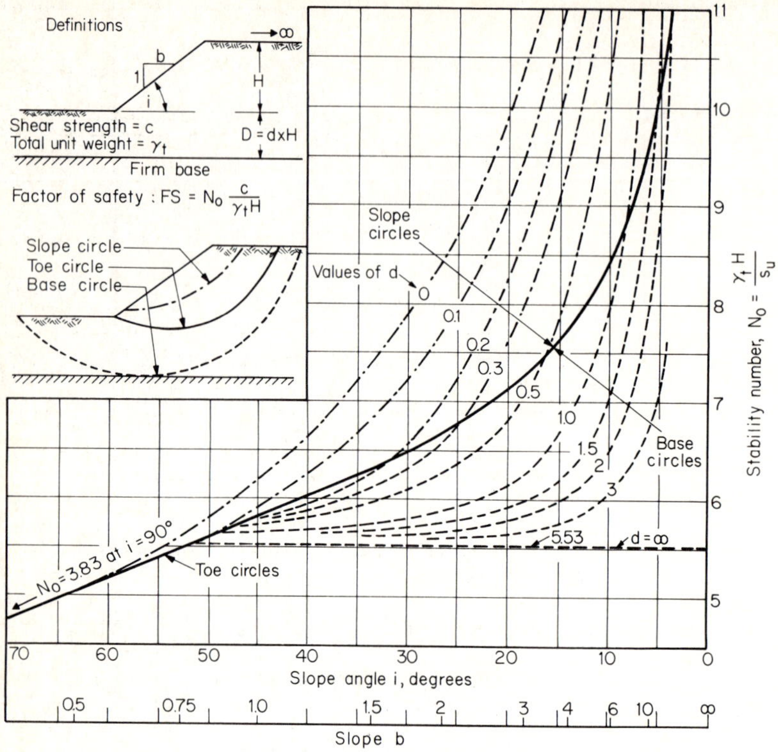

FIG. 8.43 Simplified method of stability analysis for cut slopes in cohesive soils (ϕ = 0). [From NAVFAC (1982),[42] after Janbu (1954).[11]]

Using Eq. 8.51 with FS = 1, the critical height to which a slope may be cut can be found.

Slope angle i, depth factor d_f, and the stability number N_o are related on the chart. If no strong stratum provides influence, the failure circle location is controlled by the slope angle. For slopes inclined steeper than 53° the circle describes a *toe failure*; for slopes with $i < 53°$ the circle passes beneath the toe, describing a base failure. (Note that for $i = 90°$, from Eq. 8.11, $N_o = 3.83$.) Usually it is the depth to a strong stratum, expressed by the depth factor d_f, that governs. For the chart given in Fig. 8.43 it is defined as the ratio of the depth below the toe of slope to the strong stratum, D, to the slope height H, or

$$d_f = D/H \tag{8.54}$$

If the strong stratum is within the slope, the critical circle will also be within the slope (slope circle); if it is relatively deep, the critical circle may pass below the toe (base failure); and if it is only moderately deep below the toe, the failure will be a toe circle. The location of the critical circle may be found from Fig. 8.44. Embankment stability may also be evaluated with these charts, as will be discussed.

Cousins' charts (*see Art. 8.2.3*) may also be used for the $\phi = 0$ case by plotting $\lambda_{c\phi} = 0$ and slope angle i on the chart for $r_u = 0$.

EXAMPLE OF SOLUTION WITH JANBU'S CHART Given $\gamma_t = 115$ pcf, $c = s_u = 600$ psf, $i = 35°$, $\phi = 0$, $H = 25$ ft, and the depth D to a firm base = 12.5 ft. Compute $d_f = D/H = 0.5$ and enter the chart at $i = 35°$ to find $N_o = 6.1$. From Eq. 8.53 calculate FS = (6.1)(600)/(115)(25) = 1.27. Coordinates of the critical circle are Y = (1.6)(25) = 40 ft and X = (0.76)(25) = 19 ft. Solution with Cousins' chart gives FS = 1.25; using the ordinary method of slices gives FS = 1.28.

Slopes with ϕ_u and c_u Acting

For a given value of ϕ the critical height of a slope which fails along a circular arc can also be given in terms of the stability number. Using the

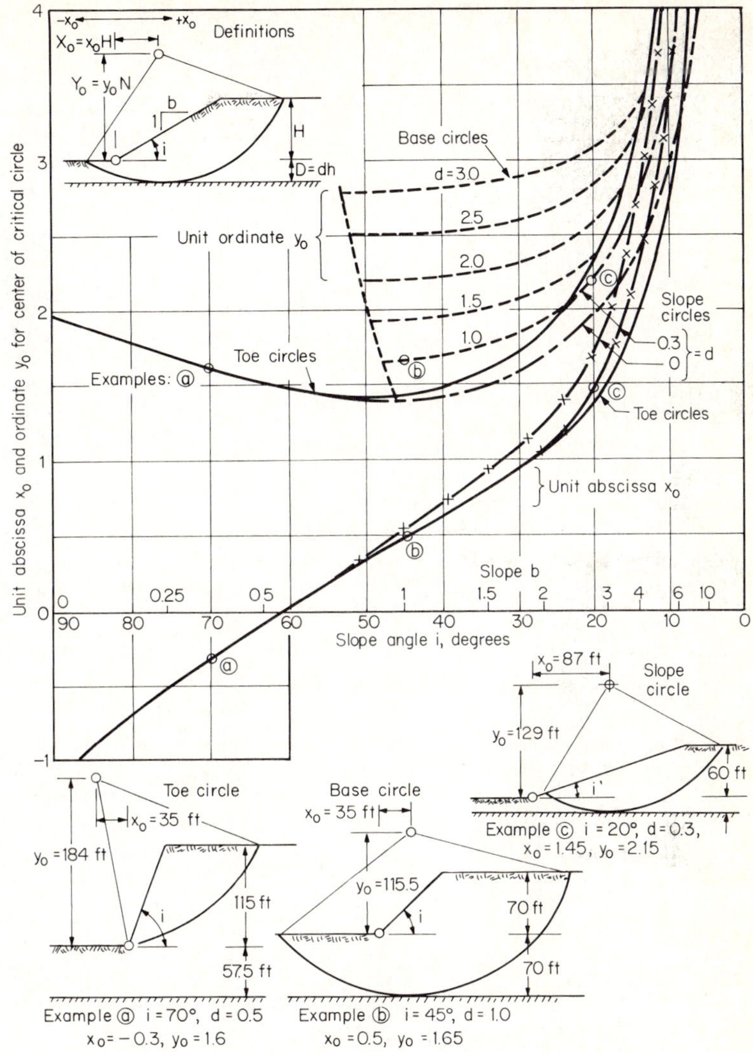

FIG. 8.44 Location of center of critical circle: $\phi = 0$ analysis. [From NAVFAC (1982)[42]; after Janbu (1954).[11]

friction circle concept, Taylor (1937)[27] developed charts relating N_s, i, and ϕ, which were subsequently modified by Janbu (1954)[11] and Cousins (1978).[35] Cousins' charts *(see Art. 8.2.3)* for the case $r_u = 0$ are applicable to total stress analysis and are useful for preliminary analysis of homogeneous dam embankments. They may be used to find FS for the conditions of a submerged slope or sudden drawdown. The steady seepage case is evaluated by effective stress analysis.

EXAMPLE OF CHART SOLUTION For the slope illustrated in Fig. 8.38, $H = 43$ ft and $i = 40°$. Undrained parameters selected are $c_u = 600$ psf and $\phi_u = 10°$ *(see discussion in Art. 3.5.3, Total Stress Analysis)*, and $\gamma_t = 125$ pcf. Compute $\lambda_{c\phi} = (125)(43)(\tan 10°)/600 = 1.58$. Enter the chart in Fig. 8.33*a* for $r_u = 0$ with $i = 40°$ to find $N_f = 9.5$. With no seepage forces, FS $= (9.5)(600)/(125)(43) = 1.06$. (The ordinary method for these conditions yields FS $= 1.07$.) The critical height may be found from Eq. 8.42 as $H_c = N_f c/\gamma_t = (9.5)(600)/(125) = 45.6$ ft.

For a *totally submerged slope* use γ_b in Eq. 8.42 to find FS $= 2.12$. For the *sudden drawdown con-*

dition, the friction angle ϕ is reduced by the ratio of the soil buoyant weight to the total soil weight, or

$$\phi_b = \frac{\gamma_b}{\gamma_t}\phi_t \tag{8.55}$$

In the foregoing problem, ϕ_b is found $\approx 5°$ and a new value computed for $\lambda_{c\phi} = 0.78$. From the chart the value $N_f = 8.0$ is found, and calculations yield FS = 0.9 (failure). The sudden drawdown case is slightly more *unstable* than the steady seepage case, as is shown in Fig. 8.38.

Consideration of Other Stability Factors

Charts have been prepared for other factors, including a surcharge load on the slope, submergence of the slope, and tension cracks, with or without full hydrostatic pressure in the cracks. The full expression for FS is given as

$$FS = \frac{U_q U_w U_t c N_s}{H + q - \gamma_w H_w} \tag{8.56}$$

where U_q, U_w, and U_t are reduction factors for surcharge, water outside the slope, and tension cracks; q is surcharge amount; and H_w is height of water outside the slope. Solutions for these conditions can be found in NAVFAC (1982),[42] based on Janbu's charts.

Validity

Application of the charts given herein is based on a number of assumptions, as follows:

There is no surcharge, no tension cracks, and no open water outside the slope.

Soil is homogeneous with strength constant with depth to a strong stratum. (If strength values vary along the failure surface, they are averaged to obtain working values, which must be done with judgment and caution.)

Failure occurs as rotation along a circular arc as all shear strength is mobilized simultaneously.

The failure surface depth is limited by a strong stratum at a depth expressed by a depth factor or by the slope angle.

Other Chart Solutions

Hunter and Schuster (1968),[43] (1971)[44] developed charts for NC unfissured clay with $\phi = 0$ which take into account the depth to GWL and a depth factor. The charts listed in Table 8.8 under Huang (1975),[36] (1977),[32] (1978),[33] (1979),[36] and (1980)[37] also are applicable to total stress analysis. Taylor's charts (1937),[27] (1948)[1] for the c, ϕ case are solved by iteration to find the lowest value for FS, a somewhat laborious process.

Nonhomogeneous Slopes

Nonhomogeneous slopes may be analyzed for total stresses with slice methods by employing parameters c_u and ϕ_u or, for the $\phi = 0$ case, s_u. For the $\phi = 0$ case, Eq. 8.46 becomes

$$FS = \frac{\Sigma s_u \, \Delta L_i}{\Sigma W_i \sin \theta_i} \tag{8.57}$$

For example, the slope given in Table 8.9 and $\phi = 0$, an average undrained strength of 800 psf is required for stability (*see discussion of Embankments over Soft Ground*).

Embankments over Soft Ground

General

Embankments placed over soft clays or organic soils may fail during construction as some critical height is reached, or immediately after construction is completed, but thereafter consolidation results in strength increase and failures are unusual. As discussed in Art. 8.2.1, where the supporting soils are homogeneous and deep, the failure surface will approach a circular arc, but if the soft stratum is relatively thin and underlain by very strong material, a planar failure surface is likely. Planar failure occurs also where the soft stratum contains a very weak zone or a thin lens of fine sand or silt in which high pore pressures can develop.

Construction of fills over soft ground is discussed in Art. 5.3.2. Prior to construction, stability analysis is performed based on total stress and the undrained strength, but during construction stability should be checked by effective stress analysis employing measured pore pressures as discussed in Art. 8.2.1. Total stress analysis of circular failure surfaces are covered in this section; planar failure surfaces are covered in the following article. These discussions pertain also to stability analysis for waterfront structures (*see Art. 9.4*).

Uncertainties in Predictions

Prediction of failure is often uncertain, and many fills have failed after rigorous analysis has indicated they would be stable. Usually, therefore, a safety factor of 1.5 is employed and the fill provided with gentle slopes. Reduced FS values require close monitoring with instrumentation during construction, and perhaps other provisions to improve stability, as will be discussed.

The major cause for uncertainty is in the values used for the undrained shear strength. As discussed above, an undetected thin seam of very weak material can be the seat of failure. In moderately to highly plastic clays, the vane shear strength corrected for the time loading rate factor (Fig. 3.34) is normally used. In other materials, anisotropy may be evaluated by obtaining strengths in triaxial compression and extension and by direct shear tests (Fig. 3.56*b*). In analysis, the strength of the fill may be ignored, since it often fails in tension. As an alternative it may be replaced with an ideal clay and assigned a value for s_u equal to that computed from Eq. 3.11 for clean sand, for example, or it may be assigned a ϕ value and analyzed based on total stresses. Failures have been known under both conditions, and which one will prevail is essentially unpredictable; therefore, it is prudent to evaluate both conditions.

Analysis

Four approaches to total stress analysis of embankments over soft ground include the use of stability charts, the ordinary method of slices, the translational failure method (*see Art. 8.2.6*), and the bearing-capacity equation. The last usually is the most conservative.

Analysis based on circular failure begins by preparing a scale drawing of the proposed fill and the soil section as illustrated in Fig. 8.45. The center coordinates of the failure arc are estimated from the chart given as Fig. 8.44. The arc is drawn with its radius tangent to the bottom of the weakest stratum and the potential failure zone divided into slices.

CHART SOLUTION To find the center coordinates, Fig. 8.44 is entered with $d_f = 66/45 = 1.5$ and $i = 40°$ and $y_o = 1.9$ and $x_o = 0.5$ are obtained. Therefore, $Y = 86$ ft and $X = 22.5$ ft. For $d_f = 1.5$ and $i = 40°$, Fig. 8.43 is entered to obtain stability number $N_s = 5.6$. Values for $s_u\,\Delta L$ are obtained for each slice, summed, and divided by the total length to find s_u (average) = 0.87 ksf. (For this example the fill material was considered to be replaced with an ideal clay and an estimated value of s_u was assigned.) From Eq. 8.53,

$$FS = \frac{s_u(\text{avg})\,N_o}{\gamma_t H} = \frac{(0.87)(5.6)}{(0.119)(45)} = 0.91$$

SLICE METHODS For a solution with the ordinary method, the weight of each slice is determined and the sum of the $W \sin\theta$ values obtained using buoyant weights below GWL. The sum of the values for $s_u\,\Delta L$ computed for the chart solution is entered into Eq. 8.57 to find FS, or

$$FS = \frac{\Sigma s_u\,\Delta L}{\Sigma W \sin\theta} = \frac{305}{368} = 0.83$$

A computer run using the simplified Bishop method, employing 20 slices and assigning values of $\phi = 25°$ for the fill and sand layer (total stress analysis) yields FS = 1.02 for circle coordinates $x = 22.5$ ft, $y = 86$ ft.

BEARING CAPACITY EQUATION In Art. 6.2.4, the bearing capacity of a surface strip footing over soft soil for the undrained, or $\phi = 0$ case, is given by Eq. 6.18 as

$$q_{ult} = 5.14\, s_u/FS$$

For the foregoing problem, $q_{ult} = 5.14(0.87)/1.0 = 4.5$ kips, or for $\gamma_t = 0.119$, the critical height $H = 38$ ft. FS is expressed as

$$FS = \frac{N_c s_u}{\gamma_t H} \tag{8.58}$$

from which FS for $H = 45$ ft is found = 0.84.

CASE HISTORY The example given in Fig. 8.45 was analyzed from data provided by Leonards (1982),[21] where he described the failure of a test embankment for Highway I-95, constructed over Boston blue clay in Massachusetts. Initially constructed to a height of 35 ft and fully instrumented, the embankment failed suddenly at a height of 53.7 ft while the elevation was being raised at the rate of 1 to 2 ft/day five years later.

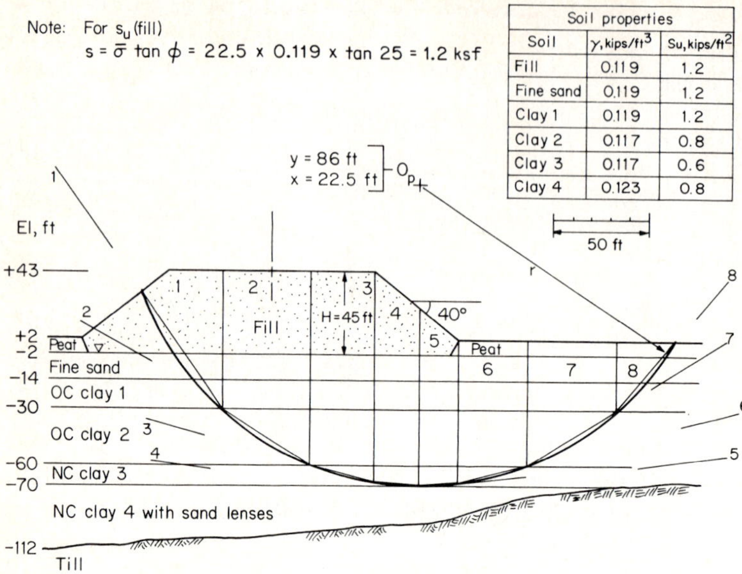

Slice	s_u, ksf	ΔL, ft	$S_u \Delta L$	θ, degrees	W, kips*	$W \sin \theta$
1	1.2	76	91.2	56	203.39	168.62
2	0.8	58	46.4	32	359.73	190.63
3	0.6	34	20.4	17	399.74	116.87
4	0.6	24	14.4	5	187.15	16.31
5	0.6	21	12.6	−5	117.60	−10.25
6	0.6	39	23.4	−14	128.46	−31.08
7	0.8	58	46.4	−32	113.40	−60.09
8	1.2	42	50.4	−51	30.24	−23.50
		$L = 352$	$\Sigma = 305.2$			$\Sigma = 367.5$

$$FS = \frac{\Sigma S_u \Delta L}{\Sigma W \sin \theta}$$

$$FS = \frac{305.2}{367.5} = 0.83$$

*For weight use γ_b or below GWL. $\gamma_e = 0.119 - 0.062 = 0.057$ ksf.

FIG. 8.45 Embankment over soft clay: stability analysis for $\phi = 0$ case using ordinary method of slices.

Failure occurred simultaneously on both sides of the embankment. Inclinometers showed the failure surface to be planar along a weak seam at a depth of about 63 ft.

Improving Stability

Three methods for improving stability are:

1. Construct the embankment slowly enough to permit consolidation and subsequent strength increase. Vertical drains not only decrease the construction time required, but in addition serve to control excessive pore-pressure increase in thin layers of fine sands and silts. Monitor pore pressures and lateral and vertical displacements and perform effective stress analysis.
2. Design the embankment with slopes with low inclinations. The slope in Fig. 8.45 was for a

test embankment and is excessively steep for a functional embankment. Slopes may be 3H:1V or flatter.

3. Construct a counterberm to resist the overturning moment.

Counterberm design may be performed with the aid of Fig. 8.46. The objective is to add weight to the area beyond the toe of the embankment to increase resistance to overturning. From the example given in Fig. 8.45, b_1 = 155 ft, $P_1 = \gamma_t H$ = 5.36 kips/ft, D = 66 ft, and the desired FS selected as 1.2 for the counterberm. $c_{all} = c_{avg}/FS$ = 0.87/1.2 = 0.73. The counterberm weight $P_2 = P_1 - 5.5\ c_{all}$ = 5.36 − (5.5)(0.73) = 1.34, and the required height H_2 = 1.34/0.119 = 11.5 ft, say 12 ft.

Compute b_1/D = 155/66 = 2.35 and P_1/P_2 = 5.36/1.34 = 4.0 and enter the upper-left-hand portion of Fig. 8.46, which indicates a Case I failure. Use the Case I figure to find b_2/D = 2.15 or b_2 = 2.15(66) = 142 ft and find x/b_2 = 0.75 (b_2 is the counterberm width). From x/b_2 is found dimension x = 0.75(142) = 106 ft. Dimension A = $(P_2/P_1)\ b_2$ = 37 ft, and x + A = 143 ft, which is the radius of the failure arc for the fill with the counterberm, located as shown in Fig. 8.47. The newly defined failure mass is divided into slices and analysis is performed using the ordinary

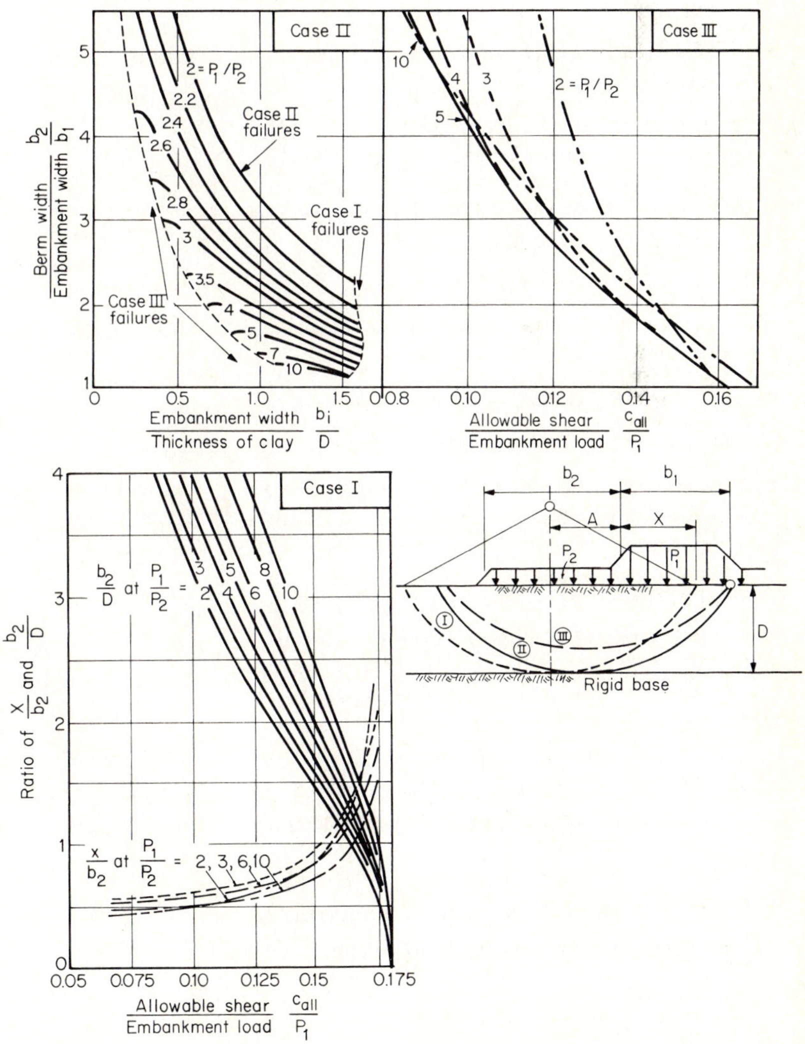

FIG. 8.46 Counterberm design: embankments on soft soils. [From NAVFAC (1982)[42]; after Jakobsen (1948).[45]]
Case I: Embankment width large compared to soft soil thickness (>1.5 times).
Case II: Embankment width ranges from 0.5 to 1.5 times soft soil thickness.
Case III: Embankment width small compared to soft soil thickness (<0.5 times).

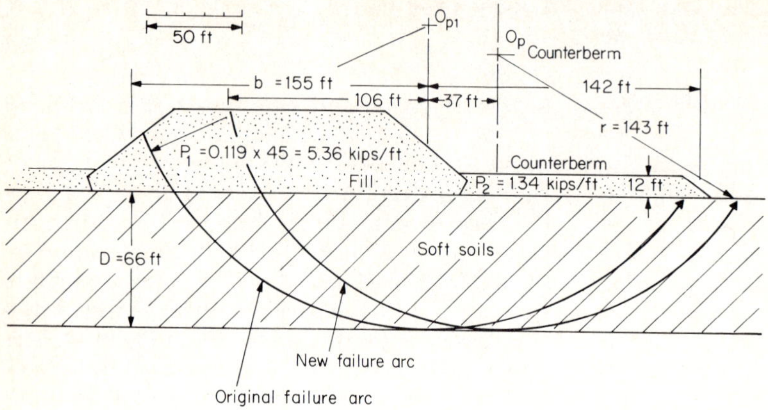

FIG. 8.47 Counterberm to provide stability for embankment shown in Fig. 8.45; FS = 1.2.

method to check the counterberm design. It is found that FS = 1.2.

The above procedure is from Jakobsen (1948),[45] who, rather than using the bearing-capacity factor 5.14, uses the more conservative factor 5.5 for the d_f = infinity case given in Fig. 8.43. He gives the critical height (from Eq. 8.15) as

$$H_c = \frac{(5.5)(s_u)}{\gamma_t\,(\text{FS})} \tag{8.59}$$

8.2.5 IRREGULAR FAILURE SURFACES

Morgenstern and Price's Method

Based on the force system of the complete Bishop method (Fig. 8.40), the method of Morgenstern and Price (1965)[10] is suitable for the analysis of any shape of failure surface, but requires a computer for solution. In analysis, assumptions are made regarding the orientation of the interslice forces between each slice and judgment is required to determine if the computed quantities are reasonable.

Whitman and Bailey (1967)[46] describe the method and compare the results with those obtained by the Fellenius and simplified Bishop methods. They report that the error resulting from the simplified Bishop method is generally 7% or less (and usually 2% or less), but that the Fellenius solution can be seriously in error, although usually on the conservative side of safety.

Janbu's Method

Relationships

Janbu developed a slice method based on the simplified Bishop method that is suitable for analysis of a failure surface of any shape [Janbu et al. (1956),[41] Janbu (1973)[12]]. Currently favored by many practitioners for analysis of soil or rock slopes where the failure surface can be reasonably assumed, in application the method has two major variations from the simplified Bishop method:

- Assumption of the failure surface rather than a series of analyses to locate the critical circle
- Introduction of a correction factor f_o to represent the influence of interslice force distribution on FS

The nomenclature and parameters of the Janbu method are illustrated in Fig. 8.48. Factor f_o depends on the geometric ratio d/L where d and L are as defined in Fig. 8.48a. In homogeneous or nearly homogeneous soils in which a potential failure surface is not apparent by inspection, the critical failure surface (lowest FS) is found by varying d. The expression for Janbu's analysis often is found written as

$$\text{FS} = f_o\,(\Sigma\{[\bar{c}\,b + (W - ub)\tan\bar{\phi}] \times [1/\cos\theta\,M(\theta)]\}/\Sigma W \tan\theta + V) \tag{8.60}$$

where $M(\theta)$ is as defined in Eq. 8.49 and given in Fig. 8.42 and V = water pressure in a tension

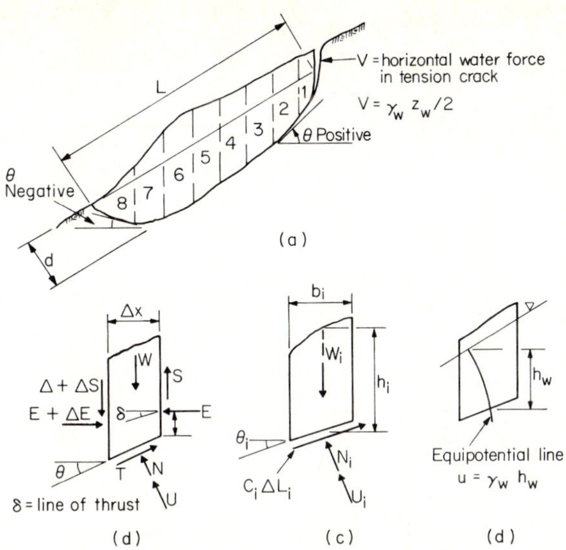

FIG. 8.48 Nomenclature for Janbu's method of analysis for noncircular failure surfaces: (*a*) failure mass geometry; (*b*) slice force system; (*c*) slice parameters for analysis; and (*d*) determination of pore pressure *u*.

crack, used more commonly in rock-mechanic problems. The correction factor f_o depends on the soil type and may be obtained from Fig. 8.49 or computed from the following:

- For $\bar{c} > \bar{0}$ and $\bar{\phi} > 0$:

$$f_o = 1 + 0.5\,[d/L - 1.4(d/L)^2] \quad (8.61)$$

- For $\bar{c} = 0$:

$$f_o = 1 + 0.31\,[d/L - 1.4(d/L)^2] \quad (8.62)$$

Procedure

Procedure for analysis is similar to that of the simplified Bishop method except that a failure surface is assumed and values for f_o calculated or obtained from the chart (Fig. 8.49). An example of a real problem is given in Fig. 8.50. Illustrated is one of the more critical sections of a proposed cut slope to be constructed during excavation of residual soils and sound gneiss in a long, narrow hill. The section shown is taken from the side slopes of the hill where the rock surface is inclined. It is assumed that the potential failure surface is along the rock surface at a point where the residual soils contact with more steeply inclined decomposed rock from where it extends up to the top of the cut as shown in the figure.

A table for data input is prepared (Table 8.11), a value for FS_1 is assumed, and calculations are made to find FS_2 for Trial 1. Iteration is performed to find $FS_1 = FS_2$ as shown in Table 8.10 for the Bishop method. Also shown in Table 8.11 are the results of parameteric studies to evaluate the effect on FS of varying values for $\bar{c}$, $\bar{\phi}$, and *u*. The problem given in Fig. 8.50 also is suitable for wedge analysis (*see Art. 8.2.6*) and the results are given in Table 8.11.

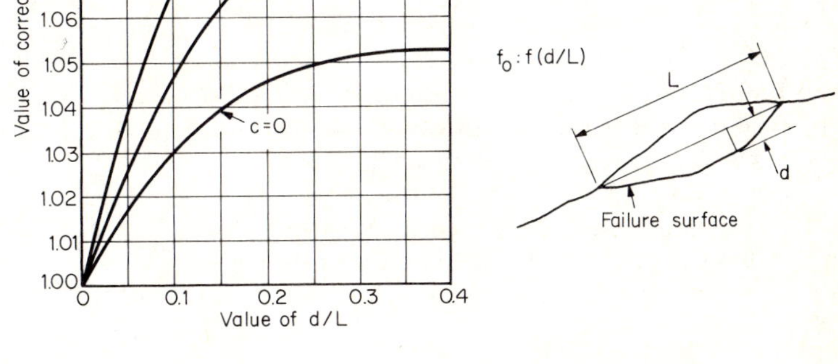

FIG. 8.49 Correction factor f_o to allow for interslice forces in Janbu's noncircular failure analysis. [From Hoek and Bray (1977)[3].]

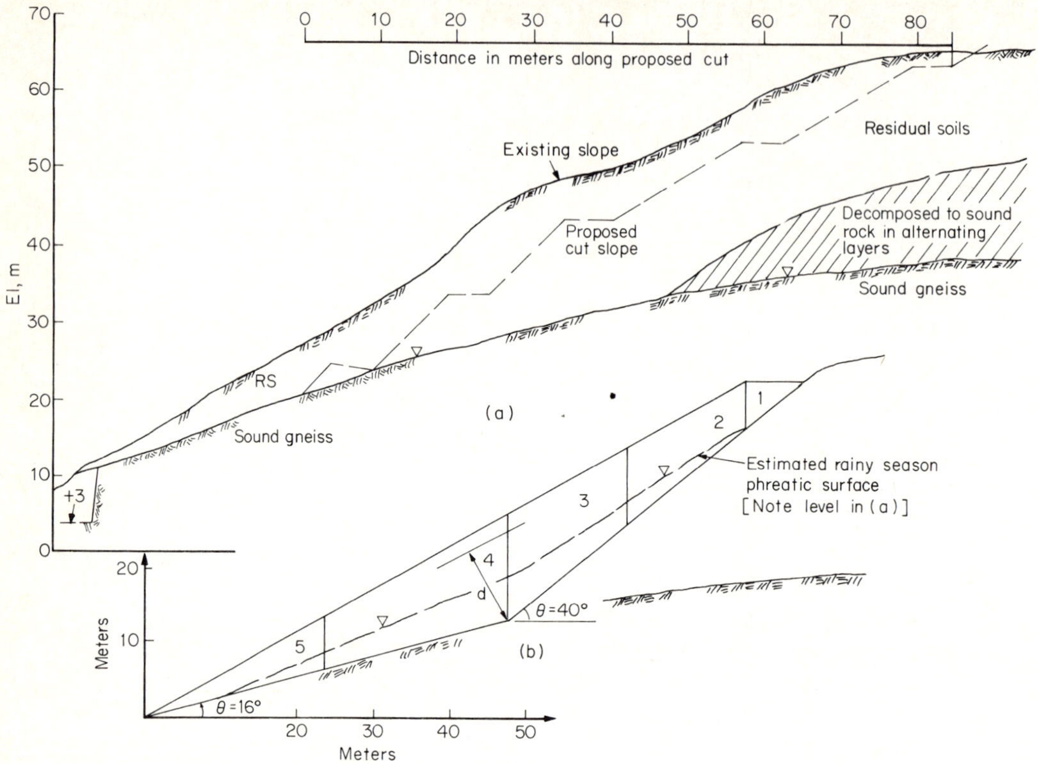

FIG. 8.50 Application of the Janbu method of stability analysis: (*a*) existing slope conditions and proposed cut slope and (*b*) geometry for a preliminary analysis.

8.2.6 PLANAR FAILURE SURFACES, WEDGES AND BLOCKS

General

Applications

In many actual situations the failure surface is not cylindrical, but rather is planar and can be approximated by one or several straight lines which divide the mass into wedges or blocks. Such cases exist

- In natural slopes where there are weak strata within or beneath the slope or where the soil strata rest on a strong stratum as in the case of residual or colluvial soil over sloping rock
- In embankments on soft ground with weak strata
- In earth dams with zoned embankments or overlying weak strata
- In sidehill fills such as for roadways or created by mine waste disposal

Various Methods

- The *infinite slope problem (see Art. 8.2.2)* involves a single planar surface long in extent compared to its depth. It is often applicable to weathered clay shale slopes (Fig. 8.18*b*).
- The *Culmann or simple wedge method (see Art. 8.1.3)* involves a single, planar failure surface of limited extent in a clay slope in which the failure surface location is unknown. The method is generally applicable to slopes steeper than 75 to 80°, and has little application except for analysis of slopes retained by anchored curtain walls *(see Art. 9.2.5)*.
- The *simple wedge method: failure surface assumed.* The force system of the Culmann method is applicable to slopes in which the fail-

TABLE 8.11
SOLUTION FOR SIMPLIFIED JANBU (EXAMPLE: FIG. 8.50)

Slice no.	$b(\gamma_h - u)\tan\phi + \bar{c}$: γ_h, t/m²	u, t/m²	ϕ, degrees	$\bar{c}$, t/m²	b, m	Result, tons	$\frac{\tan\bar{\phi}}{FS_1}$	$M_i(\theta)$*	$\cos\theta\, M_i(\theta)$	$\frac{\text{Result}}{\cos\theta\, M_{i(\theta)}}$ (R)	W, tΔ	θ, degrees	W tan θ, tons (D)	f_o†	f_o (R/D) (FS_2)
1	5.60	0	30	0	8.0	25.87	0.481	1.14	0.824	31.40	44.8	40*	37.59		
2	14.88	2.0	30	0	15.0	111.54	0.481	1.10	0.824	135.41	223.13	40	187.23		
3	21.53	3.0	30	0	15.0	160.48	0.481	1.14	0.824	194.81	322.88	40	270.93		
4	21.00	4.0	30	0	24.0	235.56	0.481	1.12	1.054	224.02	504.0	16	140.52		
5	6.48	0	30	0	24.0	89.79	0.481	1.12	1.054	85.39	155.40	16	44.56*		
										Σ 671.03			Σ 684.83	1.028	1.012

Trial 1:

$$FS_1 = 1.20 \qquad FS_2 = f_o \frac{\sum \frac{b(\gamma h - u)\tan\bar{\phi} + \bar{c}}{\cos M_i(\theta)}}{\Sigma W \tan\theta} = f_o \frac{\Sigma R}{\Sigma D} = 1.01$$

Trial 2: $FS_1 = 1.0$ $FS_2 = 0.97$, and problem is solved with FS = 1.0

Soil parameters:
$\gamma_t = 1.75\ \text{t/m}^3$ $\bar{c} = 0\ \text{t/m}^2$ $\phi = 30°$ $d = 9.5$ $L = 97.5.$

Parametric study

Case	$\bar{\phi}$, degrees	$\bar{c}$, t/m²	U, t/m²	Σ Resistance R	Σ Drive D	f_o	FS_1	FS_2	Wedge method FS
1	30	0	Yes	671.03	684.83	1.028	1.0	0.97	1.01
2	30	2	Yes	872.54	684.83	1.046	1.33	1.32	1.27
3	30	2	No	1007.82	684.83	1.046	1.55	1.55	1.50
4	30	0	No	772.16	684.83	1.028	1.17	1.17	1.18

*See Fig. 8.42 for values for $M_{i(\theta)}$.

†See Fig. 8.49 for values for f_o, or Eqs. 8.61 or 8.62.

ure surface is known or can be assumed from the soil stratigraphy. The extent of the potential failure surface may be controlled by a tension crack.

- *Multiple blocks or wedges.* Force systems acting on blocks are assumed to be those acting on a single block except that interblock forces are either evaluated by empirical relationships or by earth pressure theory. Solutions vary from graphical to numerical for two, three, or more blocks. The Janbu method *(see Art. 8.2.5)* is also suitable for evaluating multiple block conditions.

Common Features

In most cases the potential failure surface is assumed from an evaluation of the slope geometry and geologic stratification. Exceptions are certain cases of the infinite slope problem. Evaluation of pore pressures acting along the potential failure surface is possible in all solutions.

In all methods it is assumed that the free body or bodies are acted upon by gravity and external forces and normal and shear stresses distributed along the failure surface. Except for the infinite slope case, the interaction between adjacent blocks is considered in analysis.

Simple Wedge or Single Block

No Tension Crack

The force system acting on a simple wedge with no tension crack is illustrated in Fig. 8.51*a*. The mass is considered as a wedge-shaped free body containing release surfaces as lateral boundaries. FS is expressed as

$$FS = \frac{C + (N - U)\tan\phi}{T} = \frac{cL + (W\cos\theta - U)\tan\phi}{W\sin\theta} \qquad (8.8)$$

Tension Cracks and Other Factors

Generated by small shear displacements in the slope, tension cracks should be considered as an indication of potential instability. They have the effect of shortening the potential failure surface and providing openings for the entrance of water. They are more commonly considered in

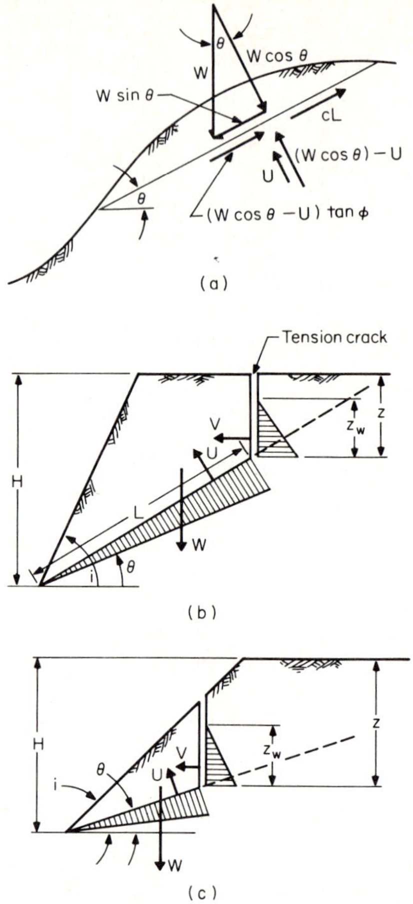

FIG. 8.51 Planar failure, simple wedges: (*a*) simple wedge with no tension crack, (*b*) tension crack in upper slope surface, and (*c*) tension crack in slope face. [From Hoek and Bray (1977).[3]]

analysis of rock slopes than soil slopes, unless in the latter case deformations have occurred. The factor of safety can reach a minimum value for a particular crack depth or location.

As shown in Fig. 8.51*b* and *c*, tension cracks can form along the slope or behind the crest changing the failure wedge geometry. Hoek and Bray (1977)[3] present a detailed discussion of plane failure analysis including the influence of tension cracks that normally develop on slopes during excavations in rock masses which are also applicable to some soil slopes. The general expression for FS with water in a tension crack and pore pressure acting along the potential failure surface is

$$FS = \frac{cL + (W \cos \theta - U - V \sin \theta) \tan \phi}{W \sin \theta + V \cos \theta} \tag{8.63}$$

where
$$L = (H - z) \csc \theta$$
$$U = \tfrac{1}{2}\gamma_w z_w (H - z) \csc \theta$$
$$V = \tfrac{1}{2}\gamma_w z_w^2$$

and, for tension crack in upper slope surface,

$$W = \tfrac{1}{2}\gamma H^2 \{[1 - (z/H)^2] \cot \theta - \cot i\}$$

or, for tension crack in the slope face,

$$W = \tfrac{1}{2}\gamma H^2 [(1 - z/H)^2 \cot \theta (\cot \theta \tan i - 1)]$$

Earthquake forces, considering the seismic coefficient *k* (*see Art. 8.2.7*), are provided for in the expression

$$FS = \{cL + [W(\cos \theta - k \sin \theta) - U - V \sin \theta] \tan \phi\}/[W(\sin \theta + k \cos \theta) + V \cos \theta] \tag{8.64}$$

Figure 8.52 shows an example of the application of the simple wedge solution with a tension crack along the slope face developing in residual soil overlying weathered rock.

Two Blocks: Graphical Analysis

Conditions

A graphical method useful for cases involving two blocks and nonhorizontal stratification, such as often exist in earth dams (Fig. 8.26), was developed by Seed and Sultan (1967).[4] The geometry of a dam section and the free body forces are illustrated in Fig. 8.53. It is assumed that contact force *P* is inclined at some predetermined angle ϕ_d to plane *OB* between the blocks, where block *A* is a passive block and block *B* is an active block. From model tests and experimental data Seed and Sultan determined that a reasonable assumption was that angle ϕ_d was a function of ϕ and FS expressed as

$$\tan \phi_d = \tan \phi/\text{FS} \tag{8.65}$$

or

$$\phi_d = (\tan^{-1})(\tan \phi/\text{FS}) \tag{8.66}$$

Although often assumed vertical in analysis, boundary *OB* was found in the studies of Seed and Sultan to yield lower values for FS when inclined upslope. They also advised that the best results were obtained when ϕ for the shell material corresponds to plain-strain-deformation conditions ($\phi = \phi_{ps}$) as measured in the laboratory.

Procedure

A trial value for FS_1 is assumed and ϕ_d calculated, and the areas scaled and block weights calculated. A force polygon is constructed for block 1 to find magnitudes for the interblock force *P* and the resultant R_1 which acts normal to *CA* at angle ϕ_d. *P* acts normal to *OB* at angle ϕ_d.

A force polygon for block 2 is then constructed by laying off T_2 at angle θ_2 and $\overline{N}_2$ perpendicular to it. Extending $\overline{N}_2$ to *P* provides closure and the value for T_2 is determined, as shown in Fig. 8.53*c*.

$$FS_2 = S_c/T_2 \tag{8.67}$$

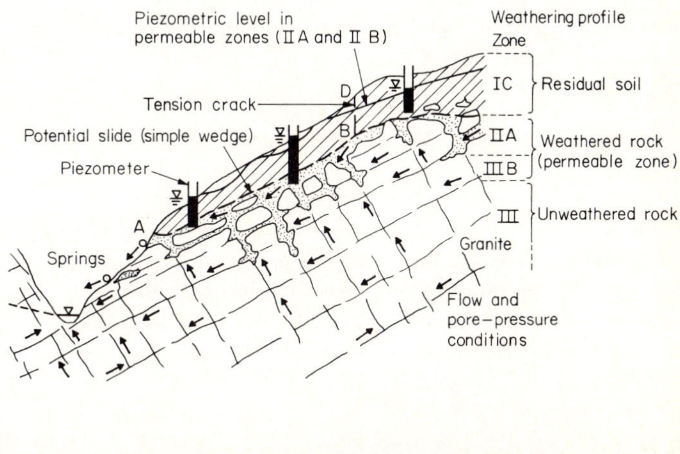

FIG. 8.52 Development of simple wedge failure in residual soil over rock. [After Patton and Hendron (1974).[15]]

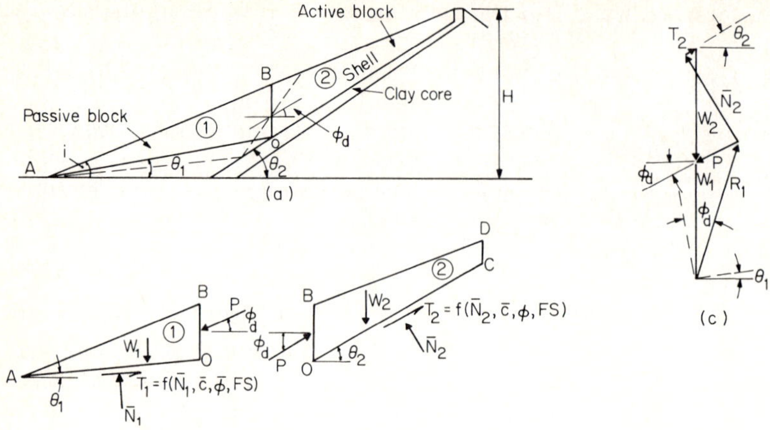

FIG. 8.53 Graphical analysis of wedge failure in sloping core dam: (*a*) dam section; (*b*) wedges and forces; and (*c*) force polygons for solution. [After Seed and Sultan (1967).[4]]

where S_c = available shear strength along line *OC*. If $FS_1 = FS_2$, the solution is the correct one; otherwise, a new value for FS_1 is assumed to find a new ϕ_d and the operation is repeated.

Example

Seed and Sultan (1967)[4] provide an example with the following data. Soil parameters: shell material, $\bar{\phi} = 40°$, $\gamma = 110$ psf; core material, $c = 2000$ psf and $cL = 2(868 \text{ ft}) = 1740$ kips/ft. $W_1 = 10{,}200$ kips/ft, $W_2 = 10{,}400$ kips/ft. $\theta_1 = 10°$ and $\theta_2 = 31.5°$, $i = 21.8°$, $AB = 975$ ft, $BD = 763$ ft, and $H = 650$ ft.

Assuming $FS_1 = 1.65$ and $\phi_d = 27°$, from the force polygons is found $FS_2 = cL/T_2 = 1740/1050 = 1.65$, and $FS_1 = FS_2$. Therefore, the correct safety factor for line *OB* vertical is FS = 1.65. They found, however, that if line *OB* was inclined upslope at an angle of 35°, a minimum value was found for FS = 1.52. Analysis with the numerical method for two blocks given in the following section yields FS = 1.61 for the case of line *OB* vertical.

Two or Three Blocks: Numerical Analysis

Two Blocks

The forces acting on two free bodies forming a potentially unstable slope are illustrated in Fig. 8.54 and are essentially those given for the graphical solution described in the foregoing problem, except for the inclusion of the pseudostatic earthquake factor *k*. As in the graphical example, the four unknowns N_1, N_2, P, and FS are found by four equilibrium equations, two for each block. The forces acting on each block are summed in the vertical and horizontal directions and the sums are equal to zero. Initially, assumptions are made that

$$\phi_d = \tan^{-1}(\tan\phi/\text{FS}) \tag{8.66}$$

and that the shear forces are

$$T_1 = (\bar{c}_1 L_1 + \bar{N}_1 \tan\bar{\phi}_1)/\text{FS} \tag{8.68}$$

$$T_2 = (\bar{c}_2 L_2 + \bar{N}_2 \tan\bar{\phi}_2)/\text{FS} \tag{8.69}$$

Solving the equations for $\Sigma H = 0$ and $\Sigma V = 0$ and then using Eq. 8.68 to find $\bar{N}_1$ and P_1 results in:

$$\bar{N}_1 = \{W_1[\cos\phi_d - r_u\cos\theta_1\cos(\phi_d - \theta_1) - k\sin\phi_d] + [\bar{c}_1 L_1 \sin(\phi_d - \theta_1)]/\text{FS}\}/\{\cos(\phi_d - \theta_1) - [\tan\phi_1 \sin(\phi_d - \theta_1)]/\text{FS}\} \tag{8.70}$$

$$P_1 = (T_1\cos\theta_1 - \bar{N}_1\sin\theta_1 - r_u W_1\sin\theta_1\cos\theta_1 - kW_1)/\cos\phi_d \tag{8.71}$$

Solving the equations for $\Sigma H = 0$ and $\Sigma V = 0$ and then using Eq. 8.69 for block 2 results in

$$\bar{N}_2 = \{W_2[\cos\phi_d - r_u\cos\theta_2\cos(\phi_d - \theta_2) - k\sin\phi_d] + [\bar{c}_2 L_2\sin(\phi_d - \theta_2)]/\text{FS}\}/\{\cos(\phi_d - \theta_2) - [\tan\phi_2\sin(\phi_d - \theta_2)]/\text{FS}\} \tag{8.72}$$

$$P_2 = (\bar{N}_2\sin\theta_2 - T_2\cos\theta_2 + r_u W_2\cos\theta_2\sin\theta_2 + kW_2)/\cos\phi_d \tag{8.73}$$

To solve the problem, a value for FS is selected and Eqs. 8.68 through 8.73 are solved to find values for P_1 and P_2. The correct value for FS is found when $P_1 = P_2$. The results obtained from the two-wedge method for the problem given in Fig. 8.50 are summarized in Table 8.10 and those obtained for the problem in Fig. 8.53 have been summarized in the foregoing section. In both cases the values obtained are of the same order of magnitude.

Three Blocks

The case of the three-block problem is illustrated in Fig. 8.54b. The equations developed for the two-block problem apply to the upper and lower blocks and it remains to develop equations for the middle block. From summing the vertical and horizontal forces and solving for the unknowns it is found that:

$$\bar{N}_3 = \{W_3(1 - r_u \cos^2 \theta_3) + (P_2 - P_1) \sin \phi_d - (\bar{c}_3 L_3 \sin \theta_3)/\text{FS}\}/[\cos \theta_3 + (\tan \phi_3 \sin \theta_3)/\text{FS}] \qquad (8.74)$$

$$\text{FCT(FS)} = (P_1 - P_2) \cos \phi_d - \bar{N}_3 \sin \theta_3 - r_u W \cos \theta_3 \sin \theta_3 + T \cos \theta_3 - kW_3 = 0 \qquad (8.75)$$

and

$$T_3 = (\bar{c}_3 L_3 + \bar{N}_3 \tan \phi_3)/\text{FS} \qquad (8.76)$$

For the solution, a value for FS is selected and Eqs. 8.68 through 8.76 are then solved to find FCT(FS). The correct value for FS is found when FCT = 0. The reference for the foregoing discussion on two- and three-block analysis is Huang (1983).[47] The concept of active and passive earth

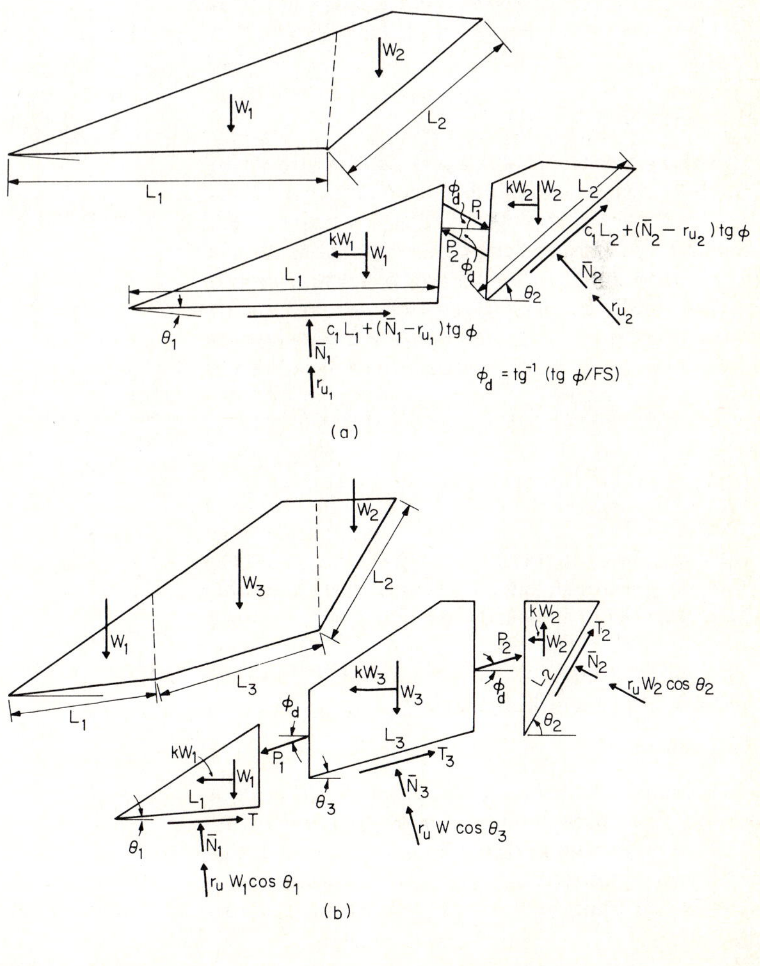

FIG. 8.54 Complex wedge systems: (*a*) free-body diagrams for two blocks and (*b*) free-body diagrams for three blocks. [After Huang (1983).[47]]

pressures is employed and when $\phi_d = 0$, $P_1 = P_P$ and $P_2 = P_A$. Shear resistance along the failure surface is reduced by a value for FS, which is not the case for some approaches using earth-pressure theory which result in overly large values for FS. Huang (1983)[47] has applied the foregoing solutions to the evaluation of the stability of mine spoil deposits placed on hillsides which often results in irregular shapes suitable for division into blocks.

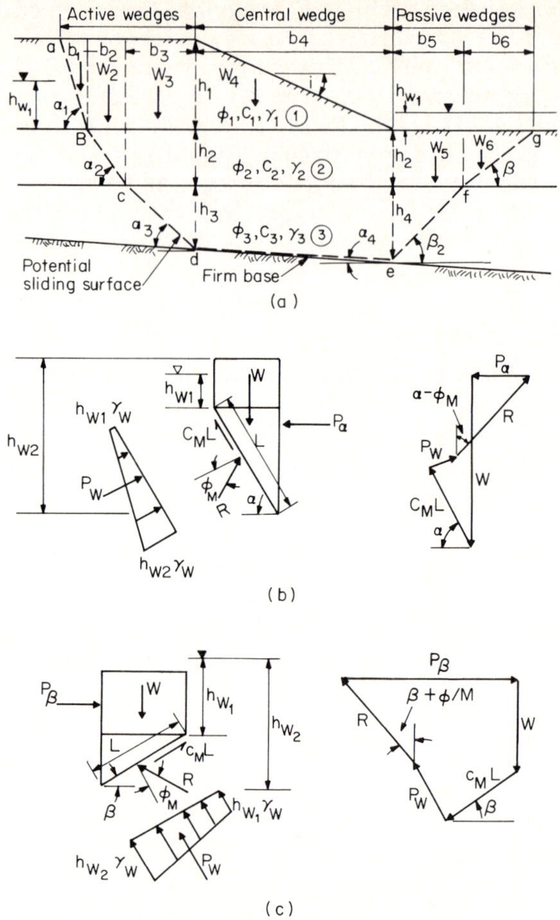

FIG. 8.55 Stability analysis of translational failure (*a*). The resultant horizontal force for a wedge sliding along surface *abcde* is P_α (*b*), and for sliding along surface *efg* is P_β (*c*). [From NAVFAC (1982).[42]]

Multiple Blocks: Earth-Pressure or Translational Failure Method

The translational failure method based on earth pressures and suitable for multiple blocks is given in NAVFAC (1982).[42] It is particularly useful where soil conditions consist of several masses with different parameters such as in the case shown for the earth dam (Fig. 8.26a).

Force Systems

The slope with a layered soil profile shown in Fig. 8.55 has been divided into a number of active and passive wedges and a central wedge. The failure surface of the central wedge is planar and controlled by either a weak stratum or the interface between two strata. In accordance with earth-pressure theory (*see Art. 9.2.3*) the failure surfaces of the passive wedges are inclined at $45° - \phi/2$ and those of the active wedges at $45° + \phi/2$. Each wedge has a weight force which is resisted by shear strength mobilized along the failure surface as shown in the free body diagrams (Fig. 8.55*b* and *c*).

Stability is evaluated by comparing the destabilizing forces of the active wedges with the stabilizing forces of the passive wedges plus the shear strength along the base of the central wedge. Analysis proceeds by trial and error to find the value for FS at equilibrium ($\Sigma P_A = \Sigma P_P$).

The terms used in Fig. 8.55 are defined as follows:

- P_α and P_β are force resultants for the active, central, and passive wedges. Depending on stratigraphy and soil strength, the central wedge may act to maintain or to upset equilibrium. If the base slopes downward, use α; if it slopes upward, use β.
- W is the total soil weight in a wedge.
- R is the resultant of normal and tangential forces.
- P_w is the resultant due to pore-water pressure on the failure surface:

$$P_w = \left(\frac{hw_1 + hw_2}{2}\right) L\gamma_w \tag{8.77}$$

- hw is the depth below the phreatic surface at the wedge boundary.

Analysis

1. Except for the central wedge where α is dictated by stratigraphy, for estimating the failure surface inclination use $\alpha = 45° + \phi/2$ and $\beta = 45° - \phi/2$.
2. For each wedge compute W and solve for P_α and P_β in terms of FS, which is applied to soil-strength values $\tan \phi$ and C as follows:

$$\phi_m = \tan^{-1}(\tan \phi/\mathrm{FS})$$

$$C_m = C/\mathrm{FS}$$

$$P_\alpha = [W - C_m L \sin \alpha - P_w \cos \alpha]\,[\tan(\alpha - \phi_m)] - [C_m L \cos \alpha - P_w \sin \alpha] \quad (8.78)$$

$$P_\beta = [W + C_m L \sin \beta - P_w \cos \beta]\,[\tan(\beta + \phi_m)] + [C_m L \cos \beta + P_w \sin \beta] \quad (8.79)$$

where

$$\tan(\alpha - \phi_m) = \tan[\alpha - \tan^{-1}(\tan \phi/\mathrm{FS})] \quad (8.80)$$

$$\tan(\beta - \phi_m) = \tan[\beta - \tan^{-1}(\tan \phi/\mathrm{FS})] \quad (8.81)$$

3. For the case of $\phi = 0$,

$$P_\alpha = W \tan \alpha - \frac{C_m L}{\cos \alpha} \quad (8.82)$$

and

$$P_\beta = W \tan \beta + \frac{C_m L}{\cos \beta} \quad (8.83)$$

4. For equilibrium, $\Sigma P_\alpha = \Sigma P_\beta$. Sum P_α and P_β forces in terms of FS, select a trial FS, calculate ΣP_α and ΣP_β. If $\Sigma P_\alpha \neq \Sigma P_\beta$, repeat with new value for FS until $\Sigma P_\alpha = \Sigma P_\beta$.

8.2.7 EARTHQUAKE FORCES

Slope Failures

Causes

Cyclic loads induced by earthquakes decrease the stability of a slope by increasing shear stresses, increasing pore pressures (pore air or pore water), and decreasing soil strength. In the extreme case, increase in pore pressures can lead to liquefaction.

Occurrence

Natural slopes composed of low-sensitivity plastic clay, dense granular soils either above or below the water table, or loose coarse-grained granular soils below the water table generally are stable during strong ground shaking. Sensitive clays and loose fine-grained granular soils above or below the water table, and metastable soils such as loess even when dry, are very susceptible to failure under cyclic loading. The presence of even a very thin layer of saturated fine-grained soil, such as silt or clayey silt, can lead quickly to instability.

Embankments over fine-grained saturated soils are particularly susceptible to failure, especially in coastal areas where lateral restraint is limited.

Earth-dam embankments can withstand moderate to strong shaking when well-built by modern standards [Seed et al. (1978)[48]]. The greatest risk of damage or failure lies with dams constructed of saturated cohesionless materials that may be subjected to strong ground shaking. The general effect of ground shaking on granular fills is slope bulging and crest settlement.

A general discussion of the effects of earthquakes on slopes and embankments is given in Art. 11.3.4 of Hunt (1984).[69]

Analysis

There are two general approaches to evaluating the effects of earthquake loadings on slopes: the conventional or pseudostatic approach and the more recent dynamic-analysis approach.

Pseudostatic Analysis

Force System

In the conventional approach, the stability of the potential sliding mass is determined as for static loading conditions and the effects of an earthquake are accounted for by including an equivalent horizontal force acting on the mass. The horizontal force (see Fig. 8.54) is expressed as the product of the weight and a *seismic coefficient* k which is related to induced accelerations. The effects on pore pressures are not considered, and a decrease in soil strength is accounted for only indirectly.

Various applications of the coefficient k can be found in the literature. It seems reasonable to increase the driving force by adding the term ($k \cos \theta$) and to decrease $\overline{N}$ in the resisting force by either ($kW \sin \theta$) or, considering that the vertical acceleration of gravity is less than the horizontal acceleration [Art. 11.4.6 of Hunt (1984)[69]], decrease $\overline{N}$ by ($0.67\ kW \sin \theta$). Equation 8.48 for the simplified Bishop method, therefore, becomes

$$FS_2 = \frac{\sum \left\{ \dfrac{b[(\gamma h - 0.67k\gamma h - u)\tan \bar{\phi} + \bar{c}]}{\cos \theta \left(1 + \dfrac{\tan \theta \tan \phi}{FS_1}\right)} \right\}}{\Sigma W(\sin \theta + k \cos \theta)} \tag{8.84}$$

Other equations including the "quake" factor k are Eq. 8.64 for the simple wedge and Eqs. 8.70 to 8.73 and 8.75 for the two- or three-block stability problem.

Seismic Coefficient k

The selection of the seismic coefficient is empirical. Terzaghi (1950)[49] suggested using the following: severe earthquakes, I = IX and $k = 0.1$; violent earthquakes, I = X and $k = 0.25$; and for catastrophic events, $k = 0.5$. For earth-dam design in the United States, values for k traditionally were selected based on the regional seismicity of the dam location, and in California values ranged from a lower limit of 0.05 to an upper limit of 0.15, although there appears to be no published justification for using these values [Seed (1967)[50]]. They do fall within the range of Terzaghi's values for intensities common to California earthquakes. Design coefficients commonly used in Japan range from 0.12 to 0.25 depending upon dam location, foundation type, and failure risk (downstream effects).

For the case of an infinitely long dam with a symmetrical and homogeneous cross section resting on a rigid foundation, k was found to vary with the depth within the dam, and therefore with dam height [Seed and Martin (1966)[51]]. Ambraseys (1960)[52] presents relationships for estimating k as a function of depth as given in Fig. 8.56.

There is not much data in the literature on the selection of k values. As discussed below, the San Fernando dam failures occurred during an event with I = XIII − IX and the computed k values at failure ranged from 0.22 to 0.55. Considering these data and the foregoing discussion, suggested values for k are summarized in terms of earthquake intensity I in Table 8.12. Based on

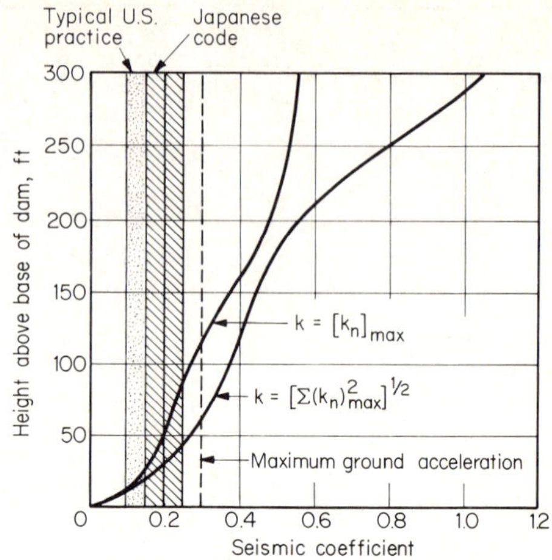

FIG. 8.56 Seismic coefficient *k* suggested for use in pseudostatic analysis of earth dam slope stability developed by Ambraseys (1960).[52] [From Seed (1967).[50]]

TABLE 8.12
SEISMIC COEFFICIENT *k*, SUGGESTED VALUES

k	Zone	Intensity *I*
0.0	0	I–IV
0.05	1	V–VI
0.10	2	VII
0.15	3	VIII–IX
0.25	3	X or greater
0.50	3	VIII or greater and high risk

studies for the Alaskan pipeline, Hall and Newmark (1977)[53] developed design accelerations for ground motion for slope stability studies as related to magnitude which are given in Table 8.13. There is generally good agreement with those given in Table 8.12.

San Fernando Dam Failures

During the San Fernando earthquake of February 9, 1971 (I = XIII − IX), the two San Fernando dams suffered major slides. Prior to the event, pseudostatic analysis employing the upper limit for k used in California determined that these dams, constructed over 50 years ago, were safe against earthquake-induced loads.

TABLE 8.13
HORIZONTAL GROUND MOTIONS: SEISMIC DESIGN CRITERIA*

Magnitude†	Design accelerations‡,§
8.0	0.60
7.5	0.45
7.0	0.30
5.5	0.12

*From Hall and Newmark (1977).[53]

†Magnitudes are considered as the maximum design earthquake developed for four seismic zones along the Alaskan pipeline.

‡Peak values which may affect slope stability and liquefaction.

§For vertical acceleration multiply by two-thirds.

Each dam was constructed with a clay core on both sides of which was placed a hydraulic sand fill. Very high pore pressures induced by the earthquake caused the sand fill to either liquefy or to reduce greatly in strength. In an analysis of the actual failures, Seed et al. (1975)[54] found that k values ranging from about 0.22 to 0.34 would have yielded FS = 1 in a pseudostatic analysis of the lower dam and in the range of 0.43 to 0.55 for the upper dam.

Time-History Dependence

The pseudostatic approach to analysis is based on a peak acceleration producing dynamic inertial forces in an embankment or slope, but studies of failures [Seed et al. (1978)[48]] indicate that the magnitude of the deformations depends on the time history of the inertial forces as well as their magnitude. Peak accelerations may be sufficiently large to drop FS below unity for brief time intervals during which displacements occur, but the movement ceases when the magnitude of the accelerations drops or is reversed.

The time element of the accelerations is much more important because of the increasing pore pressures induced by cyclic or repetitive loading. If the dynamic loads are applied in cycles with small periods, the soil will not have time to drain between loadings and the pore pressures will continue to increase, and if loadings are over a sufficiently long time interval, failure or liquefaction may occur. [See, for example, the discussion of the Turnagain Heights landslide during the Alaskan quake of 1964 in Art. 9.2.6 of Hunt (1984).[69]]

Dynamic Analysis

Seed et al. (1975)[55] evaluated the stability of the San Fernando dams, which is reviewed in some detail in Art. 11.3.4 of Hunt (1984),[69] using dynamic analysis techniques [Newmark (1965),[56] Seed (1966)[57]]. In summary, they proposed the following procedure for embankments of saturated cohesionless materials to evaluate cyclic forces more accurately than is possible with pseudostatic analysis.

1. Determine static stresses in the embankment prior to an earthquake.
2. Determine the characteristics of ground motion in the materials underlying the embankment during the design earthquake. These characteristics are given as acceleration of gravity for the location by accelerographs or are determined by seismicity study.
3. Evaluate embankment response to base excitation and compute the dynamic stresses induced in representative elements of the embankment. Analysis is most conveniently performed by using finite element techniques which incorporate nonlinear dynamic material properties by using a strain-dependent shear modulus and damping values *(see Art. 4.4.5)*.
4. Representative samples of the embankment material are subjected in the laboratory to combinations of pre-earthquake stress conditions and superimposed dynamic stress applications to determine the effects of earthquake-induced stresses on soil elements in the embankment *(see Art. 3.3.5)*. These effects will include any evidence of soil liquefaction and the magnitude of the deformations induced by the earthquake loading.
5. From a knowledge of the deformations induced in individual soil elements in the embankment, evaluate the overall deformations and the stability of the embankment section.

Summary

Pseudostatic analysis now is generally recognized as being inadequate to predict behavior during earthquakes, but may be used as a first approach. For high embankments whose failure may cause loss of life or major damage, dynamic analysis is recommended by the Committee on Earthquakes of the International Commission on Large Dams (1975).[58] Design proceeds first by conventional methods followed by dynamic analysis to investigate any deficiencies which may exist in the pseudostatic analysis.

A simplified procedure for estimating dam and embankment earthquake-induced deformations is given by Makdisi and Seed (1978).[59] Its application, however, is restricted to compacted clayey embankments and dry or dense cohesionless soils that experience very little reduction in strength due to cyclic loading.

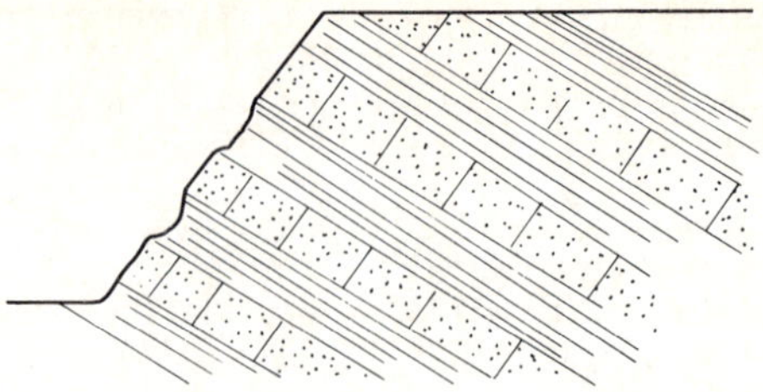

FIG. 8.57 Rock strata dipping into hillside. Stability depends on strength of weakest member and joint orientation and continuity.

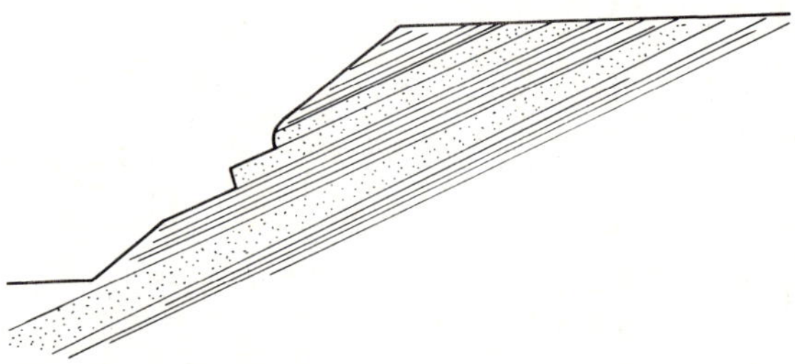

FIG. 8.58 Rock strata dipping out of slope in a potentially unstable condition. Stability depends upon strength along bedding planes.

8.3 STABILITY ANALYSIS: ROCK SLOPES

8.3.1 FACTORS TO CONSIDER

General

The major factors to consider in the stability analysis of a rock slope are the geometric configurations of the discontinuities, the degree of decomposition, the strength parameters of blocks and discontinuities, the seepage conditions, and the purpose of the slope. The latter factor establishes the degree of safety against failure required. Discontinuity orientations as related to the slope face, particularly those that provide major weakness planes, govern the analytical method selected.

Geologic Factors

Rock Structure

Discontinuities or weakness planes (Table 3.15) divide the rock mass into variously shaped blocks and in most cases control the failure surface location. If, for example, the major discontinuities dip into the slope as shown in Fig. 8.57, slope stability depends primarily on the intact strength of the weakest rock member as well as on joint orientation, although in general the joints will play a minor role in overall stability.

If, however, the planes dip out of the slope face or are parallel to it, as shown in Fig. 8.58, stability will be governed by their properties and such slopes can be expected to have much flatter stable inclinations than those exhibiting the conditions of Fig. 8.57 in the same rock material. Some of the most catastrophic failures of natural rock slopes involving enormous quantities have occurred where bedding planes were parallel or near parallel to the slope, as in the Vaiont reservoir slide in Italy in 1963; the Golden, Switzerland, slide of 1806; and the Gros Ventre, Wyoming, slide of 1925 [see Art. 9.3.3 of Hunt (1984)[69]]. Dipping sedimentary beds separated by a thin clay seam were the common factor. Major weakness planes or surfaces may also include faults, slickensides, and foliation or mylonite shear zones, as well as bedding and unconformities. In hard, jointed masses stability is controlled by the intensity, orientation, and continuity of the joints in relation to the slope face and inclination.

Rock Type

Massive jointed rock, typical of the hard crystalline rocks such as granite, rhyolite, diorite, quartzite, and marble, are divided by continuous random joints into irregularly fitting blocks locally interconnected. Natural slopes normally are subjected to falls and toppling failures resulting from seepage pressures and frost wedging. Granite is subject to slab slides due to exfoliation. Cut slopes will fail along irregular surfaces formed by the joint patterns unless failure is controlled by a major discontinuity. Terzaghi (1962)[60] gave the critical slope angle as 70° for substantial heights in hard, massive rock with random joint patterns, providing that seepage is not acting on the joint walls.

In *foliated metamorphic rocks,* such as gneiss, schist, and phyllite, foliations and slickensides cause the weakness planes, although gneiss is also commonly found jointed. Of significance in these rocks is the deep zone of decomposition that develops, particularly in warm, moist climates.

In *interbedded sedimentary rocks,* bedding planes represent weakness surfaces and individual beds usually have joints normal to the bedding planes (Fig. 8.57), referred to as cross joints. These joints have almost no effective strength. Mylonite zones and clay layers in shales and interbedded shales represent major weakness planes, as to a lesser extent do cleavage planes in hard shales and slates.

Massive clay shales are weak, intensely fractured materials *(see Art. 8.2.1),* often considered as soil mechanics problems because of low strengths. They may be subjected to progressive failure or fail in the form of the infinite slope along the weathered rock contact, as shown in Fig. 8.18*b*.

Decomposed Rock

Conditions in decomposed rock masses intermediate between sound, relatively fresh rock and soil can be extremely complex and variable. Failure surfaces can be cylindrical, irregular, or planar, varying with the rock type, degree of decomposition, and homogeneity of the mass.

In sedimentary rocks (except marine shales) decomposition may be relatively uniform depending upon the orientation of the beds to the slope surface. In igneous and metamorphic rocks, however, decomposition usually is differential leaving irregular surfaces of relatively strong rock, with transition gradual from a mature residual soil to saprolite to weathered rock with fractures to sound rock. In the mature residual soil and the saprolite, weakness planes represented by relict joints can control failure surfaces.

Typical slope conditions in decomposed shale and granite are given in Figs. 8.59 and 8.60, respectively. In both cases potential failure surfaces exist along relict joints in the shallow residual soils and along slickensides. Seams of bentonite and other clays, as well as shale my-

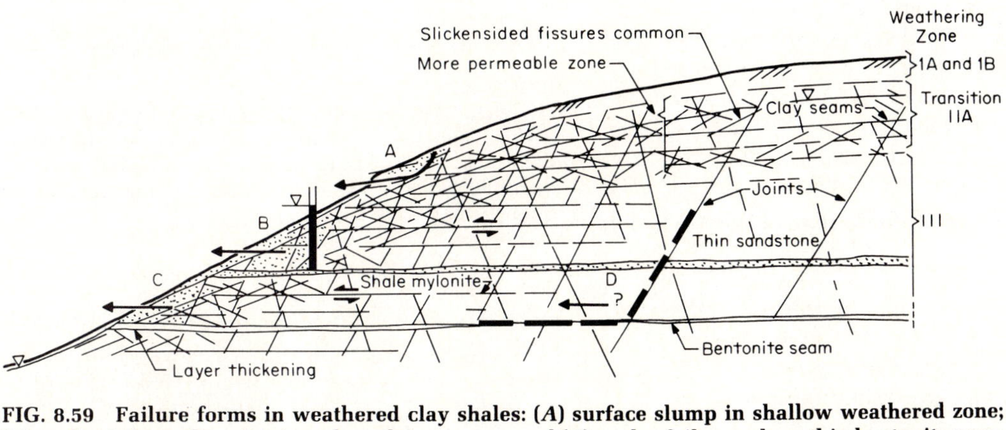

FIG. 8.59 Failure forms in weathered clay shales: (*A*) surface slump in shallow weathered zone; (*B*) wedge failure along joints and sandstone seam; and (*C*) wedge failures along thin bentonite seam may develop into large progressive failure to *D* or beyond. [From Deere and Patton (1971).[61]]

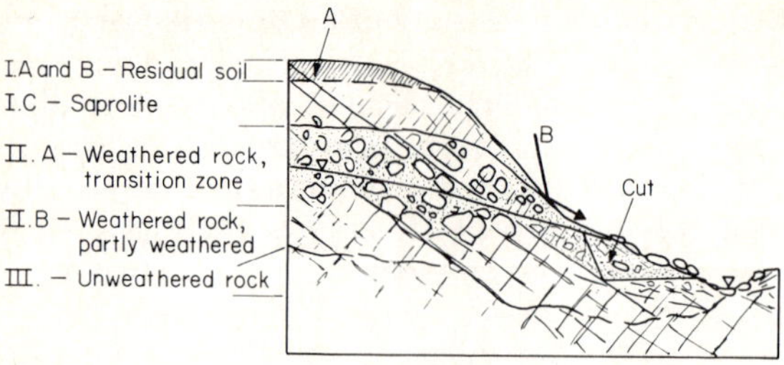

FIG. 8.60 Typical slope in decomposed granite: (*A*) potential major failure surface along relict joints and (*B*) potential slide in residual soil. [From Deere and Patton (1971).[61]]

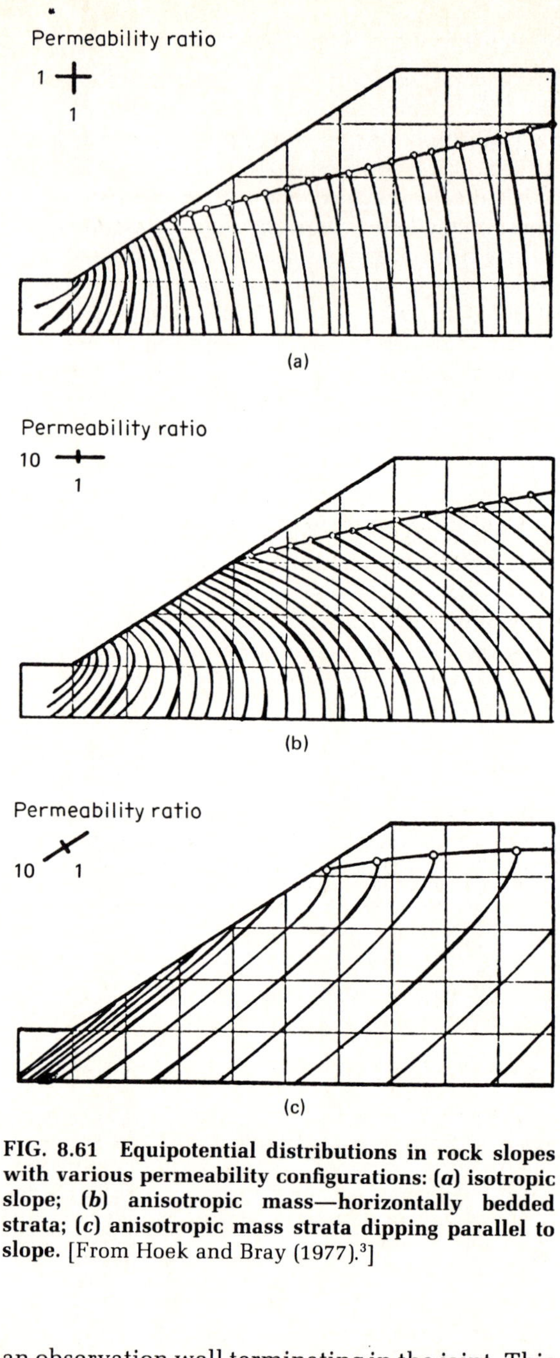

FIG. 8.61 Equipotential distributions in rock slopes with various permeability configurations: (*a*) isotropic slope; (*b*) anisotropic mass—horizontally bedded strata; (*c*) anisotropic mass strata dipping parallel to slope. [From Hoek and Bray (1977).[3]]

lonite, constitute weakness planes in the shale. The weathering zone classifications are those of Deere and Patton (1971).[61] [See Table 6.15 in Hunt (1984).[69]]

Seepage Forces

In rock slopes it is the water pressure, not the quantity of water, that causes instability. The water table often is very irregular as it is affected by variations in the fracture openings which provide "secondary permeability," as well as by seasonal variations in precipitation. Some variations in equipotential distributions in rock slopes are illustrated in Fig. 8.61.

In a zone immediately behind the slope face, rock fractures usually are more intensely distributed and much more open than those deeper in the mass and have substantially higher permeabilities. During dry periods, evaporation can cause a slope to appear dry when in reality significant pressures may exist behind the face. During wet seasons the water table may rise much more rapidly in the zone behind the face than it does deeper in the slope, resulting in a substantial increase in seepage forces in the frontal zone. If ice is blocking the flow exit paths, the water pressures will be even higher, as the water table increases in height. Either case often results in a slope failure beginning with an outward movement at the toe.

Water occupying joints exerts onto their walls a pressure equal to the product of the unit weight of water and the height to which it would rise in an observation well terminating in the joint. This pressure, corresponding to pore-water pressure in soils, was termed by Terzaghi (1962)[60] the *cleft-water pressure*. It reduces the normal force and thereby the frictional resistance along the joint walls (U in Eq. 8.86), and if a joint is steeply inclined it tends to displace the rock laterally toward the slope face exerting a pressure V (Eq. 8.86) (Fig. 8.62).

Seepage pressures are of great importance in rock slopes and their presence may reduce the stable angle by as much as 20 to 25° in comparison with dry conditions. Wherever the safe angle for an open pit is found to be less than 30°, the presence of water pressure should be suspected [Jennings (1971)[62]]. Seepage forces acting along major discontinuities such as dipping beds of sedimentary rock are almost always the cause of failure.

Strength Parameter Selection

Peak Strength

As discussed in Art. 3.4, fresh to moderately weathered rock masses normally obtain their resistance to failure by mobilization of strength along discontinuities, usually referred to as joints. Clean, smooth joints have a basic friction angle ϕ_u, but surface irregularities or asperities will increase the basic friction angle by a value j *(see Arts. 3.4.2 and 3.4.3)*, and the total shearing resistance becomes $\phi_a = \phi_u + j$. Joints with fillings may be measured to have parameters c and ϕ, although generally it is assumed that $c = 0$ for weak zones since it cannot always be counted upon. Patton and Hendron (1974)[15] and Deere (1976)[63] advise not to use a cohesive resistance for most cases where the potential failure surface consists of bedding in rock or clay zones, or foliation planes or shear zones in metamorphic rocks, because past movements are likely to have reduced strengths toward residual values. In competent rock they suggest using the effective angle of shearing resistance $\bar{\phi} = \bar{\phi}_r + j$.

A summary of typical strength values for rock-mass discontinuities for use in preliminary analysis is given in Table 3.17. In a sense joints respond to the effective stress principle in that water pressure directly counteracts the strengthening effect of the applied normal stress. This action is a major cause of failure in rock slopes.

Residual Strength

If displacements have occurred, the consolidated-drained strength ϕ_r prevails. In moderately sound rock, although displacements have occurred, some joint roughness may remain along the failure surface which increases the available strength by $\phi_r + j$ as described above. Large displacements and smooth surfaces will not add a significant value for j, but displacements under low stress levels could permit irregularities to remain and would provide j values in the order of 5 to 10° [Patton and Hendron (1974)[15]]. A value for j is obtained either by estimation from observed field conditions and the use, for example, of Table 3.17, or by field and laboratory testing *(see Arts. 3.4.2 and 3.4.3)*.

Values for ϕ_r are lower than values for $\bar{\phi}$, generally by amounts ranging from 5 to 10° or more, as indicated by values obtained from back analysis of failed slopes. The cohesion intercept should seldom be used in analyzing existing landslides, as the analysis is likely to indicate strengths unrealistically high and on the unsafe side. Analysis is based on ϕ_r or some value between ϕ and ϕ_r.

Problem Categories

The factor of safety FS desirable in a given situation is dependent upon the degree of risk or danger to lives and structures posed by a potential slope failure and is very much related to the problem category.

Natural slopes may become unstable from the effects of weathering and erosion from the development of high pore- or cleft-water pressures, or from frost wedging. Tectonic movements also have been thought to be a contributory factor in some major failures. In general, acceptable values for FS for slopes are much lower than those for foundation stability *(see Art. 6.2.4)*, generally in the order of FS = 1.5 or less, depending upon the failure risk. When lower values for FS are obtained and there is danger to lives and structures, remedial measures should be employed.

Open-pit-mine slopes often require deep excavations in relatively soft rock, and economy dictates that the slope be the steepest possible. Excavation of material from a slope that is flatter than necessary is costly, but so is the removal of landslide material which may include lost mineral rock. Therefore, the optimum balance between stability and instability is desired and low values for FS are normally accepted. A factor to consider is the loss of strength with time that occurs over the lifespan of a mine, often 30 years or more.

Hillside cuts for roadways and other construction are, in general, much shallower than open-pit mines, but require a higher value for FS against failure due to higher risk to lives and structures.

Excavations for structures are two-sided vertical slope problems and, therefore, somewhat different from the other categories, since bracing is possible to aid in slope retention. Because the cut is made in a confined rock mass, residual stresses can present significant problems *(see Art. 9.3.5)*.

Analytical Approaches

Potential Failure Surfaces

Weakness planes control the failure surface location in most cases and analysis normally is performed by assuming planar failure surfaces and employing limiting equilibrium methods. Circular failure surfaces are rare. Failure surfaces usually are three-dimensional, but in many cases they may be approximated by two-dimensional surfaces. Stability analysis for rock slopes, therefore, may be divided into two broad categories, two-dimensional analysis and three-dimensional analysis. The finite element method is used on occasion to predict potentially weak zones.

Two-Dimensional Analysis

Most rock slopes fall into one of two categories:

1. Sliding on a single, continuous plane, which is the simplest case. It is represented by a mass resting on a discontinuity or unconformity that is more or less continuous and planar through the slope (Fig. 8.58). In highly weathered rock the sliding surface often is gently curved, but still may be considered as planar for analysis.
2. Sliding on several intersecting planes, represented by combinations of planes such as a slickensided surface intersecting a soil-rock interface, several blocks providing two continuous intersecting planes, or composite surfaces of two discontinuous joint sets. In all cases no forces are considered perpendicular to the *x-y* plane.

Three-Dimensional Analysis

Four- or five-sided rock wedges are formed by inclined intersecting joint sets, and forces act in the *x-y-z* planes which require a three-dimensional approach to analysis. Such failures are less common than the two-dimensional forms, and generally involve a smaller volume of material. They are particularly troublesome in open-pit-mining operations, however, due to the lower quality of rock often associated with many commercial grade minerals.

8.3.2 ANALYTICAL METHODS

Sliding on One Continuous Plane

Simple Sliding Block

The simplest form of failure is a single block sliding along a continuous planar surface. In Fig. 8.62*a*, there are no seepage forces and the driving

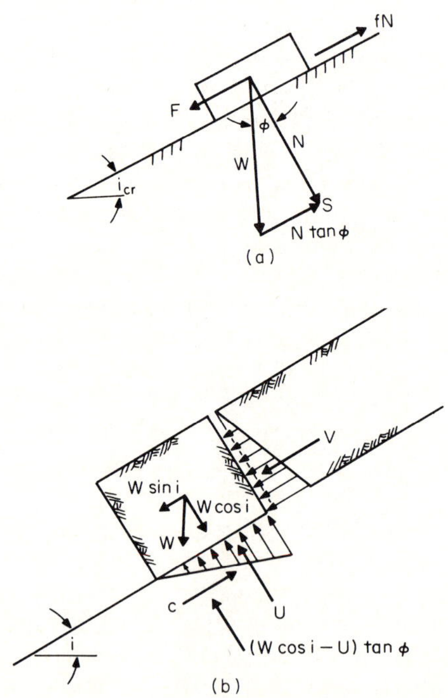

FIG. 8.62 Simple sliding blocks: (*a*) only friction acting and (*b*) cohesion and cleft-water pressures acting.

component equals the block weight component ($W \sin i$) and the resisting force $S = N \tan \phi = (W \cos i) \tan \phi$. At equilibrium

$$FS = \frac{\text{Resisting forces}}{\text{Driving forces}} = \frac{(W \cos i) \tan \phi}{W \sin i} \qquad (8.85)$$

and FS = 1 when $i = \phi$ or at i_{cr}, the critical slope inclination.

For the case of cohesion acting along the failure surface and cleft-water pressures U and V acting on the block (Fig. 8.62*b*), Eq. 8.85 becomes

$$FS = \frac{cA + (W \cos i - U) \tan \phi}{W \sin i + V} \qquad (8.86)$$

where c = cohesion acting over block area A
U = base water pressure
V = joint water pressure acting as an additional driving force

Simple Wedge

In the case of a simple wedge, the mass is resting on a continuous weakness plane that intersects the slope. As illustrated in Fig. 8.51, there are three general conditions: no tension crack, a tension crack beyond the slope face, and a tension crack within the slope face. The general equation for these conditions, including the seismic coefficient k, is given as Eq. 8.64. The tension crack location affects the calculation of the weight term and the failure surface length.

It is assumed that there are release surfaces on each side of the wedge permitting it to fail and their effect is ignored in analysis. Plane failure is valid if the strike of the failure surface is within 20° of the slope strike; otherwise, three-dimensional wedge analysis applies [Hoek and Bray (1977)[3]]. The expression for determining the force to be resisted by rock anchors is given as Eq. 8.90 (*see Art. 8.4.4*).

Sliding on Multiple Intersecting Planes: Two-Dimensional Analysis

Two or Three Blocks

A graphical procedure for the analysis of two blocks also applicable to rock slopes is given in Art. 8.2.6. Morgenstern (1968)[5] also presents a similar graphical procedure for rock slopes applicable to several blocks in which the resultant water pressure between blocks is considered. The numerical method for two or three blocks given in Art. 8.2.6 (Fig. 8.54) is also applicable to rock slopes.

In Morgenstern's method the potential sliding mass is divided into wedges and the equilibrium of each wedge considered in turn (Fig. 8.63). The known forces are the magnitudes and direction of W_1 and W_2; the directions of T_1, T_2, N_1, and N_2; the magnitudes and direction of U_1 and U_2; the magnitudes of the cohesion portion of cL of T_1 and T_2; and the direction and magnitude of the resultant water pressure P_{w12} at the interface of the wedges.

To permit construction of the polygon, assumptions are made for the inclination δ of force P transmitted across the interface of any two wedges, and for FS to obtain a value for tan ϕ/

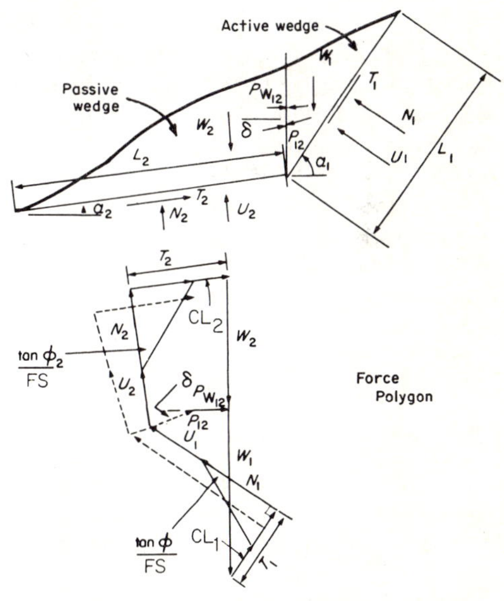

W_1, W_2 weight of a wedge
U_1, U_2 resultant water pressure acting on the base of the wedge
N_1, N_2 effective force normal to the base
T_1, T_2 shear force acting along the base of the wedge
L_1, L_2 length of the base
α_1, α_2 inclination of the base to the horizontal
P_{w12} resultant water pressure at the interface
P_{12} effective force at the interface
δ inclination of P_{12} to the horizontal

FIG. 8.63 Forces acting on two wedges: one active, one passive. [From Morgenstern and Sangrey (1978).[64] Used with permission of the National Academy of Sciences.]

FS. FS is sensitive to the assumed value for δ and conservative values result from setting $\delta = 0$. Experience indicates that values for $\delta = 10$ to $15°$ usually gives reasonable results [Morgenstern and Sangrey (1978)[64]], although Patton and Hendron (1974)[15] suggest taking $\delta = 30$ to $40°$ for less conservative results. In the procedure for solution a value for FS is assumed and the first force polygon constructed to yield the dashed lines in the figure. A value for P_{12} is obtained and the force polygon for wedge 2 constructed. If the polygons do not close, a new value for FS is selected and the procedure repeated until closure is obtained. The procedure is applicable to additional wedges.

Joint Sets Intersecting in the Same Plane

Joint sets or other fractures may be oriented in a slope so as to permit the formation of a single, but irregular failure surface intersecting the joints, as shown in Fig. 8.64. Two approaches to analysis are shown in the figure. One is to consider sliding along combinations of single joint sets and the opening of tension cracks on a secondary joint set forming a simple wedge *DCT*. It would be assumed that all stresses were the result of the weight of the wedge and seepage forces. This is the most pessimistic failure mode.

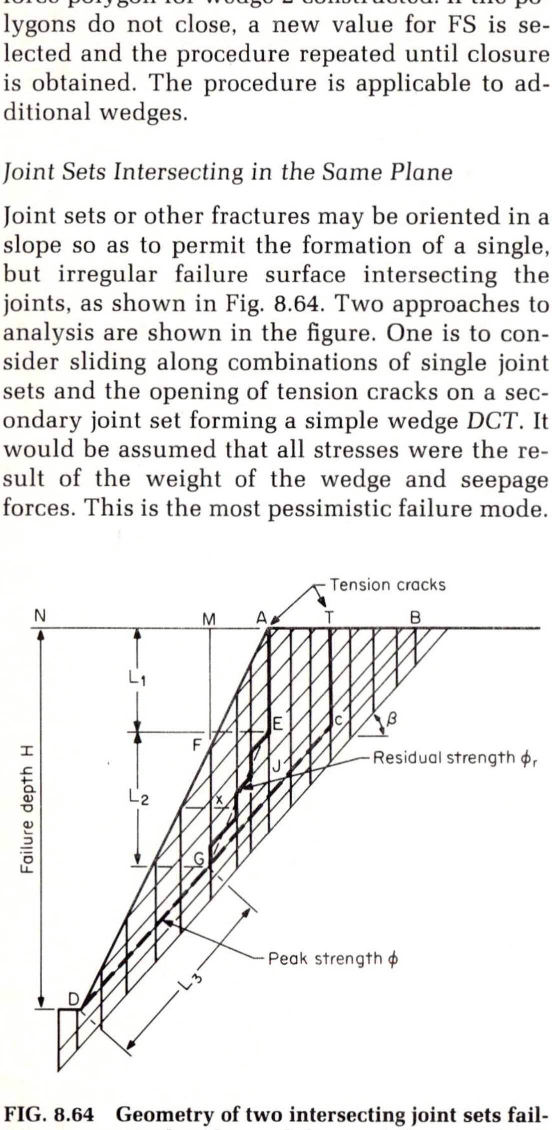

FIG. 8.64 Geometry of two intersecting joint sets failing progressively (the multilinear failure surface). [From Barton (1971).[65]]

To reduce the conservatism inherent in the above approach, Barton (1971)[65] introduced the concept of progressive failure, translational in form, and considered that the failure surface would form along a series of steps (surfaces *DGEA*). Analysis requires complete information for all joint factors including attitude, length and width, spacing, shear strength (roughness and filling), and water pressures. Joints are plotted on diagrams as stereonets [see Art. 6.1.3 of Hunt (1984)[69]].

For analysis it is assumed that no strength is available from the vertical joints and that all resistance to movement is mobilized by the strength parameters along the dipping joints. As the excavation depth increases, so does the overburden stress until it exceeds the strength available along the dipping joint surface. The limiting height is that at which the slope remains stable. As excavation proceeds the stresses on the joint surfaces beneath each of the vertical slices in the figure are increased, and if *shear displacements* can occur then the overstressing will increase progressively on the lower blocks. It cannot be assumed, therefore, that the peak strength will be mobilized simultaneously along the entire potential failure surface. Eventually, with increasing excavation depth, tension cracks begin to open at the top of the slope; if water enters the cracks, transient water pressures develop further decreasing slope stability.

Shear-strength parameters acting along the joints are selected based on two conditions: (1) peak strengths mobilized after relatively small displacements and (2) residual strength that develops with continuing displacement. The worst case would assume only ϕ_r acting in simple plane failure, but this would lead to overly conservative results. Prefailure shear displacements and increased weathering over joints cause a progressive failure mechanism that leads to the possible stepped portion of the failure surface.

For analysis the failure surface is divided into three sections as shown in Fig. 8.64: (1) a vertical part passing through the crest of the slope along which no strength is mobilized, (2) the stepped portion along which the vertical joints open and strength is given as ϕ_r because of the progressive nature of the failure, and (3) finally shear failure on the inclined joint passing through the toe along which peak values for ϕ are used.

In this approach the progressive failure mode reduces the failure depth considerably. The most pessimistic assumption is for full-tension crack depth and steady seepage reducing to zero exit pressure at the toe. The method is particularly suited to analysis of open-pit-mine slopes where overconservatism is undesirable.

Three-Dimensional Wedge Analysis

Occurrence

Three-dimensional wedge failure is common in open-pit mines on heights of one or two benches (20 to 30 m) but becomes progressively less prevalent as slope height increases. Failure seems to be associated with weakness planes that are of the same order of size as the slope height involved, with the planes representing three or more intersecting joint sets and the "free blocks" approaching tetrahedrons in shape (Fig. 8.65*a*). Sliding occurs with more frequency on the shallower slopes because weakness planes with lengths in the order of 10 to 15 m are much more common than in the order of 10 bench heights, or 100 to 150 m. Normally the slope will be cut to depth D (Fig. 8.65*b*) or some other point just above it, prior to failure.

Analysis

General wedge failure geometry is given in Fig. 8.65*b* and *c*, and a stereoplot of the failure planes in Fig. 8.65*d*. Wittke (1967)[66] proposed six basic failure forms ranging from simple sliding along a direction parallel to the line of intersection of the two planes (with both planes remaining continuously intact) through to rotation and sliding of various forms along either plane of weakness.

If both planes are investigated separately as cases of two-dimensional failure, then it will be found that the more adversely jointed will give a value for FS lower than that found for the combined wedge failure. Because many such planar features tend to occur as repeated parallel surfaces, however, consideration should be given to two-dimensional failure except for those cases where limiting values of slope angles are sought

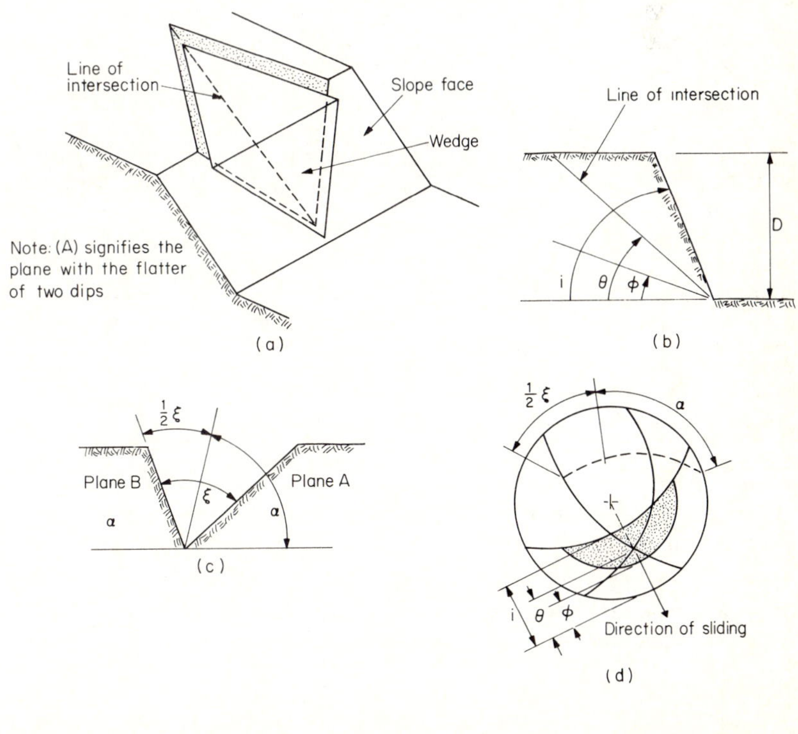

FIG. 8.65 Triangular wedge failure: (*a*) pictorial view; (*b*) view at right angles to line of intersection; (*c*) view along line of intersection; and (*d*) stereoplot of wedge failure geometry. [From Hoek and Bray (1977).[3]]

[Jennings (1971)[62]]. In the three-dimensional analysis the forces tending to disturb equilibrium are added vectorially to give the resulting driving force. These motivating forces include wedge weight, any applied external force, and cleft-water forces acting on the various faces. The failure kinematics depend on the orientation of the motivating force in relation to the orientation of the supporting planes [Morgenstern and Sangrey (1978)[68]].

Detailed discussion of the problem is given in Patton and Hendron (1974),[15] Hoek and Bray (1977),[3] and Hendron et al. (1980).[67] Two general forms of solutions are available: (1) one which calculates FS only and (2) the more complex one, which calculates the force to be resisted by anchors and provides for a tension crack.

Solution to Find FS Only

This "short solution" from Hoek and Bray (1977),[3] for a wedge with a horizontal slope crest, does not account for rotational slip or toppling or the influence of tension cracks, nor does it provide for the analysis of anchor forces required for stability. (Solutions for these conditions can be found in the foregoing references.) Cases in which there are different strength parameters and water pressures on the two weakness planes are provided for, and it is assumed that the upper slope surface either is horizontal or dips in the same direction as the slope face.

When a pair of discontinuities is selected at random from a set of field data, it is not known whether

- the planes could form a wedge (the line of intersection may plunge too steeply to daylight in the slope face, or it may be too flat to intersect the upper slope surface), or
- one of the planes overlies the other, which affects the calculation of the normal reaction on the planes, or
- one of the planes lies to the right or the left of the other plane when viewed from the bottom of the slope.

The "short solution" developed by Hoek and Bray provides for one plane overlying the other and provides a check on whether or not the two planes do form a wedge. Provision is made also for the possibility that contact may be lost on either plane due to wedge geometry or water pressure.

The geometry of the problem is illustrated in Fig. 8.66 in which 1 and 2 denote the potential failure surfaces, 3 the upper slope surface, and 4 the slope face. In the analysis, if the slope face overhangs the slope toe, the index n is assigned the value -1, and if it does not overhang, $+1$.

DATA REQUIRED Geometric data required for solution include the height H of the slope crest above point O, and the dip angle ψ and the dip direction α for each of the four planes. Rock properties required are unit weight γ, and ϕ and c acting along each of the discontinuities. To estimate the average water pressure U acting on each of the potential failure planes, Hoek and Bray suggest that if it is assumed that the discon-

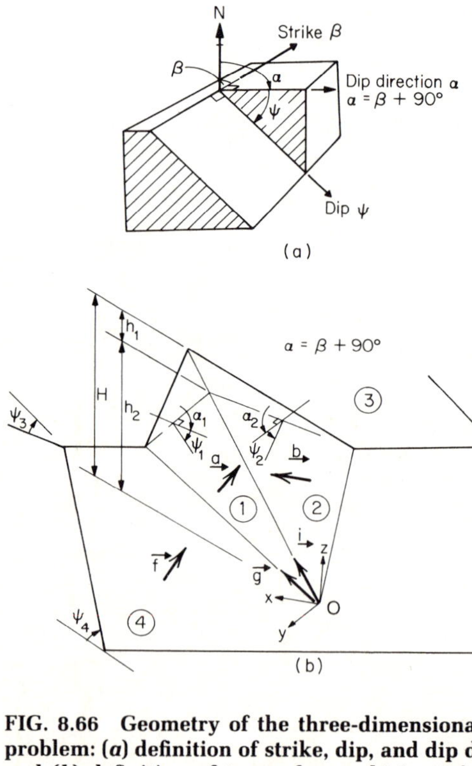

FIG. 8.66 Geometry of the three-dimensional wedge problem: (*a*) definition of strike, dip, and dip direction and (*b*) definition of terms for analysis. [After Hoek and Bray (1977).[3]]

tinuities are completely filled with water, U varies from zero at the free faces to a maximum at some point along the line of intersection, then $U_1 = U_2 = \gamma_w H_w/6$ where H_w is the overall wedge height.

Terminology

FS = factor of safety against wedge sliding calculated as the ratio of the resisting to the actuating forces

A = area of a face of the wedge

W = wedge weight (it does not enter directly into this solution but is necessary for the analysis of anchor forces)

N = effective normal reaction on a plane

S = actuating shear force on a plane

x,y,z = coordinate axes with origin at O; the z axis is directed vertically upwards, the y axis is in the dip direction of plane 2

a = unit vector in the direction normal to plane 1, with components (a_x,a_y,a_z)

b = unit vector in the direction normal to plane 2, with components (b_x,b_y,b_z)

f = unit vector in the direction normal to plane 4, with components (f_x,f_y,f_z)

g = vector in the direction of the line of intersection of planes 1 and 4, with components (g_x,g_y,g_z)

i = vector in the direction of the line of intersection of planes 1 and 2, with components (i_x,i_y,i_z)

$i = -i_z$

q = component of **g** in the direction of **b**

r = component of **a** in the direction of **b**

$k = |\mathbf{i}|^2 = i_x^2 + i_y^2 + i_z^2$

$Z = W/A_2$

$P = A_1/A_2$

$n_1 = N_1/A_2$, $n_2 = N_2/A_2$, $|Zi|/\sqrt{k} = SA_2$ (assuming contact on both planes)

$m_1 = N_1/A_2$ (denominator of FS = S_1/A_2; contact on plane 1 only)

$m_2 = N_2/A_2$ (denominator of FS = S_2/A_2; contact on plane 2 only)

Factor of Safety: Various Conditions

CASE 1: If $n_1 > 0$ and $n_2 > 0$, there is contact on both planes and

$$FS = (n_1 \tan \phi_1 + n_2 \tan \phi_2 + |p|c_1 + c_2)\sqrt{k}/|Zi| \tag{8.87}$$

CASE 2: If $n_2 < 0$ and $m_1 > 0$, there is contact only on plane 1 and

$$FS = \frac{m_1 \tan \phi_1 + |p|c_1}{[Z^2(1 - a_z^2) + kU_2^2 + 2(ra_z - b_z)ZU_2]^{1/2}} \tag{8.88}$$

CASE 3: If $n_1 < 0$ and $m_2 > 0$, there is contact only on plane 2 and

$$FS = \frac{m_2 \tan \phi_2 + c_2}{[Z^2b_y^2 + kp^2U_1^2 + 2(rb_z - s_z)pZU_1]^{1/2}} \tag{8.89}$$

CASE 4: If $m_1 < 0$ and $m_2 < 0$, water pressure causes a loss in contact on both planes and FS = 0.

Calculation sequence and a sample problem are given in Table 8.14. The analysis is suitable for a programmable calculator with which the solution is rapid.

Finite Element Approach

The effect of seepage pressures on instability and toppling of a steep 130-m-high slope in which a cut was made for the Trans-Canada Highway was analyzed by Kalkani and Piteau (1976)[68] employing the finite element method. Mass discontinuities in the granodiorite were primarily two systems of faults oriented as shown in Fig. 8.67.

The configuration of the finite element mesh is given in Fig. 8.68. The analysis considered both dry slope and high groundwater conditions, the latter producing regions of high tensile stresses as shown in Fig. 8.69, which were found to be about 3 times larger than for dry slope conditions. For stabilization the slope was unloaded by excavation and the potentially weak zones evidenced by the high tensile stresses were shotcreted.

TABLE 8.14
THREE-DIMENSIONAL WEDGE ANALYSIS: CALCULATION SEQUENCE AND EXAMPLE*

Calculation sequence	Example
A. DATA INPUT	
1. Dip ψ	$\psi_1 = 45°, \alpha_1 = 105°$
	$\psi_2 = 70°, \alpha_2 = 235°$
2. Dip direction α	$\psi_3 = 12°, \alpha_3 = 195°$
	$\psi_4 = 65°, \alpha_4 = 185°$
3. Overhang (+1, No; −1, Yes)	+1
4. Height H	130 ft
5. Unit weight γ	160 pcf
6. Frictional resistance ϕ_1, ϕ_2	$\phi_1 = 30°, \phi_2 = 35°$
7. Cohesion c_1, c_2	$c_1 = 0, c_2 = 0$
8. Water pressure U_1, U_2	$U_1 = U_2 = 1352$ psf
B. CALCULATIONS	
1. $(a_x, a_y, a_z) = \sin\psi_1 \sin(\alpha_1 - \alpha_2), \sin\psi_1 \cos(\alpha_1 - \alpha_2), \cos\psi_1$	$a_x = -0.542, f_x = -0.694$
	$a_y = -0.455, f_y = 0.583$
2. $(f_x, f_y, f_z) = \sin\psi_4 \sin(\alpha_4 - \alpha_2), \sin\psi_4 \cos(\alpha_4 - \alpha_2), \cos\psi_4$	$a_z = 0.707, f_z = 0.423$
3. $b_y = \sin\psi_2$	0.93969
4. $b_z = \cos\psi_2$	0.34202
5. $i = a_x b_y$	−0.50901
6. $g_z = f_x a_y - f_y a_x$	0.63112
7. $q = b_y(f_z a_x - f_x a_z) - b_z g_z$	0.46206
8. If $nq/i > 0$, or if $\tan\psi_3 > \sqrt{1 - f_z^2}/(f_z - nq/i)$, and $\alpha_3 = n\alpha_4$, no wedge is formed and calculation is terminated	−0.90776
9. $r = a_y b_y + a_z b_z$	−0.18256
10. $k = 1 - r_2$	0.96568
11. $Z = (\gamma H q)/3g_z$	5076.054
12. $p = -b_y f_x / g_z$	1.03372
13. $n_1 = \{(Z/k)(a_z - rb_z) - pU_i\}\, p/\lvert p \rvert$	2652.37
14. $n_2 = \{(Z/k)(b_z - ra_z) - U_2\}$	1134.42
15. $m_1 = (Za_z - rU_2 - pU_1)\, p/\lvert p \rvert$	2442.20
16. $m_2 = Zb_z - rpU_1 - U_2$	643.04
17. Since $n_1 > 0$ and $n_2 > 0$, Case 1 (Eq. 8.87) applies	FS = 0.88
18. For the case of $U_1 = U_2 = 0$	FS = 1.55

*After Hoek and Bray (1977).[3]

8.4 TREATMENTS TO PROVIDE STABILITY

8.4.1 GENERAL CONSIDERATIONS

Hazard and Risk Philosophy

Selection of a treatment to provide stability is based on an assessment of the degree of the hazard presented by the existing conditions or the proposed construction and the degree of risk involved with the occurrence of failure.

The degree of hazard relates to the potential failure itself in terms of its possible *magnitude*, such as a small volume of material with small displacements or a large volume of material with large displacements, and the *probability* of occurrence, such as unlikely, likely, or certain. *The degree of risk* relates to the consequences of failure on construction such as a relatively low risk from a small volume of material partially covering a roadway and not endangering lives to the high risk from the failure of an earth dam resulting in the loss of many lives and much destruction and damage. Safe but economical construction is the desired result, but the degree of safety acceptable varies with the degree of the hazard and the risk.

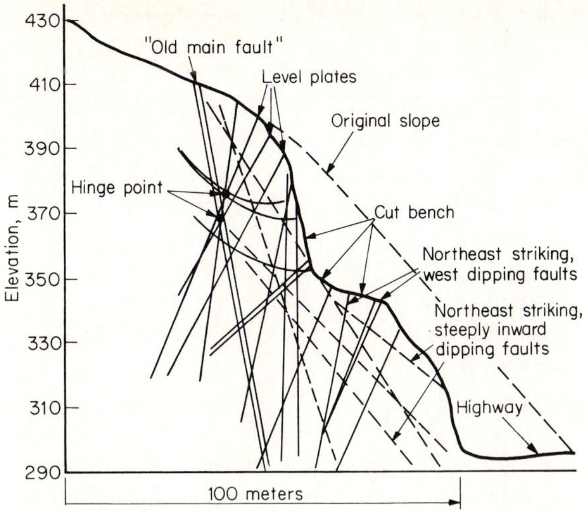

FIG. 8.67 Slope section at Hell's Gate Bluffs showing orientation of steeply dipping faults and related hinge points around which toppling is kinematically possible. [From Kalkani and Piteau (1976).[68]]

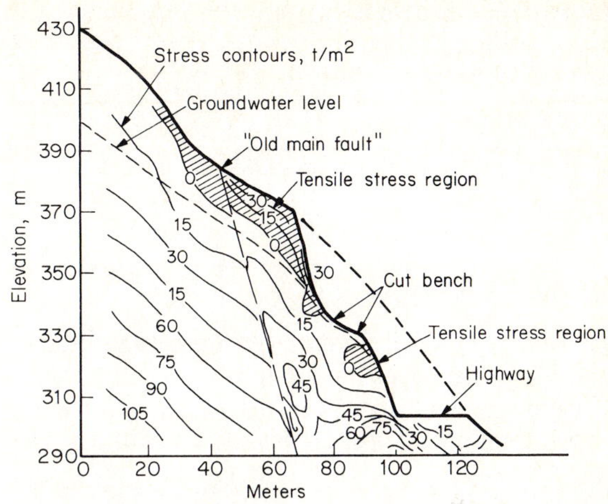

FIG. 8.69 Contours of minimum principal stresses due to excavation at Hell's Gate Bluffs slope for high groundwater condition; tensile stress regions represent zones of potential instability. [From Kalkani and Piteau (1976).[68]]

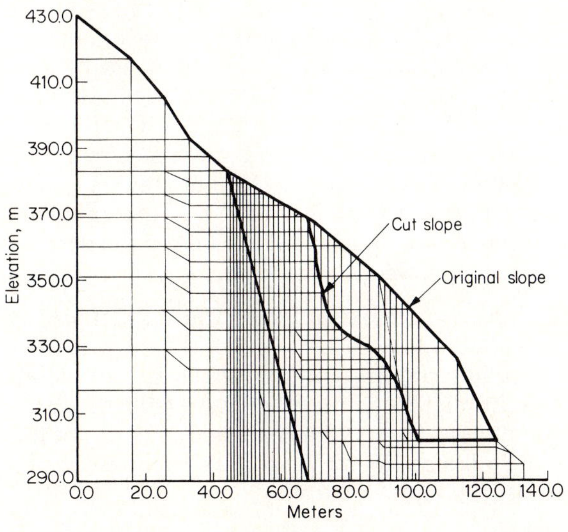

FIG. 8.68 Configuration of finite element mesh for slope at Hell's Gate Bluffs. [From Kalkani and Piteau (1976).[68]]

Natural Slopes

There are three general cases involving evaluation of natural slopes:

1. Existing slopes undergoing deformation and requiring stabilization
2. Slopes to be cut which require stability
3. Sidehill fills to be constructed which require stability

Stabilization methods can be placed in three general categories:

1. Changing slope geometry, either inclination or height, to reduce the motivating forces
2. Improving external or internal drainage to control seepage forces
3. Providing retention to increase the resisting forces

Embankments

Earth-dam design and construction involve four general areas for assessment, aside from hydrological considerations:

1. Adequate support for the embankment from the foundation materials
2. Adequate embankment strength to contain the proposed water load
3. Seepage control beneath, through, and around the embankment
4. Potential water loss and slope instability in the reservoir area

Embankments over soft ground, such as for roadways or landfill, mainly require consideration of possible foundation failure.

8.4.2 NATURAL SLOPES: ASSESSMENT FACTORS

Treatment Selection

The purposes of slope treatment can be placed in two general categories:

1. Preventive: for potentially unstable natural slopes, cut slopes, and sidehill fills
2. Remedial or corrective: for existing unstable or moving slopes or slopes totally failed

Slope failure potential usually can be established with some certainty by an experienced practitioner, although in many cases the actual occurrence of a failure is not predictable because of many natural transient factors such as weather conditions. In addition, there are many forms of slope failure that cannot be prevented. In the selection of treatment, therefore, consideration is given not only to the geologic conditions, but also to an assessment of the magnitude of the failure hazard and the risk involved, which is based on slope failure prediction. The treatment options include avoiding the hazard, accepting the risk, and reducing or eliminating the hazard.

Slope Failure Prediction

The objectives of slope failure prediction are to forecast the form of failure, the volume of material involved, and the degree of the hazard. Slope failure prediction is discussed in some detail in Art. 9.3 of Hunt (1984)[69] and is summarized briefly here. The form of failure refers to falls or toppling, slides, avalanches, or flows and is the major factor in evaluating the degree of the hazard. The key factors to assess in slope failure prediction are:

- History of local landslide activity resulting from construction, weather conditions, or seismic activity, in terms of failure forms and magnitudes.
- Geologic conditions and related potential failure forms and their suitability for analysis as related to slope material and structure.
- Shear-strength factors and their consistency, variability, and potential for change or liquefaction.
- Slope geometry, including the influence of inclination, height, and shape on potential seepage forces, runoff, and failure volume. (Runoff is an important factor in the cause of many avalanches and flows.)
- Surficial indications of instability such as creep, scars, seepage points, and tension cracks.
- Degree of existing slope activity which may be categorized as stable with no movements occurring, initial movements with tension cracks forming, partial failure with large displacements, or total failure of large volume.
- Weather, including the relationship between recent weather history and long-term conditions (less severe, average, more severe) in view of present slope activity, stability of existing cut slopes, groundwater levels, and slope seepage.

Rating the Failure Hazard and the Risk

The failure *hazard degree* is rated on the basis of the potential magnitude of the failure and its probability. It can vary from *no hazard*, where a slope is unlikely to undergo failure under any foreseeable circumstances, to *high hazard* where a slope is almost certain to undergo total failure in the near future under normal adverse conditions and will involve a large volume of material or where a slope may fail under severe conditions (moderate probability) but the potential volume and area affected are enormous and the movement velocity very high.

The *degree of risk* is rated on the basis of project type and failure consequences. It varies from *no risk* where failure will not affect human activities through *moderate risk* where failures are inconvenient but can be corrected and do not endanger lives to *high risk* where structures may be destroyed and lives lost.

Slope Condition

The slope condition refers to stable versus unstable and natural versus cut slopes, as related to hazard and risk. In potentially unstable natural slopes, the failure potential generally is recognizable but the actual occurrence of an event is unpredictable; prevention often is impossible where geologic conditions consist of slopes in mountainous regions subjected to enormous planar failures, avalanches, and flows or in regions where failures by lateral spreading and flow are possible. Most slopes subject to falls or slides are predictable and treatable.

As unstable natural slopes undergo failure, treatment may or may not be necessary, but stabilization may not be possible due to the magnitude of the volume involved and the form of failure. When unstable cut slopes undergo failure, treatment usually is necessary but also may not be practical because of the magnitude of volume involved and the failure form. New slopes to be formed by cutting or filling require treatment.

Treatment Options

Avoid the High-Risk Hazard

Where failure is essentially unpredictable and unpreventable, the consequences potentially disastrous, and the hazard high, it is prudent to avoid development along the slope or near its base. Roadways and railroads, for example, should be relocated to areas of lower hazard where stability is feasible or routed through tunnels.

Accept the Failure Hazard

Where the hazard degree is judged to be low to moderate as, for example, causing the temporary partial closure of a roadway or a failure in an open-pit mine where failure is predictable but prevention is considered uneconomical, it may be accepted. Failure is permitted to occur and the debris is removed.

In many cases failures are self-correcting and eventually a slope may reach a stable condition or work back to where failures do not affect construction, as often occurs with small slumps or planar rock slides. The true economics of this approach depend on a knowledgeable assessment of the form and magnitude of the potential failure and assurance that the risk is low to moderate (structures upslope cause a high risk condition). In addition, conditions may be such that small failures will evolve into very large ones, or that continuous and costly maintenance will be required. Public opinion regarding small but frequent failures of the nuisance type (such as rock falls) must also be considered.

A recent innovation being used in Southern California to protect homes against debris slides, avalanches, and flows is the "A" wall [Hollingsworth and Kovacs (1981)[68a]]. Two grade beams are constructed as walls about the property to be protected, joining at a point upslope. A gravity or cantilever wall joins the two walls and separates a level yard area from the existing grade upslope, providing an A-shaped wall system as shown in Fig. 8.70. The purpose of the outside walls is to deflect slope failures away from buildings. A similar approach involves the impact walls constructed in Rio de Janeiro to deflect sliding and falling rock blocks [see Fig. 9.13 of Hunt (1984)[69]].

Eliminate or Reduce the Hazard by Slope Stabilization

Where failure is essentially predictable and preventable or is occurring (or has occurred) and is adaptable to treatment, and where the risk is not acceptable, the hazard is reduced or eliminated by stabilization methods. For low- to moderate-risk conditions, the approach can be either to eliminate or to reduce the hazard depending upon comparative economics. For high-risk conditions the hazard should be eliminated. Typical preventive and remedial measures for various failure forms are summarized in Table 8.15, and a summary of slope treatment methods for stabilization is given in Table 8.16, some of which are illustrated in Fig. 8.71. As given in Art. 8.2.1, the generally acceptable factors of safety determined by stability analysis can be taken as: FS = 1.2 (low risk), FS = 1.3 (significant risk), and FS = 1.4 (high risk).

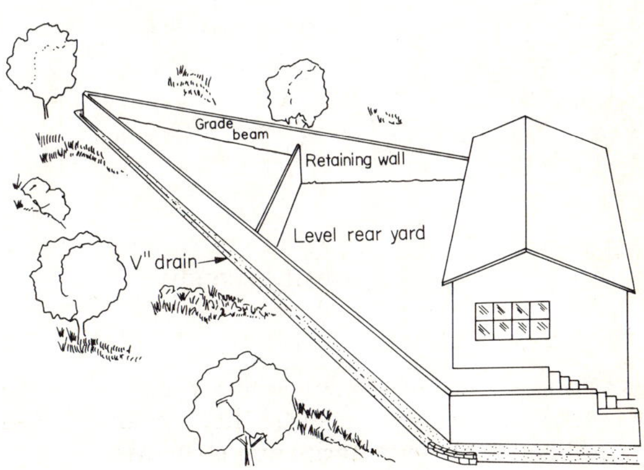

FIG. 8.70 Typical A wall layout to deflect debris slides, avalanches, and flows. [From Hollingsworth and Kovacs (1981).[68a]]

TABLE 8.15
SLOPE FAILURE FORMS: TYPICAL PREVENTIVE AND REMEDIAL MEASURES*

Failure form	Prevention during construction	Remedial measures
Rock fall	Base erosion protection. Controlled blasting excavation. Rock bolts and straps, or cables. Concrete supports, large masses. Remove loose blocks. Shotcrete weak strata.	Permit fall, clean roadway. Rock bolts and straps. Concrete supports. Remove loose blocks. Impact walls.
Soil fall	Base erosion protection.	Retention.
Planar rock slide	Small volume: remove or bolt. Moderate volume: provide stable inclination or bolt to retain. Large volume: install internal drainage or relocate to avoid.	Permit slide, clean roadway. Remove to stable inclination or bolt. Install internal drainage or relocate to avoid.
Rotational rock slide	Provide stable inclination and surface drainage system. Install internal drainage.	Remove to stable inclination. Provide surface drainage. Install internal drains.
Planar (debris) slides	Provide stable inclination and surface drainage control. Retention for small to moderate volumes. Large volumes: relocate.	Allow failure and clean roadway. Use preventive measures.
Rotational soil slides	Provide stable inclination and surface drainage control, or retain.	Permit failure, clean roadway. Remove to stable inclination, provide surface drainage, or retain. Subhorizontal drains for large volumes.
Failure by lateral spreading	Small scale: retain. Large scale: avoid and relocate, prevention difficult.	Small scale: retain. Large scale: avoid.
Debris avalanche	Prediction and prevention difficult. Treat as debris slide. Avoid high-hazard areas.	Permit failure, clean roadway; eventually self-correcting. Otherwise relocate. Small scale: retain or remove.
Flows	Prediction and prevention difficult. Avoid susceptible areas.	Small scale: remove. Large scale: relocate.

*From Hunt (1984).[69] Used with permission of McGraw-Hill Book Company.

TABLE 8.16
SUMMARY OF SLOPE TREATMENT METHODS FOR STABILIZATION†

Treatment	Conditions	General purpose (preventive or remedial)
CHANGE SLOPE GEOMETRY		
Reduce height	Rotational slides	Prevent/treat during early stages
Reduce inclination	All soil/rock	Prevent/treat during early stages
Add weight to toe	Soils	Treat during early stages
CONTROL SURFACE WATER		
Vegetation	Soils	Prevent
Seal cracks	Soil/rock	Prevent/treat during early stages
Drainage system	Soil/decomposing rock	Prevent/treat during early stages
CONTROL INTERNAL SEEPAGE		
Deep wells	Rock masses	Temporary treatment
Vertical gravity drains	Soil/rock	Prevent/treat during early stages
Subhorizontal drains	Soil/rock	Prevent/treat—early to intermediate stages
Galleries	Rock/strong soils	Prevent/treat during early stages
Relief wells or toe trenches	Soils	Treat during early stages
Interceptor trench drains	Soils (cuts/fills)	Prevent/treat during early stages
Blanket drains	Soils (fills)	Prevent
Electroosmosis*	Soils (silts)	Prevent/treat during early stages: temporarily
Chemicals*	Soils (clays)	Prevent/treat during early stages

Treatment	Conditions	General purpose (preventive or remedial)
RETENTION		
Concrete pedestals	Rock overhang	Prevent
Rock bolts	Jointed or sheared rock	Prevent/treat sliding slabs
Concrete straps and bolts	Heavily jointed or soft rock	Prevent
Cable anchors	Dipping rock beds	Prevent/treat early stages
Wire meshes	Steep rock slopes	Contain falls
Concrete impact walls	Moderate slopes	Contain sliding or rolling blocks
Shotcrete	Soft or jointed rock	Prevent
Rock-filled buttress	Strong soils/soft rock	Prevent/treat during early stages
Gabion wall	Strong soils/soft rock	Prevent/treat during early stages
Crib wall	Moderately strong soils	Prevent
Reinforced earth wall	Soils/decomposing rock	Prevent
Concrete gravity walls	Soils to rock	Prevent
Anchored concrete curtain walls	Soils/decomposing rock	Prevent/treat—early to intermediate stages
Bored or root piles	Soils/decomposing rock	Prevent/treat—early stages

*Provides strength increase.

†From Hunt (1984).[69] Used with permission of McGraw-Hill Book Company.

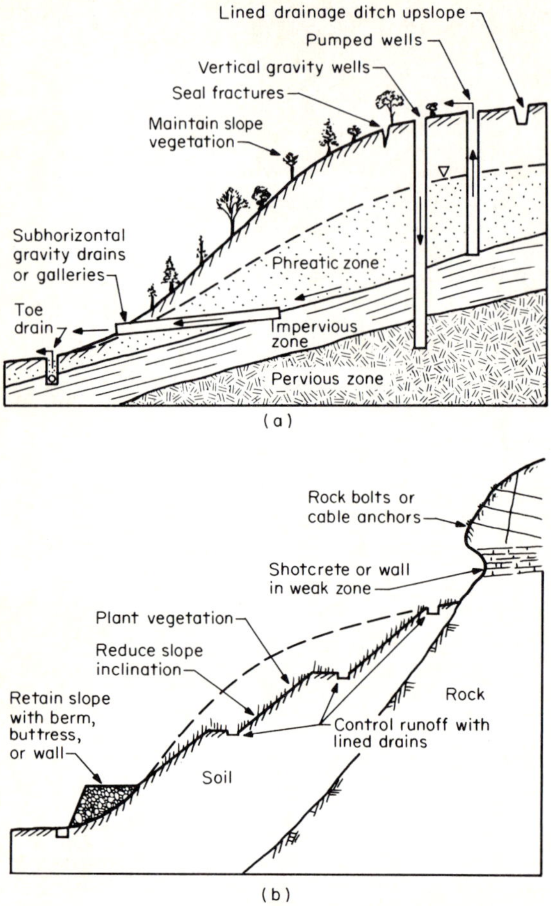

FIG. 8.71 The general methods of slope stabilization: (*a*) control of seepage forces, (*b*) reducing the driving forces and increasing the resisting forces. [From Hunt (1984).[69] Used with permission of McGraw-Hill Book Company.]

Investigation and Monitoring

Slope investigation to collect data for the preparation of sections and to collect samples for the measurement of properties for analysis is described in Hunt (1984)[69] particularly in Art. 9.5. It should be noted that for cut slopes and sidehill fills it is seldom sufficient to study only the immediate area of the cut or fill and to determine the "stable" cut angle, particularly in the case of sidehill cuts. The entire upslope area must be considered to arrive at adequate procedures to provide stability.

An important consideration in slope studies is the installation of instrumentation for monitoring movements and groundwater conditions. It provides information for evaluating unstable slopes and serves as an early warning to possible slope failure in stable or moving cut or natural slopes. The many techniques are described in Chap. 4 of Hunt (1984)[69] and include survey nets, inclinometers, deflectometers, piezometers, shear-strip indicators, etc., as illustrated in Fig. 8.72.

8.4.3 STABILIZATION OF SOIL SLOPES

Cut Slopes

Cut slopes are designed to be stable; therefore, many options exist with selection based on economic considerations.

Geometry

The stable slope angle is determined by analysis and usually is achieved by providing benches at various levels along the slope, especially for cuts greater than 8 to 10 m in height.

In general, typical cut slope angles for most soils are 27° ($1V{:}2H$) and for strong residual soils and decomposed rock 45 to 63° ($1V{:}1H$ to $2V{:}1H$). A typical profile in residual soils derived from igneous or metamorphic rock is illustrated in Fig. 8.73*a*. Clay shales often are cut to about 10° ($1V{:}6H$) without benches, and soils over a sloping rock surface are removed unless the thickness exceeds a few meters, in which case retention probably is necessary.

Surface Drainage and Erosion Protection

Lined ditches are provided upslope of the cut area and along the benches and slope face to collect runoff and direct it away from the slope face, as shown in Fig. 8.73*b*. Vegetation is planted immediately to control erosion and stabilize shallow slope portions against sloughing. The installation of wattling bundles is an effective treatment against erosion [see Art. 8.2.2 of Hunt (1984)[69]].

Internal Subhorizontal Drains

Internal subhorizontal drains (Fig. 8.71) are most useful in reducing seepage forces, especially in slopes where soil strata dip in the slope direction

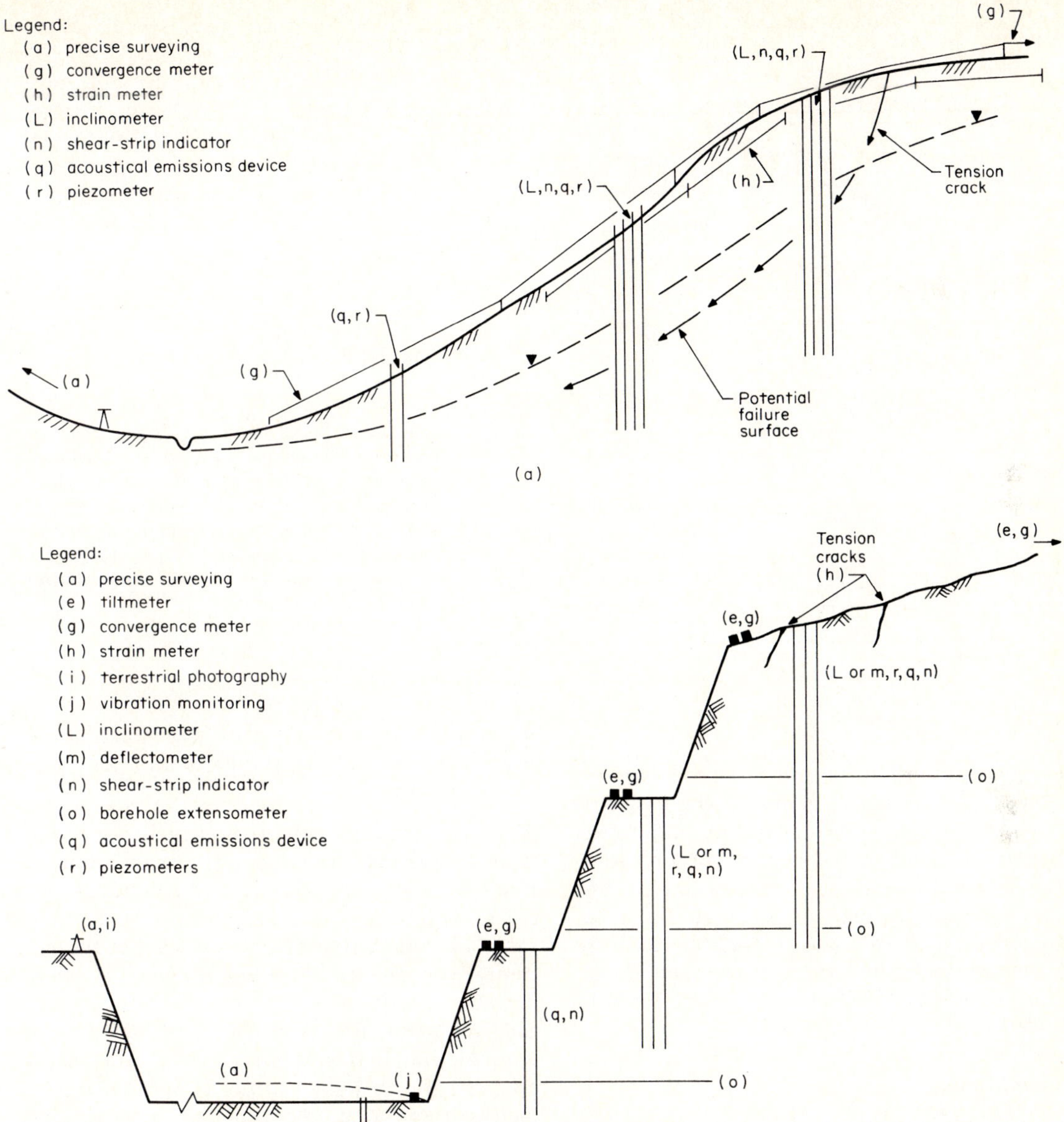

FIG. 8.72 Slope instrumentation and monitoring: (*a*) potentially unstable soil slope and (*b*) rock cut. [From Hunt (1984).[69] Used with permission of McGraw-Hill Book Company.]

or in colluvial or residual soils overlying sloping rock surfaces. Installed at a slight angle upslope to penetrate a potential phreatic zone and provide for gravity flow, they usually consist of perforated pipe, 2 in diameter or larger, forced into a predrilled hole of slightly larger diameter than the pipe. Spacing depends on the type of materials being drained; fine-grained soils may require spacing as close as 3 to 8 m, whereas for more permeable materials, 8 to 15 m may suffice.

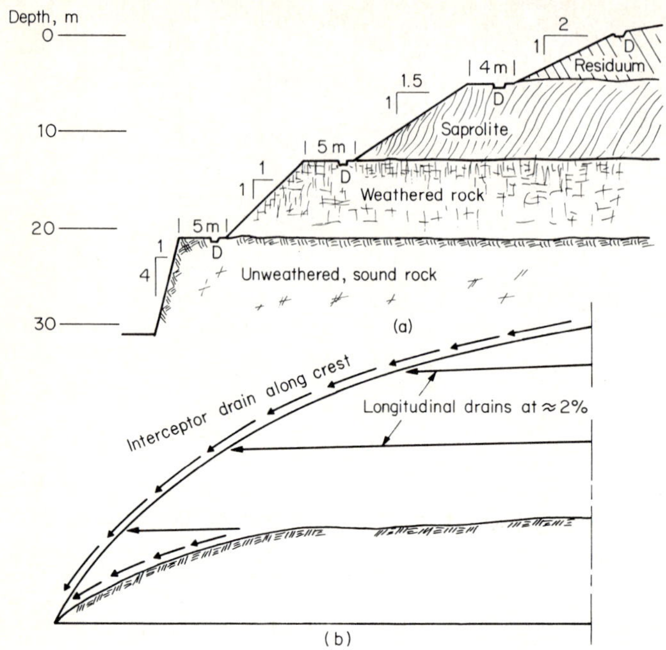

FIG. 8.73 Benching a cut slope. (*a*) Typical section with drains located at *D*. Slope in weathered rock varies with rock quality; in saprolite, varies with orientation of foliations. Attempt is made to place benches on contact between material types. (*b*) General scheme to control runnoff and drainage shown in face view.

Retaining Structures

Retaining structures (Chap. 9) are used to support slopes where adjacent construction upslope does not provide adequate space for flattening and benching, to support slopes where their toe-to-crest length and steepness would require an excessively large volume of excavation, or to provide more positive stability such as for clay shales, overconsolidated clays, metastable soils, sensitive clays, and soils over a sloping rock surface.

Unstable Slopes

Existing unstable slopes impose a limitation on selection because of slope movements, the possibility of total failure, and the time to total failure. Walls, for example, take months to build and time may not be available for their completion before total failure occurs. When a practitioner is called in to provide advice about a slope that is moving or has recently moved, as evidenced by fresh tension cracks or major deformations, time may not be available for a thorough investigation. The practitioner must rely on experience and judgment as to what remedial measures can be taken and how much time is available for their installation and for proper investigation.

As many masses fail progressively in the so-called slip-stick phenomenon, cracks and movement are not necessarily an indication that total failure is imminent. Total failure is likely during rainy seasons when slope movements are relatively large and accelerating. It appears from experience that many sliding masses can suffer displacements in the order of 2 cm/day without approaching final, sudden failure as long as they do not suffer a heavy rainfall. Accelerating velocities above 2 cm/day during a rainy period or season, however, should be considered as warning that total failure may be approaching.

Changing Geometry

Changing slope geometry may be the quickest and most practical approach if the size of the moving mass is not too large. Material is removed from the head to reduce the driving forces and, if space is available, it is placed at the toe to increase the resisting forces, as shown in Fig. 8.74.

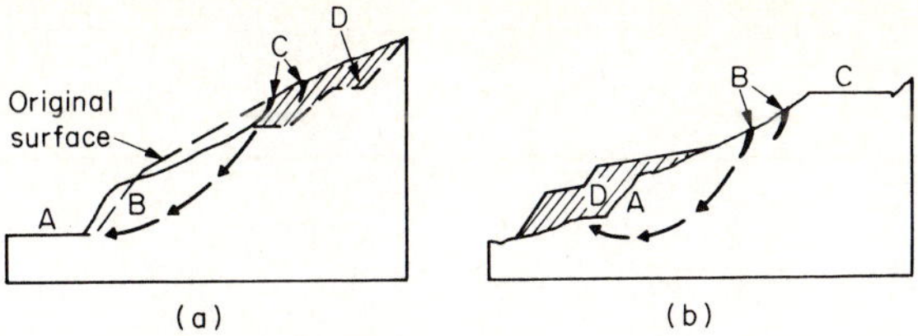

FIG. 8.74 Changing slope geometry for stabilization. (*a*) Cut for roadway *A* causes large slump *B* and tension cracks *C*. Soil removal from *D* may quickly provide short-term stability. Long-term stability will require major regrading and drainage improvements or retention. (*b*) Cut at slope base *A* causes deformation and tension cracks *B*, indicating danger to roadway *C* stabilized by adding berm *D*. Volume larger than original cut is required as deformation causes strength reduction.

Internal Drainage Improvement

Improvement of internal drainage to reduce seepage pressures is often the most practical method of stabilizing large moving masses, particularly colluvial or residual soils of substantial thickness overlying sloping rock surfaces. For most cases, subhorizontal drains are installed in soil slopes for gravity drainage (Fig. 8.75*a*). If seepage is exiting at the slope toe the installation of *toe drains* or *trenches* may be helpful in reducing pressures (Fig. 8.75*b*). Installation is made in sections and the trenches quickly backfilled with stone to avoid causing a total failure.

Vertical gravity drains are useful for perched water-table conditions overlying a nonsaturated free-draining material (Fig. 8.76*a*). Where gravity drainage is not possible, seepage forces can be reduced with pumped vertical wells (Fig. 8.76*b*) which are not only relatively costly, but require continuous or periodic operation. In all cases, slope movements break off drains, which then require reinstallation.

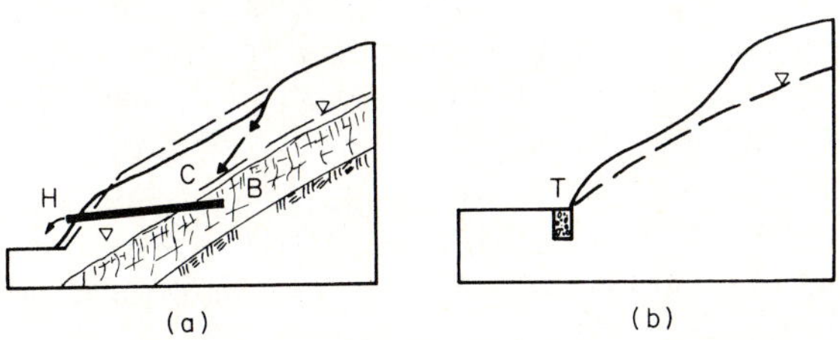

FIG. 8.75 Improving internal drainage for slope stabilization. (*a*) Subhorizontal drain *H* installed to intercept seepage and reduce forces in fractured rock *B* underlying residual soils *C* which have slumped. Drains will break up with continuing slope movements. (*b*) Rock-filled relief trench *T* reduces forces from toe seepage to provide temporary relief prior to reconstructing slope to provide long-term stability.

Electroosmosis has been used to stabilize slopes in fine-grained silty soils (*see Art. 5.4.3*) but also has the disadvantages of high cost and requiring continuous operation. In large thicknesses of these soils, however, it is the most effective method of stabilization through pore-water treatment.

External Drainage Improvement

Improvement of external drainage is not too effective with slopes undergoing movement unless failure is in the early stages and only a few shallow tension cracks have opened. Cracks are sealed with asphaltic compounds, soil-cement, or compacted soil, and interceptor trenches installed around the slide area to minimize runoff flows. Grading a large moving area will help increase stability by closing tension cracks and improving surface runoff, but since it is internal seepage forces that cause instability in the normal situation, all measures of improving surface drainage are only temporary, as the slope will continue to move and tension cracks will continue to form.

Permanent Correction

The foregoing discussion pertains generally to the temporary stabilization of slopes in the ini-

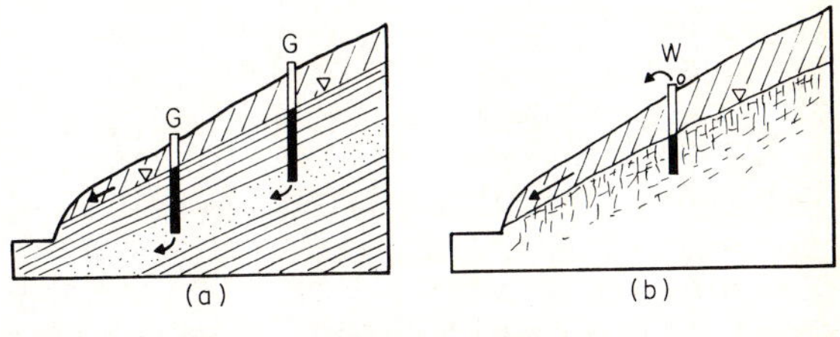

FIG. 8.76 Improving internal drainage to stabilize large masses moving slowly: (*a*) Gravity drains *G* terminating in free-draining sandstone or cavernous limestone reduce perched phreatic surface ∇ causing instability. (*b*) Pumped well *W* terminating in highly fractured rock lowers phreatic surface ∇.

tial phases of failure as evidence by slumping and the formation of tension cracks, although the construction of a berm or the installation of subhorizontal or gravity drains may prove to be satisfactory in the long run. Chemical injection with quicklime slurry into predrilled holes has, on occasion, arrested slope movements as a result of the increase in strength from chemical reaction with clay soils [Handy and Williams (1967),[70] Broms and Boman (1979)[71]]. Strength increase in saltwater clays, however, was found to be low.

Other long-term corrective measures must consider that displacements have reduced shear strength toward ϕ_r and may include the following approaches:

- Permit failure to proceed, removing material from the toe until the failing mass no longer encroaches on the roadway or other construction *(Caution: see Art. 8.4.2, Accept the Failure Hazard).*
- Regrade the slope and install an external and internal drainage system. Substantially flatter slopes than those existing originally will be required.
- Provide retention for a failing slope. Except by the construction of a berm, this is seldom a practical solution unless the slope is first treated by other methods and unless construction is planned to proceed during the dry season when total failure potential is lowest.

Slope movements should be monitored to forewarn against a major failure, particularly where the potential volume is large and during the wet season, until the permanent corrective measures are complete. Movements should be checked periodically until the slope is assuredly stable.

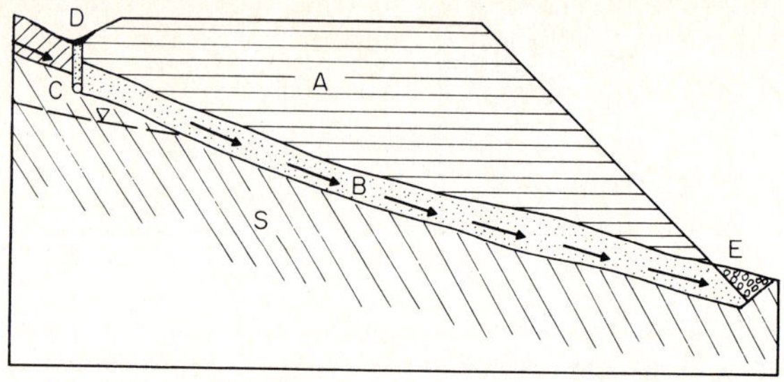

FIG. 8.77 Proper drainage provisions for a sidehill fill *A* constructed of relatively impervious materials over relatively impervious natural soils *S* subject to surface creep. Upper soils are stripped and replaced with free-draining blanket *B*. Interceptor trench drain *C* installed and sealed with lined ditch *D*. Groundwater discharge collected at low point *E* and directed downslope.

Sidehill Fills

The most important consideration in the design and construction of sidehill fills is the provision for adequate drainage. *Interceptor trench* drains (Fig. 8.77) are installed along the upslope side of the fill to intercept groundwater and direct drainage away from the fill area. Surface runoff is collected in paved ditches. A *free-draining blanket* is placed beneath the fill to relieve seepage pressures where the fill or the natural soils are slow-draining; *transverse drains* extending upslope and connected to the interceptor drain may be adequate where low seepage flows are anticipated. If the upper natural soils are potentially unstable, as evidenced by creep, they should be stripped and replaced with free-draining materials. Stepped excavation, or benching, prior to fill placement often is used to improve stability of fills covering large areas, such as for building developments. On steep slopes that continue for some distance beyond the fill, *retaining structures* may be necessary to provide stability.

Culverts beneath roadway fills must have adequate capacity to handle runoff or the fill will act as a dam. Continuous maintenance of culverts, and all other drainage methods, to remove brush and other debris which will obstruct flow, is a necessity.

Failed fills will require reconstruction with provisions for proper drainage, or retained (for example, see Fig. 9.7). On large fills, subhorizontal drains may provide temporary stability.

8.4.4 STABILIZATION OF ROCK SLOPES

Cut Slopes

Geometry

Stable slope inclination depends on rock quality, including the degree of weathering and the intensity and orientation of the discontinuities as related to the slope face. Some typical conditions are illustrated in Fig. 8.78. Typical inclinations are

- 76° (4*V*:1*H*) for hard masses of crystalline rocks, widely jointed; hard sedimentary rocks with bedding dipping vertically and perpendicular to the slope face, or dipping into the face; and horizontally interbedded hard sandstones and shales. Deterioration of expansive shales by weathering may be retarded by treatment with shotcrete or gunite.
- Follow the dip in hard masses with joints or bedding representing major discontinuities and dipping downslope. If the dip is too shallow for economical excavation, retain the blocks with rock bolts (Fig. 8.78*b*).
- 63 to 45° (2*V*:1*H* to 1*V*:1*H*) for weathered masses or those with closely spaced joints (except marine shales and dipping beds), depending on rock quality.
- 14 to 10° (1*V*:4*H* to 1*V*:6*H*) often is necessary to achieve stable slopes in marine shales, since they weather rapidly upon exposure.

Benching has been common practice in high cuts although there is some disagreement among practitioners as to its value. Some consider benches undesirable because they provide takeoff points for falling blocks [Chassie and Goughnor (1975)[72]]. To provide suitable storage they must be wide. Storage space for falling blocks should always be provided at the slope toe to protect a roadway or other works.

Blasting

Blasting for excavation must be performed carefully to minimize the formation of new cracks and the opening of existing fissures which result in a decrease in stability, as discussed in Art. 6.3.4. Bench excavation is particularly critical and is the area where rock quality is most often lowered.

Two types of controlled blasting, i.e., presplitting and smooth-wall blasting, generally are used, both providing the advantages of smoother rock faces, stronger rock surfaces requiring less reinforcement, less overbreak, less loss in the event of effluent seepage, and less water penetration.

Drainage

In zones where water pressures can build up and decrease stability, such as heavily fractured zones, soil-rock interfaces, or along bedding planes, subhorizontal drains may be installed to relieve seepage pressures.

Retention will be discussed below.

Unstable Slopes

Solutions selected for the stabilization of unstable rock slopes depend upon the volume involved, rock types, and intensity and orientation of discontinuities. Small-scale failures involving blocks, for example, are corrected by either removing the blocks, wherever practical, or retaining them with bolts.

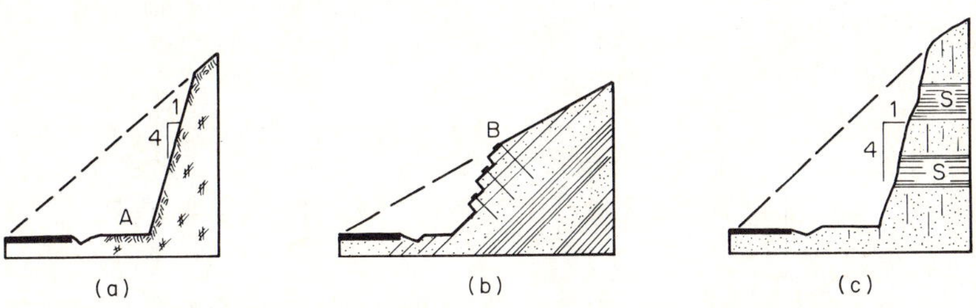

FIG. 8.78 Typical cut slopes in hard rock for a roadway. Leave space at *A* for storage of block falls and topples and scale slope of loose blocks. (*a*) Closely jointed, competent rock. (*b*) Dipping beds; follow the dip unless excessively flat, in which case retain with bolts *B*. (*c*) Horizontally bedded sandstones and shales; apply gunite to the shales *S* if subjected to differential weathering.

Drainage Improvement

Improvement of internal drainage often stabilizes large masses, especially those of dipping sedimentary rocks. Where penetration into the phreatic surface is not too great (generally less than 100 m), subhorizontal drains are installed, as discussed under *Soil Slopes*. Deep pumped wells have been used but require continuous to periodic operation to reduce seepage forces. Gravity wells are a useful solution where perched water-table conditions are causing the failure and the perched zone is underlain by a free-draining, nonsaturated rock, such as sandstone or fractured or cavernous limestone.

Drainage galleries have been used in rock masses to improve stability in both new cut slopes and moving slopes. Vertical holes drilled into the galleries from above provide for drainage from the failure zone into the gallery.

Open-Pit-Mine Slopes

Deep vertical wells have been used to relieve artesian pressures below open-pit-mine floors where heave was occurring, arresting both heave and associated slope instability [Brawner (1974)[73]]. Horizontal galleries and boreholes as long as 150 m installed in the toe zone of slowly moving masses arrested movements even where failures were very large in volume. In some cases vacuum pumps were installed to place galleries under negative pressures. Horizontal drains of slotted pipe installed in boreholes are common procedure in jointed rock masses to relieve cleft-water pressures.

Reinforcement and Retention

Various methods of reinforcing or retaining rock slopes are discussed in Hunt (1984),[69] illustrated in Fig. 8.79, and summarized as follows:

- *Concrete pedestals* are used to support overhangs remaining after lower blocks break loose, and are common in massive granite, especially in tropical climates.
- *Rock bolts* are used to reinforce jointed rock masses, dipping slabs of exfoliating granite, or beds of sedimentary rock.
- *Concrete straps and rock bolts* are used to support loose or soft rock zones or to reduce the number of bolts.

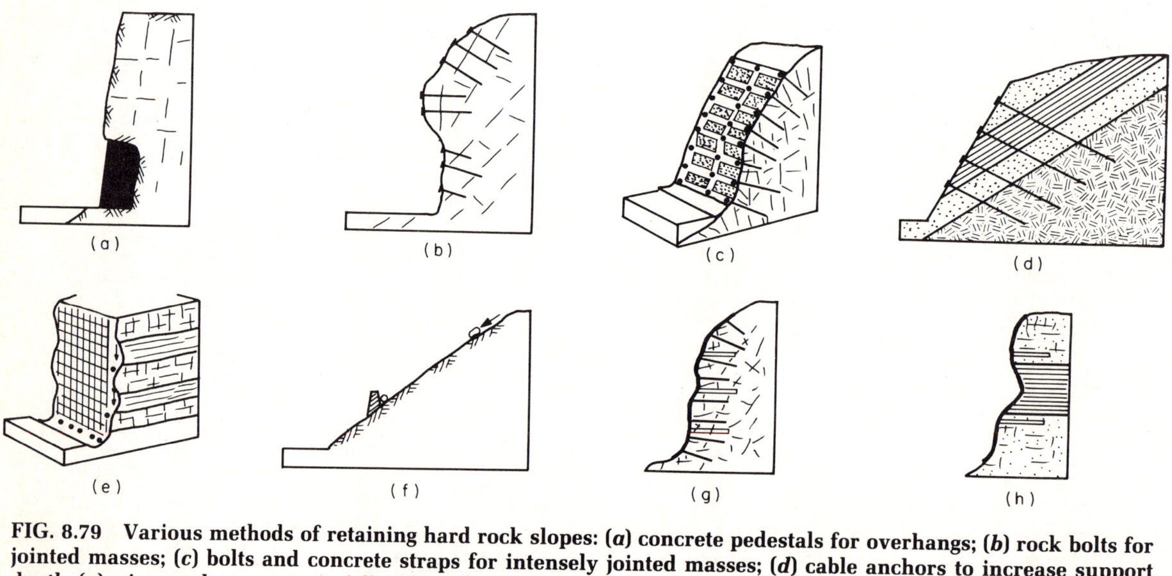

FIG. 8.79 Various methods of retaining hard rock slopes: (*a*) concrete pedestals for overhangs; (*b*) rock bolts for jointed masses; (*c*) bolts and concrete straps for intensely jointed masses; (*d*) cable anchors to increase support depth; (*e*) wire mesh to constrain falls; (*f*) impact walls to deflect or contain rolling blocks; (*g*) shotcrete to reinforce loose rock, with bolts and drains; (*h*) shotcrete to retard weathering and slaking of shales. [From Hunt (1984).[69] Used with permission of McGraw-Hill Book Company.]

- *Cable anchors* are used to reinforce thick masses, particularly dipping sedimentary beds.
- *Wire meshes*, hung on a slope, restrict falling blocks to movement along the face.
- *Concrete impact walls* are constructed along lower slopes to contain falling blocks or deflect them away from structures.
- *Shotcrete* is used to reinforce loose fractured rock or to prevent weathering and slaking of shales, especially where interbedded with more resistant rocks (*see Arts. 5.5.3 and 10.3.4*).
- *Gunite* is similar to shotcrete, except that the aggregate is smaller (*see Art. 10.3.4*).

Rock Bolts

The most common method of reinforcing rock masses is with rock bolts. The various types are described in Art. 5.5.3 and the design aspects of determining capacities are covered in Art. 7.7.4. For a single block the total reinforcing force F required to be stabilized by rock bolts or cables is found by modifying Eq. 8.63 as follows:

$$FS = [cL + (W \cos\theta - U - V \sin\theta + F \cos\beta)\tan\phi]/(W \sin\theta + V \cos\theta - F \sin\beta) \tag{8.90}$$

where β = inclination of force F normal to the failure surface as shown in Fig. 8.80 [Hoek and Bray (1977)[3]].

A procedure for determining force F to be resisted by anchors is given by Hoek and Bray (1977)[3] and Hendron et al. (1980)[67] for the case of a three-dimensional wedge with tension crack (tetrahedron). A solution for the programmable calculator HP-65 that calculates the anchor force for a given value of FS, but which requires a separate calculation for wedge weight, is given by Kovani and Fritz (1977).[74]

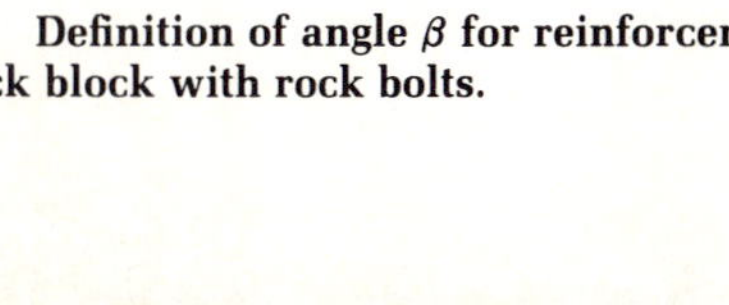

FIG. 8.80 Definition of angle β for reinforcement of a single rock block with rock bolts.

Walls

In weaker rocks, such as shales or those highly decomposed, a retaining structure may be more economical than cutting back to a stable slope angle. Rock bolts and shotcrete are seldom used in these materials. Where hillsides are steep and high and subjected to heavy rainfall, loads to be retained are high, and gravity and counterfort walls generally will not provide sufficient support. Anchored curtain walls can be economically constructed to contain high loads, and in recent years "reinforced earth" walls have been used in many locations, at times in combination with other solutions (Chap. 9).

8.4.5 EMBANKMENTS

Earth Dams

Hazard and Risk

Improperly designed and constructed earth dams can result in hazardous conditions. Inadequate embankment compaction or weak foundations can result in settlement reducing the freeboard or causing internal cracking, providing seepage flow paths. Uncontrolled internal and external seepage can result in erosion by piping and high uplift pressures in the foundation. All of these can lead to a complete failure.

Because the risk of total failure almost always involves the potential for loss of life and damage to and destruction of property, earth dams should be designed and constructed to be stable in such a manner that the hazards are eliminated.

Embankment Section

The selection of the embankment section depends on a number of factors. The availability of natural materials influences whether the embankment is homogeneous or zoned; the latter may include a clay core with a granular shell or a rock fill with impervious shells or an internal cutoff. Clay cores are common to many dams and may be constructed vertical (Fig. 8.83*b*) or slightly sloping (Fig. 8.83*c*). The stable slopes and therefore the volume in the dam are also related to the materials used for construction. All portions of the embankment must be properly compacted as a controlled fill, as discussed in Art. 5.2. Riprap is required on the upstream slope for protection against wave action.

Embankment Height

The embankment height is a function of the reservoir pool elevation required. The pool elevation determines the lateral forces to be retained by the embankment and the seepage forces through and around the embankment. The amount of seepage force in the embankment foundation is related directly to the pool elevation and may affect the embankment width.

Foundation Materials

The deformation and strength characteristics of the foundation materials influence the stable embankment slope angles. Relatively weak materials may require flatter embankment slopes than would be required for strong foundation materials.

Seepage Control

A most critical aspect of design is seepage control through, beneath, or around the embankment. Seepage through the embankment is controlled by the section design which will include various types of drains for homogeneous sections and a core zone or some other type of cutoff for nonhomogeneous sections, as illustrated in Fig. 8.81. Toe drains are a common requirement, and two forms are shown on Fig. 8.82. Filters are a necessity between materials of significantly differing permeability to control piping and internal erosion [see Art. 8.4.5 of Hunt (1984)[69]]. A dam section during construction is shown in Fig. 5.3.

Seepage through pervious materials beneath the embankment is controlled by some form of cutoff or upstream blanket, as illustrated in Fig. 8.83 and as discussed in Arts. 8.3.4 and 8.4 of Hunt (1984)[69] (see also Art. 5.5.2 of this volume for a discussion of uplift and seepage control in rock masses). Rock joints between the interface of the fill and the foundation are treated by such means

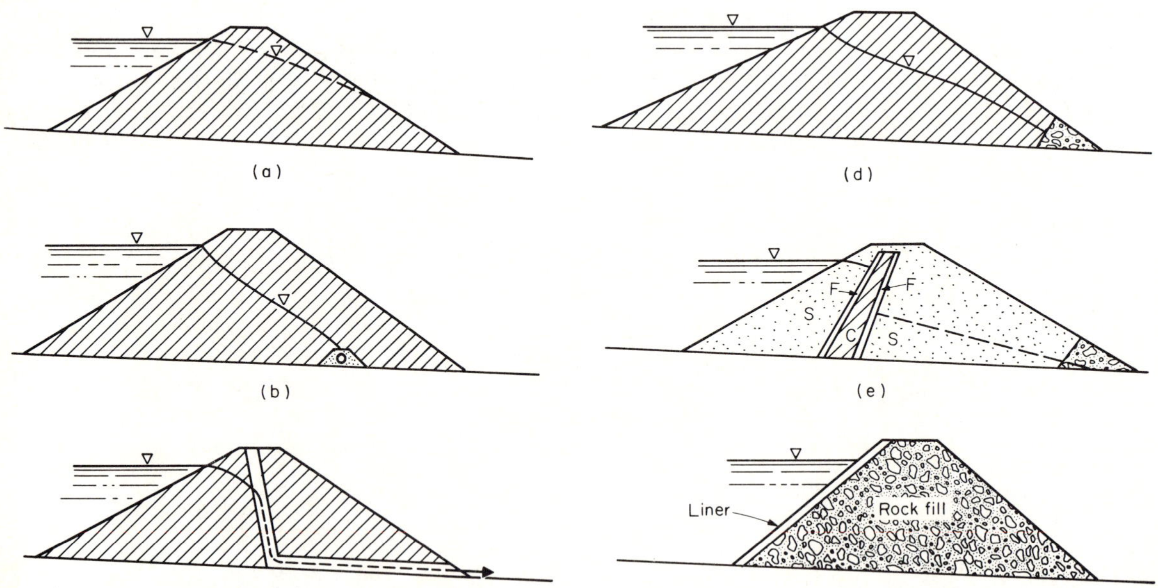

FIG. 8.81 Control of seepage through earth and rock-fill dams: (*a*) embankment without seepage control (seepage exits on downstream slope); (*b*) longitudinal drain in homogeneous embankment; (*c*) chimney and blanket drain in homogeneous embankment; (*d*) toe drain in homogeneous embankment; (*e*) zoned dam to control seepage, there being any number of combinations of pervious and impervious zones; (*f*) rock-filled dam with upstream liner. [From Hunt (1984).[69] Used with permission of McGraw-Hill Book Company.]

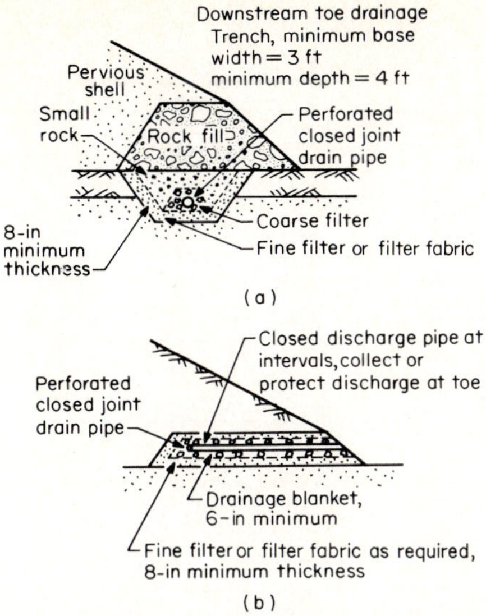

FIG. 8.82 Examples of toe drains for earthdams: (*a*) impervious surface layer and (*b*) homogeneous pervious foundation. [From NAVFAC (1982).[42]]

as slush grouting and dental concrete. Seepage through abutments is controlled by cutoffs.

Other Factors

The deterioration of the rock in the foundation can, at times, affect stability, as discussed in Art. 5.5.4.

Monitoring of deformations and pore pressures during and after construction has become standard practice for most earth-dam embankments. Instrumentation for monitoring is illustrated in Fig. 8.84.

Embankments over Soft Ground

Hazard and Risk

The degree of hazard may be considered high since failures of embankments constructed over soft ground are common. Failure results in loss of fill and a remolding and reduction in strength of the foundation soils, causing construction delays and increased costs, which compared to the costs of proper treatment can be relatively large. Because failures normally occur prior to embankment completion the public risk is not great, except for the case of earthquakes causing failure at some future date.

A major additional consideration is the effect of long-term settlement on the fill and on the structures supported by it (see Chaps. 4, 5, and 6). Design and construction, therefore, proceed on the basis of avoiding failure during construction by providing adequate stability and limiting postconstruction settlement to desirable amounts.

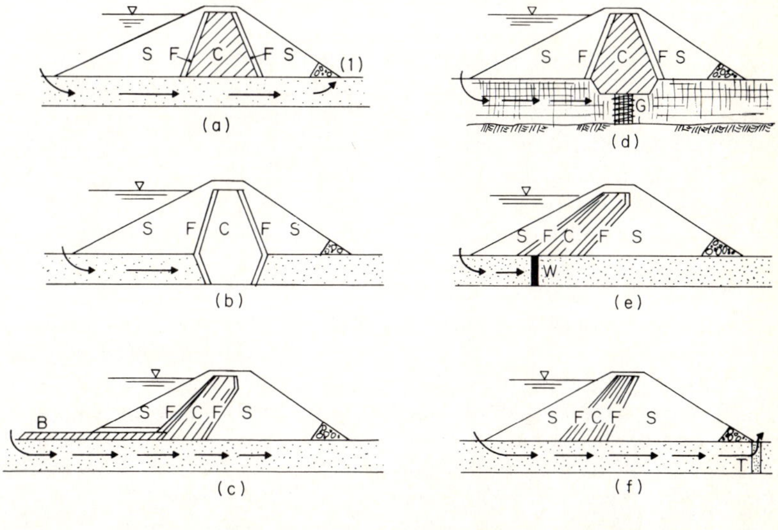

FIG. 8.83 Controlling seepage in pervious materials beneath dams: (*a*) uncontrolled foundation seepage causes water loss and uplift at (1); (*b*) extended core cutoff; (*c*) upstream blanket to reduce head; (*d*) grout curtain in fractured rock; (*e*) concrete wall or slurry trench cutoff; (*f*) downstream relief wells or trench. (NOTE: ***S*** **= shell material;** ***F*** **= filter;** ***C*** **= core material.)**

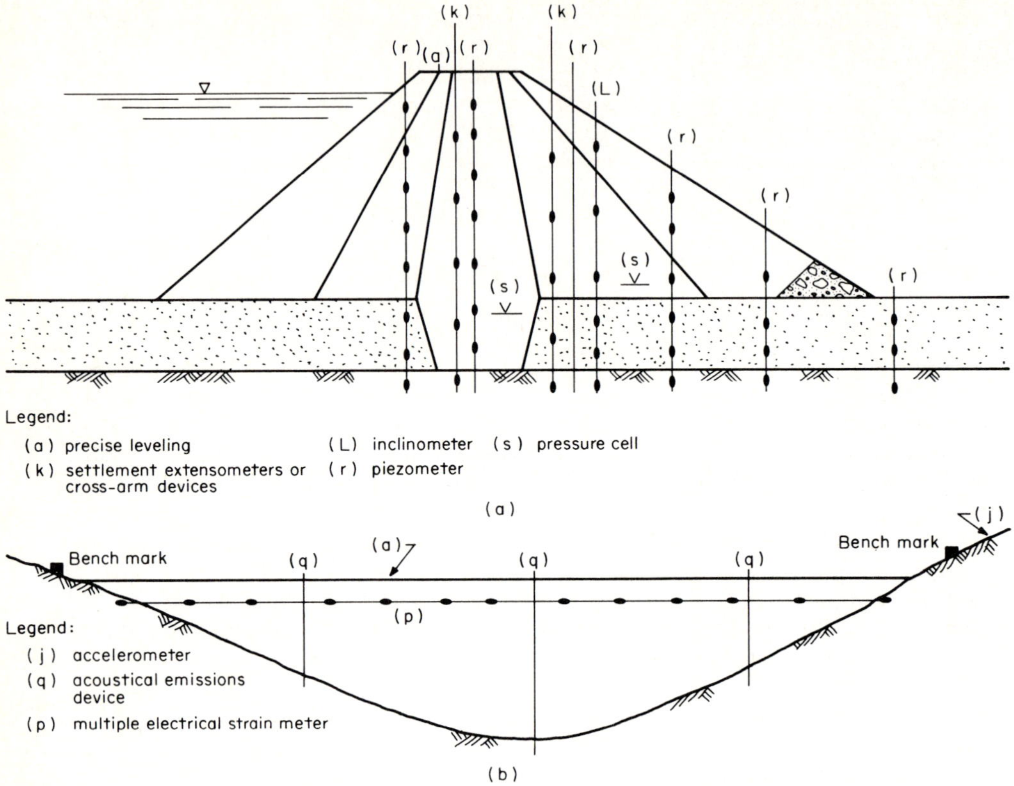

FIG. 8.84 Instrumentation for an earth dam: (*a*) cross section and (*b*) longitudinal section. [From Hunt (1984).[69] Used with permission of McGraw-Hill Book Company.]

Improving Stability and Decreasing Postconstruction Settlement

Stage construction, with proper instrumentation installed to monitor deformations and porewater pressures (Fig. 5.27*a*), provides time intervals between placement of lift thicknesses to permit consolidation and increase of strength. The installation of vertical drains *(see Art. 5.3.3)* substantially decreases the time required between lift placements, and a surcharge *(see Art. 5.3.2)* decreases postconstruction settlement. Counterberms may be required to improve stability *(see Art. 8.2.4)*. Jetties and breakwaters constructed of large rock blocks should be placed over a sand blanket (Fig. 8.27*d*) to prevent infiltration of the blocks into the soft materials and to provide stability *(see Art. 8.2.1)*.

REFERENCES

1. Taylor, D. W. (1948) *Fundamentals of Soil Mechanics*, John Wiley & Sons, New York.
2. Culmann, K. (1866) *Die graphische Statik*, Berlin.
3. Hoek, E. and Bray, J. W. (1977) *Rock Slope Engineering*, Inst. of Min. and Metal., London.
4. Seed, H. B. and Sultan, H. A. (1967) "Stability Analysis for a Sloping Core Embankment," *Proc. ASCE, J. Soil Mech. and Found. Engrg. Div.*, Vol. 93, No. SM4, pp. 69–84.
5. Morgenstern, N. R. (1968) "Ultimate Behavior of Rock Structures," *Rock Mechanics in Engineering Practice*, Stag and Zienkiewicz, eds., John Wiley & Sons, New York, Chap. 10.
6. Fellenius, W. (1927) *Erdstatische Berechnungen* (Calculation of Stability of Slopes), W. Ernst und Sohn, Berlin (rev. ed., 1939).

7. Bishop, A. W. (1955) "The Use of the Slip Circle in the Stability Analysis of Earth Slopes," *Geotechnique*, London, Vol. 5, No. 1, pp. 7–17.

8. Spencer, E. (1967) "A Method of Analysis of the Stability of Embankments Assuming Parallel Inter-Slice Forces," *Geotechnique*, London, Vol. 17, No. 1, pp. 11–26.

9. Spencer, E. (1981) "Slip Circles and Critical Shear Planes," *Proc. ASCE, J. Geotech. Engrg. Div.*, Vol. 107, No. GT7, July, pp. 929–942.

10. Morgenstern, N. R. and Price, V. E. (1965) "The Analysis of the Stability of General Slip Surfaces," *Geotechnique*, London, Vol. 15, No. 1, pp. 79–93.

11. Janbu, N. (1954) "Stability Analysis of Slopes with Dimensionless Parameters," *Harvard Soil Mech. Ser. No. 46*, Harvard University, Cambridge, Mass.

12. Janbu, N. (1973) "Slope Stability Computations," *Embankment Dam Engineering*, Hirschfield and Poulos, eds., John Wiley & Sons, New York, pp. 47–86.

13. Terzaghi, K. (1943) *Theoretical Soil Mechanics*, John Wiley & Sons, New York.

14. Fellenius, W. (1936) "Calculation of the Stability of Earth Dams," *Trans. 2d Cong. on Large Dams*, Vol. 4, Washington, D.C.

15. Patton, F. D. and Hendron, A. J., Jr. (1974) "General Report on Mass Movements," *Proc. 2d Intl. Cong. Intl. Assoc. Engrg. Geol.*, São Paulo, Brazil,, pp. V-CR 1.

16. Lambe, T. W. and Whitman, R. V. (1969) *Soil Mechanics*, John Wiley & Sons, New York.

17. Lumb, P. (1962) "Effect of Rainstorms on Slope Stability," *Proc. Symp. on Hong Kong Soils*, Hong Kong, pp. 73–87.

18. Lumb, P. (1975) "Slope Failures in Hong Kong," *Quarterly J. Engrg. Geol.*, Vol. 8, pp. 31–65.

19. Brand, E. W. (1982) "Analysis and Design in Residual Soils," *Proc. ASCE, Specialty Conf. Engrg. and Constr. in Tropical and Residual Soils*, Hawaii, January, pp. 89–143.

20. Bjerrum, L. (1973) "Problems of Soil Mechanics and Construction on Soft Clays and Structurally Unstable Soils," *Proc. 8th Intl. Conf. Soil Mech. and Found. Engrg.*, Moscow, Ses. 4, Vol. 3, pp. 111–159.

21. Leonards, G. A. (1982) "Investigation of Failures," *Proc. ASCE, J. Geotech. Engrg. Div.*, Vol. 108, No. GT2, February, pp. 187–246.

22. Bjerrum, L. (1967) "Progressive Failures in Slopes of Overconsolidated Plastic Clay and Clay Shales," *Terzaghi Lectures, ASCE* (1974), New York, pp. 139–189.

23. Sherard, J. L., Woodward, R. J., Gizienski, S. F., and Clevenger, W. A. (1963) *Earth and Earth-Rock Dams*, John Wiley & Sons, New York.

24. Lowe, J. III (1967) "Stability Analysis of Embankments," *Proc. ASCE, J. Soil Mech. and Found. Engrg. Div.*, Vol. 93, No. SM4, July.

25. Sherard, J. L. (1981) "Piezometers in Earth Dam Impervious Sections," *Proc. ASCE, Recent Developments in Geotechnical Engineering for Hydro Projects*, May, pp. 125–165.

26. Wilson, S. D. and Marsal, R. J. (1979) *Current Trends in Design and Construction of Embankments Dams*, ASCE, New York.

27. Taylor, D.W. (1937) "Stability of Earth Slopes," *Contributions to Soil Mechanics, 1925 to 1940*, Boston Soc. of Civ. Eng.

28. Bishop, A. W. and Morgenstern, N. R. (1960) "Stability Coefficients of Earth Slopes, *Geotechnique*, London, Vol. 10, No.4, pp. 129–150.

29. Morgenstern, N. R. (1963) "Stability Charts for Earth Slopes During Rapid Drawdown," *Geotechnique*, London, Vol. 13, No. 2, pp. 121–131.

30. Janbu, N. (1967) Discussion of "Dimensionless Parameters for Homogeneous Earth Slopes," by J. M. Bell, *Proc. ASCE, J. Soil Mech. and Found. Engrg. Div.*, Vol. 93, No. SM6, Proc. Paper 5545, November, pp. 367–374.

31. Huang, Y. H. (1975) "Stability Charts for Earth Embankments," *TRR 548*, Transportation Research Board, Washington, D.C., pp. 1–12.

32. Huang, Y. H. (1977) "Stability Coefficients for Sidehill Benches," *Proc. ASCE, J. Geotech. Engrg. Div.*, Vol. 103, No. GT5, May, pp. 467–481.

33. Huang, Y. H. (1978) "Stability Charts for Sidehill Fills," *Proc. ASCE, J. Geotech. Engrg. Div.*, Vol. 104, No. GT5, pp. 659–663.

34. Huang, Y. H. (1978) "Stability of Spoil Banks and Hollow Fills Created by Surface Mining," *Research Report IMMR 34-RRR1-78*, Inst. for Min. and Min. Research, University of Kentucky, Lexington, Ky.

35. Cousins, B. F. (1978) "Stability Charts for Simple Earth Slopes," *Proc. ASCE, J. Geotech. Engrg. Div.*, Vol. 104, No. GT2, February, pp. 267–279.

36. Huang, Y. H. (1979) "Stability Charts for Refuse Dams," *Proc. 5th Kentucky Coal Refuse Disposal and Utilization Seminar and Stability Analysis of Refuse Dams Workshop*, University of Kentucky, Lexington, Ky.

37. Huang, Y. H. (1980) "Stability Charts for Effective Stress Analysis of Nonhomogeneous Embankments," *TRR 749*, Transportation Research Board, Washington, D.C., pp. 72–74.

38. McCarthy, D. E. (1982) *Essentials of Soil Mechanics and Foundations*, 2d ed., Reston Publishing Co., Reston, Va.

39. Tschebotarioff, G. P. (1973) *Foundations, Retaining, and Earth Structures*, McGraw-Hill Book Co., New York.

40. Johnson, S. J. (1974) "Analysis and Design Relating to Embankments," *Proc. ASCE, Conf. Anal. and Des. in Geotech. Engrg.*, Austin, Tex., Vol. II, pp. 1–48.

41. Janbu, N. Bjerrum, L., and Kjaernsli, B. (1956) "Soil Mechanics Applied to Some Engineering Problems" (in Norwegian with English summary), Norwegian Geotechnical Inst., *Pub. 16*, Oslo.

42. NAVFAC (1982) *Design Manual DM-7.1, Soil Mechanics*, Department of the Navy, Naval Facilities Engrg. Comm., Alexandria, Va.

43. Hunter, J. H. and Schuster, R. L. (1968) "Stability of Simple Cutting in Normally Consolidated Clays," *Geotechnique*, London, Vol. 18, No. 3, pp. 372–378.

44. Hunter, J. H. and Schuster, R. L. (1971) "Chart Solutions for Analysis of Earth Slopes," *HRB No. 345*, Highway Research Board, pp. 77–89.

45. Jakobsen, R. E. (1948) "The Design of Embankments on Soft Clays," *Geotechnique*, London.

46. Whitman, R. V. and Bailey, W. A. (1967), "Use of Computers for Slope Stability Analysis," *Proc. ASCE, Soil Mech. and Found. Engrg. Div.*, Vol. 93, No. SM4, pp. 475–498.

47. Huang, Y. H. (1983) *Stability Analysis of Earth Slopes*, Van Nostrand-Reinhold, New York.

48. Seed, H. B., Makdisi, F. I., and DeAlba, P. (1978) "Performance of Earthdams During Earthquakes," *Proc. ASCE, J. Geotech. Engrg. Div.*, Vol. 104, No. GT7, July, pp. 967–994.

49. Terzaghi, K. (1950) "Mechanism of Landslides," *Engineering Geology (Berkey Volume)*; The Geologic Society of America, pp. 83–123.

50. Seed, H. B. (1967) "Slope Stability Analysis During Earthquakes," *Proc. ASCE, J. Soil Mech. and Found. Engrg. Div.*, Vol. 93, No. SM4, July, pp. 299–323.

51. Seed, H. B. and Martin, G. R. (1966) "The Seismic Design Coefficient in Earth Dam Design," *Proc. ASCE, J. Soil Mech. and Found. Engrg. Div.*, Vol. 92, No. SM3, May, pp. 25–58.

52. Ambraseys, N. N. (1960) "The Seismic Stability of Earth Dams," *Proc. 2d World Conf. on Earthquake Engrg.*, Japan, Vol. II.

53. Hall, W. J. and Newmark, N. M. (1977) "Seismic Design Criteria for Pipelines and Facilities," *The Current State of Knowledge of Lifeline Earthquake Engineering, Proc. ASCE*, New York, pp. 18–34.

54. Seed, H. B., Lee, K. L., Idriss, I. M., and Makdisi, F. I. (1975) "The Slides in the San Fernando Dam During Earthquake of February 9, 1971," *Proc. ASCE, J. Geotech. Engrg. Div.*, Vol. 101, No. GT7, July, pp. 651–688.

55. Seed, H. B., Idriss, I. M., Lee, K. L., and Makdisi, F. I. (1975) "Dynamic Analysis of the Slide in the Lower San Fernando Dam During the Earthquake of February 9, 1971," *Proc. ASCE, J. Geotech. Engrg. Div.*, Vol. 101, NO. GT9, September, pp. 889–911.

56. Newmark, N. M. (1965) "Effects of Earthquake on Dams and Embankments," *Geotechnique*, London, Vol. 15, No. 2, June.

57. Seed, H. B. (1966) "A Method for Earthquake Resistant Design of Earthdams," *Proc. ASCE, J. Soil Mech. and Found. Engrg. Div.*, Vol. 92, No. SM1, January, pp. 13–41.

58. ICOLD (1975) "A Review of Earthquake Resistant Design of Earth Dams," *Bul. 27*, Intl. Comm on Large Dams, March.

59. Makdisi, F. I. and Seed, H. B. (1978) "Simplified Procedure for Estimating Dam and Embankment Earthquake Induced Deformations," *Proc. ASCE, J. Geotech. Engrg. Div.*, Vol. 104, No. GT7, July, pp. 849–868.

60. Terzaghi, K. (1962) "Stability of Steep Slopes on Hard, Unweathered Rock," *Geotechnique*, London, Vol. 12, No. 12, pp. 251–270.

61. Deere, D. U. and Patton, F. D. (1971) "Slope Stability in Residual Soils," *Proc. 4th Panamerican Conf. Soil Mech. and Found. Engrg.*, San Juan, pp. 87–170.

62. Jennings, J. E. (1971) "An Approach to the Stability of Rock Slopes Based on the Theory of Limiting Equilibrium with a Material Exhibiting Anisotropic Shear Strength," *Proc. ASCE, 13th Symp. on Rock Mech.*, Urbana, Ill. (1972), pp. 269–302.

63. Deere, D. U. (1976) "Dams on Rock Foundations—Some Design Questions," *Rock Engineering for Foundations and Slopes, Proc. ASCE*, New York, Vol. 2, pp. 55–85.

64. Morgenstern, N. R. and Sangrey, D. A. (1978) "Methods of Stability Analysis," *Landslides: Analysis and Control*, Schuster and Krizek, eds.,

Spec. Report 176, National Academy of Sciences, Washington, D.C., Chap. 7, pp. 155–172.

65. Barton, N. (1971) "Progressive Failure of Excavated Rock Slopes," *Stability of Rock Slopes, Proc. ASCE, 13th Symp. on Rock Mech.*, Urbana, Ill. (1972), pp. 139–170.

66. Wittke, W. W. (1965) "Method to Analyze the Stability of Rock Slopes with and without Additional Loading" (in German), *Felsmechanik und Ingenieurgeologie*, Supp. 11, Vol. 30, pp. 52–79.

67. Hendron, A. J., Jr., Cording, E. J., and Aiyer, A. K. (1980) "Analytical and Graphical Methods for the Analysis of Slopes in Rock Masses," *Tech. Report GL-80-2*, U.S. Army Engineer Waterways Experiment Station, Vicksburg, Miss.

68. Kalkani, E. C. and Piteau, D. R. (1976) "Finite Element Analysis of Toppling Failure at Hell's Gate Bluffs, British Columbia," *Bul. Assoc. Engrg Geol.*, Vol. 13, No. 4, Fall.

68*a*. Hollingsworth, R. and Kovacs, G. S. (1981) "Soil Slumps and Debris Flows: Prediction and Protection," *Bul. Assoc. Engrg. Geol.*, Vol. 18, No. 1, February, pp. 17–28.

69. Hunt, R. E. (1984) *Geotechnical Engineering Investigation Manual*, McGraw-Hill Book Co., New York.

70. Handy, R. L. and Williams, W. W. (1967) "Chemical Stabilization of an Active Landslide," *Civil Engineering*, ASCE, August, pp. 62–65.

71. Broms, B. B. and Boman, P. (1979) "Lime Columns: A New Foundation Method," *Proc. ASCE, J. Geotech. Engrg. Div.*, Vol. 105, No. GT4, April, pp. 539–556.

72. Chassie, R. G. and Goughnor, R. D. (1976) "States Intensifying Efforts to Reduce Highway Landslides," *Civil Engineering*, ASCE, New York, April, p. 65.

73. Brawner, C. O. (1974) "Rock Mechanics in Open Pit Mining," *Proc. 3d Intl. Cong. Soc. for Rock Mech.*, Denver, Colo., Vol. 1A, pp. 755–773.

74. Kovari, K. and Fritz, P. (1977) "Stability Analysis of Rock Slopes for Plane Failure with the Aid of a Programmable Calculator," *Design Methods in Rock Mechanics, ASCE*, New York, pp. 25–29.

CHAPTER NINE

SURFACE RETAINING STRUCTURES

9.1 INTRODUCTION

9.1.1 BASIC FACTORS

Purpose

The purpose of a surface retaining structure generally is to provide support to a soil or rock mass that has been cut or has been placed by filling. Adequate support requires that there be no danger of the structure itself failing by rupture, overturning, or base failure and that the displacement of the structure is limited to some defined amount.

Categories of Structures

This chapter covers three general categories of surface retaining structures, each serving a different purpose and each requiring somewhat different approaches to evaluation and analysis.

Soil-Slope Retention

Where space and other factors limit slope inclinations, hillside cuts are retained by walls for which there are a large number of types. They may be placed in two general groups, rigid and semirigid, and in most cases provide resistance to lateral earth pressures by gravity forces resulting from the weight of the wall system. An exception is the anchored curtain wall. Also included in this category are walls for bridge approach fills and some cases of basement walls for buildings.

Open-Excavation Support Systems

Open excavations are made from the surface for buildings, sewer trenches, subway cuts, and other construction works. Support of the excavation sides is required where the construction area is limited or ground conditions are poor; otherwise, excavation sides may be considered as a slope stability problem and designed at some stable angle without special support. The support system should enable construction to proceed in the "dry" and should limit backslope subsidence to some tolerable amount. In most cases lateral earth pressures are resisted by a wall system which includes the wall itself, which may vary from one that is flexible to one that is rigid, and wall support which may be provided by bracing or anchors (tiebacks). In a few cases, where lateral pressures are low, support may be provided by a cantilever wall.

Waterfront Structures

Waterfront structures may serve as docks, quays, or wharfs; support for land reclamation; or to permit construction in the dry. Except for the case where a water barrier is required, such as provided by a cofferdam, support usually is provided by an anchored sheet-pile bulkhead or a relieving platform. Other permanent structures that may be used include concrete gravity walls, precast concrete floating boxes, or gravity rock dikes.

9.1.2 CRITERIA FOR SATISFACTORY PERFORMANCE

General

The satisfactory performance of a retaining structure requires adequate resistance to lateral forces and, in many cases, limited deformations. At times such structures must also serve as water barriers. In all categories of retaining structures there are a number of choices, and the selection is based on economic comparisons. In the selection process, evaluations are made of geologic conditions, the nature of the overall construction program, and the purpose of the retaining structure. The last two factors influence the height of the retaining structure and whether it must function as a water barrier. They also affect the lateral forces to be contained; the desired lifespan, which affects the type of materials to be selected for the retention system; and the deflection tolerances, which affect the system's rigidity. The following is a general review of some of the more important evaluation factors.

Evaluation Factors

Geologic Conditions

As with all planned construction, a thorough and comprehensive geotechnical investigation is required to define soil, rock, and groundwater conditions. Detailed ground topography is very important for soil-slope retention design, and information on conditions at the bottom of the body of water is necessary for the design of waterfront structures.

Slope retention is addressed in a manner similar to that employed in any slope stability problem with particular attention given to the base support of the structure, to seepage forces, and in some cases to the stability of the immediate area behind the wall as well as to the overall stability of the slope. In most cases of gravity walls, however, a portion of the slope is excavated and replaced with a free-draining granular material which significantly reduces the problem of determining the lateral forces to be resisted.

For the case of open excavations in particular, detailed information on soil stratification is required to permit accurate assessment of potential seepage flows into the excavation. In addition to the lateral forces to be contained, assessment of the stability of the excavation bottom is required, particularly where the soils are soft clays. Open excavations in rock masses may encounter high residual stresses, and these need to be defined.

With waterfront structures, the more difficult problems to be evaluated arise from the common condition of soft organic or clay soils. The strength and compressibility of these materials require close definition.

Forces Acting

The forces to be resisted by a retaining structure are imposed by *lateral earth pressures* for the most part, although many retaining structures support a vertical load in addition to their own weight. The lateral forces result not only from the earth mass or water body to be contained, but in the case of earth support the forces may be increased by some form of surcharge loading behind the structure such as from fills, railways, crane loads, or existing building foundations. In addition to backfill loads, waterfront structures are subjected to current, wave, and ice forces as well as ship breasting and docking forces. Since the distribution of the lateral forces relates to a number of factors, such as wall flexibility, the forces are not determined until a retaining system is selected.

Retaining System Selection

Prior to analysis of lateral earth pressures, a tentative retaining system is selected and dimensioned based on consideration of such factors as the purpose and general capacity of the wall and whether it is to be permanent or temporary, impervious or pervious, flexible or rigid, and deflectable or nondeflectable. Required structure capacity is determined primarily by wall height and the magnitude of the forces to be retained. Wall deflectability relates primarily to the problem of potential backslope subsidence and the subsequent settlement of structures located in the area behind the wall. This problem is common during the construction of many open excavations in urban areas.

Earth Pressure and Stability Factors Evaluated

The magnitude of the earth pressures imposed against the structure relates to the soil type to be retained, groundwater conditions, and whether or not the wall will be subject to displacement. Basement walls, for example, undergo essentially no movement upon completion of construction, and design is based on at-rest pressures. Most other walls will undergo at least a slight deflection and design is based on active earth pressures or in some cases, such as anchored bulkheads, on active and passive earth pressures. Wall flexibility controls the distribution of the earth pressures and determines whether evaluations may be based on classical theory or on empirical relationships.

Gravity retaining walls also require evaluation against sliding, overturning, and base bearing-capacity failure and settlement. Open excavations require evaluation of potential bottom failures by heave or piping, or possible foundation failure, as well as overturning. Waterfront structures may require evaluation of foundation support, sliding, or overturning, depending upon the type of structure proposed.

Retaining System Designed

Once earth pressures are determined, walls to retain soil slopes may be designed with adequate weight to resist the lateral forces or they may employ anchors. The wall itself will require a section of adequate thickness or reinforcing to provide resistance against bending and shear. Walls for open excavations and some waterfront structures will require a section modulus adequate to resist the bending that results from the lateral earth pressures. Wall supports are designed to resist the necessary reactions. Pile foundations may be provided beneath some slope retaining walls and bridge abutments, for example, and are an integral part of relieving platforms.

In general, walls to retain sidehill cuts are designed with positive drainage; temporary walls, such as those for open excavations, may permit some drainage, but it is controlled to within tolerable limits. Permanent walls for building basements are designed to be impermeable.

Construction Aspects

Construction of a retaining system requires particularly close attention to details. If a wall retains a sidehill slope, slope stability during construction must be maintained, and the installation of an adequate system to provide permanent wall drainage is an important consideration. In open excavations, the construction method used and the sequence of operations play a major role in the distribution of earth pressures, which can vary substantially from those calculated for the design. They also are a major cause of excessive backslope subsidence resulting commonly from loss of ground due to inadequate seepage control or excessive wall deflection. Monitoring of wall performance during construction is an important consideration and permits economical contingency designs. With waterfront structures, the sequence of construction operations plays a significant role. For example, the removal of soft organic soils prior to wall construction reduces substantially the forces to be retained, as does backfilling from the seaward side toward the land, rather than the reverse.

9.2 RETENTION OF SOIL SLOPES

9.2.1 EVALUATION PROCEDURE

Wall Selection

If a cut is to be made into a soil slope and analysis has determined that a retaining structure is required, if a roadway embankment is to be constructed and a bridge abutment is required, or if walls are required due to space limitations or other reasons, it is first necessary to select a wall type and its preliminary dimensions.

There are a large variety of walls to choose from and a number of factors govern the selection. The most important criteria for selection are the wall height required and the earth pressures to be retained, since each type has limits to its application. In addition to wall deflections, earth pressures relate very closely to the soil type to be retained. Compacted embankments, for example, have high strengths and will impose relatively low earth pressures when cut for wall construction, whereas cohesive soils, especially where saturated, impose relatively high lateral earth pressures. For the construction of many walls, cohesive soils are removed and replaced with a free-draining granular backfill wherever practical, depending upon the availability of materials.

Other factors to consider include the space available for construction; the availability of materials, such as stone for buttresses; the stability of the slope during cutting; the settlement tolerances of the wall where the foundation materials may be somewhat compressible; and the necessity to limit wall deflections if the wall will support a structure that exists or will be constructed upslope. The time required for construction can also be a factor, especially with failing or already failed slopes requiring correction. The governing factor in all cases is economy of construction, which also implies that the wall will provide positive slope retention for some prescribed interval of time.

Analysis

Data Collection

Topographic sections are required of the entire slope to be retained. Geologic conditions to be defined include the distribution and physical properties of the materials in the slope, as well as in the wall foundation, and the availability of materials for backfill for which clean granular soils are preferred. Groundwater conditions are determined and the possibility for a substantial rise in the phreatic surface evaluated, particularly in view of existing slope stability or potential instability during cutting.

Earth-Pressure Calculations

Earth pressures provide the forces acting to cause a wall to deflect, overturn, or slide. Their distribution depends on wall flexibility. With the relatively rigid retaining structures used for slopes, they normally are considered to be hydrostatic in distribution. Their magnitude depends on the type of materials to be retained, wall movement, groundwater conditions, and imposed surcharges.

Analyses are performed using the limit equilibrium approach, which employs Rankine or Coulomb earth-pressure theories for gravity walls to determine the force system acting on the wall and the resistance for stability as provided by the weight of the wall or soil-wall system. For the case of anchored curtain walls, analyses are done by considering the problem as one of slope stability and employing the Culmann analysis to determine the force system to be resisted by anchors.

For gravity walls, the magnitude, location, and direction of the resultant of the earth-pressure forces are determined which are influenced by

- Slope inclination and height.
- Location and magnitude of surcharge loads.
- Wall type, configuration and dimensions, embedment depth, and magnitude and direction of movements. If no movement will occur, at-rest pressures prevail; if the wall moves away from the slope, active pressures govern; and if the wall pushes into the soil, passive earth pressures prevail.
- Soil parameters for natural soils or borrow backfill, including cohesive or granular soil characteristics, seepage conditions and pore-water pressures, whether drained or undrained conditions, and wall friction.

Wall-Stability Evaluations

The wall, wall-soil, or wall-anchor system must provide adequate resistance to the soil thrust either through weight or through an anchor system. Overturning, sliding along the base, and foundation bearing failure and settlement are evaluated. Overall slope stability is evaluated to ensure that a failure surface will not develop beneath the wall.

Structural Design

Structural design of the final wall is made after all of the forces acting and other factors have been considered. The wall is sized to resist overturning, and resistance to bending and shear is provided by either adequate wall thickness or reinforcing. Deep foundations may be necessary for support where inadequate bearing capacity or excessive settlement is predicted. Drainage provisions to eliminate or reduce seepage forces are an important consideration.

9.2.2 WALL TYPES

Categories

Gravity-type walls are the most common, providing slope retention by their weight which can consist of a concrete mass, concrete in combination with soil weight, or the weight of an earth mass alone. They are classed as rigid or flexible, but all are free to deflect at the top, thereby mobilizing active earth pressures.

Nongravity walls are restrained from movement and include basement walls, many bridge abutments, and anchored walls. *Anchored concrete curtain walls* are installed in a manner which provides slope retention during construction.

Soil reinforcement wall systems include bored piles (bored piers) and "root piles" which have special applications.

Rigid Concrete Gravity Walls

Concrete gravity walls normally provide moderately high capacities, but assurance that the potential failure surface will fall within a free-draining granular backfill (the most desirable condition) may require a substantial volume of excavation and backfill (Fig. 9.24). In any event, positive drainage must be provided and there is always the possibility that drain holes will clog, the prevention of which requires periodic maintenance. Types of concrete gravity walls, their common proportions and load diagrams are given in Fig. 9.1.

Simple gravity walls constructed of a mass of plain concrete can take no tensile stresses and are not economical for high walls.

Semigravity walls contain a small amount of reinforcing steel to reduce concrete volume and provide capacity for greater heights. They have been constructed to heights of 32 m [Kulhawy (1974)[2]].

Cantilever walls of reinforced concrete are generally economical for heights in the order of about 8 m (27 ft) because of the necessity to provide adequate strength in the stem-to-base connection. The weight of the earth acting on the heel is added to the concrete weight to provide resistance against active thrust. These walls often are designed for K_o earth pressures rather than K_a because the amount of bending required to cause K_a to act may cause wall cracking, especially where the wall is supported by strong foundation materials. Failure of a base-anchored cantilever wall is illustrated in Fig. 1.3.

Counterfort walls are cantilever walls strengthened with counterforts, which are generally more economical when heights greater than 8 to 12 m are required. Proportions are similar to those of the cantilever wall, and the counterforts are spaced at ⅓ to ⅔H, depending upon wall height. For walls greater than H = 10 m, spacing may be in the order of ½H. Added resistance against overturning and sliding may be obtained by installing anchors through the base.

Buttress walls are similar to counterfort walls except that the vertical braces are placed on the wall face rather than along the back.

Flexible Gravity Walls

Flexible gravity walls are often economical where earth pressures are not great, although reinforced earth walls can be constructed to have high capacity. Flexible walls have the inherent advantages of providing positive slope drainage and a tolerance to settlement.

A *rock-filled buttress* (Fig. 9.2) normally is constructed of large pieces of rock rubble, although compacted masses of earth have been used to stabilize moving slopes where space at the toe permits.

Gabion walls (Fig. 9.3) are constructed by filling wire baskets about 50 cm (20 in) on a side, with broken stone in the order of 10 to 15 cm (4 to 6 in) across. Retention is obtained from the stone weight and its interlocking and frictional strength. The wall face is battered at about 6° from the vertical and the maximum height is in the order of 10 m. They are constructed either with a stepped face or a stepped back. The latter condition is illustrated in the Figure 9.3.

Crib walls (Fig. 9.4), constructed by forming interconnected boxes from timber, precast con-

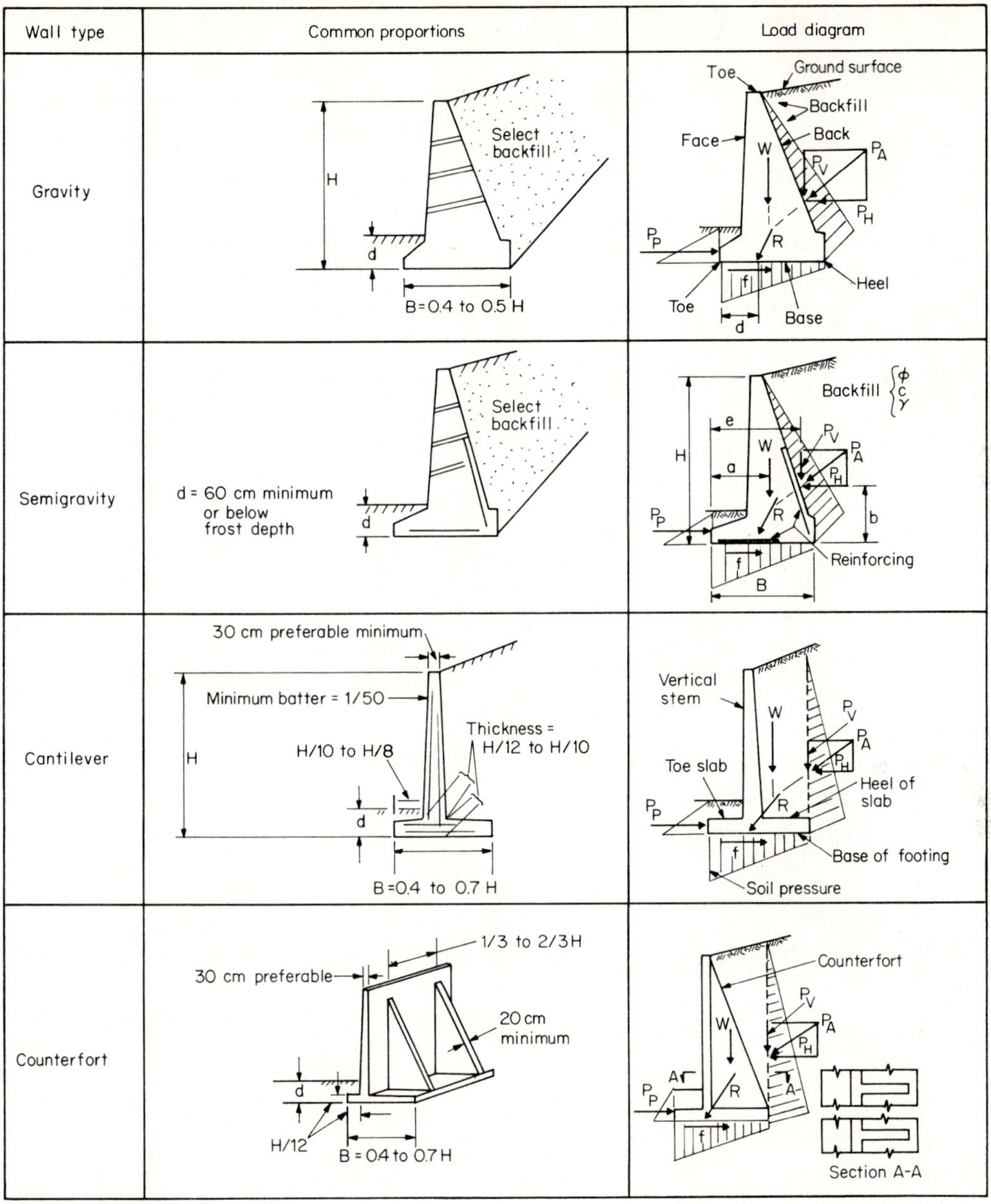

FIG. 9.1 Types of concrete gravity walls, their common proportions and load diagrams. [After NAVFAC (1982).[1]]

crete, or metal members and then filling the boxes with crushed stone or other coarse granular materials, are commonly used with compacted embankments on level ground. Precast members usually are in the order of 2 m (6 ft) in length, and the wall height is limited to an amount equal to twice the member length. Wall heights are increased by doubling box sections in depth (interlocking twin-celled crib walls), but high walls are very sensitive to transverse differ-

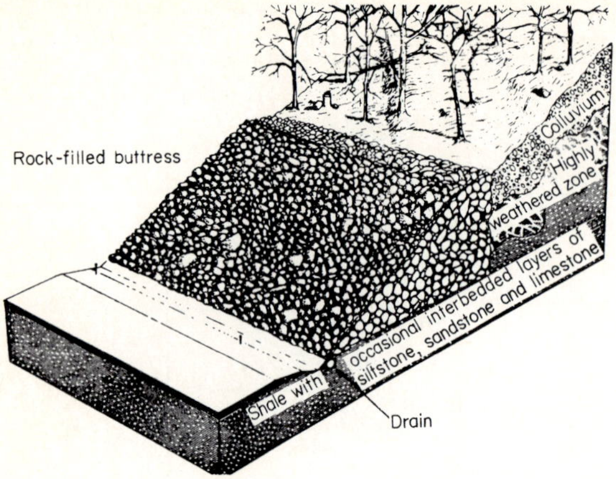

FIG. 9.2 Stabilization of cut in colluvium with rock-filled buttress. [From Royster (1973).[3]]

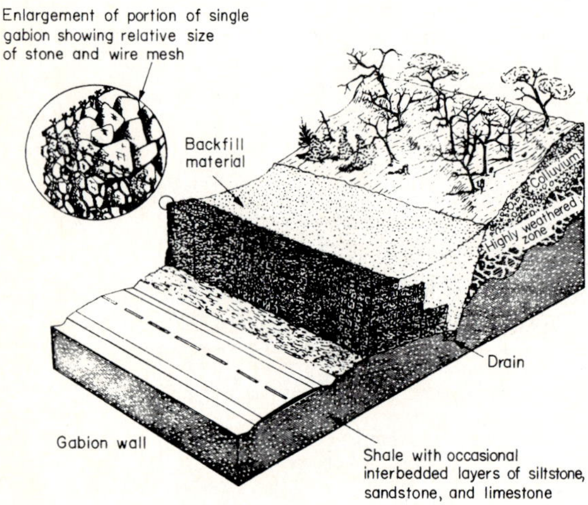

FIG. 9.3 Stabilization of cut in colluvium with gabion wall. [From Royster (1973).[3]]

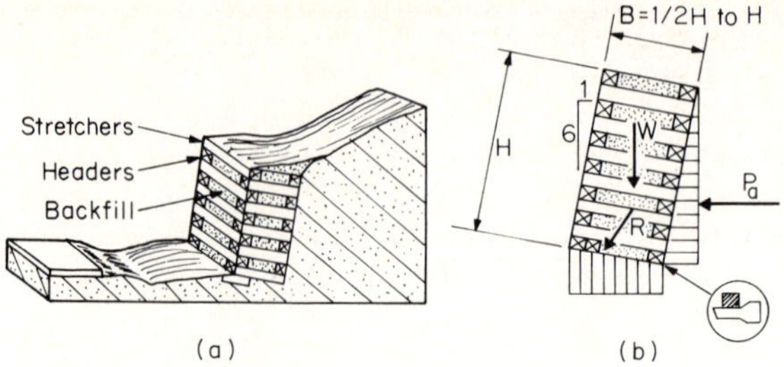

FIG. 9.4 Crib wall: (*a*) general view and (*b*) wall proportions and load diagram.

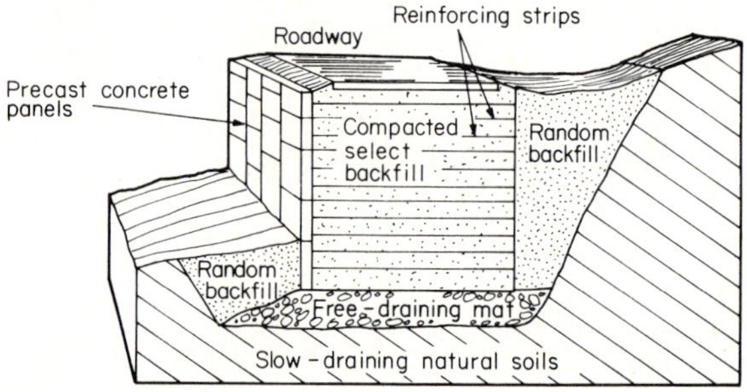

FIG. 9.5 General elements of a reinforced earth wall.

ential settlement and the weakness of the cross members precludes support of high surcharge loads. The load diagram is given in Fig. 9.4; the section resisting overturning is taken as a rectangle of dimension $H \times B$.

Reinforced earth walls (Fig. 9.5) are constructed of a compacted backfill into which strips or ties, usually of galvanized steel, are embedded to adsorb tensile forces developing in the fill. The strips are attached to a thin outer skin to retain the face, which is often made of precast concrete panels for durability and aesthetic reasons, and must have sufficient strength at all points, including connections, to resist lateral earth pressures. Ties must also be of adequate length and width to develop sufficient soil-tie friction. Lengths are usually 0.8 to 1.2 times the structure height, and to mobilize adequate friction requires backfill materials with a minimum ϕ = 25° and a maximum of 25% passing the no. 200 sieve [Gebney and McKittrick (1975)[4]]. Heights to 15 m (50 ft) or more are not uncommon. A reinforced earth wall supported on stone columns is shown in Fig. 5.38, and its use in conjunction with an earth buttress to stabilize a landslide is illustrated in Fig. 9.119 of Hunt (1984).[5] A state-of-the-art reference is *Symposium on Earth Reinforcement* (1978)[6] published by ASCE in New York.

Anchored Walls

Anchored concrete curtain walls consist of a thin "curtain" of reinforced concrete in the order of 20 to 30 cm (8 to 12 in) thick, which is constructed from the top down as the slope is cut in benches, and tied back with anchors penetrating into strong materials. They may also be constructed in the same manner as a normal wall and backfilled. They are useful for deep cuts where high retention capacity is required (Fig. 9.6), open excavations for structures *(see Art. 9.3.4)*, retention of sidehill fills (Fig. 9.7), and retention of embankments. Construction from the top down is a major advantage since the slope is always provided with support, whereas with gravity walls excavation into a slope for construction from the bottom up leaves the slope unsupported unless a bracing system or some other means of support is provided. Another advantage is that capacity can be increased even after the wall is completed by the installation of additional anchors. Anchored concrete curtain walls are discussed in detail in Art. 9.2.5.

Anchored precast concrete panels (Figs. 1.10, 9.8, and 9.32) are a recent innovation. They are installed along benches in a slope cut steeply and they readily conform to curved slopes.

Steel sheet-pile walls are formed by driving steel sheets into a slope or excavation *(see Art. 9.3.2)* and tying the sheets back with anchors. They are seldom used for sidehill slope retention because of their tendency to deflect and corrode and their relatively high costs. An application to retain a cut slope in colluvium in conjunction with other slope stabilization procedures is described by D'Appolonia et al. (1967)[7] [see also Art. 9.4.4 of Hunt (1984)[5]].

Soil Reinforcement

Bored piles or piers have been used on occasion to stabilize failed slopes during the initial stages of movement and cut slopes. Capacity is limited by the pile strength in bending as the pile acts as a cantilever unless it is anchored at the top. Unless the holes are hand-excavated, a large drilling machine is required for installation and access to the site may pose difficulties.

A *root pile system* consists of a three-dimensional lattice of small-diameter piles of precast or cast-in-place reinforced concrete, which are installed closely spaced to reinforce the earth mass. An advantage is that excavation is not required, but the system is relatively costly for the retention capacity obtained. Site access normally is required for large installation equipment. Some applications are described in Goughnour and DiMaggio (1978).[8]

9.2.3 EARTH PRESSURES

General Considerations

Forces Acting on a Rigid Gravity Wall

The total lateral thrust against a rigid gravity retaining wall is composed of thrust exerted by soil grains, thrust exerted by a pore water, and thrust exerted by the soil skeleton as a result of a surcharge. Two general theories are used to evaluate the forces acting on a wall: Coulomb's, dating from about 1776, and Rankine's, dating from 1857.

In Coulomb's concept (Fig. 9.9*a*), a wedge of earth with weight W_s is on the verge of sliding along a potential failure surface. It exerts an active thrust P_a on the wall and a force P_h on the sliding surface. (Surcharge and seepage forces would add to the driving forces.) Resistance to the active thrust is provided by wall weight W_w, a passive wedge of soil in front of the toe P_p, sliding resistance F and bearing resistance q along the base, and in most cases, friction δ acting between the wall back and the soil.

In Rankine's concept (Fig. 9.9*b*), the thrust against the wall of a wedge formed by a slip line results in a hydrostatic distribution of earth pressure P_a which is a function of the earth pressure coefficient K_a *(see Art. 3.2.3)*. It is resisted by the wall weight, passive earth pressure at the toe, sliding along the base, and base bearing capacity. Wall friction is not considered.

Either concept is used in practical analysis. Although the value for P_a computed from Coulomb's theory is the same as the value obtained from the Rankine theory for conditions of a horizontal backfill and $\delta = 0$, the Coulomb concept provides a slightly less conservative wall design when stability is evaluated employing wall friction. Both concepts have been modified and indeed, in certain instances, such as flexible sheet-pile walls, one may find analysis incorporating both, "Rankine" for active thrust and "Coulomb" for passive restraint.

FIG. 9.6 Anchored concrete curtain wall under construction from the top down. Maximum height in the photo is 15 m; final height was 25 m. João Monlevade, M. G., Brazil. Designed and constructed by Tecnosolo, S. A.

FIG. 9.7 Anchored concrete curtain wall, 10 m in height, pile-supported in rock, retaining fill over unstable colluvium. Highway BR 277, Parana, Brazil. Designed and constructed by Tecnosolo, S. A. [From Hunt (1984).[5] Used with permission of McGraw-Hill Book Company.]

FIG. 9.8 Slope stabilization with anchored premolded concrete plates which conform to the slope shape. New headquarters building for Banco Nacional Desenvolvimento Economico (BNDE), Rio de Janeiro. Designed and constructed by Tecnosolo, S. A. [From Hunt (1984).[5] Used with permission of McGraw-Hill Book Company.]

Earth-Pressure Stress States

As discussed in Art. 3.2.3, *at-rest earth pressures* (coefficient K_o) exist when the wall-earth system is rigid; i.e., no movement is occurring. The *active state of stress* exists when the top of the wall tilts even a slight amount and moves away from the soil, the soil stretches, pressures decrease, and as movement continues, eventually a shear failure occurs and a wedge is formed. Additional deformation does not reduce the earth pressure any further and it reaches a minimum value (coefficient K_a). The *passive state of stress* exists if the wall is forced into the soil. The force necessary to start movement is greater than K_o, and eventually a failure state is reached where a sliding wedge is formed. Additional deformation does not increase earth pressure any further and it is at its maximum value (coefficient K_p). Substantial deformation is required to mobilize the passive state.

Earth-Pressure Factors

Rankine's theory assumes that the slip lines that form as the soil is stretched or compressed are straight lines (Fig. 9.10*a*), which is correct if there is no adhesion between the wall and the soil. This occurs when the wall is perfectly smooth or if the wall and the soil move downward at the same rate, as may occur during settlement. Shearing stresses usually develop, however, resulting in the mobilization of *wall friction* which causes the slip lines to bend (Fig. 9.10*b*). The zone in which the slip lines approach log-spirals is termed the Prandtl zone. These curved slip lines usually have very little effect on active pressures but can have a very substantial effect on passive pressures.

The *magnitude* of the earth pressure is a function of the weight of the mass in the failing wedges and the strength mobilized along the failure surface. It depends mainly on the magnitude of the lateral strain and the direction of the force. Normally in the design of a gravity retaining wall it is standard practice to assume that the

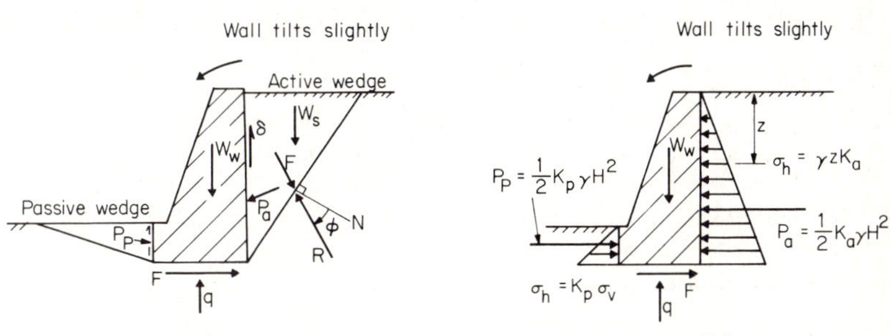

FIG. 9.9 Concepts of earth pressures acting against a rigid gravity retaining wall: (*a*) Coulomb wedges and (*b*) Rankine distribution.

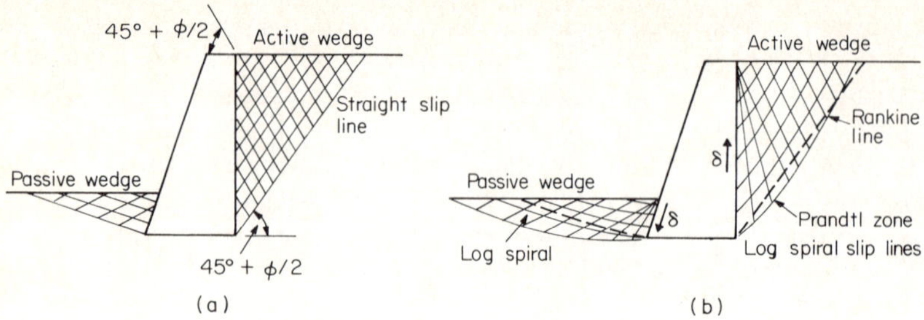

FIG. 9.10 Effect of wall friction on slip lines and the failure surface: (*a*) no wall friction and (*b*) wall friction. The effect is most significant in the passive wedge.

active state prevails because wall movements (yield) of only about 0.001H are enough to mobilize active pressures. If a rigid wall is unyielding, however, as is common in basement walls at the end of construction when they are supported structurally at top and bottom, the at-rest pressure prevails.

The earth-pressure coefficient K_a is normally about 0.3 for sands, and K_o is in the range of 0.4 to 0.5 *(see Art. 3.2.3)*; therefore, it is apparent that earth pressures for the at-rest state are significantly higher than for the active state. Earth pressures are not only selected based on wall movement. Active pressures for creeping soils are based usually on K_o, and for heavily compacted clays or swelling clays, K_o or higher. K_o for clay soils has been correlated with the overconsolidation ratio *(OCR, see Art. 3.2.3)*. Some typical values are given in Table 9.1 and Fig. 3.20.

Earth-pressure distribution, in general, does not affect the magnitude of the pressure, but its determination is important for certain types of structures, particularly those that are flexible, where distribution is not hydrostatic. Pressure distribution is related to the manner in which a wall moves or deflects. A rigid wall tilting or rotating about the bottom (Fig. 9.11*a*) has a hydrostatic distribution. If the wall is prevented from outward movement at the top but yields at the base, the lateral pressure near the top increases, and that at the bottom decreases as the wall moves (Fig. 9.11*b*). This condition is common with braced open cuts with top bracing only. If the wall translates, the pressure distribution approaches that shown in Fig. 9.11*c*.

Strength Parameter Selection

Cohesionless soils, such as clean sands with high permeability, have minimal excess hydrostatic pressures if the wall is well-drained, and effective stress parameters are used for all conditions. It is common practice, therefore, wherever practical, to excavate the entire potential failure zone behind a wall and backfill it with clean granular materials.

With *cohesive soils*, the selection of strength parameters for retaining structures in general depends on the time element as it does with slope

TABLE 9.1
COEFFICIENTS OF EARTH PRESSURE AT REST K_o*

Soil type	K_o
Normally consolidated clay	$(1 - \sin\phi)$
Compacted clay, hand-tamped	1.0–2.0
Compacted clay, machine-tamped (entire backfill)	2.0–6.0
Clay, overconsolidated	1.0–4.0
Sand, loosely dumped	0.5
Sand, compacted	1.0–1.5

*See also Fig. 3.20.

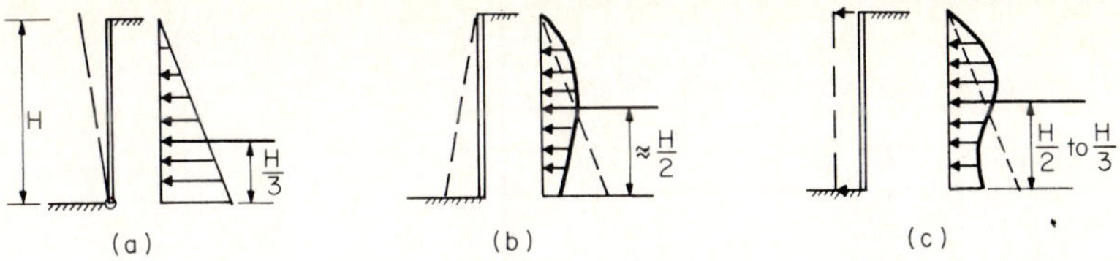

FIG. 9.11 Effect of wall movement on earth-pressure distribution in cohesionless soils: (*a*) rotation about base; (*b*) rotation about top; and (*c*) translation.

stability problems (*see Art. 8.2.1*). For walls supporting cut slopes, however, it is unusual for the wall to be in direct contact with natural soil. It is likely that the cohesive soil is placed as a compacted backfill, in which case design is based on K_o earth pressures, as will be discussed.

Seepage Forces

The thrust from seepage forces can be much larger than the soil alone, and therefore walls should be backfilled with a free-draining material or other means of drainage should be provided. Seepage pressures acting on the failure surface or against the wall are best analyzed graphically for some cases. Analysis may be based on either of two force systems which give equivalent results: (1) total weight, boundary pore pressures, and boundary effective stresses (usually the most convenient) or (2) buoyant weight, seepage force, and boundary effective stresses (*see Art. 2.3.2, Fig. 2.8*).

Rankine Concepts

Assumptions

Applications of the Rankine concept are based on the following assumptions:

- Ground surface is planar and either the wall back is vertical or full wedge can form in the backfill.
- No changes in shearing stresses between wall and soil (no wall friction).
- Failure surface develops along straight lines.
- Mohr's strength theory (*see Art. 3.2.1*) is valid and for the active state the horizontal stress

$$\sigma_h = \sigma_v N_\phi - 2c\left(\frac{1}{\sqrt{N_\phi}}\right)$$

where $N_\phi = K_p = \tan^2(45 + \phi/2)$ (Eq. 3.26, coefficient of passive stress) and $K_a = 1/K_p = \tan^2(45 - \phi/2)$ (Eq. 3.27, coefficient of active stress).

Total Lateral Earth Pressures: The Simple Case

One can find in the literature a large variety of expressions for Rankine earth pressures acting on rigid walls that provide for inclined backfills, surcharges, saturation, and cohesive soils as summarized in Table 9.3. The simple, but not uncommon, case of a wall with a vertical back has a horizontal backfill, no surcharge, and dry, free-draining granular backfill soils. For this case the horizontal stress at depth z (Fig. 9.9*b*) equals:

- At-rest condition

$$\sigma_h = \gamma z K_o \tag{3.3}$$

- Active state

$$\sigma_h = \gamma z K_a \tag{9.1}$$

- Passive state

$$\sigma_h = \gamma z K_p \tag{9.2}$$

The triangles in which the soil is in the active or passive state are known as the Rankine zones, in which the earth pressures have a hydrostatic distribution, often referred to as equivalent fluid pressures. In these zones the total earth pressures are equal to the product of the total soil weight and the earth-pressure coefficient, as follows:

$$P_o = 1/2H^2\gamma K_o \tag{9.3}$$

$$P_a = 1/2H^2\gamma K_a \tag{9.4}$$

$$P_p = 1/2H^2\gamma K_p \tag{9.5}$$

In the simple case the pressure acts at the centroid of the triangle, or $H/3$ (Table 9.3, No. 1), and represents the pressure exerted by the soil skeleton from its own weight. Some Rankine

TABLE 9.2
RANKINE EARTH-PRESSURE COEFFICIENTS FOR LEVEL BACKFILL

ϕ	10°	15°	20°	25°	30°	35°	40°
K_a	0.70	0.59	0.49	0.41	0.33	0.27	0.22
K_p	1.42	1.70	2.04	2.46	3.00	3.69	4.60

earth-pressure coefficients for a level backfill are given in Table 9.2.

Sloping Backfill, Surcharge, Submergence, and Cohesive Soils

The *sloping backfill case* (Table 9.3, No. 2) represents a geometric change from the simple case for P_a and P_p. K_a and K_p become functions of the slope angle β, expressed as K_{sa} and K_{sp} (Eqs. 9.6 and 9.7). P_a acts parallel to the slope, located at $y = H/3$, as expressed by Eq. 9.8.

The *uniform surcharge* q_s (Table 9.3, No. 3) acts on the soil skeleton with a thrust equal to q_sHK_a, which is added to the thrust from the soil skeleton due to its own weight (Eqs. 9.10 and 9.11). P_a is located at the centroid of the earth-pressure trapezoid.

Submergence (Table 9.3, Nos. 4 and 5) is considered by using the buoyant weight of soil γ_b for calculating P_a and adding to P_a the water pressure from the surface of the water table to the wall bottom (Eqs. 9.12, 9.13, and 9.14). The surcharge effect is added as $P_a + P_w + P_s$ where $P_s = q_sK_aH$.

Cohesive soils are given in Table 9.3, Nos. 6, 7, 8, and 9. Because of capillary forces and slow drainage conditions, cohesive soils normally are considered to be saturated, or nearly so, when a wall is placed directly against a cut face. The shear strength, and therefore the earth pressure, vary with time. Undrained conditions exist during and shortly after construction (short term), and drained conditions exist for some time after construction (long term) unless heavy rains occur. Values for the coefficients are given in Art. 3.2.3: drained conditions—K_p (Eq. 3.31) and K_a (Eq. 3.32); undrained conditions—K_p (Eq. 3.33) and K_a (Eq. 3.34). The case of a clay backfill is discussed in subsequent articles.

Tension cracks form in cohesive soils as a result of wall deflection, along which the soil has no shearing resistance and $P_a = 0$. For the drained state, the height of the tension crack is found from Eq. 9.1 where K_a is from Eq. 3.22, and setting $\sigma_h = 0$, as

$$H_1 = \frac{2\bar{c}}{\gamma_b\sqrt{K_a}} \tag{9.16}$$

and for the undrained state is found as

$$H_1 = 2s_u/\gamma_t \tag{9.18}$$

Theoretically, the wedge of soil formed by the tension crack adds no pressure to the wall and therefore would be subtracted from the pressure in the wedge defined by H as given by Eqs. 9.15 and 9.17. This condition, causing reduced earth pressures, might be valid over a short term but should not be considered for the long term. Not only is it probable that the cracks will fill with water exerting hydrostatic pressures, but it is also significant that for pressures corresponding to the active case to remain in effect for clays against a retaining wall, continuous wall movement must occur. Since continuous movement should not be permitted, it is prudent, therefore, when the normally undesirable condition of a clay backfill is required, to design the wall for at-rest pressures in accordance with OCR relationships and Eq. 9.3.

For granular soils with cohesion, such as a sand-clay mixture, in which a tension crack might develop, P_a may be determined as in Table 9.3, No. 1, taking a value for $K_a = 0.5$, and ignoring the tension crack.

Tension cracks do not act in the passive state, and the passive resistance is increased by cohesion along the failure surface as given in Eqs. 9.19 and 9.20.

TABLE 9.3
SUMMARY OF RANKINE EXPRESSIONS FOR EARTH PRESSURES FOR VARIOUS CONDITIONS

Conditions	Pressure distribution	Pressure at depth Z	Total pressure on wall	
1. Cohesionless dry soils, horizontal backfill	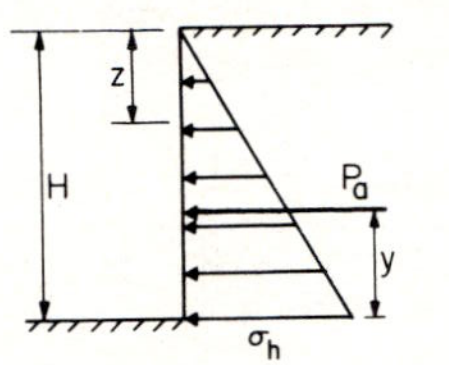	$\sigma_h = \gamma z K_a$ $\sigma_h = \gamma z K_p$ $\sigma_h = \gamma z K_o$ $y = H/3$ γ = unit soil weight $K_a = \tan^2(45 - \phi/2)$ (Eq. 3.27) $K_p = \tan^2(45 + \phi/2)$ (Eq. 3.26)	$P_a = \frac{1}{2}\gamma H^2 K_a$ $P_p = \frac{1}{2}\gamma H^2 K_p$ $P_o = \frac{1}{2}\gamma H^2 K_o$	(9.4) (9.5) (9.3)
2. Cohesionless dry soils, sloping backfill	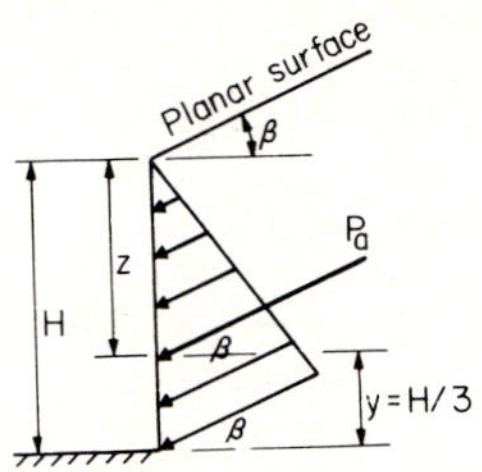	$\sigma_h = \gamma z K_{sa}$ $\sigma_h = \gamma z K_{sp}$ where $k_{sa} = \cos\beta \dfrac{\cos\beta - \sqrt{\cos^2\beta - \cos^2\phi}}{\cos\beta + \sqrt{\cos^2\beta - \cos^2\phi}}$ $k_{sp} = \cos\beta \dfrac{\cos\beta + \sqrt{\cos^2\beta - \cos^2\phi}}{\cos\beta - \sqrt{\cos^2\beta - \cos^2\phi}}$	$P_a = \frac{1}{2}\gamma H^2 K_{sa}$ $P_p = \frac{1}{2}\gamma H^2 K_{sp}$	(9.8) (9.9) (9.6) (9.7)
3. Cohesionless dry soils, horizontal backfill with surcharge	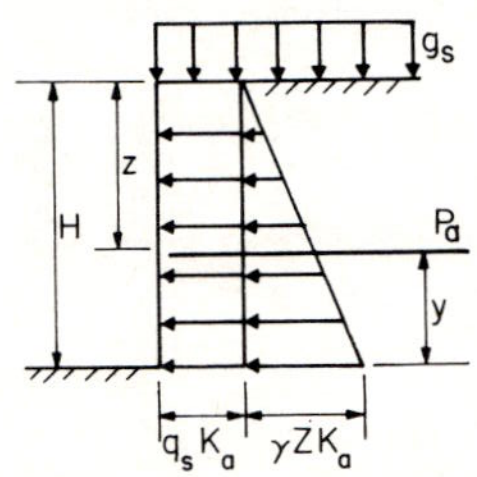	$\sigma_h = \gamma z K_a + q_s K_a$ $\sigma_h = \gamma z K_p + q_s K_p$ $y = \dfrac{H}{3}\left[\dfrac{2(q_s K_a) + (q_s K_a + \gamma H K_a)}{(q_s K_a) + (q_s K_a + \gamma H K_a)}\right]$ (centroid of trapezoid)	$P_a = \frac{1}{2}\gamma H^2 K_a + q_s H K_a$ $P_p = \frac{1}{2}\gamma H^2 K_p + q_s H K_p$	(9.10) (9.11)

(continued)

TABLE 9.3
SUMMARY OF RANKINE EXPRESSIONS FOR EARTH PRESSURES FOR VARIOUS CONDITIONS (continued)

Conditions	Pressure distribution	Pressure at depth Z	Total pressure on wall
4. Groundwater at surface, cohesionless soil	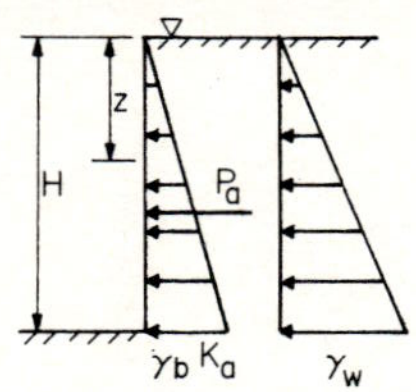	$\sigma_h = \gamma_b z K_a + \gamma_w z$ where γ_b = buoyant weight of soil γ_w = unit weight of water $\gamma_t = \gamma_b + \gamma_w$	$P_a = \frac{1}{2}\gamma_b H^2 K_a + \frac{1}{2}\gamma_w H^2$ (9.12) $P_p = \frac{1}{2}\gamma_b H^2 K_p + \frac{1}{2}\gamma_w H^2$ (9.13)
5. Groundwater at depth H_1, cohesionless soil	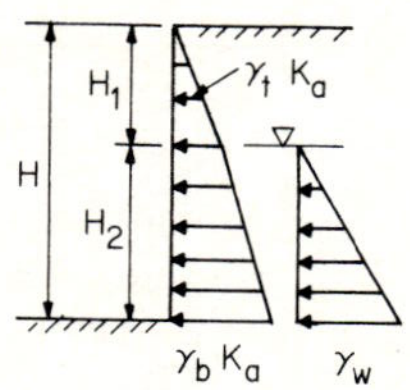	σ_h varies with GWL $\sigma_h = \gamma_t z K_a$ above GWL $\sigma_h = \gamma_b z K_a + \gamma_w z$ below GWL	$P_a = \frac{1}{2}\gamma_t H_1^2 K_a + \frac{1}{2}(H_1\gamma_t + H_2\gamma_b)K_a + \frac{1}{2}\gamma_w H_2^2$ (9.14)
6. Cohesive soil, drained conditions, active state	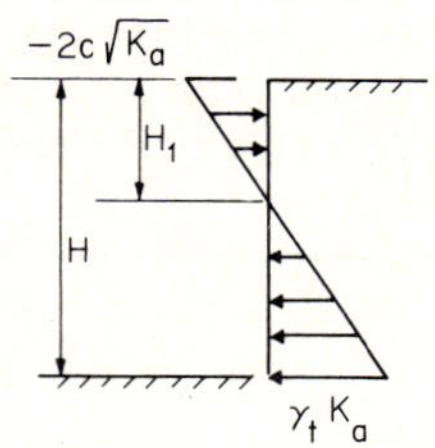	$\sigma_h = \gamma_t z K_a - 2\bar{c}\sqrt{K_a}$ where $K_a = \tan^2(45 - \bar{\phi}/2)$ (Eq. 3.27)	$P_a = \frac{1}{2}\gamma_t H^2 K_a$ (9.15) s $2\bar{c}\left(H - \frac{H_1}{2}\right)\sqrt{K_a}$ where H_1 = tension crack depth $H_1 = 2\bar{c}/\sqrt{K_a}\gamma$ (9.16) (Note: Eq. 9.15 is theoretical pressure. In practice, design for K_o and Eq. 9.3. See text.)

7. Cohesive soil, undrained conditions, $\phi = 0$, active state	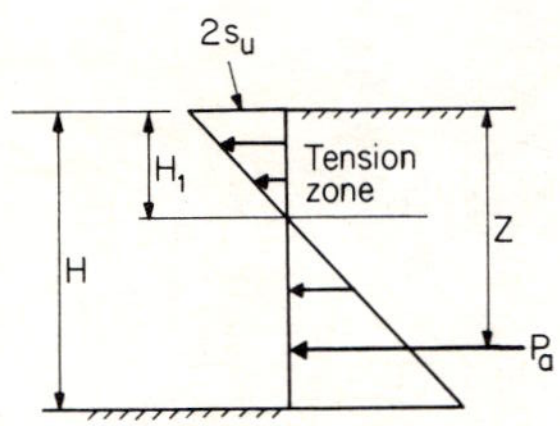	$\sigma_h = \gamma z - 2s_u$ where s_u = undrained shear strength $K_a = 1 - 2s_u/\gamma h$ (Eq. 3.34) (9.18) $H_1 = 2s_u/\gamma t$	$P_a = \frac{1}{2}\gamma_t H^2 - 2s_u H_1$ (9.17) (Note: Eq. 9.17 is theoretical pressure. In practice design for K_o and Eq. 9.3. See text.)
8. Cohesive soil, drained conditions, passive state	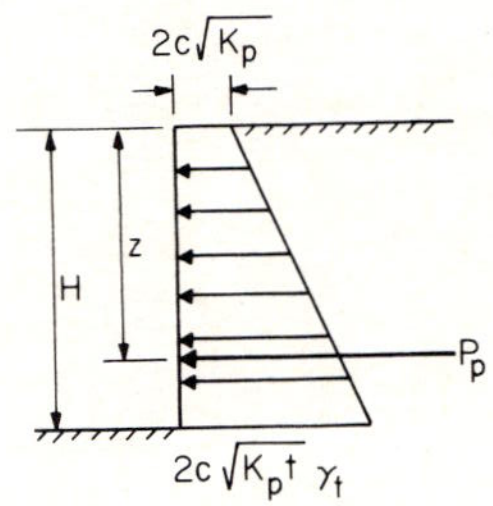	$\sigma_h = \gamma_t z K_p + 2\bar{c}\sqrt{K_p}$ where $K_p = \tan^2 (45 + \phi/2)$ (Eq. 3.26)	$P_p = \frac{1}{2}\gamma_t H^2 K_p + 2\bar{c}\sqrt{K_p}H$ (9.19)
9. Cohesive soil, undrained conditions, $\phi = 0$, passive state	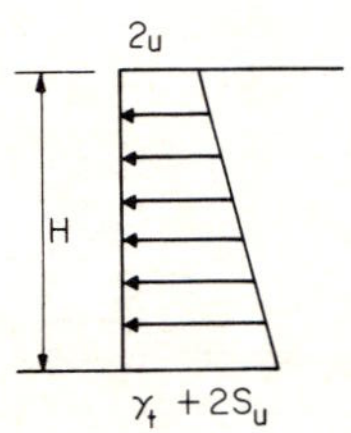	$\sigma_h = \gamma_t + 2s_u$ where $K_p = 1 + \dfrac{2s_u}{\gamma h}$ (Eq. 3.33)	$P_p = \frac{1}{2}\gamma_t H^2 + 2s_u H$ (9.20)

Irregular surfaces on walls or slopes are best analyzed by graphical solutions, as discussed in the following section.

Coulomb or Wedge Concept

Considerations

In Coulomb's classical earth-pressure theory, dating from about 1776, the limit equilibrium of an entire wedge of soil behind the wall is considered. The wedge is bounded by the wall and a planar surface passing through the heel of the wall and the resultant force on the wall which was chosen arbitrarily by Coulomb to act at an oblique angle now considered as the angle of *wall friction* (see following section). Figure 9.12 illustrates the forces acting on the wedge and the force polygon. Values for wall friction for various dissimilar materials are given on Table 9.4.

Solutions: Cohesionless Soils and Planar Surfaces

The wedge is retained satisfactorily by the wall when the forces acting on the wedge (W_s, P_a, and R in Fig. 9.12) are in equilibrium. Assumptions are made for the wall friction angle δ and soil friction ϕ; these determine the lines of action for R and P_a. The weight of an assumed wedge is calculated and the force polygon constructed to find the magnitude of P_a. Trial-and-error solutions are required for differing angles of θ to find the maximum value for P_a, corresponding to the *critical wedge*.

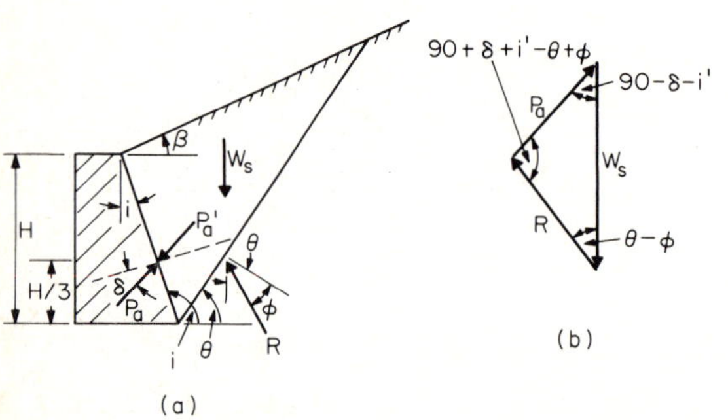

FIG. 9.12 Coulomb's conditions: (*a*) forces acting on an active wedge and (*b*) force polygon for cohesionless soil.

For the case of a uniform backslope (horizontal or inclined) and a planar wall back (vertical or inclined), P_a for the critical wedge is found from

$$P_a = \tfrac{1}{2}\gamma H^2 \times \left[\frac{\sin(i-\phi)\left(\dfrac{1}{\sin i}\right)}{\sqrt{\sin(i+\delta)} + \sqrt{\dfrac{\sin(\phi+\delta)\sin(\phi-\beta)}{\sin(i-\beta)}}}\right]^2 \tag{9.21}$$

where the portion of the expression in the brackets is often given as K_a. For the passive case,

$$P_p = \tfrac{1}{2}\gamma H^2 \times \left[\frac{\sin(i+\phi)\left(\dfrac{1}{\sin i}\right)}{\sqrt{\sin(i-\delta)} - \sqrt{\dfrac{\sin(\phi+\delta)\sin(\phi+\beta)}{\sin(i-\beta)}}}\right]^2 \tag{9.22}$$

where the portion of the expression in the brackets is often given as K_p. It can be seen that K_a and K_p are functions of slope inclination β, wall back inclination i, wall friction δ, and the soil friction angle ϕ. A number of variations of Eq. 9.21 can be found in the literature, usually resulting from differing nomenclature and definitions for the angle between the failure surface and the wall back. For the case of a horizontal backfill, vertical wall back, and wall friction = 0, the values for K_a computed from Eq. 9.21 are the same as those given in Table 9.2 for the Rankine solution. For example, if δ = 30° for these conditions, K_a = 0.297 for ϕ = 30° as compared with K_a = 0.33 from Table 9.2, resulting in a slight reduction in K_a. If the slope of the wall back is increased 20° outward, K_a increases from 0.297 to 0.501.

The *critical failure surface* for the case of a sloping backfill, vertical wall back, and wall friction = 0 can be found from

$$\cot(90-\theta) = \tan\phi + \sqrt{1+\tan^2\phi - \frac{\tan\beta}{\sin\phi\cos\phi}} \tag{9.23}$$

for either the active or passive case. In the active case, slope angle i is positive (sloping upward);

TABLE 9.4
ULTIMATE FRICTION FACTORS AND ADHESION FOR DISSIMILAR MATERIALS*

Structural material	Rock or soil material†	Friction factor tan δ	Friction angle δ, degrees
Mass concrete or masonry	Clean sound rock	0.70	35
	GW, GP	0.55–0.60	29–31
	GM, GC, SW, SP	0.45–0.55	24–29
	SM, SC	0.35–0.45	19–24
	ML (nonplastic)	0.30–0.35	17–19
	CL (stiff to hard)	0.40–0.50	22–26
	CL (medium stiff to stiff)	0.30–0.35	17–19
Steel sheet piles	Well-graded rock fill with spalls	0.40	22
	GW, GP, SW, SP	0.40	22
	Hard rock fill (single size), GM, SW	0.30	17
	GM, GC, SM	0.25	14
	ML (nonplastic)	0.20	11
Formed concrete or concrete sheet piling	Well-graded rock fill with spalls	0.40–0.50	22–26
	GW, GP, SW, SP	0.40–0.50	22–26
	Hard rock fill (single size), GW, SW	0.30–0.40	17–22
	GM, GC, SM	0.30	17
	ML (nonplastic)	0.25	14
Various	Masonry on masonry: crystalline rocks		
	Dressed soft rock on dressed soft rock	0.70	35
	Dressed hard rock on dressed soft rock	0.65	33
	Dressed hard rock on dressed hard rock	0.55	29
	Masonry on wood (cross grain)	0.50	26
	Steel on steel at sheet-pile interlocks	0.30	17

Cohesive soils s_u, tsf	Adhesion s_a, tsf
Very soft (0–0.125)	0–0.125
Soft (0.125–0.250)	0.125–0.250
Medium stiff (firm) (0.25–0.50)	0.25–0.375
Stiff (0.50–1.0)	0.375–0.475
Very stiff (1.0–2.0)	0.475–0.650

*After NAVFAC (1982).[1]

†See Table 2.5.

for the passive case, *i* is negative (sloping downward), and the right-hand portion of the expression begins with a negative value ($-\tan\phi$).

Soils with Cohesion and Irregular Surfaces

For soils with cohesion and friction acting, solution is by trial and error to find the critical wedge. The forces acting on a trial wedge and the force polygon are shown in Fig. 9.13, where a tension crack reduces the length of the failure surface. Usually the effect of the tension crack is ignored, as discussed in the previous section *(see also discussion in Art. 8.1.3)*. The total cohesion acting along the failure surface inclined at θ is equal to cL, and a total value for adhesion a acting along the wall back is taken as $A = aH$. Values for adhesion can be estimated from Table 9.4. (There is a large degree of uncertainty when analyzing walls with cohesive backfills with either the Rankine or Coulomb methods, and cohesive backfills should be avoided whenever possible, or generous values for FS employed.)

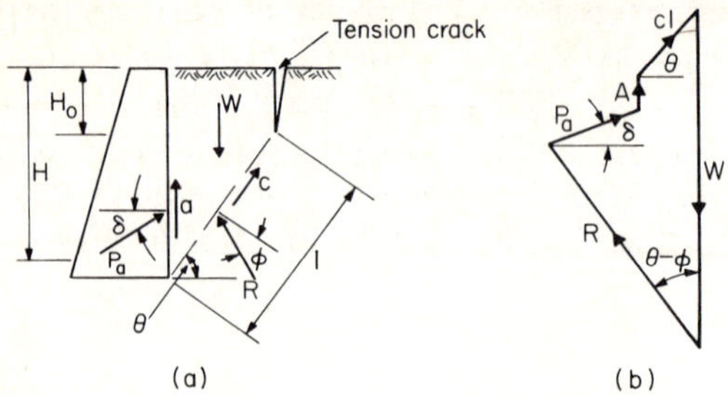

FIG. 9.13 Active forces from a wedge of soil with cohesion acting on a (*a*) simple retaining wall and (*b*) force polygon.

Irregular shapes of a wall back or the backfill slope are best analyzed by graphical solutions, a number of which have been developed for various conditions. These can be found in Taylor (1948)[9] (the Poncelet construction), Terzaghi and Peck (1967)[10] (Culmann's construction), Teng (1962),[11] and Kezdi (1975).[12]

Comments

Coulomb solutions consider straight slip lines and therefore do not truly satisfy the geometry of the failure surface (Fig. 9.10). Their use is satisfactory, however, for active pressures but can be grossly in error for passive pressures as discussed in the next section. In all straight-line solutions the location of P or the distribution of pressures cannot be found. Laboratory model tests and experience have shown that the pressure magnitudes are representative for design. With Coulomb's method, results are not more than 12% on the unsafe side for $-15° < i < +15°$ and $\theta < \beta < \phi$. The Caquot-Kerisel factors given in the following section are more nearly correct as they are based on curved slip surfaces.

Wall Friction

Description

As the wall in Fig. 9.12 moves outward, the active wedge tends to move downward, causing tangential stresses that cause the resultant force on the wall to be inclined at an angle normal to the wall termed the angle of wall friction δ. The back of the wall is no longer a principal plane and the slip lines in the vicinity of the wall are curved as log-spirals (Fig. 9.10*b*) causing different states of stress in different portions of the backfill.

The *coefficient of wall friction*, the value of angle δ between concrete and soil, usually is taken as equal to ϕ, or at times $\frac{2}{3}\phi$, depending upon the roughness of the wall and the degree of conservatism desired; it is used with Eq. 9.21 or 9.22. Ultimate values for wall friction and adhesion for dissimiliar materials are given in Table 9.4.

Effect on Active Pressure

In general, the effect of wall friction on active pressure is a slight reduction, as mentioned in the preceding section, and to ignore it is on the safe side of stability in most cases. If, however, the wall and the backfill settle together, δ could approach zero; if the wall settles more than the backfill, δ could even be negative, in which case K_a increases, resulting in unconservative analysis if the negative wall friction is ignored.

Effect on Passive Resistance

The theoretical predictions regarding passive resistance as a function of wall friction are not as well confirmed by experiment as are those regarding active thrust, and they must be used with less confidence. In fact, P_p cannot be counted upon at the toe of a rigid retaining wall and is usually ignored in design, although in flexible walls it is an important element of earth pressure. To fully mobilize passive resistance, a large amount of displacement is required, ranging from about 2% strain for dense sands to as much as 15% strain for loose sands [Lambe and Whitman (1969)[13]].

In the passive state in Fig. 9.10*b*, as the structure moves in the soil, the soil moves upward, causing substantial wall friction, and curved slip lines form against the wall. As seen in the figure, the log-spiral surface provides a much longer failure surface than does the Rankine wedge; therefore, it would seem that resistance for curved surfaces would be higher. Actually, resistance will be much smaller than it appears because the principal stresses change direction with the curved lines so as to act nearly vertically on an assumed straight slip line [Tschebotarioff (1973)[14]]. For the passive case, using straight-line failure surfaces results in resistance being significantly *overestimated* and the difference increases with increasing wall friction. It is most significant for δ greater than about 14°, or δ greater than about $\phi/3$.

Chart Solution for K_a, K_p, and Wall Friction

The charts developed by Caquot and Kerisel (1949)[15] provide values for K_a and K_p modified for values of δ to conform to log-spiral surfaces and are particularly useful in finding values for K_p for analysis of flexible retaining structures. For the case of a vertical wall, values for K_a and K_p for various values of ϕ and β are given in Fig. 9.14. These values are used in the Rankine equations, Eqs. 9.4 and 9.5. The horizontal component of earth pressure P_h can be found from $K_a \cos \delta$ and $P_v = K_a \sin \delta$.

It is to be noted that the curves in the figure are for the case of $\delta/\phi = -1.0$. The reduction factor R for other ratios is also given in the figure. Charts for walls with inclined backs can be found in Caquot and Kerisel (1949),[15] NAVFAC (1982),[1] and Tschebotarioff (1973).[14] Values for K_p with wall friction, $\beta = 0$, and $i = 0$, interpreted from Fig. 9.14, are given in Table 9.5.

Seepage Pressures

Significance

As in the case of slopes, seepage pressures are the most common cause of retaining-wall failures. Terzaghi (1936)[16] described the effect of a heavy rainfall on wall stability and noted that most failures of walls with fine-grained backfill occurred under these conditions. He observed that a wall with a clean sand backfill would remain stable as would a wall with an impervious backfill. In the latter case there would be insignificant penetration of the rainwater into the backfill.

Groundwater affects stability as follows:

- Weight of soil behind the wall is increased by saturation or partial saturation, thereby increasing earth pressures.
- Uplift on the sliding surface reduces sliding resistance.
- Uplift on the wall base reduces the effective wall weight (or the effective vertical stress at the wall base) and active thrust resistance.
- Passive pressure reduction at the toe results from reduction of effective soil weight on account of uplift pressures, and in the extreme case can be lost entirely due to piping (*see Art. 2.3.1*).

Analysis

Analysis of seepage forces is performed with flow nets (*see Art. 2.3.2*). In Fig. 9.15*a* the soil is shown saturated from the surface of the backfill to the surface at the wall toe. The equipotential lines give the hydrostatic pressure increases in 10 increments between the two points. The resultant pore-water force U at point a on the failure surface is equal to $(\frac{3}{10})h\gamma_w$. It is an added force on N (Fig. 9.16*b*) but, because it is a pore pressure, it produces no frictional resistance. The uplift pressure on the wall base also can be found from the flow net.

Pore pressures are zero throughout the backfill if a *sloping drain* is installed to carry off water seeping through the backfill, and the wall has weep holes to permit drainage (Fig. 9.15*b*). For this case, flow-net analysis is not necessary and earth pressures are determined by the Rankine or Coulomb solutions. The backfill must be considered saturated for a steady rainfall condition, however, and the pertinent soil unit weight employed.

Vertical back drains (Fig. 9.15*c*) result in a thrust against the wall greater than that for sloping drain conditions, and pore pressures that develop under steady rainfall conditions must be analyzed by flow-net construction. Various positions are assumed for a trial wedge and the pore pressures along each potential failure surface are determined to find the largest thrust. In the analysis of stresses, drained conditions are assumed and effective stress parameters used.

The determination of active pressures with pore pressures acting on a potential failure surface is illustrated in Fig. 9.16. The flow net is drawn to scale about the wall in Fig. 9.16*a*, the head drops h_p at the intersection of each equipotential line and the failure surface are scaled, and the average for each head drop $[(h_{p2} - h_{p3})/2$, for

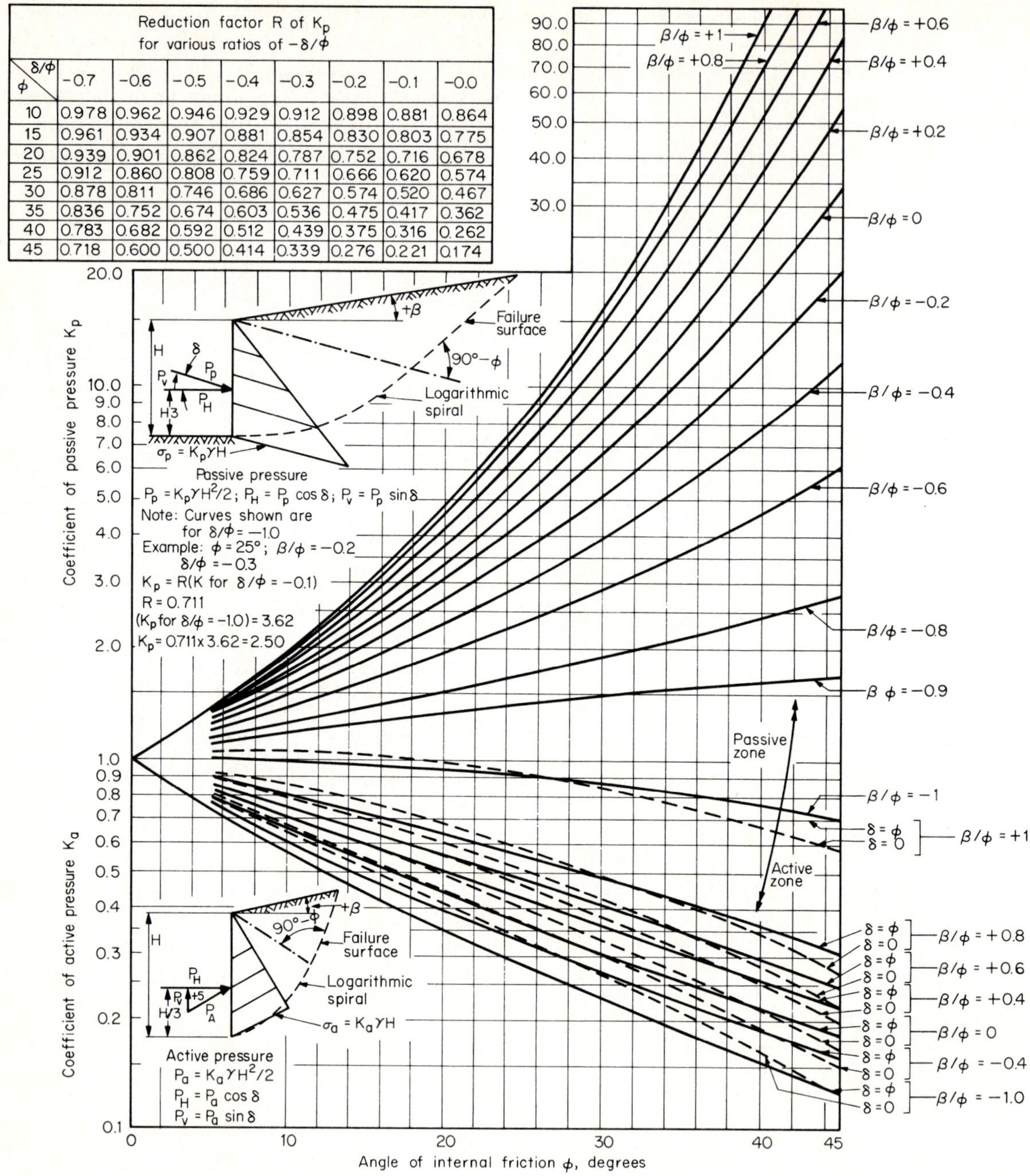

Reduction factor R of K_p for various ratios of $-\delta/\phi$

ϕ \ δ/ϕ	−0.7	−0.6	−0.5	−0.4	−0.3	−0.2	−0.1	−0.0
10	0.978	0.962	0.946	0.929	0.912	0.898	0.881	0.864
15	0.961	0.934	0.907	0.881	0.854	0.830	0.803	0.775
20	0.939	0.901	0.862	0.824	0.787	0.752	0.716	0.678
25	0.912	0.860	0.808	0.759	0.711	0.666	0.620	0.574
30	0.878	0.811	0.746	0.686	0.627	0.574	0.520	0.467
35	0.836	0.752	0.674	0.603	0.536	0.475	0.417	0.362
40	0.783	0.682	0.592	0.512	0.439	0.375	0.316	0.262
45	0.718	0.600	0.500	0.414	0.339	0.276	0.221	0.174

FIG. 9.14 Active and passive coefficients with wall friction for vertical wall and sloping backfill. Curves are for $-\delta/\phi = -1.0$. Use reduction factors for other ratios of δ/ϕ. See also Table 9.5 for $\beta = 0$, $i = 0$. [After Caquot and Kerisel (1949)[15]; from NAVFAC (1982).[1]]

TABLE 9.5
VALUES FOR K_p WITH WALL FRICTION δ ($\beta = 0$, $i = 0$)*

δ	ϕ = 20	21	22	23	24	25	26	27	28	29	30	31	32	33	34	35	36	37	38
0	2.04	2.12	2.20	2.28	2.37	2.46	2.56	2.66	2.77	2.88	3.00	3.12	3.25	3.39	3.54	3.70	3.86	4.02	4.18
11	2.65	2.78	2.91	3.06	3.20	3.40	3.59	3.75	3.91	4.16	4.33	4.58	4.82	5.09	5.36	5.73	6.09	6.45	6.81
14	2.83	2.99	3.15	3.30	3.44	3.66	3.88	4.06	4.24	4.52	4.72	5.04	5.28	5.59	5.90	6.33	6.75	7.18	7.61
17					3.67	3.93	4.18	4.39	4.59	4.90	5.13	5.45	5.76	6.05	6.34	6.97	7.46	7.94	8.42
18							4.28	4.50	4.71	5.03	5.27	5.60	5.94	6.31	6.68	7.19	7.70	8.20	8.70
19								4.61	4.83	5.17	5.42	5.76	6.11	6.50	6.89	7.42	7.95	8.48	9.01
20	3.01									5.27	5.56	5.92	6.29	6.69	7.10	7.67	8.22	8.78	9.34
21		3.25									5.71	6.09	6.47	6.89	7.31	7.90	8.48	9.06	9.64
22			3.50										6.65	7.09	7.54	8.15	8.76	9.36	9.96
23				3.75										7.29	7.75	8.40	9.04	9.67	10.30
24					4.00				Above $-\delta/\phi = -0.7$						7.93	8.65	9.32	9.98	10.64
25						4.35											9.61	10.31	11.01
26							4.75											10.59	11.38
27			$-\delta/\phi = -1.0$					5.13											
28									5.50										
29										6.00									
30											6.50								
31												7.10							
32													7.80						
33														8.60					
34															9.40				
35																10.50			
36																	11.70		
37																		13.00	
38																			14.30

*From Fig. 9.14.

example] is multiplied by the distance over which it acts along the failure surface ΔL. The products of the $h_{p(avg)} \times \Delta L$ for all head drops from h_{p1} to h_{p5} are summed and multiplied by γ_w to obtain the resultant pore-water force U. The forces acting on an active wedge are shown in Fig. 9.16*b*. To construct the force polygon (Fig. 9.16*c*), a value for θ is selected, and the forces with known magnitudes and directions, W and U, are plotted to some convenient scale. The directions of forces P_a, N, and τ are known, permitting the closure of the polygon and the determination of the magnitude of P_a. The entire procedure is repeated for other values of θ, a scaled plot of P_a for various values of θ is prepared as in Fig. 9.16*d*, and the maximum value for P_a is found.

For the case of a vertical drain behind a wall resting on an *impervious stratum*, the seepage pressure U acting on the failure surface may be obtained from the diagram in Fig. 9.17. U is found from factor f which is a function of the angle of the failure surface with the wall (α) times the water height in the wedge [Gray (1958)[17]].

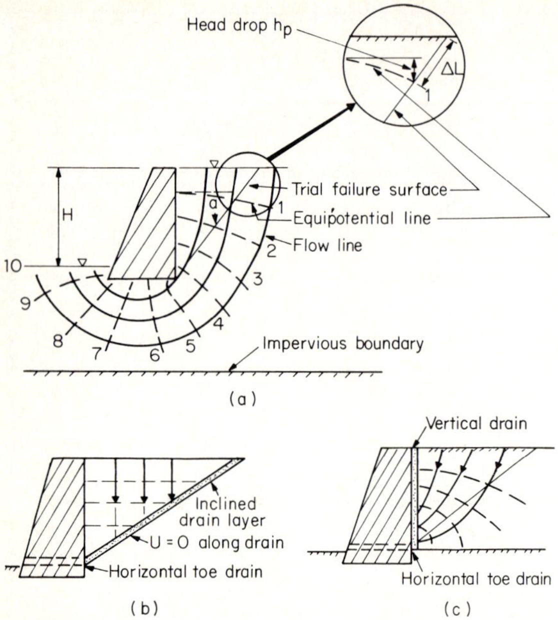

FIG. 9.15 Seepage conditions around a retaining wall during steady rainfall: (*a*) seepage around wall without drains; (*b*) seepage with drainage layer, no pore pressures within backfill; and (*c*) seepage with vertical drain.

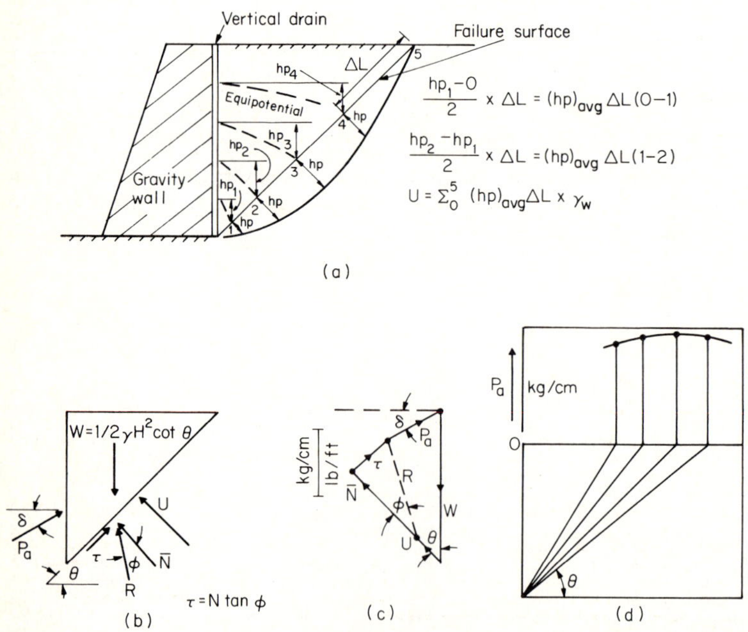

FIG. 9.16 Determination of active pressures by force polygons with pore-water pressure acting on failure surface: (*a*) determination of pore pressures with flow-net analysis, (*b*) forces acting on active wedge, (*c*) force polygon to determine active pressure, (*d*) determination of maximum P_a from several trial wedges. [From Lambe and Whitman (1969).[13] Used with permission of John Wiley & Sons, Inc.]

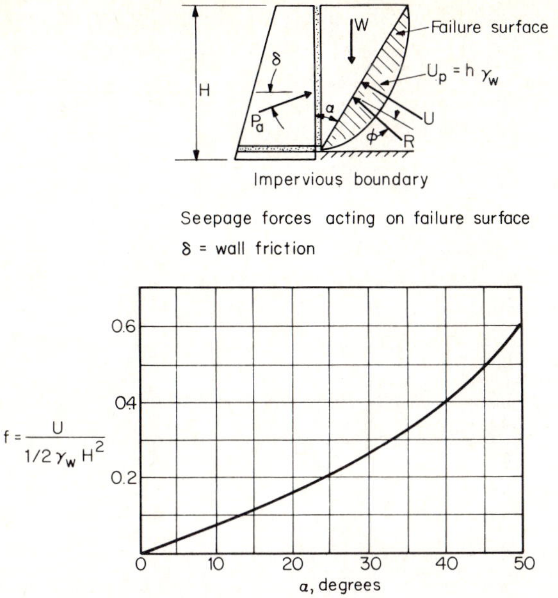

FIG. 9.17 Diagram for estimating seepage pressures on failure surface behind wall with vertical back resting on impervious stratum with vertical drain behind wall. Note that angle α is angle between wall and failure surface. [After Gray (1958).[17]]

Control

Seepage pressures are controlled by a free-draining backfill (Fig. 9.24) and drains through and behind the wall, or by backfill drains which may be either sloping drains installed in the backfill base or vertical drains installed along the back of the wall. Several schemes are illustrated in Fig. 9.18. Consideration should be given to the possibility of high lateral forces developing from frost thrust in cold climates. Pockets of coarse gravel at the weep hole intake provide partial protection against ice lenses.

Weep holes should be at least 3 in (8 cm) in diameter and provided with a filter of granular soil or fabric at the wall back to prevent backfill erosion. Horizontal spacing normally ranges from 4 to 10 ft (1.5 to 3 m), and vertical spacing is about 5 ft (1.5 m).

Surcharge Loadings

Sources

Some sources of surcharge loadings are illustrated in Fig. 9.19. Construction loads may be imposed by nearby equipment traffic and fills, and future loads can, for example, include roadway embankments and buildings.

Analysis

Determination of lateral earth pressures under surcharge loadings usually is based on elastic theory, for which the Boussinesq stresses have been modified by experiment by various workers [Terzaghi (1954)[18]]. Surcharges are analyzed as uniform loads, line or strip loads, and point loads, and results are considered to be reasonably accurate. Surcharge pressures are added to backfill and hydrostatic pressures (Fig. 9.19*d*).

Uniform loading can take the form of an earth surcharge or a large embankment and is analyzed using Rankine procedures (q_sK_a, Eq. 9.10, Table 9.3).

Line loads may be caused by highway or railroad embankments of relatively narrow width or con-

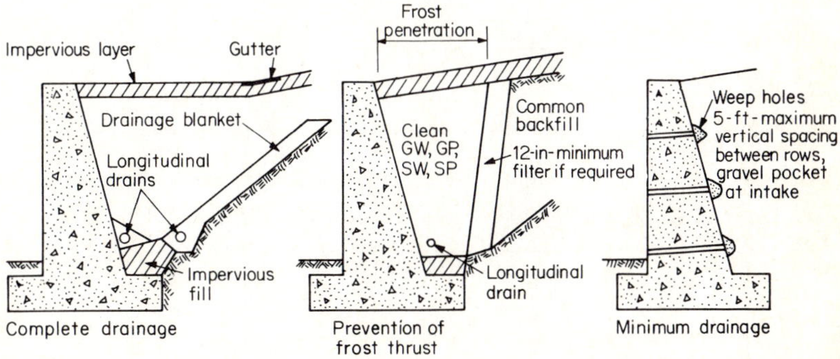

FIG. 9.18 Drainage schemes for concrete gravity retaining walls.

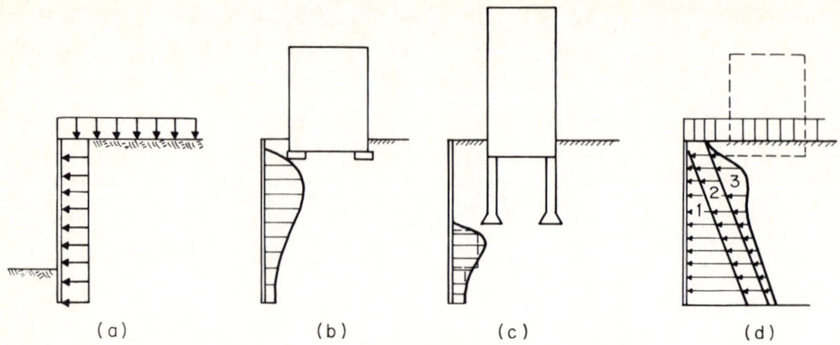

FIG. 9.19 Some forms of surcharge that increase lateral earth pressures: (*a*) excavated earth or uniform surcharge; (*b*) building on shallow foundations; (*c*) building on deep foundations; and (*d*) addition of lateral pressures—(1) backfill, (2) uniform surcharge, and (3) load from building strip footing.

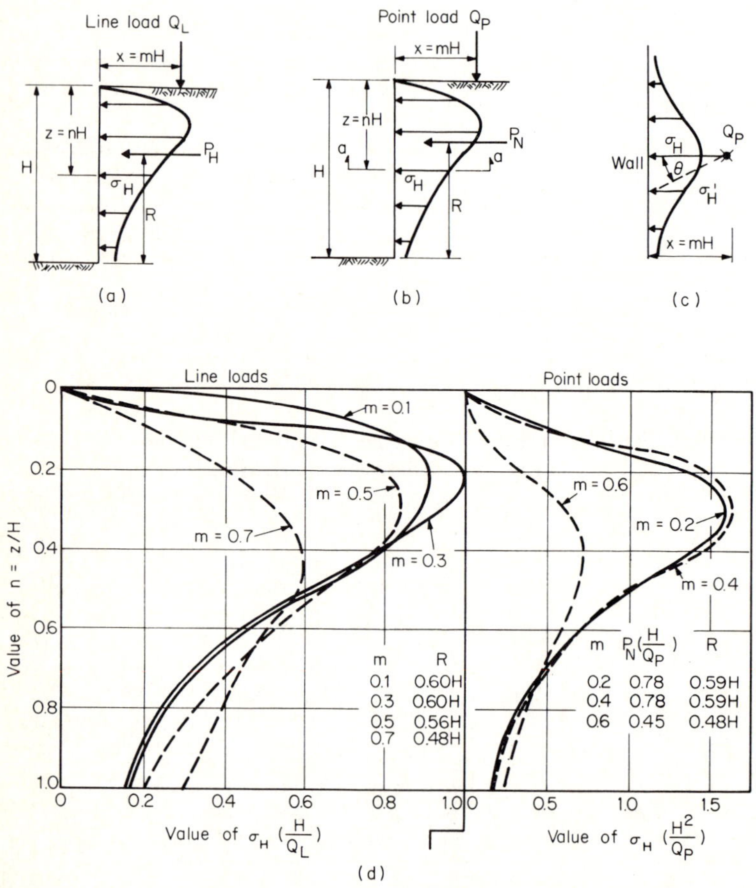

FIG. 9.20 Lateral earth pressures on walls due to line and point loads based on Boussinesq equations modified by experiment: (*a*) distribution from line load Q_L; (*b*) distribution from point load Q_p, elevation view; (*c*) distribution from point load Q_p, plan view; and (*d*) parameters *n* and *m*. [After Terzaghi, (1954)[18] from NAVFAC (1982).[1]]

tinuous building footings, resulting generally in lateral pressure distribution ranging from zero at the surface to a maximum at some particular depth as shown in Fig. 9.20*a*. For a line load of intensity Q_L (lb/ft, kN/m), the lateral pressure along the wall at depth nH can be closely approximated from

$$\sigma_h\left(\frac{H}{Q}\right) = \frac{0.20n}{(0.16 + n^2)^2} \qquad \text{for } m \leqq 0.4 \qquad (9.24)$$

$$\sigma_h\left(\frac{H}{Q}\right) = \frac{1.28m^2n}{(m^2 + n^2)^2} \qquad \text{for } m > 0.4 \qquad (9.25)$$

where n and m are distance factors as shown in Fig. 9.20*a*, values for which are given in Fig. 9.20*d*. The resultant pressure P_H is found from

$$P_H = 0.55Q_L \qquad \text{for } m \leqq 0.4 \qquad (9.26)$$

$$P_H = \frac{0.64Q_L}{m^2 + 1} \qquad \text{for } m > 0.4 \qquad (9.27)$$

and acts at a distance R above the base. Values for R are given in Fig. 9.20*d*.

Point loads are imposed by widely spread footings or piers, for example, or by crane-wheel loads. In the latter case they can be mobile. An elevation view of the pressure distribution is given in Fig. 9.20*b*, and a plan view is given in Fig. 9.20*c*. Values for the distance factors n, m, and R are given in Fig. 9.20*d*. For a point load of intensity Q_p (tons, kN), the lateral pressure along the wall at depth nH is found from

$$\sigma_h\left(\frac{H^2}{Q}\right) = \frac{0.28n^2}{(0.16 + n^2)^3} \qquad \text{for } m \leqq 0.4 \qquad (9.28)$$

$$\sigma_h\left(\frac{H^2}{Q}\right) = \frac{1.77m^2n^2}{(m^2 + n^2)^3} \qquad \text{for } m > 0.4 \qquad (9.29)$$

$$\sigma_h' = \sigma_h \cos^2 (1.1\ \theta) \qquad (9.30)$$

As shown in Fig. 9.20*d*, values for P_H are as follows:

$$P_H = 0.78\ (Q/H) \qquad \text{for } m \leqq 0.4 \qquad (9.31)$$

$$P_H = 0.45\ (Q/H) \qquad \text{for } m = 0.6 \qquad (9.32)$$

A sketch illustrating the addition of a uniform surcharge load and a line load to the backfill load is given as Fig. 9.19*d*.

In general, it is common to design for a distributed surface-load surcharge in the order of 300 psf (1.5 t/m²) for storage of construction materials and equipment applied over a work area extending about 20 to 30 ft (6 to 9 m) from the wall. This surcharge also accounts for heavy equipment loads from concrete trucks, cranes, etc., located more than about 20 ft from the wall. For closer locations, equipment loads may be accounted for as line loads. Examples of the consideration of a line load in analysis are given in Figs. 9.64 and 9.83, and of point loads from deep foundations in Fig. 9.64.

Buried loads, as imposed by adjacent foundations, are commonly a problem to be considered in the design of braced excavations, but analysis of their influence on earth pressures is not well defined. Some practitioners use the stress distribution from either the line or the point load to simulate a continuous or spread footing but simplify the analysis by dividing the curved pressure distribution into two rectangular areas as shown in Fig. 9.19*c*.

A simplified approach for a mat foundation and a continuous footing as given by Xanthakos (1979)[40] is presented in Figs. 9.21 and 9.22. Two approaches which yield different values are given for continuous footings. Xanthakos points out that whereas the magnitude of the lateral stress is influenced by the position of the external load and its proximity to the wall, its distribution hardly follows an elastic or limit theory and is decisively governed by the excavation and bracing process.

9.2.4 ANALYSIS AND DESIGN FACTORS: GRAVITY WALLS

Geotechnical Considerations

The calculation of the components of lateral earth pressure provides the data for design of the wall and analysis of its stability. As shown in Fig. 9.23, wall stability involves possible overturning (moment M_o), sliding along the base resisted by force F, possible base bearing-capacity failure at q_{max}, and possible excessive settlement of wall and backfill, S. The wall is dimensioned to resist the moment and shear forces, such dimensions relating to the type of materials used in the wall which may include earth alone, reinforced earth, earth and concrete, mass concrete, or reinforced concrete, as described in Art. 9.2.2. Overall slope stability is also assessed (Chap. 8).

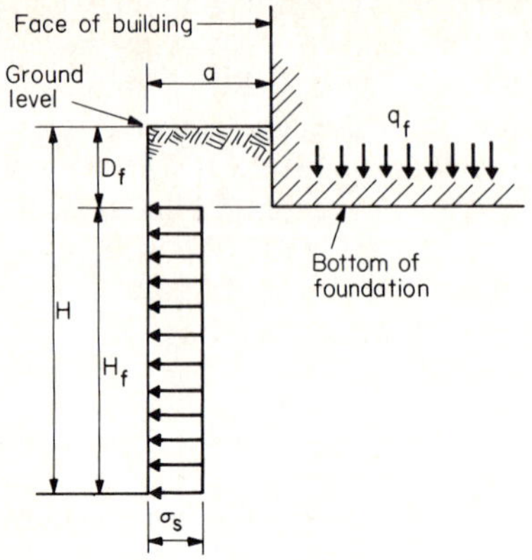

q_f = total live- and dead-load foundation pressure, psf

$$= \frac{\text{building load in width } H_f}{\text{area of width } H_f}$$

$$q_n = q_f - \text{weight of overburden} = q_f - \gamma D_f$$

$$\sigma_s = \begin{cases} 0.5q_n\left(1 - \dfrac{a}{1.5H_f}\right) & \text{for } 0 < \dfrac{a}{H_f} < 1.5 \\ 0 & \text{for } \dfrac{a}{H_f} > 1.5 \end{cases}$$

FIG. 9.21 Lateral load distribution due to building mat foundation. [From Xanthakos (1979).[40] Used with permission of McGraw-Hill Book Company.]

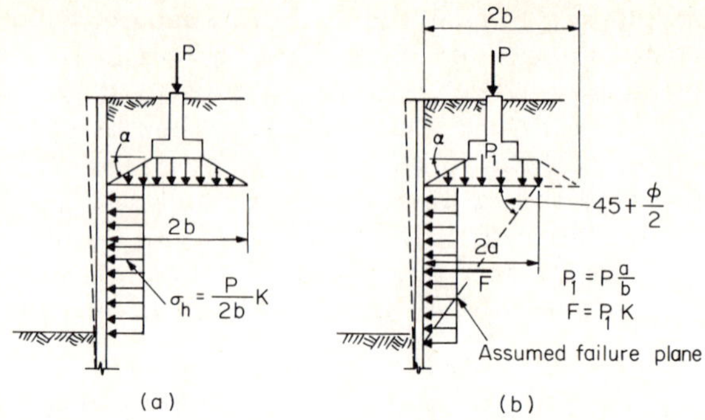

FIG. 9.22 Lateral stress distribution due to continuous footing adjacent to wall: (*a*) vertical and horizontal distribution according to an inclination angle and (*b*) distribution taking into account the zone of influence of the failure wedge. [From Xanthakos (1979).[40] Used with permission of McGraw-Hill Book Company.]

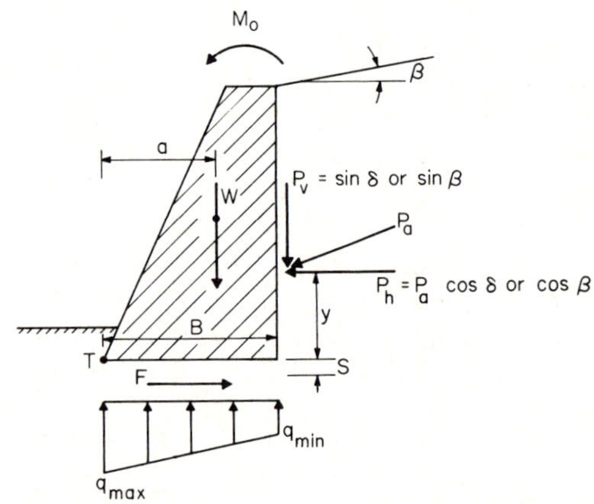

FIG. 9.23 Factors involved with overall wall stability.

Earth Pressures and Backfills

K_a vs. K_o Selection

Earth pressures are a function of backfill material weight and strength properties, groundwater conditions, slope geometry and surcharge conditions, and wall deformations. Surcharge loads can include highway and railroad embankments, buildings and other structures, and earthquake forces.

Gravity walls normally are designed for active pressure which requires wall deformations. Deflections in the order of $0.001H$ and $0.05H$ are required to develop active and passive pressures respectively in cohesionless soils and in the order of $0.004H$ to develop the active state in cohesive soils [Wu (1975)[19]]. Movements required to develop passive resistance in cohesive soils are not well defined. Unless a gravity wall is founded at a substantial depth below grade, passive resistance is usually ignored, which is on the side of safety.

Walls that are restrained at the top and are not free to tilt, such as basement walls restrained by floor slabs and bridge abutments fixed at the top by the bridge deck, do not permit soil deformation. Rigid walls founded on rock or very strong soils also may not develop enough deflection to mobilize active pressures, an important consideration in the design of cantilever walls due to their inherent weakness in the stem. Tied-back walls also permit very little movement at the wall top.

For the simple case of a vertical wall back, horizontal backfill, uniform surcharge, and clean granular backfill, K_a is found from Eq. 3.27 [$K_a = \tan^2(45 - \phi/2)$], and for similar conditions with a planar inclined backfill from Eq. 9.6, which are the Rankine expressions. For the case of an inclined wall back, K_a is found from the expression in the brackets of Eq. 9.21. Rankine expressions for cohesive soils are given in Table 9.3.

With no wall movement, soil shear strength cannot develop and the earth pressures are based on K values selected from data such as that given in Table 9.1 and Fig. 3.20, or for uncompacted sands or NC clays computed from Eq. 3.24 ($K_o = 1.0 - \sin\phi$). K_o pressures are significantly higher than K_a pressures.

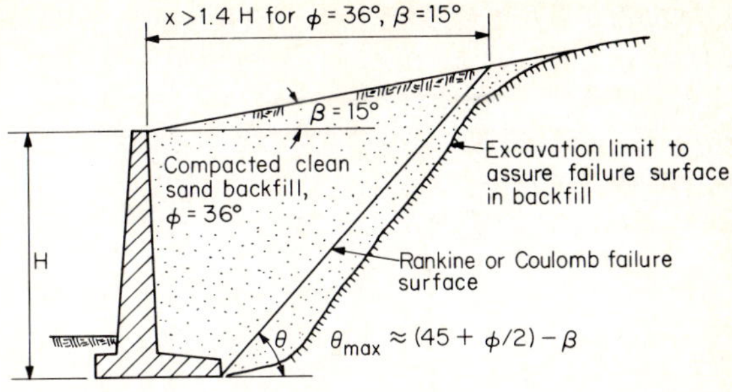

FIG. 9.24 The extent of backfill necessary to achieve active earth pressure in granular, well-drained soils. [After Bowles (1974).[23] Used with permission of McGraw-Hill Book Company.]

Granular Backfills

Because many walls are constructed by excavation into a slope, earth pressures can be controlled by the type of backfill used behind the wall. Where clean, granular soils are available they should be used because the calculated active pressures generally can be relied upon, and if drainage is adequate there will be no pore-pressure buildup, as discussed in the previous article. In clean, granular soils, such as sand, $c = 0$, and ϕ ranges between 30 and 35° ($K_a = 0.33$ to 0.27 for a horizontal backfill). In comparing wall types with respect to economics, the volume of backfill required should be considered since it can be quite significant, as shown in Fig. 9.24.

Hydrostatic distribution of earth pressure behind a wall is a simplifying assumption, and real distributions are not precisely known. For this reason, walls are designed with adequate values for safety, with FS = 1.5 as applied to overturning, sliding, or bearing capacity in granular soils. Peck et al. (1948)[20] concluded that most retaining-wall failures occur where walls are retained or founded on cohesive soils; therefore, for these materials FS = 2 is normally used.

Cohesive Backfills

In the Rankine formulas given in Table 9.3, fully consolidated clay or sand-clay mixtures in backfills should not be expected to exceed values of $K_a = 0.5$ as long as they are not heavily compacted and do not have strong swelling tendencies [Tschebotarioff (1973)[14]]. Swelling clays should be avoided as backfill materials. Tension cracks should be anticipated and evaluated as described in Art. 9.2.3 and both drained and undrained conditions considered.

Finite element analysis of the stability of a very large gravity wall to be constructed to bear on rock and support the cohesive core of a dam is described by Kulhawy (1974).[2] Design initially was based on Rankine and Coulomb methods, but the subsequent FEM analysis indicated that

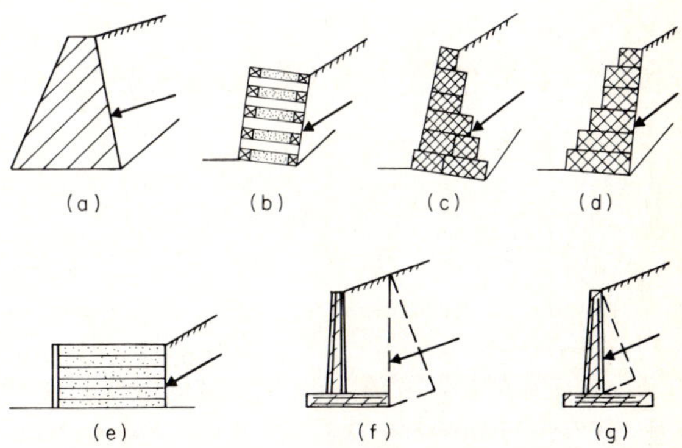

FIG. 9.25 Selection of analytical method. Use wedge concepts for walls with sloping backs (*a, b, c, d*) and wedge or Rankine concepts for walls with straight backs (*e*). Use Rankine pressures for cantilever walls since shear zone for wedge is interrupted by wall stem, applied at two locations as shown in (*f*) for sliding analysis and (*g*) for stem-pressure analysis.

the displacement of the 31.7-m (106-ft)-high wall that could develop might not be sufficient to fully develop the active state and the design was unsafe. The analysis demonstrated that the backfill would remain in the at-rest state because of insufficient compression of the rock foundation to permit tilting. The rock consisted of 6 m of "firm limestone, soft sandstone" and weathered granite overlying "firm granite." The analysis showed that the pressures developed by a "soft" core backfill would be 50% higher than from a "stiff" core backfill. The decision was made to use a stiff core backfill, to carry the foundations to the firm granite so as to limit deflections, and to design the wall for K_o pressures.

Wall Stability

Earth-Pressure Calculations

Calculations are made to determine the horizontal and vertical components of earth pressure. Applications of either the Rankine or Coulomb method depend upon the type and proportion of the wall, ground-surface configurations, and surcharges. As shown in Fig. 9.25, the Coulomb method is readily applicable to walls with sloping backs and some walls with vertical backs, and the Rankine method is applicable to walls with vertical backs. Theoretically, the Rankine approach is invalid if the wall in any way interferes with the development of the slip planes to

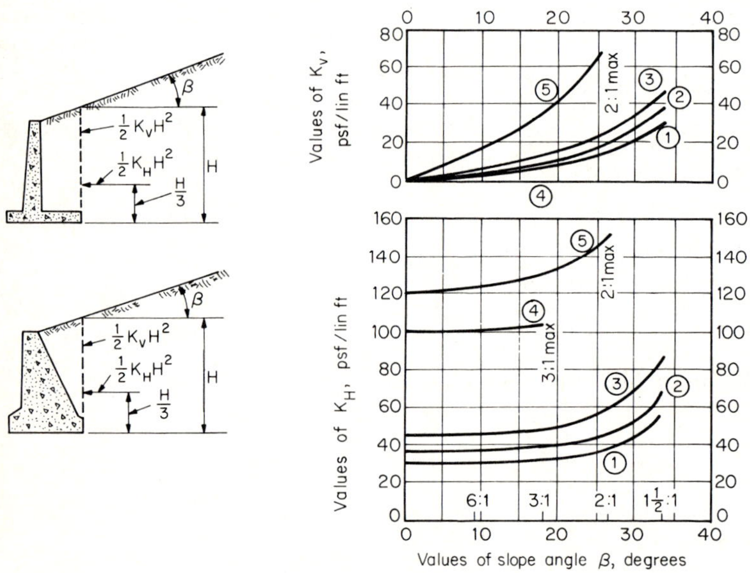

Circled numbers indicate the following soil types:
1. Clean sand and gravel: GW, GP, SW, SP
2. Dirty sand and gravel of restricted permeability: GM, GM-GP, SW, SM-SP
3. Stiff residual silts and clays, silty fine sands, clayey sands and gravels: CL, ML, CH, MM, SM, SC, GC
4. Very soft to soft clay, silty clay, organic silt and clay: CL, ML, OL, CM, MM, OH
5. Medium to stiff clay deposited in chunks and protected from infiltration: CL, CH

For type 5 material, H is reduced by 4 ft. Resultant acts at a height of $(H - 4)/3$ above base.

FIG. 9.26 Design loads for low retaining walls with straight slope backfill. *H* is less than about 20 ft (7 m). [From NAVFAC (1982),[1] after Terzaghi and Peck (1967).[10] Adapted with permission of John Wiley & Sons, Inc.]

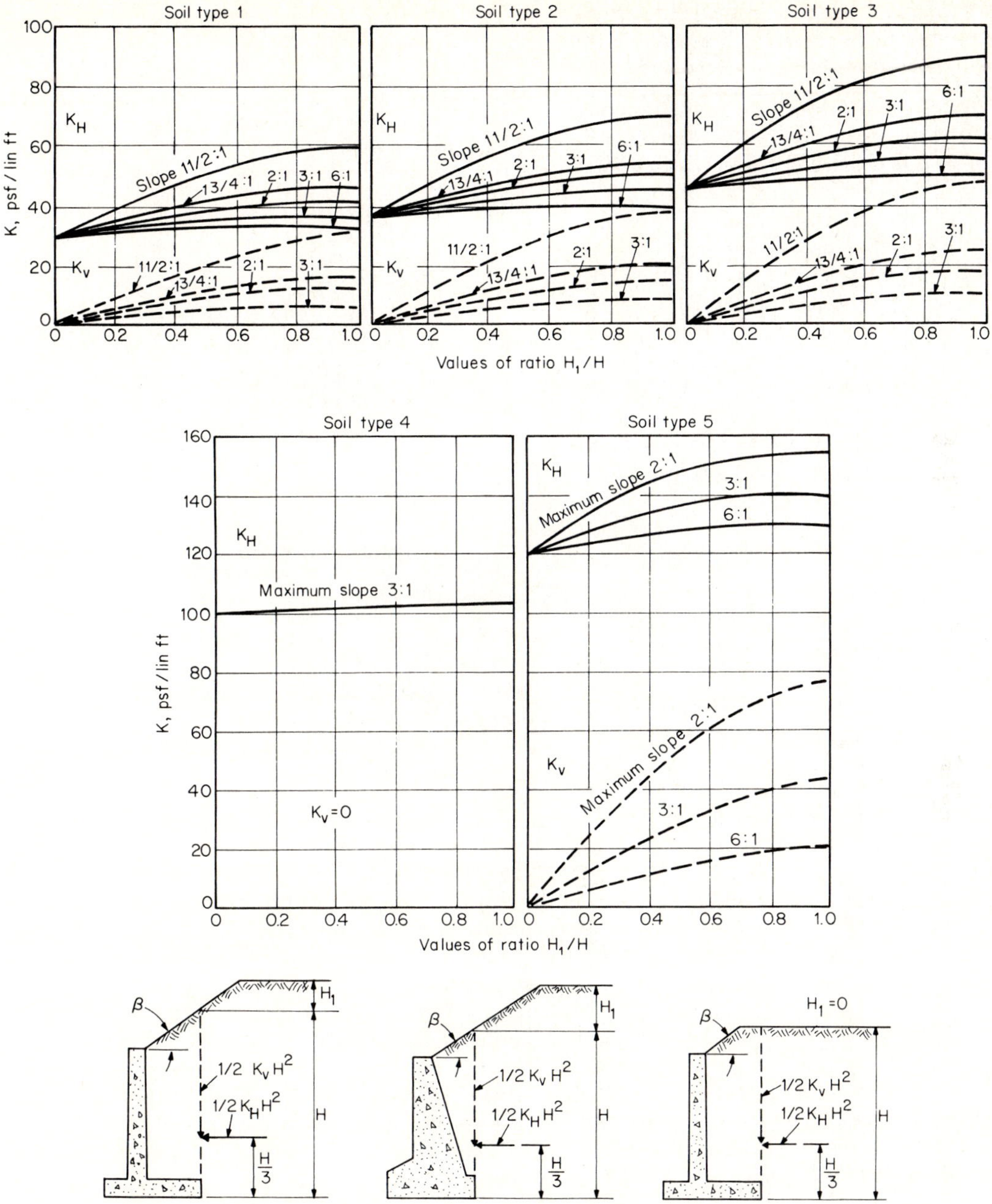

For type 5 materal H is reduced by 4 ft, resultant acts at a height of (H−4)/3 above base.
For description of soil type see Fig. 9.24.

FIG. 9.27 Design loads for low retaining walls with broken slope backfill. *H* is less than about 20 ft (7 m). [From NAVFAC (1982),[1] after Tezaghi and Peck (1967).[10] Adapted with permission of John Wiley & Sons, Inc.]

reach a state of plastic equilibrium. This situation is often ignored in practice, however, with apparently little adverse effect [Bowles (1974)[23]].

Computations of earth pressures are made using the procedures given in Art. 9.2.3, and an example of the application of the Rankine method as applied to a cantilever wall is given in Fig. 9.28. It is to be noted that the location of P and the earth-pressure distribution cannot be found with the graphical solutions. For simple cases location y is taken as $H/3$ for triangular distributions, or at the centroid of the trapezoid for uniform surcharge conditions. Conservative design loads for concrete retaining walls with H less than approximately 20 ft (7 m) can be estimated from the charts given as Fig. 9.26 for a straight slope backfill and Fig. 9.27 for a broken slope backfill.

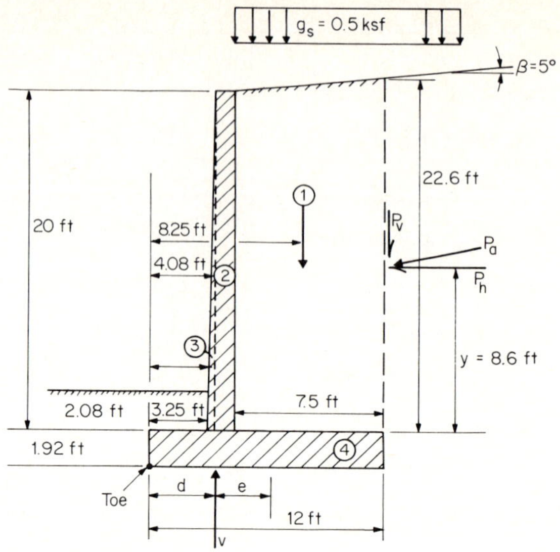

FIG. 9.28 Stability analysis of a cantilever wall. [After Bowles (1974).[23] Adapted with permission of McGraw-Hill Book Company.]

Overturning

Base width B (Fig. 9.23) must be such that the resisting moment of the wall weight (Wa) plus the moment of the vertical component of P_a (P_vB) acting on the toe at T is greater by a factor of at least 1.5 FS than the overturning moment of the horizontal component of earth pressure (P_hy) acting on the toe. An example for a cantilever wall is given in Fig. 9.28, for which the soil weight overlying the heel, a most significant factor, is also included in the computations. Note that on the figure the earth pressure is assumed to be acting along a vertical line drawn from the heel of the base slab. If the backfill or foundation soils are cohesive, FS = 2 is desired. FS (overturning) is expressed as

$$\mathrm{FS} = \frac{Wa + P_vB}{P_hy} \tag{9.33}$$

The resultant V, located at d, must fall within the middle third of the base to assure that the wall heel will always impose pressure on the foundation materials; d is found from

$$d = \frac{\Sigma M_{\text{toe}}}{\Sigma V} \tag{9.34}$$

The eccentricity e through which V acts is found from

$$e = (B/2) - d \tag{9.35}$$

A suitable base dimension is found by trial, and a reasonable first approximation can be made from Fig. 9.1. The B/H value increases with decreasing foundation and backfill strength, increasing backfill slope, and surcharge loads.

Sliding

Soil shearing resistance along the base must be greater than the horizontal component of earth pressure P_h to prevent sliding. Some practitioners add passive resistance to P_h, but others consider that it may not be counted on due to construction backfilling practices, erosion or scour, or weakening due to frost and thaw activity. In addition, some practitioners add a shear key beneath the base to increase passive resistance, while others consider that the benefits from the key are more than offset by the disturbance associated with construction (personal communication, Dr. Ralph Peck).

Sliding along the base is not uncommon in clays, and a value of at least FS = 2 is required. For a granular base support, a value of FS = 1.5 or greater is required. Sliding resistance is expressed in terms of FS as

$$\mathrm{FS} = \frac{(W + P_v)\tan\bar{\phi}' + c'L}{P_h} \tag{9.36}$$

DATA: Backfill and wall well-drained

BACKFILL: $\gamma_1 = 112$ pcf $\quad \phi = 34°$ $\quad K_{sa} = 0.286$ (from Eq. 9.6)

FOUNDATION SOIL: $\gamma_2 = 120$ pcf $\quad \phi_2 = 32°$ $\quad c = 800$ psf

CONCRETE: $\gamma_c = 150$ pcf (reinforced)

COMPUTE EARTH PRESSURES (Eqs. 9.8 and 9.10):

$$\begin{aligned} P_a &= 0.5\,\gamma H^2 K_{sa} + q_s H K_{sa} \\ &= (0.5 \times 0.112 \times 22.6 \times 22.6 \times 0.286) \\ &\quad + (0.5 \times 22.6 \times 0.286) \\ &= 11.40 \text{ kips/ft} \end{aligned}$$

$$\begin{aligned} P_h &= P_a \cos\beta = 11.35 \text{ kips/ft} \\ P_v &= P_a \sin\beta = 0.99 \text{ kips/ft} \\ y &= 8.6 \text{ [from Table 9.3 (3)]} \end{aligned}$$

1. *Check overturning* (ΣM_{toe})

Part	W (kips) + P_v	M(arm)	Moment, ft·kips
1	$(21.3 \times 7.5 \times 0.112) + (0.5 \times 7.5) = 21.64$	8.25	178.53
2	$(0.83 \times 20 \times 0.15) = 2.50$	4.08	10.20
3	$(0.21 \times 20 \times 0.15) = 0.63$	3.53	2.22
4	$(1.92 \times 12 \times 0.15) = 3.46$	6.0	20.76
5	$(P_v) = 0.99$	12.0	11.88
	$\Sigma V = 29.22$		$\Sigma M = 223.59$

$$M_{\text{overturning}} = P_h y = (11.35)(8.6) = 97.61 \text{ ft·kips}$$

$$\text{FS} = M_{\text{resist}}/M_{\text{overt.}} = 223.59/97.61 = 2.29, \text{ OK}$$

2. *Check sliding resistance* (use $\phi' = 2/3\phi$; $c' = 0.67c$ or 0.5 (max)

$$\text{FS} = \frac{(w + P_v)\tan\phi' + cL}{P_h} = \frac{(29.22 \times 0.39) + (12 \times 0.5)}{11.35} = \frac{17.47}{11.35} = 1.54$$

FS = 1.54 OK since passive resistance at toe is ignored. Passive resistance would add 14.43 kips to numerator and FS = 2.81.

3. *Check bearing capacity*

   a. Check eccentricity

$$M_a = 223.59 - 97.61 = 125.98 \text{ ft·kips} \qquad V = 29.22 \text{ kips}$$

$$d = \Sigma M_a/\Sigma V = 4.31 \text{ ft} \qquad e = (B/2) - d = 1.69 \text{ ft}$$

$e_{\max} = L/6 = 2.0$, e is OK since within middle third

   b. Pressure distribution beneath base

$$q = \frac{V}{B}\left(1 \pm \frac{6e}{B}\right) \qquad q_1 = 4.49 \text{ k} \qquad q_2 = 0.38 \text{ kip}$$

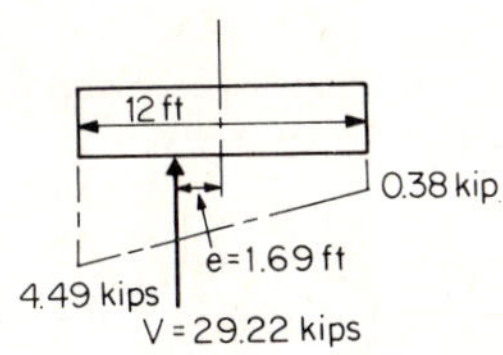

   c. From Eq. 6.16

$$q_{\text{ult}} = cN_c s_c i_c + qN_q s_q i_q + 0.5\gamma B N_\gamma s_\gamma i_\gamma$$

For strip of base slab 1 ft × L (calling B as L), $s_c = s_q = s_\gamma = 1$. i factors are found from Eqs. 6.27, 6.29, and 6.30. Use $L' = L - 2e$ for L for eccentricity. Input $\phi = 32$, $c = 0.8$, $\gamma = 0.12$, $P_o = 0.48$, $e = 1.69$, $c_a = 0$, $Q = V = 29.22$, and $P_h = 11.35$. Found are: $N_c = 35.49$, $N_q = 23.18$, $N_\gamma = 30.21$, $i_c = 0.57$, $i_q = 0.59$, $i_\gamma = 0.36$, $q_{\text{ult}} = 23.4$ ksf, $q_a = 23.4/3 = 7.8$ ksf which is substantially higher than q_1; therefore, bearing capacity is OK.

4. *Check settlements (see Art. 6.2.3. for procedures)*

5. *Design structure*

   a. Determine shear stresses in the toe and heel of the stem faces and the moment in the base slab toe and heel.

   b. Select final wall dimensions and design required reinforcing steel to resist shears and moments in the stem and base. [See Bowles (1974),[23] for example.]

FIG. 9.29 ***(continued)***

There is disagreement among practitioners as to what values to use for $\bar{\phi}'$ and c' in Eq. 9.36. Tan $\bar{\phi}'$ is often found expressed as f, for which Terzaghi and Peck (1967)[10] suggest $f = 0.55$ for a coarse-grained soil with no silt or clay and $f = 0.45$ for a coarse-grained soil containing silt. Special considerations are required where the foundation soils are clays. Even overexcavation of the clay and replacement with a well-compacted clean granular backfill may not assure that sliding will not occur between the backfill and the supporting clays. Where the supporting natural soils are cohesionless, Bowles (1974)[23] suggests using $\phi' = 0.67\phi$ to ϕ and for soil mixtures with cohesion, $c' = 0.67c$. Some practitioners use a limit of c' (max) = 0.5 tsf.

In poorly drained soils there is a danger of high pore pressures developing (Fig. 9.15*a*), thereby reducing effective stresses (*see also discussion in Art. 6.3.3, Concrete Dams: Stability Analysis*). In any event, if the value for FS cannot be raised to at least 1.5, consideration is given to wall support with a batter pile foundation. [See Peck et al. (1974)[24] for design example of pile foundation.]

Base Bearing Capacity

Toe pressure q resulting from the vertical forces (V in Fig. 9.28) must not exceed the soil bearing capacity and is found from an expression relating vertical forces V, base width B, and eccentricity e, given as

$$q = \frac{V}{B}\left(1 \pm \frac{6e}{B}\right) \qquad (6.53, 6.54)$$

in which adding the second term provides the toe pressure and subtracting the second term provides the heel pressure.

q_{max} must be less than the allowable soil pressure q_a, or q_{ult}/FS where q_{ult} is found from Eq. 6.16, for which the inclination factors i providing for inclined or eccentric loadings are found from Eqs. 6.27 through 6.30. For analysis a strip with dimensions of 1 ft × L, where B is termed L, is assumed and $s_c = s_q = s_\gamma = 1$, and $L' = L - 2e$. An example is given in Fig. 9.28. FS = 2 for granular soils and 3 for soils with cohesion. If the bearing capacity is not adequate, then pile foundations are considered.

Settlement and Slope Stability

Analysis is performed to evaluate the magnitude of the settlement to be anticipated from compression of the supporting soils under the weight of the wall, backfill, and surcharges, using the procedures described in Art. 6.2.3.

The possibility of the development of a failure surface passing beneath the wall must be evaluated using procedures described in Chap. 8.

Earthquake Loadings

Current practice for gravity-retaining-wall design generally is based on rules suggested by Seed and Whitman (1970)[25] [Richards and Elms (1979)[26]]. These rules employ the "Mononobe-Okabe" analysis dating from 1929, which is an extension of the Rankine-Coulomb wedge theory in which the earthquake forces are accounted for by the addition of horizontal and vertical inertial forces (*see Art. 8.2.7*). Wall displacements are required to mobilize soil strength for the solution to be valid. The resultant dynamic active thrust P_{ae} is given as

$$P_{ae} = 0.5\gamma H^2(1 - k_v)K_{ae} \qquad (9.37)$$

in which

$$K_{ae} = [\cos^2(\phi - \psi - i')] \Big/ \left\{ [\cos\psi \cos^2 i' \cos(\delta + i' - \psi)] \times \left[1 + \sqrt{\frac{\sin(\phi + \delta)\sin(\phi - \psi - \beta)}{\cos(\delta + i' + \psi)\cos(\beta - i')}}\right]^2 \right\} \qquad (9.38)$$

where $i' = i - 90°$ (inclination of wall back from the vertical)
$\psi = \tan^{-1}[k_h/(1 - k_v)]$
k_v = vertical acceleration of backfill
k_h = horizontal acceleration of backfill

Structural Design

Concrete Structures

Structural design of the wall proceeds after all of the forces acting have been considered in view of wall stability. Evaluations are made of the shears and moments developing in the stem and base of a cantilever wall, or in the body of a massive wall; the final wall dimensions are selected and the reinforcing steel required to resist shears and moments designed [see Bowles (1974),[23] Peck et al. (1974)[24] for a cantilever-wall example].

A cantilever wall that failed at the base-stem connection is illustrated in Fig. 1.3. The wall base, tied by a single row of anchors along the toe, was stable. The slope behind the wall consisted of residual soils rising at about a 40° inclination for a height of over 100 m.

Reinforced Earth Walls

Results of FEM analysis indicate that reinforced earth walls are rigid self-supporting units, although the rigidity in the lateral direction is considerably smaller than in the vertical direction, and the magnitude of the forces on the ties is affected by the boundary geometry of the fill and the compressibility of the foundation materials [Romstad et al. (1976),[27] Shen et al. (1976)[28]]. Under normal conditions the stress states in the backfill behind the wall approach the active state, but within the wall the soil state is approximately in the at-rest condition. Tie-force measurements on a 17-m-high wall along Highway 39 in Los Angeles County [Fig. 9.119 of Hunt (1984)[5]] suggest that a K_o distribution may be appropriate for static loading, and because of low measured residual soil-tie friction angles, FS = 4.0 was recommended to resist a pullout failure [Beaton et al. (1974)[29]]. Design procedures are given in NAVFAC (1982).[1]

Gabion Walls

Gabion walls are generally considered as low walls with design loads selected on the basis of Fig. 9.26 [Maccaferri (1979)[30]]. Wall weight is based on the specific gravity of the stone and an average porosity of 0.30. Walls are constructed either with a plane face battered at $1H{:}6V$ or a stepped face. For horizontal slopes and heights over about 10 ft the base width is about $0.5H$ and for inclined slopes the base to height ratio is about 60%. Counterforts are recommended to provide added resistance to clay slopes. They are constructed as headers and should extend from the front of the wall to a point at least one gabion length beyond the location of a possible failure surface. Spacing along the bank varies from 13 to 30 ft (4 to 9 m) depending upon the clay consistency.

9.2.5 ANCHORED CONCRETE CURTAIN WALLS

Applications

Anchored concrete curtain walls now are commonly used for retention of open excavations, but they also have application in the support of cut slopes, particularly where cuts are high, slopes are steep, and high retention capacity is required. In addition, anchors can be installed after construction is complete to increase capacity if needed, or the method can be used to strengthen existing unstable walls of other types.

Permanent anchored curtain walls have been used in Brazil since 1957 to retain cut slopes under severe natural conditions. Slopes are often as steep as 40° or more in the coastal mountains and composed of deep deposits of residual or colluvial soils over rock, which when combined with heavy rainfalls result in very high earth pressures. Conventional gravity walls usually will not provide adequate retention for these conditions.

Analysis and Design

Slope Stability Problem

Anchored concrete wall design relies heavily on experience and judgment to determine the maximum loading conditions. Because a vertical or near vertical wall will be used to retain a cut slope, the classical Culmann method of wedge analysis (see Art. 8.1.3) is considered to have valid application in locating the potential failure surface and evaluating the force to be contained in a thick residual soil formation. These walls, therefore, are often designed on the basis of slope stability methods rather than the determination of earth pressures.

Critical Failure Surface

In a relatively homogeneous formation, the critical failure surface θ_{cr} is found from Eq. 8.6 [$\theta_{cr} = (i + \phi)/2$] used in the Culmann method. Careful field investigation is necessary, however, to determine if some potential failure surface other than that found from Eq. 8.6 is present, such as a weakness plane resulting from slickensides or a soil-rock interface.

Shear-Strength Parameters and Pore Pressures

Strength parameters normally are measured in the laboratory in the direct shear device with specimens oriented to permit failure along weakness planes such as relict foliations in residual soils originating from gneiss, joints, or slickensides, if it has been judged that the in situ orientation of these surfaces may control the potential failure surface. As soil properties in residual soil slopes often cover a wide range of values, Brazilian experience has shown it to be prudent to measure strength with a large number of direct shear tests rather than a few relatively costly CU triaxial tests with pore-pressure measurements. In addition, since specimens taken from block samples are common, orientation of the potential failure surface is facilitated in the direct shear device.

Because the normal water table in steep slopes is in or near the fractured rock zone (Fig. 8.52), the overburden soils usually are unsaturated during the time of investigation. During wet seasons and heavy rains, however, the upper surficial zone may become saturated, and for evaluation shear tests are made on specimens both at natural water content and soaked. The latter case usually yields significantly lower values, particularly for cohesion. Typical values are $\phi = 30°$, $c = 2.5\ \text{t/m}^2$ (0.25 tsf), and $\gamma = 1.8\ \text{t/m}^3$ (112 pcf).

The pore pressures that might develop along the potential failure surface, and the length of surface that might begin failure, which is often progressive, are essentially indeterminate. Design normally is based, therefore, on the saturated direct shear strength mobilized along the entire potential failure surface. Because the walls are not provided with a granular backfill, "windows" are left in the wall or horizontal drains are installed for drainage.

Brazilian Analytical Method

The analytical method used in Brazil is described by Costa Nunes and Velloso (1963)[31] and Costa Nunes et al. (1972)[32] and illustrated in Fig. 9.29. It is based on finding the minimum value of FS existing for a Culmann wedge when a vertical or near vertical cut is made into a slope. From Eq. 8.7, for the critical failure surface θ_{cr}

$$\text{FS} = \frac{\dfrac{c}{\gamma H}}{\dfrac{1 - \cos(i - \phi)}{4 \sin i \cos \phi}} \qquad (8.7)$$

Substitution of Eq. 8.5 for wedge weight W yields

$$\text{FS} = \frac{cL \cos \phi}{W \sin(\theta_{cr} - \phi)} \qquad (9.39)$$

The total force F to be retained by the wall is assumed to be carried by earth or rock anchors, which must extend beyond any potential failure surface. F is found from

$$F = \frac{\lambda - 1}{\lambda} W \left[\frac{\sin(\theta_{cr} - \phi)}{\cos(\theta_{cr} - \phi + \alpha)}\right] \qquad (9.40)$$

where λ = the ratio of the working value for wall design FS, FS_d, to the minimum slope FS_s or

$\lambda = \text{FS}_d/\text{FS}_s$

α = the angle of the anchors from the horizontal

Vertical anchor spacing, usually about 2 m, is selected and divided into the wall height to find the number of rows. The product of a selected anchor working load and the number of rows provides the total capacity per running meter, which is divided by the force F to find the required horizontal anchor spacing.

In calculating the anchor working loads (*see Art. 7.7.4*), values for FS ranging from 1.5 to 2.0 are used to compensate for possible capacity reduction with time resulting from pore-pressure development on the failure surface and the loss of prestress from anchor relaxation. Slow deformations caused by soil compression behind the wall, anchor-rod stretching, and anchor-bulb movement cause relaxation [Costa Nunes (1974)[33]]. Anchor ties in residual soils usually are 1¼-in (3.15-cm) rods designed to carry test loads in the order of 30 to 40 tons metric; anchors in rock, consisting of a cylindrical group of smaller bars or cables, can carry 100 tons metric or more. The anchor type and working load are selected on the basis of economic comparisons.

Anchor lengths must be adequate to penetrate beyond any potential failure surface. In a thick

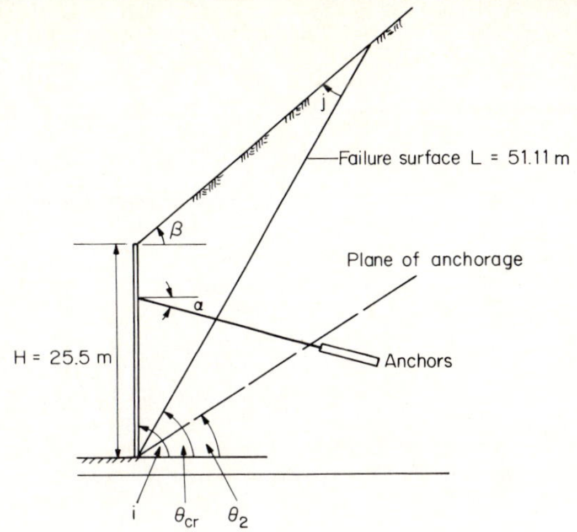

1. *Data input for soil*

$$\phi = 30° \qquad c = 2.5\ t/m^2 \qquad \gamma = 1.8\ t/m^3$$

2. *Geometry and weight*

$H = 25.5$ m $\quad i = 90° \quad \beta = 40° \quad$ assume $FS_d = 1.5$

$\theta_{cr} = (90 + 30)/2 = 60°$

From law of sines:

$$L = (H/\sin j)(\sin 90 + \beta) = 57.11 \text{ m}$$

$$\text{where } j = 180 - (90 + \beta) - (i - \theta_{cr})$$

$$W = 0.5\gamma HL\left[\frac{\sin(i - \theta_{cr})}{\sin i}\right] = 655.38 \text{ tons}$$

3. *FS of unsupported slope and factor λ*

$$FS_s = \frac{cL\cos\phi}{W\sin(\theta_{cr} - \phi)} = 0.38$$

$$\lambda = FS_d/FS_s = 3.95$$

4. *Determine anchor force F*

$$F = \frac{\lambda - 1}{\lambda} W\left[\frac{\sin(\theta_{cr} - \phi)}{\cos(\theta_{cr} - \phi + \alpha)}\right]$$

For $\alpha = 15°$, $F = 346.84$ tons per running meter

5. *Design anchor system*

Vertical spacing = 2 m/25.5 = 13 rows

Working load/anchor = 38.5 tons

Total capacity/row = 13 × 38.5 = 500.5 tons/m

Horizontal anchor spacing = 500.5/346.8 = 1.44 m

6. *Determine θ_2, plane of anchorage*

$$FS_p = 1.5 = \frac{2c/\gamma H(\sin i\cos\phi)}{\sin(i - \theta_2)\sin(\theta_2 - \phi)}$$

θ_2	FS_p	
40	0.71	
33	2.15	
34	1.63	
34.3	1.52	OK, use 34.3°

NOTE: At this location (Fig. 9.30), anchors were test-loaded to 60 t.

FIG. 9.29 Analysis of an anchored curtain wall with the Brazilian method. [After Costa Nunes et al. (1972).[32]]

deposit of residual soil, the flattest potential failure surface θ_2 is assigned FS = 1.5 and found by iteration of

$$FS_p = 1.5 = \frac{2c/\gamma H(\sin i\cos\phi)}{\sin(i - \theta_2)\sin(\theta_2 - \phi)} \qquad (9.41)$$

Wall thickness is a function of anchor spacing and is designed for practical purposes on the basis that all loads will be carried by the anchors. The wall essentially is not free to move at the top.

Alternate Analytical Procedures

Alternate analytical procedures are based either on a system of sliding wedges or on earth pressures as discussed in Art. 9.3.4. They are usually employed for analysis of permanent walls for braced excavations.

Foundations

High walls may require special foundation provisions as discussed under the following section.

Construction

Procedure

A sketch of the geologic conditions and a section through an anchored curtain wall 25.5 m in height constructed to retain a deep excavation in a slope with a history of sliding is given as Fig. 9.30. The wall, completed to a height of about 15 m, is shown in Fig. 9.6. Its maximum thickness is 50 cm.

With the "Brazilian method," construction typically proceeds from the top of the wall downward by sequential excavation of levels, anchor installation, and wall forming as illustrated in Fig. 9.31, which provides for continuous slope support. The height of each level depends on anchor spacing. In the case shown, an initial 6-ft-high bench is excavated into the slope at the wall top, a row of anchors installed, the wall formed and poured, and the anchors prestressed, using the wall as a reaction. In weak, unstable slopes,

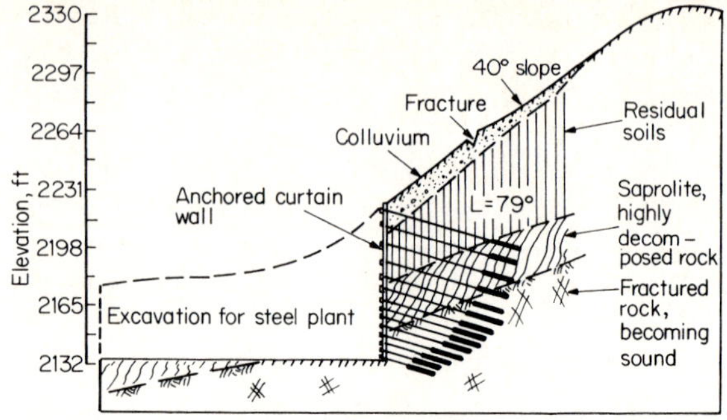

FIG. 9.30 Section through a 25.5-m (85-ft)-high anchored curtain wall constructed to retain a steep slope with a history of landslides (see also Fig. 9.6). João Monlevade, M. G., Brazil. [From Hunt and Costa Nunes (1978).[34]]

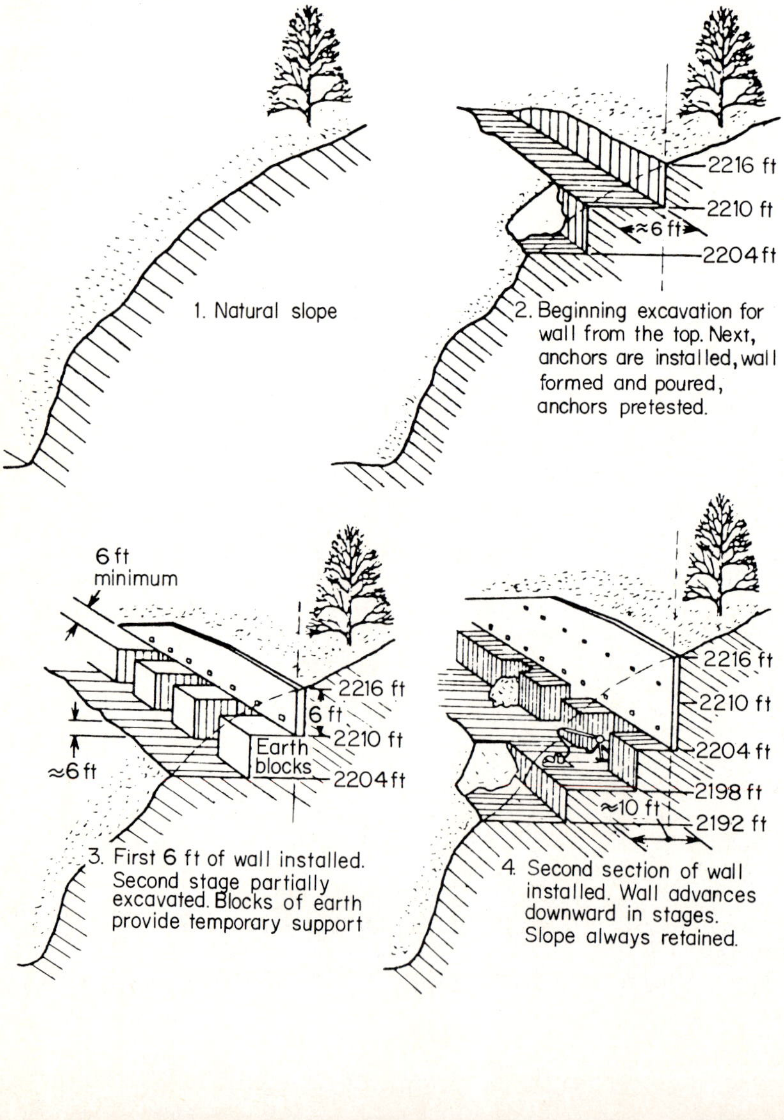

FIG. 9.31 The "Brazilian method" of construction of an anchored curtain wall from the top down. [From Hunt and Costa Nunes (1978).[34]]

the following sections proceed by excavation of alternate earth blocks along the bench, leaving blocks in place to provide temporary support. Every other anchor is installed, the remaining blocks removed and the anchors installed, and the wall section formed. In stable slopes the wall is constructed using only benches which are excavated by bulldozer where access is practical, or otherwise by hand.

Drainage

In strong, predominantly clayey soils, "windows" are left in the wall to permit drainage and evaporation; otherwise, drainage is affected by the installation of subhorizontal drains.

Foundations

Heavy walls, such as those greater than about 15 m in height, or those greater than 30 cm in thickness and supported by cable anchors, may begin to settle during construction due to wall weight and the vertical component of the anchors where installed at relatively steep inclinations. Underpinning is required until the wall reaches its final depth, which is usually in strong materials with adequate support. A pile-supported wall is illustrated in Fig. 9.7.

Anchored Premolded Plates for Slope Stabilization

Description

A novel approach to slope stabilization being used in Brazil since about 1977 involves the use of premolded concrete plates set in a hillside in rows at various levels. In the case shown in Fig. 9.32, the plates were interlocked and anchors installed in every other one. The wall illustrated was installed to stabilize a small slide that began when the slope was cut for an access road. Construction to conform with the bowl-shaped hillside was simplified by the method, thus avoiding a large amount of excavation as shown in Fig. 9.33. The stabilization of a large slope is illustrated in Figs. 1.10 and 9.8.

Analysis

For the case illustrated in Figs. 9.32 and 9.33, the anchors were designed to retain a mass bounded by a cylindrical failure surface located by the Bishop slice method *(see Art. 8.2.3)*. The geometry and parameters for the general analysis of an inclined slope are shown in Fig. 9.34.

In analysis [Tecnosolo (1978)[35]], the anchor loads and locations are selected and the lengths deter-

FIG. 9.32 Installation of anchored premolded concrete plates to stabilize a curving hillside.

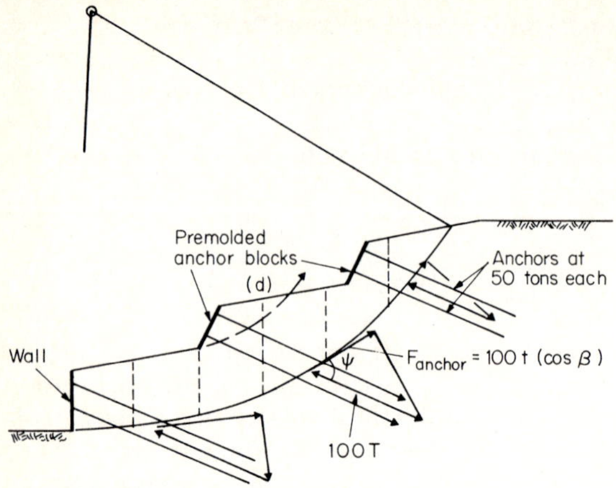

FIG. 9.33 Section illustrating design slope, anchors, and potential failure surface for slope in Fig. 9.32. [From files of Tecnosolo, S. A.]

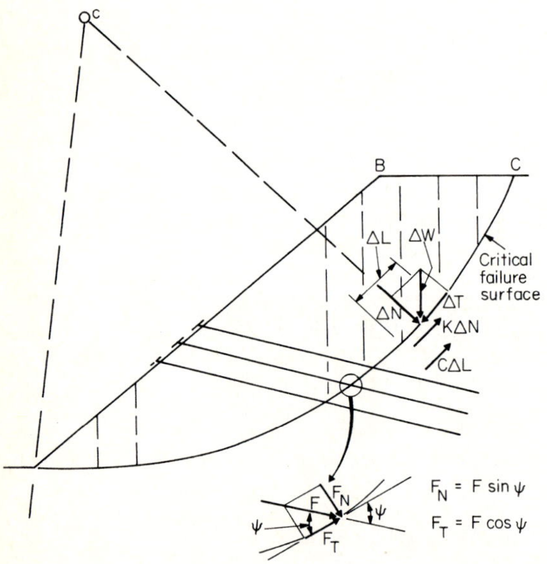

FIG. 9.34 Geometry and parameters for analysis of a slope to be stabilized by anchored plates. [From Tecnosolo, S. A. (1978).[35]]

mined to pass beyond the potential failure surface. The force F to be retained by the anchors is found from

$$F = \frac{FS_p(\Sigma \Delta T) - \Sigma K \Delta N - \Sigma c \Delta L}{FS_p \cos \psi - K \sin \psi} \tag{9.42}$$

where FS_p = FS desired from anchors
K = coefficient of friction angle ϕ (tan ϕ)
ψ = angle of force resultant from anchors with failure surface
N = normal force which may be reduced by pore pressures U, or $N - U$

Where benches are constructed, the possibility of local failure at d in Fig. 9.33 must be evaluated to ensure adequate resistance to the applied anchor forces.

Load Testing

During installation it is important that individual plates do not press into the slope because of local failure during pretensioning, since if they do the required test load will not be achieved. This may be avoided by bracing against several plates for load testing. Plate deflection after construction with subsequent anchor load reduction is not serious, since if slope movements begin, active pressures are imposed on the anchors through the plates.

Advantages of Anchored Curtain Walls

Anchored concrete curtain walls offer a number of advantages over conventional walls:

- Walls can be anchored at as many points as desirable, thereby providing high retention capacity with a relatively thin wall, usually in the order of 20 to 30 cm (8 to 12 in) thick.
- Walls can be constructed from the top downward, avoiding the necessity (in the case of conventional walls) of excavating into an open slope, which may result in failure, and then backfilling.
- Where structures exist near the top of the excavation, underpinning is not necessary, but permission is required from adjacent owners for anchor installation beneath their property line.
- When used for foundation excavations, shoring and bracing are not required and the excavation remains free of obstructions *(see Art. 9.3.2)*.
- Prestressing of anchors normally eliminates wall deformations occurring during excavation. Strain gages and other devices can be mounted

on anchor bars to monitor stress changes as a safety precaution, and additional anchors can be installed as necessary to reinforce the wall if deformations become objectionable.

- Existing conventional walls suffering distress from objectionable deflections can be reinforced by anchored curtain walls.

9.3 OPEN EXCAVATIONS: SUPPORT SYSTEMS

9.3.1 INTRODUCTION

Excavation Walls

Purposes

Open excavations are made from the surface for buildings, sewer trenches, subway cuts, and other construction works. Support of the excavation sides is required where the construction area is limited or ground conditions are poor; otherwise, the excavation sides may be considered as a slope stability problem and designed at some stable angle without special support. Support systems should also enable construction to proceed in the "dry," and subsidence of the adjacent ground in built-up areas requires control to avoid detrimental movements of nearby structures.

Wall Systems

Wall sections for soils may be soldier piles and lagging (Berlin walls), sheet piles (timber, steel, or other material), or concrete diaphragms. *Wall support* is provided by cross-lot bracing, rakers and berms, and anchors (tiebacks), or by wall rigidity and earth pressures alone (cantilever wall), as illustrated in Fig. 9.35.

Rock-mass support is provided by bolts, bolts and straps, bolts and shotcrete, or tendons.

Earth Pressures

Flexible structures result from these wall systems and they deform under earth pressures. The degree of flexibility is a function of the wall and support system. Earth-pressure magnitude and distribution are not a direct function of ground conditions as they are with rigid walls, but rather are a function primarily of structural deformation, and therefore highly indeterminate. Experience from model studies, field measurements and observations, and finite element studies indicates that the flexibility and subsequent wall deformation cause earth pressures to redistribute and vary significantly from the classical distributions. *Arching (see Art. 3.2.3)* causes the redistribution which reduces earth pressures where major yielding occurs and increases them over less yielding parts, such as where a wall is braced or anchored or where it is embedded and mobilizes passive resistance.

In rock masses the concern normally is with slippage along joints and other discontinuities, although in some cases high residual stresses may be a cause for concern.

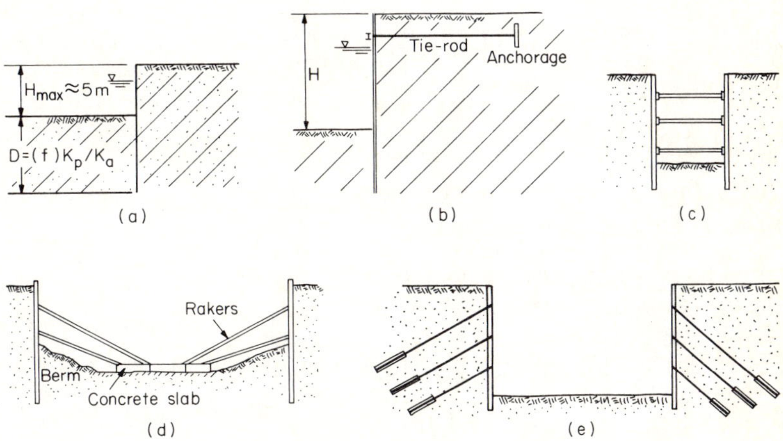

FIG. 9.35 Various types of support for flexible retaining structures: (*a*) cantilever sheet-pile wall; (*b*) anchored sheet-pile bulkhead; (*c*) strut-braced excavation; (*d*) raker and berm braced excavation; and (*e*) wall anchored with tiebacks.

Design Objectives

The design objectives are

- To determine the maximum moment M_{max} in the wall resulting from all of the earth-pressure forces to permit the selection of an adequately stiff wall section
- To calculate the forces on the wall supports (struts, rakers, anchors) from the earth pressures and select the support size
- To restrict wall movements and adjacent ground subsidence where damage to nearby structures may result

Evaluation Procedure

Wall Selection

Geologic conditions to consider include soil type, stratification, and groundwater. Significant aspects are groundwater control and seepage in granular soils and high earth pressures and bottom heave in clays; or in rock masses, rock quality, groundwater conditions, and high residual stresses.

Excavation dimensions to consider include depth, width, and shape. Depth influences the economy of the system selection, and width and shape the support system type most suitable.

Wall life refers to permanent versus temporary walls, the choice of which affects economy. Permanent walls usually are part of the structure and therefore should be completely impervious over the long term.

Backslope subsidence control requires a relatively rigid system and very careful construction execution.

Calculate Earth Pressures

The basis for earth-pressure calculations is the classical theories, but total pressures and distributions are modified by semiempirical or empirical relationships, although there are cases where the classical theories are used directly. One can find a variation in approaches to earth-pressure calculations in the literature; the methods commonly used for various types of support systems are summarized as follows:

Strut-braced walls	Semiempirical relationships
Raker-braced walls	Semiempirical relationships
Anchored walls	Semiempirical relationships
Permanent walls with top support	Rankine with K_o
Cantilever walls *(see Art. 9.4.4)*	Rankine
Anchored bulkheads *(see Art. 9.4.3)*	Rankine with semiempirical relationships

Note that in some cases the Rankine method is used for the active pressures but wall friction is considered for passive pressures.

Design Wall Elements

Calculate wall moments resulting from earth pressures, select the section modulus required to resist the bending forces, and design the supports to provide the necessary reactions.

Backslope Subsidence

Estimate the potential magnitude of backslope subsidence from published performance data and, where excessive movements are anticipated, either underpin adjacent structures, select the most rigid wall system possible, or consider mitigating the effects by chemical or compaction grouting. Another alternative is to evaluate deflections by experience, empirical data, or finite element analysis; select an economical wall system; devise contingency plans for excessive movements; and monitor performance during construction. Only the most critical cases are underpinned.

Geologic Conditions and Wall Performance

Subsurface investigations for excavation walls can require even more care than studies for slope retention as even small variations in conditions from those anticipated can result in substantial problems during construction. Soil stratification often is a critical factor because of its effects on seepage control. Strength and compressibility parameters require definition as do groundwater conditions. Geologic conditions directly influence lateral earth pressures, seepage, and the backslope subsidence potential.

Dewatering

Dewatering [see Art. 8.4.3 of Hunt (1984)[5]] is necessary for pervious walls such as Berlin or sheet-pile walls and at times for diaphragm walls in granular soils below the water table to

- Permit excavation in dry conditions
- Reduce water pressures acting on the system
- Minimize disturbance and bottom heave from piping
- Control seepage through the wall and subsequent loss of ground
- Increase passive resistance of interior berms or of the excavation bottom to wall movement
- Facilitate anchor installation and excavation for pits or permanent underdrain systems

Evaluation is necessary to ensure that dewatering will not result in excessive subsidence [Art. 10.2.1 of Hunt (1984)[5]]. A recharge system or some other method, such as an impervious wall system, may be required. Where dewatering will affect the consolidation of organic or soft clayey soils, it is extremely difficult to prevent ground settlement, and the most satisfactory solution is to avoid dewatering by using methods such as slurry wall construction.

Backslope Subsidence

Backslope subsidence results from excessive wall deflections, heave or piping of the excavation bottom, raveling of soils through openings in the wall, or lowering of the groundwater table, as illustrated on Fig. 9.36. Raveling can result in severe loss of adjacent ground and occurs through wall openings made for anchor holes, through gaps in sheeting, or between poorly driven sheet piling.

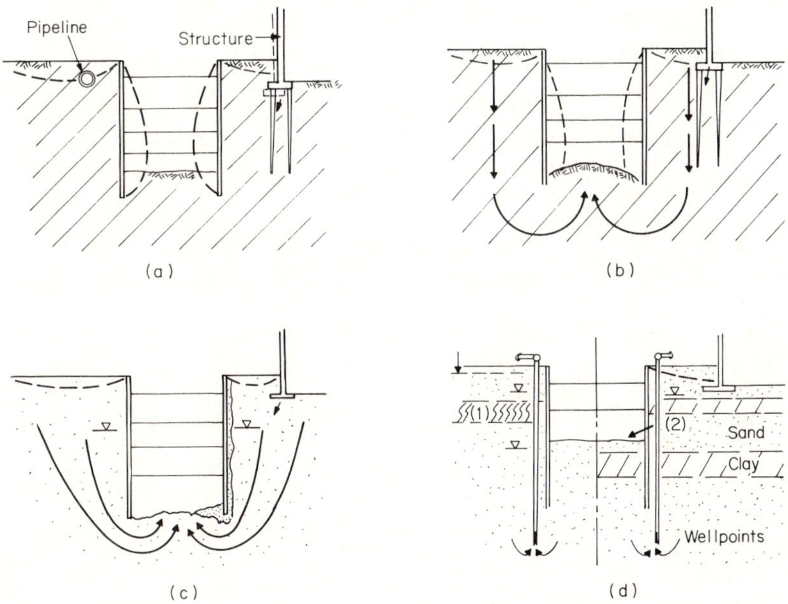

FIG. 9.36 Causes of subsidence of adjacent ground during construction of braced excavation: (*a*) lateral movement during excavation; (*b*) bottom heave in soft clay; (*c*) piping in sands below the water table; and (*d*) dewatering stratified soils, which can (1) increase overburden pressure causing consolidation of weak organic or clay soils or (2) result in migration through sheeting and loss of ground.

Normally, subsidence is related directly to the lateral deflection of the wall during construction and depends on the retention and bracing methods, construction quality, and soil type. Soil characteristics, not wall stiffness, however, are the most important variable determining the amount of movement, summarized by Peck (1969)[36] as described below.

Soft to medium saturated plastic clays usually present the most difficult conditions. Excavation reduces the load on the soil beneath the cut, and the soil in the cut tends to move upward while the soil adjacent to the cut tends to move laterally and downward before the bracing can be installed and the adjacent ground subsides. Observational data indicate that the volume of subsidence is equal to the lateral movement volume, which in turn is related to the excavation bottom heave. Adjacent subsidence, therefore, is controlled by restriction of lateral deflection and bottom heave.

Base failure may occur in the excavation bottom when the surcharge load of the adjacent soil is great enough to cause a bearing-capacity failure. Occurrence is at some critical depth that depends on the strength and depth of the clay below the excavation *(see Art. 9.3.3)*. Soft clays and organic soils consolidate readily when subjected to dewatering and increased overburden pressures, resulting in surface settlement.

In other soil types, the subsidence and lateral movement volumes may not be equal, but settlement is most effectively reduced by limiting lateral deflections.

Cohesionless sands are associated with inconsequential adjacent subsidence if the wall is properly designed and constructed and dewatering is adequate. If dewatering is not under full control, however, sand migration through some opening in the wall can result in substantial subsidence. Inadequate sheeting penetration can result in piping and boiling of the excavation bottom from excessive seepage pressures, resulting in severe loss of ground.

Cohesive granular soils usually are strong materials. Although observational data are meager, they indicate that lateral movement and subsidence are not a problem. Even small amounts of cohesion reduce sensitivity to seepage pressures and limit raveling through sheeting.

In general for *stiff, plastic clays*, the greater the stiffness, the less the lateral movements and adjacent ground settlement.

Stratified and lenticular deposits can be troublesome because adequate dewatering of sand layers below the water table may be difficult to achieve. Unless the dewatering system is designed and installed carefully, the sand layers remain saturated and when uncovered during excavation tend to migrate quickly through any opening. Once migration begins, it is difficult to arrest.

Conclusions

In granular soils below the water table, excavation can be accompanied with large inflows of water unless the water table is lowered by wellpoints or controlled by other means. Depending upon geologic conditions, dewatering can develop backslope subsidence, and the potential for raveling and piping requires careful consideration.

In soft to firm clays, even with proper design, objectionable deflections can be expected unless workmanship is excellent. Subsidence is difficult to prevent in any event for most wall systems.

In silts below the water table, dewatering may be impractical, except by electroosmosis or carefully installed vacuum wellpoints, and base instability and piping may develop.

9.3.2 WALL SYSTEMS FOR SOILS

General

Components

Wall sections can consist of soldier piles and lagging (Berlin walls), sheeting (wood, steel, precast concrete planks, premolded composite materials), or cast-in-place concrete (formed walls, diaphragm walls). Support can be provided by cross-lot bracing (struts), rakers and berms, anchors (tiebacks), or the earth and wall only (cantilevers). A summary of wall type, support system, conditions most suitable, and disadvantages is given in Table 9.6.

TABLE 9.6
WALL AND SUPPORT SYSTEMS: SUITABILITY

Wall type	Conditions most suitable	Disadvantages
Soldier piles and lagging	Overconsolidated clays Sands with cohesion Sands adequately dewatered	Cannot extend wall below excavation bottom. Effective dewatering required. Permits large movements in weak soils.
Sheet piles	Easy driving conditions in clays, sands, and clay-sand mixtures	May permit large movements in weak soils. Cannot penetrate obstacles or strong materials. Effective dewatering required.
Diaphragm walls	Soft clays and loose sands below water table where the need is to control lateral movements	Relatively costly. Penetration in strong soils by slurry trench method slow and difficult.
Support system	**Conditions most suitable**	**Disadvantages**
Cross-lot bracing	Narrow excavations	Permits large wall movements in weak soils.
Rakers and berms	Wide excavations	Permits large wall movements in weak soils.
Anchors (tiebacks)	Any excavation width where movement control is needed	Need material suitable for anchorage in reachable distance. Vertical component imposed on wall.

Lateral Movement Factors

The *magnitude* of lateral movement is directly related to the stiffness of the retaining system. As all long, relatively thin walls are flexible (even concrete diaphragm walls are considered only semirigid), the lateral movements depend far more on support spacing and stiffness and on the depth of the excavation relative to the placement of supports than on the inherent wall stiffness.

Prestressing of supports is important to reduce lateral movements, as inelastic deflections and the inherent slack in supports are removed from the system. Elastic deformations can be especially significant in tiebacks, which are usually long, thin tendons.

Shallow Trenches

The relatively small, shallow trenches required for sewers and other utilities generally are overlooked in the literature, which is primarily concerned with large, deep, complex excavations. From the aspect of danger to workers and lives lost each years from cave-ins [Civil Engineering (1977)[37]], sewer trench retention systems require as close attention as any excavation. Retention systems should be provided where there is any danger of cave-ins, and many codes require trench retention for any depth over 4 ft (1.2 m). A careful subsurface investigation should always be made.

Shallow trenches in moderately strong to strong soils usually are retained with "skeleton" sheeting; i.e., discontinuous walls composed usually of timber planks with cross-bracing. In sands with low cohesion and other relatively weak materials, sheeting will be continuous, as for example tongue and groove timber with adequate bracing. Prefabricated metal boxes are used commonly in soils where adhesion to the box is low enough to permit removal and reuse. Piping and softening of the trench bottom resulting from inadequate dewatering are common causes of sewer pipe settlement and consequent breakage. Support of shallow trenches is discussed in some detail in NAVFAC (1982).[1]

Wall Sections

Soldier Piles and Lagging (Berlin Walls)

Berlin walls are constructed of wide-flange steel H sections spaced about 2 to 3 m (6 to 10 ft) apart, driven prior to excavation. As excavation proceeds, horizontal timber sheeting (lagging) is inserted behind the H-pile flanges as shown in Fig. 9.37. The horizontal earth pressures are concentrated on the soldier piles because of their relative rigidity as compared to the lagging. Soil movement and subsidence are minimized by maintaining the lagging in firm contact with the

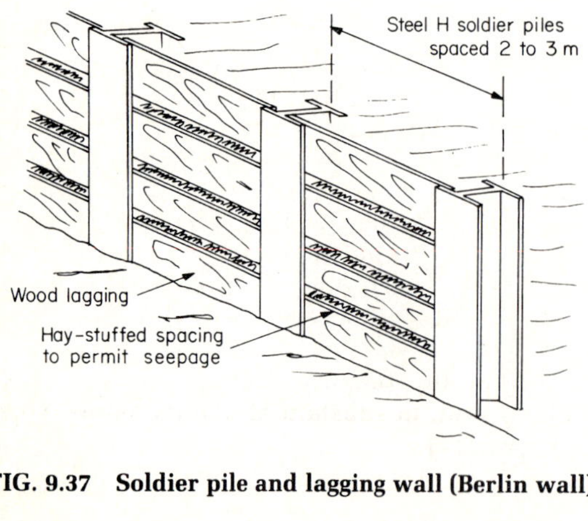

FIG. 9.37 Soldier pile and lagging wall (Berlin wall).

soil. In cuts below the water table in drainable soils, a space about 2.5 cm (1 in) is left between boards to permit drainage. The opening is stuffed with hay or filter fabric to prevent soil migration and loss of material during seepage.

Suitable conditions for Berlin walls prevail where well-constructed walls will not result in significant ground subsidence. Such conditions are provided by clays with at least some overconsolidation that will not consolidate objectionably during construction, in all soils above the water table if they have at least some cohesion, and in homogeneous, free-draining soils that can be effectively dewatered.

Unsuitable conditions include squeezing soils such as soft clays and running soils such as loose sands. Because of the necessity of leaving wall portions unsupported while trimming and lagging, significant loss of ground and adjacent ground subsidence occurs.

For excavations below groundwater level in clean granular soils, dewatering is required since leakage with Berlin walls is inevitable. The most common cause of ground loss is by material flow through the lagging and careful construction is required, particularly in stratified soils. Monitoring of the contractor's progress is required to ensure that the excavation rate does not exceed the rate at which the phreatic level is dropping. Stratified soils present particularly dangerous conditions since dewatering often is inefficient and they remain saturated and subject to sudden flow into the excavation when encountered.

In hard driving conditions caused by boulders or other obstructions, soldier piles can be predrilled if necessary or relocated to avoid the obstruction, which is an advantage over sheet-pile walls. Predrilling often is used in urban areas to avoid the noise of pile driving. Installation of H sections in predrilled holes below GWL can lead to loss of ground by inflow of cohesionless soil. The problem is resolved by drilling the holes with bentonite, which is replaced by weak concrete after the H sections are inserted.

LAGGING METHODS Lateral displacements are caused by lagging deflection resulting from improper trimming and placement. Four methods of transferring earth pressures from lagging to

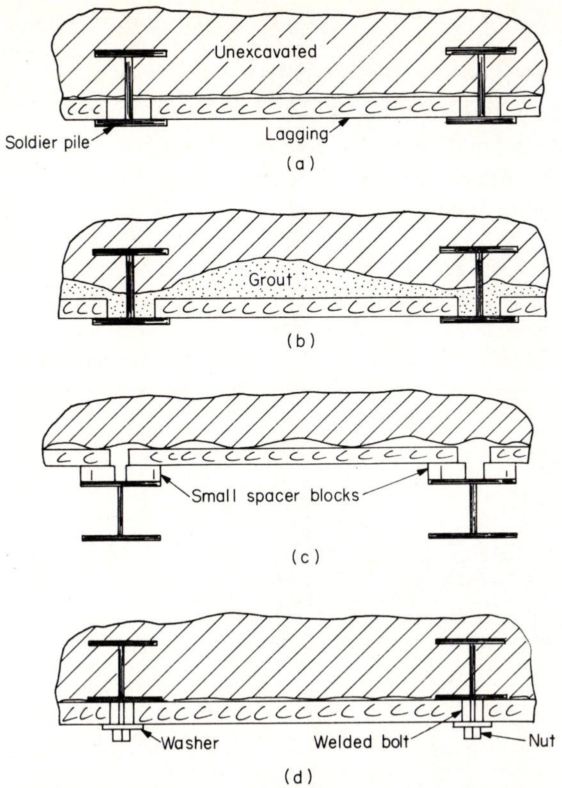

FIG. 9.38 Methods for transferring earth pressure from lagging to soldier piles: (*a*) lagging wedged against inside flanges of soldier piles; (*b*) overexcavation of (*a*) filled with grout or mortar; (*c*) lagging set between spacers, H piles, and soil; and (*d*) contact sheeting. [After Peck (1969).[36]]

soldier piles are shown in Fig. 9.38 and described as follows:

1. Lagging is wedged against the inside flanges. The clay yields as it is trimmed to make room for the lagging. Most of the earth pressure is transferred through the clay to the pile and very little through the lagging to the pile. The remaining void permits soil movement.
2. If large voids exist between the clay and the lagging in 1, they can be filled with grout or mortar to provide full contact.
3. Lagging set between spacers, the soldier piles, and the soil provides the advantage that the piles can be included in permanent reinforcements of the structural wall. Earth pressure is transferred to the piles as the excavation is carried to the face of the piles, but the heavily loaded soil behind the piles is then cut away to provide room for the spacers and the lagging and the earth wall move into the excavation, which can result in substantial surface settlement. Peck (1969)[36] observed that settlements adjacent to a wall constructed with this procedure were 3 times as great as those occurring when procedure 1 was used by the same contractor in the same conditions.
4. *Contact sheeting* is placed against the flanges of the soldier piles on the wall of the cut, permitting a tight fit because the exposed soil can be trimmed cleanly to the edge of the flange. It is the most satisfactory method to reduce lateral movements and also can serve as a form for concrete for the permanent wall.

Sheet-Pile Walls

Driven timber or steel sheets normally are used for excavation walls. For bulkhead walls, where corrosion and rot are a problem in waterfront conditions, sheets may be precast concrete, composite materials, creosoted timber, or special steel.

Timber sheets of the tongue and groove type frequently are used in sewer excavations where maximum depths are in the order of 4 to 5 m (13 to 17 ft). They cannot be driven to much penetration in strong soils such as glacial till or residual soils, but are useful in weaker materials such as moderately compact silts and sands or soft to firm clays. When driven carefully they are relatively watertight, and if their tips penetrate into soil with at least slight cohesion, groundwater control may require only sump pumping from the trench bottom. Care is required to prevent softening of the trench bottom soils which probably is the major cause of sewer pipe postconstruction settlement, except for weak soils and improper foundations *(see Art. 9.3.3).*

Steel sheets (Fig. 9.39), used in deep excavations, are most economical where retention of the higher earth pressures of soft soils is required. They have the important advantage that they can be driven to substantial depths below excavation bottom to provide control of heaving in soft clays or piping in saturated sands, which is

FIG. 9.39 Z-type sheeting installed as a temporary wall for the cooling-water structure of a nuclear power plant. Top of sheets indicates the bedrock surface irregularity. Second row of anchors is being installed as excavation in fine-grained alluvium proceeds.

not possible with the Berlin wall. They are more costly than Berlin walls, however, and are less adaptable to hard driving conditions caused by boulders, irregular rock surfaces, or other obstructions.

Less permeable than Berlin walls, they serve as *relatively* impervious walls in soils of high permeability, but do not provide a watertight boundary. Leakage can be severe if the sheets are damaged during driving, and seepage through interlocks can be sufficient to cause consolidation of highly and rapidly compressible materials such as organic soils and soft silty clays. In sandy soils raveling will not occur if the interlocks are tight, but driving sheets in loose sandy soils can result in adjacent ground settlement. Also, when sheets terminate in these soils below GWL, upward flow into the excavation with accompanying loss of support and erosion at the toe is a common occurrence.

Concrete Diaphragm Walls

Constructed as normal cast-in-place walls with supports, concrete diaphragm walls provide a watertight barrier, can be installed with a minimum of backslope subsidence, and become part of the permanent structure. Their use is becoming very popular in many countries.

Concrete walls now are commonly constructed with the slurry trench method which involves excavation of alternating panels along the proposed wall using a bentonite slurry to prevent collapse of the excavation sides. Slurry trenches over 20 m (67 ft) deep installed through thick deposits of soft organic silt, such as for the New York World Trade Center [Kapp (1969)[38]], have been found to be stable. The slurry density, usually about 1.06, cannot account totally for stability; much restraint is provided by the narrow panel widths and arching. [Slurry walls are discussed in Art. 8.4.2 of Hunt (1984).[5]]

The construction sequence is shown in Fig. 9.40. Panels are in the order of 50 to 100 cm thick (20 to 40 in) and up to 7 m (23 ft) in length, extending to the excavation bottom. Installation begins with the construction of shallow concrete or prefabricated steel guide walls. Excavation is then made by special thin-grab clamshell or other special equipment, slurry is pumped in, a guide tube placed at one end, and a prefabricated rein-

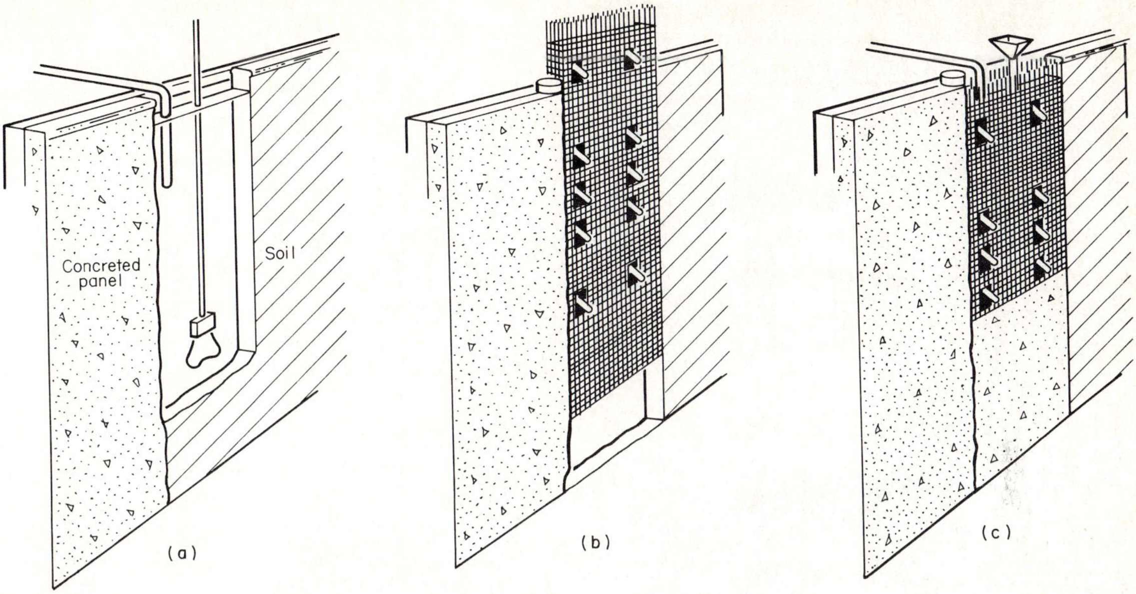

FIG. 9.40 Sequence of slurry and diaphragm-wall construction for World Trade Center, New York City. (*a*) Panel is excavated as slurry is pumped in. (*b*) Prefabricated reinforcing cage is lowered into the excavation. Openings are for anchors. (*c*) Bentonite slurry is replaced by tremie concrete. [From Kapp (1969).[38]]

forcing cage (Fig. 9.41) lowered into the excavation. The bentonite slurry is replaced by tremie concrete, the guide tube is withdrawn after initial set, and construction then proceeds to the next panel.

Because diaphragm walls are impervious, dewatering of granular soils is often not done, but the wall must be free of any openings such as "windows" or joints. Joints between panels do occur and during excavation careful observation is required to allow immediate application of corrective measures such as injection grouting, should openings be encountered. Large openings at substantial depths below GWL in soft ground, encountered suddenly without warning or preparation, can be disastrous from the aspect of sudden loss of ground.

Because they are considered semirigid, concrete diaphragm walls of shallow depths in moderate ground conditions are sometimes left unsupported. In deep excavations, support is required to restrict lateral deflections. Skin friction along the wall tends to reduce settlement of adjacent soils below amounts common to other systems such as soldier piles or steel sheeting. Johnson (1976)[39] reports vertical settlement appears to be in the order of 0.25% of excavation depth for diaphragm walls versus generally less than 1.0% for other systems.

A *soldier pile tremie concrete (SPTC) wall* is a modification of the slurry wall to increase lateral support, consisting of a tremie concrete wall in which steel soldier piles are embedded for reinforcement. In the SPTC procedure, slurry-filled holes are drilled along the wall (spaced about two wall widths apart) and steel beams (whose flanges are spaced slightly farther apart than the holes) are placed in them. The slot between the beams is then excavated while the slot is filled with slurry, followed by the placement of the tremie concrete.

In deep excavations in soft ground, these walls may alleviate the potential subsidence problem, but they still require some support system, which is much reduced, however, because of the relative wall rigidity compared to any other wall type. Where weak soils exist at foundation level and provide inadequate wall support, the steel

FIG. 9.41 Reinforcing cage to be lowered into slurry trench in preparation of basement walls for apartment building to be supported on mat foundation bearing on medium compact sands below GWL. The adjacent building, 10 stories high and supported on shallow spread footings, was not disturbed by the 12-m-deep excavation made within the diaphragm wall.

beams can be driven to penetrate into strong soils, but the wall construction need proceed only to excavation bottom or to a depth adequate to prevent heaving.

Complete coverage of slurry wall design and construction aspects is provided by Xanthakos (1979).[40]

Wall Support

Cross-Lot Bracing

A system of struts and walers to provide cross-lot bracing is common in trench excavations and other excavations of limited width (Figs. 9.42 and 9.43). In large excavations it is being used less frequently and is being replaced by other systems because heavily braced excavations are difficult and costly to work in.

Walers are horizontal members placed parallel to and against soldier piles or sheeting to which *struts* or *braces* are attached, connecting the excavation walls and providing support. They are made of steel or timber, depending upon horizontal loads and the excavation width. Long steel struts can be subjected to substantial expansion and contraction from temperature changes.

After the wall section is installed, excavation proceeds until the first level of bracing is reached. A horizontal wale is welded to the wall and the strut tightly welded against it. Horizontal spacing usually is a multiple of the soldier pile spacing. Vertical spacing may be somewhat smaller for the upper row than for the lower ones. To control deflections, construction design details regarding stiffness of the webs of the wales at the points of strut reaction are important, and the connections between the wales and the piles must be made carefully to avoid deflections and buckling. Struts usually are prestressed by jacks to about 50% of the expected load, and to provide space for working and soil removal, they usually are placed at a minimum vertical distance of 2.5 m (8 ft).

CONSTRUCTION IN SOFT GROUND Lateral movement of braced excavation walls in soft ground is common and has two components: cantilever deflection of the upper wall before placing the first brace and inward movement of the wall as the excavation is deepened. The first condition results in settlement immediately adjacent to the wall and the second in the spread of settlement away from the wall. Excessive movement occurs if the excavation is made too deep before the uppermost strut is placed. Peck (1943)[41] and Ward (1955)[42] recommend that in soft clay the top strut should be installed before the cut reaches a depth $d = 2s_u/\gamma$.

The most common cause of subsidence is overexcavation before placing each strut, and experience shows that it is even more significant than excessive strut spacing in causing lateral move-

FIG. 9.42 Cross-lot bracing for excavation in soft varved clays for power-plant cooling-water intake structure. (Courtesy of Converse Consultants.)

FIG. 9.43 Interior view of excavation in Fig. 9.42 while excavating for the third level of struts. (Courtesy of Converse Consultants.)

ments. Strut removal is also a critical function, and movements are minimized if "soft" elements are not left or introduced between the side of the excavation and the permanent wall. The relationship between construction progress, wall deflection, bottom heave, and ground subsidence in a well-constructed, braced excavation is shown in Fig. 9.44. When the excavation reached a depth of almost 9 m in the soft clays, the maximum lateral displacement was about 25 cm (10 in) and the maximum ground subsidence was about 22 cm at a distance of 5 m from the excavation.

TRENCH METHOD One of the earliest systems of cross-lot bracing used for excavations in soft clays in Boston, Chicago, and Detroit, the trench method is no longer used in developed countries because of the large amount of costly hand labor it requires but may be useful in areas of low-cost labor. A narrow trench is excavated around the building periphery, sheeted and braced, and then reinforcement is placed and the trench concreted with the sheeting left in place. Simultaneously, cross-trenches are excavated along column lines, sheeted and braced, and a concrete strut is formed in the trench bottom which later becomes part of the basement floor. The building columns are established at cross-strut intersections, and then the beams for the first floor are constructed to serve as struts. The procedure

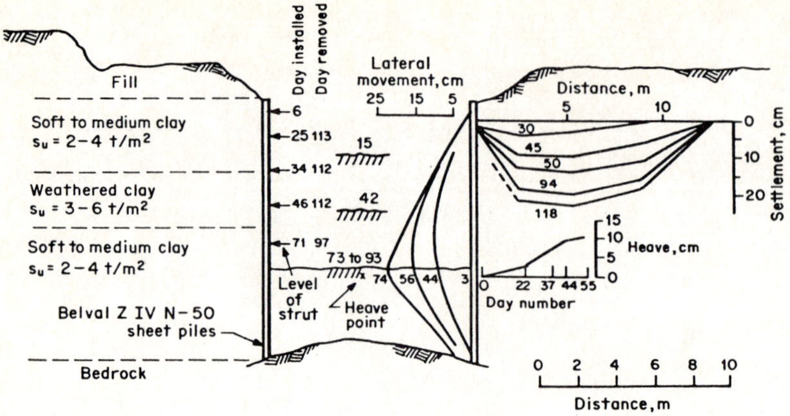

FIG. 9.44 **Relation of construction progress to lateral movement, settlement, and bottom heave from open cut in soft clay in Oslo (Vaterland I).** [From NGI (1962–1966).[43]]

provides a system of cross-lot bracing while most of the soil in the basement area remains in place, thus minimizing deflections. When the struts, walls, and columns are ready the excavation is completed.

Rakers and Berms

Inclined braces or rakers are used to support walls where the distance is too large for convenient strut support. Following wall installation, the soil in the central excavation area is removed to foundation level leaving a berm to support the wall. The center portion of the basement slab, or a "kicker" block, is formed and the upper portion of the wall braced with a raker placed against the kicker block as shown in Fig. 9.45. To minimize wall deflection, the raker should be placed in a shallow trench excavated in the berm. The berm is then lowered and the next level of rakers installed.

Movements in excess of those occurring with cross-lot bracing usually occur with the raker and berm system because of the inherent construction procedure. Stability of the berm and the kicker blocks requires analysis, which is particularly important in soft clays. Damage to a berm-wall system can occur as a result of insufficient passive resistance of the berm, slope failure in the berm, or a deep-seated heave movement beneath the berm, as shown in Fig. 9.45. These can lead to total collapse of the system, but more commonly result in large deformations of the area retained by the wall. The kicker block must be analyzed as a footing with inclined loads and should be located at a distance greater than 1.5*D* (Fig. 9.45*b*).

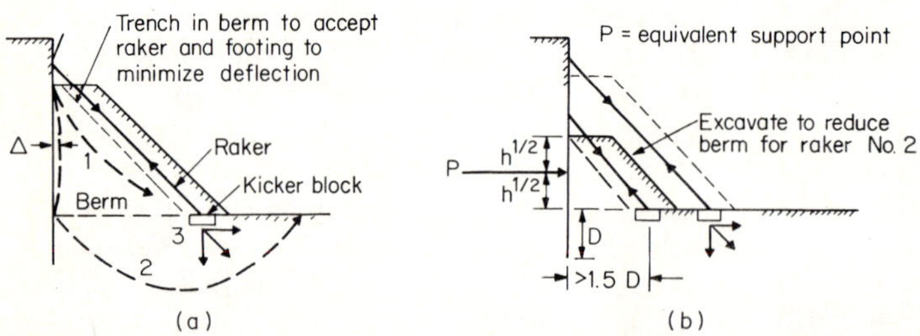

FIG. 9.45 Raker and berm method of wall support. (*a*) First-level raker installed. (*b*) Second-level raker installed. In addition to the raker forces, stability analysis is required for possible slope failures at (1) and (2) and kicker capacity at (3), as well as berm deformation itself. [After CGS (1978).[44]]

Anchors (Tiebacks)

Tiebacks *(see Art. 7.7)* are being used with increasing frequency in deep excavations for steel or concrete wall support wherever material suitable for embedment is available at some reasonable depth. They offer many advantages, including high wall capacity, maximum wall deflection control, facility of contingency designs (anchors are added as needed), and an excavation open and free of obstructions.

Following the installation of either soldier piles, steel sheet piles, or a concrete diaphragm wall, a small excavation is made along the wall and the first row of anchors installed as shown in Fig. 9.46. The trench is made only wide enough for the anchor-installation machine, and the excavation-anchor sequence is repeated until the bottom is reached. Anchors usually are proof-tested to 125 to 150% of design load; then the load is reduced and locked off at 75% or more of design load when the excavation support is temporary [Goldberg (1976)[45]]. Loads are locked off at 100% design load for permanent walls and to minimize backslope subsidence. Care is required, however, so as not to overstress soft clays.

Well-constructed anchored walls undergo less lateral deflection than braced walls and therefore provide better control of backslope subsidence, primarily because of excavation procedures and prestressing. Anchor installation requires only a small excavation to support level for equipment access, whereas for the installation of braces there is a tendency to overexcavate to below support level. Anchored walls are always prestressed to remove the "slack" from the system. Reduction in wall movement can be gained by increasing the prestress load and wall and anchor stiffness. Unless creep or other deformation occurs, the anchors maintain their load throughout the excavation sequence; with bracing, however, load and deformation typically increase as the excavation depth increases. In addition, anchors place the entire soil mass between the anchors and the wall in compression, resulting effectively in a large gravity wall. The soil-wall system is less deformable than a braced wall acting only on a cut face.

Two possible causes of loss of ground and subsidence for well-designed anchor systems are caving of the anchor holes prior to grouting and flow of cohesionless materials into the excavation through the wall openings made for anchor installation. An additional disadvantage is the possible downward movement of the wall resulting from the vertical component of the anchor forces. Because of large anchor forces it is not uncommon for an anchored wall to deflect away from the excavation into the backslope [Johnson (1976)[39]].

Cantilevers

Cantilever sheet-pile walls *(see Art. 9.4.4)* are used for temporary excavations of moderate depth, but because of large earth pressures and deflections that develop, they are seldom used to retain excavations over 3 to 4 m (10 to 13 ft) deep. Because support is derived totally from passive pressure exerted on the embedded portion, required penetration depths are relatively large. An advantage is that sheets can be pulled and reused, and salvage value is high.

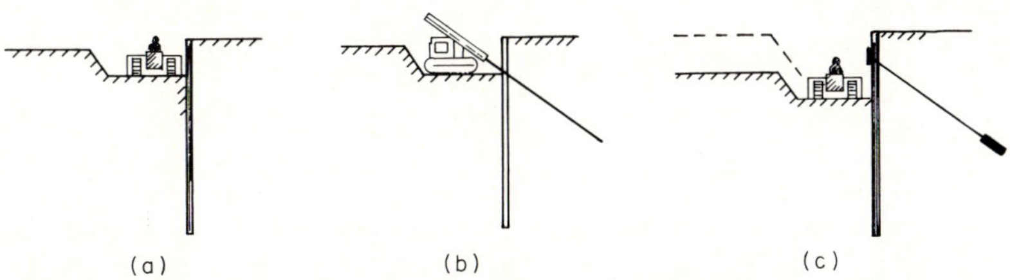

FIG. 9.46 General procedure for anchor installation for support of a sheet-pile or concrete diaphragm wall: *(a)* trench excavated after wall installed; *(b)* first anchor row installed; and *(c)* trench excavated for next anchor row and excavation lowered.

9.3.3 DESIGN CONSIDERATIONS

General

Design proceeds after the wall and dewatering systems have been selected and requires estimation of the wall penetration depth, calculations of earth pressures, selection of the axial loads for braces or anchors, selection of the wall section to resist bending moments, and, if necessary, estimates of wall deflections and backslope subsidence.

Design Approaches

Braced and anchored walls (and tunnel supports) conventionally are designed by empirical methods based on general concepts and experience because of the relative complexity of soil-structure interaction. In addition, empirical approaches can account for problems created by construction difficulties and environmental factors not considered in analytical techniques. More analytical approaches traditionally are used for systems such as bulkheads and retaining walls.

Walls, such as diaphragms and continuous sheet piles which approach plane strain conditions, have been analyzed by the FEM since it yields information on earth pressures, structural bending moments and shears, and soil and wall movements. The retention system is treated as a whole and construction sequences accounted for, which is especially pertinent to the determination of wall deflections and backslope subsidence.

Failure Modes

Braced walls usually fail into the excavation due to heave-type movements inside the excavation or from overloading of the support system. Bottom heaving results from overstressing of soft clays or piping of granular soils as discussed in the following section. Both conditions may be controlled by extending the wall to deeper penetration, although in the case of soft clays, firm soils must be reached. Drainage of granular soils is effective.

Anchored walls fail externally due to movements behind the anchorages, and their overall stability requires careful evaluation since failure can occur totally and much more rapidly than in a braced system. Possible failure modes are illustrated in Fig. 9.47. Modes *a*, *b*, and *c* are in the form of slope stability failures and *d* in the form of overturning, all of which result from either an inadequate anchor penetration beyond any potential failure surface or inadequate anchor capacity. Mode *e* is settlement or penetration of the wall into weak soils caused by the vertical component of steeply inclined anchors. Vertical movement substantially reduces anchor capacity. In addition, the anchors can rupture (*f*) or the wall can yield (*g*). In Sweden, anchor-wall failures caused by *frost action* have been reported by Broms and Stille (1976).[46] To avoid this failure

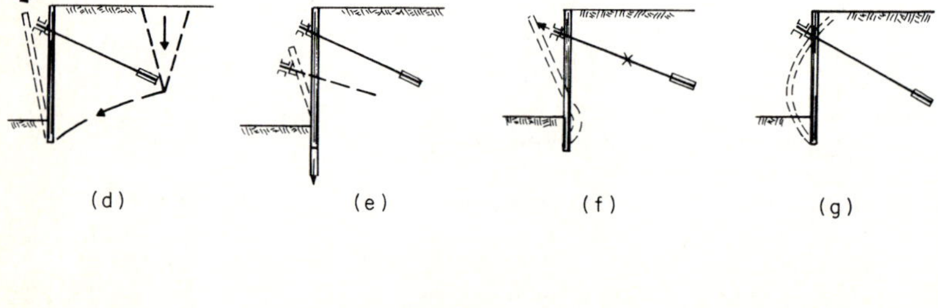

FIG. 9.47 Failure modes of anchored walls: (*a*) slope failure; (*b*) toe failure; (*c*) wedge failure; (*d*) overturning failure; (*e*) penetration failure or settlement; (*f*) anchor failure; and (*g*) wall failure.

mode, walls require either good drainage behind the wall or insulation or heating during winter.

Wall Depth Factors

General Conditions

STRONG SOILS In medium compact sands with at least some cohesion or stronger soils, sheeting or soldier piles are driven only a meter or so below the excavation bottom to mobilize passive resistance at the wall toe. The required penetration is checked by considering the equilibrium of the free-end span *(see Art. 9.4.3)* below the final brace. The allowable moment for this span is determined by summing the moments of P_a and P_p about the lower brace.

TROUBLESOME CONDITIONS Soft clays or loose sands and silts below the water table can result in bottom heave, piping, or loss of wall foundation support and require careful evaluation. Control may be effectively accomplished by increasing sheeting depth, but in severe cases in soft soils where backslope subsidence is of concern, semirigid walls such as diaphragm walls or the SPTC wall should be considered. Sands may require a deep dewatering system, especially in large excavations.

Heave in Clay

Where an excavation bottom is in *soft clay* and undrained conditions prevail, there is danger of a base failure and subsequent bottom heave when the surcharge load of the adjacent soil is great enough to cause a *bearing-capacity failure* as shown in Fig. 9.48*a*. Bottom heaves of substantial magnitudes have been observed in stiff clays. Such expansion is not related directly to shear strength, but rather is a function of stress release in overstressed or heavily preconsolidated clays, including those that are fissured or slickensided.

ESTIMATING APPROACHING FAILURE The approach to the failure state can be judged by the dimensionless number or *stability factor* N_b where

$$N_b = \gamma H/s_u \qquad (9.43)$$

where s_u = undrained strength below base level [Peck (1969)[36]]. If the strength of the underlying soil is ignored, theoretical studies show that a plastic zone would start to form at the lower corners of the cut (in an infinitely long trench) when N_b reaches 3.14. The zone would spread with increasing values for N_b until base failure occurred, which is the critical state, and N_{cb} = 5.14 *(see Art. 6.2.4)*. Cuts are not infinite in length, however, and the strength of the surcharge material is not negligible. Values for N_{cb} for ordinary rectangular-shaped cuts have been found to be in the order of 6.5 to 7.5 (rather than 5.14), and the values for the beginning of plastic zone deformation are similarly greater than 3.14. Peck (1969)[36] suggests that in soft clays the stability factor must be kept to less than about 6.

A simple relationship for estimating FS against bottom heave for variously shaped excavations is given by Bjerrum and Eide (1956)[47] as

$$FS = \frac{N_c s_u}{\gamma H + q} \qquad (9.44)$$

where q = uniform surface surcharge around the excavation. Values for N_c as a function of ex-

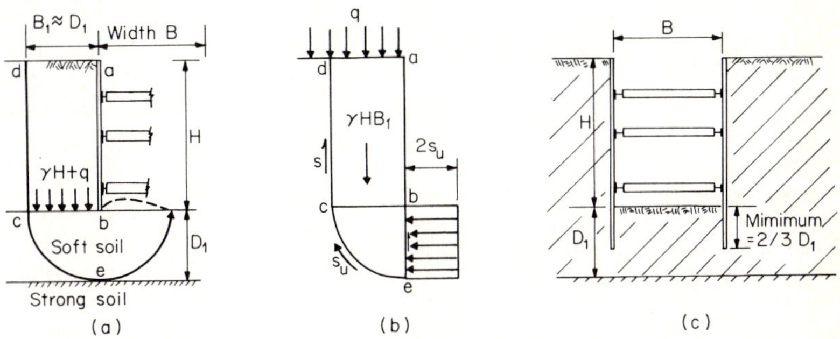

FIG. 9.48 Bottom heaving in soft clay: (*a*) braced wall in soft clay with bottom heaving; (*b*) free-body diagram; and (*c*) possible preventative measure if hard stratum is not economically reachable.

cavation width B and depth H for ratios of B/L (L = length) are given in Fig. 9.49.

CONTROL Where the potential for base failure exists the solution may be to drive sheet piles to greater depths to penetrate strong soils, which theoretically increases passive resistance against the failure surface (Fig. 9.48*b*). In deposits of soft clay extending for substantial depths below excavation bottom, Teng (1962)[11] recommends driving the sheeting to a depth below excavation bottom of at least two-thirds the soft stratum thickness below the bottom, as shown in Fig. 9.48*c*. The beneficial results of deep-driving sheeting are often overestimated, however, because of its flexibility [Peck (1969)[36]]. Bottom heave is best controlled with more rigid walls such as the concrete diaphragm wall or the SPTC wall, but they require adequate foundation support. An alternative is to excavate around the cut to reduce its height.

Piping in Granular Soils

In sands and silts below the water table, upward seepage pressures must be kept under control to avoid a quick condition and piping which develop from excess hydrostatic pressures (*see Art. 2.3.2*). In layered soils, the hydrostatic pressures in the pervious zones must be controlled to prevent bottom blowout. In either case the result is loosening of the bottom soils, which may cause collapse of the sheeting and loss of the supporting capacity for building foundations. Analysis

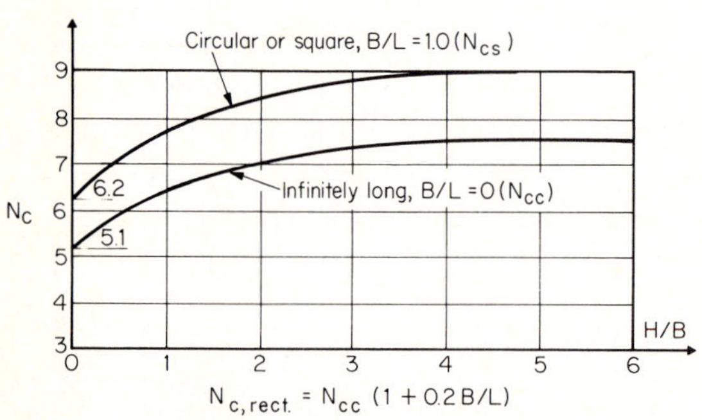

FIG. 9.49 Values of stability number N_c for heave analysis of braced excavations. FS = $(N_c s_u)/(\gamma H + q)$, where $N_{c,\text{rect.}} = N_{cc}(1 + 0.2B/L)$. [After Bjerrum and Eide (1956),[47] from NAVFAC (1982).[1]]

may be performed by flow-net construction as illustrated in Fig. 9.50.

CONTROL The most effective control is dewatering, but in deep excavations it may be more economical to penetrate the wall to a depth sufficient to intercept the flow lines with high heads (Fig. 9.50). The penetration depth required to prevent piping is a function of the pressure head between the water table and the excavation bottom (net head), the ratio of the excavation width to net head, and the soil conditions below excavation bottom. Charts have been prepared for estimating depths for various conditions including homogeneous sands of infinite depth, homogeneous sands with an underlying impervious layer, a coarse layer underlying a fine layer, and a fine layer overlying a homogeneous deposit. The latter can represent a very dangerous condition if the fine layer is below the excavation bottom but higher than a depth below the wall equal to or greater than the excavation width. The more common conditions and criteria for estimating sheeting penetration to provide FS = 1.5 against piping are given in Fig. 9.51, a modification of information provided in NAVFAC (1982).[48] As a general approximation, the wall should be extended below excavation bottom in loose sands to a depth at least equal to the net head.

Foundation Support

BRACED EXCAVATIONS AND LIGHT WALLS Except for failures resulting from heave or piping, foundation failures are seldom a problem for these support systems.

CONCRETE DIAPHRAGM WALLS Heavy permanent walls in soft ground should be provided with adequate foundation support although some support is derived from side friction along the wall. In very soft ground the use of the SPTC wall and extending the soldier piles into strong materials provide good wall support.

ANCHORED WALLS Inclined tiebacks impose a vertical component of the anchor force which may cause wall penetration. Downward movement is resisted by wall skin friction and reaction to the base. If the anchor is designed on the basis of K_a

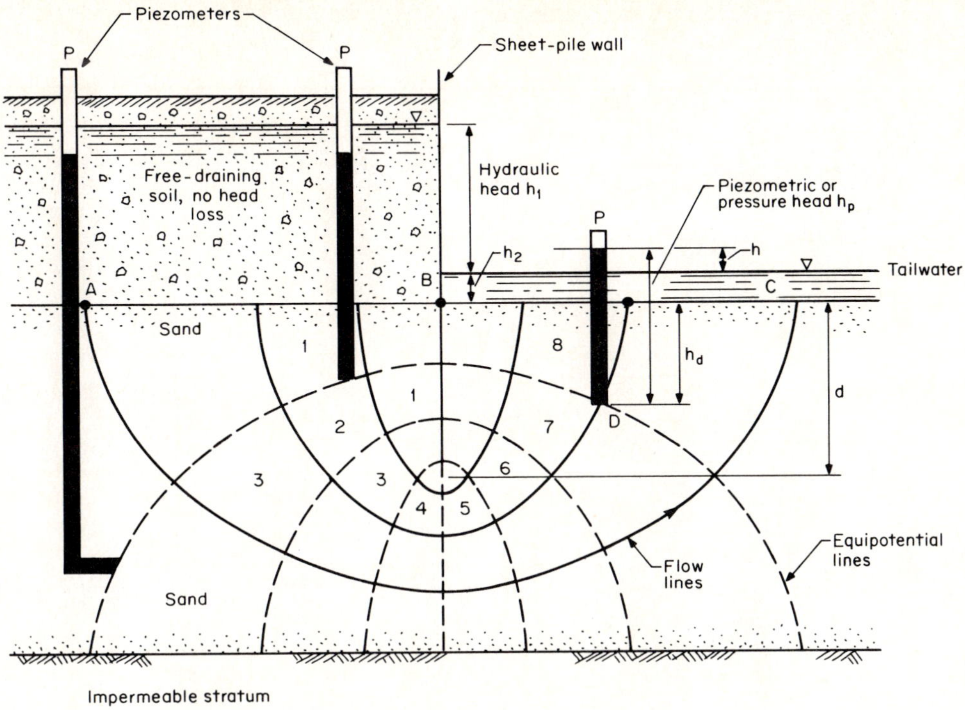

1. Find *seepage quantity beneath sheet piling.*
   Given: 0.001 m/min, h_1 = 8 m, h_2 = 2 m, h_d = 6 m

   $$L = \text{length of sheet pile wall along excavation} = 30 \text{ m}$$

   Solution: From the flow net, N_t = 4, N_e = 8. Quantity

   $$q = l(N_f/N_e)h_1 \qquad \text{(Eq. 2.12)}$$

   $$q(\text{per meter of wall}) = 0.001 \times 4/8(8) = 0.004 \text{ m}^3/\text{min}$$

   $$Q(\text{per total wall length}) = 0.004 \times 30 = 0.12 \text{ m}^3/\text{min}$$

2. The *pore pressure at point D* = $u_w = (h_1 + h_2 + h_d - 7/8h_1)\gamma_w = 9 \text{ T/m}^2$. That part, $h_1 - 7/8h_1)\gamma_w = h\gamma_w$, due only to the flow of water, is the *excess hydrostatic pressure.*

3. *Liquefaction* occurs when the exit gradient i is approximately 1; i can be expected to be highest on the outside of the wall at point B. $i = \Delta h/1$, where $\Delta h = h_1/N_e$, and l = the length of the side of the square where exit occurs, or 3 m. Therefore, $i = 8/8/3 = 0.33$, which is a marginal value in terms of desired values for FS (3 to 5). Terzaghi demonstrated that heave from boiling occurs within a distance from the wall of about $d/2$.

NOTE: FS = $i_{cr}/i = 1/0.33 = 3$.

FIG. 9.50 An example of a confined flow net beneath sheet piling. [From Hunt (1984).[5] Used with permission of McGraw-Hill Book Company.]

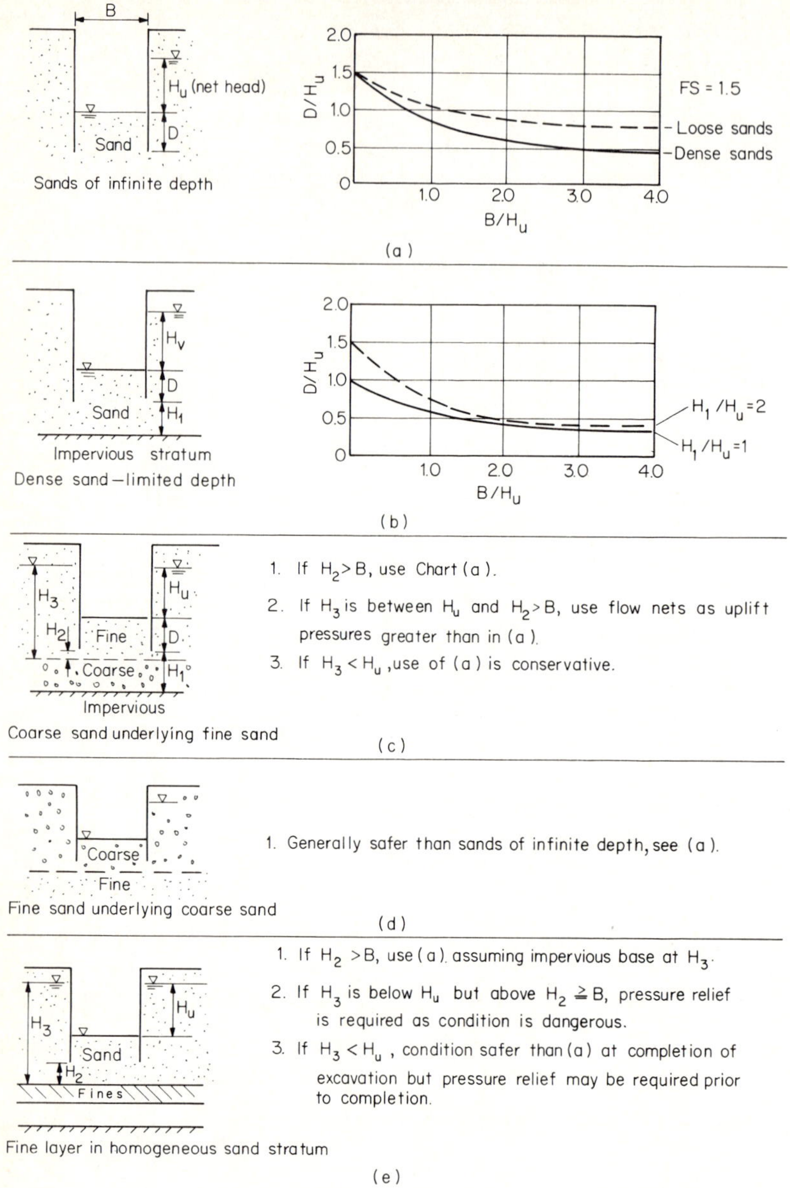

FIG. 9.51 Minimum depth of sheeting penetration to prevent piping for various conditions for FS = 1.5. [After NAVFAC (1982).[48]]

pressures, skin friction is taken as zero in both cohesionless and cohesive soils. If the anchor design basis is ½ $(K_a + K_o)$, or K_o, a skin friction value of ⅔ϕ in cohesionless soils and $c_a = s_u$ in clay soils may be assigned [CGS (1978)[44]]. For earth-pressure assumptions and wall friction see Art. 9.2.3, and for anchor designs based on earth pressures see Art. 9.3.4.

Anchored walls bearing in sands can settle when displacement piles are driven in the excavation because pile driving causes a temporary increase in pore pressures accompanied by loss of support. Broms and Stille (1976)[46] report on a number of failures of anchored walls in Sweden due to inadequate foundation support or pile driving adjacent to walls. Base capacity of a steel WF or

H-section soldier pile in sands can be increased markedly by placing it in a clean prebored hole filled with concrete.

Earth Pressures

Empirical Basis

In braced or anchored retaining walls, earth-pressure distribution during construction is a function of the support system and construction procedures. Due to its complexity, it is commonly determined by semiempirical rules that were developed for strut-supported walls but now are applied to raker-braced and anchored walls by some practitioners. The total maximum pressures are essentially fictitious but give the computed strut loads that most likely will not be exceeded. Permanent walls may be designed on the basis of Rankine pressures but usually are designed for the at-rest condition.

Earth-Pressure Variations during Construction

After the wall is installed to some predetermined depth, excavation is made to enable placement of the first row of supports, at which point the excavation is shallow, earth pressures low, and deflections small. As the excavation proceeds, however, prior to the installation of subsequent supports, earth pressures increase and the wall deflects before the supports can be placed. The exaggerated deformation line at excavation completion of a strut-braced wall is shown in Fig. 9.52 at (1).

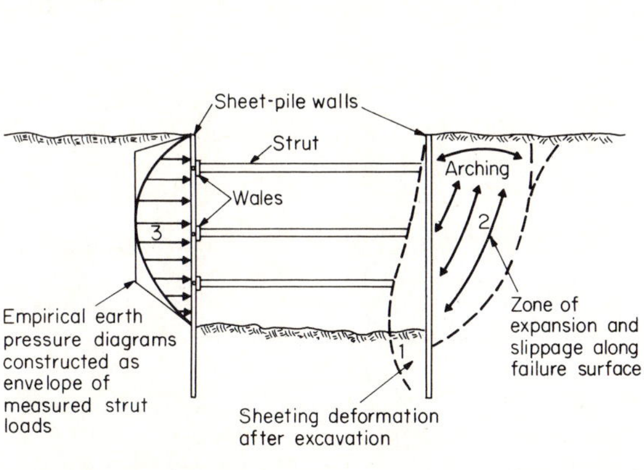

FIG. 9.52 Deformations of sheeting and soil and the resulting pressure distribution in braced cuts in sand. [After Tschebotarioff (1973).[14] Used with permission of McGraw-Hill Book Company.]

As the excavation proceeds and deflections occur, active pressures are mobilized and the distribution is approximately that of Rankine. If the wall is relatively rigid and the struts are placed carefully to restrict deflections, earth pressures against the wall assume higher values than the Rankine at the upper shoring levels. The real pressure distribution depends on the actual yielding permitted during strut placement at the various levels. Model tests have shown that yielding near the lower strut causes a slippage zone, as in Fig. 9.52 at (2), but that in granular soils arching increases lateral pressures in the upper zone and reduces them in the lower zone. In any event, the *final* resulting pressure distribution is in marked disagreement with the Coulomb or Rankine theories as shown in Fig. 9.52 at (3).

Apparent Earth Pressures

Empirical rules have been developed for design earth-pressure diagrams for horizontally braced cuts based on field and laboratory measurements. For determining magnitude and distribution, the most widely used rules are those developed by Peck [Terzaghi and Peck (1967),[10] Peck (1969)[36]], which *apply to end-of-construction conditions.* In reality, the stresses at each support level should be considered in sequence.

SANDS The apparent pressure diagram for sands is given in Fig. 9.53 which was developed to cover a scattering of test data at *dewatered* sites. It is seen as a rectangular distribution defined by

$$\sigma_h = 0.65 K_a \gamma H \qquad (9.45)$$

to which hydrostatic and surcharge pressures must be added separately.

CLAYS The pressure diagrams for clays (Fig. 9.54) are empirically based on total unit weight without regard to pore-pressure distribution and there is no need to add water pressures to the diagram, although surcharge pressure is added separately. The behavior of the soil and the bracing system depend on the stability number $N_c = \gamma H/s_u$ (Eq. 9.43) where s_u is the undrained strength of the clay beneath the cut that would be involved in a general shear failure (Fig. 9.48). It serves as a danger signal; when $N_c > 6$ caution is indicated, especially in sensitive clays.

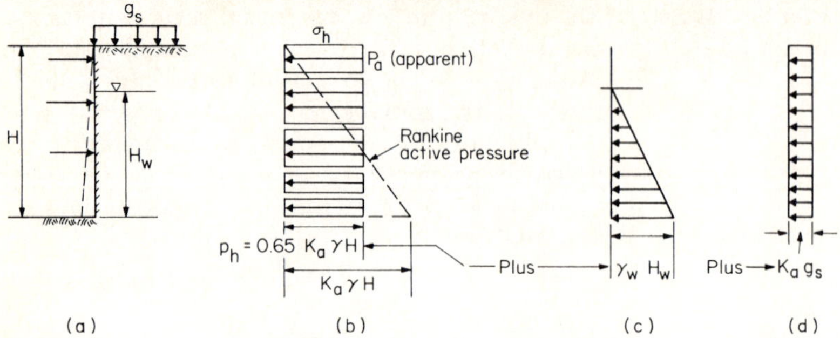

FIG. 9.53 Apparent earth pressures from sands for strut loads in braced excavations: (*a*) braced wall; (*b*) apparent pressure diagram for $p_h = 0.65\,K_a\gamma H$ compared with Rankine active distribution; (*c*) water pressure is added to apparent pressure; and (*d*) uniform surcharge pressure is added to apparent pressure. [Part (*b*) from Terzaghi and Peck (1967).[10] Used with permission of John Wiley & Sons, Inc.]

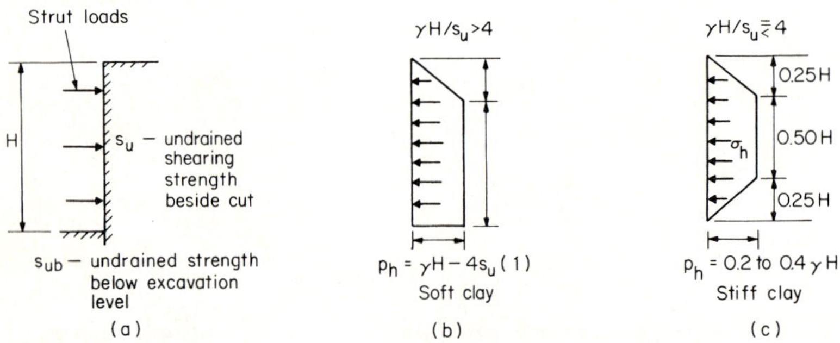

FIG. 9.54 Empirical rules for apparent pressure diagrams for calculating strut loads for braced cuts in clays: (*a*) braced wall; (*b*) diagram for $\gamma\,H/s_u >$ 4 provided that $\gamma\,H/s_{ub} < \approx 4$; and (*c*) diagram for $\gamma H/s_u \leq 4$. (NOTE: (1) If soft clay extends for considerable depth below bottom of sheeting, use $\gamma\,H - m4s_u$ where $m = 0.4$ or $\gamma H - 1.6s_u$.) [Diagrams from Peck et al. (1974).[24] Used with permission of John Wiley & Sons, Inc.]

Strength s_u applies to clays for some distance beneath the cut bottom; for cuts less than about 20 ft (6 m) deep with $N_c = 3$ to 4, no special investigations are necessary.

Peck considers that the value for p_h calculated from the apparent pressure diagram (Fig. 9.54) is reasonable for $N_c = 5.14$ for an excavation with a length considerably greater than its width, but for larger values an increment of K_a should be added (personnel communication, 1984). The total earth pressure against a cut in saturated clay is expressed as

$$P_a = 0.5\gamma H^2(K_a + \Delta K) \tag{9.46}$$

where K_a is the Rankine value

$$K_a = 1 - (4c/\gamma H) \tag{9.47}$$

and ΔK is obtained from Fig. 9.55. Employing the principle of virtual work, Henkel (1971)[49a] investigated the influence of a base failure on the earth pressure against bracing and developed the relationships between ΔK and N_c in terms of d/H where d = the depth to a strong layer and H = excavation depth.

LAYERED SOILS These soils are difficult to evaluate. Thinly layered or stratified deposits can be solved approximately by using the apparent

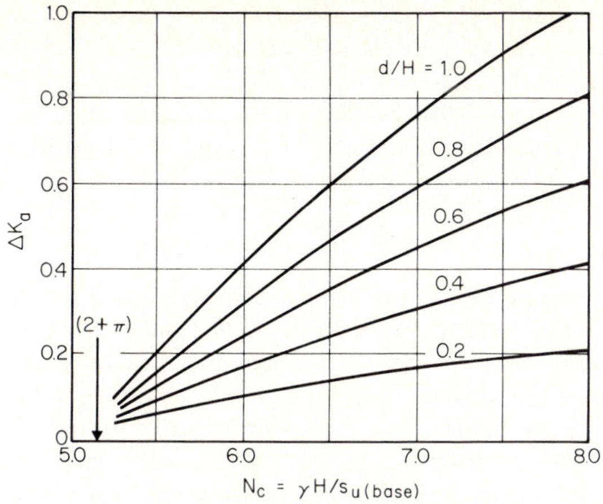

FIG. 9.55 Increment of ΔK of earth-pressure coefficient for values of base stability factor greater than $N_c = (2 + \pi)$. [After Henkel (1971).[49a]]

pressure diagrams for the predominant characteristic (sand or clay) if there is not a marked difference in material properties. Pressure diagrams for strongly layered formations can be approximated by superimposing the lateral stress diagram for each layer or by computing the Rankine distribution to find the total active Rankine pressure, multiplying by 1.2, and distributing it trapezoidally (personal communication from Dr. Peck, 1982). Excavations in sand-clay mixtures where drainage occurs can be approximated by using the drained value for ϕ and assuming $c = 0$.

Actual Magnitude and Distribution of Pressure

In a state-of-the-art report, Bjerrum et al. (1972)[49] pointed out that the *total pressure* on an excavation bracing system generally is controlled by soil shear strength, but the pressure *distribution* is determined by arching effects. The magnitude of the total active earth pressure exerted by the soil behind and beneath the wall includes the soil mass extending down to a depth along the wall where the soil movements are no longer significant. These observations were based on monitored excavations.

Soils with high modulus values, such as dense sands or soft rock, were concluded to yield a *total sum* for strut loads nearly equal to the classical value, but the *distribution* would approach the empirical trapezoidal shape as shown in Fig. 9.56*a*. This holds true also for excavations with a width that is small compared to its depth, such as trenches, and soils with lower modulus values.

Soils with low modulus values, such as soft clays, medium stiff clays, silt, and loose sand, were concluded to yield a sum of the strut loads considerably greater than the classical values computed for a depth equal to excavation depth as shown in Fig. 9.56*b*, which is due to the greater amount of deformation and greater magnitude of the arching effect in the weaker soils. The excavation causes a considerable amount of deformation below excavation depth which results in a substantial reduction in earth pressure below excavation level and an equivalent increase in earth pressure above excavation level. The reduction is felt to a depth comparable to the excavation width but is influenced strongly by the depth to a strong stratum.

Bjerrum et al. (1972)[49] commented that Peck's parameter $N_c = \gamma H/s_u$ is introduced to provide for the larger strut loads observed in soft clay due to bottom movements, but there was yet to be developed a relationship that considers the effect of problem geometry, which is significant. An approach to estimate earth pressures, however, based on bearing capacity for the conditions of either a relatively shallow strong stratum or local unloading adjacent to the excavation is

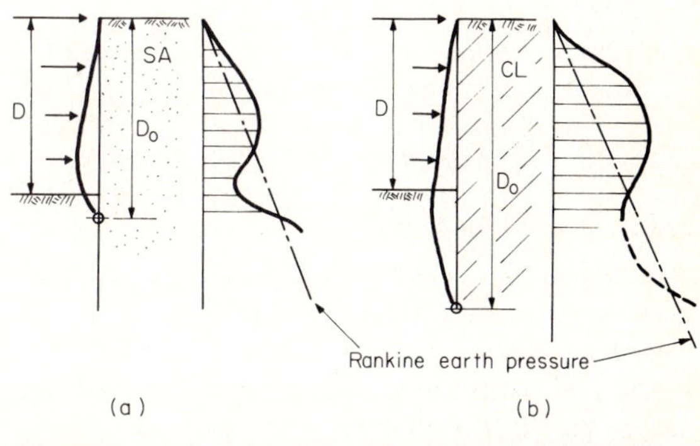

FIG. 9.56 Wall deflection and redistribution of earth pressure below excavation level in braced excavation: (*a*) insignificant deformations in dense sand and (*b*) significant deformation in soft clay. [From Bjerrum et al. (1972).[49]]

given by Henkel (1971).[49a] (See discussion for Fig. 9.55.)

Surcharge and Earthquake Loadings

As with rigid walls, other forces to be considered include surcharge and earthquake loadings, as discussed in Arts. 9.2.3 and 9.2.4 respectively. Because procedures for estimating the effects of surcharges provide only approximations and are generally conservative, especially for the case of a buried foundation, critical situations are best evaluated in terms of anticipated deflections employing the FEM as will be discussed (personal communication, Dr. D. J. Murphy).

Lateral earth pressures against flexible retaining structures in nuclear power plants under dynamic loadings are discussed by Lucks (1976).[50]

9.3.4 WALL SYSTEMS ANALYSIS

After the wall and support system is selected and the wall depth determined, the earth pressures are calculated, followed by the bending moments in the wall section. Then a section of adequate thickness is selected, strut or anchor loads are calculated, and their dimensions are determined. Increasing wall stiffness permits increasing the vertical spacing of supports but to a lesser extent than often is assumed (personal communication from Dr. Ralph Peck).

Strut-Braced Wall

Earth-Pressure Determination

The major problem to be resolved in the selection of strut dimensions is the maximum load which can be expected to be put on the individual struts. Semiempirical rules, such as Peck's, are not intended to represent the actual earth pressure or its distribution with depth but rather to constitute a hypothetical pressure diagram from which can be derived the strut loads which may be approached, but not exceeded.

The actual sum of the strut loads can be close to or greater than the theoretical Rankine value, depending on deformation magnitudes, which are very much related to construction practices. The portion of the total earth pressure to be carried by the struts is a function of the deformation, which depends also on such geologic factors as Young's modulus and stratification, as well as geometric factors such as the excavation width and depth. The actual earth-pressure distribution over the struts can be triangular, trapezoidal, rectangular, or very irregular.

Procedure with Apparent Pressure Diagram

The apparent pressure diagram is completed for the wall using Fig. 9.53 or 9.54. Analysis is performed on the wall, and its supports and the components are sized. Wall sections are selected based on the maximum moment anticipated, wales are designed to resist horizontal reactions from the soldier piles or sheeting, and the struts are designed as compression members.

The first step is to select the sheeting penetration depth and evaluate bottom conditions of potential piping or heave, as discussed in Art. 9.3.3. In clays, base stability is checked to determine if $\gamma H/s_{ub} < 6$. Experience has shown that for values less than about 6, movements of the bracing system and clay heave below base level are small. Between values of 6 and 8, movements of even a well-designed bracing system become intolerably large, and at values exceeding about 8 bracing collapse is likely. Alternately, FS against base heave is calculated with Eq. 9.44; FS $\geq$ 1.5 is satisfactory.

If the base stability is adequate for all depths of excavation, the strut loads can be determined from the apparent pressure diagrams given in Fig. 9.54 where *b* applies generally to soft to medium clays and *c* to stiff clays. The difference in the envelope shapes reflects the behavior of the clay near the bottom of the cut, which in the case of the softer clays is essentially inelastic. In practice, the earth pressures for both cases are computed and the higher value used for design regardless of the value for $\gamma H/s_u$. In most instances, the width of the apparent pressure envelope given in Fig. 9.54*c* can be considered equal to $0.3\gamma H$. The general solution to find strut loads in a braced excavation is given in Fig. 9.57.

The *wall section modulus* is selected after the maximum apparent earth pressure has been calculated and the strut spacing determined. Braced sheet piling generally is designed as a continuous member supported at strut levels, with a fic-

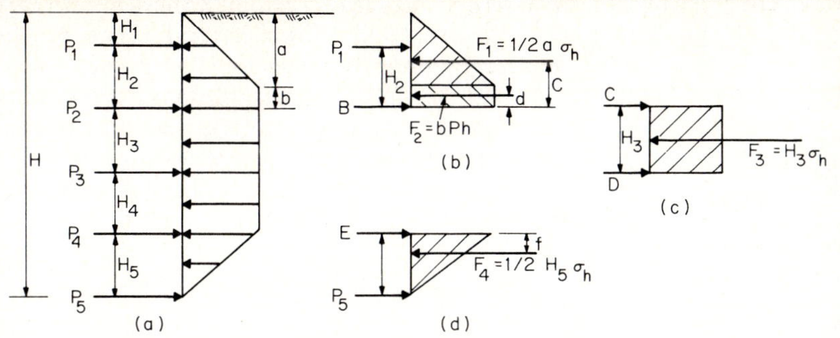

1. Compute maximum earth pressure (Fig. 9.53 or 9.54)

2. Find P_1 ($\Sigma M_B = 0$) as shown in part (*b*)

$$P_1(H_2) = (F_1c) + (F_2d)$$

3. Find $B = F_1 + F_2 - P_1$

4. In part (*c*), find $C = D = \frac{1}{2}F_3$

5. Find P_5 ($\Sigma M_E = 0$) as shown in part (*d*)

$$P_5(H_5) = F_4f$$

6. Find $E = D - P_5$

7. Find load per unit of wall width

$$P_1 = p_1 \text{ as computed}$$
$$P_2 = B + C$$
$$P_3 = C + D$$
$$P_4 = D + E$$
$$P_5 = P_5 \text{ as computed}$$

8. Strut loads = load per unit of wall width × spacing

FIG. 9.57 General solution to find strut loads in a braced excavation.

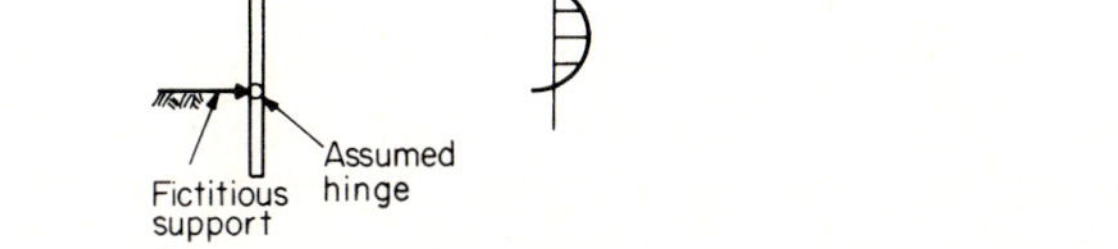

FIG. 9.58 Assumption of continuous span for the design of sheeting or soldier piles.

titious support introduced at the bottom of the excavation where the sheet piling is assumed to be hinged, as shown in Fig. 9.58. Soldier piles also are designed as fully continuous members. The maximum moment between supports is computed from Eq. 9.48, in which H_{max} is the maximum distance between supports (often governed by the construction stage) and p_h is the average earth pressure on the wall over the longest span. Arching between supports is accounted for by a reduction of 80% to obtain M_{design}. Above the excavation level for the first brace, the maximum moment is computed as for a cantilever. The required section modulus is found from

$$S = M_{max}/F_s \qquad \text{cm}^3/\text{m (in}^3/\text{ft) of width} \qquad (9.49)$$

where F_s = allowable stress in the steel. A summary analysis of a strutted excavation is given in Fig. 9.59.

If piping or heave is found not to be of concern, the sheeting may penetrate to a nominal depth below the excavation bottom. If there are no supports near the excavation bottom, however, the depth of penetration provided should be at least 1.5 times the depth required for moment equilibrium about the lowest strut. The resistance provided to the portion of the wall penetrating below the excavation bottom is computed using

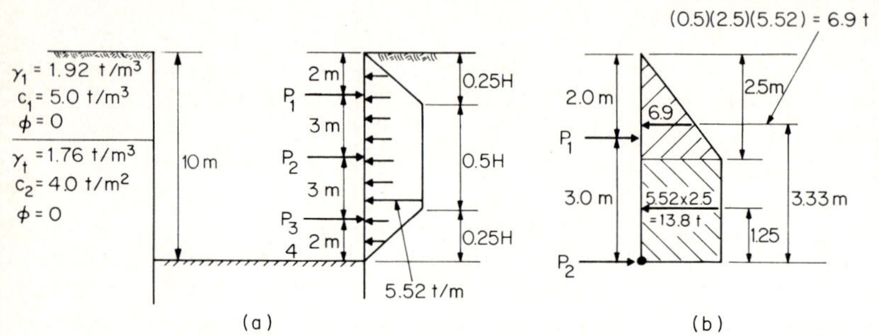

1. *Check base stability:*

γ_{avg} = 1.84 t/m³ c_b = 4.0 t/m² $\gamma H/c_b = (1.84)(10)/4 = 4.6$ OK < 6

2. *Compute maximum probable earth pressure p_h:*

$$c_{avg} = 4.5\ \text{t/m}^3 \qquad \gamma H/c = (1.84)(10)/4.5 = 4.1$$

From Fig. 9.52*b*:

$$p_h = \gamma H - 4s_u = (1.84)(10) - (4)(4.5) = 0.4\ \text{t/m}^2$$

From Fig. 9.52*c*:

$$p_h = 0.3\gamma H = (0.3)(1.84)(10) = 5.52\ \text{t/m}^2\ \text{(use)}$$

3. *Compute moment for sheet piles:*

a. Cantilever moment at (1)

$$M_{max} = [p_h(H)^2]/6 = [(4.42)(2)^2]/6 = 2.95\ \text{mt/m}$$

b. Moment at (2), use continuous span

$$M_{max} = [p_h(H)^2]/10 = [(5.52)(3)^2]/10 = 4.97\ \text{mt/m}$$

For design $M = (0.8)(4.97) = 3.97$ mt/m

Section modulus $S = M_{max}/F_s$ where F_s = allowable stress in steel

4. *Compute waler and strut loads:* To find P_1 sum moments about P_2 [see part (b)]:

$$P_1(3.0) = (6.9)(3.33) + (13.8)(1.25) = 13.4\ \text{t/m}$$
$$P_2 = (5.52)(3) = 16.56\ \text{t/m}$$
$$P_3 = P_1 = 13.4\ \text{t/m}$$

Struts located with 5-m spacing: Axial load $P_1 = P_3 = 67$ tons; $P_2 = 83$ tons. Moment in waler at (2) $[(5.52)(5)^2]/10 = 13.8$ mt

FIG. 9.59 Analysis of a strut-braced sheet-pile wall for conditions given in (*a*).

the passive resistance and ignoring wall friction [CGS (1978)[44]], as discussed for anchored walls.

Rankine Analysis

Some practitioners prefer to first compute strut loads and bending moments for each stage of construction (each support level) using Rankine pressures, then at the final stage compute strut loads and M_{max} using the apparent pressure diagram. The maximum values for strut loads and bending moments are then selected for design [NAVFAC (1982)[1]]. This procedure is described under *Anchored or Tied-Back Walls* and in Fig. 9.62 and is particularly applicable to the more rigid diaphragm walls.

Rakers and Berms

Rakers and Sheeting

It is usual to design rakers and their connections in the same manner as horizontal struts, i.e., using the empirical rules and apparent earth pressures. To analyze the flexural stresses in sheeting, the interim construction conditions are checked by determining the effective pile span using the equivalent support point shown in Fig. 9.45*b*.

Kicker Blocks

The stability of kicker blocks is analyzed as a footing with inclined loads as in the example in Fig. 9.28. In soft to medium clays, however, even if FS = 3 or more is selected, progressive movement of the blocks can be expected, and provisions are required to ensure that load is maintained in the inclined bracing. Even in good soil conditions FS should be at least 2.5 to restrict movements. In addition, adequate sheeting penetration below excavation bottom is required to resist the uplift force resulting from the kicker block reaction on the wall.

Berm Stability and Deformation

Analyses are required for the potential for a berm slope failure and for a general deep-seated failure beneath the berm, particularly in soft to firm clays as shown in Fig. 9.45*a*.

Evaluating the effect of berm size and strength and its effectiveness in providing wall restraint is a difficult task in poor soil conditions. Clough and Denby (1977)[51] used the FEM to evaluate berm wall systems in NC to slightly OC clays that undergo undrained shear during excavation; they presented their results in nondimensionalized chart form to permit quick evaluation for berm design. Study parameters included final excavation depth and width, depth to a rigid soil base, wall and raker stiffness, wall end conditions (fixed or free), number of rakers, depth to first raker, wall penetration below excavation bottom, undrained shear strength of the clay and initial tangent modulus (soil stiffness), coefficient of earth pressure at rest K_o, and the berm slope. Two cases were studied: MZ-27 sheet piles and a 0.6-m-thick slurry wall, in excavation depths varying from 9 to 17 m.

Anchored or Tied-Back Walls

Earth Pressures

The selection of the earth-pressure distributions depends on the type of wall to be installed when anchors or tiebacks are to be used for support. Empirical or apparent pressure distributions, as given for braced cuts, are used to determine total earth pressures against soldier beams and lag-

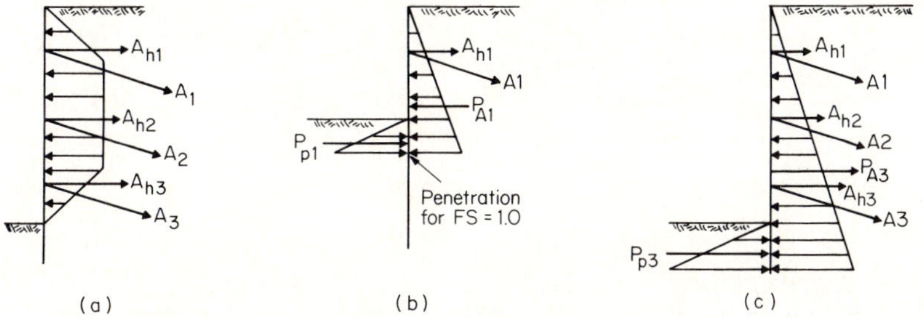

FIG. 9.60 Earth pressures acting on anchored walls. (*a*) Apparent pressures. [From Peck et al. (1974).[24] Used with permission of John Wiley & Sons, Inc.] **(*b*) Rankine pressures.** [From CGS (1978).[44]]

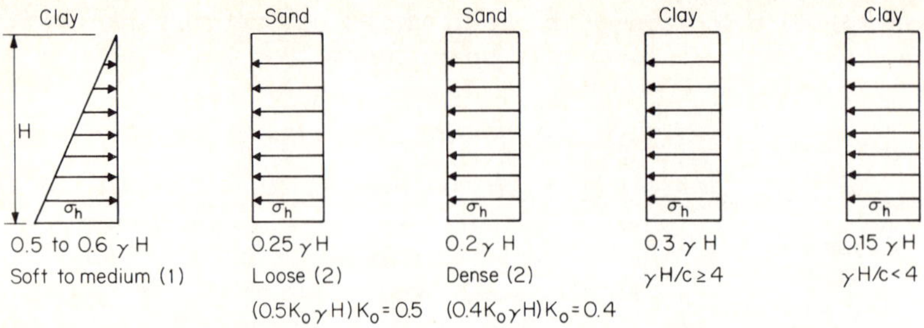

NOTE: (1) Overstressing soft clays will cause consolidation. (2) Use $0.2\gamma H$ and $0.25\gamma H$ where deformations are critical. Where deformations are not critical, use $K_{avg} = (K_o + K)/2$.

FIG. 9.61 Pressure distributions for tied-back walls for at-rest conditions. [After Goldberg et al. (1976).[45a]]

ging, or against sheet-pile walls as shown in Fig. 9.60*a* [Clough and Tsui (1974),[52] Murphy et al. (1975),[53] Peck et al. (1974)[24]]. In soft Swedish clays [Broms and Stille (1976)[46]] the earth-pressure distribution is assumed trapezoidal with a maximum intensity equal to

$$P_{max} = 1.6p_a/H \tag{9.50}$$

As there is very little, if any, lateral deflection with properly installed anchors, earth pressures can be expected to approach or equal the at-rest condition [Goldberg (1976)[45]]. Where it is necessary to minimize displacements adjacent to excavations, loads are computed based on K_o pressures and the anchor load is locked off at 100% of design load. Where displacements are not critical, loads are computed based on K_a pressures or, as some practitioners prefer, $K_{avg} = (K_a + K_o)/2$ [Goldberg et al. (1976)[45a]]. Apparent pressure distributions for tied-back walls for at-rest con-

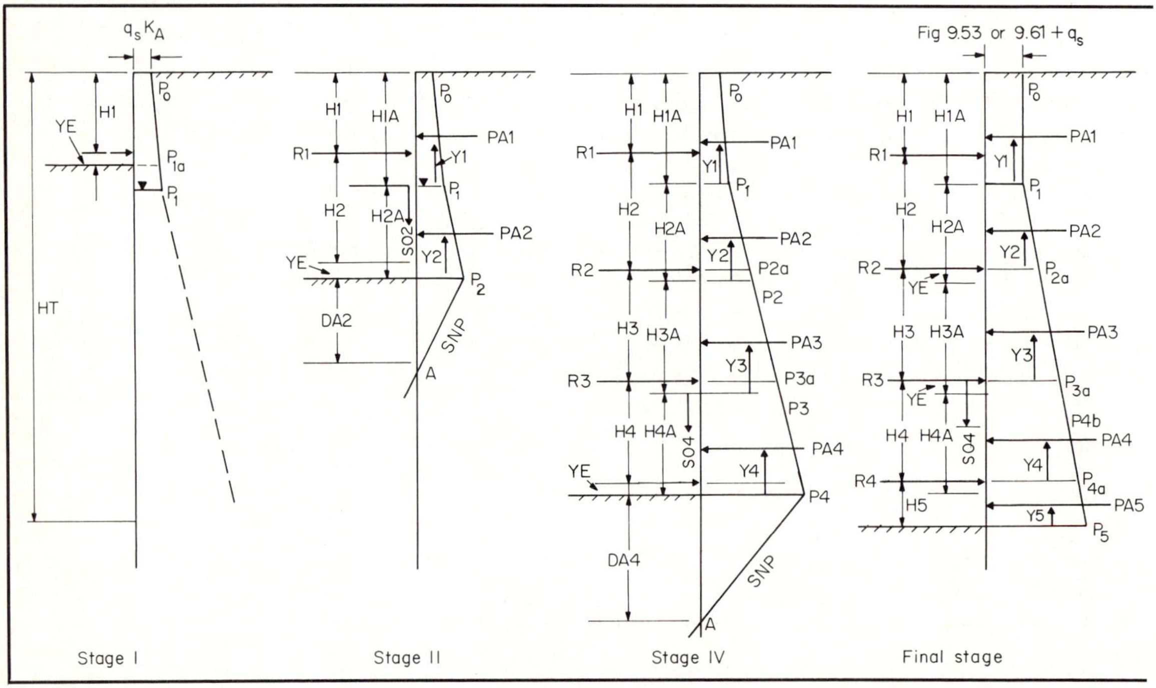

FIG. 9.62 Analysis of braced or tied-back excavation with uniform surcharge and GWL between reactions 1 and 2 based on Rankine pressures during construction stages. [After NAVFAC (1982).[1]]

1. Compute Rankine pressures for construction stages I through IV: $p_o = q_s K$ where

$$K = K_A \text{ or } K_o \text{ or } (K_A + K_o)/2$$

$$p_1 = p_o + (H1A + \gamma_t K) - 2cK$$

$$p_2 = p_1 + (k\gamma_b + \gamma_w)H2A - 2cK, \text{ etc.}$$

2. Stage I:

$$M_{max} = p_{1a}[(H1 + YE)^2]/6$$

where YE = excavation below reaction level

3. Stage II:

a. Find point of zero net pressure at A from slope of net pressure diagram:

$$SNP = (K_p/FS - K_A)\gamma_s \text{ and } DA2 = p_2/SNP$$

b. Find reaction per linear foot of wall for brace of horizontal anchor load by assuming hinge at A and summing moments about A:

$$R1 = \left\{[PA1(DA2 + H2A + Y1)] + [PA2(DA2 + Y2)] + \left[\frac{p_2(DA2)^2}{3}\right]\right\} \Big/ (DA2 + YE + H2)$$

c. Find M_{max} at point of zero shear, SO2 by solving quadratic equation:

$$R1 - PA1 - [SO2(p_1)] - \left[\frac{p_2 - p_1}{H2A}\left(\frac{SO2^2}{2}\right)\right] = 0$$

$$M2 = \left\{[R1(SO2 + H1A - H1)] - [PA1(SO2 + Y1)] - \left[\frac{p_1}{2}(SO2^2)\right] - \left[\frac{SO2^3}{6}\left(\frac{p_2 - p_1}{H2A}\right)\right]\right\} 0.80$$

4. Stage III and IV:

a. Proceed as for stage II. Find new points of zero net pressure at $DA3$ and $DA4$ and compute $R2$ and $R3$.

b. Find new points of zero shear at SO3 and SO4 and compute $M3$ and $M4$.

5. Final stage:

a. Prepare apparent pressure diagram as from Fig. 9.53 or 9.61 and provide for q_s and γ_s (below GWL).

b. Find each reaction per linear foot of wall by assuming hinge at adjacent lower strut and summing moments as for $R1$, for example:

$$R1 = \{[PA1(H2A - YE + Y1)] + [PA2(Y2)]\}/H2$$

NOTE: Some practitioners prefer the simple area method of Fig. 9.57 for the final stage reactions since moments are not used in determining the apparent pressure diagrams (personal communication, Dr. Ralph Peck).

c. M_{max} is likely to occur between $R3$ and $R4$: Find point of zero shear SO4 and p_{4b} (p_h at SO4)

{Store 01}

$$\left\{\left[p_{3a}\left(\frac{H4^2}{2}\right)\right] + \left[\frac{p_{4a} - p_{3a}}{2}\left(\frac{H4^2}{3}\right)\right]\frac{1}{H4}\right\} - [p_{3a}(SO4)] - \left[\frac{p_{4a} - p_{3a}}{2}\left(\frac{SO4^2}{H4}\right)\right] = 0$$

$$M_{max} = \left\{[(\text{Recall } 01)(SO4)] - \left[\frac{p_{3a}}{2}(SO4^2)\right] - \left[\frac{p_{4b} - p_{3a}}{6}(SO4^2)\right]\right\} 0.80$$

6. For design select the higher values for $R1$, $R2$, $R3$, and $R4$, and $M1$, $M2$, $M3$, $M4$, and M_{max}.

7. Check depth requirement to control bottom stability and piping.

FIG. 9.62 ***(continued)***

ditions are given in Fig. 9.61. Surcharge effects must also be added as discussed in Art. 9.2.3 and below. For the more rigid diaphragm walls, Rankine distributions or modifications thereof are commonly used, as shown in Fig. 9.60*b*. Passive pressures, which restrain the wall portion below the excavation bottom, are computed based on a reduced passive coefficient in accordance with

$$K'_p = K_p/\text{FS} \tag{9.51}$$

where FS is at least 1.5, although some practitioners prefer 2.0.

Procedures

Where the apparent pressure distribution is used, the horizontal anchor forces are determined in a manner similar to that described for braced cuts. A horizontal spacing is selected, the total horizontal force determined, an anchor inclination selected, and the force per anchor computed accordingly.

For the active Rankine distribution, a pressure diagram is drawn for each anchor level, as shown in Fig. 9.62, which is based on the assumption that the highest load on each anchor occurs just before it is installed. At each level except the lowest, the pressure diagram provides for a passive resistance mobilized along a depth of wall penetration adequate to establish FS = 1 against rotation, and accounts for the design forces in the previously installed anchors. The horizontal anchor load at each level is determined in a manner similar to that described for anchored bulkheads (Art. 9.4.3 and Fig. 9.83), i.e., equilibrium of all horizontal forces. For the next to lowest anchor a check is required to determine if the required penetration depth is actually available.

For the lowest anchor, the load is calculated from the horizontal force equilibrium, using the bottom penetration at the proposed design depth. In general, where the lowest anchor is more than 1 m from the wall base, the wall should penetrate below the cut bottom to the depth at which the computed resultant earth pressure is zero to avoid development of excessive wall bending moments which will increase the load on the lowest anchor. The bending moments that will develop in the wall immediately before installation of each anchor are checked against the wall design moments, because they can be significantly higher than those at the completion of the excavation.

General Example

An approach to various loading conditions assumed for analysis is given in Fig. 9.63. One aspect of the approach is to transfer the Rankine pressure triangles or trapezoid into an equivalent rectangle as shown in Fig. 9.64*a*, which presents the pressure diagrams for a 12.7-m (42-ft)-deep, anchored diaphragm wall-supported excavation for the basement of a 45-story hotel building. Active pressures were used on the street side where wall movements were not considered critical, and at-rest pressures were used on the side where an existing building was located. The sandy soils below a high GWL were dewatered.

To prepare the diagram in Fig. 9.64*a* for Level 1, for example, a trapezoidal distribution was prepared for the active pressures with the uniform surcharge based on $\sigma_h = \gamma z K_a + q_s K_a$ as shown in No. 3 of Table 9.3, and adjacent to it a triangular distribution was prepared for the line load of 5 t/m^2. Converting the triangle and trapezoid into equivalent rectangles yields the pressure diagram given for Level 1, and Levels 2 and 3 are prepared similarly. The pressure diagrams in Fig. 9.64*b* are based on at-rest pressures and provide for the stresses imposed by the adjacent building as determined from Fig. 9.20. The numbers in parentheses in Fig. 9.64*b* represent the pressure change in the lateral direction from point loads from deep foundations along the existing building line. The anchors that support the wall were installed inclined at about 35° with about 1.5 m spacing. Design load in the sandy soils was 45 tons (metric) and the anchors were all load-tested to 67.5 tonne-force. Precise leveling detected a maximum settlement of the adjacent building during construction of 5.7 mm (¼ in).

Various types of anchors, methods for estimating capacities, installation, and load-testing procedures are described in Art. 7.7.

Anchor Inclination and Length Considerations

As anchor inclination increases from the horizontal, its efficiency decreases and the vertical

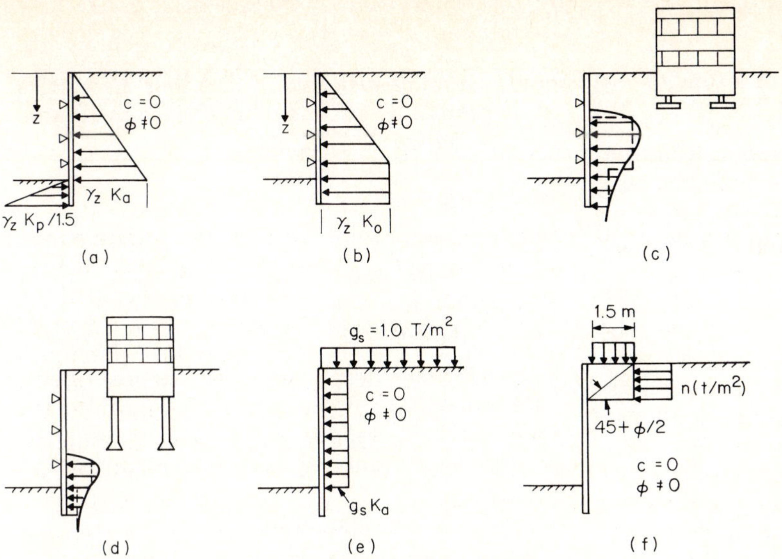

NOTE: For analysis of surcharge loads, see Art. 9.2.3.

FIG. 9.63 Various loading conditions imposed on anchored diaphragm walls: (*a*) active and passive pressures; (*b*) at-rest pressures; (*c*) surcharge from shallow building foundations; or (*d*) deep building foundations; (*e*) overall surcharge from backfill, traffic, etc.; and (*f*) surcharge from heavy construction equipment. [From Tecnosolo (1978).[35]]

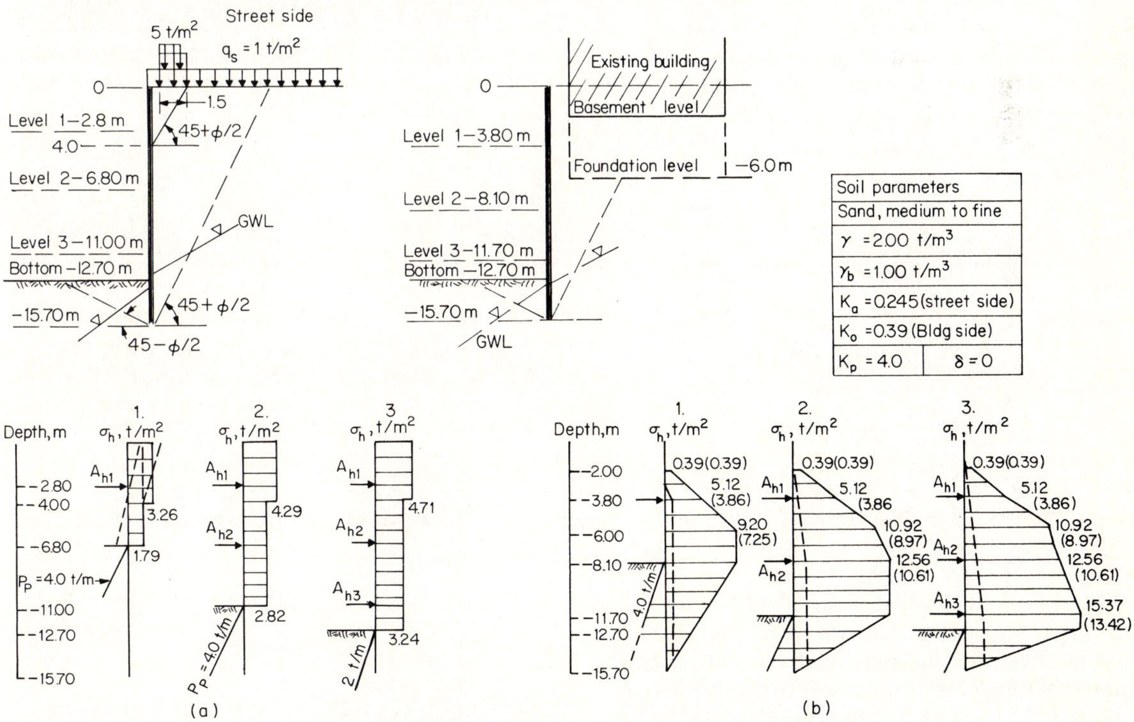

Soil parameters	
Sand, medium to fine	
γ = 2.00 t/m^3	
γ_b = 1.00 t/m^3	
K_a = 0.245 (street side)	
K_o = 0.39 (Bldg side)	
K_p = 4.0	δ = 0

FIG. 9.64 Conditions and earth-pressure diagrams used for analysis of anchored diaphragm wall, Meridien Hotel, Rio de Janeiro: (*a*) street side and (*b*) side with existing 15-story apartment building. [After Tecnosolo (1974).[54]]

component increases, although its length to reach suitable bearing decreases.

The potential for wall-base penetration resulting from the wall weight and the vertical force component from the anchors must be considered and can be especially critical for steeply inclined anchors. Vertical wall movements result in reduced anchor loads and consequently outward movement of the face. Downward movement is resisted by wall friction (Fig. 9.65) and reaction at the base. For diaphragm walls wall friction is ignored for the active case but can be significant for the at-rest case and may be taken as given on Table 9.4 or as $\delta = 0.67\phi$ for cohesionless soils and $\delta = s_u$ for cohesive soils [CGS (1974)[44]]. Inclinations of anchors for various purposes and required values for FS for load-tested anchors are given in Fig. 9.65.

The first requirement for a suitable anchor length is penetration into soil of adequate strength, but the fixed portion of all anchors must be located beyond a potential wedge-failure surface. Peck et al. (1974)[24] suggest locating this wedge at an angle of $45° - \phi/2$ from the base of the wall plus a distance of $H/5$ as shown in Fig. 9.66. The minimum depth below the surface of the fixed portion should be 5 m (17 ft) in soil and 1.5 m (5 ft) in sound rock. Anchors should not be more closely spaced in either the vertical or horizontal directions than 4 times the diameter B, as shown in Fig. 9.66.

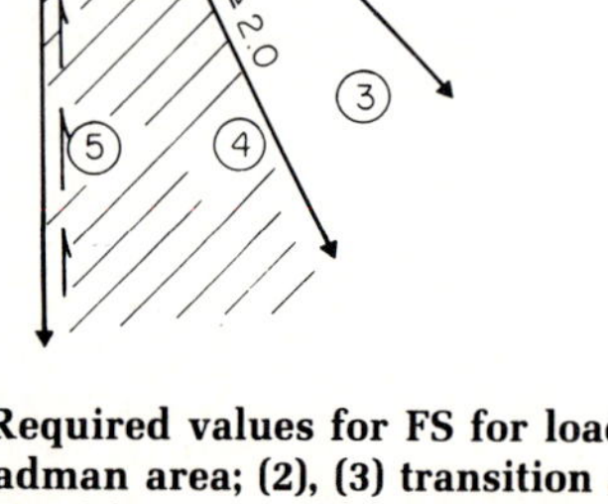

FIG. 9.65 Required values for FS for load-tested anchor: (1) deadman area; (2), (3) transition area for inclined anchors; (4) usual tension-pile area; and (5) skin friction along concrete wall resists vertical component. [After CGS (1978).[44]]

Wall Design

Flexible sheet-pile walls may be designed as flexible bulkheads with a free-earth support [D'Appolonia et al. (1967)[55] and Art. 9.4.3] or as a braced excavation.

Concrete walls may be analyzed as a continuous beam-on-an-elastic foundation to determine moments and shears *(see Art. 4.4.4)* because of the relatively close anchor spacing [Clough (1974)[56]].

Overall Wall System Stability

Overall wall stability against a general failure along a potential external rupture surface or potential internal rupture surfaces, as shown in Fig. 9.67, should be analyzed. If the earth blocks

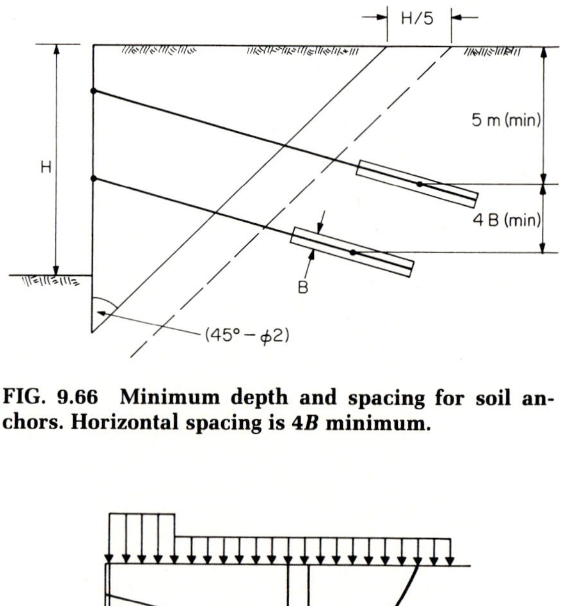

FIG. 9.66 Minimum depth and spacing for soil anchors. Horizontal spacing is 4*B* minimum.

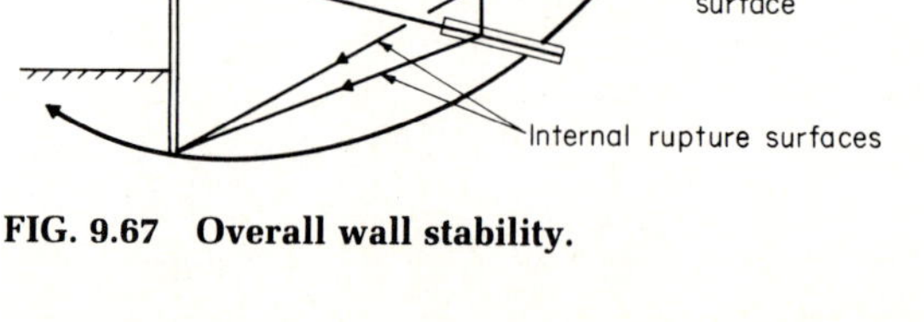

FIG. 9.67 Overall wall stability.

shown in Fig. 9.67 are found to be stable, a less conservative selection of anchor lengths than provided by the approach given in Fig. 9.66 may be possible. Only those anchors located *above* the *base* of the wall need be subjected to block failure analysis.

Graphical analysis of the stability of an anchored soil block based on a system of wedges and Rankine pressures has been presented by Ranke and Ostermayer (1968).[57] Used commonly in Europe and South America, a modified version is given in Fig. 9.68. For analysis it is assumed that a rupture surface will form from the anchor center to the base of the wall, and a vertical tension crack will exist from the anchor center to the ground surface providing an "equivalent" wall. This equivalent wall is one side of an active failure wedge with the other side inclined at $45° - \phi/2$ to the vertical. An active wedge also is assumed to form behind the anchored wall. The possible magnitude of anchor force F_A is found from a vector diagram. The magnitudes and directions of P_1, W_1, C_1, R_A, and C_A are known, as are the directions of the anchor and R_1. The vector diagram for a single soil system is given in Fig. 9.68*a* and that for a two-layer system in Fig. 9.68*b*.

For systems with multiple anchors, various possible locations for the potential lower failure surface exist: it may pass in front of or behind anchors adjacent to the one being analyzed, or it may pass through the center of one or more adjacent anchors located at other levels. Analysis usually begins at the top level, and at each subsequent level the required anchor force is the sum of all anchor forces F_A above the relevant failure surface. For example, for the second level the possible anchor magnitude is $F_{A1} + F_{A2}$.

Wall Deflection and Backslope Subsidence

Conventional Approach

Determinations of wall deflections into the excavation and backslope subsidence cannot be made with the classical theories or the com-

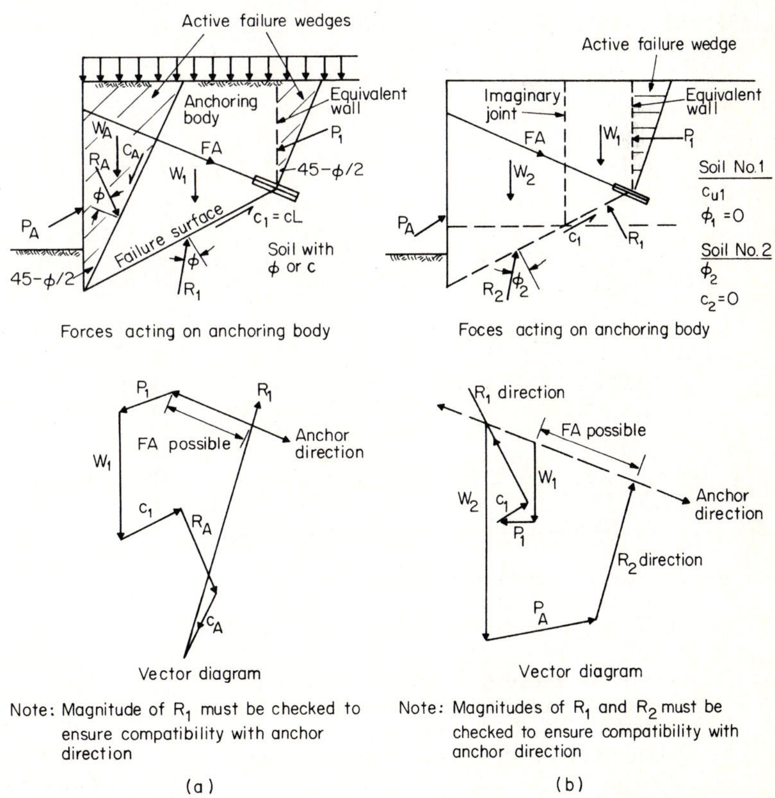

FIG. 9.68 Graphical analysis of stability of an anchored wall: (*a*) single layer and (*b*) two-layer soil profile. Water pressures are not included in the calculations of P_A and P_1. [From CGS (1978).[44]]

monly used empirical methods which are based on limit analysis. The conventional approach to the problem is to estimate settlement based on observational data from previous excavations in similar soil conditions. *Underpinning* of adjacent structures is resorted to if it is considered that they will suffer objectionable settlement during excavation. The general criteria for underpinning as a function of distance from the excavation are given in Fig. 9.69.

The chart given as Fig. 9.70 may be used for preliminary estimates of backslope subsidence. It was derived by Peck (1969)[36] from data obtained from a number of instrumented excavations in a variety of soil conditions. It can be seen that for soft clays settlement as great as 0.2% of excavation depth may be encountered at distances from the excavation as great as 3 to 4 times its depth. The susceptibility of various soil types to backslope subsidence is discussed in Art. 9.3.1.

Anchored walls in soft clays can be particularly troublesome as anchor overstressing can cause clay consolidation resulting in wall deflections away from the excavation. This results in a loss of anchor load, and it may be necessary to institute frequent and regular reloading over a period of several months until no significant loss in load is measured. Substantial overstressing should be avoided as it can cause passive failure behind the wall.

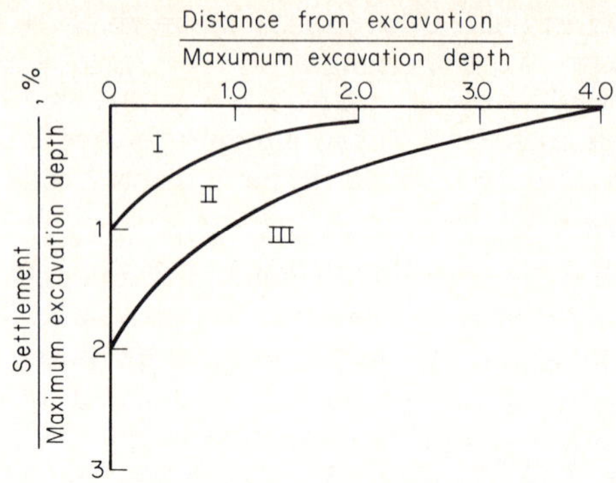

FIG. 9.70 Summary of settlement adjacent to open cuts in various conditions as a function of distance from excavation edge retained by soldier piles or sheet piles braced with either cross bracing or tiebacks. [From Peck (1969).[36]]
Zone I: Sand and soft to hard clay, average workmanship.
Zone II: Very soft to soft clay: (1) limited depth of clay below excavation bottom and (2) significant depth of clay below excavation bottom, but $N_b < N_{cb}$ *(see Art. 9.3.3)*. Settlement affected by construction activities.
Zone III: Very soft to soft clay to a significant depth below excavation bottom and with $N_b \geq N_{cb}$.

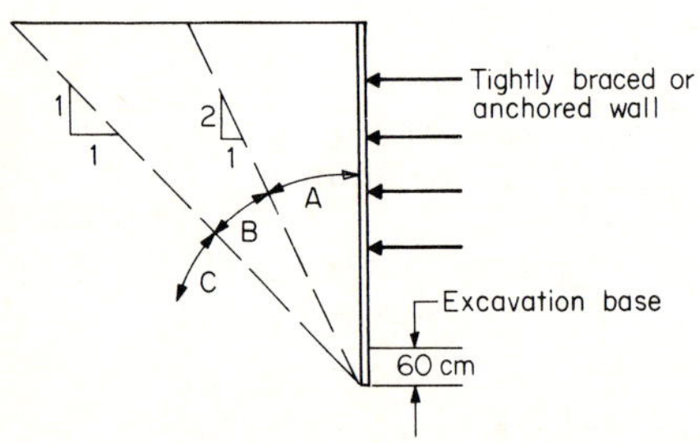

FIG. 9.69 General criteria for underpinning. [From NAVFAC (1982).[1]]
Zone A: Foundations within zone generally require underpinning.*
Zone B: Foundations within zone generally do not require underpinning.*
Zone C: Underpinning for structures must be founded in this zone.†
NOTES: ***Horizontal and vertical pressures on excavation wall of nonunderpinned foundations must be considered in Zones *A* and *B*.**
†Pressures from underpinning generally need not be considered in Zone *C*.

Finite Element Approach

Experience in recent years indicates that reasonably close estimates of wall deflections and backslope subsidence can be made with the FEM. Its application permits more efficient use of less flexible wall systems such as anchored diaphragms to replace underpinning. Clough and Mana (1976)[58] present a review of FEM applications to temporary excavations in soft clay and conclude that when used properly it can be expected to yield useful design information and relatively accurate performance predictions.

The objectives are to predict deflections and subsidence for a given design, monitor construction procedures and movements, and make design changes if necessary to control the satisfactory performance of the excavation. The FEM can simulate almost all of the complex factors of the braced or tied-back wall except random variables such as contractor error or geologic surprises.

ASSUMPTIONS Two simplifying assumptions usually are made in the analysis of braced excavations [Tsui and Clough (1974)[59]]:

1. Continuous planar walls can simulate the behavior of discontinuous walls such as the H-beam and lagging wall.
2. Plane-strain loading conditions reasonably represent the three-dimensional loading by braces and anchors. Plane-strain loading conditions assume that the wall is continuous and that discontinuous braces or ties can be simulated by planar supports placed at the same elevations as the real supports, as shown in Fig. 4.57.

ESTIMATING APPLICABILITY There are errors involved with the assumptions and plane-strain conditions cannot be arbitrarily extrapolated to many practical problems, particularly those that involve Berlin walls or light sheet-pile walls. Diaphragm walls, however, and possibly continuous sheet-pile walls approximate plane-strain conditions [Tsui and Clough (1974)[60]]. To estimate the applicability of the FEM, Tsui and Clough concluded from model studies that a definition of wall flexibility in terms of degree of consistency of behavior with plane-strain behavior was desirable.

They applied the slab-on-an-elastic-foundation theory *(see Art. 4.4.4)* to describe the model walls and concluded that the flexibility of the soil-slab system may be conveniently defined in terms of its *characteristic length* L_o, which accounts for the relative effects of both soil and slab stiffness and may be compared to the spacing of prestress loads. Their study demonstrated that horizontally closely spaced ties approximated plane-strain conditions but deviation increased with increased spacing. For excavations in which prestressed supports are used, the chart in Fig. 9.71 can be employed to quickly estimate the error or applicability of plane-strain simulation for any combination of wall rigidity, soil rigidity, and horizontal brace or tie spacing. The chart presents the maximum percent deviation of actual

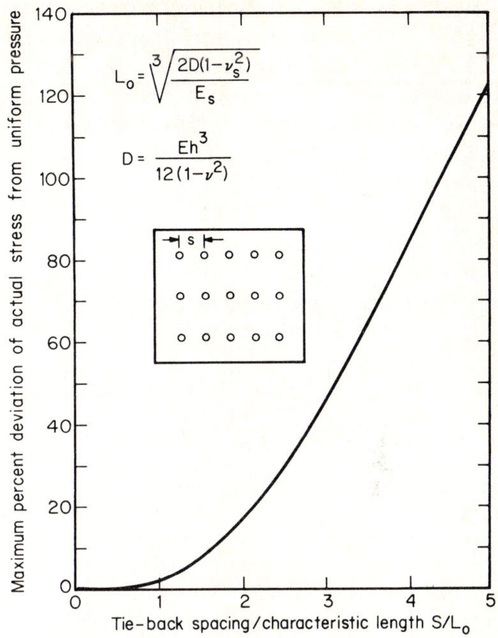

FIG. 9.71 Chart for estimating applicability of the FEM for wall-deflection analysis in terms of the maximum percent deviation of actual pressure from uniform planar pressure distribution for ratios of tied-back load spacing to characteristic length L_o. [From Tsui and Clough (1974).[59]]

stress from the uniform pressure as a function of the ratio of the tieback spacing S to the characteristic length L_o. As shown in the chart, L_o is a function of soil modulus E_s, slab modulus E, Poisson's ratio for the slab ν, Poisson's ratio for the soil ν_s, and the slab thickness h.

FEM Analysis

STRUCTURAL MODEL Flexible walls are often analyzed using the computer program "SOIL-STRUT," a general-purpose program for analyzing soil-structure interaction problems [Tsui (1974),[60] Tsui and Clough (1974)[59]]. Different element types are employed in the program to represent the different elements of a braced or anchored wall system. Springs of one-dimensional bar elements are used to model braces or tiebacks, and a linear strain isoparametric element is used to model two-dimensional structural elements.

EXCAVATION MODEL A construction sequence is selected as it and construction operations significantly affect behavior. Unfortunately, construction operations usually are subject to change after analysis is complete, and reanalysis during construction may be necessary. The actual sequence should be modeled as closely as possible. A step-by-step or incremental loading sequence is used, and for each excavation step a force system is applied to the excavation boundary so as to reduce stresses on the boundary to zero. Elements in the area excavated either have their stiffness reduced to zero or are removed altogether. Almost all FEM schemes available use this general approach, but they differ in the means of determining the force system to be applied to the excavation boundary. Clough and Mana (1976)[58] concluded that direct calculation of excavation forces from stresses of Gaussian integration points in the higher-order elements, as described by Mana (1976),[61] seems to be the most accurate technique and requires a minimum amount of input data to satisfy the excavation sequence.

SOIL MODEL Data are required on soil parameters, groundwater conditions, and stratification. The most critical parameters used in analysis are soil strength and deformation characteristics for which reliable values are difficult to obtain. Two general modes have been used in excavation studies to describe soil stress-strain behavior. The *nonlinear elastic model* allows the soil modulus to be adjusted consistently with the shear-stress levels in each element. The form of the stress-strain curve is assumed to be hyperbolic up to failure, at which point the shear modulus is reduced to zero. In this case the initial tangent modulus values are needed to define the initial slope of the stress-strain curve and subsequent modulus values are defined by obtaining tangents to the hyperbolic polynomial assumed to represent the stress-strain curve until yielding is reached. The *elastoplastic model* assumes elastic behavior until yield is reached and then plastic yielding occurs with further loading. The model uses a straight-line function for stress-strain until yield; then a secant modulus value must be chosen to represent the stress-strain curve up to yielding. Using either a nonlinear elastic model or an elastoplastic model, however, results in similar predictions of excavation behavior.

Soil Parameters

SOFT, SATURATED CLAY, UNDRAINED LOADING The nonlinear model requires the following parameters: k_o, s_u, E_i, E_{ur} (unload-reload modulus) and R_f (a curve-fitting parameter). R_f is usually an easily defined parameter, and E_i and E_{ur} are about equal in undrained loading. The parameter selection then reduces to the values k_o, s_u, and E_i, with the latter two presenting the greatest difficulties. Using laboratory and field-testing methods to obtain shear strength usually results in a wide scatter of values both by variations in the deposit itself and by disturbance during sampling and test preparation. The greatest effect of disturbance is on E_i which results in values much lower than reality [Clough and Mana (1976)[58]].

INFLUENCE ON PREDICTIONS The importance of the influence of soil parameters on deflection predictions is illustrated by the FEM analysis performed for a two-tiered retaining system in soft to firm varved clays in Hartford, Conn. [Murphy et al. (1975)[53]]. The conditions for the excavation retained by soldier pile and lagging walls anchored with tiebacks are illustrated in Fig. 9.72. The FEM analysis was performed with the SOIL-STRUT program to determine probable wall movements and to evaluate the potential for a deep-seated failure. The undrained strength from laboratory CU tests and field vane tests yielded values for s_u ranging from 0.2 to 0.6 tsf. Analysis based on s_u = 0.2 tsf resulted in predicted deflections over 3 times greater than were actually measured in the field during construction with an inclinometer. The field measurements agreed closely with movements predicted for a value of s_u = 0.4 tsf.

Conclusions

Clough and Mana (1976)[58] concluded that the use of higher values for shear strength and modulus than obtained from standard laboratory testing (undrained triaxial test) usually is justifiable or the predictions of excavation movements will be much too large. The best data for arriving at reasonable values come from experience in similar

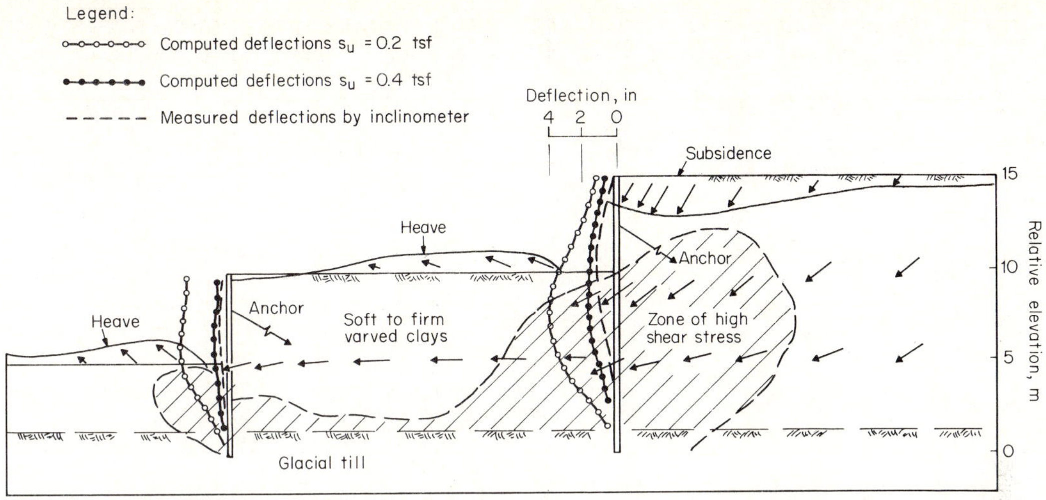

FIG. 9.72 FEM analysis of two-tier excavation in soft to firm clays supported by anchored sheet-pile walls and comparison of computed and measured deflections. Site location: Hartford, Conn. [After Murphy et al. (1975).[53]]

soil conditions. The use of "high" values of soil parameters in design to eliminate overconservatism is justified when the excavation is monitored for performance. Revaluations can be made, and if necessary contingency measures can be invoked during construction.

Performance Monitoring

Contingency Design

Wall system design is basically empirical, and the engineer often is not certain of the real pressures and deflections that will occur. Uncertainties arise in soil property measurements, and construction changes are common which lead to unanticipated performance, such as is caused by a change in the sequence of procedures. Examples include replacing a grouted anchor with a deadman at the first level, placing surcharge loads adjacent to the excavation, or overexcavating the temporary platform before placing the support. To provide for all contingencies often causes a wall to be unnecessarily costly.

The nature of wall construction provides the opportunity to strengthen the wall by placing additional braces or anchors. Initial design, therefore, can be based on relatively low safety factors but with provisions for contingency measures. This approach can proceed only if construction is monitored with early warning systems which enable reanalysis at any point.

Program Elements

The field observations and instrumentation program to monitor construction should consider measurements of the following elements:

- Movements of the retained material, the interior subgrade, and the support system
- Movement in adjacent structures
- Groundwater levels and water quality and quantity from the dewatering system
- Strains and loads in the support system
- Vibrations from blasting, pile driving, or traffic

Observations and Instrumentation Program

Various forms of instrumentation for an excavation-retaining structure in soils are illustrated in Fig. 9.73 and described in Art. 4.5.3 of Hunt (1984).[5] Piezometers are installed to monitor groundwater variations, and inclinometers are

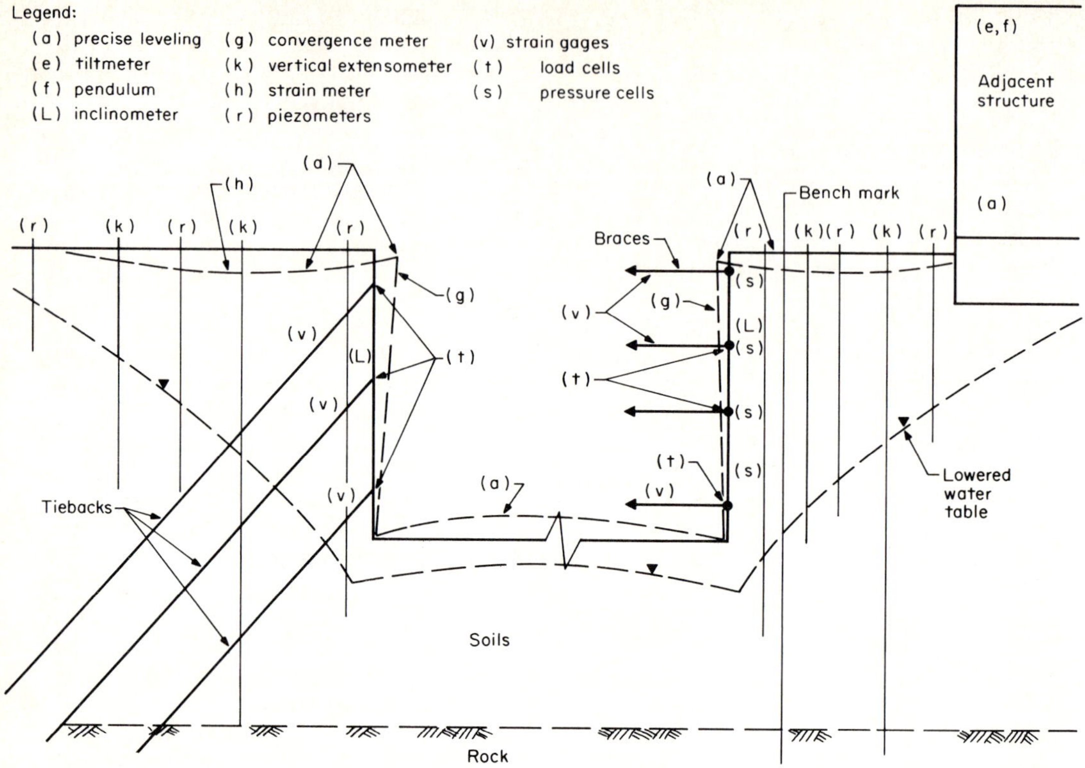

FIG. 9.73 Instrumentation of excavation retaining structure constructed in soils. [From Hunt (1984).[5] Used with permission of McGraw-Hill Book Company.]

installed immediately behind the wall to measure horizontal wall deflections. Vertical deflections of the wall and adjacent structures usually are measured optically, although tiltmeters may be placed on the wall and in adjacent structures and even attached to alarms. Anchor and bracing loads are monitored with strain gages and load cells. Strain gages must be protected from temperature changes and moisture, or corrections made. A benchmark, preferably founded in rock, is installed to permit reliable measurement of deflections of adjacent ground, structures, and the excavation bottom.

9.3.5 ROCK EXCAVATIONS FOR STRUCTURES

General

Normal Conditions

Relatively shallow, open excavations into fresh to moderately fresh rock normally do not present difficult support problems since the opening is temporary and any unstable rock can be easily trimmed, bolted, or braced as described in Art. 5.5.3. Blasting, however, must be performed carefully as described in Art. 6.3.5. (Shafts and other underground openings in rock are described in Art. 10.3.)

Troublesome Conditions

Troublesome conditions may be encountered in poor-quality rock characterized by a high degree of fracturing or weathering, and treatments such as those for soils may be necessary.

Swelling or heaving rocks can be particularly troublesome. Marine shales exhibit heave from stress release during excavation, often combined with swelling [see Art. 10.6.3 of Hunt (1984)[5]]. In these materials the best solution is to minimize shale exposure to weathering by making excavations in sections as small as practical, prevent water infiltration, and immeditely cover the bottom with foundation concrete, cyclopean concrete, or compacted earth.

High in situ residual stresses [Art. 6.6 of Hunt (1984)[5]] can be extremely troublesome as excavation causes stress relief which is manifested by bottom heave and squeezing of the walls. When one is planning construction in marine or clay shales, problem conditions are expected, but the problems associated with high residual stresses very frequently are unexpected because of the vagaries of their occurrence, unless special investigation is undertaken prior to construction. Although most frequently found in fresh crystalline rocks, they also are found in relatively flat-lying sedimentary formations composed of strata with relatively high modulus values under regional compression such as the rocks of upper New York State and adjoining Ontario.

High Residual Stresses: A Case Study

Initial Construction Phase

A case where excavation was made into intensely folded gneiss with high lateral stresses in the direction of fold compression is reported by Ward (1972).[62] The excavation was 100 ft (30 m) and 65 × 100 ft (19.5 × 30 m) in plan dimensions. Construction had proceeded with the installation of rock bolts 10 ft in length, metal straps, and shotcrete to contain slabbing in the vertically foliated gneiss. Lateral displacements, which eventually reached 20 in (51 cm), caused the failure of numerous rock bolts.

After the wall movements and bolt failures had occurred, flat-jack tests were made in the excavation which indicated that prior to construction lateral stresses as high as 900 psi (63 kg/cm^2) probably existed. Simple gravity stresses at test depth were in the order of 90 psi (6 kg/cm^2) vertical and 30 psi (2 kg/cm^2) horizontal. Eight months after the initial movements occurred, additional movements were still being observed.

FEM Analysis

Application of the FEM was made to determine the distribution of stresses behind the rock wall and the possible magnitude of wall displacement. The stress distribution provides the data for estimating rock-bolt spacing and lengths required to retain the residual stresses. The FEM net subdivides the rock mass into a finite number of elements, interconnected only at nodal points. At each point forces are applied which permit evaluation of stress, strain, force, or displacement. Anisotropy is incorporated by adjusting nodal stiffness at each point. The mesh used in the problem is given as Fig. 9.74.

General analyses were performed using different sequences of excavation, incorporating the in situ stress measurement values. It was assumed that these stresses caused the vertical slabbing of the rock in tensile failure resulting in progressive movements as stresses were released on sequential slabs. E_r for intact rock in compression was assumed to be 1.5 × 10^5 tsf and the tensile modulus 1.5 × 10^4 tsf. Joint stiffness was assumed to be 200 tsf for initial computer runs. In subsequent runs the value was modified to account for the possibility of joint failure in shear, and a joint friction value of ϕ = 40° was assumed.

Conclusions

The results of the tangential stress analysis given in Fig. 9.75 shows that the initial rock bolts of only 10 ft (3 m) length were in a high-stress zone and that bolts of 20 ft length would extend into a low-stress zone. The contractor installed bolts 40 ft long and movements were arrested. With the inclusion of creep data from long-term laboratory testing into the analysis, the displacements from the FEM analysis were found to approximate the actual displacements. Application of the FEM to the problem permitted a qualitative assessment of stress and displacement, the need for and the effectiveness of rock bolts as temporary supports, and the criteria for estimating their required minimum lengths.

9.4 WATERFRONT RETAINING STRUCTURES

9.4.1 INTRODUCTION

Purposes

Docks, quays, and wharves may be constructed with anchored sheet-pile bulkheads, relieving

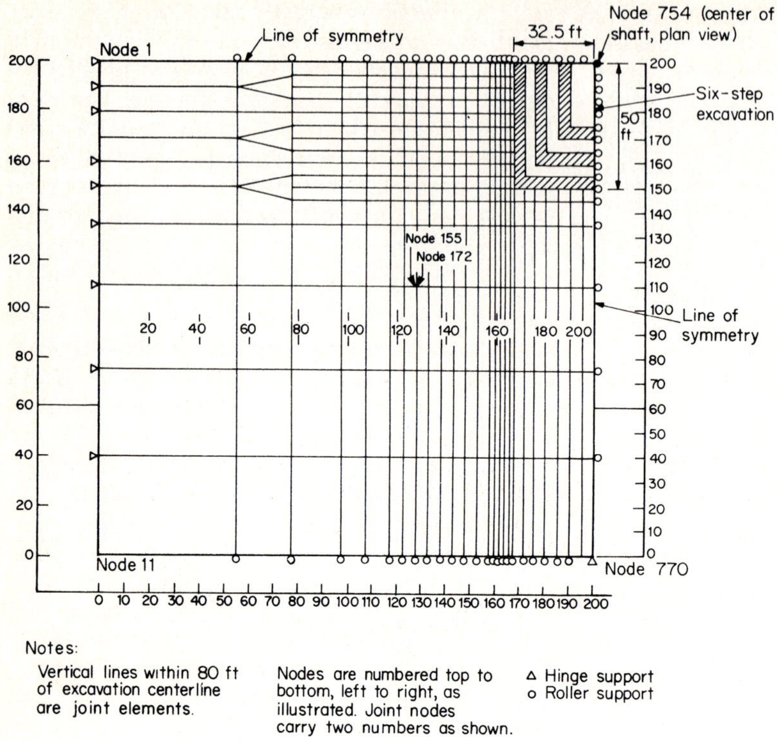

FIG. 9.74 Finite element mesh for excavation in folded gneiss with high residual stresses. [From Ward (1972).[62]]

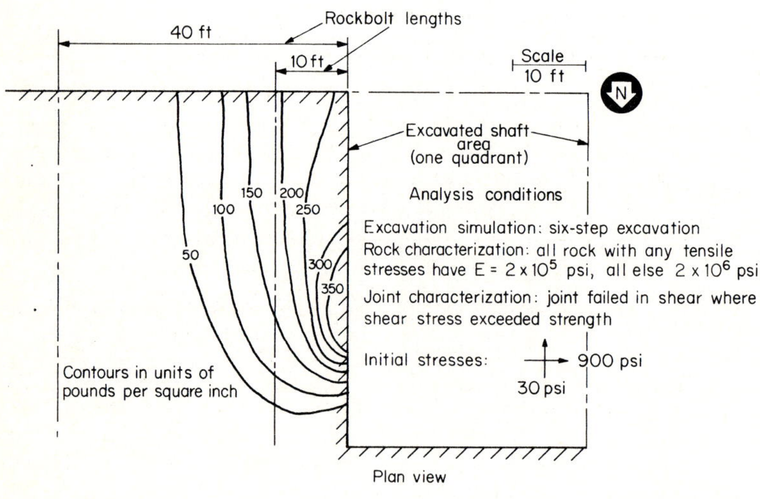

FIG. 9.75 Distribution of tensile stresses behind excavation wall analyzed with the FEM mesh of Fig. 9.74. [From Ward (1972).[62]]

platforms, cellular cofferdams, concrete gravity walls, or precast concrete floating boxes.

Land reclamation may be retained by cantilever or anchored sheet-pile bulkheads, relieving platforms, gravity rock dikes, or concrete gravity or precast boxes.

Construction in the dry may be accomplished with cellular or double-wall cofferdams.

Selection Factors

General

In the selection of the type of retaining structure for a particular project, consideration is required of the maximum water depth, height of structure above the mud line, bottom soil conditions and foundation stability, structure use and loads to be supported, and outboard conditions, i.e., wind, wave and current forces, tidal lag, and ice loads. Selection depends primarily on the life and use of the structure and then on its feasibility and comparative economics as they are influenced by natural conditions.

Materials

Materials to be used vary with loading conditions, purpose of the structure, desired life, and availability, and may include timber for relatively small structures; steel or concrete with or without stone for larger structures; or in some cases stone alone. Local conditions must be considered in selection. Concrete can be more costly than steel in developed countries with high labor costs and readily available steel sections, but in underdeveloped countries with low labor costs and only imported steel, concrete often is more economical.

Concrete structures usually have the longest life in marine conditions, except for stone. Concrete surfaces are subject to deterioration from the abrasive action of moving water carrying suspended materials, from the wedging of debris and ice, or from repeated freezing and thawing of water absorbed in the concrete. For exposed concrete in marine structures, good mix and proper curing and placement are essential. Cracks must be minimized to avoid corrosion of reinforcing steel by limiting deflections and tension in concrete members. In severe environments, coatings such as epoxy have been used to provide protection.

Steel is subject to corrosion which is particularly active in seawater, greatest in the tidal or splash zones, and severest near mean low water. Corrosion losses can be insignificant below the mud line in materials of low permeability (clay, silt, and fine sand) but can be serious near the ground surface where the steel is in contact with coarse rubble, fill, ashes, and boulders into which oxygen can penetrate. Where concrete cover is thin or quality is poor, these conditions also can lead to concrete reinforcing deterioration. Corrosion resistance is provided by cathodic protection, the use of special steel such as "mariners steel," painting, or a protective cover such as concrete or epoxy in the splash zone.

Timber is subject to rot in the zone above permanent submergence and to attack by marine borers. Creosoted timber provides protection against marine borers in mild climates but is not required in cold northern waters or in waters heavily polluted. In tropical to subtropical seas, however, even creosoted timber will suffer severe damage in a period of about 15 years from marine-borer attack [Tschebotarioff (1973)[14]].

Fabricated fiberboard is resistant to corrosion, rot, and marine-borer attack and has been used to construct lightly loaded bulkheads, particularly in shallow, quiet tidal zones.

Stone, when consisting of fresh, unweathered, sound rock blocks of materials such as basalt, granite, or diorite, provides structures that are the most resistant to deterioration.

Loading Conditions

The height of the structure and the loads to be supported impose restrictions on the selection. Cantilever sheet-pile walls, for example, are suitable only for low heights, in the order of 4 m (15 ft) maximum, and low loads. Anchored steel sheet-pile bulkheads generally are limited to about 11-m (37-ft) water depths because of restrictions of bending-moment capacity.

LANDWARD In most cases, waterfront retaining structures are constructed in the water and then the land is reclaimed by backfilling either with hydraulic fill or a combination of barge-dumped and land-dumped fill. After construction of the wall system the waterfront area is dredged to its final depth and the wall must retain all of the material placed behind it. If soft organic soils are present and not removed, the result will be very high lateral earth pressures. Surcharge loads are imposed by cranes, buildings, containerized cargo, or ore stockpiles, for example. The most severe conditions occur during sudden and large drops in the outboard water level during heavy rains which leave the backfill essentially saturated.

SEAWARD Loads are imposed by docking, breasting, and mooring forces from vessels and from ice, wind, and currents.

Foundation Conditions

Poor soil conditions are characteristic of waterfront areas where loose granular soils and soft organic materials are common. Foundation stability requires careful evaluation for many types of structures. Current forces cause scour and erosion of the foundations and for some structures, such as dikes and relieving platforms, erosion of the structure itself. *Shallow, hard rock*, especially with an irregular surface, can be very troublesome, requiring that piles be installed in grouted rock sockets to provide for stability.

Analysis and Design Factors

Prior to proceeding with analysis and design, in which lateral forces, overall stability, and construction procedures must be considered, required data include complete information on geologic conditions, soil properties, and the forces imposed by nature and the proposed facility.

Lateral Forces

Lateral forces to be considered include

- Active and passive earth pressures and effects of surcharge loadings
- Unbalanced water pressures from river flood stages and tidal ranges, especially during heavy rains
- Ship breasting forces and mooring pulls
- Wave pressure, current forces and ice loads, and effects of scour
- Earthquake forces

Stability

Stability against overturning is of concern with all types of structures. Base stability against rotational shear failure in weak soils *(see Art. 6.2.4)* requires consideration of all types, and sliding is an important consideration for gravity structures such as cofferdams and concrete walls. Slope stability can be an important factor in the design of relieving platforms and dikes.

Construction

The possible effect of construction sequence and procedures on the integrity of the structure should be considered, and those procedures that may have adverse effects on the structure should be prohibited.

Dredging removal of soft organic soils to some distance behind a bulkhead prior to its installation and then replacement with sand or stone will substantially reduce lateral earth pressures and improve overall conditions. Replacement of weak soils also is necessary to provide adequate support for gravity structures unless piles are used. Backfilling should proceed from the seaward side toward the land. If it is started from the landward side the result is a buildup of soft soils against the wall, even after dredging, since seldom can all of the soft soils be removed. Lateral pressures also increase in excess of design assumptions.

Driving piles for a relieving platform should start from the wall and proceed landward. If piles are driven seaward from the land after the wall has been installed, the resulting densification of granular soils can cause a significant increase in lateral pressure over the values assumed for design.

9.4.2 TYPES OF STRUCTURES

Flexible Walls

Anchored Sheet-Pile Bulkheads

Anchored sheet-pile bulkheads (Fig. 9.76*a*) are economical in poor soil conditions where installation by driving or jetting is possible and obstructions are not anticipated. Steel sheets are limited generally to heights above the dredge line of about 11 m (37 ft) because of the large bending moments that develop from earth pressures, especially in soft clays and organic soils. Special steel or prestressed reinforced concrete sections and stronger soil conditions permit greater heights. Steel rapidly corrodes in seawater unless provided with special protection, as

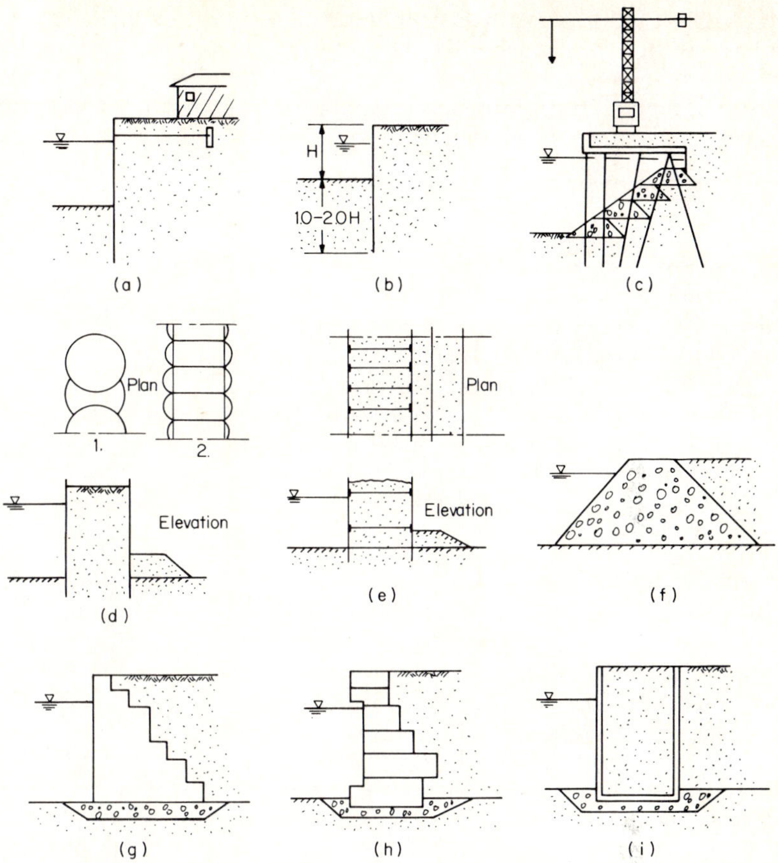

FIG. 9.76 Types of waterfront structures: (*a*) anchored bulkhead; (*b*) cantilever bulkhead; (*c*) relieving platform; (*d*) cellular cofferdam: (1) circular and (2) diaphragm; (*e*) double-wall cofferdam; (*f*) gravity rock dike; (*g*) concrete gravity wall; (*h*) concrete block wall; and (*i*) precast-concrete floating box.

discussed previously. Timber sections, concrete planks, or composite materials commonly are used to retain dredged lagoons for housing developments and small boat basins where water depths seldom exceed 3 to 4 m (10 to 13 ft).

Cantilever Bulkheads

Cantilever bulkheads (Fig. 9.76*b*) usually are constructed with steel sheets or concrete planks. Materials with low strength, such as wood and fabricated fiber board, normally require support with anchors. Steel sheet-pile walls are subject to large deflections and are limited to heights above the dredge line in the order of 4.5 m (15 ft). They are suitable only for support of granular soils or stiff clays. Capacity is severely sensitive to scour because their restraint is developed by passive resistance below the dredge line.

Relieving Platforms

Relieving platforms (Fig. 9.76*c*) are pile-supported concrete platforms which relieve the vertical loads on the underlying soils and are constructed to withstand relatively high lateral loads resulting from great water depth, thick deposits of weak soils, or high surcharge loads from settlement-sensitive structures and cranes. They have a medium first cost and may be of medium durabilty and maintenance cost depending upon pile types used.

Platform Variations

A high-level platform is located on the surface, whereas a low-level platform is buried under a thickness of fill. Support is provided either totally by piles or partially by piles and partially by a sheet-pile wall. Anchorage is provided by raking piles. Platform width must be sufficient to provide total or almost total shielding of the sheet piles from the effects of loads above and behind the platform. Various schemes are illustrated in Fig. 9.92.

Backslope Retention

Sheet piles may be placed in front of or behind the platform. Unless the sheeting is in front of the platform, it is necessary to provide a stable slope for the soils beneath the platform to prevent failure from inboard loads and to provide scour protection. This is normally accomplished with short inboard sheeting and stone berms. There are many variations in relieving platform designs.

Gravity Structures

Cellular Cofferdam

Cellular cofferdams, consisting of circular walls constructed of interlocking steel sheet piles driven to form a cell, are suitable for moderate to large heights. The two main types are the circular cofferdam, where the cells are linked by arcs (Fig. 9.76*d.1*), and the diaphragm type shown in Fig. 9.76*d.2*. The cells are filled with sand, stone, concrete or some other deformation-resistant material and must be provided with adequate foundation support. They often provide a satisfactory and economical retaining bulkhead when topped by a small outboard seawall as they have a low to moderate initial cost, but are subject to damage from barge or ship impact and corrosion. Their basic purpose is to permit construction in the dry by providing an impervious water barrier.

Double-Wall Cofferdam

Double-wall cofferdams (Fig. 9.76*e*) are constructed of two sheet-pile walls, supported internally with tie-rods and externally on one side with a berm, and filled with granular materials. They are suitable for moderate heights, and are used mainly as a dam to permit construction in the dry.

Gravity Dike or Embankment

Gravity dikes (Fig. 9.76*f*) are constructed to enable land reclamation. They usually involve a lowest first cost where proper quarry stone is available, but they encroach upon and reduce the land area available for development and are subject to erosion. Usually they are constructed of crusher-run quarry stone and are protected in the wave zone with riprap. A filter layer may be needed between the rock dike and the landfill. Strong underlying soils are required for support. When used as a cofferdam a clay core or blanket is required, although at times steel sheeting or grouting is used as a cutoff.

Gravity Concrete Walls

Gravity concrete walls can be constructed as one massive block (Fig. 9.76g) or by stacking a number of precast elements (Fig. 9.76*h*). The latter type is limited to relatively shallow water. Requiring adequate support on strong soils or piles, these wall types are used for reclamation for which they involve the highest first cost but have the lowest maintenance and the longest life and provide a maximum usable land area.

Precast Concrete Floating Boxes

Precast concrete floating boxes (Fig. 9.76*i*) avoid the problems associated with constructing a massive concrete wall in situ in waterfront conditions, and have been used where long life and low maintenance costs are important. They are constructed of costly, high-quality concrete in a graving dock or basin, floated or towed to the site, sunk to bear on a prepared stone bed, and then filled with granular material. A graving dock is not necessary if they are constructed of precast concrete segments lowered into place to form boxes.

Selection of marine gravity retaining structures is discussed by Johnston et al. (1976).[63]

9.4.3 ANCHORED SHEET-PILE BULKHEADS

Analysis and Design Considerations

Support

Support of an anchored sheet-pile bulkhead is provided by passive resistance on the front of the wall embedded below the dredge line and from anchors installed near the top of the wall. A sheet-pile member with a section modulus adequate to resist earth pressure forces is selected and the anchorage designed. Overall stability and the stresses in the members depend mainly on the relative piling stiffness, piling penetration depth, soil strength and compressibility, and amount of anchor yield. Anchorage usually is provided by ties connected to deadmen placed at a distance behind the wall beyond any possible rupture surface.

Methods

In the past, penetration depth has been estimated and forces analyzed with one of two systems: the free-earth-support method or the less conservative fixed-earth-support method. General diagrams for earth pressures, bending moments, and deflections for each method are given in Fig. 9.77. The currently favored free-earth-support method is based on the assumption that the maximum possible passive resistance is mobilized below the mud line. The fixed-earth-support method, now seldom used, is based on the assumption that a point of contraflexure (bending moment = 0) occurs at some depth x below the mud line.

Excavation and Backfill

Thick soft soil deposits result in high lateral forces and backfill settlement which severely affect the performance of the bulkhead and anchor system. They are usually removed, therefore, from some distance behind the bulkhead prior to its installation and replaced with granular backfill. Backfill should proceed from the seaward to the landward side to avoid buildup of weak materials along the bulkhead. Several schemes to reduce lateral pressures and backslope subsidence are shown in Fig. 9.78.

Stability Factors

Anchored bulkheads commonly fail in one of several modes including general foundation failure, failure of the anchorage, or failure of the embedded section.

Bearing-capacity failure in soft clays may occur due to the unbalanced weight of the material behind the wall. Failure takes the form of rotational movement and may be analyzed by methods given in Art. 8.2.4. Resisting forces should include the water weight W_w and lateral pressure P_w acting on the front of the bulkhead [Peck et al. (1974)[24]].

Anchorage failure may result from underestimation of anchor forces, overestimation of anchor resistance, location of anchor too close to the wall, or ground settlement imposing fill loads on the anchor ties.

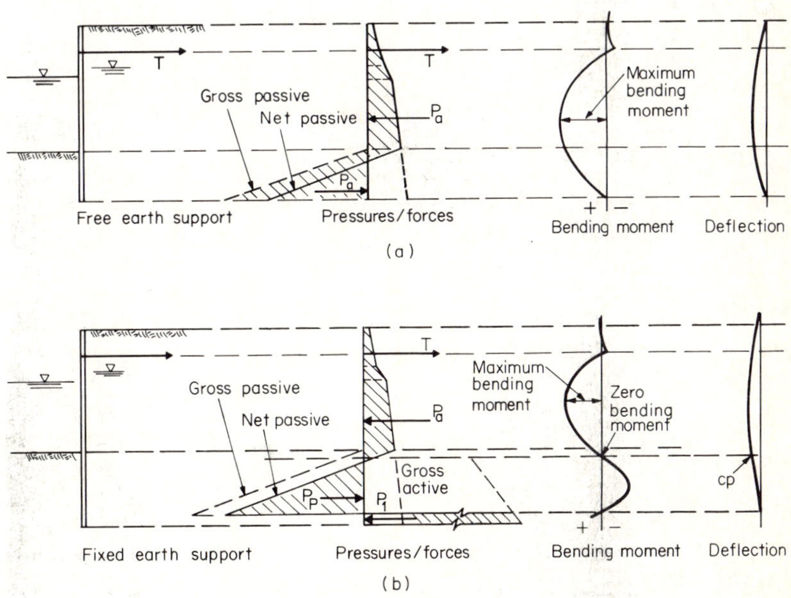

FIG. 9.77 Conventional soil pressure, bending moment, and deflection diagrams for the free-earth support *(a)* and fixed-earth support or equivalent beam *(b)* methods used in analysis of anchored sheet-pile structures.

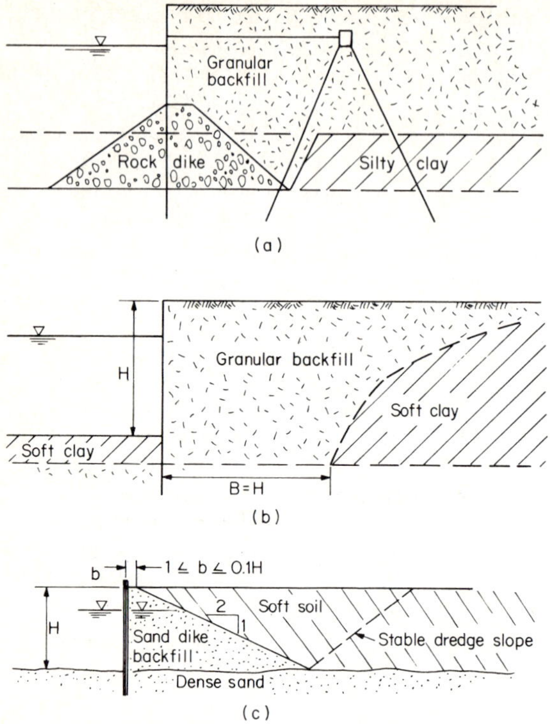

FIG. 9.78 Methods to reduce lateral earth pressures and backslope subsidence in soft soils: (*a*) excavate soft soil, install bulkhead, place rock dike and backfill; (*b*) excavate soft soil, install bulkhead, place granular backfill from bulkhead landward; and (*c*) excavation-backfill scheme where granular backfill is scarce. [Part (c) from NAFVAC (1982).[1]]

Toe failure of the embedded portion may occur due to inadequate penetration permitting excessive movement or even ground rupture or to high upward seepage forces caused by an unbalanced hydraulic head behind the bulkhead resulting in piping and loosening of toe materials.

DESIGN DETAIL It is prudent to place the wales on the seaward side; otherwise, the connections between the piles and the wales are required to resist in tension all force on the anchor rods.

Earth Pressures

Distribution

Model studies, field observations, and FEM studies indicate that sheet-piling flexibility causes earth pressures to redistribute and vary significantly from those predicted by classical theories. The resulting bending moments are much smaller than classical theory would indicate, and the bending moment decreases with increasing sheeting flexibility [Bjerrum et al. (1972)[49]]. Redistribution around the wall as it deflects results from arching, which reduces the earth pressures on some wall parts and increases them on others. There is a major reduction in earth pressure between the anchor and the dredge level where major yielding occurs, and an increase in pressures on the less yielding parts at the anchor and below the dredge line as shown in Fig. 9.79*a*. This reduction in bending moment should be accounted for in design to obtain a lower and more economical section modulus. In modern design practice this is accounted for with Rowe's reduction factors, as will be discussed, but only for the case of embedment in sands denser than the loose state [NAVFAC (1982)[1]].

The results of an FEM analysis compared with actual wall deflections in loose sands measured by inclinometer is presented by Bjerrum et al. (1972).[49] Earth pressures calculated by the FEM and by the free-end support method are given in Fig. 9.79*a*. The bending moments calculated by the FEM and the free-end-support method and by using Rowe's reduction factor are given in Fig. 9.79*b*. The agreement between the FEM study and analysis by Rowe's method is apparent, as is the large reduction in moment obtained from classical theory.

Backfill Materials

Studies of a number of existing bulkheads [Casagrande (1973)[21]] demonstrated the importance of the type of backfill material and of anchorage rigidity on bulkhead performance. Bulkheads backfilled with coarse material such as crushed stone or sand and gravel, and with unyielding anchors, show good agreement between observed tie-rod stresses and bending moments with those predicted using Rowe's reductions. Where backfill materials finer than coarse sand were used, however, earth pressures approaching the at-rest condition developed after several years. Casagrande explained this as due to the loss of arching that occurs from tidal fluctuations and wave and current action. The result is

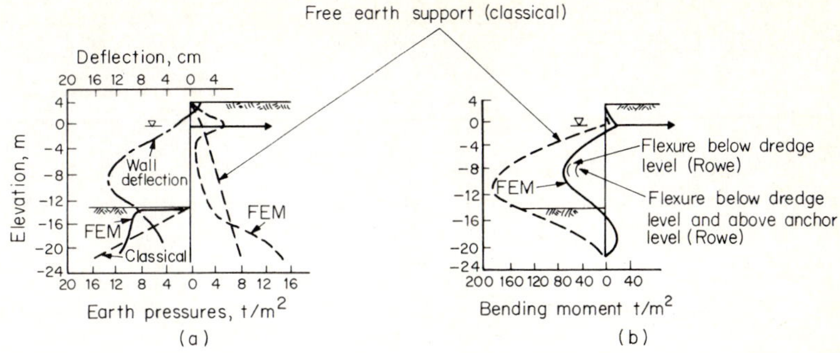

FIG. 9.79 Comparison of earth pressures (*a*) and bending moments (*b*) from measured wall deflections, FEM analysis, free-earth-support method, and Rowe's reductions. [After Bjerrum et al. (1972).[49]]

higher anchor forces and higher bending moments than anticipated using Rowe's reduction.

Where conditions are uncertain either the value for FS should be increased to 2.5 or the conventional design assumptions should be modified to consider the type of backfill material. The use of coarse, granular materials such as gravel, crushed stone, or slag assures continued minimum active earth pressures, provided that the backfill is not compacted. Even light compaction may permanently increase the earth pressures into the passive range. The finer the backfill, the less is the assurance that active pressure will be effective for an extended period. Plastic deformation, vibrations, water-level fluctuations, temperature changes, and other effects may cause the materials to develop at-rest earth pressures, even against flexible bulkheads.

Backfills often are placed hydraulically and consist of relatively fine-grained materials. According to NAVFAC (1982),[1] active pressure coefficients for loose hydraulic fill range from about 0.35 for clean sands to about 0.50 for silty fine sand. Hydraulic fill should be placed by procedures which permit runoff of wash water to prevent buildup of large hydraulic pressures wherever practical.

Maximum Pressure Conditions

The maximum earth pressure normally exists when the water level is at its lowest stage, but it can be increased substantially if a heavy rainstorm or rapidly receding high water level produces a saturated or near-saturated condition behind the wall and above the outside water level. The lag in the water level depends on the backfill material, ranging from negligible in coarse sand and gravel to a meter or so in fine or silty sand to full hydrostatic pressure between water levels for a silt or clay backfill. A high head differential also can cause flow beneath the wall and piping at the toe, resulting in a loss of passive resistance as already mentioned. Analysis may be made by flow net, as shown in Fig. 9.50.

Earth-Pressure Determinations

Three general methods have been used to determine earth pressures behind anchored bulkheads, as illustrated in Fig. 9.80 [Casagrande (1973)[21]]. *Classical approaches* generally employ Rankine active pressures and Coulomb passive pressures with wall friction in granular soils or Rankine pressures for the undrained case in cohesive soils. As indicated in the foregoing discussion, the classical approach has a number of weaknesses, but is currently the favored method, using Rowe's reduction factors. For soils behind the bulkhead consisting predominantly of grain sizes smaller than coarse sand, the earth pressure that will develop several years after completion may be appreciably greater than the minimum active pressure. This can lead to a much greater anchor pull than designed for. Depending upon the flexibility of the sheet piles, arching may temporarily or permanently reduce the maximum bending moment M_{max}. Since reduction in M_{max} is automatically associated with an increase in earth pressure in the anchor and toe areas, appreciable stresses may be thrown temporarily or permanently into the anchor, depending on the type of backfill and the sheeting

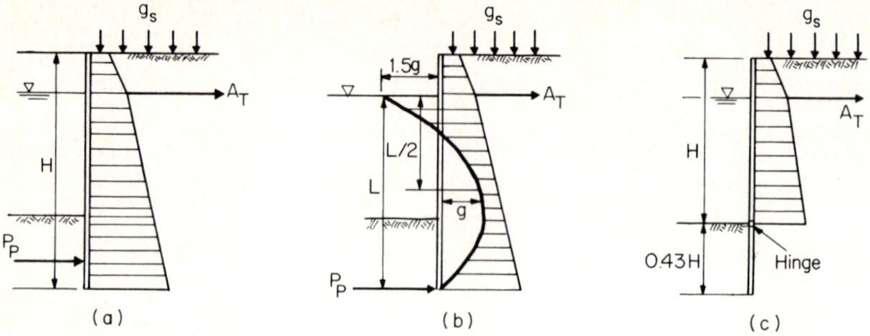

FIG. 9.80 Basic assumptions for three widely used methods of anchored bulkhead design: (*a*) classical; (*b*) Danish rules; and (*c*) Tschebotarioff. [After Casagrande (1973).[21]]

flexibility. Consideration of the type of backfill, therefore, is extremely important in the selection of earth pressure coefficients for design.

Danish rules, developed empirically in the early 1900s, combine active earth pressures with an appreciable reduction in M_{max}. There is an increase in anchor pull, the magnitude of which depends on the reduction in P_a, given by the expression

$$q = k \frac{10h + 2L}{10h + 3L} P_a \tag{9.52}$$

where P_a = average active earth pressure over bulkhead height
h = height of submerged soil equivalent to load above anchor level
k = empirical coefficient ≈ 0.8
L = distance between anchor and bottom of sheeting

The method has proved satisfactory for coarse granular backfill, but for finer-grained soils the reduction in M_{max} may only be temporary, which would result in a bulging of the wall.

Tschebotarioff's method (1951),[64] (1973)[14] involves the assumption of a hinge at dredge level. M_{max} is smaller than that resulting from the classical method, but larger than that of the Danish method. It is a simplified equivalent beam based on the fixed-earth-support method.

Sowers and Sowers (1967)[65] analyzed a number of bulkhead failures and concluded that the failures were not the result of inadequacies in earth-pressure theories, but rather were caused by neglect of load restrictions, poor design, or deterioration. Casagrande (1973)[21] points to the problem of anticipating what the real earth pressures will be, as discussed previously.

Free-Earth-Support Method

Design Basis

In this discussion it is assumed that all weak soils have been removed to the mud line and replaced with granular backfill, and embedment is in either relatively strong clays or in sands. Pressure distribution is assumed to be in accordance with the classical theories and some fraction of the maximum possible passive resistance (usually P_p/FS, Fig. 9.81) is mobilized below the mud line (dredge line). Distribution depends on embedment depth, which must be adequate to mobilize sufficient passive pressure to provide limiting equilibrium.

Conventional pressure distribution, bending moment, and deflection diagrams for the free-earth-support method are given in Fig. 9.77*a*. The bulkhead is stable (limiting equilibrium) when the sum of the moments of P_a and P_p/FS taken about the anchor equal zero. The anchor force is found from the condition that the sum of the horizontal forces equal zero, or $T = P_a - P_p$.

Loading conditions and resulting pressure distributions against the bulkhead are illustrated in Fig. 9.81. P_1 represents the resultant horizontal pressure due to soil weight behind the bulkhead and is based on total unit weight above GWL plus submerged unit weight below GWL times the active earth-pressure coefficient K_a, employing K_{a1} for the backfill material above the mud line, and where embedment is in sands, K_{a2} for the material below the mud line. To these pressures are added the unbalanced water pressure P_2, which should provide for sudden drawdown conditions, and surcharge pressures including uniform surcharge P_3 and line or concentrated surcharge loads P_4. P_5 represents the mobilized

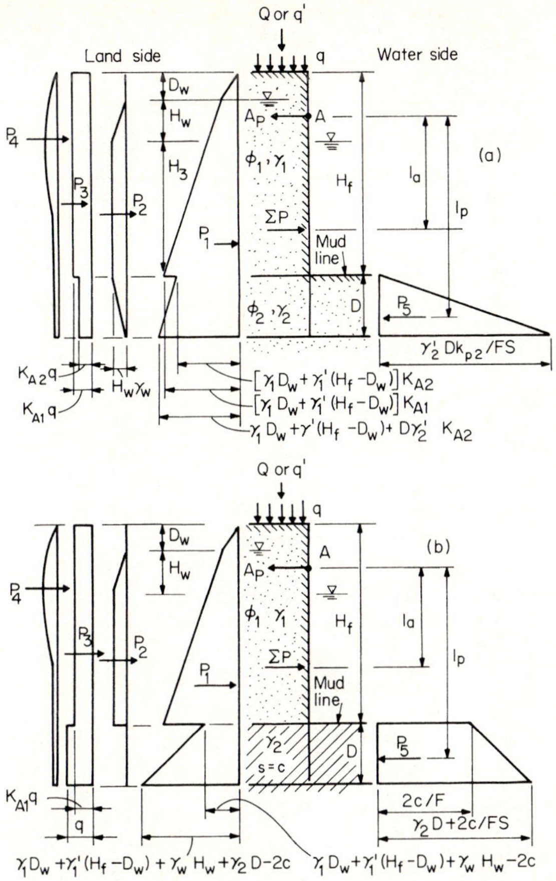

FIG. 9.81 Loads against anchored bulkhead retaining sand backfill: (*a*) sand embedment and (*b*) plastic clay embedment. [From Peck et al. (1974).[24] Used with permission of John Wiley & Sons, Inc.]

portion of the passive pressure of the soil in front of the embedded portion of the bulkhead which is a function of the passive coefficient for embedment in sands or of the undrained shear strength for embedment in clays. Earth pressures for a soft clay stratum retained by the bulkhead are given in Fig. 9.82.

Because active pressures are mobilized with very little wall deflection their coefficients are used directly in analysis, whereas with passive resistance substantial movement is required for their mobilization. Therefore, a value for FS is applied to either the passive coefficient or the undrained strength.

Procedures

An example of the analysis of an anchored bulkhead with a granular backfill is given in Fig. 9.83 for two conditions of embedment: clay and sand. The analytical procedures are described as follows.

CASE A Clay embedment involves the following steps:

Step 1. Check that the bearing capacity of the clay is satisfactory. Use Eq. 9.44.

Step 2. Compute the earth pressures. The active earth pressures are computed to the mud line using the equation from Table 9.3, No. 3, that includes the uniform surcharge q_s. The net water pressure is added for the unbalanced water condition. At the mud line, the active pressure

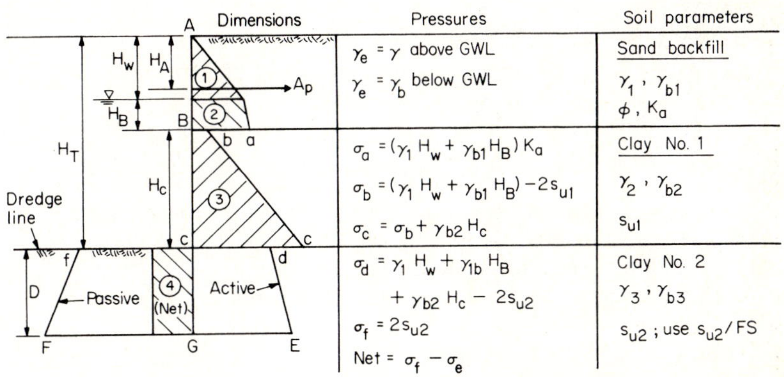

FIG. 9.82 Earth pressures against anchored bulkhead retaining sand backfill and soft clay, with embedment in clay. [After USS (1974).[70]]

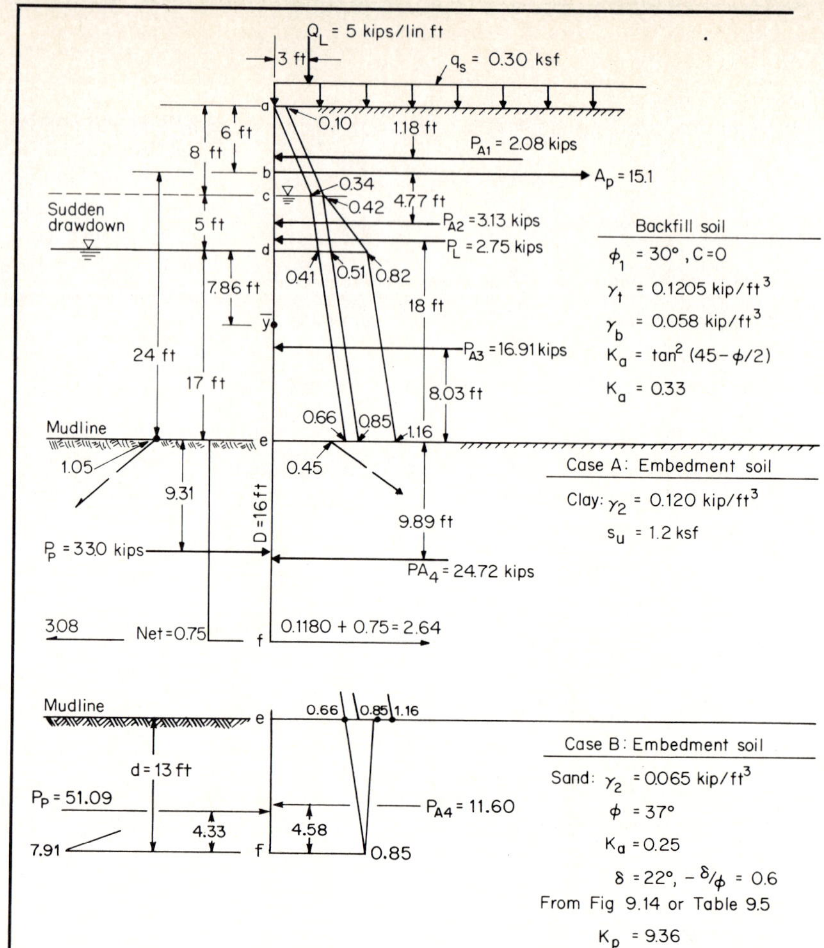

CASE A: Clay Embedment

1. *Check bearing capacity* (Eq. 9.44):

$$\text{FS} = \frac{5.14 s_u}{q + \gamma H} = \frac{5.14(1.2)}{0.30 + 30(0.1205) - 17(0.0624)} = 2.16, \text{ OK}$$

2. *Compute earth pressures*

2.1 Active pressures including uniform surcharge q_s. From Table 9.3(3):

$$\sigma_h = \gamma s K_a + q_s K_a$$

At (a) $\sigma_h = 0.3(0.33) = 0.10$ ksf

(b) $\sigma_h = 0.1205(6.0)(0.33) = 0.24 + 0.10 = 0.34$ ksf

(c) $\sigma_h = 0.1205(8.0)(0.33) = 0.32 + 0.10 = 0.42$ ksf

(d) $\sigma_h = [0.1205(8.0) + 0.058(5.0)](0.33) = 0.41 + 0.10 = 0.51$ ksf

add net water pressure $0.0624(5.0) = 0.31$

0.82 ksf

(e) $\sigma_h = [0.1205(8.0) + 0.058(22.0)](0.33) = 0.75 + 0.10 = 0.85$ ksf

add net water pressure $0.0624(5.0) = 0.31$

1.16 ksf(fill)

FIG. 9.83 Analysis of an anchored sheet-pile bulkhead. Case A: Embedment in clay. Case B: Embedment in sand.

CASE A: Clay Embedment

(e) Clay embedment σ_h =

surcharge load	= 0.30
0.1205(8.0) + 0.058(22.0)	= 2.24
Net water pressure (0.0624)(5)	= 0.31
	2.85 ksf
(−) $2s_u$	2.40
	0.45 ksf (clay)

$(f)\sigma_h = \gamma_2 D - \sigma_{h(e)} = 0.118D - 0.45$ ksf

2.2 Passive pressures

At (e) $2s_u/FS = 1.2$ ksf

(f) $\gamma D - 2s_u/FS = 0.118D - 1.20$

$(-)0.118D - 0.45$

Net pressure (e) to (f) = 0.75 ksf

2.3 Line load surcharge pressure (Fig. 9.20 and Eq. 9.26):

$m = x/H = 3/30 = 0.1$

$P_L = 0.55Q_L = 0.55(5.0) = 2.75$ ksf, acting at $H_L = 0.60H = 0.6(30) = 18$ ft

3. *Evaluate wall equilibrium:* Forces and moments about (b) and find D.

Location	*Force*	*Arm*	*Moment*
$(abc)PA_1 = \frac{0.10 + 0.42}{2}(8) =$	2.08	(−)1.18	(−)2.46
$(cd)PA_2 = \frac{0.42 + 0.82}{2}(5) =$	3.13	4.77	14.94
$PL =$	2.75	6.00	16.50
$(de)PA_3 = \frac{0.82 + 1.16}{2}(17) =$	16.91	15.97	270.05
	24.88 k		301.49

P_{net} (below embedment) $= (-)0.75D \quad 24 + \frac{D}{2} \quad (-)(18.0D + 0.38D^2)$

To find D,

$\Sigma M_b = 0 = 301.49 - 18.0D + 0.38D^2$, and $\underline{D = 13.1 \text{ ft}}$

For FS add 20% or

$D = 13.1 \times 1.2 = 15.7$ ft, say $\underline{D = 16 \text{ ft}}$

4. *Compute anchor pull:*

$\Sigma H = 0 = 24.88 - 0.75D - A_p$

$= 24.88 - 0.75(13.1) - A_p$, and $\underline{A_p = 15.1 \text{ kips}}$

5. *Find M_{max} for sheet piles.* Let point of contraflexure be at the mud line and let V = 0 occur at y ft below the water line:

$$A_p = PA_1 + PA_2 + P_L + (0.82)y + \frac{1}{2}\left(\frac{1.16 - 0.82}{17}\right)y^2$$

$$15.1 = 2.08 + 3.13 + 2.75 + 0.82y + 0.01y^2$$

$$0.01y^2 + 0.82y - 7.12 = 0, \text{ and } \underline{y = 7.86 \text{ ft}}$$

Find M_{max} at V = 0

$$M_{max} = (15.1)(14.86) - (2.08)(16.04) - (3.13)(10.09) - (2.75)(8.86) - \frac{(0.82)(7.86)^2}{2} - \frac{(0.01)(7.86)^2}{2}\left(\frac{7.86}{3}\right) = \underline{108.13} \text{ ft·kips}$$

Required section modulus $S = M_{max}/f_s = 108.13(12)/32 = 40.55$ in³/ft for steel of $f_s = 32{,}000$ psi (see Case B).

FIG. 9.83 ***(continued)***

CASE B: Sand Embedment

1. *Compute earth pressures:*

(a) through (d) as computed for Case A
(e) at mud line

$$\sigma_h = [0.1205(8) + (0.058)(22)](0.25) = 0.560$$

$$\sigma_h = (\text{net water pressure } (0.0624)(5) = 0.312$$

$$\sigma_h = (\text{surcharge})(0.30)(0.25) = \underline{0.075}$$

$$0.947 \text{ ksf}$$

(f) Select Trial D, say, $D = 13$

$$\sigma_h = [0.1205(8) + (0.058)(22) + (0.065)(13)](0.25) = 0.77$$
$$\sigma_h = \text{surcharge} = \underline{0.075}$$
$$0.846 \text{ ksf}$$

(g) Compute passive pressure

$$\sigma_h(e) = 0$$
$$\sigma_h(f) = (0.065)(13)(9.36) = 7.91 \text{ ksf}$$

2. *Solve for toe failure FS (should be about 2.5)*

Pressure	Force	Arm	Moment
PA_1	2.08	(−)1.18	(−)2.46
PA_2	3.13	4.77	14.94
PL	2.75	6.0	16.50
PA_3	16.91	15.97	270.05
PA_4	11.60	30.38	353.83
PP	51.09	32.67	1669.05

$$FS = \Sigma M_P/\Sigma M_A = 1669.05/652.9 = 2.55, \text{ OK}$$

(NOTE: If FS is too low or too high, then try new trial D. Changes of D of only 1 ft result in significant changes in moments.)

3. *Compute anchor pull:*

$$A_p = \Sigma P_A - \Sigma P_p = (2.08 + 3.13 + 2.75 + 16.91 + 11.60) - (51.09/2.5) = 16.0 \text{ kips}$$

4. *Find M_{max} for sheet piles: Find y for V = 0 (see Case A):*

$$16.0 = 2.08 + 3.13 + 2.75 + 0.83\,y + 0.01\,y^2$$

$$0.01y^2 + 0.83y - 7.54 = 0, \text{ and } \underline{y = 8.25 \text{ ft}}$$

$$M_{max} = (16.0)(15.24) - (2.08)(16.42) - (3.13)(10.48) - (2.75)(9.25) - 28.23 - 1.81 = \underline{121.4 \text{ ft}\cdot\text{kips}}$$

5. *Use Rowe's moment reduction to find M_{design}:*

$$\rho = \frac{(M + D)^4}{EI} = \frac{[(30 + 13)12]^4}{(30 \times 10^6)I} = \frac{2363}{I}$$

Evaluate various sections

Section	I, in⁴	S, in³	ρ, in²/lb	Reduction (from Fig. 9.84)
PZ 38	280.8	46.8	8.4	0.87
PZ 32	220.4	38.3	10.7	0.77
PZ 27	184.2	30.2	12.8	0.72

Try PZ 32 with $f_s = 29{,}000$ psi (Mariner steel)

$$M_{design} = M_{max}(0.75) = 121.4(0.77) = \underline{93.5 \text{ ft}\cdot\text{kips}}$$

$$f_s = M_{design}/S = \frac{93.5(1000)(12)}{38.3} = 29{,}288 < 29{,}000 \text{ psi, OK}$$

FIG. 9.83 ***(continued)***

in the clay is equal to the sum of the uniform surcharge, the soil weight, and the net water pressure minus $2s_u$ (Eq. 3.34). (The net water pressure $H_w\gamma_w$ is used since the full water pressure $H_3\gamma_w$ may be subtracted from both the active and passive pressures.) The active pressure at the toe of the wall equals the product of the soil total unit weight and the embedment depth D plus $2s_u$.

The passive pressure in the clay at the mud line is $2s_u$, to which is normally applied a value of FS = 2 where one is confident of strength measurements. At the wall toe it is a function of the total soil weight, penetration depth D, and $2s_u$/FS (Eq. 3.33). The active pressure at the toe is subtracted from the passive pressure to find the *net pressure* acting along depth D.

The resultant pressures from the line and concentrated surcharge loads, and their location in reference to the mud line, are found using the procedures given in Art. 9.2.3.

Step 3. Evaluate wall equilibrium. To find the required depth of embedment D for equilibrium, the resultant horizontal forces of each element of earth pressure are computed and their moments found about the point of application of the anchor pull and summed. The moment for the net pressure below the mud line is given in terms of D and results in a quadratic equation which is solved to find D_1. The final depth is found by increasing D by 20 to 40% to increase FS.

Step 4. Compute anchor pull A_p by summing the horizontal forces.

Step 5. Find M_{max} for the sheet piles. The maximum moment (point of zero shear) normally is assumed to be located at a distance y below the water line. The horizontal forces are summed in terms of y, resulting in a quadratic equation which is solved for y. In relatively strong soils the real point of contraflexure of the sheeting is assumed to be the mud line. If the material at the mud line is loose or otherwise weak, the point of contraflexure is taken at 1 to 2 ft below the dredge line, but if the sheeting is driven into hard material at shallow depths overlain by weak soils the point of contraflexure is taken at the top of the hard material. Either case raises the point of application of the passive resistance and reduces the span length of the sheeting, H_p, which significantly influences the moments between the anchorage and the dredge line since they are a function of H_p^3.

If strong soils occur at the mud line, M_{max} is found by summing moments about the water line. If the point of contraflexure is below the mud line, the horizontal force and moment resulting from depth d is subtracted from the forces and moments for above the mud line that were found under Step 3. A new reaction is found at depth d which is subtracted from the anchor pull to find AP' (although the original anchor pull is used in anchorage computations). The depth y below the outside water level is then computed for the new value AP' and M_{max} subsequently found for this y [Peck et al. (1974)[20]].

An allowable stress f_s for the steel is selected and the required section modulus S for the sheeting found from

$$S = M_{max}/f_s \qquad (9.53)$$

The bulkhead anchorage is then designed.

CASE B Sand embedment involves the following steps:

Step 1. Compute earth pressures. The active earth pressures above the mud line are computed as in Case A, Step 2. The active pressure in the sand embedment at the mud line is equal to the sum of the soil weights, net water pressure, and uniform surcharge multiplied by K_{a2}. At depth D_1 below the mud line the active pressure is equal to the sum of the soil unit weights and the uniform surcharge times $Ka2$. The passive pressure at depth D is equal to the overburden pressure times K_p. The value for K_p is found from Fig. 9.14 in which a value for wall friction is applied.

Step 2. One approach to evaluate wall equilibrium to find the required penetration depth D in a manner similar to that described under Case A. The sum of the moments results in a cubic equation which is solved by trial and error. D is found by summing y and D_1. Using the final value for D selected, the equilibrium of the wall is checked with computing

$$FS = M_p/\Sigma M_a \qquad (9.54)$$

An alternate and simpler method is to select a trial value for D (usually between 10 and 16 ft depending upon outboard height), compute the earth pressures at (e) and (f) and find the value for FS from Eq. 9.54. FS should fall between 2.0 and 3.0 for coarse-grained soils and between 1.5 and 2.0 for fine-grained soils. If the computed value for FS is either too high or too low, another value is selected and the computations repeated until the desired FS is obtained.

Step 3. Compute anchor pull by summing the horizontal forces.

Step 4. Find M_{max} for the sheet piles in a manner similiar to that described for Case A. Rowe's moment reduction is applied to find the design value for M, as described in the following section, and the anchorage then designed.

Rowe's Reduction Factor

A relationship between the degree of flexibility of an anchored bulkhead and the reduction of the actual bending moment M, as compared to the free-earth-support value M_{max} has been developed by Rowe (1952),[66] (1955),[67] (1956),[68] (1957),[69] and referred to as *Rowe's reduction factor*. The ratio M_{design}/M_{max} is expressed as a function of the flexibility coefficient ρ for granular soils as

$$\rho = (H + D)^4/EI \qquad (9.55)$$

where $H + D$ = total sheeting length
E = modulus of elasticity for steel
I = moment of inertia

A value for ρ is calculated and entered into Fig. 9.84 to find the reduction factor to be applied to M_{max}. M_{design} is obtained by successive trials of sheeting size until the maximum bending stress f_s in the sheeting equals or is slightly less than the allowable bending stress for the section selected. No reduction in M_{max} is permitted for penetration in fine-grained, loose, or very loose granular soils [NAVFAC (1982)[1]].

Anchorage Design

Types of Anchorage

In relatively strong upper soils, anchor reaction may be provided by buried concrete blocks, continuous concrete beams, or sheet-pile walls. Resistance is derived from passive pressures, as shown in Fig. 9.85a. In weak upper soils, or

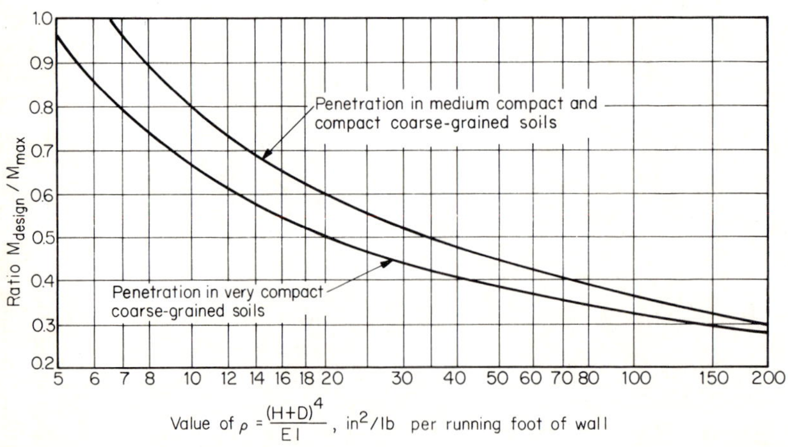

FIG. 9.84 Rowe's moment-reduction curves for granular soils. [From NAVFAC (1982),[1] after Rowe (1952).[66]]

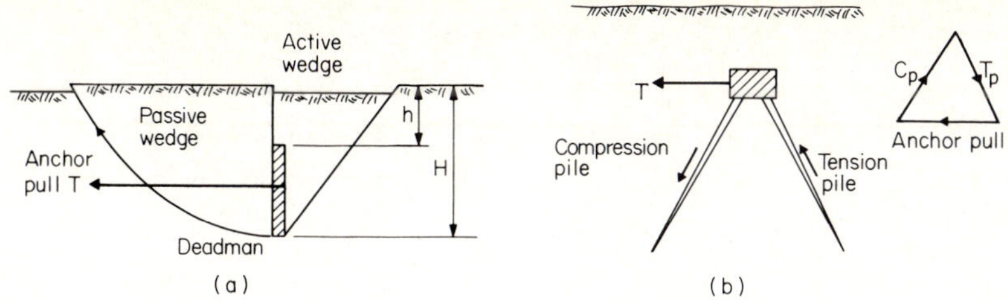

FIG. 9.85 Anchorages for sheet-pile bulkheads: (*a*) continuous deadman near ground surface and (*b*) tension and compression piles.

where there is a need to minimize wall deflections, anchor reaction is provided by braced piles driven to form an A frame, as shown in Fig. 9.85*b*.

Tiebacks, as used for open excavations, have also been used for bulkhead anchorage and are particularly useful in stabilizing existing bulkheads where surface excavation for some other anchor system would be difficult or not feasible due to existing structures. The vertical force component on the wall, however, must be considered.

Stability of Deadman Anchorages

In designing the anchorage, the anchor pull found from the bulkhead analysis is usually increased by a factor of 1.2 for safety. A concrete block or other similiar form of deadman anchorage must be located beyond any potential failure surface. As illustrated in Fig. 9.86, location in Zone 1 provides no resistance, in Zone 2 partial resistance, and in Zone 3 full resistance.

The stability analysis for a deadman anchor is illustrated in Figs. 9.87 and 9.88. If a deadman must be positioned close to the wall, the anchorage resistance is decreased and an additional passive reaction is required for stability at the wall base. Soils within the passive wedge (Fig. 9.87*b*) should be compacted to not less than 90% standard Proctor density and fine-grained soils of medium to high plasticity should not be used at the anchorage. A safety factor of 2 is applied to the ultimate values found for anchor force A_{pc} or A_p, and the tie-rod location to the anchorage is made at the location of the resultant earth pressures acting on the vertical face of the anchorage.

Anchor ties are protected against corrosion by coatings and wrappings or by encasement. They also require protection against deflections caused by consolidation of the underlying soils due to fill and surcharge loads. This is accomplished by placing the tendon in a rigid pipe with a diameter larger than the anticipated settlement, and in extreme cases vertical support may be required.

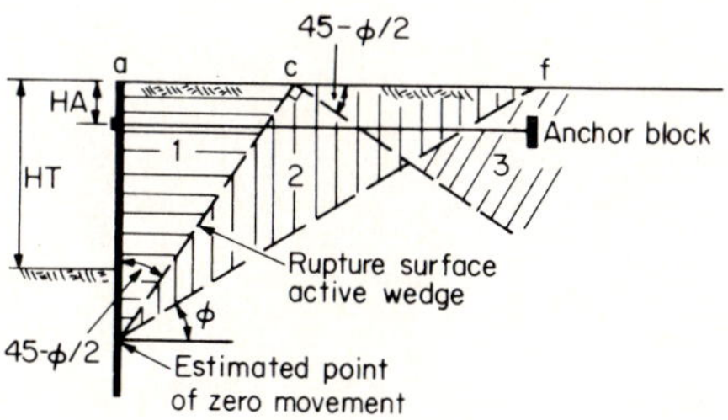

Anchor block position:

Zone 1—provides no resistance

Zone 2—provides partial resistance; ΔP_p is transferred to wall

Zone 3—provides full resistance

FIG. 9.86 Effects of anchor location relative to wall on anchor resistance.
[From USS (1974).[70]]

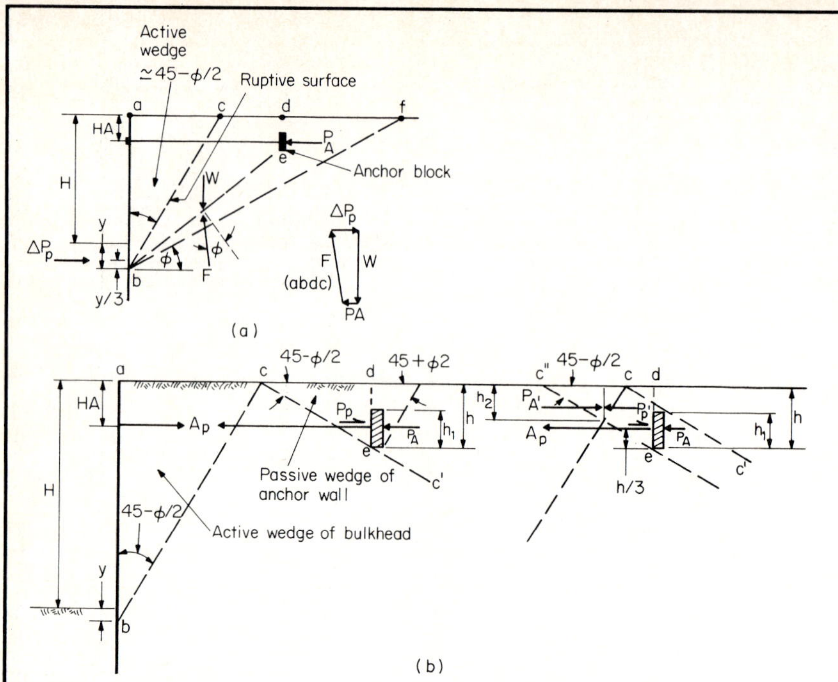

Forces per linear foot of anchor wall

Anchor wall right of cc'	Anchor wall left of cc'
For $h_1 \geq h/2$ $P_A = 0.5K_a\gamma h^2$ $P_p = 0.5K_p\gamma h^2$ K_p from Fig. 9.14 Using $-\delta/\phi = 0.5$	For $h_1 \geq h_2$ $P_A = 0.5K_a\gamma h^2$ $P_p = 0.5K_p\gamma h^2 - (P_p' - P_A')$ $= 0.5K_p\gamma h^2 - (0.5K_p\gamma h_2 - 0.5K_a\gamma h_2^2)$ K_a from Eq. 9.21

FIG. 9.87 Design criteria for deadman anchorage: (*a*) forces acting with anchor block in zone 2 and (*b*) continuous anchor wall located between rupture surface and slope of friction angle. Load ΔP_p transferred to base of wall. [From NAVFAC (1982).[1]]

Continuous wall:

1. $h_1 \geq h/2$
   Ultimate $A_{pc}/d = P_p - P_a$
   (P_p and P_a taken per linear foot of wall)

2. $h_1 < h/2$

Individual anchors:

1. $h \geq h/2$
   a. If $d > b + h$, $A_p(\text{ult}) = b(P_p - P_a) + 2P_o \tan\phi$ where
   P = resultant force of soil at rest on vertical area cde or c''de
   b. If $d = h + b$, $A_p/d = 70\%\ A_{pc}/d$, and $L = L' = h$
   c. If $d < h + b$, $A_p/d = A_{pc}/d - L/L'(0.3A_{pc}/d)$, and $L' = h$

2. $h_1 < h/2$

Ultimate A_{pc}/d or A_p/d = bearing capacity of strip footing of $B = h_1$ and surcharge load $\gamma(h - h_1/2)$ (see Art. 6.2.4, Eq. 6.14). Use ϕ where $\tan\bar{\phi}' = 0.6\tan\phi$

FIG. 9.88 Design criteria for deadman anchorage. Effect of depth and spacing of anchor blocks on ultimate anchor resistance R_{pc} and A_p. [After NAVFAC (1982).[1]]

9.4.4 CANTILEVER BULKHEADS

Earth Pressures

Support Source

Sheet piles driven to an adequate penetration depth into the ground become fixed as a vertical cantilever. Support to resist lateral earth pressures is derived from structural rigidity and passive resistance mobilized by the embedded portion. Substantial lateral deflections can be anticipated at the unsupported top.

Pressure Distributions

Rankine pressures normally are assumed for cantilever bulkhead analysis. When force P in Fig. 9.89 is applied to the top of the wall, rotation occurs about pivot point O, mobilizing passive pressures above and below point O. As both active and passive pressures are exerting force on the wall, the *net* passive pressure is the passive pressure minus the active pressure and is equal to zero at the pivot point.

Studies have shown the pressure distribution to be curvilinear, but for design purposes the curved-shaped areas are replaced by triangles with point d located so as to place the wall in a state of static equilibrium.

The conventional assumed pressure diagram for a cantilever sheet-pile wall with granular backfill and embedded in granular soils is given in Fig. 9.90. The case of clay above and below the mud line is given in Fig. 9.91*a*, and the case for a granular backfill overlying clay embedment is given in Fig. 9.91*b*.

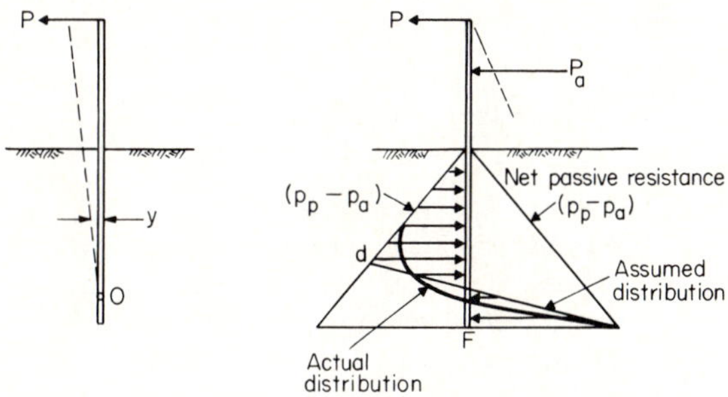

FIG. 9.89 Earth pressure on cantilever sheet piling.

Analysis

Granular Soil Backfill and Embedment

Step 1. Calculate the resultant earth pressures σ_{A1}, σ_{A2}, σ_E, and σ_J in terms of depth D as shown in Fig. 9.90*a*. Coefficients K_a and K_p are found in the same manner as described for the anchored bulkhead. Either a value for FS is applied to K_p or

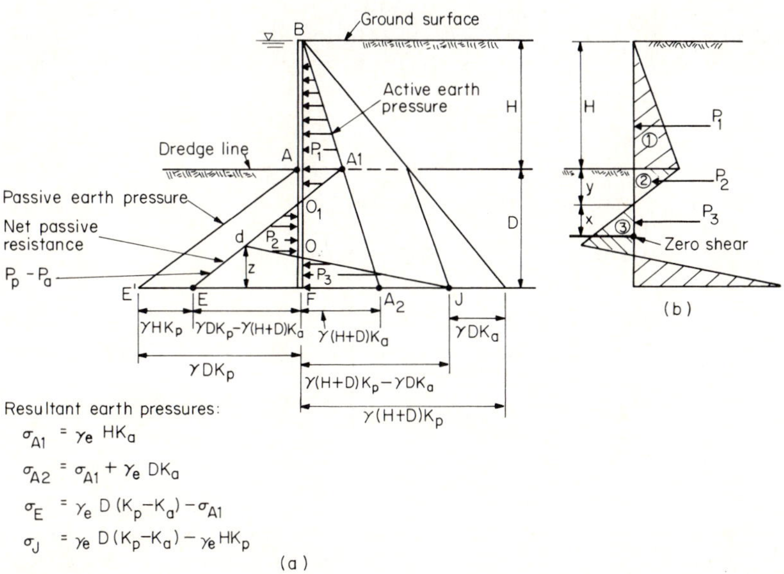

FIG. 9.90 Cantilever sheet-pile wall in granular backfill and embedment soils: (*a*) assumed net earth-pressure diagram and (*b*) location of zero shear. [From USS (1974).[70]]

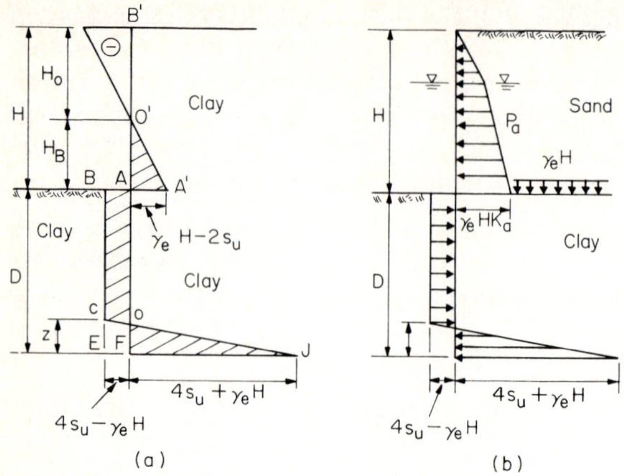

FIG. 9.91 Assumed net-pressure diagrams for cantilever sheet-pile wall in cohesive soils: *(a)* clay above and below mud line and *(b)* granular backfill over clay. [After USS (1974).[70]]

the penetration depth D is increased by 20 to 40% later in the computations. Effective soil weights γ_e are used, i.e., total unit weight above GWL and the submerged unit weight below, and sudden drawdown conditions need to be considered, as well as the occurrence of surcharge loads.

Step 2. From statics the sum of the horizontal forces must equal zero and the sum of the moments about any point must equal zero. The unknown distances for the determination of the moments are Z and D. The $\Sigma F_H = 0$ may be written in terms of areas:

$$\text{Area }(BAA_1) + \text{area }(AA_1A_2F) + \text{area }(ECJ) - \text{area }(EA_1A_2) = 0$$

or

$$\frac{1}{2}(H)\sigma_{A1} + (\sigma_{A1} + \sigma_{A2})\frac{D}{2} + (\sigma_E + \sigma_J)\frac{Z}{2} - (\sigma_E + \sigma_{A1})\frac{D}{2} = 0$$

Solving for Z,

$$Z = \frac{(\sigma_E - \sigma_{A1})D - H\sigma_{A1}}{\sigma_E + \sigma_J} \tag{9.56}$$

Taking $\Sigma M_F = 0$,

$$M_F = \frac{1}{2}(H)\sigma_{A1}\left(D + \frac{H}{3}\right) + (\sigma_{A1})\frac{D^2}{2} + (\sigma_E + \sigma_J)\frac{Z^2}{6} - (\sigma_E + \sigma_{A2})\frac{D^2}{6} + (\sigma_{A2} - \sigma_{A1})\frac{D^2}{6} = 0 \tag{9.57}$$

A value for Z is found from Eq. 9.56 by assuming a trial value for D, for which Table 9.7 is useful. Z is then substituted into Eq. 9.56 to check if $\Sigma M_F = 0$, or approximately so. If the sum of the moments about F is found not to equal zero, a new trial value for D is assumed and entered into Eqs. 9.56 and 9.57, and the computations are continued until the correct solution is obtained. If a value for FS has not been applied to K_p then the depth D is increased as discussed previously.

TABLE 9.7
ESTIMATION OF PENETRATION DEPTH: CANTILEVER SHEET-PILE WALL*

SPT value N, blows/ft	Relative density D_R	Penetration depth, H above mud line
0–4	Very loose	2.0H
5–10	Loose	1.5H
11–30	Medium dense	1.25H
31–50	Dense	1.0H
>50	Very dense	0.75H

*From USS (1974).[70]

Step 3. To find the maximum moment and the required size of the sheet piling requires finding the point of zero shear, located as shown in Fig. 9.90*b*. Distance *y* is found from

$$y = \sigma_{A1}/\gamma_b(K_p - K_a)$$

Distance x is found from

$$\tfrac{1}{2}\gamma_b(K_p - K_a)x^2 = P_1 + P_2$$

or

$$x = \sqrt{\frac{2(P_1 + P_2)}{\gamma_b(K_p - K_a)}}$$

where $P_1 = \tfrac{1}{2}\sigma_{A1}(H)$
$P_2 = \tfrac{1}{2}\sigma_{A1}(y)$

and $P_3 = \tfrac{1}{2}\gamma_b(K_p - K_a)x^2$

$$M_{max} = P_1L_1 + P_2L_2 + P_3L_3$$

where $L_1 = (H/3) + y + x$
$L_2 = (2y/3) + x$
$L_3 = x/3$

The steel sheeting is then selected as described for anchored bulkheads.

Clay Backfill and Clay Embedment

Step 1. The bearing capacity, or critical height, is checked with the procedure given for the anchored bulkhead using Eq. 9.44. Earth pressures are calculated as shown in Fig. 9.91*a* (see also Table 9.3) and surcharge effects added if necessary. Equilibrium conditions existing immediately after the sheet piling is driven (short-term conditions) are analyzed based on $\phi = 0$, and s_u prevails (in the diagrams $s_u = \tfrac{1}{2}U_c$). FS = 1.5 to 2.0 is applied, depending upon the reliability of the strength data. Long-term stability is analyzed based on an assumed strength increase with mobilization of ϕ due to backfill and surcharge pressures, but with a reduction or elimination of s_u [USS (1974)[70]]. A final value for ϕ = 20 to 30° is suggested, and the procedures described for granular soils are used for analysis.

Step 2. Distance *Z* and depth *D* are found in a manner similar to that described for granular soils, although it is first required to find the point of zero pressure *O′* acting at depth H_o, found from

$$H_o = 2s_u/\gamma$$

$\Sigma F_H = 0$ in terms of areas is

$$\text{Area }(O'A'A) + \text{area }(CJE) - \text{area }(BAEF) = 0$$

or

$$\frac{1}{2}(\gamma H - 2s_u)H_B + \frac{8s_u(Z)}{2} - (4s_u - \gamma H)D = 0$$

and

$$Z = \frac{2D(4s_u - \gamma H) - (\gamma H - 2s_u)H_B}{8s_u}$$

which is solved in terms of *D*. Summing the moments about *F*,

$$\Sigma M_F = \frac{1}{2}(\gamma H - 2s_u)(H_B)\left(D + \frac{H_B}{3}\right) + \frac{8s_uZ^2}{6} - (4s_u - \gamma H)\frac{D^2}{2} = 0$$

and substituting the equation found for *Z*, which includes *D*, results in a quadratic equation from which *D* and then *Z* can be found. Since a value for FS was applied to s_u, it need not be applied to increase *D*.

9.4.5 OTHER SYSTEMS: ANALYSIS AND DESIGN FACTORS

Relieving Platforms

Loads and Earth Pressures

Relieving platforms carry the weight of surcharge loads and most of the weight of backfill above water level, transmitting the loads through piles to underlying strong strata. Depending upon the scheme selected, the system either eliminates or substantially reduces lateral earth pressures. Raking piles provide the stability.

Support Schemes

Various support schemes are illustrated in Fig. 9.92.

A *pile-supported deck with no sheeting* (Fig. 9.92*a*) is supporting a light structure. The fill be-

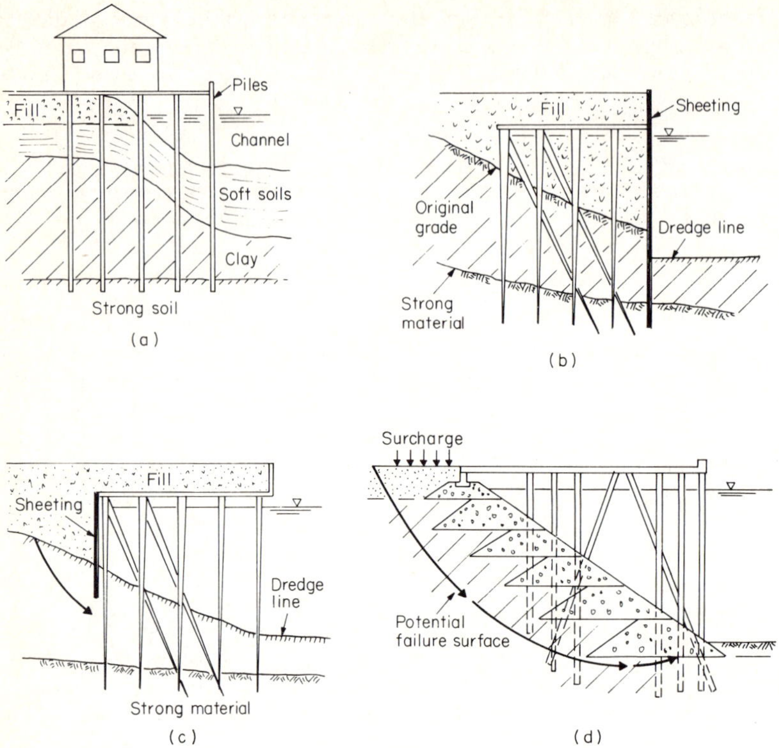

FIG. 9.92 Various schemes for relieving platforms: (*a*) no sheeting; (*b*) sheeting on water side; (*c*) sheeting on landward side; and (*d*) slope support with rock dikes.

neath the platform is placed on sufficiently shallow slopes to avoid shear failure of the soft organic soils which would result in a mud wave into the channel and loss of the fill. With sufficiently shallow slopes a retaining structure is not required. The piles are designed with consideration given to lateral earth pressures *(see Art. 7.6.5)*.

Sheeting on the outboard side of a low-level platform (Fig. 9.92*b*) retains the backfill. Although most of the surcharge loads will be carried by the platform, dredging will result in lateral earth pressures on the sheeting. The screening effect of the piles will reduce earth pressures and must be considered in analysis [Tschebotarioff (1973)[14]]. Suction dredging can result in piping of fine granular soil through gaps in the sheeting, causing loss of fill and subsidence to the rear of the platform.

Sheeting on the landward side (Fig. 9.92*c*) requires shorter lengths to retain the backfill and is advantageous in deeper water. Consideration must be given to the possibility of erosion of the soil slope in front of the sheeting from current and eddies and the overall slope stability under backfill loads.

Small rock dikes in series (Fig. 9.92*d*) provide slope support and erosion protection. High inland fills placed over weak soils require analysis regarding potential slope failures and high lateral forces on the pile foundations.

High- versus low-level platform: Cases *b* and *c* may be designed as high-level platforms where the deck is exposed on the surface and supports no fill loads. The main advantage of the low-level platform is to reduce the pile lengths. It is usually situated at a depth not less than one-fourth of the wall height [Cornfield (1975)[71]]. The platform width should be adequate to provide almost total shielding of the sheet piling or other slope-protection system from surcharge effects.

Cellular Cofferdams

Loading Conditions

Permanent cellular cofferdams are installed as waterfront structures to support heavy loads, such as cranes, and to provide docking and mooring facilities. *Temporary* cofferdams are used to permit dry excavation such as for bridge piers, or locks and dams, and therefore are required to retain a high head of water on one side. They can be cellular types, double-wall types, or earth- or rock-filled dikes with an impervious cutoff. Double-wall cofferdams are similar to braced excavation systems (*see Art. 9.3*) except that the soil is within the sheeting and the outside is open.

This article treats in a general manner only cellular cofferdams. They are essentially gravity structures that acquire their stability largely from the shear resistance of the cell fill. There are many aspects of an adequate design, and since failure often may result in disastrous consequences, they should be designed only by engineers who have substantial experience with cellular cofferdams.

Geometric Considerations

Height of the structure is selected based on providing protection against the maximum highest flood level anticipated during the life of the structure, when their purpose is to permit construction in the dry. Cofferdams as high as 30 m (100 ft) have been built, and in the case of high structures earth berms (Fig. 9.76*d*) may be placed on the inboard side to provide added lateral resistance against water pressures. *Width* is determined after the height is estimated giving consideration to overturning forces, resistance to sliding, foundation conditions, etc.

Sheeting design depends largely on the type of fill to be placed in the cells. The most critical aspect is the estimation of lateral earth pressures and the resulting tensions on the sheeting interlocks. *Penetration depth* is selected based on consideration of the protection required against scour of the foundation materials, the capacity of the foundation soils to support the structure, and protection against seepage uplift forces.

Tilting Stability and Cell Backfill

The structure in Fig. 9.93*a* is bearing on rock and resisting water pressure P_W. The resistance to failure by tilting is provided by the shearing resistance of the fill and the friction in the inter-

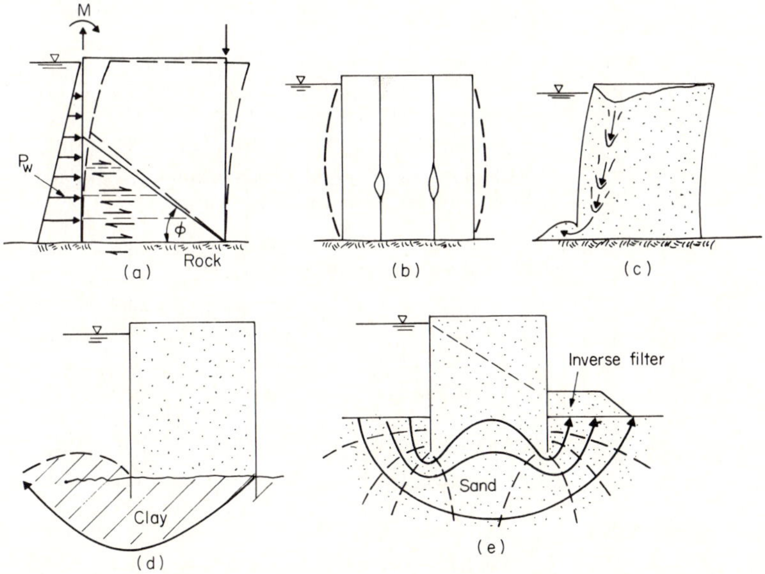

FIG. 9.93 Factors influencing stability of cellular cofferdams: (*a*) internal stability and horizontal shear; (*b*) bulging and interlock failure; (*c*) pullout of sheeting and escape of fill; (*d*) bearing-capacity failure in clay foundation; and (*e*) seepage in sand foundation and control against piping with inverse filter.

locks. [The concept of internal horizontal shear was introduced by Cummings (1957).[72]] If a berm is used, the resisting moment due to its passive resistance is added to the fill resistance and interlock friction. The horizontal frictional resistance is a function of the soil weight and its ϕ value.

Interlock tension is a function of the internal cell pressure which depends upon P_a within the cell, which in turn depends on local conditions and the construction method. If a cofferdam is constructed by driving sheet piles through a substantial thickness of sand and gravel to bedrock and then excavating the material inside the cell, the use of K_a may be acceptable and values ranging from 0.4 to 0.5 are commonly used. If the cells, however, are filled to a major portion of their height by dumping in water or by pumping into a dry cell, the lateral pressures may exceed K_o [Casagrande (1973)[21]]. In seismically active areas consideration should be given to the possibility of liquefaction of loose granular backfill in the cells and subsequent loss of internal friction and support. It is prudent, especially for permanent structures, to compact the backfill by vibrating or some other means and design the structure for the higher earth pressures that will result. Although some bulging of the cells is normal, excessive interlock pressures will result in sheeting separation (Fig. 9.93*b*), loss of backfill, and possibly cell collapse. Rock fill should be limited to pieces not exceeding 30 cm (12 in) as large pieces can exert concentrated stresses at points along the sheeting and cause failure.

Other Stability Considerations

On the water side, *uplift* resulting from inadequate penetration causes pullout of the sheeting which permits the fill to escape (Fig. 9.93*c*). *Foundation failure* by base rupture (Fig. 9.93*d*) may occur in weak soils. In these conditions the procedure is either to excavate the weak soils to strong materials and construct the cell, or to excavate and backfill with strong materials and then construct the cell.

Piping and loss of support at the toe (Fig. 9.93*e*) occurs in granular soils from seepage beneath the cell. Preventive measures are adequate sheeting penetration and the construction of a berm designed as an inverted filter to reduce inboard seepage forces. *Sliding* must always be considered, particularly where the cell rests on a bedrock surface and there is little or no sheeting penetration. Resistance may be obtained by increasing the wall width.

Design procedures are discussed in Tschebotarioff (1973),[14] Dismuke (1975),[73] Lacroix et al. (1970),[74] Schroder and Maitland (1979),[75] and NAVFAC (1982).[1]

Other Gravity Structures

Gravity Rock Dikes

Gravity rock dikes, constructed of crusher-run quarry stone and protected in the wave zone with riprap, are designed in the same manner as any embankment. A filter layer may be needed between the rock dike and the landfill. When used as a cofferdam, a clay core is required, or steel sheeting or grouting may be considered as alternate methods of cutoff.

Strong underlying soils are required since resistance to lateral load is derived almost entirely from frictional resistance between the dike fill and the dredged or prepared bottom. Excavation for a shear key is somewhat wider than for other gravity types.

Concrete Gravity Walls

Massive walls, block wall composition, and precast boxes forming concrete gravity walls are designed in the same manner as gravity retaining structures, as discussed in Art. 9.2.4. They are supported either on strong soils or on pile foundations. Stability against overturning, sliding, and foundation failure must be considered.

REFERENCES

1. NAVFAC (1982) *Design Manual DM 7.2, Foundations and Earth Structures*, Department of the Navy, Naval Facilities Engrg. Comm., Alexandria, Va.
2. Kulhawy, F. H. (1974) "Analysis of a High Gravity Retaining Wall," *Proc. ASCE, Conf. on Analysis and Design in Geotech. Engrg.*, University of Texas, Austin, Vol. I, pp. 159–171.

3. Royster, D. C. (1973) "Highway Landslide Problems Along the Cumberland Plateau, Tennessee," *Bul. Intl. Assoc. Engrg. Geol.*, Vol. 10, No. 4.

4. Gebney, D. S. and McKittrick, D. P. (1975) "Reinforced Earth: A New Alternative for Earth Retention Structures," *Civil Engineering*, ASCE, October, pp. 58–61.

5. Hunt, R. E. (1984) *Geotechnical Engineering Investigation Manual*, McGraw-Hill Book Co., New York.

6. *Symposium on Earth Reinforcement (1978), Proc. ASCE*, ASCE Annual Conv., Pittsburgh, Pa.

7. D'Appolonia, E. D., Alperstein, R. A., and D'Appolonia, D. J. (1967) "Behavior of a Colluvial Soil Slope," *Proc. ASCE, J. Soil Mech. and Found. Engrg. Div.*, Vol. 93, No. SM4, pp. 447–473.

8. Goughnor, R. D. and DiMaggio, J. A. (1978) "Soil-Reinforcement Methods on Highway Projects," *Symp. Earth Reinforcement, Proc. ASCE*, New York, pp. 371–399.

9. Taylor, D. W. (1948) *Fundamentals of Soil Mechanics*, John Wiley & Sons, New York.

10. Terzaghi, K. and Peck, R. B. (1967) *Soil Mechanics in Engineering Practice*, 2d ed., John Wiley & Sons, New York.

11. Teng, W. C. (1962) *Foundation Design*, Prentice-Hall, Englewood Cliffs, N.J.

12. Kezdi, A. (1975) "Lateral Earth Pressure," *Foundation Engineering Handbook*, Winterkorn and Fang, eds., Van Nostrand-Reinhold, New York, Chap. 5.

13. Lambe, T. W. and Whitman, R. V. (1969) *Soil Mechanics*, John Wiley & Sons, New York.

14. Tschebotarioff, G. P. (1973) *Foundations, Retaining, and Earth Structures*, McGraw-Hill Book Co., New York.

15. Caquot, A. and Kérisel, J. (1949) *Traite de Mechanique des Sols*, Gauthier-Villars, Paris.

16. Terzaghi, K. (1936) "Effect of the Type of Drainage of Retaining Walls on the Earth Pressures," *1st Conf. on Soil Mech. and Found. Engrg.*, Harvard University, Cambridge, Mass., No. J4, pp. 215–218.

17. Gray, H. (1958) "Contribution of the Analysis of Seepage Effects in Backfills," *Geotechnique*, London, Vol. 8, No. 4, Inst. of Civ. Engr., p. 166.

18. Terzaghi, K. (1954) "Anchored Bulkheads," *Trans. ASCE*, Paper 2720, Vol. 119, pp. 1243–1324.

19. Wu, T. H. (1975) "Retaining Walls," *Foundation Engineering Handbook*, Winterkorn and Fang, eds., Van Nostrand-Reinhold Co., New York, Chap. 12.

20. Peck, R. B., Ireland, H. O., and Teng, C. Y. (1948) "A Study of Retaining Wall Failures," *Proc. 2d Intl. Conf. Soil Mech. and Found. Engrg.*, Rotterdam, Vol. 3, pp. 296–299.

21. Casagrande, L. (1973) "Comments on Conventional Design of Retaining Wall Structures," *Proc. ASCE, J. Soil Mech. and Found. Engrg. Div.*, Vol. 99, No. SM2, February, pp. 181–198.

22. Coyle, H. M. and Bartoskewitz, R. E. (1976) "Earth Pressure on Precast Panel Retaining Wall," *Proc. ASCE J. Geotech. Engrg. Div.*, Vol. 102, No. GT5, May, pp. 441–456.

23. Bowles, J. E. (1974) *Analytical and Computer Methods in Foundation Engineering*, McGraw-Hill Book Co., New York.

24. Peck, R. B., Hanson, W. E., and Thornburn, T. H. (1974) *Foundation Engineering*, 2d ed., John Wiley & Sons, New York.

25. Seed, H. B. and Whitman, R. V. (1970) "Design of Earth Retaining Structures for Dynamic Loads," *Lateral Stresses in the Ground and the Design of Earth Retaining Structures, Proc. ASCE*, New York, pp. 103–147.

26. Richards, R., Jr. and Elms, D. G. (1979) "Seismic Behavior of Gravity Retaining Walls," *Proc. ASCE, J. Geotech. Engrg. Div.*, Vol. 105, No. GT4, April, pp. 449–464.

27. Romstad, K. M., Herrman, L. R. and Shen, C. K. (1976) "Integrated Study of Reinforced Earth—I: Theoretical Formulation," *Proc. ASCE, J. Geotech. Engrg. Div.*, Vol. 102, No. GT5, May, pp. 457–472.

28. Shen, C. K., Romstad, K. M., and Kerrman, L. R. (1976) "Integrated Study of Reinforced Earth—II: Behavior and Design," *Proc. ASCE, J. Geotech. Engrg. Div.*, Vol. 102, No. GT5, June, pp. 577–590.

29. Beaton, J. L., Forsyth, R. A., and Chang, J. C. (1974) "Design and Field Behavior of the Reinforced Earth Embankment—Road 39," *Report*, California Department of Transportation, Division of Highways, Sacramento, presented at the January Meeting, ASCE National Water Resources Engineering, Los Angeles.

30. Maccaferri (1979) "Gabions Structural Design Data," Maccaferri Gabions of America, New York.

31. Costa Nunes, A. J. da and Velloso, D. A. (1963) "Estabilização de Taludes em Capas Residuais de Origem Granito-Gnáissica," *2d Panamer. Conf. Soil Mech. and Found. Engrg.*, Brazil, Vol. 2, pp. 383–394.

32. Costa Nunes, A. J. da, Cerqueira, C. A. G., and Novaes, J. L. M. (1972) "Approveitamento de Terrenos Acidentados com Cortinas Anchoradas," *I*

Exposição E Encontro Nacional da Construção, São Paulo, Brazil.

33. Costa Nunes, A. J. da (1974) "Tied-Back Walls," *Proc. ASCE, Conf. Analysis and Design in Geotech.* Engrg., University of Texas, Austin, Vol. 2, pp. 146–149.

34. Hunt, R. E. and Costa Nunes, A. J. da (1978) "Retaining Walls: Taking It from the Top," *Civil Engineering*, ASCE, May, pp. 73–75.

35. Tecnosolo (1978) "Estabilização de Taludes com Anchoragem," *Spec. Pub. 29*, Tecnosolo S. A., Rio de Janeiro, Brazil.

36. Peck, R. B. (1969) "State-of-the-Art Report: Deep Excavations and Tunneling in Soft Ground," *Proc. 7th Intl. Conf. Soil Mech. and Found. Engrg.*, Mexico City, pp. 225–290.

37. Civil Engineering (1977) "Can Trench Cave-in Deaths Be Cut?" *Civil Engineering*, ASCE, September, pp. 82–86.

38. Kapp, M. S. (1969) "Slurry Trench Construction for Basement Wall of World Trade Center," *Civil Engineering*, ASCE, April.

39. Johnson, E. G. (1976) "The Sixty State Street Building," *Geotechnical Lecture Series on Lateral Earth Pressure*, Boston Soc. Civ. Engr.

40. Xanthakos, P. P. (1979) *Slurry Walls*, McGraw-Hill Book Co., New York.

41. Peck, R. B. (1943) "Earth Pressure Measurements in Open Cuts, Chicago Subway," *Trans. ASCE*, No. 108, pp. 1009–1036.

42. Ward, W. H. (1955) "Experience with Some Sheet-Pile Coefferdams at Tilbury," Corresp. to *Geotechnique*, London, Vol. 6, No. 4, pp. 327–330.

43. NGI (1962–1966) "Measurements of a Strutted Excavation," *Norw. Geotech. Inst. Tech. Report 1–9*, Oslo.

44. CGS (1978), "Excavations and Retaining Structures," Part 4, *Canadian Foundation Engineering Manual*, Canadian Geotech. Soc., Montreal, March.

45. Goldberg, D. T. (1976) "Lateral Support Systems," *Geotechnical Lecture Series on Lateral Earth Pressure*, Boston Soc. Civ. Engr., Boston.

45a. Goldberg, D. T., Jaworski, W. E., and Gordon, M. D. (1976) *Lateral Support Systems and Underpinning*, Reports No. FHWA-RD-75-128, 129, and 130 (Vols. I, II, III), prepared for Federal Highway Administration, Washington, D.C.

46. Broms, B. and Stille, H. (1976) "Failure of Anchored Sheet Pile Walls," *Proc. ASCE., J. Geotech. Engrg. Div.*, Vol. 102, No. GT3, March, pp. 235–251.

47. Bjerrum, L. and Eide, O. (1956) "Stability of Strutted Excavations in Clay," *Geotechnique*, London, Vol. 6, No. 1, pp. 32–47.

48. NAVFAC (1982) *Design Manual DM 7.1, Soil Mechanics*, Department of the Navy, Naval Facilities Engrg. Comm., Alexandria, Va.

49. Bjerrum, L., Clausen, C. J. F., and Duncan, J. M. (1972) "Earth Pressures on Flexible Structures: A State-of-the-Art Report," *5th European Conf. on Soil Mech. and Found. Engrg.*, Madrid, Vol. 2, pp. 169–196.

49a. Henkel, D. (1971) "The Calculation of Earth Pressure in Open Cuts in Soft Clays," *The Arup Journal*, December.

50. Lucks, A. S. (1976) "Lateral Earth Pressures—Nuclear Power Plant Design," *Geotechnic Lecture Series on Lateral Earth Pressures*, Boston Soc. Civ. Engr., Boston.

51. Clough, G. W. and Denby, G. M. (1977) "Stabilizing Berm Design for Temporary Walls in Clay," *Proc. ASCE, J. Geotech. Engrg. Div.*, Vol. 103, No. GT2, pp. 75–90.

52. Clough, G. W. and Tsui, Y. (1974) "Performance of Tied-back Walls in Clay," *Proc. ASCE, J. Geotech. Engrg. Div.*, Vol. 100, No. GT12, December, pp. 1259–1274.

53. Murphy, D. M., Clough, G. W., and Woolworth, R. S. (1975) "Temporary Excavation in Varved Clay," *Proc. ASCE, J. Geotech. Engrg. Div.*, Vol. 101, No. GT3, March, pp. 279–295.

54. Tecnosolo (1974) "Atirantamento de Paredes Diaframa, Hotel Meridien, Copacabana," *Topicas de Geomecânica Tecnosolo*, Pub. No. 15, Tecnosolo, S. A., Rio de Janeiro.

55. D'Appolonia, E., Alperstein, R. A., and D'Appolonia, D. J. (1969) "Behavior of a Colluvial Soil Slope," *Proc. ASCE, Stability and Performance of Slopes and Embankments*, pp. 489–518.

56. Clough, G. E. (1974) "Tied-back Walls," *Proc. ASCE, Conf. on Analysis and Design in Geotech.* Engrg., University of Texas, Austin, Vol. 2, pp. 100–102.

57. Ranke, A. and Ostermayer, H. (1968) "Beitrag Zur Stabilitatuntersuchung—Marfach Verankerter Bahgruben-umschleissungen," *Die Bautechnik*, Vol. 10.

58. Clough, G. W. and Mana, A. (1976) "Lessons Learned in Finite Element Analysis of Temporary Excavations in Soft Clay," *Numerical Methods in Geotechnical Engineering, ASCE*, New York, Vol. I, pp. 496–510.

59. Tsui, Y. and Clough, G. W. (1974) "Plane Strain

Approximations in Finite Element Analysis of Temporary Walls," *Proc. ASCE, Conf. Analysis and Design in Geotech. Engrg.*, University of Texas, Austin, Vol. 1, pp. 173–197.

60. Tsui, Y. (1974) "A Fundamental Study of Tied-back Wall Behavior," Ph.D. dissertation, Duke University, Durham, N.C.

61. Mana, A. (1976) "Finite Element Analysis of Deep Excavation Behavior," Ph.D. thesis, Stanford University, Stanford, Calif.

62. Ward (1972) "Bedrock Can Be a Hazard—Wall Movements of a Deep Rock Excavation Analyzed by Finite Element Method," *Soils*, Vol. 1, Joseph S. Ward & Assocs., Caldwell, N.J., February.

63. Johnston, R. C., Barnheimer, M., and Zarrella, D. A. (1976) "Optimizing Selection of Marine Retaining Walls," *Proc. ASCE, J. Geotech. Engrg. Div.*, Vol. 102, No. GT12, December, pp. 1209–1224.

64. Tschebotarioff, G. P. (1951) *Foundations, Retaining, and Earth Structures*, 1st ed., McGraw-Hill Book Co., New York.

65. Sowers, G. B. and Sowers, G. F. (1967) "Failures of Bulkhead and Excavation Bracing," *Civil Engineering*, ASCE, January, pp. 72–77.

66. Rowe, P. W. (1952) "Anchored Sheet Pile Walls," *Proc. Inst. of Civ. Engr.*, London, Vol. 1, pp. 27–70, January, and discussion 1, September, pp. 616-647.

67. Rowe, P. W. (1955) "A Theoretical and Experimental Analysis of Sheet Pile Walls," *Proc. Inst. of Civ. Engr.*, London, Vol. 4, January, pp. 32–87, and discussion 6, November, pp. 825–855.

68. Rowe, P. W. (1956) "Sheet Pile Walls at Failure," *Proc. Inst. of Civ. Engr.*, London, Vol. 5, May, pp. 267–315, and discussion 6, February, pp. 347–361.

69. Rowe, P. W. (1957) "Limit Design of Flexible Walls," *Proc. Midland Soil Mech. and Found. Engrg. Soc.*, Great Britain, Vol. 1, February, pp. 29–40.

70. USS (1974) *Steel Sheet Pile Design Manual*, U.S. Steel Corp., Pittsburgh, Pa.

71. Cornfield, G. M. (1975) "Sheet Pile Structures," *Foundation Engineering Handbook*, Winterkorn and Fang, eds., Van Nostrand-Reinhold, New York, Chap. 13.

72. Cummings, E. M. (1957) "Cellular Cofferdams and Locks," *Proc. ASCE, J. Waterways and Harbour Div.*, Paper 1366, September.

73. Dismuke, T. D. (1975) "Cellular Structures and Braced Excavations," *Foundation Engineering Handbook*, Winterkorn and Fang, eds., Van Nostrand-Reinhold, New York, Chap. 14.

74. Lacroix, Y., Esrig, M. I., and Luscher, U. (1978) "Design, Construction and Performance of Cellular Cofferdams," *ASCE Spec. Conf. on Lateral Stresses and Earth Retaining Structures*," Cornell University, Ithaca, N.Y., pp. 271–328.

75. Schroeder, W. L. and Maitland, J. K. (1979) "Cellular Bulkheads and Cofferdams," *Proc. ASCE, J. Geotech. Engrg. Div.*, Vol. 105, No. GT7, July, pp. 823–837.

CHAPTER TEN

UNDERGROUND OPENINGS

10.1 INTRODUCTION

10.1.1 GENERAL CONSIDERATIONS

Categories of Underground Openings

Tunnels are constructed in soil or rock for transportation systems, water supply, and sewage disposal.

Caverns and chambers are constructed in rock for underground power stations, gas and petroleum storage, and various other uses.

Mines are constructed in rock and can take many forms including tunnels, shafts, and horizontal slots.

Basic Study Elements

The basic study elements for any underground construction can be considered to be

Excavation: The methods required for the removal of material, generally related to its hardness and strength.

Retention of the opening: Support requirements to prevent closure during and after construction.

Ground improvement: Treatments required to improve unfavorable or hazardous conditions that may endanger workers, cause delays and extra construction costs, affect construction performance, and cause surface subsidence.

Other structures: The effect of underground construction on overlying surface structures from ground subsidence and the effect on adjacent structures from blasting or ground displacement.

Geology and Risk

Geologic conditions are the controlling factors in underground construction, which for a given project can provide a range of variation to a much greater degree than other categories of construction, especially in the case of long tunnels. Adequate identification of all geologic conditions which may be encountered is usually a difficult and costly undertaking, and seldom will all significant conditions be known prior to construction.

Risks in underground construction are many and should be shared between the owner and the contractor to minimize delays and necessary extra construction costs resulting from unanticipated conditions.

Typical subsurface excavations and some significant factors are illustrated in Fig. 10.1.

10.1.2 THE BASIC FACTORS

Ground Conditions

Ground conditions have been classified in terms of how the ground reacts to an underground opening, as given in Table 10.1. Two general groupings are recognized: favorable conditions presenting no unusual difficulties to well-exe-

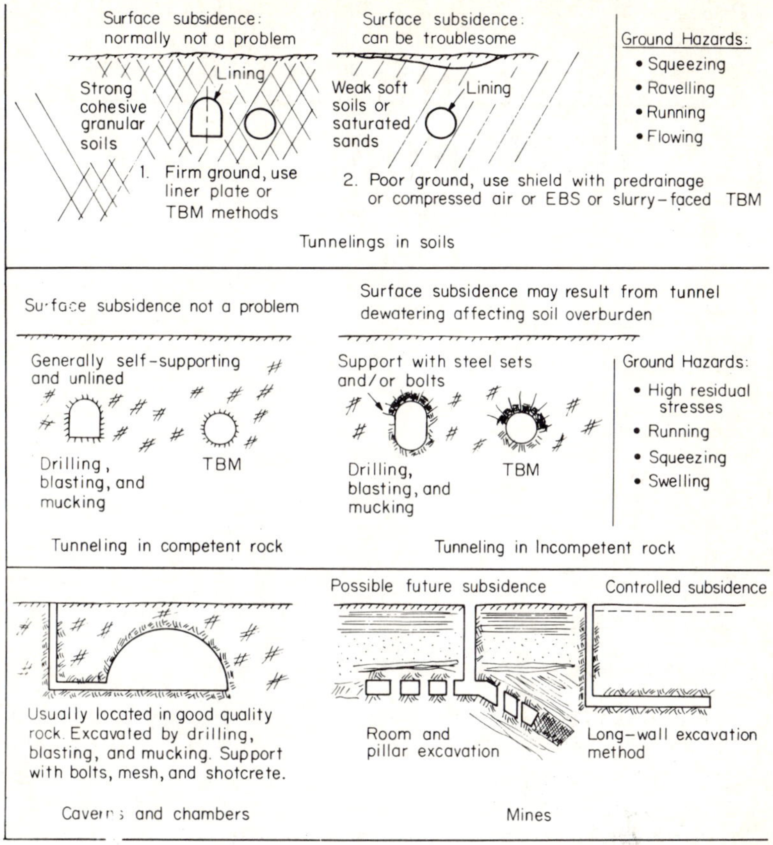

FIG. 10.1 Typical subsurface excavations and some significant factors.

cuted construction and unfavorable or hazardous conditions. Unfavorable conditions can endanger workers, cause construction delays and cost extras, affect future performance of the excavation, and result in excessive surface subsidence. They either require treatment prior to excavation or are provided for with proper supports, depending upon their nature.

Rock Masses

Competent rock is hard intact rock which does not require support, but its hardness is a factor to consider when evaluating removal techniques. *Incompetent rock* contains discontinuities such as joints, foliation planes, bedding planes, and faults which divide the mass into blocks. These can vary from hard to soft rock depending upon their mineralogy and degree of decomposition. Discontinuities can vary from widely spaced where intact rock is approached to closely spaced forming a crushed mass. *Troublesome* conditions are caused by highly crushed rock, high residual stresses, and flowing, squeezing, or swelling ground.

Soils

"Firm ground" provides favorable conditions since a tunnel roof can be left unsupported for several days without inducing perceptible movement of the surrounding ground. Troublesome ground can ravel, run, flow, squeeze, or swell and without proper treatment can fill an opening immediately or gradually depending on various conditions.

Water

The most serious problem encountered during excavation often is the sudden inflow of large

TABLE 10.1
CLASSIFICATION OF GROUND CONDITIONS SUMMARIZED

Condition	Associations	Performance
Competent rock	Hard, intact rock with tightly closed and widely spaced joints.	Self-supporting, remains open indefinitely even in large excavations.
Incompetent rock	Contains discontinuities: joints, foliation planes, bedding planes and faults, etc. (See Table 10.2, Classes II, III, IV, and V.)	Depending upon discontinuity spacing and opening width, blocks fall into the opening.
High residual stresses	Stresses higher than overburden remaining from folding, metamorphism, etc.	Results in popping, slabbing, rock bursts, and closure.
Flowing ground (rock)	Saturated, cohesionless, highly fragmented, pervious material associated with faults and other shear zones.	Saturated debris flows as a slurry into opening, often under high initial pressures and quantities, filling the opening.
Squeezing ground (rock)	Relatively weak, clay-rich plastic material with high saturation percentage associated generally with fault gouge, closely sheared and altered micaceous rocks, and soft shales.	Material moves into opening under pressure immediately upon exposure.
Swelling ground (rock)	Soft shales, fault gouge, altered pyroclastic and micaceous rocks containing montmorillonite.	Material expands upon contact with moisture and after a delay moves into the opening or exerts high pressures on supports.
Firm ground (soil)	Strong cemented or cohesive granular soils, stiff to hard clays (nonswelling).	Opening can be left unsupported for several days without inducing perceptible ground movement.
Raveling ground (soil)	Slightly cohesive granular soils with relict structure such as loess and residuals, or stiff fissured clays. Friable soils.	Material breaks into chunks, flakes and angular fragments which drop from the roof, enlarging the cavity.
Running ground (soil)	Dry, cohesionless soils.	Discharges immediately into opening to form a cone with sides about 34°.
Flowing ground (soil)	Cohesionless soils or soils with low cohesion below GWL.	Invades tunnels as a stream extending for large distances.
Squeezing ground (soil)	Soft to firm clays. Tendency is probable if $(p_z - p_a)/s_u > 5$.*	All exposed, unsupported faces advance into the opening slowly and continuously.
Swelling ground (soil)	Stiff clays.	See Swelling ground: (Rock).

*Peck (1969).[1]

quantities of water. Groundwater and seepage pressures are a major factor in raveling, flowing, and squeezing ground. The possibly high corrosive effects of water on linings also must be considered.

Gases

Gases can be both lethal and explosive and include "dead air" (air with oxygen removed), carbon dioxide, methane, and sulfur dioxide.

Temperature

Ground temperature usually is not a concern unless the excavation is more than 150 m below the surface; then high temperatures can begin to affect labor.

Earthquakes

Ground shaking is a factor to be considered in the design of linings and support systems as with any structure located in active areas, although ground motions are much less severe than for surface structures [Dowding and Rozen (1978)[2]]. Fault-zone crossings are potentially hazardous locations since displacements may shear the tunnel or other opening.

Ground conditions are discussed in more detail in subsequent articles.

Ground Response to Openings

Stress Field Changes

Underground excavation results in stress relief which causes a redistribution of the natural stress fields about the opening and new strains and deformations in the surrounding materials. Stress distribution depends on the nature of the geologic formation and can be highly complex. It may lead to high compressive forces in the vertical direction while the mass tends to expand under reduced compressive forces in the horizontal direction. It also depends very much on the shape, such as circular or slot, and the size of the opening. Size is significantly related to geologic conditions in that the larger the opening, the more variable are the conditions likely to be encountered.

Time is an important consideration since in many situations, particularly in raveling soils and fractured rock, *arching (see Art. 3.2.3)* often results and stresses diminish with time as deformation occurs. However, excessive deformation results in high stresses around the excavation and, in some cases, in surface subsidence. The construction method and execution techniques have a very significant influence on performance with respect to the earth pressures to be supported and ground subsidence.

Arching and Stand-Up Time

Many geologic formations, such as raveling soil and blocky or seamy rock, do not react at once to the stress changes produced by underground excavation. The natural roof support is removed by excavation and the loosened zone above the opening yields to form a so-called roof tension arch *(see Art. 10.3.3)*. The major share of the total overburden weight in the yielding mass above the opening is transferred by frictional forces to the unyielding ground on both sides of the opening. In actuality, arching action occurs along a tunnel axis, whereas dome or half-dome action occurs over the newly excavated portion in front of the lined portion. The body of soil or rock which transfers the load is referred to as the *ground arch*.

If the arch or half-dome is left unsupported, a few blocks drop from the roof leaving a small gap, then more blocks drop out widening the gap, and eventually the mass constituting the half-dome drops into the opening to form a new half-dome. The process continues until the section adjoining the working face is filled with soil or rock debris. In time a cavity may even reach the surface.

Stand-up time refers to the time interval during which the soil or rock mass can remain without support after the opening is made. It is largely unpredictable in many instances as it is a function not only of material strength properties but also of formation characteristics such as relict structures and stratification in soils and discontinuities in rock masses, seepage forces acting on the face and arch, and the size of the unsupported opening. Stand-up time can range from zero in running sands to several days in firm ground and highly fractured rock to essentially permanent in competent rock.

Both arching and stand-up time have been considered by practitioners, such as Terzaghi, in the development of empirical relationships for estimating loads on tunnel supports.

Support Requirements

Support requirements for underground excavations are placed in two categories: temporary support during construction and permanent support which may include a lining. Competent rock, including the massive crystallines and some hard sedimentary rock, is self-supporting during and after construction, and the necessity for a permanent lining depends on the proposed use of the facility. Most other materials require support during construction and a permanent lining. During design stages, consideration is given to incorporating the temporary support into the permanent lining. Earth pressures for design are estimated with the aid of analytical or empirical relationships and can range from full overburden pressures in soft to firm clays to zero in competent rock through a wide range of values in between for other geologic conditions.

Rock Masses

Competent rock is usually self-supporting both during and after construction. Incompetent rock is supported temporarily with steel ribs and timber lagging; liner plates; rock bolts, wire mesh, and shotcrete; or concrete in varying combinations. Bolts, mesh, and shotcrete may provide the permanent lining in moderate to fair quality rock, whereas rock of poor quality usually requires a concrete lining. In a mine excavated as a slot, roof support is provided by pillars and perhaps some rock bolting.

Soils

During excavation in soils, support is required at the face of the opening and, depending upon ground conditions, may be provided by breasting boards, an excavating machine, a slurry wall, or air pressure. The arch and sides may be supported temporarily by liner plates or a shield. Permanent lining is required and is provided by cast-in-place concrete or by concrete or steel segments.

A summary of construction support methods for the roof arch is given in Table 10.2. It is seen that there are many alternatives and selection is closely related to ground conditions. For estimating project costs and preparing budgets, the designer needs to forecast not only support requirements but also construction methods.

Construction Aspects

During the planning and design stages all aspects of construction require consideration, including temporary support provisions as mentioned above, excavation methods, ground-improvement techniques, and in urban areas the need to control surface subsidence, all of which are closely related.

Excavation Methods

Rock masses are excavated with conventional drilling, blasting, and mucking techniques; by roadheader operating with or without a shield; or by tunnel boring machine (TBM).

In soils, if liner plates and breasting boards are used, material at the face is removed by hand or by a machine such as a backhoe. Hydraulically advanced shields with either a closed or partially closed face displace soft soils; with an open face the soil is removed by hand or machine as the shield advances. Tunnel boring machines with or without a slurry face or an earth pressure balanced (EPB) shield excavate material at the face, where it is removed by conveyor system.

Ground Improvement

Various techniques are used to improve unfavorable ground with the result, for example, that flowing or running ground is changed to raveling ground or even firm ground and rock masses are made more competent. *Air pressure* is used in squeezing soils and soils with a tendency to flow. *Dewatering* with wellpoints, deep wells, or water collectors is used in free-draining soils or fractured rock masses below GWL. *Grouting* is used in granular soils or pervious rock masses below GWL. *Electroosmosis* or *freezing* is used in fine-grained saturated soils, the latter particularly where they are stratified and dewatering or

TABLE 10.2
GROUND CONDITIONS AND SUPPORT METHODS: GENERAL RELATIONSHIPS*†

Class		Ground conditions	Construction support of roof arch
I	R	Hard, intact rock; massive crystallines (granite, quartzite)	None necessary
II	R	Hard stratified rock (sandstones) or schistose (gneiss) with only occasional joints	None necessary, gunite (2- to 5-cm layer), or bolts (1- to 2-m spacing)
III	R	Massive, moderately jointed rock (granites, sandstones, limestones, dolomites)	Gunite (2- to 7-cm layer) or bolts (0.7- to 2.5-cm spacing) with mesh and gunite (2 to 3 cm)
IV	R	Moderately blocky and seamy rock (stratified sandstones, schists, crystallines)	Shotcrete (10 to 15 cm), bolts (0.5- to 1.5-m spacing) with mesh and gunite (3 to 5 cm), or steel ribs and blocking
V	R	Completely crushed, unweathered rock; soft rocks with medium strength; decomposed rocks (Note: saturated or flow zones, improve by dewatering or grouting)	Shotcrete (7 to 20 cm) over mesh and steel ribs, shotcrete and bolt system (or continuously resin-cemented anchors), steel ribs and blocking, or cast concrete arches
VI	R	Squeezing or swelling rock	Heavy application of mesh reinforced concrete, circular section, or liner plate and steel ribs and cast concrete arches
VII	S	Strong soils: residuals, stiff to hard clays and clayey sands, cemented dense sands	Shotcrete (10 to 20 cm) over mesh and metal arches or steel ribs and liners with cast concrete
VIII	S	Moderately strong soils; clayey sands (firm ground)	Steel ribs and liners with cast concrete or jacked in-place steel or concrete segments
IX	S	Medium dense to loose sands, saturated, below GWL; loose fine sands (dry and running)	Stabilize with air pressure, dewatering, grouting or freezing, then support as for VIII, or use TBM and slurry-faced method
X	S	Soft soils; organics, clays, silts below GWL (flowing and squeezing ground)	Use shield tunneling with air pressure or slurry-face methods and line with grouted steel or concrete segments

*After ABGE (1980)[3]; from AFTES (1974).[4]

†In many cases the size of the opening is an important factor in selection.

grouting is less effective than in homogeneous formations.

Surface Subsidence

Stress changes in the ground caused by tunneling will necessarily result in deformations. Excavation causes settlement at the crown which may migrate through the overburden to be finally manifested as surface subsidence which is generally differential. Dewatering also can result in subsidence [see Art. 10.2.5 of Hunt (1984)[5]]. One of the more difficult aspects of tunnel engineering is evaluation of the effects of deformation on the surface and the potential for damage to overlying structures. If subsidence is anticipated, consideration is given to using the tunneling method which will result in the least subsidence, to ground improvement, or to underpinning, which is relatively costly and not always successful.

Subsidence magnitude depends very much on geologic conditions and on excavation techniques. It is usually negligible in firm soil and most rock masses when good construction practice is followed. It is potentially significant over very large areas when tunneling in clays or when excavating long-wall mines, or some time after a room-and-pillar form of mine is closed; over relatively small areas in other ground conditions if tunneling techniques are poor so that loss of ground is permitted; or in certain soil formations overlying rock masses which are extensively dewatered.

10.1.3 EVALUATION PROCEDURES

Exploration and Testing

Reconnaissance Phase

Geologic reconnaissance employing literature review, interpretation of remote-sensing imagery, and field reconnaissance is important on any project, but it may be the most important element of exploration where underground excavations in rock masses are involved over a large area and at substantial depths. Faults and joint patterns are identified and evaluations are made of rock types and quality existing at the anticipated construction depth.

Borings, Adits, and Geophysics

Test borings are standard procedure for soil and rock tunnels, but in the latter case, if the excavation is located far below the surface and in difficult terrain, cost and time will limit the number of borings made. For tunnels it would be ideal to drill horizontal borings along the alignment, but present drilling technology does not enable drilling holes that remain essentially horizontal for much more than about 100 m. Normal boring spacing for a project in soils is much closer than for one in rock, unless the terrain is level, the proposed construction not excessively deep, or the area not large (such as that of a chamber). In tunnels in rock, test borings often are made from the tunnel heading during construction, which is good practice to follow where hazardous ground conditions may exist.

Pilot tunnels and *adits* often are used to explore rock conditions and to allow the performance of in situ testing for measuring rock properties.

Geophysical methods are necessary for investigations in rock masses to supplement the information from the test borings and adits. Seismic refraction is the method used most commonly; reflection methods may have application where the excavation will be at depths below 300 m or so.

Testing to measure material properties is performed in the laboratory on soil and rock specimens recovered during test boring, as well as in situ. Information is required on strength and deformation properties and, in rock masses, on residual stresses. Permeabilities often are estimated from published data, but may be measured by in situ testing, particularly in rock masses.

Design

Earth pressures and support loads are determined analytically or empirically, depending upon ground conditions, and the support and lining requirements for the excavation selected.

The selection of construction methods is influenced by their close relationship with design and performance. Consideration is given to

- Incorporation of temporary supports into permanent linings
- Control of surface subsidence where it may be detrimental to overlying structures
- Control of effects of blasting or ground displacement on adjacent structures
- Impact of the construction method on the permanent support system, including excavation, ground improvement, and support

Contract Documents

Owner-Contractor Relationships

In tunnel construction, particularly in rock, there is a great amount of uncertainty and risk because explorations cannot be expected to disclose all significant geologic features and hazards. To have the contractor assume all of the risk can be unnecessarily costly to the owner [Deere et al. (1974)[6]]. Forms of contracts that enable the owner's engineer and the contractor to work together to select suitable support methods for given ground conditions, wherein both share the risk for unknown conditions, should result in the most economical and safely completed project.

Specifications and Tender Documents

Except under arrangements where the work is carried out under a design-construct contract, the trend in tunnel construction is for the owner to provide as much geologic information to tendering contractors as practical and even to define acceptable tunneling procedures in the specifications [Deere et al. (1974)[6]]. These can include, for example, the need for freezing, predraining, grouting, or compressed air; whether the opening can be driven by full-face methods or multiple-drift methods are required, or if a TBM is suitable or desired (with or without the slurry face); blasting control in terms of kilograms of explosives per cubic meter of rock, overbreak, and length of drill round; and support-method requirements.

Construction Monitoring

The installation of instrumentation and the use of other techniques to monitor the response of the surrounding ground and structures adjacent to the underground excavation is becoming standard practice for most projects, and is particularly important for those located in urban areas.

Surface Subsidence

For evaluating the effects of excavation on overlying structures, it is important to measure ground movements throughout the mass around the tunnel as well as subsidence at the surface. Lateral deflection can be as important as settlement in causing damage to overlying structures, and there can be a significant time lag between disturbance and loss of ground around the tunnel prior to surface or near-surface manifestations. Surface effects are monitored optically by precise leveling of monuments on the surface and in overlying structures. Tiltmeters and pendulums may be installed in structures. Piezometers monitor changing groundwater levels, inclinometers measure lateral movements in soils, and borehole extensometers measure strains in rock or soil.

A survey of the condition of overlying structures should be performed prior to construction. Cracks or other signs of excessive distortion should be documented, as well as the types and depths of foundations providing support. All nearby underground utilities must be precisely located.

Excavations

Closure of the opening is measured with borehole extensometers and convergence meters. Roof and wall pressures to be contained by the support system may be measured with in situ stress meters. Loads and stresses in the lining or in anchors are monitored with load or pressure cells and strain gages. Collapse warning in large excavations such as caverns or mines may be provided by the acoustical-emission device. Blasting vibrations are monitored when blasting occurs adjacent to an existing tunnel or near overlying structures.

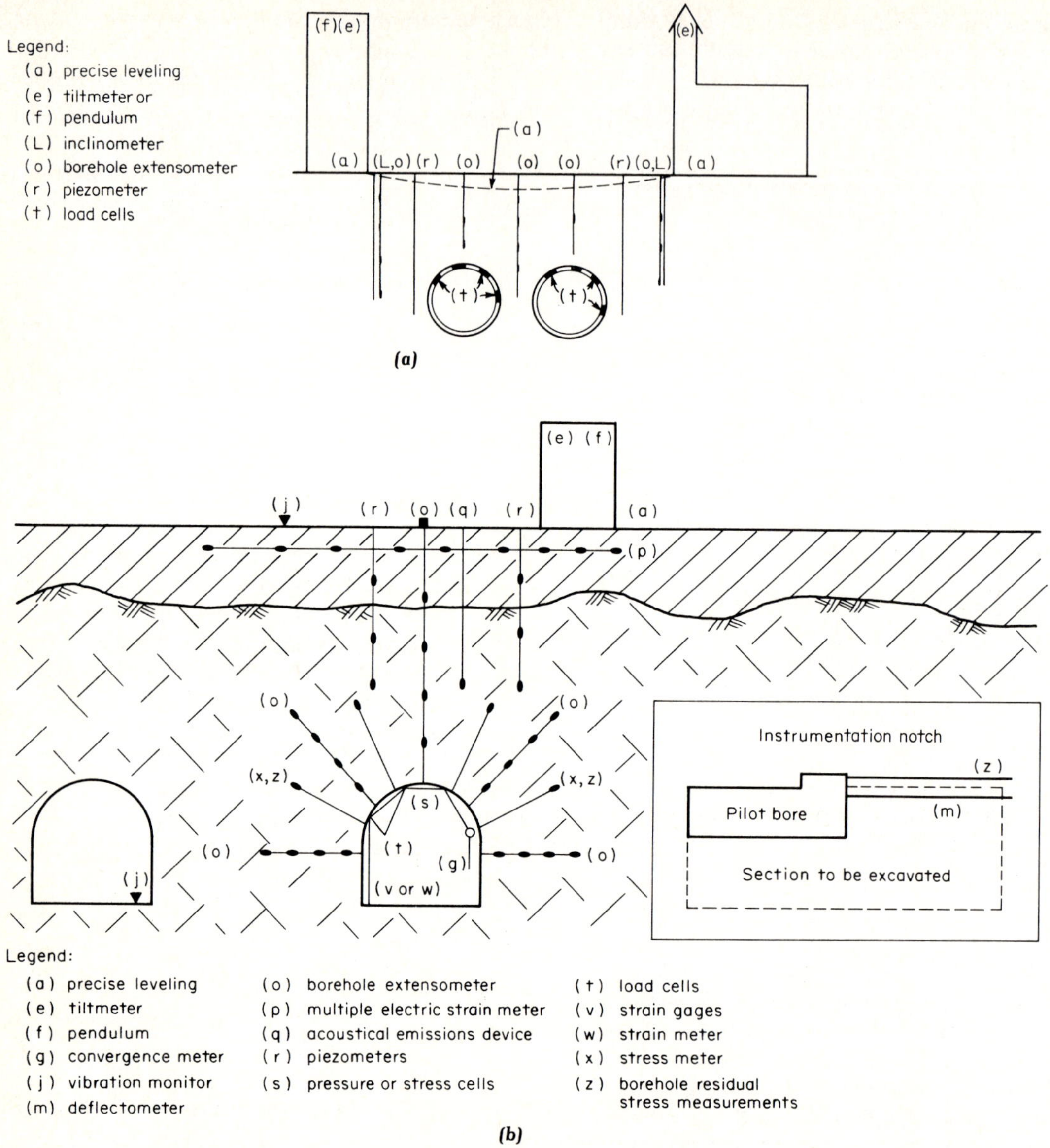

FIG. 10.2 Monitoring conditions associated with underground excavation with instrumentation: (*a*) tunnels in soil in urban areas and (*b*) tunnels, caverns, and mines in rock. [From Hunt (1984).[5] Used with permission of McGraw-Hill Book Company.]

Instrumentation and monitoring for underground excavations are discussed in Art. 4.5.5 of Hunt (1984)[5] and illustrated in Fig. 10.2. Detailed coverage of instrumentation for tunnels is given by Cording et al. (1975).[7]

10.2 TUNNELS IN SOILS

10.2.1 GENERAL

Installation

Tunnels in soil are also referred to as earth tunnels or soft-ground tunnels and are constructed for water supply, sewage disposal, and rail and roadway systems. Subaqueous tunnels are a subcategory.

Tunneling proceeds by removal of soil at the tunnel face as a lining is advanced to maintain the roof, sides, and floor. Under favorable conditions, temporary support in front of the lining is provided by ribs and lagging or ribs and liner plates, whereas in unfavorable conditions temporary support usually is provided by a shield behind which is placed temporary support in the form of liner plates or permanent segments. Temporary and permanent linings, the latter usually concrete, are designed to resist deformation under earth pressures.

Troublesome ground conditions are improved to facilitate installation and improve tunnel performance by using air pressure, dewatering, grouting, or freezing, depending upon soil and water conditions.

Performance

Adequacy of performance results from two major factors: the ability of the lining to resist deformations under earth pressures and ground surface subsidence; both are very dependent upon the method of construction and the quality of its execution. Ground subsidence is the result primarily of soil deformations occurring near the advancing tunnel heading and is largely dependent upon the construction method, advance rate, and presence of internal pressure for a given set of ground conditions [Peck (1969)[1]]. Surface manifestation is a dish-shaped trough along the tunnel alignment which can cause differential deflections in overlying buildings, utility lines, and other structures, unless adequately controlled.

Ground Conditions

Described in terms of soil types and their potential performance characteristics as firm, raveling, running, flowing, squeezing, or swelling ground (Table 10.1), ground conditions are related to suitable excavation and support methods and the surface subsidence potential. A general summary of these relationships is given in Table 10.3.

10.2.2 DESIGN FACTORS

Lining Stresses

Soil tunnels are lined with concrete, steel, cast iron, or timber and designed to resist deformation under earth pressure. Lining performance is a complex problem of soil-structure interaction which cannot be solved precisely by analysis. Ideally, performance depends on lining rigidity and soil type and deformation properties, as these factors influence the earth pressures to be supported. The disturbance that occurs necessarily during installation, however, substantially affects earth pressures and their distribution.

Flexible Lining

In the ideal case, a circular, completely flexible lining inserted into the ground without disturbing the ground inside or out would be subjected to a vertical pressure $p_z = \gamma z$ and a horizontal pressure $p_h = K_o p_z$, causing an elliptical pressure distribution about the lining as illustrated at a on Fig. 10.3.

Removal of soil from within the lining causes it to deflect to adjust to the pressures until it is slightly elliptical and support is provided by its *ring stress* capacity. The radial stress is virtually equal in all directions, expressed as $\frac{1}{2}(1 + K_o)p_z$ (b on Fig. 10.3). As the lining deflects, the overlying soil deforms, develops shearing stresses causing arching, and redistributes the earth pressure on the lining. If an elliptical lining were

TABLE 10.3
SOIL CONDITIONS, GENERAL EXCAVATION METHODS, AND SUBSIDENCE

Soil conditions	Excavation method*	Construction hazard†	Treatment‡	Subsidence potential§
Strong cemented or cohesive granular soils	Liner plate, ribs and lagging, TBM	Usually none	None needed	Not a problem with good construction practice
Stiff to hard clays¶	Liner plate, ribs and lagging, shield, TBM	Possible squeezing	Careful excavation	Can be significant with liner plates, negligible with TBM
Slightly cohesive fine-grained granular soils with relict structure (loess, residuals)	Liner plate, ribs and lagging, or TBM	Raveling ground	Provide support promptly, predrain, advance face in small-size drifts	Small to negligible with proper construction, can result in sinkholes if uncontrolled
Cohesionless granular soil above GWL	Liner plate or TBM with slurry face	Running ground, dry flows	Chemical grout	Local surface effects and sinkholes, negligible with TBM and slurry face
Cohesionless granular soil below GWL (dense and coarse-grained)	Liner plate or TBM with slurry face	Possible flowing ground	Air pressure, dewatering	Local surface effects and sinkholes, neligible with TBM and slurry face
Cohesionless fine-grained granular below GWL (loose to medium compact)	Shield or TBM with slurry face	Flowing ground	Air pressure, dewatering, chemical grout, freezing	Local surface effects and sinkholes, negligible with TBM and slurry face
Very soft to medium clays and organic soils	Shield or TBM with slurry face	Squeezing ground	Air pressure	Large-area surface effects difficult to prevent but reduced by TBM with slurry face

*See Table 10.9.

†See Table 10.1.

‡See Table 10.6.

§See Table 10.11.

¶Since swelling is a time-delay phenomenon, the rapid placement of supports usually prevents movement, but the possibility of the future development of high pressures on the tunnel lining must be considered.

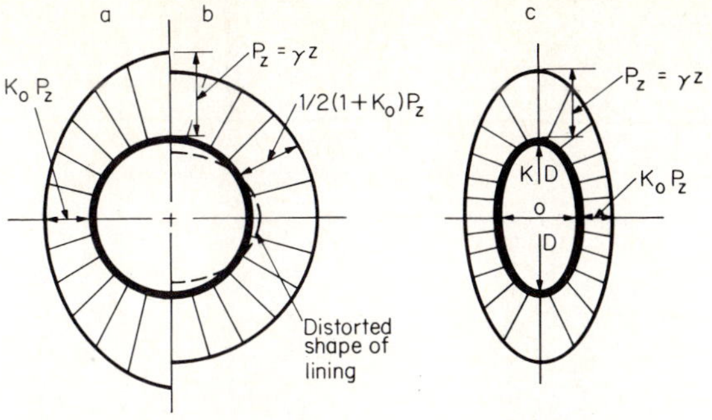

FIG. 10.3 Earth pressures acting and deflections occurring in hypothetical flexible tunnel linings. [From Peck (1969).[1]]

installed with its shape conforming to the equilibriumpolygon for the normal forces acting on the lining, removal of the soil from within would cause no stress redistribution or deflection of the lining, as shown at c in Fig. 10.3.

In design, the lining is selected to carry the ring stress associated with the radial pressures and to prevent local buckling. The deformations that occur are considered in view of damage to architectural finishes, the amount of which is related to soil stress-strain characteristics, time-dependent soil deformation characteristics, and tunnel depth and dimensions.

Rigid Linings and Reality

In a perfectly rigid circular lining inserted without any disturbance, no deformations occur upon removal of the soil, and therefore there is no redistribution of earth pressure. The differences between the vertical and horizontal earth pressures result in bending moments imposed on the lining. In reality, flexural rigidities of linings are intermediate between perfect flexibility and perfect rigidity. Additional complexities are introduced by horseshoe-shaped tunnels or tunnels with a roof arch and straight sides.

For additional discussion on design see Peck (1969)[1] and Cording et al. (1975).[7]

Earth Pressures and Support Loads

General Concepts

The concepts developed by Terzaghi, published in 1946 [Proctor and White (1977)[8]], are still applied by most practitioners for estimating earth pressures and tunnel support loads. In *sands* in general, and in *raveling soils*, he determined that the load on the roof of tunnel supports does not exceed a small fraction of the overburden weight, unless the cover is very shallow, and the loads do not increase significantly with time. Due to *arching* action (*see Art. 10.1.2*), if the overburden thickness is greater than about 1.5 times the combined tunnel width and height, the load is practically independent of depth. (In circular tunnels, the tunnel height is taken as zero.)

The concept for *sands* is illustrated in Fig. 10.4*a*. While excavating the tunnel and installing supports, frictional forces transfer most of the overburden weight from height H onto the ground on both sides of the tunnel. The balance, equivalent to height H_p, is carried by the roof support. The width of the ground arch B_1 is equal to the tunnel width B plus the tunnel height H_t and its thickness over the tunnel is about 1.5 B_1. Above the ground arch, pressure conditions in the sand remain almost unaffected by the tunneling operation. H_p increases almost in direct proportion to B_1 modified by factors for the relative density of the soil and the deflection of the ground arch, and for design is given as H_d in terms of feet of earth, as will be discussed in the subsequent section. The total load on a unit area of roof is the product of soil unit weight and H_d.

The concept for *homogeneous clays* (the clay located above the tunnel is not significantly stiffer than the clay along the sides or beneath the bottom) is illustrated in Fig. 10.4*b*. A block of soil around and above the opening tends to deflect, resisted by shear strength mobilized along the sides of the block. Based on actual measurements of radial loads on liners, the initial loads during construction may be as low as 20% of overburden, but in time may approach an all-around pressure equal to full overburden pressure p_z [Peck (1969)[1]]. In swelling clays and those similar to London clays, the all-around pressure may in time exceed overburden pressure.

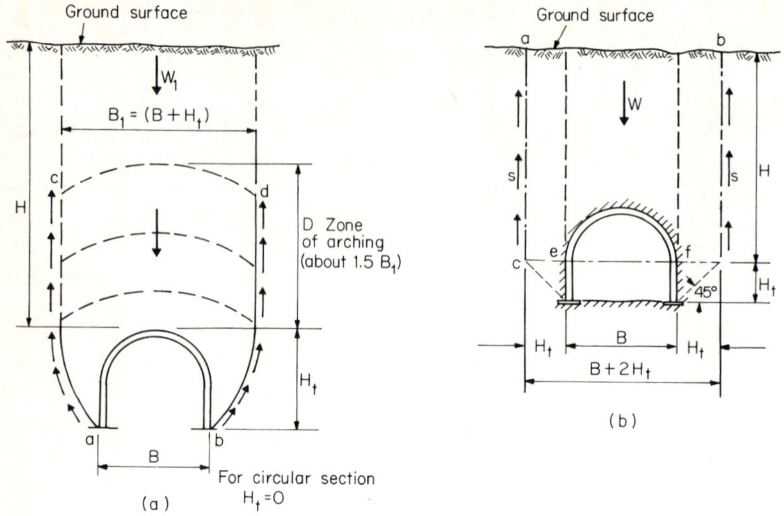

FIG. 10.4 Load action on underground openings in earth; (*a*) granular material and (*b*) homogeneous clay. [From Proctor and White (1977).[8]]

Peck (1969)[1] suggests that the ring load for a single tunnel lining be designed for an all-around pressure $p_z = \gamma z$, except for heavily overconsolidated clays for which it "might be preferable to provide for a ring load corresponding to the pressure $\frac{1}{2}(1 + K_o)p_z$. Because of the influence of construction procedures and other factors, the state-of-the-art is such that further refinement is not yet warranted." This statement apparently holds true today.

Estimating Loads on Temporary Supports

Criteria developed by Terzaghi for estimating design loads for temporary supports for soil tunnels at depths greater than $1.5(B + H_t)$ are summarized in Table 10.4. There are two major groupings: soils that develop arching action (including running, flowing, and raveling ground) and clays (classed as squeezing or swelling ground). The design load is given as H_d in feet of earth, and the unit load on the roof is given as

$$p_r = \gamma H_d \quad \text{or} \quad wH_d \qquad (10.1)$$

and the total load per unit of tunnel length is given as

$$P_r = \gamma H_d B \qquad (10.2)$$

For circular tunnels the width of the arching zone is greatly reduced by the absence of vertical walls and by the arching of the roof, so the roof load is estimated by assuming $H_t = 0$. The conditions under *Remarks* in Table 10.4 are discussed in Art. 10.2.3.

For running ground, H_d is a function of $B + H_t$ modified by a factor for relative density; and for flowing ground of H or $B + H_t$. Raveling ground employs H_d found for running ground modified by time factors, and in the case of the tunnel below GWL using compressed air, air pressure p_a. T is the elapsed time between mining and completion of the permanent structure in minutes, and t is the stand-up time in minutes of an unsupported strip of roof of width b.

Values for t are estimated from the equation

$$t = R_s/B \qquad (10.3)$$

where R_s = unit stand-up time of the ground
B = tunnel width in ft

R_s is estimated by excavating a pocket 1 ft square in the wall of a test pit in soil equal to the crown of the future tunnel. The time that elapses from the start of the excavation until the first ground drops out of the pocket roof is the unit stand-up time.

For example, if R_s is found to be equal to 30 days and the tunnel width B = 10 ft, t = 3 days. If H_d has been found to be equal to 10 ft, then for an elapsed time T = 4 days, the design load for raveling ground above GWL = $H_d(T - t)/T$ = 10(4

TABLE 10.4
DESIGN LOADS FOR TEMPORARY SUPPORTS IN EARTH TUNNELS AT DEPTHS MORE THAN 1.5 $(B + H_t)$*

Type of ground	Ground condition	Design load H_d	Remarks	Eq.
Running ground above water table	Loose	$0.50\,(B + H_t)$		(10.4)
	Medium	$0.40\,(B + H_t)$		(10.5)
	Dense	$0.30\,(B + H_t)$		(10.6)
Running ground in compressed-air tunnel		Disregard air pressure; H_d equal to that for running ground, above water table with equal density; circular lining or floor indicated.		
Flowing ground in free-air tunnel		H or $2\,(B + H_t)$, whichever is smaller.	Circular lining or stiff floor required.	(10.7)
Raveling ground	Above water table	$\frac{T - t}{T} H_d$ (running)		(10.8)
	Below water table (free air)	$\frac{T - t}{T} 2H_d$ (running)	Circular lining or floor indicated.	(10.9)
	Below water table (compressed air)	$\frac{T - t}{T} 2H_d - \frac{P_a}{w}$	Circular lining or floor indicated.	(10.10)
Squeezing ground	Homogeneous	$H - \frac{p_a}{w} - \frac{Hq_u}{2w(B + 2H_t)}$	After complete blowout, $p_a = 0$. In this event the stress in ribs should not exceed yield point of steel.	(10.11)
	Soft roof, stiff sides	$H - \frac{p_a}{w} - \frac{Hq_u}{2wB}$		(10.12)
	Stiff roof, soft sides	$H - \frac{p_a}{w} - \frac{Hq_u}{2w(B + 6H_t)}$		(10.13)
Swelling ground	Intact	Very small.	Permanent roof support should be completed within a few days after mining or else compressible ribs should be used.	
	Fissured, above water table; Fissured, below water table, free-air tunnel	H_d equal to that for raveling ground with same stand-up time H.		

*From Proctor and White (1977).[8]

NOTE:

B = tunnel width, ft.

H = vertical distance between tunnel roof and ground surface, ft.

H_t = tunnel height, ft; for circular tunnels = 0.

H_d = design load, ft of earth.

p_a = air pressure, psf = $144p_c$ when p_c is given in psi.

p_r = load on projected roof area, psf = wH_d.

q_u = unconfined compressive strength of ground above roof, psf.

T = time elapsed between mining and completion of permanent structures, min.

t = stand-up time of unsupported strip of roof width b, min.

w = average total unit weight of soil above tunnel roof, pcf (γ).

− 3)/4 = 2.5 ft of overburden. If the tunnel is filled with compressed air under a pressure p_a (psf), the compressed air carries the weight of a soil layer with a height p_a/γ which reduces the roof load accordingly.

The expressions for estimating support loads in soft to firm clays (squeezing ground) are given in terms of the tunnel dimensions, the overburden thickness and weight, the unconfined compressive strength of the clay, and air pressure, with modifications for a soft roof and stiff sides and for a stiff roof with soft sides. Swelling clays are stiff to hard, but the degree of swelling depends on factors besides consistency. As with "firm ground," intact swelling clays are unlikely to develop significant support loads for at least a week or so after the opening is made; therefore, prompt installment of the permanent lining is important. Otherwise, pressure steadily increases and supports may be crushed. Fissured swelling clays start to swell soon after the clay is exposed and supports should be installed and back-packed as rapidly as possible. Loads may increase to full overburden pressure or higher prior to installation of the lining.

Foundations

Horseshoe-shaped tunnels supported by ribs and liner plates or lagging require consideration of the bearing capacity of the wall footings and of bottom heave (*see Art. 9.3.3*). The critical condition occurs during construction, while the tunnel is being excavated and before the floor is finally in place. Excessive footing settlement will endanger the tunnel integrity and result in surface subsidence as well. Determination of the allowable bearing value and required footing width usually determines the applicability of the ribs and lagging versus the shield method for tunneling.

Allowable Bearing Values

Ribs are commonly supported on closely spaced rectangular footings, corresponding to strip footings, with their tops almost flush with the tunnel bottom. The ultimate bearing capacity is determined by procedures given in Art. 6.2.4 (Eq. 6.10), where the load per footing is $Q/2$ from the roof load above. Typical values for q_{all} for various ground conditions are given in Table 10.5. In general, for soils with cohesion and friction, q_{all} should not exceed $2/3q_{ult}$. The values for sands are based on the load-settlement curves presented in Terzaghi and Peck (1948)[9] and are selected to restrict settlement to within about ½ in. The values given for clays are based on the same settlement criteria. With footings in clay, should a blowout occur, the load increases by $p_aB/2$, where p_a is the air pressure. Where footings designed for q_{all} are over 3 ft wide, consideration is given to flooring the tunnel as a mat or to providing a circular cross section.

TABLE 10.5
ALLOWABLE SOIL PRESSURES FOR STRIP FOOTINGS UNDER RIBS*

Ground condition	q_{all}†
Firm ground and stiff clay	10–12 ksf
Running ground, very loose	2.4 ksf
Running ground, medium dense	13.2 ksf
Running ground, very dense	24.0 ksf
Flowing ground (see text)	Use stiff floor or circular tunnel
Raveling ground, dry or moist	Use values for running ground
Raveling ground, free-air tunnel below GWL (see text)	Use stiff floor or circular tunnel
Clays, soft to firm‡	$1.43U_c$

*After Terzaghi (1946) in Proctor and White (1977).[8]

†Generally to limit settlements to about ½ in.

‡During blowout load on footing, increases by $p_aB/2$.

For some conditions in flowing or raveling ground special considerations are required. Ground will flow in free-air tunnels below GWL or if the air pressure is not high enough to dry the tunnel bottom. Installation of steel ribs with footing support is not possible because ground flows out from beneath the footings into the tunnel, causing the footings to sink. In raveling ground, if soil permeability is relatively high, softening of the bottom will occur along with raveling of the roof and sides; footing support will be lost unless the tunnel is dried with compressed air or the soil adequately drained. If neither is feasible, a stiff floor or a circular lining is required.

10.2.3 CONSTRUCTION ASPECTS

Tunneling Methods

Selection

Selection of the tunneling method to be employed is based on consideration of ground conditions, the dimensions of the drift or heading to be excavated, and roof and face-opening stability during excavation and prior to support placement. Basic soil tunneling methods include ribs and lagging, liner plates, shields, and machine excavators. Machines are referred to as "moles," digger shields, or tunnel boring machines (TBM). Ground conditions are considered from the aspect of transforming troublesome conditions (running, flowing, or squeezing ground) into more favorable conditions (firm or raveling ground).

Excavation Procedure

Excavation proceeds in cycles, each consisting of excavating the working face ahead a certain distance and providing support. Unless some form of shield is used, structural members of H- or I-beam steel, termed *ribs* or *sets*, are erected to support the sides and roof. *Lagging*, consisting of wooden boards or steel plates, are placed longitudinally between the ribs for additional support. They can be installed tight against each other or with spacing in between, depending on ground conditions. Alternatively, steel *liner plates*, which can be corrugated to increase stiffness, are installed bolted together with or without ribs. The distance advanced at each cycle depends largely on the ground conditions and usually does not extend beyond the spacing of the steel sets or the width of one course of liner plates.

A *shield* functions as a mobile support for the earth which would otherwise be exposed while mining one cycle. Lining in shield-driven or shield-excavated tunnels often is constructed of steel or concrete sections installed behind the shield to form the ring as it advances. Ribs and lagging and cast-in-place concrete also are often used with shields.

Ground Conditions and Improvement

Tunnel Performance

Soil conditions have been grouped in Art. 10.1.2 (Table 10.1) into classifications based on their performance during excavation. Ground conditions, associated soil types, and general effects on tunneling and surface subsidence are illustrated in Fig. 10.5 and described in a general manner, including a brief statement on treatments to improve conditions, in Table 10.6. In addition to advancing a face in small drifts rather than full-face, as will be discussed in a following section, ground treatment is provided by air pressure, dewatering, grouting, or freezing.

Air Pressure

Compressed air is used to improve all types of unstable ground and is the only method for stabilizing very soft squeezing ground, except for tunneling with the slurry-face TBM or earth-balance shield methods. In pervious soils the compressed air (usually taken as $p_a = \gamma_w z$, where z = tunnel depth below GWL) drives the water from the soil, drying it and increasing its strength. When the boundary between the air-dried soil and saturated soil reaches GWL, *air leakage* occurs, the amount depending upon soil permeability. The greatest danger of blowouts is in pervious soils. In soft clays, air pressures are set to control ground loss. The tendency for squeezing is high if $(p_z - p_a)/s_u > 5$ [Peck (1969)[1]], where p_z is vertical pressure to the tunnel center and p_a is air pressure above atmospheric.

Pressures up to 16 psi above normal atmospheric pressure are common and require a bulkhead to

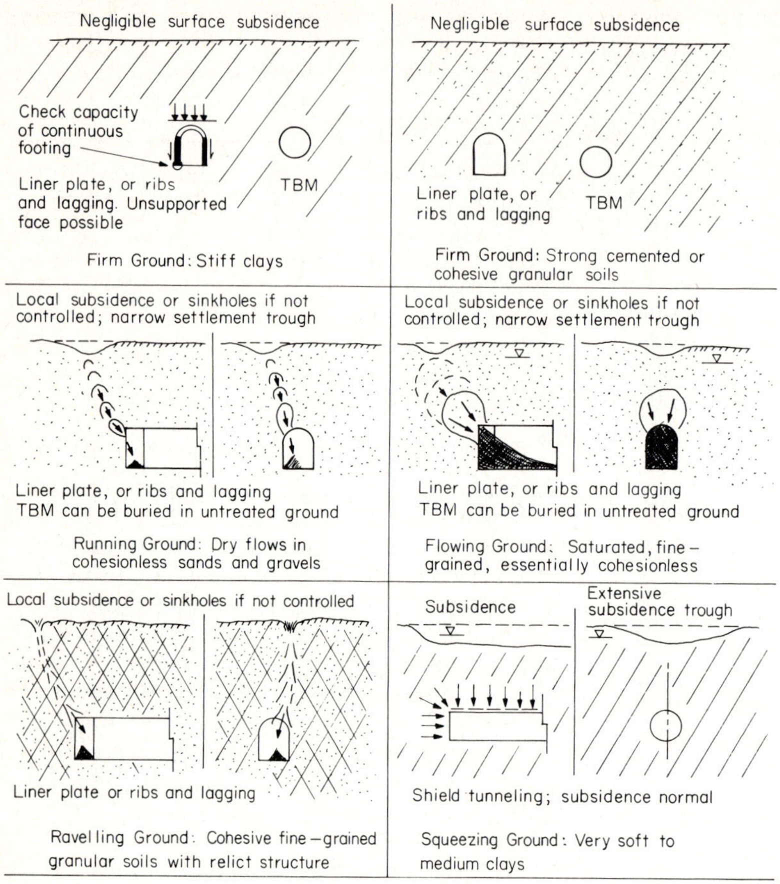

FIG. 10.5 Ground conditions, associated soil types, and general effects on tunneling.

seal the tunnel and air locks to permit entry and exit. Allowable working time and decompression time are given in Table 7.4. The maximum depth for work under compressed air is 100 ft below GWL. In severe cases at substantial depths below GWL, air pressure is supplemented with drainage, grouting, freezing, or electroosmosis.

Dewatering or Predrainage

Predrainage can transform raveling or flowing ground, which is the result in most cases of seepage forces, into more favorable ground. It is accomplished with wellpoints, vacuum points, eductor wells, deep wells, electroosmosis, or water collectors. The latter are used where the base of the water-bearing stratum is only a few feet below the tunnel bottom and wellpoint spacing would be very small. Water collection involves pumping from a shaft penetrating the pervious stratum in which radial collector tubes have been installed. In thicker strata, wellpoints can be jetted down from the tunnel bottom. Dewatering also is discussed in Arts. 2.3.3 and 5.4.3.

Penetration Grouting

Chemical grouting is limited primarily to use in silty to coarse sandy soils and ideally it should penetrate the pores and not split the ground. The most commonly used grouting solutions are silica based (40 to 70% silicate) with gel times that can be set from seconds to hours [Clough

TABLE 10.6
GROUND CONDITIONS AND TUNNELING OPERATIONS IN SOILS*

Condition	Soil types	Effects on tunneling	Improvement treatments
Firm ground	Cemented or cohesive granular.	Stand-up time sufficient to allow excavation to proceed a meter or so without immediate support and without collapse danger.	In plastic clays with squeezing tendencies use air pressure.
	Stiff clays.	Tunneling difficulties may occur when $(p_z - p_a)/s >$ 5 due to squeezing.†	
Raveling ground	Slightly cohesive fine-grained granular, with apparent cohesion. Residuals with relict structure.	Unstable in roof and upper face resulting in spalling and falls. Can be progressive and lead to cavities above tunnel, possibly extending to surface causing sinkholes. Can continue after lining installed, resulting in nonuniform loadings and even lining failure.	Provide support promptly. Seepage forces acting toward face are common cause. Predrainage can transform raveling ground into firm ground. Advance face in small drifts rather than full-face. Backfill annular space between shield and lining carefully.
Running ground	Dry sands or gravels, no cohesion, weak cementation.	Dry flow from unsupported face, or even through small openings, builds a soil pile until angle of repose is reached. Can destroy arching effect resulting in higher lining pressures than anticipated.	Chemical grouting; effective in extensive formations.
Flowing ground	Saturated fine-grained soils and high seepage pressures.	Thick, slurry-like mass can fill tunnel. Destroys arching effect and results in high lining pressures. The most dangerous condition.	Transform into raveling or firm ground by adequate drainage, chemical grouts, air pressure, or freezing.
Squeezing ground†	Very soft to medium clays.	Ground extrudes plastically into tunnel. Must be contained; surface subsidence control difficult.	Performance controlled by tunneling methods: shield tunneling with air pressure or slurry TBM.

*From Peck (1969).[1]

†Squeezing tendency: probable occurrence if $(p_z - p_a)/s_u > 5$ where p_z = total vertical pressure at depth z in tunnel center and p_a = air pressure above atmospheric.

(1981)[10]]. In tunnel construction, grouting most commonly is used to control ground movements and groundwater flow in critical urban areas or to reduce or eliminate the need for air pressure where groundwater pressures are high. Various types of grouts and penetration grouting are discussed in Art 5.4.2. The success of grouting decreases rapidly as the silt content of the sand increases.

Freezing

Ground freezing *(see Art. 5.4.3)* also serves to minimize the use of compressed air and to control loss of ground, and in some instances is more effective in stratified soil formations than is grouting or dewatering. Freezing with nitrogen is quicker than using brine, although it is more costly.

TABLE 10.7
GROUND IMPROVEMENT METHODS

Method	Applications	Comments
Air pressure	Plastic stiff clays with squeezing tendencies. Soft to medium clays. Saturated fine-grained soils with high seepage pressures and flow tendencies, and saturated cohesionless soils.	Reduces amount and rate of inflow into space between shield skin and soil and reduces squeezing into face during shield tunneling. Reduces or eliminates hydraulic gradient toward tunnel. Blowout danger when overburden thin. In very pervious soils, escaping air may dry the soils eliminating apparent cohesion resulting in running.
Dewatering with wellpoints or deep wells*	Homogeneous granular soils below GWL with $k >$ 10^{-4} cm/sec (silty sands). Stratified deposits with free-draining layers.	May cause ground surface subsidence where overlying soils are soft clays or organics. Wellpoints are limited to depths of about 4.5 m. Greater depths require jet-eductor system or deep wells.
Chemical grouting*	Dry cohesionless sands with a tendency to run. Saturated fine-grained soils and high seepage pressures and tendency to flow (most suitable method for these soils unless excavation is with slurry mole).	Chemical grouts preferred over other types because of control over set times, superior penetration, and relatively high tensile strength. Resins are suitable for soils with k values as low as 10^{-6} cm/sec. Grout wall may be discontinuous and care is required during excavation. Below GWL always use drainage with grouting to avoid high seepage pressures causing flow through small openings in grout curtain.
Freezing*	Saturated fine-grained granular soils. Best application is in stratified formations and fine sands and silts which may be difficult to grout or dewater [Jones and Brown (1978)[11]].	Time-consuming and costly. Careful installation required to ensure thick and continuous wall. Surface heave may occur during freezing followed by collapse upon thawing. May cause cracking of completed adjacent concrete lining from ground temperature changes.

*For further discussion see Arts. 8.4.2 and 8.4.3 of Hunt (1984).[5]

Ground-improvement methods are summarized in Table 10.7.

Ribs and Lagging and Liner-Plate Tunneling

General Procedures

Ribs and lagging and liner-plate methods, often used to construct horseshoe-shaped tunnels, are most suitable in firm or swelling ground. They can be used economically in squeezing ground if the squeezing rate is not too excessive to permit support with bracing; suitability depends on soil shear strength, tunnel width, and overburden pressure on the crown.

Tunneling proceeds by excavating either full face or in benches, depending on ground conditions. Earth removal is accomplished by hand or by small machines. In *unitary excavation*, small pockets are removed in succession starting at the crown and proceeding along the arch toward the spring line (Fig. 10.6). The roof of each pocket is immediately supported by a unit of lagging, usually steel liner plates. *Full-arch excavation* is an alternative to unitary excavation, wherein the entire portion of tunnel above the spring line is excavated for the width of one or more courses before support is erected. Support may be achieved with liner plates or with steel ribs and lagging.

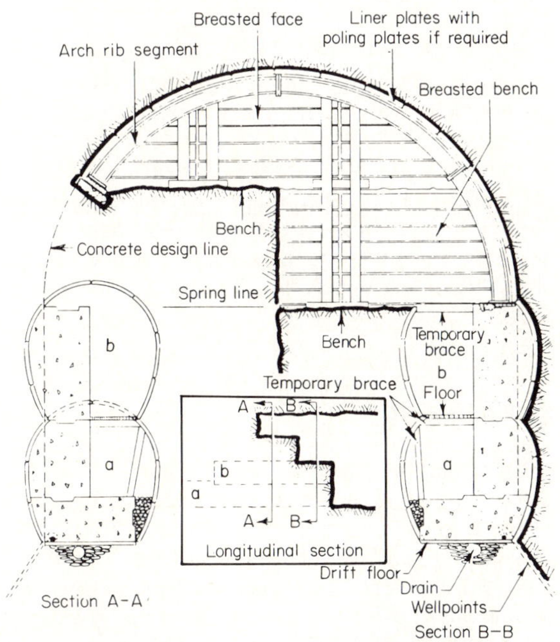

FIG. 10.6 Side-drift method in earth with vertical faces. [From Proctor and White (1977).[8]]

With liner plates, steel ribs can be installed after the plates for additional support. With ribs and lagging, the ribs are installed first with the lags of timber or steel resting on the outer flanges of the ribs. For the rib and lagging method, therefore, longer stand-up time is required. In full-arch excavation the lags usually are installed from the spring line.

The *side drift method* with vertical faces used for large tunnels is illustrated in Fig. 10.6. Section A-A is located a few feet back from the face; Section B-B is taken through the spring line bench. Drift tunnels (a) are driven first and if below GWL compressed air may be used, and subdrains and wellpoints may be installed. As the drift advances, invert and side-wall concrete is placed. Drift tunnels (b) are then driven above drifts (a) in free air, if possible, and the side-wall concrete is brought up to the spring line. As the upper drifts advance, the crown liner plates of the lower drifts are removed and temporary horizontal braces are installed and floored over. When below GWL, drifts (a) are driven much further ahead of the general excavation (dashed lines in the longitudinal section) to properly drain the ground ahead.

Arch excavation proceeds with a crown bench followed by a spring line bench. The core face can be vertical and breasted as shown, or sloped at the angle of repose of loose ground.

Ground Conditions and Support Requirements

FIRM GROUND Usually, perceptible ground movements do not occur for several days; therefore, temporary support is not required if the permanent lining is placed promptly. Customary advance length is about 5 ft.

RAVELING GROUND Raveling is the most common form of ground behavior and can be considered as running ground with some binder such as cohesion or cement. The ground breaks up into chunks, flakes, and angular fragments which drop from the roof. Stand-up time is adequate to permit the installation of at least one support unit, such as a rib or liner plate, and increases with decreasing width of the unsupported section. The spacing between roof supports is decreased from the customary 5 ft with the spacing dependent upon the stand-up time. All cavities above the roof support are immediately backpacked with pea gravel or grout to prevent raveling and loosening of the materials over the roof.

STOPING This refers to the progressive disintegration of the roof which can continue over a period of years until a sinkhole forms at the surface unless back-packing is adequate. The bigger the tunnel, the more extensive the stoping can be. In fast-raveling ground, with stand-up time less than about 7 min, liner plates are used with or without ribs, and breasting, forepoling, and spiling may be necessary at the face. *Spiles* are small sections of wood or steel with a sharpened end driven into the soil at the face along the arch to provide support during unitary excavation in ground with a very low stand-up time. *Poling plates* are steel sections that can be jacked ahead of a tunnel face to provide temporary support in the form of a partial shield in poor ground.

RUNNING GROUND Stand-up time is zero. Breasting boards at the face and tightly closed roof supports are required, although a certain amount of material will run into the tunnel as boards are being placed. The leading edges of the roof and side supports are kept buried at all times in the ground ahead of the excavation.

FLOWING GROUND Water and soil move into the tunnel from all sides as well as the bottom. Behavior depends on cohesion, permeability, and hydrostatic head. It occurs only in free-air tunnels below GWL, or in tunnels where compressed air pressure is inadequate to dry the bottom. To change flowing ground to running ground (cohesionless soils) or to raveling ground (slight cohesion) or to squeezing ground (soft organic silts), the air pressure should be equal to the pressure in the pore water at the level of the tunnel bottom ($p_a = \gamma_w z$), where z = depth below GWL. Predrainage is an alternative or a supplementary measure in free-draining soils.

SQUEEZING GROUND Soft to medium clays and organic silts in the working face and unsupported roof advance toward the tunnel opening slowly but continuously. Stand-up time is not a factor since the clay neither spalls nor ravels, and adequate time is always available for installation of supports by hand mining and the full-arch process using breasting boards. The major factor is the large volume loss and surface subsidence that accompanies squeezing. With hand mining, squeezing is reduced by air pressure but surface settlement may still be undesirable. Shield tunneling permits a lower air pressure and may reduce loss of ground. Ribs and lagging or liner plates are heavier than for raveling ground, and spacing along the axis usually is small to limit soil pressures on footings and subsequent settlement.

SWELLING GROUND Occurs in heavily overconsolidated clays, particularly in shallow tunnels (less than 30 m depth), and pressures on supports and linings can exceed overburden. In *intact clays*, swelling is rarely perceptible within the first week of exposure. They are classed as firm ground, and the permanent lining may be installed without temporary supports, but they should be provided within a few days of mining. *Fissured swelling clays* are classed as raveling soils since they tend to break into fragments along joints or slickensides, and swelling tends to increase the raveling process. Temporary supports and backpacking should be provided promptly.

Shield Tunneling

Applications

Shields offer advantages which may offset their much higher costs under certain conditions. For example, in soils requiring breasting, the rate of progress is increased; under watercourses, the tunneling hazard is reduced; through flowing or squeezing ground, there is a reduction in loss of ground and an increase in tunneling rate and, in general, a reduction in ground surface settlement. Tunneling conditions justifying the use of a shield are summarized in Table 10.8.

Procedures

The method employs an outer shell, usually cylindrical, that is jacked ahead of the excavation to provide protection to workers and machinery. The general components are illustrated in Fig. 10.7. After or during the forward thrust into the soil, the soil is removed either along the entire face or in smaller sections, depending upon ground conditions. In some cases, modern open-face shields employ a digger arm or backhoe and conveyor system for soil removal and metal orange-peel breasting boards installed in the upper portion of the face. They also may have articulation joints along the length of the shield. In very soft river silts it is sometimes practical to push the shield into the materials and displace them, a method used in the construction of some subaqueous tunnels. An alternate procedure is to leave a small opening in the shield face for soil

TABLE 10.8
TUNNELING CONDITIONS JUSTIFYING THE USE OF A SHIELD*

	Land tunnels					
	In open country		In cities		Subaqueous tunnels	
Tunneling conditions	Small	Large	Small	Large	Small	Large
Firm ground	—	—	—	—	—	H-opt.
Slowly raveling	—	—	—	—	—	H-opt.
Rapidly raveling and swelling	—	—	—	—	—	H-adv.
Rapidly raveling, above water table	—	B-opt.	—	B-opt.	—	—
Rapidly raveling below water table, compressed air	—	B-opt.	—	B-opt.	H-adv.	BH-comp.
Running or cohesive running above water table	—	B-opt.	—	B-opt.		
Running or cohesive running below water table, compressed air	—	B-opt.	S-opt.	SB-adv.	H-comp.	BH-comp.
Flowing (quicksand or soft silt) below water table, free-air or inadequate compressed-air pressure	H-opt.	HB-comp.	SH-comp.	SBH-comp.	H-comp.	BH-comp.
Rapidly squeezing (very soft clay), compressed air	E-opt.	E-opt.	ES-adv.	ES-comp.	EH-adv.	EH-comp.
Slowly squeezing (soft or medium clay) free air	—	—	—	S-adv.	—	H-opt.
Slowly squeezing (soft or medium clay) compressed air	—	—	—	—	—	H-adv.
Erratic ground	—	BH-adv.	S-opt.	SBH-adv.	H-adv.	BH-comp.

*From Proctor and White (1977).[8]

Incentive for using shield:
S—to avoid excessive settlement in city tunnels
H—to minimize hazards to miners
B—to facilitate breasting, thereby making faster progress
E—to avoid loss of ground caused by rapid squeeze and make faster progress

Degrees of necessity for using shield:
comp.—compulsory
adv.—advisable
opt.—optional
dash—unnessary

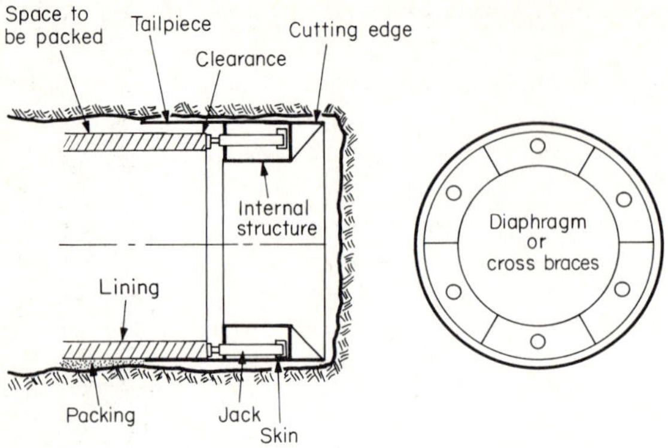

FIG. 10.7 Heading of a shield tunnel and essential elements of a shield. [From Proctor and White (1977).[8]]

to enter the tunnel where it is removed. In displacement tunneling the effect on nearby installations must be evaluated.

After a shove, the jacks are retracted and a ring consisting of lining segment is installed inside the tail piece of the cylinder. The ring is not exposed to the soil until the shield is again jacked ahead. Precast concrete liners have been used extensively in Europe and Japan but are only recently finding application in the United States as they become competitive with steel liner plates. Conventional precast segments have indented grooves for rubberized waterproofing strips and are bolted. In the stiff London clays where groundwater problems are minimal, tapered segments forced into place have been installed without bolts.

The annular space occupied by the tailpiece and the necessary clearances between the tailpiece and the lining are about 6 to 10 cm. If the space is not filled, the surrounding soil can move into it, resulting in loss of ground equal to the total volume of the annular ring. To minimize ground loss and surface subsidence, the void should be filled as quickly as possible with pea gravel or grout or the lining should be expanded into the void.

Compressed Air

In plastic clays, shield tilting in a downward direction may tend to occur during advance and may become unmanageable in clays where values of $(p_z - p_a)/s_u$ approach 7 [Peck (1969)[1]]. The proper use of compressed air and adequate clearance between the lining and the tailpiece are necessary to relieve the problem. The clearance permits the shield to be realigned by steering with the jacks.

Compressed air is used also to control inflow of soil and water into the tunnel, and to control infiltration p_a should correspond to the water head at the tube bottom. This causes excess air pressure at the tube top, creating blowout danger where soil cover is thin and overburden pressures inadequately low. *Subaqueous tunneling* in sand requires careful removal of materials from small openings in the face. The costs and danger involved may make the bentonite slurry or sunken tube techniques more economical.

Machine Excavation (TBM)

Procedures

A tunnel boring machine, also referred to as a TBM or mole, consists of a rotating or oscillating cutter that chisels or shaves the soil from the circular face as the machine is advanced. It can be mounted inside a cylindrical shell which serves most of the functions of the conventional shield in providing protection against unstable ground. TBMs advance by jacking against grips, which is not convenient with ribs and lagging.

In stable or firm ground, such as stiff to hard clays, the working face may be left entirely unsupported, without a protective cylindrical tailpiece. If a cast-in-place concrete lining is to be installed, temporary support is provided by steel ribs expanded against lagging in contact with the soil. The alternate, more common procedure, is to place segments of steel, cast iron, or precast concrete jacked or wedged into contact with the soil, as permanent lining. The rapid completion of the ring effectively stops subsidence that can result from squeezing ground.

Slowly raveling ground is controlled readily by providing the mole with a shield because of the rapid rate of advance. In more rapidly raveling ground, considerable protection is provided by a machine with the cutter face consisting of a disk or flat cone interrupted by occasional slots for the cutting knives and material entrance.

Troublesome Ground

Flowing or running ground can be more troublesome to machine tunneling than to conventional methods because of the rapid advance rate and the difficulties inherent in recognizing poor ground conditions. Flowing ground and even rapidly raveling ground have buried machines and resulted in substantial loss of ground. It is even more imperative in machine tunneling to transform running, flowing, or rapidly raveling ground into firm ground, or at least into slowly raveling ground. Investigations for machine tunneling require even more care than for conventional tunneling because of the rapid advance rates.

The feasibility of machine tunneling also is dependent on soil stratigraphy and the occurrence of boulders and other obstacles. For efficiency, different types of materials require different types of cutting tools and, if the apparatus will encounter material changes requiring frequent tool changes, machine excavation will be inefficient. In particular, stratified deposits presenting a wide range of material in the same face are often difficult to remove with any given tool. Boulders or other obstacles also may cause progress to be inefficiently slow [ENR (1980)[12]].

Slurry-Shield or Slurry-Face TBM

Method

Machine tunneling in combination with a bentonite or ordinary clay slurry is a recent development that is very effective in saturated granular and soft clay soils, even without using air pressure [NCE (1979),[13,14] Clough (1981)[10]]. Although originally developed in Britain, as of 1983 Japan is by far the leading country in the world in slurry-shield tunneling. Schematic diagrams of the slurry-shield operation are given in Fig. 10.8. The slurry shield (Fig. 10.8*b*) consists of a steel cylinder with an open-head cutting wheel located behind a slurry that supports the tunnel face. The cylinder (shield) contains a watertight rubber tail seal around the inner perimeter. A tunnel lining of concrete segments is installed with a system that enables grouting to be kept within a few meters of the face.

Operations

Slurry pressure to the face varies generally between 1 to 3 kN/m^2 higher than groundwater pressure and must be altered for different soil conditions. An agitator mixes the excavated muck with the slurry, which is then pumped through a return line to a treatment and disposal location. Large particles such as cobbles are screened from the slurry.

Initially bentonite suspension was used, but experience and technical developments have made possible the use of other suspensions, which vary with ground conditions. In clayey soil the excavated muck is diluted and used as slurry; in sandy soils composites of clay and bentonite, clay and CMC (sodium carboxylmethylcellulose), or clay and a polymer are used.

In large-diameter machines, airlocks located near the cutting wheel supply compressed air to the chamber permitting access to the face for removing obstructions or changing the cutting picks. Backfilling between the erected segments and the excavated tunnel walls is accomplished with a grout sufficiently liquid to fill the space completely and designed to gel quickly. Mortar usually is used, and gel time control improved by adding liquid silicic acid.

Directional control may be a problem as it can be affected by the slurry penetrating the space around the tunnel and causing the shield to rise, or it can be affected by the specific gravity of the backfill grout which is greater than that of the slurry. Before placing a shield in the launching shaft, it is necessary to strengthen the ground around the shaft by chemical grouting or freezing [NCE (1979)[14]].

Surface subsidence appears from experience to date to be effectively minimized because of the ability inherent with the operation to keep earth pressures in balance during tunneling. There can be, however, as much loss of ground into the annular void behind the tailpiece with either the slurry face or EPB shields as with any other shield if improper procedures are followed.

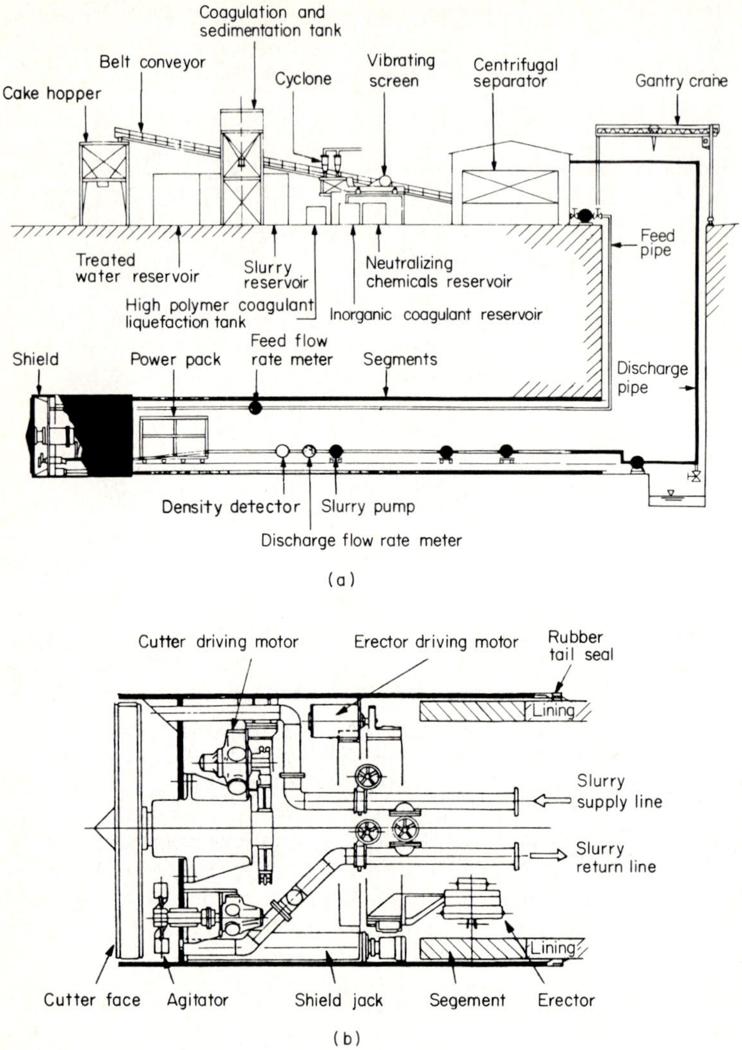

FIG. 10.8 The slurry-shield or slurry-face TBM method of tunneling. (*a*) General system of the slurry-shield method in 1980. (Courtesy of Mitsubishi.) **(*b*) Major elements of the slurry shield and cutter face.** [From NCE (1979).[14]]

Earth-Pressure Balance (EPB) Shield

Method

Introduced in Japan in about 1976, the basic principle behind the EPB shield is that the face of the tunnel is stabilized by keeping excavated soil in contact with the face at all times under sufficient pressure to prevent soil movement, without the use of air pressure. No slurry or additives are mixed with the excavated soil and the system is primarily applicable to soils above the water table or in soils of relatively low permeabilities below GWL [Clough (1981)[10]].

Operations

The shield (Fig. 10.9) is advanced by jacks reacting against the in-place liner, carried as a steady forward movement to keep constant pressure on the soil face as it is being excavated. Jacking pressure can be adjusted to control yawing and pitching or to guide the machine into a curve, as with all shields. The excavated soil fills the plenum behind the cutting face from where it is removed by the conveyor screw. To balance the earth pressures, the excavated and discharged volumes are balanced by synchronizing the ro-

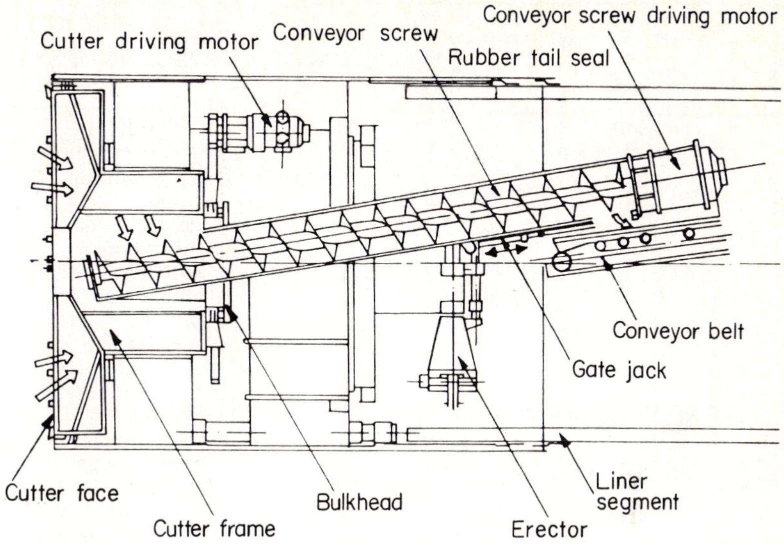

FIG. 10.9 The earth-pressure balance shield or EPB shield method of tunneling in soil. [From NCE (1979).[14]]

tation of the screw conveyor with the shield advancement speed. Water is added at the cutting face to turn the soil into pliable muck, while the cutting head rotates alternately in two directions, usually reversing after each shove to maintain the alignment against torque deflection. At the end of each shove the muck is removed in cars and a hydraulic erector mounted at the machine bulkhead is activated to install segments of liner plate, fitted with neoprene gaskets for waterproofing.

In its first application in the United States, a Mitsubishi-built machine averaged 30 ft per day on a 12-ft-diameter, 3057-ft-long sewage tunnel constructed in San Francisco Bay muds [ENR (1981)[15]; Clough et al. (1983)[16]]. The path chosen along the city waterfront required cutting through about 90 timber piles, which the machine managed without very much difficulty. Surface subsidence was on the order of 1½ in. A major cost advantage over the slurry-shield method is the elimination of extensive aboveground installations for separating out and recycling slurry (Fig. 10.8*a*).

Cut and Cover Tunneling

Applications

Cut and cover tunneling is generally suitable where tunnels are located near the surface and there is danger of blow-out, as for river crossings, in built-up urban areas or where ground conditions are poor and the danger of surface subsidence and detrimental settlement of nearby structures is severe.

River Crossings (Sunken Tube Method)

For river crossings, a broad trench is excavated, a stone foundation bed is prepared, and precast concrete units are floated to the site and sunk onto the prepared bed. The excavation is tightly backfilled and a soil cover is placed over the tunnel to provide for uplift protection [Schmidt and Grantz (1979)[17]].

Urban Areas

In urban areas, construction proceeds as for a normal open excavation (Chap. 9) and the slurry-wall method is used frequently. The excavation sides are installed in segments, material is removed from within the walls, the floors are poured, followed by the roof, and finally a backfill is placed over the completed tunnel.

Summary

A summary of soil tunneling methods, suitable ground conditions, and associated problems is given in Table 10.9.

TABLE 10.9
SOIL TUNNELING METHODS SUMMARIZED

Method	Description	Most suitable conditions	Comments
Liner plate, ribs, and lagging	Excavate face either full-face or in drifts. Install light liner plates and steel-rib supports. Place forms and concrete lining.	Firm ground or squeezing ground with movements sufficiently slow to permit bracing installation.	Often used to construct horseshoe-shaped tunnels. In clays wall footing bearing capacity, settlement, and bottom heave require analysis. Capacity controls applicability of method. In squeezing ground may result in surface subsidence. Use air pressure below GWL.
Shield tunneling	Outer cylindrical shield shell jacked ahead of excavation. Soil is excavated from partially open face under air pressure, or displaced by closed face when very soft. Jack pressure is relieved, jacks withdrawn, and steel segments placed within cylinder tailpiece which has annular space. Segments are exposed to soil when shell jacked ahead. Digger shield has a small backhoe mounted on the shield framework.	Soft to medium clays, organic soils, and soft silts. Stiff clays. Subaqueous tunnels in sands.	Downward tilting may occur in plastic clays; relieved by air pressure. In soft river silts compressed air controls inflow of soil and water. Blowout danger exists where soil cover is thin. Displacement method may have detrimental effects on adjacent structures. In sands careful removal of materials from small openings in face can be costly and dangerous. In soft cohesive soils, usually results in soil disturbance and surface subsidence.
Machine excavation (mole, TBM)	Rotating or oscillating cutter chisels or shaves soil from a circular face as machine advances. Can be mounted inside a shield. Metal or concrete segments are wedged or jacked into soil contact providing permanent lining.	Firm ground or slowly raveling ground with shield.	Rapid advance rate may result in machine burial if flowing, running, or rapidly raveling ground encountered. Ground improvement is necessary. Different cutting heads are used for different soils; therefore application may be inefficient in stratified soils, rapidly changing conditions or where boulders are present. In stiff to hard clays, surface subsidence negligible.
Slurry-faced TBM	Large steel cylinder with open cutting head sits behind a slurry which supports tunnel face. Contains a watertight tail seal in form of rubber gasket around inner perimeter, and a slurry pressure regulator. Tunnel lining installed in concrete segments with a system permitting grouting to be kept within a few meters of the face.	Firm ground. Raveling ground. Saturated granular and soft clay soils.	Very effective in poor ground conditions. Minimizes surface subsidence by maintaining balance of earth pressures. Slurries are formed with bentonite mixtures or with clay-soil muck.

TABLE 10.9 (continued)

Method	Description	Most suitable conditions	Comments
Earth pressure balance shield (EBP shield)	TBM utilizes excavated soil in contact with the tunnel face at all times and under pressure to prevent soil movements. Linings installed in concrete segments.	Soils above GWL. Soils with relatively low permeabilities below GWL.	No slurry or additives are mixed with the excavated soil. A recent method undergoing development to increase applicability to various soil conditions.
Sunken tube for subaqueous	For river crossings, open excavation made, foundation mat placed, precast sections floated to site and sunk in place, and excavation backfilled and fill placed over tunnel.	River crossings where excavation is relatively shallow.	More economical and safer than tunneling for shallow river crossings.
Cut and cover on land	In urban areas construction proceeds as for a normal open excavation usually employing the slurry-wall method. The tunnel is formed and backfilled.	Urban areas with many underground utilities, poor ground conditions, and surface subsidence problems.	More efficient than tunneling in urban areas when depths are relatively shallow.

10.2.4 GROUND SUBSIDENCE

Magnitude and Distribution

Deformations

Normally the deformations around a tunnel result in trough-like depressions more or less symmetrical around the vertical axis of the tunnel and paralleling its alignment. Field observations [Peck (1969)[1]] and model tests [Atkinson and Potts (1977)[18]] indicate that the trough width and depth are a function of soil type, tunnel diameter, depth below the surface, and construction method. As related to deflections of surface structures, the problem lies in differential subsidence. On the assumption that proper construction methods and procedures will be used, the vertical and horizontal extent of surface subsidence can be predicted with reasonable accuracy. Poor construction procedures can result in disaster, even in moderately stable ground.

Table 10.10 lists a number of case histories for which normal and maximum settlement, construction method, and tunnel depths and diameters for five soil groupings are compared. Normal settlement occurs over a tunnel from routine operations. Settlements that occur from nonroutine events such as runs, blowouts, and collapses are excluded. The number of case histories is small, but the data provide an idea of settlement that might be anticipated. Settlements resulting from dewatering also are excluded.

Trough Characteristics

Schmidt observed, as reported by Peck (1969),[1] that the settlement curves resemble an error function or Gaussian possibility curve (Fig. 10.10) described by the equation

$$s = S_{max}[\exp - (x^2/2i^2)] \qquad (10.14)$$

where S_{max} = maximum settlement over the crown
x = horizontal distance
i = horizontal distance to inflection point P

TABLE 10.10
DOCUMENTED SETTLEMENT OVER TUNNELS: NORMAL CONDITIONS*

Soil conditions	Construction method	Settlement, cm: First tunnel	Settlement, cm: Second tunnel	Depth, m	Diameter, m	Distance between tunnels, m
1. Dense sands, above water table	Hand-mined, no shield	8.4	14.7 (total)	10.3–13.3	5.3	7.9
2. Dense sand, below water table	Hand-mined, with shield, no air	1.8	6.7	10.6	5.3	6.7
	Digger shield, with air, dewatered	0.9 (normal) 2.1 (max)	1.2 (normal) 3.0 (max)	10.9	5.3	10.0
3. Cohesive granular						
a Cohesive fine sand, dense, dewatered	Digger shield with air	0.3 (normal) 0.9 (max)	0.3	10.9	5.3	10.0
b. Residual, saprolite	Hand-mined horseshoe	6.1	—	30.3	10.0	—
4. Nonswelling, stiff to hard clay	Shield, hand-mined	0.9 (max)	—	13.0	10.9	—
	Hand-mined horseshoe	3.7 (normal) 8.2 (max)	—	23.3	7.3	—
5. Soft to stiff saturated clay						
a. Sensitive	Shield with air hand-mined	73.2 (max)	—	17.0–21.8	7.9	—
b. Moderately sensitive	Shield, with air breasted face	6.1 (max) 4.6 (normal)	—	17.9	5.5	—
c. Varved clay	Hand-mined, liner plates, with air	1.8 (normal) 3.0 (max)	6.1 (normal) 10.7 (max)	11.8	6.1	8.5
d. Sensitive	Digger shield with air	0.6 (max)	—	18.2	3.0	—

*After Peck (1969).[1]

The properties of the curve are related to tunnel depth z and diameter $2r$. Peck presented a relationship between $z/2r$ and i/r for various soil types. Cording and Hansmire (1975)[19] suggest that this relationship can be expressed in terms of angle β.

The relationship between settlement trough width $2w$ and $z/2r$ for various soil types, with boundaries given by β, is given in Fig. 10.11, as is the relationship between $z/2r$ and i/r. The curves are used to estimate the differential deflection to be anticipated as a function of distance from the tunnel centerline, as well as the total trough width. It may be noted that the trough for soft clays is several times the width of that for hard clays or dry sands.

Effects of a *second tunnel* driven adjacent to the first are substantially greater than those resulting from the first tunnel because of the disturbance caused by the first. For estimating the effects of the second tunnel use Fig. 10.11, but substitute i/r' for i/r where $r' = r + d/2$ and d = distance between the two tunnels.

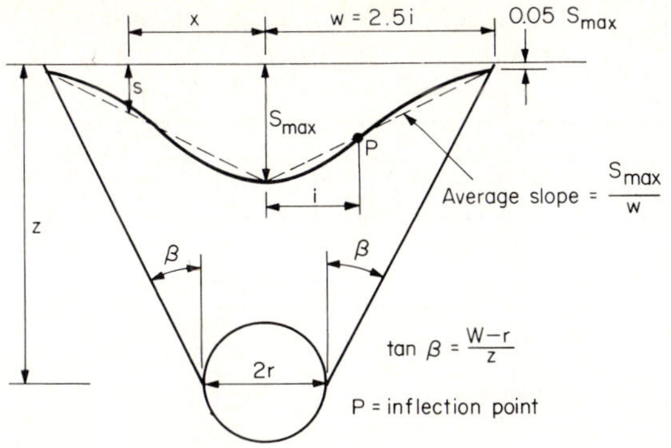

FIG. 10.10 Settlement trough characteristics. [After Peck (1969),[1] Cording and Hansmire (1975),[19] Ghaboussi et al. (1978).[20]]

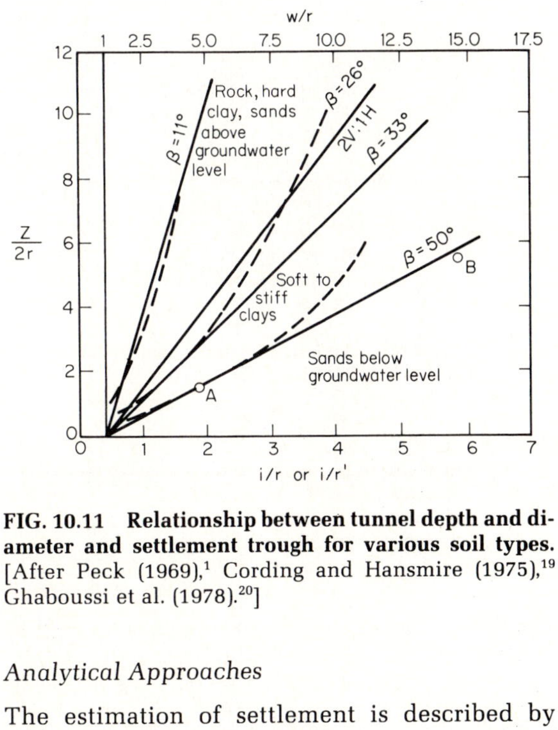

FIG. 10.11 Relationship between tunnel depth and diameter and settlement trough for various soil types. [After Peck (1969),[1] Cording and Hansmire (1975),[19] Ghaboussi et al. (1978).[20]]

Analytical Approaches

The estimation of settlement is described by Cording et al. (1975).[7] In Fig. 10.10 the point of inflection P, or settlement ordinate at distance i, according to the properties of the probability curve, is equal to $0.61S_{max}$. The area under the curve, or settlement volume, is

$$V_s = 2.5iS_{max} \tag{10.15}$$

Lost ground volume and stability problems in the tunnel are closely related and are a function of the soil type and strength, and construction procedures. In general, it has been found that where the tunnel depth–shear strength ratio $(p_z - p_a)/s_u < 4$ with good construction techniques, the surface volume may be less than 1% of the amount excavated. For $(p_z - p_a)/s_u > 4$, in the order of 6 to 7, surface settlement volumes of 2 to 5% might be expected. In sands, with good construction techniques, a tunnel above GWL may incur a maximum loss of ground of 0.5 to 1.0% of the excavated volume; but if construction difficulties are encountered or poor procedures employed, settlement volumes may reach 2.5 to 5% of excavated volume for a single tunnel.

Cording et al. (1975)[7] suggest the following procedure for estimating settlement at the ground surface:

1. Estimate volume of lost ground V_s as a percent of excavated volume as described above. (In sands use $V_s = 1$, assuming no unusual problems, or increase the anticipated volume of lost ground if tunneling through loose soils, a poor construction technique is anticipated, or dewatering is incomplete.)
2. Estimate width parameter i for the settlement trough from Fig. 10.11.
3. Calculate $S_{max} = V_s/2.5i$ (Eq. 10.15).
4. For a second tunnel, assume the additional volume of lost ground is the same as for the first tunnel, center the total settlement trough over the midpoint between the tunnels, and using the tunnel radius plus half the distance between the tunnel centerlines (r′), estimate i and S_{max} as in Steps 2 and 3.

Analysis of surface subsidence employing the FEM is described by Ghaboussi et al. (1978).[20] The value of analytical solutions, in general, is limited by the necessary assumptions concerning soil behavior and tunnel deformations. The analyses do not account sufficiently for construction methods and procedures and the heading advance rate.

Settlement estimated with the foregoing procedure can be compared with the permissible deflections in structures required to avoid excessive distortions and damage as given in Fig. 6.9 and Art. 6.1.2, and a rational decision can be made as to the necessity for underpinning or ground treatment, particularly where a structure's importance and sensitivity to distortion and the given ground conditions are marginal. Some form of treatment is always considered for critical situations, and monitoring of deformations and deflections should be performed.

Ground Treatments and Subsidence Control

Tunneling Operations

Good tunneling operations are required in any ground conditions to minimize the effects of ground subsidence. Subsidence potential and the treatment to reduce or eliminate it during tun-

TABLE 10.11
GROUND CONDITIONS AND SURFACE SUBSIDENCE*

Soil conditions	Subsidence potential	Treatment
Cohesionless silt, sand and gravel (potentially running or flowing ground)	In dense soils, negligible with proper construction techniques. Poor construction, particularly in loose soils, results in dry soil runs or saturated soil flows causing cavities and large surface subsidence (relatively narrow trough), and possibly sinkholes.	Above GWL: Rapid installation of liner plates; grout loose soils. Below GWL: Dewatering, perhaps with grouting; or use TBM with slurry face. Air pressure reduces seepage pressures but escaping air may dry permeable soils causing running. Grout always with dewatering to reduce seepage forces through small openings in grout curtain.
Cohesive granular (clayey sand, silt, loess, most residuals; often with relict structure) (potentially raveling to firm ground)	Small to negligible with proper construction techniques. Poor techniques permit overbreak, raveling, and spalling possibly resulting in surface sinkholes.	Raveling soils: Advance face in small-size drifts rather than full face. Below GWL: Predrain, especially stratified soils with permeable layers. Air pressure aids to reduce hydraulic gradient. Shield tunnels: Carefully backfill space between shield and lining to prevent future collapse.
Stiff to hard nonswelling clays (firm to squeezing ground)	Negligible with shield or mole. Raveling occurs with high seepage pressures and secondary structure. Liner-plate method can result in ground loss as clay squeezes inward before permanent lining installed. Subsidence negligible if $p_z/s_u > 4$ with good construction techniques.	Use shield or mole. With liner plates, air pressure p_a reduces ground loss to an extent corresponding to a decrease in overburden pressure from p_z to $p_z - p_a$.
Soft to stiff saturated clays ($s_u \approx 0.1$ to 1.0 tsf) (squeezing ground)	Disturbed zone about tunnel results in subsidence several times that expected in cohesive granular soils. (Results in squeezing into openings as tunneling proceeds rather than raveling, running, or flowing.)	Quickly backfill space between shield and lining (frictional forces between shield skin and soil always tend to pull clay into annular space, however). Air pressure tends to reduce amount and rate of inflow. TBM with slurry face greatly reduces subsidence potential.

*From Peck (1969).[1]

neling for various soil groups and ground conditions are summarized on Table 10.11.

Where structures overlie the alignment, and particularly where they are important or sensitive to settlement, the traditional approaches have been to use air pressure during tunneling and underpinning for the structures. Where prevention of differential settlement is critical and the supporting soils are granular and relatively pervious, both penetration grouting (*see Art. 10.2.3*) and compaction grouting have been demonstrated to be effective and generally more economical than underpinning or compressed air.

Penetration Grouting

A successful application of penetration grouting at a location where four tunnels of the Washington Metro Transportation System were to pass beneath a freeway and a large overpass structure is cited by Clough (1981).[10] The crowns of the upper two tunnels were in sands and sands and gravels and passed within 2 m of the bridge foundation. A silica grout was injected into the granular stratum forming a strengthened, impermeabilized zone between the roadway, the bridge foundations, and the tunnels. At the bridge location the four tunnels were completed with a maximum of 20 mm of settlement and no sand runs, whereas settlement in untreated regions ranged from 40 to 120 mm.

As used for construction of many European subways, the grouted zone is in the order of 2.3 to 3 m thick, extending over the proposed tunnel and down to about invert elevation where the tunnel is entirely in granular soil, or to a clay stratum if the tunnel passes partially through sand underlain by clay. A procedure for selecting the size, shape, and strength of the zone to be stabilized is presented by Tan and Clough (1980).[21]

Compaction Grouting

Originally developed in the United States for leveling slabs and improving the support of foundations undergoing settlement (*see Art. 5.4.1*), compaction grouting shows promise for replacing underpinning where tunnels will pass beneath relatively brittle, lightly loaded structures. Clough (1981)[10] cites a case where a portion of the Baltimore Region Rapid Transit System was planned to pass at a crown depth of 15 m below two- and three-story dwellings and stores with brick bearing walls. Crowns were about 1.5 to 3 m below GWL in dense sands with silt and clay lenses.

The grout, consisting of a highly viscous sand-mortar mix, was planned to be injected under high pressure immediately above the proposed tunnel locations in the zone which loosens as the ground displaces into the tail void forming behind a shield. Grout pipes were installed prior to construction but grouting was planned to be used only if surface settlement approached 13 mm. In actuality, grouting of each pipe was started once the tail of the shield passed beneath the grout pipe and significant subsurface settlement was measured. Use of the grouting procedure held settlement to less than 13 mm.

Monitoring

Deformations of the soil surrounding the tunnel, of the surface, and in overlying structures should always be monitored, beginning before and continuing after construction as described in Art. 10.1.3. Permissible deflections in structures to avoid excessive distortion and damage are given in Fig. 6.9 and Art. 6.1.2.

10.3 UNDERGROUND OPENINGS IN ROCK MASSES

10.3.1 GENERAL

Installation

Underground openings in rock can take the form of tunnels, caverns or chambers, and horizontal slots. Excavation is made by drilling, blasting, and mucking; by roadheaders; or by TBMs. Progress rates depend on rock hardness, rock quality in terms of discontinuities, and the presence of hazardous conditions.

Support Requirements

Opening support requirements depend on rock quality. Hard, massive, or competent rock usually is self-supporting. In the United States, lesser quality rock traditionally has been supported temporarily with rock bolts or with steel sets and lagging, although in recent years combinations of rock bolts, wire mesh, and shotcrete, as used in many other countries, have been finding increasing usage. This latter system, an inherent part of the new Austrian tunneling method (NATM), often also serves as permanent lining, as an alternative to reinforced concrete. In mines, pillars are left for support and roof support may be improved with rock bolts.

Support methods as related to rock quality are summarized on Table 10.2, and general rock tunnel nomenclature is given in Fig. 10.12.

Performance

Performance depends on the ability of lined or unlined openings to resist deformations under external pressures in a manner similar to soil tunnels. *Surface heave*, however, may result where pressure tunnels are driven close to the surface and the pressure imbalance causes uplifting of the overburden. Leaks from such tunnels where located adjacent to slopes may also induce landslides.

Subsidence normally is not a problem associated with underground excavation in rock except for some soft rocks, for excavations in rock overlain by soil formations where extensive dewatering is induced, or for mining operations, particularly coal and other soft rock mining.

Ground Conditions

Normal conditions of concern are rock hardness, which impacts on excavation rates and costs,

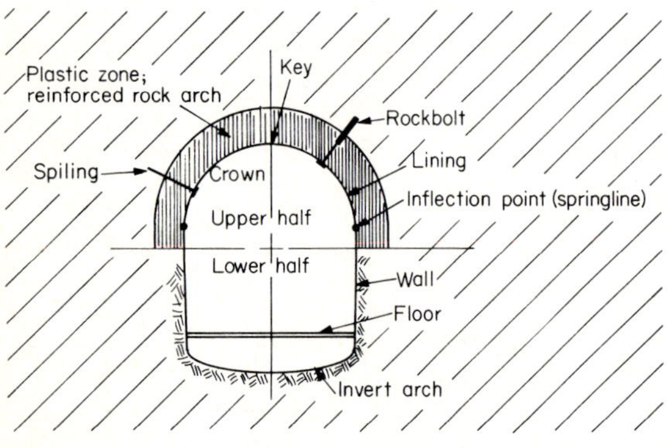

FIG. 10.12 Rock tunnel nomenclature.

and jointing and other defects, which impact on support pressures and methods.

Hazardous ground can include rock with high residual stresses which cause opening closure, "popping," and even "rock bursts"; running or flowing ground wherein tremendous quantities of slurry-like material suddenly flow into the excavation as the result of high seepage forces in a crushed-rock zone; and squeezing and swelling ground, which causes heaving of walls and the deflection and crushing of roof supports. These conditions endanger workers and equipment, cause substantial delays and extra costs, and require extensive treatment for prevention or correction.

Various treatments are used for hazardous ground conditions such as rock bolts to contain high residual stresses, dewatering and grouting in running and flowing ground, and slip-joint supports in squeezing and swelling ground.

10.3.2 ROCK-MASS CONDITIONS

Conditions requiring assessment when considering underground excavations in rock include rock quality (competent versus incompetent rock), hazardous ground (high residual stresses; running, squeezing, or swelling ground), water (pressures, quantities, and corrosive properties), gases (dangers to workers), and temperature (worker comfort).

Rock Quality

Competent Rock

Competent rock is hard, massive, intact rock with few defects that essentially has a permanent stand-up time. Joints are tight and generally spaced wider than 1 m apart. Associated rock types include most igneous rocks forming batholiths or laccoliths, massive metamorphics such as quartzite and marble, and thickly bedded sedimentaries including sandstone, limestone, and dolomite.

Excavation may be difficult due to hardness, and careful blasting techniques are required to maintain rock quality. In hard, brittle rock containing high residual stresses, "popping" or even "rock bursts" may occur, as discussed under *Hazardous Ground*.

Incompetent Rock

Incompetent rock contains defects or discontinuities such as joints, foliations, faults, bedding planes, or cleavage which provide weakness planes in the mass, dividing it into blocks. Soft materials such as shales, schists, or decomposed rock also are considered as incompetent.

Stand-up time and support loads *(see Art. 10.3.3)* and *overbreak*, the amount of excavation beyond the payline, will depend on defect type, intensity, and orientation with the opening, as well as the strength of intact blocks. The effect on overbreak of the orientation of the major joint set for various conditions is illustrated in Fig. 10.13, and the effect of joint strike and dip orientation is given in Table 10.12. (Note that engineering geologists often refer to all defects as joints, except for faults where they present a large zone of low-quality materials.) In all cases the amount of overbreak is related to the size of the opening and the joint spacing and orientation with the opening.

In horizontally bedded rock (Fig. 10.13*a* and *b*), blocks break loose along joints and the opening proceeds upward until it is bridged by a slab. In Fig. 10.13*c* the tunnel is parallel to subparallel to a predominant steep to vertical weakness; wall slabbing and overbreak occur, and large wedges

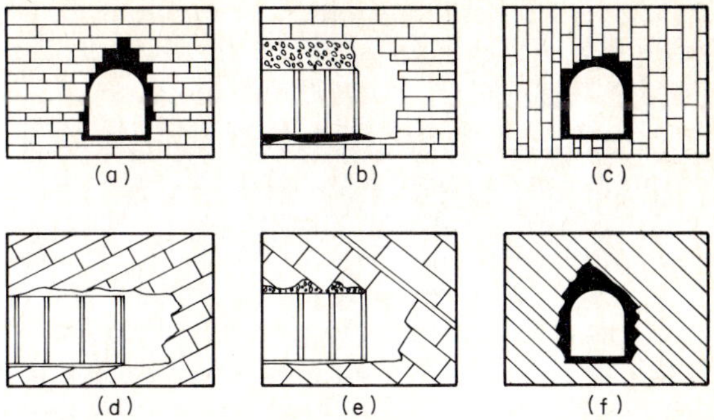

FIG. 10.13 Effect of major joint set orientation on overbreak: (*a*) and (*b*) horizontal strata left unsupported; (*c*) vertical strata; (*d*) strata dipping into tunnel overbreaks mainly in face; (*e*) large overbreak in roof caused by weak seam; and (*f*) dipping strata overbreak to form peak.

may loosen in the crown. In Fig. 10.13*d* the weakness planes dip into the face and most of the overbreak occurs at the face, leaving it irregular. Subhorizontal weaknesses dip in the tunnel direction in Fig. 10.13*e*, and the roof tends to break to the weakness which forms a peak where the strata dip into the alignment (Fig. 10.13*f*). In jointed rock the direction effect is generally overbreak, but if the weakness is a small fault, foliation shears or similar zones, or poor

TABLE 10.12
EFFECT OF JOINT STRIKE AND DIP ORIENTATIONS ON TUNNELING*

Strike perpendicular to tunnel axis				**Strike parallel to tunnel axis**	
Drive with dip		**Drive against dip**			
Dip = 45–90°	**Dip = 20–45°**	**Dip = 45–90°**	**Dip = 20–45°**	**Dip = 45–90°**	**Dip = 20–45°**
Very favorable	Favorable	Fair	Unfavorable	Very unfavorable	Fair
Dip 0°–20°: Unfavorable, irrespective of strike.					

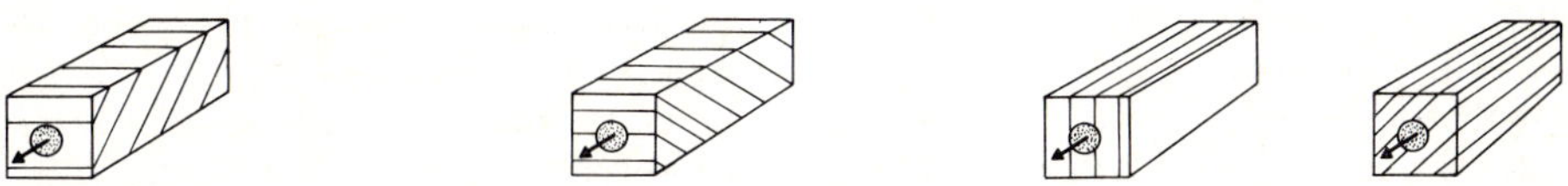

*After Wickham et al. (1972).[22]

rock intercepting the tunnel, the problem can be much more severe, leading to large fallouts and block movements.

Terzaghi (1946)[23] placed rock into several classes of quality for which he derived estimated support loads which practitioners use for estimating support requirements, as given in Table 10.2. *Intact rock* contains neither joints nor hair cracks and breaks only across sound rock. Blasting may result in spalling from the roof several hours or days later, and high residual stresses may result in "popping" of slabs from sides or roof. *Stratified rock* consists of individual strata separated by very weak planes. The strata may or may not contain transverse joints, and spalling is common. *Moderately jointed rock* contains joints and hairline cracks, but the blocks have adequate interlock to enable vertical walls to stand without support. Spalling and popping may occur. *Blocky and seamy rock* consists of chemically intact or almost intact fragments entirely separated from each other and imperfectly interlocked, and vertical walls may require support. *Crushed* but chemically intact rock has the character of a crusher run, and where the fragments are small it exhibits the properties of a water-bearing sand when below GWL. *Squeezing rock* slowly advances into the tunnel without perceptible volume increase, and *swelling rock* advances into the tunnel chiefly due to expansion (see *Hazardous Ground* below).

Hazardous Ground

High Residual Stresses

Rock stresses, often many times the overburden stress, remain from folding deformation, metamorphism, slow cooling at great depths, and the removal of overburden [for discussion, see Art. 6.6 of Hunt (1984)[5]]. High horizontal stresses relative to vertical stresses may also be encountered around valley bottoms due to stress concentrations produced by erosion of the valley notch.

High residual stresses cause slabbing or spalling of walls and roof, closure of an opening, popping and rock bursts, the latter two conditions resulting in slabs separating suddenly and explosively. *Popping* refers to thin slabs of rock detaching suddenly after excavation, and normally occurs only in hard, brittle, intact rock in a state of intense elastic deformation. Preceded by an inward bending of slabs, the phenomenon commonly involves only rock breakage in the immediate vicinity of the walls and is contained by rock bolts, ribs, or other support. If not contained, a progressive general slabbing may occur extending to the roof.

Rock bursts can be highly explosive, involving relatively large volumes. Some conditions where they might occur include:

- Massive granite or gneiss, with few joints, at considerable depths
- Geologically complex conditions characterized by rock fracture and anisotropy such as exist where highly competent, brittle rocks are interbedded with less competent, somewhat plastic rocks
- Stress concentrations associated with or near large faults

High residual stresses are difficult to detect during investigation, since the simplest methods employing borehole techniques [Art. 4.4.4 of Hunt (1984)[5]] are limited to relatively short boreholes. Hydraulic fracturing can be used to obtain approximate estimates of in situ stresses in deep boreholes. The effect of high residual stresses as evidenced by even minor slabbing and fracturing is best observed during construction of a shaft or pilot tunnel. High horizontal stresses tend to create new fractures and slabs primarily in the crown and floor of the excavation, whereas if vertical stresses are higher than horizontal stresses, slabbiness is most likely to be first exhibited on the side walls [Cording et al. (1975)[7]]. Measurements are then made in the excavation walls using borehole stress relief techniques. The construction of a shaft in gneiss with high residual stresses is described in Art. 9.3.5.

Running or Flowing Ground

Crushed rock below GWL can act as running or flowing ground. It is commonly associated with faults and other shear zones, particularly with the highly fractured rock of the foot wall or hanging wall of a fault where the water level in the downthrown block through which the tunnel is advancing is much lower than the water level in the adjacent block, as illustrated in Fig.

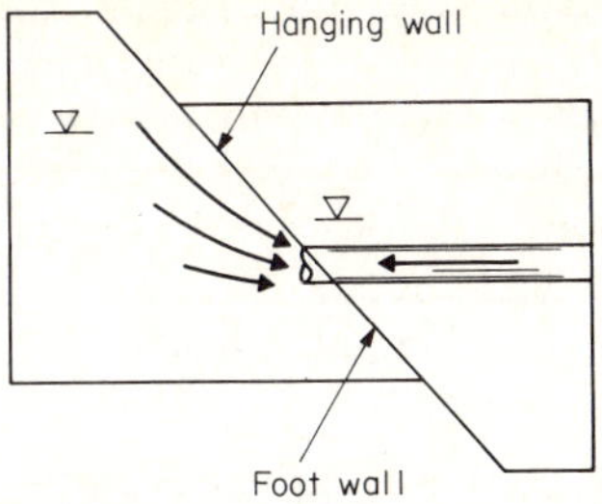

FIG. 10.14 Differential water pressures along the sides of a normal or dip-slip fault can be extremely hazardous to tunnels approaching the hanging wall side.

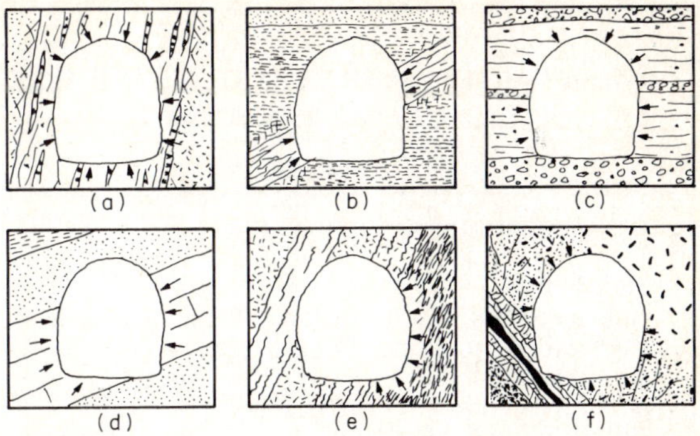

FIG. 10.15 Some conditions promoting ground squeezing in tunnels. Arrows indicate direction of squeezing ground into tunnel opening. (*a*) Wide shear zone in granite; (*b*) fault in brittle shale; (*c*) layer of altered volcanic ash (bentonite); (*d*) soft, poorly compacted mudstone layer; (*e*) closely sheared, altered biotite gneiss; (*f*) extensively altered fine-grained porphyry. [From Walhstrom (1973).[25]]

10.14. When the foot wall is encountered in excavation, the saturated debris, lacking in cohesive binder, flows as a slurry into the opening, often under high initial pressures and quantities, filling the opening.

Squeezing Ground

In squeezing ground, relatively weak, clay-rich, plastic material with high saturation percentages moves into an opening under pressure from surrounding rocks immediately upon exposure. Pressures are often high enough to crush and deform tunnel linings if support is not adequate.

The squeezing rate depends on material stiffness. Under the lower confining pressures existing at shallow depths, squeezing can be anticipated from fault gouge, closely sheared and altered micaceous rocks, and soft shales. Shales at moderate to great depths generally have adequate strength from precompression to resist squeezing. At substantial depths, squeezing is most likely to be associated with fault gouge and closely sheared and altered rocks containing clays and micaceous materials. Some conditions for squeezing ground are given in Fig. 10.15.

In Fig. 10.15*a*, the tunnel is in a wide shear zone in granite which has been thoroughly crushed and consists of a gouge matrix, essentially impermeable but damp, containing lenses of angular rock fragments. The flat fault in brittle shale in Fig. 10.15*b* has produced a gouge containing a braided pattern of shears which results in localized swelling at the fault exposure. Squeezing in Fig. 10.15*c* is caused by overburden pressures on the layer of altered volcanic ash containing montmorillonite, and in Fig. 10.15*d* by overburden pressures acting on the soft mudstone layer. Squeezing occurs in Fig. 10.15*e* in a closely sheared and altered biotite gneiss containing abundant chlorite and clay minerals, and in Fig. 10.15*f* in an extensively altered fine-grained porphyry adjacent to a vein.

Semple et al. (1973)[24] present analytical procedures for evaluating support pressures that develop with time in squeezing ground.

Swelling Ground

Swelling ground results from material expanding upon contact with moisture and either enters the opening or causes high pressures on linings. The major difference between swelling and squeezing ground, which does not involve volume change, is the time delay involved for swelling to occur. Caused by the moisture that is available in normally wet tunnel conditions or groundwater, it may continue for months or years until equilibrium conditions are reached. Invert heave up to 25 cm/yr has been observed [Einstein and Bischoff (1975)[26]].

Swelling is most pronounced in clay gouge in fault zones and clay-rich rocks including shales, altered pyroclastics, chemically weathered fer-

TABLE 10.13
SWELLING PROPERTIES OF GOUGE AND ALTERED ROCK, ROBERTS TUNNEL, COLORADO*

Description	Final heave, % vol. change	Final swell pressure, psi
Gouge consisting predominantly of crushed wall rock with minor to moderate clay-mineral alteration.	0.5–10.0	300–4250
Gouge containing abundant clay minerals, with predominant montmorillonite, and smaller amounts of kaolinite and sericite. Clay minerals are of hydrothermal origin.	10.25–18.5	4,500–11,800
Hydrothermally altered wall rock with minor to moderate amounts of clay minerals. Some samples show extensive hydrothermal silification.	9.0–10.0	3,900–4,250
Hydrothermally altered wall rock with abundant clay minerals, predominantly montmorillonite, but with some kaolinite, sericite, and chlorite.	13.75–25.0	8,000–12,000

*From Walhstrom (1973).[25]

romagnesian rocks, and hydrothermally altered rocks, particularly when the clay mineral is montmorillonite. Anhydrite, found associated with reddish shales, siltstones, halite, and at times, limestone, is converted to gypsum when exposed to water and can generate considerable swelling pressures. Examples of swelling pressures and the percent volume change that can occur are given on Table 10.13.

Water

Pressures and Quantities

The sudden inflow of huge quantities of water under high pressures is the most serious problem to be anticipated in underground excavations in rock. Water pressure equal to full hydrostatic pressure can be encountered which can cause bursting of the roof, floor or heading, even in hard rock. High pressures are the cause of running ground and can be part of the driving force behind squeezing ground.

Sharp et al. (1972)[27] cite a case where tunneling for the San Jacinto tunnel in California encountered maximum flows of 16,000 gal/min (60,500 L/min) with pressures as high as 600 psi (4.1 MPa) associated with a fault zone filled with clay gouge. On the hanging side of the fault, piezometric pressures were relatively low and controllable by draining, whereas on the foot-wall side the piezometric level was high and perched as shown in Fig. 10.14. When the tunnel penetrated the clay gouge, 3000 m^3 of crushed material and water were released suddenly into the tunnel.

Quantities also can be very high in highly porous rock such as vesiculated lavas or cavernous limestone, but pressures are not necessarily high.

Corrosive Water

Deleterious to concrete linings, corrosive water may result from water flowing over anhydrite, gypsum, or other sulfates and becoming weak solutions containing sulfuric salts [see also Art. 8.3.4 of Hunt (1984)[5]].

Gases

Gases encountered in deep closed excavations can be both lethal and explosive, and the air should be monitored continuously during construction to detect undesirable gases and to determine the necessary ventilation provided for their removal.

Dead Air

Dead air has the same composition as surface air except that the oxygen has been removed by oxidation from reaction with iron compounds and sulfides, and it is deleterious to workers.

Carbon Dioxide

With a density of about 1.5 times that of air, carbon dioxide tends to collect in the lowest locations. In amounts in excess of 15 to 25% of the tunnel atmosphere it is lethal, and combined with water it produces carbonic acid, mildly corrosive to metals and concrete lining. It results from the oxidation of carbonaceous materials and is a common abundant constituent of warm waters of deep origin within the earth. Large amounts can be expected in regions of recent volcanic activity.

Methane (Marsh Gas)

A simple, nontoxic hydrocarbon, with a density about one-half that of air, marsh gas forms a highly explosive mixture in concentrations above about 5%. It has caused explosions and fires in excavations in shales and in coal-mining operations where it had formed by the slow decay of bituminous materials of organic origin. When confined by impervious barriers, the gas can move great distances through joints and other openings to accumulate in unexpected locations.

Sulfur Dioxide and Hydrogen Sulfide

Both sulfur dioxide and hydrogen sulfide have the strong, distinctive odor of rotten eggs. Sulfur dioxide results from oxidation of sulfur or sulfides in sedimentary rocks or in hydrothermal mineral deposits containing sulfides. Hydrogen sulfide is a by-product of sulfur-bearing organic materials and sulfides of organic origin; it is a constituent of many thermal waters of deep-seated, presumably igneous origin. Concentrations of less than 1% are lethal, but both gases are easily detectable because of their strong odors.

Temperature

Temperature becomes a concern generally when excavations are greater than about 150 m below the surface since high temperatures can affect laborers. Even in shallow openings, high temperatures can result from poor ventilation and high humidity when working under air pressure. Normal groundwater temperature at shallow depths is in the order of 12°C (54°F) in temperate to cool climates.

The *geothermal step* or *gradient* at a particular location is given as the change of temperature of 1°C for some depth in meters. Gradients for some European tunnels ranging in depth from 1600 to 2100 m varied from 36 to 58 m respectively [Krynine and Judd (1957)[28]]. In the Simplon tunnel, water with a temperature of 56°C (132°F) was encountered at a depth of 2100 m. Walhstrom (1973)[25] gives some general values of geothermal gradients in terms of increase in temperature per degree Celsius as a function of depth as follows: tectonically active areas, 60 to 80 m; volcanically active areas with igneous activity, 10 to 15 m; and worldwide average, 30 to 35 m.

10.3.3 DESIGN FACTORS

Stress Around an Opening

Support of an underground excavation in rock generally is considered a two-step problem: temporary support during excavation and permanent support. Temporary support, however, often is incorporated into permanent support. Design requires an assessment of rock pressures which may be addressed analytically or empirically. The problem of stress distribution around a hole, shaft, or slot formed in an anistropic, nonhomogeneous, discontinuous medium, as is common in rock masses, can be extremely complex, and in most instances support selection is based on experience and empirical relationships.

Theoretical Considerations

Tunnel support loads depend generally on the stress state in the rock prior to excavation, such loads being considered in terms of a *vertical load* on some horizontal section equal to the weight of the overlying rock and a *horizontal pressure* acting at the elevation of the section which may range between wide limits. In most rock masses the horizontal pressure is considerably lower than the vertical load at a given point, but in masses containing residual stresses, the horizontal pressure can be many times the vertical load and can be close to the unconfined compressive strength of the rock at any depth.

Where excavating a tunnel in *intact rock*, theory shows that the effect on the state of stress decreases rapidly with increasing distance from the tunnel walls. At the walls the normal radial stress is zero and the normal circumferential or tangential stress is roughly equal to twice the stress which acted at the same point prior to tunneling. With increasing distance from the walls the radial stress increases, the circumferential stress decreases, and at a distance equal to about the tunnel diameter, the state of stress in the rock is practically unaltered. At the walls the state of stress is similar to that in a rock specimen subject to an unconfined compression test, and the rock in the wall, if intact, does not fail until the circumferential stress equals the unconfined compressive strength. Where the rock is not acted on by horizontal forces the circumferential stress does not exceed about twice the overburden pressure [Terzaghi (1946)[23]].

If the rock modulus is sufficiently low, the tunnel will deform under the vertical load or compressive stress and begin to assume an elliptical shape. Stresses will be concentrated on the walls and blocks will be expelled into the cavity, thereby increasing the tunnel width and provoking a rise of the floor and downward deflection of the roof, which eventually folds into the cavity.

The magnitudes of the external forces acting about a cavity are determined by the conditions of primary stresses and the physical properties of the rock mass. The direction of the forces is determined by a process of readjustment during and after construction of the cavity and is very much related to rock heterogeneity and residual stresses. With the process of readjustment, lateral forces with much higher intensity than the vertical forces can develop. Such is the case, for example, with phyllites where the schistosity is vertical and parallel to the tunnel axis.

Stress Relief Effects

Figure 10.16*a* illustrates the mechanical theory of the process of relief of stresses about a lined cavity after suffering a radial deformation. If the lining has sufficient capacity p_r^0, displacement will not occur, nor will a plastic zone develop.

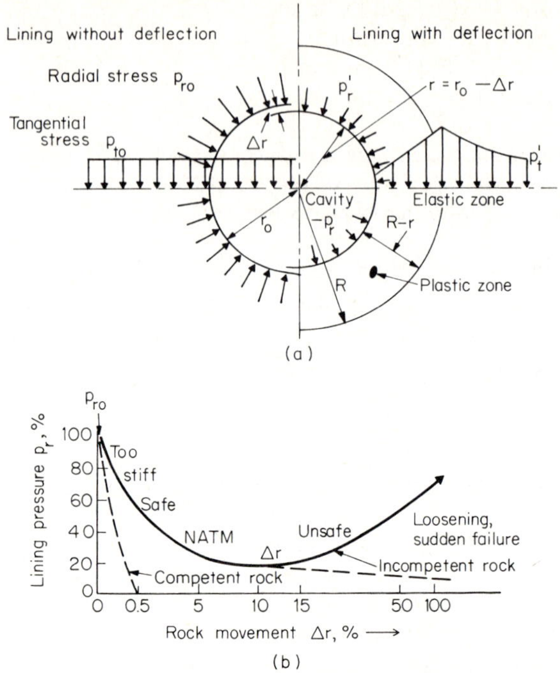

FIG. 10.16 Rock-mass stress distribution before and after tunneling, and rock movements, pressures, and stability: (*a*) stresses around tunnel lining with and without deflection and (*b*) lining pressure, rock movement, and stability. [From Rabcewicz (1975).[30]]

The ratio $R/r = 1$ and the tangential stress p_{to} correspond to the larger stress about the lining. If the lining is permitted to deform by some value Δr, the capacity required for support reduces by p_r^1 as does the corresponding tangential stress along the lining, which becomes p_t^1. A plastic zone develops with a radius R with tangential stresses a maximum at its border.

The relationship between lining pressure given as a percentage of p_r and rock movement Δr is given in Fig. 10.16*b*. It shows that the lower the lining pressure, the greater is the rock movement until a point where loosening and failure of the rock occur and lining pressures increase significantly. Linings that are too rigid and permit no movement will carry uneconomically high loads, whereas in poor rock, linings that are too flexible will permit such magnitudes of movement as to be unsafe. The objective of the NATM (*see Art. 10.3.4*) is to provide a temporary semirigid lining that is stressed by a moderate rock load just above its theoretical minimum value [Rabcewicz (1975)[29]]. In competent rock, stable equilibrium is reached in time under very small strains, often in the order of 10 to 25 mm, and a temporary lining is not required. In incompetent rock deformation continues, followed by eventual collapse unless support is provided.

Geometric Factors

Peripheral stress distribution is dependent upon the shape, but not on the size of the opening in the ideal case. The *radial distance* affected by the opening, however, is a function of the opening's size and shape as well as the magnitude and distribution of the primary stresses existing in the rock [Kruse (1970)[31]]. The stress concentration decreases rapidly with distance from the opening, and measurements with modern instrumentation have shown that in tunnels it generally dissipates within a meter to a few meters, depending on rock conditions [Walhstrom (1973)[25]].

Large stress concentrations will occur around openings with sharp corners, and such conditions should be avoided to maintain rock quality. In other than competent rock, it is becoming common practice to excavate an *invert arch* (Fig. 10.12) to avoid stress concentrations in horseshoe-shaped tunnels. For a hydrostatic stress field in viscous or plastic materials (squeezing ground) a circular section is preferred, in part because no tensile stresses develop.

Other Stress Factors

In addition to gravity-induced loads, which may change in time because of creep or swelling ground, there can also be residual or tectonic stresses which can be of large magnitude in elastic rocks at substantial depths, as already mentioned.

Roof Tension Arch or Bridge Action

Relaxation

Natural rock-mass equilibrium is disturbed by excavation; the mass over the opening loosens and compressive stresses decrease while stresses are redistributed to the wall rock as compressive stresses. The rock displacement as it deflects into the opening is often referred to as *relaxation*, which is dependent on the direction and magnitude of the existing primary rock stresses. Relaxations normal to the axis of the opening are considered for analysis to be analogous to the displacement of a thin plate subjected to biaxial stress.

Arching and Stand-Up Time

Moderate-sized openings, even in incompetent rock that is highly fractured or altered, often will tend to stabilize by *arching* within a few days of excavation, although supports are necessary to minimize caving. The loosened zone above the opening often is referred to as the *roof tension arch* and depending upon geologic conditions it may remain stable or it may collapse.

Blocky or seamy rock commonly does not react at once to the stress changes produced by tunnel excavation. Following blasting, which removes the natural roof support and may form a half-dome, some blocks drop from the roof, leaving a small gap. If left unsupported, more blocks drop out; this action can be aggravated by machinery vibrations or groundwater seepage, with the result that the gap widens and eventually the mass constituting the half-dome drops into the tunnel and a new half-dome is formed. The process continues until the tunnel section adjoining the working face is filled with rock debris. The rate at which this progressive deterioration or raveling occurs depends on block sizes and shapes, joint widths and fillings, and the distance between the new working face and the last support. The time element for collapse to occur is termed the *stand-up time* and, as described, is related to rock-mass quality and span distance. An approximation of stand-up time in terms of rock quality and span distance as measured by the RQD is given in Table 10.14.

Stand-up time and the thickness of the roof tension arch are factors to consider in determining support requirements. Bolts, for example, must penetrate through the roof tension zone to be adequate.

Estimating Support Pressures

Criteria of Terzaghi (1946)[23]

Support pressures are related directly to rock quality, and a number of practitioners have de-

TABLE 10.14
STAND-UP TIME VERSUS ROCK QUALITY (RDQ MEASUREMENT)*

Class	Description	RDQ	Average stand-up time	Caveability
I	Very good rock	100–90	10 yr for 5-m span	Very poor
II	Good rock	90–70	6 months for 4-m span	Occasional large blocks
III	Fair rock	70–50	1 week for 2-m span	Fair
IV	Poor rock	50–25	5 hr for 1.5-m span	Caves readily; good fragmentation
V	Very poor rock	< 25	10 min for 0.5-m span	Very good

*After Bieniawski (1974).[32]

NOTE: For complete table, see Table 5.25 in Hunt (1984).[5]

veloped various systems for describing rock quality to enable a qualitative assessment of rock conditions from which judgments are made regarding support pressures and systems requirements. Many tunnel geologists and engineers still use the empirical criteria developed by Terzaghi (1946)[23] from a study of data obtained from existing tunnels. His system, relating support pressures and rock load estimates to nine classes of rock conditions, is given in Table 10.15. The logic follows that given for sands in Art. 10.22. The total load P per foot of tunnel length is found from Eq. 10.2.

Rock loads estimated by Terzaghi were for steel supports and timber blocking, a system that permits some loosening of rock blocks about the tunnel, and the more loosening that occurs the greater are the loads that ultimately require support. Rock loads can be much smaller where other support systems such as rock bolts, shotcrete and bolts, shotcreted steel ribs, and cast-in-place linings are placed near the heading, since rock movements are limited and large loosened zones will not develop. Rock loads for these support systems are described below.

Criteria of Barton et al. (1974)[35]

Based on a study of 200 existing tunnels and caverns, mostly located in Scandinavia, Barton, Lien, and Lunde (1974)[35] developed an expression for rock-mass quality Q in terms of six factors describing various aspects of rock-mass conditions in a quantitative manner and related Q to permanent support pressure.

$$Q = (\mathrm{RQD}/J_n)(J_r/J_a)(J_w/\mathrm{SRF}) \qquad (10.16)$$

where RQD = rock quality designation [Deere (1963),[36] (1969)[37]]

J_n = joint set number (describes number and distribution of joints)

J_r = joint roughness number (describes joint surface condition)

J_a = joint alteration number (describes joint fillings)

J_w = joint water reduction factor (describes quantity and pressure of joint water flow)

SRF = stress reduction factor (provides for weakness zones which may cause rock-mass loosening during excavation, residual stresses, and squeezing and swelling rock)

The subdivisions and the weighted value for each factor can be found in Barton et al. (1974)[35] or in Table 5.26 of Hunt (1984).[5]

The range in possible Q values (approximately 0.001 to 1000) covers conditions ranging from heavy squeezing ground to sound, unjointed rock. From their study, Barton et al. developed an empirical equation relating permanent support pressure P to Q as follows:

$$P_{\mathrm{roof}} = (2.0/J_r)Q^{-1/3} \qquad (10.17)$$

The relationship between rock-mass quality and permanent support pressure is given in Fig.

TABLE 10.15
OVERBURDEN ROCK LOAD CARRIED BY ROOF SUPPORT*

Rock conditions	Rock load H_p, ft	Remarks
1. Hard and intact	Zero	Sometimes spalling or popping occurs
2. Hard stratified or schistose	0 to 0.5*B*	Light pressures
3. Massive, moderately jointed	0 to 0.25*B*	Load may change erratically from point to point
4. Moderately blocky and seamy	$0.25B$ to $0.35(B + H_t)$	No side pressure
5. Very blocky and seamy	0.35 to $1.10(B + H_t)$	Little or no side pressure
6. Completely crushed but chemically intact	$1.10(B + H_t)$	Considerable side pressure; softening effect of seepage towards bottom of tunnel
7. Squeezing rock, moderate depth	$(1.10$ to $2.10)(B + H_t)$	Heavy side pressure
8. Squeezing rock, great depth	$(2.10$ to $4.50)(B + H_t)$	Heavy side pressure
9. Swelling rock	Up to 250 ft irrespective of value of $(B + H_t)$	Very heavy pressures

*From NAVFAC (1982)[33]; after Proctor and White (1977).[34]

NOTES:

1. Above values apply to tunnels at depth greater than $1.5(B + H_t)$.
2. The roof of the tunnel is assumed to be located below the water table. If it is located permanently above the water table, the values given for rock conditions 4 to 6 can be reduced by 50%.
3. Some very dense clays which have not yet acquired properties of shale rock may behave as squeezing or swelling rock.
4. Where sandstone or limestone contain horizontal layers of immature shale, roof pressures will correspond to rock condition "very blocky and seamy."

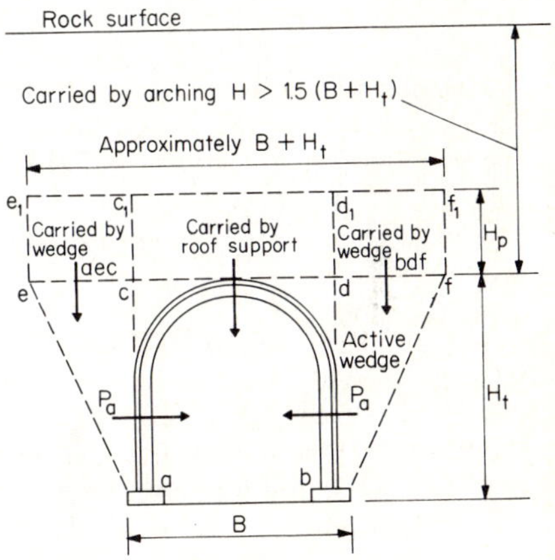

10.17. In the figure, the shaded portion is their estimate of the range to be expected in practice from their evaluation of the records. The system relies heavily on the RQD [see Art. 2.4.5 of Hunt (1984)[5]] and it must be considered that in their study the supports used were perhaps not always the supports needed, but the system is intended to permit estimation of actual support methods (see *Art. 10.3.4*). A comparison of various methods of predicting rock loads in terms of rock quality and groundwater is given in Fig. 10.18. It serves as a useful summary of anticipated rock load ranges.

Caverns

Barton et al. studied roof-support pressures measured by load cells installed between supports and the rock face for 26 large caverns excavated in hard rock with various degrees of jointing. They concluded that there did not appear to be any trend or necessity to increase the support

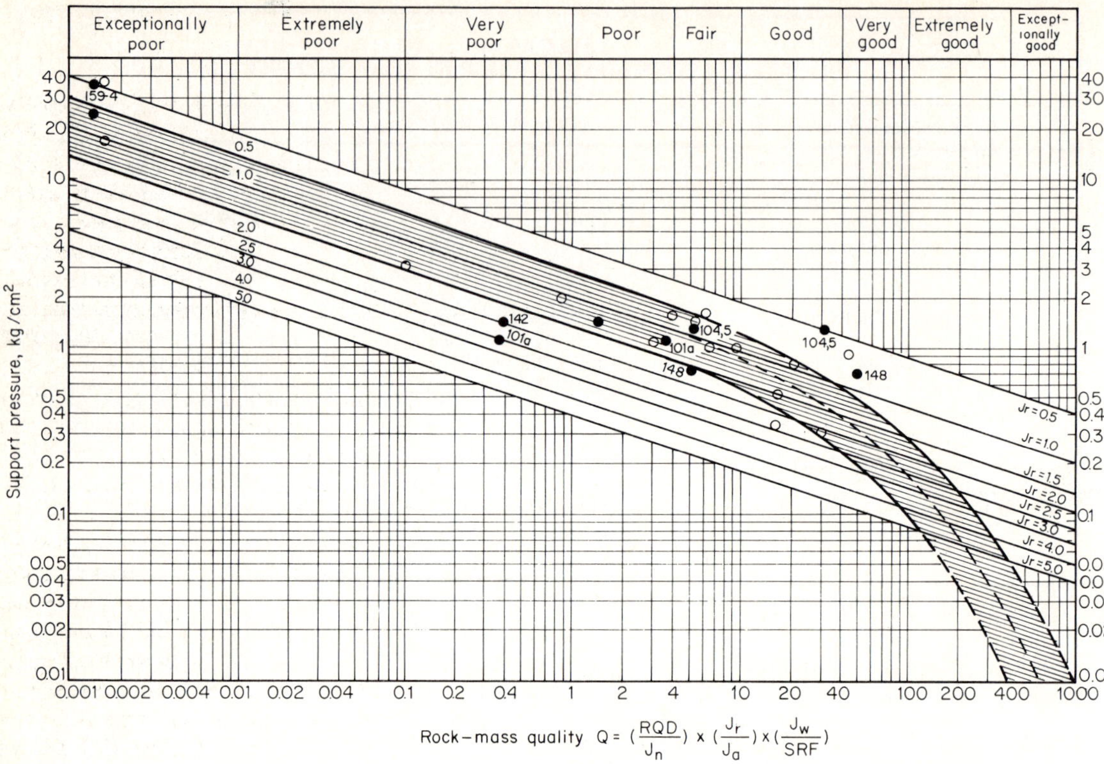

FIG. 10.17 Relationship between rock-mass quality Q and permanent support pressure. [From Barton et al. (1974).[35]]

pressure with increasing cavern dimensions. Roof support pressures were found generally to range from 0.5 to 1.5 kg/cm² and wall pressures from approximately 0.3 to 0.7 kg/cm². These pressures are in the range of one-half the values based on Terzaghi's criteria.

Analytical Methods: Tunnels and Caverns

Classical Approach

For tunnels, a circular opening is assumed in an elastic, semi-infinite plate, loaded on the remote vertical boundaries by externally applied pressures. Mathematical solutions are employed, based on the assumption of linear elasticity. Discussion regarding tunnels and other shapes can be found in Cording et al. (1975),[7] Lee and Lo (1976),[39] and Goodman (1980).[40]

Finite Element Method

When applying the FEM the initial stresses along the circumference of a tunnel are removed by applying a system of loads equal and opposite to the initial stresses. Computed stress changes are then superimposed on to the initial stresses in the rock mass to determine the new stresses, their distributions, and the resulting deformations.

Limitations and Applications

Analysis requires information on the in situ rock-mass stress condition and deformation moduli, and for the FEM, detailed information as well on the orientation and concentration of all discontinuities. In situ stresses are measured by flat-jack tests, borehole stress relief tests, or hydraulic fracturing [Art. 4.4 of Hunt (1984)[5]]. De-

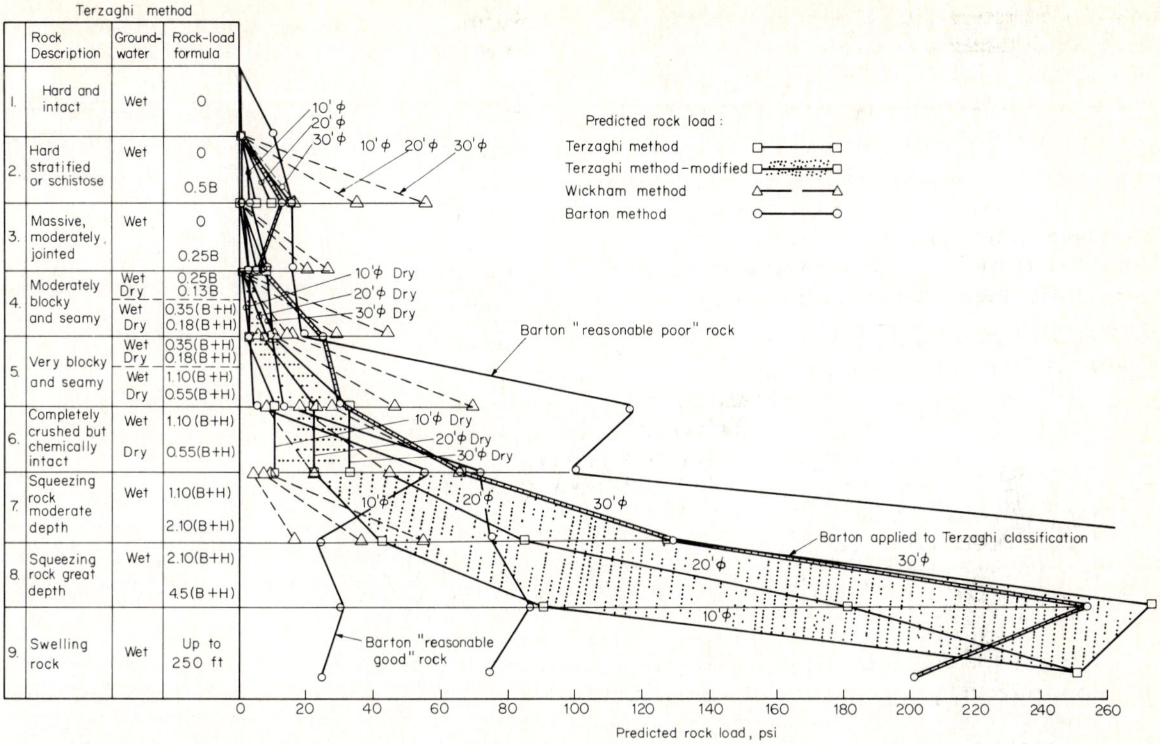

FIG. 10.18 Comparison of various methods of predicting rock load in terms of rock quality and groundwater. [From Rose et al. (1981).[38]]

formation moduli are measured by plate loading or radial jacking tests, by the dilatometer used in a borehole, or by extensometer measurements when in situ stresses are measured [Art. 3.5.3 of Hunt (1984)[5]; also, see the summary in Art. 4.1.4 in this volume].

In all in situ testing, the orientation of the applied load to the discontinuities in the rock and scale effects are most important. The geometry of flat jacks prohibits load applications and deformations radial to a tunnel axis, whereas plate and radial jacking tests do not. During tunnel driving, radial stresses diminish to zero at the opening edge while tangential stresses become concentrated [Kruse (1970)[31]], and joints with nearly tangential orientation have the greatest tendency to open. This condition influences borehole extensometer measurements as well as the magnitude of compression generated by plate-bearing tests. Rock moduli computed from radial deflections, therefore, are lower than those measured by flat-jack tests. Part of the difference is due also to scale effects, since a radial test normally stresses a larger area than does a plate-jack test, and large plate-jack tests have been found to give lower values for moduli than smaller diameter plates in the same rock conditions.

Procuring representative data on in situ rock properties is costly and difficult, and the poorer the rock quality the less representative the data are likely to be. Analytical methods based on in situ testing, therefore, generally are limited to rock of "good" to "exceptional" quality with relatively narrow ranges of variation, as applied to large tunnels or caverns. In these cases, testing costs relative to excavation costs are low, and there is a need to evaluate progressive deformation and the possibility of full-scale roof failure, rather than partial or confined failure such as falling blocks or running ground. Rational design based on in situ measurements is described by Abel (1967)[41] and Rabcewicz and Golser (1973).[42]

The FEM is practical for a two-dimensional problem, but often becomes impractical when the problem is three-dimensional. In caverns and tunnels excavated in benches, or with side galleries, etc., the method has severe limitations at present [Peck and Deere (1977)[43]].

Mine Openings

General

Normally self-supporting structures, mine openings usually are rectangular rather than cylindrical or elliptical, especially in the case of coal mines. Stresses are carried by walls, pillars, and other unexcavated portions, rather than by linings or steel supports.

In stratified rock the roof stratum deflects, forming a pressure arch which carries little or no load from the overlying strata. The overburden loads are transmitted around the opening to the walls and pillars. The roof stratum deflection is a function of span dimensions, stratum weight and thickness, tensile strength, and joint characteristics. Heavily jointed rock may be pattern bolted to form a composite beam or slab [Lang (1972)[44]].

When the roof does not contain vertical joints or it has been reinforced by bolting, roof stability may be evaluated by elementary beam theory with reasonable results. The condition of vertical joints can be evaluated by the FEM [Barker and Hatt (1973)[45]]. Pillar support capacity will be a function of the pillar material and the pillar base material, although punching through the roof is also possible. Some soft rocks, such as halite, undergo creep deformation over long time intervals and under relatively low stress levels [Hendron (1969)[46]].

Pillar Support

The capability of a pillar to support the roof is a function of the compressive strength of the rock, the cross-sectional area of the pillar, the roof load, and the strength of the floor and roof strata against punching. The cross-sectional area of the pillar may be reduced in time by weathering and spalling of its walls. [Some typical strength properties of coal are given in Art. 10.3.4 of Hunt (1984).[5]] Assuming that the roof and floor support are stronger than the pillar, the FS against failure is given by the ratio of pillar strength s to average pillar stress σ_v

$$\text{FS} = s/\bar{\sigma}_v \tag{10.18}$$

in which the average pillar stress given by the contributary area is

$$\bar{\sigma}_v = \frac{A_t}{A_p}\sigma_v \tag{10.19}$$

where A_t = roof area supported by pillar
A_p = pillar area
σ_v = vertical stress at roof level from overburden column of width equal to one-half the distance to adjacent pillars, ignoring strength of overburden

Pillar strength often is not merely the unconfined compressive strength of the pillar material because of shape and size effects on strength measurements of rock specimens [Bieniawski (1968)[47]]. For a pillar shaped as a rectangular prism (height $H >$ width W), Goodman (1980)[40] provides the following expression for estimating strength, corrected for size and shape effects and assuming that $H > h_{crit} \geq h$:

$$s = \left(0.875 + 0.250\frac{W}{H}\right)\left(\frac{h}{h_{crit}}\right)^{1/2}(U_c) \tag{10.20}$$

where U_c = unconfined compressive strength of the pillar material
h = height of tested specimen = $2d$
h_{crit} = maximum height of a cubical specimen such that an increase in dimension will produce no further reduction in strength, given as 1 m by Bieniawski (1968)[47] for coal

Pillar strength controls the relative separation of the rooms, whereas roof stability generally controls room widths.

Roof Stability

Roof stability can be critical in horizontally bedded rock. Idealized collapse conditions of continuous (unjointed) slabs are illustrated in Fig.

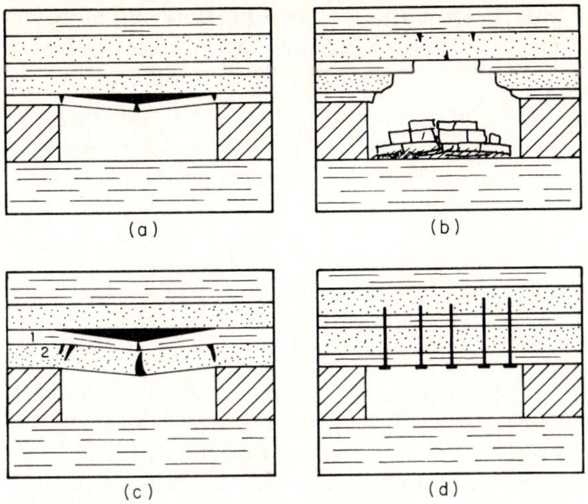

FIG. 10.19 Idealized mine-roof-collapse conditions: (*a*) thin slab deflects away from thicker overlying slab; (*b*) collapse extends upward, eventually forming a stable arch; (*c*) thicker lower slab deflects together with thinner overlying slab; and (*d*) rock bolts increase effective roof thickness and stability. [After Goodman (1980).[40] Used with permission of John Wiley & Sons, Inc.]

10.19. Model tests [Goodman (1980)[40]] have shown that a relatively thin roof slab under excessive load separates from a thicker upper layer, flexes downward, and cracks at the ends of its upper surface and at the middle of its lower surface (Fig. 10.19*a*). (Alternatively, cracks occur at slab defects.) Finally collapse occurs, leaving short cantilevers remaining at the opening sides. The adjacent slab flexes, cracks, and collapses, also leaving a short cantilever; as subsequent slabs break loose, a stable arch (Fig. 10.19*b*) results as long as there is still competent rock overlying the opening. If it extends high enough and the overburden is sufficiently weak, ground subsidence occurs.

Assuming that the roof slab acts as a clamped beam, and that the horizontal compressive stress in the beam is zero (a conservative assumption since beams are greatly strengthened by compressive stress), the *maximum tensile stress* σ_t at the top surface near the ends as given by Goodman (1980)[40] is

$$\sigma_{t(\max)} = \frac{\gamma L^2}{2t} \tag{10.21}$$

where γ = rock weight per unit volume
L = span
t = beam thickness

In the *beam center*, the maximum tensile stress is one-half the value near the ends. FS against failure is then the ratio of tensile strength (*see Art. 3.4.2*) to the maximum tensile stress. The *maximum deflection* of a clamped elastic beam is given by

$$\rho = \frac{\gamma L^4}{32Et^2} \tag{10.22}$$

where E is Young's modulus.

Where a thin beam (1) overlies a thick beam (2) (Fig. 10.19*c*), it will transfer its load to the lower beam, and the stresses and deflections of the lower beam can be calculated from Eqs. 10.21 and 10.22 by assigning it a new beam unit weight given by

$$\gamma_a = \frac{E_2t_2^2(\gamma_2t_2 + \gamma_1t_1)}{E_2t_2^3 + E_1t_1^3} \tag{10.23}$$

In the case of a thin-under-thick roof slab (Fig. 10.19*d*), rock bolts may be used to increase the effective roof thickness and decrease the maximum tensile stress. The bolts are designed to supply a force per unit area Δq, given as

$$\Delta q = \frac{\gamma_2t_2E_1t_1^3 - \gamma_1t_1E_2t_2^3}{E_1t_1^3 + E_2t_2^3} \tag{10.24}$$

and the tensile stresses in the two layers are given by

$$\sigma_{\max} = \frac{(\gamma t \pm \Delta q)L^2}{2t^2} \tag{10.25}$$

Measurements of the tensile strength of intact rock are described in Art. 3.4.2.

Floor Support

Coals are usually found in a sequence of strata in which the coals are underlain by clays. Failure of a mine floor supporting coal pillars frequently occurs when the subclay strata include layers of relatively soft clays or other low-strength material. Manifestations of failure range from floor heave to pillar collapse which result from a bearing-capacity failure of the supporting materials. The results of a study of several existing coal mines by Rockaway and Stephenson

(1982)[48] indicated that the foundation failure of coal pillars can be evaluated by Terzaghi's bearing-capacity equation (Eq. 6.11) modified by an approach defined by Vesić (1975)[49] which provides for a two-layer system as follows:

$$q_{ult} = c_1 N_m + q \qquad (6.34)$$

where q = surcharge

c_1 = undrained strength of the soft stratum

N_m = modified bearing-capacity factor which is a function of the ratio of the cohesive strength of two layers, $k = c_2/c_1$, the thickness of the soft stratum relative to the footing width and the foundation shape

Vesić's formulation of N_m assumes a modified local shear failure which is similar to that observed by Rockaway and Stephenson. It considers the widest range of factors which interact to alter predicted bearing capacity and is also the most conservative of the bearing-capacity equations for a two-layered problem. N_m for the case of a stiff clay layer over a soft clay layer is given in Art. 6.2.4 *(BEARING CAPACITY ANALYSIS: Two-Layered Soil Profiles)*. The expression for N_m for the case of a soft layer over a stiff layer can be found in Vesić (1975).[49] In general, where the thickness of the soft upper layer is smaller than the foundation width, the presence of a rigid underlying layer results in an increase in bearing capacity. If a normal analysis is performed, therefore, based on local shear, the results will be conservative, and in any event in the presence of supporting soft clays pillar design should be based on consideration of estimated settlement to avoid large deflections.

10.3.4 SUPPORT OF ROCK OPENINGS

Until recent years tunnels constructed for engineering purposes were lined with concrete almost as a standard procedure, and design and construction considered the problem in two steps: temporary support during excavation and permanent support. Modern practice generally incorporates the construction support system into the permanent support requirements, and reinforced concrete is becoming less common for rock masses. Excavation size and rock quality are the factors controlling the selection of the support system, and a summary of support systems for various rock classes is given in Table 10.2.

Caverns and mines are seldom lined. Caverns generally are constructed in good quality rock requiring minimal supports; mine support generally is provided by pillars with additional roof support provided by rock bolts where necessary.

Methods

Little or No Support Required

Moderate-sized excavations in competent rock or rock with joints irregular and discontinuous with surfaces tight and unaltered, with RQD values ranging from 90 to 100, usually require little or no auxiliary support. Tunnels greater than 8 m in diameter and chambers usually will be pattern bolted [Deere et al. (1974)[6]]. The exception is when high residual stresses cause spalling and popping. Mild popping can be contained with rock bolts and wire mesh or with steel straps, more severe conditions with shotcrete and rock bolts or with steel sets and close lagging.

Rock Bolts and Cables

In jointed, laminated, or otherwise unstable rock the plastic or relaxation zone around the arch is kept in compression with rock bolts or cables *(see Art. 5.5.3)*.

Rock bolts can be installed by several methods: driven in and expanded, driven in and wedged, or installed in drill holes and grouted with morter or resins (Figs. 5.43 and 5.44). Bolts fully grouted with resins are becoming increasingly common because they are relatively easy to install, can develop capacity within minutes of installation, and provide better support than mechanically tensioned, cement-grouted bolts. They are particularly useful as *spiling* at the heading of a rapidly advancing tunnel. *Prestressed cables*, consisting of a number of wires, are anchored in place with a special cement grout and provide higher capacities than is possible with bolts. Types of rock bolts and analytical procedures for their design are described by Jaeger (1972)[50] and Lang (1972).[44]

Rock-bolt lengths typically range from one-fourth to one-third the opening width in rock of fair to excellent quality or in large openings greater than about 15 m in width, whereas in poor rock lengths may range up to one-half the opening width. Extensometer measurements provide data on the relaxation depth and bolts must penetrate an adequate distance beyond the tensile zone. Spacing usually ranges from 1 to 3 m, depending on rock quality (see Table 10.2), the length of the tunnel section excavated, and the intermediate support such as mesh or shotcrete placed with the bolts [Deere et al. (1974)[6]]. Wire mesh and shotcrete are used to contain finely jointed rocks, thereby allowing increased bolt spacing. Bolts and steel straps are used in mines and open shafts.

Loads on bolts are more accurately determined by load cells mounted behind the bolt than by calibrated torque wrenches that are greatly influenced by thread conditions.

Shotcrete and Gunite

Shotcrete (see Art. 5.5.3) is a rapid-setting, wet-mix concrete containing coarse aggregate (12 mm) pneumatically sprayed onto the rock face at high velocity, establishing a firm bond with the rock and acting as reinforcement. Normally sprayed on in 50-mm-thick layers immediately after blasting, it can be used in combination with other support methods, as the sole support for small tunnels, or for the final support for a large opening if it is placed as a continuous arch. A thick layer acts as an arch in compression, but a thin layer acts in tension. The tensile strength is significantly increased by the addition of 25-mm-long wire fibers included in the concrete mix, by the installation of a wire mesh between layers, or by the use of microsilica shotcrete.

It is used to reinforce and stabilize fractured rock of poor to very poor quality, slaking shales, and other poor ground conditions. As it is semirigid, desirable small deflections are permitted which mobilizes strength and reduces squeezing loads *(see Art. 10.3.3)*.

Application of shotcrete requires care, particularly on the part of the worker at the nozzle. Brekke (1972)[51] describes shotcrete and its application to poor ground conditions. Clay seams require special treatment for stabilization which is best achieved by cleaning exposed clay from the rock surface wherever possible.

Gunite is a cement-sand-water mixture similar to shotcrete except that the maximum size of the aggregate is smaller. It is applied to reinforce moderately jointed rock.

New Austrian Tunneling Method (NATM)

The new Austrian tunneling method employs shotcrete and rock bolts, often with wire mesh, to contain unstable rock (Fig. 10.20). It also includes detailed rock classification and routine extensive instrumentation. Rock classification, each with its own particular support requirement, is determined round by round by the contractor and engineer jointly. Instrumentation, consisting principally of Gloetzl pressure cells and extensometers, is used to indicate when strains have stopped, and hence no more layers of shotcrete or other supports are needed.

Rabcewicz (1975)[30] describes the application of the NATM to provide support for a tunnel in the Austrian Alps driven through difficult conditions including talus of cohesionless rock debris and boulders and intensely folded phyllites. The suitability of the method is given in Fig. 10.16*b*, and the general construction procedure is illustrated in Fig. 10.22.

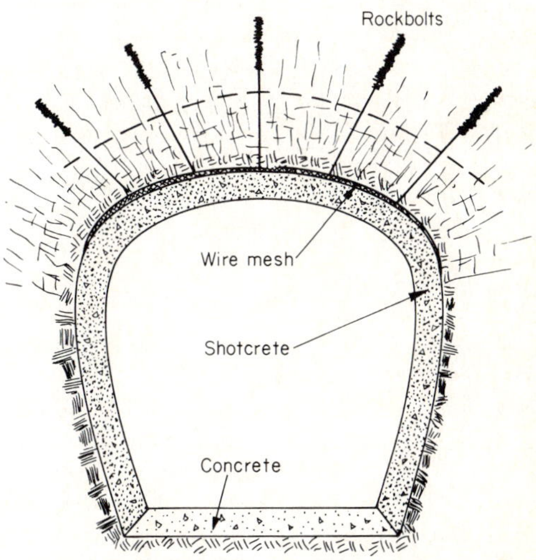

FIG. 10.20 Support of unstable ground with rock bolts, wire mesh, and shotcrete, as used in the New Austrian Tunneling Method (NATM).

Steel Ribs and Timber Blocking

The traditional method used in the United States for many years consists of structural steel bent to conform to the required tunnel cross section and blocked with timber to provide contact between the rib and the roof, as illustrated in Fig. 10.21. Time delay involved with installation can allow loosening of the rock mass, resulting in high support loads, reduced stand-up time, and more difficulty in placing supports in the adjacent section.

As steel ribs are a visual support, loading distortions can be observed and additional support placed before collapse occurs. Failure of steel-rib supports usually is associated with inadequate blocking of the rib to the rock, particularly in the vicinity of the side arches, above the spring line.

Concrete Linings

Barton et al. (1974)[35] recommend concrete linings only for transportation tunnels in the poorest rock classes where the effectiveness of other methods is uncertain. It is usually cast in place and steel reinforced and used in conjunction with rock bolts or shotcrete, or with steel ribs and timber blocking as shown in Fig. 10.21. Water tunnels are normally lined except in rock of very high quality where erosion, spalling, and falls are not anticipated. Precast concrete liners are used usually for primary and final support in good quality rock excavated by a TBM.

Back-packing should be carefully performed immediately after installation of the lining to fill all spaces and to prevent further rock roof displacement with the attendant increase in pressures which, in time, can become considerable.

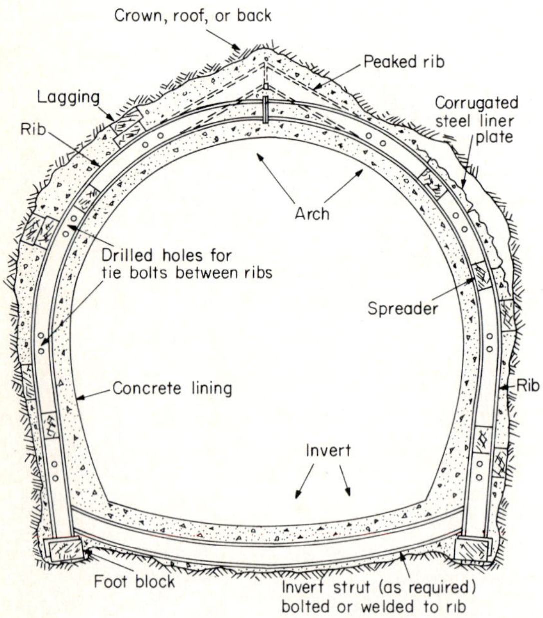

FIG. 10.21 Conventional support with steel ribs and timber blocking. Peaked ribs provide space for pumping concrete into forms for lining. [From Walhstrom (1973).[25]]

Selection

Tunnels and Caverns

Support methods for underground openings can be determined qualitatively by such criteria as those of Terzaghi (1946)[23] or Barton et al. (1974),[35] or quantitatively by using rock-mass properties determined from in situ testing together with qualitative information.

Qualitative determinations are based on rock-mass quality and experience and can incorporate in situ test data on material properties. Barton et al. provide an essentially qualitative procedure for selecting support systems for caverns and tunnels based on their evaluation of 200 case records *(see Art. 10.3.3)*. They developed a relationship between rock quality Q, the excavation dimensions, and the support actually used. A summary of support systems in terms of rock quality is given in Table 10.16.

Support methods have been divided by Barton et al. into 38 categories as given in Table 10.17, which can be used for evaluating roof or wall support. The excavation support ratio (ESR), which modifies roof or wall dimensions (SPAN), reflects construction practice in that the degree of safety and support required is a function of the excavation purpose and its use for machinery and personnel. Approximate ESR values for various types of excavations are given in Table 10.18. The support recommendations given are for permanent support; for temporary support, where reductions are possible, they recommend

TABLE 10.16
ROCK QUALITY AND SUPPORT SYSTEMS SUMMARIZED

Rock quality	Q	Terzaghi's class*	Support system†
Exceptional to extremely good	1000–100	Hard and intact	No support to untensioned, grouted rock bolts even in spans to 100 m
Very good	100–40	Massive, moderately jointed	Widely spaced (to 3 m) pattern bolting
Good	40–10	Hard, stratified schistose and massive, moderately jointed	More closely spaced pattern bolting, and for larger spans, bolting with mesh reinforced concrete
Fair and poor	10–4 4–1	Moderately blocky and seamy	Rock-bolt spacing decreases and shotcrete thickness increases generally in relationship to the SPAN and ESR‡
Very poor	1.0–0.1	Very blocky and seamy	Range from closely spaced bolting and shotcrete to bolting and steel ribs and concrete arches, depending on SPAN/ESR value and the presence of swelling clays or squeezing rock
Extremely poor	0.1–0.01	Squeezing rock, moderate depth	Heavy applications of mesh-reinforced shotcrete or steel ribs and cast concrete arches for support of all ratios of SPAN/ESR
Exceptionally poor	0.01–0.001	Squeezing rock great depth, swelling rock	

*Correlations of Q values with Terzaghi's classification made by Barton et al. (1974).[35]

†Summary of systems given in Table 10.17 for various Q values.

‡SPAN = excavation width or height; ESR = excavation support ratio; see Table 10.18.

increasing the ESR value to 1.5 and increasing Q to $5Q$.

The support systems given are based on normal excavation practice; if drilling and blasting quality are better or worse than normal, then the support will be either conservative or inadequate, respectively. As most of the cases studied by Barton et al. were from projects in Scandinavia, the range of rock types was limited, and support methods included in their analysis were primarily bolting and shotcrete common to European practice, rather than the steel-rib support, free-span concrete arch roofs, and precast sectional linings common to North America.

Pressure Tunnels

Water-pressure tunnel design requires an evaluation of the effect of tunnel pressure on overburden pressure to prevent blowout in competent rock and on leakage that would cause instability. In incompetent rock the problem is the prevention of rupture or excessive deflection of the adjacent rock. Liner thickness for pressure tunnels required to resist deflection and cracking

TABLE 10.17
TUNNEL SUPPORT MEASURES FOR ROCK MASSES*

Support category	Q†	Conditional factors: RQD/*Jn*	Conditional factors: *Jr*/*Jn*	Conditional factors: SPAN/ESR, m	*P*, kg/cm² (approx.)	SPAN/ESR, m	Type of support	Remarks (see corresponding list at end)
A. ROCK QUALITY RANGE: 1000–10								
Support measures for rock masses of "exceptional," "extremely good," "very good," and "good" quality								
1‡	1000–400	—	—	—	<0.01	20–40	sb (utg)§	—
2‡	1000–400	—	—	—	<0.01	30–60	sb (utg)	—
3‡	1000–400	—	—	—	<0.01	46–80	sb (utg)	—
4‡	1000–400	—	—	—	<0.01	65–100	sb (utg)	—
5‡	400–400	—	—	—	0.05	12–30	sb (utg)	—
6‡	400–100	—	—	—	0.05	19–45	sb (utg)	—
7‡	400–100	—	—	—	0.05	30–65	sb (utg)	—
8‡	400–100	—	—	—	0.05	48–88	sb (utg)	—
9	100–40	≧20	—	—	0.25	8.5–19	sb (utg)	—
		<20	—	—			B (utg) 2.5–3 m	—
10	100–40	≧30	—	—	0.25	14–30	B (utg 2–3 m	—
		<30	—	—			B (utg) 1.5–2 m +clm	—
11‡	100–40	≧30	—	—	0.25	23–48	B (tg) 2–3 m	—
		<30	—	—			B (tg) 1.5–2 m +clm	—
12‡	100–40	≧30	—	—	0.25	40–72	B (tg) 2–3 m	—
		<30	—	—			B (tg) 1.5–2 m +clm	—
13	40–10	≧10	≧1.5	—	0.5	5–14	sb (utg)	I
		≧10	<1.5	—			B (utg) 1.5–2 m	I
		<10	≧1.5	—			B (utg) 1.5–2 m	I
		<10	<1.5	—			B (utg) 1.5–2 m +S 2–3 cm	I

14	40–10	≧10	—	≧15	0.5	9–23	B (tg) 1.5–2 m +clm	I, II
		<10	—	≧15			B (tg) 1.5–2 m +S (mr) 5–10 cm	I, II
		—	—	<15			B (utg) 1.5–2 m +clm	I, III
15	40–10	>10	—	—	0.5	15–40	B (tg) 1.5–2 m +clm	I, II, IV
		≦10	—	—			B (tg) 1.5–2 m +S (mr) 5–10 cm	I, II, IV
16‡	40–10	<15	—	—	0.5	30–65	B (tg) 1.5–2 m +clm	I, V, VI
See remark XII		≦15	—	—			+S (mr) 10–15 cm	
B. ROCK QUALITY RANGE: 10–1 **Support measures for rock masses of "fair" to "poor" quality**								
17	10–4	>30	—	—	1.0	3.5–9	sb (utg)	I
		≧10, ≦30	—	—			B (utg) 1–1.5 m	I
		<10	—	≧6 m			B (utg) 1–1.5 m +S 2–3 cm	I
		<10	—	<6 m			S 2–3 cm	I
18	10–4	>5	—	≧10 m	1.0	7–15	B (tg) 1–1.5 m +clm	I, III
		>5	—	<10 m			B (utg) 1–1.5 m +clm	I
		≦5	—	≧10 m			B (tg) 1–1.5 m +S 2–3 cm	I, III
		≦5	—	<10 m			B (utg) 1–1.5 m +S 2–3 cm	I
19	10–4	—	—	≧20 m	1.0	12–29	B (tg) 1–2 m +S (mr) 10–15 cm	I, II, IV
		—	—	<20 m			B (tg) 1–1.5 m +S (mr) 5–10 cm	I, II

TABLE 10.17
TUNNEL SUPPORT MEASURES FOR ROCK MASSES* (continued)

Support category	Q†	Conditional factors: RQD/Jn	Conditional factors: Jr/Jn	Conditional factors: SPAN/ESR, m	P, kg/cm² (approx.)	SPAN/ESR, m	Type of support	Remarks (see corresponding list at end)
B. ROCK QUALITY RANGE: 10–1								
Support measures for rock masses of "fair" to "poor" quality								
20‡ See remark XII	10–4	—	—	≧35 m	1.0	24–52	B (tg) 1–2 m +S (mr) 20–25 cm	I, V, VI
		—	—	<35 m			B (tg) 1–2 m +S (mr) 10–20 cm	I, II, IV
21	4–1	≧12.5	≦0.75	—	1.5	2.1–6.5	B (utg) 1 m +S 2–3 cm	I
		<12.5	≦0.75	—			S 2.5–5 cm	I
		—	>0.75	—			B (utg) 1m	I
22	4–1	>10, <30	>1.0	—	1.5	4.5–11.5	B (utg) 1m + clm	I
		≦10	>1.0	—			S 2.5–7.5 cm	I
		<30	≦1.0	—			B (utg) 1 m +S (mr) 2.5–5 cm	I
		≧30	—	—			B (utg) 1 m	I
23	4–1	—	—	≧15 m	1.5	8–24	B (tg) 1–1.5 m +S (mr) 10–15 cm	I, II, IV, VII
		—	—	<15 m			B (utg) 1–1.5 m +S (mr) 5–10 m	I
24‡ See remark XII	4–1	—	—	≧30 m	1.5	18–46	B (tg) 1–1.5 m +S (mr) 15–30 cm	I, V, VI
		—	—	<30 m			B (tg) 1–1.5 m +S (mr) 10–15 cm	I, II, IV
C. ROCK QUALITY RANGE: 1.0–0.1								
Support measures for rock masses of "very poor" quality								
25	1.0–0.4	>10	>0.5	—	2.25	1.5–4.2	B (utg) 1m + mr or clm	I
		≦10	>0.5	—			B (utg) 1 m + S (mr) 5 cm	I
		—	≦0.5	—			B (utg) 1m + S (mr) 5 cm	I

26	1.0–0.4	—	—	—	2.25	3.2–7.5	B (tg) 1 m +S (mr) 5–7.5 cm	VIII, X, XI
		—	—	—			B (utg) 1 m + S 2.5–5 cm	I, IX
27	1.0–0.4	—	—	≧12 m	2.25	6–18	B (tg) 1 m + S (mr) 7.5–10 cm	I, IX
		—	—	<12 m			B (utg) 1 m +S (mr) 5–7.5 cm	I, IX
		—	—	>12 m			CCA 20–40 cm +B (tg) 1 m	VIII, X, XI
		—	—	<12 m			S (mr) 10–20 cm +B (tg) 1 m	VIII, X, XI
28‡ See remark XII	1.0–0.4	—	—	≧30 m	2.25	15–38	B (tg) 1 m +S (mr) 30–40 cm	I, IV, V, IX
		—	—	≧20, <30			B (tg) 1 m +S (mr) 20–30 cm	I, II, IV, IX
		—	—	<20 m			B (gt) 1 m +S (mr) 15–20 cm	I, II, IX
		—	—	—			CCA *sr) 30–100 cm +B (tg) 1 m	IV, VIII, X, XI
29‡	0.4–0.1	>5	>0.25	—	3.0	1.0–3.1	B (utg) 1 m + S 2–3 cm	—
		≦5	>0.25	—			B (utg) 1 m + S (mr) 5 cm	—
		—	≦0.25	—			B (tg) 1 m + S (mr) 5 cm	—
30	0.4–0.1	≧5	—	—	3.0	2.2–6	B (tg) 1 m + S 2.5–5 cm	IX
		<5	—	—			S (mr) 5–7.5 cm	IX
		—	—	—			B (tg) 1 m +S (mr) 5–7.5 cm	VIII, X, XI

TABLE 10.17
TUNNEL SUPPORT MEASURES FOR ROCK MASSES* (continued)

Support category	Q†	Conditional factors: RQD/*Jn*	Conditional factors: *Jr*/*Jn*	Conditional factors: SPAN/ESR, m	*P*, kg/cm² (approx.)	SPAN/ESR, m	Type of support	Remarks (see corresponding list at end)
31	0.4–0.1	>4	—	—	3.0	4–14.5	B (tg) 1 m +S (mr) 5–12.5 cm	IX
		≦4, ≧1.5	—	—			S (mr) 7.5–25 cm	IX
		<1.5	—	—			CCA 20–40 cm +B (tg) 1 m	IX, XI
		—	—	—			CCA (sr) 30–50 cm +B (tg) 1 m	VIII, X, XI
32 See remark XII	0.4–0.1	—	—	≧20 m	3.0	11–34	B (tg) 1 m +S (mr) 40–60 cm	II, IV, IX, XI
		—	—	<20 m			B (tg) 1 m +S (mr) 20–40 cm	III, IV, IX, XI
		—	—	—			CCA (sr) 40–120 cm +B (tg) 1 m	IV, VIII, X, XI
D. ROCK QUALITY RANGE: 0.1–0.001								
Support measures for rock masses of "extremely poor" and "exceptionally poor" quality								
33‡	0.1–0.01	≧2	—	—	6	1.0–3.9	B (tg) 1 m +S (mr) 2.5–5 cm	IX
		<2	—	—			S (mr) 5–10 cm	IX
		—	—	—			S (mr) 7.5–15 cm	VIII, X
34	0.1–0.01	≧2	≧0.25	—	6	2.0–11	B (tg) 1 m +S (mr) 5–7.5 cm	IX
		<2	≧0.25	—			S (mr) 7.5–15 cm	IX
		—	<0.25	—			S (mr) 15–25 cm	IX
		—	—	—			CCA (sr) 20–60 cm +B (tg) 1 m	VIII, X, XI

35	0.1–0.01	—	—	≧15 m	6	6.5–28	B (tg) 1 m +S (mr) 30–100 cm	II, IX, XI
See remark XII		—	—	≧15 m			CCA (sr) 60–200 cm +B (tg) 1 m	VIII, X, XI, II
		—	—	<15 m			B (tg) 1 m +S (mr) 20–75 cm	IX, XI, III
		—	—	<15 m			CCA (sr) 40–150 cm +B (tg) 1 m	VIII, X, XI, III
36†	0.01–0.001	—	—	—	12	1.0–2.0	S (mr) 10–20 cm	IX
		—	—	—			S (mr) 10–20 cm +B (tg) 0.5–1.0 m	VIII, X, XI
37	0.01–0.001	—	—	—	12	1.0–6.5	S (mr) 20–60 m	IX
		—	—	—			S (mr) 20–60 cm +B (tg) 0.5–1.0 cm	VIII, X, XI
38	0.01–0.001	—	—	≧10 m	12	4.0–20	CCA (sr) 100–300 cm	IX
See remark XIII		—	—	≧10 m			CCA (sr) 100–300 cm +B (tg) 1 m	VIII, X, II, XI
		—	—	<10 m			S (mr) 70–200 cm	IX
		—	—	<10 m			S (mr) 70–200 cm +B (tg) 1 m	VIII, X, III, XI

REMARKS

I. For cases of heavy rock bursting or "popping," tensioned bolts with enlarged bearing plates often used, with spacing of about 1 m (occasionally down to 0.8 m). Final support when "popping" activity ceases.
II. Several bolt lengths often used in same excavations, i.e., 3, 5, and 7 m.
III. Several bolt lengths often used in same excavation, i.e., 2, 3, and 4 m.
IV. Tensioned cable anchors often used to supplement bolt support pressures. Typical spacing 2–4 m.
V. Several bolt lengths often used in some excavations, i.e., 6, 8, and 10 m.
VI. Tensioned cable anchors often used to supplement bolt-support pressures. Typical spacing 4–6 m.
VII. Several of the older generation power stations in this category employ systematic or spot bolting with areas of chain link mesh, and a free span concrete arch roof (25–40 cm) as permanent support.
VIII. Cases involving swelling, for instance, montmorillonite clay (with access of water). Room for expansion behind the support is used in cases of heavy swelling. Drainage measures are used where possible.

(Continued)

TABLE 10.17
TUNNEL SUPPORT MEASURES FOR ROCK MASSES* (continued)

IX. Cases not involving swelling clay or squeezing rock.

X. Cases involving squeezing rock. Heavy rigid support is generally used as permanent support.

XI. According to the experience of Barton et al., in cases of swelling or squeezing, the temporary support required before concrete (or shotcrete) arches are formed may consist of bolting (tensioned shell-expansion type) if the value of RQD/J_n is sufficiently high (i.e., > 1.5), possibly combined with shotcrete. If the rock mass is very heavily jointed or crushed (i.e., $RQD/J_n < 1.5$, for example a "sugar cube" shear zone in quartzite), then the temporary support may consist of up to several applications of shotcrete. Systematic bolting (tensioned) may be added after casting the concrete (or shotcrete) arch to reduce the uneven loading on the concrete, but it may not be effective when $RQD/J_n < 1.5$, or when a lot of clay is present, unless the bolts are grouted before tensioning. A sufficient length of anchored bolt might also be obtained using quick setting resin anchors in these extremely poor quality rock masses. Serious occurrences of swelling and/or squeezing rock may require that the concrete arches are taken right up to the face, possibly using a shield as temporary shuttering. Temporary support of the working face may also be required in these cases.

XII. For reasons of safety the multiple-drift method will often be needed during excavation and supporting of roof arch. Categories 16, 20, 24, 28, 32, 35 (SPAN/ESR > 15 m only).

XIII. Multiple drift method usually needed during excavation and support of arch, walls and floor in cases of heavy squeezing. Category 38 (SPAN/ESR > 10 m only).

*From Barton, Lien, and Lunde (1974).[35]

†$Q = (RQD/J_n)(J_r/J_a)(J_w/SRF)$. See Table 5.26 of Hunt (1984).[5]

‡Estimates of support by Barton et al. Insufficient case records available for reliable estimation of support requirements.

The type of support to be used in categories 1 to 8 will depend on the blasting technique. Smooth-wall blasting and thorough barring-down may remove the need for support. Rough-wall blasting may result in the need for single applications of shotcrete, especially where the excavation height is > 25 m. Future case records should differentiate categories 1 to 8.

§*Key to support tables:*

sb = spot bolting
B = systematic bolting
(utg) = untensioned, grouted
(tg) = tensioned (expanding shell type for competent rock masses, grouted posttensioned in very poor quality rock masses; see remark XI)
S = shotcrete
(mr) = mesh reinforced
clm = chain link mesh
CCA = cast concrete arch
(sr) = steel reinforced

Bolt spacings are given in meters. Shotcrete or cast concrete arch thickness is given in centimeters.

TABLE 10.18
THE EXCAVATION SUPPORT RATIO (ESR) APPROPRIATE TO A VARIETY OF UNDERGROUND EXCAVATIONS*

Type of excavation	ESR	No. of cases
Temporary mine openings, etc.	ca. 3–5?	2
Vertical shafts:	ca. 2.5?	0
(i) circular section	ca. 2.0?	0
(ii) rectangular/square section		
Permanent mine openings, water tunnels for hydropower (exclude high-pressure penstocks), pilot tunnels, drifts and headings for large excavations, etc.	1.6	83
Storage rooms, water-treatment plants, minor road and railway tunnels, surge chambers, access tunnels, etc. (cylindrical caverns?)	1.3	25
Power stations, major road and railway tunnels, civil defense chambers, portals, intersections, etc.	1.0	79
Underground nuclear power stations, railway stations, sports and public facilities, factories, etc.	ca. 0.8?	2

*From Barton, Lien, and Lunde (1974).[35]

can be used based on the rock modulus as measured by in situ testing or to resist existing high residual stresses. Pressure-tunnel design is discussed by Kruse (1970)[3] and Jaeger (1972).[50]

Incompetent rock can be troublesome. Under internal pressure the concrete lining expands and compresses the surrounding rock. If the rock modulus is low, the concrete lining may crack irrespective of the magnitude of the overburden pressure, and water may then escape into pervious zones. Reinforcement in some instances may be inadequate to control the cracking, and steel liners may be necessary, particularly in penstocks near slopes.

10.3.5 CONSTRUCTION ASPECTS

Tunneling Conditions Reviewed

Deere et al. (1974)[6] placed rock tunneling conditions and associated problems in three general categories: good, average to difficult, and very difficult to hazardous.

Good Tunneling Ground

In good tunneling ground the rock is hard, massive, and tightly jointed. Very little support is required except for large-diameter tunnels and chambers which are usually pattern bolted (Q = 1000 to 40). Adverse conditions may include residual stresses or very hard rock, making excavation difficult.

Average to Difficult Tunneling Ground

Most conditions encountered in tunneling are represented by average to difficult ground (Q = 40 to 0.1). Three important considerations are tunnel size effect, directional effect of rock structure, and the "support appropriateness" effect.

A small tunnel driven in poor rock may present few support problems, whereas a large-diameter tunnel could encounter severe problems where shear zones or seams intersect the arch, sidewalls, or face. Large blocks could move into a larger opening which would remain stable at a smaller opening. The orientation of rock structure related to tunnel direction becomes important when there is one major joint or other weakness plane or zone (anisotropic conditions), whereas it may not be very important in rock with several joint sets, none of which is dominant (isotropic conditions). Some effects of joint orientation are given in Art. 10.3.2. "Support appropriateness" refers to the use of reasonable methods of support for reasonable rates of advance. Advancing too rapidly may require overly costly supports, or the contractor may use supports for a given situation that are more costly than required (for example, heavy steel ribs rather than shotcrete and rock bolts).

Very Difficult to Hazardous Tunneling Ground

Very difficult to hazardous tunneling conditions usually are caused by major faulting or very high residual stresses. Faulting can result in wet, seamy, heavily jointed or sheared rock; squeezing ground under high pressures; and high water pressures and quantities (Q = 0.1 to 0.001).

When such conditions are encountered, supports must be applied as quickly as possible to prevent cave-ins, pressure increase with time must be anticipated, and adequately strong supports provided. If drainholes do not relieve water pressures, then grouting is required. These conditions often are difficult to define during predesign investigations and provide the risks attendant to tunneling and other underground excavations. Test borings drilled from the tunnel heading are imperative if there is a chance for the presence of very difficult or hazardous conditions.

Excavation

Conventional Methods

Cycles of drilling, blasting, mucking, and installation of supports compose conventional excavation methods and in competent rock when the opening is of moderate size proceeds against the full face of the heading. Large openings are drilled from one or several tiers of platforms mounting the drills for the blasting holes to excavate the arch and walls, as illustrated in Fig. 10.22, where the opening is being made in three lifts. In this case the rock mass at the heading is being presupported by rock bolts, and as the drift is opened it is supported by wire mesh and shotcrete (as in the NATM). Steel ribs and blocking could be used as an alternative to the mesh and shotcrete.

Multidrift methods open the heading in smaller sections and are used in large excavations or in incompetent rock to provide better control, as shown in Fig. 10.23.

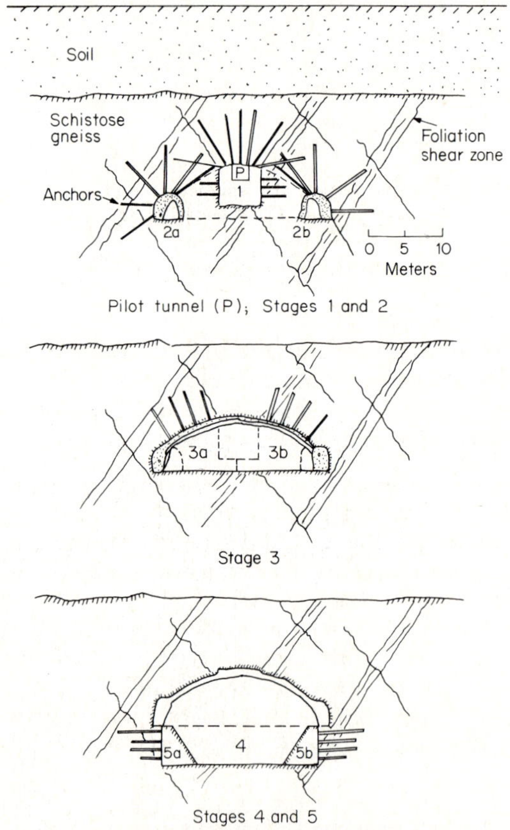

FIG. 10.23 Multidrift method of tunneling showing excavation and support sequences, DuPont Circle Station, Washington, D.C. [From Deere et al. (1974).[6]]

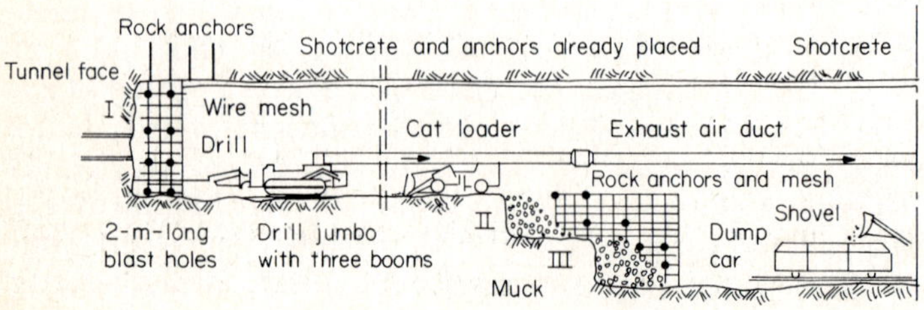

FIG. 10.22 Conventional NATM tunneling methods in rock. [From Rabcewicz (1975).[30]]

Excavation sequences and initial support procedures are often the most critical aspects of tunneling. Initial support procedures in the heading must be designed to prevent fallouts or collapse which would endanger workers or stop the progress of the excavation, or which would affect the amount of overbreak, the degree of block loosening around the perimeter, and the final loads to be supported. Proper blasting control is very important to minimize overbreak, maintain rock quality, and avoid damage to adjacent structures. Blasting procedures, effect, and control are described in a general manner in Arts. 5.6.3 and 6.3.4.

Roadheaders

Roadheaders, developed in the coal mining industry, are being used for soft rock excavation and, in some cases, in harder, fractured rock. The machine, which may be positioned within a shield, contains a boom on which is mounted a rotating ball with picks which cut the rock. Irregularly shaped openings are readily excavated and if hard materials are encountered, requiring blasting, the machine is easily removed from the face area.

Tunnel Boring Machines

A modern development that permits excavation without blasting, TBMs contain a cutter head (Fig. 10.24), and means to carry cuttings from the tunnel face to the rear of the machine. Machines have been designed for excavation in soft rocks such as shale, or for hard rocks such as granite and quartzite with unconfined compressive strengths in excess of 20,000 psi (1400 bars).

Where conditions are favorable the TBM has a number of advantages. Blasting effects on nearby structures are eliminated, and the rock remains intact, cut in a smooth circular cross section. This provides maximum stability and therefore requires less support and minimizes overbreak. The machines are readily adaptable to lining

FIG. 10.24 Atlas Copco Jarva MK 30 TBM, diameter 9.83 m (32 ft 3 in), weight 900 tonnes. On the Chicago TARP project this machine averaged 20.6 m/day drilling in dolomite with a compressive strength of 550 to 1880 bars. (Courtesy of Atlas Copco.)

with precast concrete segments which are backpacked and grouted after application. Economies realized substantially offset the initial cost of the machine in favorable geologic conditions. In suitable rock the progress rate is much greater than with conventional mining techniques, ranging from over 120 m/day in soft rock to about 30 m/day in moderately hard limestones and sandstones.

Rock hardness and advance rates have been found to be related [Tarkoy (1975)[52]]. A combination of hardness tests based on rebound (Schmidt L-hammer and the shore scleroscope) and abrasion (modified Taber abrasion test) were studied to evaluate their usefulness to predict TBM advance rates, presented for various rock types in terms of "total hardness" [Fig. 3.1 of Hunt of (1984)[5]]. A relationship between total rock hardness and TBM penetration rates developed from actual projects is given in Fig. 10.25.

Disadvantages of the TBM include its high initial cost, which causes it to be generally impractical for short tunnels unless a machine is already available for the desired diameter, and a possible limit in the tunnel diameter, although machines have been used for diameters to 12 m. They can be uneconomical in very hard rock, and yield poor performance where hard rocks alternate with soft and in extensive zones of highly altered and fractured rocks, as well as in squeezing, swelling, and running ground. TBM technology is advancing rapidly, however, and improvements are being made constantly.

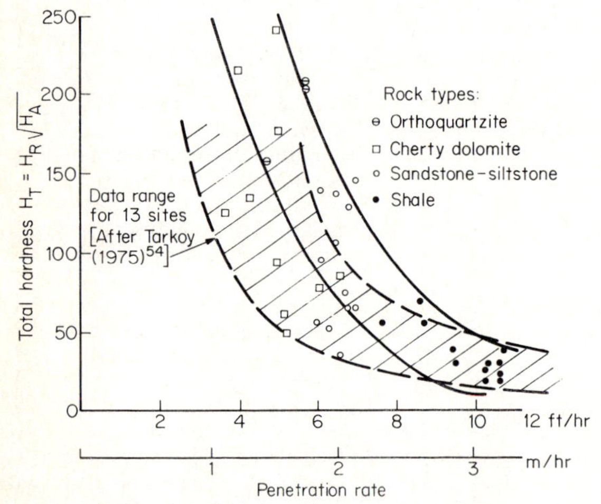

FIG. 10.25 Total rock hardness versus penetration rate of TBM. (H_R = Schmidt hardness, H_A = abrasion test hardness). [From Merritt (1974).[53]]

Ground Treatment

With respect to treatment of the ground to cope with or improve conditions, rock masses can be classed as competent rock not requiring treatment, incompetent rock, highly stressed rock, water flows and running ground, and squeezing and swelling ground. The effects on tunneling for various ground conditions and the general treatments are summarized in Table 10.19.

Incompetent Rock, Heavily Jointed

Heavily jointed rock can be prereinforced in advance of the face with *spiles;* untensioned steel members, such as reinforcing bars, are placed in holes predrilled in a regular pattern and fully grouted. Installed about the periphery, they are angled out about the face in fan form. A field performance study is described by Korbin and Brekke (1978).[55]

Highly Stressed Rock

In tunnels and caverns, popping and closure from high residual stresses are controlled with bolts which must penetrate beyond the tensile zone, often used together with mesh and straps or with steel sets. Rock bursts in mines at depths of thousands of meters below the surface are a serious problem. When anticipated, the potential bursting zone has been covered with a rock-block fill to contain the burst. A reduction in release energy to decrease the hazard is accomplished by unstressing or modifying the excavation sequence [Board and Crouch (1975)[56]]. Acoustical emission devices monitoring rock stress changes may provide warning [Art. 4.3.5 of Hunt (1984)[5]].

Water and Running Ground

Predraining for groundwater control is one of the most important aspects of underground excava-

TABLE 10.19
ROCK CONDITIONS AND UNDERGROUND EXCAVATIONS

Condition	Associations	Effects on opening	Treatments
Competent rock	Hard massive, intact (igneous in batholiths or laccoliths, quartzite, gneiss, sandstone)	Resistant to excavation. Seldom requires support. Permanent stand-up time if opening not excessively large.	Careful blasting techniques required to maintain quality.
Incompetent rock	Contains discontinuities (joints, foliations, bedding, faults) or soft rocks (shales, schists, decomposed)	Overbreak and stand-up time depends on defect intensity and orientation with tunnel alignment and strength of intact blocks.	Careful excavation techniques so as not to lower quality. Prereinforce with spiles from the heading. Predrain with pilot holes and grout crushed or other weak zones.
Residual stresses	Adjacent to fault zones, massive rock at great depth, folded rock	Closure of opening, "popping," and in extreme cases, rock bursts.	Install rock bolts beyond the tensile zone, mesh, and straps. When forewarned, fill to contain burst.
Running ground	Saturated, crushed, or sheared zones, particularly along faults	Saturated debris flows into tunnel, often under high initial pressures and large quantities. Can close tunnel; occurs suddenly and is highly dangerous.	Predrainage and grouting (several grout types and installation phases may be required).
Squeezing ground	Wet, weak plastic materials (soft shales, fault gouge, altered pyroclastic and micaceous rock)	Material moves into opening immediately upon exposure under pressure from surrounding rocks often high enough to crush and deform lining.	Overexcavate and rapidly apply thick mesh-reinforced concrete or liner plates with cast concrete or steel arches or provide temporary openings in the lining to allow stress release and pressure reduction with time prior to placing final lining. Use steel sets with slip joints.
Swelling rock	Soft shales, fault gouge, altered pyroclastic and micaceous rocks containing montmorillonite	Expansion occurs upon moisture contact causing high pressures on linings. Usually a time delay before noticed and continues for months or years.	As for squeezing ground; also excavate invert arch and bolt slab. Use crushable back-packing to allow pressure release. Grouting and predrainage may control source water from rock adjacent to swelling rock.

tion in rock. Since the presence of water often is unpredictable, it is prudent to drill pilot feeler holes in front of the heading which should be spaced to explore at angles from the heading as well as directly in front, since water-bearing zones can exist at any orientation. In addition to serving for detection, the holes also predrain high-pressure water trapped by a geologic structure, such as fault zone.

Grouting is required if flows persist to seal off the water flow as well as strengthen the rock mass. Grout mixtures follow flow channels and ordinarily do not penetrate clay or gouge-filled fractures; therefore, grouting is not particularly effective in sealing slow, seeping water or strengthening soft, impermeable materials. Materials injected can vary from cement grout, to cement-chemical mixtures to chemicals, depending upon the severity of conditions (*see Art. 5.5.3*). Cement grouts are substantially lower in cost than chemical grouts, but the latter have the important advantages of superior penetration over cement in fine openings, and additives can permit very short set times, in the order of minutes. Injection pressures vary depending upon conditions and care is required to maintain pressures greater than water pressure but less than that required to cause new fractures or to open existing fractures, in most cases. Pressures of 7 tsf are usual, although pressures as high as 100 tsf have been used to compress rock masses.

Very difficult conditions are caused by saturated, crushed rock zones associated with faulting and folding. When these are encountered or anticipated, the heading may be sealed with a concrete bulkhead and grout injected to stabilize the mass. Lancaster-Jones (1969)[57] cites a case where four phases of treatment were carried out to stabilize a wet crushed quartzite sand before tunneling could proceed. A preliminary injection was made of sodium silicate to assist penetration of a subsequent cement grout which was pumped under pressure up to 1400 psi (100 kg/cm^2) to fissure the ground and compress the sandy mass. A strong network of cement was formed in the fractured ground and the main water pressures were sealed. A third treatment was made using chemical grouts injected under pressures of 140 psi, designed to set in 15 min. This treatment tended to bind the sand grains together, increasing the material's strength. A final injection of cement was made to compress the mass futher. The final result was so satisfactory that explosives were necessary for excavation.

Squeezing and Swelling Ground

If the anticipated squeezing or swelling pressures are moderate, the material can be overexcavated to remove zones moistened by tunneling operations and a thick mesh-reinforced concrete lining or liner plates with cast concrete or steel arches can be quickly installed. An alternate procedure, preferred by some practitioners, is to permit swelling to occur to relieve pressures by providing temporary openings in the lining [Walhstrom (1973)[25]].

Einstein and Bischoff (1975)[26] recommend several procedures as follows. Excavate an invert arch, since the main swelling zone is often the floor area. Place an anchored or bolted invert slab (Fig. 10.21) with bolting extending preferably below the main swelling zone. Use frangible back-packing or compressible supports which will deflect and relieve swelling pressures. Grouting may be used to seal off sources supplying water to the swelling rock such as adjacent porous rocks or fractured zones. Predrainage of these zones will aid in reducing the problem if the water supply is limited.

In squeezing ground, the use of steel sets with slip joints permits movement to occur under controlled conditions to reduce squeezing pressures and support requirements [Yagi (1979)[58]]. At an optimum level of displacement, based on field measurements, bolts on the slip joints are tightened so that further inward movement results in loading of the sets.

Ground Subsidence

Ground subsidence resulting from underground excavations in rock is normally not a concern of the same magnitude as it is for tunneling in soils. There are, however, some notable exceptions, most of which involve tunneling or mining in soft rocks when the opening closes, causing strains in the overlying rocks resulting in surface subsidence. In certain situations, soils overlying rock may consolidate either as a result of predrainage or because of the natural drainage occurring in the rock during excavation.

Mining

Extraction of materials such as coal, salt, sulfur, and gypsum from soft rocks may result in surface subsidence during mining operations, or subsidence may occur many years after operations have ceased. Coal mining is the most noticeable cause of surface subsidence and varies in occurrence, in a general manner, with the extraction method. With the room and pillar method, mine collapse caused by pillar failure may occur years after operations terminate, whereas with the long-wall method of extraction collapse occurs in a somewhat controlled manner during extraction. In either case, surface deflections can be significant.

Ground subsidence due to mining operations is discussed in Art. 10.3 of Hunt (1984),[5] including a description of the extraction operations, mine-collapse mechanisms, investigation, analysis, prevention, and control. Mathematical modeling by FEM techniques to evaluate mine subsidence phenomena is described by Mabry (1973)[59] and Dahl and Choi (1975).[60]

Soft Rock Tunneling

During construction of the Garrison Dam test tunnel in a clay shale, a maximum surface subsidence of 4 cm was reported [Peck (1969)[1]]. The tunnel, which had a diameter of 10.9 m, was 37 m below the surface and was excavated by blasting and supported by ribs and lagging. Unconfined compressive strength of the shale was in the order of 10 tsf (1000 kPa).

Subway tunnel construction in the soft Keuper sandstone underlying Nuremberg, Germany, caused concern regarding potential damage to some of the historic structures in the city. Gartung and Bauernfiend (1977)[61] reported on a study made for the St. Lorenz church supported on shallow foundations bearing on the sandstone. The subway tunnel, with a diameter of approximately 9 m and a crown depth of 10 m, passes very near and below the church foundations. Although without significant joints or faults, the sandstone has a low peak strength, in the order of 20 kg/cm^2 with a confining pressure of about 1 kg/cm^2 and an unconfined compressive strength of about 6 kg/cm^2.

The FEM was used to simulate the influence of the tunnel excavation on the ground stress state and to compute settlements. Analysis of existing settlement data from small tunnel construction in the Keuper led to the selection of a value for K_o = 0.3. The computed settlements of 1 cm were considered objectionable so a tied-back concrete wall was installed to provide constraint. Computed settlements with the wall in place were in the order of 0.5 cm. During construction using the NATM, movements were monitored with extensometers, precise surface leveling, deflectometers for lateral deformations, and a sensitive inverse pendulum installed in the church tower. The measured settlements were slightly less than those computed by the FEM analysis; the church corner immediately adjacent to the tunnel settled 2.9 mm and had been computed to settle 4.3 mm.

Soils Overlying Rock

Scandinavian cities such as Stockholm and Goteborg [Broms (1978)[62]] and Oslo [Karlsrud and Sander (1978)[63]] suffer surface subsidence, often detrimental to structures, during tunneling in the underlying rocks. In Oslo the rocks vary from clay-shales and calcareous shales to limestone, and in Sweden they are granite and gneiss. The bedrock surface is highly irregular due to glaciation and in both localities is overlain by soft glacial clays with a stiff crust. In the deeper depressions in the rock surface, often the location of a weak shear zone, a sand layer is sandwiched between the rock and the soft clay as shown in Fig. 10.26.

During excavation, leakage into the tunnel occurs along rock fractures and sheared zones which, connected to the sand stratum, lower the pore pressures, increasing overburden stresses on the soft clays resulting in consolidation and surface subsidence. Even after lining is installed, leakage may follow the fractured blasting zone around the tunnel to an adjacent groundwater reservoir with a lower pressure head.

Grouting the zone about the tunnel generally reduces leakage, but it is impractical to grout every fracture, and even a few open fractures are ade-

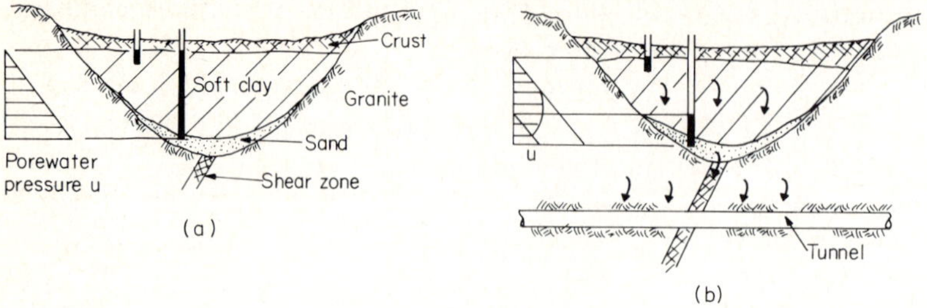

FIG. 10.26 Cause of subsurface subsidence from tunneling in Stockholm: (*a*) initial groundwater conditions and (*b*) groundwater level in sand lowers, causing clay to consolidate. [After Broms (1978).[62]]

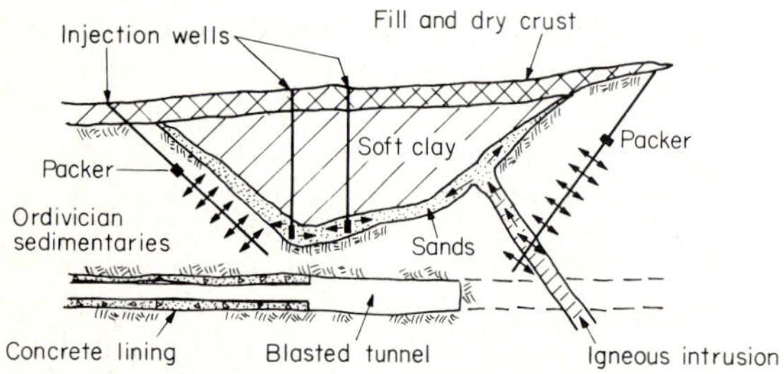

FIG. 10.27 Artificial groundwater recharge to reduce clay consolidation and surface subsidence in Oslo. [After Karlsrud and Sander (1978).[63]]

quate to permit drainage in the sand layer resulting in clay consolidation. Artificial recharge by injecting water into the sand stratum and the rock, as illustrated in Fig. 10.27, helps to maintain pore-pressure equilibrium in the sand, and although some drainage from the sand usually occurs, settlement is reduced substantially. Excavation with the TBM should result in less rock fracture than occurs during blasting; therefore, fully lined tunnels should serve to reduce surface subsidence, which is not always the case with lined blast-excavated tunnels.

REFERENCES

1. Peck, R. B. (1969) "State-of-the-Art Report: Deep Excavations and Tunneling in Soft Ground," *Proc. 7th Intl. Conf. Soil Mech. and Found. Engrg.*, Mexico City, Vol. 1, pp. 225–290.
2. Dowding, C. H. and Rozen, A. (1978) "Damage to Rock Tunnels from Earthquake Shaking," *Proc. ASCE, J. Geotech. Engrg. Div.*, Vol. 104, No. GT2, February, pp. 175–191.
3. ABGE (1980) *Sustentação e Revestimento em Tuneis e Galerias*, Assoc. Brasileira de Geologia de Engenharia, São Paulo, *Tradução No. 10*, August.
4. AFTES (1974) "Tunnels and Underground Works," Assoc. Francesa de Obras Subterrâneas (AFTES), Paris, No. 1, January/February.
5. Hunt, R. E. (1984) *Geotechnical Engineering Investigation Manual*, McGraw-Hill Book Co., New York.
6. Deere, D. U., Merritt, A. H., and Cording, E. S. (1974) "Engineering Geology and Underground Construction," *Proc. 2d Intl. Cong., Intl. Assoc. Engrg. Geol.*, São Paulo, Vol. 7, pp. GR1–27.
7. Cording, E. J., Hendron, A. J., Jr., Hansmire, W. H., Mahar, J. W., MacPherson, H. H., Jones, R. A., and O'Rourke, T. D. (1975) *Methods for Geotechnical Observations and Instrumentation in Tunneling*, University of Illinois at Urbana-Champaign, National Technical Information Service Pub. 252 585 (Vol. 1) and 252 586 (Vol. 2, Appendices), U.S. Department of Commerce, Springfield, Va.
8. Proctor, R. V. and White, T. L. (1977) *Earth Tunneling with Steel Supports*, Commercial Shearing Inc., Youngstown, Ohio (reprint 2d ed., 1968).
9. Terzaghi, K. and Peck, R. B. (1948) *Soil Mechanics in Engineering Practice*, John Wiley & Sons, New York.
10. Clough, G. W. (1981) "Innovations in Tunnel Construction and Support Techniques," *Bul. Assoc. Engrg. Geol.*, Vol. 18, No. 2, May, pp. 151–168.

11. Jones, J. S. and Brown, R. R. (1978) "Temporary Tunnel Support by Artificial Ground Freezing," *Proc. ASCE, J. Geotech. Engrg. Div.*, Vol. 104, No. GT10, October, pp. 1257–1276.

12. ENR (1980) "Miners Muscle Soft Ground Maze," *Engineering News-Record*, May, pp. 26–30.

13. NCE (1979) "Tunneling," *New Civil Engineer International*, Inst. of Civ. Engr., London, February, pp. 28–31.

14. NCE (1979) "Facing Up to the Future—Tunneling Supplement," *New Civil International*, Inst. of Civ. Engr., London, March.

15. ENR (1981) "Kudos for Japanese on U.S. Bore," *Engineering News-Record*, July 16, pp. 56–61.

16. Clough, G. W., Sweeney, B. P., and Finno, R. J. (1983) "Measured Soil Response to EPB Shield Tunneling," *Proc. ASCE, J. Geotech. Engrg. Div.*, Vol. 109, No. GT2, February, pp. 131–149.

17. Schmidt, B. and Grantz, W. C. (1979) "Settlements of Immersed Tunnels," *Proc. ASCE, J. Geotech. Engrg. Div.*, Vol. 105, No. GT9, September, pp. 1031–1047.

18. Atkinson, J. H. and Potts, D. M. (1977) "Subsidence Above Shallow Tunnels in Soft Ground," *Proc. ASCE, J. Geotech. Engrg. Div.*, Vol. 103, No. GT4, April, pp. 307–325.

19. Cording, E. J. and Hansmire, W. H. (1975) "Displacements Around Soft Ground Tunnels," *Proc. 5th Pan Amer. Cong. Soil Mech. and Found. Engrg.*, Buenos Aires, November, General Report, Ses. 4.

20. Ghaboussi, J., Ranken, R. E., and Karshenas, M. (1978) "Analysis of Subsidence over Soft-Ground Tunnels," *Proc. Intl. Conf. Evaluation and Prediction of Subsidence*, Pensacola Beach, Fla., ASCE, New York, pp. 182–196.

21. Tan, D. Y. and Clough, G. W. (1980) "Ground Control for Shallow Tunnels by Soil Grouting," *Proc. ASCE, J. Geotech, Engrg, Div.*, Vol. 106, No. GT9, September, pp. 1037–1057.

22. Wickham, G. E., Tiedmann, H. R., and Skinner, E. H. (1972) "Support Determinations Based on Geologic Conditions," *Proc. 1st North American Tunneling Conf.*, AIME, New York, pp. 43–64.

23. Terzaghi, K. (1949) "Introduction to Tunnel Geology," *Rock Tunneling with Steel Supports*, R. V. Proctor and L. White, eds. (revised, 1977), Commercial Shearing Inc., Youngstown, Ohio (reprint 2d ed., 1968).

24. Semple, R. M., Hendron, A. J., and Mesri, G. (1973) "The Effect of Time Dependent Properties of Altered Rock on Tunnel Support Requirements," U.S. Department of Transportation, Federal Railway Administration, Cont. No. DOT FR 30022, Washington, D.C.

25. Walhstrom, E. E. (1973) *Tunneling in Rock*, Elsevier Scientific Pub. Co., Amsterdam, The Netherlands.

26. Einstein, H. H. and Bischoff, N. (1975) "Design of Tunnels in Swelling Rock," *Design Methods in Rock Mechanics, Proc. 16th Symp. Rock Mech.*, University of Minnesota, Minneapolis, ASCE, New York (1977), pp. 185–195.

27. Sharp. J. C., Maini, Y. N., and Brekke, T. L. (1973) "Evaluation of the Hydraulic Properties of Rock Masses," *New Horizons in Rock Mechanics, Proc. 14th Symp. on Rock Mech.*, University Park, Penn., ASCE, pp. 481–500.

28. Krynine, D. and Judd, W. R. (1957) *Principles of Engineering Geology and Geotechnics*, McGraw-Hill Book Co., New York.

29. Rabcewicz, L. V. (1975) "Princípios e Modos de Aplicação do Novo Método Austríaco de Abertura de Túneis com Atenção Particular às Condições Geotécnicas e Topográficas Brasileiras," Lecture given to Associação Brasileira de Geologia de Engenharia, *ABGE Tradução No. 8*, November 1979.

30. Rabcewicz, L. V. (1975) "Tunnel Under the Alps Uses New Cost-Saving Lining Method," *Civil Engineering*, ASCE, New York, October, p. 66.

31. Kruse, G. H. (1970) "Deformability of Rock Structures, California State Water Project," *Determination of the In Situ Modulus of Deformation of Rock*, *Spec. Tech. Pub. 477*, ASTM, Philadelphia, pp. 58–88.

32. Bieniawski, Z. T. (1974) "Geomechanics Classification of Rock Masses and Its Application to Tunneling," *Proc. 3d Intl. Cong. Rock Mech.*, Denver, Vol. IIA, pp. 27–32.

33. NAVFAC (1982), *Soil Mechanics Design Manual 7.1*, Department of the Navy, Naval Facilities Engineering Command, Alexandria, Va, May.

34. Proctor, R. V. and White L. (1977) *Rock Tunneling with Steel Supports*, Commercial Shearing Inc., Youngstown, Ohio (reprint 2d ed., 1968).

35. Barton, N., Lien, R., and Lunde, J. (1974) "Analysis of Rock Mass Quality and Support Practice in Tunneling: A Guide for Estimating Support Requirements," *NGI Internal Report 54206*, Norwegian Geotech. Inst., Oslo.

36. Deere, D. U. (1963) "Technical Description of Rock Cores for Engineering Purposes," *Rock Mech. Engrg. Geol. 1*, pp. 18–22.

37. Deere, D. U. (1969) "Geologic Considerations," *Rock Mechanics in Engineering Practice*, Stagg and

Zienkiewicz, eds., John Wiley & Sons, New York, Chap. 1.

38. Rose, D., Kaboli, P., and Mayes, R. (1980) "Influence of Geologic Logs and Descriptions on Tunnel Designs and Costs," *AEG Newsletter*, Assoc. of Engrg. Geol., Vol. 24, No. 1, January, pp. 24–26.

39. Lee, C. F. and Lo, K. Y. (1976) "Finite Element Study of a Shallow Tunnel in Rock Exhibiting Clay-Induced Anistropy," *Numerical Methods in Geomechanics*, ASCE, New York, Vol. II, pp. 834–846.

40. Goodman, R. E. (1980) *Introduction to Rock Mechanics*, John Wiley & Sons, New York.

41. Abel, J. F. (1967) "Tunnel Mechanics," *Quarterly*, Colorado School of Mines, Boulder, Colo., Vol. 62, No 2, April.

42. Rabcewicz, L. V. and Golser, J. (1973) "Principles in Dimensioning the Supporting System for the New Austrian Tunneling Method," *Water Power*, March, p. 88.

43. Peck, R. B. and Deere, D. U. (1977) "The Role of Theory in Geotechnical Field Observations," *Structural and Geotechnical Mechanics*, W. J. Hall, ed., Prentice-Hall, Englewood Cliffs, N.J. (The Newmark Vol.), pp. 204–219.

44. Lang, T. A. (1972) "Rock Reinforcement," *Bul. Assoc. Engrg. Geol.*, Vol. IX, No. 3, Summer, pp. 215–240.

45. Barker, R. M. and Hatt, F. (1973) "Joint Effects in Bedded Formation Roof Control," *New Horizons in Rock Mechanics, Proc. ASCE, 14th Symp. on Rock Mech.* (1972), University Park, Pa., pp. 247–262.

46. Hendron, A. J., Jr. (1969) "Mechanical Properties of Rock," *Rock Mechanics in Engineering Practice*, Stagg and Zienkiewicz, eds., John Wiley & Sons, New York, Chap. 2.

47. Bieniawski, Z. T. (1968) "The Effect of Specimen Size on Compressive Strength of Rock," *Intl. J. Rock Mech. Min. Sci.*, Vol. 5, pp. 325–335.

48. Rockaway, J. D. and Stephenson, R. W. (1982) "Geotechnical Evaluation of the Support of Coal Pillars in Underground Coal Mines," *Bul. Assoc. Engrg. Geol.*, Vol. XIX, No. 1, February, pp. 5–14.

49. Vesić, A. S. (1975) "Bearing Capacity of Shallow Foundations," *Foundation Engineering Handbook*, Winterkorn and Fang, eds., Van Nostrand-Reinhold Co., New York, Chap. 3.

50. Jaeger, C. (1972) *Rock Mechanics and Engineering*, Cambridge University Press, Great Britain.

51. Brekke, T. L. (1972) "Shotcrete in Hard-Rock Tunneling," *Bul. Assoc. Engrg. Geol.*, Vol. IX, No. 3, Summer, pp. 215–240.

52. Tarkoy, P. J. (1975) "A Study of Rock Properties and Tunnel Boring Machines Advance Rates in Two Mica Schist Formations," *Applications of Rock Mechanics, Proc. ASCE 15th Symp. on Rock Mech.* (1973), Custer State Park, North Dakota, pp. 415–447.

53. Merritt, A. H. (1974) "Tunnel Boring Machines—Geologic Control," *Proc. 2d Intl. Cong. Intl. Assoc. Engrg. Geol.*, São Paulo, Vol. 2, pp. VII-PC-2.1 to 2.7.

54. Tarkoy, P. J. (1975) *Rock Hardness Index Properties and Geotechnical Parameters for Predicting TBM Performance*, Ph.D. thesis (unpublished), University of Illinois, Urbana.

55. Korbin, G. E. and Brekke, T. L. (1978) "Field Study of Tunnel Reinforcement," *Proc. ASCE, J. Geotech. Engrg. Div.*, Vol. 104, No. GT8, August, pp. 1091–1108.

56. Board, M. P. and Crouch, S. L. (1975) "Mine Planning to Control Rock Bursts in Cut and Fill Excavations," *Design Methods in Rock Mechanics, Proc. 16th Symp. Rock Mech.*, University of Minnesota, Minneapolis, ASCE (1977), New York, pp. 249–255.

57. Lancaster-Jones, P. F. F. (1969) "Methods of Improving Properties of Rock Masses," *Rock Mechanics in Engineering Practice*, Stagg and Zienkiewicz, eds., John Wiley & Sons, New York, Chap. 12.

58. Yagi, Y. (1979) "Application of the NATM to the Nakayama Tunnel," *Tunneling under Difficult Conditions*, Kitamure, ed., Pergamon Press, Oxford, England, pp. 193–198.

59. Mabry, R. E. (1973) "An Evaluation of Mine Subsidence Potential," *New Horizons in Rock Mechanics, Proc. ASCE, 14th Symp. on Rock Mech.*, University Park, Pa., pp. 263–288.

60. Dahl, H. D. and Choi, D. S. (1975) "Some Case Studies of Mine Subsidence and Its Mathematical Modeling," *Applications of Rock Mechanics, Proc. ASCE, 15th Symp. on Rock Mech.* (1973), Custer State Park, North Dakota, pp. 1–22.

61. Gartung, I. and Bauernfiend, P. (1977) "Subway Tunnel at Nurnberg—Predicted and Measured Deformations," *Intl. Symp. on Field Measurements in Rock Mechanics*, Zurich, April, Fed. Inst. of Tech., Zurich, pp. 473–483.

62. Broms, B. (1978) "Subsidence from Lowering of the Groundwater Table," *Proc. Intl. Conf. on Evaluation and Prediction of Subsidence*, Pensacola Beach, Fla., January, ASCE, New York, pp. 441–458.

63. Karlsrud, K. and Sander, L. (1978) "Subsidence Problems Caused by Rock Tunneling in Oslo," *Proc. Intl. Conf. on Evaluation and Prediction of Subsidence*, Pensacola Beach, Fla., January, ASCE, New York, pp. 197–213.

APPENDIX A

CONVERSION TABLES

ENGLISH TO METRIC TO THE INTERNATIONAL SYSTEM*

Units†	English	Metric		SI
Length	1 mi = 1760 yds = 5280 ft	1.609 km	or	1609 m
	1 ft = 12 in	0.3048 m	or	30.48 cm
	1 in	2.54 cm	or	25.4 mm
Area	1 mi^2 = 640 acres	2.59 km^2	or	2.59 × 10^6 m^2
	1 acre = 43,560 sq ft^2	0.4047 ha	or	4047 m^2
	1 ft^2 = 144 in^2	0.0929 m^2	or	929.0 cm^2
	1 in^2	6.452 cm^2		6.452 cm^2
Volume	1 acre·ft = 43,560 ft^3	1233.49 m^3		1233.49 m^3
	1 yd^3 = 27 ft^3 = 1728 in^3	0.7646 m^3		0.7646 m^3
	1 ft^3 = 7.48 gal	0.0283 cu m	or	28.32 liters
	1 U.K. gal	4546 cm^3	or	4.546 liters
	1 U.S. gal = 231 in^3	3785 cm^3	or	3.785 liters
	1 in^3	16.387 cm^3		16.387 cm^3
Mass	1 ton (short) = 2000 lb	0.9072 tonne	or	907.18 kg
	1 ton (long) = 2240 lb	1.016 tonnes	or	1016 kg
	1 lb = 16 oz	0.4536 kg	or	453.6 g
	1 oz	28.352 k		28.352 g
Density	1 pcf	16.019 kg/m^3		0.157 kN/m^3
Force	1 tonf (short)	0.9072 tonnef		8.897 kN
	1 tonf (long)	1.016 tonnesf		9.964 kN
	1 lbf	0.4536 kgf		4.448 N
Pressure or stress	1 tsf (short) = 2000 psf	9.764 t/m^2		95.76 kPa
	1 tsf (short) = 13.89 psi	0.9764 kg/cm^2		95.76 kPa
	1 tsf (long) = 2240 psf	10.936 t/m^2		107.3 kPa
	1 psf = 0.00694 psi	0.000488 kgf/cm^2		0.04788 kPa
	1 psi =	0.0703 kgf/cm^2		6.895 kPa
	1 bar =	1.02 kg/cm^2		100 Pa
	1 atm = 33.9 ft of water (39.2°F)	76 mm mercury at 0°C		
	1 atm = 1.058 tsf = 14.7 psi	1.033 kgf/cm^2		1.0133 bars

(continued)

ENGLISH TO METRIC TO THE INTERNATIONAL SYSTEM* (*Continued*)			
Units†	**English**	**Metric**	**SI**
Velocity	1 ft/year = 1.9025×10^{-6} ft/min	0.9659×10^{-6} cm/s	0.9659×10^{-8} m/s
Acceleration of gravity	1 ft/s^2 32 ft/s^2	0.3048 m/s^2 980 cm/s^2	0.3048 m/s^2 980 gals
Flow	1 gal/min = 192 ft^3/day	0.0038 m^3/min	3.8 liters/minute
Temperature	t°F = 1.8 t°C + 32	t°C = (t°F − 32)/1.8	
Water, unit weight	62.4 pcf	1 g/cm^3 = 1 t/m^3	9.81 kN/m^3

METRIC TO ENGLISH TO SI			
Units	**Metric**	**English**	**SI**
Length	1 km = 1000 m 1 m = 100 cm 1 cm = 10 mm	0.6214 mi 3.28 ft = 39.37 in 0.3937 in	1 km 1 m 1 cm
Area	1 km^2 = 100 ha 1 ha = 10,000 m^2 1 m^2 = 10,000 cm^2	0.386 mi^2 2.47 acres 10.764 ft^2 = 1550 in^2	1 km^2 1 ha 1 m^2
Volume	1 m^3 = 1000 liter 1 liter = 1000 cm^3	1.31 yd^3 = 35.315 ft^3 0.264 gal = 61 in^3	1 m^3 1 liter
Mass	1 tonne = 1000 kg 1 kg = 1000 g	2204.6 lb 2.205 lb	1 tonne 1 kg
Density	1 t/m^3 = 1 g/cm^3	62.428 pcf	9.807 kN/m^3
Force	1 tonnef = 1000 kgf 1 kgf = 1 kilopond = 1000 g	1.10 tonf (short) 2.205 lbf	9.807 kN 9.807 N
Unit weight	1 t/m^3 = 1 g/cm^3	62.428 pcf	9.807 kN/m^3
Pressure or stress	1 t/m^2 = 0.1 kg/cm^2 1 kg/cm^2 1 kg/cm^2 1.02 kg/cm^2	0.1024 tsf (short) 1.024 tsf = 14.223 psi 0.9678 atm 14.495 psi	9.807 kPa 98.07 kPa 1 bar (100 kPa)
Velocity	1 cm/s	1.9685 ft/min	1 cm/s
Acceleration of gravity	1 cm/s^2 = 0.01 m/s^2	0.3048 ft/s^2	1 gal
Flow	1 m^3/s = 1000 liters/second 1 liter/second	15,800 gal/min 15.8 gal/min = 0.263 gal/s	1 m^3/s

*Metric and SI units are generally the same except for force and pressure; therefore under the SI column some additional useful metric conversions are given.

†For abbreviations see *Symbols* at end of table.

SYMBOLS			
Time	**English**	**Metric**	**SI**
mo—month hr—hour min—minute s—second	in—inch ft—foot yd—yard mi—mile oz—ounce lb—pound t—ton gal—gallon psi—lb/in^2 tsf—tons/ft^2 pcf—lbs/ft^3	mm—millimeters cm—centimeters m—meters km—kilometers ha—hectares g—grams kg—kilograms t—tonnes f—force	N—newton (force) Pa—pascal = N/m^2 (pressure) 1 N = 0.2248 lb = 0.102 kg = 102 g 1 kPa = 0.0104 tsf (short) = 0.145 psi = 0.0102 kgf/cm^2 1 bar = 100 kPa
Pressure			
atm—atmosphere			

NOTE: M = mega (10^6), k = kilo (10^3), h = hecto $(10)^2$, da = deka = (10), d = deci (10^{-1}), c = centi (10^{-2}), m = milli (10^{-3}), μ = micro (10^{-6}).

SUMMARY TABLE OF CONVERSIONS							
		System of units			Mutual proportion		
Name	**Symbol**	**SI**	**Metric**	**Imperial**	**SI**	**Metric**	**Imperial**
Length	l	mm m	mm m	in ft	1 1	1 1	0.03937 3.281
Area	A	mm^2 m^2	mm^2 m^2	in^2 ft^2	1 1	1 1	1.55×10^3 10.764
Volume	V	milliliter m^3	cm^3 m^3	in^3 ft^3	1 1	1 1	0.061 35.315
Velocity	v	mm/s m/s	mm/s m/s	in/s ft/s	1 1	1 1	0.03937 3.281
Rate of flow	Q	milliliter/second m^3/s	cm^3/s m^3/s	ft^3/s ft^3/s	1 1	1 1	3.531×10^{-3} 35.315
Mass	m	g t	g t	lb lb	1 1	1 1	2.205×10^{-3} 2205
Density	ρ	kg/m^3 t/m^3	kg/m^3 t/m^3	lb/ft^3 lb/ft^3	1 1	1 1	0.06243 62.4
Force	P, F	N kN MN	kgf kgf tf	lbf lbf kips	1 1 1	0.101971 101.971 101.971	0.22481 224.81 224.81
Stress pressure	σ, p	kN/m^2 MN/m^2	kgf/cm^2 tf/m^2	lbf/in^2 tf/ft^2	1 1	0.0101971 101.971	14.504 9.3238
Unit weight	γ	kN/m^3	gf/cm^3	lbf/ft^3	1	0.10197	6.3657

APPENDIX B

SYMBOLS

SYMBOLS

Symbol	Represents	Art.	Symbol	Represents	Art.
A	Amplitude	5.6.2	C_R	Gradation range	2.3.2
A	Area	—	C_r	Consolidation recompression curve	4.3.4
A	Pore-pressure parameter	3.2.2	C_u	Uniformity coefficient	2.2.4
A_p	Anchor force	9.4.3	CD	Consolidated-drained test	3.3.2
A_p	Pile tip area	7.3.2	CU	Consolidated-undrained test	3.3.3
A_s	Pile shaft area	7.3.2	CBR	California bearing ratio	5.2.3
a	Acceleration	5.6.2	CPT	Cone penetration test	3.3.2
a	Area	—	C_α	Secondary compression coefficient	4.3.4
a_v	Compressibility coefficient	4.1.3	C_τ	Soil suction index	4.3.6
B	Bulk modulus	4.1.3	c	Cohesion	3.1.1
B	Foundation width	4.3.1	$\bar{c}$, c'	Effective stress cohesion	3.2.1
B	Pore-pressure parameter	3.2.2	c_a	Apparent cohesion	3.3.2
B	Tunnel width	10.2.2	c_a	Footing-soil adhesion	6.2.4
b	Foundation width	4.2.2	c_a	Pile-soil adhesion	7.3.2
b_c, b_q, b_γ	Base tilt factors	6.2.4	c_h	Horizontal consolidation coefficient	5.3.3
C	Compressibility factors	4.3.3	c_u	Undrained cohesion	3.3.2
C	Pore-pressure parameter	3.2.2	c_v	Vertical consolidation coefficient	4.3.4
C	Total cohesion	3.3.2			
C_c	Compression index	4.3.4			
C_N	SPT depth correction factor	6.2.2			

(Continued)

SYMBOLS

Symbol	Represents	Art.
D	Constrained modulus	4.1.3
D	Damping ratio	3.3.5
D	Diameter	—
D_f	Foundation depth	6.2.2
D_R	Relative density	2.2.4
D_{max}	Maximum density	2.3.2
D_{min}	Minimum density	2.3.2
D_{10}	Diameter at which 10% of soil is finer	2.2.4
d	Depth	8.2.2
d	Diameter	—
d_w	Well diameter	5.3.3
E	Energy, pile driving	7.5.1
E	Moment arm	8.1.3
E	Young's modulus	4.1.3
ER	Energy ratio	5.6.3
E_b	Beam modulus	4.4.1
E_c	Compression modulus	4.1.4
E_d	Dynamic Young's modulus	4.1.3
E_i	Initial tangent modulus	4.1.3
E_p	Pile modulus	7.4.1
E_r	In situ static rock modulus	4.1.3
E_s	In situ static soil modulus	4.1.3
E_{se}	Secant modulus	4.1.3
E_{seis}	Dynamic field modulus	4.1.4
E_t	Tangent modulus	4.1.3
E_{ur}	Unload-reload modulus	9.3.4
e	Eccentricity	6.2.4
e	Void ratio	2.2.4
e_0	Initial void ratio	4.3.4
e_f	Pile-hammer efficiency	7.5.1
F	Force	3.2.3
FS	Factor of safety	—
F_ω	Pile taper correction factor	7.3.4
f	Coefficient of friction	3.1.1
f	Frequency	5.6.2
f	Function of	8.1.3
f_c	Settlement coefficient	4.3.2
f_c	Working load, concrete	7.3.2
f'_c	Ultimate strength, concrete	7.3.2
f_L	Local sleeve friction, CPT	7.3.4
f_n	Natural frequency	6.2.7
f_o	Slope factor, Janbu	8.2.5
f_s	Allowable stress, steel	9.4.3
f_s	Shaft friction	7.3.2
G	Shear modulus	4.1.3
G_d	Dynamic shear modulus	3.3.5
G_s	Specific gravity, solids	2.2.2
G_w	Specific gravity, water	2.2.2
GWL	Groundwater level, static	3.1.2
g	Acceleration of gravity	5.6.2
g_c, g_q, q_γ	Ground slope factors	6.2.4
H	Height	—
H	Horizontal force	8.2.6
H	Hydraulic head	2.3.2
H	Stratum thickness	4.3.4
H_c	Capillary rise	2.3.1
H_c	Critical height	7.1.3
H_A	Abrasion test hardness	10.3.5
H_R	Schmidt hardness	10.3.5
H_T	Total hardness	10.3.5
h	Hydraulic head	2.3.2
h	Sample thickness	4.3.4
h_e	Elevation head	2.3.2
h_p	Pressure head	2.3.2
h_t	Total hydraulic head	2.3.2
I, I_o	Epicentral intensity	8.2.7
I	Moment of inertia	4.4.1
I	Stress influence factors	4.2.2
I_B	Boussinesq stress influence factor	4.2.2
I_p	Pile tip influence factor	4.2.5
I_s	Point-load index	3.4.2
I_s	Settlement influence factor	4.2.5
I_w	Westergaard stress influence factor	4.2.2
I_z	Strain influence factor	4.3.3
i	Hydraulic gradient	2.3.2
i_{cr}	Critical hydraulic gradient	2.3.2
i	Inclination of wall back	9.2.3
i	Slope inclination with horizontal	8.1.3
i_c, i_q, i_γ	Foundation inclination factors	6.2.4
J	Damping factor, wave equation	7.5.2

SYMBOLS

Symbol	Represents	Art.
J	Soil restraint coefficient	4.4.3
JRC	Joint roughness coefficient	3.4.3
j	Joint roughness angle	3.4.3
j	Seepage force per unit volume	2.3.2
K	Dynamic bulk modulus	4.1.3
K	Pressuremeter constant	4.1.4
K	Spring constant	4.4.5
K	Stiffness factor	4.4.2
K_a	Earth pressure, active coefficient	3.2.3
K_H	Earth pressure, wall backfill coefficient	9.2.4
K_o	Earth pressure, at rest coefficient	3.2.3
K_p	Earth pressure, passive coefficient	3.2.3
K_s	Lateral earth pressure factor	7.3.2
K_V	Earth pressure, wall backfill coefficient	9.2.4
k	Permeability coefficient	2.3.1
k	Pseudoseismic coefficient	8.2.7
k, k_s	Joint stiffness	—
k, k_y	Spring constant	6.2.7
k_g	Pressuremeter bearing capacity factor	6.2.4
k_h	Horizontal acceleration	9.2.4
k_h	Horizontal permeability	2.3.2
k_h	Horizontal subgrade reaction modulus	4.4.3
k_{mean}	Average permeability	2.3.2
k_n	Normal joint stiffness	—
k_r	Subgrade reaction modulus, mat	4.4.1
k_s	Subgrade reaction modulus	4.4.1
k_{sv}	Vertical subgrade reaction modulus	4.4.1
k_t	Subgrade reaction modulus, test value	4.4.1
k_v	Vertical acceleration	9.2.4
k_v	Vertical permeability	2.3.2
L	Length	—
L	Long waves	5.6.2
L_o	Characteristic length	4.4.4
LI	Liquidity index	2.2.4
LL	Liquid limit	2.2.4
l	Length	6.1.2
M	Moment	—
M_m	Maximum imposed moment	7.6.3
M_y	Pile yield moment	7.6.3
m	Mass	5.6.2
m_v	Volume change coefficient	4.1.3
N	Normal force component	3.1.1
N	Standard penetration resistance (SPT)	4.1.4
$\overline{N}$	Effective normal force component	8.2.2
N', N_1	SPT corrected for $\overline{p}_o$	6.2.2
N_b	Stability factor, excavations	9.3.3
N_c	Bearing-capacity factor, cohesion	6.2.4
N_c	Stability number, excavations	9.3.3
N_{cq}, $N_{\gamma q}$	Modified bearing-capacity factors, slopes	6.2.5
N_e	Number of equipotential drops	2.3.2
N_f	Number of flow channels	2.3.2
N_M	Bearing-capacity factor, modified	6.2.4
N_p	Laterally loaded pile resistance factor	4.4.3
N_q	Bearing-capacity factor, foundation depth	6.2.4
N_s, N_f, N_o	Slope stability numbers	8.1.3
N_γ	Bearing-capacity factor, foundation width	6.2.4
N_ϕ	$\tan^2(45 + \phi/2)$	3.2.3
NC	Normally consolidated	3.3.1
n	Porosity	2.2.2
n_e	Effective porosity	2.3.2
n_h	Horizontal subgrade reaction modulus constant	4.4.3
O_c	Center of gravity	8.1.2
O_p	Origin	3.1.3

SYMBOLS

Symbol	Represents	Art.
OC	Overconsolidated	3.3.1
OCR	Overconsolidation ratio	3.2.3
P	Force, load, pressure	—
P	Primary waves	5.6.2
P_a	Active pressure	3.2.3
P_a	Atmospheric or air pressure	—
P_h	Horizontal pressure	9.2.3
PI	Plasticity index	2.2.4
P_{oh}	At-rest horizontal pressure	3.3.3
PL	Plastic limit	2.2.4
P_L	Limiting pressure, pressure meter test	4.1.4
P_p	Passive pressure	3.2.3
P_u	Ultimate lateral capacity, pile	4.4.3
P_v	Vertical pressure	9.2.3
$P(t)$	Ground acceleration input	4.4.5
p	Unit pressure	—
p_c	Preconsolidation pressure	3.2.3
p_f	Foundation pressure	4.3.4
p_o	Overburden pressure	3.2.3
$\bar{p}_o$, p'_o	Effective overburden pressure	3.2.3
p_s	Seepage pressure	2.3.2
Q	Flow rate	2.3.2
Q	Load	—
Q	Love waves	5.6.2
Q	Rock-mass quality	10.3.3
Q	Quake factor, wave equation	7.5.2
Q	Quick or UU triaxial test	3.3.3
Q_a	Allowable bearing value	6.2.2
Q_b	Total ultimate capacity	6.2.4
Q_L	Line load	9.2.3
Q_p	Point load	9.2.3
Q_p	Ultimate pile tip capacity	7.3.2
Q_s	Ultimate pile shaft capacity	7.3.2
Q_u	Total ultimate capacity	7.3.2
q	Flow rate per unit area	2.3.2
q	Flow quantity per unit of time	2.3.1
q	Mean normal stress at footing base	6.2.4
q_a	Allowable unit capacity	6.2.2
q_b	Ultimate unit bearing capacity	6.2.4
q_c	Cone point resistance (CPT)	3.3.2
q_c	Uniaxial compressive strength	3.4.2
q_h	Lateral unit soil bearing capacity	7.6.3
q_p	Ultimate unit pile tip capacity	7.3.2
q_s	Surcharge loading	6.2.4
q_s	Unit foundation pressure	4.2.2
q_{ult}	Ultimate unit bearing capacity	6.2.4
R	Radial distance	5.6.3
R	Radius	—
R	Rayleigh waves	5.6.2
R	Resultant force	8.2.6
R	Schmidt rebound hammer value	3.4.3
R	CU triaxial test	3.3.2
R_u	Ultimate soild resistance, wave equation	7.5.2
RQD	Rock-quality designation	2.2.2
r	Radius	—
r_u	Pore-pressure ratio	8.2.1
S	Beam deflection slope	4.4.1
S	Saturation percent	2.2.4
S	Section modulus	9.3.4
S	Settlement	7.4.1
S	Set, pile driving	7.5.1
S	Shear waves	5.6.2
S	Shearing resistance, strength	3.4.2
S	Slow or CD triaxial test	3.3.2
$\mathcal{S}$	Shape factor, flow nets	2.3.2
SG	Specific gravity	2.2.2
SL	Shrinkage limit	2.2.4
SR	Cyclic stress ratio	3.3.5
SPT	Standard penetration test	3.3.2
SSI	Soil-structure interaction	4.4.5
S_{max}	Maximum shearing strength	3.1.1
S_t	Sensitivity	3.3.4
S_t	Tensile strength	3.4.2
s_c, s_q, s_γ	Footing shape factors	6.2.4

SYMBOLS

Symbol	Represents	Art.
s_r	Remolded or ultimate undrained strength	3.3.4
s_u	Undrained shear strength	3.3.3
T	Period	5.6.2
T	Pile-stiffness factor	4.4.3
T	Shear force	3.1.1
T	Tension	7.7.4
T_h	Theoretical time factor, horizontal	5.3.3
T_{MR}	Flexural strength	3.4.2
T_v	Theoretical time factor, vertical	4.3.4
TBM	Tunnel boring machine	10.2.3
t	Time	4.3.4
t	Load transfer	7.4.1
U	Average consolidation ratio	4.3.4
U	Total pore pressure or seepage force	8.2.2
UC	Underconsolidated	3.3.1
UU	Unconsolidated, undrained triaxial test	3.3.3
U_c	Unconfined or uniaxial compressive strength	3.3.3
U_j	Uniaxial compressive strength along joint	3.4.3
u	Unit pore pressure	2.3.2
u_a	Unit pore-air pressure	3.3.2
u_w	Unit pore-water pressure	2.3.2
V	Beam shear	4.4.1
V	Joint or tension crack water pressure	8.2.5
V	Velocity	5.6.2
V	Vertical force	6.3.3
V	Volume	—
V_a	Volume of air or gas	2.2.4
V_F	Field V_p	2.2.2
V_L	Laboratory V_p	2.2.2
V_p	Compression wave velocity	5.6.2
V_s	Shear-wave velocity	5.6.2
V_s	Volume of solids	2.2.4
V_v	Volume of voids	2.2.4
V_w	Volume of water	2.2.4
v	Particle or vibration velocity	5.6.2
v_d	Discharge velocity	2.3.2
v_s	Seepage velocity	2.3.2
W	Weight	2.2.2
W_s	Weight of solids	2.2.4
W_t	Total sample weight	2.2.4
W_w	Weight of water	2.2.4
w	Water content	2.2.4
w	Average unit weight, material over tunnel roof	10.2.2
w_n	Natural water content	
x	Distance along x axis	4.4.1
y	Distance along y axis	4.4.1
y	Unit deflection	4.4.1
Z, z	Depth	3.1.2
z	Displacement	7.4.1
z_b	Bearing depth	7.3.3
z_c	Critical depth	7.3.2
z_o	Tension crack depth	7.4.1
z_w	Depth to GWL	2.3.2
α	Angle of obliquity	3.1.1
α	Compressibility factor	4.3.6
α	Compression modulus factor	4.3.3
α	CPT compression coefficient	6.2.2
α	Force angle inclination	9.2.5
α	Friction ratio, adhesion	7.3.4
β	Modulus reduction factor	4.1.4
β	Shaft friction factor	7.3.4
β	Slope inclination behind wall	9.2.3
γ	Soil unit weight	2.2.2
γ_b	Buoyant unit weight	2.2.4
γ_d	Dry unit weight	2.2.4
γ_D	Maximum density	2.2.4
γ_L	Minimum density	2.2.4
γ_N	Natural density	2.2.4
γ_s	Saturated unit weight	2.2.4
γ_t	Total unit weight	2.2.4
γ_w	Unit weight, water	2.3.2
Δ, δ	Change or increment	6.1.2
δ	Soil-pile friction angle	7.3.2
δ	Soil-wall friction angle	9.2.3
ϵ	Strain	4.1.2
ϵ_c	Cyclic axial strain	3.3.5

SYMBOLS

Symbol	Represents	Art.
ζ	Unit shear strain	3.3.5
θ	Angle between failure surface and horizontal	8.1.3
θ	Angle between normal stress and major principal stress	3.1.3
θ_c, θ_{cr}	Slope of critical failure surface	3.1.3
λ	Damping ratio	3.3.5
λ	Pile shaft capacity coefficient	7.3.4
λ	Pressuremeter shape factor	4.3.3
λ	Wavelength	5.6.2
μ	Loading-time-rate correction factor, vane shear	3.3.3
ν	Poisson's ratio	4.1.2
ρ	Consolidation settlement	4.3.1
ρ	Mass or bulk density	2.2.2
ρ	Rowe's reduction coefficent	9.4.3
Σ	Sum	—
σ	Stress	4.1.2
$\bar{\sigma}$, σ'	Effective stress	3.2.2
σ_d	Deviator stress	3.3.2
σ_{dc}	Cyclic deviator stress	3.3.5
σ_h	Horizontal stress	3.1.2
σ_n	Normal stress	3.1.3
$\bar{\sigma}_n$	Effective normal stress	3.2.2
$\bar{\sigma}_{nf}$	Effective normal stress at failure	3.1.3
σ_t	Uniaxial tensile strength	3.4.2
σ_v	Vertical stress	2.3.2
$\bar{\sigma}_v$, $\bar{\sigma}_o$	Effective vertical overburden stress	3.2.3
σ_z	Vertical stress at depth z	4.2.2
σ_1, σ_2, σ_3	Principal stresses	3.1.3
τ	Joint shear strength	3.4.3
τ	Shear stress	3.1.3
τ_m	Matrix soil suction	4.3.6
τ_{max}	Maximum shear stress	3.1.3
ϕ	Friction angle	3.1.1
$\bar{\phi}$, ϕ'	Effective friction angle	3.2.2
ϕ_a, ϕ_b	Apparent friction angle	3.4.2
ϕ_d	Drained friction angle	3.4.3
ϕ_r	Residual drained strength	3.3.4
ϕ_u	Undrained friction angle	3.3.3
ω	Circular frequency	5.6.2

NOTE: Not included are symbols for minerals or chemical elements, classification systems, or dimensions; and some occasional, less significant symbols that appear only once, at which point they are defined.

INDEX

ABOUT THE AUTHOR

Roy E. Hunt received his B.S. in Geology from Upsala College in 1952 and his M.A. in Soil Mechanics and Foundation Engineering from Columbia University in 1956. He is a fellow of both the American Society of Civil Engineers and the American Institute of Consulting Engineers.

Since 1952 he has worked for a number of consulting firms specializing in geotechnical engineering. He has been a project geologist, chief of a soil mechanics laboratory, a project engineer, and the principal of a firm. His experience ranges in many directions, including work on such diverse projects as earth and concrete dams, highways, airfields, foundations for structures, tunnels, development of large land areas, environmental conservation, retaining structures, and slope failure. These projects have taken him across the United States and around the world, to France, Israel, Brazil, Bolivia, Indonesia, and several other countries.

Mr. Hunt has published widely, with articles on soil and foundation engineering, slope failures, retaining structures, and applications of remote-sensing interpretation and terrain analysis. He is the author of the *Geotechnical Engineering Investigation Manual*, also published by McGraw-Hill, which received the Burwell Award from the Geologic Society of America, Engineering Division, and the Claire P. Holdredge Award from the Association of Engineering Geologists.